CLINICALLY
ORIENTED
Anatomy

THIRD EDITION

KEITH L. MOORE

B.A., M.Sc., Ph.D., F.I.A.C., F.R.S.M.
Professor of Anatomy and Cell Biology
Faculty of Medicine, University of Toronto
Toronto, Ontario, Canada
Visiting Professor of Clinical Anatomy
Department of Anatomy, Faculty of Medicine
University of Manitoba
Winnipeg, Manitoba, Canada

ILLUSTRATED PRIMARILY BY

Dorothy Chubb, Elizabeth Blackstock, Jean Calder, Dorothy Irwin, Nancy Joy, Nina Kilpatrick, David Mazierski, Angela Cluer, Stephen Mader, Bart Vallecoccia, Sari O'Sullivan, and Kam Yu

PHOTOGRAPHY BY

John Kozie and Paul Schwartz

SANS TACHE

Williams & Wilkins

BALTIMORE • PHILADELPHIA • HONG KONG
LONDON • MUNICH • SYDNEY • TOKYO

A WAVERLY COMPANY

Editor: Timothy S. Satterfield
Managing Editor: Linda Napora
Copy Editor: Kathy Lumpkin
Designer: Norman W. Och
Illustration Planner: Lorraine C. Wrzosek
Production Coordinator: Anne Stewart Seitz
Indexer: Deborah Tourtlotte

Accurate indications, adverse reactions, and dosage schedules for drugs are provided in this book, but it is possible that they may change. The reader is urged to review the package information data of the manufacturers of the medications mentioned.

PRINTED IN CANADA

First Edition, 1980
Second Edition, 1985

Library of Congress Cataloging-in-Publication Data

Moore, Keith L.
    Clinically oriented anatomy / Keith L. Moore.—3rd ed. / illustrated primarily by Dorothy Chubb . . . [et al.]; photography by John Kozie and Paul Schwartz.
       p. cm.
    Includes bibliographical references and index.
    ISBN 0-683-06133-X
    1. Human anatomy.   I. Title.
    [DNLM: 1. Anatomy.   QS M822c]
    QM23.2.M67 1992
    611—dc20
    DNLM/DLC
    for Library of Congress   91-30649
                                                     CIP

Dedicated to the memory of
Professor J. C. Boileau Grant
1886–1973
M.C., M.B., Ch.B., Hon. D.SC. (Man.), F.R.C.S. (Edin.)

Late Professor Emeritus of Anatomy in the University of Toronto, Professor Grant was Professor and Head of Anatomy in the University of Manitoba from 1919–1930 before he went to Toronto. He retired as Professor and Chairman of Anatomy in Toronto in 1959 and for several years was Visiting Professor of Anatomy in the University of California, Los Angeles. The above photograph was taken in 1956 when Professor Grant was awarded an honorary degree by the University of Manitoba.

Grant's Atlas, Grant's Method, Grant's Dissector and the *J.C.B. Grant Museum* in the University of Toronto are living memorials to this renowned Canadian anatomist.

Throughout the text, liberal use has been made of illustrations from the following Williams & Wilkins publications, which the author acknowledges with sincere thanks.

Agur AMR: *Grant's Atlas of Anatomy*, ed 9, 1991.

Basmajian JV: *Primary Anatomy*, ed 8, 1982.
Figs. 17, 20, 21, 1.16, 2.59, 5.8, 5.105, 6.46, 6.48, 6.61, 6.108, 6.111, 6.112, 6.113, 7.12, 7.47, and 7.82.

Basmajian JV, Slonecker CE: *Grant's Method of Anatomy*, ed 11, 1989.
Figs. 14, 138*A* and *B*, 141*A* and *B*, 2.8, 2.16, 2.21, 2.52, 2.70, 2.76, 2.84, 2.101, 3.2, 3.3*A*, 3.9, 3.21, 3.36, 3.55, 3.57, 5.33, 5.74, 4.4, 4.5, 6.12, 6.19, 7.54*A*, 7.57, 7.64, 7.74, 7.100, 8.13, 8.18, and 8.24.

Carpenter MB: *Core Text of Neuroanatomy*, ed 3, 1985.
Figs. 4.49*A*, 4.50, 4.52, 7.36, 7.44, 7.45, and 7.52.

Dilts PV, Jr, Greene JW, Jr, Roddick JW, Jr: *Core Studies in Obstetrics and Gynecology*, ed 2, 1977.
Figs. 3.5, 3.6, 3.15, and 3.62.

Salter RD: *Textbook of Disorders and Injuries of the Musculoskeletal System*, ed 2, 1983.
Figs. 5.94, 5.103, 6.95*B*, 6.101, 6.123, 6.125, 6.127, and 6.132.

The enthusiastic endorsement of the first two editions of *Clinically Oriented Anatomy* by students and professors in Canada, the United States, and overseas has been gratifying. I am particularly pleased that students have sent letters of appreciation, some of which contained suggestions for improvement. As I prepared this edition, I talked with first year students at every university I visited. All comments received careful consideration, with the result that this edition should be more helpful to students. Many students requested that a separate chapter giving a *summary of the cranial nerves* appear in this edition, similar to the one that was in the first edition. This has resulted in repetition of some facts and clinical comments, but these summaries emphasize the main points of clinical interest in the cranial nerves.

The preparation of a gross anatomy book of general suitability is difficult because the teaching of gross anatomy varies according to the objectives of the course. However, this book appears to meet the requirements of most students in acquiring a basic knowledge of the structure of the human body, and some appreciation of the practical applications of this knowledge. Because *textbooks are now an integral part of teaching programs*, I have attempted to simplify, correlate, explain, integrate, and describe the material in such a way that beginning students will understand it.

Because the recital of anatomical facts means little to students unless the practical relevance is made apparent, a *deliberate attempt has been made to use the clinical sciences to facilitate the understanding of anatomy*. The structures described in this book are those that are likely to be of importance to the general practitioner of medicine, dentistry, or other health professions. The references to the clinical significance of various facts were inserted after much consideration and in realization of the fact that there is room for considerable difference of opinion in these areas. The information presented in the clinical comments (screened in blue) and patient oriented problems is true and complete to the best of my knowledge, but the book is not intended as a guide to medical treatment. The reader is urged to seek the advice of his/her physician or clinical instructors concerning clinical subjects. The clinical materials will be helpful to students in their tutorials and in preparing for problem-based learning sessions.

The success that this book has had is clearly related to the quality of the illustrations, particularly those from *Grant's Atlas*. These superb drawings are often much more efficient than text material in conveying complex anatomical information. They also aid in the recognition of structures in dissected specimens and in living persons during surgery. *In this edition, there are many new illustrations* and others have been improved. The most noticeable change is the color enhancement of the illustrations from *Grant's Atlas*. Photoretouch dyes were used on photographic prints of the halftone illustrations, making it possible to add colors without changing the original drawings. *More emphasis has been placed on sectional anatomy* with the inclusion of more magnetic resonance images (MRIs) and computed tomographic (CT) scans.

Photographs of living persons are also used extensively because they aid the learning of **living anatomy** and serve as *a bridge to physical diagnosis*, the branch of medicine that uses this aspect of anatomy during the examination of patients. It was not surprising to find that students were particularly enthusiastic about the clinical comments in this book because *nothing stimulates students more than the correlation of the anatomy taught with patient oriented problems*. Care has been taken to make the clinical vignettes concise and accurate. The enthusiasm for this book demonstrated by anatomy teachers and clinicians is an honor. My clinical orientation resulted from a 35-year association with many colleagues who assisted me with the medical considerations of the various anatomical regions. It will be obvious to specialists that what is presented is only a small portion of the clinical material that could have been included. This material has been limited to keep the book to a reasonable size. Omission of anatomical facts of importance to certain specialists does not imply that they are not of anatomical significance.

The Third Edition has fewer pages and illustrations (reduced from 1100 to approximately 900) owing to the careful selection of material based on its use for 12 years. The extensive culling of illustrations and revision of the text includes factual changes, as well as those designed to improve clarity and readability. Facts of secondary importance, or subjects not usually covered in Gross Anatomy, are set in intermediate type. In accordance with requests from many students, *colored line drawings face the tables* to illustrate the chief attachments of muscles. The terminology in this book adheres to the internationally accepted *Nomina Anatomica* (6th ed., Churchill Livingstone, 1989) approved by the Twelfth International Congress of Anatomists held in London, England in 1985. In accordance with international agreement, the terminology departs from strict Latin, in some cases by anglicizing terms, or by using direct English translations. Eponyms commonly used clinically appear in parentheses, *e.g.*, **sternal angle** (angle of Louis), to assist students in translating the clinical language that is often used in hospitals and patients' charts. In most cases, the official term is printed in **boldface** as in the above example.

University of Toronto   K.L.M.

Acknowledgments

A book of this size and complexity cannot be accomplished without much help. I renew my sincere thanks to several colleagues in my department, especially: Dr. J.W.A. Duckworth, Dr. C. Smith, and Dr. E.G. Bertram, Professors Emeriti of Anatomy; Dr. D.L. McRae, late Professor Emeritus of Radiology; Dr. B. Liebgott and Dr. I.M. Taylor, Professors of Anatomy and Cell Biology; Dr. K. McCuaig, Dr. P. Stewart, and Dr. M.J. Wiley, Associate Professors of Anatomy and Cell Biology; and Anne Agur, Assistant Professor of Anatomy and Cell Biology; Dr. Ming Lee, Senior Tutor in Anatomy and Cell Biology, and Steve Toussaint, Chief Technician and Prosector.

For their enthusiastic support in providing most of the diagnostic imaging material, I thank Drs. E.L. Lansdown, J. Heslin, W. Kucharzyck, E. Becker, A.M. Arenson, J. Campbell, and D. Armstrong. Of the many external colleagues who made constructive suggestions and contributions for the Third Edition, I thank Dr. T.V.N. Persaud, University of Manitoba; Dr. William Forrest, Queen's University; Dr. Eugene Daniels, McGill University; Dr. Margaret Hines, Ohio State University; Dr. Julian J. Baumel, Creighton University; Dr. David Page, Tufts University; Dr. Tom White, University of Tennessee; Dr. Edmund Crelin, Yale University School of Medicine; Dr. Robb Chase, Stanford University School of Medicine; Dr. Barry T. Hinton, University of Virginia; Dr. Raymond F. Gasser, Louisiana State University; Dr. David Peck, University of Kentucky; Dr. Arthur Dalley, II, Creighton University School of Medicine; Dr. Julian Dwornik, University of South Florida; Dr. Todd Olson, Albert Einstein College of Medicine; Dr. Raphael Poritsky, Case Western Reserve; and Dr. Duane Haines, The University of Mississippi Medical Center. To all these people and to any I may have failed to mention, *I extend my sincere thanks*.

My appreciation is extended again to the medical illustrators and photographers who prepared the illustrations for this book. My sincere thanks to them is expressed by including their names on the title page.

To my secretary and wife, Marion, I extend grateful appreciation for her friendly help and patience with the computer preparation of this edition. Marion also spent many hours discussing the manuscript with me and proofreading it. Her understanding, support, and friendship are unique.

Throughout the text, liberal use has been made of illustrations from other Williams & Wilkins publications, especially *Grant's Atlas*, ed. 9, 1991, which the author and publisher acknowledge with thanks. Illustrations from other texts were also requested and used; these are acknowledged in the text. I am grateful to all these authors.

It has been my pleasure to work with the staff of Williams & Wilkins for 15 years, and for this edition I have had the pleasure of working with: Sara Finnegan, President; John Gardner, Vice President, International Division; Tim Satterfield, Editor-in-Chief, Textbooks and Allied Health Sciences; Vicki Vaughan and Linda Napora, Managing Editors; Kathy Lumpkin, Copy Editor; Deborah Tourtlotte, Indexer; Anne Stewart Seitz, Production Coordinator; Bob Och, Art Director; and Lorraine Wrzosek, Illustration Planner. They have spared no effort in making this edition better than the first two. I should like to thank them for their patient help during the preparation and publication of my present work.

K.L.M.

Contents

Anatomy is the oldest basic medical science. As far as we know, it was first studied formally in Egypt (ca. 500 B.C.). Later it was taught in Greece by **Hippocrates** (460–377 B.C.), who is regarded by many as the *Father of Medicine* and a founder of the science of anatomy. In addition to the *Hippocratic Oath* attributed to him, he wrote several books on anatomy. In one he stated, "The nature of the body is the beginning of medical science." Another Greek physician and scientist **Aristotle** (384–322 B.C.) is credited with being the first person to use the term *anatome*, a Greek word meaning "cutting up or taking apart." The Latin word *dissecare* has a similar meaning. If interested in the history of anatomy, consult Persaud (1984).

Anatomy is the science of the structure and function of the body. *Clinically oriented anatomy* emphasizes structure and function as it relates to the practice of medicine and the other health sciences. For example, a hard fall on the elbow may fracture it and injure the ulnar nerve causing numbness in the fifth digit (little finger). *Clinical anatomy is exciting to learn* because it is of practical importance. Furthermore, there are very few structures that cannot be visualized in living patients by one or another of the relatively new imaging techniques, *e.g.*, ultrasound, computed tomography (CT), and magnetic resonance imaging (MRI).

Cadaver anatomy is the foundation of our knowledge of anatomy. It is the best means we have for reaching our main objective of acquiring knowledge of the structure of the *living body*. Learn to make and trust your own observations. During dissection observe, feel, move, and dissect the parts of the body. Dissection is a well-established method of investigation, and it is a stimulating way of learning anatomy if you know about the clinical importance of the structures you are dissecting. To learn this, read about the structures in your text *before* you go to the laboratory.

Anatomy is concerned with the study of living human beings and cannot be learned completely by studying the bodies of dead persons. Much can be learned by observing the surface of the body. The clinical use of **surface anatomy and living anatomy** begins when the physician, dentist, or other health professional first examines a patient.

The aim of surface anatomy is the visualization in the "mind's eye" of structures that lie beneath the skin and are hidden by it. For example, in patients with stab or **gunshot wounds,** the physician must visualize the structures that might have been injured beneath the wound. Surface anatomy is the basis for the physical examination of the body that forms a part of physical diagnosis. A thorough study of a patient's body, with emphasis on the area of complaint, is most helpful in making the correct diagnosis of the anatomical basis for the complaint, *e.g.*, chest pain or loss of sensation in the digits (fingers).

Palpation (examining the body with the hands) is a clinical technique you will use when studying living anatomy. Palpation of arterial pulses is part of every routine evaluation of a patient. You will learn to use various instruments to observe parts of the body (*e.g.*, the eyes using an *ophthalmoscope*) or to listen to the functioning parts of the body (*e.g.*, the heart and lungs using a *stethoscope*). You will also learn to use a *reflex hammer* for examining the functional state of nerves and muscles. Try hard to associate living anatomy with the anatomy that you learn in the laboratory. If you do this, your own body will provide useful memory aids.

When x-rays were discovered in 1895 by Wilhelm **Roëntgen**, observations were made on the living structure of the human skeletal system. Bones and joints were readily visualized on radiographs (roentgenograms or x-ray films). Later, by introducing radiopaque or radiolucent substances, the form and function of various organs and cavities could be studied (*e.g.*, the esophagus during swallowing). **Radiological anatomy** is the study of the structure and function of the body using radiographical techniques. It is an important part of anatomy and is the anatomical basis of radiology, the branch of medical science dealing with the use of radiant energy in the diagnosis and treatment of disease. The sooner you learn to identify normal structures on radiographs, the easier it will be for you to recognize and understand the changes visible on radiographs that are caused by disease and injury.

Developmental anatomy, which includes *embryology*, is the study of human growth and development. Much can be learned about the structure and function of adults by studying changes that occur during development (Moore, 1988a).

In most gross anatomy courses the body is examined regionally because dissections are best performed in this way. The regions usually designated are: the *thorax* (chest), *abdomen*, *pelvis* and *perineum*, *back*, *lower limb*, *upper limb*, *head*, and *neck*. However, from the functional standpoint, it is more helpful to describe the parts and organs of the body by systems.

The systems of the body are usually grouped together and described as follows:

1. *The integumentary system*: the skin and its appendages (*e.g.*, hair and nails).
2. *The skeletal system*: composed of bones and certain cartilaginous parts (*e.g.*, in the chest and nose).
3. *The articular system*: consisting of joints and their associated bones and ligaments.
4. *The muscular system*: comprising the muscles. Sometimes the muscular and skeletal systems are considered together as the *musculoskeletal system*. As the primary function of this system is locomotion, the muscular system is sometimes considered to be a subdivision of a larger system called the *locomotor system*.
5. *The nervous system*: consisting of the brain and spinal cord (*central nervous system*) and the nerves arising from them (*peripheral nervous system*). Parts of the nervous system concerned

chiefly with the regulation of visceral activity (*e.g.*, of the heart and stomach) are referred to as the *autonomic nervous system*. Because of their close developmental and functional associations, the *special sense organs* (sight, hearing, taste, and smell) are usually described with the nervous system (Barr and Kiernan, 1988). The nervous system is the master system that controls and coordinates the activities of all other systems.

6. *The circulatory system* or vascular system: comprising the heart and blood vessels (arteries, veins, and capillaries), and including the *lymphatic system* composed of lymph nodes and vessels. The cardiovascular system refers to the heart and blood vessels.

7. *The digestive system* (alimentary system): composed of the oral cavity, pharynx, and gut. It extends from the mouth to the anus. Associated with it are glands (*e.g.*, the pancreas and liver). This system is concerned with the assimilation of food.

8. *The respiratory system*: comprising the lungs and a system of tubes by which the air reaches them. It is concerned with the exchange of oxygen and carbon dioxide.

9. *The urinary system*: composed of the kidneys, urinary bladder, and excretory passages. It is concerned with the elimination of waste material in the urine.

10. *The reproductive system* (genital system): comprising various organs that are concerned with reproduction. Because of their close association, especially in the adult male, the urinary and genital systems are often referred to as the *urogenital system*.

11. *The endocrine system*: consisting of ductless glands (*e.g.*, the hypophysis cerebri or pituitary gland), which produce secretions called *hormones* that are carried by the circulatory system to all parts of the body.

For details about the histological structure of these systems, see Cormack (1987).

Medical and Anatomical Terminology

Although you are familiar with the common terms for many parts and regions of the body, you should learn to use the internationally adopted anatomical nomenclature, *e.g.*, use the word *axilla* instead of "armpit" and *clavicle* rather than "collar bone." Despite this, you must learn what the common terms refer to so that you can understand the words your patients may use when they describe their complaints. In addition, you must use terms that they can understand when you explain their medical or dental problems to them.

The terminology used in this book is recommended by the internationally accepted collection of Latin terms called *Nomina Anatomica* (Warwick, 1989). Some terms have been translated into English. Unfortunately the language and jargon used most commonly in hospitals sometimes differs from that used in anatomy courses (Moore, 1988b). As this is often a source of confusion between students and clinicians, commonly confused terms are clarified in this text by putting the clinical designations in parentheses, *e.g.*, *auditory tube* (eustachian tube), *uterine tube* (fallopian tube), and *suprarenal gland* (adrenal gland).

Anatomical terminology is important because it introduces you to a large part of medical terminology. To be understood, you must express yourself clearly, using the proper terms in the correct way. Because most terms are derived from Latin and Greek, *medical language can be difficult at first*, but as you learn the origin of medical terms (Skinner, 1961; Squires, 1986), the words make sense. For example the term *decidua*, used to describe the lining or endometrium of the pregnant uterus, is derived from Latin and means "a falling off." This is appropriate because the endometrium "falls off" or is shed, after the baby is born, just as the leaves of deciduous trees fall after summer.

Clear communication is fundamental in clinical medicine, dentistry, and the allied health sciences. To describe the body clearly and to indicate the position of its parts and organs relative to each other, anatomists around the world have agreed to use the same descriptive terms of position and direction. As clinicians also use these terms, it is important to learn them well. Practice using them so that it will be clear what you mean when you describe parts of the body in patients' histories or during discussions of patients with clinicians.

Eponyms are not used in anatomy because they give no clue as to the type or location of the structure involved. Who would ever guess that Wharton's duct is the submandibular duct (L. *sub*, under, + *mandibula*, jaw), the duct of the *submandibular gland*, one of the salivary glands. In addition, some eponyms are historically inaccurate, *e.g.*, Poupart was not the first anatomist to describe the inguinal ligament (Poupart's ligament). However, because some eponyms are so commonly used by clinicians, they are clarified in this text by putting the eponyms in parentheses, *e.g.*, *submandibular duct* (Wharton's duct).

The Anatomical Position

All descriptions in human anatomy are expressed in relation to the anatomical position (Figs. 1 and 2). This position is adopted

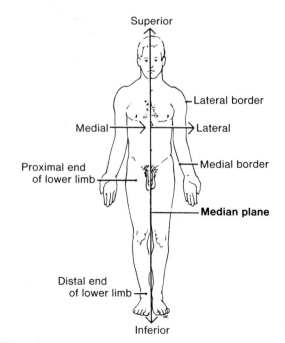

Figure 1. The anatomical position and some descriptive terms.

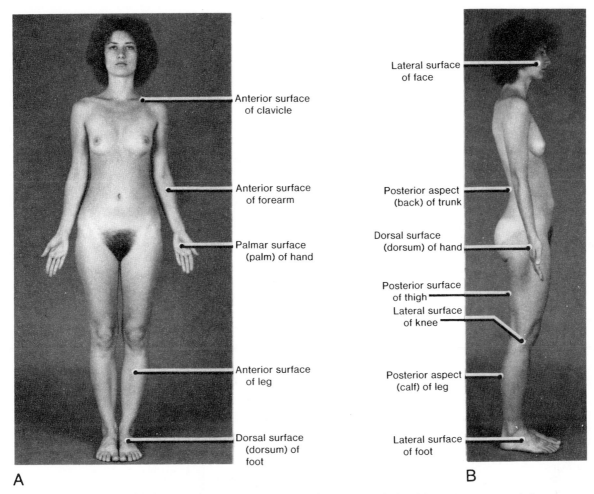

Figure 2. Photographs of a 27–year-old woman demonstrating the anatomical position and some anatomical terms. *A*, Anterior view. *B*, Lateral view.

worldwide for giving anatomical descriptions and must be understood. By using the anatomical position, any part of the body can be related to any other part of it. A person in the anatomical position is standing erect (or lying supine as if erect) with the head, eyes, and toes directed anteriorly (forward), the upper limbs by the sides with the palms facing anteriorly, and the lower limbs together with the digits (toes) pointing anteriorly. *You must always visualize the anatomical position in your "mind's eye" when describing patients* (or cadavers) lying on their backs (the supine position), sides, or fronts (the prone position). Always describe the body *as if* it were in the anatomical position, otherwise confusion as to the meaning of your description may exist and serious consequences could result.

The Anatomical Planes

Anatomical descriptions are also based on four imaginary planes (median, sagittal, coronal, and horizontal) that *pass through the body in the anatomical position* (Fig. 3).

The Median Plane (Figs. 1 and 3).   This is the imaginary vertical plane passing longitudinally through the body from front to back, dividing it into right and left halves. The median plane intersects the surface of the front and back of the body at the *anterior and posterior median lines*, respectively.

The Sagittal Planes (Fig. 3).   These are imaginary vertical planes passing through the body *parallel to the median plane*. These planes are named after the *sagittal suture of the skull*, with which they are parallel. The sagittal plane that passes through the median plane of the body is often referred to as the *median sagittal plane*, or the *midsagittal plane*, but it is preferable to use the official terminology. The sagittal planes that divide the body into right and left portions, but do not pass through the median plane, are sometimes referred to as *parasagittal planes* (G. *para*, beside). However, this terminology is redundant because any plane that is parallel with a sagittal plane is still sagittal. For descriptive purposes, it is helpful to give a point of reference, *e.g.*, a sagittal plane through the midpoint of the clavicle (collar bone).

The Coronal Planes (Figs. 3 and 15).   These are imaginary vertical planes passing through the body *at right angles to the median plane*, dividing it into anterior (front) and posterior (back) portions. These planes are named after the *coronal suture of the skull*, which is in a coronal plane. The coronal plane is also referred to as the *frontal plane*, probably because several of them pass through the forehead, within which is the frontal bone.

The Horizontal Planes (Fig. 3).   These are imaginary planes passing through the body at right angles to both the median and coronal planes. It may help to remember that they are *parallel*

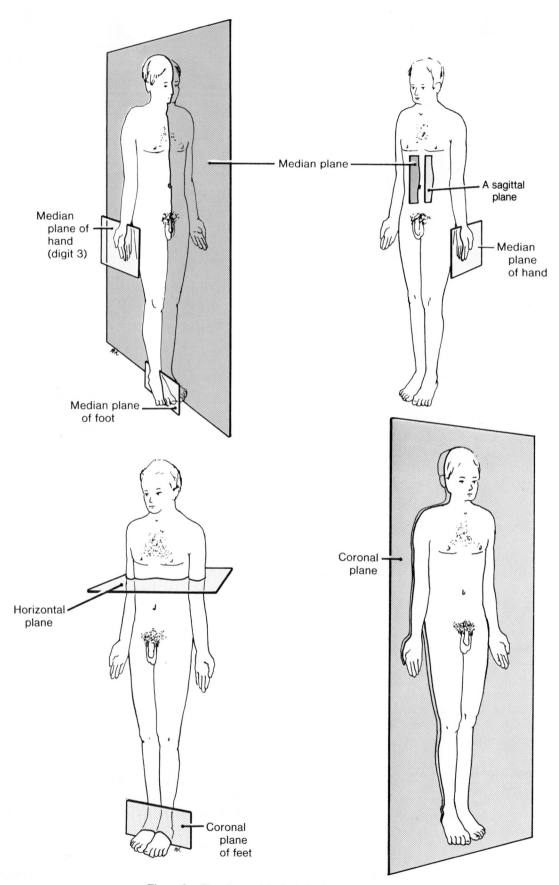

Figure 3.   The planes of the body in the anatomical position.

to the horizon. A horizontal plane divides the body into superior (upper) and inferior (lower) parts. Again it is best to give a reference point, *e.g.*, a horizontal plane through the *umbilicus* (navel). The horizontal plane is also referred to as the *transverse plane*, but such a reference may be erroneous. For example, a transverse section of the hand is in a horizontal plane, but a transverse section of the foot is in the coronal plane.

Sections of the Body

To describe and display many internal structures, slices or sections of the body and its parts are cut in various planes. *Longitudinal sections* (L. *longitudo*, length) run lengthwise in the direction of the long axis of the body, or any of its parts, and they are applicable regardless of the position of the body. Longitudinal sections may be cut in median, sagittal, or coronal planes. *Vertical sections* are the same as longitudinal sagittal sections, except that they are taken through the body, or a part of it, in the anatomical position. *Transverse sections* (Fig. 4*B*) or cross-sections are slices of the body or its parts which are cut at right angles to the longitudinal axis of the body or its parts. *Oblique sections* (Fig. 4*A*) are sections of the body or any of its parts that are not cut in one of the anatomical planes of the body, *i.e.*, they slant or deviate from these planes.

Terms of Relationship

Various adjectives are used to describe the relationship of parts of the body in the anatomical position (Figs. 1 and 2; Table 1).

Anterior means "nearer to the front" of the body, *e.g.*, the nipples and umbilicus are on the anterior surface of the body. Usually the anterior surface of the hand is called the *palmar surface*, or palm (Fig. 2*A*), and the inferior surface of the foot

Table 1.
Commonly Used Terms of Relationship and Comparison

Term	Meaning of the term	Example of its use
Superior (Cranial)	Nearer to the head	The heart is superior to the stomach.
Inferior (Caudal)	Nearer to the feet	The stomach is inferior to the heart.
Anterior (Ventral)	Nearer to the front	The sternum is anterior to the heart.
Posterior (Dorsal)	Nearer to the back	The kidneys are posterior to the intestine.
Medial	Nearer to the median plane	The 5th digit (little finger) is on the medial side of the hand.
Lateral	Farther from the median plane	The 1st digit (thumb) is on the lateral side of the hand.
Proximal	Nearer to the trunk or point of origin (*e.g.*, of a vessel)	The elbow is proximal to the wrist, and the proximal part of an artery is its beginning.
Distal	Farther from the trunk or point of origin (*e.g.*, of a nerve)	The wrist is distal to the elbow and the distal part of the lower limb is the foot.
Superficial	Nearer to or on the surface	The muscles of the arm are superficial to its bone (humerus).
Deep	Farther from the surface	The humerus is deep to the arm muscles.
External (Outer)	Toward or on the exterior	The auricle or pinna is external to the middle ear.
Internal (Inner)	Toward or in the interior	The spiral organ concerned with hearing is internal to the middle ear.
Central	Nearer to or toward the center	The spinal cord is part of the CNS.
Peripheral	Farther or away from the center	The spinal nerves leaving the spinal cord are part of the PNS.
Parietal	Pertaining to the external wall of a body cavity	The parietal pleura forms the external wall of the pleural cavity.
Visceral	Pertaining to the covering of an organ	The visceral pleura covers the external surface of a lung.

is usually called the *plantar surface*, or sole. *Ventral is interchangeable with anterior* and is commonly used in neuroanatomy where it is advantageous because it is equally applicable to humans and the animals that are commonly used in research (Barr and Kiernan, 1988).

Posterior means "nearer to the back" of the body, *e.g.*, the gluteal region (buttock) is on the posterior surface (Fig. 2*B*). *Dorsal is interchangeable with posterior.* When describing the posterior or dorsal surface of the hand or foot, the term *dorsum* is commonly used (Fig. 2*A*).

Superior means "nearer to the head" (Fig. 1), *e.g.*, the heart is superior to the diaphragm. *Cranial* and *cephalic* are interchangeable adjectives. *Rostral* is often used instead of anterior; when describing parts of the brain, it means "nearer to the front

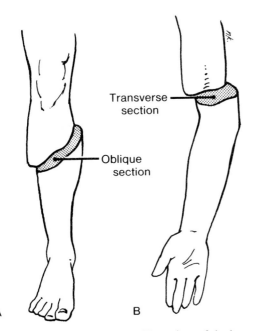

Transverse section

Oblique section

A   B

Figure 4. Oblique (*A*) and transverse (*B*) sections of the lower and upper limbs, respectively.

end,'' *e.g.*, the frontal lobe is rostral to the cerebellum (see Fig. 7-36).

Inferior means "toward the feet" or lower part of the body, *e.g.*, the diaphragm is inferior to the heart. The term *caudal* is a Latin word meaning "tail." It is interchangeable with *inferior* but is commonly used only in descriptions of embryos.

Medial means "toward the median plane" of the body (Fig. 1), *e.g.*, the anterior nares (nostrils) are medial to the eyes. In dentistry the term *mesial* (G. *mesos*, middle) is equivalent to *medial* and means "toward the midline of the dental arch."

Lateral means "farther away from the median plane" of the body (Fig. 1). Hence the fifth digit (little toe) is lateral to the first digit (great toe) and the fifth digit (little finger) is medial to the first digit (thumb).

Intermediate means "between two structures," one of which is medial and the other lateral, *e.g.*, the fourth digit (ring finger) is intermediate between the fifth digit (little) and third digit (middle finger).

Combined Terms. Often terms are combined to indicate a direction; *e.g.*, inferomedial means toward the feet and the median plane. In a posteroanterior (PA) radiograph of the chest (the routine one), a beam of x-rays passes through the thorax from posterior to anterior, *i.e.*, the x-ray tube is posterior to the patient and the x-ray film is anterior.

Terms of Comparison

These terms compare the relative position of two structures with each other.

Proximal means "nearest the trunk" (*e.g.*, a limb) or "point of origin" of a vessel, nerve, or organ. In the limbs, proximal (L. *proximus*, next) is used to indicate positions nearer to the attached end of a limb, *e.g.*, the thigh is at the proximal end of the lower limb.

Distal means "farthest from the trunk" (*e.g.*, a limb) or "point of origin" of a vessel, nerve, or organ. In the limbs, distal (L. *distans*, distant) is used to indicate positions farther from the attached end of a limb, *e.g.*, the foot is at the distal end of the lower limb. *In dentistry, distal* means "farther from the midline of the dental arch" (*i.e.*, farther from the midpoint between the central incisor teeth).

Superficial means "nearer to or on the surface," *e.g.*, the superficial fascia is closer to the skin or surface of the body than the deep fascia. Similarly the scalp is superficial to the calvaria (skullcap).

Deep means "farther from the surface," *e.g.*, in the arm, the bone (humerus) is deep to the muscles, fasciae, and skin.

Internal means "toward or in the interior of an organ or cavity," *e.g.*, the internal surface of the urinary bladder is the interior of the organ. The term is also used to describe structures that pass from the anterior to the posterior surface of the body or that enclose other structures. Hence, the internal surface of a rib is the surface toward the interior, and the internal carotid artery passes to the interior of the skull.

External means "toward or on the exterior of an organ or cavity," *e.g.*, the external surface of the urinary bladder is the exterior of the organ. Similarly, the external surface of a rib is the surface closer to the skin and the external carotid artery passes to the exterior of the skull.

Ipsilateral means "on the same side of the body" (L. *ipse*, same, + *lateralis*, side), *e.g.*, the right thumb and great (big) toe are ipsilateral.

Contralateral means "on the opposite side of the body" (L. *contra*, opposite), *e.g.*, the right hand and left hand are contralateral.

Ambiguous Terms.   Do not use the terms on, over, and under because it may not always be clear what you mean, *e.g.*, if you say a structure passes over another one, do you mean anterior, superior, or superficial to it? Terms such as front, back, in front of, behind, forward, backward, upper, lower, above, below, upward, and downward may be ambiguous and are best avoided. They are not ambiguous if they are used only when referring to the body in the anatomical position. In fact we usually use "upper limb" and "lower limb" instead of "superior" and "inferior" limbs because there is no way anyone could be confused about which limb is meant.

Terms of Movement

Anatomy is concerned with the living body. Therefore various terms are used to describe the different movements of the limbs and other parts of the body (Table 2). Movements take place at joints where two or more bones meet or articulate with one another.

Flexion indicates "bending, or making a decreasing angle" between the bones or parts of the body (Fig. 5). Usually the

Table 2.
Commonly Used Terms of Movement

Term	Explanation of term and example of its use
Flexion	*Bending* or decreasing the angle between body parts, *e.g.*, flexing the elbow joint (Fig. 5*C*)
Extension	*Straightening* or increasing the angle between body parts, *e.g.*, extending the knee joint (Fig. 5*D*)
Abduction	*Moving away from the median plane*, *e.g.*, abducting the upper limb (Fig. 6*A*)
Adduction	*Moving toward the median plane*, *e.g.*, adducting the lower limb (Fig. 6*B*)
Rotation	*Moving around the long axis*, *e.g.*, medial and lateral rotation of the lower limb (Fig. 5*F*)
Circumduction	*Circular movement* combining flexion, extension, abduction, and adduction, *e.g.*, circumducting the upper limb (Fig. 7)
Eversion	*Moving the sole of the foot away from the median plane*, *e.g.*, when the lateral surface of the foot is raised
Inversion	*Moving the sole of the foot toward the median plane*, *e.g.*, when you examine the sole of your foot to remove a splinter
Supination	*Rotating the forearm and hand laterally* so that the palm faces anteriorly (Fig. 8), *e.g.*, when a person extends a hand to beg
Pronation	*Rotating the forearm and hand medially* so that the palm faces posteriorly (Fig. 8), *e.g.*, when a person pats a child on the head
Protrusion	*Moving anteriorly*, *e.g.*, sticking the chin out
Retrusion	*Moving posteriorly*, *e.g.*, tucking the chin in

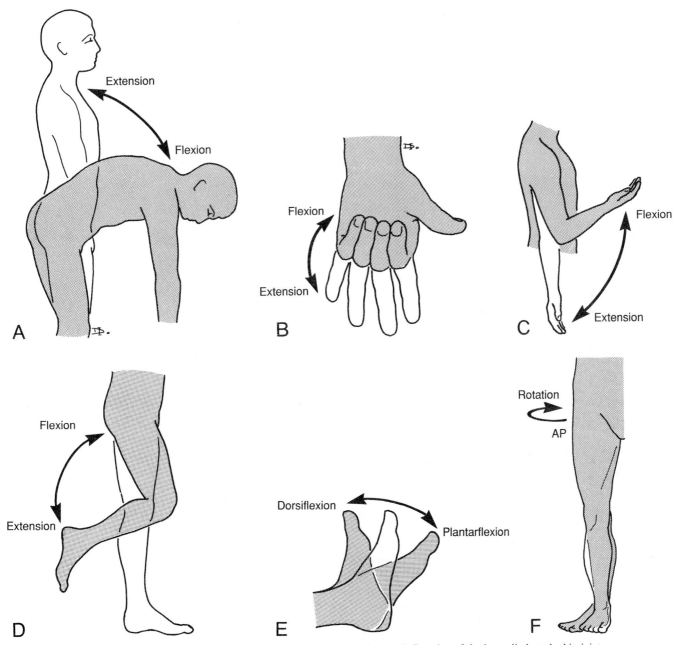

Figure 5. *A* to *E*, Flexion and extension of various parts of the body. *F*, Rotation of the lower limb at the hip joint.

movement is an anterior bending in a sagittal plane, *e.g.*, when bowing (Fig. 5*A*). Understand that flexion of the forearm at the elbow joint is an anterior bending of the limb (Fig. 5*C*), whereas flexion of the leg at the knee joint is a posterior bending of the limb (Fig. 5*D*). To describe flexion of the foot at the ankle joint, as in walking uphill, the term *dorsiflexion* is used (Fig. 5*E*).

Extension indicates ''straightening of a bent part, or making an increasing angle'' between the bones or parts of the body (Fig. 5). Extension usually occurs in the posterior direction, but extension of the leg at the knee joint is in an anterior direction (Fig. 5*D*). Extension of a limb or part beyond the normal limit is referred to as *hyperextension* (overextension). This may cause injury (*e.g.*, hyperextension of the neck, the so-called *whiplash injury*, may occur during a rear end automobile collision). Un-

derstand that when the lower limb is completely extended, the ankle joint is plantarflexed (Fig. 5*E*), *e.g.*, when standing on the tiptoes.

Abduction means ''moving away from the median plane in the coronal plane'' (Fig. 6*A*), *i.e.*, drawing away laterally, *e.g.*, when moving the upper limb away from the body. The Latin prefix *ab-* means ''off'' or ''away from.'' In abduction of the digits (the fingers or toes), the term means spreading them apart, *i.e.*, moving them away from the third digit (middle finger) or median plane or axial line of the hand (see Fig. 6-108).

Adduction is opposite to abduction (Fig. 6*B*), *i.e.*, it means ''moving toward the median plane in a coronal plane,'' *e.g.*, when moving the upper limb toward the body. The Latin prefix *ad-* means ''to'' or ''toward.'' In adduction of the digits (the

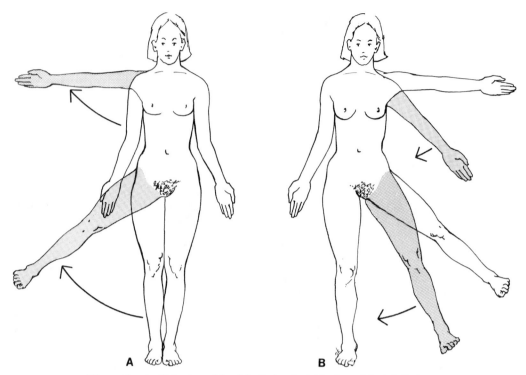

Figure 6. *A*, Abduction of the right limbs. *B*, Adduction of the left limbs.

fingers), the term means moving them toward the median plane or axial line of the hand, *i.e.*, moving the digits toward the third digit (see Fig. 6-108).

Opposition is the "movement by which the first digit (thumb) pad is brought to another digit (finger) pad." We use this movement to hold a pen or to pinch something.

Reposition is the term used to describe the movement of the first digit from the position of opposition back to its anatomical position.

Protrusion is "a movement made anteriorly" (forward) as occurs when protruding the mandible (sticking the chin out). Drawing the shoulders anteriorly is called *protraction*.

Retrusion is "a movement made posteriorly" (backward) as occurs in drawing the mandible posteriorly (tucking the chin in). Drawing the shoulders posteriorly, *e.g.*, squaring the shoulders, is called *retraction*.

Elevation means "raising or moving a part superiorly," *e.g.*, elevating the shoulders as occurs when shrugging them or in raising the upper limbs superior to the shoulders.

Depression means "lowering or moving of a part inferiorly," *e.g.*, when depressing or lowering the shoulders as occurs when standing at ease.

Circumduction (Fig. 7) is a term that is derived from the Latin words *circum*, meaning around, and *duco*, to draw. Circumduction therefore refers to a *circular movement* that is a combination of flexion, extension, abduction, and adduction occurring in such a way that the distal end of the part being moved describes a circle. It can occur at any joint at which the above mentioned four types of movement are possible, *e.g.*, the hip joint.

Rotation involves "turning or revolving a part of the body around its long axis," *e.g.*, rotation of the humerus (arm bone)

at the shoulder joint. *Medial rotation* brings the anterior surface of the limb closer to the median plane, whereas *lateral rotation* takes the anterior surface away from the median plane (Fig. 5F).

Eversion of the foot "moves the sole away from the median plane," *i.e.*, the sole faces inferolaterally.

Inversion of the foot "moves the sole toward the median plane," *i.e.*, the sole faces inferomedially. Eversion and inversion involve three movements at the subtalar joint which are discussed in Chapter 5.

Pronation is the movement of the forearm and hand that rotates the radius medially around its long axis so that the palm of the hand faces posteriorly and its dorsum faces anteriorly (Fig. 8A). When the elbow is flexed at a 90 degree angle, pronation moves the forearm and hand so that the palm of the hand faces inferiorly (*i.e.*, turned downward). During pronation of the forearm and hand, the radius of the forearm crosses the ulna diagonally, moving the thumb medially.

Supination is the movement of the forearm and hand that rotates the radius laterally around its long axis so that the dorsum of the hand faces posteriorly and the palm faces anteriorly (Fig. 8B). The forearms and hands are supinated in the anatomical position (Figs. 1 and 2). When the elbow is flexed at a 90 degree angle, supination moves the forearm and hand so that the palm is turned superiorly (*i.e.*, faces upward).

The movement called "pronation of the foot" is not recognized officially and therefore is not commonly used. When it is, the movement refers to eversion and abduction of the foot which causes a lowering of its medial border.

The Meaning of Terms. As mentioned on p. 2, the official list of terms (*Nomina Anatomica*) is in Latin, but some terms in this book are translated into English, *e.g.*, "muscle" instead of "musculus." However, it is senseless to use English equivalents for some terms, *e.g.*, foramen magnum, which means large opening when anglicized.

Anatomy is a descriptive science, consequently many anatomical terms indicate the shape, size, location, function, or resemblance of a structure to something. Some muscles are given descriptive terms to indicate their main characteristics, *e.g.*, the *deltoid* is triangular-shaped like *delta*, the fourth letter of the Greek alphabet. The suffix *-oid* means "like something." Therefore deltoid means "like delta." Biceps means *two headed* (L. *bi*, two, + *caput*, head) and triceps (L. *tri*, three) *three headed*. Some muscles are named according to their shape; *e.g.*, the *piriformis* has a pear-like form (L. *pirum*, pear). Other muscles are named according to their location; *e.g.*, the *temporalis* muscle is in the *temporal region* of the skull. In some cases actions are used to describe muscles, *e.g.*, the *levator scapulae* elevates the scapula. Thus there are good reasons for the names given to muscles and other parts of the body (Squires, 1986), and if you learn their meanings and think about them as you read and dissect, you should have less difficulty remembering their names.

The Meaning of Abbreviations.   Abbreviations are used for brevity in this book and in clinical charts. The following ones are commonly used. They are easy to understand if you know the words being abbreviated, *e.g.*, " fl. carpi ulnaris m." refers to the flexor carpi ulnaris muscle in the forearm. When it is obvious that the structure is a muscle, the abbreviation is omitted. Frequently you will also see abbreviations on illustrations for arteries and ligaments (*e.g.*, "post. circumflex humeral a." refers to the posterior circumflex humeral artery, and "sup. glenohumeral lig." refers to the superior glenohumeral ligament). As "a." is used for artery, the abbreviation for arteries is "aa." Similarly "v." indicates a vein and "vv." veins; "n." indicates nerve and "nn." nerves. Vertebrae and spinal nerves are designated by region and

number, *e.g.*, the *first lumbar vertebra* is abbreviated to *L1 vertebra*.

Anatomical Variations

The structure of people varies considerably. The bones of the skeleton vary amongst themselves, not only in their basic shape, but also in lesser details of surface structure. There is also a wide variation in the size, shape, and form of the attachment of muscles. Similarly, there is considerable variation in the method of division of nerves and arteries.

In most descriptions of structures you will see the words usually or normally used; this implies that *there is a normal range of variation*. The frequency of variations mentioned in this and other standard textbooks is based on studies of North American people. The frequency of variations often differs among various human groups, and variations collected in one population cannot be assumed to apply to members of another population (Tobias et al., 1988). For example, bifid (forked) ribs occur in 1% to 2% of most populations, but Martin (1960) found this abnormality in radiographs of 8.4% of Samoans. Some variations, such as those in the abdomen, are clinically important and any surgeon operating without a good knowledge of them is sure to have problems.

Textbooks describe the structure of the body that is observed in most people (60% to 70%). Clinically significant variations are mentioned in the "Clinically Oriented Comments" that are printed in intermediate type and screened in blue. For a discussion of the meaning of "normal," see Moore (1989). Bergman et al. (1988) have written a book on human variations.

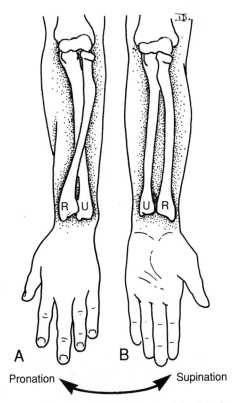

Figure 8.   Pronation (*A*) and supination (*B*) of the right forearm and hand. Note that the positions of the radius (*R*), ulna (*U*), and hands change during pronation.

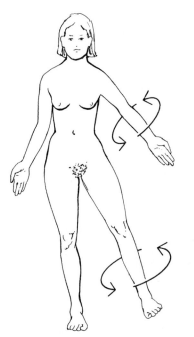

Figure 7.   Circumduction of the left limbs.

Diagnostic Imaging

Anatomical knowledge is essential for interpreting radiographs (''x-rays''), and radiological imaging often provides dynamic information about the structure and function of the body. During your first year and when you begin to practice your profession, you will examine the anatomy of the body in radiographs or other images, *e.g.*, CT (computed tomographic) scans or MRIs (magnetic resonance images), nearly every day. You will see structures much more frequently this way than you will see them displayed during an operation or autopsy. Familiarity with normal radiographs and other images will enable you to recognize abnormalities, *e.g.*, congenital malformations, variations, tumors, and fractures.

Initially, each image on a normal radiograph should be identified on a skeleton or in a dissection. The essence of a radiological examination is that a highly penetrating beam of x-rays ''transilluminates'' the patient, showing tissues of differing densities within the body as images of differing densities on x-ray film (Fig. 9). A tissue or organ that is relatively dense (*e.g.*, compact bone) absorbs more x-rays than a less dense tissue such as fat (Table 3). Consequently, a dense tissue or organ produces a relatively transparent area (erroneously called a white area) on the x-ray film because relatively fewer x-rays reach the silver salt/gelatin emulsion in the film. Therefore relatively fewer grains of silver are developed at this area when the film is processed. A very dense substance is said to be *radiopaque*, whereas a substance of less density is said to be *radiolucent*.

Computerized Tomography (CT) images show sections of the body that resemble transverse sections (Fig. 10). During this process, a small beam of x-rays is passed through the body as the x-ray tube moves in an arc or a circle around the body. The amount of radiation absorbed by each different volume element of the chosen body plane varies with the amount of fat, water density of the tissue, and bone in each element. A multitude of linear energy absorptions is measured and put into a computer. It matches the many linear energy absorptions to each point within the section or plane that is scanned and displays the CT image on a print-out or a cathode ray tube. The CT display of anatomy relates very well to the radiographs, in that areas of great absorption are relatively transparent and those with little absorption are black (Fig. 10). CT has provided greatly improved sensitivity of measurement of x-ray absorption differences. Furthermore, owing to the magnetic storage of CT images, they can be manipulated after they are obtained.

Magnetic Resonance Imaging (MRI) also shows sections of the body that resemble anatomical cross-sections very closely (see Fig. 1-65*A*). *MRIs are obtained without x-rays*. A patient is put in a scanner with a strong magnetic field and is pulsed with radiowaves. The signals emitted from the patient are stored in a computer and reconstructed into various images of the body. In addition to horizontal images, median, sagittal, and coronal images can be produced.

Basic radiological nomenclature should be learned early in your studies. The standard view of the thorax or chest is a **PA**

Table 3.
The Basic Principles of X-ray Image Formation

Most radiolucent	Air	Least radiodense
↑	Fat	↑
	Water and most tissues[1]	
	Cancellous bone	
↓		↓
Least radiolucent	Compact bone	Most radiodense

[1]Includes cytoplasm and uncalcified intercellular substances.

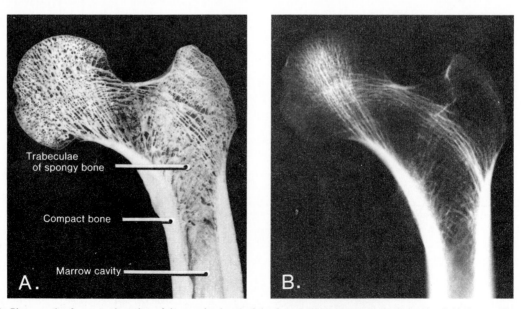

Figure 9.   *A*, Photograph of a coronal section of the proximal end of the femur (thigh bone). *B*, Radiograph of this bone. Observe the tension and pressure lines related to the weight bearing function of this bone.

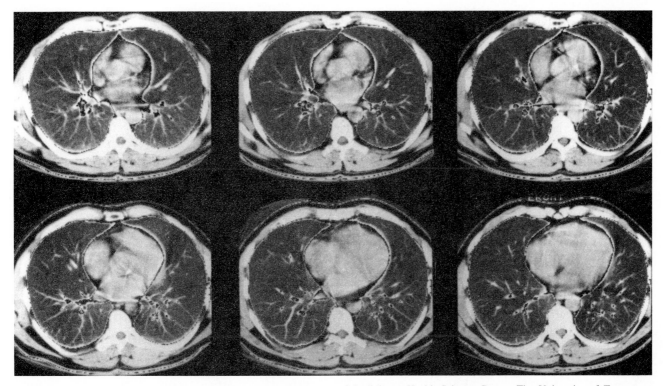

Figure 10. CT (computed tomographic) scans of the thorax showing the heart, pulmonary arteries, and lungs. The thoracic vertebrae and aorta are also clearly shown. (Courtesy of Dr. Tom White, Department of Radiology, Health Science Center, The University of Tennessee, Memphis, TN.)

projection (see Fig. 1-49). It is one in which the x-rays traverse the patient from posterior (*P*) to anterior (*A*). Consequently the x-ray tube is posterior to the patient and the x-ray film is anterior. An *AP projection* is one in which the x-rays traverse the patient from anterior (*A*) to posterior (*P*), indicating that the x-ray tube is anterior to the patient and the x-ray film is posterior. A *lateral projection* (Fig. 13*B*) is made with the patient's body part (*e.g.*, the wrist) close to the x-ray film.

(*e.g.*, by a kick in the shin), bleed when fractured, and change considerably as we get older. Like other organs, bones have blood vessels, lymph vessels, and nerves, and they may become diseased. When broken or fractured, a bone heals. Unused bones, *e.g.*, in a paralyzed limb, *atrophy* (become thinner and weaker). Bone may be absorbed, as occurs in the mandible (jaw) following the extraction of teeth. *Bones also hypertrophy* (become thicker and stronger) when they have increased weight to support for a long period.

Bones and the Skeletal System

Bone, a rigid form of connective tissue, forms most of the skeleton and is the chief supporting tissue of the body. The adult skeletal system or **skeleton** (G. dried) consists of many bones (205 in case your lay friends inquire). A few cartilages are included in it (*e.g.*, the *costal cartilages* connecting the anterior ends of the ribs to the sternum or breastbone). The skeletal system consists of two main parts: (1) the *axial skeleton* consisting of the skull, vertebral column, sternum, and ribs, and (2) the *appendicular skeleton* consisting of the bones of the pectoral (shoulder) and pelvic (hip) girdles and the limbs.

The study of bones is called **osteology** (G. *osteon*, bone, + *logos*, study). Although the bones studied in the laboratory are lifeless and dry, owing to the removal of protein from them, *the bones in our bodies are living organs* that hurt when injured

Types of Bone

There are two main *types of bone*, **spongy bone** (cancellous) and **compact bone** (dense), but there are no sharp boundaries between the two types because the differences between them depend on the relative amount of solid matter and the number and size of the spaces in each of them (Fig. 11). For details on the structure of bone, see Cormack (1987). All bones have an outer shell of compact bone around a central mass of spongy bone, except where the latter is replaced by marrow or a *medullary cavity* (Fig. 11), or an air space, *e.g.*, a *maxillary sinus* in the face (see Fig. 7-14). In the adult there are two types of bone marrow, red and yellow. *Red marrow* is active in blood formation (**hematopoiesis**, G. *hemato*, blood, + *poiein*, to form), whereas yellow marrow is mainly inert and fatty. In most long

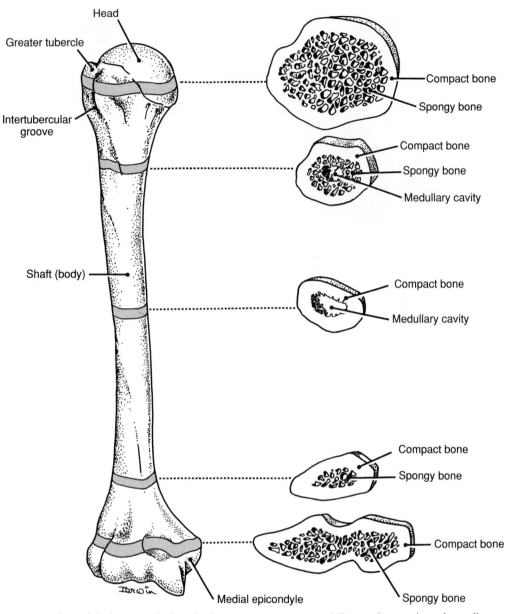

Figure 11. Transverse sections of the humerus, the bone in the arm. Observe that its body (shaft) is a tube of compact bone. During life the marrow or medullary cavity contains red or yellow marrow or a combination of both.

bones there is a medullary cavity in the body (shaft) of the bone that contains *yellow marrow*; in adult life most of the hematopoietic tissue in yellow marrow has been replaced by fat.

Compact bone appears solid except for microscopic spaces (Fig. 11). Its crystalline structure gives it hardness and rigidity and makes it opaque to x-rays (Fig. 9*B*; Table 3).

Classification of Bones

Bones are classified as *axial bones* (skull, vertebrae, ribs, and sternum) or *appendicular bones* (limb bones). They are also classified according to their shape.

1. **Long bones** such as those in our limbs (Figs. 8, 9, and 11) are tubular in shape and have a body (shaft) and two ends, which are either concave or convex. Usually the body is hollow. The length of long bones is greater than their breadth, even though some are short (*e.g.*, in our digits). The ends of long bones are expanded for articulation with other bones.

2. **Short bones** are cuboidal in shape and are found only in the foot and wrist.

3. **Flat bones** or squamous bones usually serve protective or reinforcement functions (*e.g.*, the flat bones of the skull). They consist of two plates of compact bone with spongy bone and marrow between them, *e.g.*, the bones of the *calvaria* (skullcap) and sternum (breastbone). Most flat bones help to form the walls of cavities (*e.g.*, the thoracic and cranial cavities); hence, most of them are gently curved like the ribs.

4. Irregular bones have various shapes (*e.g.*, the facial bones and vertebrae). The bodies of vertebrae have some features of long bones.

5. Sesamoid bones are round or oval nodules that develop in certain tendons (*e.g.*, the *patella*, or kneecap). They were called sesamoid bones because of their resemblance to *sesame seeds* (G. *sesamodes*, like sesame). They are commonly found where tendons cross the ends of long bones in the limbs. They protect the tendon from excessive wear, and they change the angle of the tendon as it passes to its distal attachment. This results in a greater mechanical advantage at the joint.

6. Accessory bones or supernumerary bones develop when additional ossification centers appear and give rise to extra bones. Many bones develop from several centers of ossification and the separate parts normally fuse. Sometimes one of these centers fails to fuse with the main bone, giving the appearance of an extra bone, but careful study can usually show that the apparent extra bone is the missing part of the main bone. Circumscribed areas of bone are often seen in the suture lines of the skull where the flat bones come together (see Fig. 7-3). These wormlike bones are called **sutural bones** (Wormian bones). *Accessory bones are common in the foot*, and it is important to remember this so that they will not be mistaken for bone chips in radiographs and other images. Bones sometimes form in soft tissues where they are not normally present (*e.g.*, in scars). They are called *heterotopic bones*. In the old days cavalrymen often developed these bones in their buttocks and thighs, probably due to hemorrhagic (bloody) areas that underwent calcification and eventual ossification.

Bone Markings and Other Features

The surfaces of bones are not smooth and glossy, except over articular areas such as their heads, which are covered by cartilage (Fig. 11), and where tendons, blood vessels, and nerves move in grooves. Bones display a variety of prominences (bumps), depressions, and foramina (openings). Markings appear wherever tendons, ligaments, and fascia are attached, but the attachment of the fleshy fibers of a muscle make no markings on a bone (Fig. 20). Bone markings become prominent during puberty (around 12 to 15 years of age) and become progressively more marked as adulthood occurs. For descriptive purposes, the various markings and features of bones are given names to distinguish them.

Elevations (Fig. 11). A *linear elevation* is referred to as a *line* or a *ridge*. Prominent ridges are called *crests*. A rounded elevation is called: (1) a *tubercle* (small raised eminence); (2) a *protuberance* (swelling or knob); (3) a *trochanter* (large blunt elevation); (4) a *tuberosity* (large round elevation, *e.g.*, the frontal tuberosity, Fig. 15); and (5) a *malleolus* (a hammerhead-like elevation). A sharp elevation or projecting part is called a *spine* (L. *spina*, a thorn) or a *spinous process* (like those of the vertebrae in the furrow of your back).

Facets (F. little faces). These are small, smooth flat areas or surfaces of a bone, especially where it articulates with another bone. *Articular facets* are covered with hyaline cartilage (*e.g.*, the facets of a vertebra, see Fig. 1-2).

Other Features of Bones (Fig. 11). A rounded articular area of a bone is called a *head* (*e.g.*, the head of the humerus),

or a *condyle* (G. knuckle). An *epicondyle* (G. *epi*, upon) is a prominent process just proximal to a condyle. Small hollows in bones are described as *fossae* (L. pits), whereas long narrow depressions are referred to as *grooves*. An indentation at the edge of a bone is called a *notch*. When a notch is bridged by a ligament or bone it is called a *foramen* and opens into a pit or a *canal* which has an orifice at each end. A *meatus* (L. passage) is a canal that enters a structure but does not pass through it, *e.g.*, the *external acoustic meatus* (ear canal).

Living bones are plastic tissues containing organic and inorganic components. They consist essentially of intercellular material impregnated with mineral substances, mainly hydrated calcium phosphate. The salts, representing about 60% of the weight of a bone, are deposited in the matrix of collagenic fibers. The fibers in the intercellular material give the bones resilience and toughness, whereas the salt crystals give them hardness and some rigidity. Even a bone that is burned retains its shape, but its fibrous tissue is destroyed. As a result, the bone becomes brittle, inelastic, and crumbles easily.

The relative amount of organic to inorganic matter in bones varies with age. Organic matter is greatest in childhood; hence, the bones of children will bend somewhat. In some metabolic disturbances, such as **rickets** and *osteomalacia*, there is inadequate calcification of the matrix of bones. Because calcium provides hardness to bones, the uncalcified areas bend, particularly if they are weight-bearing bones. This results in progressive deformities such as *knock-knees* and *bowlegs*.

Fractures are more common in children than in adults owing to the combination of their slender bones and carefree activities. Fortunately many of these breaks are hairline and *green-stick fractures*. In these fractures, the bone breaks like a willow bough. Fortunately, fractures heal more rapidly in children than in adults. For example, a femoral fracture occurring at birth is united in 3 weeks, whereas union takes up to 20 weeks in persons 20 years of age and older (Salter, 1983).

During old age, especially in women, both the organic and inorganic components of bone decrease, producing a condition called **osteoporosis** (G. *osteon*, bone, + *poros*, pore). There is a *reduction in the quantity of bone* (atrophy of skeletal tissue); as a result, the bones of old people lose their elasticity and fracture easily. Fractures of the neck of the femur (''hip fractures'') are especially common in elderly women because osteoporosis is more severe in them.

During a *sternal puncture* a wide bore needle is inserted through the thin cortical layer of bone of the sternum (breastbone) into the spongy bone. A *sample of red marrow* is aspirated with a syringe for laboratory examination. Study of bone marrow provides valuable information for evaluating many hematological diseases. **Bone marrow transplantation** is sometimes performed in the treatment of leukemia.

Trauma to a bone (*e.g.*, during an accident) may cause it to break. For the fracture to heal properly, the broken ends must be brought together in their normal position. This is called *reduction*. During bone healing, the surrounding fibroblasts (connective tissue cells) proliferate and secrete collagen around the fracture. As this *collar of callus* calcifies, it holds the bones together. Remodeling of bone occurs in the fracture area and the callus calcifies. Eventually the callus is resorbed and replaced by bone. After several months, little evidence of the fracture remains, especially in young people.

Development of Bones

Bones develop from condensations of *mesenchyme* (embryonic connective tissue). The mesenchymal model of a bone which forms during the embryonic period may undergo direct ossification, called *intramembranous ossification* (membranous bone formation), or it may be chondrified (transformed into a cartilaginous bone model). This model later becomes ossified by *intracartilaginous ossification* (endochondral bone formation). Consequently, bone forms by replacing membrane or cartilage. The process of ossification is similar in each case and the final histological structure of the bone is identical (Cormack, 1987).

Ossification of Long Bones (Fig. 12). The first indication of ossification in the cartilaginous model of a long bone is visible near the center of the future body (shaft). This is called the *primary ossification center*. Primary centers appear at different times in different developing bones, but most primary centers appear between the 7th and 12th weeks of prenatal life. Virtually all primary centers are present at birth. By this time, ossification from the primary center has almost reached the ends of the cartilage model of the long bone. The body, which is ossified from a primary center, is called the *diaphysis*.

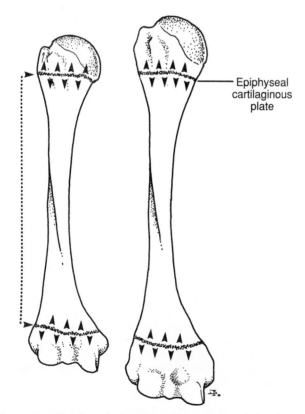

Figure 12. Sketches of a long bone, illustrating how it lengthens at each end. Growth occurs on both sides of the epiphyseal cartilaginous plates, as indicated by the arrows.

Epiphyseal cartilaginous plate

Most *secondary ossification centers* appear after birth. The parts of a bone ossified from these centers are called *epiphyses* (G. *epi*, upon, + *physis*, growth). The epiphyses of the bones at the knee are the first to appear; they may be present at birth. The cartilaginous epiphyses undergo much the same changes that occur in the diaphysis (Fig. 12). As a result, the body of the bone becomes capped at each end by the epiphyses. The part of the diaphysis nearest the epiphysis is referred to as the *metaphysis*. The diaphysis grows in length by proliferation of cartilage at the metaphysis. To enable growth in length to continue until the adult length of a bone is attained, the bone formed from the primary center in the diaphysis does not fuse with that formed from the secondary centers in the epiphyses until the adult size of the bone is reached. During the growth of a bone, a plate of cartilage known as the **epiphyseal cartilaginous plate** (epiphyseal plate or disc) intervenes between the diaphysis and epiphyses.

The diaphysis consists of a hollow tube of compact bone surrounding the medullary cavity, whereas the epiphyses and metaphyses consist of spongy bone covered by a thin layer of compact bone (Fig. 11). During the first two postnatal years, secondary ossification centers appear in the epiphyses that are exposed to weight and pressure (*e.g.*, at the knee and hip joints). These centers, referred to as *pressure epiphyses*, are located at the ends of long bones where they are subjected to pressure from opposing bones. Some centers ossify parts of bone associated with the attachment of muscles and tendons. These centers, subjected to traction rather than pressure, are referred to as *traction epiphyses*, *e.g.*, the tubercles of the humerus (Fig. 11).

The epiphyseal cartilaginous plates are eventually replaced by bone at each of its two sides, diaphyseal and epiphyseal. When this occurs, growth of the bone ceases and the diaphysis fuses with the epiphyses, a process called *synostosis* (G. *syn*, with, + *osteon*, bone, + *osis*, condition). The bone formed at the site of this plate is particularly dense and is recognizable on a radiograph as an *epiphyseal line* (Fig. 14). There are tables in textbooks of radiology and pediatrics that show the expected times of appearance and fusion of various ossification centers (*e.g.*, Behrman 1992).

Ossification of Short Bones. The development of short bones is similar to that of the primary ossification center of long bones, and only one bone, the calcaneus (heel bone), develops a secondary center of ossification.

Blood and Lymph Vessels of Bones

Arterial Supply of Bones (Fig. 14). Bones are richly supplied with arteries. The vessels pass into them from the *periosteum*, the fibrous connective tissue membrane investing them. *Periosteal arteries* enter the body of a bone at numerous points and supply the compact bone; they are responsible for its nourishment. Consequently, a bone from which the periosteum has been removed will die. Near the center of the body, a *nutrient artery* passes obliquely through the compact bone and divides into longitudinally directed branches. They supply the spongy bone and bone marrow; blood vessels are especially rich in those parts that contain *red bone marrow*.

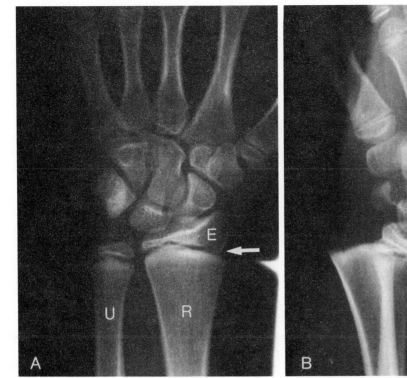
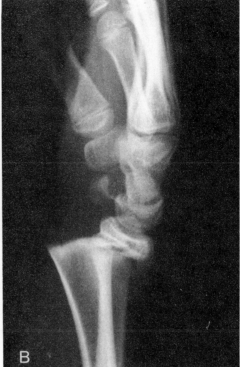

Figure 13. Radiographs of the wrist of a child. *R*, indicates the radius; *U*, ulna; *E*, epiphysis at the distal end of the radius. The *arrow* indicates the site of the *epiphyseal cartilaginous plate*. *A*, Frontal projection of the left wrist showing the normal position of the epiphyses at the distal ends of the radius and ulna. *B*, Lateral projection of the right wrist showing dorsal displacement of the epiphysis at the distal end of the radius.

Metaphyseal and epiphyseal arteries supply the ends of bones. They arise mainly from the arteries that supply the associated joint. In growing bones these arteries are separated by the epiphyseal cartilaginous plates. These arteries are important because they supply these *growth plates*. Consequently any prolonged disturbance of blood flow in them affects bone growth. Examination of the articular ends of dried bones reveals many *vascular foramina* through which the arteries pass.

Venous and Lymphatic Drainage of Bones. Veins accompany the arteries, and many of the large ones leave through foramina near the articular ends of the bones. Long bones and others containing red bone marrow have numerous large veins. *Lymph vessels* are abundant in the periosteum.

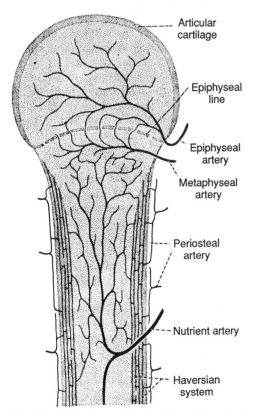

Articular cartilage

Epiphyseal line

Epiphyseal artery

Metaphyseal artery

Periosteal artery

Nutrient artery

Haversian system

Figure 14. The proximal part of a long bone illustrating its arterial supply. Note that the epiphysis and metaphysis have their own arteries.

The developmental changes in bones are clinically important. A general index of growth status during infancy, childhood, and adolescence is indicated by **bone age**, as determined by radiographs. Radiologists can determine the age of a person by studying the *ossification centers*. Two main criteria are used: (1) the appearance of calcified material in the diaphysis and/or epiphyses (the time of which is specific for each epiphysis and diaphysis of each bone for each sex), and (2) the disappearance of the dark line representing the epiphyseal cartilaginous plate. Loss of this line indicates that the epiphysis has fused with the diaphysis; it occurs at specific times for each epiphysis. Fusion of the epiphyses with the diaphysis occurs 1 to 2 years earlier in girls than in boys. *Determination of bone age* is also used in establishing the approximate age of human skeletal remains in medicolegal cases.

Some diseases speed up ossification times as compared to the chronological age of the individual; other diseases slow them down. The growing skeleton is sensitive to relatively slight and transient illnesses and to periods of malnutrition. Proliferation of cartilage at the metaphysis slows down during starvation and illness, but degeneration of cartilage cells in the columns continues, producing a dense line of provisional calcification. These lines later become bone with thickened trabeculae; they are called *lines of arrested growth*.

Without a basic knowledge of bone growth and the appearance of bones on radiographs and in other images at various ages, *an epiphyseal cartilaginous plate could be mistaken for a fracture*, and separation of an epiphysis could be interpreted as a displaced piece of bone (Fig. 13). If you know the age of a patient and the location of the epiphyses, these errors can be avoided, especially if you note that the edges of the diaphysis and epiphysis are smoothly curved in the region of the epiphyseal cartilaginous plate. **Bone fractures** always leave a sharp, often uneven edge of bone. An injury that causes a fracture in an adult usually causes displacement of an epiphysis in a child.

The loss of the arterial supply to an epiphysis or other parts of a bone results in death of bone tissue, a condition referred to as **avascular necrosis** (ischemic or aseptic necrosis). After every fracture, small areas of adjacent bone undergo necrosis. In some fractures, there may be avascular necrosis of a large fragment of bone. A number of clinical disorders of epiphyses in children result from avascular necrosis of unknown etiology (cause). They are referred to as *osteochondrosis* (Salter, 1983).

Innervation of Bones

Many nerve fibers accompany the blood vessels supplying bones. The periosteum of bones is rich in sensory nerves, called periosteal nerves. Some of these sensory fibers are pain fibers. **Vasomotor nerves** are ones which cause constriction or dilation of vessels.

The periosteum is especially sensitive to tearing or tension; this explains why *pain from broken bones is usually severe.* The nerves that accompany the arteries inside the bones are mostly vasomotor nerves. Drilling into compact bone without anesthesia (*e.g.*, for the insertion of screws to hold a metal plate) may cause dull pain (an aching sensation), but drilling into spongy bone is usually very painful.

Architecture of Bones

The structure of a bone varies according to its function. In long bones designed for rigidity, weight bearing, and attachment of muscles and ligaments, the amount of compact bone is relatively greatest near the middle of the body (Fig. 11) where it is liable to buckle. Compact bone provides strength for weight bearing. In addition, as described previously, long bones have elevations (lines, ridges, crests, tubercles, and tuberosities) that serve as buttresses where heavy muscles attach. Living bones have some elasticity (flexibility) and great rigidity (hardness).

Functions of Bones

The five main functions of bones are to provide:

1. *Protection of vital structures*, *i.e.*, by forming rigid walls of cavities that contain vital structures (*e.g.*, the ribs in the thoracic wall that protect the heart and lungs).
2. *Support for the body, e.g.*, the vertebral column forms the structural framework for the trunk.
3. *A mechanical basis for movement*, *i.e.*, by providing attachments for muscles and providing levers for other bones that produce the movements (*e.g.*, the transverse and spinous processes on vertebrae; p. 35).
4. *Blood cells*, *i.e.*, the red bone marrow in the ends of long bones, in the sternum and ribs, in the vertebrae, and in the diploë of the flat bones of the skull (Figs. 11 and 15) are sites for the *development of red blood cells*, some lymphocytes, granulocytic white cells, and platelets of the blood. For more information, see Cormack (1987).
5. *Storage for salts*, *i.e.*, the calcium, phosphorus, and magnesium salts in bones provide a mineral reservoir for the body.

Joints (Articulations)

The articular system consists of the joints (articulations) and their associated bones and ligaments. The study of joints is called **arthrology** (G. *arthros*, joint, + *logos*, study). *Arthritis* (G. *itis*) means inflammation of joints, and *arthroscopy* is the examination of their interiors.

Classification of Joints

Joints are classified according to the type of material uniting or binding the articulating bones. There are three types of joint: *fibrous joints*, united by fibrous tissue; *cartilaginous joints*, united by cartilage or a combination of cartilage and fibrous tissue; and *synovial joints*, united by cartilage with a synovial membrane enclosing a joint cavity.

Fibrous Joints

As stated, the bones in these articulations are united by fibrous tissue. The amount of movement occurring at a fibrous joint (*synarthrosis*) depends on the length of the fibers uniting the bones.

Sutures (Figs. 15 and 16*A*).   Although separated, the bones are held together by several layers of strong connective tissue. The union of the articulating surfaces is extremely tight and there is little movement between the bones. *Sutures occur only in the skull*; hence, they are sometimes referred to as "skull-type" joints. The edges of the bones may overlap (*squamous-type*) or interlock in a jigsaw fashion (*serrate-type*). In the skull of a newborn infant, the bones of the *calvaria* (skullcap) do not make full contact with each other. At these places, the sutures form wide areas of fibrous tissue ("soft spots") called *fontanelles* or fonticuli. The **anterior fontanelle** is the most prominent; most lay people call it the baby's "*soft spot.*" Fusion of the bones across the suture lines (*synostosis*) begins on the internal aspect of the calvaria (calvarium is incorrect) during a person's early

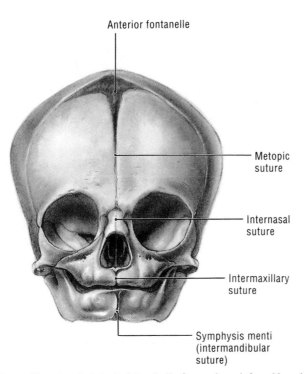

Anterior fontanelle

Metopic
suture

Internasal
suture

Intermaxillary
suture

Symphysis menti
(intermandibular
suture)

Figure 15. Anterior view of the skull of a newborn infant. Note the small size of the face compared with the calvaria that houses the brain. The orbits (eye sockets) are also large at this stage of development.

20s and progresses throughout life. Nearly all evidence of the skull sutures is obliterated in very old people.

Syndesmosis (G. *syndesmos*, ligament). In this type of fibrous joint, the bones are united by a sheet of fibrous tissue. The tissue may be a ligament or a fibrous membrane; *e.g.*, an *interosseous membrane* connects the interosseous border of the radius to this border of the ulna (Fig. 16*B*). At a syndesmosis, slight to considerable movement can be achieved. The degree of movement depends on the distance between the bones and the degree of flexibility of the uniting fibrous tissue. The interosseous membrane between the bones in the forearm is broad enough and sufficiently flexible to allow considerable movement, such as occurs during pronation and supination of the forearm and hand (Fig. 8).

Gomphosis (G. *gomphos*, bolt, + *osis*, condition). This is a unique joint between a tooth and the bone in its alveolus (socket) where the fibrous tissue of the *periodontal ligament* firmly anchors the tooth. Movement at this joint (*i.e.*, a loose tooth) is a pathological condition.

Cartilaginous Joints

There are two types of cartilaginous joint; bones are united by either hyaline cartilage or fibrocartilage.

Primary Cartilaginous Joints (Synchondroses, Hyaline Cartilaginous Joints). The bones are united by hyaline cartilage, which permits slight bending during early life. This type of joint is usually a temporary union, *e.g.*, during the development of a long bone. As described on p. 14, an *epiphyseal cartilaginous plate* separates the ends (epiphyses) and the body

(diaphysis) of a bone (Fig. 16*C*, top). Primary cartilaginous joints permit growth in the length of a bone (Fig. 12). When full growth is achieved, the cartilage is converted into bone and the epiphyses fuse with the diaphysis. Some synchondroses are permanent, *e.g.*, where the *costal cartilage* of the first rib attaches to the sternum (Fig. 16*C*, middle).

Secondary Cartilaginous Joints (Symphyses, Fibrocartilaginous Joints). The surfaces of the articulating bones are covered with hyaline cartilage and the bones are united by strong fibrous tissue and/or fibrocartilage. These are strong, slightly movable joints. The joints between the vertebral bodies in the vertebral column (spine), where the bones are united by *fibrocartilaginous intervertebral discs*, are secondary cartilaginous joints (Figs. 16*C* and 1-23*A*). They are designed for strength and shock absorption. Cumulatively, these joints give considerable flexibility to the vertebral column (Fig. 5*A*). Another example of a secondary cartilaginous joint is the *pubic symphysis* between the bodies of the pubic bones (see Fig. 3-1). The *manubriosternal joint* between the manubrium and body of the sternum is also a secondary cartilaginous joint (see Fig. 1–6). In the newborn infant there is a secondary cartilaginous joint between the right and left halves of the mandible (lower jaw), called the *mandibular symphysis* (symphysis menti). It disappears when the two parts of the mandible (lower jaw) ossify and fuse to form one bone during infancy. Its former site is visible in adults (see Fig. 7-2).

Synovial Joints

Synovial joints are the most common and important type functionally (Fig. 17). They normally provide free movement between the bones they join, and they are typical of nearly all the joints of the limbs (*e.g.*, the shoulder and hip joints). They are called synovial joints because they contain a lubricating substance called *synovial fluid* and are lined with a *synovial membrane* or capsule. Injury to synovial joints (*e.g.*, torn ligaments and cartilages) is common in sports at all ages and in both sexes (Griffith, 1986).

The Three Distinguishing Features of a Synovial Joint (Fig. 17). Synovial joints have: (1) a *joint cavity*, (2) an *articular cartilage*, and (3) an *articular capsule* (fibrous capsule lined with a synovial membrane). They are usually reinforced by *accessory ligaments* which are either separate or attached to the articular capsule. Friction between the bones is reduced to a minimum in synovial joints because the articular surfaces are covered with a thin layer of *articular cartilage*, which is lubricated by viscous *synovial fluid*.

The Articular Cartilage (Fig. 17). This cartilage is usually of the hyaline type, although the matrix contains many collagenous fibers. This cartilage has no nerves or blood vessels; it is nourished by the synovial fluid covering its free surface. The nutrients in the synovial fluid come from the capillaries in the synovial membrane.

The Articular Capsule (Fig. 17). The articular capsule envelops the articulation or joint. It consists of two parts: a **fibrous capsule** and a **synovial membrane** (capsule). When the term *joint capsule* is used, it is usually its fibrous part that is meant. The synovial membrane is a vascular connective tissue

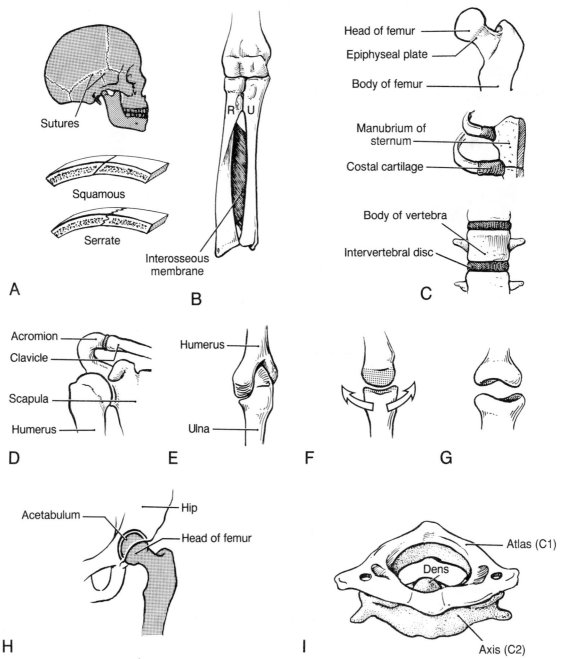

Figure 16. The various types of joint. *A* and *B*, Fibrous joints. *C*, Cartilaginous joints. The upper two are primary cartilaginous joints (synchondroses) and the lower one is a secondary cartilaginous joint (symphysis). *D*, A plane joint. *E*, A hinge joint. *F*, A condyloid joint. *G*, A saddle joint. *H*, A ball and socket joint. *I*, A pivot joint.

membrane that lines the entire joint cavity, although it does not cover the articular cartilage. The *synovial membrane produces synovial fluid* that lubricates the joint. It *regenerates if damaged.* For details on the histological structure of the synovial membrane, see Cormack (1987). Articular capsules are usually strengthened by accessory ligaments which are either part of their fibrous capsules (*intrinsic ligaments*) or are separated from them (*extrinsic ligaments*). These ligaments limit the movements of the joints in undesirable directions so that they will not be damaged. The articular capsule and its accessory ligaments are important in maintaining the normal relationship between the articulating bones. Severe trauma to a joint results in strained or *torn ligaments*, a common knee injury in contact sports such as football. This produces pain on the medial and/or lateral sides of the knee (Griffith, 1986).

Some synovial joints have other distinguishing features besides the three listed on page 17. The following three additional characteristics are relatively common. *Articular discs* are present in some synovial joints where the articulating surfaces are incongruous (*e.g.*, the articular disc of the wrist joint; see Fig. 6-128). They are usually *fibrocartilaginous pads* that help to protect and hold the bones together. In some cases, however, they

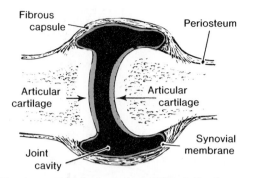

Figure 17. A synovial joint (*e.g.*, the knee joint). For demonstration, the bones have been pulled apart and the articular capsule has been inflated. Normally the joint cavity is a potential space that contains a small amount of fluid. The periosteum, a vascular connective tissue membrane, invests the bones and is closely associated with the fibrous capsule (the fibrous part of the articular capsule). The other part of the capsule is the synovial membrane (shown in red).

are attached to only one of the bones (*e.g.*, the fibrocartilaginous discs or *menisci in the knee joint*; see Fig. 5-92). Articular discs have no nerves except at their attached margins. Some synovial joints have a *fibrocartilaginous ring*, called a *labrum* (L. lip), which deepens the articular surface for one of the bones, *e.g.*, the *glenoid labrum* in the shoulder joint (see Fig. 6-117). In other synovial joints a tendon passes within the capsule of the joint, *e.g.*, the tendon of the long head of the biceps brachii muscle runs within the capsule of the shoulder joint (see Fig. 6-114).

Types of Synovial Joint. There are *six types* of synovial joint. They are classified according to the shape of the articulating surfaces and/or the type of movement they permit.

Plane Joints (Fig. 16*D*). These articulations are numerous and are nearly always small. They permit gliding or sliding movements, *e.g.*, the joint between the acromion of the scapula and the clavicle. The opposed surfaces of the bones are flat or almost flat. Most plane joints move in only one axis; hence, they are called *uniaxial joints*. Movement of plane joints is limited by their tight articular capsules. Injuries to these joints are common (*e.g.*, "**separated shoulder**").

Hinge Joints (Ginglymus). These articulations also move in only one axis (*uniaxial joints*) at right angles to the bones involved (*e.g.*, the elbow, Fig. 16*E*). Hinge joints *permit flexion and extension only*. The articular capsule of these joints is thin and lax where the movement occurs, but the bones are *joined by strong collateral ligaments*.

Condyloid Joints (Fig. 16*F*). These are *biaxial joints* which allow movement in two directions. They have two axes at right angles to each other. Condyloid joints (G. knucklelike) permit flexion and extension, abduction and adduction, and circumduction.

Saddle Joints (Fig. 16*G*). These *biaxial joints* are appropriately named because the opposing surfaces of the bones are shaped like a saddle, *i.e.*, they are concave and convex where they articulate with each other. The carpometacarpal (knuckle) joint of the first digit (thumb) is a good example of a saddle joint.

Ball and Socket Joints (Fig. 16*H*). These articulations are

multiaxial joints that move in several axes. In these highly movable joints, the spheroidal surface of one bone moves within the socket of another (*e.g.*, the hip and shoulder joints). Flexion and extension, abduction and adduction, medial and lateral rotation, and circumduction can occur at ball and socket joints. See Table 2 for an explanation of these movements.

Pivot Joints (Fig. 16*I*). These articulations are uniaxial joints that allow rotation. In these joints, a rounded process of bone rotates within a sleeve or ring (see Fig. 4-34). In the atlanto-axial joint, the fingerlike dens (odontoid process) of the *axis* (C2 vertebra) rotates in a collar formed by the anterior arch of *atlas* (C1 vertebra) and the transverse ligament.

Innervation of Joints

Joints have a rich nerve supply. The nerve endings are located in the articular capsule, both in the fibrous capsule and synovial membrane. The *articular nerves* supplying a joint are branches of the ones that supply the overlying skin and the muscles that move the joint. **Hilton's law** states that the nerves supplying a joint also supply the muscles moving the joint and the skin covering the attachments of these muscles. The main type of sensation from a joint is **proprioception** (L. *proprius*, one's own, + *receptor*, receiver), which provides information concerning the movement and position of the parts of the body. Impulses pass from nerve endings in the articular capsule to the spinal cord and brain (Fig. 18) and act in reflexes concerned with the control of the muscles acting on the joints (Fig. 19). *Pain fibers are numerous in the fibrous capsule* and in its associated ligaments. These sensory endings respond to twisting and stretching such as occurs with distention of the joint with fluid (*e.g.*, "water on the knee" owing to *synovitis* or inflammation of the synovial membrane).

Arterial Supply and Venous Drainage of Joints

Numerous *articular arteries* supply joints. They arise from vessels around the joint (*e.g.*, the *epiphyseal arteries*, Fig. 14). The arteries often communicate or anastomose to form networks, *e.g.*, the anastomoses around the elbow joint (see Fig. 6-55). Diffusion occurs readily from these arteries into the joint cavity. *Veins accompany the arteries* and, similar to them, are present in the articular capsule, especially in the synovial membrane.

Synovial joints are well designed to withstand wear, but heavy use over several years can cause degenerative changes (*e.g.*, in the articular discs). Some destruction is inevitable during normal activities (*e.g.*, jogging), which wear away the articular cartilages and sometimes erode the underlying articulating surfaces of the bones. These degenerative changes usually result in bone and joint inflammation (osteoarthritis) that is often accompanied by stiffness, discomfort, or pain. **Osteoarthritis** is common in old people and usually affects their hip and knee joints, which support the weight of their bodies.

Excessive stretching and twisting of the articular capsule is very painful. *The fibrous capsule is highly sensitive*, but the synovial membrane is relatively insensitive. Compared with pain arising in the skin, joint pain is poorly localized and may be referred to the overlying skin or muscle. There

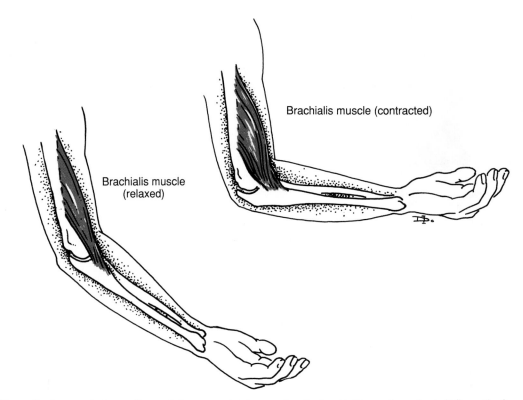

Brachialis muscle (contracted)

Brachialis muscle (relaxed)

Figure 18. The pull of a muscle is usually exerted across a joint, which decreases the angle between the articulating bones. For example, when the brachialis muscle contracts it flexes the forearm at the elbow joint.

may be visceral disturbances (*e.g.*, nausea) associated with joint pain. It is generally believed that changes in temperature, humidity, or pressure cause joints to be more sensitive or painful. Some arthritic people use this phenomenon to predict the weather. It may be that environmental changes reflexively alter the blood flow to the joints.

Most substances in the bloodstream, whether normal or pathological, easily enter the joint cavity. Similarly, traumatic infection of a joint may be followed by **arthritis** (G. *arthron*, joint, + *itis*, inflammation) and *septicemia* (''blood poisoning''). The joint cavity of a synovial joint can be examined by inserting a cannula and a small telescope, called an *arthroscope*, into it. This surgical procedure, called **arthroscopy**, enables an orthopedic surgeon to examine the joint for abnormalities, *e.g.*, torn articular discs. Some surgical procedures can also be performed during arthroscopy (*e.g.*, by inserting instruments through small puncture incisions; Gross, 1989). Because the opening made in the articular capsule for inserting the arthroscope is small, healing is usually more rapid after this procedure than after traditional surgery.

Muscle Tissue and the Muscular System

Contractility is highly developed in muscle tissue. Muscle cells or fibers produce contractions which move parts of the body,

and the associated connective tissue controls the contractions and conveys nerve fibers and capillaries to the muscle cells. Because they are long and narrow when relaxed, muscle cells are commonly called muscle fibers. The muscular system is responsible for movement. *Skeletal muscle* moves bones, *cardiac muscle* pumps blood from the heart, and *smooth muscle* moves substances through hollow viscera (*e.g.*, the intestines) and blood vessels (*e.g.*, the aorta). The muscular system also gives form to the body and provides heat.

Types of Muscle

There are three varieties of muscle: (1) *skeletal muscle* (voluntary) mainly produces movements of the skeleton; (2) *cardiac muscle* (involuntary) forms most of the wall of the heart; and (3) *smooth muscle* (involuntary) forms part of the walls of most vessels and hollow organs (*e.g.*, the stomach).

Skeletal Muscle

This type of muscle is what most people refer to as muscle (''flesh''). They are commonly called *skeletal* because most of them are attached, at least one end, to some part of the skeleton. We have over 600 muscles, and they constitute about 40% of our body weight.

Skeletal muscles are often called voluntary muscles because we can control many of them voluntarily, *e.g.*, we can flex (bend) our elbows at will (Figs. 5*C* and 18). However, many actions of these muscles are automatic, such as those that occur during *stretch reflexes* (Fig. 19). Skeletal muscles produce movement by shortening; *i.e.*, they pull and never push (Fig. 18). *Usually the pull of a muscle is across a joint and draws the articulating bones together*; *i.e.*, it reduces the angle between the bones. The end of a skeletal muscle blends with the structure on which it pulls, *e.g.*, bone or cartilage (Fig. 20). In some cases the muscle tapers into a long tendon of collagenous fibers (*e.g.*, the tendo calcaneus or *Achilles tendon* (see Fig. 5-66), which inserts into the calcaneus (heel bone). In certain regions, muscles are attached by sheet-like tendons called *aponeuroses*; these are membranous expansions of muscle such as those of the flat muscles of the abdomen (see Fig. 2-8). The fleshy part of a muscle is called its belly (Figs. 18 and 20).

Attachments of Skeletal Muscle.   A skeletal muscle has at least two attachments. Some muscles (*e.g.*, facial muscles) are attached to the skin, whereas other muscles (*e.g.*, those of the tongue) are attached to the mucous membrane. A few muscles are attached to *fascia* (a sheet of fibrous tissue), and other muscles form circular bands (*e.g.*, the external anal sphincter; see Fig. 3-63). For purposes of description, most muscles are commonly described as having an *origin* and an *insertion*. However, some muscles may act in both directions under different circumstances. It is therefore best in most cases to use the terms *proximal attachment* and *distal attachment*, instead of origin and insertion. In all colored illustrations of bones in this book, proximal attachments of muscles are shown in red and distal attachments are shown in blue. Skeletal muscles frequently form part of a lever system (Fig. 18), where the power exerted by a muscle depends in part on its angle of pull. The optimum angle of pull is a right angle, and as its angle of pull increases, the power of its pull decreases.

Architecture of Skeletal Muscle.   In some muscles most fibers run parallel to the long axis of the muscle, and a few

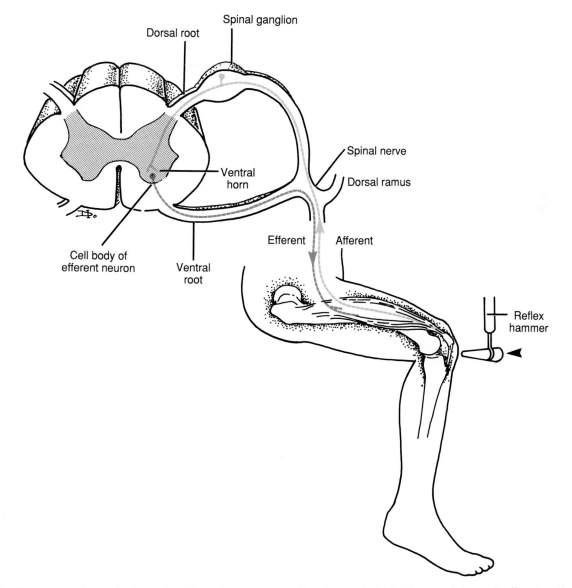

Figure 19.   The stretch reflex, a simple one involving only two neurons. The first neuron is afferent and has its cell body in a spinal ganglion (dorsal root ganglion). The second neuron is efferent and has its cell body in the ventral horn of the spinal cord.

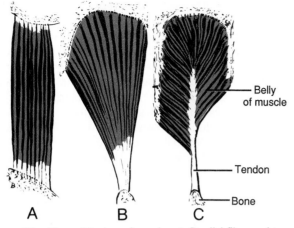

Figure 20. The architecture of muscles. *A,* Parallel fibers end to end. *B,* Nearly parallel fibers (fan-shaped). *C,* Pennate fibers converging on a central tendon. The chief components of a striated skeletal muscle fiber are its fleshy belly (*red*) and fibrous tendon (*white*).

of them run through its whole length. Muscles with their fibers arranged in parallel, or nearly so, can lift a weight through a long distance (Figs. 18 and 20*A*). In other muscles the fibers are oblique to the long axis of the muscle. Because of their resemblance to feathers (Fig. 20*C*), they are called *pennate muscles* (L. *pennatus,* feather). They are known as: (1) *unipennate muscles* when their fibers have a linear or narrow origin resembling half a feather; (2) *bipennate muscles* when their fibers arise from a broad surface resembling a whole feather; (3) *multipennate muscles* when septa extend into the attachments of the muscles, dividing them into several featherlike portions; and (4) *circumpennate muscles* when the fibers converge on a tendon extending into their substance.

In other muscles the fibers converge from a wide attachment to a fibrous apex (Fig. 20*B*). This type of muscle is described as a fan-shaped or *triangular muscle.* On contraction, its fleshy muscle fibers shorten by a third to a half of their resting length. Hence, the muscle appears to "swell," as illustrated by your brachialis muscle when you flex your forearm (Figs. 5*C* and 18).

Actions of Skeletal Muscle. The *structural unit* of a muscle is a muscle cell or fiber. The *functional unit,* consisting of a motor neuron and the muscle fibers it controls, is called a **motor unit**. The number of muscle fibers in a motor unit varies from one to several hundred, but usually there are about 100. The number of fibers varies according to the size and function of the muscle. Large motor units, where one neuron supplies several hundred muscle fibers, are found in the large trunk and thigh muscles (Fig. 19), whereas in small eye and hand muscles, where precision movements are required, the motor units contain only a few muscle fibers.

When a nerve impulse reaches a motor neuron in the spinal cord (Fig. 19), another nerve impulse is initiated, which makes all the muscle fibers supplied by that motor unit contract simultaneously. Movements result from an increasing number of motor units being put into action while antagonistic muscles are relaxing. During maintenance of a given position or posture, the muscles involved are in a *state of reflex contraction*. **Tone** is the state of excitability of the nervous system controlling or

influencing skeletal muscles. The maintenance of tone depends on impulses reaching the brain and spinal cord from sensory endings in the muscles, tendons, and joints (Fig. 19).

During movements of the body, the main muscles are called into action. These muscles, called *prime movers* or *agonists* (G. *agon,* a contest), contract actively to produce the desired movement. A muscle that opposes the action of a prime mover is called an *antagonist*. As a prime mover contracts, the antagonist progressively relaxes; this coordination produces a smooth movement. The antagonists are also called into action at the end of a violent movement to protect the joints from injury (*e.g.,* the hip, knee, and ankle joints during kicking a football).

When a prime mover passes over more than one joint, certain muscles prevent movement of the intervening joint. These kinds of muscle are called *synergists* (G. *syn,* together, + *ergon,* work). Thus synergists complement the action of the prime mover. Other muscles, called *fixators,* steady the proximal parts of a limb (*e.g.,* the forearm) while movements are occurring in distal parts (*e.g.,* the hand). The same muscle may act as a prime mover, antagonist, synergist, or fixator under different conditions.

Cardiac Muscle

Cardiac muscle (**heart muscle**) or *myocardium* forms most of the walls of the heart. Although cardiac muscle is composed of striated fibers (cells), contractions of cardiac muscle are not under voluntary control. The heart rate is regulated by a *pacemaker* composed of special cardiac muscle fibers, which are innervated by the *autonomic nervous system* (Fig. 26; see also Fig. 1-63).

Smooth Muscle

Smooth muscle forms the muscular layers in the *walls of blood vessels* and the digestive tract. Like cardiac muscle, smooth muscle is innervated by the autonomic nervous system (Fig. 26). Hence it is *involuntary muscle* that can undergo partial contraction for long periods. This is important in regulating the size of the lumen of tubular structures (*e.g.,* the intestine and blood vessels). In the walls of the digestive tract, uterine tubes, and ureters, the smooth muscle cells undergo *peristaltic waves* or rhythmic contractions. This process, known as *peristalsis,* propels the contents along tubular structures (*e.g.,* food in the intestine).

Tone is abolished by anesthesia. Consequently, dislocated joints or fractured bones are more easily reduced (placed in their normal positions) when the patient is given an anesthetic agent, a compound that reversibly depresses neuronal function. In some diseases of the nervous system, tone is increased (*spasticity*) or decreased (*flaccidity*).

Muscles depend on a rich supply of blood for oxygen and nutrients and on nerves for carrying impulses to induce contraction. If a muscle is deprived of its innervation, it is said to be denervated, and eventually it will *atrophy* (waste away) as muscle fibers decrease in size. A muscle can also enlarge or *hypertrophy* from constant use (*e.g.,* in weight lifting). In these cases the muscle fibers grow, but the number of fibers remains the same. *Skeletal muscle has limited powers of re-*

generation. When there is massive muscle damage, the muscle tissue is eventually replaced by fibrous scar tissue.

Muscle actions can be tested in various ways. **Muscle testing** is usually performed when nerve injuries are suspected. There are two common methods of determining the status of motor function: (1) the patient performs certain movements against resistance produced by the examiner, and (2) the examiner performs certain movements against resistance produced by the patient. For example, when testing flexion of the forearm (Figs. 5C and 18), the patient is asked to flex their forearm while the examiner resists the effort. The other method is to ask the patient to keep the forearm flexed while the examiner attempts to extend it. The latter technique enables the examiner to gauge the power of the movement.

Electromyography is another method for testing the action of muscles (Basmajian and DeLuca, 1985). Electrodes are inserted in a muscle and the patient is asked to perform certain movements. The differences in electrical action potentials of the muscles are amplified and recorded. A resting normal muscle shows no activity. Using this technique, it is possible to analyze the activity of an individual muscle during different movements. *Electrical stimulation of muscles* may also be used as part of the treatment program for restoring the action of muscles. Heart muscle responds to increased demands by increasing the size of its fibers; this is called *compensatory hypertrophy*. No new cardiac muscle fibers are formed because these cells are unable to divide (*i.e.*, undergo mitosis). When the myocardium (heart muscle) is damaged (*e.g.*, during a **heart attack**), fibrous scar tissue is formed. When this tissue becomes *necrotic* (*i.e.*, dies), the lesion is called a *myocardial infarction* (MI).

Similar to skeletal and cardiac muscle, smooth muscle cells undergo compensatory hypertrophy in response to increased demands. During pregnancy the smooth muscle cells in the wall of the uterus not only increase in size (*hypertrophy*), but they increase in number (*hyperplasia*). Consequently, smooth muscle has a remarkable regenerative capacity.

Blood Vessels and the Cardiovascular System

There are *three types* of blood vessel: *arteries*, *veins*, and *capillaries*.

Arteries

Arteries carry blood away from the heart and distribute it to the body. There are three types of artery: arterioles, muscular arteries, and elastic arteries.

Arterioles (Fig. 21).   These are the *smallest type of artery*. They have relatively narrow lumina (lumens, cavities) and thick muscular walls. The degree of pressure within the arterial system (*arterial pressure*) is mainly regulated by the degree of tonus in the smooth muscle in the walls of the arterioles. If the tonus of

this muscle is increased above normal, *hypertension* (high blood pressure) results.

Muscular Arteries (Fig. 22).   These arteries distribute blood to various parts of the body; because of this, they are often referred to as *distributing arteries*. Their walls consist chiefly of circularly disposed smooth muscle fibers, which constrict their lumina when they contract. As a result, they regulate the flow of blood to different parts of the body as required (*e.g.*, the blood flow to the skeletal muscles in the limbs increases during exercise).

Elastic Arteries (Fig. 23A).   These are the *largest type of artery* in the body (*e.g.*, the aorta has a diameter of about 2 cm). Their walls consist chiefly of *elastin* (yellow, elastic, fibrous mucoprotein). The maintenance of blood pressure within the arterial system between contractions of the heart results from the elasticity of these arteries. This quality allows them to expand when the heart contracts and to return to normal between cardiac contractions. The outermost layer of arteries, known as the *tunica adventitia*, is supplied with oxygen by tiny arteries called *vasa vasorum* (L. vessels of vessels).

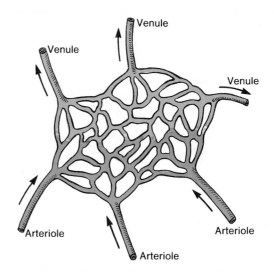

Figure 21.   A capillary bed or network.

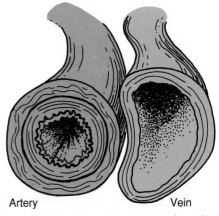

Figure 22.   A muscular artery and its accompanying vein (vena comitans). Note that the wall of the artery is much thicker than the vein and that the lumen of the vein is wider than the artery.

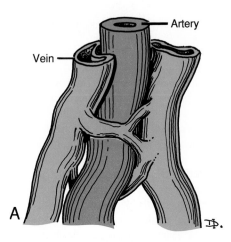

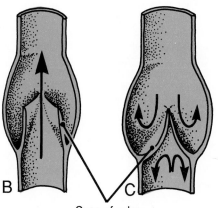

Cusp of valve

Figure 23. *A*, An artery and its accompanying veins (venae comitantes). *B*, A bicuspid valve (two cusps) in the open position. *C*, Valve closed by the back pressure of blood. The *arrows* indicate the direction of blood flow.

The most commonly acquired abnormality of the arteries is an *atherosclerotic lesion*. It results in **atherosclerosis** associated with fibrosis and calcification, a condition called "hardening of the arteries" by many people. Expansion of an atherosclerotic lesion in the tunica intima (innermost coat) of muscular and elastic arteries may result in **thrombosis** (clotting of blood) and occlusion of muscular arteries. The complications of atherosclerosis are *ischemic heart disease* (poor blood supply to the heart), *myocardial infarction* (*MI*), stroke, and *gangrene* (death of tissue), *e.g.*, in parts of the limbs. For a discussion of the origin (pathogenesis) of atherosclerotic lesions, see Cormack (1987).

Veins

Veins return blood to the heart from the capillary beds (Fig. 21); their walls are thinner than those of arteries. Contracting skeletal muscles (*e.g.*, in the "calf" of the leg) compress the veins, "milking" the blood superiorly along the vessels toward the heart. When standing, the venous return from the legs depends largely on this "calf pump" produced by the muscular activity of the posterior leg muscles, commonly called the *calf muscles*. The smallest veins are called **venules** (Fig. 21). These tributaries unite to form larger veins which commonly join to form *venous plexuses* or networks (*e.g.*, the dorsal venous network of the hand, see Fig. 6-87). *Veins tend to be double or multiple*. The veins that accompany muscular arteries, one on each side (Fig. 23), are called *venae comitantes* (L. accompanying veins). Systemic veins are more variable than arteries, and *anastomoses* (communications) occur more often between them.

Many veins contain valves (Fig. 23*B* and *C*) that prevent the backflow of blood. The valves support the columns of blood superior to them (*e.g.*, in the legs) and divide these columns into smaller amounts of less weight. This facilitates the return of blood to the heart and prevents pooling of blood. The muscular actions in active limbs (*e.g.*, in the lower limbs during jogging), assist the venous valves in keeping the blood moving toward the heart. Veins are also supplied by *vasa vasorum* (defined on p. 23). As with arteries, these small *nutrient vessels* supply oxygen to the veins. Veins are also generously supplied with lymph vessels (Fig. 24).

When the walls of veins lose their elasticity, they are weak. A weakened, varicose vein tends to dilate under the pressure of supporting a column of blood against gravity. **Varicose veins** have a caliber greater than normal and their valve cusps do not meet (Fig. 23*C*) or have been destroyed by inflammation. These veins are said to have *incompetent valves* because they are unable to perform their function competently. Varicose veins (dilated, lengthened, and tortuous) are common in the posterior and medial parts of the lower limbs, particularly in the "calves" of the legs.

Capillaries

Capillaries are *simple endothelial tubes* that connect the arterial and venous sides of the circulation (Fig. 21). They are generally arranged in communicating networks called **capillary beds**. The blood flowing through a capillary bed is brought to it by arterioles and is carried away from it by venules. The volume of blood flowing through a capillary bed is under the control of the autonomic nervous system (Fig. 26). As blood is forced through the capillary bed by the hydrostatic pressure in the arterioles, an interchange of materials with the surrounding tissue occurs (*e.g.*, oxygen passes out of the bed and carbon dioxide enters it).

In some regions (*e.g.*, in the digits), there are direct connections between arteries and veins (*i.e.*, there are no capillaries between them). The sites of such communications, called *arteriovenous anastomoses* (**AV shunts**), permit blood to pass directly from the arterial to the venous side of the circulation without having to pass through capillaries. AV shunts are numerous in the skin, where they have an important role in conserving body heat.

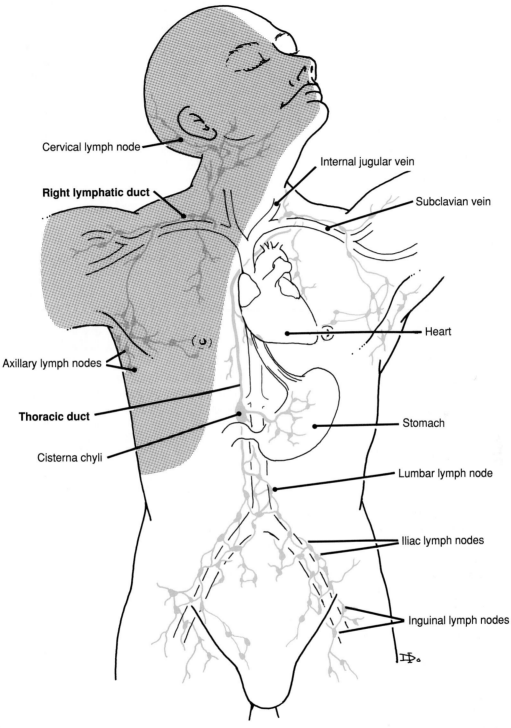

Cervical lymph node

Right lymphatic duct

Internal jugular vein

Subclavian vein

Axillary lymph nodes

Heart

Thoracic duct

Stomach

Cisterna chyli

Lumbar lymph node

Iliac lymph nodes

Inguinal lymph nodes

Figure 24. The lymphatic system. Observe that: (1) the lymphatics of the upper limb enter the *axillary lymph nodes* in the axilla; (2) the lymphatics of the head and neck drain into *cervical lymph nodes* that are mainly grouped around the vessels of the neck (L. *cervix*). All areas except the shaded one are drained by the thoracic duct. It is drained by the right lymphatic duct.

Lymphatics and the Lymphatic System

The lymphatic system is part of the circulatory system (Fig. 24). It consists of a vast network of vessels, called *lymphatics*, that are connected with masses of lymphatic tissue called lymphatic or *lymph nodes* (glands). The lymphatic system also includes *lymphatic organs* (*e.g.*, the spleen). The lymphatic system drains excess fluid from the tissues and provides defense mechanisms for the body. When tissue fluid enters a lymph vessel it is called **lymph**. Excess tissue fluid is filtered through the lymph nodes and returned to the bloodstream.

The lymphatic system consists of: (1) *lymph plexuses*, or networks of very small lymph vessels, called *lymph capillaries*, which begin in the intercellular spaces of most tissues; (2) *lymph nodes* (Fig. 24) composed of small masses of lymphatic tissue arranged along the course of lymph vessels through which lymph passes on its way to the venous system; (3) *aggregations of lymphoid tissue* located in the walls of the alimentary canal (*e.g.*, the tonsils) and in the spleen and thymus; and (4) *circulating lymphocytes*, which are formed in (a) *lymphoid tissue* located throughout the body (*e.g.*, in the lymph nodes and spleen) and (b) in *myeloid tissue* located in red bone marrow.

Lymph Vessels

Most lymph vessels are not visible in dissections, but they can be demonstrated *in vivo* (L. in living persons). *Lymph capillaries* are small vessels that begin blindly in most tissues. They drain lymph from the tissues and are in turn drained by small lymph vessels. They join to form larger and larger collecting vessels or trunks that pass to *regional lymph nodes* (Fig. 24). As a general rule, lymph traverses one or more lymph nodes before it enters the bloodstream.

The lymphatics carrying lymph to lymph nodes are called *afferent lymph vessels*, whereas those leaving a lymph node are called *efferent lymph vessels*. An efferent lymph vessel of one lymph node becomes an afferent lymph vessel of another lymph node in a chain (Fig. 24). After traversing one or more lymph nodes, the lymph enters larger lymph vessels called *lymph trunks*, which unite to form either (1) the **thoracic duct**, which enters the junction of the left internal jugular and left subclavian veins (Fig. 24), or (2) the *right lymphatic duct*, which enters the junction of the right internal jugular and right subclavian veins. In general, the thoracic duct drains lymph from the entire body, except for (1) the right side of the head and neck, (2) the right upper limb, and (3) the right half of the thoracic cavity. The **cisterna chyli** is a dilated sac at the inferior end of the thoracic duct into which the intestinal and lumbar (aortic) lymphatic trunks empty.

Superficial lymph vessels are located in the skin and deep to it, *i.e.*, in the *superficial fascia*. The lymph capillaries run parallel to the superficial vessels of the skin and then join to form slightly larger vessels. These vessels also form a network on the deep surface of epithelia, which line cavities. The superficial lymph vessels eventually drain into deep lymph vessels.

Deep lymph vessels are located in the deep fascia between the muscles and the superficial fascia and accompany the major blood vessels of the region concerned. They have thick walls containing connective tissue and smooth muscle; they also have *valves*.

Lymph Nodes

Lymph nodes are flattened, oval or kidney-shaped structures that can be easily palpated when they are swollen. Many lymph nodes are located in the axilla (armpit) and inguinal region (groin), but there are numerous lymph nodes distributed along the large vessels of the neck and trunk. Lymph nodes consist of *aggregations of lymphatic tissue* and vary from about the size of a pinhead to 2.5 cm or more in diameter. Two of the principal functions of lymph nodes are the production of lymphocytes and antibodies.

Lymph

Lymph (L. *lympha*, clear water) is usually a clear watery fluid, but it may be faintly yellow or slightly opalescent. It is collected by the lymph capillaries from the interstitial fluid compartment (intercellular spaces) in various tissues. *Lymph has the same constituents as blood plasma* (*e.g.*, protein and many lymphocytes). Lymph coming from the intestine also contains fat, fatty acids, glycerol, amino acids, glucose, and other substances (*e.g.*, chemotherapeutic drugs given to a patient for the treatment of cancer).

Functions of the Lymphatics

Drainage of Tissue Fluid. Lymph capillaries are primarily involved with the collection of plasma from the tissue spaces and the transport of plasma to the venous system. On its way, lymph passes through the lymph nodes where particulate matter (*e.g.*, dust inhaled into the lungs) is largely filtered out by phagocytes (G. *phagein*, to eat) or scavenger cells called *macrophages*. In a similar way, bacteria and other microorganisms drained from an infected area are trapped and ingested by the *phagocytes*, thereby preventing them from entering the blood.

Absorption and Transport of Fat. The small lymphatic capillaries, called *lacteals*, which drain the intestine contain considerable amounts of emulsified fat after a fatty meal (see Fig. 2-59). This creamy lymph, called *chyle* (G. *chylos*, juice), after passing through various lymph vessels, is conveyed by the thoracic duct to the left subclavian vein where it enters the blood stream (Fig. 24).

Defense Mechanism of the Body. The lymphatic system provides an important *immune mechanism* for the body. When minute amounts of foreign protein are drained from an infected area by lymphatic capillaries, immunologically competent cells produce a specific **antibody** to the foreign protein, and/or lymphocytes are dispatched to the infected area. The antibody is carried to the area via the bloodstream and the tissue fluid. When foreign tissue is transplanted, it is the **lymphocytes** that reject the graft.

> The lymph vessels and lymph nodes draining an infected area often become inflamed. This is called *lymphangitis* (G. *angeion*, vessel), whereas inflammation of a lymph node is called *lymphadenitis* (G. *aden*, gland). Lay people often refer to inflamed lymph nodes as "swollen glands."
>
> The extensive *spread of cancer cells* via the lymphatic system and their aggregation may block lymph vessels, resulting in *lymphedema* (G. *oidema*, swelling), an accumulation of fluid in the affected region. Malignant cells may also accumulate in the lymph nodes, which causes them to enlarge. Consequently, the lymphatic system is commonly involved in the spread (metastasis) of cancer cells. This is referred to as **lymphogenous dissemination of cancer cells.** Cancer can spread by permeating the lymph vessels as solid cell growths from which minute cellular emboli (G. *embolus*, a plug) may break free and pass to regional lymph nodes (Fig. 24).
>
> Radiological study of lymph vessels and lymph nodes, called *lymphography*, is possible after the cannulation of appropriate peripheral lymph vessels and the injection of a radiopaque contrast material.

Nervous Tissue and the Nervous System

The nervous system is composed of nervous tissue and relatively small amounts of connective tissue. There are basically two types of cells (1) *neurons* (nerve cells), which transmit neural impulses, and (2) *neuroglial cells,* which are supporting cells that provide assistance and nourishment to the neurons. The nervous system provides the mechanism which enables the body to react to continuous changes in its external and internal environments. It also controls and integrates the activities of various parts of the body (*e.g.,* the heart and lungs). For descriptive purposes, the nervous system is divided into central and peripheral parts (Fig. 25).

The Central Nervous System

The central nervous system (**CNS**) consists of the *brain and spinal cord*, which are enclosed in the cranium and vertebral column, respectively. *Cerebrospinal fluid* (**CSF**) and *meninges* (L. membranes) surround the CNS and provide additional protection (Bertram and Moore, 1982). The principal roles of the CNS are to integrate and coordinate incoming and outgoing neural signals and to carry out the higher mental functions such as thinking and learning. A *bundle of nerve fibers* or axons in the CNS, connecting neighboring or distant parts, is called a *tract* (Fig. 25). A collection of nerve cell bodies in the CNS is called a *nucleus.*

The Peripheral Nervous System

The peripheral nervous system (**PNS**) is the peripheral extension of the CNS. It is anatomically and operationally continuous with the brain and spinal cord. *Cranial nerves* arise from the brain and *spinal nerves* emerge from the spinal cord (Fig. 25). The PNS conveys neural impulses to the CNS from the sense organs (*e.g.,* the eyes and ears), and from sensory receptors in various parts of the body (*e.g.,* in muscles, Fig. 19). The PNS also conveys neural impulses from the CNS to the muscles and glands. A bundle of nerve fibers or axons in the PNS is called a **nerve**. Nerves are strong, whitish cords in living persons that are arranged in *fascicles* (L. bundles), which are held together by a connective tissue sheath (Fig. 25C). A network of nerves is called a *nerve plexus* (*e.g.,* the celiac or solar plexus in the superior part of the abdomen), and a collection of nerve cells or *neurons* is called a *ganglion, e.g.,* the spinal ganglion (Fig. 25C).

A peripheral nerve fiber consists of an *axon,* a *myelin sheath* (in some cases), and a *neurilemmal sheath* (of Schwann). The components of all but the smallest peripheral nerves are arranged in bundles or *nerve fasciculi* (Fig. 25C). Peripheral nerves are fairly strong and resilient because the delicate nerve fibers in them are strengthened and protected by three connective tissue coverings. The entire nerve is surrounded by a thick sheath of loose connective tissue, called the *epineurium*. It contains fatty tissue, blood vessels, and lymphatics. A more delicate connective tissue sheath enclosing a fasciculus of nerve fibers, called the *perineurium*, provides an effective barrier to the penetration of foreign substances in or out of the nerve fibers. Individual nerve fibers are surrounded by a delicate covering of connective tissue called the *endoneurium*.

The regions of attachment between neurons or between neurons and effector organs in the CNS and PNS are called **synapses** (G. *synapto*, to join). Synapses are the sites of contact between neurons at which one neuron is excited or inhibited by another neuron. *A typical spinal nerve* arises from the spinal cord by two roots. The *ventral root* contains motor or *efferent fibers* from motor neurons in the ventral horn of the spinal cord (Fig. 25C). The *dorsal root* carries sensory or *afferent fibers* from cells in the spinal (dorsal root) ganglion. The dorsal and ventral roots unite to form a *spinal nerve*, which divides into two branches (L. rami), a dorsal primary ramus and a ventral primary ramus (Fig. 27). The dorsal rami supply nerve fibers to the back, whereas the ventral rami supply nerve fibers to the limbs and anterolateral regions of the trunk. The dorsal and ventral *primary rami of spinal nerves* contain: (1) motor or efferent fibers from ventral horn cells of the spinal cord (Fig. 19); (2) sensory or afferent fibers of spinal ganglion cells (Fig. 25); and (3) autonomic fibers (Fig. 27).

The Autonomic Nervous System

The autonomic nervous system (**ANS**) is a system of nerves and ganglia concerned with the distribution of impulses to the heart, smooth muscle, and glands (Figs. 25B, 26; Table 4). It also receives afferent impulses from these parts of the body. The

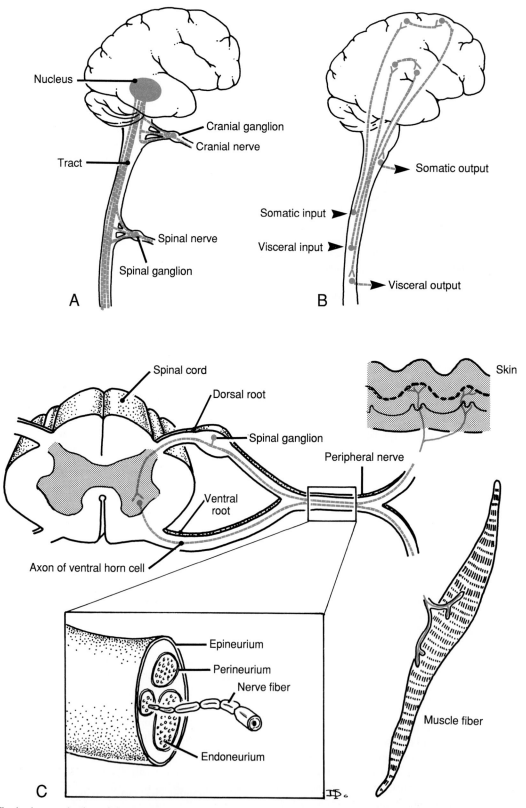

Figure 25. The basic organization of the nervous system. *A*, Shows a *spinal tract* (a collection of nerve fibers) arising from a *nucleus* (a collection of nerve cells) in the brain and passing into the *spinal cord*. The cranial and spinal nerves arise from the brain and spinal cord, respectively. *B*, Somatic and visceral sensory input on the right side and somatic and autonomic output on the left side. *C*, The organization of the peripheral nervous system (PNS).

Table 4.
Functions of the Autonomic Nervous System (ANS)

Tissue/Organ	Sympathetic stimulation	Parasympathetic stimulation
Heart	Increases rate and strength of contraction; heart beats quickly; dilates the coronary arteries, increasing the blood supply to cardiac muscle	Decreases rate and strength of contraction; heart beats slowly; constricts the coronary arteries, decreasing the blood supply to cardiac muscle
Blood vessels		
Skin	Constricts cutaneous vessels	Does not innervate most cutaneous vessels
Skeletal muscle	Dilates vessels in this type of muscle	Vessels in this type of muscle are not innervated
Viscera (except heart and lungs)	Constricts vessels in most viscera	No innervation of vessels in most viscera
Lungs	Dilates the bronchi	Constricts the bronchi
Arrector pili muscles	Contracts these smooth muscles causing the hairs to ''stand on end''	No innervation of these bundles of smooth muscle
Eyes		
Iris	Contracts dilator pupillae muscle and dilates pupil	Contracts smooth muscle sphincter and constricts the pupil
Ciliary muscle	No innervation of these smooth muscle cells	Contracts ciliary muscle and accommodates lens for near vision
Glands		
Sweat	Contraction of myoepithelial cells and expulsion of sweat	No innervation of these glands
Lacrimal	Inhibits secretion of tears	Promotes secretion of tears
Salivary	Decreases the secretion of saliva	Stimulates the secretion of saliva
Gastric	Inhibits the secretion of gastric juices	Stimulates the secretion of gastric juices
Intestinal	Inhibits the secretion of digestive juices	Stimulates the secretion of digestive juices
Suprarenal (adrenal) medulla	Promotes secretion of epinephrine and norepinephrine	No innervation of the suprarenal medulla
Cortex	Promotes glucocorticoid secretion	No innervation of the suprarenal cortex
Liver	Promotes glycogenolysis and decreases secretion of bile	Promotes glycogenesis and increases secretion of bile
Stomach	Decreases motility	Increases motility
Intestine	Decreases motility	Increases motility producing peristalsis
Pancreas	Inhibits the secretion of enzymes	Promotes the secretion of enzymes
Spleen	Contracts spleen causing the discharge of blood into the circulation	No innervation of the spleen
Kidney	Constricts it vessels resulting in decreased urine formation	No innervation of its blood vessels
Urinary bladder	Relaxes smooth muscle in its wall and contracts internal sphincter, resulting in retention of urine	Contracts smooth muscle in its wall and relaxes internal sphincter, resulting in passage of urine
Uterus	Inhibits contraction of the pregnant uterus	Has minimal effect
Other genital organs	Constricts epididymis, ductus deferens, seminal vesicle, and prostate, which results in ejaculation of semen	Dilation of blood vessels resulting in erection of penis and clitoris Causes secretion of fluid from bulbourethral glands in men and greater vestibular glands in women (see Ch. 3)

ANS consists of two parts (Fig. 26): (1) the *sympathetic system* and (2) the *parasympathetic system*, both of which act on structures such as smooth muscle, glands, and blood vessels. The sympathetic system stimulates activities that are performed during emergency and stress situations (*i.e.*, fight, fright, and flight situations) when the heart beats quickly and the blood pressure rises. The parasympathetic system stimulates activities that conserve and restore body resources (*e.g.*, the heart beats slowly). The sympathetic system has connections with the thoracic and lumbar regions of the spinal cord from T1 to L2 or L3 segments. The parasympathetic system has cranial and sacral parts which

are connected with the brain through cranial nerves III, VII, IX, and X and with the spinal cord through spinal nerves S2 to S4 (sometimes S2 is absent).

Some neurons in the sympathetic system correspond to the efferent or motor neurons of the PNS. These efferent neurons are located in: (1) *paravertebral ganglia* or ganglia of the sympathetic trunk (Figs. 26 and 27); (2) *prevertebral ganglia* or visceral ganglia (*e.g.*, the celiac ganglion); and (3) the medulla of the suprarenal gland (adrenal gland). In Figure 27, observe that the axon of a sympathetic neuron sends a preganglionic fiber via the ventral root and the *white ramus communicans* (con-

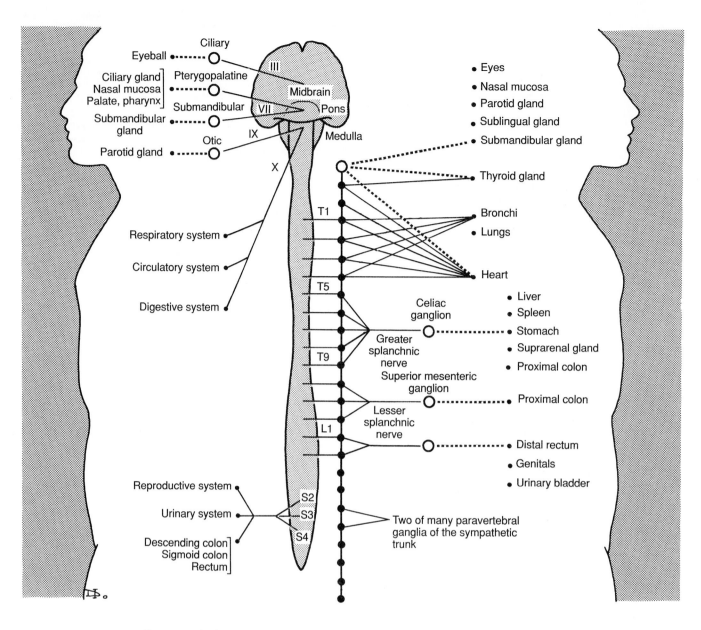

Parasympathetic

Sympathetic

Figure 26. The general plan of the autonomic nervous system (ANS). Note that it consists of two parts, *sympathetic* and *parasympathetic*.

These parts are anatomically separate and are usually functionally reciprocal.

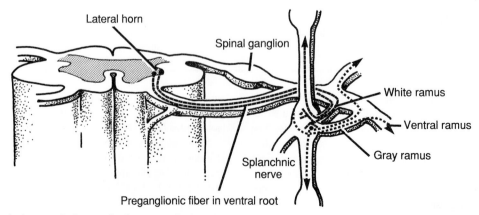

Figure 27. The typical sympathetic contribution to a spinal nerve. Note that the neurons in the lateral horn of gray matter, extending from segments T1 through L2 or L3, are the source of the preganglionic

sympathetic fibers. Observe that these fibers reach the sympathetic trunk by way of the ventral root and the white ramus communicans.

necting branch) to a paravertebral ganglion. Here, it may synapse with excitor neurons, but some *preganglionic fibers* ascend or descend in the sympathetic trunk to synapse at other levels. Other preganglionic fibers pass through the paravertebral ganglia without synapsing to form *splanchnic nerves* to the viscera (*e.g.*, the intestine).

Postganglionic fibers or axons of excitor neurons pass through the *gray ramus communicans* (Fig. 27) and pass from the sympathetic ganglion into the spinal nerve. They run to blood vessels, sweat glands, and the arrector pili muscles associated with hairs. Some *postganglionic fibers* pass superiorly to the cervical sympathetic ganglia and supply structures in the head. Others pass inferiorly to supply the lower limbs. Hence, the *sympathetic trunks* are composed of ascending and descending fibers (preganglionic efferent, postganglionic efferent, and afferent fibers).

SUGGESTED READINGS

Barr ML, Kiernan JA: *The Human Nervous System: An Anatomical Viewpoint*, ed 5. Philadelphia, JB Lippincott, 1988.

Basmajian JV, De Luca CJ: *Muscles Alive: Their Functions Revealed By Electromyography*, ed 5. Baltimore, Williams & Wilkins, 1985.

Behrman RE: *Nelson Textbook of Pediatrics*, ed 14. Philadelphia, WB Saunders, 1992.

Bergman RA, Thompson SA, Afifi AK, Saadeh FA: *Compendium of Human Anatomic Variation: Text, Atlas, and World Literature*. Baltimore, Urban & Schwarzenberg, 1988.

Bertram EG, Moore KL: *An Atlas of the Human Brain and Spinal Cord*. Baltimore, Williams & Wilkins, 1982.

Cormack DH: *Introduction to Histology*. Philadelphia, JB Lippincott, 1984.

Cormack DH: *Ham's Histology*, ed 9. Philadephia, JB Lippincott, 1987.

Griffith HW: *Complete Guide To Sports Injuries*. Los Angeles, Price Stern Sloan, 1986.

Gross AE: Orthopedic surgery: Adult. In Gross A, Gross P, Langer B (Eds): *A Complete Guide for Patients and Their Families*. Toronto, Harper & Collins, 1989.

Martin EJ: Incidence of bifidity and related rib abnormalities in Samoans. *Am J Phys Anthropol*, 18: 179–187, 1960.

Moore KL: *The Developing Human: Clinically Oriented Embryology*, ed 4. Philadelphia, WB Saunders, 1988a.

Moore KL: Anatomical terminology/clinical terminology. Clin Anat 1:7–13, 1988b.

Moore KL: Meaning of ''Normal.'' Clin Anat 2:235–239, 1989.

Persaud TVN: *Early History of Human Anatomy From Antiquity to the Beginning of the Modern Era*. Springfield, Charles C Thomas, 1984.

Salter RB: *Textbook of Disorders and Injuries of the Musculoskeletal System*, ed 2. Baltimore, Williams & Wilkins, 1983.

Skinner HA: *The Origin of Medical Terms*, ed 2. Baltimore, Williams & Wilkins, 1961.

Squires BP: *Basic Terms of Anatomy and Physiology*, ed 2. Toronto, WB Saunders, 1986.

Tobias PV: Some changes in anatomical nomenclature. Clin Anat 3:79, 1990.

Tobias PV, Arnold M, Allan JC: *Man's Anatomy*, ed 4. Johannesburg, Witwatersrand University Press, 1988.

Warwick R: *Nomina Anatomica*, ed 6. Edinburgh, Churchill Livingstone, 1989.

The thorax (chest), located between the neck and abdomen, is the superior part of the trunk. It lodges and provides protection for the heart and lungs and some abdominal organs (*e.g.*, the liver and spleen). The muscles covering the *osteocartilaginous thoracic cage* (Fig. 1-1) are described in other chapters. For example, the pectoral muscles (L. *pectus*, chest) are described with the upper limb (Chap. 6) because they act from the thorax to move the upper limbs. **Chest pain is common**, the significance of which varies from negligible to very serious. A person experiencing a *heart attack* usually describes a crushing substernal pain that does not disappear with rest. The evaluation of a patient with chest pain is largely concerned with discriminating between serious conditions and the many minor causes of pain. To perform a clinical examination of the chest, a good knowledge of its vital organs is required. The heart, trachea, lungs, and pleurae are the most important structures in the chest. As these organs are constantly moving in living persons, the thorax is one of the most dynamic regions of the body.

The Thoracic Wall

This wall is mainly composed of bones (vertebrae, ribs, and sternum) and muscles. It is constructed so that the volume of the thoracic cavity can be varied during respiration. The thoracic wall of infants is thin owing to their underdeveloped muscles, and their thoracic cages are soft and pliable.

Skin and Fascia of the Thoracic Wall

The skin of the thoracic area, as elsewhere, consists of a superficial layer, the *epidermis*, and a deep connective tissue layer, the *dermis*. The fasciae are deep to the skin. The **superficial fascia** (also called tela subcutanea or hypodermis) is composed of *loose connective tissue*. It contains a variable amount of fat. The superficial fascia, extending between the dermis and the underlying deep fascia, also contains sweat glands, blood and lymphatic vessels and nerves. The **deep fascia** is thin but is usually dense and loosely attached to the superficial fascia. It forms an envelope, deep to the superficial fascia, which is adherent to the underlying muscles and constitutes their covering, called *epimysium*. The deep fascia can be separated only by sharp dissection because the epimysium sends septa into the muscles. Deep fascia covers the muscles up to their attachment to bone (*e.g.*, the sternum and ribs) and is itself *attached to the periosteum* of bones. Not only does the deep fascia hold the parts of the thorax together, but it presents a barrier to infection.

Bones of the Thoracic Wall

The *osteocartilaginous thoracic cage* (Fig. 1-1) is formed by part of the vertebral column (12 *thoracic vertebrae* and *intervertebral discs*); 12 pairs of *ribs* and *costal cartilages*, and the *sternum*. The ribs and costal cartilages form the largest part of the thoracic cage.

The Thoracic Vertebrae

These 12 bones are described in detail with the vertebral column in Chapter 4 (p. 331). Their special features related to the thoracic cage are: (1) they have *facets on their bodies* for articulation with the heads of ribs (Fig. 1-2D); (2) there are *facets on their transverse processes* for articulation with the tubercles of ribs (Fig. 1-2B), except for the inferior two or three ribs, and (3) they have *long spinous processes* (spines).

The Costal Facets (Figs. 1-2 and 1-3). Thoracic vertebrae are unique in that they have facets on their bodies and transverse processes for articulation with the ribs (T11 and T12 are exceptions). Two demifacets are located laterally on the bodies of T2 to T9 (Fig. 1-2D). The superior demifacet articulates with the head of its own rib and the inferior demifacet articulates with the head of the rib inferior to it. The costal facets of other vertebrae vary somewhat. T1 has a single costal facet for the head of the first rib and a small demifacet for the cranial part of the second rib (Fig. 1-2A). T10 has only one costal facet which is partly on its body and partly on its pedicle. T11 and T12 have only a single costal facet on their pedicles (Fig. 1-2C).

The Spinous Processes (Fig. 1-2). These *bony projections* from the vertebral arches of T5 to T8 are nearly vertical and overlap the vertebrae like roof shingles. This arrangement covers the small intervals between the laminae of adjacent vertebrae, thereby preventing sharp objects such as a knife from penetrating the vertebral canal and injuring the spinal cord. The spinous processes of T1, T2, T11 and T12 are horizontal, and those of T3, T4, T9 and T10 slope inferiorly.

Surface Anatomy of the Thoracic Vertebrae. The spinous processes of all the thoracic vertebrae are usually palpable in the *posterior median line* located at the bottom of the median furrow of the back (see Figs. 4-8 and 4-9). When the neck and trunk are flexed, the first spinous process that is visible is C7; for this reason it is called the *vertebra prominens*. The spinous process of T1 may also be prominent. The spinous processes of the other thoracic vertebrae are usually not as conspicuous when the trunk is flexed, but they are palpable. The spinous process of T10 (or T11) is often shorter than the others; this produces a slight depression over it. These surface anatomical facts are helpful when counting or manipulating the thoracic vertebrae.

Movements between adjacent vertebrae are relatively small in the thoracic region. This limited movement protects the

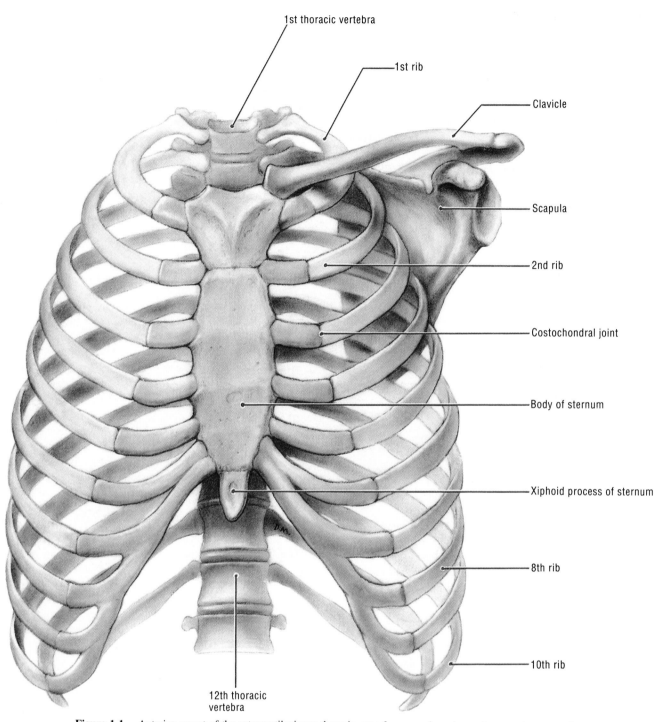

Figure 1-1. Anterior aspect of the osteocartilaginous thoracic cage from a male cadaver; its posterior aspect is shown in Fig. 4-2.

heart and lungs. The bodies of T5 to T8 are related to the thoracic aorta, which often flattens their left sides (Fig. 1-18). When the aorta develops an **aneurysm** (G. localized dilation), the bodies of these vertebrae may be partly eroded by pressure from the aneurysm (see Fig. 2-104). In some cases these bony changes are visible on radiographs and other images (*e.g.*, CT) of the thorax.

The Ribs

There are usually 12 of these *elongated flat bones* on each side of the thorax. They form the largest part of the osteocartilaginous thoracic cage. The ribs curve anteriorly and inferiorly from the thoracic vertebrae. Each typical rib has a head, neck, tubercle, and shaft (Figs. 1-1 and 1-3).

The head of a rib is wedge-shaped and presents two articular facets for articulation with the numerically corresponding vertebra and the vertebra superior to it. These facets are separated by the *crest of the head.*

The neck of a rib is the stout, flattened part located between the head and the tubercle. The neck lies anterior to the transverse process of the corresponding vertebra. Its superior border, called the *crest of the neck,* is sharp, whereas its inferior border is rounded.

The tubercle of a rib is on the posterior surface at the junction of its neck and shaft and is most prominent on the superior ribs. The tubercles of most ribs have a smooth convex facet, which articulates with the corresponding transverse process of the vertebra, and a rough nonarticular part for attachment of the *lateral costotransverse ligament* (Fig. 1-21). The tubercles of the 8th to 10th ribs have flat facets for articulation with similar facets on the transverse processes of the vertebrae.

The shaft (body) of a rib is thin, flat, and curved. Forming its largest part, the shaft has external and internal surfaces, thick and rounded superior borders, and thin, sharp inferior borders (Fig. 1-3). A short distance beyond the tubercle, the shaft ceases to pass posteriorly and swings sharply anteriorly. The point of greatest change in curvature is called the *angle of the rib.* Here the rib is both curved and twisted. The thorax of a person lying on his/her back is supported by the angles of the ribs and the spinous processes of the vertebrae. The **costal groove** (sulcus) and flange formed by the inferior border of the rib protect the

intercostal nerve and vessels that accompany the rib (Figs. 1-18, 1-19, and 1-23*C*).

True Ribs (Fig. 1-1). The *first seven* (and sometimes the eighth) pairs of ribs are called true or *vertebrosternal ribs* because they are *connected to the sternum* by their costal cartilages.

False Ribs (Fig. 1-1). The *8th to 12th* pairs of ribs are false or *vertebrochondral ribs.* Each of the 8th to 10th ribs is connected by its costal cartilage to the cartilage of the rib superior to it. The 11th and 12th pairs of false ribs are often called vertebral or *floating ribs* because they are unattached anteriorly. They end in the muscles of the anterior abdominal wall. Although these ribs are free, *they are not floating*; they articulate with the body of their own vertebra.

Typical Ribs (Figs. 1-1 and 1-3). The 3rd to 9th ribs are typical; *i.e.,* they have the features described on p. 34. However, typical ribs vary slightly in length and other characteristics.

Atypical Ribs (Figs. 1-1 and 1-4). The 1st, 2nd, and 10th to 12th pairs of ribs are atypical.

The First Rib (Figs. 1-1, 1-4, and 1-5). This is the *broadest and most curved* of all the ribs. It is also the shortest of the true ribs. The first rib is clinically important because so many structures cross or attach to it. It is flat and has a prominent **scalene tubercle** on the internal border of its superior surface for the attachment of the *scalenus anterior* muscle (Fig. 1-20). The *subclavian vein* crosses the first rib anterior to the scalene tubercle and subclavian artery. The inferior trunk of the *brachial*

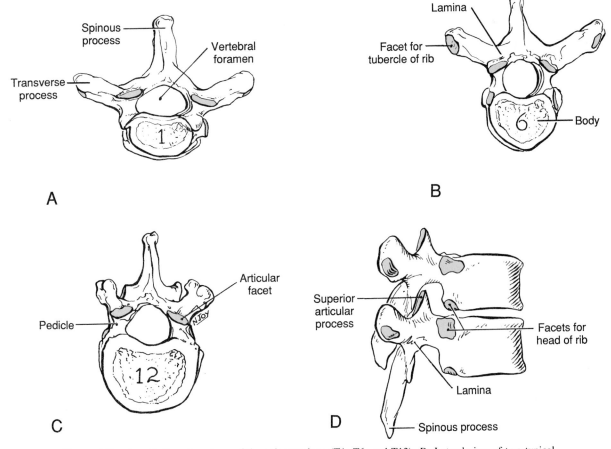

Figure 1-2. *A to C,* Superior views of thoracic vertebrae (T1, T6, and T12). *D,* Lateral view of two typical thoracic vertebrae (T6 and T7).

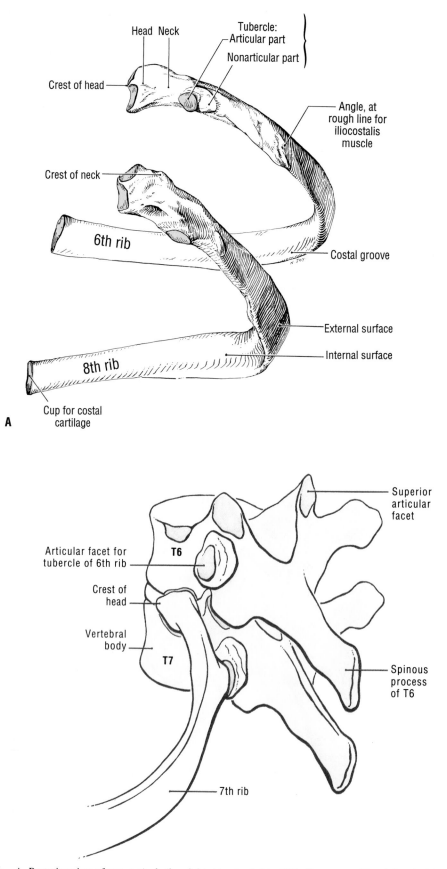

Figure 1-3. *A*, Posterior view of two *typical ribs*, right side; the 3rd to 10th ribs are considered "typical."
B, Posterolateral view of a costovertebral joint.

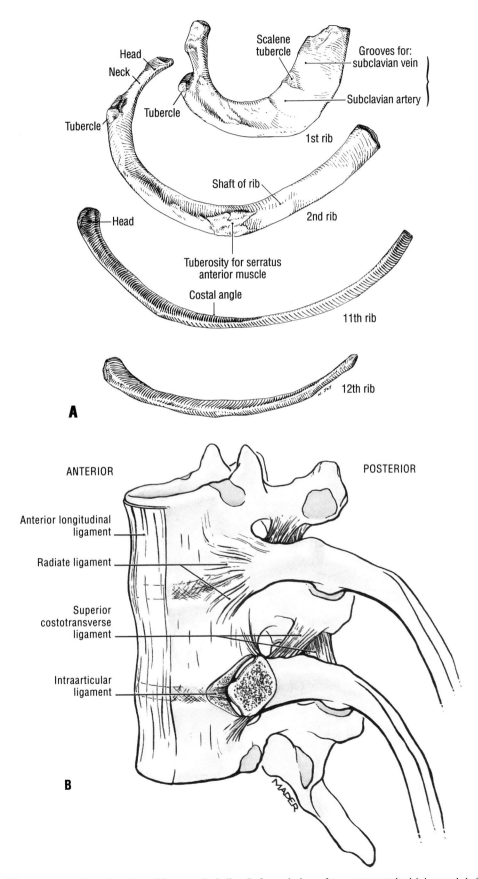

Figure 1-4. *A*, Superior view of four atypical ribs. *B*, Lateral view of two costovertebral joints and their associated ligaments.

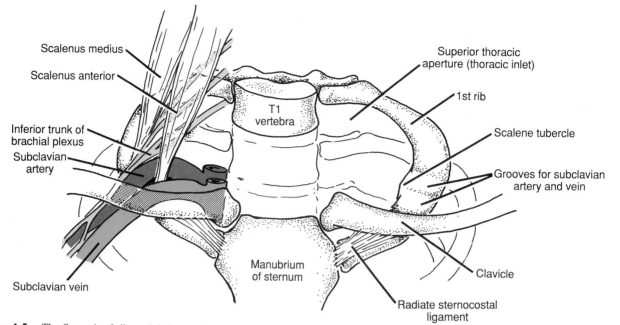

Figure 1-5. The first pair of ribs and their costal cartilages, showing their articulation with T1 vertebra and the manubrium of the sternum. On the right, the structures crossing the first rib are shown; two muscle attachments are shown on the left rib. Observe the superior thoracic aperture (thoracic inlet). During life it is filled with the structures entering and leaving the thorax.

plexus, a network of nerves on their way to the upper limb, passes posterior to it (Fig. 1-19). A distinct groove is formed by the subclavian vessels and brachial plexus on the superior surface of the first rib. This rib articulates with the body of the first thoracic vertebra. The prominent tubercle of the first rib articulates with the transverse process of this vertebra (Fig. 1-2A). *The first rib is difficult to palpate* because of the clavicle, but the first intercostal space can be felt just inferior to this "collar" bone.

The Second Rib (Figs. 1-1 and 1-4). This rib has a curvature similar to the first rib, but it is thinner, much less curved, and is about twice as long as the first rib. It can easily be distinguished from the first rib by the presence of a broad, rough eminence, which is the **tuberosity for the serratus anterior muscle.**

The 10th Rib (Fig. 1-1). This rib usually articulates with T10 vertebra only.

The 11th and 12th ribs (Figs. 1-1 and 1-4). These ribs are short, especially the 12th pair. They are capped with a small costal cartilage, have a single facet on their heads, and *have no neck or tubercle.* The 11th rib has an ill-defined angle and a shallow costal groove. The 12th rib has neither of these features and may be shorter than the first rib.

Although there are usually 12 pairs of ribs, the number may be increased by the development of *cervical ribs* (p. 524) or lumbar ribs, or the number may be decreased by failure of the 12th pair to form (*agenesis*). Persons with trisomy 21 (Down syndrome) may have only 11 pairs of ribs. It is extremely rare to find an additional rib or pair of ribs in both the cervical and lumbar regions of the same person. *Lumbar ribs* are more common than cervical ribs, but they are usually quite small (see Fig. 4-44). The 12th rib may be absent on one side and shortened on the other side. *Lumbar ribs have clinical significance* in that they may confuse the identification of vertebral levels in radiographs and other images. In addition, a lumbar rib may be erroneously interpreted as a fractured transverse process of the L1 vertebra. For a description of other variations of ribs, see Bergman et al., 1988.

Variations in the sternocostal junctions are not uncommon. The first seven costal cartilages nearly always join the sternum (Fig. 1-1), but in some people, only six cartilages articulate with it. In other people, the eighth costal cartilage, especially on the right, joins the sternum (Fig. 1-19). The 10th rib also "floats" in some people.

Bifid ribs (forked ribs) occur in 1 to 2% of most populations, but this abnormality is present in 8.4% of Samoans (Martin, 1960). This possibility must be considered when a rib count seems to indicate that the cartilages of the superior eight ribs articulate with the sternum. Usually the condition is unilateral; hence, unilaterality of eight apparent true ribs favors the presence of a bifid rib. In some cases, however, eight normal ribs articulate with the sternum (Fig. 1-19). *Fused ribs* are uncommon and they are often associated with a vertebral malformation, *e.g.*, *hemivertebra* (a half vertebra resulting from agenesis of one half).

The angles of the ribs are their weakest parts. Despite their ability to bend under stress, they may be fractured by direct violence or indirectly by crushing injuries that commonly occur in automobile accidents. Because the ribs of infants and children are mainly cartilaginous, they have the property of returning to their original shape after compression (*i.e.*, they are elastic and do not often fracture). However, the ribs of adults are breakable and ribs have been fractured by muscular men hugging frail women too vigorously. The first two pairs of ribs, which are protected by the clavicle, and the last two pairs, which are unattached anteriorly and free to swing, are not commonly fractured.

Fractured ribs, commonly referred to as "cracked ribs," are painful because the fractured parts move during respiration, coughing, and sneezing. The middle ribs are the ones most commonly fractured. Crushing injuries tend to break the ribs at their weakest point, *i.e.*, just anterior to their costal angles (Fig. 1-3). *Direct injuries may fracture a rib anywhere*, and the broken ends may be driven inwardly and injure the internal organs (*e.g.*, the heart, lungs and/or spleen). Careful palpation along a broken rib often reveals local tenderness, even when a rib fracture may not be visible on a radiograph. Most rib fractures are accompanied by strain or *tearing of intercostal muscle fibers* (Fig. 1-19). These fractures are relatively common in athletes who compete in body contact sports (*e.g.*, quarterbacks in football).

Flail chest ("stove-in chest") occurs when a sizable segment of the anterior and/or lateral thoracic wall is freely movable owing to multiple rib fractures. This allows the loose segment of the wall to move paradoxically (*i.e.*, inward on inspiration and outward on expiration). This impairs ventilation and thereby affects oxygenation of the blood; if severe, death results. During treatment, the loose segment is often fixed by hooks and/or wires so that it cannot move.

Sometimes a piece of rib is used for **autogenous bone grafting** (Fig. 1-19), *e.g.*, for reconstruction of the mandible following excision of a tumor. Following the operation, the missing piece of rib regenerates from the remaining periosteum, but it never returns to its original form.

Coarctation of the aorta (narrowing of the distal part of the aortic arch) results in enlargement of the intercostal and other arteries as a means of providing a collateral circulation to inferior parts of the body. These enlarged intercostal arteries erode or notch the inferior borders of the corresponding ribs. Radiographic demonstration of *notching of the ribs* is very useful in confirming a suspected case of this congenital malformation of the aorta.

The Costal Cartilages (Figs. 1-1 and 1-19).   These *hyaline cartilage bars*, more rounded than the ribs, extend from the anterior ends of the ribs. The first seven pairs of costal cartilages (and sometimes the eighth) are connected with the sternum. The 8th to 10th pairs of costal cartilages articulate with the inferior border of the cartilage of the preceding rib, and the costal cartilages of the 11th and 12th ribs are pointed and end in the musculature of the anterior abdominal wall. The costal cartilages contribute significantly to the elasticity and mobility of the ribs.

The Costal Margins (Figs. 1-1, 1-7, and 1-8).   The medial ends of the seventh to tenth costal cartilages join to form a cartilaginous costal margin on each side. Together these margins form the *costal arch*. The angle formed where the right and left costal margins converge at the inferior end of the body of the sternum is called the *infrasternal angle*. It is located at the *xiphisternal joint*, where the body of the sternum articulates with the xiphoid process.

The costal cartilages of young people provide resilience to the thoracic cage, preventing many crushing injuries or direct blows from fracturing the ribs and/or the sternum. In older adults the costal cartilages often undergo superficial calcification. As a result, they lose some of their elasticity and become brittle. This *calcification makes them radiopaque* (*i.e.*, visible on radiographs), which is confusing at first to inexperienced observers. In some very old people the costal cartilages may ossify, resulting in a rigid anterior thoracic wall. As a result *cardiopulmonary resuscitation* (p. 106) may be painful.

The Costovertebral Joints

A typical rib articulates with the vertebral column at two joints (Figs. 1-3*B* and 1-4*B*): (1) the joints of the heads of the ribs and (2) the costotransverse joints. Typically the head of a rib articulates with the sides of the bodies of two thoracic vertebrae, and the tubercle articulates with the tip of a transverse process. The costovertebral joints are the *plane type of synovial joint* (p. 19), which allows for gliding or sliding movements.

Joints of the Heads of the Ribs (Figs. 1-2 to 1-4 and 1-22). The head of each typical rib articulates with the demifacets of two adjacent vertebrae and the intervertebral disc between them. The head articulates with the superior part of the corresponding vertebra, the inferior part of the vertebra superior to it, and the adjacent intervertebral disc. For example, the head of the sixth rib articulates with the superior part of the body of T6 vertebra, the inferior part of T5, and the intervertebral disc between these bones. The *crest of the head* is attached to the intervertebral disc by an *intraarticular ligament*. It is located within the joint and divides it into two synovial cavities. There are exceptions to the general arrangement just described. The heads of the 1st, sometimes the 10th, and usually the 11th and 12th ribs articulate with only their own vertebral bodies. In these cases there are no intraarticular ligaments and the joint cavities are not divided.

An *articular capsule* surrounds each joint and connects the head of the rib with the circumference of the joint cavity. The capsule is strongest anteriorly where a *radiate ligament* (Fig. 1-4*B*) fans out from the anterior margin of the head of the rib to the sides of the bodies of two vertebrae and the intervertebral disc between them. The heads of the ribs are connected so closely to the vertebral bodies that only slight gliding movements occur at the joints of the heads of the ribs.

The Costotransverse Joints (Figs. 1-2 to 1-4, 1-21, and 1-22).   The tubercle of a typical rib articulates with the facet on the tip of the transverse process of its own vertebra to form a *synovial joint*. These small joints are surrounded by thin articular capsules, which are attached to the edges of the articular facets. A *lateral costotransverse ligament*, passing from the tubercle of the rib to the tip of the transverse process, strengthens the joint on each side. In addition, a costotransverse ligament unites the posterior surface of the neck of the rib to the anterior surface of the transverse process. A *superior costotransverse ligament* joins the crest of the neck of the rib to the transverse process superior to it. The aperture between this ligament and the vertebral column permits passage of the spinal nerve and the dorsal branch of the intercostal artery (Figs. 1-4*B*, 1-18, and 1-22). The 11th and 12th ribs do not articulate with transverse processes and have freer movements as a result. The strong costotransverse ligaments binding these joints limit their movements to slight gliding. However, the articular surfaces on the tubercles of the superior six ribs are convex and fit into concavities on the transverse processes. As a result, some superior and inferior movements of the tubercles are associated with rotation of the ribs.

The Sternocostal Joints

The 1st to 7th ribs articulate via their costal cartilages with the lateral borders of the sternum (Figs. 1-1, 1-6, and 1-8). The first pair of costal cartilages articulates with the sternum at *primary cartilaginous joints* (synchondroses). The costal cartilages are united directly to the hyaline cartilage in the depressions located at the superolateral margins of the manubrium of the sternum.

The 2nd to 7th pairs of costal cartilages articulate with the sternum at *synovial joints*, but joint cavities are often absent in the inferior ones. The thin, weak articular capsules of these joints are strengthened anteriorly and posteriorly by *radiate sternocostal ligaments* (Fig. 1-5). These thin, broad membranous bands pass from the costal cartilages to the anterior and posterior surfaces of the sternum, forming a feltlike covering for the sternum.

In Fig. 1-6 observe the following: the first pair of costal cartilages articulates with the manubrium only; the second pair of costal cartilages articulates with the manubrium and the first segment (sternebra) of the sternum; the third to fifth

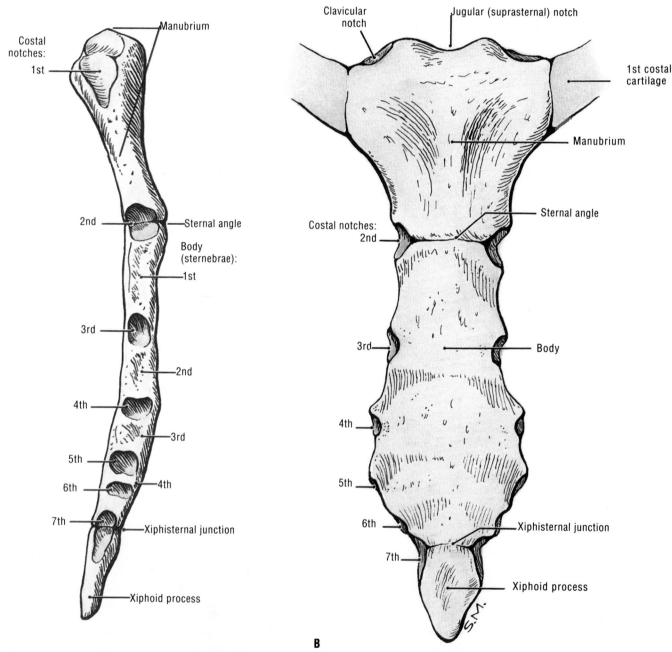

Figure 1-6. The sternum from a female cadaver. *A*, Lateral view. *B*, Anterior view. Observe the sternal angle at the junction of the manubrium and body (manubriosternal joint).

pairs of costal cartilages articulate with two segments of the sternum; the sixth pair of costal cartilages articulates with the fourth segment of the sternum only; and the seventh pair of costal cartilages articulates with the fourth segment of the sternum and the xiphoid process.

The Costochondral Joints

Each rib has a cup-shaped depression in its anterior end into which its costal cartilage fits (Figs. 1-1, 1-3*A*, and 1-4). The rib and its costal cartilage are firmly bound together by the continuity of the periosteum of the rib with the perichondrium of the costal cartilage. No movement normally occurs at these joints.

The Interchondral Joints

The articulations between the adjacent borders of the sixth and seventh, seventh and eighth, and eighth and ninth costal cartilages are *plane synovial joints*. Each of these articulations is enclosed within an articular capsule that is lined with a synovial membrane. The joints are strengthened by *interchondral ligaments*. The articulation between the ninth and tenth costal cartilages is a fibrous joint (*i.e.*, the cartilages are joined by fibrous tissue).

A *rib dislocation* usually refers to displacement of the costal cartilage, usually the second to seventh, from the sternum; *i.e.*, there is dislocation of the synovial joint. This causes severe pain at the time of injury and during respiratory movements. The injury produces a lumplike deformity at the dislocation site. Dislocation of the costovertebral joints may also occur, but this is less common owing to their strong articular capsules and radiate ligaments (Fig. 1-22). *Rib dislocations are common in body contact sports*, especially football, hockey, wrestling, and baseball. Possible complications are pressure on or damage to nearby nerves, blood vessels, and muscles.

A *rib separation* usually refers to a dislocation of the costochondral junction between the rib and its costal cartilage—and not to a lack of continuity in the shaft of the rib that may be detected in a radiograph of a fractured rib. In separations of the 3rd to 10th ribs, tearing of the perichondrium and/or periosteum occurs, and one of the costal cartilages (usually the 10th) separates from the inferior border of the costal cartilage superior to it. Consequently, the cartilage of the ''slipping rib'' can move superiorly and override the one superior to it, causing pain.

Rickets (p. 13) causes the ends of the ribs (usually the second to eighth) to enlarge where they join their costal cartilages. The beadlike bony enlargements are sometimes referred to as the ''rachitis rosary.''

The Sternum

The sternum (breastbone) is an *elongated flat bone* that resembles a short broadsword or dagger. It forms the middle part of the anterior wall of the thorax (Figs. 1-1 and 1-6 to 1-8). The sternum (G. *sternon*, chest) *consists of three parts*: manubrium, *body*, and *xiphoid process*.

The Manubrium (Figs. 1-1 and 1-6 to 1-9). The manu-brium (L. handle), the superior part of the sternum, is located anterior to T3 and T4 vertebrae. It is wider and thicker than the other two parts of the sternum. Its narrow inferior end gives it a somewhat triangular shape. Broad and thick superiorly, the manubrium slopes inferoanteriorly. The superior surface of the manubrium is indented by the *jugular notch* (suprasternal notch), which can easily be palpated. On each side of this notch there is an oval articular facet, called the *clavicular notch*, which articulates with the medial end of the clavicle. Just inferior to the clavicular notch, the costal cartilage of the first rib is fused with the lateral margin of the manubrium. This is a flexible but strong primary cartilaginous joint (p. 17). The inferior border of the manubrium is oval and rough where it articulates with the body of the sternum at the *manubriosternal joint*. The manubrium and body lie in slightly different planes; hence, their junction forms a projecting **sternal angle** (angle of Louis). This bony landmark is located opposite the second pair of costal cartilages. *The sternal angle is an important clinical guide to the accurate numbering of the ribs* (p. 44).

The Body (Figs. 1-1 and 1-6 to 1-9). The body of the sternum, the longest of its three parts, is located anterior to T5 to T9 vertebrae. It is longer, thinner, and narrower than the manubrium, but its width varies owing to the scalloping of its lateral borders by the *costal notches*. The body is broadest at the level of the fifth pair of sternocostal joints and then gradually tapers inferiorly. The *anterior surface* of the body is slightly concave from side to side. In young people four *sternebrae* are obvious. They articulate with each other at primary cartilaginous joints. The sternebrae begin to fuse from the inferior end between puberty and 25 years of age. In adults the sternum is marked by three transverse ridges, representing lines of fusion of the four originally separate sternebrae. These ridges are usually absent in the sterna of elderly people. The *posterior surface* of the body is slightly concave, and transverse ridges marking the lines of fusion of the sternebrae may be visible in a sternum from a young adult, but they are less distinct than the ridges on the anterior surface.

The Xiphoid Process (Figs. 1-1 and 1-6 to 1-8). This thin, sword-shaped process (G. *xiphos*, sword) is the smallest and most *variable part of the sternum*. Although it is often pointed, the xiphoid process (xiphisternum) may be blunt, bifid, curved, or deflected to one side or anteriorly. The xiphoid is *cartilaginous at birth* and remains so until early childhood. It may begin to ossify during the third year of life and then it consists of a bony core surrounded by hyaline cartilage. However, ossification usually does not begin until much later. The xiphoid usually ossifies and unites with the body of the sternum around 40 years of age, but it may not be united even in very old people. *The xiphoid process is an important landmark in the median plane* for the following reasons: (1) its junction with the body of the sternum at the xiphisternal joint indicates the inferior limit of the thoracic cavity anteriorly and the site of the infrasternal angle; and (2) it is a midline pointer to the diaphragmatic surface of the liver, diaphragm, and inferior border of the heart (Fig. 1-47; see also Fig. 2-36).

The fetal halves of the sternum may fail to unite owing to defective ossification. This uncommon condition is referred

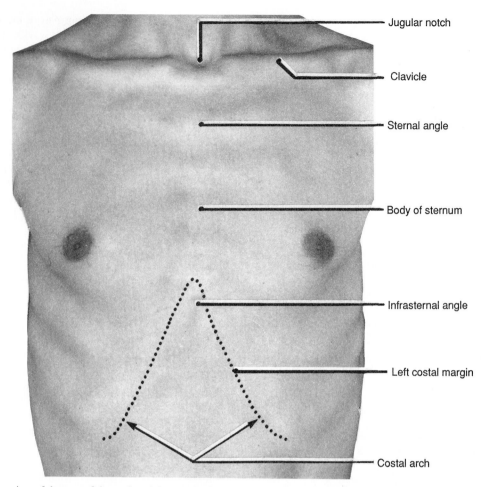

Jugular notch

Clavicle

Sternal angle

Body of sternum

Infrasternal angle

Left costal margin

Costal arch

Figure 1-7.   Anterior view of the root of the neck and thorax of a 27-year-old man, showing the surface features of the clavicles, ribs, and sternum. Observe that the costal margins form the costal arch and the infrasternal angle is located where they converge at the xiphisternal joint.

to as *cleft sternum*. Severe cases of cleft sternum are often associated with *ectopia cordis*, in which the heart is outside the thorax (Moore, 1988). Sometimes there is an opening in the body of the sternum called the *sternal foramen*. This common defect results from abnormal ossification and is of no clinical significance, except that its presence should be identified so that it will not be misinterpreted on a radiograph as a bullet hole. Although the xiphoid process is commonly perforated in elderly persons owing to incomplete ossification, this observation is of no clinical significance. In infants it is not unusual to see the tip of the xiphoid process protruding anteriorly beneath the skin. Sometimes this condition persists.

The body of the sternum is usually significantly shorter in females than in males and is less than twice the length of the manubrium (Figs. 1-1 and 1-8). This *sex difference in the sternum* may be useful in sexing human skeletal remains. Not uncommonly, men in their early forties suddenly detect their ossified xiphoid processes and consult their physician about the "hard lump" in the "pit of their stomach" (epigastric fossa; see Fig. 2-5A). Never having felt their xiphoid process before, they fear they have developed stomach cancer.

The ribs are more commonly broken than the sternum; however, **sternal fractures** are common following traumatic compression of the thorax (*e.g.*, in automobile accidents when the driver's chest is driven into the steering wheel). The body

of the sternum is commonly fractured near the sternal angle, and it is often a *comminuted fracture* (*i.e.*, broken into pieces). Fortunately the ligaments covering the sternum usually confine the fragments so that a compound fracture does not usually occur (*i.e.*, there is not an open wound leading to the broken pieces). In severe accidents, the body of the sternum separates from the manubrium and is driven posteriorly. This may *rupture the aorta* and/or injure the heart and liver, resulting in sufficient loss of blood and/or damage to the heart muscle to cause the patient to die.

The sternum is important to the hematologist because it has a readily accessible medullary (marrow) cavity. During a procedure called **sternal puncture**, a large bore needle is inserted through the cortex of the sternum into the *red bone marrow*, a sample of which is aspirated for laboratory evaluation. Although the skin is anesthetized, it is still an uncomfortable procedure.

To gain access to the *anterior mediastinum* for surgical operations on the heart and great vessels (Figs. 1-40 and 1-45), it is often necessary to divide the sternum in the median plane. Similarly, in operations on the *thymus*, it is usually necessary to divide the manubrium. For a description of this gland, see p. 107.

In some people the body of the sternum projects infero-posteriorly and presses on the heart. This widens it, making it appear large on AP chest radiographs. A lateral chest radiograph reveals the true cause of the heart's abnormal shape. This rare condition, usually congenital, is called *pectus excavatum* (L. funnel chest). In other people, the chest is flattened on each side and the sternum projects anteriorly. This uncommon keel-like condition, called *pectus carinatum* or "pigeon chest," was given its name because the person's chest looks somewhat like the keel of a boat (L. *carina*, keel).

The Sternal Joints

The joints of the sternum are between its parts. There are two articulations: the manubriosternal (sternomanubrial) and xiphisternal (xiphosternal) joints.

The Manubriosternal Joint (Figs. 1-1 and 1-6 to 1-9). This articulation is between the manubrium and body of the sternum. The *sternal angle* indicates the manubriosternal joint. In adults this is a *secondary cartilaginous joint* (symphysis). The bony articular surfaces are covered with hyaline cartilage, and the articulating bones are connected by a fibrocartilaginous disc. In about 30% of people, the central part of the cartilaginous disc undergoes absorption, forming a cavity, but *this is not a joint cavity*. The manubriosternal joint is strengthened by anterior and posterior fibrous ligaments, which extend across the joint from the manubrium to the body. In most people the manubriosternal joint moves slightly during respiration.

In newborn infants the articulating bones forming the manubriosternal joint are joined by collagenous and elastic fibers.

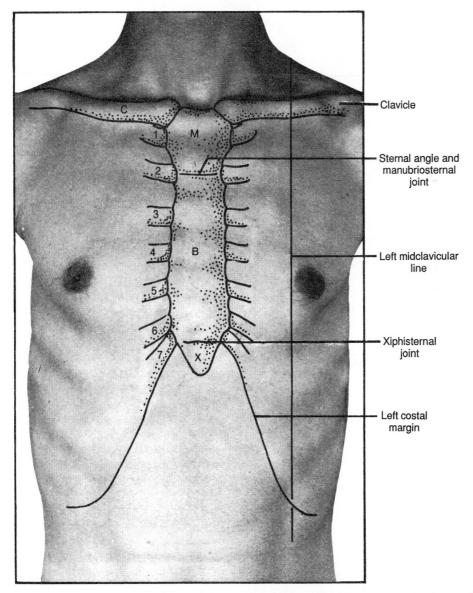

Figure 1-8. Anterior view of the thorax of the same man shown in Fig. 1-7. Outlines of the clavicle (*C*), costal cartilages (numbered), and parts of the sternum have been drawn. *M*, indicates manubrium; *B*, body; and *X*, xiphoid process. Note that the sternal angle is adjacent to the second pair of costal cartilages. Counting of ribs starts here and proceeds inferolaterally.

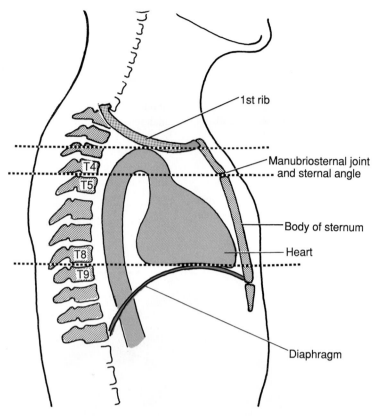

Figure 1-9. Median section of the osteocartilaginous thoracic cage, indicating its surface markings and vertebral levels. Note that the man-ubrium is anterior to T3 and T4 vertebrae and that the sternal angle and manubriosternal joint are at the T4 to T5 vertebral level.

Later the bones are united by hyaline cartilage, forming *a primary cartilaginous joint*. The plate of cartilage is later converted into fibrocartilage, forming *a secondary cartilaginous joint*. In about 10% of people over 30 years of age, the superficial part of this cartilage becomes ossified. In very old people, fusion of the manubrium and body occurs (*synostosis*).

The Xiphisternal Joint. This articulation between the xiphoid process and body of the sternum is a *primary cartilaginous joint* (synchondrosis); these bones are united by hyaline cartilage. By 40 years of age, the xiphoid and this cartilage have usually ossified. In most elderly people, the xiphisternal (xiphosternal) joint is ossified and the xiphoid is fused with the body of the sternum.

Surface Anatomy of the Anterior Thoracic Wall

The clavicle lies subcutaneously at the junction of the thorax and neck (Figs. 1-7 and 1-8). It can be easily palpated, especially at its medial end where it articulates with the manubrium of the sternum. The *sternum* also lies subcutaneously in the middle part of the thorax and is palpable throughout its length. The smooth superior border of the manubrium is easily felt because it has a shallow concavity that forms the floor of the *jugular notch*. Put your second digit (index finger) in your jugular notch and then slide it slowly down the median plane of your manubrium until

you feel a horizontal ridge. This is the *sternal angle*, which is located at the junction of the manubrium and body (Figs. 1-6 to 1-8). Because fat does not accumulate here and because the manubriosternal joint moves during respiration, this angle is palpable and may be visible.

The sternal angle is located about 5 cm inferior to the jugular notch. It lies at the level of the second pair of costal cartilages (Figs. 1-1 and 1-8), and when traced laterally, it directs the palpating finger to the second costal cartilage, the starting point from which the ribs are counted. The small inferior part of the sternum, the *xiphoid process*, lies in the depression where the converging costal margins form the *infrasternal angle* (Figs. 1-1 and 1-7). The xiphoid projects over the left lobe of the liver into the *epigastric region* of the abdomen (see Figs. 2-5A and 2-36). In young people, the xiphoid is usually movable at the *xiphisternal joint*. On each side of the sternum, the ribs can be palpated as they approach the sternum. The first rib is difficult to feel because it lies deep to the clavicle and is covered by muscles (Fig. 1-19).

The costal margins are palpable with ease, extending inferolaterally from the xiphisternal joint (Figs. 1-1, 1-7, and 1-8). The superior part of the costal margin is usually formed by the 7th costal cartilage, and its inferior part is usually formed by the 8th to 10th costal cartilages. The degree of prominence of the costal margins varies; they are obvious in thin, physically fit persons when they inspire (Fig. 1-17). The *infrasternal angle* also varies in size from person to person and increases during inspiration. This angle is used in *cardiopulmonary resuscitation* (*CPR*) for locating the proper hand position on the body of the sternum (p. 106).

> Often, a need to count the ribs arises (*e.g.*, to determine which rib is diseased or injured). First, the sternal angle is located and then the palpating digit is passed directly laterally from it to the second costal cartilage. The ribs are then counted inferolaterally from this point to avoid confusion with the fused inferior costal cartilages. Very uncommonly, the sternal angle occurs at the level of the third costal cartilage. This variation should be suspected if the manubrium is longer than usual and the sternal angle is more than 5 cm inferior to the jugular notch.

The Breasts

Both men and women have breasts; normally they are well-developed only in women. These glands are usually rudimentary throughout life in men and consist of only a few small ducts. The breasts (L. mammae) are the most prominent superficial structures of the anterior thoracic wall, especially in women. They are situated on the anterior surface of the thorax, overlying the *pectoral muscles* (pectoralis major and serratus anterior, Figs. 1-10 to 1-12). The mammary glands are accessory organs of the female reproductive system. They secrete milk for the nourishment of the infant, a process known as *lactation*. They often extend toward the axillae (armpits) forming *axillary tails*. The amount of fat surrounding the glandular tissue determines the size of the breasts.

The Female Breasts

During puberty (usually 12 to 15 years of age), the female breasts normally grow, and the pink circular areas of skin around the nipples, called *areolae*, enlarge (Fig. 1-10). The lactiferous ducts give rise to buds that form 15 to 20 lobules of glandular tissue, which constitute the *mammary gland*. Each lobule is drained by a *lactiferous duct* (Fig. 1-11), which opens on the nipple. These ducts extend from the nipple in a manner similar to spokes of a wheel. Deep to the areola, each duct has a dilated portion called the *lactiferous sinus*, in which milk accumulates during lactation.

The areolae contain numerous *sebaceous glands*, which enlarge during pregnancy and secrete an oily substance that provides a protective lubricant for the areola and nipple. The areolae, variable in size, are pink in white *nulliparous women* (those who have not borne children; Fig. 1-10). There is no fat beneath the areolae. During the first pregnancy the areolae of white women change permanently to brown; the depth of color depends on the woman's complexion.

The nipples are conical or cylindrical prominences that are located in the center of the areolae. There is no fat in the nipples. In nulliparous women they are usually located at the level of the fourth intercostal spaces. However, *the position of the nipple varies considerably* and therefore cannot be used as a guide to the fourth intercostal space. The tip of the nipple is fissured and contains the openings of the lactiferous ducts (Fig. 1-11). The nipples are composed mostly of circularly arranged smooth muscle fibers that compress the lactiferous ducts and erect the nipples when they contract.

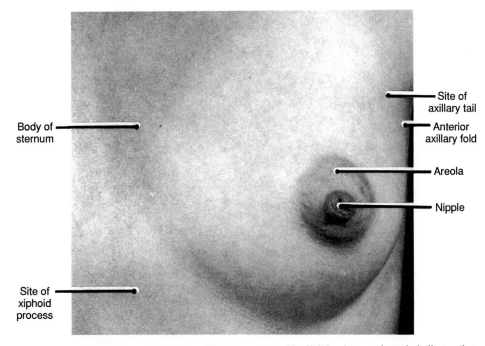

Figure 1-10.   Breast of a 27-year-old nulliparous woman. Her lightly pigmented areola indicates that she has not borne a child.

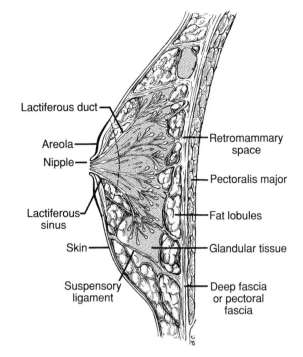

Lactiferous duct

Areola

Nipple

Lactiferous sinus

Skin

Suspensory ligament

Retromammary space

Pectoralis major

Fat lobules

Glandular tissue

Deep fascia or pectoral fascia

Figure 1-11. Sagittal section of a woman's breast. Observe that it consists of glandular, fibrous, and fat tissues.

The mammary gland is a modified sweat gland; this explains why it has no special capsule or sheath. It lies in the *superficial fascia*, anterior to the thorax (Fig. 1-11). The deep aspect of the breast is separated from the pectoral muscles by the *deep fascia*. Between the breast and deep fascia, there is an area of loose connective tissue that contains little fat. This zone, referred to as the *retromammary space* (bursa), allows the breast to move freely on the deep fascia covering the pectoralis major muscle. Although easily separated from the deep fascia, the mammary gland is firmly attached to the skin of the breast by *suspensory ligaments* (Cooper's ligaments). These fibrous bands, which support the breast, run between the skin and the deep fascia (Fig. 1-11). The rounded contour and most of the bulk of the breasts are produced by fat lobules, except during pregnancy and lactation when the mammary glands enlarge. The shape of the breast varies considerably in different persons and races and in the same person at different ages.

During **puberty** the lactiferous ducts undergo branching and there is an increased deposition of fat. As a result, progressive enlargement of the breasts occurs. In some women the breasts enlarge slightly during their menstrual cycles owing to the increase in gonadotropic hormones (follicle-stimulating hormone [FSH] and luteinizing hormone [LH]). During pregnancy the breasts enlarge greatly owing to the formation of new glandular tissue. The *milk-secreting cells*, referred to as *alveoli*, are arranged in grapelike clusters or lobules. Although the mammary glands are prepared for secretion by midpregnancy, milk is not secreted until after delivery of the fetus and placenta. *Colostrum*, a creamy white to yellowish *premilk fluid*, may be expressed from the nipples during the last trimester of pregnancy. Near term, colostrum often leaks from the nipples.

In *multiparous women* (those who have borne several children), the breasts may become large and pendulous. In elderly women the breasts are small and wrinkled owing to the decrease in adipose and glandular tissue. Although breasts vary markedly in size, their roughly circular bases are fairly constant and have the following limits in well-developed females: vertically from the *second to sixth ribs* and laterally from the edge of the *sternum to the midaxillary line* (Fig. 1-17). Two-thirds of the breast rest on the pectoralis major muscle; one-third covers the serratus anterior muscle (Fig. 1-11). Its inferior border overlaps the superior part of the *rectus sheath* (see Fig. 2-7).

Arterial Supply of the Breast (Figs. 1-19 and 1-20).   There is an abundant blood supply to the breast. The arteries are mainly from the *internal thoracic artery* (formerly the internal mammary) via its perforating branches, which pierce the second to fourth intercostal spaces. The breast also receives several branches from the *axillary artery*, mainly from its *lateral thoracic and thoracoacromial branches* (see Fig. 6-16) and lateral and anterior cutaneous branches from the *intercostal arteries* (in the third to fifth intercostal spaces).

Venous Drainage of the Breast (Figs. 1-12 and 1-19). Veins from the breast drain into the axillary, internal thoracic, lateral thoracic, and intercostal veins. The chief venous drainage is to the *axillary vein* (see Fig. 6-33).

Lymphatic Drainage of the Breast (Figs. 1-12, 1-19, and 1-20).   Most lymph passes from the mammary gland along interlobular lymphatic vessels to a *subareolar plexus*. From here and other parts of the breast, most lymph vessels follow the veins of the breast to the axilla. Most of the lymphatic drainage (about 75%) is to the **axillary lymph nodes**, mainly the *pectoral group* of nodes. They are located along the inferior border of the triangular pectoralis minor muscle, which lies deep to the pectoralis major. From the deep surface of the breast, the lymphatics pass through the pectoralis major muscle and drain into the *apical group* of axillary lymph nodes. Lymph from the medial part of the breast drains into the *parasternal lymph nodes*, which are located within the thorax along the internal thoracic vessels. Lymph from the skin of the breast may pass to the abdominal wall and the opposite breast.

Innervation of the Breast (Figs. 1-18 to 1-20).   The breast is supplied by lateral and anterior cutaneous branches of the second to sixth intercostal nerves. These nerves include both sensory and sympathetic fibers, which supply the skin, smooth muscle of the areolae and nipples, blood vessels, and mammary glands.

The breasts are usually equal in size, but if one is larger and more inferior, it is usually the right one. The *superolateral quadrant* of the breast contains a large amount of glandular tissue and is *where most breast cancers develop*. This tissue may extend superior to the clavicle and/or inferiorly into the epigastrium (see Fig. 2-5*A*). Similarly, glandular tissue from one breast may cross the median plane. The vascularity of the breasts begins to increase early in pregnancy. The increased blood supply dilates the vessels, often making them visible. The connections of the intercostal veins with the *vertebral venous plexuses* (see Fig. 4-53 and p. 365) provide a route for spread of cancer cells from the breast to the vertebrae and then to the skull and brain. When carcinoma or

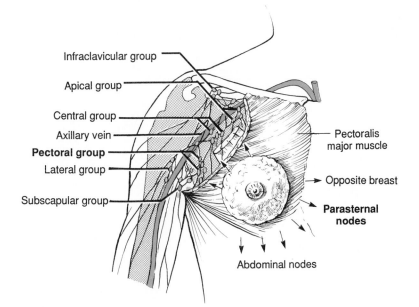

Figure 1-12. Lymphatics of the breast and axilla. Most lymph vessels drain into the pectoral group of axillary lymph nodes. Lymph from the medial part of the breast enters the parasternal lymph nodes (Fig. 1-19).

CA of the breast invades the *retromammary space* (Fig. 1-11) and attaches to or invades the deep fascia covering the pectoralis major muscle, contraction of this muscle causes the breast to move superiorly. This is a clinical sign of advanced malignant disease of the breast (Healey and Hodge, 1990).

The lymphatic and venous drainage of the breasts is of highest importance in the spread of CA of the breast (**breast cancer**), one of the two most common types of cancer in women (Fig. 1-13*B*). Cancer cells are carried from the breast by lymph vessels to lymph nodes, chiefly those in the axilla (Fig. 1-12). The cells lodge in these nodes where they produce nests of tumor cells called *metastases* (G. *meta*, beyond + *stasis*, a placing). As there is free communication between lymph nodes inferior and superior to the clavicle and between the axillary and cervical (neck) lymph nodes, metastases from the breast may develop in the supraclavicular lymph nodes, the opposite breast, or in the abdomen.

The axillary lymph nodes are the most common site of metastases from CA of the breast. Enlargement of these nodes in a woman therefore suggests the possibility of breast cancer. Cancerous nodes tend to be hard, but they are not usually tender. The absence of enlarged axillary lymph nodes is no guarantee that metastasis from a breast cancer has not occurred. Often there is dimpling and a thickening of the skin over the site of a CA of the breast (Fig. 1-13*B*), giving it the appearance of an orange peel; this skin change is called *peau d'orange* (Fr. orange skin). Interference with the lymphatic drainage of the breast produces the leathery thickening, whereas dimpling of the skin is mainly caused by infiltration of cancer cells along the suspensory ligaments (Figs. 1-11 and 1-13). This invasion shortens the ligaments and causes the skin to invaginate (dimple). *Subareolar CA* may cause inversion of the nipple by the same mechanism.

Mastectomy (excision of a breast) is not as common an operation as it once was. In simple *modified radical mastectomy* the breast is removed down to the deep fascia (Fig. 1-

11). *Radical mastectomy* is a more extensive surgical operation during which the breast, pectoral muscles, fat, fascia, and all lymph nodes in the axilla and pectoral region are excised. Radical mastectomies are uncommon now (Stone, 1989). Often only the tumor and surrounding tissues, such as the one visible in Figure 1-13*B*, are removed. This surgical operation is called a **lumpectomy** or a wide local excision. During mastectomy care is taken to preserve the *long thoracic nerve* (see Fig. 6-23). Cutting this nerve results in paralysis of the serratus anterior muscle and a *winged scapula* (p. 525).

Mammography (radiographic examination of the breast) is one of the radiographic techniques used to detect breast masses. Mammographs are also used by surgeons to guide them during the removal of breast tumors, cysts, and abscesses. Breast tumors emit more heat than normal breast tissue, consequently *thermography* is used as a method of measuring and recording heat radiation emitted by the breast. It is sometimes used in conjunction with mammography. *Computerized tomography* (*CT*) is also combined with mammography for detection of breast cancer. Before CT scans are taken, an iodide-contrast material is given intravenously to the patient. Breast cancer cells have an unusual affinity for iodide and so become recognizable.

Accessory breasts (*polymastia*) or nipples (*polythelia*) may occur superior or inferior to the normal breasts. Usually supernumerary breasts consist of a nipple and areola that may be mistaken for a mole or nevus (birthmark). They may appear anywhere along a line extending from the axilla to the groin; this is the location of the *embryonic mammary ridge* (Moore, 1988).

About 1% of breast cancers occur in males. A malignant breast tumor in males is often hard and tends to infiltrate the deep fascia, pectoralis major muscle, and the apical group of axillary lymph nodes (Fig. 1-12). *CA of the breast is uncommon in males*, but the consequences are serious because they are often not detected until invasion of deep tissues has occurred. A transient increase in the size of the breasts of boys

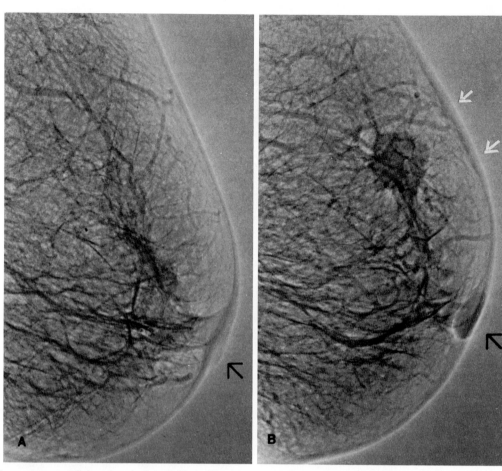

Figure 1-13. Mammograms; mediolateral projection. *A*, Normal breast showing the suspensory ligaments between the lobules of fat and glandular tissue. Veins may also be seen. The skin thickness is uniform except at the nipple (*arrow*). *B*, A breast with a carcinoma; it appears as a jagged density. Note the overlying skin thickening (*white arrows*) resulting from impaired lymphatic drainage. The black arrow indicates the nipple. (Courtesy of Dr. T. Connor, Women's College Hospital, Toronto, Ontario, Canada.)

commonly occurs during puberty; however, in some boys with the *Klinefelter syndrome* (over 50%), the breasts enlarge during puberty and usually increase slowly in size over a period of years. Enlargement of the breasts (*gynecomastia*) is often the presenting symptom in young men with this syndrome, who usually have an XXY sex chromosome complex (Moore, 1988).

Movements of the Thoracic Wall

During inspiration, movements of the thoracic wall and diaphragm result in an increase in all diameters of the thorax (Figs. 1-9 and 1-14 to 1-16). These movements, resulting from muscular action, increase the intrathoracic volume. The resulting changes in pressure result in air being alternately drawn into the lungs (*inspiration*) through the nose, mouth, larynx, and trachea and expelled from the lungs (*expiration*) through the same passages (Fig. 1-14). The thorax increases in diameter in three dimensions to increase its volume: vertically, transversely, and anteroposteriorly.

The Vertical Diameter of the Thorax (Figs. 1-14, 1-49, and 1-50). *During inspiration* the vertical diameter of the thorax is increased as the diaphragm is lowered by contraction. During vigorous inspiration the domes of the diaphragm descend 5 to 10 cm, forcing the abdominal viscera inferiorly. *During expiration* the vertical diameter is returned to normal by the subatmospheric pressure produced in the pleural cavities by the elastic recoil of the lungs. As a result, the domes move superiorly, diminishing the vertical diameter of the thorax. During deep expiration, forced contraction of the abdominal muscles aids in expelling air from the lungs. In addition, when in the supine position, the weight of the abdominal viscera helps to move the diaphragm superiorly, especially when the lower body is raised.

The Transverse Diameter of the Thorax (Fig. 1-15). This dimension of the thorax is increased when the intercostal muscles contract, raising the ribs. As the ribs ascend, they move laterally during the so-called *bucket-handle movement*. When the handle is raised, its convexity moves laterally. In a similar way, a rib (attached anteriorly and posteriorly) moves superolaterally when the intercostal muscles contract.

The Anteroposterior Diameter of the Thorax (Fig. 1-16). This dimension of the thorax is also increased when the intercostal muscles contract, raising the ribs. Movement of the ribs at the costovertebral joints through the axes of their necks causes their anterior ends and the sternum to rise—the so-called

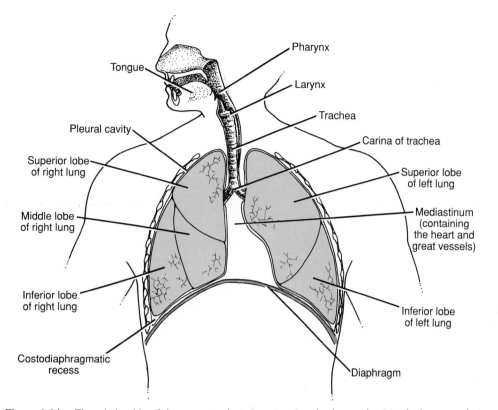

Figure 1-14.   The relationship of the upper respiratory system (conducting portions) to the lower respiratory system (respiratory portions).

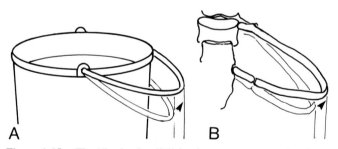

Figure 1-15.   The "bucket-handle" inspiratory movement. *A*, when the pail handle is raised, its convexity moves laterally, away from its attachments. *B*, similarly, when the intercostal muscles contract the ribs move superolaterally, increasing the transverse diameter of the thorax.

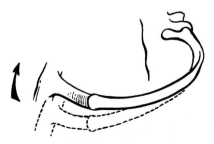

Figure 1-16.   The "pump-handle" inspiratory movement. Anterior parts of the ribs move anteriorly like a pump handle. This action moves the sternum up and down, increasing and decreasing the anteroposterior diameter of the thorax.

pump-handle movement. The second to sixth ribs are primarily involved. Because they slope inferiorly, any elevation of them during inspiration results in movement of the sternum at the manubriosternal joint and an increase in the anteroposterior diameter of the thorax. The sternum not only moves anteriorly but changes to an oblique position, straightening the sternal angle.

During expiration the elastic recoil of the lungs produces a subatmospheric pressure in the pleural cavities. This and the weight of the thoracic walls cause the transverse and anteroposterior diameters of the thorax to return to normal.

> *Paralysis of half of the diaphragm*, resulting from injury or operative division of its motor supply from one of the *phrenic nerves*, does not affect the other half because each half has a separate nerve supply (Fig. 1-42). One can detect paralysis of the diaphragm radiologically by noting its paradoxical movement; instead of descending on inspiration it is pushed superiorly by pressure from the abdominal viscera.

The Thoracic Apertures

The bony thorax has two apertures or openings (Fig. 1-1). The superior thoracic aperture is often referred to as the *thoracic inlet* (Fig. 1-5), and the inferior thoracic aperture is sometimes called the *thoracic outlet*.

The Superior Thoracic Aperture

The head, neck, and limbs communicate with the thoracic cavity through the **thoracic inlet** or superior thoracic aperture

(Figs. 1-1 and 1-5). Through this relatively small, *kidney-shaped opening* (about 5 cm anteroposteriorly and 11 cm transversely) pass structures entering and leaving the thorax, such as the trachea (windpipe), esophagus (gullet), and the great arteries and veins that supply and drain the head, neck, and upper limbs. This aperture is limited by the body of the *first thoracic vertebra* posteriorly, the *first pair of ribs and their costal cartilages* anterolaterally, and the superior end of the *manubrium of the sternum* anteriorly. As the margin of the aperture slopes inferoanteriorly, the apex of each lung and its covering of pleura (pleural cupula) project superiorly through the lateral parts of the thoracic inlet (Figs. 1-26 and 1-33).

The Inferior Thoracic Aperture

At the **thoracic outlet** or inferior thoracic aperture, the thoracic cavity is separated from the abdominal cavity by the musculotendinous *thoracic diaphragm* (Figs. 1-14 and 1-20). Most structures that pass from the thorax to the abdomen, or vice versa, go through openings in the diaphragm (*e.g.*, the inferior vena cava, aorta, and esophagus; Fig. 1-29). The thoracic outlet is uneven and is much larger than the superior thoracic aperture. The inferior thoracic aperture, which slopes inferoposteriorly, is limited by the *12th thoracic vertebra* posteriorly, *the 12th pair of ribs and costal margins* anterolaterally, and the *xiphisternal joint* anteriorly (Figs. 1-1 and 1-9).

On its way to the upper limb, the subclavian artery crosses the first rib, producing a distinct groove (Figs. 1-4 and 1-5). This artery may be compressed where it passes over this rib, producing *vascular symptoms*, *e.g.*, pallor, coldness, and *cyanosis* (blue color) of the hands. Less frequently, nerve pressure symptoms (numbness and tingling) in the digits result from pressure on the inferior trunk of the brachial plexus (Fig. 1-5). These conditions have been described under several different terms (*e.g.*, the *thoracic inlet syndrome*), depending on what the author thought was the cause of the symptoms. Usually the condition is called the **neurovascular compression syndrome**. The designation thoracic inlet syndrome has also been used to describe the signs and symptoms resulting from multiple enlarged lymph nodes that constrict the superior thoracic aperture or thoracic inlet. These nodes usually enlarge as the result of infiltration of them by malignant cells from a *lymphosarcoma*, a lymphatic tumor. As a result, blood does not drain normally from the head, neck, and upper limbs and they become congested with blood and appear swollen.

Muscles of the Thorax

Many muscles are attached to the ribs, such as the anterolateral muscles of the abdomen (Figs. 1-19 and 1-20) and some back muscles (see Chap. 4). The *pectoral muscles*, covering the anterior thoracic wall (see Fig. 6-5), usually act on the upper limbs, but the *pectoralis major* can also function as an *accessory muscle of respiration* to expand the thoracic cavity when inspiration is deep and forceful (*e.g.*, after a 100 m dash). In addition, the *serratus anterior muscle*, which protrudes the scapula (p. 507), runs around the anterolateral surface of the thorax from the scapula (Figs. 1-17 and 1-19). It is also an accessory muscle of respiration.

The *accessory muscles of respiration* (pectoralis major and serratus anterior) are often used by patients with respiratory problems or heart failure when they struggle to breathe. They hold onto a table to fix their pectoral girdles (clavicles and scapulae) so these muscles can act on their attachments to the ribs (see Figs. 6-8 and 6-9).

Muscles of the Thorax Proper

The serratus posterior, levatores costarum, intercostal and subcostal muscles, and transversus thoracis are the muscles of the thorax proper.

The Serratus Posterior Muscles (see Figs. 4-40 and 6-40). These flat *inspiratory muscles* run from the vertebrae to the ribs and are innervated by the intercostal nerves. The *serratus posterior superior* muscle lies at the junction of the neck and back. It arises from the inferior part of the *ligamentum nuchae* in the neck, and from the spinous processes of C7 and T1 to T3 vertebrae. This muscle runs inferolaterally to insert into the superior borders of the second to fourth (or fifth) ribs. The *serratus posterior inferior* muscle lies at the junction of the thoracic and lumbar regions. It arises from the spinous processes of the last two thoracic and the first two lumbar spinous processes. It runs superolaterally and attaches to the inferior borders of the inferior three or four ribs near their angles. The serratus posterior superior elevates the superior four ribs, increasing the diameter of the thorax and raising the sternum, whereas the serratus posterior inferior muscle depresses the inferior ribs, preventing them from being pulled superiorly by the diaphragm.

The Levator Costarum Muscles (Fig. 1-21). These 12 fan-shaped muscles are attached to the transverse processes of C7 and T1 to T11 vertebrae and pass inferolaterally to attach to the ribs inferiorly, close to their tubercles. As their name indicates, they *elevate the ribs*.

The Intercostal Muscles

Typically, the spaces between the ribs, called *intercostal spaces*, contain three layers of muscle (Figs. 1-18 to 1-23). The superficial layer is the *external intercostal muscle*, the middle layer is the *internal intercostal muscle*, and the deepest layer is the *innermost intercostal muscle*.

The External Intercostal Muscles (Figs. 1-18 to 1-23). Each of the *11 pairs* of muscles occupy the intercostal spaces from the tubercles of the ribs posteriorly to the costochondral junctions anteriorly. Anteriorly the muscle fibers are replaced by the *external intercostal membranes*. The muscles run inferoanteriorly from the rib above to the rib below. Each muscle is attached superiorly to the inferior border of the rib and inferiorly to the superior border of the rib below. The external intercostal muscles are *continuous inferiorly with the external oblique* muscles of the anterolateral abdominal wall.

The Internal Intercostal Muscles (Figs. 1-18 to 1-21 and 1-23). The *11 pairs* of muscles run deep to and at right angles to the external intercostal muscles. Their fibers run inferoposteriorly from the floors of the costal grooves to the superior borders of the ribs inferior to them. The internal intercostal

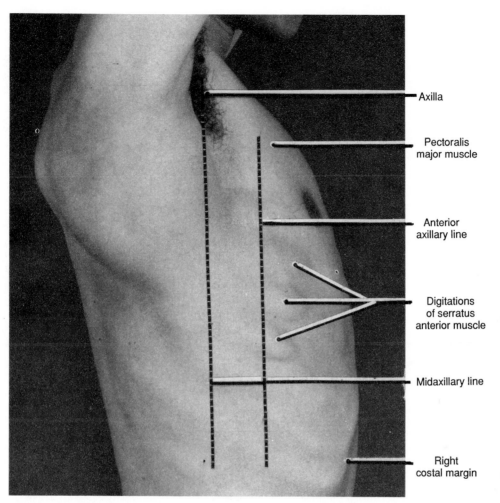

Axilla

Pectoralis
major muscle

Anterior
axillary line

Digitations
of serratus
anterior muscle

Midaxillary line

Right
costal margin

Figure 1-17. Right lateral view of a 27-year-old man with his limbs raised over his head. Most of the contour of his thorax is formed by his pectoralis major muscles, whereas in females the breasts produce varying contours depending on their development.

muscles attach to the shafts of the ribs and their costal cartilages as far anteriorly as the sternum and as far posteriorly as the angles of the ribs. Between the ribs posteriorly, the internal intercostal muscles are replaced by the *internal intercostal membranes*. The inferior internal intercostal muscles are *continuous with the internal oblique* muscles of the anterolateral abdominal wall.

The Innermost Intercostal Muscles[1] (Figs. 1-18 to 1-23). The deepest intercostal muscles are similar to the internal intercostal muscles and are really deep portions of them. The innermost intercostal muscles are separated from the internal intercostal muscles by the intercostal nerves and vessels. These muscles pass between the internal surfaces of adjacent ribs and occupy the middle part of the intercostal spaces.

The Subcostal Muscles (Fig. 1-22). Variable in size and shape, these muscles are thin muscular slips that extend from the internal surface of the angle of one rib to the internal surface of the rib inferior to it. Crossing one or two intercostal spaces, they run in the same direction as the internal intercostal muscles and lie internal to them.

The Transversus Thoracis Muscles (Figs. 1-18 to 1-20). These thin muscles consist of four or five slips that are attached posteriorly to the xiphoid process, the inferior part of the body of the sternum, and the adjacent costal cartilages. They pass superolaterally and are attached to the second to sixth costal cartilages. The transversus thoracis muscles are *continuous inferiorly with the transversus abdominis* muscle. The internal thoracic vessels run anterior to these muscles, between them and the costal cartilages and internal intercostal muscles.

Actions of the Intercostal Muscles (Figs. 1-15 and 1-16). All these inspiratory muscles *elevate the ribs*. This movement expands the thoracic cavity in the transverse and antero-posterior diameters (p. 48). All three layers of intercostal muscles *keep the intercostal spaces rigid*, thereby preventing them from bulging out during expiration and from being drawn in during inspiration.

The Intercostal Spaces

These spaces between the ribs are deeper anteriorly than posteriorly and deeper between the superior than the inferior ribs (Figs. 1-18 to 1-23). The intercostal spaces *widen on inspiration* (Fig. 1-49). Each space contains three muscles and a **neuro-**

[1]These muscles are also called intercostalis intimi or intimal intercostal muscles.

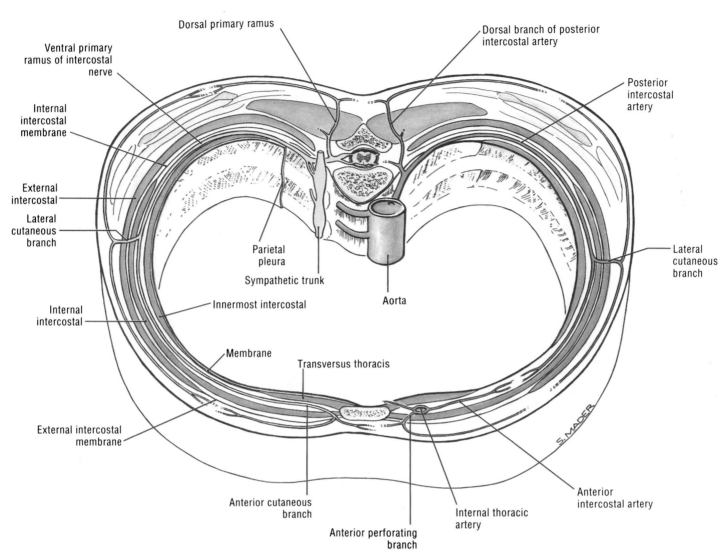

Figure 1-18. Transverse section of the thorax showing the contents of an intercostal space. For purposes of illustration, nerves are shown on the right and arteries on the left. Examine the intercostal vessels and nerves running in the plane between the middle and innermost layers of muscle. The inferior intercostal vessels and nerves occupy the corresponding morphological plane in the abdominal wall.

vascular bundle (vein, artery, and nerve; often referred to as **VAN**).

The Intercostal Nerves

There are *12 pairs* of thoracic nerves. As soon as they pass through the intervertebral foramina, the nerves divide into ventral and dorsal primary rami (Fig. 1-18). The ventral rami of *T1 to T11 are called intercostal nerves because they enter the intercostal spaces* (Figs. 1-18 to 1-20 and 1-23). The ventral ramus of T12, being inferior to the 12th rib, is not in an intercostal space; it is therefore called the *subcostal nerve*. The dorsal rami pass posteriorly, immediately lateral to the articular processes of the vertebrae, to supply the muscles, bones, joints, and skin of the back.

Typical Intercostal Nerves

A typical intercostal nerve (*third to sixth*) enters the intercostal space posteriorly, between the parietal pleura and the internal intercostal membrane (Figs. 1-21 and 1-23*B*). At first it runs near the middle of the intercostal space, across the internal surface of the internal intercostal membrane and muscle (Fig. 1-18). Near the angle of the rib, it passes between the internal intercostal and the innermost intercostal muscles. Here *the intercostal nerve enters and is sheltered by the costal groove* (Fig. 1-3), where it lies just inferior to the intercostal artery (Figs. 1-22 and 1-23*B*). It continues anteriorly between the internal and innermost intercostal muscles, giving branches to these and other muscles and a lateral cutaneous branch (Fig. 1-18). Anteriorly the nerve appears on the internal surface of the internal intercostal muscle (Fig. 1-19), external to the transversus thoracis muscle and internal thoracic vessels. Near the sternum the intercostal nerve turns anteriorly and ends as an anterior cutaneous branch (Fig. 1-18). The intercostal nerves supply successive segments of the thoracoabdominal wall. The band of skin supplied by each pair of intercostal nerves is known as a *dermatome* (p. 57). The group of muscles supplied by a pair of intercostal nerves is known as a *myotome*.

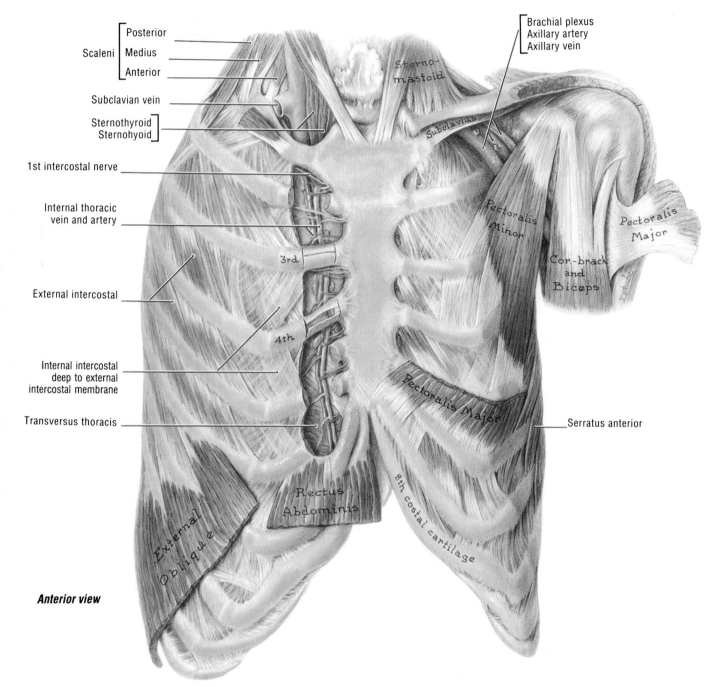

Anterior view

Figure 1-19. Dissection of the anterior thoracic wall. Observe the internal thoracic vessels running about 1 cm lateral to the border of the sternum. Examine the parasternal lymph nodes (*green*), which receive lymph vessels from the intercostal spaces, costal pleura, diaphragm, and breasts. Although the seventh costal cartilage is usually the last one to attach to the sternum, it is not uncommon (as in this specimen), for the eighth to do so. The H-shaped cut through the perichondrium of the third and fourth costal cartilages on the right side was used to shell out segments of cartilage as an illustration of one surgical approach to the thoracic cavity.

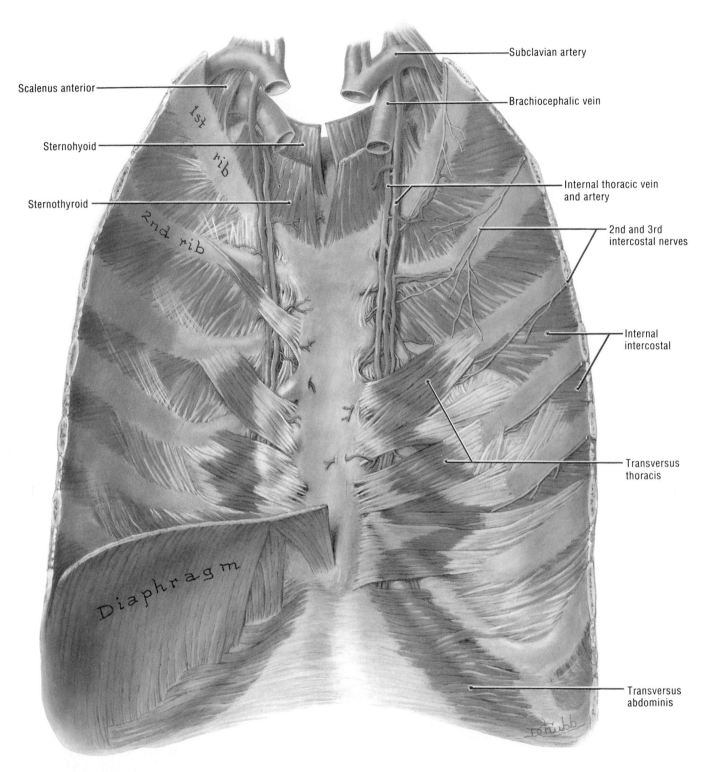

Subclavian artery

Brachiocephalic vein

Internal thoracic vein and artery

2nd and 3rd intercostal nerves

Internal intercostal

Transversus thoracis

Transversus abdominis

Scalenus anterior

Sternohyoid

Sternothyroid

1st rib

2nd rib

Diaphragm

Posterior view

Figure 1-20.   Dissection of the anterior thoracic wall. Observe the internal thoracic arteries arising from the subclavian arteries and the internal thoracic veins, which are tributaries of the brachiocephalic veins.

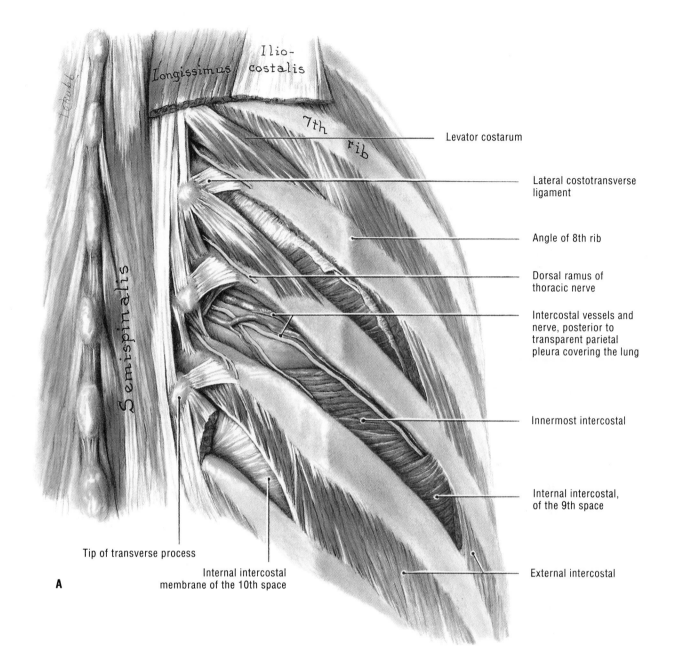

Levator costarum

Lateral costotransverse ligament

Angle of 8th rib

Dorsal ramus of thoracic nerve

Intercostal vessels and nerve, posterior to transparent parietal pleura covering the lung

Innermost intercostal

Internal intercostal, of the 9th space

External intercostal

Tip of transverse process

Internal intercostal membrane of the 10th space

A

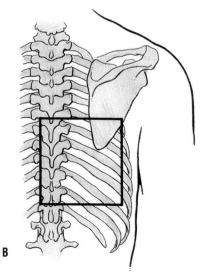

Figure 1-21.   *A*, Dissection of the posterior end of an inferior intercostal space. Medial to the angles of the ribs, the iliocostalis and longissimus muscles have been removed to expose the levator costarum muscle. The internal intercostal membrane (Fig. 1-23*A*) has been removed to show the intercostal nerves and vessels. *B*, Diagram indicating the area shown in *A*.

B

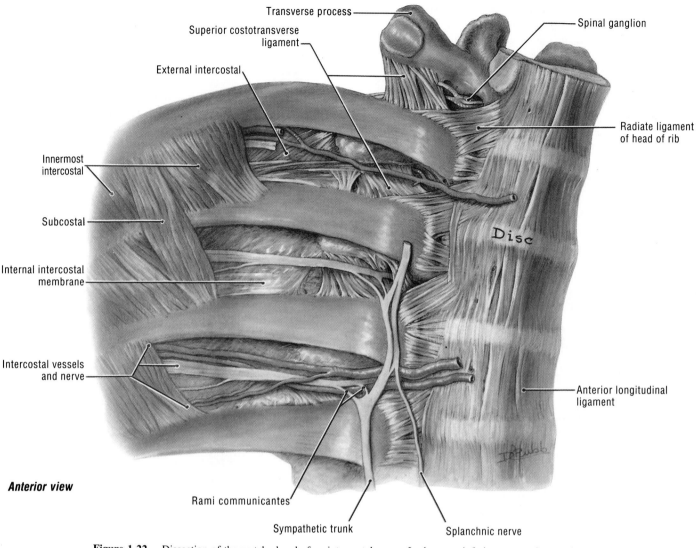

Transverse process
Superior costotransverse ligament
External intercostal
Innermost intercostal
Subcostal
Internal intercostal membrane
Intercostal vessels and nerve
Anterior view
Rami communicantes
Sympathetic trunk
Splanchnic nerve
Spinal ganglion
Radiate ligament of head of rib
Disc
Anterior longitudinal ligament

Figure 1-22. Dissection of the vertebral end of an intercostal space. In the most inferior space, observe the order of the structures (intercostal vein, artery, and nerve). Note their collateral branches.

Branches of a Typical Intercostal Nerve (Figs. 1-18 to 1-23).

1. *Rami communicantes* or communicating branches connect each intercostal nerve to a sympathetic trunk. The nerve sends a white ramus communicans to a ganglion of the *sympathetic trunk* and receives a gray ramus communicans from a ganglion of the trunk. *Sympathetic nerve fibers* are distributed through all branches of the intercostal nerve to blood vessels, sweat glands, and smooth muscle.

2. *Collateral branches* arise near the angles of the ribs and supply the intercostal muscles.

3. *Lateral cutaneous branches* arise beyond the angles of the ribs and pierce the internal and external intercostal muscles about halfway around the thorax (Fig. 1-18). The cutaneous branches divide into anterior and posterior branches, which supply the skin on the lateral aspect of the thoracic and abdominal walls.

4. *Anterior cutaneous branches* supply the skin on the anterior aspect of the thorax and abdomen.

5. *Muscular branches* supply the subcostal, transversus thoracis, levatores costarum, and serratus posterior muscles.

The thoracoabdominal nerves (inferior five intercostal nerves; T7 to T11) supply the thoracic and abdominal walls. Communicating branches link these nerves to one another posteriorly as they traverse the anterior abdominal wall. As soon as they enter the superficial fascia they are known as the *anterior cutaneous nerves of the thorax* and abdomen (Fig. 1-18; see also Figs. 2-7 and 2-13).

The subcostal nerve (see Figs. 2-82 and 2-99) follows the inferior border of the 12th rib and passes into the abdominal wall. It is distributed to the abdominal wall and skin in the gluteal region (buttocks).

Atypical Intercostal Nerves

In the first part of their course, the first and second intercostal nerves pass on the internal surfaces of the first and second ribs.

The first intercostal nerve has no anterior cutaneous branch and usually has no lateral cutaneous branch. It divides into a large superior and a small inferior part. The superior part joins the *brachial plexus*, the nerve plexus supplying the upper limb (Fig. 1-19; see also Figs. 6-17 and 6-18). The inferior part becomes the first intercostal nerve. The second intercostal nerve may also contribute a small branch to the brachial plexus. The lateral cutaneous branch of the second intercostal nerve is called the *intercostobrachial nerve* because it supplies the floor of the axilla

and then communicates with the medial brachial cutaneous nerve to supply the medial side of the upper limb as far as the elbow region.

The Dermatomes

Through the cutaneous branches of the dorsal and ventral rami (Fig. 1-18), each spinal nerve supplies a continuous strip of skin, which extends from the posterior midline to the anterior

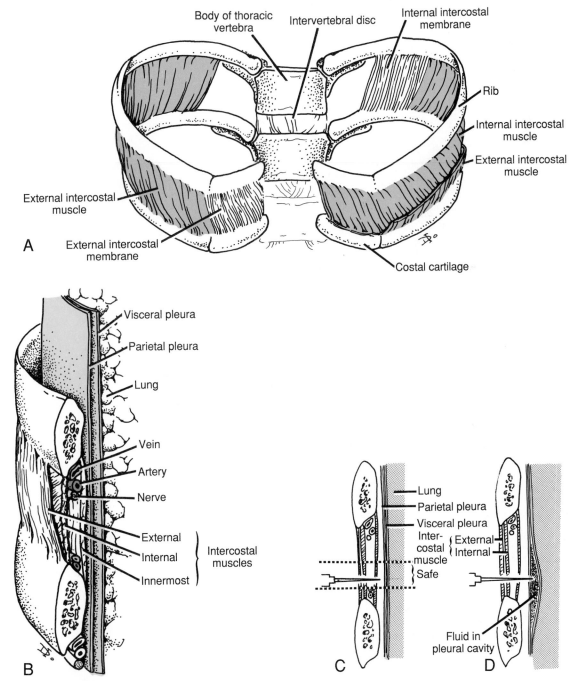

Figure 1-23. The intercostal spaces. *A*, the external and internal intercostal muscles and membranes, *B*, the position of the *neurovascular bundle* (VAN) between the intercostal muscles, *C*, a needle is passed through an intercostal space superior to the rib to avoid the neurovascular bundle, *D*, a needle is passed through an intercostal space to withdraw fluid from the pleural cavity.

midline. These *strips of skin* are called **dermatomes**. This term is derived from Greek and means *skin slices*. As illustrated in Figure 1-24, the thoracic and abdominal dermatomes are arranged in a segmental fashion because the thoracoabdominal nerves arise from segments of the spinal cord. *There is considerable overlapping of contiguous or closely related dermatomes;* *i.e.*, each segmental nerve overlaps the territories of its neighbors (see Fig. 6-109). The first two intercostal nerves (T1 and T2) supply the upper limbs in addition to supplying the thoracic wall (Fig. 1-24). Also note that the inferior five intercostal nerves (T7 to T11) supply the abdominal wall, as well as the thoracic wall (Figs. 1-18 and 1-24; see also Fig. 2-13).

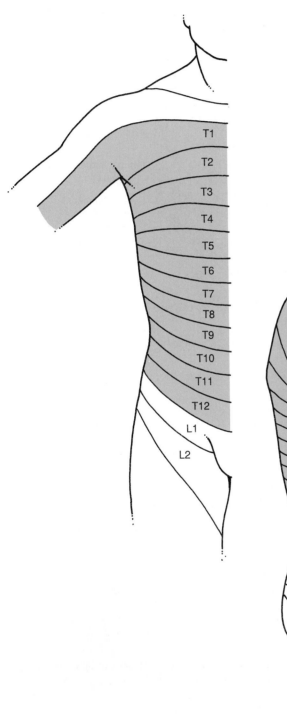
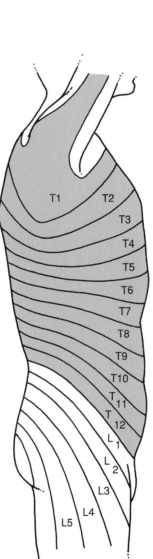
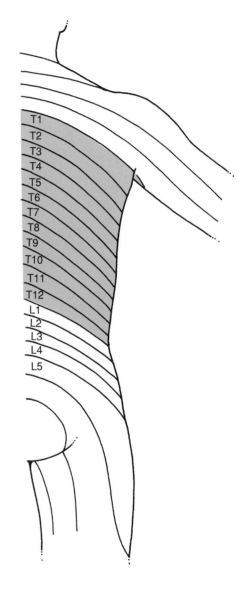

Figure 1-24.   The dermatomes of the trunk and proximal parts of the limbs. The thoracic dermatomes are shown in red.

Herpes zoster (''shingles'') is a viral disease of the spinal ganglia. The term literally means a ''creeping, *girdle-shaped cutaneous eruption.*'' When the virus invades the spinal ganglia, a sharp burning pain is produced in the area of skin supplied by the nerves involved. A few days later, the involved dermatomes become red and vesicular eruptions appear.

It is sometimes necessary to insert a hypodermic needle through an intercostal space to obtain a sample of pleural fluid or to remove blood or pus from the pleural cavity (Fig. 1-23*D*). Sometimes a tube is inserted through an intercostal space to drain fluid or air from the pleural cavity. For more about this procedure called *thoracotomy*, see p. 64. The site chosen for aspiration (removal by suction) of the pleural cavity depends on where the abnormal fluid is located. Usually, however, the tube is passed through the intercostal space just lateral to the angles of the ribs. In other cases it may be necessary to *anesthetize an intercostal nerve*. To avoid damage to the main intercostal vessels and nerve, the needle is inserted superior to the rib, just high enough to avoid the collateral branches (Fig. 1-23*C*). *Local anesthesia of an intercostal space* may be produced by injecting an anesthetic agent around the origin of an intercostal nerve, just lateral to the vertebra concerned (Figs. 1-18 and 1-22). This procedure, involving infiltration of a local anesthetic agent around the intercostal nerve trunk and its collateral branch, is known as an **intercostal nerve block**. The term *block* indicates that the nerve endings (pain receptors) in the skin and the transmission of impulses in the sensory nerves carrying information about pain are interrupted (blocked) before they can reach the spinal cord and brain. Because there is considerable overlapping of contiguous dermatomes, complete anesthesia results only when two or more consecutive intercostal nerves are anesthetized.

Pus (a product of inflammation), originating in the region of the vertebral column, tends to pass around the thorax along the course of the neurovascular bundles (Figs. 1-21 to 1-23), and to escape on the surface where the cutaneous branches of the intercostal nerves pierce the muscles to supply the skin (Fig. 1-18; see also Fig. 2-13).

The Intercostal Arteries

Each intercostal space is supplied by three arteries, a large posterior intercostal artery and a small pair of anterior intercostal arteries (Figs. 1-18 and 1-23*B*).

The Posterior Intercostal Arteries

The first two posterior intercostal arteries arise from the *superior intercostal artery*, a branch of the costocervical trunk of the subclavian artery (Figs. 1-20 and 1-75). Nine pairs of posterior intercostal arteries and one pair of subcostal arteries arise posteriorly from the thoracic aorta. Each posterior intercostal artery gives off a *posterior branch*, which accompanies the dorsal ramus of the spinal nerve (Fig. 1-18) to supply the spinal cord, vertebral column, back muscles, and skin (see Fig. 4-52*B*). Each artery also gives off a small *collateral branch* that crosses the intercostal space and runs along the superior border of the rib inferior to the space (Fig. 1-21). The terminal branches of the posterior intercostal artery anastomose anteriorly with the anterior intercostal artery (Fig. 1-18). The posterior intercostal

artery accompanies the intercostal nerve through the intercostal space. Close to the angle of the rib it enters the costal groove, where it lies between the intercostal vein and nerve (Fig. 1-22). At first the artery runs between the pleura and the internal intercostal membrane (Figs. 1-18 and 1-21); it then runs between the innermost intercostal and internal intercostal muscles.

The Anterior Intercostal Arteries

The anterior intercostal arteries supplying the *superior six intercostal spaces* are derived from the *internal thoracic arteries* (Figs. 1-18 to 1-20). There are two anterior intercostal arteries for each intercostal space. The arteries supplying the *seventh to ninth intercostal spaces* are derived from the *musculophrenic arteries*, branches of the internal thoracic arteries. These arteries pass laterally, one near the inferior margin of the superior rib and the other near the superior margin of the inferior rib. At their origins, the first two arteries lie between the pleura and the internal intercostal muscles, whereas the next four arteries are separated from the pleura by the transversus thoracis muscle (Fig. 1-20). The anterior intercostal arteries supply the intercostal muscles and send branches through them to the pectoral muscles, breasts, and skin. There are no anterior intercostal arteries in the inferior two intercostal spaces. These spaces are supplied by the posterior intercostal arteries and their collateral branches.

The Intercostal Veins

The intercostal veins accompany the intercostal arteries and nerves; they lie deepest in the costal grooves (Figs. 1-19 to 1-23). There are *eleven posterior intercostal veins* and one *subcostal vein* on each side. They receive lateral cutaneous, muscular, intervertebral, and posterior tributaries. *The intercostal veins contain valves*, which direct the blood posteriorly. The posterior intercostal veins anastomose with the *anterior intercostal veins*, which are tributaries of the internal thoracic veins. Most intercostal veins end in the *azygos vein* (p. 115), which conveys venous blood to the *superior vena cava* (Fig. 1-76). The superior intercostal veins drain into the superior vena cava (Fig. 1-72).

The Internal Thoracic Vessels

The Internal Thoracic Artery (Figs. 1-19 and 1-20). This vessel arises in the root of the neck from the inferior surface of the first part of the *subclavian artery* at the medial border of the scalenus anterior muscle. The internal thoracic artery (formerly the internal mammary) descends into the thorax posterior to the clavicle and first costal cartilage. It runs on the internal surface of the thorax a little lateral to the sternum. It lies on the pleura posteriorly and is *crossed by the phrenic nerve*, the nerve to the diaphragm. The internal thoracic artery runs inferiorly in the thorax posterior to the superior six costal cartilages and the intervening intercostal muscles. At the level of the third costal cartilage, it continues inferiorly, anterior to the transverse thoracis muscle, to end in the sixth intercostal space where it *divides into the superior epigastric and musculophrenic arteries.*

The Internal Thoracic Veins (Fig. 1-20).   These vessels are the venae comitantes of the internal thoracic artery. In the region of the first to third intercostal space, they usually unite to form a trunk (single vein) that empties into the corresponding *brachiocephalic vein*; however, the right internal thoracic trunk may empty into the superior vena cava.

> Because the internal thoracic vessels descend about 2.5 cm lateral to the sternum, it is usually possible to insert a needle between them and the sternum when removing fluid from the pericardial cavity around the heart (Fig. 1-41), a procedure called *pericardiocentesis* (p. 87). However, these vessels are not always in the usual position; therefore, there is a possibility of piercing them in some patients.

The Thoracic Cavity

The thoracic cavity is divided into three divisions: *two pleural cavities* and the *mediastinum*. The pleural cavities are completely separate from each other, and with the lungs, they occupy most of the thoracic cavity (Fig. 1-14). The lungs and their covering membranes, called *pleurae*, occupy the lateral parts of the thoracic cavity. The space between the lungs and pleurae is called the *mediastinum*. It extends from the superior thoracic aperture to the diaphragm inferiorly (p. 79). The mediastinum contains the heart, thoracic parts of the great vessels, and other important structures (*e.g.*, thoracic parts of the trachea, esophagus, thymus, autonomic nervous system, and lymphatic system).

Pleurae and Pleural Cavities

Each lung is surrounded by a **pleural sac** (Fig. 1-25). *A parietal layer* of pleura forms its external wall and a *visceral layer* invests the lung. As the lung expands during inspiration, the visceral pleura is brought into contact with the parietal pleura, reducing the *pleural cavity* to a very thin space. The pleural

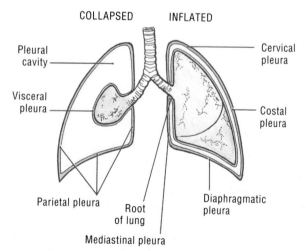

Figure 1-25. The lungs, pleural cavities, and pleurae. In the inflated lung, note that the parietal pleura is applied to the visceral pleura but is separated from it by a narrow "space," the pleural cavity.

cavity is a separate and closed potential space. Normally, contains only a capillary layer of serous fluid, which is secreted by the pleura. This *pleural fluid* lubricates the pleural surfaces and reduces friction between the parietal and visceral layers of pleura. Hence, the apposed surfaces of parietal and visceral pleura slide smoothly against each other during respiration. As a result, respiratory movements occur without interference.

The Visceral Pleura

The moist and shiny visceral (pulmonary) pleura *closely covers the lung* and is adherent to all its surfaces (Figs. 1-25 and 1-38A). It provides the lung with a smooth, slippery surface that enables it to move freely on the parietal pleura. The visceral pleura dips into the fissures of the lungs so that the lobes are also covered with it (Fig. 1-30). In these fissures the visceral pleura of adjacent lobes is in contact. The visceral pleura is continuous with the parietal pleura at the *root of the lung* (Figs. 1-25 and 1-27), where structures enter and leave it (p. 66).

The Parietal Pleura

The external wall of the pleural cavity, formed by parietal pleura, is adherent to the thoracic wall and diaphragm by connective tissue. Consequently it moves as these structures change their positions during respiration. The parietal pleura is also attached to the *pericardium* or pericardial sac, the fibroserous sac that encloses the heart (Figs. 1-30A, 1-41, and 1-42). The parietal pleura is given different names according to the parts with which it is associated: *costal pleura* (ribs and sternum); *mediastinal pleura* (mediastinum); *diaphragmatic pleura* (diaphragm); and pleural cupula (apex of the lung).

The Costal Pleura (Figs. 1-25 to 1-27 and 1-76).   This part of the parietal pleura covers the internal surfaces of the sternum, costal cartilages, ribs, intercostal muscles, and the sides of the thoracic vertebrae. It is separated from these structures by a thin layer of loose connective tissue, called *endothoracic fascia*, which provides a natural *cleavage plane* for the surgical separation of the costal pleura from the thoracic wall.

The Mediastinal Pleura (Figs. 1-14, 1-25 to 1-27, and 1-40).   This part of the parietal pleura *covers the mediastinum*, the space between the two pleural sacs. It is continuous with the costal pleura anteriorly and posteriorly; with the diaphragmatic pleura inferiorly; and with the cupula or cervical pleura superiorly. Superior to the root of the lung, the mediastinal pleura is a continuous sheet between the sternum and the vertebral column. At the *root of the lung*, the mediastinal pleura passes laterally, where it encloses the structures in the root (Figs. 1-31 and 1-32), and becomes continuous with the visceral pleura. Inferior to the root of the lung, the mediastinal pleura passes laterally as a double layer from the esophagus to the lung, where it is continuous with the visceral pleura. This double layer of pleura is called the *pulmonary ligament*. It is continuous superiorly with the mediastinal pleura enclosing the lung root and ends inferiorly in a free border (Figs. 1-31 and 1-32). The parietal layers of mediastinal pleura come into close apposition between the aorta and esophagus, just superior to the diaphragm (Fig. 1-29).

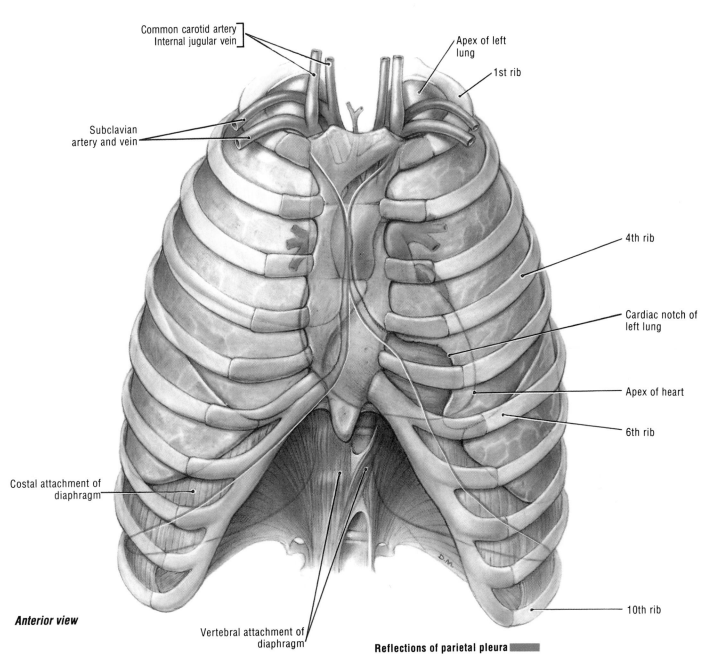

Common carotid artery
Internal jugular vein

Apex of left lung

1st rib

Subclavian artery and vein

4th rib

Cardiac notch of left lung

Apex of heart

6th rib

Costal attachment of diaphragm

10th rib

Anterior view

Vertebral attachment of diaphragm

Reflections of parietal pleura

Figure 1-26. The contents of the thorax and the sternal and costal lines of pleural reflection. Observe that the pleural sacs do not extend to the anterior ends of the seventh, eighth, and ninth intercostal spaces (*i.e.*, they do not reach the costal margins). The reason for this is that the diaphragm is attached to the internal surfaces of these costal cartilages.

The Diaphragmatic Pleura (Figs. 1-14, 1-25, 1-27, and 1-29).   This part of the parietal pleura covers the superior surface of the diaphragm, lateral to the mediastinum.

The Pleural Cupula or Cervical Pleura (Figs. 1-25 to 1-27 and 1-33).   The pleural cupula (cupola) is the dome-shaped apex of the pleural sac that is formed by the cervical pleura. The pleural cupula is the continuation of the costal and mediastinal layers of pleura that *covers the apex of the lung*. It extends superiorly posterior to the clavicle and first rib and passes through the superior thoracic aperture into the root of the neck. Its summit is nearly 3 cm superior to the level of the medial third of the clavicle at the level of the neck of the first rib. The pleural cupulae, well seen in radiographs and other images of the lungs, result from the obliquity of the superior thoracic aperture or thoracic inlet (Fig. 1-5). The cupulae are separated by the trachea and esophagus and the blood vessels passing to and from the neck (Fig. 1-30A). Each cupula is strengthened by a layer of dense fascia, called the *suprapleural membrane*, which is attached to the internal border of the first rib and the anterior border of the transverse process of C7 vertebra. This membrane usually contains some fibers of the *scalenus minimus muscle*, which give it added strength (see Fig. 8-30).

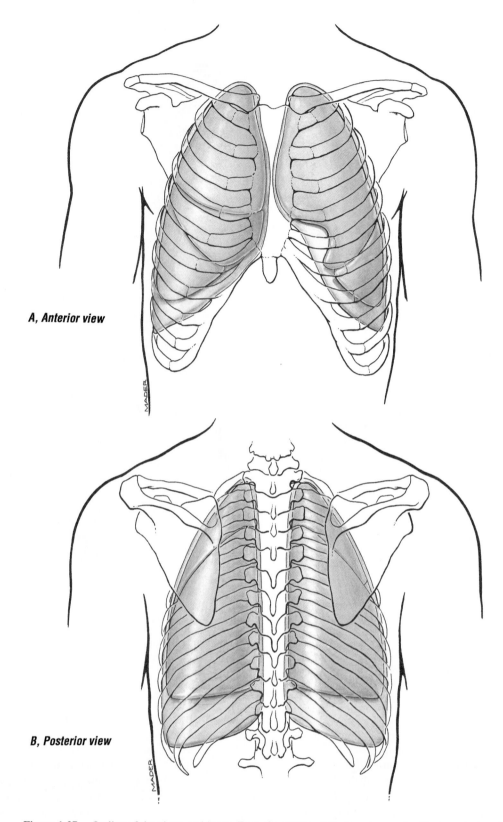

A, Anterior view

B, Posterior view

Figure 1-27.   Outline of the pleura and lungs. Trace the outline of the lung (covered with visceral pleura) and the parietal pleura as observed in quiet respiration.

Because of the inferior slope of the first pair of ribs, the cupulae of the pleura and the apices of the lung project superiorly through the thoracic inlet into the neck, posterior to the sternocleidomastoid muscles (Figs. 1-25, 1-33, and 1-48). Consequently, *the pleural cupulae and apices of the lungs may be injured in wounds of the neck*. In such cases air will enter the pleural cavity, producing *pneumothorax* (p. 64). Because the parietal pleura is innervated by intercostal nerves, inflammation of the pleura (**pleuritis** or pleurisy) results in pain being referred to the cutaneous distribution of these nerves, *i.e.*, to the thoracic and abdominal walls (Fig. 1-24).

Pleural Reflections

The relatively abrupt lines along which the parietal pleura folds back or changes direction from one wall of the pleural cavity to another are known as *pleural reflections*. They occur where the costal pleura becomes continuous with the mediastinal pleura anteriorly and posteriorly and with the diaphragmatic pleura inferiorly (Figs. 1-26 and 1-27). The sternal and costal lines of pleural reflection are important landmarks in clinical medicine.

The sternal line of pleural reflection (anterior costomediastinal reflection) is where the costal pleura is continuous with the mediastinal pleura posterior to the sternum (Figs. 1-26 and 1-27). The right and left sternal reflections are indicated by lines that pass inferomedially from the sternoclavicular joints to the anterior median line at the level of the sternal angle. Here the two pleural sacs come in contact and may slightly overlap each other. *On the right side*, the sternal line of reflection passes inferiorly in the median plane to the posterior aspect of the xiphoid process, where it turns laterally. *On the left side*, the sternal line of pleural reflection passes inferiorly in the median plane to the level of the fourth costal cartilage. Here it passes to the left margin of the sternum and then continues inferiorly to the sixth costal cartilage (Fig. 1-26).

The costal line of pleural reflection is where the costal pleura is continuous with the diaphragmatic pleura. It passes obliquely across the *8th rib* in the midclavicular line, the *10th rib* in the midaxillary line, and the *12th rib* at its neck or inferior to it.

The fact that the pleural sacs are separate is important clinically because it makes it possible to collapse a lung on one side without collapsing the one on the other side. *The pleurae descend inferior to the costal margin in three regions* where an abdominal incision might inadvertently enter a pleural sac: (1) the right part of the infrasternal angle (Figs. 1-26 and 1-33) and (2) the right and left costovertebral angles (Fig. 1-27). In persons whose 12th ribs are very short, the pleurae lie inferior to the posterior costal margin after crossing the 11th rib and are therefore in danger during surgery in this region.

The parietal pleura is visible radiographically in certain regions only. The visceral pleura is usually not visible, except in certain views that demonstrate the fissures of the lung. Any disease process that causes *thickening of the pleurae* is likely to make them more visible in radiographs.

The Pleural Recesses

During deep inspiration the lungs fill the pleural cavities (Figs. 1-25 [left lung] and 1-49), but during quiet respiration parts of these cavities are not occupied by the lungs. Here, portions of the parietal pleura are in contact. These *potential spaces* are called pleural recesses.[2]

The Costodiaphragmatic Recesses (Figs. 1-14, 1-29, and 1-76; see also Fig. 2-25). These slitlike intervals are between the costal and diaphragmatic pleurae on each side. The layers of parietal pleura in these areas are separated by only a capillary layer of fluid, and they become alternately smaller and larger as the lungs move in and out of them during inspiration and expiration.

The Costomediastinal Recesses (Figs. 1-26 and 1-29). These potential spaces lie along the anterior margin of the pleura; here the costal and mediastinal parts of the parietal pleura come into contact. The left recess is larger because of the presence of a semicircular deficiency, the *cardiac notch*, in the left lung anterior to the pericardium (Figs. 1-30B and 1-32). The costomediastinal recesses lie at the anterior ends of the fourth and fifth intercostal spaces. During inspiration and expiration, a thin tonguelike edge of the left lung, called the *lingula* (L. dim. of *lingua*, tongue), slides in and out of the left costomediastinal recess.

The pleural cavities are potential spaces. They are filled with a thin film of pleural fluid, which facilitates movement of the lungs. Obliteration of a pleural cavity by disease or during surgery (*pleurectomy*), however, does not cause appreciable functional consequences. In other surgical procedures (*pleural poudrage*), adherence of the visceral and parietal layers of pleura is purposely induced by covering the apposing pleural surfaces with a slightly irritating powder. This surgical operation may be performed to prevent recurring *spontaneous pneumothorax* (entry of air in the pleural cavity) resulting from disease of the lung.

During inspiration and expiration the normally moist, smooth pleurae make no detectable sound during *auscultation* (listening to the sounds made by the thoracic viscera); however, inflammation of the pleurae (*pleuritis*) makes the surfaces of the lung rough. The resulting friction rub (*pleural rub*) may be heard with a stethoscope. Irritation of the parietal pleura causes pain that is referred to the thoracoabdominal wall (innervated by intercostal nerves) or to the shoulder. Pleuritis may also cause *pleural adhesions* to form between the parietal and visceral layers of pleura.

The accumulation of significant amounts of fluid in the pleural cavity (*hydrothorax*) may result from a variety of causes. In advanced cases of pleuritis, serum from the inflamed pleurae may exude or effuse from the blood vessels of the pleurae into the pleural cavity, forming a *pleural exudate*. As fluid accumulates, the negative or subatmospheric pressure is lessened, allowing the lung to retract. When the lung is completely retracted, additional fluid will displace the heart and mediastinum toward the opposite side. If there is a **chest wound**, blood may also enter the pleural cavity (*hemothorax*). In unusual cases, *chyle* (lymph and emulsified fat)

---

[2]A pleural recess is a deep hollow created by a pleural reflection, a "folding back" of the pleura.

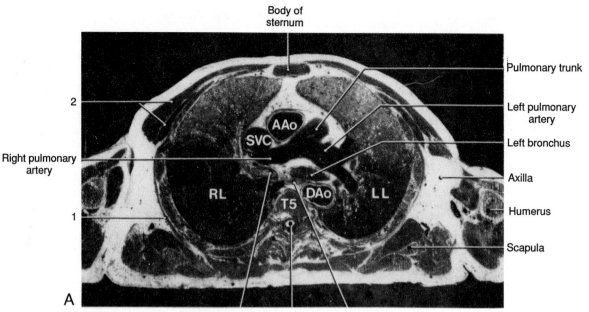

Body of
sternum

Pulmonary trunk

Left pulmonary
artery

Left bronchus

Axilla

Humerus

Scapula

Right pulmonary
artery

2

1

A

Right bronchus   Spinal cord   Esophagus

Figure 1-28.   *A*, Transverse section of the thorax of an adult female cadaver at the level of T5 vertebra. In keeping with radiographic conventions, as in CT scans and MRI, the section is viewed inferiorly. AAo indicates ascending aorta; DAo, descending aorta; SVC, superior vena cava; RL, right lung; LL, left lung; T5, body of fifth thoracic; 1, serratus anterior muscle; 2, pectoralis major and minor muscles. *B*, CT scans of the thorax at progressively lower levels showing the heart and lungs. (Courtesy of Dr. Tom White, Department of Radiology, College of Medicine, The Health Science Center, the University of Tennessee, Memphis, TN.)

may pass into the pleural cavity from a *ruptured thoracic duct* (Fig. 1-39). This condition is called *chylothorax*.

Entry of air into a pleural cavity (**pneumothorax**), following a penetrating wound or rupture of a lung, results in partial collapse of the lung. Fractured ribs often produce a pneumothorax, but the most common type is *spontaneous pneumothorax*. It usually results from rupture of bullae (blebs) on the surface of the lungs. As mentioned previously (p. 63), the pleural cupulae (Fig. 1-33) are vulnerable to stab wounds in the root of the neck. These injuries may penetrate the pleural cavity and lung, producing an *open pneumothorax*. In these cases, there is communication between the atmosphere and the pleural cavity. This condition is sometimes referred to as a ''blowing wound'' or a *sucking pneumothorax*.

Most cases of uncomplicated pneumothorax are not dangerous, but if the opening in the visceral pleura has a flap over it, air can enter the pleural cavity on inspiration but cannot leave it during expiration. In such cases, the amount of air in the pleural cavity increases (*positive pressure pneumothorax*), pushing the mediastinum to the other side, compressing the opposite lung, and creating a medical emergency. Air or fluid (blood, pus, chyle) can be removed from the pleural cavity with a wide-bore needle (Fig. 1-23*C*), but if a large amount is present, it may be necessary to insert a draining tube through the intercostal space. The other end of the tube is connected to an underwater seal in a bottle. This procedure is called a *closed thoracotomy*.

The pleura may also be injured by a needle when a *stellate ganglion block* (p. 815) or a nerve block of the brachial plexus is performed (p. 794). The anesthetist's needle tears the pleura as the lungs move during respiration. Because the pleura crosses the 12th rib (Fig. 1-27), it may also be injured during removal of a kidney (*nephrectomy*). When the 12th rib is very

short, the 11th rib may be mistaken for it; consequently a posterior incision reaching it would result in opening of the pleural cavity to the atmosphere (*open pneumothorax*).

Arterial Supply of the Pleurae

The arterial supply of the *parietal pleura* (Figs. 1-18 to 1-22) is from the arteries that supply the thoracic wall (intercostal, internal thoracic, and musculophrenic arteries). The arterial supply of the *visceral pleura* is from the **bronchial arteries**, which are branches of the thoracic aorta (Figs. 1-38, 1-69, and 1-75).

Venous and Lymphatic Drainage of the Pleurae

The veins from the parietal pleura join the systemic veins in adjacent parts of the thoracic wall. Similarly, its *lymphatic vessels* drain into adjacent lymph nodes of the thoracic wall (intercostal, parasternal, posterior mediastinal, and diaphragmatic). These nodes may in turn drain into the axillary lymph nodes (Figs. 1-12 and 1-19). *The veins from the visceral pleura* drain into the pulmonary veins and its numerous lymphatic vessels drain into nodes at the hila of the lungs (Figs. 1-31 and 1-32).

The nerve supply of the pleurae is described with the lungs on p. 77.

The Lungs

The lungs (L. *pulmones*) are the essential organs of respiration (Figs. 1-28 to 1-32). Their main function is to *oxygenate the mixed venous blood*. Within them, the inspired air is brought

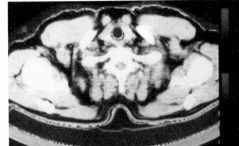
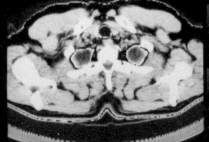
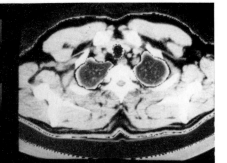
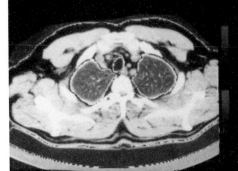
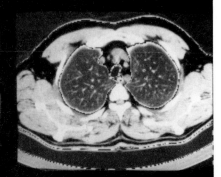
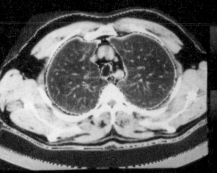
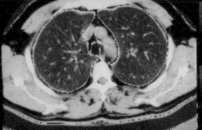
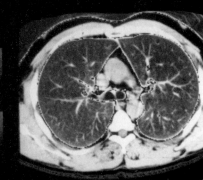
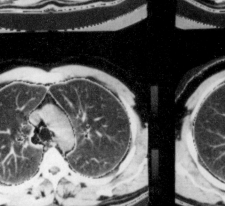
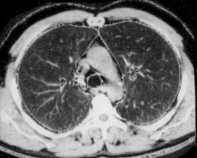
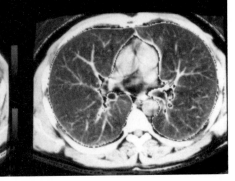

Figure 1-28*B*

65

into close relation to the blood in the pulmonary capillaries (Fig. 1-38). Although cadaveric lungs are shrunken, hard and discolored, *healthy lungs during life are normally light, soft, and spongy*. They are very elastic and will recoil to about one-third their size when the thoracic cavity is opened (Fig. 1-25). Each lung fills its side of the thoracic cavity and is *radiolucent* (Fig. 1-37).

In healthy people who live in a clean environment, the lungs are light pink, but in city dwellers they are often dark and mottled owing to the accumulation of inhaled dust particles, which become trapped in their fixed pulmonary *phagocytes* (scavenger cells). Each lung is conical in shape and is contained in its own pleural sac (Figs. 1-25 and 1-33). The lungs are separated from each other by the heart and great vessels in the middle mediastinum (Figs. 1-14 and 1-30A). The lungs are attached to the heart and trachea by the structures in the *roots of the lungs* (pulmonary arteries, pulmonary veins, and main bronchi) and to the pericardium by the *pulmonary ligaments*[3] (Figs. 1-30 to 1-32). The hardened lungs in an embalmed cadaver have surface impressions which were formed by the structures adjacent to them, *e.g.*, the ribs and aorta (Figs. 1-31 and 1-32). These markings provide clues to the relationships of the lungs, but they are not visible during surgery or in fresh postmortem specimens. Each lung has an *apex, base, root*, and *hilum* (hilus).

The Apex of the Lung (Figs. 1-26 to 1-28, 1-30 to 1-32, and 1-48). The rounded, tapered superior end or apex of the lung extends through the superior thoracic aperture into the root of the neck. Here, it lies in close contact with the dome formed by cervical pleura, called the *cupula of the pleura*. Owing to the obliquity of the superior thoracic aperture (thoracic inlet), the apex of the lung extends up to 3 cm superior to the anterior end of the first rib and its costal cartilage and the medial end of the clavicle. These bony structures afford some protection to the apex, but its most superior part is protected only by soft tissues. *The apex of the lung is crossed by the subclavian artery* (Figs. 1-5 and 1-62), which produces a groove in the mediastinal surface of the embalmed lung. However, the artery is separated from the cupula by the *suprapleural membrane* (p. 61).

The Base of the Lung (Figs. 1-29, 1-31, 1-32, 1-39, and 1-50). This is the concave diaphragmatic surface of the lung, which is *related to the dome of the diaphragm*. The base of the right lung is deeper because the right dome rises to a more superior level. Its inferior border is thin and sharp where it *enters the costodiaphragmatic recess* (Fig. 1-29; see also Fig. 2-25).

The Root of the Lung (Figs. 1-31 and 1-32). The root serves as *the attachment of the lung* and is the "highway" for transmission of the structures entering and leaving the lung at the hilum. It connects the medial surface of the lung to the heart and trachea and is surrounded by the reflection of parietal to visceral pleura (see also p. 60).

The Hilum of the Lung[4] (Figs. 1-31 and 1-32). This is where the root is attached to the lung. It contains the main

bronchus, pulmonary vessels (one artery and two veins), bronchial vessels, lymph vessels, and nerves entering and leaving the lung.

The Main Differences Between the Right and Left Lungs (Figs. 1-30 to 1-32). The right lung has three lobes, the left lung two. *The right lung is larger and heavier* than the left lung, but it is shorter and wider because the right dome of the diaphragm is higher and the heart and pericardium bulge more to the left. The anterior margin of the right lung is straight, whereas this margin of the *left lung has a deep cardiac notch*.

Fresh healthy lungs always contain some air, consequently pulmonary tissue removed from them will float in water. A diseased lung partly filled with fluid may not float. Fetal lungs and those from a stillborn infant have never expanded, therefore they will sink when placed in water. The lungs of a liveborn infant will float. These observations are of medicolegal significance in determining whether a dead infant was stillborn (*i.e.*, has never inspired air) or whether it was born alive and had started to breathe.

Auscultation of the lungs (listening to their sounds with a stethoscope) and tapping the chest with a finger (*percussion of the lungs*) include the root of the neck, so that sounds in its apices may be heard. As mentioned previously, a stab or other wound superior to the medial part of the clavicle may pierce both the cupula of the pleura and the apex of the lung, causing serious breathing problems (p. 63).

Lobes and Fissures of the Lungs

The lungs are divided into lobes by fissures (Figs. 1-30 to 1-32). The right lung has *horizontal and oblique fissures*, whereas the left lung has only one, the *oblique fissure*.

The left lung is divided into superior (upper) and inferior (lower) lobes by a long deep *oblique fissure*, which extends from its costal to its medial surface. The *superior lobe* has a wide **cardiac notch** on its anterior border, where the lung is deficient owing to the bulge of the heart. This leaves part of the anterior aspect of the pericardium or pericardial sac uncovered by lung tissue (Figs. 1-30A and 1-42). The anteroinferior part of the superior lobe has a small tonguelike projection called the *lingula* (L. tongue). The *inferior lobe* of the left lung is larger than the superior lobe and lies inferoposterior to the oblique fissure.

The right lung is divided into superior (upper), middle, and inferior (lower) lobes by *horizontal and oblique fissures* (Figs. 1-30 and 1-31). The horizontal fissure separates the superior and middle lobes, and the oblique fissure separates the inferior lobe from the middle and superior lobes. The superior lobe is smaller than in the left lung, and the middle lobe is wedge-shaped.

Occasionally an extra fissure divides a lung, or a fissure is absent. For example, the left lung sometimes has three lobes and the right lung may have only two. A *lobule of the azygos vein* appears in the right lung in about 1% of people. It develops when the *apical part of the apicoposterior bronchus* (Fig. 1-35) grows superiorly, medial to the arch of the azygos vein, instead of lateral to it. As a result, the azygos vein lies at the bottom of a deep fissure in the superior lobe. This fissure with the azygos vein at its inferior end produces a linear marking on a radiograph of the thorax, which sep-

---

[3]The pulmonary ligaments are not true ligaments; they are folds of the two layers of pleura. Hence they are not fibrous ligaments.

[4]The hilum (L. "a small thing") is a general term for a pit or depression at the part of an organ where the vessels and nerves enter. *Hilum and hilus are synonymous.* Hilum is recommended by Nomina Anatomica (1989).

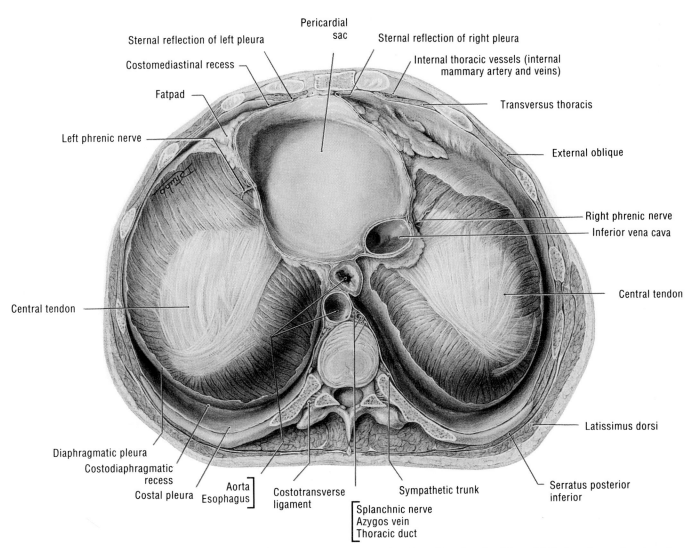

Sternal reflection of left pleura

Pericardial sac

Sternal reflection of right pleura

Costomediastinal recess

Internal thoracic vessels (internal mammary artery and veins)

Fatpad

Transversus thoracis

Left phrenic nerve

External oblique

Right phrenic nerve

Inferior vena cava

Central tendon

Central tendon

Latissimus dorsi

Diaphragmatic pleura

Costodiaphragmatic recess

Aorta
Esophagus

Costotransverse ligament

Sympathetic trunk

Serratus posterior inferior

Costal pleura

Splanchnic nerve
Azygos vein
Thoracic duct

Figure 1-29. Superior view of a dissection of the diaphragm, mediastinum, and pericardial sac. Most of the diaphragmatic pleura has been removed. Observe the sternal reflection of the left pleural sac, which fails to meet that of the right sac in the median plane, ventral to the pericardium. Note that the right and left pleural sacs almost meet between the esophagus and aorta and that the costal pleura on reaching the vertebral column becomes the mediastinal pleura.

arates the apex of the lung from the remainder of the superior lobe.

Surfaces of the Lungs

Each lung has three surfaces (costal, mediastinal, and diaphragmatic), which are named according to their relationships.

The Costal Surface of the Lung (Figs. 1-21, 1-27, 1-30, and 1-33). This surface is large, smooth, and convex. It is *related to the costal pleura*, which separates it from the ribs, their costal cartilages, and the innermost intercostal muscles. The posterior part of this surface is related to the thoracic vertebrae; because of this, this area of the lung is sometimes referred to as the *vertebral part of the costal surface*.

The Mediastinal Surface of the Lung (Figs. 1-14, 1-27, 1-31, 1-32, 1-34, and 1-40). This *medial surface* is concave because it is related to the middle mediastinum containing the pericardium and heart. Because two-thirds of the heart are to the left, the *pericardial concavity* is understandably deeper in the left lung. The mediastinal surface of the embalmed lung shows a *cardiac impression* produced by the heart and great vessels. This surface also contains the *root of the lung*, around which the pleura forms a "sleeve" or covering. The *pulmonary ligament* hangs inferiorly from the *pleural sleeve* around the root of the lung.

Some people have difficulty visualizing the root of the lung, its pleural sleeve, and the pulmonary ligament hanging inferiorly from it. Tobias et al. (1988) offer a good analogy that has been somewhat modified here. Put on an extra large lab coat and abduct your upper limb and digits. The *root of the lung* is comparable to your arm and forearm; your coat sleeve represents the *pleural sleeve* of visceral pleura; the *pulmonary ligament* is comparable to the slack of your sleeve

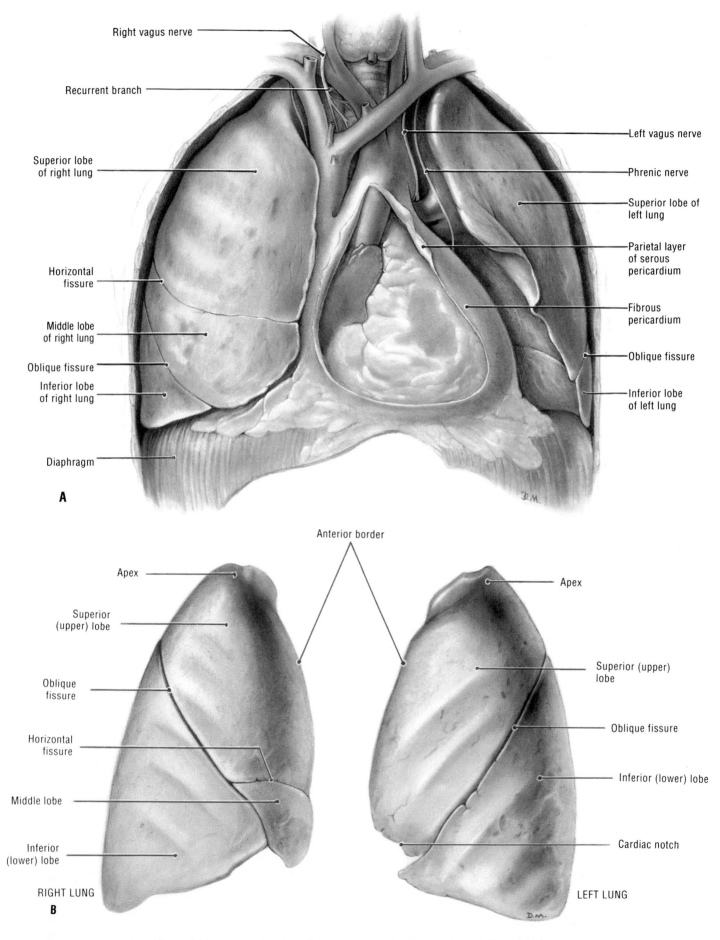

Figure 1-30. *A*, Anterior view of the lungs and pericardium. *B*, Lateral view of the lungs.

hanging from your wrist, and your abducted digits represent the *branches of the main bronchus* and the pulmonary vessels.

The Diaphragmatic Surface of the Lung (Figs. 1-29 to 1-33 and 1-39).

This deeply concave surface, often referred to as the *base of the lung*, rests on the convex dome of the diaphragm. The concavity is deeper in the right lung because of the higher position of the right dome. Laterally and posteriorly, the diaphragmatic surface is bounded by a thin sharp margin that projects into the *costodiaphragmatic recess* of the pleura.

> When clinicians refer to the base of the lung, they are not usually referring to its diaphragmatic surface or anatomical base that is related to the diaphragm. They are speaking about the inferior limit of the posterior surface of the inferior (lower) lobe (Fig. 1-30*A*). To auscultate this area, the clinician applies a stethoscope to the inferoposterior aspect of the thoracic wall at the level of the 10th thoracic vertebra (Fig. 1-27*B*).

Borders of the Lungs

Each lung has three borders: anterior, posterior, and inferior.

The Anterior Border of the Lung (Figs. 1-29 to 1-32 and 1-34).

This border is thin and sharp and *overlaps the pericardium*. There is an indentation in the anterior border of the left lung, called the *cardiac notch*. In each lung the anterior border *separates the costal surface from the mediastinal surface*, and it more or less corresponds to the anterior border of the pleura. During deep inspiration the anterior border of the lung *projects into the costomediastinal recess* of the pleura.

The Posterior Border of the Lung (Figs. 1-27, 1-31, and 1-32).

This border is broad and rounded and lies in the deep concavity at the side of the thoracic region of the vertebral column, called the *paravertebral gutter*.

The Inferior Border of the Lung (Figs. 1-29 to 1-32).

This border circumscribes the diaphragmatic surface of the lung and separates the diaphragmatic surface from the costal surface. It is thin and sharp where it *projects into the costodiaphragmatic recess* of the pleura during all phases of respiration, but it is blunt and rounded medially where it separates the diaphragmatic surface from the mediastinal surface.

Surface Markings of the Lungs

The anterior borders of the lungs follow the sternal lines of pleural reflection, except at the cardiac notch inferior to the level of the fourth costal cartilage. The location of the lungs can be outlined on the surface of the thorax using the following clinical landmarks.

The apex of the lung is represented by a line drawn superolaterally from the sternoclavicular joint to a point 2.5 cm superior to the medial third of the clavicle and then inferolaterally to the junction of the middle and lateral thirds of the clavicle.

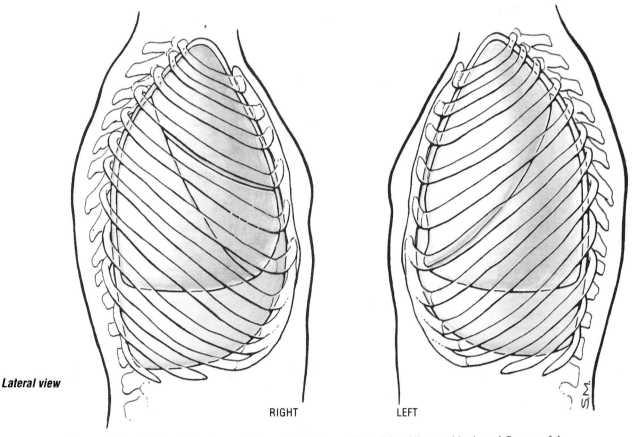

Lateral view

RIGHT   LEFT

Figure 1-33.   Outline of the pleura and lungs. Follow the outlines of the oblique and horizontal fissures of the right lung and the oblique fissure of the left lung.

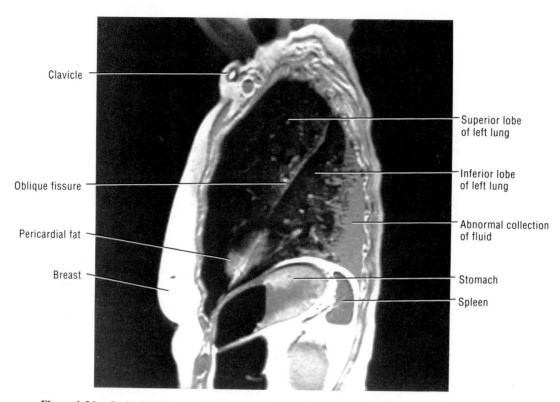

Clavicle

Oblique fissure

Pericardial fat

Breast

Superior lobe of left lung

Inferior lobe of left lung

Abnormal collection of fluid

Stomach

Spleen

Figure 1-34. Sagittal MRI (magnetic resonance image) of the left lung. (Courtesy of Dr. W. Kucharczyk, Clinical Director of Tri-Hospital Magnetic Resonance Centre, Toronto, Ontario, Canada.)

The anterior border of the right lung essentially corresponds to the anterior border of the right pleura. Between the levels of the second and fourth cartilages, its anterior border is near the median plane (Fig. 1-27A). Inferior to the fourth costal cartilage, the surface of the right lung gradually diverges from this plane and leaves the sternum posterior to the sixth costal cartilage.

The anterior border of the left lung essentially corresponds to the anterior border of the left pleura as far as the level of the fourth costal cartilage (Fig. 1-27A). Here, the anterior border deviates laterally to a point about 2.5 cm lateral to the left edge of the sternum to form the *cardiac notch*. It then turns inferiorly and slightly medially to the sixth left costal cartilage.

The inferior borders of the lungs are indicated by a line drawn from the inferior end of the line representing the anterior border that crosses the 6th rib in the midclavicular line, the 8th rib in the midaxillary line, and the 10th rib in the midscapular line. These borders end about 2.5 cm lateral to the spinous process of T10 vertebra. They *lie two ribs superior to the pleura* on each of the three vertical lines just mentioned. In children the inferior borders of the lung are about one rib more superior than in adults. *Remember*, the levels of the inferior borders of the lungs vary according to the phase of respiration. The markings just described apply to quiet respiration.

Surface Markings of the Lung Fissures (Figs. 1-26, 1-27, and 1-33). The projection of the *oblique fissure* is essentially the same for each lung and is indicated by a line that runs from the spinous process of T2 vertebra around the thorax to the sixth costochondral junction. The projection of the *horizontal fissure* in the right lung is indicated by a line that runs

from the anterior border of the lung along the fourth costal cartilage to the oblique fissure.

The Bronchi, Roots, and Bronchopulmonary Segments of the Lungs

The principal or **main bronchi**, one to each lung, pass inferolaterally from the bifurcation of the trachea at the level of the sternal angle to the roots of the lungs (Figs. 1-14, 1-31, 1-32, and 1-69). Like the trachea, *the bronchial walls are supported by C-shaped rings of cartilage.* The bronchus accompanies the pulmonary artery into the somewhat wedge-shaped *hilum* (hilus) of the lung, where it subdivides. Each main bronchus has a characteristic branching pattern, referred to as the **bronchial tree** (Figs. 1-35 and 1-37).

The right main bronchus is wider, shorter, and *more vertical than the left one* (Figs. 1-35 and 1-69). About 2.5 cm long, it passes directly to the root of the lung. *The left main bronchus* is about 5 cm long and passes inferolaterally, inferior to the arch of the aorta and anterior to the esophagus and descending aorta. Within each lung the bronchi divide in a constant fashion and in constant directions so that each branch supplies a clearly defined sector of the lung (Figs. 1-35 to 1-38). Each main bronchus divides into *secondary bronchi* or **lobar bronchi** (two on the left and three on the right), each of which supplies a lobe of the lung. Each lobar bronchus divides into *tertiary bronchi* or **segmental bronchi**, which supply specific segments of the lung called *bronchopulmonary segments* (Figs. 1-36 and 1-38A).

The Roots of the Lungs (Figs. 1-30 to 1-32, 1-72, and 1-

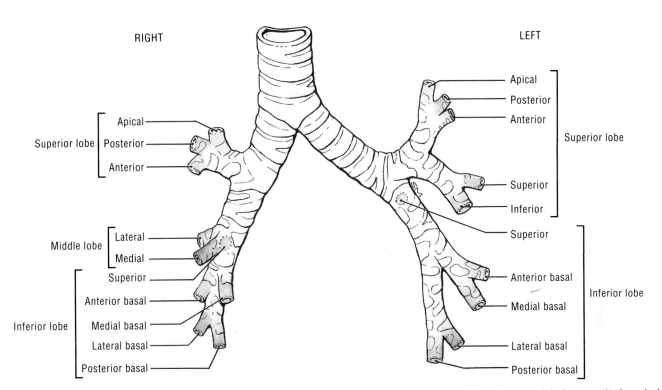

Figure 1-35. The segmental bronchi, correlating with the bronchopulmonary segments shown in Fig. 1-36. Each bronchopulmonary segment takes its name from the segmental bronchus supplying it. Observe that there are 10 segmental bronchi on the right and 8 on the left. The left lung has two segments fewer than the right because (1) the apical and posterior bronchi have a common stem and (2) the medial basal bronchus is usually lacking.

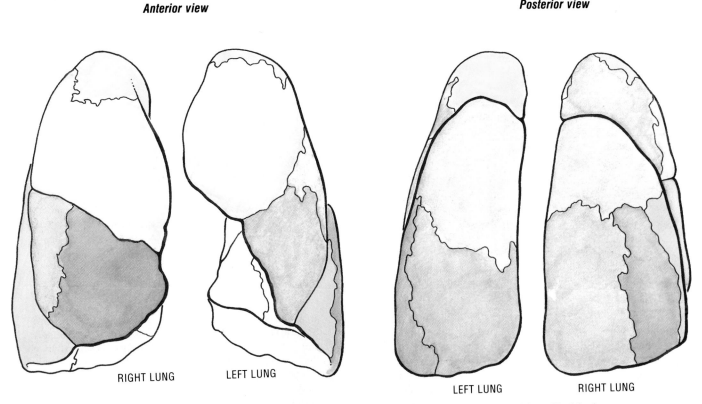

Figure 1-36. The bronchopulmonary segments. To prepare these specimens, the segmental bronchi of fresh lungs were isolated at the hilum and injected with latex of various colors.

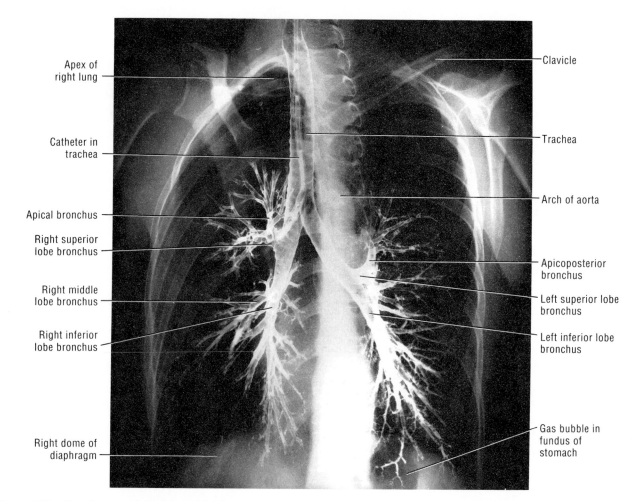

Apex of
right lung

Catheter in
trachea

Apical bronchus

Right superior
lobe bronchus

Right middle
lobe bronchus

Right inferior
lobe bronchus

Right dome of
diaphragm

Clavicle

Trachea

Arch of aorta

Apicoposterior
bronchus

Left superior lobe
bronchus

Left inferior lobe
bronchus

Gas bubble in
fundus of
stomach

Figure 1-37.   Bronchogram of the right and left bronchial tree. This is a slightly oblique, posteroanterior view. (Courtesy of Dr. D.E. Sanders and Dr. E.L. Lansdown, Department of Radiology, University of Toronto, Toronto, Ontario, Canada.)

73). These roots are formed by the structures entering and leaving the lungs at the hila. They attach the mediastinal surfaces of the lungs to the heart and trachea. The *main structures in the roots of the lungs are the main bronchi* and *pulmonary vessels.* Other structures, all embedded in connective tissue, are the bronchial arteries and veins, nerves, lymphatic vessels, and lymph nodes. Each pulmonary artery passes transversely into the hilum, anterior to its bronchus. The two *pulmonary veins* on each side (superior and inferior) ascend from the hilum to the left atrium of the heart (Fig. 1-44).

The Bronchopulmonary Segments (Figs. 1-35 to 1-38).   The segments of a lung *supplied by segmental bronchi* are called bronchopulmonary segments. Within each segment there is further branching of the bronchi. Each segment is pyramidal in shape with its apex facing the root of the lung and its base on the pleural surface. The names and arrangement of the bronchopulmonary segments are shown in Figures 1-35 and 1-36. Each segment is named according to the segmental bronchus that supplies it. The left superior lobe has a *lingular bronchopulmonary segment* (Fig. 1-36). Each bronchopulmonary segment has its own segmental bronchus, artery, and vein.

When the bronchi are examined with a *bronchoscope* (an instrument inserted via the mouth for examining the trachea and bronchi), a keel-like ridge called the **carina** (L. keel) is observed between the orifices of the main bronchi (Fig. 1-14). Normally the carina is nearly in a sagittal plane and has a fairly definite edge. If the *tracheobronchial lymph nodes* (Fig. 1-38) in the angle between the main bronchi enlarge (*e.g.,* owing to the lymphogenous spread of cancer cells from a *bronchogenic carcinoma*), the carina becomes distorted, widened posteriorly, and immobile. Hence, morphological *changes in the carina are important diagnostic signs* for the bronchoscopist in assisting with the differential diagnosis of disease of the respiratory system. The mucous membrane at the carina, one of the most sensitive areas of the tracheobronchial tree, is associated with the *cough reflex.* For example, when a child aspirates a peanut he/she chokes and coughs. Once the peanut passes the carina, coughing usually stops, but the resulting *chemical bronchitis* (inflammation of the bronchus) caused by substances released from the peanut and the collapse of the lung (*atelectasis*) distal to the foreign body soon cause difficult breathing (*dyspnea*). In addition, during postural drainage of the lungs (p. 76), lung secretions

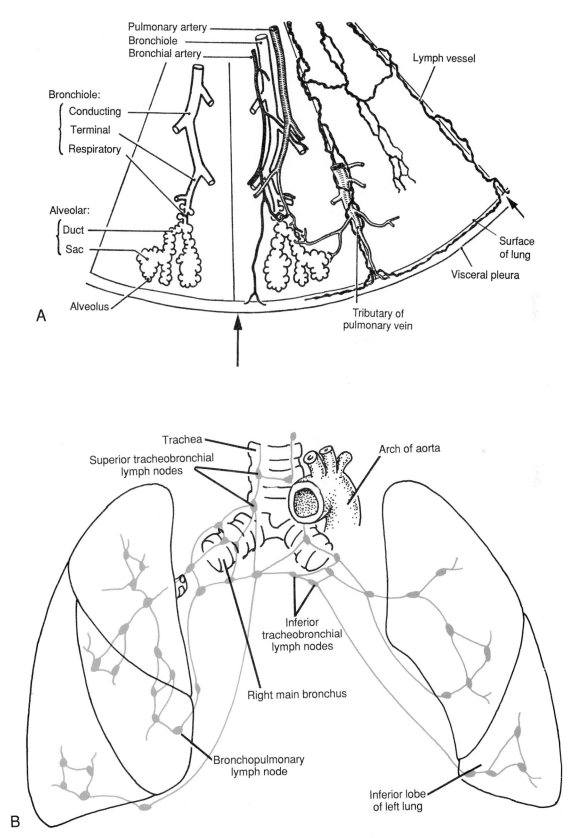

Figure 1-38. *A*, the structure of part of a lung. The base of a bronchopulmonary segment is between the arrows. Within the bronchopulmonary segment, the segmental bronchi divide to form bronchioles. *B*, the lymphatic drainage of the lungs. Observe that the entire right lung drains into the tracheobronchial lymph nodes and that the superior part of the left lung drains similarly into nodes on the left side. Note, however, that the inferior part of the left lung drains mainly to the right side.

pass to the carina and usually cause coughing, which helps to expel them.

Because the right bronchus is wider and shorter and runs more vertically than the left bronchus (Figs. 1-35 and 1-69), foreign bodies and aspirated material are more likely to enter and lodge in the right middle or inferior bronchi than in the left bronchus. A common hazard encountered by dentists is an aspirated foreign body, such as a piece of a tooth or filling material. To avoid this possible occurrence, some dentists insert a thin rubber dam in the mouth.

Knowledge of the normal anatomy of the bronchopulmonary segments is essential for the accurate interpretation of chest radiographs and for *surgical resection of diseased segments*. Bronchial and pulmonary disorders (*e.g.*, a tumor or an abscess) often localize in a bronchopulmonary segment, and these may be surgically resected without seriously disrupting surrounding lung tissue. *Cardiothoracic surgeons* often refer to the medial basal segment of the lung as the *cardiac segment* (Fig. 1-35). During surgical treatment of **lung cancer** (Ginsberg, 1989), the surgeon may remove a whole lung (*pneumonectomy*), a lobe (*lobectomy*), or one or more bronchopulmonary segments (*segmentectomy* or wedge resection).

Although each bronchopulmonary segment is supplied by its own nerve, artery, and vein, the planes between them are crossed by branches of the pulmonary veins and sometimes by small branches of the pulmonary arteries. The fact that the segments are marked by these *intersegmental veins* forms a landmark during the surgical resection of a segment (Figs. 1-36 and 1-38). Each bronchopulmonary segment is surrounded by connective tissue that is continuous with the visceral pleura (Fig. 1-38A). The connective tissue septa separating the segments prevent air from passing between segments. Therefore air in a bronchopulmonary segment whose segmental bronchus is obstructed is absorbed by the bloodstream. This causes *segmental atelectasis* or collapse of the affected segment of lung tissue.

Knowledge of the branching of the bronchial tree is essential for determining the appropriate posture for draining an infected area of the lung. For example, when a patient with *bronchiectasis* (dilation of bronchi) is positioned in bed on his/her left side, secretions from the right lung and bronchi flow toward the carina of the trachea. As this is a very sensitive area, the cough reflex is stimulated and the patient brings up *purulent sputum* (*i.e.*, sputum containing pus), clearing the right bronchial tree. Alternatively, a person with *bronchiectasis of the lingula* of the left superior lobe drains it by lying on his/her right side.

Because oropharyngeal and nasopharyngeal contents containing bacteria may be aspirated into the lungs and cause inflammation of the lungs (*pneumonitis* or **pneumonia**) or a lung abscess in one or more bronchopulmonary segments, the position of very ill and unconscious patients is changed frequently to promote good aeration of their lungs. The usual *supine position* one assumes in bed with the head slightly elevated by a pillow is poor for lung drainage.

A tumor may block a segmental bronchus and cause collapse of its bronchopulmonary segment, owing to absorption of the air in the segment by the blood that is still circulating through it. The collapsed portion of lung can be determined radiographically using a technique known as *bronchography* (Fig. 1-37). Only one side of the bronchial tree is usually injected at a time because the contrast medium partly obstructs the flow of air to the segmental bronchi that are being visualized. Impaired oxygenation of the blood may occur in ill people if both sides are filled with contrast medium at one time. Bronchography is not commonly performed now; it has largely been replaced by CT and *endoscopy* (inspection of the interior of the bronchi) using a *bronchoscope* (p. 74).

Arterial Supply of the Lungs

The pulmonary arteries arise from the pulmonary trunk (Figs. 1-43 and 1-44) and distribute deoxygenated blood (low in oxygen but not without it) to the lungs for *aeration* (changing venous blood into arterial blood). The right and left pulmonary arteries pass to the corresponding root of the lung and give off a branch to the superior lobe before entering the hilum (Figs. 1-31 and 1-32). Within the lung each pulmonary artery descends posterolateral to the main bronchus and sends branches to the lobar and segmental bronchi on their posterior surfaces. Hence, there is a branch to each lobe, bronchopulmonary segment, and lobule of the lung (Figs. 1-35 to 1-38). The terminal branches of the pulmonary arteries divide into capillaries in the walls of the *alveoli*, which are the air sacs where gaseous exchange takes place between the blood and air.

The bronchial arteries supply blood to the connective tissue of the bronchial tree (Figs. 1-38A and 1-69). These small vessels pass along the posterior aspects of the bronchi to supply them as far distally as the respiratory bronchioles. The *two left bronchial arteries* arise from the superior part of the *thoracic aorta* (Fig. 1-75), superior and inferior to the left main bronchus (Fig. 1-69). The *single right bronchial artery* often arises as a common trunk with the third (or fifth) *posterior intercostal artery* (Fig. 1-75) or from the superior left bronchial artery.

Pulmonary thromboembolism (**PTE**) is a common cause of morbidity (sickness) and mortality (death). An *embolus* (G. plug) forms when a *thrombus* (blood clot), fat globules, or air bubbles are carried from a distant site (*e.g.*, from the leg veins following a fracture). The thrombus passes through the right side of the heart and is carried to a lung via a pulmonary artery. The immediate result of PTE is complete or partial obstruction of the arterial blood flow to the lung.

Embolic obstruction of a pulmonary artery produces a sector of lung that is ventilated but not perfused. A very large embolus may occlude the pulmonary trunk or one of the main pulmonary arteries; in such cases *the patient suffers acute respiratory distress* and may die in a few minutes. A medium-sized embolus may block an artery to a bronchopulmonary segment and may produce an *infarct* (area of dead tissue). In healthy people, a collateral circulation (*i.e.*, an indirect blood supply) often develops so that infarction does not occur. There are abundant anastomoses with branches of the bronchial arteries in the region of the terminal bronchioles (Fig. 1-38A). In ill people in whom circulation in the lung is impaired (*e.g.*, in a person with *chronic congestion*), PTE commonly results in infarction of the lung. Because an area of pleura is also deprived of blood, it becomes inflamed (*pleuritis*), which results in pain. The pain is referred to the cutaneous distribution of the intercostal nerves (*i.e.*, to the thoracic wall or, in the case of the inferior nerves, to the anterior abdominal wall; Fig. 1-24; see also Fig. 2-13).

Venous Drainage of the Lungs

The pulmonary veins carry oxygenated blood from the lungs to the left atrium of the heart. Beginning in the pulmonary capillaries, the veins unite into larger and larger vessels that run mainly in the interlobular septa (Fig. 1-38A). *A main vein drains each bronchopulmonary segment*, usually on the anterior surface of the corresponding bronchus. The two pulmonary veins on each side, superior and inferior ones, open into the posterior aspect of the left atrium (Fig. 1-44). The superior right pulmonary vein drains the superior and middle lobes of the right lung, and the superior left pulmonary vein drains the superior lobe of the left lung. The right and left inferior pulmonary veins drain the respective inferior lobes.

The bronchial veins drain the large subdivisions of the bronchi, but they drain only part of the blood delivered by the bronchial arteries to the bronchial tree; some of this blood is drained by the pulmonary veins. The right bronchial vein drains into the *azygos vein* and the left bronchial vein drains into the *accessory hemiazygos vein* or into the *left superior intercostal vein* (Figs. 1-39 and 1-76).

Innervation of the Lungs and Pleurae

The lungs and visceral pleura are innervated by the anterior and posterior *pulmonary plexuses*, which are located anterior and posterior to the roots of the lungs (Figs. 1-63, 1-72, and 1-73). They are mixed plexuses containing vagal (parasympathetic) and sympathetic fibers. These networks of nerves are formed by the *vagus nerves* and *sympathetic trunks*. The parasympathetic ganglion cells are located in the pulmonary plexuses and along the branches of the bronchial tree.

The efferent fibers of the vagus nerve (parasympathetic) are: motor to the smooth muscle of the bronchial tree (*bronchoconstrictor*); inhibitor to the pulmonary vessels (*vasodilator*); and *secretor* to the glands of the bronchial tree (*secretomotor*). *The afferent fibers of the vagus nerve* are sensory to the respiratory epithelium (touch and pain) and to the branches of the bronchial tree (stretch). *The efferent sympathetic fibers are:* inhibitor to the bronchial muscle (*bronchodilator*); motor to the pulmonary vessels (*vasoconstrictor*); and *inhibitor* to the glands of the bronchial tree. Afferent sympathetic fibers are also present in the bronchial tree, but their function is unknown.

Innervation of the Parietal Pleura (Figs. 1-29, 1-68, 1-72, and 1-73). The *costal pleura* and the peripheral part of the *diaphragmatic pleura* are supplied by the **intercostal nerves**. They mediate the sensations of touch and pain. The central part of the diaphragmatic pleura and the *mediastinal pleura* are supplied by the **phrenic nerves**.

The visceral pleura is insensitive to pain or contact because it receives no nerves of general sensation, but *the parietal pleura is very sensitive to pain*, particularly its costal part. Irritation of the costal and peripheral parts of the diaphragmatic pleura results in local pain and in referred pain along

the intercostal nerves to the thoracic and abdominal walls (Fig. 1-24; see also Figs. 2-6 and 2-13). Irritation of the mediastinal and central diaphragmatic areas of the parietal pleura results in pain that is referred to the root of the neck and over the shoulder. The explanation for this is that these areas are supplied by the same segments of the spinal cord (C3 to C5) that give origin to the phrenic nerve (Fig. 1-24).

Lymphatic Drainage of the Lungs

There are two lymphatic plexuses or networks of lymph vessels that communicate freely. There are superficial and deep plexuses.

The **superficial lymphatic plexus** lies deep to the visceral pleura (Fig. 1-38A), and lymph vessels from it drain into the *bronchopulmonary lymph nodes* that are located in the hilum of the lung. From them, lymph drains to the superior and inferior *tracheobronchial lymph nodes*, which are located superior and inferior to the bifurcation of the trachea, respectively (Figs. 1-38B, 1-63, and 1-70). These lymph vessels drain the lung and visceral pleura.

The **deep lymphatic plexus** is located in the submucosa of the bronchi and in the peribronchial connective tissue. There are no lymph vessels in the walls of the alveoli. Lymph vessels from the deep lymphatic plexus drain into the *pulmonary lymph nodes*, which are located in the lung along the large branches of the main bronchi. Lymph vessels from these nodes follow the bronchi and pulmonary vessels to the hilum of the lung, where they drain into the *bronchopulmonary lymph nodes* (Fig. 1-38B). Lymph vessels then pass to *tracheobronchial lymph nodes* around the trachea and main bronchi (Figs. 1-63 and 1-70). The lymph then passes to the right and left *bronchomediastinal lymph trunks* (Fig. 1-39), which are formed by the junction of the efferent lymph vessels from the parasternal, tracheobronchial, and anterior *mediastinal lymph nodes* (Figs. 1-19, 1-38B, and 1-39). These trunks usually terminate on each side at the junction of the subclavian and internal jugular veins. The left bronchomediastinal trunk may terminate in the *thoracic duct*.

Lymph from the lungs carries *phagocytes* which contain ingested carbon particles from inspired air. In many people, especially cigarette smokers and/or city dwellers, these particles give the surface of the lungs a mottled gray to black appearance (Fig. 1-31). The carbon particles are also carried to the lymph nodes in the hila of the lungs and mediastinum, giving them a similar appearance. If the parietal and visceral layers of pleura adhere (*pleural adhesions*), the lymphatics in the lung and visceral pleura may drain into the *axillary lymph nodes* (Fig. 1-12). The presence of carbon particles in these nodes is presumptive evidence of pleural adhesions.

Bronchogenic carcinoma (*CA of a bronchus*) is the most common cancer in men and is responsible for about 30% of malignancies. The major causative factors for this type of cancer are cigarette smoking and urban living. Because of the arrangement of the lymphatics, these tumors may metastasize to the pleurae, hila of the lungs, and mediastinum and from them to other parts of the body (*e.g.*, the skull, including the brain). Tumorous involvement of a *phrenic nerve* (Figs. 1-29 and 1-42) may result in paralysis of half of the dia-

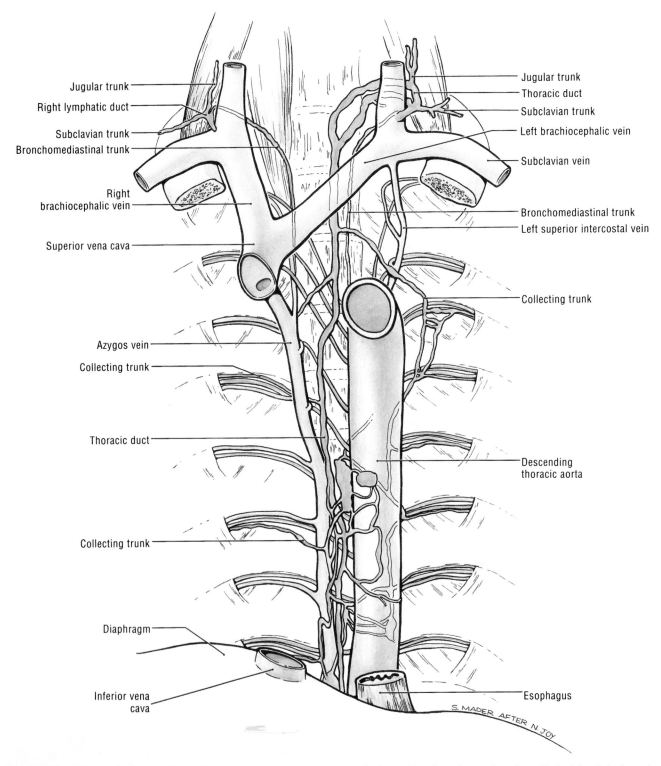

Jugular trunk
Right lymphatic duct
Subclavian trunk
Bronchomediastinal trunk
Right brachiocephalic vein
Superior vena cava
Azygos vein
Collecting trunk
Thoracic duct
Collecting trunk
Diaphragm
Inferior vena cava

Jugular trunk
Thoracic duct
Subclavian trunk
Left brachiocephalic vein
Subclavian vein
Bronchomediastinal trunk
Left superior intercostal vein
Collecting trunk
Descending thoracic aorta
Esophagus

S. MADER AFTER N. JOY

Figure 1-39.   Structures in the posterior mediastinum. Note the thoracic duct, right lymphatic duct, and azygos vein. Observe the collecting trunks of the thoracic duct, and note that this main lymphatic duct opens into the internal jugular vein near its union with the left subclavian vein. Note that the right lymphatic duct is very short and is formed by the union of the right jugular, subclavian, and bronchomediastinal trunks.

phragm. Because of the *intimate relationship of the recurrent laryngeal nerve* to the *apex of the lung* (Fig. 1-62), this nerve may be involved in **apical lung cancers**. This usually results in *hoarseness*, owing to paralysis of a vocal fold (cord), because this nerve supplies all but one of the laryngeal muscles.

Involvement of the hilar and mediastinal lymph nodes occurs by *lymphogenous dissemination of cancer cells*. Lymph from the entire right lung drains into the *tracheobronchial lymph nodes* on the right side, and most lymph from the left lung drains into these nodes on the left side (Fig. 1-38*B*). Lymph from the lungs drains into the venous system via the right and left bronchomediastinal trunks (Fig. 1-39). This lymph may therefore carry cancer cells into the venous system and to the right atrium of the heart. After passing through the pulmonary circulation, the blood and cancer cells return to the heart and are distributed to the entire body.

Common sites of hematogenous metastasis of cancer cells (spreading via the blood) from a bronchogenic carcinoma are the brain, bones, lungs, and suprarenal (adrenal) glands. Tumor cells probably enter the systemic circulation by invading the wall of a sinusoid or venule in the lung (Fig. 1-38*A*), and are transported via the pulmonary veins, left heart, and aorta. Often the lymph nodes superior to the clavicle (*supraclavicular lymph nodes*) are enlarged and hard when there is CA of a bronchus owing to metastases from the primary tumor. Consequently, the *supraclavicular lymph nodes* are commonly referred to as **sentinel lymph nodes** because their enlargement alerts the examiner to the possibility of malignant disease in the thoracic and/or abdominal organs.

The Mediastinum

The space between the two pleural sacs is called the *thoracic mediastinum* (L. middle septum). It is a thick partition of tissue in and on each side of the median plane (Figs. 1-14 and 1-40). It contains all the structures in the thorax (chest) except the lungs and pleurae. It extends from the *superior thoracic aperture* (p. 49) to the diaphragm inferiorly and from the sternum and costal cartilages anteriorly to the bodies of the thoracic vertebrae posteriorly. Structures in the mediastinum (*e.g.*, the heart and great vessels) are surrounded by loose connective tissue, nerves, blood and lymph vessels, lymph nodes, and fat. In a cadaver this tissue is rigid, but in living persons the looseness of the connective tissue and fat and the elasticity of the lungs and pleura enable the mediastinum to accommodate movement and volume changes in the thoracic cavity (*e.g.*, movements of the trachea, bronchi, and lungs during respiration, pulsations of the heart and great vessels, and volume changes in the esophagus during swallowing).

Subdivisions of the Mediastinum

For descriptive purposes, the mediastinum is divided into superior and inferior parts. In the supine position the *superior mediastinum* extends inferiorly from the superior thoracic ap-

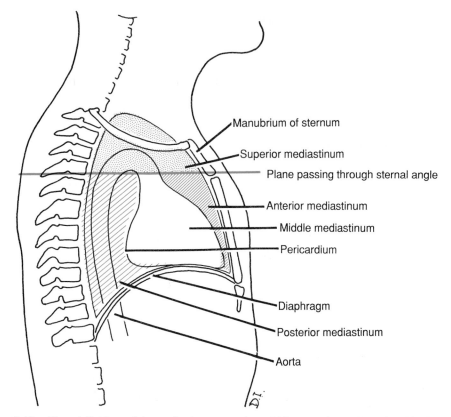

Figure 1-40.   The subdivisions of the mediastinum: superior, middle, posterior, and anterior. The last three divisions constitute the inferior mediastinum.

Manubrium of sternum
Superior mediastinum
Plane passing through sternal angle
Anterior mediastinum
Middle mediastinum
Pericardium
Diaphragm
Posterior mediastinum
Aorta

erture (thoracic inlet) to a horizontal plane passing through the sternal angle and the inferior border of T4 vertebra (Fig. 1-40). The *inferior mediastinum*, between this plane and the diaphragm, is subdivided by the pericardium into three parts: anterior, middle, and posterior. Some structures pass through the mediastinum (*e.g.*, the esophagus, vagus and phrenic nerves, and thoracic duct) and therefore lie in more than one subdivision.

The Superior Mediastinum (Figs. 1-40, 1-42, and 1-45).   As its name indicates, this subdivision is superior to the inferior mediastinum and *superior to the manubriosternal plane*, passing through the sternal angle to the intervertebral disc between T4 and T5 vertebrae. Structures in the superior mediastinum are: the *thymus* (or its remains) anteriorly; the *great vessels related to the heart* in the middle; and the *esophagus, trachea and thoracic duct* posteriorly (p. 115).

The Anterior Mediastinum (Figs. 1-20, 1-40, 1-42, and 1-65).   This is the smallest subdivision of the mediastinum. It is located *anterior to the pericardium* or pericardial sac and *posterior to the sternum* and transversus thoracis muscles. Although usually small in the adult, it is relatively large during infancy and childhood because the *thymus* extends into it from the superior mediastinum. During infancy the image of the thymus is as wide or wider than that of the heart in radiographs and other images of the thorax (p. 89).

The Middle Mediastinum (Figs. 1-31, 1-32, 1-40, and 1-45).   This subdivision is of the highest importance clinically because it *contains the pericardium and heart* and the immediately adjacent parts of the *great arteries*, *phrenic nerves*, *main bronchi*, and other structures in the roots of the lungs (p. 74).

The Posterior Mediastinum (Figs. 1-39 and 1-40).   This subdivision is located *posterior to the pericardium* and diaphragm and anterior to the bodies of the inferior eight thoracic vertebrae. Its main contents are the *esophagus* and descending *thoracic aorta*, which have descended into it from the superior mediastinum (Fig. 1-66 and p. 114).

When a person is standing or sitting erect, the levels of the subdivisions of the mediastinum are slightly more inferior than shown in Figure 1-40, and the plane between the superior and inferior mediastina is oblique. This occurs because the structures in the mediastinum, especially the pericardium and its contents (the heart and great vessels) move inferiorly. However, when a person is supine as is common in bed or on the operating table, the levels are similar to those shown in Figure 1-40 because the abdominal viscera push the mediastinal structures superiorly.

Much of the mediastinum can be visualized and certain minor surgical procedures can be carried out with a tubular, lighted instrument called a *mediastinoscope*. This instrument is inserted through a small incision in the neck, just superior to the manubrium (Ginsberg, 1989). *Mediastinoscopy* is commonly done to obtain tissue from the mediastinal and hilar lymph nodes (*e.g.*, to determine if cancer cells have metastasized to them from a *bronchogenic carcinoma*). The mediastinum can also be explored and biopsies can be taken by removing part of a costal cartilage, often the third one as illustrated in Figure 1-19.

The Pericardium

The pericardium (G. around the heart) is a *double-walled fibroserous sac* that encloses the heart (G. *kardia*) and the roots

of its great vessels (Figs. 1-29 and 1-40 to 1-42). This conical *pericardial sac* is located in the middle mediastinum, posterior to the body of the sternum and the second to sixth costal cartilages and anterior to T5 to T8 vertebrae. The pericardium or pericardial sac, containing the heart and *pericardial cavity*, consists of **two parts** (Fig. 1-41; Table 1-1): (1) a strong external layer composed of tough fibrous tissue, called the *fibrous pericardium*, and (2) an internal double-layered sac composed of a transparent membrane called the *serous pericardium*.

The Fibrous Pericardium

The fibrous pericardium is the tough, more or less conical outer sac of pericardium, which *protects the heart against sudden overfilling* (Figs. 1-40 to 1-42). Its truncated *apex* is pierced by the aorta, pulmonary trunk, and superior vena cava (Figs. 1-45 and 1-46). The *ascending aorta* carries the pericardium superiorly beyond the heart to the level of the *sternal angle*. The external surface of the fibrous pericardium has a dull appearance, and *its base rests on and is fused with the central tendon of the diaphragm* (Figs. 1-29, 1-30*A*, 1-42, and 1-45). Anteriorly the fibrous pericardium is attached to the posterior surface of the sternum by condensations of connective tissue called the *sternopericardial ligaments*. The fibrous pericardium is also fused with the fibroelastic external coat (tunica adventitia) of the great vessels entering and leaving the heart (Fig. 1-45). Thus *the pericardial sac is influenced by movements of the sternum, diaphragm, and heart*. The central tendon of the diaphragm and the pericardium are pierced on the right side by the *inferior vena cava* posteriorly (Figs. 1-29, 1-43, and 1-46).

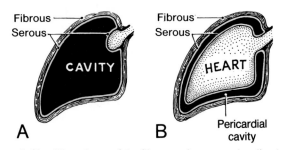

Figure 1-41.   The scheme of the fibrous and serous pericardia. *A*, The embryonic heart beginning to invaginate the pericardial cavity. *B*, After invagination is complete, the pericardial cavity is greatly reduced. Observe that the pericardium forms a double-walled sac that encloses the heart. See also Table 1-1.

Table 1-1.
Layers of the Pericardium and Heart

Pericardium[1]	Fibrous	
	Serous ————{	Parietal
		Visceral[2]
Heart	Epicardium	
	Myocardium	
	Endocardium	

[1]The *pericardium* consists of an external sac, the fibrous pericardium, and an internal double-layered sac, the serous pericardium (Fig. 1-41). The *pericardial sac* is composed of the two layers of pericardium.
[2]The visceral layer of the serous pericardium is reflected onto the heart as the epicardium.

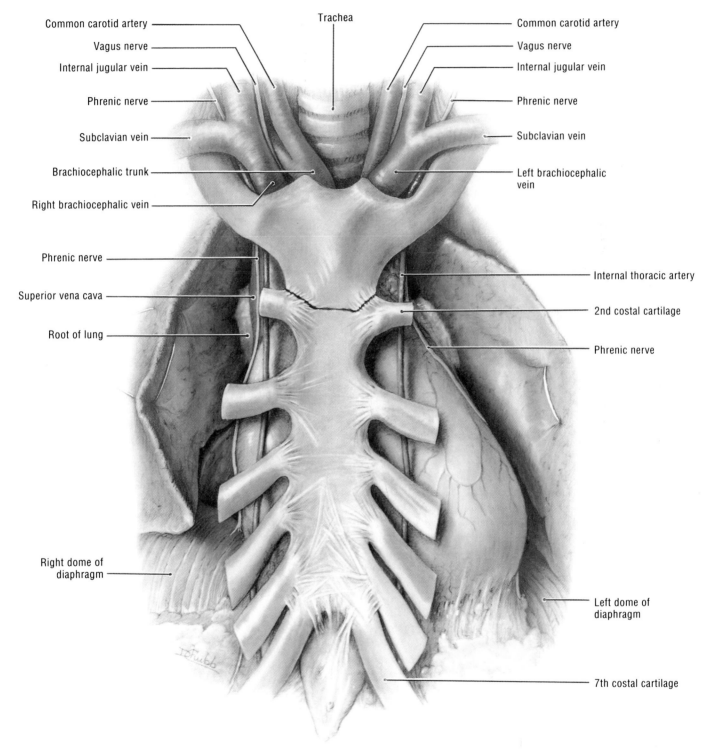

Common carotid artery

Vagus nerve

Internal jugular vein

Phrenic nerve

Subclavian vein

Brachiocephalic trunk

Right brachiocephalic vein

Trachea

Common carotid artery

Vagus nerve

Internal jugular vein

Phrenic nerve

Subclavian vein

Left brachiocephalic vein

Phrenic nerve

Superior vena cava

Root of lung

Internal thoracic artery

2nd costal cartilage

Phrenic nerve

Right dome of diaphragm

Left dome of diaphragm

7th costal cartilage

Figure 1-42.   Dissection of the thorax showing the relationship of the pericardial sac and heart to the sternum. Note that about one-third of the heart lies to the right of its median plane and two-thirds are to the left.

The fibrous pericardium extends 1 to 1.5 cm to the right of the sternum and 5 to 7.5 cm to the left of the median plane at the level of the fifth intercostal space (Figs. 1-42, 1-48, and 1-49). It is separated from the sternum and costal cartilages of the second to sixth ribs by the lungs and pleurae (Fig. 1-30A), except where it is attached to the posterior surface of the sternum by the sternopericardial ligaments and where the bulge of the heart intervenes.

The Serous Pericardium

The serous pericardium consists of **two layers**: a *parietal pericardium* and a *visceral pericardium* (Fig. 1-41; Table 1-1). The parietal pericardium is fused to the internal surface of the *fibrous pericardium* and is so closely adherent to it that they are difficult to separate. The *visceral pericardium is reflected onto the heart* where it forms the **epicardium**, the external layer of the heart wall. The potential space between the parietal and visceral layers of serous pericardium is called the **pericardial cavity** (Fig. 1-41). It normally contains a thin film of serous fluid that enables the heart to move and beat in a frictionless environment. The visceral pericardium is reflected from the heart and great vessels to become continuous with the parietal pericardium where the aorta and pulmonary trunk leave the heart and the superior and inferior venae cavae and pulmonary veins enter the heart (Figs. 1-45 and 1-46).

The Pericardial Sinuses

The sinuses of the pericardium develop during folding of the embryonic heart, a process which produces reflections of the pericardium (Moore, 1988). *The aorta and pulmonary trunk are enclosed by a common sheath of visceral pericardium.* When the pericardial sac is opened anteriorly (Figs. 1-45 and 1-46), a digit can be inserted into a recess posterior to the ascending aorta and pulmonary trunk and anterior to the superior vena cava. This *recess*, known as the **transverse pericardial sinus**, lies posterior to the ascending aorta and pulmonary trunk (Figs. 1-46 and 1-53). At its left and right ends, it communicates with the main part of the pericardial cavity. The atria are posteroinferior to the transverse pericardial sinus, and the great arteries (aorta and pulmonary trunk) are within the same sheath of serous pericardium (Figs. 1-53 and 1-60). This situation exists because they both developed from the same part of the embryonic heart, the *truncus arteriosus* (Moore, 1988).

As the pulmonary veins and inferior vena cava penetrate the fibrous pericardium to enter the heart, they bulge into the pericardial cavity. Consequently, they are partly covered by serous pericardium, which forms a somewhat inverted, U-shaped blind recess, called the **oblique pericardial sinus** (Figs. 1-46 and 1-60). It is a wide, slitlike recess posterior to the heart that extends posterior to the left atrium. The thoracic part of the *inferior vena cava* is to the right of this sinus (Fig. 1-46). This large vein enters the pericardium inferiorly, immediately after it passes through the *vena caval foramen* in the central tendon of the diaphragm (Fig. 1-29; see also Fig. 2-98). The oblique pericardial sinus, which can be entered inferiorly, will admit several digits; however, they cannot pass around any of the vessels because this sinus is a *blind recess* (cul-de-sac).

In Fig. 1-46 note the close relationship of the transverse and oblique pericardial sinuses. Only two layers of serous pericardium separate them. The heart shown in Fig. 1-60 was removed from the pericardial sac shown in Fig. 1-46. If you can imagine how they were related *in situ*, you can visualize the oblique pericardial sinus.

The transverse pericardial sinus is especially important to cardiac surgeons. After the pericardial sac has been opened, a digit and a ligature can be passed through the transverse pericardial sinus between the great arteries and the pulmonary veins (Figs. 1-46 and 1-53). By tightening the ligature, the surgeon can control and/or stop the circulation through the great arteries while surgery is carried out on the ascending aorta or the pulmonary trunk. Because the *inferior vena cava* passes through the pericardium immediately after passing through the diaphragm (Fig. 1-46), its entire thoracic part (about 2 cm) is within the pericardium. Consequently, to expose this large vein, the pericardial sac has to be opened. The same is true for the terminal part of the *superior vena cava*, which is partly inside and partly outside the pericardium (Fig. 1-46).

Vessels and Nerves of the Pericardium

Arterial Supply of the Pericardium (Figs. 1-18 to 1-20, 1-42, 1-45, 1-69, and 1-72). Its main arteries are the *pericardiacophrenic and musculophrenic arteries*, which are branches of the internal thoracic arteries. It also receives pericardial branches from the bronchial, esophageal, and superior phrenic arteries. The visceral layer of the serous pericardium or *epicardium*, is supplied by the *coronary arteries* (p. 98).

Venous Drainage of the Pericardium (Figs. 1-39, 1-72, and 1-76). Its veins are tributaries of the *azygos system* of veins. Pericardiacophrenic veins also enter the *internal thoracic veins*.

Innervation of the Pericardium (Figs. 1-42, 1-68, 1-72, and 1-73). The nerves of the pericardium are derived from the vagus and phrenic nerves and the sympathetic trunks.

Pericarditis (inflammation of the pericardium) *causes substernal pain*, frequently severe, and *produces pericardial effusion* (passage of fluid from the pericardial capillaries into the pericardial cavity). If the effusion is extensive, the excess fluid in the pericardial cavity may interfere with the action of the heart by compressing the pulmonary veins as they cross the pericardial sac and atria (Figs. 1-44 and 1-45). This compression, resulting in a condition called **cardiac tamponade**, occurs because the fibrous pericardium is inelastic.

Pericardial friction rub is an important physical sign of acute pericarditis. Normally the small amount of serous fluid in the pericardial sac prevents friction between the epicardium of the beating heart and the pericardium. Usually the moist, moving layers of serous pericardium make no detectable sound during *auscultation*. However, inflammation of the pericardium causes the surfaces to become rough and the resulting friction, called **friction rub**, *sounds like the rustle of silk*, especially during forced expiration when the patient is bending anteriorly.

Penetrating wounds of the pericardium (e.g., from a knife) commonly pierce the heart and bleeding occurs into the pericardial cavity. As the blood accumulates, the heart is com-

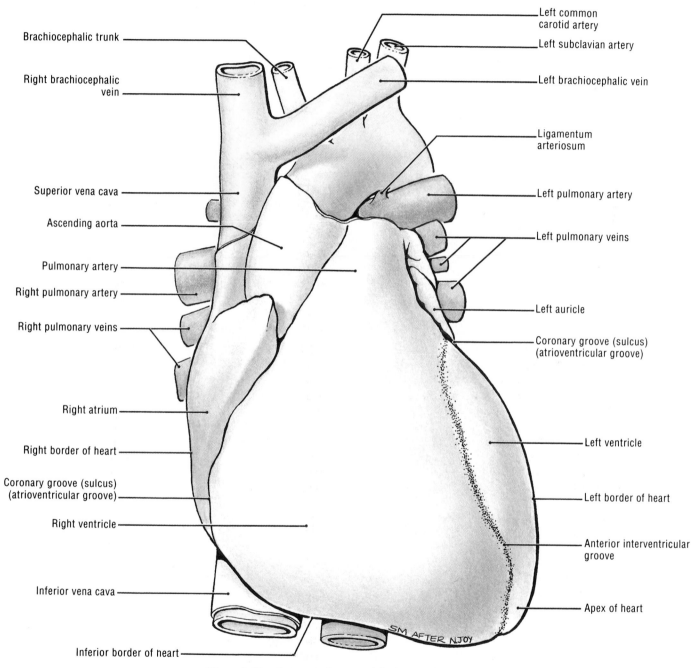

Brachiocephalic trunk

Right brachiocephalic vein

Superior vena cava

Ascending aorta

Pulmonary artery

Right pulmonary artery

Right pulmonary veins

Right atrium

Right border of heart

Coronary groove (sulcus) (atrioventricular groove)

Right ventricle

Inferior vena cava

Inferior border of heart

Left common carotid artery

Left subclavian artery

Left brachiocephalic vein

Ligamentum arteriosum

Left pulmonary artery

Left pulmonary veins

Left auricle

Coronary groove (sulcus) (atrioventricular groove)

Left ventricle

Left border of heart

Anterior interventricular groove

Apex of heart

SM AFTER NJOY

Figure 1-43. Sternocostal aspect of the heart and great vessels.

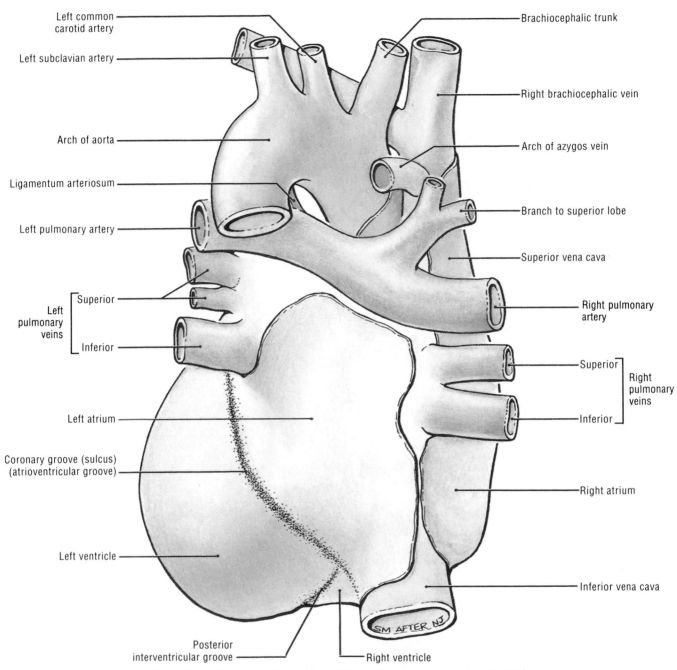

Left common carotid artery

Left subclavian artery

Arch of aorta

Ligamentum arteriosum

Left pulmonary artery

Left pulmonary veins
  Superior
  Inferior

Left atrium

Coronary groove (sulcus) (atrioventricular groove)

Left ventricle

Posterior interventricular groove

Brachiocephalic trunk

Right brachiocephalic vein

Arch of azygos vein

Branch to superior lobe

Superior vena cava

Right pulmonary artery

Right pulmonary veins
  Superior
  Inferior

Right atrium

Inferior vena cava

Right ventricle

SM AFTER NJ

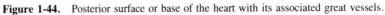

Figure 1-44. Posterior surface or base of the heart with its associated great vessels.

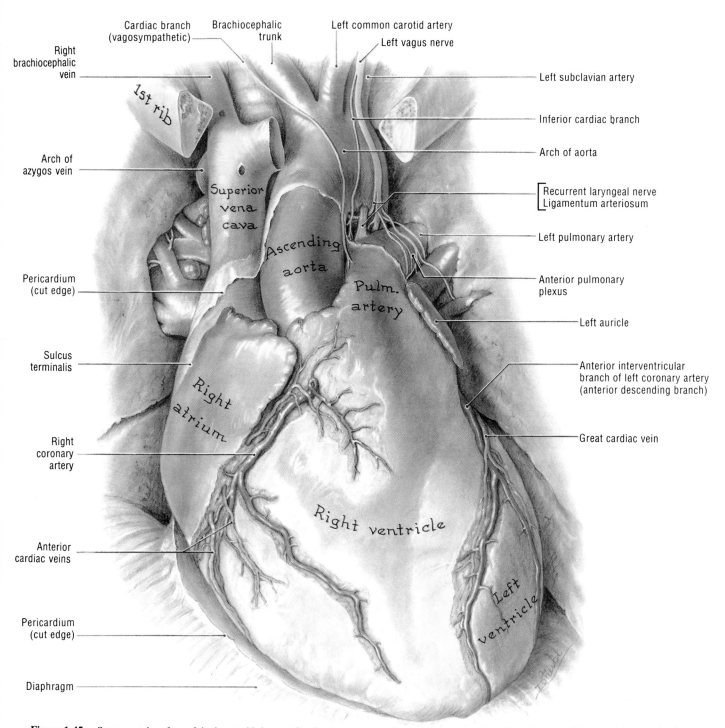

Cardiac branch (vagosympathetic)

Brachiocephalic trunk

Left common carotid artery

Left vagus nerve

Right brachiocephalic vein

1st rib

Arch of azygos vein

Superior vena cava

Ascending aorta

Pericardium (cut edge)

Pulm. artery

Sulcus terminalis

Right atrium

Right coronary artery

Anterior cardiac veins

Right ventricle

Pericardium (cut edge)

Left ventricle

Diaphragm

Left subclavian artery

Inferior cardiac branch

Arch of aorta

Recurrent laryngeal nerve
Ligamentum arteriosum

Left pulmonary artery

Anterior pulmonary plexus

Left auricle

Anterior interventricular branch of left coronary artery (anterior descending branch)

Great cardiac vein

Figure 1-45. Sternocostal surface of the heart with its associated nerves and great vessels, in situ. The pericardium is opened. Note that when the heart is in its normal position, the term *right ventricle* is misleading because the right ventricle is clearly anterior.

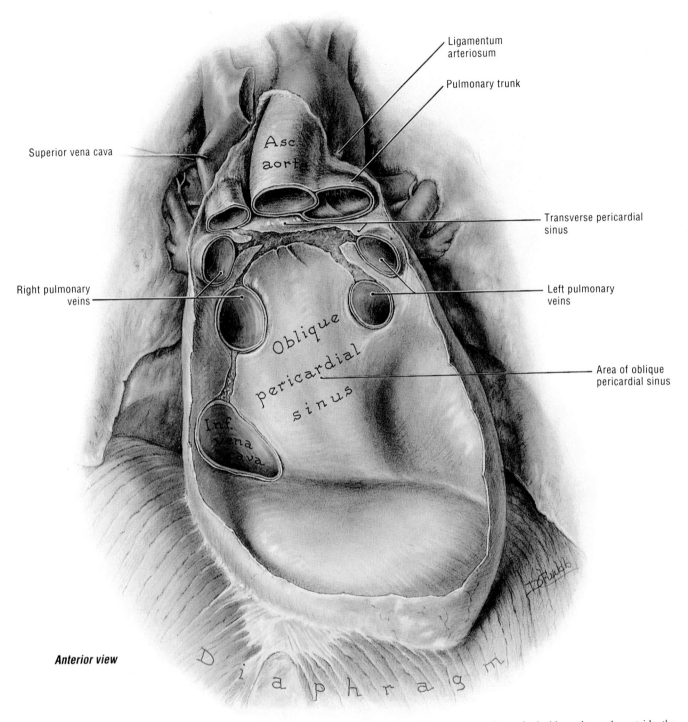

Anterior view

Figure 1-46.   Interior of the pericardial sac. To remove the heart shown in Fig. 1-60, the eight vessels piercing it were severed. Observe that the oblique pericardial sinus is circumscribed by five veins. Note that the superior vena cava is partly inside and mostly outside the pericardium.

pressed and circulation fails. The veins of the face and neck become engorged owing to compression of the superior vena cava as it enters the inelastic pericardium (Fig. 1-45).

Pericardiocentesis (peri- + G. *kentesis*, puncture) or paracentesis of the pericardium (a tapping or drainage of fluid from the pericardial cavity) is sometimes necessary to relieve the pressure of accumulated fluid on the heart. To remove the excess fluid, a wide-bore needle may be inserted through the left fifth or sixth intercostal space near the sternum (Fig. 1-42). However, care is taken not to puncture the *internal thoracic artery* (Fig. 1-19). This approach to the pericardial sac is possible because the cardiac notch in the left lung leaves part of this sac exposed (Fig. 1-30*A*). This area is also used as a route for *intracardiac injection* (*e.g.*, of **heparin**, an anticoagulant). The pericardial sac may also be reached by entering the left part of the infrasternal angle and passing the needle superoposteriorly (Figs. 1-29 and 1-42). At this site and angle, the needle should avoid the lung and pleurae and enter the pericardial cavity.

The Heart

The heart, which is slightly larger than a clenched fist, is a *double self-adjusting muscular pump*, the two parts of which normally work in unison. It propels the blood through the blood vessels to various parts of the body. The right side of the heart receives deoxygenated blood (low in oxygen but not without it) from the body and pumps it to the lungs, whereas the left side receives oxygenated blood from the lungs and pumps it into the aorta for distribution to the body.

The heart's weight varies from 280 to 340 gm in men and from 230 to 280 gm in women (Williams et al., 1989).

The heart has four chambers. Each side consists of an *atrium* (L. antechamber), a receiving area that pumps blood into a *ventricle* (L. little belly), a discharging chamber. The wall of each chamber consists of three layers (Table 1-1): an internal layer or *endocardium*; a middle layer or *myocardium* composed of cardiac muscle; and an external layer or *epicardium*. The myocardium forms the main mass of the heart.

The heart and the roots of the great vessels occupy the pericardium (pericardial sac), which is located in the middle mediastinum (Figs. 1-40 and 1-41; Table 1-1). These structures are related anteriorly to the sternum, the costal cartilages, and the medial ends of the third to fifth ribs on the left side (Figs. 1-42, 1-45, and 1-47). *The heart is situated obliquely in the middle mediastinum* (Fig. 1-40). The heart is not in the median plane; it is located about two-thirds to the left and one-third to the right of it (Fig. 1-42). The heart has a *base* (posterior aspect), *apex* (inferolateral end), *three surfaces* (sternocostal, diaphragmatic, and pulmonary), and *four borders* (right, inferior, left, and superior).

The Base of the Heart (Figs. 1-40, 1-43, 1-44, and 1-60). The base is located posteriorly and is *formed mainly by the left atrium*. It lies opposite T5 to T8 (supine position) and T6 to T9 vertebrae (erect position) and faces superiorly, posteriorly, and toward the right shoulder. The base or *posterior aspect of the heart* is quadrilateral in shape and is its most superior part,

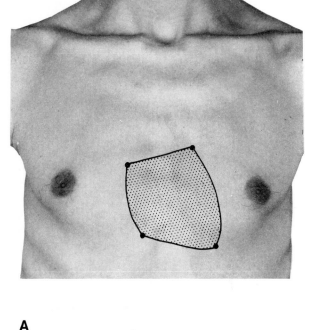

A

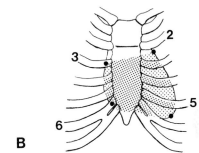

B

Figure 1-47. *A*, Thorax of a 27-year-old man showing the surface markings of the sternocostal surface of his heart. *B*, Anterior part of the thoracic cage showing the landmarks used when drawing an outline of the heart on the chest.

from which the ascending aorta and pulmonary trunk emerge and into which the superior vena cava enters. The base is separated from the diaphragmatic surface of the heart by the posterior part of the *coronary groove* (L. sulcus). *The heart does not rest on its base*; the term "base" derives from the somewhat conical shape of the heart, the base being opposite the apex.

The Apex of the Heart (Figs. 1-43, 1-45, and 1-47 to 1-50). The blunt apex is formed by the left ventricle, which points inferolaterally. The apex is *located posterior to the left fifth intercostal space in adults*, 7 to 9 cm from the median plane and just medial to the left midclavicular line. However, its location varies slightly with the person's position and the phase of respiration. The **apex beat** ("heart beat") is an impulse imparted by the heart; it is *its point of maximal pulsation*.

The Sternocostal (Anterior) Surface of the Heart (Figs. 1-43, 1-45, and 1-48 to 1-50). This surface of the heart is *mainly formed by the right ventricle* and is visible in PA radiographs of the thorax.

The Diaphragmatic (Inferior) Surface of the Heart (Figs. 1-29, 1-44, and 1-60). This surface of the heart is usually

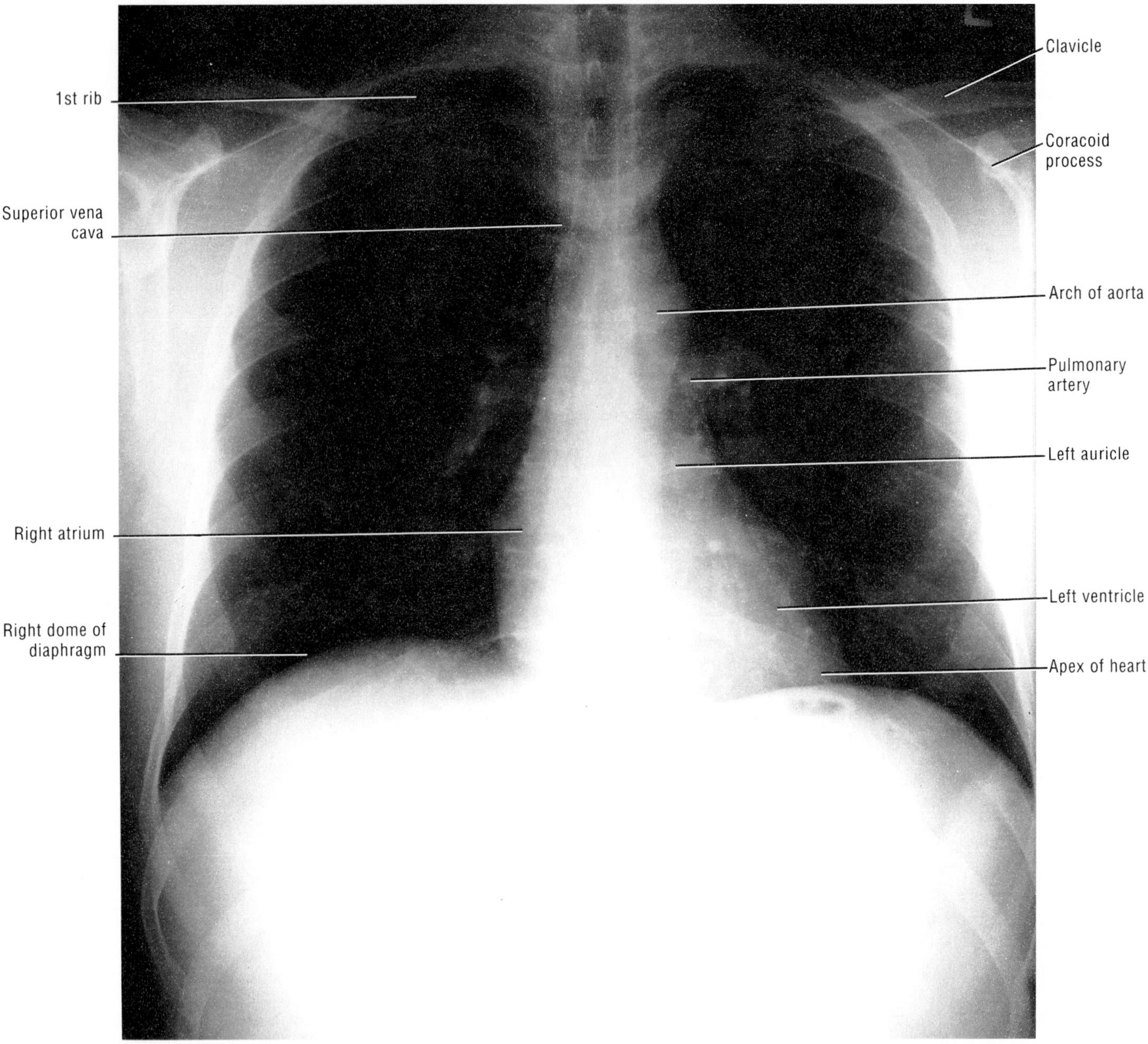
1st rib
Superior vena cava
Right atrium
Right dome of diaphragm
Clavicle
Coracoid process
Arch of aorta
Pulmonary artery
Left auricle
Left ventricle
Apex of heart

Figure 1-48. PA radiograph of the thorax, primarily to show the cardiovascular silhouette (cardiac shadow). Note that the dome of the diaphragm is somewhat higher on the right. Observe that the right mediastinal border of the heart is formed by the right atrium. The lesser convexity superior to this is produced by the superior vena cava. Note that the left mediastinal border of the heart is formed by the arch of the aorta, pulmonary trunk, left auricle, and left ventricle. (Courtesy of Dr. E.L. Lansdown, Professor of Radiology, University of Toronto, Toronto, Ontario, Canada.)

horizontal or slightly concave. It is *formed by both ventricles,* mainly the left one. As its name implies, this surface is related to the central tendon of the diaphragm. The *posterior interventricular groove* (sulcus) divides this surface into a right one-third and a left two-thirds.

The Pulmonary (Left) Surface of the Heart (Figs. 1-30*A*, 1-43, and 1-45). This surface of the heart is *mainly formed by the left ventricle* and occupies the *cardiac notch* of the left lung.

The Borders of the Heart (Figs. 1-43, 1-45, and 1-48 to 1-50). The heart has four borders: right, inferior, left, and superior. Actually these borders are the borders of its sternocostal (anterior) surface. The *right border* is formed by the *right atrium*. It is slightly convex and is almost in line with the superior and inferior venae cavae. The *inferior border*, which is sharp and thin, is nearly horizontal. It is formed *mainly by the right ventricle* and only slightly by the left ventricle. The *left border* is formed *mainly by the left ventricle* and only slightly by the left auricle (atrial appendage). The *superior border* is where the great vessels enter and leave the heart. It is formed by the *right and left auricles* with the superior conical portion of the right ventricle, the *conus arteriosus* (infundibulum), between them. The pulmonary trunk arises from the conus arteriosus.

Surface Anatomy of the Heart

The surface markings of the four borders of the sternocostal surface of the heart are variable because they depend partly on the physique of the individual (below). However, the outline of the average heart can be traced on the anterior surface of the chest by using the following guidelines (Fig. 1-47).

The superior border of the heart corresponds to a line connecting the *inferior margin of the second left costal cartilage* (3 cm to the left of the median plane) to the *superior border of the third right costal cartilage* (2 cm from the median plane). **The right border of the heart** corresponds to a line drawn from the *third right costal cartilage* (2 cm from the median plane) to the *sixth right costal cartilage* (2 cm from the median plane). This line is slightly convex to the right. **The inferior border of the heart** corresponds to a line drawn from the *inferior end of the right border* to a point in the *fifth intercostal space* close to the midclavicular line. The left end of this line corresponds to the location of the apex beat. **The left border of the heart** corresponds to a line connecting the left ends of the lines representing the superior and inferior borders. *The apex beat of the heart* in adults can be felt or heard in the fifth left intercostal space, just medial to the midclavicular line (7 to 9 cm from the median plane). In men and women, the apex beat is usually slightly inferomedial to the nipple.

The surface markings just described apply to the average adult heart. The surface anatomy of the heart is modified by age, sex, body size and build, respiration, position, and disease of the heart or lungs. A tall and thin or *ectomorphic person* has a long narrow thorax that contains a long narrow or **vertical heart**, similar to the shape seen during deep inspiration (Fig. 1-49). The broad, heavy-set, and stout or *mesomorphic person* has a broad thorax that contains a broad or **transverse heart** similar to the shape seen after deep expiration (Fig. 1-50). This type of heart is also associated with infancy, pregnancy, and obesity. Most hearts are between vertical and transverse and are called **oblique hearts**.

The location of the nipple is not a reliable guide to the position of the apex beat in mature females owing to the variation in the size and pendulousness of the breast, especially in women who have borne children. In infants and children, the apex beat is more superior and further laterally than in adults. The site of the apex beat almost corresponds to the position of the apex of the heart. *The apex beat in a newborn infant may be palpated in the fourth left intercostal space*, in or just lateral to the midclavicular line. After 2 years of age, the apex beat is usually detected in the fifth intercostal space, slightly medial to the midclavicular line. In many people the pulsations of the apex are visible.

Percussion is a commonly used diagnostic procedure for determining the *density of the heart*. The character of the sound changes as different areas of the chest are tapped. Place the middle digit of your left hand approximately parallel and to the left of the left border of your heart. Now tap with the middle digit of your other hand. While percussing, move your digit to the right and note how the sound changes as you

move from over your lung to your heart. Owing to the *cardiac notch* in the left lung, part of the fibrous pericardium is uncovered by lung tissue (Fig. 1-30*A*). This area of the left chest is known as the *area of superficial cardiac dullness*. Being devoid of overlying lung tissue, it gives a dull note during percussion. By noting the areas of cardiac dullness, you can determine the rough outline of your heart.

Radiological Anatomy of the Heart

The heart and great vessels and the blood within them are of the same order of density. Hence, AP chest films show the contour of the heart and great vessels in the middle mediastinum (Fig. 1-48). This is called the **cardiovascular silhouette** or cardiac shadow. Because the heart and great vessels are full of blood, the silhouette stands out in contrast to the clearer areas occupied by the air-filled lungs. Because the fibrous pericardium is attached to the diaphragm, *the cardiovascular silhouette becomes longer and narrower during inspiration* (Fig. 1-49) and shorter and broader during expiration (Fig. 1-50). In PA radiographs of the thorax, the borders of the cardiovascular silhouette (superior to inferior) are formed by the heart and great vessels (Fig. 1-48). *The right border of the cardiovascular silhouette* is formed by: (1) the superior vena cava; (2) the right atrium; and (3) the inferior vena cava. *The left border of the cardiovascular silhouette* is formed by: (1) the arch of the aorta, which produces a characteristic prominence known as the *aortic knob*; (2) the pulmonary trunk; (3) the left auricle; and (4) the left ventricle. The radiographic appearance of the heart varies in different people because the outline of the heart is variable. Its appearance depends partly on the physique of the person.

Physicians know the parts of the heart and great vessels that form the cardiovascular silhouette so they can detect heart abnormalities in radiographs and other images (CT scans and MRIs).[5] **Enlargement of the heart** may result from *hypertrophy*, especially in cases of high blood pressure. In these cases, the walls of the heart grow thicker by increasing the number and size of the cardiac muscle fibers. However, *enlargement of the heart usually results from dilation*. When blood regurgitates from the aorta into the left ventricle, this chamber dilates to accommodate the extra blood. Similarly, when blood regurgitates through a *diseased mitral valve* (left atrioventricular valve) into the left atrium from the left ventricle, the left atrium dilates to accommodate the additional blood.

A complication of pericarditis may be *hydropericardium* (p. 82), in which the pericardial sac becomes distended with fluid and interferes with the return of blood to the heart. Severe pericardial effusion causes an enlargement of the cardiovascular silhouette, and differentiation of it from heart enlargement can be determined by *ultrasonography* or by studying CT scans or MRIs.

In the usual chest radiograph, the separate cavities of the heart are not distinguishable, but *angiocardiography* can be used to study the cavities of the heart and great vessels. When a suitable contrast medium is injected via a catheter, the course of the blood can be followed fluoroscopically by either

---

[5]CT scan indicates computerized tomographic scan and MRIs, magnetic resonance images.

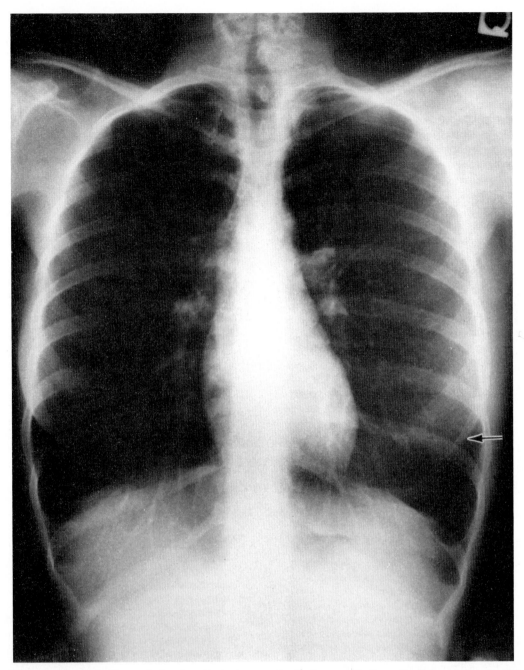

Figure 1-49. PA radiograph of the thorax during deep inspiration. Observe that the ribs are widely separated. Note the low position of the diaphragm and the recesses of the pleural cavity that are not filled by the lungs. During inspiration the heart appears more vertical because the diaphragm pulls the pericardial sac inferiorly. Note the difference between the relation of the left half of the diaphragm and the anterior end of the seventh rib (arrow).

cineradiography or serial radiographic films in various projections. Most congenital **cardiac defects** can be diagnosed with the aid of cardiac catheterization. *Right cardiac catheterization* consists of passing a radiopaque catheter under sterile conditions into a peripheral vein (*e.g.*, the external iliac) and guiding it with the aid of fluoroscopy into the great veins (*e.g.*, the inferior vena cava), the right heart chambers, or the pulmonary artery. When a measured dose of contrast material is injected into the right ventricle, it traverses the pulmonary circulation and enters the left side of the heart. *Left cardiac catheterization* consists of inserting a catheter into a peripheral artery (*e.g.*, the femoral artery in the thigh) and guiding it by fluoroscopy through parts of the aorta to the left ventricle.

Chambers of the Heart

The heart has *four chambers*, two atria and two ventricles. The **coronary groove** (sulcus) encircles most of the superior part of the heart and *separates the atria from the ventricles* (Figs. 1-

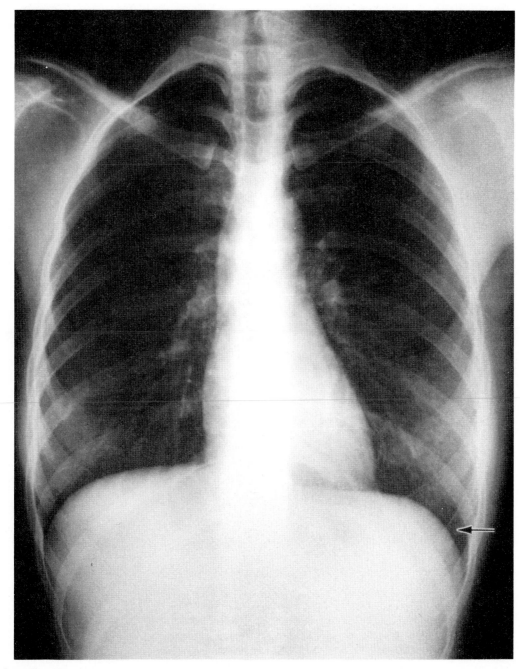

Figure 1-50.   PA radiograph of the thorax shown in Fig. 1-49 during expiration. Observe the high position of the diaphragm and lungs, which are not as translucent because they contain less air. The arrow indicates the anterior part of the seventh rib. Note that the diaphragm has risen and casts dome-shaped shadows on each side (slightly higher on the right). Observe the characteristic ''aortic knob'' produced by the arch of the aorta (see also Fig. 1-48).

43 and 1-44). Similarly, division of the ventricles is indicated by the anterior and posterior *interventricular grooves* (sulci).

The Right Atrium (Figs. 1-43 to 1-45, 1-48 to 1-52, and 1-60).   This chamber forms the right border of the heart between the superior and inferior venae cavae. It receives venous blood from these large vessels and the coronary sinus. The **coronary sinus**, lying in the posterior part of the *coronary groove*, receives blood from the veins of the heart and opens into the right atrium. The internal wall of the right atrium consists of: (1) a smooth posterior part, called the *sinus venarum* (sinus of the venae cavae), which receives the venae cavae and coronary sinus, and (2) a

rough anterior part, which has internal muscular ridges (*musculi pectinati*) that resemble the teeth of a comb (L. *pectin*).

The right auricle (atrial appendage) is a small, conical muscular pouch that projects to the left from the right atrium and overlaps the ascending aorta (Figs. 1-43, 1-45, and 1-51). The two distinct parts of the right atrium are separated externally by a shallow vertical groove on the posterior aspect of the right atrium, called the *sulcus terminalis*, and internally by a vertical crest or ridge called the *crista terminalis*. The crista extends between the two vena caval orifices.

The interatrial septum forms the thin posteromedial wall of

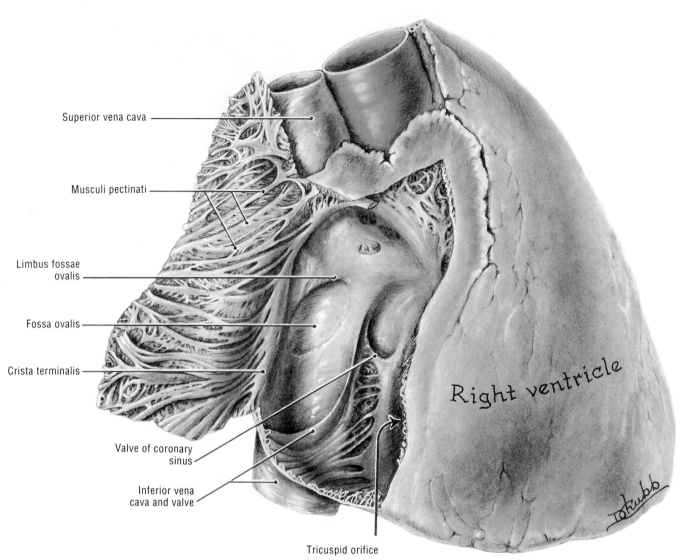

Superior vena cava

Musculi pectinati

Limbus fossae
ovalis

Fossa ovalis

Crista terminalis

Valve of coronary
sinus

Inferior vena
cava and valve

Right ventricle

Tricuspid orifice

Figure 1-51. Anterolateral view of the interior of the right atrium.

the right atrium. A prominent feature of this dividing septum, between the caval orifices, is the thumbprint-sized *fossa ovalis* (Fig. 1-51), a large, shallow translucent *oval depression*. The fossa ovalis has an incomplete sharp margin known as the *limbus fossae ovalis*. The small *opening of the coronary sinus* is located between the right atrioventricular orifice and the orifice of the inferior vena cava. This opening of the coronary sinus is guarded by a thin semicircular valve, called the *valve of the coronary sinus*, which is located along the inferior right margin of the orifice of the coronary sinus. This valve closes the orifice during contraction of the right atrium, thereby preventing regurgitation of blood.

The superior vena cava (Figs. 1-42 to 1-44 and 1-54), returning blood from the superior half of the body, opens into the superoposterior part of the right atrium at about the level of the right third costal cartilage. Its orifice, usually valveless, is directed inferoanteriorly.

The inferior vena cava (Figs. 1-43, 1-44, and 1-51), returning blood from the inferior half of the body, opens into the

inferior part of the right atrium, almost in line with the smaller superior vena cava. The rudimentary *valve of the inferior vena cava* is a thin fold of variable size that connects the anterior margin of its orifice to the anterior part of the *limbus fossae ovalis*. This valve is nonfunctional after birth.

Without knowledge of the development of the right atrium, its adult anatomy is difficult to understand and remember. The primitive right atrium was enlarged by the incorporation of most of the *sinus venosus*, a venous part of the embryonic heart (Moore, 1988). The primitive right atrium is represented in the adult by the right auricle. The *coronary sinus* is also derived from the sinus venosus. In the adult right atrium, the incorporated part of the sinus venosus is represented by the *sinus venarum*. The separation between the primitive atrium, represented by the auricle, and the sinus venarum, derived from the sinus venosus, is indicated externally by the *sulcus terminalis* and internally by the *crista terminalis*. Before birth the *valve of the inferior vena cava* directed the flow of oxygenated blood from the placenta, via the *umbilical vein*, through an opening in the interatrial septum called the *foramen ovale* and into the left atrium. After birth the foramen ovale normally closes and is represented in the adult interatrial septum by the *fossa ovalis*, the margins of which are called the *limbus fossae ovalis* (L. *limbus*, border). The valve of the inferior vena cava has no function after birth. It varies considerably in size and is occasionally absent.

A probe-sized *atrial septal defect* (ASD) appears in the superior part of the fossa ovalis in up to 25% of people. *Before birth*, every fetus has such a communication (the foramen ovale) between the right and left atria, but it normally closes *after birth*. A small ASD is usually of no clinical significance (Behrman, 1992), but large ASDs are clinically significant because they allow recently oxygenated blood from the lungs in the left atrium to be shunted through the defect into the right atrium (Moore, 1988). This overloads the pul-

monary system and causes enlargement of the right atrium and ventricle and dilation of the pulmonary trunk.

The Right Ventricle (Figs. 1-43, 1-45, 1-51, 1-52, and 1-55). This chamber forms the largest part of the sternocostal (anterior) surface of the heart, a small part of the diaphragmatic surface, and almost the entire inferior border of the heart. Its superior left angle tapers into a cone-shaped pouch, called the infundibulum or **conus arteriosus** (L. *infundibulum*, funnel), which leads into the pulmonary trunk. Internally the conus arteriosus has a funnel-shaped appearance. Its internal wall is smooth, whereas the rest of the right ventricular wall is roughened by a number of irregular muscle bundles (*papillary muscles*) and muscular ridges and bridges. The irregular muscular elevations, called *trabeculae carneae*, project from the ventricular wall giving it

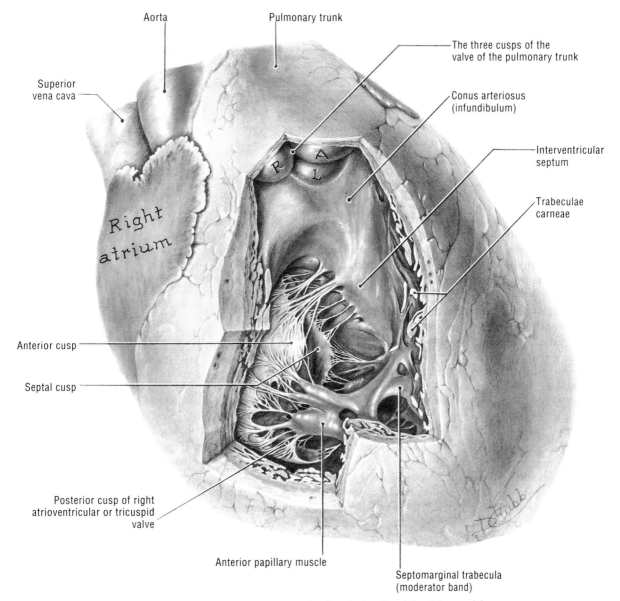

Figure 1-52.   Dissection of the heart showing the interior of the right ventricle.

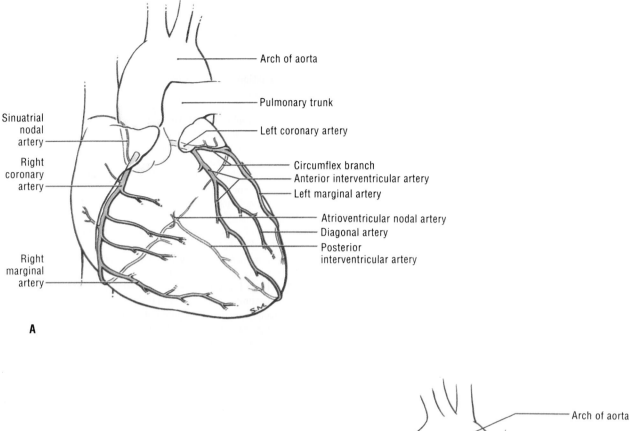

Arch of aorta

Pulmonary trunk

Left coronary artery

Circumflex branch
Anterior interventricular artery
Left marginal artery

Atrioventricular nodal artery
Diagonal artery
Posterior interventricular artery

Sinuatrial nodal artery

Right coronary artery

Right marginal artery

A

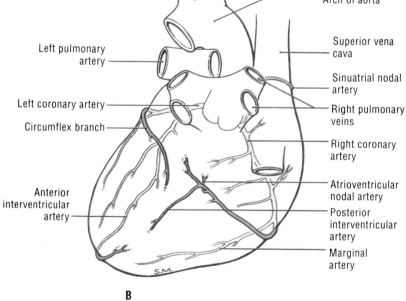

Arch of aorta

Superior vena cava

Sinuatrial nodal artery

Right pulmonary veins

Right coronary artery

Atrioventricular nodal artery

Posterior interventricular artery

Marginal artery

Left pulmonary artery

Left coronary artery

Circumflex branch

Anterior interventricular artery

B

Figure 1-58. *A*, Anterior and *B*, Posteroinferior views of the heart showing the coronary arteries. *C*, Diagram explaining the arteriogram shown in *D*, which is through the courtesy of Dr. I Morrow, Department of Radiology, Health Sciences Centre, University of Manitoba, Winnipeg, Manitoba, Canada.

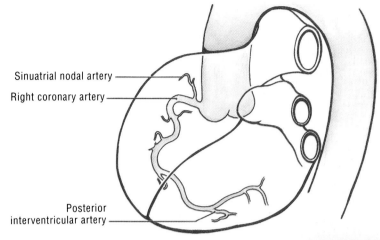

Sinuatrial nodal artery

Right coronary artery

Posterior
interventricular artery

C

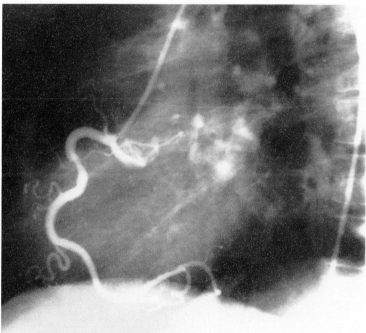

D

Figure 1-58C–D

When the supply of oxygen to the myocardium is cut off owing to coronary occlusion, the area of muscle concerned undergoes necrosis and infarction. The pain resulting from **MI** is often more severe than with angina pectoris, and *the pain does not disappear after 1 or 2 min of rest*. MI may also follow excessive exertion by a person with stenotic coronary arteries. The straining heart muscle demands more oxygen than the narrow arteries can provide; as a result, the ischemic area of myocardium undergoes infarction and a **heart attack** occurs. Coronary occlusion of any but the smallest branches of an artery usually results in death of the cardiac muscle fibers it supplies. The damaged muscle is replaced by fibrous tissue and a scar forms. If parts of the conducting system are affected by the blockage (*e.g.*, stenosis of the AV nodal artery), a *heart block* may occur. In this case the ventricles may continue to contract independently at their own rate.

The coronary arteries can be visualized by **coronary angiography**. A long narrow catheter is passed into the ascending aorta via the femoral (thigh) or brachial (arm) arteries. Under fluoroscopic control, the tip of the catheter is placed just inside the opening of a coronary artery (Figs. 1-55 and 1-57). A small injection of radiopaque contrast material is then made, and radiographs or cineradiographs are taken to show the lumen of the artery and its branches, as well as any stenotic areas that may be present.

The three most **common sites of coronary occlusion** are in: (1) the anterior interventricular branch of the left coronary artery; (2) the right coronary artery; and (3) the circumflex branch of the left coronary artery. In some patients with obstruction of the coronary circulation and severe angina pectoris, a *coronary bypass graft* operation is carried out (Goldman, 1989). A segment of a vein, *e.g.*, the internal thoracic vein or great saphenous vein (Fig. 1-20; see also Fig. 2-7) is

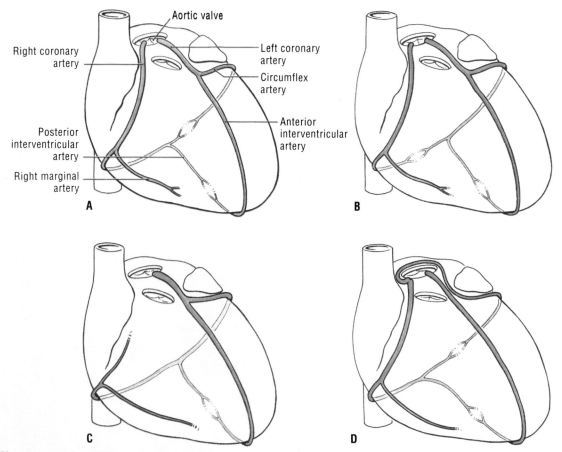

Figure 1-59. Anterior views of hearts showing variations of the coronary arteries. *A*, The coronary arteries share equally in the blood supply of the heart. *B*, The left coronary artery is supplying part of the usual right coronary artery territory. Note that the posterior interventricular branch comes off the circumflex branch. *C*, There is only one coronary artery. *D*, the circumflex branch arises from the right aortic sinus.

connected to the ascending aorta or to the proximal part of a coronary artery and then to the coronary artery distal to the stenosis. A **coronary bypass graft** shunts blood from the aorta or a coronary artery to a branch of a coronary artery to increase the flow distal to the obstruction. Simply stated, it provides a detour around the narrow area (arterial stenosis) or blockage (arterial atresia). *Revascularization of the myocardium* may also be achieved by surgically anastomosing an internal thoracic artery with a coronary artery. In selected patients, a technique called *percutaneous transluminal coronary angioplasty* is used. A catheter with a small inflatable balloon attached to its tip is passed into the obstructed coronary artery. When the obstruction is reached, the balloon is inflated and the vessel is opened. In other cases *thrombokinase* is injected via the catheter; this enzyme dissolves the blood clot.

Venous Drainage of the Heart

The heart is drained mainly by veins that empty into the **coronary sinus** (Figs. 1-45, 1-51, and 1-60), and partly by small veins (*venae cordis minimae* and *anterior cardiac veins*) that open directly into the chambers of the heart, principally those on the right side. *The coronary sinus is the main vein of the heart*. This short wide venous channel (about 2 cm long) runs from left to right in the posterior part of the *coronary groove*. The coronary sinus receives the great cardiac vein at its left end and the middle and small cardiac veins at its right end.

The coronary sinus drains all the venous blood from the heart, except that carried by the anterior cardiac veins and the venae cordis minimae. It opens into the right atrium, immediately to the left of the inferior vena cava and posterior to the right atrioventricular orifice (Fig. 1-51). The small *valve of the coronary sinus*, which is variable in size and form, lies to the right of its orifice. This valve, a remnant of the valve of the sinus venosus (a part of the embryonic heart), appears to have *no function postnatally*. Usually the veins accompany the coronary arteries and their branches, but they have not been given the same names.

The **great cardiac vein** is the *main tributary of the coronary sinus* (Figs. 1-45 and 1-60). It begins near the apex of the heart and ascends in the anterior interventricular groove with the anterior interventricular artery. Here it *enters the left end of the coronary sinus*. It has a valve at its entrance into this sinus. The great cardiac vein drains the areas of the heart supplied by the left coronary artery (p. 99). The *middle cardiac vein* also begins at the apex of the heart but ascends in the posterior interventricular groove with the posterior interventricular artery (Fig. 1-60). It enters the right side of the coronary sinus. The *small cardiac vein* runs in the coronary groove (with the marginal

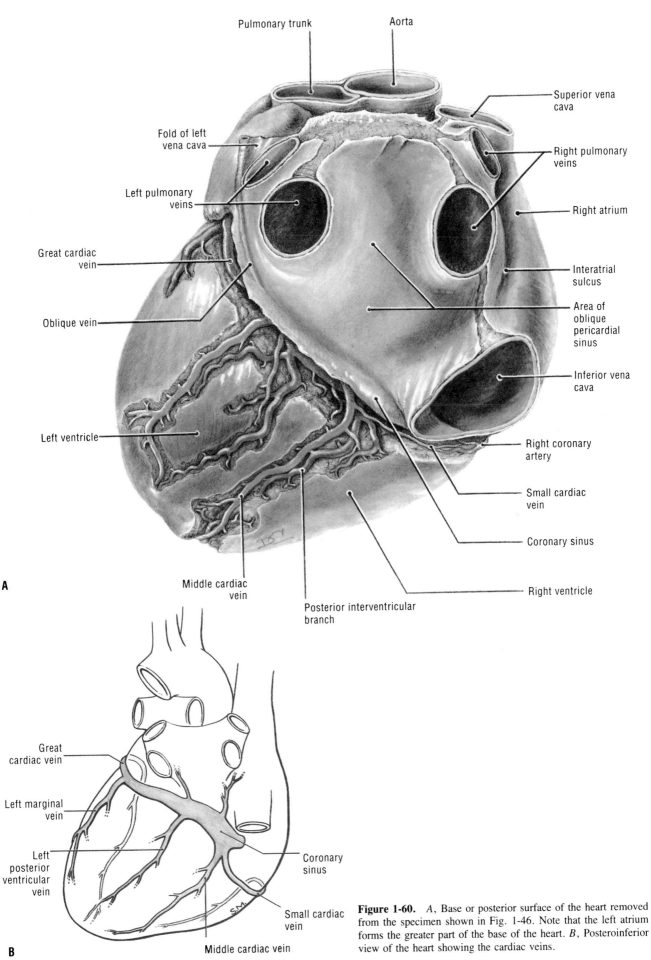

Figure 1-60. *A*, Base or posterior surface of the heart removed from the specimen shown in Fig. 1-46. Note that the left atrium forms the greater part of the base of the heart. *B*, Posteroinferior view of the heart showing the cardiac veins.

103

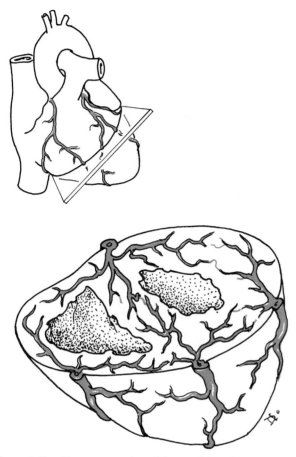

Figure 1-61. Transverse section of the ventricles of a heart showing the branches of the coronary arteries penetrating the myocardium. Note the anastomoses between the vessels in the interventricular septum.

branch of the right coronary artery). It enters the coronary sinus to the right of the middle cardiac vein. Although it usually terminates in the coronary sinus, it may open directly into the right atrium. The middle and small cardiac veins drain most of the area of the heart supplied by the right coronary artery (p. 99). *The posterior vein of the left ventricle* passes along its inferior surface (Fig. 1-60) and empties near the middle of the coronary sinus. It is frequently paired with terminal branches of the circumflex branch of the left coronary artery.

> The *oblique vein of the left atrium* is a small, relatively unimportant vessel, that runs over the posterior wall of the left atrium (Fig. 1-60A) and enters the left end of the coronary sinus. There are several small *anterior cardiac veins* (Fig. 1-45) that begin over the anterior surface of the right ventricle, cross over the coronary groove, and usually *end directly in the right atrium*; sometimes they enter the small cardiac vein. The *venae cordis minimae* or smallest cardiac veins (once called Thebesian veins) are minute vessels that begin in the myocardium and open directly into the chambers of the heart, chiefly the atria. Although called veins, they may also carry blood to the myocardium.

Lymphatic Drainage of the Heart

The lymph vessels from the myocardium and subendocardial connective tissue pass to the *subepicardial lymph plexus*. Vessels from this plexus pass to the coronary groove and follow the coronary arteries (Fig. 1-61). A single lymph vessel, formed by the union of various vessels from the heart, ascends between the pulmonary trunk and left atrium and ends in the inferior *tracheobronchial lymph nodes*, usually on the right side (Figs. 1-38B, 1-62, and 1-70).

Conducting System of the Heart

This system consists of cardiac muscle cells and conducting fibers (not nervous tissue) that are *specialized for initiating impulses and conducting them rapidly through the heart* (Fig. 1-63). They initiate the normal heart beat and coordinate the contractions of the four heart chambers. Both atria contract together, as do the ventricles, but atrial contraction occurs first. The conducting system gives the heart its automatic rhythmic beat. For the heart to pump efficiently and the systemic and pulmonary circulations to operate in synchrony, there must be coordination of the events in the *cardiac cycle*.

The Sinuatrial Node (SA Node)[6] (Fig. 1-63). This collection of specialized cardiac muscle fibers (*nodal tissue*) in the wall of the right atrium initiates the impulses for contraction. It is the natural **pacemaker of the heart**. The SA node is located at the superior end of the *crista terminalis* at the junction of the anteromedial aspect of the superior vena cava and the right auricle (Figs. 1-51 and 1-58). It may be located by following the *nodal artery* from the right coronary artery (occasionally from the left coronary). The SA node is *supplied by both divisions of the ANS* (autonomic nervous system). It gives off an impulse about 70 times/min in most people. The rate at which the node produces impulses can be altered by nervous stimulation: Sympathetic stimulation speeds it up and parasympathetic stimulation slows it down.

The Atrioventricular Node (AV node) (Figs. 1-51 and 1-63). This much smaller collection of nodal tissue is also composed of specialized cardiac muscle cells. It is *located in the interatrial septum*, on the ventricular side of the orifice of the coronary sinus. Impulses from the cardiac muscle fibers of both atria converge on the AV node, which distributes them to the ventricles via the atrioventricular bundle. The AV node conducts the impulses slowly, but sympathetic stimulation speeds up conduction and parasympathetic stimulation slows it down.

The Atrioventricular Bundle (AV Bundle) (Fig. 1-63). This collection of specialized conducting muscle fibers, often called *Purkinje fibers*, originates in the AV node and runs through the membranous part of the interventricular septum. Here, it lies just inferior to the septal cusp of the tricuspid valve. *The AV bundle is the only bridge between the atrial and ventricular myocardium.* At the junction between the membranous and muscular parts of the interventricular septum, the AV bundle divides

---

[6]Its name, sinuatrial node (sinoatrial node), serves as a reminder that it was in the wall of the sinus venosus during embryonic development and was absorbed into the right atrium with the sinus venosus (Moore, 1988).

into right and left limbs, commonly called *right* and *left branches*. Each branch proceeds deep to the endocardium into the walls of the ventricles. The *right branch of the AV bundle* stimulates the anterior papillary muscle, and the wall of the right ventricle. The *left branch of the AV bundle* stimulates the papillary muscles, and the wall of the left ventricle.

Summary of the Conducting System of the Heart (Fig. 1-63). The *SA node initiates an impulse*, which is rapidly conducted to the muscle fibers of the atria, causing them to contract. *The impulse enters the AV node and is distributed through the AV bundle and its branches*, which pass through the interventricular septum and the papillary muscles and walls of the ventricles. The papillary muscles contract first, tightening the chordae tendineae and drawing the cusps of the atrioventricular valves together. Next, contraction of the ventricular muscle occurs.

> When there is an *atrial septal defect* (*ASD*), the AV bundle usually lies in the margin of the defect. Obviously this vital part of the impulse-conducting system must be preserved during *ASD surgery* because destruction of it would cut the only

physiological link between the atrial and ventricular musculature.

The passage of impulses over the heart from the SA node can be amplified and recorded as an *electrocardiogram* (**ECG**). As many heart problems involve abnormal functioning of the impulse-conducting system, ECGs are of considerable clinical importance in detecting the cause of irregularities of the heart beat. Patients with a *massive myocardial infarction* (**MI**) usually experience crushing substernal chest pain and have an abnormal ECG. The nerve impulses from the heart, which are responsible for producing the pain of MI, enter the spinal cord through the superior thoracic ganglia of the sympathetic trunk (Fig. 1-22).

Artificial pacemakers produce electrical impulses that initiate ventricular contractions at a predetermined rate. The battery-powered pacemaker, about the size of a pocket watch, is implanted for permanent pacing of the heart. An electrode with a catheter connected to it is inserted into a vein and followed with a fluoroscope. The terminal of the electrode is passed to the right atrium and through the tricuspid valve to the right ventricle, where it is firmly fixed to the trabeculae

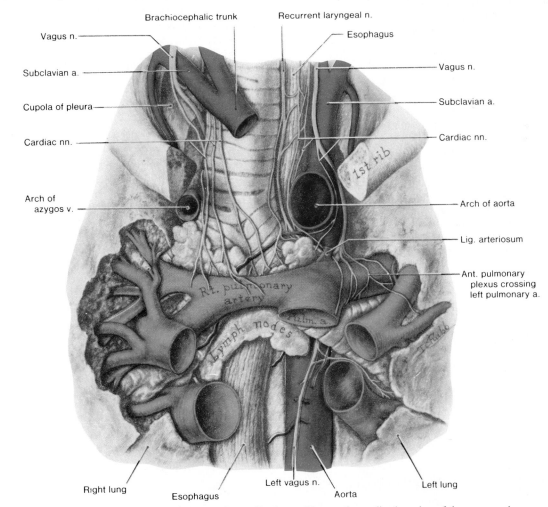

Figure 1-62. Dissection of the superior mediastinum. Observe the cardiac branches of the vagus and sympathetic nerves streaming down the sides of the trachea and forming the cardiac plexus.

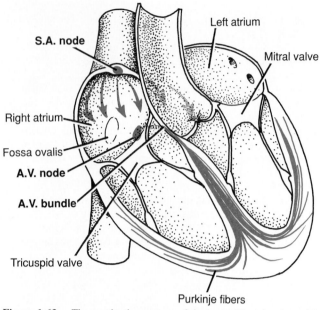

Figure 1-63.   The conducting system of the heart. Note the sinuatrial node (SA node) at the superior end of the crista terminalis and the atrioventricular node (AV node) in the inferior part of the interatrial septum. The AV bundle begins at the AV node and divides into right and left limbs at the junction of the membranous and muscular parts of the interventricular septum.

carneae in its walls (Fig. 1-52). Here, it is in contact with the endocardium.

When the heart stops beating, attempts are made to start it beating again. This can be done by trained health professionals using **external heart massage**, or *cardiopulmonary resuscitation* (*CPR*). During CPR, firm pressure is applied to the chest vertically downward over the inferior part of the sternum (Fig. 1-42). The sternum moves posteriorly 4 to 5 cm, forcing blood out of the heart into the great arteries.

Fibrillation of the heart refers to multiple, rapid, circuitous contractions or twitchings of cardiac muscular fibers. In *atrial fibrillation* the normal regular rhythmical contractions of the atria are replaced by rapid irregular twitchings of different parts of their walls simultaneously. The ventricles respond at irregular intervals to the *dysrhythmic impulses* received from the atria, but usually a satisfactory circulation is maintained. In *ventricular fibrillation*, the normal ventricular contractions are replaced by rapid, irregular, twitching movements that do not pump (*i.e.*, they do not maintain the systemic circulation, including the coronary circulation). The damaged conducting system of the heart does not function normally. As a result, an irregular pattern of contractions occurs in all areas of the ventricles simultaneously, except in those that have been infarcted. Ventricular fibrillation is the most disorganized of all *dysrhythmias*, and in its presence no effective cardiac output occurs. The condition is fatal if allowed to persist.

Brain anoxia (lack of oxygen to the brain) and *brain death* usually occur before the abnormal heart movements cease. To *defibrillate the heart*, an electric shock may be given to the heart through the thoracic wall via large electrodes (paddles). This shock causes cessation of all cardiac movements

and a few minutes later the heart may begin to beat more normally. As a result, pumping of the heart is re-established and some degree of systemic (including coronary) circulation results.

Innervation of the Heart

The heart is supplied by autonomic nerve fibers from the *vagus nerves* and the *sympathetic trunks*. Branches of both vagus nerves and both sympathetic trunks form **the cardiac plexus** (Figs. 1-62, 1-68, 1-72, and 1-73). It lies anterior to the bifurcation of the trachea, posterior to the arch of the aorta, and superior to the bifurcation of the pulmonary trunk. *Stimulation of the sympathetic nerves increases the heart rate* and the force of the heart beat causing dilation of the coronary arteries that supply the heart. This results in the supply of more oxygen and nutrients to the myocardium. *Stimulation of the parasympathetic nerves* slows the heart rate, reduces the force of the heart beat, and constricts the coronary arteries. The CNS[7] via the cardiac plexus, exercises control over the action of the heart and monitors blood pressure and respiration. This plexus also transmits afferent fibers to the vagus nerves from the great vessels and lungs. These sensory fibers transmit impulses from *pressure receptors* in the arch of the aorta, superior vena cava, and elsewhere.

The pain of angina pectoris and myocardial infarction commonly radiates from the substernal region and left pectoral region to the left shoulder and the medial aspect of the left arm (Fig. 1-64). This is known as **cardiac referred pain**. Less commonly the pain radiates to the right shoulder and arm, with or without concomitant pain on the left side. These cutaneous zones of reference for cardiac referred pain coincide with the segmental distribution of the sensory fibers that enter the same spinal cord segments as the fibers coming from the heart (Fig. 1-24).

The heart is insensitive to touch, cutting, cold, and heat, but ischemia and the resulting accumulation of metabolic products stimulate pain endings in the myocardium. The af-

---

[7]Central nervous system.

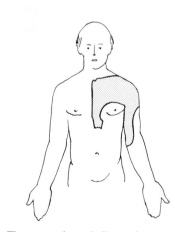

Figure 1-64.   The screened area indicates the common site of referred pain from the heart during a heart attack.

ferent pain fibers run centrally in the middle and inferior cervical branches and in the thoracic cardiac branches of the sympathetic trunk (Figs. 1-72 and 1-73). The axons of these primary sensory neurons enter spinal cord segments T1 to T4 or T5 on the left side. Cardiac pain is therefore referred to the left side of the chest and the medial aspect of the left arm (Figs. 1-24 and 1-64). Synaptic contacts may also be made with commissural neurons (connector neurons) that conduct impulses to neurons on the right side of comparable areas of the cord. This explains why pain of cardiac origin, although usually referred to the left side, may be referred to the right side or both sides.

The Superior Mediastinum

This subdivision of the mediastinum is located superior to the pericardium and the horizontal plane passing through the *sternal angle* to the intervertebral disc between T4 and T5 vertebrae (Fig. 1-40). From anterior to posterior, **the main contents of the superior mediastinum** are (Figs. 1-42 and 1-65 to 1-70): (1) the *thymus* (or its remains); (2) the *great vessels* related to the heart and pericardium; (3) the *phrenic and vagus nerves*; (4) the *cardiac plexus* of nerves; (5) the *trachea*; (6) the *left recurrent laryngeal nerve*; (7) the *esophagus*; (8) the *thoracic duct*; and (9) the *prevertebral muscles* (*e.g.*, the longus colli muscle; see p. 807 and figures 8-26 and 8-29).

The Thymus

The thymus (or its remains) is located in the anterior part of the superior mediastinum (Figs. 1-40 and 1-65). Its two flat flask-shaped lobes are a prominent feature of this region *during infancy* and childhood. In vivo, it has a pink, lobulated appearance, lies immediately posterior to the manubrium, and extends into the anterior mediastinum anterior to the pericardium. In some newborn infants the thymus may also extend superiorly through the superior thoracic aperture into the neck. *The thymus plays an important role in the development and maintenance of the immune system.* As puberty is reached, the thymus begins to diminish in relative size, and by adulthood, it is composed largely of adipose tissue and is often scarcely recognizable. However, it continues to produce *T-lymphocytes* (Cormack, 1987).

The Arterial Supply of the Thymus (Figs. 1-20, 1-42, and 1-67). There is a rich blood supply to the thymus. It is derived from the *inferior thyroid, anterior intercostal*, and *internal thoracic arteries*.

Venous and Lymphatic Drainage of the Thymus (Figs. 1-19, 1-38B, 1-63, 1-70, and 1-73). The *veins of the thymus* end in the left brachiocephalic, internal thoracic, and inferior thyroid veins. The *lymph vessels of the thymus* end in the parasternal, brachiocephalic, and tracheobronchial lymph nodes.

The thymus grows steadily during childhood and reaches its greatest size by puberty. Thereafter it begins to involute and much of it is replaced by fibrous tissue and fat. Most of the thymus is usually located in the superior mediastinum posterior to the manubrium and in the superior half of the anterior mediastinum, but it may extend as far inferiorly as the xiphoid process. *Thymomas* (usually benign tumors originating from thymic tissue) are uncommon, but they may be responsible for vague *retrosternal pain* (*i.e.*, posterior to the sternum), coughing, and *dyspnea* (shortness of breath) owing to pressure on the trachea (Fig. 1-65). A **thymoma** may also compress the superior vena cava and cause engorgement of the neck veins.

The Brachiocephalic Veins

The brachiocephalic veins (formerly called innominate veins) are *located in the superior mediastinum* (Figs. 1-40, 1-65, 1-68, 1-72, and 1-73). Each vein is formed posterior to the medial end of the clavicle by the union of the internal jugular and subclavian veins. This is a union of the veins from the arm (L. *brachium*), head (G. *kephale*), and neck. They have no valves. At the level of the inferior border of the first right costal cartilage, *the two brachiocephalic veins unite to form the superior vena cava*. Each brachiocephalic vein receives the internal thoracic, vertebral, inferior thyroid, and most superior (highest) intercostal veins.

The Right Brachiocephalic Vein (Figs. 1-42, 1-68, and 1-72). This short, almost vertical vein forms posterior to the right sternoclavicular joint by the *union of the right internal jugular and subclavian veins*. It descends in the superior mediastinum, posterior to the manubrium and lateral to the brachiocephalic trunk. The *right vagus nerve (CN X)* lies between these vessels, and the *right phrenic nerve* lies posterolateral to the right brachiocephalic vein. This vein joins the left brachiocephalic vein to form the *superior vena cava* close to the sternum and the inferior border of the first costal cartilage. The right brachiocephalic vein *receives the right lymphatic duct* (Fig. 1-39).

The Left Brachiocephalic Vein (Figs. 1-39, 1-42, 1-65, 1-68, and 1-73). This large vein forms posterior to the left sternoclavicular joint by the *union of the left internal jugular and subclavian veins*. It passes to the right and inferiorly, posterior to the manubrium, where it *unites with the right brachiocephalic vein to form the superior vena cava*. The left brachiocephalic vein is over twice as long as the right brachiocephalic vein because it passes from the left to right side, and shunts blood from the head and neck and left upper limb to the right atrium. During its descent, it crosses the left common carotid artery, the brachiocephalic trunk, the left vagus nerve, and the left phrenic nerve. The left brachiocephalic vein is separated from the manubrium by the thymus (or its remains), and the origins of the sternohyoid and sternothyroid muscles (Fig. 1-20). In addition to the tributaries common to both brachiocephalic veins mentioned previously, *the left brachiocephalic vein receives the thoracic duct*, the largest lymph vessel in the body (Fig. 1-69) and the *left superior intercostal vein* (Fig. 1-76).

The left brachiocephalic vein in children may ascend from the superior mediastinum, superior to the jugular notch, and enter the root of the neck. This occurs because their necks are relatively short. The possible high location of the left brachiocephalic vein must be kept in mind when doing a *tracheostomy* in children. For a discussion of this procedure for creating an opening in the trachea, see p. 817.

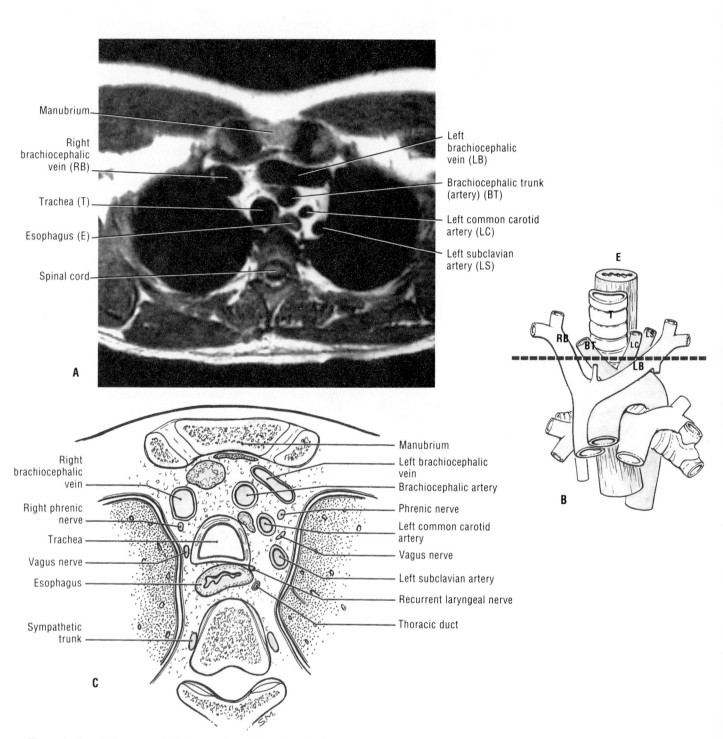

Figure 1-65.   *A*, Transverse MRI (magnetic resonance image) of the superior mediastinum superior to the arch of the aorta. (Courtesy of Dr. W. Kucharczyk, Clinical Director of Tri-Hospital Magnetic Resonance Centre, Toronto, Ontario, Canada.) *B*, Diagram showing the level of the MRI. *C*, Diagram of transverse section of superior mediastinum at same level as the MRI scan.

The Superior Vena Cava

The superior vena cava (**SVC**), the great vein draining blood from the head and neck, is located in the superior mediastinum (Figs. 1-51, 1-52, 1-67, 1-68, and 1-72). About 7 cm long, this large vessel enters the right atrium of the heart vertically from its superior aspect. The SVC returns blood from all structures superior to the diaphragm, except the lungs and heart (*i.e.*, the head, neck, upper limbs, and thoracic wall). The SVC forms posterior to the first right costal cartilage by the union of the right and left brachiocephalic veins. It passes inferiorly and ends at the level of the third costal cartilage, where it *enters the right atrium*. The SVC lies in the right side of the superior mediastinum (Figs. 1-67 and 1-72), anterolateral to the trachea and posterolateral to the ascending aorta. The right phrenic nerve lies between the SVC and the mediastinal pleura, which partly surrounds the right surface of this vein (Fig. 1-72). *The terminal half of the SVC is in the middle mediastinum*, where it lies beside the ascending aorta in the pericardium (Figs. 1-44 and 1-45). The shadow cast by the SVC is often seen in radiographs of the chest (Fig. 1-48).

The Arch of the Aorta

The arch of the aorta or aortic arch is the part of the aorta that is *located in the superior mediastinum* (Figs. 1-62 and 1-65). It is the curved continuation of the ascending aorta, which begins posterior to the second right sternocostal joint at the level of the sternal angle. It then arches superoposteriorly and to the left, but its main direction is posterior. It passes anterior to the trachea to reach the left side of the trachea and esophagus. It then arches over the root of the left lung as it passes inferiorly on the left side of the body of T4 vertebra (Figs. 1-66 and 1-73). It ends by becoming the *descending thoracic aorta* posterior to the second left sternocostal joint (Figs. 1-70 and 1-73). The terminal part of the arch of the aorta is visible in radiographs of the chest (Figs. 1-48 to 1-50). The shadow it casts is often called the *aortic knob* (knuckle). The **ligamentum arteriosum** passes from the root of the left pulmonary artery to the inferior concave surface of the arch of the aorta (Figs. 1-45, 1-62, 1-73, and 1-75). *The left recurrent laryngeal nerve* hooks around the arch, posterior to the ligamentum arteriosum, and then ascends between the trachea and esophagus (Figs. 1-68 and 1-69).

> The ligamentum arteriosum is the *remnant of the ductus arteriosus*, an embryonic vessel that shunted blood from the left pulmonary artery to the aorta in order to bypass the nonfunctioning lungs (Moore, 1988).

Branches of the Arch of the Aorta (Figs. 1-67, 1-70, and 1-73).   Typically there are **three branches** of the arch of the aorta: the *brachiocephalic trunk*, the *left common carotid artery*, and the *left subclavian artery*. These arteries arise from the superior aspect of the arch and supply the head, neck, upper limbs, and part of the body wall.

The Brachiocephalic Trunk (Figs. 1-42, 1-67, 1-68, and 1-70).   This is the first and largest of the three branches of the arch of the aorta. It arises posterior to the manubrium of the sternum in or a little to the left of its median plane. Here, it is

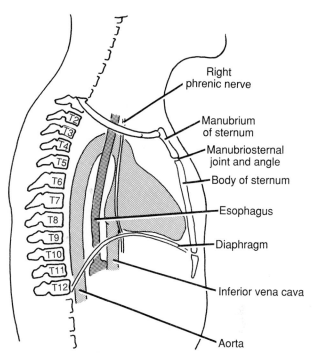

Figure 1-66.   Sagittal section of the thorax showing the relationships of the ascending aorta, arch of the aorta, and descending thoracic aorta.

anterior to the trachea and posterior to the left brachiocephalic vein. The brachiocephalic trunk ascends superolaterally to reach the right side of the trachea and the right sternoclavicular joint, where it *divides into the right common carotid and right subclavian arteries.*

The Left Common Carotid Artery (Figs. 1-67, 1-68, 1-70, and 1-73).   The second branch of the arch of the aorta is the left common carotid artery, which has a short course in the superior mediastinum. It arises from the arch slightly posterior and to the left of the brachiocephalic trunk and posterior to the manubrium. It ascends anterior to the left subclavian artery and is at first anterior to the trachea, and then is to its left. It enters the neck by passing posterior to the left sternoclavicular joint.

The Left Subclavian Artery (Figs. 1-63, 1-67, 1-68, 1-70, and 1-73).   This artery arises from the posterior part of the arch of the aorta, just behind the left common carotid artery. It ascends with this vessel through the superior mediastinum and lies against the left lung and pleura laterally. As it leaves the thorax and enters the root of the neck, the left subclavian artery passes posterior to the left sternoclavicular joint. It has no branches in the mediastinum.

> In advanced cases of arterial disease associated with an **aneurysm of the arch of the aorta**, pressure may be exerted on the trachea and esophagus, which lie on its right side (Fig. 1-63), causing difficulty in breathing and swallowing.
>
> The most superior part of the arch of the aorta is usually about 2.5 cm inferior to the superior border of the manubrium, but it may be more superior or inferior than this. Sometimes the arch curves over the root of the right lung and passes inferiorly on the right side. This congenital malformation is called a *right aortic arch*. In some cases this abnormal arch, after passing over the root of the right lung, passes posterior

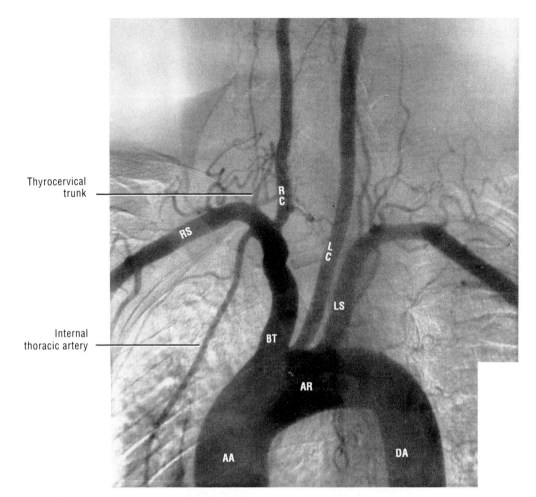

Thyrocervical trunk

Internal thoracic artery

RC

RS

LC

LS

BT

AR

AA

DA

Figure 1-67. Aortic angiogram. Observe the ascending aorta (*AA*), the arch of the aorta (*AR*), the descending aorta (*DA*), the brachiocephalic trunk (*BT*) branching into the right subclavian (*RS*) and right common carotid (*RC*) arteries, the left subclavian (*LS*) and left common carotid (*LC*) arteries arising directly from the aorta. (Courtesy of Dr. E.L. Lansdown, Professor of Radiology, University of Toronto, Toronto, Ontario, Canada.)

to the esophagus to reach its usual position on the left side (Moore, 1988). Less frequently, a *double aortic arch* forms a **vascular ring** around the esophagus and trachea. If the trachea is compressed enough to affect breathing, surgical interruption of the ring may be required. Variations in the origins of the branches of the arch may also occur (Fig. 1-71). A *retroesophageal right subclavian artery* crosses posterior to the esophagus to reach the upper limb. In unusual cases this abnormal right subclavian artery may compress the esophagus and cause difficulty in swallowing (*dysphagia*). Occasionally additional arteries arise from the arch of the aorta: usually the right or left vertebral artery. Less commonly, there is an accessory artery to the thyroid gland, called the *thyroid ima* (p. 820).

A *patent ductus arteriosus* results from failure of the ductus arteriosus to close after birth and become the *ligamentum arteriosum*. Usually this transformation is completed by the end of the third month after birth. **Coarctation of the aorta** is a congenital abnormality in which the aorta has a purse-string-like constriction (Moore, 1988). As a result, the aortic lumen is constricted (stenotic). When the coarctation is inferior to the ligamentum arteriosum, a good collateral cir-

culation usually develops between the proximal and distal parts of the aorta by way of the intercostal and internal thoracic arteries. This *postductal type of coarctation* is compatible with many years of life because the collateral circulation carries blood to the descending aorta inferior to the stenosis.

The Vagus Nerves

The vagus nerves (*Xth cranial nerves*) arise from the medulla of the brain (Chap. 9). The thoracic parts of these nerves descend from the neck, posterolateral to the common carotid arteries. Each nerve enters the superior mediastinum posterior to the appropriate sternoclavicular joint and brachiocephalic vein.

The Right Vagus Nerve (Figs. 1-42, 1-62, 1-68, and 1-72). This nerve enters the thorax anterior to the right subclavian artery and runs posteroinferiorly through the superior mediastinum on the right side of the trachea. It then passes posterior to the right brachiocephalic vein and the *superior* vena cava, where it breaks up into a number of branches that contribute to the *right pulmonary plexus*, located posterior to the root of the right lung. This plexus supplies the bronchi and lungs. The vagus

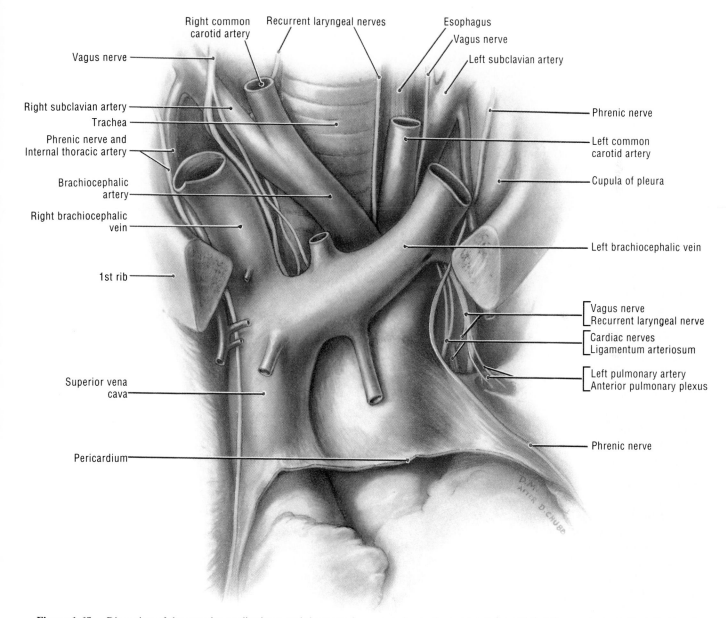

Figure 1-68. Dissection of the superior mediastinum and the root of the neck after removal of the thymus. Observe that the great veins are anterior to the great arteries and that the nerves cross the left side of the arch of the aorta.

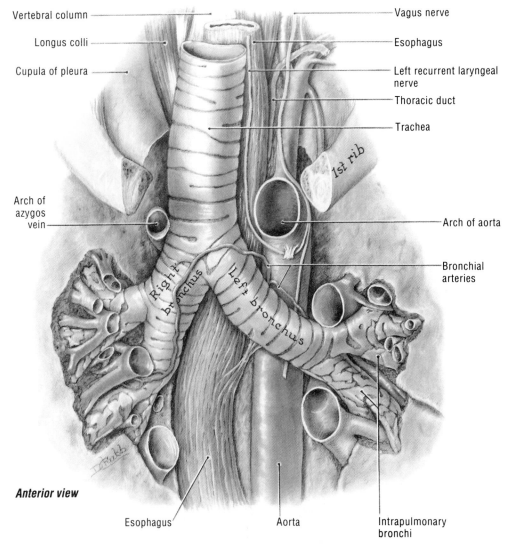

Vertebral column

Longus colli

Cupula of pleura

Arch of azygos vein

Anterior view

Esophagus

Vagus nerve

Esophagus

Left recurrent laryngeal nerve

Thoracic duct

Trachea

1st rib

Arch of aorta

Bronchial arteries

Aorta

Intrapulmonary bronchi

Right bronchus

Left bronchus

Figure 1-69. Deep dissection of the superior mediastinum showing four parallel structures: trachea, esophagus, left recurrent laryngeal nerve, and thoracic duct. Note that the right bronchus is more vertical than the left bronchus. The course of the right bronchial artery is abnormal. Usually it passes posterior to the bronchus.

nerve leaves this plexus as a single nerve and passes to the esophagus, where it again breaks up and contributes fibers to the *esophageal plexus*. This plexus supplies the esophagus, pericardium, and pleura. The right vagus nerve also gives rise to *cardiac nerves*, which contribute to the *cardiac plexus* supplying the heart (p. 106). This plexus is located between the arch of the aorta and the bifurcation of the trachea. After passing anterior to the subclavian artery, the right vagus nerve gives rise to the **right recurrent laryngeal nerve,** which hooks around the right subclavian artery and ascends in the neck between the trachea and esophagus to supply the *larynx* or speech organ (see also Fig. 8-64).

The Left Vagus Nerve (Figs. 1-42, 1-68, and 1-73). This nerve descends in the neck posterior to the left common carotid artery, between it and the left subclavian artery. When it reaches the left side of the arch of the aorta, it diverges posteriorly from the left phrenic nerve. It is separated laterally from the phrenic nerve by the left superior intercostal vein. The left vagus nerve

passes posterior to the root of the left lung, where it breaks up into a number of branches that contribute to the *left pulmonary plexus*. This plexus supplies the bronchi and lungs. The left vagus nerve leaves this plexus as a single trunk and passes to the esophagus, where it joins fibers from the right vagus in the *esophageal plexus*. This plexus supplies the esophagus, pericardium, and pleura. Before contributing to these plexuses, the left vagus nerve curves medially at the inferior border of the arch of the aorta and gives off the *left recurrent laryngeal nerve*. This nerve hooks around the ligamentum arteriosum (Figs. 1-62 and 1-69), and passes superiorly on the right side of the aortic arch. The left recurrent laryngeal nerve ascends in the neck to the larynx in a groove between the trachea and esophagus.

The recurrent laryngeal nerves supply nearly all the intrinsic muscles of the larynx (see Fig. 8-63). Consequently, any investigative procedure (*e.g., mediastinotomy*) or disease process in the superior mediastinum may involve these nerves

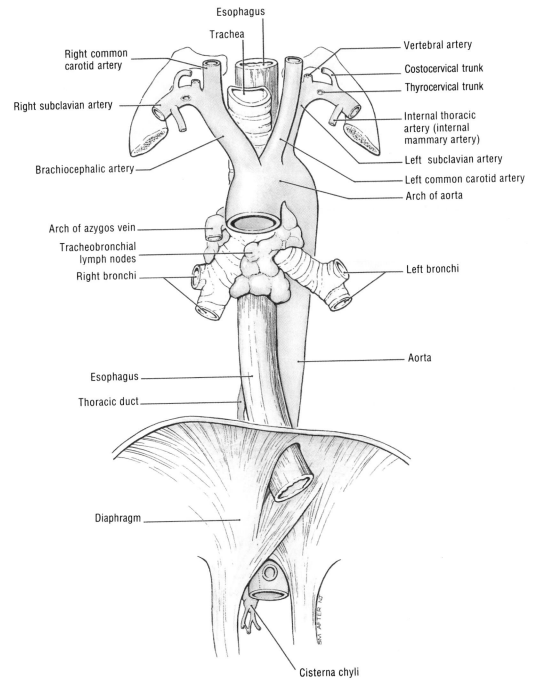

Figure 1-70. Anterior view of the esophagus, trachea, bronchi, and aorta. Observe that the arch of the aorta curves posteriorly on the left side of the trachea and esophagus. Note that the right crus of the diaphragm is larger and longer than the left one, and that it splits to enclose the esophagus.

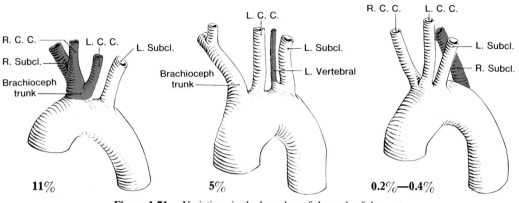

Figure 1-71.   Variations in the branches of the arch of the aorta.

and affect the voice. Because the left recurrent laryngeal nerve winds around the arch of the aorta (Fig. 1-69) and ascends between the trachea and esophagus, it may be involved when there is: (1) a bronchial or *esophageal carcinoma*; (2) enlargement of mediastinal lymph nodes; or (3) an *aneurysm of the arch of the aorta*. In the latter condition the nerve may be stretched by the dilated aorta.

The Phrenic Nerves

Each phrenic nerve enters the superior mediastinum between the subclavian artery and the origin of the brachiocephalic vein (Figs. 1-42, 1-72, and 1-73). *The phrenic nerves are the sole motor supply to the diaphragm*; about one-third of their fibers are sensory to the diaphragm. They arise from the ventral rami of C3 to C5 nerves (principally C4).

The Right Phrenic Nerve (Figs. 1-63, 1-68, and 1-72).   This nerve is in contact with venous structures throughout its course. It passes along the right side of the right brachiocephalic vein, the superior vena cava, and the pericardium over the right atrium. The right phrenic nerve passes anterior to the root of the right lung and descends on the right side of the inferior vena cava to the diaphragm, which it pierces near the inferior vena caval opening (Fig. 1-29).

The Left Phrenic Nerve (Figs. 1-29, 1-42, 1-68, 1-71, and 1-73).   This nerve descends between the left subclavian and left common carotid arteries. *It crosses the left surface of the arch of the aorta*, anterior to the left vagus nerve, and passes over the left superior intercostal vein. It then descends anterior to the root of the left lung and runs along the pericardium, superficial to the left auricle and ventricle of the heart, where it pierces the diaphragm to the left of the pericardium.

The Trachea

The trachea (windpipe) is a *wide fibrocartilaginous tube*, the superior half of which is in the neck. It descends anterior to the esophagus (Figs. 1-68 and 1-69). As it enters the superior mediastinum, it inclines a little to the right of the median plane. *The posterior surface of the trachea is flat* where it is applied to the esophagus. It is kept patent by a series of C-shaped *tracheal*

cartilages. The thoracic part of the trachea is 5 to 6 cm in length; it ends at the level of the sternal angle by dividing into the right and left main bronchi (Fig. 1-69).

The arch of the aorta is at first anterior to the trachea and then is on its left side (Figs. 1-67 to 1-70, 1-72, and 1-73). Superior to the arch, the brachiocephalic trunk and the left common carotid artery are at first anterior and then on the right and left sides of the trachea, respectively. These vessels separate the trachea from the left brachiocephalic vein. The posterior surface of the trachea lies anterior and a little to the right of the esophagus and the *left recurrent laryngeal nerve*. This nerve sends branches to the esophagus and trachea.

Numerous tracheal and *tracheobronchial lymph nodes* are associated with the trachea (Figs. 1-38*B*, 1-63, and 1-70). They are clinically important because lymph from the lungs drains into them. Consequently they become enlarged when *bronchogenic carcinoma* develops (p. 77). Widening or *distortion of the carina* (Fig. 1-14), results from invasion and enlargement of these nodes by cancer cells. These changes can be observed with a *bronchoscope* (p. 74).

The Esophagus

The esophagus (gullet) is a *narrow fibromuscular tube*, which is usually flattened anteroposteriorly; it connects the pharynx (throat) with the stomach (Figs. 1-65 to 1-70 and 1-72). It is 25 to 30 cm in length and has cervical, thoracic, and abdominal parts. The esophagus enters the superior mediastinum between the trachea and vertebral column. It lies anterior to T1 to T4 vertebrae. Initially it inclines to the left, but it is moved by the arch of the aorta to the median plane opposite the roots of the lungs. Inferior to the arch, however, the esophagus again inclines to the left as it approaches and passes through the diaphragm (Fig. 1-70). In the superior mediastinum, the thoracic duct lies on the left side of the esophagus, deep to the arch of the aorta (Figs. 1-67 and 1-69).

The Posterior Mediastinum

This part of the mediastinum is located anterior to T5 to T12 vertebrae and posterior to the pericardium and diaphragm (Figs.

1-39, 1-40, and 1-66). *The posterior mediastinum contains*: (1) several longitudinal structures (thoracic aorta, thoracic duct, azygos and hemiazygos veins, esophagus, and esophageal plexus) and (2) several transverse structures (posterior intercostal arteries, thoracic duct as it passes from right to left, certain intercostal veins, and parts of the hemiazygos veins).

The Thoracic Aorta

The descending aorta is divisible into thoracic and abdominal portions. The thoracic aorta is located in the posterior mediastinum (Figs. 1-40, 1-66, 1-70, and 1-74). As the *continuation of the arch of the aorta*, it begins on the left side of the inferior border of the body of T4 vertebra and descends in the posterior mediastinum on the left sides of T5 to T12 vertebrae. This large artery, somewhat in excess of 2 cm in diameter, commonly flattens the vertebral bodies (Fig. 1-18). As it descends, it approaches the median plane and displaces the esophagus to the right (Figs. 1-69 and 1-70). The thoracic aorta descends through the posterior mediastinum against the left pleura, with the thoracic duct and azygos vein to its right (Figs. 1-72 to 1-74). A fine autonomic plexus, called the *thoracic aortic plexus*, surrounds it. The thoracic aorta lies posterior to the root of the left lung, the pericardium and esophagus. *The thoracic aorta terminates anterior to the vertebral column at the inferior border of T12 vertebra*. It enters the abdomen through the most posterior opening in the diaphragm, called the *aortic hiatus* (Figs. 1-29 and 1-70). The *thoracic duct* and *azygos vein* descend on its right side and accompany it through this hiatus (Figs. 1-73 and 1-76).

Branches of the Thoracic Aorta (Figs. 1-69 and 1-75). The thoracic aorta has the following branches: bronchial, esophageal, pericardial, mediastinal, posterior intercostal, subcostal, and superior phrenic. The small **bronchial arteries** consist of one right and two left bronchial arteries (Fig. 1-75). Only the left bronchial arteries arise from the aorta. Usually, two *esophageal branches* supply the middle third of the esophagus. The *pericardial branches* send twigs to the pericardium and the *mediastinal branches* supply lymph nodes and other tissues in the posterior mediastinum. The thoracic aorta also gives rise to nine pairs of **posterior intercostal arteries**, which pass to the 3rd to 11th intercostal spaces, inclusively. The *subcostal arteries* also arise from the thoracic aorta (Fig. 1-75) and the superior phrenic arteries arise from its most inferior part. The *phrenic arteries* pass to the posterior surface of the diaphragm, where they anastomose with the musculophrenic and pericardiacophrenic arteries, branches of the internal thoracic artery.

The Thoracic Duct

This *main lymphatic duct* lies on the bodies of the inferior seven thoracic vertebrae (Fig. 1-39). It *conveys most of the lymph of the body to the venous system* (Figs. 1-70 and 1-73; see also p. 26). In addition to receiving lymph vessels from the posterior mediastinal lymph nodes and intercostal lymph nodes, it receives all the lymphatic vessels inferior to the diaphragm. It drains the **cisterna chyli** (the expanded inferior end of the thoracic duct), which lies anterior to the T12 vertebra and posterior to the right

of the aorta (Fig. 1-70). The thoracic duct passes superiorly from the cisterna chyli through the *aortic hiatus* in the diaphragm (Figs. 1-29, 1-39, and 1-70). It is usually thin-walled and dull white in color; often it is beaded owing to its numerous valves. The thoracic duct ascends in the posterior mediastinum between the thoracic aorta and the azygos vein (Figs. 1-39 and 1-74). It reaches the right side of the esophagus and passes superiorly, anterior to the origins of the right posterior intercostal arteries (Fig. 1-73). At the level of T4, T5 or T6 vertebrae, *the thoracic duct crosses to the left, posterior to the esophagus* (Figs. 1-65 and 1-67), and ascends to the superior mediastinum (p. 80). The thoracic duct usually empties into the venous system near the union of the left internal jugular and subclavian veins (Fig. 1-39). It normally divides into two or three separate branches, all of which open at the angle between these two veins.

Because the thoracic duct is thin-walled and may be colorless or white, depending on how much fat is in the chyle, it may not be easily seen in a cadaver or a patient during surgery unless digestion was in an active phase before death or surgery, respectively. Because the thoracic duct is not easily identified, it may be inadvertently injured during investigative and/or surgical procedures in the posterior mediastinum. **Laceration of the thoracic duct** during an accident or during lung surgery, results in *chyle* (L. *chylus*, juice) escaping from it. The chyle may enter the pleural cavity producing *chylothorax* (p. 64). This fluid may be removed by a needle tap (p. 59), but in some cases it may be necessary to tie off the thoracic duct. The lymph then returns to the venous system by other lymph channels, which join the duct superior to the ligature (Fig. 1-39). *Variations of the thoracic duct are common* (Bergman et al., 1989) because the superior part of the duct represents the original left member of a pair of vessels in the embryo (Moore, 1988). Sometimes two thoracic ducts are present for a short distance.

The Azygos System of Veins

The azygos system consists of veins on each side of the vertebral column that drain the back and thoracic and abdominal walls (Figs. 1-72 and 1-76). It exhibits much variation, not only in its origin, but in its course, tributaries, anastomoses, and termination. The *azygos vein* and its main tributary, the *hemiazygos vein*, usually arise from the posterior aspect of the inferior vena cava and the renal vein, respectively (Fig. 1-76). The azygos and hemiazygos veins provide another means of venous drainage from the abdomen and thorax.

Azygos means unpaired and is used to describe the asymmetry of the longitudinal veins in the azygos system, which is located on both sides of the vertebral column (Fig. 1-76). Its variable origin from the posterior aspect of the inferior vena cava, at or inferior to the level of the renal veins, results from its complex developmental origin (Moore, 1988).

The Azygos Vein (Figs. 1-29, 1-39, 1-70, 1-72, 1-74, and 1-76). *The azygos vein connects the superior and inferior venae cavae*, either directly by joining the inferior vena vena or indirectly by the hemiazygos and accessory hemiazygos veins. The azygos vein drains blood from the posterior walls of the thorax

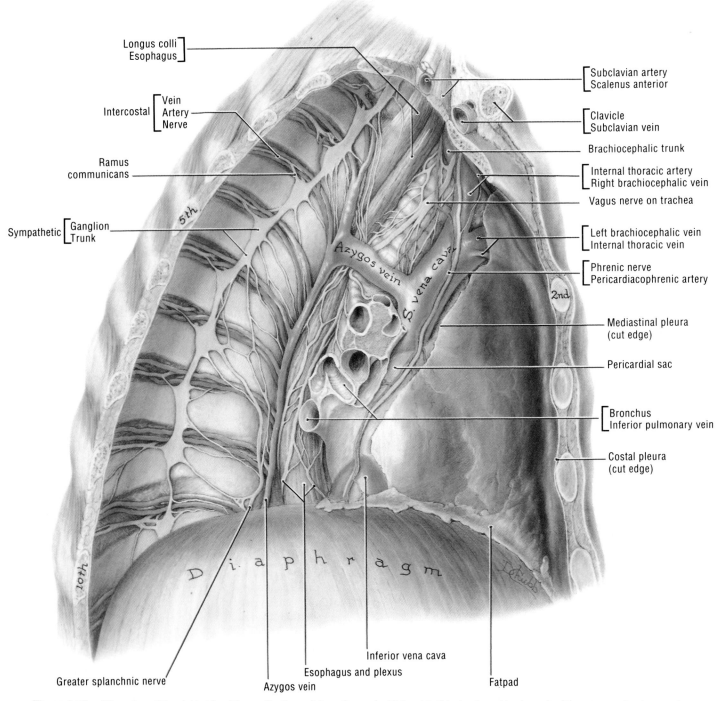

Longus colli
Esophagus

Intercostal Vein Artery Nerve

Ramus communicans

Sympathetic Ganglion Trunk

Subclavian artery
Scalenus anterior

Clavicle
Subclavian vein

Brachiocephalic trunk

Internal thoracic artery
Right brachiocephalic vein

Vagus nerve on trachea

Left brachiocephalic vein
Internal thoracic vein

Phrenic nerve
Pericardiacophrenic artery

Mediastinal pleura
(cut edge)

Pericardial sac

Bronchus
Inferior pulmonary vein

Costal pleura
(cut edge)

Greater splanchnic nerve

Azygos vein

Esophagus and plexus

Inferior vena cava

Fatpad

Figure 1-72.   Dissection of the right side of the mediastinum. Most of the costal and mediastinal pleurae has been removed to expose the underlying structures. Observe that the right side of the mediastinum, the "*blue side*," is dominated by the arch of the azygos vein, the superior vena cava, and the right atrium.

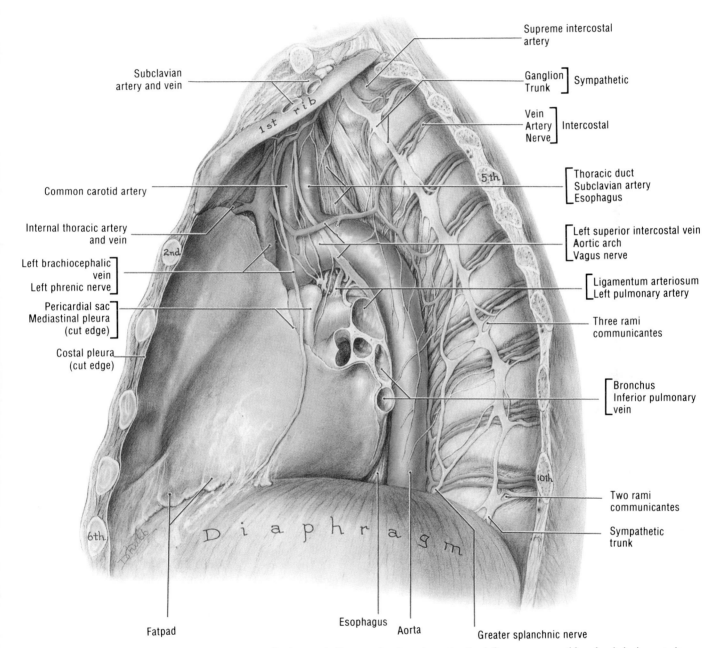

Subclavian
artery and vein

Common carotid artery

Internal thoracic artery
and vein

Left brachiocephalic
vein
Left phrenic nerve

Pericardial sac
Mediastinal pleura
(cut edge)

Costal pleura
(cut edge)

Supreme intercostal
artery

Ganglion
Trunk } Sympathetic

Vein
Artery } Intercostal
Nerve

Thoracic duct
Subclavian artery
Esophagus

Left superior intercostal vein
Aortic arch
Vagus nerve

Ligamentum arteriosum
Left pulmonary artery

Three rami
communicantes

Bronchus
Inferior pulmonary
vein

Two rami
communicantes

Sympathetic
trunk

1st rib

2nd

5th

10th

6th

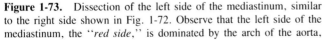

Fatpad

Esophagus   Aorta

Greater splanchnic nerve

Figure 1-73.   Dissection of the left side of the mediastinum, similar to the right side shown in Fig. 1-72. Observe that the left side of the mediastinum, the "*red side*," is dominated by the arch of the aorta, the thoracic aorta, the left common carotid and subclavian arteries. Observe that the sympathetic trunk is attached to intercostal nerves by connecting branches called rami communicantes.

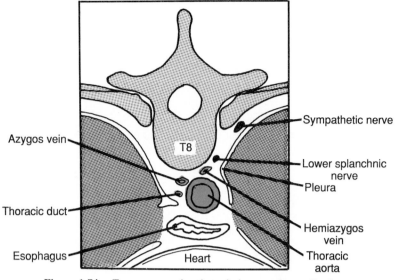

Figure 1-74. Transverse section through the posterior mediastinum.

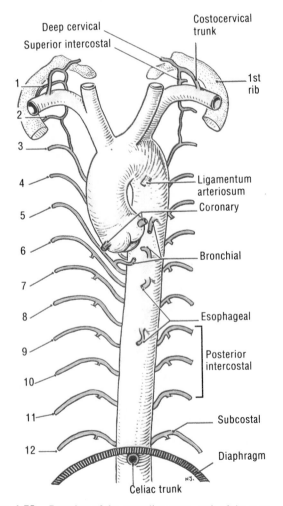

Figure 1-75. Branches of the ascending aorta, arch of the aorta, and thoracic aorta.

and abdomen. It ascends in the posterior mediastinum, passing close to the right sides of the bodies of the inferior eight thoracic vertebrae. The azygos vein is covered anteriorly by the esophagus as it passes posterior to the root of the right lung. It then arches over the superior aspect of this root to join the superior vena cava. In addition to the posterior intercostal veins, the azygos vein communicates with the *vertebral venous plexuses* (see Fig. 4-53), which drain the back, vertebrae, and structures in the vertebral canal. The azygos vein also receives the mediastinal, esophageal, and bronchial veins.

The Hemiazygos Vein (Figs. 1-74 and 1-76). This vein arises on the left side by the junction of the left subcostal and ascending lumbar veins. It ascends on the left side of the vertebral column, posterior to the thoracic aorta, as far as T9 vertebra. Here it crosses to the right, posterior to the aorta, thoracic duct, and esophagus, and *joins the azygos vein*. The hemiazygos vein receives the inferior three posterior intercostal veins, the inferior esophageal veins, and several small mediastinal veins.

The Accessory Hemiazygos Vein (Fig. 1-76). This vein begins at the medial end of the fourth or fifth intercostal space and descends on the left side of the vertebral column from T5 to T8. It receives tributaries from veins in the fourth to eighth intercostal spaces and sometimes from the left bronchial veins. It crosses over T7 or T8 vertebrae, posterior to the thoracic aorta and thoracic duct, where it *joins the azygos vein*. Sometimes the accessory hemiazygos joins the hemiazygos vein and opens with it into the azygos vein. The accessory hemiazygos is frequently connected to the left superior intercostal vein, as in Fig. 1-76.

The azygos, hemiazygos and accessory hemiazygos veins vary markedly from one person to another (Bergman et al., 1988); however, this system offers an **alternate means of**

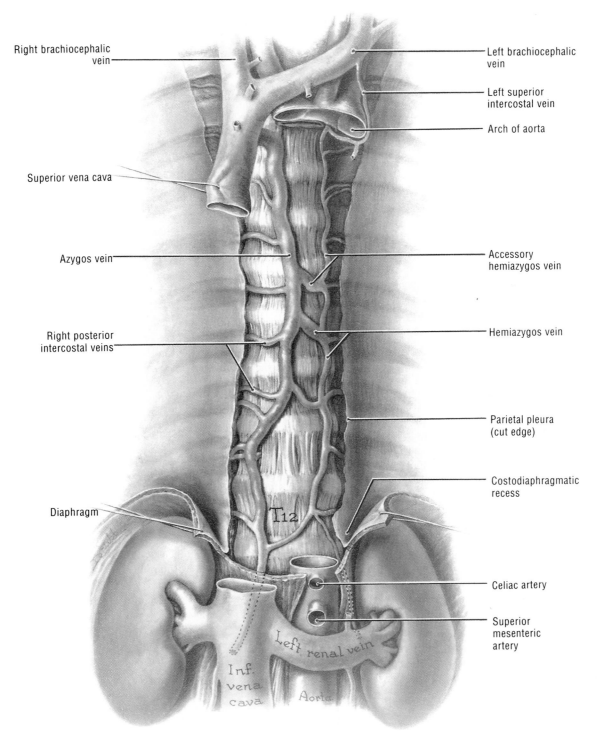

Figure 1-76. The *azygos system of veins*. Note that the azygos vein forms a direct connection between the inferior and superior venae cavae.

venous drainage from the thoracic, abdominal, and back regions when there is *obstruction of the inferior vena cava*. In some people, an accessory azygos vein parallels the main azygos vein. In other people, there is no hemiazygos system of veins. A clinically important variation, although uncommon, is when the azygos system receives all the blood from the inferior vena cava, except that from the liver. In these people, the azygos system drains nearly all the blood inferior to the diaphragm except from the digestive tract.

The Esophagus

The course and relationships of the esophagus in the superior mediastinum are described on p. 114. It passes from the superior mediastinum into the posterior mediastinum, descending posterior and to the right of the arch of the aorta and posterior to the pericardium and left atrium (Figs. 1-40, 1-69, and 1-70). It then deviates to the left and *passes through the esophageal hiatus in the posterior part of the diaphragm*, anterior to the descending thoracic aorta (Figs. 1-29, 1-66, 1-70, 1-72, and 1-73). *The anterior relations of the esophagus* in the posterior mediastinum are: the left main bronchus; the tracheobronchial lymph nodes; the pericardium and left atrium; and the diaphragm inferiorly. *The posterior relations of the esophagus* are: the vertebral bodies of T5 to T12; the thoracic duct; the azygos, hemiazygos, and accessory hemiazygos veins; and the right intercostal arteries (Figs. 1-66, 1-69, and 1-73 to 1-76). At its inferior end, the descending aorta lies between the esophagus and the vertebrae.

The right side of the esophagus is close to the mediastinal pleura and the lung, except where it is crossed by the azygos vein (Figs. 1-72 and 1-74). The left side of the esophagus is also close to the mediastinal pleura superior to the arch of the aorta, except where the thoracic duct and left subclavian artery intervene. The arch of the aorta and descending aorta lie to the left of the esophagus to the level of T7 vertebra (Figs. 1-66, 1-69, and 1-70). The esophagus is surrounded by the *esophageal plexus*, which is largely formed by esophageal branches of both vagus nerves. It also receives sympathetic branches through the *splanchnic nerves* (Figs. 1-72 and 1-73).

Impressions of the Esophagus (Figs. 1-69 and 1-70). The esophagus may have three impressions or "constrictions" in its thoracic part. These may be observed as narrowings of the lumen in oblique radiographs of the thorax, which are taken as barium is swallowed. The esophagus is compressed by: (1) the *arch of the aorta*; (2) the point at which it is crossed by the *left main bronchus*; and (3) the diaphragm as it passes through the esophageal hiatus (see also Fig. 2-30). There are no constrictions in the empty esophagus; however, as it expands during filling, the above structures compress its walls.

The impressions that are produced in the esophagus by adjacent structures are of clinical interest because of the slower passage of substances at these sites. The impressions indicate where swallowed foreign objects are most likely to lodge and where a stricture or narrowing may develop following the accidental drinking of caustic liquids (*e.g.*, lye).

The most common congenital malformation of the esophagus is **esophageal atresia**, which is commonly associated with *tracheoesophageal fistula*. This malformation, occurring in about one in 2500 newborn infants, results from incomplete division of the foregut into respiratory and digestive portions (Moore, 1988). In most cases (90%), the superior portion of the esophagus ends as a blind pouch and a fistula (an abnormal canal) connects the inferior portion of the esophagus to the trachea. Because the esophagus is blocked, food, saliva, and mucous secretions overflow into the larynx, causing coughing and gagging. Some of this material enters the trachea and bronchi (Behrman, 1992). If attempts are made to feed the infant, milk fills the esophageal pouch and overflows into the larynx and trachea, causing more coughing and gagging.

The Anterior Mediastinum

This small subdivision of the mediastinum lies between the body of the sternum and the transversus thoracis muscles anteriorly and the pericardium posteriorly (Figs. 1-20, 1-40, and 1-66). It is continuous with the superior mediastinum at the sternal angle and is limited inferiorly by the diaphragm. Superior to the level of the fourth costal cartilages, *the anterior mediastinum is a very narrow space* owing to the closeness of the right and left pleurae (Figs. 1-27, 1-33, and 1-34). It consists of loose connective tissue, fat, lymph vessels, some lymph nodes, a few branches of the internal thoracic artery, and (inferiorly) the *sternopericardial ligaments*. These ligaments attach the pericardium to the superior end of the xiphoid process. In infants and children the anterior mediastinum contains the inferior part of the thymus. In unusual cases, this gland may extend as far inferiorly as the fourth costal cartilages.

PATIENT ORIENTED PROBLEMS

Case 1-1

While having a heated argument with a client, a 48-year-old lawyer experienced a sudden, *crushing substernal pain in her chest* that radiated along the medial aspect of her left arm. The client helped her to the chesterfield where the lawyer attempted to relieve the pain by squirming, stretching, and belching. When her secretary noted that she was pale, perspiring, and writhing in pain, she called a physician and an ambulance. The ambulance attendants administered oxygen and rushed her to the hospital, where she was admitted to the intensive care unit (*ICU*). She was placed under observation with *ECG monitoring* for detection of potential fatal *arrhythmias*. Her blood pressure was low (a sign of **shock**).

On questioning, the resident learned that the patient had had previous attacks of substernal discomfort during stress, which she was reluctant to describe as pain. She said that this discomfort always passed when she rested. When the resident asked the patient to describe her present chest pain, she said that it was the worst pain she had ever felt, and clenched her fist to demonstrate its viselike nature. She said that when the pain struck, she had a feeling of weakness and nausea. On *auscultation* the resident detected an occasional arrhythmia. The ECG was also abnormal.

Diagnosis. Acute myocardial infarction (**MI**) owing to coronary atherosclerosis.

> **Problems.** Define acute MI and coronary atherosclerosis. Explain the anatomical basis of the referred pain from the heart to the left side of the chest, shoulder, and medial aspect of the arm. These problems are discussed on p. 123.

Case 1-2

A 58-year-old man who had lived in an industrial area all his life consulted his physician because he was coughing up blood (*hemoptysis*), and was experiencing shortness of breath (*dyspnea*) during exertion. It was learned that he had been a heavy cigarette smoker for over 40 years and had had a "*smoker's cough*" for several years. He stated that his shortness of breath and cough had been getting worse for the last few months. He first noticed that his *sputum was blood-streaked* about 3 weeks earlier and stated that he had experienced vague chest pain on the left side at that time. Physical examination revealed that his left medial *supraclavicular lymph nodes were slightly enlarged* and more firm than usual. His breath sounds and resonance were diminished on the left side compared with the right side. The physician requested chest films.

Radiologist's Report. There is an obscuration of the hilum of the left lung by a mass. Normal left mediastinal contours superior to the hilum could not be recognized, and there was slight radiolucency of the remainder of the left lung. The mediastinum was shifted slightly to the left. The radiologist suggested that these changes were most likely caused by a tumor in the left superior lobe bronchus with metastases to the left hilar lymph nodes.

Endoscopy. On examination of the interior of the main bronchi under local anesthesia with a bronchoscope, the *otolaryngologist*[8] observed a growth that was partly obstructing the origin of the left superior lobe bronchus. Through the bronchoscope, he obtained a biopsy of the tumor. The enlarged supraclavicular lymph nodes were also biopsied for microscopic examination.

Mediastinoscopy. Examination of the mediastinum with a mediastinoscope inserted through a suprasternal incision under anesthesia revealed enlarged tracheobronchial lymph nodes. Through the mediastinoscope, the surgeon biopsied these nodes.

Pathologist's Report. Bronchogenic carcinoma was detected in the bronchial biopsy. The supraclavicular lymph nodes did not show definite tumor involvement. The mediastinal lymph nodes showed the presence of many malignant tumor cells.

Diagnosis. Bronchogenic carcinoma.

> **Problems.** Using your knowledge of the anatomical relations of the bronchi and lungs, state which structures are likely to be involved by direct extension of a malignant tumor of the bronchus. Where would you expect tumor cells to spread via the lymph and blood? What is unusual about the lymph drainage of the inferior lobe of the left lung? Explain the probable anatomical basis for metastasis of tumor cells

from a bronchogenic carcinoma to the brain. These problems are discussed on p. 123.

Case 1-3

During an argument with his wife, a 44-year-old inebriated man was stabbed with a paring knife, the blade of which was 9 cm long. The knife, penetrated the fourth intercostal space along the left sternal border. By the time he was taken to the emergency room of a hospital, the patient was semiconscious, in shock, and gasping for breath. In a few moments he became unconscious and died. The coroner (medical examiner) arranged for an autopsy.

Autopsy Report. Death was caused by an excessive loss of blood and *cardiac tamponade* resulting from the stab wound in the thorax.

> **Problems.** Using your knowledge of the surface anatomy of the thorax, what organ(s) would you expect to be punctured by the knife? Where would the blood likely accumulate? Discuss the probable cause of death. These problems are discussed on p. 124.

Case 1-4

A short, thin man with spindly limbs (*gracile habitus*), 42 years of age, complained about recent difficulties in breathing during exercise (*exertional dyspnea*) and extreme fatigue. He stated that other than being physically underdeveloped, he had been well most of his life, until the last year or so when he had had several respiratory infections. Physical examination revealed a prominent right ventricular cardiac impulse. A moderately loud *midsystolic murmur* was heard over the second and third intercostal spaces along the inferior left sternal border. An ECG showed changes suggestive of *right ventricular hypertrophy*. Radiographs of the thorax and angiocardiographic studies were requested.

Radiologist's Report. Enlargement of the right side of the heart, especially of the right outflow tract, a small aortic knob, dilation of the pulmonary artery and its major branches, and increased pulmonary vascular markings were observed. During *right cardiac catheterization*, the catheter easily passed from the right atrium into the left atrium. Serial samples of blood for determination of oxygen saturation were taken as the catheter was withdrawn from the left atrium into the right atrium and then through the inferior vena cava (IVC). These studies revealed increased oxygen saturation of the right atrial blood compared with blood in the IVC. Serial determinations of pressures showed unequal pressures in the atria (slightly higher in the left atrium).

Diagnosis. Atrial septal defect (ASD). It was classified as a secundum type (ASD) with a left to right atrial shunt of blood.

> **Problems.** Was this man born with this ASD? Where else may defects occur in the interatrial septum? What additional complications do you think might occur in this patient in view of the left to right shunt of blood? These problems are discussed on p. 124.

---

[8]A physician who specializes in the treatment of diseases of the ears and upper respiratory tract.

body. Common sites of hematogenous metastasis from bronchogenic carcinoma are the brain, bones, lungs, and suprarenal glands. Often the medial supraclavicular lymph nodes, particularly on the left side, are enlarged and hard because they contain tumor cells.

The anatomical basis for involvement of the supraclavicular or sentinel lymph nodes is that lymph passes cranially from the thoracic and abdominal viscera via the bronchomediastinal trunks and the thoracic duct to reach the venous system. Backflow of lymph from the thoracic duct can pass into the deep supraclavicular nodes, posterior to the sternocleidomastoid muscles. This is probably the reason why nodes on the left side are most commonly involved. The brain is a common site for hematogenous spread of bronchogenic carcinoma. Tumor cells probably enter the blood through the wall of a capillary or venule in the lung and are transported to the brain via the internal carotid artery and vertebral artery systems. Once in the cranium, the tumor cells probably pass between the endothelial cells of the capillaries and enter the brain.

Although most cancer cells from the lung are probably transported to the brain via the arterial system, others may be carried by the venous system. It has been suggested that constant coughing and enlarged mediastinal lymph nodes compress the superior and inferior venae cavae, causing the blood draining the bronchi to reverse its flow and pass via the bronchial veins into the azygos venous system (Fig. 1-76). From here, blood and tumor cells pass to the *extradural vertebral venous plexuses* around the spinal dura mater (see Fig. 4-53). As this plexus communicates with the cranial *dural venous sinuses*, tumor cells can be transported to the brain (see Fig. 7-30). The passage of tumor cells to the veins of the vertebral column also explains the frequency of metastases to vertebrae.

Case 1-3

The knife, entering the fourth intercostal space at the left sternal border, did not penetrate the left lung owing to the cardiac notch in its anterior border, which begins at the fourth costal cartilage. The knife probably nicked the parietal layer of pleura of the left lung and then passed through the conus arteriosus of the right ventricle and the aortic vestibule of the left ventricle, immediately inferior to the aortic orifice.

Blood passed from the wounds in both ventricles into the pericardial sac. As blood accumulated in the pericardial cavity, severe compression of the heart (*cardiac tamponade*) and great veins occurred. This pressure increased until it exceeded the pressure in these large veins, preventing normal venous return to the heart and outflow of blood from the heart to the lungs. This explains the patient's shock and gasping for breath before his death.

Case 1-4

ASD is a congenital malformation because it is an imperfection in the interatrial septum that developed during embryonic life; thus it was present at birth (L. *congenitus*, born with). The common form of heart defect is the secundum type of ASD, so classified because it results from abnormal development of the foramen ovale and septum secundum (Moore, 1988). The septum primum and septum secundum normally fuse in such a way that no opening remains between the right and left atrium. The site of the prenatal opening, and where most defects occur, is represented by the fossa ovalis in the adult right atrium (Fig. 1-51).

The left to right shunt of blood occurs because left atrial pressure exceeds that in the right atrium during the major part of the cardiac cycle. Some of the patient's blood therefore makes two circuits through the lungs. As a result of this shunting, the workload of the right ventricle increases and its muscular wall hypertrophies. The cavities of the right atrium, right ventricle, and pulmonary artery dilate to accommodate the excess amount of blood in them.

During the first 40 years, a majority of ASD patients have a moderate to good exercise tolerance, despite the fact that the opening in the interatrial septum is often 2 to 4 cm in diameter. Usually patients with ASDs do not exhibit *cyanosis* (dark bluish coloration of the skin). *Pulmonary vascular disease* (arteriosclerosis) is likely to develop with increased pulmonary artery pressure, particularly if recurrent respiratory infections occur. Severe pulmonary hypertension may eventually result in higher pressure in the right atrium than in the left, reversing the shunt and causing cyanosis, severe disability, and heart failure.

Case 1-5

As the right main bronchus is wider, shorter, and more vertical than the left main bronchus, foreign bodies more frequently pass into the right than into the left bronchus. Foreign bodies that are commonly found are nuts, hardware, pins, crayons, and dental material (*e.g.*, part of a tooth). The right middle and inferior lobes of the right lung are usually involved because (1) the right inferior lobe bronchus is in line with the right main bronchus and (2) the foreign body often lodges in the inferior lobe bronchus, superior to the origin of the middle lobe bronchus.

When there is complete obstruction of a main bronchus, the entire lung eventually collapses, becoming a nonaerated or *atelectatic lung*. Airlessness of the lung owing to absorption of air from the alveoli, as in the present case, is called *atelectasis*. Collapse of the lung occurs when the air (*i.e.*, oxygen and nitrogen) in the trapped part of the lung is absorbed into the blood perfusing the lung. Depending on the site of obstruction, collapse may involve an entire lung, a lobe, or a bronchopulmonary segment.

Because a collapsed lung is of soft tissue density, atelectatic lungs, lobes, or segments appear as homogeneous dense shadows on radiographs, in contrast to normal air-filled lung, which is relatively lucent and appears dark on films. When atelectasis of a sizable segment of lung occurs, the heart and mediastinum are drawn toward the obstructed side and remain there during inspiration and expiration. The diaphragm on the normal side moves normally, whereas on the opposite side it moves much less.

Case 1-6

The pleurae are continuous with each other around and inferior to the root of the lung. Normally, these layers are in contact during all phases of respiration. The potential space between them (*pleural cavity*) contains a capillary film of fluid. The visceral pleura normally slides smoothly on the parietal pleura during respiration, facilitating movement of the lung. When the pleurae are inflamed (*pleuritis*), the pleural surfaces become rough and the rubbing of their surfaces produces friction that is audible as a *pleural rub* during auscultation.

The pleural cavity is not usually visible on radiographs, but when air, fluid, pus, or blood collect between the visceral and parietal layers of pleura, the pleural cavity becomes apparent. If the inflammatory process in the present case had not been treated, an effusion or exudation of serum would have occurred from the blood vessels supplying the pleura. *Pleural exudate* collects in the pleural cavity and is visible on radiographs as a more or less homogeneous density that obscures the normal markings of the lung. Large pleural effusions are associated with a shift of the heart and other mediastinal structures to the opposite side. If the inflamed pleurae become infected, pus accumulates in the pleural cavity (*empyema*). A small empyema may be drained by thoracentesis during which a widebore needle is inserted posteriorly through the seventh intercostal space, along the superior border of the eighth rib. The insertion of a needle close to the superior border of the rib avoids injury to the intercostal nerves and vessels (Fig. 1-23*D*).

The parietal pleura, particularly its costal part, is very sensitive to pain, whereas the visceral pleura is insensitive. Afferent fibers from pain endings in the costal pleura and the pleura on the peripheral part of the diaphragm are conveyed through the thoracic wall as fine twigs of the intercostal nerves. Irritation of nerve endings in the pleura resulting from rubbing of the inflamed pleurae together, particularly during inspiration, produces stabbing pain. Referred pain from the pleura is felt in the thoracic and abdominal walls, the areas of skin innervated by the intercostal nerves. Pain around the umbilicus is explained by the fact that the 10th intercostal nerve supplies the band of skin that includes the umbilicus (Fig. 1-24; see also Fig. 2-13*A*).

The mediastinal pleura and the pleura on the central part of the diaphragm are supplied by sensory fibers from the phrenic nerve (C3, **C4**, and C5). Irritation of these areas of pleura, as in the present case, stimulates nerve endings in the pleura, resulting in pain being referred to the root of the neck and over the shoulder. These areas of skin are supplied by the supraclavicular nerves (C3 and C4), which are derived from two of the segments of the spinal cord that give origin to the phrenic nerve (Fig. 1-24). The heart and mediastinal structures shift to the affected side where there is pneumonitis and occupy the space created by the slight loss of volume of the consolidated lung tissue, resulting from the loss of air from alveolus. If pleural effusion or empyema occurs, the heart and mediastinal structures are pushed toward the opposite side by the accumulated serum or pus, respectively.

Case 1-7

Pulmonary thromboembolism (*PTE*) is an important cause of morbidity (sickness) and mortality (death) in patients confined to bed, pregnant patients, and women who have been taking birth control pills for a long time. In the present case, it is probable that the thrombus was released from the great saphenous vein or one of its tributaries (see Fig. 5-7*B*). It was transported to the pulmonary circulation, where it caused complete or almost complete obstruction of the right pulmonary artery. This led to various respiratory and hemodynamic disturbances (*e.g.*, dyspnea, tachypnea, arrhythmia, and tachycardia).

Three factors are involved in thrombus formation or **thrombogenesis**: stasis, abnormalities of the wall of the vessel, and alterations in the blood coagulation system. Various conditions are associated with thromboembolism (*e.g.*, pregnancy, fractures of the pelvis or lower limbs, abdominal operations, and the use of oral contraceptives). Possibly the lengthy car trip with a safety belt around her abdomen and the birth control pills were contributory factors to thrombogenesis in the present case.

The radiologist probably visualized the right ventricle by *right cardiac catheterization*. A radiopaque catheter was probably inserted into the left femoral vein just inferior to the inguinal ligament. It would then have been guided with the aid of fluoroscopy into the inferior vena cava, right atrium, and right ventricle. The patient's chest discomfort probably resulted from right heart strain and distention of the left pulmonary artery and its branches. As the right main pulmonary artery was obstructed, most of the blood was going through the left pulmonary artery. When the right lung failed to receive an adequate blood supply, changes very likely resulted in the pleura and pulmonary tissue. In view of the *pleural rub*, there was likely some pleuritic chest pain caused by irritation of nerve endings of the costal pleura. The referred pain was felt in the thoracic wall, the area of skin innervated by the intercostal nerves (Fig. 1-24).

SUGGESTED READINGS

Anderson RH, Becker AE: *Cardiac Anatomy*, London, Gower, 1987.

Behrman RE: *Nelson Textbook of Pediatrics*, ed 14. Philadelphia, WB Saunders, 1992.

Bergman RA, Thompson SA, Afifi AK, Saadeh FA: *Compendium of Human Anatomic Variation: Text, Atlas and World Literature*, Baltimore, Urban & Schwarzenberg, 1988.

Cormack DH: *Ham's Histology*, ed 9. Philadelphia, JB Lippincott, 1987.

Davies MF, Anderson RH, Becker AE: *The Conduction System of the Heart*, London, Butterworths, 1983.

Ginsberg RJ: Thoracic surgery. In Gross A, Gross P, Langer B (Eds): *Surgery: A Complete Guide for Patients and Their Friends*, Toronto, Harper & Collins, 1989.

Goldman BS: Cardiovascular surgery. In Gross A, Gross P, Langer B (Eds): *Surgery: A Complete Guide for Patients and Their Friends*, Toronto, Harper & Collins, 1989.

Haagensen CD: *Diseases of the Breast*, ed 3. Philadelphia, WB Saunders, 1986.

Healey JE Jr, Hodge J: *Surgical Anatomy*, ed 2. Toronto, BC Decker, 1990.

Maden RE: Cardiopulmonary surgery. In McCredie IA, (Ed): *Basic Surgery*, New York, Macmillan Publishing, 1977.

Martin EJ: Incidence of bifidity and related rib abnormalities in Samoans, Am J Phys Anthropol, 18:179–187, 1960.

Miller WS: *The Lung*, ed 2. Springfield, Ill, Charles C Thomas, 1974.

McAlpine WA: *Heart and Coronary Arteries*, New York, Springer, 1975.

Moore KL: *The Developing Human: Clinically Oriented Embryology*, ed 4. Philadelphia, WB Saunders, 1988.

Ross RS, Lesch M, Braunwald E: Acute myocardial infarction. In Thorn GW, Adams RD, Braunwald E, Isselbacher KE, Petersdorf G (Eds): *Harrison's Principles of Internal Medicine*, ed 10. New York, McGraw-Hill, 1983.

Stone R: General surgery. In Gross A, Gross P, Langer B (Eds): *Surgery: A Complete Guide for Patients and Their Friends*, Toronto, Harper & Collins, 1989.

Swartz MA, Moore ME: *Medical Emergency Manual Differential Diagnosis and Treatment*, ed 3. Baltimore, Williams & Wilkins, 1983.

Tobias PV, Arnold M, Allan JC: *Man's Anatomy: A Study in Dissection*, ed 4. vol. 1. Johannesburg, Witwatersrand University Press, 1988.

Williams PL, Warwick R, Dyson M, Bannister LH: *Gray's Anatomy*, ed 37. New York, Churchill Livingstone, 1989.

The abdomen is the part of the trunk that lies between the thorax and pelvis. *Abdominal pain* is the usual presenting symptom associated with abdominal disease, which is common and serious. **Appendicitis** (inflammation of the appendix) ranks high on the list of diseases leading to hospitalization and is *a major cause of abdominal pain*. It is the most common reason for emergency abdominal surgery in children and adolescents (Filler, 1989). However, any abdominal organ is subject to disease or injury. If a hollow gastrointestinal (GI) organ ruptures, its irritating contents escape into the peritoneal cavity, producing **peritonitis** (inflammation of the peritoneum, the serous sac that lines the abdominal cavity and covers most of the GI organs). The abdomen is located between the thoracic diaphragm and the pelvic brim (Fig. 2-29). *The diaphragm forms the roof of the abdominal cavity*; it has no floor because it is continuous with the pelvic cavity. The abdominal cavity extends superiorly into the osteocartilaginous thoracic cage to about the fifth anterior intercostal space (Fig. 2-1) when a person is supine; consequently, parts of the liver, stomach, and spleen are protected by the thoracic cage.

Contents of the Abdomen. The abdomen contains: (1) the *peritoneal cavity*; (2) the *GI organs* comprising the inferior part of the esophagus, the stomach, the intestine, including the vermiform appendix; (3) the liver and biliary system; (4) the pancreas; (5) the suprarenal (adrenal) glands; (6) the kidneys and the superior parts of the ureters; and (7) nerves, lymphatics, and blood vessels (*e.g.*, the aorta and inferior vena cava).

The Abdominal Wall

The abdominal wall extends from the osteocartilaginous thoracic cage to the pelvis (Fig. 2-1). Its major part is muscular.

Subdivisions of the Abdominal Wall. Although the abdominal wall is continuous, it is helpful for descriptive purposes to subdivide it into: (1) the *anterior abdominal wall*; (2) the right and left *lateral walls* (loins, flanks); and (3) the *posterior abdominal wall*. The combined term *anterolateral wall* is often used because some structures (*e.g.*, the external oblique muscle and cutaneous nerves) are located in both the anterior and lateral walls.

Layers of the Abdominal Wall. The abdominal wall consists of skin, subcutaneous tissue (superficial fascia and fat), deep fascia, muscles, transversalis fascia, extraperitoneal fat, and parietal peritoneum.

The Anterior Abdominal Wall

When a patient's abdomen is examined, the anterior abdominal wall is inspected and palpated. When the abdomen is sur-

gically operated on, it is usually the anterior abdominal wall that is incised. Most of this wall consists of three muscular layers, each of which has its fibers arranged in a different direction. This three-ply structure of the muscular layers is the same as in the intercostal spaces (see Figs. 1-19 and 1-23).

Boundaries of the Anterior Abdominal Wall (Figs. 2-1 to 2-3). The anterior abdominal wall is limited or bounded *superiorly* by the right and left costal margins; *inferiorly* by a line connecting each anterior superior iliac spine to the pubic symphysis; and *on each side* by a vertical line through the anterior iliac spines.

Surface Anatomy of the Anterior Abdominal Wall

The **umbilicus** (navel, "belly-button") is the obvious feature of the anterior abdominal wall of most people. This puckered scar represents the former site of attachment of the *umbilical cord*, which contained vessels that carried oxygenated blood to the fetus and deoxygenated blood from it (Moore, 1988). Its location varies considerably in obese people, depending on whether they are in the supine or erect position. Furthermore, the umbilicus tends to be more inferior in children and old persons. *In*

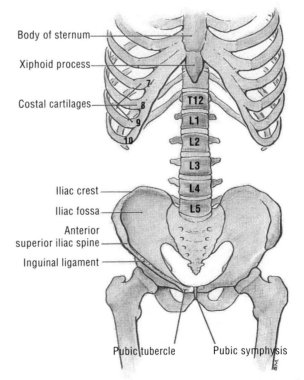

Figure 2-1. The skeleton of the abdomen, illustrating the boundaries and bony landmarks of the anterior abdominal wall. Note that the most superior point of the iliac crest is at the level of L4 vertebra.

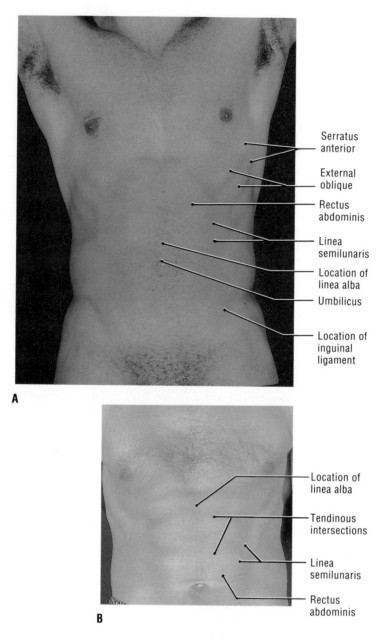

Figure 2-2.   Surface anatomy of the anterior abdominal wall of a physically fit man.

physically fit people, the umbilicus lies at the level of the intervertebral disc between L3 and L4 vertebrae (Figs. 2-1 and and 2-4*A*). This is about midway between the xiphoid process of the sternum and the pubic symphysis. A line joining the xiphoid process to the pubic symphysis indicates the position of the **linea alba** (L. *alba*, white), a median fibrous white line or band (Figs. 2-2 and 2-7). This line divides the anterior abdominal wall into right and left halves. The position of the linea alba is indicated by a vertical skin groove in the anterior median line; it is clearly visible in thin muscular persons (Fig. 2-2).

The **linea semilunaris** is a curved line or groove (convex laterally) that extends from the ninth costal cartilage to the pubic tubercle (Figs. 2-1 to 2-3). This semilunar line indicates the lateral border of the rectus abdominis muscle, which is located 5 to 8 cm from the median plane (Fig. 2-6). These lines are obvious in persons with good abdominal muscular development (Fig. 2-2). In such persons, three or more transverse grooves are also visible in the skin overlying the tendinous intersections of the rectus abdominis muscles (Figs. 2-2 and 2-6). The site of the inguinal ligament is indicated by the *inguinal groove* (Fig. 2-3). It also indicates the division between the anterior abdominal wall and the thigh. A fold of skin, called the *inguinal fold*, is visible just superior to the inguinal groove (Figs. 2-3 and 2-46). The *inguinal ligament* may be felt along its entire length.

The **pubic symphysis** (symphysis pubis) may be felt as a firm resistance in the anterior midline at the inferior end of the linea alba (Figs. 2-1 and 2-3). In some people the fat covering this secondary cartilaginous joint makes it difficult to palpate, but its superior border is usually easy to locate. The *pubic crest* (Fig. 2-1) may be felt by extending a digit (finger) laterally from

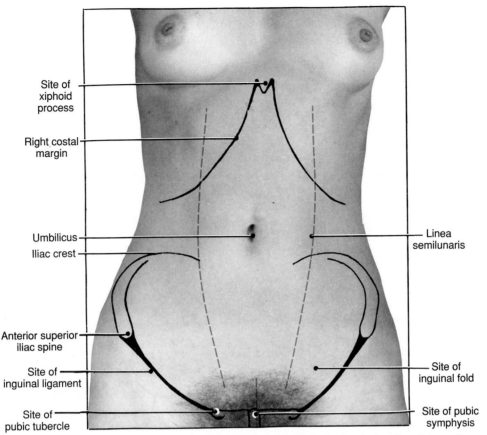

Figure 2-3. The anterior abdominal wall of a 27-year-old woman, indicating the location of the surface landmarks shown in Fig. 2-1.

the pubic symphysis for about 2.5 cm. This crest terminates at a clinically important landmark called the **pubic tubercle**. It is the prominent, rounded elevation of the body of the pubis to which the inguinal ligament attaches (Figs. 2-1 and 2-3). The **iliac crest**, extending posteriorly from the anterior superior iliac spine (Figs. 2-1 and 2-3), can be easily felt throughout its length. The *iliac tubercle* can be palpated about 6 cm posterior to the anterior superior iliac spine (Fig. 2-1; see also Fig. 5-2). The *epigastric fossa* ("pit of the stomach") is a slight depression in the epigastric region, just inferior to the xiphoid process (Fig. 2-5A). This fossa is particularly noticeable when a person is in the supine position because the viscera spread out, drawing the anterior abdominal wall posteriorly in this region.

Planes and Regions of the Abdomen

For purposes of description, the abdomen is divided by four planes (two horizontal and two sagittal) into the *nine regions* shown in Figure 2-5A. These regions are helpful clinically for describing the location of a pain, swelling or incision, or for indicating the location of viscera (*e.g.*, the duodenum, Fig. 2-4B). Of the various horizontal planes that have been described, the subcostal and transtubercular planes are the ones that are commonly used (Fig. 2-4A).

The subcostal plane (*SCP*) is a horizontal one that joins the most inferior points of the costal margins (Fig. 2-4A), which are

the inferior margins of the 10th costal cartilages (Fig. 2-1). This plane usually passes through the superior part of the body of L3 vertebra.

The transtubercular plane (*TTP*) passes through the *iliac tubercles* on the iliac crests (Fig. 2-4A). This plane usually passes through the body of L5 vertebra.

The sagittal planes are the *midclavicular lines* (Fig. 2-4A). Each line extends inferiorly from the midpoint of the clavicle to the *midinguinal point*. This is the midpoint of the line joining the anterior superior iliac spine and the pubic symphysis (Fig. 2-3).

Two other horizontal planes are also used when describing the abdomen.

The transpyloric plane (*TPP*) is a horizontal plane situated midway between the jugular notch in the manubrium of the sternum (see Figs. 1-6 and 1-7) and the pubic symphysis (Figs. 2-3 and 2-4B). It is *the key plane of the abdomen* because many abdominal viscera are related to it (*e.g.*, the stomach and duodenum). In most people this plane is approximately midway between the xiphisternal joint and the umbilicus (Fig. 2-1). The transpyloric plane passes through the inferior border of L1 vertebra, and as its name implies, it *passes through the pylorus of the stomach*, the junction between the stomach and the duodenum (Fig. 2-4B). This plane also passes through the anterior parts of the ninth costal cartilages, the duodenojejunal junction, the neck of the pancreas, and the hila of the kidneys (Fig. 2-46).

Midclavicular lines

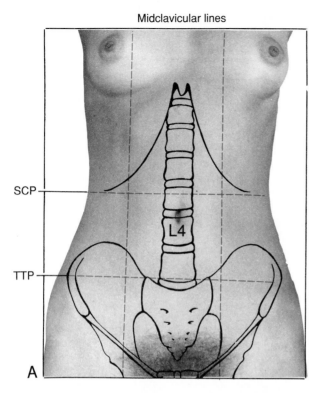

A

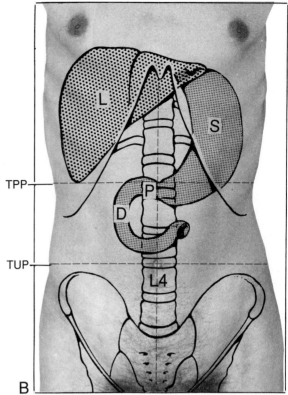

B

Figure 2-4. The commonly used horizontal and sagittal planes (*red*) and points of reference for the abdomen. *A*, The planes used for dividing the abdomen into the nine regions shown in Fig. 2-5*A*. The sagittal planes are the *midclavicular lines*, which pass through the midpoints of the clavicles. The horizontal planes are: the *subcostal plane* (*SCP*), joining the most inferior points of the costal margins and passing through L3 vertebra, and the *transtubercular plane* (*TTP*), joining the tubercles of the iliac crests and passing through L5 vertebra. *B*, Two other commonly used horizontal planes: *transpyloric* (*TPP*) and *transumbilical* (*TUP*). Liver (*L*), stomach (*S*), pylorus (*P*), and duodenum (*D*).

The transumbilical plane (*TUP*), as its name implies, passes through the umbilicus (Fig. 2-4*B*). In persons with a reasonably firm anterior abdominal wall, this plane indicates the level of the intervertebral disc between L3 and L4 vertebrae.

Quadrants of the Abdomen (Fig. 2-5*B*). For simplicity, clinicians routinely divide the abdomen into four quadrants, *right and left upper and lower quadrants*. To obtain these they use the median and transumbilical planes. These four quadrants are used to describe the general location of a pain or a tumor. For example, the pain of acute appendicitis usually starts in the periumbilical region (the region around the umbilicus) and later localizes in the right lower quadrant.

Pain in the periumbilical region followed by nausea and vomiting are early symptoms of **appendicitis**. Later the patient usually describes *right lower quadrant pain*. Hence knowledge of the abdominal regions is necessary to record these symptoms in a patient's chart.

Protuberance of the abdomen is normal in infants and young children because their GI tracts contain considerable amounts of air. In addition, their abdominal cavities are still enlarging and their abdominal muscles are gaining strength. A child's liver is also relatively large which accounts for some of the bulging. During pregnancy protrusion of the abdomen occurs as a result of the presence of a *fetus*. The abdomen also expands in both sexes following the deposition of *fat* or

the accumulation of *feces*, *fluid*, or *flatus* (gas in GI tract). Note that the five common causes of abdominal protrusion all begin with an "*f.*" During old age, abdominal muscle laxity also contributes to abdominal protuberance. Abdominal tumors and the accumulation of fluid in the peritoneal cavity (*ascites*) also produce abdominal enlargement. As it enlarges, the anterior abdominal muscles thin out, the skin grows, and the nerves and blood vessels lengthen.

Reddish elongated lines or streaks called *striae gravidarum* (L. *gravidus*, heavy) often appear in the abdominal skin of pregnant women. These striae gradually change into thin, white scarlike lines called *lineae albicantes* (white lines or "stretch marks"). These lines also appear in the skin of obese men and women. When examining the abdomen, the hands should be warm so that the patient's muscles will not become tense. To relax the abdominal wall, patients are examined in the supine position with their thighs semiflexed and a pillow under their knees. If this is not done, the deep fascia of the thighs pulls on the deep layer of superficial fascia of the abdomen and tenses the abdominal wall.

Fascia of the Anterior Abdominal Wall

The fascia (fibrous tissue) beneath the skin of the anterior abdominal wall consists of superficial and deep layers.

The Superficial Fascia (Fig. 2-6).   Over most of the wall, the superficial fascia consists of one layer that contains a variable amount of fat. In some persons the fat is several inches thick. The superficial fascia just superior to the inguinal ligament can be divided into two layers: (1) *a fatty superficial layer* (Camper's fascia) containing a variable amount of fat and (2) *a membranous deep layer* (Scarpa's fascia) containing fibrous tissue and very little fat. The superficial vessels and nerves are located between these two layers of superficial fascia. The fatty superficial layer merges with the superficial fascia of the thigh, and the membranous deep layer is continuous with the deep fascia of the thigh, called the *fascia lata* (see Fig. 5-32). The membranous deep layer is also continuous with the superficial fascia of the perineum (Colles' fascia) and with that investing the scrotum and penis and the labia majora (Fig. 2-18; see also Chap. 3). The membranous deep layer of superficial fascia fuses with the deep fascia of the abdomen.

The Deep Fascia.   Little can be said about the deep fascia of the anterior abdominal wall, except that it is a very thin, strong layer over the superficial muscles and cannot be separated easily from them.

Surgeons use the membranous deep layer of superficial fascia for holding sutures during closure of abdominal skin incisions. Between the deep layer of superficial fascia and the deep fascia of the abdomen, there is a potential space in which fluid may accumulate. For example, urine may pass through a rupture in the urethra (*extravasation of urine*, p. 320) into this space and extend superiorly into the anterior abdominal wall between the membranous deep layer of the superficial fascia and the deep fascia.

Muscles of the Anterior Abdominal Wall

Most of the abdominal wall is muscular and extends between the thoracic cage and the bony pelvis (Figs. 2-1 and 2-6 to 2-8). There are *four important paired muscles* in the anterior abdominal wall: *three flat muscles* (external oblique, internal oblique, and transversus abdominis) and *one straplike muscle* (rectus abdominis). The combination of muscles and aponeuroses (sheet-like tendons) in the anterior abdominal wall affords considerable protection to the abdominal viscera, especially when the muscles are in good physical condition. The flat muscles cross each other

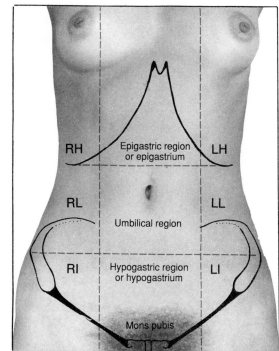

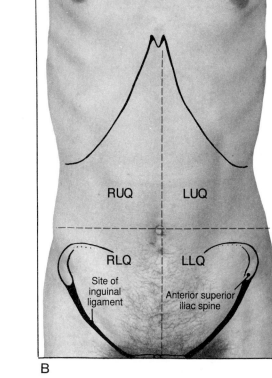

Figure 2-5.   The two commonly used clinical methods of dividing the abdomen for descriptive purposes. *A,* For specific localization of a finding, the abdomen is divided into nine regions using the subcostal and transtubercular planes and the midclavicular lines (see also Fig. 2-4). The terms used to designate the three central regions are: epigastric, umbilical or periumbilical, and pubic or hypogastric. The superior part of the epigastric region, just inferior to the xiphoid process, is often referred to as the epigastric fossa (''pit of the stomach'' by lay people). The inferior part of the pubic region, just superior to the mons pubis, is often referred to as the suprapubic area. The lateral regions are: *LH,* left hypochondriac; *RH,* right hypochondriac; *LL,* left lateral (lumbar); *RL,* right lateral (lumbar); *RI,* right inguinal; and *LI,* left inguinal. Some regions are referred to in two ways: *e.g.,* epigastric region or epigastrium and hypochondriac region or hypochondrium. *B,* For general localization the abdomen is divided into four quadrants (right upper, right lower, left upper, and left lower). The vertical line is in the median plane and the horizontal line is the transumbilical plane.

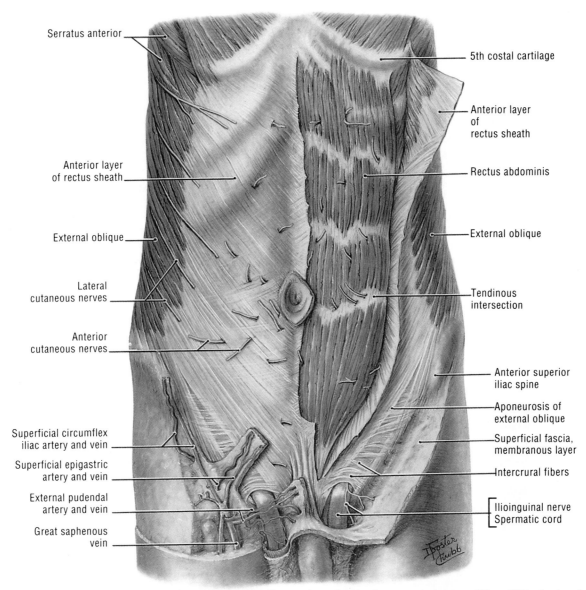

Serratus anterior

5th costal cartilage

Anterior layer
of
rectus sheath

Anterior layer
of rectus sheath

Rectus abdominis

External oblique

External oblique

Lateral
cutaneous nerves

Tendinous
intersection

Anterior
cutaneous nerves

Anterior superior
iliac spine

Aponeurosis of
external oblique

Superficial circumflex
iliac artery and vein

Superficial fascia,
membranous layer

Superficial epigastric
artery and vein

Intercrural fibers

External pudendal
artery and vein

Ilioinguinal nerve
Spermatic cord

Great saphenous
vein

Figure 2-6. Superficial dissection of the anterolateral abdominal wall. The anterior layer of the rectus sheath is reflected on the left side to show the anterior cutaneous nerves (T7 to T12) piercing the rectus abdominis muscle and rectus sheath.

in such a way (similar to a three-ply corset) that strengthens the abdominal wall and diminishes the risk of protrusion of viscera (herniation) between the muscle bundles.

In about 80% of people there is an insignificant, small triangular abdominal muscle, called the *pyramidalis*, which is located anterior to the inferior part of the rectus abdominis. It tenses the linea alba.

The anterior abdominal wall may be the site of **congenital hernias**. Most of them occur in the umbilical and inguinal regions. *Inguinal hernias* are discussed subsequently (p. 147). *Umbilical hernias are common* in low birthweight and black infants (Behrman, 1992). These hernias are usually small (1 to 5 cm) and congenital (present at birth). They *result from incomplete closure of the anterior abdominal wall* after the umbilical cord is ligated at birth. Herniation occurs through

the defect created by the degenerating umbilical vessels. Hernias may also occur through defects in the linea alba (Fig. 2-7); these are *median hernias*. Extraperitoneal fat and/or omentum (a peritoneal fold) protrude through these defects. If the hernia occurs in the epigastric region (Fig. 2-5A), it is called an *epigastric hernia*. This type tends to occur after 40 years of age and is often associated with obesity.

The External Oblique Muscle (Figs. 2-2 and 2-6 to 2-9). This is the largest and most superficial of the three flat abdominal muscles. It is located in the anterolateral part of the abdominal wall. Its fleshy part forms the anterolateral portion and its *aponeurosis* (sheet of tendon fibers) forms the anterior part. Its fibers run inferoanteriorly and medially in the same direction as do the extended digits (fingers) when they are in one's side pockets. The attachments, nerve supply, and main

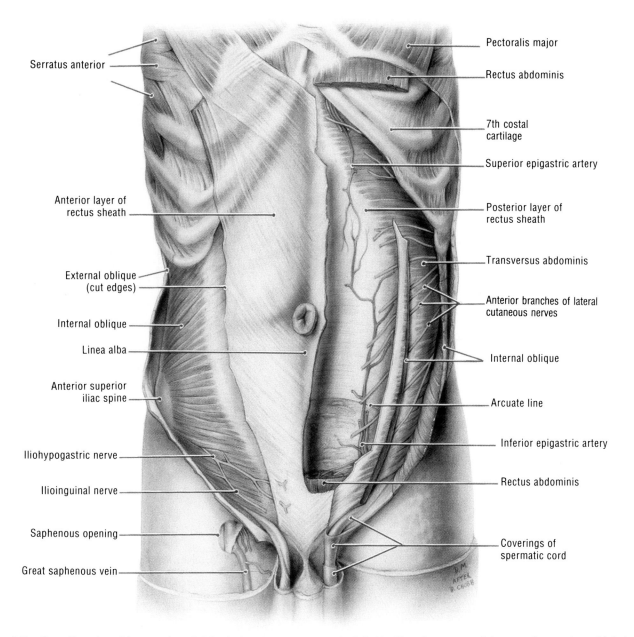

Serratus anterior

Anterior layer of
rectus sheath

External oblique
(cut edges)

Internal oblique

Linea alba

Anterior superior
iliac spine

Iliohypogastric nerve

Ilioinguinal nerve

Saphenous opening

Great saphenous vein

Pectoralis major

Rectus abdominis

7th costal
cartilage

Superior epigastric artery

Posterior layer of
rectus sheath

Transversus abdominis

Anterior branches of lateral
cutaneous nerves

Internal oblique

Arcuate line

Inferior epigastric artery

Rectus abdominis

Coverings of
spermatic cord

Figure 2-7.   Deep dissection of the anterolateral abdominal wall. Most of the external oblique muscle is excised on the right side. The rectus abdominis muscle is excised and the internal oblique muscle is divided on the left side. Note the anastomosis between the superior and inferior epigastric arteries, which indirectly unites the arteries of the upper and lower limbs.

actions of the external oblique are given in Table 2-1. As the fibers of this muscle pass medially, they become aponeurotic. This aponeurosis ends medially in the *linea alba*. Inferiorly it folds back on itself to form the *inguinal ligament* between the anterior superior iliac spine and the pubic tubercle. Medial to the pubic tubercle the external oblique aponeurosis is attached to the *pubic crest*. Some fibers of the inguinal ligament cross the linea alba and attach to the opposite pubic crest. These fibers form the *reflex inguinal ligament* (Fig. 2-9). The medial part of the inguinal ligament is reflected horizontally back and is attached to the pecten pubis as the *lacunar ligament* (Figs. 2-1 and 2-16B). Just superior to the medial part of the inguinal ligament, there is an opening in the aponeurosis called the *su-*

perficial inguinal ring (Fig. 2-9). This clinically important ring is discussed with the inguinal region (p. 143).

   The Internal Oblique Muscle (Figs. 2-7 to 2-11).   This is the intermediate of the three flat abdominal muscles. Its attachments, nerve supply, and main actions are given in Table 2-1. The fibers of the internal oblique muscle run superoanteriorly, at right angles to those of the external oblique. Its fibers also become aponeurotic and the aponeurosis splits to form a sheath for the rectus abdominis muscle (Fig. 2-8C). The inferior fibers of the aponeurosis arch over the *spermatic cord* as it lies in the inguinal canal. They then descend posterior to the *superficial inguinal ring* to attach to the pubic crest and pecten pubis (Figs. 2-1, 2-7, and 2-9). The most inferior tendinous fibers of the

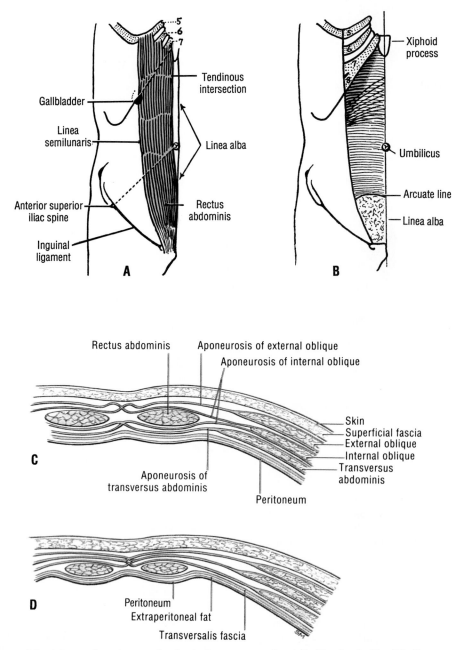

Figure 2-8. The rectus abdominis muscle and rectus sheath. *A*, Anterior view of the right rectus abdominis after removal of the anterior layer of its sheath. *B*, The rectus muscle has been removed to show the posterior wall of its sheath. *C* and *D*, Transverse sections of the anterior abdominal wall showing the interlacing fibers of the aponeuroses of the right and left oblique and transversus abdominis muscles.

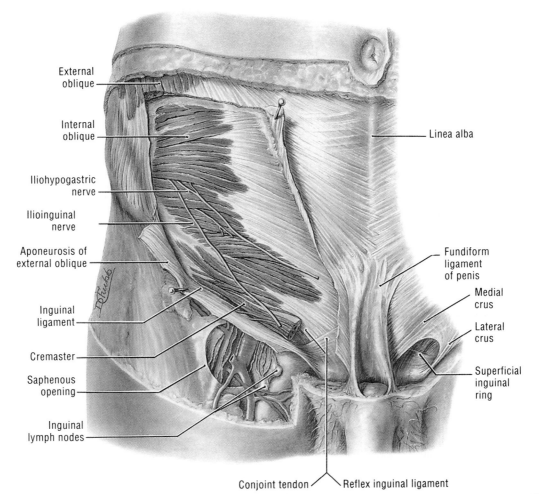

External oblique

Internal oblique

Iliohypogastric nerve

Ilioinguinal nerve

Aponeurosis of external oblique

Inguinal ligament

Cremaster

Saphenous opening

Inguinal lymph nodes

Linea alba

Fundiform ligament of penis

Medial crus

Lateral crus

Superficial inguinal ring

Conjoint tendon   Reflex inguinal ligament

Figure 2-9. Superficial dissection of the anterolateral abdominal wall, especially the inguinal region. Note that only two structures run between the external and internal oblique muscles, namely the iliohypogastric and ilioinguinal nerves.

internal oblique muscle join with aponeurotic fibers of the transversus abdominis muscle to form the *conjoint tendon*, which turns inferiorly to insert into the pubic crest and pecten pubis.

The Transversus Abdominis Muscle (Figs. 2-7, 2-8, 2-10, and 2-11).   This is the innermost of the three flat abdominal muscles. Its attachments, nerve supply, and main actions are given in Table 2-1. Its fibers run more or less horizontally, except for the most inferior ones, which pass inferiorly and run parallel to those of the internal oblique muscle. Muscle fibers of the transversus abdominis end in an aponeurosis which contributes to the formation of the rectus sheath.

Actions of the Three Flat Abdominal Muscles (Table 2-1).   The anterolateral abdominal wall is unsupported and unprotected by bone. However, the three-ply structure of its flat muscles and their extensive aponeuroses form a strong expandable support, which provides considerable *protection for the abdominal viscera*. Normally, quiet rhythmic movements of the anterolateral abdominal wall accompany respirations. When the diaphragm contracts during inspiration, its domes flatten and descend, increasing the vertical dimension of the thorax (see Chap. 1). To make room for the viscera (*e.g.*, the stomach and intestine), the anterolateral abdominal wall expands as its mus-

cles relax. When the thoracic cage and diaphragm relax during expiration (see Fig. 1-50), the anterolateral abdominal wall passively sinks in. However, in the forced expiration that occurs during coughing, sneezing, vomiting, and straining, all the anterior abdominal muscles act strongly in compressing the abdominal contents.

Acting together, the flat abdominal muscles increase the intra-abdominal pressure. When the ribs and diaphragm are fixed, compression of the viscera by the anterior abdominal muscles occurs, which raises the intra-abdominal pressure. This action produces the force required for defecation (bowel movement), micturition (urination), and parturition (childbirth).

Acting separately, the flat abdominal muscles move the trunk (see Chap. 4). If the pelvis is fixed, both external oblique muscles can flex the trunk. Acting separately, one external oblique muscle can laterally flex the trunk and rotate it to the opposite side (see Fig. 4-36). If the thorax is fixed, both external oblique muscles tilt the anterior part of the pelvis superiorly and flex the trunk. Similarly, when the pelvis is fixed, one internal oblique muscle can flex the trunk and rotate it to the same side. If the thorax is fixed, one internal oblique muscle can laterally flex the trunk and rotate the pelvis to the opposite side. See Chap. 4 for a

Table 2-1.
Principal Muscles of the Anterolateral Abdominal Wall

Muscle	Origin	Insertion	Innervation	Action(s)[1]
External oblique	External surfaces of 5th to 12th ribs	Linea alba, pubic tubercle and anterior half of iliac crest	Inferior six thoracic nn. and subcostal n.	Compress and support abdominal viscera Flex and rotate trunk
Internal oblique	Thoracolumbar fascia, anterior two-thirds of iliac crest, and lateral half of inguinal ligament	Inferior borders of 10th to 12th ribs, linea alba, and the pubis via the conjoint tendon	Ventral rami of inferior six thoracic and first lumbar nn.	
Transversus abdominis	Internal surfaces of 7th to 12th costal cartilages, thoracolumbar fascia, iliac crest, and lateral third of inguinal ligament	Linea alba with aponeurosis of internal oblique, pubic crest, and pecten pubis via conjoint tendon		Compresses and supports abdominal viscera
Rectus abdominis	Pubic symphysis and pubic crest	Xiphoid process and 5th to 7th costal cartilages	Ventral rami of inferior six thoracic nn.	Flexes trunk and compresses abdominal viscera

[1]These abdominal muscles have three main actions: (1) movement of the trunk; (2) compression of the viscera; and (3) respiration.

further discussion of these movements of the vertebral column and trunk (p. 350).

The Rectus Abdominis Muscle (Figs. 2-2, 2-3, and 2-6 to 2-8). This long, broad, strap muscle is the *principal vertical muscle* of the anterior abdominal wall. Its attachments, nerve supply, and main actions are given in Table 2-1. *The two muscles are separated by the linea alba* and lie close together inferiorly. The rectus abdominis is three times as wide superiorly as inferiorly; it is narrow and thick inferiorly and broad and thin superiorly. The lateral border of the rectus muscle and its sheath are convex and form a clinically important surface marking known as the *linea semilunaris*. Most of the rectus abdominis muscle is enclosed in the **rectus sheath**, formed by the aponeuroses of the three flat abdominal muscles (Figs. 2-6 to 2-8). The anterior layer of the rectus sheath is firmly attached to the rectus muscle at three or more *tendinous intersections* (Fig. 2-6). When this muscle is tensed in muscular persons, each stretch of muscle between the tendinous intersections bulges outward (Fig. 2-2). The location of the tendinous intersections is indicated by grooves in the skin between the muscle bulges. They are usually located at the level of the xiphoid process, umbilicus, and halfway between these structures.

Actions of the Rectus Abdominis Muscles. In addition to helping the other abdominal muscles to compress the abdominal viscera (*e.g.*, during coughing, vomiting, and defecating), these muscles depress the ribs and *stabilize the pelvis during walking*. This fixation of the pelvis enables the thigh muscles to act effectively. Similarly, during lower limb lifts from the supine position, the rectus abdominis muscles contract to prevent tilting of the pelvis by the weight of the limbs.

The Linea Alba and Rectus Sheath (Figs. 2-2 and 2-7 to 2-9). These structures have been mentioned several times and have been briefly described; however, owing to their clinical importance, a more detailed description and summary of the rectus sheath and its relationship to the linea alba follows.

The rectus sheath is the strong, incomplete fibrous compartment of the rectus abdominis muscle. It forms by the fusion and separation of the aponeuroses of the flat abdominal muscles (Fig. 2-8). At its lateral margin, the internal oblique aponeurosis splits into two layers, one passing anterior to the rectus muscle and the other passing posterior to it. The anterior layer joins with the aponeurosis of the external oblique to form the anterior wall of the rectus sheath. The posterior layer joins with the aponeurosis of the transversus abdominis muscle to form the posterior wall of the rectus sheath. The fibers of the anterior and posterior walls of the sheath interlace in the anterior median line to form a complex tendinous raphe, called the *linea alba* (Figs. 2-7 to 2-9), which is an intermixture of the aponeurotic fibers of the oblique and transverse abdominal muscles (Williams et al., 1989). It is narrow inferior to the umbilicus, but is wide superior to it. A groove is visible in the skin superficial to it in thin muscular persons (Fig. 2-2). The linea alba lies between the two parts of the rectus abdominis muscle; the umbilicus is located just inferior to its midpoint (Figs. 2-6 and 2-7).

Superior to the costal margin, the posterior wall of the rectus sheath is deficient because the transversus abdominis muscle passes internal to the costal cartilages and the internal oblique muscle is attached to the costal margin. Hence, superior to the costal margin, the rectus muscle lies directly on the thoracic wall (Fig. 2-7). The inferior one-fourth of the rectus sheath is also deficient because the internal oblique aponeurosis does not split here to enclose the rectus muscle (Fig. 2-8). The inferior limit of the posterior wall of the rectus sheath is marked by a crescentic border called the *arcuate line* (Fig. 2-7). The position of this line is usually midway between the umbilicus and the pubic crest (Fig. 2-1). Inferior to the arcuate line, the aponeuroses of the three flat muscles pass anterior to the rectus muscle to form the anterior layer of the rectus sheath (Figs. 2-7 and 2-8C).

Important structures in the rectus sheath, in addition to the rectus abdominis muscle, are the superior and inferior *epigastric vessels*, and the terminal parts of the inferior five intercostal and subcostal vessels and nerves (Fig. 2-7).

The Transversalis Fascia

This somewhat transparent internal investing layer *lines most of the abdominal wall* (Figs. 2-12 and 2-13*B* and *C*); posteriorly it fuses with the anterior lamina of the *thoracolumbar fascia* (Fig. 2-102). The transversalis fascia (fascia transversalis) covers the deep surface of the transversus abdominis muscle and its aponeurosis and is continuous from side to side, deep to the linea alba (Figs. 2-8 and 2-11 to 2-13). Each part of the transversalis fascia is named according to the structures it covers. It is called the *diaphragmatic fascia* on the diaphragm; the *iliac fascia* on the iliacus muscle; the *psoas fascia* on the psoas major muscle; and the *pelvic fascia* in the pelvis. The transversalis fascia also extends into the thigh with the iliac fascia to form the *femoral sheath* (p. 405). It also passes through the inguinal canal to form the *internal spermatic fascia*, part of the covering of the spermatic cord (Figs. 2-7, 2-12, and 2-15*A*). Internal to the transversalis fascia is the *peritoneum*, the extensive serous membrane that lines the abdominal and pelvic cavities (Fig. 2-12). The transversalis fascia is separated from the peritoneum by a variable amount of subperitoneal fat, referred to as *extraperitoneal fat*.

Nerves of the Anterior Abdominal Wall

The skin and muscles of the anterior abdominal wall are supplied mainly by the ventral rami of the *inferior six thoracic nerves* (*i.e.*, the continuation of the *inferior intercostal nerves*, T7 to T11) and the *subcostal nerves* (T12). The inferior part of the anterior wall is supplied by two branches of the ventral ramus of the first lumbar nerve via the iliohypogastric and ilioinguinal nerves (Figs. 2-7, 2-9, and 2-13). The *iliohypogastric nerve* supplies the skin over the inguinal region (groin). The *ilioinguinal nerve* runs anteroinferiorly, just superior to the iliac crest,

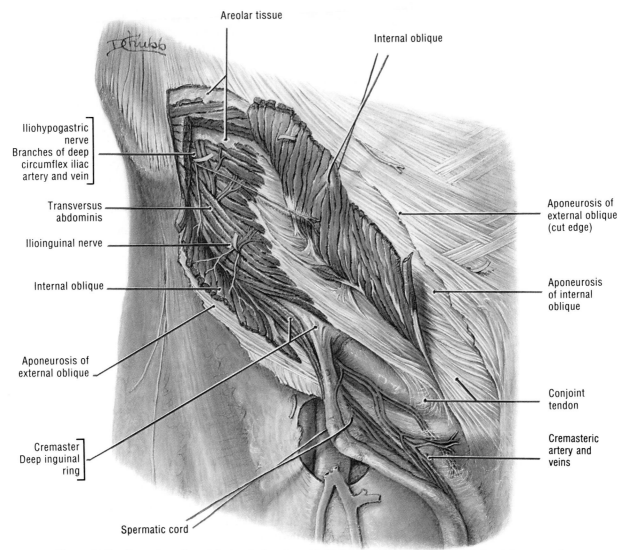

Areolar tissue

Internal oblique

Iliohypogastric nerve
Branches of deep circumflex iliac artery and vein

Transversus abdominis

Ilioinguinal nerve

Internal oblique

Aponeurosis of external oblique

Cremaster
Deep inguinal ring

Spermatic cord

Aponeurosis of external oblique (cut edge)

Aponeurosis of internal oblique

Conjoint tendon

Cremasteric artery and veins

Figure 2-10.   Deep dissection of the inguinal region with the internal oblique muscle reflected and the spermatic cord retracted.

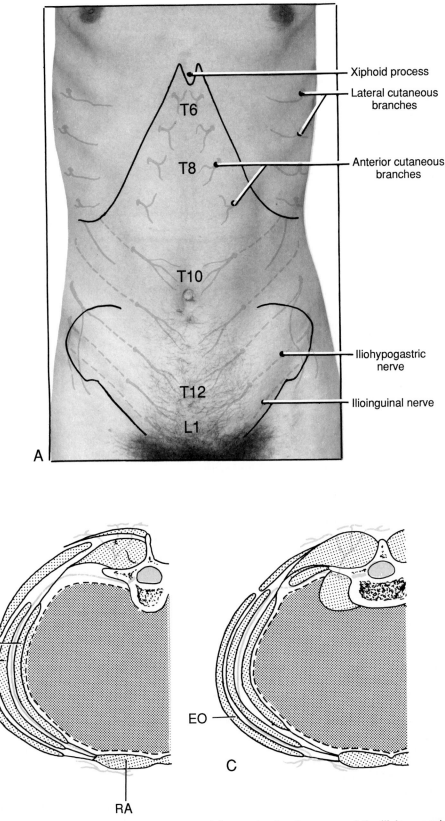

Figure 2-13. *A*, Distribution of the thoracoabdominal nerves in a 27-year-old man. *B*, Oblique section of the trunk illustrating the course of the inferior six thoracic nerves. Note that the ventral rami of these nerves run between the internal oblique (*IO*) and transversus abdominis (*TA*) muscles—*the neurovascular plane*—to the lateral margin of the rectus abdominis (*RA*) muscle. *C*, Transverse section of the lower abdomen showing the course of the iliohypogastric and ilioinguinal nerves. Note that the ventral ramus pierces the internal oblique and then runs between this muscle and the external oblique (*EO*) to pierce the external oblique aponeurosis. The broken lines (-----) in *B* and *C* indicate the transversalis fascia.

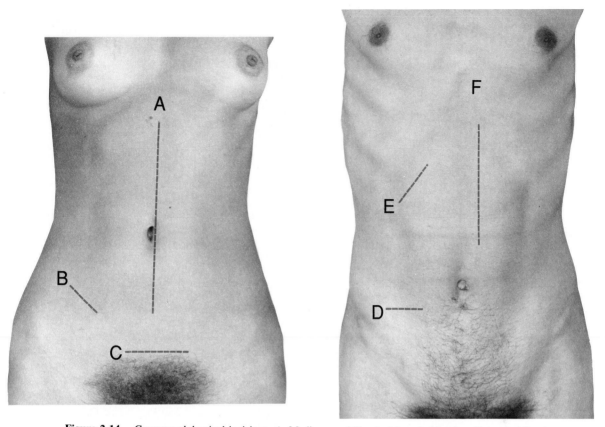

Figure 2-14. Common abdominal incisions. *A,* Median or midline incision. *B,* Muscle splitting (gridiron) incision. *C,* Suprapubic incision. *D,* Transverse incision. *E,* Subcostal incision. *F,* Paramedian incision.

to its nerve supply. The posterior layer of the rectus sheath and peritoneum are then incised to enter the peritoneal cavity.

Suprapubic incisions are made transversely (Fig. 2-14*C*), superior to the pubic symphysis, for gaining access to the pelvic organs (*e.g.,* the uterus during a *hysterectomy,* see Chap. 3).

Subcostal incisions are made to gain access to the gallbladder on the right side and the spleen on the left. The incision is made at least 2.5 cm inferior to the costal margin to avoid the seventh thoracic nerve (Figs. 2-13*A* and 2-14*E*), which runs close to this margin and crosses the abdominal muscles.

One of the common incisions is the *muscle-splitting (gridiron) incision* used for appendectomy (Fig. 2-14*B*). The common type is the oblique *McBurney incision* which is made about 2.5 cm superomedial to the anterior superior iliac spine (McBurney's point). This incision is now less popular than an almost *transverse incision* in the line of the skin crease (Fig. 2-14*D*), made 2.5 cm superomedial to the anterior superior iliac spine. In either case, the external oblique aponeurosis is incised obliquely in the direction of its fibers and retracted. The musculoaponeurotic fibers of the internal oblique and transversus abdominis muscles, lying at right angles to those of the external oblique (Fig. 2-13), are then split in the line of their fibers and retracted without dividing them. Carefully made, *the entire exposure cuts no musculoaponeurotic fibers;* therefore when the incision is closed, the muscles move together and the abdominal wall is as strong after the operation as it was before. When kept relatively small and done carefully, this incision provides good access and avoids cutting, tearing, and stretching the nerves.

An *incisional hernia* is a protrusion of an organ or tissue through a surgical incision. The surgeon who has a thorough knowledge of the anatomy of the anterolateral abdominal wall and makes incisions accordingly will only occasionally have to operate on this kind of hernia. However, if the muscular and tendinous layers of the abdomen do not heal properly, an incisional hernia can result. Infection and/or obesity are predisposing factors (Harrison, 1989).

Because the inferior intercostal nerves (T11 and T12) and the iliohypogastric and ilioinguinal nerves (L1) supply the abdominal musculature (Fig. 2-7), injury to them during surgery or an accident could result in weakening of muscles in the inguinal region, creating a predisposition to development of *direct inguinal hernia* (p. 148). The ilioinguinal nerve also gives off motor branches to the fibers of the internal oblique muscle, which are inserted into the lateral border of the conjoint tendon. Division of this nerve paralyzes these fibers and weakens the conjoint tendon; this may also result in a direct inguinal hernia.

Peritonitis (inflammation of the peritoneum) causes pain in the overlying skin and a reflex increase in the tone of the anterior abdominal muscles. Rhythmic movements of the anterior abdominal wall normally accompany respirations (*i.e.,* the abdomen moves out with the chest). If the abdomen goes in as the chest goes out (*paradoxical abdominothoracic rhythm*) and muscle rigidity is present, it is probable that peritonitis or *pneumonitis* (inflammation of the lungs) is present. A special feature of an examination of an *acute abdomen* (patient with intense abdominal pain) is the finding of spasm of the anterolateral abdominal muscles, sometimes amounting to boardlike rigidity. This palpable spasm of the muscles is called "*guarding*" or splinting.

Arterial Supply of the Anterior Abdominal Wall

Small arteries arise from anterior and collateral branches of the *posterior intercostal arteries* in the 10th and 11th intercostal spaces and from anterior branches of the *subcostal arteries* to supply the anterior abdominal wall (Figs. 2-6, 2-7, 2-10, and 2-11; see also Fig. 1-18). They anastomose with the superior epigastric arteries, the superior lumbar arteries, and with each other. The **main arteries of the anterior abdominal wall** are the *inferior epigastric* and *deep circumflex iliac* arteries, which are branches of the external iliac artery, and the *superior epigastric artery*, which is a terminal branch of the internal thoracic artery (Figs. 2-7 and 2-11; see also Fig. 1-19). The *inferior epigastric artery* runs superiorly in the transversalis fascia to reach the *arcuate line*; there it enters the rectus sheath. The *deep circumflex iliac artery* runs on the deep aspect of the anterior abdominal wall, parallel to the inguinal ligament, and along the iliac crest between the transversus abdominis and internal oblique muscles. The *superior epigastric artery* enters the rectus sheath superiorly, just inferior to the seventh costal cartilage (Fig. 2-7).

Venous and Lymphatic Drainage of the Anterior Abdominal Wall

The *superficial epigastric vein* and the *lateral thoracic vein* anastomose (Fig. 2-6), thereby uniting the veins of the superior and inferior halves of the body. The three *superficial inguinal veins* end in the great saphenous vein of the lower limb (Figs. 2-6 and 2-7).

The *superficial lymph vessels* of the anterolateral abdominal wall, superior to the umbilicus, pass to the *axillary lymph nodes* (see Fig. 1-12), whereas those inferior to the umbilicus drain into the *superficial inguinal lymph nodes* (Figs. 2-9 and 2-11). However, lymph from the anterior abdominal wall, in general, drains to the *lumbar lymph nodes* and to the common and external *iliac lymph nodes* (see Figs. 24 [p. 24] and 3-73).

Internal Surface of the Anterior Abdominal Wall

The internal or deep surface of this wall is covered with *parietal peritoneum*. It exhibits several *peritoneal folds*, some of which contain remnants of fetal vessels that carried blood to and from the placenta before birth (Fig. 2-12). *A peritoneal fold is an elevation of peritoneum with a free edge*, which is usually raised by an underlying blood vessel or its ligamentous remains. There are hollows or fossae between adjacent folds. Superior to the umbilicus there is a large median fold called the *falciform ligament* (Fig. 2-25). It passes from the deep surface of the superior half of the anterior abdominal wall to the liver and the diaphragm. The free edge of this ligament contains the *ligamentum teres*, the remnant of the *umbilical vein*.

Before birth the umbilical vein carried oxygenated blood from the placenta to the fetus (Moore, 1988). This vein is patent for some time after birth and may be used for exchange transfusions during early infancy (*e.g.*, in infants with *erythroblastosis fetalis* or hemolytic disease of the fetus, Behrman, 1992). Although reference is often made to the "obliterated" umbilical vein, it is usually patent to some extent. It often becomes very dilated if there is *portal hypertension* (raised pressure in the portal system, such as that caused by *cirrhosis of the liver* (p. 210).

Inferior to the umbilicus, five umbilical folds (two on each side and one in the median plane) pass superiorly toward the umbilicus. The *lateral umbilical folds*, formed by elevations of peritoneum covering the *inferior epigastric arteries*, run superomedially on each side (Figs. 2-7, 2-11, and 2-12). The *medial umbilical folds* are formed by elevations of peritoneum covering the *medial umbilical ligaments*, the obliterated parts of the fetal umbilical arteries. These vessels carried blood from the fetus to the placenta for oxygenation before birth (Moore, 1988). They ascend obliquely from the lateral walls of the pelvis to the umbilicus. The *median umbilical fold* is formed by an elevation of peritoneum covering the *median umbilical ligament*, the fibrous remnant of the *urachus* (Fig. 3-31*B*) that joined the fetal bladder to the umbilicus (Moore, 1988). The median umbilical ligament and fold pass from the deep aspect of the umbilicus to the apex of the urinary bladder (see Fig. 3-31). Of the five folds inferior to the umbilicus, *only the lateral umbilical folds contain vessels that carry blood*.

The Inguinal Region

The inguinal region is very important surgically because it is the site of *inguinal hernias* ("ruptures") in both sexes; however, they are much more common in males. The inguinal region is an area of weakness in the anterior abdominal wall, especially in males, owing to the prenatal penetration of the wall by the testis and spermatic cord.

The Inguinal Canal (Figs. 2-9 to 2-12, 2-16, and 2-17). This is *an oblique passage*, about 4 cm long in adults, through the inferior part of the anterior abdominal wall. It runs inferomedially, just superior and parallel to the medial half of the *inguinal ligament*. The inguinal canal has *two walls* (anterior and posterior), *two openings* (one at each end called the superficial and deep *inguinal rings*), a *roof* (superior wall), and a *floor* (inferior wall).

The anterior wall of the inguinal canal is formed mainly by the aponeurosis of the external oblique muscle. It is reinforced laterally by fibers of the internal oblique muscle; sometimes by those of the transversus abdominis muscle.

The posterior wall of the inguinal canal is formed throughout by the transversalis fascia, which is reinforced medially by the *conjoint tendon*, the common tendon of the internal oblique and transversus abdominis muscles.

The floor of the inguinal canal is formed by the superior surface of the inguinal ligament and the lacunar ligament (Fig. 2-16*B*).

The roof of the inguinal canal is formed by arching fibers of the internal oblique and transversus abdominis muscles. The *inferior epigastric artery* (Figs. 2-7 and 2-11) lies at the *medial boundary of the deep inguinal ring*; hence, its pulsations form a useful landmark during surgery for determining the location of this ring.

Owing to the obliquity of the inguinal canal, *the deep and superficial inguinal rings do not coincide*. Consequently, increases in intra-abdominal pressure act on the deep inguinal ring, forcing the posterior wall of the canal against the anterior wall. This strengthens this weak part of the anterior abdominal wall. The inguinal canal has been likened to an arcade of three arches formed by the three flat abdominal muscles. Contraction of the external oblique muscle approximates the anterior wall of the canal (formed mainly by the aponeurosis of the external oblique) to the posterior wall (formed mainly by the transversalis fascia). Contraction of the internal oblique and transversus abdominis muscles makes them taut; as a result, the roof of the canal descends and the passage is constricted. During standing, these muscles continuously contract.

During coughing and straining, the raised intra-abdominal pressure threatens to force some of the abdominal contents through the canal, producing a hernia. However, vigorous contraction of the arched fleshy fibers of the internal oblique and transversus abdominis muscles "clamp down." The action is like a half-sphincter that helps to prevent herniation (Fig. 2-16). Immedi-

ately posterior to the superficial inguinal ring is the conjoint tendon; the rectus abdominis muscle is posterior to the conjoint tendon (Figs. 2-9 and 2-10). When intra-abdominal pressure rises, the flat muscles of the abdomen contract, forcing the external oblique aponeurosis against the conjoint tendon, which then pushes against the rectus abdominis muscle. Hence, the conjoint tendon and rectus abdominis muscle reinforce the posterior surface of the superficial inguinal ring, tending to prevent herniation.

The Superficial Ring of the Inguinal Canal (Figs. 2-6, 2-9, 2-15, and 2-16A). Although it is called a ring, the superficial (external) inguinal ring is a more or less *triangular aperture* (deficiency) in the aponeurosis of the external oblique muscle. The base of this triangle is formed by the *pubic crest* and its apex is directed superolaterally. The sides of the triangle are formed by the *medial and lateral crura* (L. legs) of the superficial inguinal ring. Emerging from the superficial inguinal ring is the *spermatic cord* in the male (Fig. 2-6) and the *round ligament of the uterus* in the female (Fig. 2-18). In addition, the *ilioinguinal nerve* makes its exit through the ring to supply skin on the superomedial aspect of the thigh. The central point of the superficial inguinal ring is superior to the **pubic tubercle** (Figs. 2-1 and 2-3).

The *lateral crus of the superficial inguinal ring* is formed by the part of the external oblique aponeurosis that is attached to the pubic tubercle via the inguinal ligament. The spermatic cord rests on the inferior part of this crus (Figs. 2-6 and 2-9). The *medial crus of the superficial inguinal ring* is formed by the part of this aponeurosis that diverges to attach to the pubic bone and pubic crest, medial to the pubic tubercle. *Intercrural fibers* from the inguinal ligament arch superomedially across the superficial

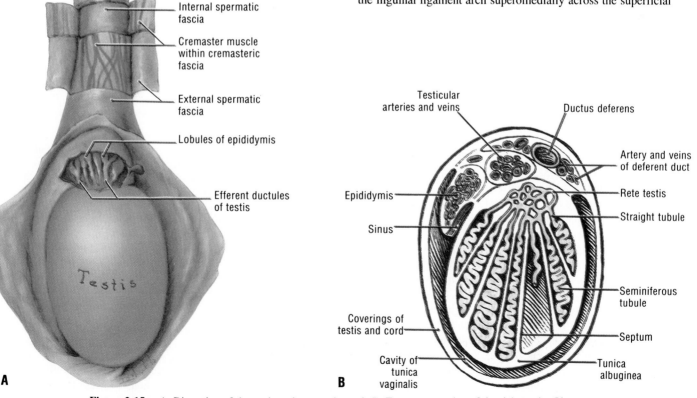

Figure 2-15. *A*, Dissection of the testis and spermatic cord. *B*, Transverse section of the right testis. Observe the cavity of the tunica vaginalis, which is artificially distended to show its visceral and parietal layers.

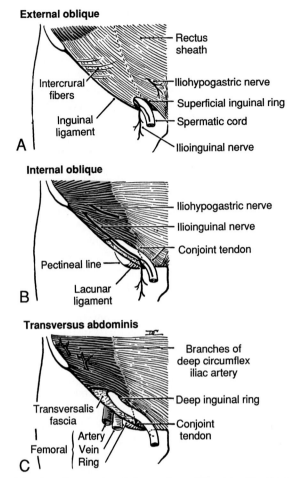

Figure 2-16. Progressive dissections of the three flat muscles of the anterior abdominal wall. Note that the anterior wall of inguinal canal is formed throughout by the external oblique aponeurosis and that its posterior wall is formed by the transversalis fascia and conjoint tendon.

inguinal ring. They prevent the crura from spreading apart (Figs. 2-6, 2-9, and 2-16A).

The superficial inguinal ring is palpable just superior and lateral to the pubic tubercle (Fig. 2-17). In men it can be examined by invaginating the skin of the scrotum with the tip of a digit (often the index finger), and probing gently superolaterally along the spermatic cord. If the ring is enlarged, it may admit the digit without causing pain. In women and children, the dimensions of the superficial inguinal ring are much less than in men and palpation of it is difficult. In male infants the superficial inguinal ring does not normally admit the tip of a digit.

The Deep Ring of the Inguinal Canal (Figs. 2-10 to 2-12, 2-15, and 2-16). This slitlike opening in the transversalis fascia is *located just lateral to the inferior epigastric artery.* This deep (internal) ring is immediately superior to the midpoint of the inguinal ligament and medial to the origin of the transversus abdominis muscle from the inguinal ligament (Table 2-1). The deep inguinal ring is the opening of a fingerlike diverticulum of the transversalis fascia (Fig. 2-16C). It formed prenatally when the *processus vaginalis* evaginated ("pushed through") the transversalis fascia (Moore, 1988). The margins of the deep ring are not sharply defined, as are those of the superficial ring. When the external oblique is reflected and the epigastric vessels are displaced, it ceases to exist as a ring; however, from the internal aspect, a dimple in the peritoneum often marks the site of the deep inguinal ring.

Descent of the Testes. To understand the inguinal canal, some knowledge of the migration and descent of the testes is essential. The testes *develop in the lumbar regions* deep to the transversalis fascia, between it and the peritoneum. They normally pass through the inguinal canals into the scrotum just before birth (Moore, 1988). The site of the inguinal canal in the fetus is first indicated by the *gubernaculum*, a ligament that extends from the testis through the anterior abdominal wall and inserts into the internal surface of the scrotum. Later, a fingerlike outpouching or diverticulum of peritoneum, called the *processus vaginalis*, follows the gubernaculum and evaginates the anterior abdominal wall to form the inguinal canal. The processus vaginalis pushes extensions of the layers of the anterior abdominal wall before it (Fig. 2-12). In males these prolongations of the layers of the anterior abdominal wall become the coverings of the spermatic cord (Figs. 2-12 and 2-15). In both sexes the opening produced by the processus vaginalis in the external oblique aponeurosis forms the *superficial inguinal ring* (Fig. 2-16A). The testes usually enter the inguinal canals just before birth and pass inferomedially through them to enter the scrotum. Normally the stalk of the processus vaginalis obliterates shortly after birth, leaving only the part surrounding the testis, which becomes the *tunica vaginalis* (Fig. 2-15*B*). The *scrotal ligament* is the adult derivative of the gubernaculum (Fig. 2-19*A*).

Maldescent of a testis (undescended testis or *cryptorchidism*) is a common abnormality. The testes are undescended in about 3% of full-term and 30% of premature infants. Undescended testes are usually located somewhere along the inguinal canal. Most undescended testes descend during the first few weeks after birth.

Descent of the Ovaries.   The ovaries also descend from their sites of origin on the posterior abdominal wall to a point just inferior to the pelvic brim (see Fig. 3-7); however, they do not normally enter the inguinal canals. The processus vaginalis normally obliterates completely and the gubernaculum attaches to the uterus, where it is divided into the *ligament of the ovary* and the round ligament of the uterus (Fig. 2-18; see also Fig. 3-44). The *round ligament of the uterus* passes through the inguinal canal and attaches to the internal surface of the labium majus (homologous to half of the scrotum). Persistence of the processus vaginalis in a female, called a *canal of Nuck* clinically, may result in an indirect inguinal hernia. Cysts in the inguinal canal and labium majus may also develop from remnants of the processus vaginalis.

Summary of the Inguinal Canal (Figs. 2-16 and 2-17).   The inguinal canal is an oblique passage through the inferior part of the anterior abdominal wall. The chief protection of the inguinal canal is muscular. Its main constituent is the *spermatic cord* in males and the *round ligament of the uterus* in females. It contains the *ilioinguinal nerve* in both sexes. It has an opening at each end, the deep and superficial inguinal rings. The *deep ring* is a slitlike opening in the transversalis fascia and the *superficial ring* is a triangular opening in the aponeurosis of the external oblique. The inguinal canal has two walls (anterior and posterior), a roof, and a floor. The *anterior wall* is formed mainly by the *aponeurosis of the external oblique muscle* and is reinforced laterally by fibers of the internal oblique. The *posterior wall* is formed mainly by the *transversalis fascia* and is reinforced medially by the conjoint tendon. The *floor* is formed by the inguinal and lacunar ligaments. The *roof* is composed of the arching fibers of the internal oblique and transversus abdominis muscles.

The Spermatic Cord

This cord, which suspends the testis in the scrotum, consists of structures running to and from the testis (Figs. 2-6, 2-7, 2-12, 2-15, and 2-19; Table 2-2). They are surrounded by protective coverings derived from the anterior abdominal wall. The spermatic cord *begins at the deep inguinal ring*, lateral to the inferior epigastric artery, where its constituents assemble, and *ends at the posterior border of the testis*.

It passes through the inguinal canal, emerges at the superficial inguinal ring, and descends within the scrotum to the testis. As the cord leaves the inguinal canal, it acquires its third covering,

the *external spermatic fascia*. It can be easily felt as a firm cord as it descends in the scrotum.

Constituents of the Spermatic Cord (Figs. 2-10, 2-11, 2-15, 2-19, 2-20, and 2-22).   Within the coverings of the cord are:

1. *The Ductus Deferens*. This large duct of the testis, formerly called the vas deferens, lies in the posterior part of the spermatic cord and is easily palpable because of its thick wall of smooth muscle.
2. *Arteries*. The testicular artery is a long slender vessel that arises from the anterior aspect of the aorta at the level of L2 vertebra (Fig. 2-39), where the testis developed in the embryo (Moore, 1988). The *testicular artery* is the main vessel supplying the testis and epididymis. The *artery of the ductus deferens* is a slender vessel that arises from the inferior vesical artery (Fig. 2-19; see also Fig. 3-24). It *accompanies the ductus deferens* throughout its course and anastomoses with the testicular artery near the testis. The *cremasteric artery* is a small vessel that arises from the inferior epigastric artery (Figs. 2-7 and 2-19). It *accompanies the spermatic cord* and supplies the cremaster muscle and other coverings of the spermatic cord. It also anastomoses with the testicular artery near the testis (Fig. 2-19*B*).
3. *Veins* (Figs. 2-15 and 2-19*C*). Up to 12 veins leaving the posterior surface of the testis anastomose to form a *pampiniform plexus*. This unusual name is derived from the Latin word *pampinus* meaning tendril, the spirally coiling organ of a climbing plant. This large *vinelike venous plexus*, forming a large part of the spermatic cord, surrounds the ductus deferens and arteries in the spermatic cord. It is located within the internal spermatic fascia and ends in the testicular vein (Fig. 2-105).
4. *Nerves*. There are sympathetic fibers on the arteries and sympathetic and parasympathetic fibers on the ductus deferens. These *autonomic sensory nerves* carry the impulses that produce deep visceral pain when the testis is squeezed or injured, producing excruciating pain and a sickening sensation. The genital branch of the *genitofemoral nerve* (Fig. 2-11) passes into the spermatic cord and supplies the cremasteric muscle.
5. *Lymph vessels* (Fig. 2-22). The lymph vessels draining the testis and immediately associated structures pass superiorly in the spermatic cord. These vessels end in the *lumbar* and *preaortic lymph nodes*, situated between the common iliac and renal veins.

Sometimes remnants of the stalk of the processus vaginalis persist, which may become swollen with fluid to form a *hydrocele of the spermatic cord*. The **pampiniform plexus of veins** sometimes becomes varicose (dilated and tortuous), producing a condition known as varicocele that feels like a ''bag of worms.'' *Varicoceles* (varicose veins of the spermatic cord) are more common on the left side and often result from defective valves in the testicular vein. The wormlike swelling disappears when the person lies down and the scrotum is elevated. Persons appearing to have two testes on one side usually have a varicocele or a testicular tumor.

Coverings of the Spermatic Cord (Figs. 2-6, 2-7, 2-9, 2-10, 2-12, and 2-15*A*).   The structures passing to and from the testis (ductus deferens and associated nerves and vessels), and constituting the *spermatic cord*, are surrounded by three layers

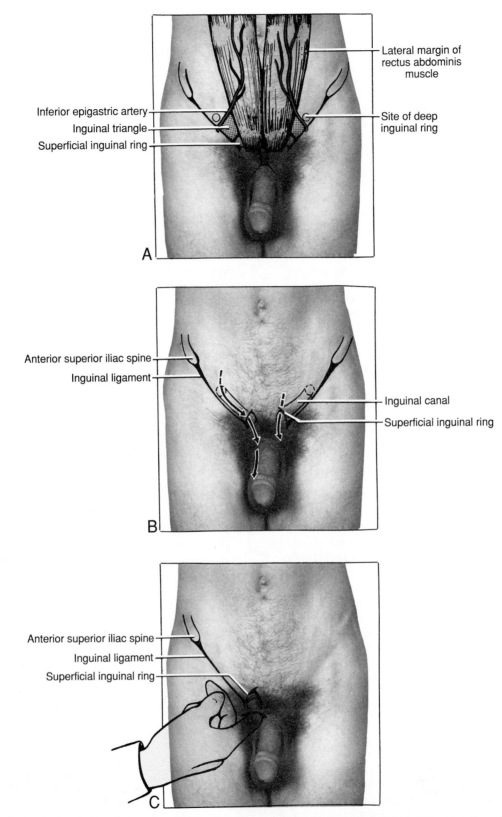

Figure 2-17.   The inguinal region, inguinal canals, and types of hernia. *A*, The boundaries of the inguinal triangle (*screened*): inguinal ligament inferiorly; inferior epigastric artery laterally; and rectus abdominis muscle medially. *B*, The course and presentation of inguinal hernias. The indirect inguinal hernia (right side) passes through the inguinal canal and enters the scrotum. The direct inguinal hernia (left side) bulges anteriorly through the inferior part of the inguinal triangle. *C*, Invagination of the scrotum during palpation of the superficial inguinal ring.

Table 2-2.
Corresponding Layers of the Anterior Abdominal Wall, Spermatic Cord, and Scrotum[1]

Layers of Anterior Abdominal Wall	Scrotum and Coverings of Testis	Coverings of Spermatic Cord
Skin	Skin	
Superficial fascia	Superficial fascia and dartos muscle	Scrotum[2]
External oblique aponeurosis	External spermatic fascia	External spermatic fascia
Internal oblique muscle	Cremaster muscle	Cremaster muscle
Fascia of internal oblique muscle	Cremasteric fascia	Cremasteric fascia
Transversus abdominis muscle		
Transversalis fascia	Internal spermatic fascia	Internal spermatic fascia
Extraperitoneal tissue		
Peritoneum	Tunica vaginalis	

[1]Also see Figures 2-12 and 2-15.
[2]The scrotum consists of skin and superficial fascia. The dartos muscle is in the superficial fascia.

of fascia derived from the anterior abdominal wall (Table 2-2). These coverings are not easily separated from one another.

The Internal Spermatic Fascia (Figs. 2-12 and 2-15*A*). As the processus vaginalis evaginated the transversalis fascia at the deep inguinal ring, it carried a thin layer of fascia before it that became the internal spermatic fascia. It constitutes the filmy innermost covering of the spermatic cord.

The Cremaster Muscle and Cremasteric Fascia (Figs. 2-9, 2-10, and 2-15*A*). As the processus vaginalis, with its covering of transversalis fascia (future internal spermatic fascia), evaginated under the edge of the internal oblique muscle, it acquired a few of this muscle's fibers and some of its investing fascia. These fibers and fascia form the cremaster muscle and cremasteric fascia, respectively. The cremasteric fascia forms the middle covering of the spermatic cord, which contains loops of the **cremaster muscle**. This muscle, which is continuous with the internal oblique, *reflexly draws the testis to a more superior position in the scrotum*, particularly in cold temperatures. Contraction of the cremaster can be produced by lightly stroking the skin on the medial aspect of the superior part of the thigh, the area supplied by the *ilioinguinal nerve* (Fig. 2-13*A*). This results in contraction of the cremaster muscle supplied by the genital branch of the genitofemoral nerve (L1 and L2). The reflex raising of the testis is called the **cremasteric reflex**. The testis, located outside the pelvis in the scrotum, is sensitive to cold; hence, the cremaster muscle draws it superiorly in the scrotum for warmth and protection against injury.

The External Spermatic Fascia (Figs. 2-12 and 2-15*A*). As the processus vaginalis evaginated the external oblique aponeurosis, forming the superficial inguinal ring, it carried an extension

of this aponeurosis before it. This layer became the external spermatic fascia, the thin outermost covering of the spermatic cord. This fascia, attached superiorly to the crura of the superficial inguinal ring, is continuous with the deep fascia covering the external oblique muscle (Figs. 2-6 and 2-7).

Testing the cremasteric reflex is part of a complete physical examination in male patients because abdominal and cremasteric reflexes may be absent in both upper and lower motor neuron disorders. The *cremasteric reflex* is much more active in children. During childhood, hyperactive cremasteric reflexes may simulate undescended testes. The reflex can be abolished by having the child sit in a cross-legged squatting position. If the testis is normal, it can be palpated in the scrotum. The cremasteric reflex is often sluggish or absent in older men. Unilateral absence of this reflex is not, in itself, indicative of neurological disease.

A hernia is a protrusion of a structure, viscus, or organ from the cavity in which it belongs. Some lay people refer to a hernia as a *rupture*, indicating that it is like the "blowout" of a tire caused by pressure of the contents on a weak area. Because the scrotum and the layers within it represent outpouchings of the anterior abdominal wall (Fig. 2-12), hernias through the abdominal wall in the inguinal region or into the scrotum are particularly common in males. The labia majora in females are homologous with the scrotum, but they consist mostly of fat (Fig. 2-18). Hence, the potential weak area of the abdominal wall is small and congenital inguinal hernia is much less common in females.

Failure of the stalk of the processus vaginalis to obliterate leads to a number of extremely common surgical conditions, of which *indirect inguinal hernia* is the most common and important. An inguinal hernia typically contains part of a viscus, most commonly the small intestine. *There are two types of inguinal hernia*: indirect and direct. Indirect inguinal hernia is the more common type at all ages (75% of cases) and in both sexes. As its name indicates, an indirect hernia takes an indirect course through the anterior abdominal wall.

An Indirect Inguinal Hernia (Fig. 2-17*B*). This hernia leaves the abdominal cavity *lateral to the inferior epigastric vessels*. It traverses the deep inguinal ring, inguinal canal, and superficial inguinal ring. Consequently, an indirect (oblique) inguinal hernia is covered by all three layers of the spermatic cord. *The hernial sac of an indirect inguinal hernia is the remains of the processus vaginalis*, which normally obliterates, except for the part that forms the tunica vaginalis (Fig. 2-15*B*). If the entire stalk of the processus vaginalis persists, the hernia is complete and extends into the scrotum. Indirect inguinal hernia is about 20 times more common in males than in females. However, if the processus vaginalis persists in a female (canal of Nuck), a hernia may enter it and follow it through the inguinal canal into the labium majus.

There are two types of indirect inguinal hernia: congenital and acquired. Although the term congenital means "present at birth," the herniation of a loop of bowel into a patent processus vaginalis may not be present at birth; the herniation may occur during infancy, childhood, or adulthood. In other words, the potential for a *congenital inguinal hernia* is present at birth if the patent processus vaginalis does not obliterate. When herniation occurs late in life, it is probable that a small remnant of the processus vaginalis was present at birth. Herniation then occurred following *lifting*, pushing, coughing, or straining during urination or defecation.

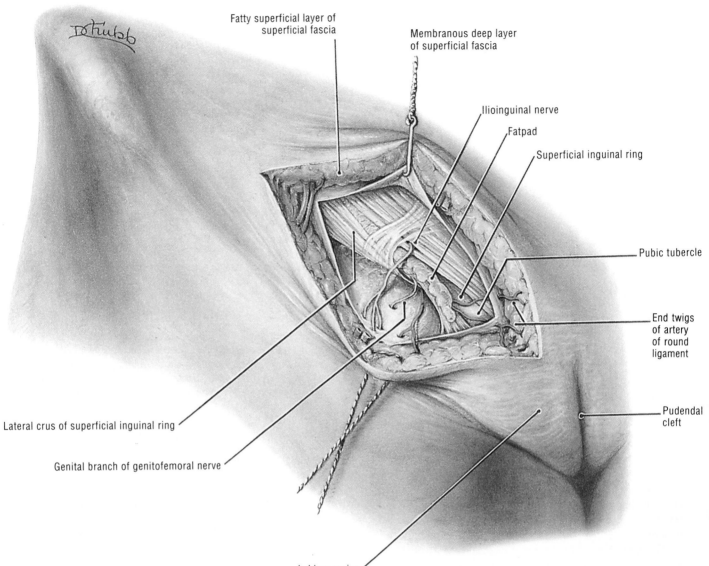

Fatty superficial layer of
superficial fascia

Membranous deep layer
of superficial fascia

Ilioinguinal nerve

Fatpad

Superficial inguinal ring

Pubic tubercle

End twigs
of artery
of round
ligament

Pudendal
cleft

Lateral crus of superficial inguinal ring

Genital branch of genitofemoral nerve

Labium majus

Figure 2-18.   Dissection of the inguinal canal in a woman. Note the structures issuing from the superficial inguinal ring: (1) the round ligament of the uterus, (2) a pad of fat, (3) the genital branch of the genitofemoral nerve, and (4) the artery of the round ligament. This artery is homologous with the cremasteric artery in the male (see also Fig. 2-19).

Direct Inguinal Hernia (Fig. 2-17).   This hernia *protrudes anteriorly* through the posterior wall of the inguinal canal and leaves the abdominal cavity *medial to the inferior epigastric vessels*. It passes through some part of the *inguinal triangle* (Hesselbach's triangle), usually its inferior part. This triangle is limited (bounded) by the inguinal ligament inferiorly, the inferior epigastric artery laterally, and the rectus abdominis muscle medially. The **inguinal triangle** lies just posterolateral to the superficial inguinal ring. It includes the area of the posterior wall of the inguinal canal that is formed by transversalis fascia only. Consequently, the inguinal triangle is a potential weak area of the anterior abdominal wall and the potential site of a direct inguinal hernia. Normally the gap between the conjoint tendon and the inguinal ligament is obliterated when the anterior abdominal muscles contract during coughing or straining. This action of the muscle is called the "shutter action."

The hernial sac of a direct inguinal hernia is composed of peritoneum. The hernia does not pass through the deep inguinal ring, but emerges through or around the *conjoint tendon* to reach the superficial inguinal ring. If the hernia passes lateral to the conjoint tendon, it pushes the peritoneum and transversalis fascia ahead of it and emerges through the superficial inguinal ring, either superior or inferior to the spermatic cord. If the hernia passes through the weakened fibers of the conjoint tendon, it is covered by peritoneum, transversalis fascia, and fibers of the conjoint tendon. Direct inguinal hernia is much less common than indirect inguinal hernia and it usually occurs in men over 40 years of age. This type of hernia is uncommon in women. *Direct inguinal hernia usually results from weakening of the anterior abdominal muscles*, which reduces their "shutter action" in closing the gap between the conjoint tendon and the inguinal ligament. When a weakened condition of the muscles exists, repeated increased intra-abdominal pressure—as occurs in lifting, pushing, coughing and straining—may induce herniation.

Some people have difficulty remembering the difference between direct and indirect inguinal hernias because they think "direct" means the hernia goes directly through the inguinal canal. This is wrong. "*Direct*" means the hernia passes directly through the anterior abdominal wall. Hence, a direct inguinal hernia bypasses the deep inguinal ring by protruding through the posterior wall of the canal. A direct hernia may produce a general bulging of the abdominal wall, or the hernial sac may emerge from the superficial inguinal ring and turn superiorly. "*Indirect*" means that the hernia passes indirectly through the anterior abdominal wall by entering the deep inguinal ring, passing through the inguinal canal, and emerging through the superficial inguinal ring.

The Scrotum and Testes

The scrotum and testes are considered here because their development is closely related to the inguinal canal.

The Scrotum (Figs. 2-12, 2-15, 2-17, and 2-19).   The scrotum develops from two cutaneous outpouchings of the anterior abdominal wall, called *labioscrotal swellings*, which normally fuse to form a pendulous cutaneous pouch (Moore, 1988). Later the *testes* and *spermatic cords* pass into it. The scrotum consists of two layers, skin and superficial fascia (Table 2-2). The thin skin is dark colored and rugose (wrinkled), especially in virile men. The scrotum is divided on its surface into right and left halves by a *scrotal raphe* (see Fig. 3-54). This cutaneous ridge indicates the bilateral origin of the scrotum from the labioscrotal swellings. The superficial fascia is devoid of fat, but it contains a thin sheet of smooth muscle called the *dartos muscle*. Because its fibers are attached to the skin, contraction of them causes the scrotum to wrinkle when cold, which helps to regulate the loss of heat through its skin. This is important because *spermatogenesis* (formation of sperms) requires a controlled temperature. The superficial fascia of the scrotum is continuous anteriorly with the membranous deep layer of the superficial fascia of the anterior abdominal wall (Fig. 2-12 and p. 131), and posteriorly with the superficial fascia of the perineum (see Chap. 3). The superficial fascia, including the dartos muscle, forms an incomplete scrotal septum that divides the scrotum into right and left halves, one for each testis.

The coverings of the testis are continuous with the coverings of the spermatic cord (Figs. 2-12 and 2-15A). The outermost covering of the testis, the *external spermatic fascia*, is continuous with this layer of the spermatic cord (Table 2-2), which is continuous with the external oblique aponeurosis at the superficial ring. Internal to this layer is the *cremaster muscle* and *cremasteric fascia*, and inside this layer is the *internal spermatic fascia*, which is continuous with this covering of the spermatic cord and the transversalis fascia (Fig. 2-12). Inside the internal spermatic fascia is the *tunica vaginalis testis* (Fig. 2-15B). This closed serous sac of peritoneum is a derivative of the *processus vaginalis* (p. 144). The **tunica vaginalis** is a peritoneal sac surrounding the testis, which was cut off from the greater peritoneal sac before birth (Fig. 2-15B). It consists of two layers: the *parietal layer* is adjacent to the internal spermatic fascia and the *visceral layer* is adherent to the testis and epididymis (Figs. 2-19 and 2-20). Laterally the visceral layer of the tunica vaginalis passes between the testis and the epididymis to form the *sinus of the epididymis* (Figs. 2-15B and 2-19A). A very small amount of fluid normally separates the visceral and parietal layers of the tunica vaginalis. This enables the testis to move freely within the scrotum.

Arterial Supply and Venous Drainage of the Scrotum (Figs. 2-6, 2-7, 2-10, and 2-11; see also Fig. 3-24).   The skin and dartos muscle of the scrotum are supplied by the perineal branch of the *internal pudendal artery* and by the external pudendal branches of the *femoral artery*. The scrotum is also supplied by the *cremasteric artery*, a branch of the inferior epigastric artery. The *scrotal veins* accompany the arteries. The external pudendal veins enter the *great saphenous vein*.

Innervation of the Scrotum (Figs. 2-6, 2-7, 2-10, 2-11, 2-13, and 2-16; see also Fig. 3-61).   Several nerves supply the scrotum. The genital branch of the *genitofemoral nerve* sends sensory branches to the anterior and lateral surfaces of the scrotum; *it also supplies the cremaster muscle*. The anterior surface of the scrotum is also supplied by scrotal branches of the *ilioinguinal nerve*. The perineal branch of the *pudendal nerve* supplies the posterior surface of the scrotum, and perineal branches of the *posterior femoral cutaneous nerve* supply the inferior surface of the scrotum.

Lymph Vessels of the Scrotum (Figs. 2-9 and 2-22).   The lymph vessels ascend in the superficial fascia of the scrotum and drain into the *superficial inguinal lymph nodes*.

The Testes and Their Ducts (Figs. 2-15 and 2-19 to 2-21).   The testes (testicles), the main male reproductive organs, are paired ovoid glands that are suspended in the scrotum by the spermatic cords. The surface of each testis is covered by the visceral layer of the tunica vaginalis, except where it is attached to the epididymis and spermatic cord. Internal to this layer is the *tunica albuginea*, the connective tissue coat of the testis. The testes produce male germ cells or *sperms* (spermatozoa) and male sex hormones called *androgens*. The sperms are formed in the *seminiferous tubules* of the testes, which join to form a network of canals known as the *rete testis* (Fig. 2-15B). Small *efferent ductules* (15 to 20) connect the rete testis to the head of the epididymis.

The Epididymis (Figs. 2-15 and 2-19 to 2-21).   This comma-shaped structure is applied to the superior and posterolateral surfaces of the testis. Its superior expanded part, called the *head*, is composed of the *lobules of the epididymis*, which are the coiled ends of *efferent ductules* of the testis. These ductules transmit the sperms from the testis to the epididymis. The *body of the epididymis* consists of the highly convoluted *duct of the epididymis*. The sperms are stored in the epididymis where they undergo the final stages of maturation. The *tail of the epididymis* is continuous with the *ductus deferens* (vas deferens), the duct which transports the sperms from the epididymis to the *ejaculatory duct* for expulsion into the *prostatic urethra* (see Figs. 3-29 and 3-30).

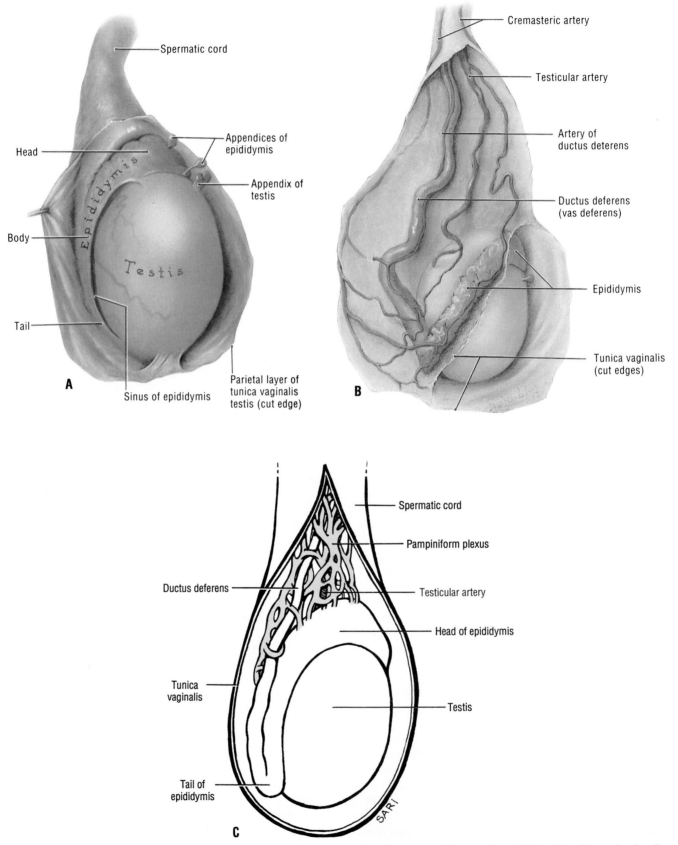

Figure 2-19. *A*, Lateral view of a dissection of the testis. The tunica vaginalis has been incised longitudinally and reflected. *B*, Dissection of the testis and spermatic cord to show the blood supply. Note the an-astomosis between the three arteries. *C*, Dissection of the testis primarily to show the pampiniform plexus of veins.

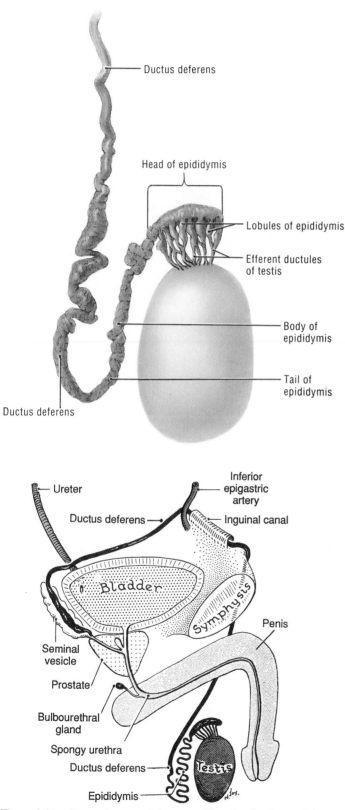

Figure 2-20. The testis, epididymis, and ductus deferens after removal of their coverings (shown in Fig. 2-19). Observe the efferent ductules that pass from the rete testis to the head of the epididymis. These ductules become coiled to form the lobules of the epididymis, which are part of the head.

Ductus deferens

Head of epididymis

Lobules of epididymis

Efferent ductules
of testis

Body of
epididymis

Tail of
epididymis

Ductus deferens

Ureter

Inferior
epigastric
artery

Ductus deferens

Inguinal canal

Bladder

Symphysis

Penis

Seminal
vesicle

Prostate

Bulbourethral
gland

Spongy urethra

Ductus deferens

Testis

Epididymis

Figure 2-21. The male urogenital system. Note that the ductus deferens passes superiorly from the testis, enters the superficial inguinal ring, and passes through the inguinal canal. Sperms (spermatozoa) are produced in the testis and enter the epididymis where they are stored and undergo maturation. They are then propelled through the ductus deferens and ejaculatory duct into the urethra.

Because the anterior third of the scrotum is supplied mainly by the L1 segment of the spinal cord via the ilioinguinal and genitofemoral nerves and the posterior two-thirds of the scrotum are supplied mainly by the S3 segment through the perineal and posterior femoral cutaneous nerves, a spinal anesthetic agent must act more superiorly to anesthetize the anterior surface of the scrotum.

The presence of excess fluid in the processus vaginalis after birth is called a *hydrocele*. The size of the hydrocele depends on how much of the processus vaginalis is patent. Infants with a normally obliterated processus vaginalis may have residual peritoneal fluid in the cavity of their tunica vaginalis (Fig. 2-15*B*), but this fluid usually is absorbed during the first year. Certain pathological conditions (*e.g.*, injury and/or inflammation of the epididymis) may result in an increase in the fluid in the tunica vaginalis, producing a *hydrocele of the testis* and scrotal enlargement. A *hematocele of the testis* is a collection of blood in the cavity of the tunica vaginalis, often resulting from trauma to the scrotum and testis. This damages the vessels around the testis (Fig. 2-19*B*) and blood enters the tunica vaginalis.

Cells from a **testicular tumor** may spread by lymphogenous dissemination to the *lumbar lymph nodes*, whereas cancer of the scrotum metastasizes to the *superficial inguinal lymph nodes* (Figs. 2-9 and 2-22). Rudimentary structures may be observed around the testis and epididymis when the tunica vaginalis is opened (Fig. 2-19). These are small vestigial remnants of the genital ducts in the embryo (Moore,

Figure 2-22. Lymphatic drainage of the scrotum and testes. *A*, The testicular lymph vessels ascend in the spermatic cord and end in the lumbar and preaortic lymph nodes. The lymph vessels of the scrotum drain into the superficial inguinal lymph nodes shown in *A* and *B*.

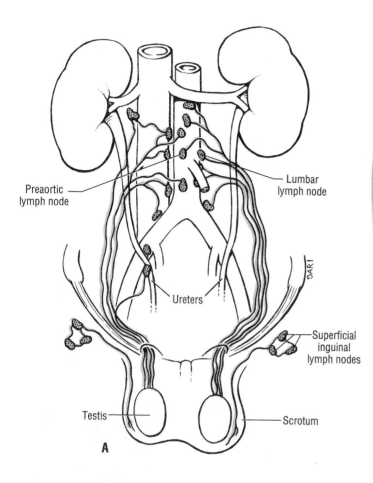

Preaortic lymph node

Lumbar lymph node

Ureters

Superficial inguinal lymph nodes

Testis

Scrotum

A

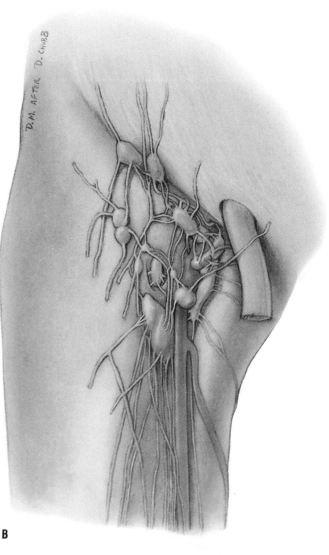

B

1988). They are rarely observed, unless pathological changes occur in them. The *appendix of the testis* is a vesicular remnant of the cranial end of the paramesonephric duct (embryonic female genital duct). It is attached to the superior pole of the testis (Fig. 2-19*A*). The *appendix of the epididymis* is the remnant of the cranial end of the mesonephric duct (embryonic male genital duct), which is attached to the head of the epididymis. This duct and the mesonephric tubules associated with it normally form the efferent ductules and epididymis. Unused parts normally degenerate but they may persist and form appendices of the testis and epididymis.

The *ductus deferens* (vas deferens) is ligated bilaterally when sterilizing a man. To perform this operation, called a **deferentectomy** or *vasectomy*, the ductus deferens is isolated on each side by incising the superior wall of the scrotum (Farrow, 1989). Following deferentectomy, sperms can no longer pass to the urethra (Fig. 2-21), so they degenerate in the epididymis and ductus deferens. However, the secretions of the auxiliary genital glands (*e.g.*, the seminal vesicles) can still be ejaculated.

The Peritoneum and Peritoneal Cavity

The peritoneum is a thin, *transparent serous membrane* that consists of two layers. The peritoneum lining the abdominal wall is called *parietal peritoneum* (Fig. 2-12), whereas the peritoneum investing the viscera is called *visceral peritoneum* (Fig. 2-23). Both types of peritoneum consist of a single layer of squamous epithelium called *mesothelium*. The parietal and visceral layers of peritoneum are separated from each other by a capillary film of *peritoneal fluid*. This serous fluid lubricates the peritoneal surfaces, enabling the viscera to move on each other without friction.

The disposition of the peritoneum in the adult appears meaningless and complex if its developmental changes and

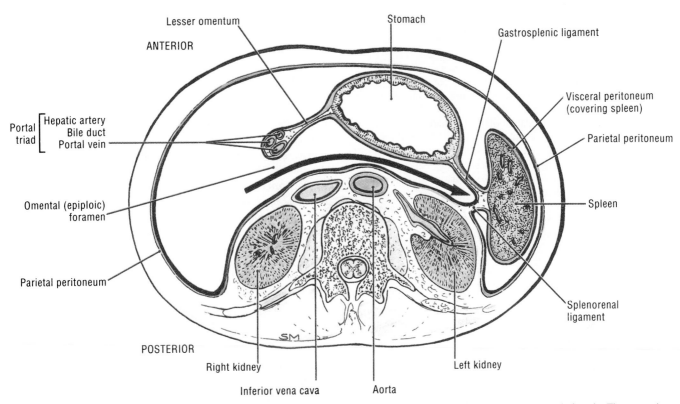

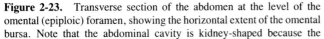

Figure 2-23. Transverse section of the abdomen at the level of the omental (epiploic) foramen, showing the horizontal extent of the omental bursa. Note that the abdominal cavity is kidney-shaped because the vertebral column and the great vessels protrude into it. The arrow indicates the path taken by a digit passed through the omental (epiploic) foramen into the omental bursa to reach the hilum of the spleen.

modifications are not considered. In the early embryo the peritoneum is a large sac that lines the walls of the abdominal cavity. The primordia (beginnings) of the viscera are located outside this sac in the extraperitoneal tissue (Table 2-2). Later in the embryonic period, as the viscera develop, they protrude into the peritoneal sac to varying degrees. Some organs such as the kidneys protrude only slightly (Fig. 2-23); they are called *retroperitoneal organs*. Other organs, such as the ascending colon, protrude further into the peritoneal sac and are covered on each side with peritoneum (Fig. 2-24C). Organs such as the stomach and jejunum protrude completely into the peritoneal sac and are completely covered with visceral peritoneum. The peritoneum forms the *serosa of the walls* of these organs. They are also attached to the posterior abdominal wall by two apposed layers of peritoneum, called a *mesentery* (Figs. 2-23 and 2-24A).

When an organ protrudes into the peritoneal sac, it takes its vessels and nerves with it. They are located between the two layers of peritoneum forming the mesentery. There is also loose connective tissue between these layers that contains a variable amount of fat cells. *Viscera with a mesentery are mobile*, the degree of which depends on the length of the mesentery. As the developing organs enlarge, they obliterate the peritoneal cavity almost completely. Hence, the peritoneal cavity after birth is only a potential space with a small amount of *peritoneal fluid* between the parietal and visceral layers. As the fetal organs assume their adult positions, the *peritoneal cavity* is divided into two peritoneal sacs, the *greater and lesser sacs* of peritoneum. A surgical incision through the anterior abdominal wall enters the *greater peritoneal sac*. The lesser sac, known as the **omental bursa**, lies posterior to the

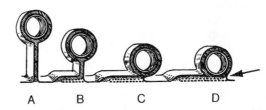

Figure 2-24. Diagrams illustrating the primitive mesentery of the large intestine in various stages of absorption. The arrow indicates the paracolic gutter where the visceral peritoneum is attached to the parietal peritoneum, and where an incision is made during mobilization of the intestine prior to surgery (*e.g.*, on the ascending colon). Sometimes the ascending and descending colon have a short mesentery like that shown in *B*.

stomach, lesser omentum, and liver (Figs. 2-23 and 2-28). The peritoneal cavity is closed in males, but in females there is communication with the exterior through the uterine tubes, uterus, and vagina.

If the mesothelium forming the peritoneum is damaged or removed in any area (*e.g.*, during surgery), the two layers of peritoneum may adhere to each other forming an **adhesion** (L. *adhaesio*, to stick to). Adhesions interfere with the normal movements of the viscera. The cutting or division of adhesions at a surgical operation is called an *adhesiotomy*. Commonly during dissection, the apposing layers of peritoneum are adherent (Fig. 2-35). These adhesions (strands of fibrous tissue) probably resulted from an inflammatory process (*e.g.*, peritonitis). You can usually break these adhesions down with your digits (fingers).

Under certain pathological conditions, the peritoneal cavity may be distended to form a large space containing several liters of fluid. **Ascites** (hydroperitoneum) is an accumulation of serous fluid in the peritoneal cavity. Widespread metastases of cancer cells cause exudation of ascitic fluid that is often blood stained and contains cancer cells. Fluid in the peritoneal cavity represents an imbalance between the rate of fluid formation and its absorption by the peritoneum. Patients with advanced liver disease (*e.g.*, *cirrhosis*) usually have ascites.

If the intestine ruptures as the result of a penetrating wound (*e.g.*, caused by a knife) or if a closed abdominal injury results from, say an automobile accident, gas and intestinal material enter the peritoneal cavity. *GI rupture* results in free air located inferior to the diaphragm, which can be detected by abdominal radiographs.

Various terms are used to describe different parts of the peritoneum and peritoneal cavity. These are: mesentery, omentum, ligament, fold, and recess.

Mesentery (Figs. 2-23 and 2-24). This is a *double layer of peritoneum* that encloses an organ and connects it to the abdominal wall. Mesenteries have a core of loose connective tissue containing a variable number of fat cells and lymph nodes, together with vessels and nerves passing to and from the viscus. The mesentery of the stomach is called the *mesogastrium* (G. *gaster*, stomach) and the mesentery of the transverse colon is called the *transverse mesocolon*. The most mobile parts of the intestine have a mesentery (*e.g.*, the transverse colon and most parts of the small intestine). Some viscera have no mesentery and are extraperitoneal or retroperitoneal (*e.g.*, the ascending colon and kidneys). These organs lie on the posterior abdominal wall and are covered by peritoneum anteriorly.

Omentum (Figs. 2-23 and 2-25). This is a *double-layered sheet or fold of peritoneum*. The lesser omentum and the greater omentum attach the stomach to the body wall or to other abdominal organs.

The Lesser Omentum (Figs. 2-26 to 2-28 and 2-67). This fold of peritoneum connects the lesser curvature of the stomach and the proximal part of the duodenum to the liver. Individually these connections are referred to as the *hepatogastric ligament* and the *hepatoduodenal ligament*. The Greek words for the liver and stomach are *hepar* and *gaster*, respectively. The lesser omentum lies posterior to the left lobe of the liver and is attached to the liver in the *fissure for the ligamentum venosum* (the remnant of the fetal ductus venosus; Moore, 1988). It is also attached to the *porta hepatis*, the transverse fissure or gate (L. *porta*) on the inferior surface of the liver through which the bile duct, vessels, and nerves enter or leave the liver.

The Greater Omentum (Figs. 2-25 to 2-28). This fat-laden fold of peritoneum hangs down from the greater curvature of the stomach and connects the stomach with the diaphragm, spleen, and transverse colon. This double-layered peritoneal fold normally fuses during the fetal period, thereby obliterating the inferior recess of the omental bursa. As a result, the apronlike greater omentum is *composed of four layers of peritoneum*. They can be partially separated in adults, which is commonly per-

formed as a surgical entrance to the omental bursa (p. 169) and the posterior aspect of the stomach. The greater omentum contains a variable amount of extraperitoneal tissue and fat. In emaciated persons, this omentum may be as thin as a piece of paper, whereas in obese persons, it is of considerable thickness and weight. It is usually long enough to reach the pelvis, but in cadavers it is often contracted so that it barely covers the intestine. After passing inferiorly as far as the pelvis, the greater omentum loops back on itself, overlying and attaching to the transverse colon, which runs across the abdomen just inferior to the stomach (Figs. 2-27 and 2-28).

Peritoneal Ligament (Fig. 2-25). A peritoneal ligament is a double layer of peritoneum that connects an organ (*e.g.*, the liver) with another organ or with the abdominal wall. Ligaments may contain blood vessels or remnants of vessels (*e.g.*, the falciform ligament contains the ligamentum teres, a remnant of the fetal umbilical vein). For descriptive purposes, the greater omentum is divided into three parts: (1) The apronlike part, called the *gastrocolic ligament*, is attached to the transverse colon; (2) the left part, called the *gastrosplenic ligament* (gastrolienal ligament), is attached to the spleen (G. *splen*) and connects the spleen to the greater curvature of the stomach; and (3) the superior part, called the *gastrophrenic ligament*, is attached to the diaphragm (G. *phren*).

Peritoneal Folds (Fig. 2-12). A peritoneal fold (L. *plica*) is a reflection of peritoneum with more or less sharp borders. Often it is formed by peritoneum that covers blood vessels, ducts, and obliterated fetal vessels (*e.g.*, the medial and lateral *umbilical folds*).

Peritoneal Recesses (Figs. 2-28 and 2-65). In certain places the peritoneum folds to form blind pouches (cul-de-sacs) or tubular cavities that are closed at one end with an opening into the peritoneal cavity. The duodenojejunal area often has two or three peritoneal recesses. In the ileocecal area there is a *retrocecal recess* where the peritoneal cavity extends superiorly, posterior to the cecum. Frequently the vermiform appendix lies in this recess.

Subdivisions of the Peritoneal Cavity

Lifting the inferior end of the apronlike greater omentum reveals that this large fold of peritoneum, together with the transverse colon and transverse mesocolon, forms a shelf that subdivides the peritoneal cavity into supracolic and infracolic compartments.

The Supracolic Compartment (Figs. 2-25, 2-28C, and 2-65). This compartment is divided into smaller parts or spaces by the *falciform ligament*. The right and left anterior *subphrenic recesses* are located between the inferior surface of the diaphragm and the superior surfaces of the right and left lobes of the liver. The *hepatorenal recess* is located between the inferior surface of the right lobe of the liver and the right kidney.

The Infracolic Compartment (Fig. 2-28C). This compartment is divided into right and left *infracolic spaces* by the mesentery of the small intestine. Owing to the obliquity of the mesentery, the right infracolic space is superior and to the right and the left infracolic space is inferior and to the left.

Related to the ascending and descending colon are longitu-

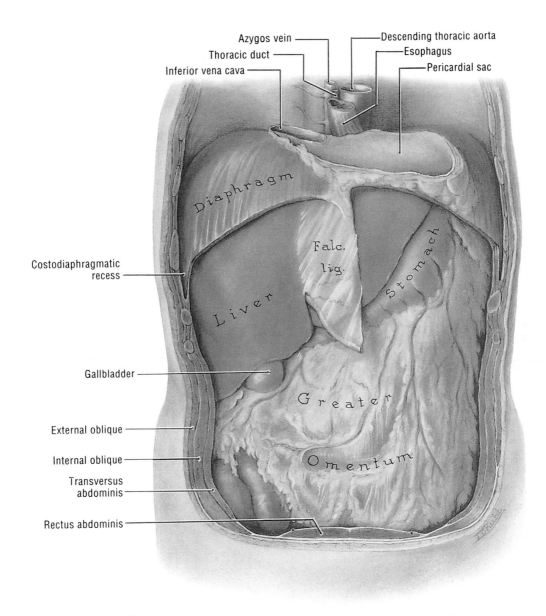

Azygos vein

Thoracic duct

Inferior vena cava

Descending thoracic aorta

Esophagus

Pericardial sac

Costodiaphragmatic recess

Diaphragm

Falc. lig.

Liver

Stomach

Gallbladder

Greater

External oblique

Internal oblique

Transversus abdominis

Omentum

Rectus abdominis

Figure 2-25.   The abdominal contents after removal of the thoracoabdominal wall.

dinal depressions or channels, called *paracolic gutters* or grooves (Figs. 2-24*D* and 2-28*C*). There are medial and lateral paracolic gutters on the right (associated with the ascending colon) and the left (associated with the descending colon). The *right lateral paracolic gutter* is of particular surgical significance because it is continuous superiorly with the *hepatorenal recess* (Fig. 2-65), and beyond this with the *omental bursa* and its superior recess (Fig. 2-28*A*). Inferiorly this gutter is continuous with the rectovesical pouch in the male (see Fig. 3-33) and the *rectouterine pouch* in the female (see Fig. 3-34). On each side of the vertebral column is a *paravertebral gutter* or groove containing a kidney, a ureter, and a part of the colon.

The paracolic gutters or grooves slope posterosuperiorly so that when fluid accumulates in the peritoneal cavity, it

follows these gutters to the superior part of the abdomen when a person is in the supine position.

The greater omentum prevents the *visceral peritoneum* covering the intestine from adhering to the *parietal peritoneum* lining the anterior abdominal wall. It has considerable mobility and can migrate to any area of the abdomen and wrap itself around an inflamed organ such as the appendix. Because it "walls off" and *protects an infected organ from other viscera*, it has been called the "policeman of the abdomen."

The division of the peritoneal cavity into peritoneal recesses is of clinical importance in connection with the spread of pathological fluids (*e.g.*, pus). The subdivisions determine the extent and direction of the spread of fluids (*e.g.*, blood), which may be present in the peritoneal cavity when an organ

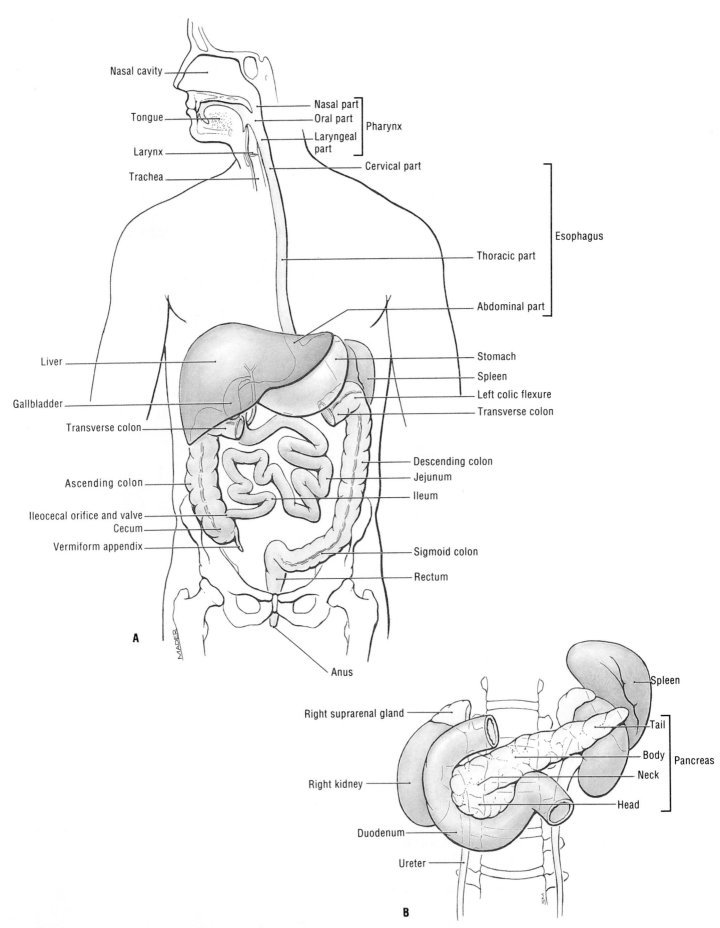

Figure 2-26. The digestive system and its relationship to the upper respiratory system. *A*, The digestive tract extends from the mouth to the anus. *B*, The spleen, pancreas, kidneys, and duodenum after removal of the stomach, liver, and colon.

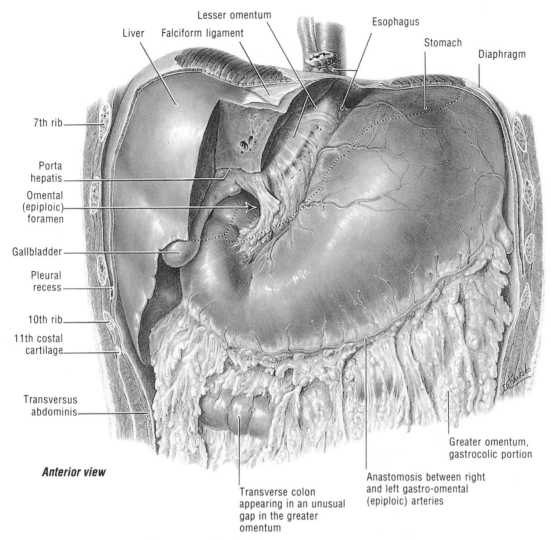

Lesser omentum

Liver   Falciform ligament

Esophagus

Stomach

Diaphragm

7th rib

Porta hepatis

Omental (epiploic) foramen

Gallbladder

Pleural recess

10th rib

11th costal cartilage

Transversus abdominis

Anterior view

Transverse colon appearing in an unusual gap in the greater omentum

Anastomosis between right and left gastro-omental (epiploic) arteries

Greater omentum, gastrocolic portion

Figure 2-27. Dissection of the stomach and omenta associated with it. The stomach is inflated with air and the left part of the liver is cut away. Observe that the first or superior part of the duodenum almost occludes the omental foramen (*arrowhead*).

is diseased or injured. Normally these recesses are in communication with each other, but they may become separated from each other by adhesions between the peritoneum and viscera (Fig. 2-35).

The Omental Bursa

The omental bursa (*lesser sac of peritoneum*) is the large compartment or recess of the peritoneal cavity that is located between the stomach and the posterior abdominal wall (Figs. 2-23, 2-27, and 2-28). The omental bursa is also located posterior to the lesser omentum and the stomach. As the anterior and posterior walls of the bursa slide smoothly on each other, the omental bursa gives considerable movement to the stomach, permitting it to slide freely on the bursa during contraction and distention. The omental bursa is an extension of the main peritoneal cavity into the invaginated right side of the dorsal mesentery of the stomach (gastric mesentery). The inferior extension

of the omental bursa, called the *inferior recess* (Fig. 2-28A), is between the duplicated layers of the gastrocolic ligament of the greater omentum. *In adults the inferior recess of the omental bursa is a potential space*; it is usually shut off from the main part of the bursa owing to adhesion of the layers of the gastrocolic ligament (Fig. 2-28B). The omental bursa also has a *superior recess* (Fig. 2-28A), which is limited superiorly by the diaphragm and the posterior layers of the *coronary ligament* (Fig. 2-48). The omental bursa is in communication with the main peritoneal cavity (greater sac of peritoneum) through the **omental foramen** (epiploic foramen or foramen of Winslow), located posterior to the free edge of the lesser omentum (Figs. 2-27 and 2-28A). The omental (epiploic) foramen is usually large enough to admit two digits.

The Latin term *omentum*, meaning "fat skin," is widely used, whereas the Greek term *epiploon* for the omentum is unfamiliar to most people. Consequently the adjective "epiploic" has been replaced by "omental" in the current list of

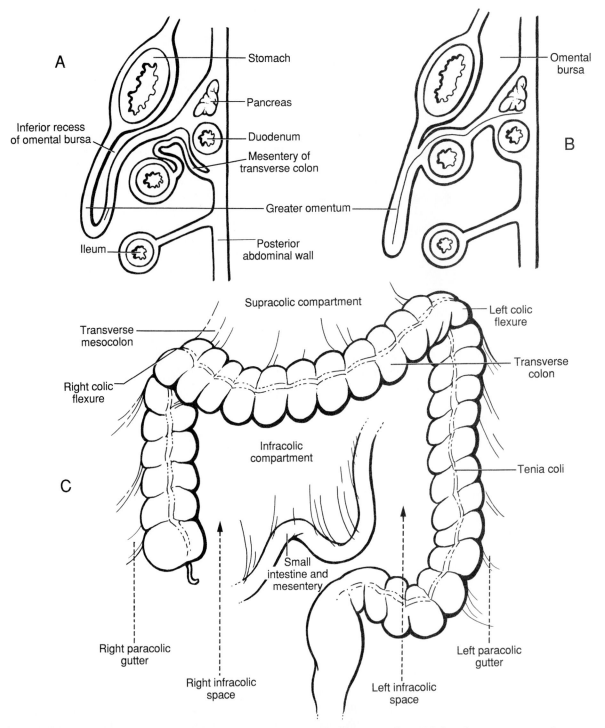

Figure 2-28. *A* and *B*, Sagittal sections of the abdomen. *A*, The vertical extent of the omental bursa. *B*, Fusion of the layers of the greater omentum and fusion of it with the mesentery of the transverse colon. *C*, The supracolic and infracolic compartments after removal of the greater omentum.

terms in Nomina Anatomica (Warwick, 1989). Therefore *omental foramen* is used instead of epiploic foramen. Similarly the gastroepiploic vessels are now known as the gastro-omental vessels. It will take time to gain acceptance of these new terms because the old ones are firmly entrenched in the minds of both basic science and clinical teachers.

Boundaries of the Omental Foramen (Figs. 2-27, 2-33, and 2-48).   The boundaries of the omental foramen are the following: *anteriorly*, the portal vein, hepatic artery, and bile duct (all in the free edge of the lesser omentum); *posteriorly*, the inferior vena cava and right crus of the diaphragm; *superiorly*, the caudate lobe of the liver; and *inferiorly*, the superior part of the duodenum, portal vein, hepatic artery, and bile duct.

> Although uncommon, loops of small intestine may pass through the omental foramen into the omental bursa and become strangulated by the edges of the foramen. As none of the boundaries of omental (epiploic) foramen can be incised owing to the presence of blood vessels, the swollen intestine is usually decompressed using a needle so it may be returned to the main part of the peritoneal cavity through the omental foramen.

> When the *cystic artery* is accidentally severed during *cholecystectomy* (removal of the gallbladder), hemorrhage from it can be controlled by compressing the hepatic artery between the second digit (index finger) in the omental foramen and the first digit (thumb) on its anterior wall.

The Abdominal Cavity

This is the larger part of the abdominopelvic cavity (Fig. 2-29). It is located superior to the pelvic inlet or superior aperture of the pelvis and is limited superiorly by the thoracic diaphragm. The abdominal cavity is *continuous inferiorly at the pelvic brim with the pelvic cavity*, the smaller part of the abdominopelvic cavity. A large part of the abdominal cavity is under the cover of the osteocartilaginous thoracic cage (Figs. 2-25 and 2-36). Parietal peritoneum lines the walls of the abdominal cavity; hence, *the peritoneum and peritoneal cavity are within the abdominal cavity*. Neither the abdominal cavity nor the peritoneal cavity is actually a cavity. The peritoneal cavity contains a capillary layer of fluid and the abdominal cavity is occupied by the peritoneum and abdominal viscera (Fig. 2-26). The abdominal cavity is kid-

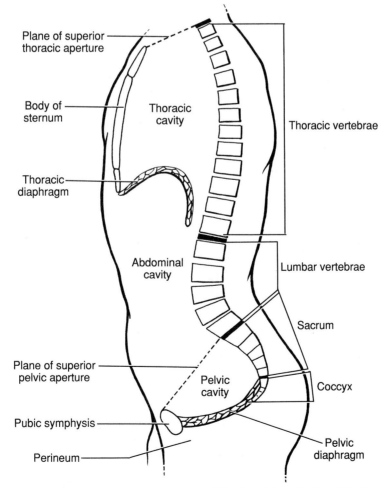

Figure 2-29.   The thoracic and abdominopelvic cavities as viewed in a median section of the body. The inferior broken line indicates the plane of the superior pelvic aperture (pelvic inlet), which divides the abdominopelvic cavity into abdominal and pelvic cavities. The thoracic diaphragm separates the thoracic and abdominal cavities. The pelvic diaphragm forms the inferior boundary of the pelvic cavity.

ney-shaped in transverse section because the vertebral column protrudes into it (Fig. 2-23).

The Abdominal Viscera

Before studying the viscera (L. soft parts or internal organs) and omenta in detail, observe their general arrangement in a cadaver (Figs. 2-25 to 2-27). Note that the liver, stomach, and spleen almost fill the domes (concavities) of the diaphragm. Observe that the *falciform ligament*, which is normally attached along a continuous line to the anterior abdominal wall as far inferiorly as the umbilicus, divides the liver into right and left parts. Also note that the fat-laden *greater omentum* conceals almost all the intestine (bowel, gut). Observe that the *gallbladder* protrudes from under the inferior border of the liver and that the C-shaped curve of the *duodenum* is closely related to the head and neck of the *pancreas*. The size of the peritoneal cavity is greatly exaggerated in Figure 2-26 because very little of the small intestine is shown. Observe the *spleen* in the left upper quadrant. Normally it is about the size of one's fist and is usually largely concealed by the stomach.

The Esophagus

The esophagus (gullet) is a fairly straight *muscular tube* (23 to 25 cm long) that extends from the pharynx to the stomach (Fig. 2-26). As food mixes with saliva, it passes down the esophagus, which propels swallowed food to the stomach to be digested. The wall of the esophagus contains a few mucous glands to supply additional lubrication. The esophagus follows the curve of the vertebral column as it descends through the neck and posterior mediastinum (see Fig. 1-66) and pierces the diaphragm just to the left of the median plane (Fig. 2-27). The short abdominal part of the esophagus (about 1 cm) forms a groove in the left lobe of the liver and *enters the stomach at its cardiac orifice*, posterior to the seventh left costal cartilage (Figs. 2-30 and 2-34). Here, the esophagus is covered anteriorly and laterally by peritoneum and is encircled by the *esophageal plexus of nerves* (see Fig. 1-73).

The right border of the abdominal part of the esophagus is continuous with the lesser curvature of the stomach, but its left border is separated from the fundus of the stomach by the *cardiac notch* (Fig. 2-34). At the inferior end of the esophagus, the esophagogastric junction, there is a physiological mechanism known as the *esophageal sphincter*, which contracts and relaxes. Radiological studies show that food stops momentarily here and that this sphincter is normally quite efficient in preventing reflux of gastric contents into the esophagus. When one is not eating, this sphincter is normally closed to prevent food or stomach juices from regurgitating into the esophagus.

Arterial Supply of the Esophagus (Figs. 2-38, 2-39, and 2-44).   The arteries supplying the abdominal part of the esophagus are derived from the left gastric branch of the *celiac artery*,

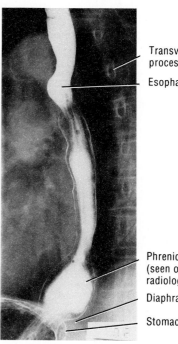

Transverse process

Esophagus

Phrenic ampulla (seen only radiologically)

Diaphragm

Stomach

Figure 2-30.   Radiograph of the esophagus after swallowing barium. (Courtesy of Dr. E.L. Lansdown, Professor of Radiology, University of Toronto, Toronto, Ontario, Canada.)

and the left *inferior phrenic artery*, a branch of the abdominal aorta.

Venous Drainage of the Esophagus (Fig. 2-44; see also Fig. 1-76).   The veins from the abdominal part of the esophagus drain into the *azygos vein* and the *left gastric vein*.

Lymphatic Drainage of the Esophagus (Figs. 2-41 and 2-44; Table 2-4; see also Fig. 1-70).   Lymph vessels from the abdominal part of the esophagus drain into the *left gastric lymph nodes*. Efferent vessels from these nodes drain into the *celiac lymph nodes*, but some vessels empty directly into the thoracic duct.

Innervation of the Esophagus (Figs. 2-33 and 2-42; see also Fig. 1-73).   The abdominal part of the esophagus is supplied by the *vagal trunks* (anterior and posterior gastric nerves), the thoracic *sympathetic trunks*, the greater and lesser *splanchnic nerves*, and the plexus of nerves around the left gastric and inferior phrenic arteries.

The incidence of **cancer of the esophagus** is low in North America. However the onset of difficulty in swallowing (*dysphagia*) in anyone over 45 years of age, especially in males, raises suspicion of esophageal cancer. A *barium swallow* (Fig. 2-30) in these patients shows a persistent filling defect owing to narrowing of the esophageal lumen by the cancer. During *esophagoscopy* a tumor may be observed and biopsied. An *esophagoscope* is a relatively thin, movable *fiberoptic tube*. Most cancer cells from a tumor of the abdominal part of the esophagus metastasize to the left gastric lymph nodes (Fig. 2-41), but some enter the thoracic duct and pass to the venous system (see Fig. 1-39). *Pyrosis* (''heartburn'') is the most common type of esophageal discomfort or pain. This is a well-known feeling of warmth (*burning sensation*) posterior to the inferior part of the sternum. *Heartburn* is often accompanied by regurgitation of small amounts of food or gastric fluid into the esophagus and pharynx. Esophageal pain often accompanies *dysphagia* (G. difficulty in swallowing) and may be severe if food is unable to pass a constricted segment of the esophagus.

The Stomach

The stomach is the greatly expanded portion of the digestive tract or *alimentary canal* between the esophagus and the small intestine (Figs. 2-25 to 2-28, 2-31, and 2-34). In the supine position, it is usually located in the left upper quadrant, where it occupies parts of the epigastric, umbilical, and left hypochondriac regions (Fig. 2-5). In most people the stomach is J-shaped and its pyloric part lies horizontally or ascends to the proximal part of the duodenum (Figs. 2-27 and 2-31). *The stomach acts as a food blender and reservoir* where gastric juices digest the food. It is a very distensible organ. The empty stomach is only of slightly larger caliber than the large intestine, but it is capable of considerable expansion and can hold 2 to 3 liters of food. A newborn infant's stomach, only the size of a lemon, can hold up to 30 ml of milk.

The stomach has two curvatures (Figs. 2-27, 2-31, and 2-34). The *lesser curvature* is continuous with the right border of the esophagus and forms the concave border of the stomach. The *angular notch* (incisura angularis) is a sharp angulation of the lesser curvature, which indicates the junction of the body and the pyloric part of the stomach (Fig. 2-34). The *greater curvature* is continuous with the left border of the esophagus and forms the *convex border of the stomach*. The greater curvature is four to five times longer than the lesser curvature.

Parts of the Stomach (Fig. 2-34). For descriptive purposes, the stomach is divided into five parts: a cardiac part, fundus, body, pyloric part, and pylorus.

The cardiac part of the stomach or cardia is a rather indefinite region around the cardiac orifice, which receives the opening of the abdominal part of the esophagus (Fig. 2-34).

The cardiac part or cardia of the stomach was given its name because it lies near the diaphragm where the pericardial sac containing the heart rests (Fig. 2-25). The Greek word *kardia* means heart.

The fundus of the stomach, a rounded vault, is its dilated portion to the left and superior to the cardiac orifice (Figs. 2-27, 2-31, and 2-34). The most superior part of the stomach, the fundus is related to the left dome of the diaphragm. It usually contains a bubble of gas, which is visible in radiographs (Figs. 2-30 and 2-31).

The body of the stomach (Figs. 2-26A, 2-27, and 2-34). This is the major portion of the stomach. It lies between the fundus and the pyloric antrum. Together the body and fundus form the most capacious area of the stomach. They are widely continuous with each other.

The pyloric part of the stomach consists of a wide portion, the *pyloric antrum*, and a narrow portion, the *pyloric canal*. The pyloric canal (1 to 2 cm long) is continuous with the markedly constricted terminal part called the *pylorus*, which separates the stomach from the duodenum.

The pylorus of the stomach (G. gatekeeper) is the distal sphincteric region that guards the *pyloric orifice* (Figs. 2-31 and 2-34). Its wall is thicker because it contains extra circular smooth muscle. The middle layer of the muscularis externa is greatly thickened to form the *pyloric sphincter*, which controls the rate of discharge of stomach contents into the duodenum. The pylorus is normally in tonic contraction (*i.e.*, it is closed except when emitting the semifluid contents of the stomach). At irregular intervals, *gastric peristalsis* passes a semifluid mass, called *chyme* (G. juice), through the pyloric canal into the small intestine for further mixing, digestion, and absorption.

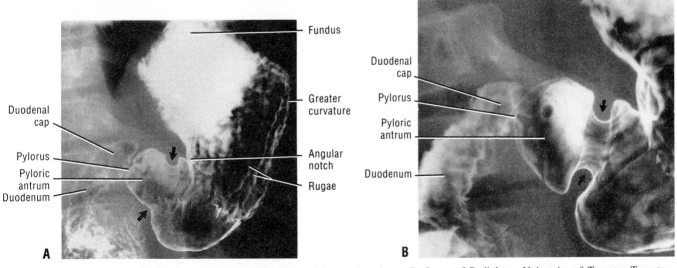

Figure 2-31. *A*, Radiograph of the stomach and small intestines following a barium meal. *B*, Radiograph of the pyloric region of the stomach and the proximal part of the duodenum. (Courtesy of Dr. E.L. Lansdown, Professor of Radiology, University of Toronto, Toronto, Ontario, Canada.)

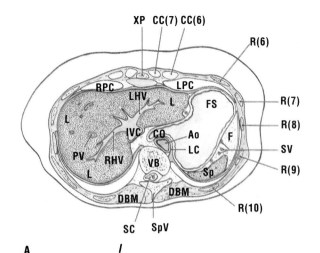

XP CC(7) CC(6)
R(6)
RPC LHV LPC
L FS
R(7)
L R(8)
IVC CO Ao
PV LC SV
RHV VB Sp R(9)
L
DBM DBM
R(10)
SC SpV

A   I

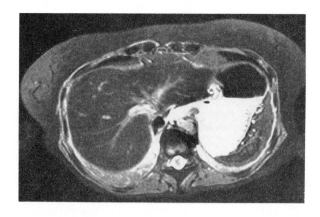

II

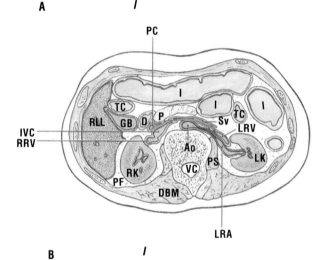

PC
I
TC I I
RLL D P Sv TC
IVC GB LRV
RRV Ao
RK VC LK
PF PS
DBM

LRA

B   I

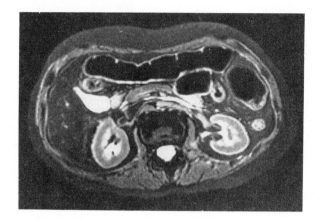

II

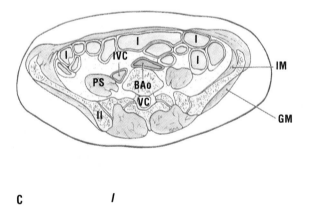

I I
I IVC I
PS BAo IM
VC
II GM

C   I

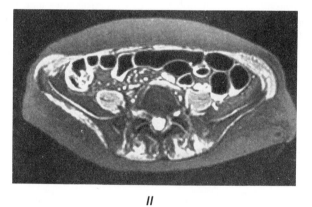

II

Figure 2-32. *A, B,* and *C,* Transverse MRIs (magnetic resonance images) of the abdomen. (Courtesy of Dr. W. Kucharzyck, Clinical Director of Tri-Hospital Resonance Centre, Toronto, Ontario, Canada.) *AO,* indicates aorta; *BAo,* bifurcation of aorta; *CC,* costal cartilage; *CO,* cardiac orifice of stomach; *D,* duodenum; *DBM,* deep back muscles; *DC,* descending colon; *F,* fat; *FS,* fundus of stomach; *GB,* gallbladder; *I,* intestine; *II,* ilium; *IM,* iliacus muscle; *IVC,* inferior vena cava; *L,* liver; *LC,* left crus; *LHV,* left hepatic vein; *LPC,* left pleural cavity; *LK,* left kidney; *LRA,* left renal artery; *P,* pancreas; *PF,* perirenal fat; *PC,* portal confluence; *PS,* psoas major muscle; *PV,* portal vein (triad); *R,* rib; *RHV,* right hepatic vein; *RK,* right kidney; *RLL,* right lobe of liver; *SC,* spinal cord; *RPC,* right pleural cavity; *SP,* spleen; *SpV,* spinous process of vertebra; *Sv,* splenic vein; *SV,* splenic vessels; *VB,* vertebral body; *VC,* vertebral canal; *X,* xiphoid process of sternum.

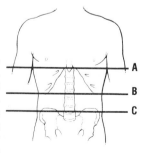

A
B
C

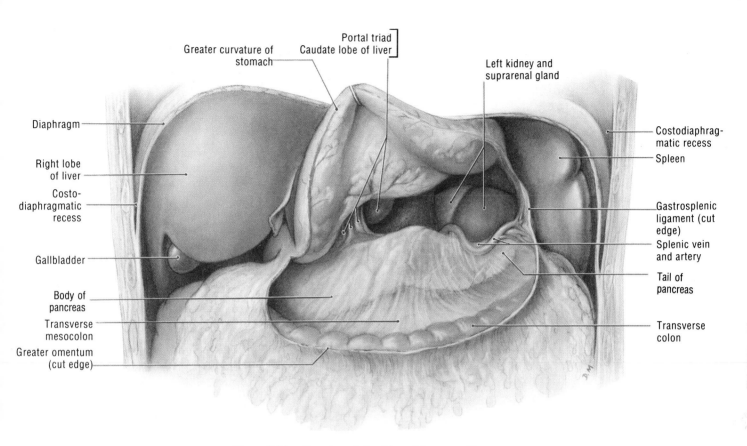

Figure 2-33.   Posterior relationships of the omental bursa.

The shape and position of the stomach vary in different persons and in the same individual as a result of movements of the diaphragm during respiration, the stomach's contents, and the position of the person (*i.e.*, whether lying down or standing). In the supine position, the stomach commonly lies in the left upper quadrant; in the erect position, it moves inferiorly (1 to 16 cm). In asthenic (G. *astheneia*, weakness) persons, the body of the stomach may extend into the pelvis major (see Fig. 3-7).

Surface Anatomy and Markings of the Stomach (Figs. 2-4*B*, 2-27, and 2-36).   The surface markings of the stomach vary greatly because, as mentioned, its size and position change under a variety of circumstances (*e.g.*, after a heavy meal). However, the *cardiac orifice* is usually located posterior to the seventh left costal cartilage, 2 to 4 cm from the median plane, at the level of T10 or T11 vertebra. The most superior part of the *fundus* is usually located posterior to the fifth left rib in the midclavicular line. The *pylorus* usually lies in the *transpyloric plane* (Fig. 2-4*B*). A normal stomach is not palpable because its walls are flat and rather flabby. *Fluoroscopic studies* (radiological examination of the internal parts of the stomach with a *fluoroscope*) show that the cardiac orifice remains fairly stationary (Fig. 2-30), but that the level of the pylorus varies from L1 to L3 vertebrae in the supine position. In the erect position, the location of the pylorus varies from L2 to L4 vertebrae. It is usually on the right side, but occasionally it is in the median plane.

Relations of the Stomach (Figs. 2-23, 2-25 to 2-28, 2-33, 2-35, and 2-36).   The stomach is covered entirely by perito-

neum, except where the blood vessels run along its curvatures and at a small bare area posterior to the cardiac orifice. The two layers of the lesser omentum surround the stomach and leave its greater curvature as the greater omentum. *The fundus of the stomach is in contact with the diaphragm*, posterior to the inferior left costal cartilages. In radiographs of the chest and esophagus, there is often a gas bubble inferior to the left dome of the diaphragm (Fig. 2-30; see also Figs. 1-48 to 1-50). The longitudinal rugae (L. wrinkles) are also outlined by gas in radiographs of the stomach (Fig. 2-31). The anterior surface of the stomach is in contact with: (1) the diaphragm in the fundic region, (2) the left lobe of the liver, and (3) the anterior abdominal wall (Figs. 2-25 to 2-27).

The Stomach Bed (Figs. 2-28, 2-33, 2-35, and 2-47).   The "bed" of the stomach is formed by the posterior wall of the omental bursa and retroperitoneal structures between it and the posterior abdominal wall (*e.g.*, the pancreas and left kidney). Superiorly the stomach bed contains part of the diaphragm, the spleen, and the left suprarenal gland. Inferiorly it consists of the body and tail of the pancreas and the transverse mesocolon.

Acquired hiatus hernias are common owing to the higher pressure in the abdominal cavity compared to the pressure in the thorax (chest). They occur most often in people after middle age, possibly owing to weakening of the muscle forming the esophageal hiatus in the diaphragm (see Figs. 1-29 and 1-70). A portion of the fundus of the stomach may herniate through the esophageal hiatus into the thorax. Very few

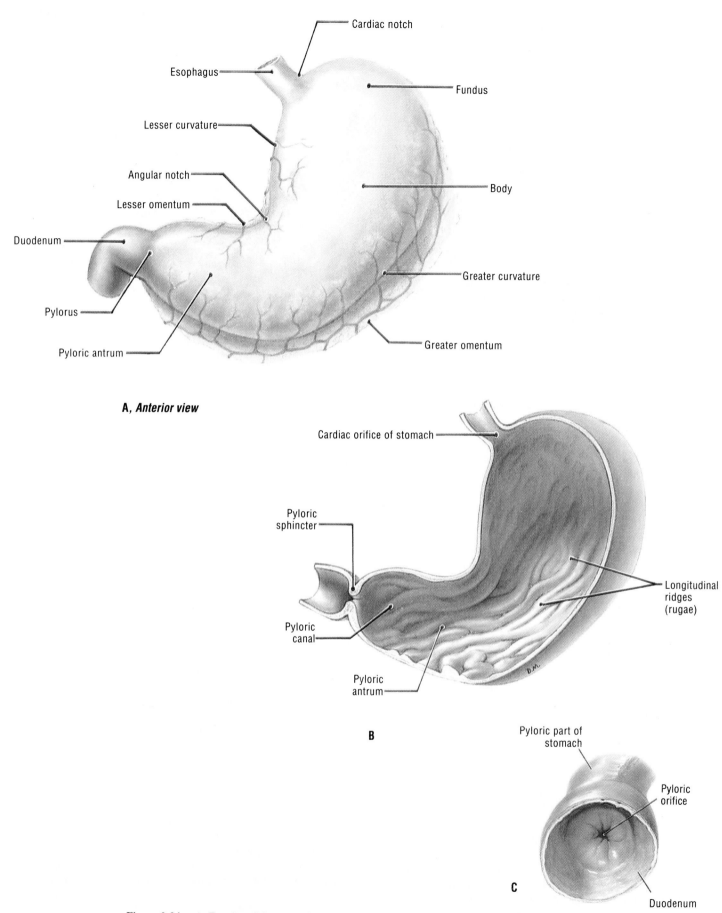

A, *Anterior view*

Figure 2-34. *A,* Exterior of the stomach. *B,* Interior of the stomach showing the mucous membrane. *C,* The pylorus or distal sphincteric region of the stomach.

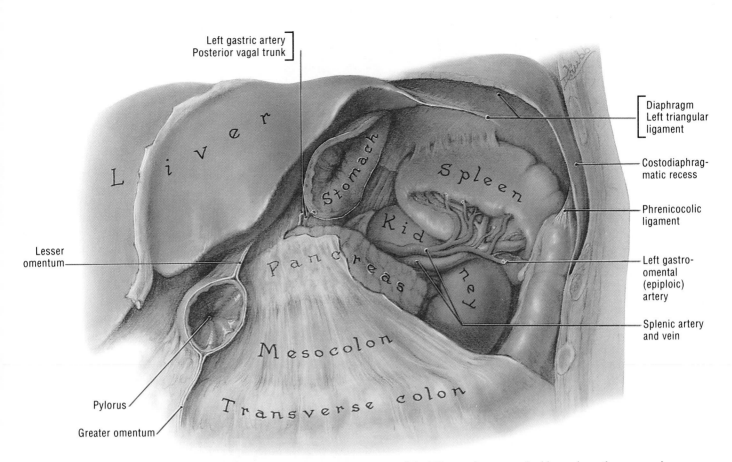

Left gastric artery
Posterior vagal trunk

Diaphragm
Left triangular
ligament

Costodiaphrag-
matic recess

Phrenicocolic
ligament

Left gastro-
omental
(epiploic)
artery

Splenic artery
and vein

Lesser
omentum

Pylorus

Greater omentum

Figure 2-35.   Dissection of the stomach bed. Most of the stomach has been excised and the peritoneum of the omental bursa covering the bed has been largely removed, as has the peritoneum covering the inferior part of the kidney and pancreas. In this specimen the pancreas is unusually short and adhesions bind the spleen to the diaphragm; adhesions are pathological features but not unusual.

patients with **hiatus hernia** experience significant symptoms. There are two main types of hiatus hernia.

In *sliding hiatus hernia* (Fig. 2-37*A*), the gastroesophageal region slides superiorly into the chest through the esophageal hiatus when the person lies down or bends over. There is often some regurgitation of stomach contents into the esophagus because the clamping action of the crura of the diaphragm on the inferior end of the esophagus is weak (see Fig. 1-70).

In *paraesophageal hiatus hernia* (Fig. 2-37*B*), which is far less common, the gastroesophageal region remains in its normal position, but a pouch of peritoneum, often containing part of the fundus of the stomach, extends through the esophageal hiatus, anterior to the esophagus. In these cases there is usually no regurgitation of gastric contents because the cardiac orifice is in its normal position.

Arterial Supply of the Stomach (Figs. 2-38, 2-39, and 2-42 to 2-45).   The stomach has a rich blood supply from all three branches of the celiac trunk: (1) *The left gastric artery*, its smallest branch, supplies five branches to the stomach as it passes along the lesser curvature of the stomach; (2) The *right gastric* and *right gastro-omental arteries* (gastroepiploic arteries) are branches of the common hepatic artery; and (3) The *left gastro-omental* and *short gastric arteries* are branches of the splenic artery.

The left gastric artery, a small branch of the *celiac trunk*, passes superiorly and to the left across the posterior wall of the omental bursa. It lies in the floor of this bursa, posterior to the parietal peritoneum (Fig. 2-42). The left gastric artery passes from the posterior abdominal wall to the cardiac part of the stomach. It then runs inferiorly between the layers of the lesser omentum (*hepatogastric ligament*) and along the lesser curvature, frequently as two branches, to the pylorus. It supplies both surfaces of the stomach and anastomoses with the right gastric artery (Fig. 2-38).

The right and left gastro-omental (gastroepiploic) arteries run along the greater curvature of the stomach, supplying both its surfaces (Fig. 2-38). These vessels *run between the layers of the greater omentum*, a short distance from its attachment to the stomach. The *right gastro-omental artery*, a branch of the gastroduodenal, runs to the left and anastomoses with the left gastro-omental artery. It sends branches to the right part of the stomach, the superior part of the duodenum, and the greater omentum. The *left gastro-omental artery*, a branch of the splenic, runs between the layers of the *gastrosplenic ligament* to the greater curvature of the stomach. It passes to the right between the layers of the greater omentum, supplying the stomach and omentum, and ends by anastomosing with the right gastro-omental artery (Fig. 2-38).

The short gastric arteries (four to five) are also branches of

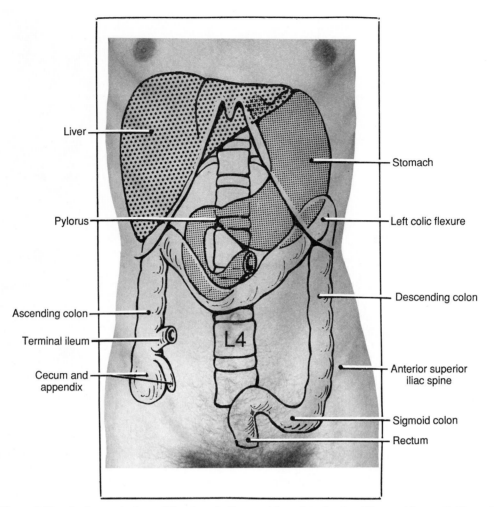

Figure 2-36. Surface projections of the stomach, liver, and large intestine in a 27-year-old man. Outlines of the vertebrae are shown for reference.

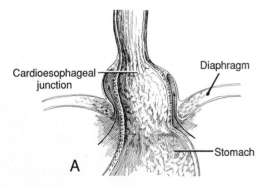

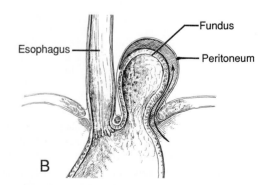

Figure 2-37. *A*, Sliding hiatus hernia showing the cardioesophageal junction situated superior to the esophageal hiatus in the diaphragm. *B*, paraesophageal hiatus hernia with this junction in its normal position. Note that the pouch of peritoneum extending through the esophageal hiatus into the chest contains part of the fundus of the stomach. (Modified and reprinted with permission from McCredie JA: *Basic Surgery*. New York: Macmillan Publishing Company, 1977.)

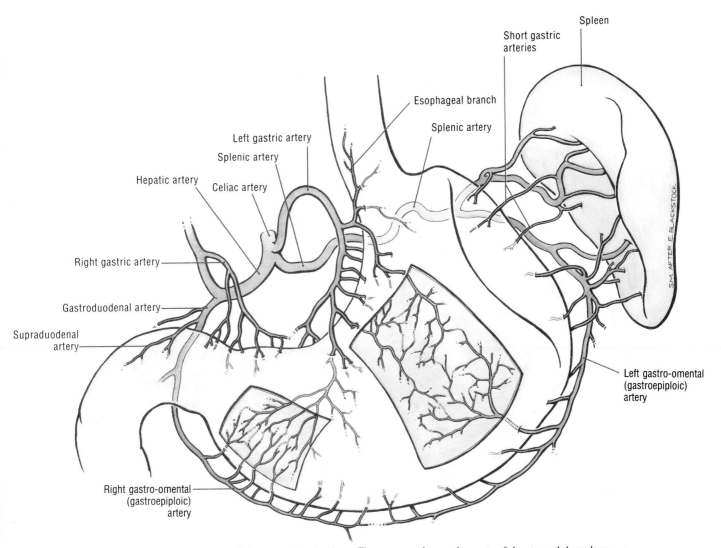

Figure 2-38. Arteries of the stomach and spleen. The serous and muscular coats of the stomach have been removed from two areas to show the anastomotic arterial networks in the submucous coat.

the splenic artery. They run between the layers of the gastro-splenic ligament to the fundus of the stomach, where they anastomose with branches of the left gastric and left gastro-omental (gastroepiploic) arteries.

Veins of the Stomach (Figs. 2-33 and 2-40). The gastric veins parallel the arteries in position and course and drain into the *portal system of veins*. There is considerable variation in the way these veins drain into this system. Usually the right and left gastric veins drain directly into this portal vein. The right gastro-omental (gastroepiploic) vein usually drains into the superior mesenteric vein, but the right gastro-omental vein may enter the portal vein directly or join the splenic vein. The left gastro-omental vein and the short gastric vein drain into the splenic vein or one of its tributaries.

Lymphatic Drainage of the Stomach (Figs. 2-41, 2-44, and 2-45). Lymph vessels from the stomach accompany the arteries along the greater and lesser curvatures. *There are four major areas of lymphatic drainage*, each of which has its own regional lymph nodes (*e.g.*, gastric and pancreaticosplenic). The

lymph vessels drain lymph from the anterior and posterior surfaces of the stomach toward its curvatures, where many of the *gastro-omental (gastroepiploic) lymph nodes* are located. The efferent vessels from these nodes accompany the large arteries to the *celiac lymph nodes*, which surround the celiac trunk.

The largest area of lymphatic drainage is from the lesser curvature, including a large part of the body of the stomach, to the *left gastric lymph nodes*, which lie along the left gastric artery. The next largest area of lymphatic drainage is from the right part of the greater curvature, including most of the pyloric part of the stomach, to the *gastro-omental (gastro-epiploic) lymph nodes* (Fig. 2-41), which lie along the right gastro-omental vessels. Many lymph vessels from this area pass directly to the **pyloric lymph nodes**, located on the anterior surface of the head of the pancreas, close to the pylorus. The third area of lymphatic drainage, smaller than the previous two, is from the left part of the greater curvature to the gastro-omental lymph nodes, which lie along the left gastro-omental vessels. Other vessels pass to the *pancreati-*

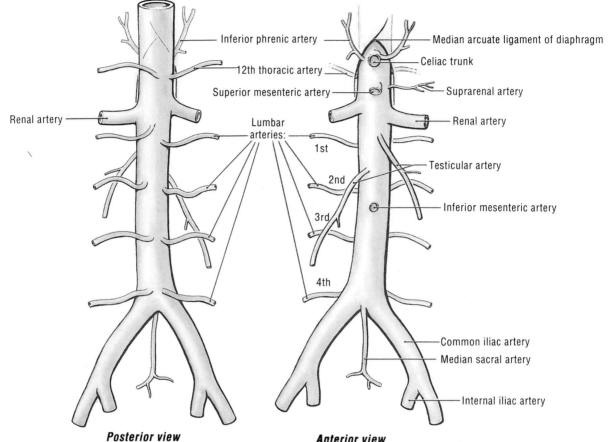

Renal artery

Inferior phrenic artery

12th thoracic artery

Superior mesenteric artery

Lumbar
arteries:

1st

2nd

3rd

4th

Median arcuate ligament of diaphragm

Celiac trunk

Suprarenal artery

Renal artery

Testicular artery

Inferior mesenteric artery

Common iliac artery

Median sacral artery

Internal iliac artery

Posterior view

Anterior view

Figure 2-39.   The abdominal aorta and its branches. Besides its several paired branches, it has three unpaired branches that supply the GI system. These are the celiac trunk and the superior and inferior mesenteric arteries.

cosplenic lymph nodes, which lie along the splenic vessels. The fourth and smallest area of lymphatic drainage is from the part of the lesser curvature related to the pyloric part of the stomach. Lymph vessels from this area pass to the *right gastric lymph nodes*, which lie along the right gastric artery. Lymph from all four major groups of lymph nodes drains into the **celiac lymph nodes** (Figs. 2-41 and 2-45), which are located around the origin of the *celiac trunk*. Lymph from these nodes passes with that from other parts of the GI tract to the *cisterna chyli* and *thoracic duct* (see Fig. 1-39).

Innervation of the Stomach (Fig. 2-42; Table 2-3; see also Fig. 8-33).   The parasympathetic nerve supply is derived from the anterior and posterior **vagal trunks** and their branches. Both trunks are often located close to where the left gastric artery reaches the stomach (Fig. 2-38). The sympathetic nerve supply is mainly from the *celiac plexus* through the plexuses around the gastric and gastro-omental (gastroepiploic) arteries. The efferent sympathetic fibers to the stomach arise from T6 to T9 segments of the spinal cord.

The anterior vagal trunk, derived mainly from the *left vagus nerve*, usually enters the abdomen as a single branch that lies on the anterior surface of the esophagus (Fig. 2-42; see also Fig. 8-33). It runs toward the lesser curvature of the stomach, where it gives off hepatic and duodenal branches that leave the stomach in the hepatoduodenal ligament. The rest of the anterior vagal

trunk continues along the lesser curvature, giving rise to anterior gastric branches.

The posterior vagal trunk, derived mainly from the *right vagus nerve*, enters the abdomen on the posterior surface of the esophagus and passes toward the lesser curvature of the stomach (Fig. 2-42; see also Fig. 8-33). It gives off a celiac branch that runs to the *celiac plexus*. It then continues along the lesser curvature, giving rise to posterior gastric branches.

Pylorospasm (spasmodic contraction of the pylorus) sometimes occurs in infants, usually between 2 to 12 weeks of age. It is characterized by failure of the smooth muscle fibers encircling the pyloric canal to relax normally. As a result, food does not pass easily from the stomach into the duodenum and the stomach becomes overly full. This usually results in vomiting.

Malformations of the stomach are uncommon, except for congenital *hypertrophic pyloric stenosis*. This marked thickening of the smooth muscle in the pylorus affects approximately 1 in every 150 male infants, and 1 in every 750 female infants (Moore, 1988). The elongated, overgrown pylorus is hard, and severe *stenosis* (narrowing) of the pyloric canal is present. The proximal part of the stomach is secondarily dilated owing to the pyloric obstruction. Although the cause of congenital hypertrophic pyloric stenosis is unknown, genetic factors appear to be involved because of its high incidence in infants of monozygotic twins.

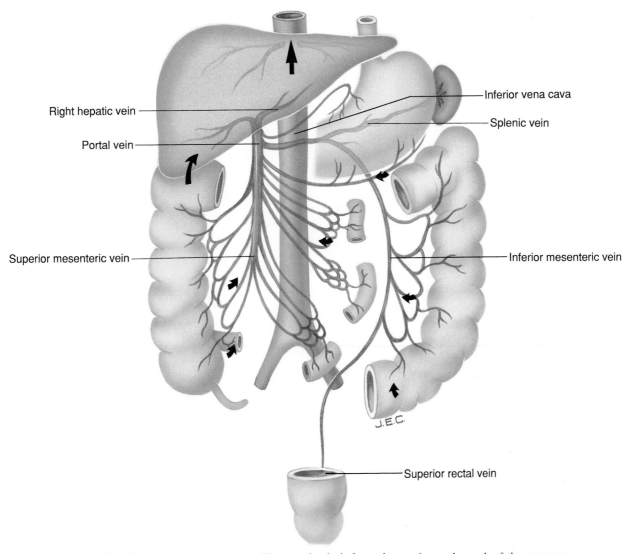

Right hepatic vein

Portal vein

Superior mesenteric vein

Inferior vena cava

Splenic vein

Inferior mesenteric vein

Superior rectal vein

J.E.C.

Figure 2-40.   The portal venous system. The portal vein is formed posterior to the neck of the pancreas by the union of the splenic and superior mesenteric veins.

When the body or the pyloric part of the stomach contains a tumor, the mass may be palpable. The incidence of *carcinoma of the stomach* is higher in certain countries (*e.g.*, Scandinavia) than in others (*e.g.*, North America). It is also more common in men than women. The etiological factors are unknown, but this malignancy appears to be related to diet. Smoked fish and spices have been implicated in areas that show a high incidence of gastric cancer. Since the development of flexible fiberendoscopes, *gastroscopy* has become common. This instrument enables physicians to observe gastric lesions and to take biopsies.

Although uncommon, part of the stomach may herniate through the diaphragm at birth owing to a congenitally large esophageal hiatus. Sometimes the stomach enters the thorax through a large posterolateral defect in the diaphragm. This type of *congenital diaphragmatic hernia* occurs about once in every 2000 newborn infants (Behrman, 1992; Moore, 1988).

Gastrectomy (removal of the entire stomach) is very uncommon. Because the anastomoses of the arteries supplying the stomach provide a good collateral circulation, one or more of its major arteries may be ligated without seriously affecting its blood supply. During a *partial gastrectomy* (*e.g.*, excision of the pyloric antrum), the greater omentum is incised inferior to the right gastro-omental (gastroepiploic) artery (Figs. 2-27 and 2-38). When all the omental branches of this artery are ligated, the greater omentum does not degenerate because the omental branches of the left gastro-omental artery are still intact. *Partial resection of the stomach* (*e.g.*, because of carcinoma [CA]), usually involves removal of all involved regional lymph nodes. The *pyloric nodes* are especially important because they receive lymph from the pyloric region of the stomach where CA is frequent. Because lymph from this region also drains into the *right gastro-omental nodes*, they are frequently involved. In advanced cases, cancer cells spread by lymphogenous dissemination to the *celiac lymph nodes* grouped around the origin of the celiac trunk (Fig. 2-44) because all gastric lymph nodes drain into them. CA of the stomach can spread to: (1) the liver, (2) the pelvis, and

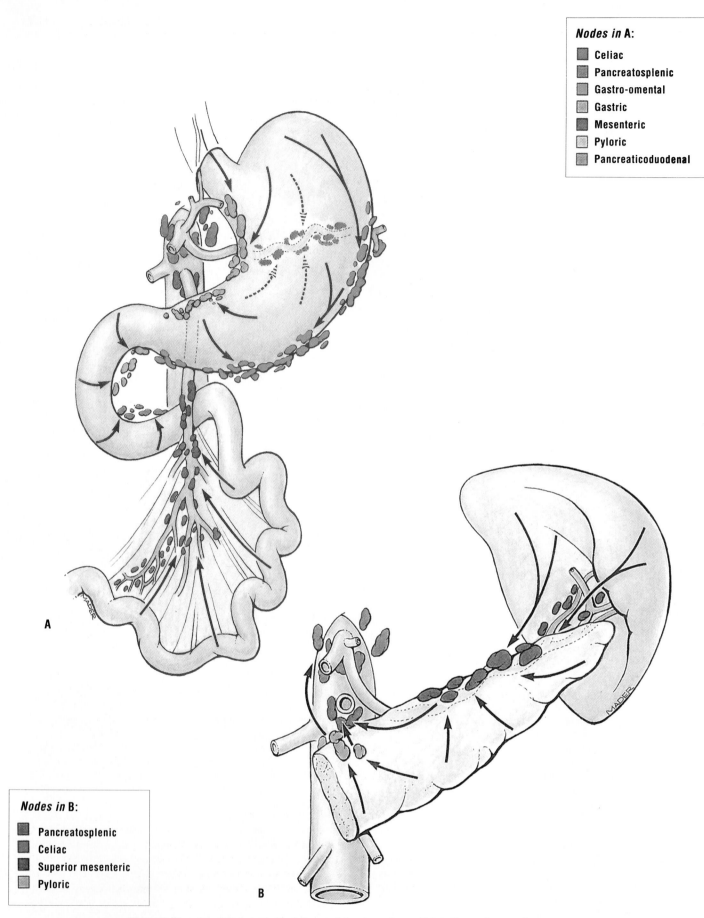

Nodes in A:
- Celiac
- Pancreatosplenic
- Gastro-omental
- Gastric
- Mesenteric
- Pyloric
- Pancreaticoduodenal

Nodes in B:
- Pancreatosplenic
- Celiac
- Superior mesenteric
- Pyloric

Figure 2-41. *A* and *B*, Lymphatic drainage of the stomach, small intestine, spleen, and pancreas.

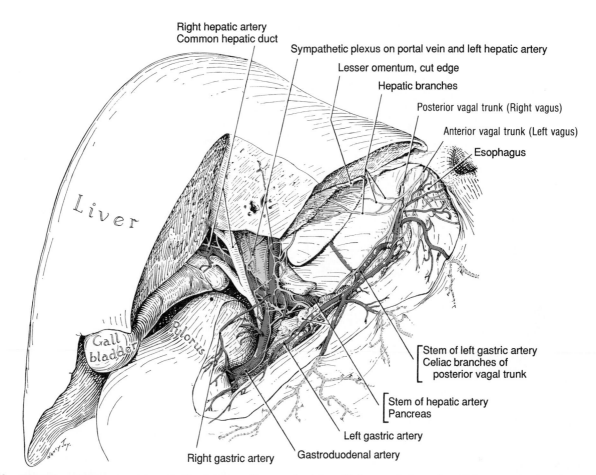

Right hepatic artery
Common hepatic duct

Sympathetic plexus on portal vein and left hepatic artery

Lesser omentum, cut edge

Hepatic branches

Posterior vagal trunk (Right vagus)

Anterior vagal trunk (Left vagus)

Esophagus

Liver

Gall bladder

Pylorus

Stem of left gastric artery
Celiac branches of
    posterior vagal trunk

Stem of hepatic artery
Pancreas

Left gastric artery

Gastroduodenal artery

Right gastric artery

Figure 2-42. Dissection of the vagus nerves in the abdomen. The nerves to the stomach are branches of the anterior and posterior vagal trunks. The parasympathetic fibers arrive via the right and left vagus nerves and the sympathetic fibers come via preganglionic fibers from the right and left sympathetic trunks, which synapse in preaortic ganglia. Both kinds of nerve mingle in the rich tangle of nerve plexuses on the anterior aspect of the aorta, especially around the celiac trunk, where they form the celiac plexus (see also Fig. 2-64).

Table 2-3.
The Three Types of Splanchnic Nerve

Name	Type	Origin
1. Thoracic (greater, lesser, and lowest splanchnic nn.)	Sympathetic	Branches of 5th to 12th thoracic sympathetic ganglia
2. Lumbar splanchnic nn.	Sympathetic	Branches of the four lumbar sympathetic ganglia
3. Pelvic splanchnic nn.	Parasympathetic	Branches of ventral rami of sacral spinal nerves, S2, S3 (S4)

and the gastric juices erode the mucosa, forming a *gastric ulcer*. Excess acid secretion is associated with these lesions. Vagotomy may be done in conjunction with resection of the ulcerated area. Often a *selective vagotomy* is performed during which only the gastric branches of the vagus are sectioned. This has the desired effect on the acid-producing parietal cells of the stomach without affecting other abdominal structures supplied by the vagus. Pain impulses from the stomach appear to be carried in the sympathetic nerves because the pain of a recurrent peptic ulcer may persist after complete vagotomy, whereas patients who have had a bilateral sympathectomy may have a perforated peptic ulcer with no pain.

The Small Intestine

(3) the rest of the body via the thoracic duct and venous system (see Fig. 1-39).

The secretion of acid by the parietal cells of the stomach is largely controlled by the vagus nerves; hence, section of the vagal trunks (*vagotomy*) is sometimes performed to reduce the production of acid in persons with *peptic ulcers* (lesions of the mucosa of the stomach or duodenum). Usually mucus covers the mucosa, forming a barrier between the acid and the cells. Sometimes, however, this protection is inadequate

The pylorus empties the contents of the stomach into the *duodenum* (Figs. 2-26 and 2-31), the first part of the small intestine (small bowel, small gut). Its other two parts are the *jejunum* and *ileum*. The coils of small intestine occupy most of the abdominal cavity (Figs. 2-25 and 2-26). Most digestion occurs in the small intestine, the length of which varies, but it averages 6 to 7 meters. Excision of up to one-third of it is compatible with a fairly normal life-style.

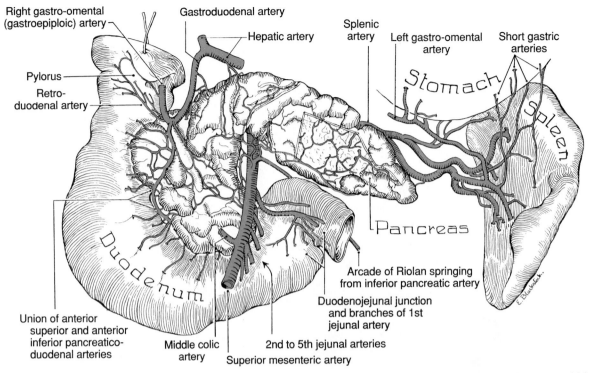

Figure 2-43. Blood supply of the pancreas, duodenum, and spleen; anterior view. A slice has been removed from the tail of the pancreas to show the internal arrangement of its vessels. Note the many arteries entering the hilum of the spleen; *these are end arteries*, which do not have significant anastomoses in the spleen.

The Duodenum

This is the *first part of the small intestine* (Figs. 2-26, 2-34, 2-44 to 2-46, and 2-52). The pancreas is its most intimate relation. The duodenum pursues a C-shaped course from the pylorus around the head and neck of the pancreas and becomes continuous with the jejunum. The duodenum (L. *duodeni*, twelve) was given its name because it is usually 12 fingerbreadths long (about 25 cm). The position of the duodenum is variable, but it begins at the pylorus on the right side (2 to 3 cm from the median plane) and ends on the left side at the *duodenojejunal junction* (2 to 3 cm from the median plane).

The duodenum is the shortest, widest, and most fixed part of the small intestine. It forms a U-shaped loop, which faces superiorly and to the left and is molded around the head and neck of the pancreas. The duodenum is particularly important because it *receives the openings of the bile and pancreatic ducts* (Figs. 2-49 and 2-51). All but the first 2.5 cm of the duodenum is posterior to the peritoneum (*i.e.*, *retroperitoneal*, Fig. 2-48). For purposes of description, **the duodenum is divided into four parts**, which are related to the vertebral column as follows (Figs. 2-45, 2-46, and 2-52): (1) a *superior or first part* anterolateral to the body of L1 vertebra; (2) a *descending or second part* to the right of the bodies of L1 to L3 vertebrae; (3) a *horizontal or third part* anterior to L3 vertebrae; and (4) an *ascending or fourth part* to the left of the body of L3 and rising as high as L2 or L1 vertebra. Hence, its superior (1st) and ascending (4th) parts are only about 5 cm apart.

The Superior (First) Part of the Duodenum (Figs. 2-26, 2-27, and 2-43 to 2-46, and 2-52). This part, about 5 cm long, is the most movable of the four parts of the duodenum. It begins at the pylorus and passes to the right, posteriorly, and slightly superiorly toward the neck of the gallbladder and the right kidney. Hence, it passes almost at a right angle to the pylorus from which the stomach contents enter the duodenum. In right anterior oblique radiographs, the superior part of the duodenum appears short because of its oblique and angled direction. Radiologists refer to the beginning or **ampulla** of the superior part of the duodenum as the *duodenal cap* or *bulb* (Fig. 2-31*B*). The proximal half of the superior part of the duodenum has a mesentery and is mobile. The greater omentum and the *hepatoduodenal ligament* are attached to this part of the duodenum (Fig. 2-27); hence, this part moves with the stomach. The distal half of the superior part of the duodenum has no mesentery and is immobile. It is fixed to the posterior abdominal wall (Fig. 2-48*A*).

The principal relations of the superior part of the duodenum are (Figs. 2-48 to 2-50 and 2-52): *anteriorly*, the peritoneum, gallbladder, and quadrate lobe of the liver; *posteriorly*, the bile duct, portal vein, inferior vena cava, and gastroduodenal artery; *superiorly*, the neck of the gallbladder; and *inferiorly*, the neck of the pancreas. Because of its close relationship to the gallbladder, the anterior surface of this part of the duodenum is commonly stained with bile in cadavers.

The Descending (Second) Part of the Duodenum (Figs. 2-26, 2-43 to 2-46, 2-48, 2-49, 2-51, and 2-52). This part of the duodenum is about 7.5 cm long and has no mesentery. It descends retroperitoneally along the right sides of L1 to L3 vertebrae. During its descent, it passes to the right of and parallel to the inferior vena cava. **The bile duct** (common bile duct) and main pancreatic duct enter the posteromedial wall of the de-

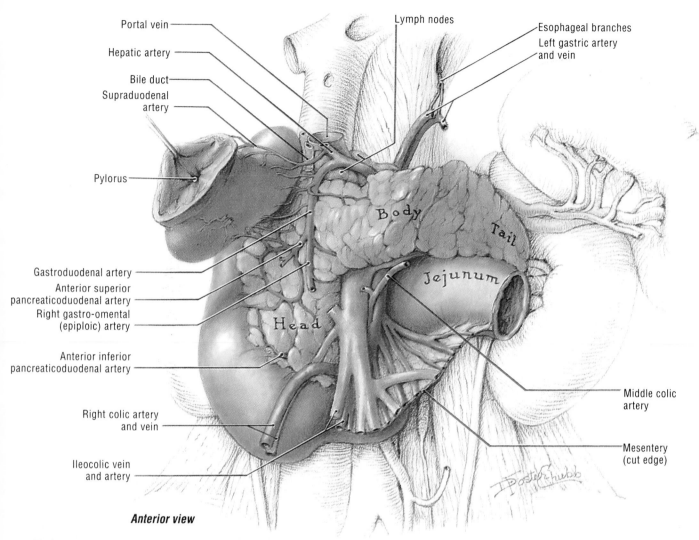

Portal vein

Hepatic artery

Bile duct

Supraduodenal artery

Pylorus

Gastroduodenal artery

Anterior superior pancreaticoduodenal artery

Right gastro-omental (epiploic) artery

Anterior inferior pancreaticoduodenal artery

Right colic artery and vein

Ileocolic vein and artery

Lymph nodes

Esophageal branches
Left gastric artery and vein

Body

Tail

Jejunum

Head

Middle colic artery

Mesentery (cut edge)

Anterior view

Figure 2-44. Anterior view of a dissection of the duodenum, pancreas, and bile duct. Observe that the duodenum embraces the head of the pancreas and that its superior or first part (retracted) is overlapping the pancreas and passing posteriorly, superiorly, and to the right. Note that the body of the pancreas is arched anteriorly where it crosses the vertebral column.

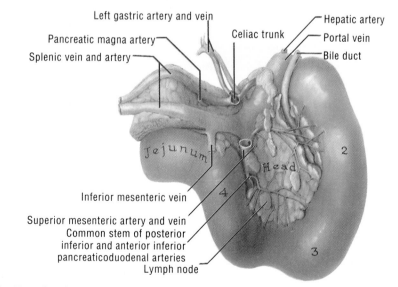

Left gastric artery and vein

Pancreatic magna artery

Splenic vein and artery

Celiac trunk

Hepatic artery

Portal vein

Bile duct

Jejunum

Head

2

4

3

Inferior mesenteric vein

Superior mesenteric artery and vein
Common stem of posterior inferior and anterior inferior pancreaticoduodenal arteries

Lymph node

Figure 2-45. Posterior view of a dissection of the duodenum, pancreas, and bile duct. Observe that the bile duct is descending in a long fissure (opened here) in the posterior part of the head of the pancreas.

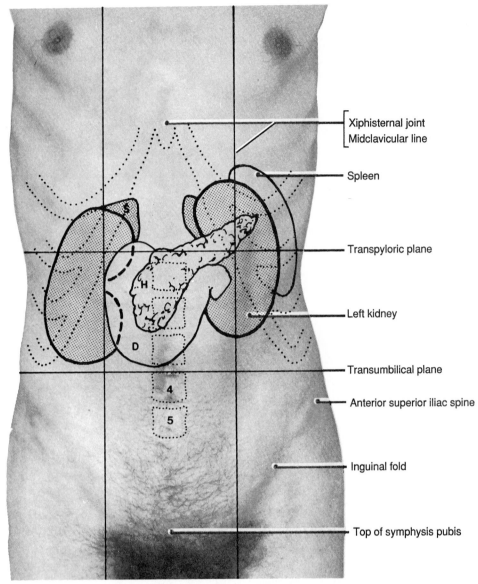

Xiphisternal joint
Midclavicular line

Spleen

Transpyloric plane

Left kidney

Transumbilical plane

Anterior superior iliac spine

Inguinal fold

Top of symphysis pubis

Figure 2-46. Surface projection of the duodenum (D) and pancreas, kidneys, spleen, and suprarenal glands (S). Note that the U-shaped duodenum (D), embracing the head of the pancreas (H), lies superior to the umbilicus and that its two ends are not far apart.

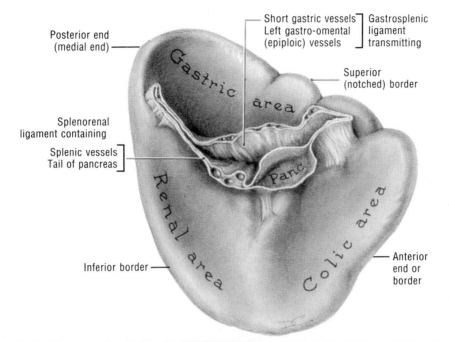

Posterior end
(medial end)

Short gastric vessels ⎤ Gastrosplenic
Left gastro-omental ⎥ ligament
(epiploic) vessels ⎦ transmitting

Superior
(notched) border

Splenorenal
ligament containing

Splenic vessels
Tail of pancreas

Inferior border

Anterior
end or
border

Figure 2-47. Visceral surface of the spleen. Note the characteristic notches in its superior border. Observe that this cadaveric spleen has impressions of the structures that were in contact with it.

scending part of the duodenum, about two-thirds of the way along its length. These ducts enter the wall obliquely, where they usually unite to form a short dilated tube known as the *hepatopancreatic ampulla*. This ampulla (formerly called the ampulla of Vater) opens on the summit of the *major duodenal papilla*, located 8 to 10 cm distal to the pylorus. The opening of this papilla is guarded by the *sphincter of the hepatopancreatic ampulla*, which is capable of constricting it and thereby of controlling the discharge of bile and pancreatic secretions into the duodenum. In some people the bile and pancreatic ducts do not join, but open separately on the major duodenal papilla.

The principal relations of the descending part of the duodenum are (Fig. 2-52): *anteriorly*, the transverse colon, transverse mesocolon, and some coils of the small intestine; *posteriorly*, the hilum (hilus) of the right kidney, renal vessels, ureter, and psoas major muscle; and *medially*, the head of the pancreas, pancreatic duct, and bile duct.

The Horizontal (Third) Part of the Duodenum (Figs. 2-26, 2-43 to 2-46, 2-48, 2-49, and 2-52). This part, about 10 cm long, runs horizontally from right to left across L3 vertebra, passing anterior to the inferior vena cava, aorta, and inferior mesenteric artery. This part is also retroperitoneal and adherent to the posterior abdominal wall.

The principal relations of the horizontal part of the duodenum are: *anteriorly*, the superior mesenteric artery and coils of the small intestine; *posteriorly*, the right psoas major muscle, inferior vena cava, aorta, and right ureter; and *superiorly*, the head of the pancreas and superior mesenteric vessels.

The Ascending (Fourth) Part of the Duodenum (Figs. 2-26, 2-43 to 2-46, 2-48, 2-52, and 2-53). This short part (about

2.5 cm) ascends on the left side of the aorta anterior to the left renal vessels to the level of the L2 vertebra. Here it meets the body of the pancreas and joins the jejunum at the *duodenojejunal flexure*. It bends abruptly (anteriorly) and becomes continuous with the jejunum. This distal end is covered with peritoneum and is movable; however, most of this part of the duodenum is retroperitoneal, immobile, and adherent to the posterior abdominal wall. The duodenojejunal flexure is supported by a fibromuscular band called the *suspensory muscle of the duodenum* (ligament of Treitz). The superior part of this slender band contains striated muscle; its intermediate part consists of elastic tissue; and its inferior part contains smooth muscle. This suspensory muscle passes from the superior surface of the ascending (fourth) part of the duodenum and the duodenojejunal flexure and divides into two parts: One part is attached to the right crus of the diaphragm, close to the esophageal opening, and the other part is attached to connective tissue around the celiac trunk. *The suspensory muscle of the duodenum supports the duodenojejunal flexure* and widens its angle, thereby facilitating movement of the intestinal contents.

The principal relations of the ascending part of the duodenum are (Fig. 2-48): *anteriorly*, the beginning of root of the mesentery and coils of jejunum; *posteriorly*, the left psoas major muscle and the left margin of aorta; *medially*, the head of pancreas; and *superiorly*, the body of the pancreas.

Peritoneal Recesses of the Duodenum (Fig. 2-53). Several peritoneal folds and recesses are related to the duodenum, particularly near the duodenojejunal junction. Most of them are inconstant. This is where the intestine changes from a retroperitoneal (duodenum) to a intraperitoneal

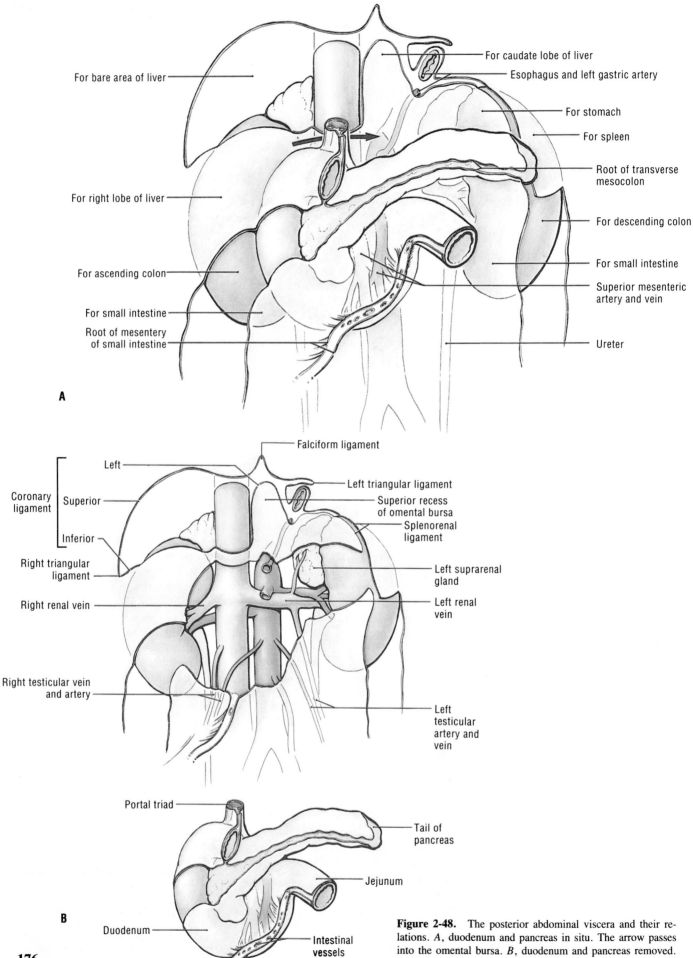

For bare area of liver

For caudate lobe of liver

Esophagus and left gastric artery

For stomach

For spleen

Root of transverse mesocolon

For right lobe of liver

For descending colon

For small intestine

Superior mesenteric artery and vein

For ascending colon

For small intestine

Root of mesentery of small intestine

Ureter

A

Falciform ligament

Left

Left triangular ligament

Coronary ligament

Superior

Superior recess of omental bursa

Splenorenal ligament

Inferior

Right triangular ligament

Left suprarenal gland

Right renal vein

Left renal vein

Right testicular vein and artery

Left testicular artery and vein

Portal triad

Tail of pancreas

Jejunum

B

Duodenum

Intestinal vessels

Figure 2-48. The posterior abdominal viscera and their relations. *A*, duodenum and pancreas in situ. The arrow passes into the omental bursa. *B*, duodenum and pancreas removed.

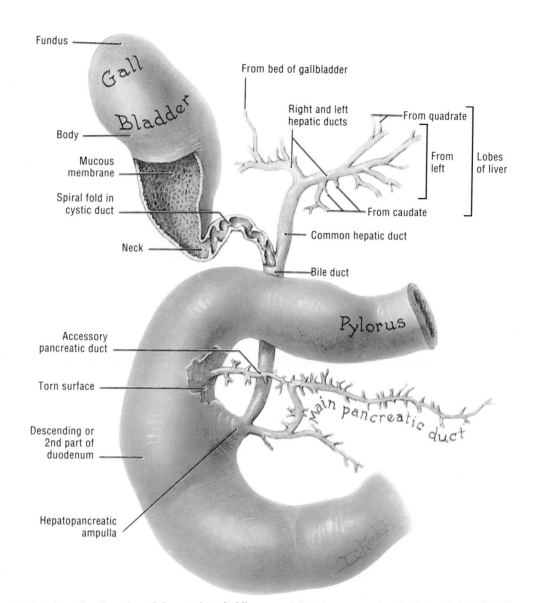

Fundus

Gall Bladder

Body

Mucous membrane

Spiral fold in cystic duct

Neck

From bed of gallbladder

Right and left hepatic ducts

From quadrate

From left

Lobes of liver

From caudate

Common hepatic duct

Bile duct

Pylorus

Accessory pancreatic duct

Torn surface

Main pancreatic duct

Descending or 2nd part of duodenum

Hepatopancreatic ampulla

Figure 2-49. Anterior view of a dissection of the extrahepatic bile passages and pancreatic ducts. Note that the cystic duct is sinuous and that its mucous membrane forms a spiral fold (spiral valve). Observe that the main pancreatic duct with its tributaries resembles a herringbone and that the common hepatic duct with its tributaries resembles a deciduous tree. Note that the accessory pancreatic duct is joined to the main pancreatic duct in this specimen.

position (jejunum). From this junction, a duodenal fold passes superiorly and to the left. In some people, a *superior duodenal recess* posterior to it opens inferiorly. An inferior duodenal fold also extends to the left from the distal part of the duodenum. In many people, this fold covers an *inferior duodenal recess*, which opens superiorly. The terminal part of the ascending part of the duodenum is covered with peritoneum on its posterior surface. As a result, there is a *retroduodenal recess* between the superior and inferior duodenal recesses. Occasionally a paraduodenal fold is raised by the inferior mesenteric vein, and in these cases there may be a *paraduodenal recess* posterior to it. It is on the left side of the duodenojejunal junction and opens to the right. All the duodenal recesses may unite to form one large recess. Often, a duodenojejunal-mesocolic recess, usually referred to as the

mesocolic recess, is located just superior to the duodenojejunal flexure.

Arterial Supply of the Duodenum (Figs. 2-43 to 2-45).   The duodenum has a rich blood supply. Because it is derived from the embryonic foregut and midgut, it is supplied by the *celiac* (foregut) and *superior mesenteric* (midgut) arteries. The main blood supply to the duodenum is from the superior and inferior *pancreaticoduodenal arteries*, which are branches of the gastroduodenal and superior mesenteric arteries, respectively. The proximal half of the duodenum is supplied by the *superior pancreaticoduodenal artery* and the distal half by the *inferior pancreaticoduodenal artery*. These vessels anastomose to form anterior and posterior *arterial arcades*, which lie in the angle between

Figure 2-50.   Relations of the gallbladder: *anteriorly*, anterior abdominal wall and visceral surface of the liver; *posteriorly*, transverse colon and superior and descending parts of the duodenum.

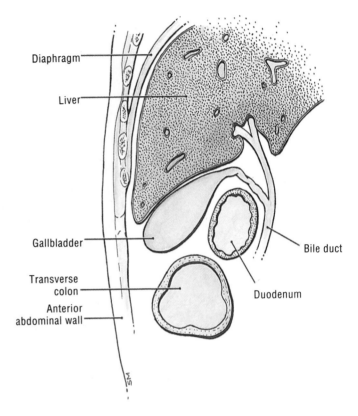

Diaphragm

Liver

Gallbladder

Transverse colon

Anterior abdominal wall

Bile duct

Duodenum

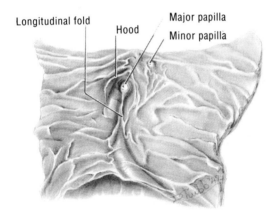

Longitudinal fold

Hood

Major papilla

Minor papilla

Figure 2-51.   Interior of the descending part of the duodenum, showing the spirally disposed folds called plica circulares. Note that the major duodenal papilla projects into this part of the duodenum. On the tip of the papilla is the orifice of the bile duct and, inferior to it, the orifice of the main pancreatic duct; usually these two ducts open together. Note that the minor duodenal papilla, into which the accessory pancreatic duct opens, is about 2 cm anterosuperior to the major duodenal papilla.

the duodenum and pancreas. The superior part of the duodenum may receive blood from: (1) the *supraduodenal artery*, (2) the *right gastric artery*, (3) the *right gastro-omental (gastroepiploic) artery*, and (4) the *gastroduodenal artery*. These vessels often anastomose with each other.

Venous Drainage of the Duodenum (Figs. 2-40, 2-44, 2-45, and 2-52).   In general, the veins follow the arteries and drain into the *portal venous system*. Most duodenal veins drain into the *superior mesenteric vein*, but some enter the portal vein

directly. There are also numerous small veins on the anterior and posterior surfaces of the superior part of the duodenum, some of which drain into the *superior pancreaticoduodenal veins*. One of the anterior veins, called the **prepyloric vein** (of Mayo), ascends anterior to the pylorus and drains into the *right gastric vein*.

> The prepyloric vein is clinically important because it is used by surgeons as a guide to the gastroduodenal junction and the site of the pyloric orifice.

Lymphatic Drainage of the Duodenum (Figs. 2-41, 2-44, 2-45, and 2-60; Table 2-4).   The lymph vessels on the anterior and posterior surfaces of the duodenum anastomose freely with each other within the wall of the duodenum. The anterior vessels follow the arteries and drain superiorly to the *pancreaticoduodenal lymph nodes* along the splenic artery and the *pyloric lymph nodes* along the gastroduodenal artery. Efferent vessels pass to the **celiac lymph nodes**. The posterior lymph vessels pass posterior to the head of the pancreas and drain inferiorly into the *superior mesenteric lymph nodes*, located around the origin of the superior mesenteric artery.

Innervation of the Duodenum (Figs. 2-42 and 2-61; see also Fig. 8-33).   The duodenum is supplied by the vagus and sympathetic nerves via the plexuses on the pancreaticoduodenal arteries.

> In duodenal ulcer, a type of *peptic ulcer* (p. 239), the mucosa is eroded to form a craterlike depression that penetrates the duodenal wall to various depths. **Duodenal ulcers** are commonly located in the duodenal ampulla or ''cap'' (Fig. 2-31). In some cases the ulcer perforates, permitting the duo-

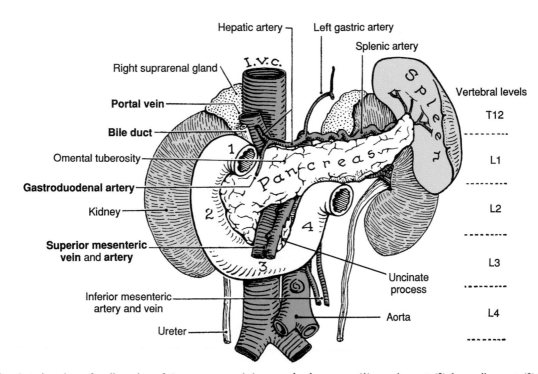

Figure 2-52. Anterior view of a dissection of the pancreas and the structures related to it. Note that the uncinate process projects medially, posterior to the superior mesenteric vessels. The numbered parts of the duodenum are: (1) superior part; (2) descending part; (3) horizontal part; and (4) ascending part.

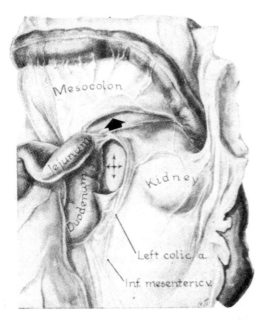

Figure 2-53. Dissection of the duodenal folds and recesses (*arrows*). The transverse colon and its mesentery have been reflected superiorly. Note their relationship to the duodenojejunal flexure and the left kidney. The paraduodenal recess (*left arrow*) is located medial and posterior to the inferior mesenteric vein, between it and the ascending part of the duodenum. Observe the superior duodenal recess (*superior arrow*), guarded by the superior duodenal fold; the inferior duodenal recess (*inferior arrow*), guarded by the inferior duodenal fold; and the retroduodenal recess (*right arrow*). The inferior duodenal fold is bloodless, but the superior duodenal fold often contains the inferior mesenteric vein. The large arrow indicates the mesocolic recess.

denal contents to enter the peritoneal cavity. This gives rise to *peritonitis*. As the superior part of the duodenum is close to the liver and gallbladder, either of them may adhere to or be ulcerated by a duodenal ulcer. The proximity of the duodenum to the gallbladder (Figs. 2-27 and 2-50) explains the frequency with which adhesions are found in persons who have had attacks of *cholecystitis* (inflammation of the gallbladder). This relationship also explains how a *gallstone* may ulcerate from the fundus of the gallbladder and enter a perforated duodenum. Because of the intimate relationship of the pancreas to the duodenum (Figs 2-44, 2-48A, and 2-52), this gland may be invaded by a posterior duodenal ulcer. *Erosion of the gastroduodenal artery* (Figs. 2-43, 2-44, and 2-52), a posterior relation of the superior part of the duodenum, by a duodenal ulcer results in *severe hemorrhage into the peritoneal cavity*.

During the early fetal period, the entire duodenum has a mesentery (Moore, 1988), but most of it becomes fused to the posterior abdominal wall owing to pressure from the overlying transverse colon (Figs. 2-28 and 2-33). Because this is a secondary attachment, the duodenum and the closely associated pancreas may be separated from the underlying retroperitoneal viscera during surgical operations involving the duodenum (Fig. 2-48B), without endangering the blood supply of the kidney or ureter.

The duodenal recesses may become sites of intraperitoneal or *internal hernias*. A loop of small intestine may enter one of these recesses and become impacted. The common hernia of this type, called a **paraduodenal hernia**, is into the *paraduodenal recess*. Because the peritoneal fold guarding the recess usually has to be cut to relieve the strangulation and

Table 2-4.
Lymphatic Drainage of Gastrointestinal Viscera

Viscera	Regional Lymph Nodes	Central Lymph Nodes
Esophagus (abdominal part)	Left gastric	Celiac
Stomach	Gastric, gastro-omental, and pancreaticosplenic	Celiac
Duodenum	Pancreaticoduodenal and pyloric	Celiac and superior mesenteric
Jejunum and ileum	Mesenteric	Superior mesenteric
Pancreas	Pancreaticosplenic and pyloric	Celiac and superior mesenteric
Liver	Hepatic (mostly) and phrenic (partly)	Celiac and parasternal
Gallbladder and bile duct	Cystic and hepatic	Celiac
Cecum, appendix, ascending and transverse colon (proximal two-thirds)	Ileocolic, paracolic, and middle colic	Superior mesenteric
Transverse colon (distal one-third), descending and sigmoid colon	Paracolic and intermediate colic	Inferior mesenteric

to remove the herniated loop of bowel, it is important to know that the *paraduodenal fold is vascularized*. It contains the inferior mesenteric vein and the superior left colic artery.

The Jejunum and Ileum

The jejunum begins at the duodenojejunal flexure (Figs. 2-48 and 2-53). Together the jejunum and ileum are 6 to 7 m long; the jejunum constitutes about two-fifths and the ileum comprises the remainder. The jejunum and ileum are the greatly coiled parts of the small intestine. They are covered to a varying extent by the *greater omentum* (Figs. 2-25 and 2-27). Although there is no clear line of demarcation between the jejunum and ileum, the character of the intestine changes gradually. *Intestinal localization is of surgical importance*, therefore the gross characteristics of the jejunum and ileum are described.

The jejunum is often empty; hence, its name (L. *jejunus*, empty). It is thicker, more vascular, and redder in living persons than is the ileum. Most of the jejunum lies in the umbilical region of the abdomen, whereas the ileum occupies much of the pubic (hypogastric) and right inguinal regions (Fig. 2-5A). The terminal part of the ileum usually lies in the pelvis major (see Fig. 3-7) and ascends over the right psoas major muscle and right iliac vessels to enter the cecum (Fig. 2-26). The circular folds (*plicae circulares*) of the mucous membrane are large and well developed in the superior part of the jejunum (Fig. 2-54), whereas they are small in the superior part of the ileum and absent in its

terminal part. When the jejunum of a living person is grasped between the forefinger and thumb, the circular folds can be felt distinctly through the wall of the superior part of the jejunum. As they are low and sparse in the superior part of the ileum and absent in its inferior part, it is possible to distinguish the superior parts from the inferior parts of the jejunum and ileum.

The Mesentery of the Jejunum and Ileum (Figs. 2-55 to 2-58). The jejunum and ileum are suspended from the posterior abdominal wall by a fan-shaped mesentery. The *root of the mesentery* (about 15 cm long) is directed obliquely, inferiorly, and to the right from the left side of L2 vertebra to the right sacroiliac joint. Between these points, *the root of the mesentery crosses* the following structures: (1) the horizontal part of the duodenum, (2) aorta, (3) inferior vena cava, (4) psoas major muscle, (5) right ureter, and (6) right testicular (or ovarian) vessels. *The jejunum and ileum have varying degrees of mobility.* The proximal part of the jejunum and the distal part of the ileum have shorter mesenteries and are therefore less mobile than other parts. The pleated, fan-shaped mesentery consists of two layers of peritoneum, between which are jejunal and ileal blood vessels, lymphatics, nerves, and extraperitoneal fatty tissue (Fig. 2-56). The jejunal mesentery contains less fat than that of the ileum; thus the arterial arcades of the jejunum are easier to observe than in the ileum (Fig. 2-57). This anatomical characteristic is another criterion surgeons use to differentiate the jejunum from the ileum.

Arterial Supply of the Jejunum and Ileum (Figs. 2-39, 2-48, and 2-55 to 2-58). The arteries to the jejunum and ileum arise from the *superior mesenteric artery*, the second of the unpaired branches of the abdominal aorta. This artery usually arises at the level of L1 vertebra, about 1 cm inferior to the celiac trunk and posterior to the body of the pancreas and the splenic vein. It descends across the left renal vein, the uncinate process of the pancreas, and the horizontal part of the duodenum to enter the mesentery. The **superior mesenteric artery** runs obliquely in the root of the mesentery to the right iliac fossa, sending many branches to the intestines. Its last ileal branch anastomoses with a branch of the ileocolic artery. The 15 to 18 jejunal and ileal branches arise from the left side of the superior mesenteric artery and pass between the two layers of the mesentery. The arteries unite to form loops or arches called *arterial arcades*, from which vasa recta (L. straight vessels) arise. The **vasa recta** do not anastomose within the mesentery. They pass from the arcades to the mesenteric border of the intestine, where they pass more or less alternately to opposite sides. There are many anastomoses of the blood vessels in the wall of the intestine. This vascularity is greater in the jejunum than in the ileum, but the arterial arcades are shorter and more complex in the ileum.

Venous Drainage of the Jejunum and Ileum (Figs. 2-40, 2-45, and 2-52). The *superior mesenteric vein* drains the jejunum and ileum. It accompanies the superior mesenteric artery, lying anterior and to its right in the root of the mesentery. The superior mesenteric vein crosses the horizontal (third) part of the duodenum and the uncinate process of the pancreas. It terminates posterior to the neck of the pancreas by uniting with the splenic vein to form the *portal vein*. The tributaries of the superior mesenteric vein have an arrangement similar to the branches of

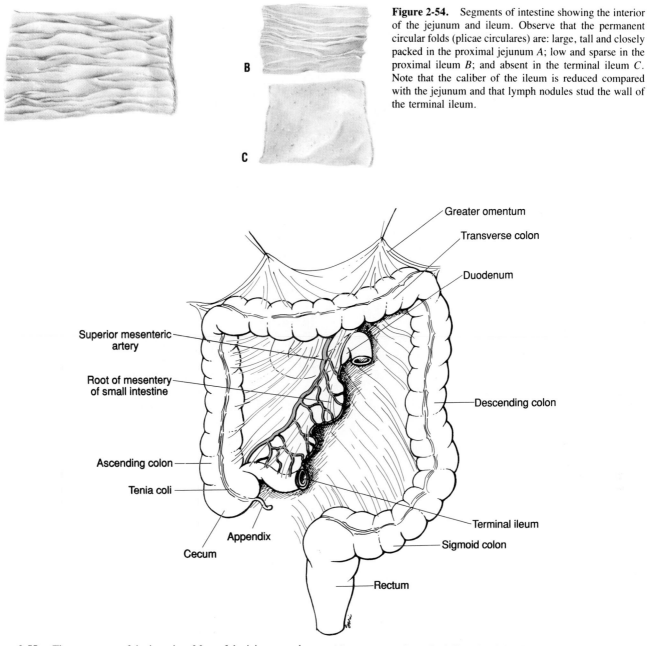

Figure 2-54. Segments of intestine showing the interior of the jejunum and ileum. Observe that the permanent circular folds (plicae circulares) are: large, tall and closely packed in the proximal jejunum *A*; low and sparse in the proximal ileum *B*; and absent in the terminal ileum *C*. Note that the caliber of the ileum is reduced compared with the jejunum and that lymph nodules stud the wall of the terminal ileum.

Figure 2-55. The mesentery of the intestine. Most of the jejunum and ileum are removed. Note that the root of the mesentery extends in an oblique manner from the left to the right side, a distance of about 15 cm in adults (see also Fig. 2-81).

the superior mesenteric artery, and they drain the same areas supplied by these arteries.

Lymphatic Drainage of the Jejunum and Ileum (Figs. 2-56, 2-59, and 2-60; Table 2-4). The lymphatics in the intestinal villi, called *lacteals* (L. *lactis*, milk), empty their milklike fluid into a plexus of lymph vessels in the walls of the jejunum and ileum. The lymph vessels then pass between the two layers of the mesentery to the **mesenteric lymph nodes**. These nodes are in three locations: (1) close to the wall of the intestine, (2) amongst the arterial arcades, and (3) along the proximal part of the superior mesenteric artery. The lymph vessels from the ter-

minal ileum follow the ileal branch of the ileocolic artery to the *ileocolic lymph nodes*. Efferent lymph vessels from all the mesenteric lymph nodes drain into the *superior mesenteric lymph nodes*.

Innervation of the Jejunum and Ileum (Figs. 2-42 and 2-61; Table 2-3). The nerves of the jejunum and ileum are derived from the *vagus* and *splanchnic nerves* through the celiac ganglion and the nerve plexuses around the superior mesenteric artery. The *superior mesenteric nerve plexus* receives its parasympathetic fibers from the celiac division of the posterior vagal trunk, and its sympathetic fibers from the superior mesenteric ganglion.

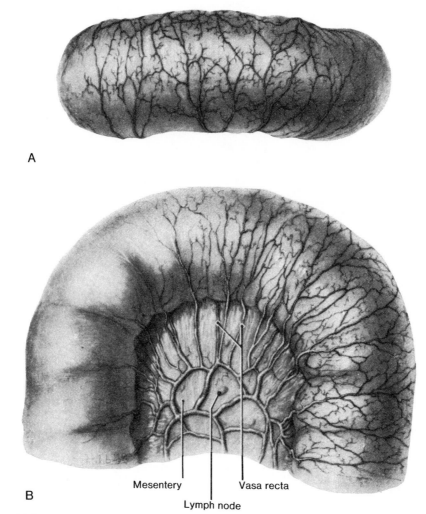

B

Mesentery   Vasa recta

Lymph node

Figure 2-56.   Arterial supply of the jejunum. *A*, Antimesenteric border. *B*, The arteries in the mesentery and wall of the intestine.

Figure 2-57.   Arteries of the jejunum and ileum. Compare the diameter, thickness of the wall, number of arterial arcades, long or short vasa recta, presence of translucent (fat-free) areas at the mesenteric border, and fat encroaching on the wall of the gut. Note that the arterial arcades in the ileum are more complex and that the vasa recta are shorter than in the jejunum.

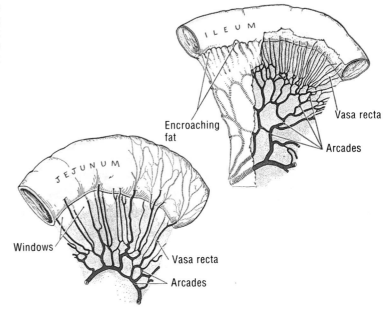

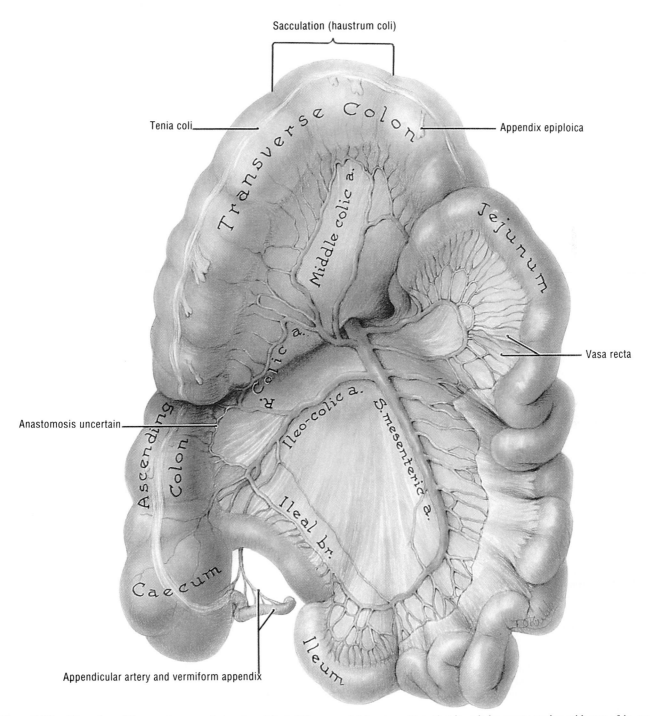

Figure 2-58. Dissection of the superior mesenteric artery. Most of the peritoneum has been stripped off. Observe the extensive field of supply of this artery. Note that it ends by anastomosing with one of its own branches, the ileal branch of the ileocolic artery.

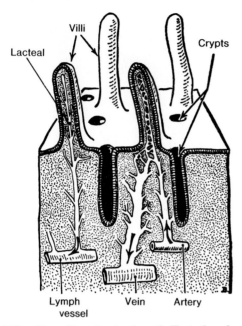

Villi

Lacteal

Crypts

Lymph vessel   Vein   Artery

Figure 2-59. Three-dimensional, schematic illustration of a section of small intestine, showing its arterial supply and venous and lymphatic drainage.

Congenital malformations resulting from *abnormal rotation of the midgut* during the fetal period are common. Sometimes the intestinal loops fail to return to the abdominal cavity from the umbilical cord during the 10th fetal week (Moore, 1988). This large hernia, called an *omphalocele*, usually contains most of the small intestine. The hernial sac is formed by the amnion that covers the umbilical cord. Sometimes the intestines return to the abdominal cavity during the fetal period, but there is incomplete closure of the anterior abdominal wall in the umbilical region. As a result, part of the omentum or an intestinal loop may herniate after birth, forming an *umbilical hernia*, which is often no larger than a cherry. This type of hernia is covered by skin and is most noticeable when the infant cries.

Abnormal rotation and poor fixation of the intestine sometimes occurs as it returns to the abdominal cavity during the 10th fetal week. Incomplete rotation usually results in *nonfixation of the mesentery*; as a result, the entire midgut loop may hang from a narrow pedicle. Twisting of the intestine and the superior mesenteric vessels often occurs. This twisting, called *volvulus* (L. *volvo*, to roll), may render the small intestine *ischemic* (deficient in blood) and it may undergo *necrosis*.

An **ileal diverticulum** (Meckel's diverticulum) is one of the most common malformations of the digestive tract (Fig. 2-62*B*). It occurs in 1 to 2% of people. This blind sac or fingerlike pouch is the remnant of the proximal part of the embryonic *yolk stalk* (Moore, 1988). It is of clinical significance because it sometimes becomes inflamed (*ileal diverticulitis*) and may cause symptoms that mimic appendicitis. The wall of an ileal diverticulum contains all layers of the ileum and it may contain patches of gastric-type of epithelium and pancreatic tissue. The gastric mucosa may secrete acid, producing ulceration and bleeding from the diverticulum. Typically, an ileal diverticulum is 3 to 6 cm long and *projects*

from the antimesenteric border of the ileum, within 50 cm of the ileocecal junction. It may be connected to the umbilicus by a fibrous cord or a *fistula*, which represents a *persistent yolk stalk*.

Occlusion of vasa recta (arteries or veins) results in poor nutrition or drainage of the part of the intestine concerned. An *embolus* in an artery or a *thrombosis* in a vein may lead to *necrosis* of the segment of bowel concerned and *ileus* of the paralytic type (G. *eileos*, intestinal colic). If the condition is diagnosed early enough (*e.g.*, using a *superior mesenteric arteriogram*), the obstructed portion of the vessel may be cleared surgically. Owing to the many anastomoses of blood vessels (Fig. 2-56), blockage of a single vessel or a small group of vessels is not usually followed by *gangrene* (death of tissue) of the intestine.

The Spleen

The spleen (G. *splen*, L. *lien*) is a large, soft vascular lymphatic organ in the left upper quadrant (Figs. 2-26, 2-38, 2-44, 2-46, and 2-47). The *largest single mass of lymphoid tissue in the body*, the spleen is located between the layers of the dorsal mesogastrium (gastric mesentery), which suspends the stomach from the posterior abdominal wall (Fig. 2-23). It is in contact with the posterior wall of the stomach and is connected to its greater curvature by the gastrosplenic (gastrolienal) ligament and to the left kidney by the splenorenal (lienorenal) ligament. During life the spleen is a soft, purplish, freely movable organ, which is considerably larger than in most cadavers. It is located in the left hypochondrium, posterior to the stomach and anterior to the superior part of the left kidney (Figs. 2-35 and 2-46). The spleen lies against the diaphragm laterally, which separates it from the pleural cavity, and is related to the 9th and 11th ribs on the left side. The diaphragmatic surface of the spleen is convexly curved to fit the concavity of the diaphragm. Its anterior and superior borders are sharp and are often notched (Figs. 2-35 and 2-47). These notches represent the remains of the lobulated fetal spleen (Moore, 1988). The posterior and inferior borders of the spleen are rounded.

The *gastrosplenic* and *splenorenal ligaments* are attached to the hilum (hilus) of the spleen on its medial aspect (Fig. 2-47), where the branches of the splenic artery enter and the tributaries of the splenic vein leave (Figs. 2-43 and 2-44). Except at the hilum, the spleen is completely surrounded by peritoneum. Recall that it develops between the layers of the dorsal mesogastrium (Moore, 1988). The hilum is usually intimately related to the tail of the pancreas (Figs. 2-26*B*, 2-46, and 2-47). The spleen varies in size and shape, but it is usually about 12 cm long and 7 cm wide and will usually fit into one's cupped hand. It normally contains a large amount of blood. Its capsule and trabeculae contain some smooth muscle, which enables it to expel its blood into the circulation. The shape of the spleen is affected by the fullness of the stomach and the transverse colon at the left colic flexure (Figs. 2-26*A* and 2-47). The distended stomach gives the spleen the shape of a segment of orange, whereas when the colon is full the spleen has a four-sided appearance.

Surface Anatomy of the Spleen (Figs. 2-26 and 2-46). The spleen is under the shelter of the osteocartilaginous thoracic cage. It normally lies deep to the 9th to 11th ribs and its external

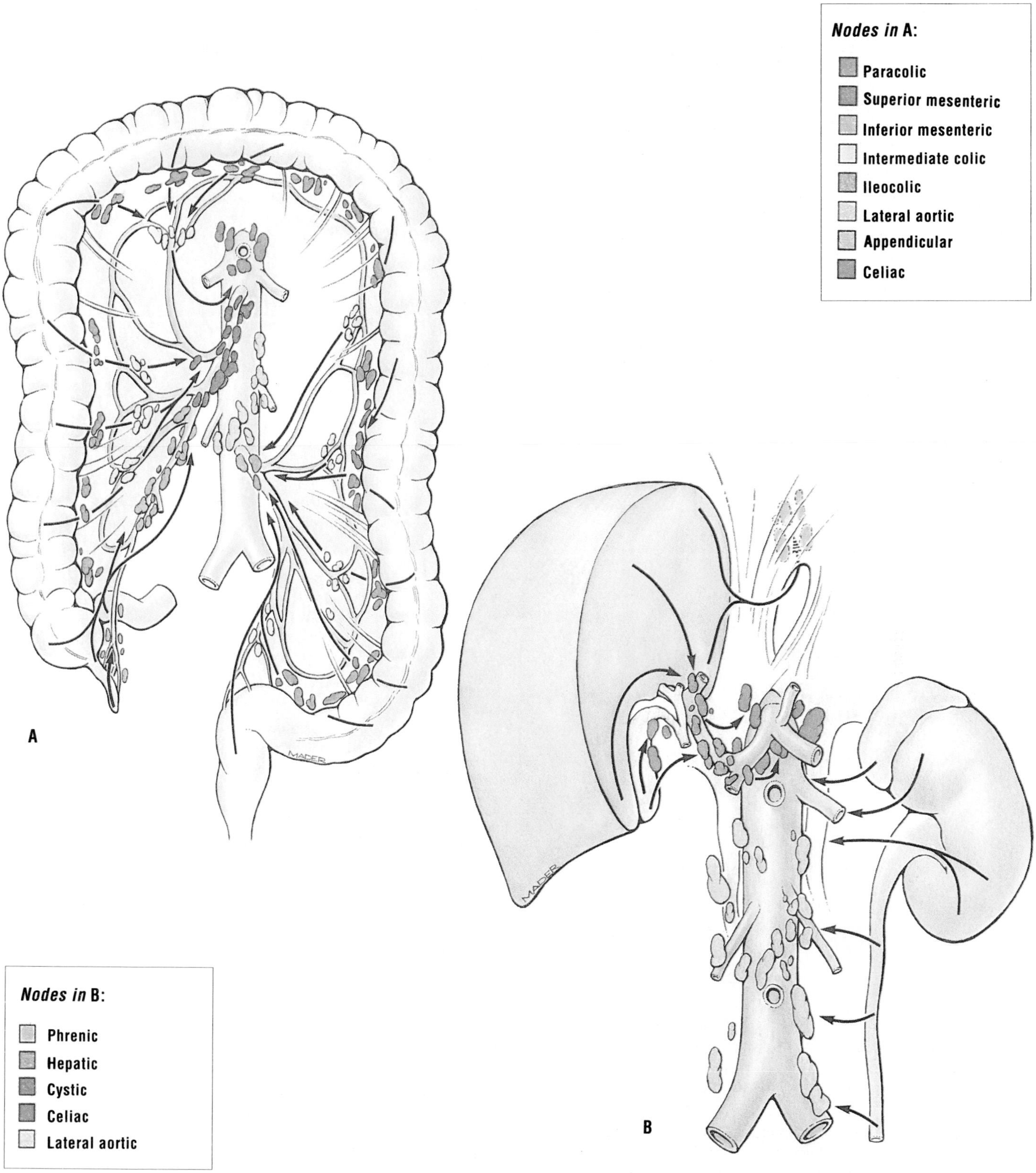

Nodes in A:

■ Paracolic
■ Superior mesenteric
■ Inferior mesenteric
□ Intermediate colic
■ Ileocolic
■ Lateral aortic
■ Appendicular
■ Celiac

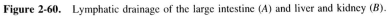

Nodes in B:

□ Phrenic
■ Hepatic
■ Cystic
■ Celiac
□ Lateral aortic

A

B

Figure 2-60.   Lymphatic drainage of the large intestine (*A*) and liver and kidney (*B*).

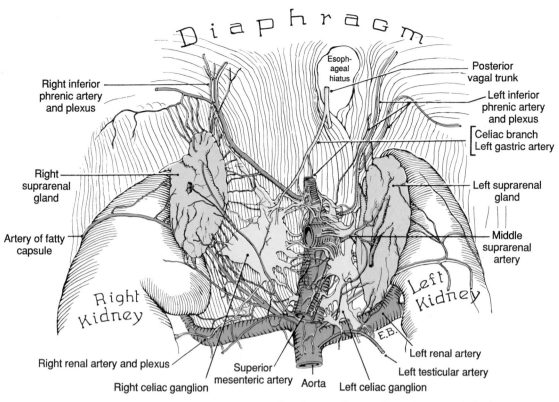

Figure 2-61. Dissection of the celiac trunk, celiac plexus, celiac ganglia, and suprarenal glands.

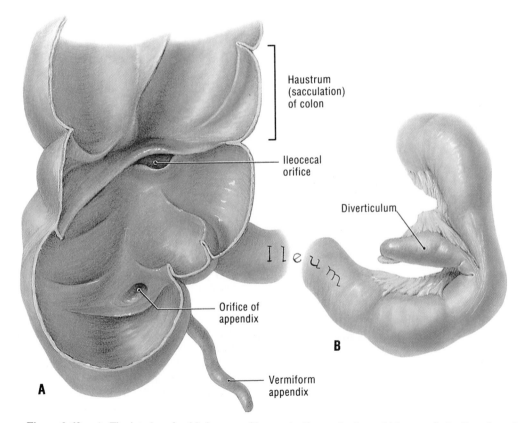

Figure 2-62. *A*, The interior of a dried cecum. Observe the ileocecal valve, which controls the flow through the ileocecal orifice. *B*, An ileal (Meckel's) diverticulum.

186

surface is convex to fit these bones. The long axis of the spleen lies in the line of the 10th rib, where it rests on the left colic flexure. Normally the spleen does not extend inferior to the left costal margin; hence, a normal spleen is seldom palpable through the anterolateral abdominal wall. The anterior tip of the spleen usually does not extend farther medially than the midclavicular line.

Arterial Supply of the Spleen (Figs. 2-33, 2-35, 2-38, 2-43 to 2-45, and 2-47). The *splenic artery* is the largest branch of the celiac trunk. It follows a tortuous course posterior to the omental bursa, anterior to the left kidney, and along the superior border of the pancreas. Between the layers of the splenorenal ligament, the **splenic artery** divides into five or more branches, which enter the hilum of the spleen. The branches of the splenic artery supply the individual elements of the spleen as *end arteries*. There are no anastomoses between the small branches of the splenic arteries. Consequently obstruction of one of them results in death of splenic tissue (*splenic infarction*).

Venous Drainage of the Spleen (Figs. 2-33, 2-35, 2-40, and 2-47). The *splenic vein* is formed by several tributaries that emerge from the hilum of the spleen. It is joined by the inferior mesenteric vein and runs posterior to the body and tail of the pancreas throughout most of its course. The splenic vein unites with the superior mesenteric vein posterior to the neck of the pancreas to form the *portal vein*.

Lymphatic Drainage of the Spleen (Figs. 2-41 and 2-45). Lymph vessels arise from the spleen and pass along the splenic vessels to the *pancreaticosplenic lymph nodes*. These nodes are related to the posterior surface and superior border of the pancreas.

Innervation of the Spleen (Figs. 2-61 and 2-82). The nerves to the spleen are derived from the *celiac plexus*. They are distributed mainly to the branches of the splenic artery and are vasomotor in function.

If ruptured (*e.g.*, by a fractured left rib), the spleen will bleed profusely because its capsule is thin and its parenchyma is soft and pulpy. Repair of a ruptured spleen is difficult but preferable to removal (**splenectomy**) in children and adolescents because it has an important role in the immune system (Cormack, 1987).

Partial removal of the spleen is followed by rapid regeneration. Even total splenectomy does not produce serious effects, especially in adults, because its functions are assumed by other reticuloendothelial organs. When the spleen is diseased it may be 10 or more times its normal size (*splenomegaly*). In some cases, the spleen may fill the left half of the abdomen. When a spleen is grossly enlarged, it projects inferior to the left costal margin and its notched superior border faces inferomedially. Its notched border is helpful when *palpating an enlarged spleen* because when the patient takes a deep breath, these notches can often be palpated as it moves inferoanteriorly.

Certain conditions (trauma, tumors, and some hematological diseases) often require *splenectomy*. During this operation, the surgeon has to be aware of the intimate relationship of the tail of the pancreas to the hilum of the spleen to avoid injury to this important digestive gland. Although well protected by the ribs (Fig. 2-46), *the spleen is the most frequently injured organ in the abdomen* when severe blows are received

to the left hypochondrium. Injury usually results from rib fractures and is partly the result of its thin capsule and friability. Sometimes football players have their spleens ruptured when they are hit hard on the left side. *Rupture of the spleen causes severe intraperitoneal hemorrhage and shock.* Splenectomy is often performed to prevent the patient from bleeding to death. The spleen may rupture spontaneously in some patients with *infectious mononucleosis*, malaria, and septicemia ("blood poisoning") because it is large and more friable under these conditions.

An accessory spleen (one or more) is usually near the spleen's hilum, but they may be embedded partly or wholly in the tail of the pancreas. They may also be found between the layers of the gastrosplenic ligament. Accessory spleens are common (10%) and usually are about 1 cm in diameter. Awareness of their possible presence is important because, if not removed during splenectomy, they may result in persistence of the symptoms that indicated removal of the spleen (*e.g.*, splenic anemia).

The relationship of the *costodiaphragmatic recess* of the pleural cavity to the spleen is clinically important (Fig. 2-35). This potential space occurs at the level of the 10th rib in the midaxillary line. Its existence must be kept in mind when doing a *splenic needle biopsy* or when injecting radiopaque material into the spleen for visualization of the portal vein (*splenoportography*). If care is not taken, this material may enter the pleural cavity and cause *pleuritis* (p. 63).

The Pancreas

The pancreas is an elongated (12 to 15 cm), soft, grayish-pink digestive gland (Figs. 2-44, 2-46, and 2-52). It is located near the transpyloric plane, more or less transversely, across the posterior abdominal wall and posterior to the stomach. The transverse mesocolon is attached to its anterior margin (Fig. 2-48A). The pancreas (G. *pankreas*, sweetbread; Fr. *pas* [pan], all + *kreas*, flesh) is both an exocrine and endocrine gland. It produces (1) an external secretion (*pancreatic juice*) that enters the duodenum via the pancreatic duct (Fig. 2-49) and (2) internal secretions (*glucagon* and *insulin*) that enter the blood.

The pancreas is located in the epigastric and left hypochondriac regions and its right part lies across the bodies of L1 to L3 vertebrae (Figs. 2-46 and 2-52). It has a *head, neck, body,* and *tail* (Fig. 2-26B) and its shape somewhat resembles an inverted, curved tobacco pipe (Tobias et al., 1988). Its right side (*head*) is usually inferior to the transpyloric plane, and its left side (*tail*) is slightly superior to this plane. The pancreas lies posterior to the omental bursa, where it forms a major part of the *stomach bed* (Fig. 2-35).

The head of the pancreas is located within the curve of the duodenum and is embraced by it (Figs. 2-43 to 2-46). It has a prolongation, called the *uncinate process* (L. hook-shaped), which extends superiorly and to the left and lies posterior to the superior mesenteric vessels; here it rests against the aorta posteriorly (Figs. 2-48 and 2-52). The head rests posteriorly on the inferior vena cava, the right renal vessels, and the left renal vein. The *bile duct*, on its way to the duodenum, lies in a groove on the posterosuperior surface of the head of the pancreas; sometimes it is embedded in it (Fig. 2-45).

The neck of the pancreas (Figs. 2-26*B*, 2-43, 2-44, 2-45, and 2-48), about 2 cm long, is continuous with the superior left portion of the head and merges imperceptibly into the body of the pancreas. The neck is grooved posteriorly by the superior mesenteric vessels. Its anterior surface is covered with peritoneum and is adjacent to the pylorus of the stomach. The superior mesenteric vein joins the splenic vein posterior to the neck of the pancreas to form the *portal vein*.

The body of the pancreas (Figs. 2-26*B*, 2-33, 2-35, and 2-52) extends slightly superiorly as it extends to the left across the aorta and superior lumbar vertebrae, posterior to the omental bursa. The body is somewhat triangular in cross-section and *has three surfaces*: anterior, posterior, and inferior. The *anterior surface* is covered with peritoneum (Fig. 2-48*A*) and forms part of the bed of the stomach, where it provides attachment for the transverse mesocolon. The *posterior surface* is devoid of peritoneum (Fig. 2-48*B*) where it is in contact with the aorta, superior mesenteric artery, left suprarenal gland, and the left kidney and its vessels. *The body of the pancreas is intimately related to the splenic vein* (Figs. 2-40 and 2-45). Where it lies anterior to the aorta, the body of the pancreas lies between the celiac trunk and the superior mesenteric artery (Figs. 2-45 and 2-48). The inferior border of the body of the pancreas separates the posterior surface from the inferior surface. The body has a small projection, the *omental tuberosity*, arising from its superior border, and contacting the lesser omentum (Fig. 2-52). This tuberosity is located immediately inferior to the celiac trunk (Fig. 2-45).

The tail of the pancreas (Figs. 2-26*B*, 2-33, 2-44, 2-47, and 2-52) is its narrow left end. It is thick and may be pointed or blunt. It passes between the two layers of the *splenorinal (lienorenal) ligament* with the splenic vessels. Its end usually contacts the hilum of the spleen.

The Pancreatic Ducts (Figs. 2-49 and 2-63).   There are two ducts, a main one and an accessory one.

The pancreas develops from two outgrowths, called dorsal and ventral *pancreatic buds* (Moore, 1988). They normally fuse in such a way that their ducts communicate. Usually the duct of the ventral pancreatic bud forms the proximal end of the *main pancreatic duct*, and the distal part of the dorsal pancreatic duct forms the remainder. Frequently the proximal (duodenal) end of the dorsal pancreatic duct also persists and forms an *accessory pancreatic duct*.

The main pancreatic duct begins at the tail of the pancreas and runs through the substance of the gland, receiving tributaries in a herringbone pattern (Fig. 2-49). The main duct is embedded rather superficially beneath the posterior surface of the pancreas. When the main duct is joined by the parts of the duct in the head and uncinate process, it becomes Y-shaped. Within the head, the main pancreatic duct turns inferiorly and comes into close relationship with the bile duct. The pancreatic and bile ducts pierce the posteromedial wall of the descending (second) part of the duodenum obliquely, near its middle (Fig. 2-49). Usually the pancreatic and bile ducts unite to form a short, dilated *hepatopancreatic ampulla* (ampulla of Vater). The ampulla opens via a common duct into the duodenum at the summit of the *major duodenal papilla* (Fig. 2-51). There is a sphincter around the terminal part of the main duct, called the *pancreatic duct sphincter*. There is also one around the hepatopancreatic ampulla called the *hepatopancreatic sphincter* (sphincter of Oddi). These sphincters control the flow of bile and pancreatic juice into the duodenum.

The *accessory pancreatic duct* drains part of the head of the

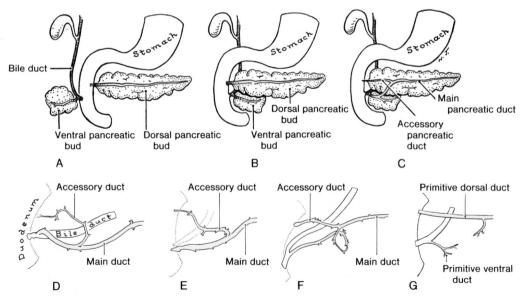

Figure 2-63.   The development of the pancreas and the variability in the anatomy of the pancreatic ducts. *A,* In the embryo the smaller ventral pancreatic bud arises with the bile duct and the larger dorsal pancreatic bud arises independently from the duodenum. *B,* The duodenum rotates laterally on its long axis, bringing the ventral pancreatic bud and bile duct posterior to the dorsal pancreatic bud. *C,* A connecting segment unites the dorsal duct to the ventral duct. In *A* to *C,* note that the ventral bud forms the head, including the uncinate process and that the dorsal bud forms the neck, body, and tail of the pancreas. *D,* Often (44%), the accessory pancreatic duct loses its connection with the duodenum. *E,* In some people (10%), the accessory pancreatic duct opens into the duodenum. *F,* In other people (20%), the proximal part of the accessory duct is small. *G,* In a few people (9%), the main pancreatic duct has no connection with the duct draining the head.

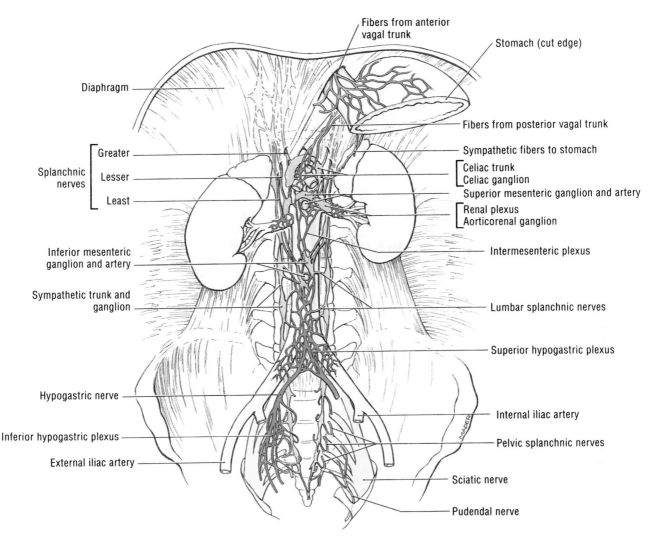

Figure 2-64. The autonomic nerve supply of the abdomen and pelvis. Both sympathetic and parasympathetic fibers form a complex tangle of nerves around the stems of the celiac, superior mesenteric, and inferior mesenteric arteries. Observe that the sympathetic fibers here synapse outside the sympathetic trunk in ganglia, some of which are small and scattered, but two of which are large, the celiac and aorticorenal ganglia.

pancreas; it is variable (Figs. 2-49 and 2-63C to G). Usually it is connected to the main pancreatic duct, but in about 9% of people it is a completely separate duct that opens into the duodenum at the *minor duodenal papilla* (Fig. 2-51).

Arterial Supply of the Pancreas (Figs. 2-43 to 2-45 and 2-52). The arteries of the pancreas are derived from the *splenic* and *pancreaticoduodenal arteries*. Up to 10 small branches of the splenic artery supply the body and tail of the pancreas. The anterior and posterior superior pancreaticoduodenal arteries, from the *gastroduodenal artery*, and the anterior and posterior inferior pancreaticoduodenal arteries, from the *superior mesenteric artery*, supply the head of the pancreas. The pancreaticoduodenal arteries anastomose freely with each other. The groove between the anterior part of the head of the pancreas and the duodenum lodges the *anterior pancreaticoduodenal arcade*, whereas the corresponding groove between the posterior part of the head and the duodenum lodges the *posterior pancreaticoduodenal arcade*.

Venous Drainage of the Pancreas (Figs. 2-40 and 2-45). The pancreatic veins drain into the portal, splenic, and superior mesenteric veins, but most of them empty into the *splenic vein*.

Lymphatic Drainage of the Pancreas (Figs. 2-41, 2-44, and 2-45; Table 2-4). The lymph vessels of the pancreas follow the blood vessels. Most of them end in the *pancreaticosplenic nodes* that lie along the splenic artery on the superior border of the pancreas, but some vessels end in the *pyloric lymph nodes*. Efferent vessels from these nodes drain to the *celiac, hepatic,* and *superior mesenteric lymph nodes*.

Innervation of the Pancreas (Figs. 2-42 and 2-64; Table 2-3, p. 171). The nerves are derived from the vagus and splanchnic nerves. The pain fibers are carried by the splanchnic nerves. The sympathetic and parasympathetic fibers reach the gland by passing along the arteries from the *celiac* and *superior mesenteric plexuses*.

Because the main pancreatic duct usually joins the bile duct to form the *hepatopancreatic ampulla* and empties via a common duct that pierces the duodenal wall, a *gallstone*

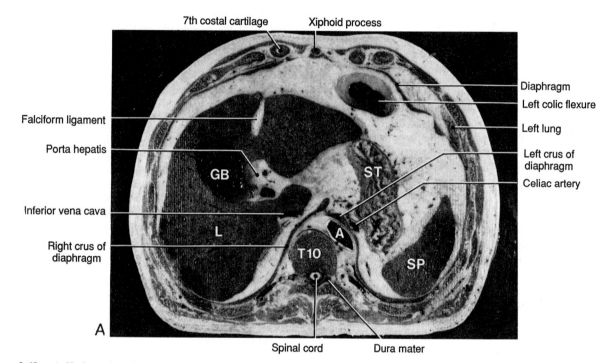

7th costal cartilage   Xiphoid process

Falciform ligament

Porta hepatis

Inferior vena cava

Right crus of
diaphragm

Diaphragm

Left colic flexure

Left lung

Left crus of
diaphragm

Celiac artery

GB   ST

L

A

T10

SP

A

Spinal cord   Dura mater

Figure 2-69.   *A*, Horizontal section of the abdomen of an adult female cadaver at the level of T10 vertebra. In keeping with radiographic convention, the section is viewed inferiorly. Hence, the right side of the body appears on the left side of the photograph. *GB*, indicates gallbladder; *ST*, stomach; *A*, descending aorta; *L*, liver; *SP*, spleen; *T10*, tenth thoracic vertebra. *B*, CT scans of the abdomen. The elongated pancreas is well shown in the middle of the third row. (Courtesy of Dr. Tom White, Department of Radiology, The Health Science Center, The University of Tennessee, Memphis, TN.) If you have difficulty recognizing any of the structures, see the drawings accompanying the MRIs on page 162.

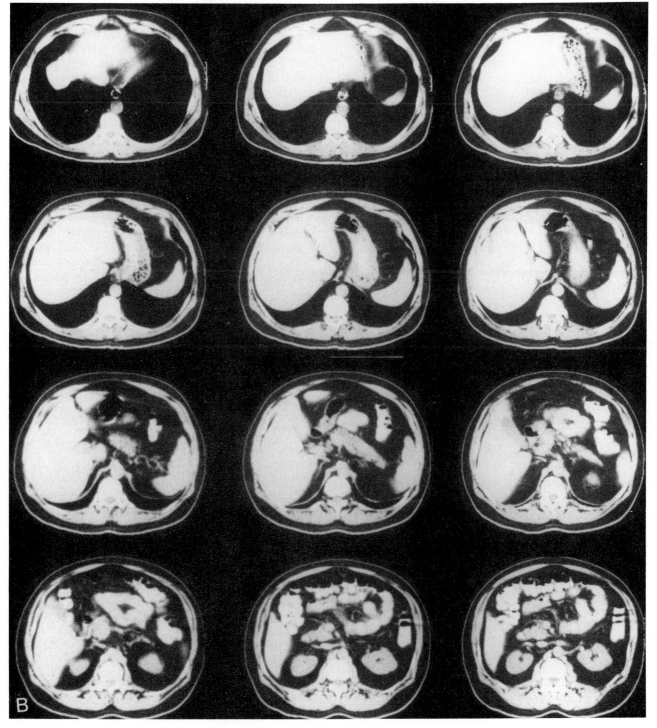

Figure 2-69*B*

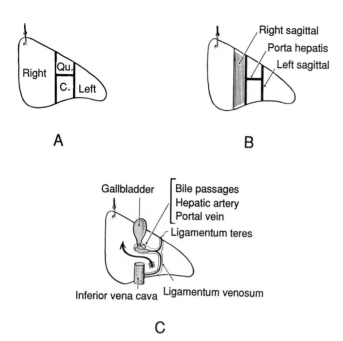

Figure 2-70. Features of the visceral surface of the liver. *A*, Its four lobes: right, quadrate (*Qu.*), caudate (*C.*), and left. Formerly the caudate and quadrate lobes were considered part of the right lobe, but now the caudate lobe and most of the quadrate lobe are considered to be part of the left lobe. *B*, The H-shaped fissures and grooves defining the lobes. *C*, The structures in the fissures and grooves. The arrow traverses the omental (epiploic) foramen and passes into the omental bursa. The division of the liver into right and left lobes runs along the plane of the gallbladder and the inferior vena cava (Fig. 2-72*A* and *B*).

On the visceral surface, the layers of the falciform ligament reflect onto the liver along the line of the fissure for the ligamentum teres, as far as the porta hepatis (Figs. 2-67 and 2-70*C*). At the superior end of the falciform ligament, its two layers separate from each other, exposing a hand-sized triangular area on the superior surface called the **bare area of the liver** because it is devoid of peritoneum (Figs. 2-67 and 2-68). Here, the two peritoneal layers diverge laterally and are reflected onto the diaphragm to form the *coronary ligament* (Fig. 2-68). Because the liver is applied directly to the diaphragm in this area, the right and left layers of the coronary ligament are spread apart and surround the bare area. The right reflection of the falciform ligament constitutes the anterior layer of the coronary ligament. It passes to the right and then bends sharply at the *right triangular ligament* to become the posterior layer of the coronary ligament. The left reflection of the falciform ligament forms the *left triangular ligament*, which becomes continuous with the posterior layer of the coronary ligament (Figs. 2-48 and 2-68).

From the *porta hepatis* the peritoneal reflections pass to the lesser curvature of the stomach and the superior part of the duodenum as the **lesser omentum** (Figs. 2-23, 2-27, and 2-67). The *left triangular ligament* (posterior layer) is continuous with the lesser omentum (Figs. 2-35 and 2-68). The portion of the lesser omentum, extending between the liver and the stomach is called the *hepatogastric ligament*, and the part between the liver and the duodenum is called the *hepatoduodenal ligament*. The two layers in the *free edge of the lesser omentum* enclose the *hepatic artery*, *portal vein*, *bile duct* (Figs. 2-23 and 2-66), a few lymph nodes and vessels, and the hepatic plexus of nerves.

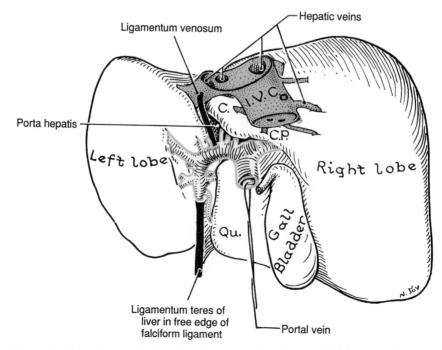

Figure 2-71. Visceral surface of the liver. Observe the branches of the portal vein (*yellow*) entering the liver and the three large and several small hepatic veins (*blue*) leaving the liver to join the inferior vena cava. *C.* indicates caudate lobe; *C.P.* caudate process; *Qu.* quadrate lobe.

Arterial Supply of the Liver (Figs. 2-33, 2-40, 2-66, 2-67, and 2-72). The liver has a *double blood supply* from the hepatic artery (30%) and the portal vein (70%). The right and left hepatic arteries carry oxygenated blood to the liver and the **portal vein** carries venous blood containing the products of digestion absorbed from the GI tract. The arterial blood is conducted to the central vein of each liver lobule. The *common hepatic artery* arises from the celiac trunk and passes anteriorly to the right in the posterior wall of the omental bursa. It runs inferior to the omental foramen to reach the superior part of the duodenum. After giving off the gastroduodenal artery, it passes between the layers of the lesser omentum as the *hepatic artery proper*. This artery ascends in the free edge of the lesser omentum anterior to the portal vein and to the left of the bile duct. Near the porta hepatis the hepatic artery proper divides into right and left terminal branches, called the *right* and *left hepatic arteries*. Variations of the hepatic arteries are common (Fig. 2-73).

The portal vein, supplying most of the blood to the liver, is formed posterior to the neck of the pancreas by the union of the superior mesenteric and splenic veins (Fig. 2-40). It runs in the free right edge of the lesser omentum, posterior to the bile duct and hepatic artery (Fig. 2-23) and anterior to the omental foramen. At the right end of the *porta hepatis*, the portal vein terminates by dividing into right and left branches, each of which supplies about half of the liver (Figs. 2-40 and 2-72*E*).

Venous Drainage of the Liver (Figs. 2-40, 2-66, and 2-71). The hepatic veins draining blood from the liver are formed by the union of the central veins in its lobules. The *hepatic veins* open into the inferior vena cava just inferior to the diaphragm. There are superior and inferior groups of veins. The superior group may consist of only right and left veins, but usually there is a middle vein from the caudate lobe. The inferior group consists of 6 to 18 small veins, which drain blood from the right lobe, including part of the caudate lobe.

Lymphatic Drainage of the Liver (Figs. 2-41, 2-44, 2-60*B*, and 2-67; Table 2-4). Most of the *deep lymph vessels* from the liver converge at the porta hepatis and end in the **hepatic lymph nodes** scattered along the hepatic vessels and ducts in the lesser omentum, such as the *cystic lymph node* near the neck of the gallbladder and the *lymph node of the omental foramen*. Efferent vessels from the hepatic lymph nodes drain into the *celiac lymph nodes*, which are clustered around the celiac trunk and proximal parts of its branches. Lymph from these nodes enters the *thoracic duct* (see Figs. 1-39 and 1-70). Some of the deep lymph vessels follow the hepatic veins to the *vena caval foramen* in the diaphragm (Fig. 2-98), and end in the middle group of **phrenic lymph nodes**. Efferent vessels from these nodes end in the *parasternal lymph nodes*. Most of the *superficial lymph vessels* from the liver join the lymph vessels in the *porta hepatis* and enter the hepatic lymph nodes, which drain into the celiac lymph nodes. Lymph vessels from the bare area on the diaphragmatic surface pass through the diaphragm (*e.g.*, via the sternocostal hiatus and vena caval foramen; Fig. 2-98) and enter the phrenic and mediastinal lymph nodes. Lymph from these nodes empty into the right lymphatic and thoracic ducts (see Fig. 24, p. 25).

Innervation of the Liver (Fig. 2-42; see also Fig. 1-73). The nerves to the liver contain both *sympathetic* and *parasympathetic fibers*. These nerves reach the liver via the *hepatic plexus*, the largest derivative of the *celiac plexus*, which also receives filaments from the left and *right vagus* and *right phrenic nerves*. The hepatic plexus of nerves accompanies the hepatic artery and portal vein and their branches and enters the liver at the porta hepatis.

Hepatic tissue for diagnostic purposes may be obtained by *liver biopsy*. The needle puncture is commonly made through the right 10th intercostal space in the midaxillary line. The biopsy is taken while the patient is holding his/her breath in full expiration to reduce the *costodiaphragmatic recess* (Figs. 2-25 and 2-27) and to lessen the possibility of damaging the lung and contaminating the pleural cavity.

The liver is a common site of metastatic carcinoma (CA) from any organ drained by the portal system of veins (Fig. 2-40). Cancer cells may also pass to the liver from the thorax, especially the breast, owing to the communications between thoracic lymph nodes and lymph vessels draining the bare area of the liver. Normally the liver is a soft mass with a jellylike consistency, almost all of which is protected by the osteocartilaginous thoracic cage (Fig. 2-36). In some normal people, especially those with a wide infrasternal angle, the inferior border and middle portions of the liver may be palpable. The right lobe may be palpable 1 to 2 cm inferior to the right costal margin; hence, a palpable inferior border of the liver does not by itself indicate liver enlargement (*hepatomegaly*). Large livers are associated with CA, *cirrhosis of the liver*, congestive heart failure, fatty infiltration, and Hodgkin's disease.

Liver transplants have been performed with good success (75%) since the addition of *cyclosporine* to the antirejection therapy (Robinette, 1989). A transplanted liver usually begins to function within minutes. A liver transplant, involving two teams of surgeons, usually takes about eight hours. For a description of the technique, see Robinette (1989). Usually the donated liver is from a clinically dead person, but recently a live-donor transplant was performed. Part of the left lobe of the liver was removed from a mother and transplanted into her infant daughter. Liver transplants from living donors are done only in emergency situations.

The liver is easily ruptured because it is large, fixed in position, and friable. The bleeding may be severe because the hepatic veins are located in rigid canals and are unable to contract. Often the liver is torn by the end of a fractured rib, which perforates the diaphragm. In **cirrhosis of the liver**, there is progressive destruction of hepatocytes and replacement of them by fibrous tissue. This tissue surrounds the intrahepatic blood vessels and biliary ducts, impeding circulation of blood through the liver. The consistency of the liver is very firm owing to the large amount of fibrous tissue. The surface of the liver has a nodular appearance, accounting for the term "*hob-nail liver.*" There is often a reddish yellow or tawny color to the liver, hence the name cirrhosis (G. *kirrhos*, tawny, orange colored + *osis* condition).

Peritonitis (*e.g.*, resulting from a ruptured appendix) may result in the formation of localized *abscesses* (collections of pus) in various parts of the peritoneal cavity. A common site for an abscess is in the *subphrenic recess* (Fig. 2-65), located between the diaphragmatic surface of the liver and the diaphragm. **Subphrenic abscesses** occur much more frequently on the right side because of the frequency of ruptured appendices, duodenal ulcers, and gallbladders. As the right subphrenic recesses is continuous with the *hepatorenal*

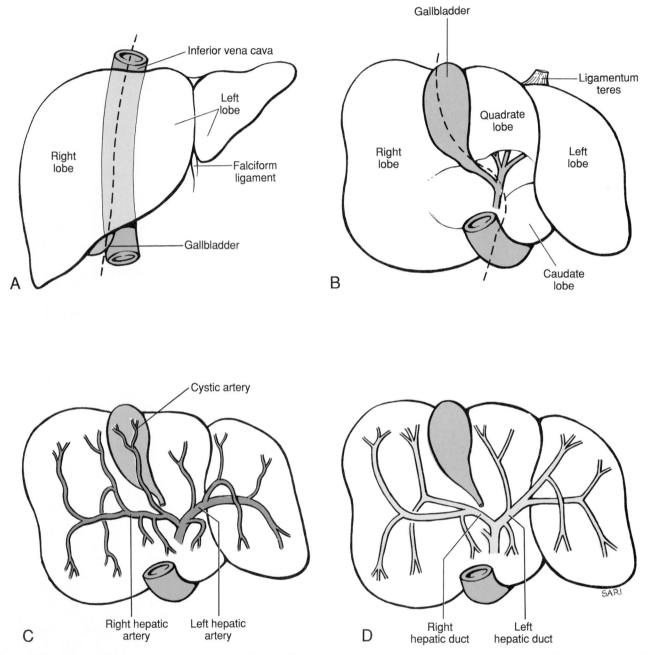

Figure 2-72. Segmental and vascular anatomy of the liver. *A*, Diaphragmatic surface of the liver. *B*, Visceral surface. The broken lines in *A* and *B* indicate the separation between the functional right and left lobes. *C* and *D*, The distribution of the hepatic arteries and biliary ducts, respectively. Note that the quadrate lobe is supplied by the left hepatic artery only, and that the caudate lobe is supplied by both arteries. *E*, The segments of the liver (II to VII). The first segment is not visible in this view. Each segment of the liver is supplied by a branch of the hepatic artery, bile duct, and portal vein.

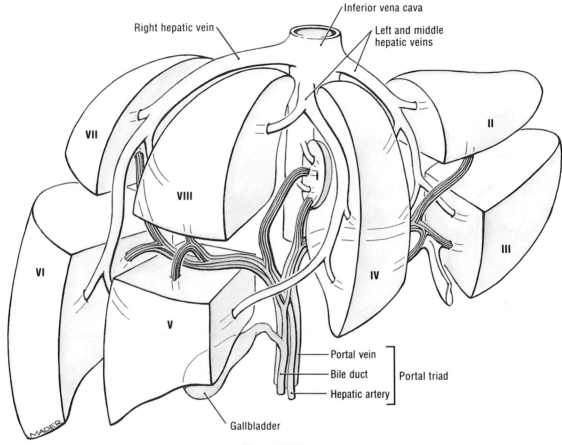

Figure 2-72E

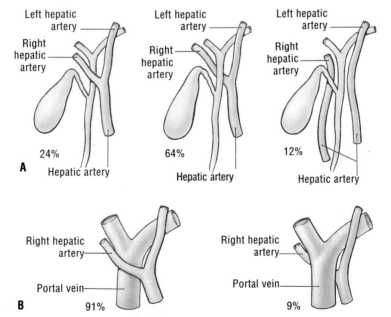

Figure 2-73.   Variations of the hepatic arteries. *A*, Right hepatic artery variants. In a study of 165 cadavers, three patterns were observed: the right hepatic artery crossed anterior to the bile passages (24%); the right hepatic artery crossed posterior to the bile passages (64%); and the aberrant artery arose from the superior mesenteric artery (12%). *B*, The right hepatic artery crossed anterior to the portal vein in 91% of 165 specimens and posterior to it in 9% of cases.

recess (Fig. 2-65), pus from a subphrenic abscess may drain into this recess when the patient is supine, because in this position the hepatorenal recess is the most posterior part of the peritoneal cavity.

The Biliary Ducts

Bile is secreted by hepatic cells into *bile canaliculi*, the smallest branches of the intrahepatic duct system. Most of the canaliculi drain into small *interlobular bile ducts*, which join with others to form progressively larger ducts. Eventually, right and left *hepatic ducts* are formed that emerge from the *porta hepatis* (Figs. 2-49, 2-50, 2-66, and 2-67). The right hepatic duct drains the right lobe of the liver and the left hepatic duct drains the left lobe. The areas of the liver drained by the hepatic ducts are the same as those supplied by the right and left branches of the hepatic artery and portal vein. Shortly after leaving the porta hepatis, the right and left hepatic ducts unite to form the *common hepatic duct*. About 4 cm in length, it passes inferiorly and to the right, between the layers of the lesser omentum. It is joined on the right side at an acute angle by the *cystic duct* from the gallbladder to form the **bile duct** (common bile duct). This large duct is 8 to 10 cm long and 5 to 6 mm in diameter (Figs. 2-49, 2-72*E*, and 2-74).

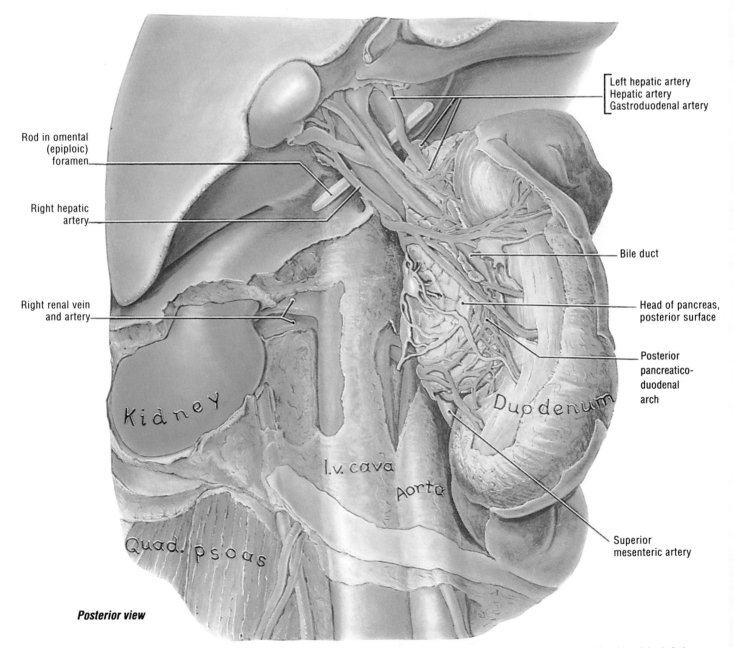

Left hepatic artery
Hepatic artery
Gastroduodenal artery

Rod in omental (epiploic) foramen

Right hepatic artery

Right renal vein and artery

Bile duct

Head of pancreas, posterior surface

Posterior pancreatico-duodenal arch

Kidney

l.v. cava

Aorta

Duodenum

Quad. psoas

Superior mesenteric artery

Posterior view

Figure 2-74.   Dissection of the bile duct and structures associated with it. The duodenum has been pulled to the left, taking the head of the pancreas with it. Observe the close posterior relationship of the inferior vena cava to the portal vein and bile duct.

The bile duct runs in the free edge of the lesser omentum with the hepatic artery and portal vein (Figs. 2-23, 2-66, and 2-74). It passes inferiorly, anterior to the omental foramen, where it is anterior to the right edge of the portal vein and on the right side of the hepatic artery (Figs. 2-27 and 2-66). The bile duct passes posterior to the superior part of the duodenum and the head of the pancreas, occupying a groove in the posterior part of the head or embedded in it (Figs. 2-45 and 2-52). As it runs posterior to the duodenum, the bile duct lies to the right of the gastroduodenal artery (Fig. 2-44). On the left side of the descending part of the duodenum, the bile duct comes into contact with the pancreatic duct and the two of them run obliquely through the wall of the duodenum, where they usually unite to form the *hepatopancreatic ampulla* (Figs. 2-49 and 2-63D). The distal constricted end of the ampulla opens into the descending part of the duodenum at the summit of the major duodenal papilla, 8 to 10 cm from the pylorus (Figs. 2-49 and 2-51).

The circular muscle around the distal end of the bile duct is thickened to form the *choledochal sphincter* (G. *choledochus*, containing bile), a muscular sheath that surrounds the bile duct just before and after it penetrates the duodenal wall (Fig. 2-49). As mentioned on p. 188, there is also a sphincter around the hepatopancreatic ampulla called the *hepatopancreatic sphincter*. This sphincter controls the discharge of both bile and pancreatic juice into the duodenum. When the choledochal sphincter contracts, bile cannot enter the ampulla and/or the duodenum; hence, it backs up and passes along the cystic duct into the gallbladder for concentration and storage of the bile (Fig. 2-49 and p. 190).

Arterial Supply of the Bile Duct (Figs. 2-45, 2-66, and 2-74). The bile duct is supplied by several arteries: (1) the distal or retroduodenal part is supplied by the *posterior superior pancreaticoduodenal artery*; (2) the middle part is supplied by the *right hepatic artery*; and (3) the proximal part is supplied by the *cystic artery*. The supraduodenal branches of the *gastroduodenal artery* also supply the bile duct as it passes posterior to the superior part of the duodenum. There is considerable variation in the arrangement of these arteries.

Venous Drainage of the Bile Duct (Figs. 2-40, 2-66, and 2-77). The veins from the proximal part of the bile duct and the hepatic ducts generally enter the liver directly. The *posterior*

superior pancreaticoduodenal vein drains the distal part of the bile duct and empties into the portal vein or one of its tributaries.

Lymphatic Drainage of the Bile Duct (Fig. 2-44; Table 2-4). Lymph passes to the *cystic lymph node* near the neck of the gallbladder, the node of the omental foramen, and the *hepatic lymph nodes* along the hepatic vessels. Efferent lymph vessels pass to the *celiac lymph nodes* (Fig. 2-60).

The Gallbladder and Cystic Duct

The Gallbladder (Figs. 2-25 to 2-27, 2-33, 2-49, 2-66, 2-67, 2-71, and 2-72). This piriform (pear-shaped) sac lies along the right edge of the quadrate lobe of the liver in a shallow depression on its visceral surface, called the *gallbladder fossa*. It hangs inferiorly like a pear with the *cystic duct* representing the stem. Its rounded *fundus* usually projects beyond the inferior margin of the liver. During life the gallbladder is a rather thin-walled bluish-green sac. It is normally covered on its posterior and inferior surfaces by peritoneum. Occasionally it is completely invested with peritoneum and may even be connected to the liver by a short mesentery. The gallbladder *concentrates the bile secreted by the liver and stores it* in the intervals between active phases of digestion. The gallbladder usually holds 30 to 60 ml of bile. For descriptive purposes, the gallbladder is divided into a fundus, body, and neck.

The fundus of the gallbladder (Figs. 2-25, 2-27, and 2-49) is its wide end which projects from the inferior border of the liver. It is usually located at the tip of the ninth costal cartilage in the midclavicular line, where the *linea semilunaris* marks the lateral edge of the rectus abdominis and meets the costal margin. It is directed inferiorly, anteriorly, and to the right where it comes into relationship with the posterior surface of the anterior abdominal wall and the descending part of the duodenum.

The body of the gallbladder (Figs. 2-49, 2-50, 2-66, 2-67, and 2-76), its main part, is directed superiorly, posteriorly, and to the left from the fundus. It lies in contact with the visceral surface of the liver to which it is attached by loose connective tissue. It also contacts the right part of the transverse colon and the superior part of the duodenum.

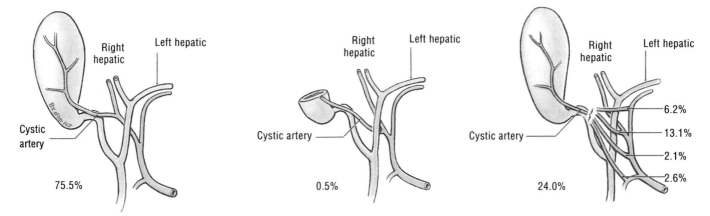

Figure 2-75. Variations in the origin and course of the cystic artery. The cystic artery usually arises from the right hepatic artery in the angle between the common hepatic duct and the cystic duct. However, when

it arises on the left of the biliary ducts, it usually crosses anterior to these ducts. The right-hand diagram is a composite of the variations of the cystic artery passing anterior to the common hepatic duct.

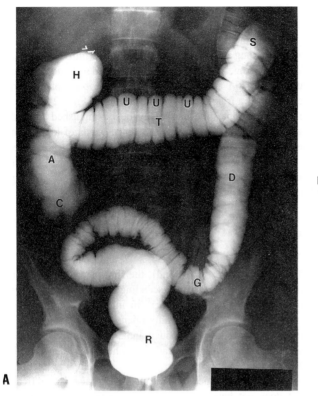

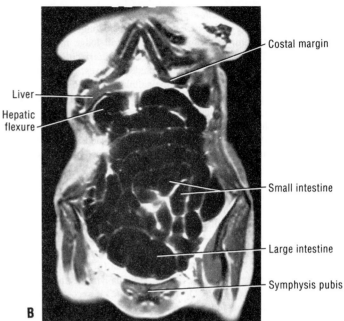

Figure 2-78. *A*, Single contrast AP radiograph of the abdomen following a barium enema. Observe the haustra (sacculations) in the wall of the colon. (Courtesy of Dr. E.L. Lansdown, Professor of Radiology, University of Toronto, Toronto, Ontario, Canada.) C - Caecum, A - Ascending colon, H - hepatic (right colic) flexure, T - Transverse colon, S - Splenic (left colic) flexure, D - Descending colon, G - Sigmoid colon, R - Rectum, U - Haustra. *B*, Coronal MRI (magnetic resonance image) of the abdomen. (Courtesy of Dr. W. Kucharzyck, Clinical Director of Tri-Hospital Resonance Centre, Toronto, Ontario, Canada.)

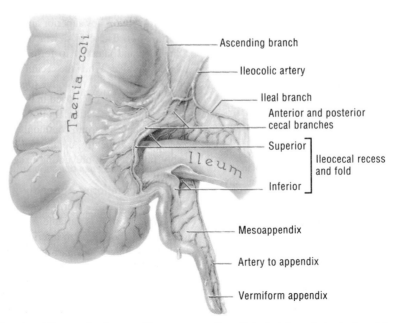

Figure 2-79. The ileocecal region. Observe that the vermiform appendix is in one free border of the mesoappendix and its artery is in the other. Note that the anterior tenia coli leads to the appendix; this is a guide to the appendix during appendectomy.

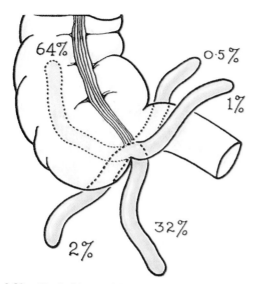

Figure 2-80. The incidence of the various locations of the vermiform appendix. In most people, the appendix is located posterior to the cecum (*i.e.*, retrocecal).

to the ascending colon). In this position, it usually has no mesentery (*i.e.*, it is retroperitoneal like the ascending colon). *The base of the appendix is fairly constant* and usually lies deep at the junction of the lateral and middle thirds of the line joining the anterior superior iliac spine and the umbilicus (*McBurney's point*; p. 141). The three teniae coli of the cecum converge at the base of the appendix and form a complete outer longitudinal muscle coat for it.

Arterial Supply of the Cecum and Vermiform Appendix (Figs. 2-58 and 2-79). The cecum is supplied by the *ileocolic artery*, a branch of the superior mesenteric artery, and the appendix is supplied by the *appendicular artery*, a branch of the ileocolic artery. It descends posterior to the terminal part of the ileum and enters the mesoappendix.

Venous Drainage of the Cecum and Vermiform Appendix (Fig. 2-40). The *ileocolic vein*, a tributary of the superior mesenteric vein, drains blood from the cecum and appendix. The *superior mesenteric vein* unites with the splenic vein to form the portal vein.

Lymphatic Drainage of the Cecum and Vermiform Appendix (Fig. 2-60; Table 2-4). Lymph vessels from the cecum and appendix pass to lymph nodes in the mesoappendix and to *ileocolic lymph nodes* that lie along the ileocolic artery. Efferent lymph vessels pass to the superior mesenteric lymph nodes.

Innervation of the Cecum and Vermiform Appendix (Figs. 2-61 and 2-82). The nerves to the cecum and appendix are derived from the *celiac* and *superior mesenteric ganglia*.

Although the human appendix is considered to be a vestigial organ, it has actually developed progressively in primates (Scott, 1980). In infants and children it has the appearance of a well-developed *lymphoid organ* and may have important immunological functions. The structure of the appendix varies with age. During old age the lymphoid tissue atrophies and is replaced largely by connective tissue.

Inflammation of the appendix (**appendicitis**) is one of the common causes of an *acute abdomen*. Close to 90,000 North American children have this operation annually. Appendicitis is usually caused by obstruction of the appendix (Filler, 1989), most often by fecal material. When its normal secretions cannot escape, it swells, obstructing its blood supply. The pain of acute appendicitis usually commences in the periumbilical region and then localizes in the right lower quadrant. In typical cases, digital pressure over McBurney's point registers the maximum abdominal tenderness, but in cases of retrocecal appendix (Fig. 2-80), the maximum tenderness may be in the right lateral region between the ribs and the iliac crest. If required, an **appendectomy** is usually performed through a muscle splitting incision in the right lower quadrant (Fig. 2-14*B* and *D*), which is centered at McBurney's point (p. 141). The cecum is delivered into the surgical wound and the mesoappendix containing the appendicular vessels is firmly ligated and divided (Fig. 2-79). The base of the appendix is tied, the appendix is excised, and its stump is usually cauterized and invaginated into the cecum.

If the appendix is not obvious during surgery, the teniae coli of the cecum are used as a guide because they converge on the base of the appendix. In unusual cases of malrotation or incomplete rotation of the cecum, the appendix is not located in the lower right quadrant. When the cecum is high (*subhepatic cecum*), the appendix is located in the right hypochondriac region and the pain in these cases is localized there, not in the lower right quadrant. Acute infection of the appendix may result in *thrombosis of the appendicular artery* and development of **gangrene** (necrosis owing to obstruction of blood supply). Rupture of an inflamed appendix results in infection of part or all of the peritoneum (*i.e.*, *general peritonitis*), increased abdominal pain, and abdominal rigidity.

The Ascending Colon

The ascending colon (G. *kolos*, large intestine) varies from 12 to 20 cm in length (Figs. 2-26, 2-58, 2-60, and 2-78 to 2-81). It ascends on the right side of the abdominal cavity from the cecum to the right lobe of the liver, where it turns to the left as the *right colic (hepatic) flexure*. It usually has no mesentery and lies retroperitoneally along the right side of the posterior abdominal wall. Some people (about 25%) have a short mesentery. The ascending colon is separated superiorly from the back muscles (*e,g.*, iliacus and quadratus lumborum; Fig. 2-82) by the kidney and inferiorly by the nerves of the posterior abdominal wall (ilioinguinal and iliohypogastric). It is usually separated from the anterior abdominal wall by coils of small intestine and the greater omentum (Fig. 2-25). The ascending colon is covered by peritoneum anteriorly and on its sides, which attaches it to the posterior abdominal wall. On the lateral side of the ascending colon, the peritoneum forms a trench or groove called the *right paracolic gutter* (Fig. 2-24*D* and p. 155). The depth of this groove depends on how much gas the ascending colon contains.

Arterial Supply of the Ascending Colon (Fig. 2-58). The ascending colon and right colic flexure are supplied by the *ileocolic* and *right colic arteries*, branches of the superior mesenteric artery.

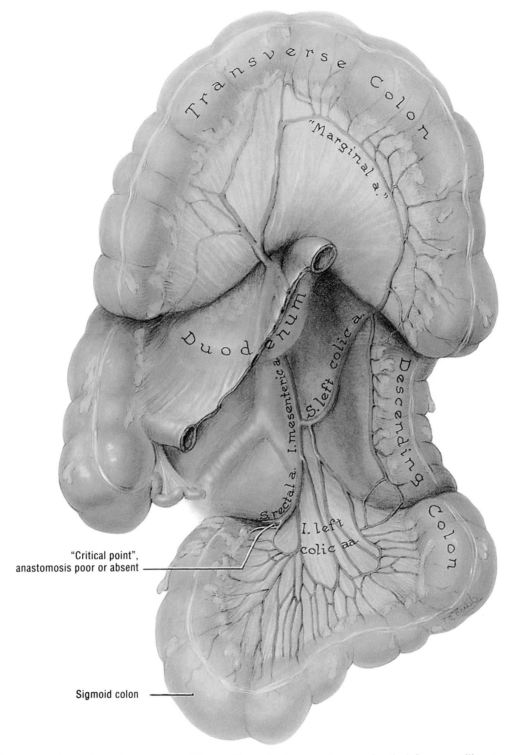

Figure 2-81. The mesenteric arteries. The mesentery of the small intestine has been cut at its root and discarded with the jejunum and ileum. Observe that the inferior mesenteric artery arises posterior to the duodenum, and on crossing the left common iliac artery, it becomes the superior rectal artery. Observe several *sigmoid arteries* (inferior left colic arteries) arising from the left side of the inferior mesenteric artery.

Venous Drainage of the Ascending Colon (Fig. 2-40).   The *ileocolic* and *right colic veins*, tributaries of the superior mesenteric vein, drain blood from the ascending colon.

Lymphatic Drainage of the Ascending Colon (Fig. 2-60*A*; Table 2-4).   The lymph vessels of the ascending colon pass to the *paracolic* and *epicolic lymph nodes* and from them to the *superior mesenteric lymph nodes*.

Innervation of the Ascending Colon (Figs. 2-61, 2-64, and 2-82).   The nerves to the ascending colon are derived from the *celiac* and *superior mesenteric ganglia*.

> When a person is in a supine position, fluid in the right *hepatorenal recess* (Fig. 2-65) passes along the right paracolic gutter (*e.g.*, pus from a ruptured appendix) to the *rectouterine* and/or *rectovesical pouch* (see Figs. 3-33 and 3-34).
>
> Prior to resection (surgical excision) of all or part of the ascending colon (*e.g.*, owing to *colonic cancer*), it has to be mobilized. The basic principle of mobilization of the colon is reconstruction of its primitive mesentery to the stage shown in Fig. 2-24*A*. An incision is made along the attachment of the visceral peritoneum to the parietal peritoneum in the right paracolic gutter (Fig. 2-24*D*), and the colon is reflected medially by cleavage of the fused layers of fascia. During mobilization of the colon, neither the vessels supplying it nor the ureter and kidney are disturbed because they lie anterior to the separated layers of fascia (Fig. 2-92).

The Transverse Colon

Despite its name, this part of the colon is not really transverse; it hangs down as a loop to a variable extent. The transverse colon, about 45 cm in length, is *the largest and most mobile part of the large intestine* (Figs. 2-26 to 2-28, 2-36, 2-58, and 2-78). It crosses the abdomen from the *right colic flexure* to the left colic flexure, where it bends inferiorly to become the descending colon. The *left colic flexure* lies on the inferior part of the left kidney and is attached to the diaphragm by the *phrenicocolic ligament*, a horizontal fold of peritoneum (Fig. 2-35). The left colic flexure is at a more superior level and in a more posterior plane than is the right colic flexure. Between these two colic flexures, the transverse colon is freely movable and forms a loop that is directed inferiorly and anteriorly. The transverse colon has a mesentery, called the *transverse mesocolon*, which is connected to the inferior border of the pancreas and to the greater omentum that covers it anteriorly (Figs. 2-27, 2-28*B*, and 2-48*A*). The mesocolon is a double layer of peritoneum that suspends the transverse colon from the posterior abdominal wall (Fig. 2-28). Because it is freely movable, the transverse colon is variable in position. It may be at the level of the transpyloric plane or it may extend inferiorly as far as the pelvic brim (see Fig. 3-7).

Arterial Supply of the Transverse Colon (Figs. 2-58 and 2-81).   The transverse colon is supplied mainly by the *middle colic artery*, a branch of the superior mesenteric artery, but it also receives blood from the *right and left colic arteries*. The left colic artery is a branch of the inferior mesenteric.

Venous Drainage of the Transverse Colon (Fig. 2-40).   Blood is drained from the transverse colon via the *superior mesenteric vein*.

Lymphatic Drainage of the Transverse Colon (Fig. 2-60*A*; Table 2-4).   Lymph from the transverse colon passes to the lymph nodes that lie along the middle colic artery. The *superior mesenteric lymph nodes* receive lymph vessels from these nodes.

Innervation of the Transverse Colon (Fig. 2-61).   The nerves of the transverse colon, which follow the right and middle colic arteries, are derived from the *superior mesenteric plexus*. They transmit sympathetic and vagal nerve fibers. The nerves that follow the left colic artery are derived from the *inferior mesenteric plexus*.

The Descending Colon

This part of the large intestine, 22 to 30 cm in length (Figs. 2-26, 2-36, 2-78, and 2-82), descends from the left colic flexure into the left iliac fossa, where it is *continuous with the sigmoid colon*. As it descends, the colon passes anterior to the lateral border of the left kidney and the transversus abdominis and quadratus lumborum muscles. The caliber of the descending colon is considerably smaller than that of the ascending colon. It usually has no mesentery and lies retroperitoneally along the left side of the posterior abdominal wall. Its posterior surface is attached to the posterior abdominal wall but the descending colon, like the ascending colon, can be mobilized surgically. In some people (about 33%), the descending colon has a mesentery. The descending colon is related to the diaphragm superiorly and the quadratus lumborum muscle. The iliohypogastric and ilioinguinal nerves intervene between it and this muscle.

Arterial Supply of the Descending Colon (Figs. 2-81 and 2-82).   The descending colon is supplied by the *left colic* and *superior sigmoid arteries*, branches of the inferior mesenteric artery.

Venous Drainage of the Descending Colon (Fig. 2-40).   The descending colon is drained by the *inferior mesenteric vein*.

Lymphatic Drainage of the Descending Colon (Fig. 2-60*A*; Table 2-4).   The lymph vessels from the descending colon pass to the *intermediate colic lymph nodes* along the left colic artery. From them the lymph passes to the *inferior mesenteric lymph nodes* around the inferior mesenteric artery. However, those from the left colic flexure also drain to the *superior mesenteric lymph nodes* by vessels that accompany the superior mesenteric artery.

Innervation of the Descending Colon (Figs. 2-64 and 2-92).   The nerve supply is derived from the sympathetic and parasympathetic systems. It receives its sympathetic supply from the lumbar part of the sympathetic trunk and the *superior hypogastric plexus* by means of plexuses on the branches of the inferior mesenteric artery. The parasympathetic supply is derived from the *pelvic splanchnic nerves*.

The Sigmoid Colon

The sigmoid colon forms a sinuous, S-shaped loop of variable length (usually about 40 cm) that reminded early anatomists of the Greek letter sigma (S). The sigmoid colon (pelvic colon) is the portion of the large intestine between the descending colon and rectum (Figs. 2-26*A*, 2-78, and 2-81). It extends from the pelvic brim to the third segment of the sacrum (see Figs. 3-7 and 3-48), where it joins the rectum. The termination of the

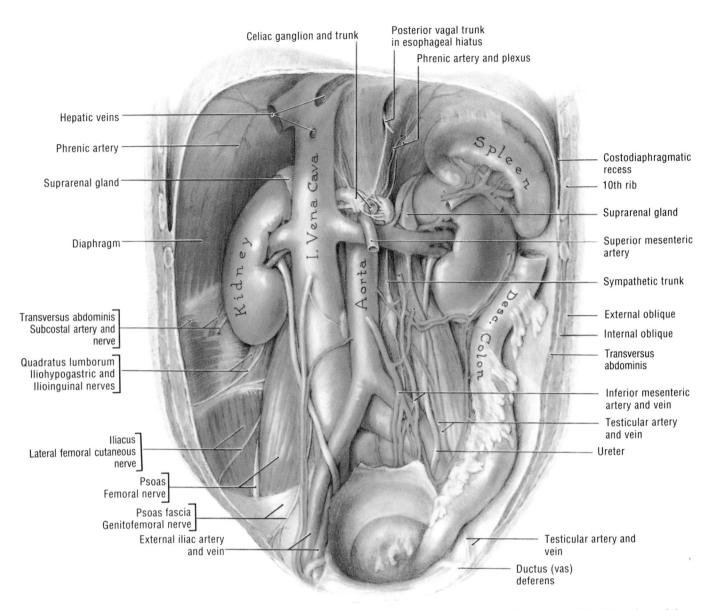

Celiac ganglion and trunk

Posterior vagal trunk in esophageal hiatus

Phrenic artery and plexus

Hepatic veins

Phrenic artery

Suprarenal gland

Diaphragm

Transversus abdominis
Subcostal artery and nerve

Quadratus lumborum
Iliohypogastric and Ilioinguinal nerves

Iliacus
Lateral femoral cutaneous nerve

Psoas
Femoral nerve

Psoas fascia
Genitofemoral nerve

External iliac artery and vein

Spleen

Kidney

I. Vena Cava

Aorta

Desc. Colon

Costodiaphragmatic recess

10th rib

Suprarenal gland

Superior mesenteric artery

Sympathetic trunk

External oblique

Internal oblique

Transversus abdominis

Inferior mesenteric artery and vein

Testicular artery and vein

Ureter

Testicular artery and vein

Ductus (vas) deferens

Figure 2-82. Dissection of the posterior abdominal wall of a man showing the great vessels, kidneys, ureters, and suprarenal glands. Most of the fascia has been removed. Observe that the ureter crosses the external iliac artery just beyond the common iliac bifurcation and that the testicular vessels cross anterior to the ureter and pass with the ductus deferens into the inguinal canal.

longitudinal bands of smooth muscle (teniae coli) indicates the beginning of the rectum (Fig. 2-60*A*). It usually has a long mesentery (*sigmoid mesocolon*) and therefore has considerable freedom of movement. The sigmoid colon usually occupies the rectovesical pouch in males (see Fig. 3-33) and the rectouterine pouch in females (see Fig. 3-34). The root of its mesentery has a V-shaped attachment, superiorly along the external iliac vessels and inferiorly from the bifurcation of the common iliac vessels to the anterior aspect of the sacrum. Posterior to the apex of the mesentery (*i.e.*, retroperitoneally) lies the left ureter and the division of the left common iliac artery (Fig. 2-82).

The *omental (epiploic) appendages (appendices epiploicae)* are very long in the sigmoid colon. These little sacs of omentum are generally distended with fat (Fig. 2-58). The *rectosigmoid junction* is about 15 cm from the anus. The shape and position

of the sigmoid colon depend on how full it is. Feces are stored in the sigmoid colon until just before defecation. Usually the sigmoid colon lies relatively free within the pelvis minor, inferior to the small intestine (Fig. 2-26*A*). Posterior to the sigmoid colon are the left external iliac vessels, the left sacral plexus and the left piriformis muscle.

Arterial Supply of the Sigmoid Colon (Figs. 2-81 and 2-82). Two to three sigmoid arteries supply the sigmoid colon. Branches of the inferior mesenteric artery, the *sigmoid arteries*, descend obliquely to the left where they divide into ascending and descending branches. The most superior sigmoid artery anastomoses with the descending branch of the left colic artery.

Venous Drainage of the Sigmoid Colon (Fig. 2-40). The inferior mesenteric vein returns blood from the sigmoid colon.

Lymphatic Drainage of the Sigmoid Colon (Fig. 2-60*A*;

Table 2-4). Lymph passes to *intermediate colic lymph nodes* on the branches of the left colic arteries, and from them to the *inferior mesenteric lymph nodes* around the inferior mesenteric artery.

Innervation of the Sigmoid Colon (Figs 2-64, 2-82, and 2-92). The nerve supply is derived from the *sympathetic and parasympathetic systems*. It derives its sympathetic supply from the lumbar part of the sympathetic trunk and the *superior hypogastric plexus* by means of plexuses on the branches of the inferior mesenteric artery. The parasympathetic supply is derived from the *pelvic splanchnic nerves*.

The Rectum

The rectum (L. *rectus*, straight) is the fixed terminal part of the large intestine (Figs. 2-26, 2-36, and 2-78; see also Figs. 3-48 to 3-50). It is only partially covered with peritoneum and has no mesentery. The inferior part of the rectum is continuous with the anal canal. Because the rectum is a pelvic organ, it is described with the other viscera in the pelvic cavity (see Chap. 3, p. 289).

The Anal Canal

The anal canal is the terminal part of the GI tract (Figs. 2-26A and 2-78). It terminates at the anus in the perineum with which region it is described (see Chap. 3, p. 299).

The interior of the sigmoid colon can be observed with fiberoptical instruments known as a *colonoscope* and a *sigmoidoscope* (Stone, 1989). The colonoscope, inserted into the colon via the anus, is a flexible tube that can be used to examine the entire colon. The *sigmoidoscope* is a rigid tube that is used to examine inferior parts of the colon. *Most tumors of the large intestine occur in the rectum*; about 12% of them appear near the rectosigmoid junction, which can be palpated during digital examination. *Sigmoidoscopy* may be followed by a radiographical study of the large intestine (Fig. 2-78). Liquid barium is put into the colon during a procedure known as a *barium enema*.

A chronic disease of the colon called *ulcerative colitis* is characterized by severe inflammation and ulceration of the colon and rectum. In some of these patients, a **total colectomy** is performed during which the terminal ileum and colon, as well as the rectum and anal canal, are removed. An *ileostomy* is then constructed to establish an opening between the ileum and the skin of the anterior abdominal wall. This creates an ''*artificial anus*.'' Colectomy is also used in the surgical treatment of some patients with **cancer of the colon** (Stone, 1989). In *subtotal colectomy*, the rectum and anal canal are preserved and the ileum is joined to the rectum by an end-to-end anastomosis. The most mobile parts of the colon are used for colostomies; thus the usual ones are: *cecostomy*, *transverse colostomy*, and *sigmoidostomy*. The type of **colostomy** constructed depends on which part of the colon has been resected. When parts of the colon are removed, an adequate blood supply to the remaining segments must be preserved. Failure to do so results in necrosis of the segment of intestine concerned.

The Portal Vein and Portal-Systemic Anastomoses

The portal vein is the main channel of the *portal system of veins* (Fig. 2-40). It collects blood from the abdominal part of the GI tract, the gallbladder, pancreas, and spleen, and carries it to the liver. There it branches to end in expanded capillaries known as *sinusoids*. From these the blood is collected by *hepatic veins*, which drain into the *inferior vena cava*. The portal vein carries blood from three major veins: the superior and inferior mesenteric veins and the splenic vein.

The portal vein is formed posterior to the neck of the pancreas by the union of the splenic and superior mesenteric veins (Fig. 2-45). The inferior mesenteric vein usually ends posterior to the pancreas by joining the splenic vein (Fig. 2-40). However, it may join the portal vein or even the superior mesenteric vein. The portal vein ascends to the liver in the free margin of the lesser omentum, posterior to the bile duct and hepatic artery (Fig. 2-74). At the *porta hepatis*—a transverse fissure through which the bile duct, vessels and nerves enter or leave the liver (Figs. 2-67 and 2-71)—the portal vein divides into right and left branches, which empty their blood into the *hepatic sinusoids*. This blood contains the products of digestion (carbohydrates, fats, and protein from the intestine) and the products of red cell destruction from the spleen.

The portal venous system communicates with the systemic venous system in several locations (Fig. 2-83). These portal-systemic anastomoses are important clinically. When the portal circulation is obstructed (*e.g.*, owing to liver disease), blood from the GI tract can still reach the right side of the heart through the inferior vena cava via a number of collateral routes. The portal vein and its tributaries have no valves; hence blood can flow from the obstructed liver to the inferior vena cava via these alternate routes. In *portal hypertension* venous pressure in the portal venous system is increased; consequently some blood in the portal venous system may reverse its direction and pass through the portal-systemic anastomoses into the systemic venous system. This causes the veins in the portal-systemic anastomotic areas to dilate and become varicose.

The main portal-systemic anastomotic areas are illustrated in Figure 2-83.

1. In the gastroesophageal region, the *esophageal tributaries of the left gastric vein anastomose with the esophageal veins*, which empty into the azygos vein (see Figs. 1-39 and 1-76).
2. In the anorectal region, the *superior rectal vein anastomoses with the middle and inferior rectal veins* (Fig. 2-83), which are tributaries of the internal iliac and internal pudendal veins, respectively.
3. In the paraumbilical region, the *paraumbilical veins in the falciform ligament anastomose with subcutaneous veins* in the anterior abdominal wall.
4. In the retroperitoneal region, *tributaries of the splenic and pancreatic veins anastomose with the left renal vein*. Short veins also connect the splenic and colic veins to the lumbar veins of the posterior abdominal wall. The veins of the bare area of the liver also communicate with the veins of the diaphragm and the right internal thoracic vein.

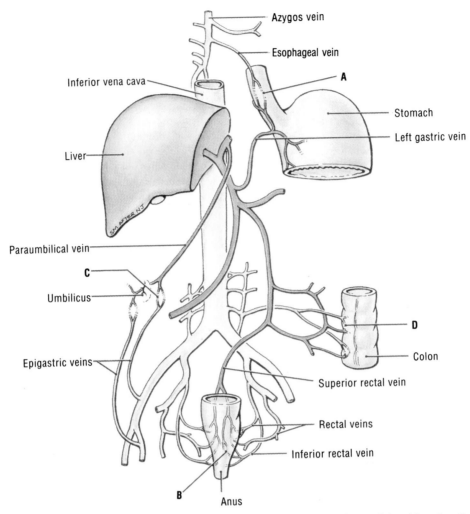

Figure 2-83. The *portal-systemic anastomoses*. These communications provide a collateral portal circulation in cases of obstruction in the liver or portal vein. In this diagram, portal tributaries are darker blue and systemic tributaries are lighter blue. *A* to *D* indicate sites of anastomoses.

In some patients with *portal hypertension*, the abnormal increase in pressure in the portal vein and its tributaries is caused by **cirrhosis of the liver**, a disease characterized by progressive destruction of hepatic parenchymal cells and replacement of them by fibrous tissue. *Portal hypertension* and the diseases causing it are serious conditions, but the symptoms are modified by the portal-systemic anastomoses, which provide alternative pathways for the blood to flow. Because there are no functionally competent valves in the portal venous system, the increase in portal pressure is reflected throughout the system. Blood tends to be diverted into the systemic venous system in regions where portal-systemic anastomoses occur (Fig. 2-83). The veins in these areas tend to become dilated and tortuous and are called **varicose veins**.

Varicose veins in the anal region are called *hemorrhoids* (piles), and those in the gastroesophageal region are called *esophageal varices*. The veins in both locations may become so dilated that their walls rupture, resulting in hemorrhage. Bleeding from esophageal varices is often severe and may be fatal. In severe cases of portal obstruction, even the *paraumbilical veins may become varicose* and look somewhat like small snakes under the skin. This condition is referred to as

caput medusae because of its resemblance to the serpents on the head of Medusa, a character in Greek mythology.

A common way of reducing portal pressure is to divert blood from the portal venous system to the systemic venous system by creating a communication between the portal vein and the inferior vena cava. This **portacaval anastomosis** may be done where these vessels lie close to each other posterior to the liver (Fig. 2-74). Another way of reducing portal pressure is to join the splenic vein to the left renal vein (*splenorenal anastomosis*), following removal of the spleen (*splenectomy*).

The Kidneys

The kidneys (L. *renes*), one on each side of the vertebral column, lie in the paravertebral gutters (grooves) at the level of T12 to L3 vertebrae (Figs. 2-23, 2-26*B*, 2-82, 2-84, 2-85, 2-86, and 2-96). Their long axes are almost parallel with the axis of the body. A ureter runs inferiorly from each kidney and passes over the pelvic brim at the bifurcation of the common iliac artery. It runs along the lateral wall of the pelvis and enters the urinary bladder. The kidneys remove excess water, salts, and the products of protein metabolism from the blood and maintain its pH. The waste products removed from the blood are conveyed in the urine to the urinary bladder by the ureters.

Position, Form, and Size of the Kidneys (Figs. 2-46, 2-82, and 2-84 to 2-90). Each kidney lies in a mass of *perirenal (perinephric) fat*, posterior to the peritoneum (*i.e.*, retroperitoneally) on the posterior abdominal wall. They lie alongside the vertebral column, against the psoas major muscles. The superior parts of the kidneys are protected by the thoracic cage and are tilted so that their superior poles are nearer the median plane than their inferior poles. Owing to the large size of the right lobe of the liver, the right kidney lies at a slightly lower level than the left kidney (Fig. 2-82).

Each kidney has anterior and posterior surfaces, medial and lateral margins (borders), and superior and inferior poles. The lateral margin is convex and the medial margin is indented or concave where the **renal sinus** and renal pelvis are located. Fresh adult kidneys are reddish-brown in color and measure about 10 cm in length, 5 cm in width, and 2.5 cm in thickness. The left kidney is often slightly longer than the right one. Each kidney is ovoid in outline, but its indented medial margin gives it a

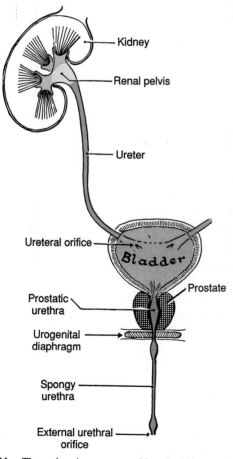

Figure 2-84. The male urinary system. Note the kidneys where urine is formed; the ureters that convey urine to the urinary bladder; the urinary bladder where urine is temporarily stored; and the urethra through which the urine passes to the exterior. The membranous urethra is surrounded by the urogenital diaphragm and the spongy urethra runs through the penis.

somewhat bean-shaped appearance. At this concave part of each kidney there is a vertical cleft, the **renal hilum** (hilus), through which the renal artery enters and the renal vein and the renal pelvis leave the kidney. The hilum leads into a space within the kidney called the *renal sinus*, which is about 2.5 cm deep (Figs. 2-87 and 2-88). The renal sinus is occupied by the renal pelvis, renal calices, renal vessels and nerves, and varying amounts of fat. *The renal vein is anterior*; the renal artery is posterior to the vein; the renal pelvis is posterior to the artery (Fig. 2-87*A*).

Surface Anatomy and Markings of the Kidneys (Figs. 2-46 and 2-86). In very muscular and/or obese people, the kidneys may be impalpable. In thin adults with poorly developed abdominal muscles, the inferior pole of the right kidney is usually palpable by bimanual examination in the right lateral region as a firm, smooth, somewhat rounded mass that descends during inspiration. The normal left kidney is usually not palpable. The levels of the kidneys change during respiration and with changes in posture. Each kidney moves about 3 cm in a vertical direction during the movement of the diaphragm that occurs with deep breathing (p. 48). The hilum of the left kidney lies in the transpyloric plane, about 5 cm from the median plane. This plane cuts through the superior part of the right kidney. As the posterior approach to the kidney is the usual surgical one, it is helpful to know that the inferior pole of the right kidney is about a fingerbreadth superior to the iliac crest and that its superior pole is superior to the 12th rib.

Gross Structure of the Kidneys (Figs. 2-87 to 2-90). The kidneys are closely invested by a *strong fibrous capsule*, which gives the fresh kidney a glistening appearance. This capsule strips easily from a normal kidney. It passes over the lips of the hilum to line the renal sinus and become continuous with the walls of the calices. The kidney and its capsule are surrounded by *pararenal fat*, but it is sparse on the anterior surface. This fat is less dense (lower specific gravity than the kidney); thus an outline of the kidney is usually visible in radiographs, CTs and MRIs.

The funnel-shaped renal pelvis is continuous inferiorly with the ureter. It is surrounded by the fat, vessels, and nerves in the renal sinus (Fig. 2-88*B*). The word *pelvis* is derived from the Greek word *pyelos*, meaning a basin. Hence a *pyelogram* is a radiograph of the renal pelvis and ureter (Fig. 2-89), and *pyelonephritis* (G. *nephros*, kidney) indicates inflammation of the renal pelvis and kidney. Within the *renal sinus*, the renal pelvis usually divides into two wide, cup-shaped *major calices* (G. flower cups). Each major calix (calyx) is subdivided into 7 to 14 *minor calices*. The urine empties into a minor calix from collecting tubules that pierce the tip of a *renal papilla* obliquely (Fig. 2-88). It then passes through the major calix, renal pelvis, and ureter to enter the urinary bladder.

Renal Fascia and Renal Fat (Figs. 2-85, 2-88, and 2-96). Each kidney, invested by a *fibrous renal capsule*, is embedded in a substantial mass of *perirenal fat* that constitutes a fatty renal capsule. Very little fatty tissue lies anterior to the kidney. The *fatty renal capsule* is in turn covered by fibroareolar tissue called the *renal fascia*. This fascia encloses the kidney, its surrounding fibrous and fatty capsules, and the suprarenal (adrenal) gland. These coverings help to maintain these organs in position. Superiorly the renal fascia is continuous with the fascia on the inferior surface of the diaphragm (*diaphragmatic*

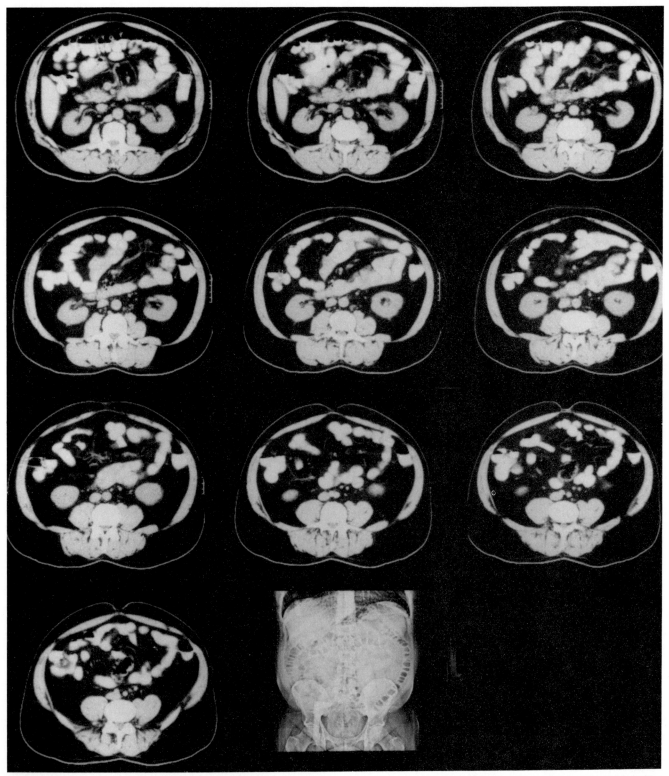

Figure 2-85.   CT scans through the abdomen in the horizontal planes of the kidneys. For the identification of structures, see Figs. 2-32*B* and 2-90. (Courtesy of Dr. Tom White, Department of Radiology, The Health Sciences Center, The University of Tennessee, Memphis, TN.)

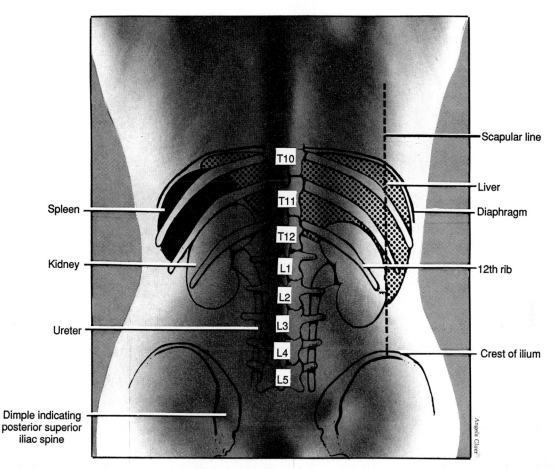

Figure 2-86.   Surface anatomy and markings on the back of a 21-year-old woman. Observe that the kidneys lie on each side of the vertebral column from T12 to L3 vertebrae, and that the right kidney is slightly inferior to the left one. Note that the superior poles of the kidneys are protected by the 11th and 12th ribs and that the inferior pole of the right kidney is about a fingerbreadth superior to the crest of the ilium.

fascia). Medially the anterior layers of fascia on the right and left sides blend with each other anterior to the abdominal aorta and inferior vena cava. The posterior layer of renal fascia fuses medially with the fascia over the psoas major muscle. The layers of renal fascia are loosely united inferiorly and may be easily separated inferior to the kidney. The encasement of the kidney in fat is an important factor in anchoring it in position. The amount of fat in the fatty capsule varies with the individual. The extraperitoneal fat outside the renal fascia *(pararenal fat)* is located between the peritoneum of the posterior abdominal wall and the renal fascia.

Relations of the Kidneys (Figs. 2-23, 2-82, 2-84 to 2-86, and 2-89 to 2-91). *Posteriorly*, each kidney lies on muscle. The posterior surface of the superior pole is related to the diaphragm, which separates it from the pleural cavity and 12th rib. More inferiorly, the kidney is related to the quadratus lumborum muscle and sometimes encroaches slightly on the *psoas major muscle* medially and the transversus abdominis muscle laterally. The subcostal nerve and vessels and the iliohypogastric and ilioinguinal nerves descend diagonally across the posterior surface of the kidney. *Anteriorly*, the relations of the kidneys differ on the two sides, except that the anterior and medial aspects of the superior pole of each kidney are covered by the corresponding suprarenal gland.

Anterior Relations of the Right Kidney (Figs. 2-48, 2-65, and 2-67).   Its superior pole is related to the inferior surface of the liver. Except for this pole, the right kidney is separated from the liver by the hepatorenal recess. More inferiorly, the descending part of the duodenum passes across the hilum of this kidney. The right colic flexure lies anterior to its lateral border and inferior pole. In Fig. 2-48, observe that the suprarenal, duodenal, and colic areas of the kidney are not covered by peritoneum. Part of the small intestine lies across the inferior pole of the right kidney; it is separated from it by a film of peritoneal fluid and peritoneum.

Anterior Relations of the Left Kidney (Figs. 2-35 and 2-48).   This kidney is related anteriorly to the suprarenal gland, stomach, spleen, pancreas, jejunum, and descending colon. The gastric, splenic, and jejunal areas are covered by peritoneum. The left kidney, along with the pancreas and spleen, is in the *stomach bed*, where it is covered by the posterior wall of the omental bursa.

The lobes of the kidney are demarcated on the surface of fetal and infantile kidneys. This external evidence of the lobes usually disappears by the end of the first year. Uncommonly, evidence of the fetal lobes may be visible in adult kidneys (Moore, 1988). Occasionally the left kidney is somewhat

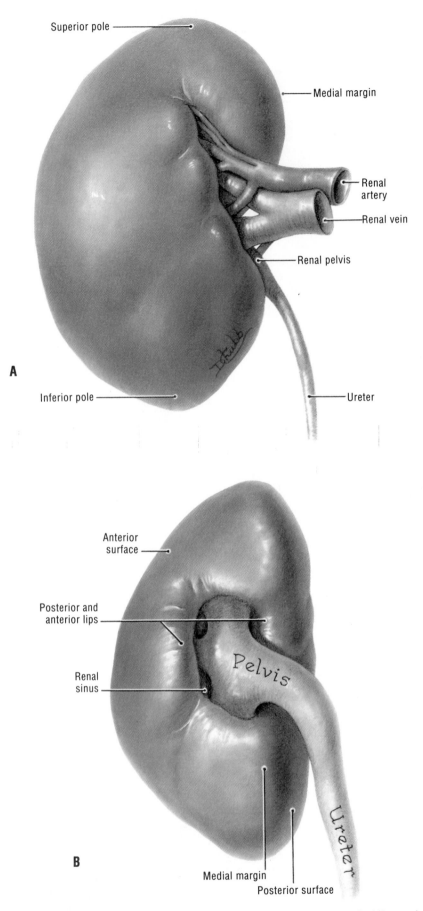

Superior pole

Medial margin

Renal
artery

Renal vein

Renal pelvis

A

Inferior pole

Ureter

Anterior
surface

Posterior and
anterior lips

Renal
sinus

Pelvis

B

Medial margin

Posterior surface

Ureter

Figure 2-87. The right kidney. *A*, Anterior view showing the order of the structures at the hilum: vein, artery, and renal pelvis. *B*, Anteromedial view after removal of the renal vessels to show the renal sinus.

214

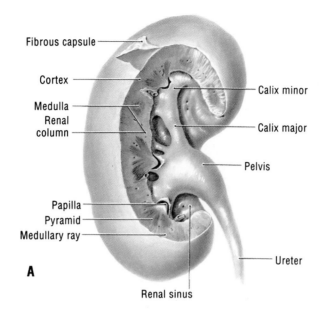

Fibrous capsule

Cortex

Medulla
Renal
column

Calix minor

Calix major

Pelvis

Papilla
Pyramid
Medullary ray

Ureter

A

Renal sinus

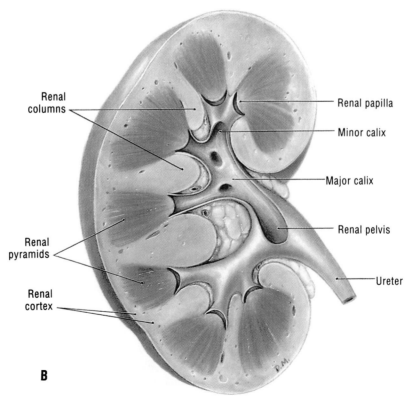

Renal
columns

Renal papilla

Minor calix

Major calix

Renal pelvis

Renal
pyramids

Ureter

Renal
cortex

B

Figure 2-88.   *A*, Right kidney from which the anterior lip of the renal sinus is cut away. *B*, Longitudinal section of the kidney. Note that each renal pyramid ends as a renal papilla on which a dozen or more large collecting tubules open. One to four papillae project into each minor calix.

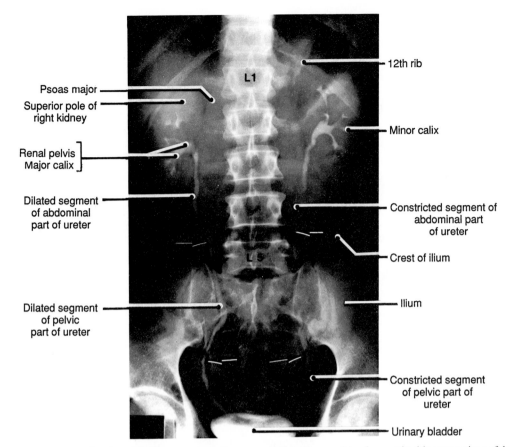

Psoas major
Superior pole of right kidney
Renal pelvis
Major calix
Dilated segment of abdominal part of ureter
Dilated segment of pelvic part of ureter

L1
L5

12th rib
Minor calix
Constricted segment of abdominal part of ureter
Crest of ilium
Ilium
Constricted segment of pelvic part of ureter
Urinary bladder

Figure 2-89. Intravenous urogram (pyelogram). The contrast medium was injected intravenously and was concentrated and excreted by the kidneys. This AP projection shows the calices, renal pelves, and ureters outlined by the contrast medium filling their lumina. Note the difference in shape and level of the renal pelves and the constrictions and dilations in the ureter resulting from peristaltic contractions of their smooth muscle walls. The arrows indicate narrowings of the lumen resulting from peristaltic contractions. (Courtesy of Dr. John Campbell, Department of Radiology, Sunnybrook Medical Centre, University of Toronto, Toronto, Ontario, Canada.)

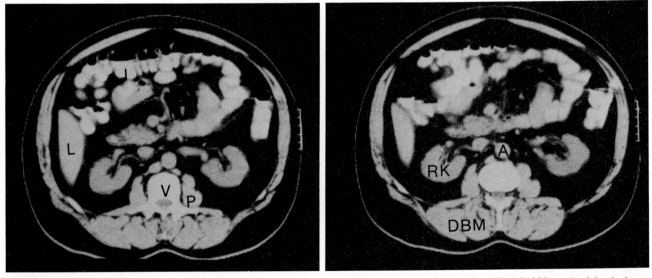

Figure 2-90. CT scans through the abdomen in the horizontal plane at the level of the kidneys. These sections are visualized as if you were looking at a transverse section of the abdomen from below, with the right side to your left. *L*, indicates the liver; *I*, intestine; *V*, lumbar vertebra; *P*, psoas major muscle; *RK*, right kidney; *A*, abdominal aorta; *DBM*, deep back muscles. (Courtesy of Dr. Tom White, Department of Radiology, The Health Sciences Center, University of Tennessee, Memphis, TN.)

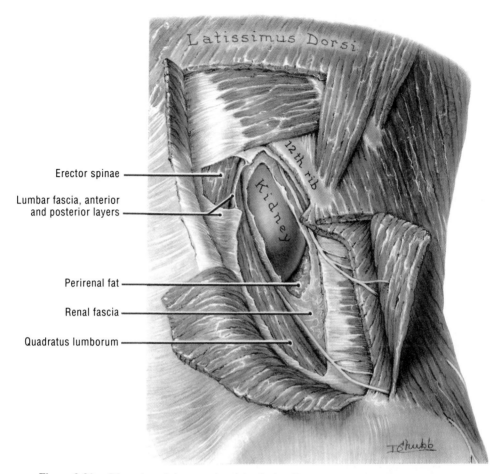

Erector spinae

Lumbar fascia, anterior
and posterior layers

Perirenal fat

Renal fascia

Quadratus lumborum

Figure 2-91. Dissection of the posterior abdominal wall on the right side, posterolateral view.

triangular in shape, probably as a result of molding by the spleen. The superior pole of the kidney is narrow and the inferior pole is wide, giving it a *humpbacked appearance.* This abnormally shaped kidney has no clinical significance.

The kidney can be removed without damaging the suprarenal gland because a weak septum of renal fascia separates the kidney from this gland. Similarly the gland can be removed without affecting the kidney. If the fatty renal capsule is thin or absent, as occurs in emaciated patients, the kidneys are difficult to examine radiologically because the perirenal fat does not produce a clear outline of the kidney. The kidneys move superiorly and inferiorly slightly during respiration because they are in contact with the diaphragm. It has been determined radiographically that the normal renal mobility is about 3 cm (roughly the height of one vertebral body). Blood from an injured kidney or pus from a *perinephric abscess* may force its way inferiorly into the pelvis and between the anterior and posterior layers of pelvic fascia.

The close relationship of the kidneys to the psoas major muscles explains why extension of the thighs may increase pain resulting from inflammation in the pararenal regions. The common surgical approach to kidney is the *retroperitoneal approach* (Fig. 2-91). Removal of a kidney (**nephrectomy**) is indicated when contamination of the peritoneal cavity is likely (*e.g.*, if there is inflammatory renal disease). During

surgery the subcostal, iliohypogastric, and ilioinguinal nerves are vulnerable to injury. The transabdominal approach to the kidney is used for surgery involving the renal vessels or ureters.

Renal transplantation is now an established operation for the treatment of selected cases of *chronic renal failure.* The transplant site for the kidney is the lower abdomen. Its renal artery and vein are joined to the external iliac artery and vein, and the ureter is sutured into the bladder. For surgical details and illustrations, see Robinette (1989). Rejection is a serious problem in transplants, but this can usually be controlled by drugs, which the recipient has to take for the rest of his/her life.

The Ureters

These thick-walled, expandable muscular ducts with narrow lumina carry urine from the kidneys to the urinary bladder (Figs. 2-82, 2-84, and 2-87 to 2-89). As the urine passes along the ureters, peristaltic waves occur in their walls. Each ureter is continuous superiorly with the funnel-shaped renal pelvis. The *abdominal part of the ureter* is about 12.5 cm long and 5 mm wide. The inferior half or pelvic part of the ureter is described in Chap. 3 (p. 267). The pale-colored abdominal ureter adheres closely to the parietal peritoneum and is retroperitoneal throughout its entire course. It descends almost vertically, *anterior to the psoas major muscle* (Fig. 2-82). As the right ureter descends, it lies closely related to the inferior vena cava, the lumbar lymph nodes, and the sympathetic trunk (Fig. 2-92). The ureter crosses the brim of the pelvis and the external iliac artery, just beyond the bifurcation of the common iliac artery (Fig. 2-82).

Arterial Supply of the Kidneys and Ureters (Figs. 2-82, 2-87, and 2-93 to 2-95).   The *renal arteries* are large vessels that usually arise at right angles from the aorta, at the level of the intervertebral disc between L1 and L2 vertebrae. The right renal artery passes posterior to the inferior vena cava. Typically each artery divides close to the hilum into five segmental arteries. Most of these vessels pass anterior to the pelvis of the kidney, but one or two may pass posterior to it. Based on arterial distribution, *renal segments* are described. Each segment is supplied

by a *segmental artery*. The initial branches of these arteries, called *lobar arteries*, divide into *interlobar arteries*.

The Arterial Supply of the Ureter (Fig. 2-93).   The arteries usually come from three sources, but they may arise from any of the following vessels (the main abdominal sources are printed in italics): *renal*, *testicular* or *ovarian*, *aorta*, common and internal iliac, and superior and/or inferior vesical (male) or uterine arteries.

Venous Drainage of the Kidneys and Ureters (Figs. 2-82, 2-87, 2-96, and 2-103).   Several veins drain the kidney and unite in a variable fashion to form the renal vein. The renal veins lie anterior to the renal arteries and *the left renal vein passes anterior to the aorta*, just inferior to the origin of the superior mesenteric artery. Each renal vein drains into the inferior vena

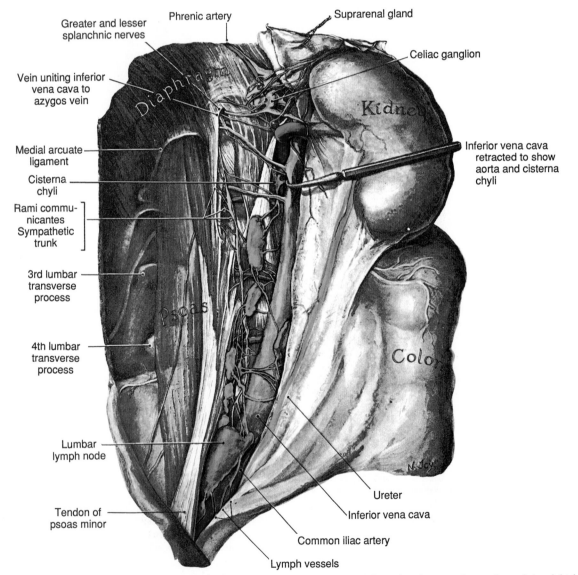

Figure 2-92.   Dissection of the posterior abdominal wall showing the right lumbar lymph nodes, celiac ganglion, splanchnic nerves, and sympathetic trunk. The right suprarenal gland, kidney, ureter, and colon are turned to the left so that the posterior surface of the right kidney is facing anteriorly. The inferior vena cava is pulled medially and the third and fourth lumbar veins are removed.

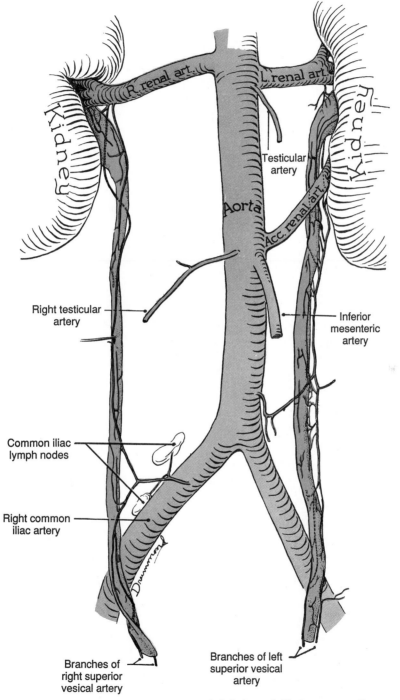

Figure 2-93. The arterial supply of the kidneys and ureters. Observe that the ureter is supplied by long arteries that come from three main sources: (1) the *renal artery* superiorly, (2) the *superior vesical artery* inferiorly, and (3) the *common iliac artery* near its middle or *aorta*. Note that these branches approach the ureter from the medial side. An accessory renal artery is also present in this specimen.

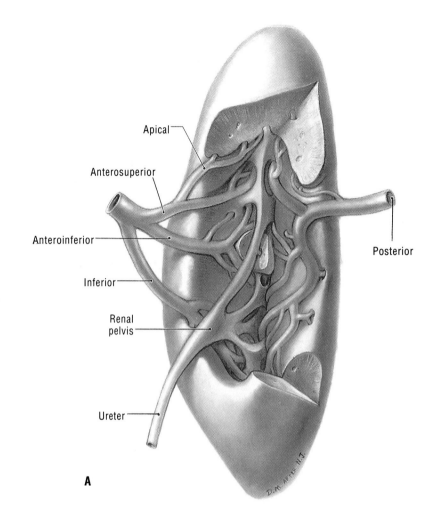

Apical

Anterosuperior

Anteroinferior

Posterior

Inferior

Renal
pelvis

Ureter

A

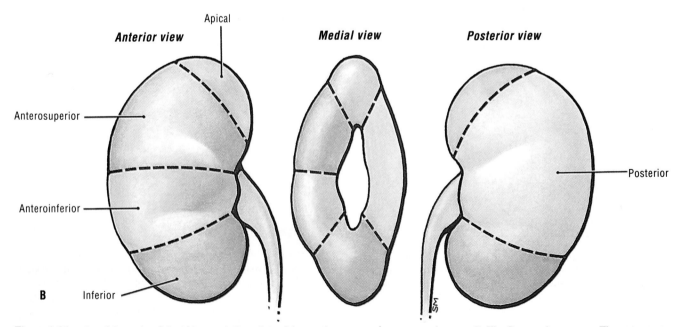

Apical

Anterior view   *Medial view*   *Posterior view*

Anterosuperior

Posterior

Anteroinferior

B   Inferior

Figure 2-94. Arterial supply of the kidneys. *A*, Branches of the renal artery in the renal sinus. The posterior lip of the renal sinus has been incised, superiorly and inferiorly, near the limits of the territory of the posterior segmental artery. *B*, The five renal segments. These segments are based on the arterial distribution shown in *A*.

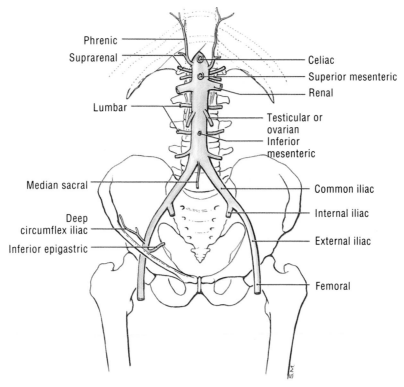

Figure 2-95. The abdominal aorta showing its relationship to the vertebral column. It is unusual for the median sacral artery to arise from the anterior surface of the aorta as here; its usual origin is from the posterior aspect.

cava. *Venous drainage from the ureters* is into the testicular or ovarian veins.

Lymphatic Drainage of the Kidneys and Ureters (Figs. 2-22, 2-60B, 2-92, and 2-93).   The lymph vessels of the kidney follow the renal vein and drain into the *lumbar (lateral aortic) lymph nodes*. Lymph vessels from the superior part of the ureter join those from the kidney or pass directly to these nodes. Lymph vessels from the middle part usually drain into the *common iliac lymph nodes*, whereas those from its inferior part drain into the common, external, or internal *iliac lymph nodes*.

Innervation of the Kidneys and Ureters (Figs. 2-61, 2-64, 2-82 and 2-96; Table 2-3).   Nerves to the kidneys and ureters come from the *renal plexus* and consist of sympathetic and parasympathetic fibers. This plexus is supplied by fibers from the lesser and lowest splanchnic nerves.

Rapid excessive distention of the ureter causes severe rhythmic pain called *ureteric colic*. Frequently the pain results from a *kidney stone* that passes into the ureter, where it is known as a **ureteric calculus** (L. pebble). These are usually composed of calcium oxalate, calcium phosphate, and/or uric acid. Calculi may be located in the calices, renal pelvis, ureter, or urinary bladder. Calcium-containing stones are radiopaque, whereas those composed of uric acid are radiolucent. Ureteric stones may cause complete or intermittent obstruction of urinary flow. The obstruction may occur superiorly at the ureteropelvic junction or anywhere along the ureter.

Ureteric colic is usually a sharp, stabbing pain that follows the course of the ureter (*i.e.*, from the lateral region or loin to the inguinal region or groin) as the stone is gradually forced down the ureter. In men the pain is frequently also referred to the scrotum, and in women, it may radiate to the labia majora (Fig. 2-18). The ureter is supplied with pain afferents that are included in the lowest splanchnic nerve (Fig. 2-64); the impulses enter via T12 and L1 segments. This explains why the spasmodic and agonizing pain is referred to the lateral and inguinal regions of the abdomen (Fig. 2-13A). The methods used to remove stones from the ureter are described with the pelvis (see Chap. 3, p. 268).

Large kidney calculi or stones, unable to enter the ureter, remain in the kidney or renal pelvis. Here, they may cause bleeding, infection, obstruction, and loss of kidney function. **Kidney stones** can be removed in three ways: open surgery, percutaneous surgery, and lithotripsy (Farrow, 1989). *Open surgery* is usually required when the stone is large or cannot be treated in other ways. An incision is made in the lumbar region and the stone is removed through the renal pelvis. In some cases the kidney has to be incised to remove the stone(s) from the calices. In *percutaneous surgery* the stones are removed with an instrument called a nephroscope that is inserted through a small incision. The surgeon can pass a small telescope through the *nephroscope* so that the stone(s) can be observed and removed. If they are too large to remove, they can be broken with an ultrasound probe or hydroelectric probe. Most kidney stones can also be treated with *lithotripsy*, a technique that focuses a shock-wave through the body that breaks up the stone. The patient can then pass the small fragments in the urine.

To study the kidneys, ureters, and urinary bladder radiologically, a contrast medium may be injected intravenously

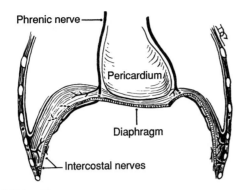

Figure 2-101.   Nerve supply to the diaphragm. Note that each phrenic nerve provides the sole motor supply to its half of the diaphragm. Observe that the peripheral part of the diaphragm receives its sensory supply from the inferior intercostal nerves.

vena cava, whereas the left inferior phrenic vein usually joins the left suprarenal vein. Some veins from the posterior curvature of the diaphragm drain into the *azygos* and *hemiazygos veins.*

Lymphatic Drainage of the Diaphragm (Fig. 2-60; see also Figs. 1-19 and 1-39).   Lymphatic plexuses on the thoracic and abdominal surfaces of the diaphragm communicate freely. The diaphragmatic lymph nodes are on the thoracic surface of the diaphragm. Lymph vessels from these nodes enter the *phrenic lymph nodes,* and from them, lymph drains into the *parasternal* and posterior *mediastinal lymph nodes.* Some lymph vessels from the abdominal surface of the diaphragm drain into the *superior lumbar (lateral aortic) lymph nodes.*

Innervation of the Diaphragm (Fig. 2-101; see also Figs. 1-72 and 1-73).   The entire motor supply to the diaphragm is from the *phrenic nerves,* which arise from the ventral rami of segments C3 to C5 of the spinal cord. The phrenic nerves also supply sensory fibers (pain and proprioception) to most of the diaphragm. Peripheral parts of the diaphragm receive their sensory supply from the inferior six or seven *intercostal nerves* and the *subcostal nerve* (Fig. 2-99).

The phrenic nerves divide into three branches on the thoracic surface of the diaphragm, or just superior to it (Fig. 2-101). The contribution to the phrenic nerve from the ventral ramus of C5 may be derived as a branch from the nerve to the subclavius muscle; this is called the *accessory phrenic nerve* (see Fig. 8-12). The superior level of origin of the phrenic nerves (C3 to C5) results from the caudal migration of the developing diaphragm relative to the vertebral column (Moore, 1988).

Actions of the Diaphragm

The diaphragm is the chief muscle of inspiration. When it contracts, its right and left domes move inferiorly so that its convexity is flattened. This descent of the domes increases the vertical diameter of the thoracic cavity (p. 48). During forced inspiration the muscular portion of the diaphragm contracts to its maximum extent (see Fig. 1-50), drawing its central tendon inferiorly at least 4 cm. As the diaphragm descends, it pushes the abdominal viscera before it. This increases the volume of

the thoracic cavity and decreases the intrathoracic pressure, resulting in air being taken into the lungs. In addition, the volume of the abdominal cavity is somewhat decreased and the intra-abdominal pressure is somewhat increased because the antero-lateral abdominal wall moves outward. During quiet inspiration the central tendon moves very little.

Diaphragmatic movements are also important in blood circulation because the increased abdominal pressure and decreased thoracic pressure accompanying contraction of the diaphragm help to return blood to the heart. When the diaphragm contracts, compressing the abdominal viscera, blood in the inferior vena cava is forced superiorly into the heart. This movement is facilitated by the enlargement of the vena caval foramen and the dilation of the inferior vena cava that occurs as the diaphragm contracts.

> The diaphragm is also an important muscle in abdominal straining. It assists the anterior abdominal muscles in raising the intra-abdominal pressure during *micturition* (urination), *defecation* (bowel movements), and *parturition* (childbirth). During these functions, people often inspire and then close their glotis (opening in the larynx). This traps air in the respiratory tract, preventing the diaphragm from rising. A grunt is produced when air escapes. The diaphragm is also used when lifting heavy weights. A weight lifter about to raise a heavy load inspires deeply to raise the intra-abdominal pressure and to give additional support to the vertebral column.

Shape and Position of the Diaphragm

It is clinically important to acquire a three-dimensional concept of the dome-shaped diaphragm. One relies on this "mind's eye view" particularly when viewing radiographs of the thorax and abdomen. Its posterior attachment is considerably more inferior than its anterior attachment (Fig. 2-100). Hence, much of it cannot be seen in PA radiographs (see Figs. 1-49 and 1-50). Normally the right dome bulges more superiorly into the thorax than does the left one (Fig. 2-101). The level of the domes varies in relation to the ribs and vertebrae according to: (1) the phase of respiration, (2) the posture assumed, and (3) the size and degree of distention of the abdominal viscera. The diaphragm is at its most superior level when a person is supine, with the upper body lowered, because the abdominal viscera push the diaphragm into the thoracic cavity. When a patient lies on one side, the hemidiaphragm next to the bed or radiography table rises to a more superior level owing to the greater push of the viscera on that side. Conversely, the diaphragm assumes an inferior level when a person is sitting or standing. This explains why patients with *dyspnea* (difficult breathing) prefer to sit up rather than lie down and to have their upper body elevated when they are recumbent.

> Section of the phrenic nerve in the neck results in complete paralysis and eventual atrophy of the muscular part of the corresponding half of the diaphragm, except in persons who have an accessory phrenic nerve (see Fig. 8-12). *Paralysis of a hemidiaphragm* can be recognized radiographically by its permanent elevation and paradoxical movement. Instead of descending on inspiration, it is forced superiorly by the

increased intra-abdominal pressure secondary to descent of the opposite unparalyzed hemidiaphragm.

Referred pain from the diaphragm is felt in two different areas owing to the difference in the sensory nerve supply of the diaphragm. Irritation of the diaphragmatic pleura or diaphragmatic peritoneum is referred to the shoulder region, the area of skin supplied by segment C4 of the spinal cord. This segment also contributes ventral rami to the phrenic nerve. Sensation from peripheral regions of the diaphragm, innervated by the inferior intercostal nerves, is referred to the skin over the costal margin of the abdominal wall (Figs. 2-7 and 2-13).

Because there are communications between lymphatic vessels on the abdominal surface of the diaphragm with those on its thoracic surface, CA of the GI organs may spread directly via lymph vessels that pass through the diaphragm, either directly or through its apertures, into the thorax.

The Posterior Abdominal Wall

The posterior abdominal wall is composed principally of muscles and fascia attached to the vertebrae, hip bones, and ribs. It also contains fat, important nerves, vessels, and lymph nodes (Figs. 2-82, 2-91, 2-92, 2-96, and 2-99). The posterior abdominal wall is composed of three types of bone (lumbar *vertebrae*, *sacrum*, and wings of *ilium*), and four different muscles (posterior part of the *diaphragm*, *psoas major*, *iliacus*, and *quadratus lumborum*). The bones of the posterior abdominal wall are discussed in the section on the back (see Chap. 4).

Muscles of the Posterior Abdominal Wall

The diaphragm has been considered, starting on p. 224. There are three paired muscles in the posterior abdominal wall that are clinically important: psoas major, iliacus, and quadratus lumborum.

The Psoas Major Muscle (Figs. 2-82, 2-92, 2-96, 2-99, and 2-102). This long, thick fusiform muscle lies lateral to the lumbar vertebrae. *Psoas is a Greek word* meaning "muscle of the loin." (Butchers refer to the psoas muscle as the tenderloin.) The psoas muscles pass inferolaterally and run deep to the inguinal ligament to reach the lesser trochanter of the femur. The attachments, nerve supply, and main actions of the psoas muscles are summarized in Table 2-5. The lumbar plexus is embedded in this muscle.

The Psoas Minor Muscle (Fig. 2-92). This small weak muscle with a short belly and a long tendon is present in 50 to 60% of people, but it may be present on only one side. It is attached superiorly to the sides of T12 and L1 vertebrae and the intervening intervertebral disc, and inferiorly to the iliopubic eminence on the pelvic brim (see Fig. 5-2). Located anterior to the psoas major, it helps the psoas major to flex the pelvis and lumbar region of vertebral column. It is innervated by the ventral ramus of L1 nerve.

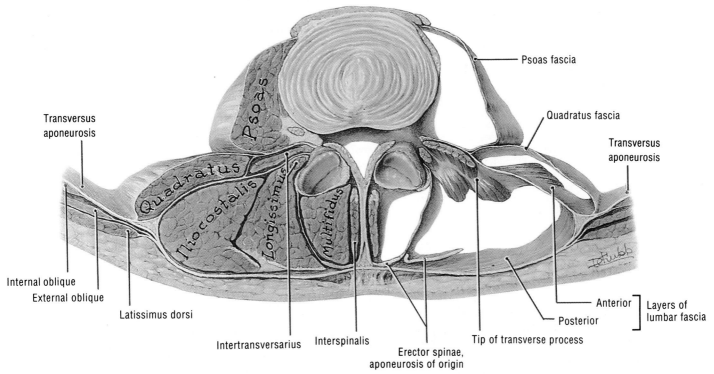

Figure 2-102. Transverse section through the superior lumbar region. The left side has been dissected to illustrate the fascia covering the muscles of the back. On the left side, observe that the transversus aponeurosis or aponeurosis of the transversus abdominis muscle splits to form the two layers of the thoracolumbar fascia.

The Iliacus Muscle (Figs. 2-82 and 2-99).   This large triangular muscle lies along the lateral side of the inferior part of the psoas major muscle. Attached to the superior two-thirds of the iliac fossa, it extends across the sacroiliac joint. Most of its fibers join the tendon of the psoas major muscle. The attachments, nerve supply, and main actions of this muscle are summarized in Table 2-5. The psoas and iliacus muscles are referred to jointly as the **iliopsoas muscle**, which is the *chief flexor of the thigh*. It is also a stabilizer of the hip joint and helps to maintain the erect posture at this joint (see Chap. 5, p. 386).

The iliopsoas muscle has extensive and clinically important relations to the kidneys, ureters, cecum, appendix, sigmoid colon, pancreas, lumbar lymph nodes, and nerves of the posterior abdominal wall (Figs. 2-82, 2-92, and 2-99). When any of these structures is diseased, movements of this muscle may cause pain. As it lies along the vertebral column and crosses the sacroiliac joint, disease of the intervertebral and sacroiliac joints may cause *spasm of the iliopsoas muscle*, a protective reflex. Owing to the relationship of the pancreas to the posterior abdominal wall (Figs. 2-48, 2-52, and 2-74), *CA of the pancreas* in advanced stages invades the muscles and nerves of the posterior abdominal wall, producing excruciating pain.

Although the prevalence of *tuberculosis* in North America has been greatly reduced, this infection still occurs, especially in the native population and may spread via the blood (*hematogenous spread*) to the vertebrae, particularly during childhood. An abscess resulting from tuberculosis in the lumbar region tends to spread from the vertebrae into the *psoas sheath* (fascia enclosing the psoas muscle) and to produce a *psoas abscess*. As a consequence, the *psoas fascia* thickens to form a strong stockinglike tube (Fig. 2-102). Pus from the abscess passes inferiorly along the psoas within this fascial tube over the pelvic brim and deep to the inguinal ligament.

The pus usually points (surfaces) in the *femoral triangle* in the superior part of the thigh (see Fig. 5-21 and p. 393). Pus can also reach the psoas sheath by tracking down from the posterior mediastinum when the thoracic vertebrae are diseased (see Figs. 1-66 and 1-74).

The inferior part of the *iliac fascia* is often tense and raises a fold that passes to the internal aspect of the iliac crest (Figs. 2-82 and 2-99). The superior part of this fascia is loose and may form a pocket, sometimes called the *fossa iliacosubfascialis*, posterior to the above mentioned fold. Part of the large intestine may become trapped in this fossa (*e.g.*, the cecum and/or appendix on the right and the sigmoid colon on the left). Considerable pain results from this condition.

The Quadratus Lumborum Muscle (Figs. 2-91, 2-96, 2-99, and 2-102).   This quadrilateral muscle (L. *quadrus*, square) forms a thick muscular sheet in the posterior abdominal wall. It lies adjacent to the transverse processes of the lumbar vertebrae and is broader inferiorly. The attachments, nerve supply, and main actions of this muscle are summarized in Table 2-5.

Fascia of the Posterior Abdominal Wall

Each muscle in the posterior abdominal wall is enclosed in fascia. The name of the fascia is derived from the muscle(s) it encloses.

The Iliac Fascia (Figs. 2-92, 2-98, 2-99, and 2-102).   The *iliopsoas fascia* covering the psoas and iliacus muscles, is usually referred to as the iliac fascia (fascia iliaca). Although thin superiorly, it thickens inferiorly as it approaches the inguinal ligament. The part of the fascia covering the psoas major muscle, called the *psoas fascia*, is attached medially to the lumbar vertebrae and pelvic brim. Superiorly this fascia is thickened to form the *medial arcuate ligament* of the diaphragm. The fascia

Table 2-5.
Principal Muscles of the Posterior Abdominal Wall

Muscle	Superior Attachments	Inferior Attachment(s)	Innervation	Actions
Psoas major[1]	Transverse processes of lumbar vertebrae; sides of bodies of T12 to L5 vertebrae and the intervening intervertebral discs	By a strong tendon to the lesser trochanter of the femur	Lumbar plexus via ventral branches of L2, L3, and L4	Acting superiorly with the iliacus m., it flexes the thigh Acting inferiorly by itself, it flexes the vertebral column laterally. It is used to balance the trunk when sitting. Acting inferiorly with its partner and the iliacus mm., it flexes the trunk
Iliacus[1]	Superior two-thirds of iliac fossa, ala of sacrum, and anterior sacroiliac ligaments	Lesser trochanter of femur and shaft of femur inferior to it	Femoral n. (L2 to L4)	Flexes thigh and stabilizes hip joint; acts with psoas major
Quadratus lumborum	Medial half of inferior border of 12th rib and tips of lumbar transverse processes	Iliolumbar ligament and internal lip of iliac crest	Ventral branches of T12 and L1 to L4	Extends and laterally flexes the vertebral column; fixes 12th rib during inspiration

[1]The psoas major and iliacus muscles are often described together as the *iliopsoas muscle* when flexion of the thigh is discussed (Chap. 5). The iliopsoas is the *chief flexor of the thigh* and, when the thigh is fixed, it is a strong flexor of the trunk (*e.g.*, during situps).

is fused laterally with the anterior layer of the *thoracolumbar fascia*. Inferior to the iliac crest, it is continuous with the part of the iliac fascia covering the iliacus muscle. The psoas fascia also blends with the fascia covering the quadratus lumborum muscle. The *iliac fascia* is continuous with the transversalis fascia (Figs. 2-11 and 2-12) and is continuous inferiorly with the fascia in the thigh. The posterior margin of the transversalis fascia is attached to the inguinal ligament and is continuous there with the iliac fascia as it passes into the thigh.

The Quadratus Lumborum Fascia (Figs. 2-91, 2-96, 2-99, and 2-102).   The fascia covering the quadratus lumborum muscle is a dense membranous layer that is continuous laterally with the anterior layer of the thoracolumbar fascia. The quadratus lumborum fascia is attached to the anterior surfaces of the transverse processes of the lumbar vertebrae, the iliac crest, 12th rib, and transversalis fascia. It is thickened superiorly to form the *lateral arcuate ligament* and inferiorly it is adherent to the *iliolumbar ligament*.

The Thoracolumbar Fascia (Figs. 2-91 and 2-102; see also Fig. 4-44).   This is an extensive sheet of fascia covering the deep muscles of the back. The lumbar part of the thoracolumbar fascia extends between the 12th rib and the iliac crest. It is attached laterally to the internal oblique and transversus abdominis muscles. *The thoracolumbar fascia splits into three layers medially.* The quadratus lumborum muscle lies between its anterior and middle layers and the deep back muscles are enclosed between its middle and posterior layers. The thin anterior layer of thoracolumbar fascia (which forms the quadratus lumborum fascia) is attached, along with the psoas fascia, to the anterior surfaces of the lumbar transverse processes. The thick middle layer of thoracolumbar fascia is attached to the tips of the transverse processes. The dense posterior layer of this fascia is attached to the spinous processes of the lumbar and sacral vertebrae, and to the *supraspinous ligament* (see Fig. 4-28).

Nerves of the Posterior Abdominal Wall

There are two types of nerve in this wall: *somatic nerves* of the lumbar plexus and its branches, and visceral or *splanchnic nerves* of the autonomic nervous system (Figs. 2-64, 2-92, and 2-99). The five *lumbar nerves* pass from the spinal cord through the intervertebral foramina, inferior to the corresponding vertebrae, where they divide into dorsal and ventral primary rami. Each ramus contains sensory and motor fibers. The dorsal primary rami pass posteriorly to supply the muscles and skin of the back (Fig. 2-91), whereas the ventral primary rami pass into the psoas major muscles. Here they are connected to the *sympathetic trunk* by rami communicantes (Figs. 2-92 and 2-99). The ventral rami give branches to the psoas major, quadratus lumborum, and iliacus muscles (Table 2-5).

The ventral rami of L1 to L3 nerves and the superior branch of L4 form the *lumbar plexus* (Fig. 2-99). The inferior branch of L4 and all of L5 form the *lumbosacral trunk*, which descends to the *sacral plexus* (see Fig. 3-16). The subcostal nerve is the ventral ramus of T12. It passes posterior to the lateral arcuate ligament of the diaphragm, about 1 cm caudal to the 12th rib (Figs. 2-82 and 2-99). Usually the *subcostal nerve* sends a branch to the ventral ramus of L1 nerve and then runs inferolaterally

across the anterior surface of the quadratus lumborum muscle. The subcostal nerve pierces the transversus abdominis muscle and runs in the anterior abdominal wall between this muscle and the internal oblique. It supplies the anterior abdominal wall inferior to the umbilicus and superior to the pubic symphysis (Fig. 2-13A).

The Lumbar Plexus

The lumbar plexus is a network of nerves *formed within the psoas major* muscle by the ventral rami of L1 to L4 nerves (Fig. 2-99). Hence the origin of the nerves contributing to the lumbar plexus can be studied only when the psoas muscle has been carefully removed. The nerves forming the lumbar plexus pass through the psoas major muscle at different levels. In many people (about 50%), there is a contribution from the *subcostal nerve* (the large ventral ramus of T12). All five of the lumbar ventral rami receive gray rami communicantes from the sympathetic trunk, and the superior two send white rami communicantes to the sympathetic trunk (Figs. 2-92 and 2-99). The largest and most important branches of the lumbar plexus are the obturator and femoral nerves: they are derived from the same spinal cord segments (L2, L3, and L4).

The Obturator Nerve (Fig. 2-99; see also Fig. 3-10).   This nerve descends through the psoas major muscle, leaving its medial border at the brim of the pelvis. It pierces the psoas fascia, crosses the sacroiliac joint, passes lateral to the internal iliac vessels and ureter, and enters the pelvis minor (p. 248). The obturator nerve leaves the pelvis through the obturator foramen and enters the thigh. In some people (about 30%), the dorsal division of L3 and L4 nerves gives off branches that unite to form an *accessory obturator nerve*. It is located along the medial border of the psoas major. The obturator is the *nerve of the adductor muscles of the thigh* (see Table 5-2).

The Femoral Nerve (Figs. 2-82 and 2-99).   This nerve also pierces the psoas major muscle and runs inferolaterally within it to emerge between the psoas major and iliacus muscles, just superior to the inguinal ligament. In the abdomen the femoral nerve supplies the psoas and iliacus muscles (iliopsoas muscle). In the thigh it is *the nerve of the extensor muscles of the knee* (see Table 5-1).

The Ilioinguinal and Iliohypogastric Nerves (Figs. 2-7, 2-13, 2-82, and 2-99).   These nerves are derived from L1 segment, often by a common stem. They enter the abdomen posterior to the *medial arcuate ligament* and pass inferolaterally, anterior to the quadratus lumborum muscle. Often these nerves do not separate until they are under the transversus abdominis muscle. They pierce this muscle near the anterior superior iliac spine and pass through the internal and external oblique muscles to supply the skin of the suprapubic and inguinal regions. Both nerves also supply branches to the abdominal musculature. The *iliohypogastric nerve* (*L1*) sends a lateral branch to the skin of the gluteal region (buttocks) and an anterior branch to the skin of the hypogastric region. The *ilioinguinal nerve* (*L1*) passes through the superficial inguinal ring and supplies the skin of the groin and scrotum or labium majus (Figs. 2-13A and 2-18).

The Genitofemoral Nerve (Figs. 2-11, 2-82, and 2-99).   This nerve arises from the ventral divisions of L1 and L2 segments and pierces the anterior surface of the psoas major muscle and the iliac fascia. It runs inferiorly and divides lateral to the common and external iliac arteries into the femoral and genital branches.

The Lateral Femoral Cutaneous Nerve (Figs. 2-82 and 2-99; see also Fig. 5-10).   This cutaneous nerve of the thigh arises from the dorsal branches of L2 and L3 ventral rami and passes through the psoas major muscle, emerging superior to the iliac crest. It runs inferolaterally on the iliacus muscle and enters the thigh posterior to the inguinal ligament, just medial to the anterior superior iliac spine. This nerve supplies the skin on its anterolateral surface of the thigh.

The Lumbosacral Trunk

The lumbosacral trunk is a large flat nerve that is formed by the union of the inferior part of the ventral ramus of L4 with the ventral ramus of L5 nerve (Fig. 2-99). Its L4 component descends through the psoas major muscle on the medial part of the transverse process of L5 vertebra and then passes over the ala (wing) of the sacrum. The lumbosacral trunk descends into the pelvis and takes part in the formation of the *sacral plexus*. It is not a branch of the lumbar plexus.

The Autonomic Nerves

The autonomic nerves in the posterior abdominal wall consist of sympathetic and parasympathetic portions (see Fig. 26 on p. 30). The efferent nerves to the viscera, part of the autonomic nervous system, emerge from the spinal cord and brain stem as fibers of certain spinal nerves (Figs. 2-82, 2-92, and 2-99). The abdominal part of the sympathetic trunk on each side enters the abdomen by passing posterior to the medial arcuate ligament. It is usually composed of four *lumbar sympathetic ganglia*, which have connecting parts. The sympathetic trunks lie in a groove along the medial border of the psoas major muscle (Figs. 2-64, 2-92, 2-96, and 2-99). Medially, the sympathetic trunk gives off *lumbar splanchnic nerves* to the aortic plexus.

The Sympathetic Trunks and Nerves (Figs. 2-61, 2-64, 2-82, 2-92, 2-96, and 2-99; Table 2-3).   *The right sympathetic trunk* lies posterior to the inferior vena cava, lumbar lymph nodes, and right ureter. This relationship is surgically important. Both sympathetic trunks pass anterior to the small lumbar vessels supplying the posterior abdominal wall, and then run posterior to the common iliac vessels as they enter the pelvis. The two trunks unite in the median *ganglion impar* on the coccyx. The sympathetic and parasympathetic nerves are distributed to the abdominal viscera by a rich tangle of nerve plexuses and ganglia along the anterior surface of the abdominal aorta. A principal part of the abdominal autonomic nervous system is the **celiac plexus** (solar plexus) and its ganglia. It is located on each side of the celiac trunk at the level of the superior part of L1 vertebra.

The thoracic splanchnic nerves are the main source of sympathetic nerves in the abdomen. The greater, lesser, and lowest splanchnic nerves are branches of the 5th to 12th thoracic sympathetic ganglia. These nerves are composed of *preganglionic fibers* which come from the spinal cord via white rami communicantes and pass through the sympathetic ganglia without

synapsing (see Figs. 26 and 27 in Overview). They end in the celiac and aorticorenal ganglia (Figs. 2-61 and 2-64), from which they are relayed as unmyelinated postganglionic fibers.

The Greater Splanchnic Nerve (Figs. 2-64, 2-92, and 2-99).   This large nerve is formed by 4 to 5 roots from the sympathetic trunk between the 6th and 10th ganglia. It runs inferiorly on the thoracic vertebral bodies medial to the sympathetic trunk and lateral to the azygos vein. It usually pierces the corresponding crus of the diaphragm and ends in the celiac ganglion.

The Lesser Splanchnic Nerve (Figs. 2-61, 2-64, and 2-92).   Usually this smaller nerve arises by two roots from the 9th and 10th sympathetic ganglia and runs inferiorly, lateral to the greater splanchnic nerve. It pierces the corresponding crus of the diaphragm and ends in the inferior part of the celiac ganglion, which is called the *aorticorenal ganglion*.

The Least Splanchnic Nerve (Fig. 2-64).   This tiny nerve is formed by branches from the 11th and/or 12th sympathetic ganglion. It usually pierces the corresponding crus of the diaphragm, near or with the lesser splanchnic nerve, and ends in the *renal plexus*. Sometimes it passes posterior to the medial arcuate ligament with the sympathetic trunk. It is often absent.

The Abdominal Autonomic Plexuses (Figs. 2-61 and 2-64).   These plexuses surround the abdominal aorta and its major branches. Collections of sympathetic nerve cells (*sympathetic ganglia*) are scattered amongst the celiac and mesenteric plexuses. The *parasympathetic ganglia* (collections of parasympathetic nerve cells) are located in the walls of the viscera, *e.g.*, the *myenteric plexus* (of Auerbach) is in the muscular coat of the stomach and intestine (Cormack, 1987).

The Intermesenteric Plexus (Fig. 2-64).   This plexus (part of the aortic plexus) consists of 4 to 12 nerves on the anterior and anterolateral aspects of the aorta, between the superior and inferior mesenteric arteries. This plexus receives contributions from the first two lumbar splanchnic nerves and gives rise to renal, testicular (or ovarian), and ureteric branches. Occasionally, it gives off branches to the duodenum, pancreas, aorta, and inferior vena cava.

The Superior Hypogastric Plexus (Fig. 2-64; see also Figs. 3-47 and 3-48).   This plexus is continuous with the intermesenteric plexus. It receives branches from the lumbar ganglia of the sympathetic trunks. It lies anterior to the inferior part of the aorta, its bifurcation, and the median sacral vessels. It receives the inferior two lumbar splanchnic nerves and divides into right and left *hypogastric nerves*, which pass to the inferior hypogastric plexus. The superior hypogastric plexus supplies *ureteric* and *testicular plexuses* and a plexus on each common iliac artery.

The Inferior Hypogastric Plexus (Fig. 2-64; see also Fig. 3-47).   This plexus is formed on each side by a hypogastric nerve from the superior hypogastric plexus. They are situated on the sides of the rectum, uterine cervix, and urinary bladder. Sympathetic ganglia within these plexuses surround the corresponding internal iliac artery. The right and left plexuses receive small branches from the superior sacral sympathetic ganglia and the sacral parasympathetic outflow from S2 to S4 (*pelvic splanchnic nerves*). Extensions of the inferior hypogastric plexus send autonomic fibers along the blood vessels, which form visceral plexuses on the walls of the pelvic viscera (*e.g.*, the rectal and vesical plexuses).

Afferent Fibers in Sympathetic Nerves (Figs. 2-64, 2-92, and 2-99). Although the sympathetic nerves are motor, they also carry some sensory fibers from sense organs in the viscera. These fibers pass toward the spinal cord in the splanchnic nerves as far as the sympathetic trunk. They then pass superiorly or inferiorly to reach the level of the spinal cord that is to convey the impulses conducted by them to the central nervous system. They then leave the sympathetic trunk in a *white ramus communicans* and enter a spinal nerve and the spinal cord via its dorsal root. The cell bodies of the visceral sensory fibers are in the *spinal ganglia*.

Afferent Fibers in Parasympathetic Nerves. Although the parasympathetic nerves are visceral efferent (*i.e.*, motor to smooth muscle and glands), the viscera supplied by them contain sense organs and afferent fibers from them pass back to the central nervous system in the parasympathetic nerves. All the sensory fibers in parasympathetic nerves have their cell bodies in the sensory ganglion of the nerve supplying the viscus with parasympathetic fibers (*e.g.*, in the sensory ganglion of the vagus nerve).

Pain arising from an abdominal viscus varies from dull to very severe, but it is poorly localized. It radiates to the part of the body supplied by somatic sensory fibers associated with the same segment of the spinal cord that receives visceral sensory fibers from the viscus concerned (see Fig. 1-24). This is called **visceral referred pain**. For example, pain from a gastric ulcer is referred to the epigastric region because the stomach is supplied by pain afferents that reach T7 and T8 segments of the spinal cord via the *greater splanchnic nerve*. The pain is interpreted by the brain as though the irritation occurred in the area of skin supplied by the dorsal roots of T7 to T9 nerves (Fig. 2-13A; see also Fig. 1-24).

Pain from an inflamed vermiform appendix passes centrally in the lesser splanchnic nerve on the right side and is initially referred to the umbilical region, which lies in the T10 dermatome (Fig. 2-13A; see also Fig. 1-24). Pain is later referred to the lower right quadrant when the parietal peritoneum in contact with the appendix becomes inflamed. Pain arising from the parietal peritoneum is the somatic type and is usually severe. It can be precisely located as to the site of its origin. The anatomical basis for this localization of pain is that the parietal peritoneum is supplied by somatic sensory fibers through the thoracic nerves, whereas the appendix is supplied by visceral sensory fibers in the lesser splanchnic nerve.

Inflamed parietal peritoneum is extremely sensitive to stretching. Hence, when pressure is applied to the anterior abdominal wall over the site of inflammation (*e.g.*, McBurney's point, p. 141) and suddenly removed, extreme localized pain is usually felt. When the inflamed parietal peritoneum is stretched by digital pressure and then rebounds, pain is produced. This is called *rebound tenderness*.

The treatment of some patients with arterial disease in the lower limbs may include the surgical removal of two or more lumbar sympathetic ganglia, with division of their rami communicantes. This operation is called a *lumbar sympathectomy*. Surgical access to the sympathetic trunks is commonly through a lateral extraperitoneal approach because the sympathetic trunks lie retroperitoneally in the extraperitoneal fatty tissue (Fig. 2-92). The muscles of the anterior abdominal wall are

split and the peritoneum is moved medially and anteriorly to expose the medial edge of the psoas major muscle, along which the sympathetic trunk lies (Figs. 2-92, 2-96, and 2-99). The left trunk is overlapped slightly by the aorta (Fig. 2-82) and, on rare occasions, by a persisting left inferior vena cava (Fig. 2-106). The right sympathetic trunk is covered by the inferior vena cava (Figs. 2-82 and 2-96). Hence, the surgeon has to retract these structures medially to expose the sympathetic trucks (Fig. 2-93). They usually lie in the groove between the psoas major muscle laterally and the lumbar vertebral bodies medially, where they are often obscured by fat and lymphatic tissue.

Identification of the sympathetic trunks is not easy (Figs. 2-92, 2-96, and 2-99). Knowing this, great care is taken not to remove inadvertently part of: (1) the genitofemoral nerve, (2) the lumbar lymphatics, or (3) the ureter. Uncommonly, pathologists see these parts in surgical specimens sent to them. The intimate relationship of the sympathetic trunks to the aorta and the inferior vena cava also makes these large vessels vulnerable to injury during a *lumbar sympathectomy*.

Arteries of the Posterior Abdominal Wall

These arteries arise from the abdominal aorta (Fig. 2-95), except for the subcostal arteries which are branches of the thoracic aorta.

The Subcostal Arteries (Figs. 2-7, 2-39, and 2-82). These arteries are the last branches of the descending thoracic aorta. They were given their name because they are located inferior to the costal margins, the intercostal arteries, and the 12th rib. Each artery runs laterally over the body of T12 vertebra and posterior to the splanchnic nerves, sympathetic trunk, pleura, and diaphragm. *They enter the abdomen posterior to the lateral arcuate ligaments* with the subcostal nerves (T12). They then run anterior to the quadratus lumborum muscle and posterior to the kidney, before piercing the aponeurosis of the origin of the transversus abdominis muscle and passing between this muscle and the internal oblique muscle. The subcostal arteries anastomose anteriorly with the inferior epigastric and inferior intercostal arteries and posteriorly with the lumbar arteries.

The Abdominal Aorta

The abdominal aorta is the continuation of the descending thoracic aorta (Figs. 2-39, 2-82, 2-93, 2-95, and 2-103). It begins at the aortic hiatus in the diaphragm at the level of the intervertebral disc between T12 and L1 vertebrae and ends around the level of L4 vertebra by dividing into two common iliac arteries. Throughout its course the aorta lies against vertebral bodies.

The Relations of the Abdominal Aorta (Figs. 2-61, 2-82, 2-95, 2-99, and 2-103; see also Fig. 1-70).

Anteriorly, the abdominal aorta is related to the *celiac trunk* and its branches, the *celiac plexus*, omental bursa, *pancreas*, left renal vein, ascending part of the *duodenum*, root of *the mesentery*, and intermesenteric plexus of nerves.

Posteriorly, the abdominal aorta descends anterior to the *bodies of L1 to L4 vertebrae*, the intervening intervertebral discs, and the corresponding part of the *anterior longitudinal ligament*. On the right, the abdominal aorta is related superiorly to the

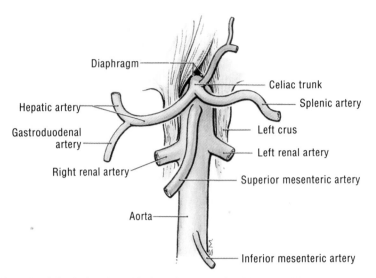

Figure 2-103. The abdominal aorta and its branches. Note that it begins where the diaphragm rests on the celiac trunk at the level of the intervertebral disc between T12 and L1 vertebrae.

cisterna chyli, thoracic duct, and right *crus of the diaphragm.* The inferior vena cava lies to the right of the aorta. *On the left,* the abdominal aorta is related superiorly to the left crus of the diaphragm and the *left celiac ganglion.* The *duodenojejunal flexure* is on its left, opposite L2 vertebra, and the *sympathetic trunk* runs along its left side.

The Surface Anatomy of the Abdominal Aorta (Figs. 2-46 and 2-95). This large artery may be represented by a broad band, about 2 cm wide, extending from a median point, about 2.5 cm superior to the *transpyloric plane*, to a point slightly inferior and to the left of the umbilicus. The latter point indicates the level of bifurcation of the aorta into the common iliac arteries. The *aortic bifurcation* is also just to the left of the midpoint of the line joining the highest points of the iliac crests. This line is very helpful when examining obese persons in whom the umbilicus is not a reliable landmark. When the anterior abdominal wall is relaxed, particularly in children and thin adults, the most inferior part of the abdominal aorta may be readily compressed against the body of L4 vertebra by firm pressure on the anterior abdominal wall, just inferior to the umbilicus. The pulsations of the abdominal aorta can be felt.

The Branches of the Abdominal Aorta (Figs. 2-39, 2-82, 2-93, 2-95, and 2-103). The branches of the aorta may be grouped into four types: (1) three unpaired visceral branches; (2) paired visceral branches; (3) paired parietal branches; and (4) an unpaired parietal branch.

The Unpaired Visceral Branches (Figs. 2-39, 2-95, and 2-103). These vessels arise from the anterior surface of the aorta. They are: the celiac trunk (*CT*), superior mesenteric artery (*SMA*), and inferior mesenteric artery (*IMA*). They arise at the following vertebral levels: CT (*T12*); SMA (*L1*); and IMA (*L3*).

The Paired Visceral Branches (Figs. 2-39, 2-61, 2-93, and 2-95). These vessels arise from the sides of the aorta at the following vertebral levels:

1. The *middle suprarenal arteries* (L1), one or more on each side, arise close to the origin of SMA. They run laterally on the crura of the diaphragm to the suprarenal glands;

2. The *renal arteries* (L1) arise just inferior to the SMA. Occa-

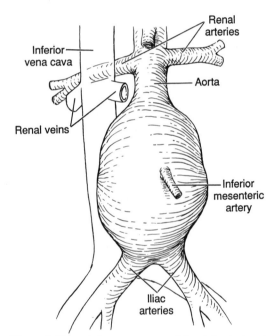

Figure 2-104. A large aneurysm of the abdominal aorta.

sionally there is an *accessory renal artery*, particularly on the left side;

3. The *gonadal arteries* (L2) are long slender vessels that arise from the aorta, a short distance inferior to the renal arteries. They pass inferiorly on the psoas major muscle. The right artery passes over the inferior vena cava. The *testicular artery* passes through the deep inguinal ring, enters the inguinal canal, and leaves it to form part of the spermatic cord (Fig. 2-19). The *ovarian artery* follows a similar course through the abdomen (see Fig. 3-26), but it crosses the proximal ends of the external iliac vessels to enter the pelvis minor, where it supplies the ovary and uterine tube.

The Paired Parietal Branches of the Aorta (Figs. 2-39, 2-61, 2-82, 2-92, 2-95, and 2-103). These vessels arise from its posterolateral surfaces. The *inferior phrenic arteries* arise just inferior to the diaphragm and pass superolaterally over the crura of the diaphragm. Each artery gives rise to several superior suprarenal arteries and then spreads out on the inferior surface of the diaphragm. The four pairs of **lumbar arteries** arise from the posterolateral surfaces of the abdominal aorta. Each pair passes around the sides of the superior four lumbar vertebrae. The lumbar arteries pass posteromedial to the sympathetic trunks and, on the right, run posterior to the inferior vena cava. The lumbar arteries divide between the transverse processes of the lumbar vertebrae into anterior and posterior branches.

The anterior branch of each lumbar artery passes deep to the quadratus lumborum muscle, and then runs around the abdominal wall between the internal oblique and transversus abdominis muscles, where it anastomoses in the posterior part of the rectus abdominis muscle with the inferior epigastric arteries (Fig. 2-7). The anterior branches supply the anterolateral walls of the inferior half of the abdomen. *The posterior branch of each lumbar artery* passes posteriorly, lateral to the articular processes of the vertebra, and supplies the spinal cord (see Fig. 4-52 and p. 362), cauda equina, spinal meninges, erector spinae muscles, and the overlying skin (see Chap. 4). The spinal arteries arise from the posterior branches of the lumbar arteries and pass through the intervertebral foramina to supply the vertebrae.

The Unpaired Parietal Artery (Figs. 2-39 and 2-95).   The median sacral artery is a tiny vessel which was the dorsal aorta in the sacral region of the embryo. It became smaller as the tail of the embryo disappeared. Typically the *median sacral artery* arises from the posterior surface of the aorta, just proximal to its bifurcation. It descends in the midline anterior to L4 and L5 vertebrae, usually giving off a small lumbar artery on each side called the *fifth lumbar artery* or arteria lumbalis ima (L. *ima*, lowest). Their distribution is similar to that of the lumbar arteries.

The abdominal aorta and its branches can be studied radiologically after injections of radiopaque contrast medium. When the distal part of the abdominal aorta and/or the external iliac artery are not occluded, a catheter can be passed into a femoral artery just inferior to the inguinal ligament (Fig. 2-95). Under fluoroscopic control the catheter can be passed to the desired level of the aorta for injection of the contrast medium, or it can be manipulated until it is in a branch of the aorta for *selective angiography* of this branch. When *atherosclerosis* (*e.g.*, narrowing of the aorta or iliac arteries owing to lipid deposits in their walls) is present, this technique may not be feasible. In some of these cases, the abdominal aorta is injected with contrast material by direct needle puncture (*translumbar aortography*). The level of puncture is from L1 to L3 vertebra (Fig. 2-95), but it is sometimes performed at T12 and L1 level because of the constancy of the position of the aorta at this level in the aortic hiatus of the diaphragm (Fig. 2-103). The *needle puncture* is commonly made 1 cm below the 12th rib, 6 to 8 cm to the left of the median plane.

Atherosclerosis of a common iliac artery just distal to the bifurcation of the aorta is common. Patients with this type of *arteriosclerotic occlusive disease* complain of posterior leg (calf), thigh, or hip pain on exertion. This *claudication* (pain) disappears when they rest. Sometimes this vascular condition is treated by *bypass graft surgery* (Gross et al., 1989). An aortobifemoral bypass graft is inserted to save a limb that is receiving an inadequate supply of blood. A graft made of Dacron is joined to the aorta through an abdominal incision.

The graft is then brought down and joined to the femoral arteries through separate groin incisions.

Because the aorta lies posterior to the pancreas and stomach (Figs. 2-52 and 2-68), a tumor of these organs may transmit the pulsations of the aorta and be mistaken for an aneurysm. *Aneurysm of the abdominal aorta* (localized enlargement) distal to the renal arteries (Fig. 2-104) may also be repaired by inserting a graft (Ameli, 1989). The aneurysm is opened and a Dacron graft is sewn into position. The wall of the aneurysmal aorta is sewn over the graft to protect it. For a description of a typical case study of **abdominal aortic aneurysm,** see Case 4-5 on p. 371.

During certain surgical procedures in the abdomen, it may be necessary to ligate or retract one or more lumbar arteries for several minutes. Prolonged pressure on or ligation of the lumbar artery giving rise to the large *radicular artery*, which supplies the inferior two-thirds of the spinal cord, leads to circulatory impairment of this part of the cord. This may result in an area of dead nervous tissue, which could result in paralysis of the lower limbs (*paraplegia*) and loss of all sensation inferior to the infarcted area. The clinically important radicular arteries are discussed further in Chap. 4 (p. 364).

Veins of the Posterior Abdominal Wall

The veins of the posterior abdominal wall are tributaries of the inferior vena cava (Figs. 2-52, 2-100, and 2-105), except for the left testicular (or ovarian) vein which enters the renal vein.

The Inferior Vena Cava

The inferior vena cava (IVC) is the *largest vein in the body*; it has no valves except for a variable, nonfunctional one that is located at its orifice (p. 92). The IVC returns blood from the lower limbs, most of the abdominal wall, and the abdominopelvic viscera. Blood from the viscera passes through the *portal system* and liver (Fig. 2-40) before entering the inferior vena cava via the *hepatic veins*. The IVC begins anterior to L5 vertebra by the union of the common iliac veins. This union occurs about 2.5 cm to the right of the median plane, inferior to the bifurcation of the aorta and posterior to the proximal part of the right common iliac artery. The IVC ascends on the right psoas major muscle to the right of the median plane and aorta. It passes through the *vena caval foramen* in the the diaphragm at the level of T8 vertebra (Figs. 2-98 and 2-100). It then pierces the fibrous pericardium and enters the inferior part of the right atrium of the heart (see Figs. 1-43 and 1-51).

Relations of the Inferior Vena Cava (Figs. 2-44, 2-52, 2-67, 2-71, 2-82, and 2-100).   Posteriorly the IVC *lies on the bodies of L3 to L5 vertebrae* to the right of the aorta. It ascends on the right psoas major muscle, right sympathetic trunk, right renal artery, right suprarenal gland, right celiac ganglion, and *the right crus of the diaphragm* as it passes to the vena caval foramen. Anteriorly the relations of the IVC are the *peritoneum, superior mesenteric vessels* in the root of the mesentery, horizontal part of the duodenum, and *head of the pancreas*, with the portal vein and bile duct intervening. Superior to the first part of the duodenum, the IVC is posterior to the omental foramen. It then enters a groove on the inferior surface of the liver, between

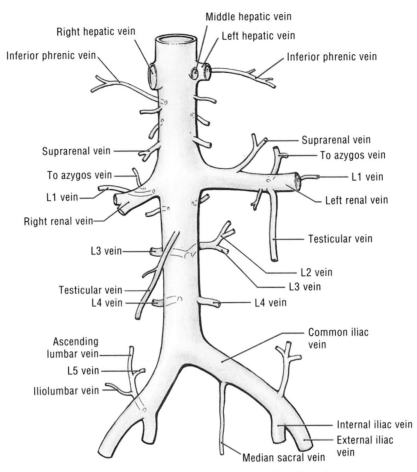

Figure 2-105. The inferior vena cava and its tributaries.

the right and caudate lobes. To the left of the IVC is the aorta. To its right are the right *ureter* and *kidney* and the descending part of the *duodenum*.

Tributaries of the Inferior Vena Cava (Fig. 2-105). These veins correspond to the branches of the aorta. The tributaries are: (1) the *common iliac veins*; (2) the third and fourth *lumbar veins*; (3) the right *testicular* or *ovarian vein*; (4) the *renal veins*; (5) the *azygos vein* (see Fig. 1-76); (6) the right *suprarenal vein*; (7) the inferior *phrenic veins*; and (8) the *hepatic veins*. The right and left *common iliac veins* are formed by the union of the external and internal iliac veins. The right testicular or ovarian vein and the right suprarenal vein usually drain into the IVC, whereas on the left side these veins usually drain into the left renal vein.

The renal veins drain into the IVC at the level of L2 vertebra (Figs. 2-52, 2-74, 2-82, and 2-105). They lie anterior to the corresponding renal artery. The right renal vein receives few if any tributaries other than those from the kidney, whereas the left renal vein also receives blood from the left suprarenal gland and the testis or ovary.

The azygos system of veins (see Figs. 1-39 and 1-76) is described in Chap. 1 (p. 115). The azygos vein connects the superior and inferior venae cavae, either directly or indirectly. It commonly arises from the posterior aspect of the IVC at the level of the renal veins, but it may begin as the continuation of the right subcostal vein or from the junction of that vein and the

right ascending lumbar vein. The azygos vein enters the thorax through the *aortic hiatus* or the right crus of the diaphragm. The inferior *hemiazygos vein* arises from the posterior surface of the left renal vein, but it may also arise from the union of the left subcostal and left ascending lumbar vein.

The right suprarenal vein is short and drains into the posterior aspect of the IVC, whereas the left suprarenal vein is long and usually drains into the left renal vein; however, it may also drain into the IVC.

The inferior phrenic veins drain blood from the abdominal surface of the diaphragm. The right inferior phrenic vein generally empties into the IVC, whereas the left one usually joins the left suprarenal vein.

The hepatic veins (Figs. 2-40 and 2-105) are short ones that open into the IVC, just as it passes through the vena caval foramen in the diaphragm. The right hepatic vein sometimes passes through this foramen before entering the IVC.

The lumbar veins consist of four or five segmental pairs (Fig. 2-105). Their dorsal branches drain the back and communicate with the *vertebral venous plexuses* (see Fig. 4-53). The mode of termination of the lumbar veins varies. They may drain separately into the IVC or the common iliac vein, but they are generally united on each side by a vertical connecting vein, the *ascending lumbar vein*. This vein lies posterior to the psoas major muscle. Each ascending lumbar vein passes posterior to the *medial arcuate ligament* to enter the thorax. The right ascending lumbar vein joins the right subcostal vein to form the azygos vein, whereas the left ascending lumbar vein unites with the left subcostal vein to form the hemiazygos vein (see Fig. 1-76).

Three collateral routes are available for venous blood to pass to the right side of the heart if the IVC is obstructed or when ligation of this vein is necessary. In such cases an extensive collateral venous circulation is soon established owing to enlargement of superficial and/or deep veins. The first route bypassing the IVC is through various anastomoses in the abdomen and pelvis that enable blood to reach the *superficial* and *inferior epigastric veins* (Figs. 2-6 and 2-7). Blood ascends in them to the thoracoepigastric veins, superior epigastric veins, and the superior vena cava. The second route bypassing the IVC is via tributaries of the IVC that anastomose with the *vertebral system of veins* (see Fig. 4-53). These veins, passing within the vertebral canal and vertebral bodies, can also provide a route for metastasis of cancer cells to the vertebral bodies, or to the brain from an abdominal or pelvic tumor. The third route bypassing the IVC is via the *lateral thoracic vein*, which connects the circumflex iliac veins with the axillary vein (Fig. 2-6).

Sometimes the IVC is ligated or plicated (folded to reduce its size) to prevent emboli following thrombosis of the veins in the pelvis or lower limbs from reaching the lungs and producing *pulmonary infarcts* (p. 76). Because of the close relationship of the IVC to the inferior part of the vertebral column (Fig. 2-96), it and the common iliac are vulnerable to injury during repair of a *herniated nucleus pulposus* of an intervertebral disc (see Fig. 4-32). Because these vessels lie anterior to the fifth intervertebral disc, they could be injured by an instrument called a *rongeur* if it is unintentionally pushed through the L4 and L5 intervertebral disc during removal of the herniated portion of its nucleus pulposus. A rongeur is a strong biting forceps used for gouging bone and removing herniated intervertebral discs.

Development of the IVC is complex (Moore, 1988). At one stage during embryonic development, there are *two inferior venae cavae*. Usually the left one degenerates, but it may persist as a small or large vessel (Fig. 2-106). This extra IVC joins the left common iliac vein to the left renal vein.

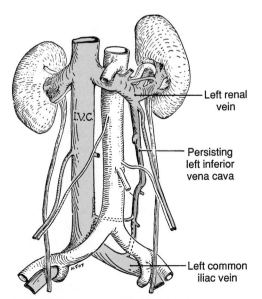

Figure 2-106.   A persistent left inferior vena cava.

Left renal vein

Persisting left inferior vena cava

Left common iliac vein

If this abnormal vein is large, it obscures the left sympathetic trunk.

Lymphatics of the Posterior Abdominal Wall

The lymph nodes on this wall lie along the aorta, inferior vena cava, and iliac vessels (Figs. 2-60, 2-92 and 2-93; see also Fig. 3-73). The *external iliac lymph nodes* are scattered along the external iliac vessels. Inferiorly, the medial lymph nodes receive lymph from the lower limbs and pelvic viscera, whereas the lateral lymph nodes receive lymph from the areas supplied by the inferior epigastric and deep circumflex iliac vessels. The *common iliac lymph nodes* are scattered along the common iliac vessels. They receive lymph from the external and internal iliac lymph nodes. The medial group of common iliac lymph nodes also drains lymph directly from the pelvis. Lymph from the common iliac lymph nodes passes to the lumbar lymph nodes.

The Lumbar (Lateral Aortic) Lymph Nodes (Fig. 2-60).   These nodes lie on both sides of the abdominal aorta and inferior vena cava. They receive lymph directly from the posterior abdominal wall, kidneys and ureters, testes or ovaries, uterus, and uterine tubes. They also receive lymph from the descending colon, pelvis, and lower limbs through the inferior mesenteric and common iliac lymph nodes (Fig. 2-60). Efferent lymph vessels from these large lymph nodes form the right and left *lumbar lymph trunks*. These trunks terminate in the cisterna chyli.

The Cisterna Chyli (see Fig. 24, p. 25, and Fig. 1-70).   This saclike expansion at the inferior end of the *thoracic duct* is about 5 cm long and 6 mm wide. It is frequently absent. When present, it is located between the origin of the abdominal aorta and azygos vein. It lies on the right sides of the bodies of L1 and L2 vertebrae and is usually *located posterior to the right crus of the diaphragm*. The cisterna chyli receives lymph from the right and left lumbar lymph trunks, the intestinal lymph trunks, and a pair of lymph vessels that descend from the inferior *intercostal lymph nodes*. Lymph from the digestive tract first passes to the lymph nodes close to the viscera concerned (*e.g.*, *paracolic lymph nodes*), and then it passes along lymph vessels that follow the major blood vessels to the mesenteric lymph nodes (Fig. 2-60). From them the lymph is carried to the *lumbar lymph nodes*.

The Thoracic Duct (see Figs. 1-29, 1-39, and 1-70).   This is the *main lymphatic duct*. It begins in the cisterna chyli and ascends through the *aortic hiatus* in the diaphragm to the thorax. It opens near or at the angle of union of the internal jugular and subclavian veins. The thoracic duct, which receives all the lymph that forms inferior to the diaphragm, is also considered in Chap. 1 (p. 115).

PATIENT ORIENTED PROBLEMS

Case 2-1

A 32-year-old accountant complained to her physician about a steadily burning pain of about 2 weeks duration in the "pit of her stomach." On careful questioning, it was revealed that the pain usually began about 2 hrs after she had eaten and then disappeared when she ate again or drank a glass of milk. Except for mild tenderness in the right upper quadrant just lateral to her xiphoid process, the physical examination was normal. Suspecting a peptic ulcer, the physician ordered plain radiographs of the patient's abdomen and upper GI studies.

Radiologist's Report. The plain radiographs were normal, but the GI studies showed the presence of a peptic ulcer in a moderately deformed duodenal cap.

Diagnosis. Active duodenal ulcer.

Treatment. Initially the patient responded well to medical treatment (*e.g.*, antacids, frequent bland feedings, and abstinence from smoking and drinking alcohol).

Follow Up. Although the patient followed the physician's instructions for 2 months, she began to work long hours again, smoke heavily, and consume excessive amounts of coffee and alcohol. Her symptoms returned and vomiting sometimes occurred when the pain was severe. One evening she developed acute upper abdominal pain, vomited, and fainted. She was rushed to hospital. Examination revealed extreme pain and rigidity of the abdomen and rebound tenderness. On questioning, the patient revealed that her ulcer had been "acting up" and that she had noticed blood in her vomitus. Emergency surgery was performed and a perforated duodenal ulcer was found. There was a generalized chemical peritonitis resulting from the escape of bile and the contents of her GI tract into the peritoneal cavity.

> **Problems.** What structures closely related to the superior part of the duodenum might be eroded by a perforated duodenal ulcer? Name the congenital malformation of the ileum in which peptic ulcers commonly develop. Explain the anatomical basis for abdominal pain in the right upper and lower quadrants. What nerves might be cut during surgical procedures to reduce the secretion of acid by the parietal cells of the stomach. These problems are discussed on p. 239.

Case 2-2

On the way home from work, a 42-year-old office worker suddenly experienced a sharp, lancinating (tearing) pain in his right side. The pain was so excruciating that he doubled up and moaned in agony. A fellow worker took him to hospital. When the physician asked him to describe the onset of pain, the patient said that he first felt a slight pain between his ribs and hip bone and then it gradually increased until it was so severe that it brought tears to his eyes. He said that this unbearable pain lasted about 30 minutes and then suddenly eased. He explained that the pain comes and goes, but it seemed to be moving toward his groin.

During the physical examination, the physician noted that there was some tenderness and guarding (muscle spasm) in the right lower quadrant, but there was no rigidity. While palpating the tender area as deeply as possible, he suddenly removed his hand. Instead of wincing, the patient seemed relieved that the probing had stopped (absence of rebound tenderness). By this time the patient reported that he felt the pain in his right groin and testis and along the medial side of his thigh. The physician noted that the right testis was unusually tender and was retracted. When asked to produce a urine sample, the patient stated that it was difficult and painful for him to urinate (*dysuria*). The nurse reported that his urine sample contained blood (*hematuria*). Although the physician was quite certain that the man had a kidney or ureteric stone, he ordered radiographic studies of the abdomen.

Radiologist's Report. A small calcified object, compatible with a *uric acid stone*, is visible in the region of the inferior end of the right ureter.

Diagnosis. Ureteric calculus (stone) in the superior end of the right ureter.

> **Problems.** What probably caused the patient's initial attack of excruciating pain? Based on the anatomy of the ureter, at what other sites do you think a ureteric calculus is likely to become lodged? Explain the intermittent exacerbation of pain and the course taken by the pain. Briefly discuss referred pain from the ureter. These problems are discussed on p. 240.

Case 2-3

A 14-year-old boy suffered pain in his right groin while attempting to lift a heavy weight. As soon as he noticed a lump in the region where he felt pain he decided to lie down. The bulge soon disappeared and he went home. On the way, he blew his nose very hard and again experienced pain and the swelling reappeared in his right groin. Fearing that he may have a "rupture," his father called the family physician.

During the physical examination, the physician inserted the tip of his digit into the boy's superficial inguinal ring. Nothing was felt until he asked the patient to cough; he then felt an impulse on the tip of his digit. When the patient was in a prone position the bulge disappeared, but when he was asked to strain (as occurs during defecation), a plum-sized bulge appeared in the right inguinal region.

Diagnosis. Indirect inguinal hernia.

> **Problems.** What is an indirect inguinal hernia? Explain the embryological basis of this kind of hernia. What layers of the spermatic cord cover the hernial sac? What structures are endangered during an operation for repair of an indirect inguinal hernia? These problems are discussed on p. 240.

Case 2-4

A 22-year-old married female medical student woke up one morning not feeling as well as usual; she was anorexic and had crampy abdominal pains. As this coincided with the time of her expected menses, she thought that these cramps were the beginning of her usual painful menstruation (*dysmenorrhea*). Because she had missed her last menstrual period, she also thought that she might be having early symptoms of a ruptured *ectopic pregnancy*. Because she had a slight fever and felt dizzy, she decided to stay in bed. The pain soon localized around her umbilicus. By evening the site of pain shifted to the right lower quadrant of her abdomen and she suspected *acute appendicitis*. As she was in considerable pain, her husband decided to take her to the hospital.

Clinical Findings. There was a slight elevation of her temperature, an increased pulse rate, and an abnormally high white blood cell count (*leukocytosis*). When asked to indicate where the pain began, she circled her umbilical area. When asked where she now felt pain, she put her finger on McBurney's point. During gentle palpation of her abdomen, the physician detected localized rigidity (*muscle spasm*) and tenderness in her right lower quadrant. When the physician suddenly removed her palpating hand from the area of McBurney's point, the patient winced in pain (*rebound tenderness*).

Diagnosis. Acute appendicitis.

> **Problems.** What type of incision would the surgeon most likely make to expose the vermiform appendix? Why is this a good incision anatomically? How would you locate McBurney's point? What part of the appendix is usually deep to this point? Based on your knowledge of dissection, how do you think the patient's appendix would be exposed? Where would her appendix most likely be located? What position of an inflamed appendix might give rise to pelvic or rectal pain? Discuss referred pain from the appendix. These problems are discussed on p. 240.

Case 2-5

A 58-year-old obese man with a history of heartburn, indigestion (*dyspepsia*), and belching after heavy meals complained about recent epigastric and retrosternal pain. He stated that the pain behind his breastbone (retrosternal) developed recently and that it was most severe after dinner, especially when he stooped down. Fearing these might be heart pains (*angina pectoris*), his wife insisted that he consult a physician. When asked if he had noticed any other abnormalities, he stated that he often brought up (regurgitates) small amounts of sour or bitter-tasting substances (*gastric reflux*), particularly when he stooped to tie his shoes. He also reported that he recently had been having bouts of hiccups and that he occasionally had difficulty swallowing (*dysphagia*). The physician ordered an electrocardiogram (ECG) and radiographic studies. The ECG showed no evidence of heart disease.

Radiologist's Report. The radiographs of the abdomen were negative, but a fluoroscopic examination of the thorax showed a round space filled with gas and fluid in the inferior part of the patient's posterior mediastinum. On swallowing a barium sulphate emulsion, the barium was seen to enter this space, which was identified as the gastroesophageal region of the stomach. There was no radiological evidence of gastric or duodenal ulcers.

Diagnosis. Sliding hiatus hernia.

> **Problems.** What is a diaphragmatic hernia? Does hiatus hernia have an embryological basis? Is it usually present at birth? What caused the patient's epigastric and retrosternal pain and hiccups? Based on your anatomical knowledge, what structures do you think would be endangered in the surgical repair of a hiatus hernia? These problems are discussed on p. 241.

Case 2-6

A fair, fat, flatulent, 40-year-old woman with five children was rushed to hospital with severe colicky pain in the right upper quadrant of her abdomen. When asked where she first felt pain, she pointed to her epigastric region. When asked where the pain was now, she ran her fingers under her right ribs (hypochondriac region) and around her right side to her back, stating that the pain was felt near the lower end of her shoulder blade (inferior angle of her scapula). On questioning, she said the sharp midline pain followed a heavy meal containing several fatty foods, after which she felt nauseated and vomited. Gradually there was an increase in pain. During gentle palpation of her abdomen, the

physician noted rigidity and tenderness in the right upper quadrant, especially during inspiration. She ordered radiographic studies.

Radiologist's Report. There is a small calculus (gallstone) in the cystic duct and the gallbladder is enlarged.

Diagnosis. Biliary colic resulting from impaction of a gallstone in the cystic duct.

> **Problems.** What is a gallstone? Explain the anatomical basis of the patient's pain in: (1) the epigastric region, (2) the right hypochondrium, and (3) the infrascapular region. Does peritoneum separate the gallbladder from the liver? What structures are endangered during cholecystectomy? These problems are discussed on p. 241.

Case 2-7

A 54-year-old mechanic was admitted to hospital because of severe epigastric pain and vomiting of blood (*hematemesis*). It was obvious that he had been drinking heavily. On examination, it was noted that the blood in his vomitus was bright red. On questioning, it was learned that the patient had exhibited upper GI bleeding on previous occasions, but never so profusely. His blood pressure was low and his pulse rate was high. The patient's skin and conjunctivae were slightly yellow (jaundiced). His eyes appeared to be slightly sunken. Spider nevi or angiomas (branching arterioles) were present on his cheeks, neck, shoulders, and arms. His abdomen was large and pendulous. Palpation of the patient's abdomen revealed some enlargement of the liver (*hepatomegaly*) and spleen (*splenomegaly*). Several bluish, dilated varicose veins radiated from his umbilicus, forming a *caput medusae*. During a proctoscopic examination, internal hemorrhoids were observed. On questioning, the patient said that he sometimes saw blood in his stools (bowel movements). At other times he said they were black and shiny.

Diagnosis. Alcoholic cirrhosis of the liver.

> **Problems.** Discuss anatomically the basis of the patient's hematemesis, hemorrhoids, bloody stools, and caput medusae. What is the likely cause of the ascites and splenomegaly? Thinking anatomically, how would you suggest that blood pressure in the portal system could be reduced? These problems are discussed on p. 242.

DISCUSSION OF PATIENTS' PROBLEMS

Case 2-1

A peptic ulcer is an ulceration of the mucous membrane of the stomach or duodenum. Ulcers are common in the stomach (gastric ulcers) and duodenum (duodenal ulcers). They are usually found within 3 cm of the pylorus and occur more commonly in males. Both gastric and duodenal ulcers tend to bleed. Sometimes organs and vessels adjacent to the duodenum, usually the pancreas, become adherent to an ulcer and are eroded; *e.g.*, a posterior penetrating ulcer may erode the gastroduodenal artery or one of its branches. This causes sudden massive hemorrhage, which may be fatal. Peptic ulcers may occur in an ileal (Meckel's) diverticulum, a remnant of the yolk stalk attached to the ileum. Gastric tissue may be present in the wall of this diverticulum, which may secrete acid that causes ulcer formation.

The pain resulting from a gastric ulcer is referred to the epigastric region because the stomach is supplied with pain afferents that reach T7 and T8 segments through the greater splanchnic branch of the sympathetic trunk. Pain resulting from a peptic ulcer is referred to the anterior abdominal wall superior to the umbilicus because both the duodenum and this area of skin are supplied by T9 and T10 nerves (Fig. 1-24). When a duodenal ulcer perforates, there may be pain throughout the abdomen. Sometimes the peritoneal gutter associated with the ascending colon may act as a watershed and direct the escaping inflammatory material into the right iliac fossa. This explains why pain from an anterior perforation of a duodenal ulcer may cause right upper and lower quadrant pain.

As the vagus nerves largely control the secretion of acid by the parietal cells of the stomach, section of the vagus nerves (*vagotomy*) as they enter the abdomen is sometimes performed to reduce acid production. Vagotomy may be performed in conjunction with *resection* of the ulcerated area and the acid-producing part of the stomach. Often only the gastric branches of the vagus nerves are cut (*selective vagotomy*), thereby avoiding adverse effects on other organs (*e.g.*, dilation of the gallbladder).

Case 2-2

The patient's initial attack of excruciating pain was almost certainly caused by passage of the kidney stone from his renal pelvis into the superior end of his right ureter. Calculi that are larger than the lumen of the ureter (3 mm) cause severe pain when they attempt to pass through it. The pain moves inferomedially as the calculus passes along the ureter. The patient very likely experienced the severe pain when the calculus was temporarily impeded owing to angulation of the ureter as it crossed the pelvic brim and, later, when it became wedged in the ureter where it passed through the wall of the urinary bladder. At the inferior end of the ureter, there is a definite narrowing of the lumen; this is a common site of obstruction. The pain ceases when it passes into the urinary bladder, although tenderness along the course of the ureter often persists for some time.

Ureteric pain results from passage of the calculus through the ureter. As the ureter is a muscular tube in which peristaltic contractions normally convey urine from the kidney to the urinary bladder, pain results from distention of the ureter by the calculus and the urine that is unable to pass by it. The smooth muscular coat of the ureter normally undergoes peristaltic contractions from its superior to its inferior end. As the peristaltic wave approaches the obstruction, forceful smooth contraction causes excessive dilation of the ureter between the wave and the stone. It is the ureteric distention that produces the lancinating or sharp pain. Exacerbation of pain occurs as distention increases.

The afferent pain fibers supplying the ureter are included in the lesser splanchnic nerve. Impulses also enter L1 and L2 segments of the spinal cord, and the pain is felt in the cutaneous areas innervated by the inferior intercostal nerves (T11 and T12), the iliohypogastric and ilioinguinal nerves (L1), and the genitofemoral nerve (L1 and L2). These are the same regions of the spinal cord that supply the ureter (T11 to L2). Consequently, the pain commences in the lateral region and radiates to the groin and scrotum. The retraction of the testis by the cremaster muscle and the pain along the medial part of the front of the thigh indicate

that the genital and femoral branches of the genitofemoral nerve (L1 and L2) were involved.

Ureteral colic is caused by distention of the ureter, which stimulates pain afferents in its wall. As there was no peritonitis, there was no rigidity and no rebound tenderness. When peritonitis is present, pressing the hand into the abdominal wall and rapidly releasing it causes pain when the abdominal musculature springs back into place, carrying the inflamed peritoneum with it. Hence, the abdominal rebound test is useful in the differentiation of ureteric colic from appendicitis and intestinal colic.

Case 2-3

A complete indirect inguinal hernia is an outpouching of the peritoneal sac that enters the deep inguinal ring, traverses the inguinal canal, and exits through the superficial inguinal ring. The embryological basis of an indirect inguinal hernia is persistence of all or part of the *processus vaginalis*, an embryonic diverticulum of the peritoneum that pushes through the abdominal wall and forms the inguinal canal. The processus vaginalis evaginates all layers of the abdominal wall before it and in males, they become the coverings of the spermatic cord.

The hernial sac (former processus vaginalis) may vary from a short one not extending beyond the superficial ring to one that extends into the scrotum (or labium majus), where it is continuous with the tunica vaginalis. A persistent processus vaginalis predisposes a person to indirect inguinal hernia by creating a weakness in the anterior abdominal wall. It also forms a hernial sac into which abdominal contents may herniate if the intra-abdominal pressure becomes very high, as occurs during straining while lifting a heavy object. Once the deep inguinal ring has been enlarged by a herniation of the intestine, coughing may cause herniation to occur again. This is the basis of the test done during physical diagnosis, where the examiner's digit is inserted through the superficial ring into the inguinal canal, and the patient is asked to cough (Fig. 2-17C).

During the surgical repair of an indirect inguinal hernia, the genital branch of the *genitofemoral nerve* is endangered because it traverses the inguinal canal in both sexes, and exits through the superficial inguinal ring. The ilioinguinal nerve may also be injured. It supplies the skin of the superomedial area of the thigh, the skin over the root of the penis, and the part of the scrotum or labia majora with sensory fibers. If this nerve is injured, anesthesia of these areas of skin will likely result. If the nerve is constricted by a suture, postoperative neuritic pain may also occur in these areas.

As the ductus deferens lies immediately posterior to the hernial sac, it may be damaged when the sac is freed, ligated, and excised. Because the hernial sac is within the spermatic cord, the pampiniform plexus of veins and the testicular artery may also be injured, resulting in impairment of circulation to the testis. Injury to the vessels of the spermatic cord may result in atrophy of the testis on that side.

Case 2-4

The type of skin incision used for an appendectomy depends on the type of patient and the certainty of the diagnosis. Usually a muscle-splitting (gridiron) incision is made, which is an oblique or almost transverse one (Fig. 2-14B and D). The center of the incision is at McBurney's point, which is at the junction of the lateral and middle thirds of the line joining the anterior superior iliac spine and the umbilicus. In most cases, this point overlies the base of the appendix. Following incision of the skin and the superficial fascia, the aponeurosis of the external oblique muscle is incised in the direction of the fibers of this muscle. The other two muscles of the anterior abdominal wall (internal oblique and transversus abdominis) are then split (not cut) in the direction of their fibers. This lessens the chances of injuring the nerves supplying them. Next the transversalis fascia and the parietal peritoneum are incised to expose the cecum. The base of the appendix is indicated by the point of convergence of the three teniae coli.

Variations in length and position of the appendix may give rise to varying signs and symptoms in appendicitis. For example, the site of maximum tenderness in cases of retrocecal appendix may be just superomedial to the anterior superior iliac spine, even as far superior as the transumbilical plane. If the appendix is long (10 to 15 cm) and extends into the pelvis minor, the site of pain in a female might suggest peritoneal irritation resulting from a ruptured ectopic pregnancy. As the appendix crosses the psoas major muscle, the patient often flexes the right thigh to relieve the pain. Thus hyperextension of the thigh (*psoas test*) causes pain because it stretches the muscle and its inflamed fascia. Tenderness on the right side during a rectal examination may indicate an inflamed pelvic appendix.

Initially, the pain of typical acute appendicitis is referred to the periumbilical region of the abdomen; later the site of pain usually shifts to the right lower quadrant. Afferent nerve fibers from the appendix are carried in the lesser splanchnic nerve and impulses enter T10 segment of the spinal cord. As impulses from the skin in the periumbilical region are also sent to this region of the spinal cord (see Fig. 1-24), the pain is interpreted as somatic rather than visceral, apparently because impulses of cutaneous origin are received more often by the brain.

The shift of pain to the right lower quadrant is caused by irritation of the parietal peritoneum, usually on the posterior abdominal wall. Afferent fibers from this region of peritoneum and skin are carried in the inferior intercostal and subcostal nerves. The pain during palpation results from stimulation of pain receptors in the skin and the peritoneum, whereas the increased tenderness detected in the right side of the *rectouterine pouch* (rectovesical pouch in a male) is caused by irritation of the parietal peritoneum in this pouch. When the abdominal wall is depressed and allowed to rebound, the patient usually winces because, as the abdominal muscles spring back, the inflamed peritoneum is carried with it.

If the patient had previously had her appendix removed, an inflamed ileal diverticulum could give rise to signs and symptoms similar to appendicitis. An ileal (*Meckel's*) diverticulum represents the remnant of the proximal portion of the yolk stalk and appears as a fingerlike projection from the antimesenteric border of the ileum.

Case 2-5

A diaphragmatic hernia is a herniation of abdominal viscera into the thoracic cavity through an opening in the diaphragm. *Hiatus hernia* is common, particularly in older people. Usually the gastroesophageal region of the stomach herniates through the *esophageal hiatus* into the inferior part of the thorax. Hiatus hernia is usually acquired, but a congenitally enlarged esophageal hiatus may be a predisposing factor. Understand that the right crus passes to the left of the midline; hence, the esophageal hiatus and hernia are to the left of the midline even though they are within the right crus.

There are two main types of hiatus hernia (1) *sliding hiatus hernia* and (2) *paraesophageal hiatus hernia*, but some hernias present features of both types and are referred to as mixed hiatus hernias. The thoracic region of the vertebral column becomes shorter with age owing to dessication of the intervertebral discs, and the abdominal fat generally increases during middle age. Both of these occurrences favor development of hiatus hernias. Most of the present patient's complaints (heartburn, belching, regurgitation, and epigastric pain) resulted from irritation of the esophageal mucosa by the reflux of gastric juice. The irritant effect of the gastric juice produces esophageal spasm, resulting in dysphagia and retrosternal pain. Pain endings in the esophagus are stimulated by the forcible contractions of the smooth muscle in esophageal wall.

Pain of gastroesophageal origin is referred to the epigastric and retrosternal regions, the cutaneous areas of reference for these regions of the viscera (see Fig. 1-24). The patient's hiccups are caused by spasmodic contractions of the diaphragm, which result from pressure created by the hernia. Probably enlargement of the esophageal hiatus stimulates fibers of the phrenic nerves supplying the diaphragm. As the esophageal hiatus also transmits the vagus nerves and esophageal branches of the left gastric vessels, these structures, as well as the esophagus, must be protected from injury during surgical repair of hiatus hernias.

Case 2-6

Obese middle-aged women who have had several children are most prone to gallbladder disease. Thus the common aphorism "forty, flatulent, and fat" describes most patients, but it does not characterize all patients with gallstones. *Gallstones are more common in women* over 20, but this is not necessarily so after 50 years of age. In about 50% of persons, gallstones are "silent" (asymptomatic). A gallstone is a concretion in the gallbladder, cystic duct, or bile duct, composed chiefly of cholesterol crystals. The pain is severe when a biliary calculus is lodged in the cystic or bile duct. The patient's sudden severe pain in the epigastric region (biliary colic) was caused by a gallstone wedged in the cystic duct.

The pain referred to the right upper quadrant and scapular region results from inflammation of the gallbladder and distention of the cystic duct. The nerve impulses pass centrally in the greater splanchnic nerve on the right side and enter the spinal cord through the dorsal roots of T7 and T8 nerves. This visceral referred pain is felt in the right upper quadrant of the abdomen and in the right infrascapular region because the source of the stimuli entering this region of the cord is wrongly interpreted as cutaneous (see Fig. 1-24). Often the inflamed gallbladder irritates the peritoneum covering the peripheral part of the diaphragm, resulting in a parietal referred pain in the inferior part of the thoracic wall. This part of the peritoneum is supplied by the inferior intercostal nerves. In other cases the peritoneum covering the diaphragm is irritated and the pain is referred to the shoulder region because this area of peritoneum is supplied by the phrenic nerve. The skin of the shoulder region is supplied by the supraclavicular nerves (C3 and C4), the same segments of the cord that receive pain afferents from the central portion of the diaphragm.

When fat enters the duodenum, *cholecystokinin* causes contraction of the gallbladder. In the present case it is very likely that the patient's gallbladder contracted vigorously after her fatty meal, squeezing a stone into her cystic duct. *Acute cholecystitis* is associated with a gallstone impacted in the cystic duct in a high percentage of cases. The impacted calculus causes sudden distention of the gallbladder, which compromises its arterial supply and its venous and lymphatic drainage.

Usually peritoneum does not separate the gallbladder from the liver. The gallbladder lies in a fossa on the inferior surface of its right lobe. The peritoneum on this surface of the liver passes over the inferior surface of the gallbladder. The abdominal rigidity detected in the present case resulted from involuntary contractions of the muscles of the anterior abdominal wall, particularly the rectus abdominis. This muscle spasm was a reflex response to stimulation of nerve endings in the peritoneum associated with the dilated gallbladder.

Anatomical variations in the gallbladder and cystic duct and in the arteries supplying them are common. Because of this, surgeons must determine the existing anatomical pattern and identify the cystic, bile, and hepatic ducts and the cystic and hepatic arteries before dividing the cystic and duct and its artery. As there may be accessory cystic branches from the hepatic arteries, unexpected hemorrhage may occur during cholecystectomy.

Case 2-7

Hepatic cirrhosis is a disease characterized by progressive destruction of hepatic parenchymal cells. These hepatic cells are replaced by fibrous tissue, which contracts and hardens. The fibrous tissue surrounds the intrahepatic blood vessels and biliary radicles (roots). As this process advances, circulation of blood through the branches of the portal vein and of bile through the biliary radicles in the liver is impeded. As pressure in the portal vein rises (*portal hypertension*), the liver becomes more dependent on the hepatic artery for its blood supply and blood pressure in the portal vein rises, reversing blood flow in the normal portacaval anastomoses. This results in portal blood entering the systemic circulation. As these anastomotic veins seldom possess valves, they can conduct blood in either direction. This causes enlargement of the veins (*varicose veins*) forming these anastomoses at the inferior end of the esophagus (*esophageal varices*), the inferior end of the rectum and anal canal (*hemorrhoids*), and around the umbilicus (*caput medusae*).

Because of pressure during swallowing and defecation, the esophageal varices and hemorrhoids, respectively, may rupture, resulting in bloody vomitus and/or bleeding from the anus and bloody stools (feces). Internal hemorrhoids are varicosities of the tributaries of the superior rectal vein. Blood may also pass in a retrograde direction in the paraumbilical veins (small tributaries of the portal vein) via the vein in the ligamentum teres. In portal hypertension the paraumbilical veins may become varicose, forming a radiating venous pattern at the umbilicus, called a *caput medusae*, owing to its resemblance to the snakes adorning the head of Medusa, a mythological character.

In cirrhosis of the liver, the ramifications of the portal vein are compressed by the contraction of the fibrous tissue in the portal canals. As a result, there is increased pressure in the splenic and superior and inferior mesenteric veins. Fluid is forced out of the capillary beds drained by these veins into the peritoneal cavity. Accumulation of fluid in the peritoneal cavity is called *ascites*. The spleen usually enlarges (*splenomegaly*) when there is hepatic cirrhosis because of increased pressure in the splenic vein. As there are no valves in the portal system, pressure in the splenic vein is equal to that in the portal vein. A common method of reducing portal pressure is by diverting blood from the portal vein to the inferior vena cava through a surgically-created anastomosis (*portacaval anastomosis*). Similarly, the splenic vein may be anastomosed to the left renal vein (*splenorenal anastomosis*).

SUGGESTED READINGS

Ameli FM: Vascular Surgery. In Gross A, Gross P, Langer B (Eds): *Surgery; A Complete Guide for Patients and Their Families*, Toronto, Harper & Collins Publishers, 1989.

Behrman RE: *Nelson Textbook of Pediatrics*, ed 14. Philadelphia, WB Saunders, 1992.

Bergman RA, Thompson SA, Afifi AK, Saadeh FA: *Compendium of Human Anatomic Variation. Text, Atlas and World Literature*. Baltimore, Urban & Schwartzenberg, 1988.

Cormack DH: *Ham's Histology*, ed 9. Philadelphia, JB Lippincott, 1987.

Ellis H: *Clinical Anatomy: A Revision and Applied Anatomy for Clinical Students*, ed 7. Oxford, Blackwell Scientific Publications, 1983.

Farrow GA: Urology. In Gross A, Gross P, Langer B (Eds): *Surgery: A Complete Guide for Patients and Their Families*, Toronto, Harper & Collins, 1989.

Filler RM: General Surgery Pediatric. In Gross A, Gross P, Langer B (Eds): *Surgery: A Complete Guide for Patients and Their Families*, Toronto, Harper & Collins, 1989.

Ger R: Surgical anatomy of hepatic venous system. Clin Anat 1:15–22, 1988.

Gorgollon P: The normal human appendix: A light and electron microscopic study. J. Anat. 126:87-101, 1978.

Gross A, Gross P, Langer B (Eds): *Surgery: A Complete Guide for Patients and Their Families*, Toronto, Harper & Collins, 1989.

Harrison AW: General Complications of Surgery. In Gross A, Gross P, Langer B (Eds): *Surgery; A Complete Guide for Patients and Their Families*, Toronto, Harper & Collins, 1989.

Healey JE, Jr, Hodge J: *Surgical Anatomy*, ed 2. Toronto, B.C. Decker, 1990.

Moore KL: *The Developing Human: Clinically Oriented Embryology*, ed 4. Philadelphia, WB Saunders, 1988.

Robinette MA: Transplant Surgery. In Gross A, Gross P, Langer B (Eds): *Surgery: A Complete Guide for Patients and Their Families*. Toronto, Harper & Collins, 1989.

Scott GBD: The primate cecum and appendix vermiformis: a comparative study, J. Anat. 131:549–563, 1980.

Skandalakis JE, Gray SW, Rowe JS: *Anatomical Complications in General Surgery*, New York, McGraw-Hill, 1983.

Stone R: General surgery. Adult. In Gross A, Gross P, Langer B (Eds): *Surgery: A Complete Guide for Patients and Their Families*, Toronto, Harper & Collins, 1989.

Tobias PV, Arnold M, Allan JC: *Man's Anatomy: A Study In Dissection*, ed 4. Johannesburg, Witwatersrand University Press, 1988, vol 1.

Warwick R (Ed): *Nomina Anatomica*, ed 6. Edinburgh, Churchill Livingstone, 1989.

Williams PL, Warwick R, Dyson M, Bannister LH: *Gray's Anatomy*, ed 37. New York, Churchill Livingstone, 1989.

Woodburne RT, Burkel WE: *Essentials of Human Anatomy*, ed 8. New York, Oxford University Press, 1988.

The pelvis (L. basin) is the region where the trunk and lower limbs meet (Fig. 3-1). *The perineum*, the area between the thighs and buttocks (Fig. 3-54), is the diamond-shaped region of the trunk inferior to the *pelvic diaphragm* (p. 295), which separates the pelvic cavity from the perineum (see Fig. 2-29).

The Pelvis

The pelvis is the inferior part of the trunk and the *pelvic cavity* is the basin-shaped inferior part of the *abdominopelvic cavity*, which is located inferior to the plane of the **pelvic brim**[1] (Figs. 3-1 and 3-7; see also Fig. 2-29). However, the abdominal and pelvic cavities are continuous across the pelvic brim. The pelvic cavity is bounded inferiorly by the *pelvic diaphragm*, which forms the *pelvic floor* and separates the pelvis from the perineum. The circumference of the **superior pelvic aperture** (pelvic inlet) is outlined on each side by the pelvic brim and extends from the superior border of the *pubic symphysis* anteriorly to the *promontory of the sacrum* posteriorly (Fig. 3-1A).

The Bony Pelvis

The bony pelvis is the skeleton of the pelvis (Figs. 3-1 to 3-4). It surrounds the pelvic cavity and forms the *pelvic girdle*[2] for attachment of the lower limbs. The bony pelvis is formed anteriorly and laterally by the two *hip bones* (ossa coxae); posteriorly by the *sacrum* and *coccyx*, and anteriorly by the meeting of the two pubic bones at the *pubic symphysis* (Figs. 3-1 to 3-4). The sacrum and coccyx, the inferior parts of the vertebral column, are interposed dorsally between the hip bones. The **hip bones** are large and irregularly-shaped; they consist of three parts: *ilium, ischium*, and *pubis*. These bones meet at the *acetabulum*, the cup-shaped cavity in the lateral surface of the hip bone into which the head of the femur fits (Figs. 3-3B and 3-4). The hip bones are described with the lower limb (Chap. 5, p. 373). The four parts of the bony pelvis are bound together by dense ligaments (Figs. 3-8 and 3-9), and are joined at four articulations: two synovial joints, the *sacroiliac joints*, and two secondary cartilaginous joints, the *pubic symphysis*, and *sacrococcygeal joint* (Figs. 3-1 and 3-4). The bones involved in the last two joints are connected by fibrocartilaginous discs, as well as by ligaments.

[1]The *pelvic brim* is formed in continuity by the pubic crest, pecten pubis (pectineal line of pubis), arcuate line of the ilium, and the alae and promontory of the sacrum (Fig. 3-1).

[2]The pelvic girdle is formed by the two hip bones. The cavity between the pelvic girdle and the sacrum is the pelvis or pelvic cavity.

The Pelvic Brim

The pelvic cavity is located inferior to the plane of the pelvic brim (Fig. 3-7; see also Fig. 2-29). This is an oblique plane that forms an angle of about 55 degrees to the horizontal. It coincides with a line joining the *sacral promontory* to the superior border of the *pubic symphysis*. This line also gives the anteroposterior (AP) diameter of the superior pelvic aperture or pelvic inlet (Fig. 3-6A). The pelvic brim is formed by the following structures (Figs. 3-1 to 3-7 and 3-15A): superior margin of *pubic symphysis, pubic crest, pecten pubis* (pectineal line of the pubis), *arcuate line* of the ilium, anterior border of the *ala* (L. wing) of the sacrum, and the *promontory* of the sacrum.

The pelvis is divided into a **pelvis major** (false pelvis), which is part of the *abdominal cavity*, and a **pelvis minor** (true pelvis, "obstetric pelvis"), which contains the pelvic cavity (Figs. 3-4 and 3-7; see also Fig. 2-29). The *pelvis major* lies between the iliac fossae, superior to the pelvic brim. The pelvis minor is located inferior to the oblique plane of the pelvic brim; it is particularly important in obstetrics and gynecology because it is an integral part of the *birth canal*.

Sex Differences Between Male and Female Pelves

(Fig. 3-1; Table 3-1). The pelves of males and females differ in several respects that are striking and easily recognized. These differences are related mainly to the heavier build and stronger muscles of men and to the adaptation of the pelvis in women for childbearing. The general structure of the male pelvis is heavier and thicker, and it usually has more prominent bone markings. *The female pelvis is wider, shallower, and has larger superior and inferior pelvic apertures.* In typical male and female pelves, observe that: (1) the hip bones are usually farther apart in females owing to the broader sacrum; this explains the relatively wider hips of women; (2) the ischial tuberosities in females are farther apart because of the wider pubic arch; (3) the sacrum in females is less curved, which increases the size of the inferior pelvic aperture and the diameter of the birth canal; and (4) the obturator foramina are round in the male and oval in the female. For a detailed account of sex differences in the pelvis, see Ellis (1983).

Although there are usually clear-cut anatomical differences between male and female pelves, the pelvis of any person may have anatomical features typical of the opposite sex. The pelvic type is based on overall architecture and not on the appearance of the superior pelvic aperture alone (Fig. 3-5 and Table 3-1). The presence of certain male characteristics in a female pelvis may present hazards to successful vaginal delivery of a fetus (p. 283).

The Pelvis Major

The *false pelvis* lies superior to the superior pelvic aperture (pelvic inlet) or pelvic brim (Figs. 3-1, 3-4, and 3-7). Its cavity is part of the abdominal cavity (see Fig. 2-29).

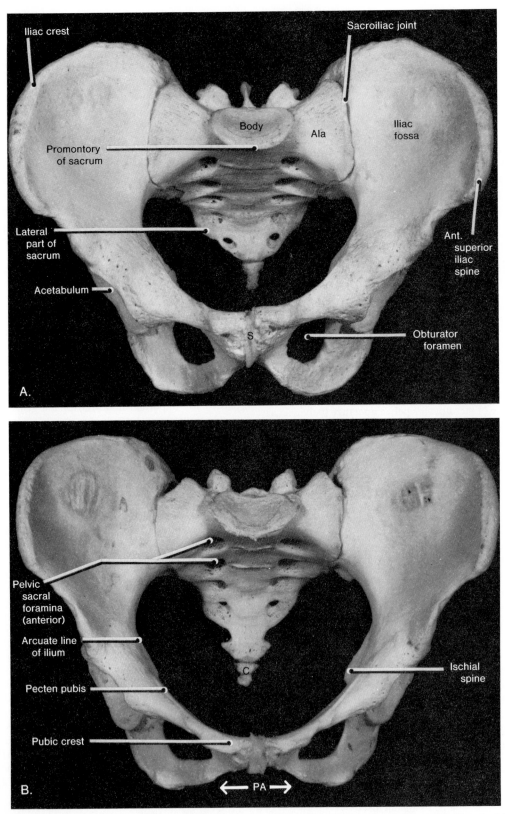

Figure 3-1.   Anterior views of male (A) and female (B) bony pelves. Note that the superior pelvic aperture (pelvic inlet) is larger in the female pelvis and that the pubic arch (PA) is much wider than in the male pelvis. The subpubic angle, located at the apex of this arch, is at the inferior border of the pubic symphysis (S). Note the thinness of the bone in the iliac fossae of the female pelvis. C, Anterior view showing the pelvis in situ.

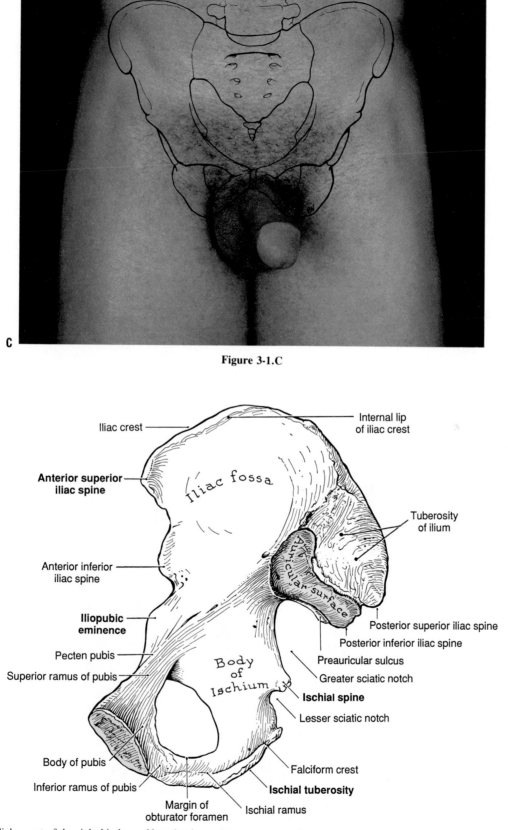

Figure 3-1.C

Figure 3-2. Medial aspect of the right hip bone. Note that it consists of three parts: ilium, ischium, and pubis and is fan-shaped. The spread of the "fan" is the ala (L. wing) of the ilium, and its broad "handle" is formed by the body of the pubis.

Labels in Figure 3-2:
Iliac crest — Internal lip of iliac crest — Iliac fossa — **Anterior superior iliac spine** — Tuberosity of ilium — Auricular surface — Anterior inferior iliac spine — Posterior superior iliac spine — Posterior inferior iliac spine — **Iliopubic eminence** — Preauricular sulcus — Pecten pubis — Body of Ischium — Greater sciatic notch — Superior ramus of pubis — **Ischial spine** — Lesser sciatic notch — Body of pubis — Falciform crest — Inferior ramus of pubis — **Ischial tuberosity** — Margin of obturator foramen — Ischial ramus

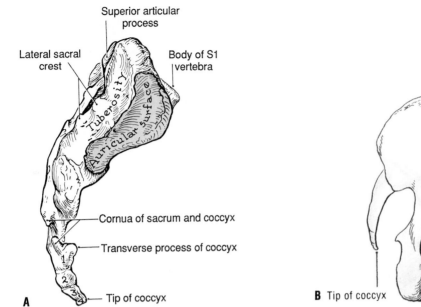

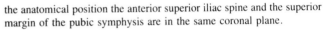

Figure 3-3. *A*, Right lateral aspect of the sacrum and coccyx. *B*, Lateral aspect of the right hip bone and coccyx, demonstrating that in

the anatomical position the anterior superior iliac spine and the superior margin of the pubic symphysis are in the same coronal plane.

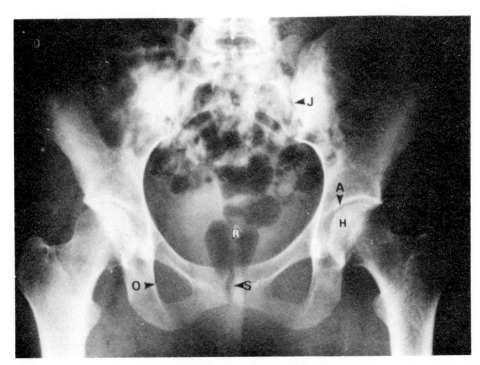

Figure 3-4. Anteroposterior (AP) radiograph of a female pelvis. Note the wide pubic arch and subpubic angle and that there is a considerable amount of air in the rectum (*R*) and the sigmoid colon superior to it. The arrows point to the pubic symphysis (*S*), the sacroiliac joint (*J*),

the superior surface of the acetabulum (*A*), and the margin of the obturator foramen (*O*). The head of the femur (*H*) is articulating with the cup-shaped acetabulum.

ANTHROPOID

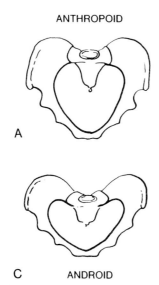

A

PLATYPELLOID

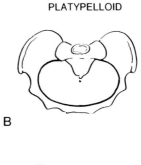

B

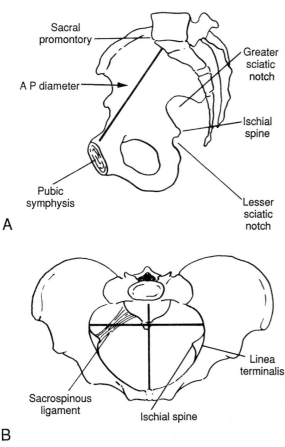

C   ANDROID

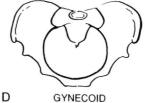

D   GYNECOID

Figure 3-5. The four types of bony pelvis. *A* and *B* are common in males, whereas those shown in *C* and *D* are common in females. *A*, Anthropoid pelvis (present in some males and about 23% of females). Note that the AP diameter of the superior pelvic aperture is greater than the transverse diameter. *B*, Platypelloid pelvis (uncommon in both sexes). *C*, Android pelvis (present in most males and in about 32% of females).

Note that the superior pelvic aperture has a wide transverse diameter, but the posterior part of the aperture is narrow. *D*, Gynecoid pelvis (present in about 43% of females). This type of pelvis is the most spacious obstetrically; hence, a woman with this type usually has an uneventful delivery of her fetus.

Figure 3-6. The typical female pelvis illustrating the anteroposterior (*AP*) and transverse diameters of the superior pelvic aperture or pelvic inlet. In *B*, observe that the transverse diameter of the superior pelvic aperture is greater than the AP diameter.

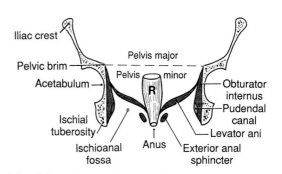

Figure 3-7. Schematic coronal section of a pelvis showing the funnel-shaped pelvic diaphragm, formed by the levator ani and coccygeus muscles. Note that the pelvis minor (true pelvis) lies inferior to the plane of the pelvic brim. *R* indicates the ampulla of the rectum which rests on and is anchored to the pelvic diaphragm.

Table 3-1.
The Main Differences Between Male and Female Pelves

	Male	Female
General structure	Thick and heavy	Thin and light
Muscle attachments	Well marked	Poorly marked
Pelvis major	Deep	Shallow
Pelvis minor	Narrow and deep	Wide and shallow
Superior pelvic aperture	Heart shaped	Oval or rounded
Inferior pelvic aperture	Comparatively small	Comparatively large
Subpubic angle	Narrow	Wide
Obturator foramen	Round	Oval
Acetabulum	Large	Small

The pelvis major **contains abdominal viscera** (*e.g.*, the sigmoid colon, Fig. 3-4; see also Fig. 2-26*A*). The pelvis major is bounded anteriorly by the abdominal wall, laterally by the iliac fossae, and posteriorly by L5 and S1 vertebrae.

The Pelvis Minor

The **true pelvis** lies inferior to the superior pelvic aperture (pelvic inlet) and pelvic brim (Figs. 3-1, 3-4, and 3-7). It is limited inferiorly by the inferior pelvic aperture (pelvic outlet), which is closed by the *pelvic diaphragm* that is composed mainly of the levator ani muscles (Fig. 3-7; see also Fig. 2-29). Its inferior boundary corresponds roughly to a line joining the tip of the coccyx to the inferior border of the pubic symphysis. *The cavity of the pelvis minor is the pelvic cavity*; it **contains pelvic viscera** (*e.g.*, the urinary bladder). The pelvic cavity, short and curved, forms the basinlike part of the abdominopelvic cavity. It is tilted anteroinferiorly when it is in the anatomical position (Figs. 3-1 and 3-4).

The Walls of the Pelvis Minor (Figs. 3-1 to 3-4). The *posterior wall*, which is notably longer than its anterior wall, is formed by the concave pelvic surface of the sacrum and coccyx.

The *anterior wall* is formed by the pubic symphysis, the body of the pubis, and the pubic rami. The *lateral walls* are formed by the pelvic aspects of the ilium and ischium.

The Superior Pelvic Aperture (Figs. 3-1 and 3-4 to 3-7). *The pelvic inlet* is variable in contour; sexual, racial, and nutritional differences influence its shape. It is heart-shaped in males and some females. *In most females this opening is larger than in males* and is rounded or oval in contour. The pelvic inlet is encroached upon by the **sacral promontory**. The periphery of the superior pelvic aperture, formed by the pelvic brim, is indicated by the *linea terminalis* (iliopectineal line). This terminal line is an oblique ridge on the internal surface of the ilium, which is continued on the pubis. It forms the inferior boundary of the iliac fossae and separates the true and false pelves.

The Inferior Pelvic Aperture (Figs. 3-1 to 3-3, 3-8, and 3-53). *The pelvic outlet* does not have a smooth contour because it is bounded posteriorly by the *sacrum* and *coccyx*, anteriorly by the *pubic symphysis*, and laterally by the *ischial tuberosities*. The plane of this aperture makes an angle of 10 to 15 degrees with the horizontal when the pelvis is in the anatomical position. The *greater and lesser sciatic notches* between the sacrum and coccyx and the ischial tuberosities are divided into *greater and*

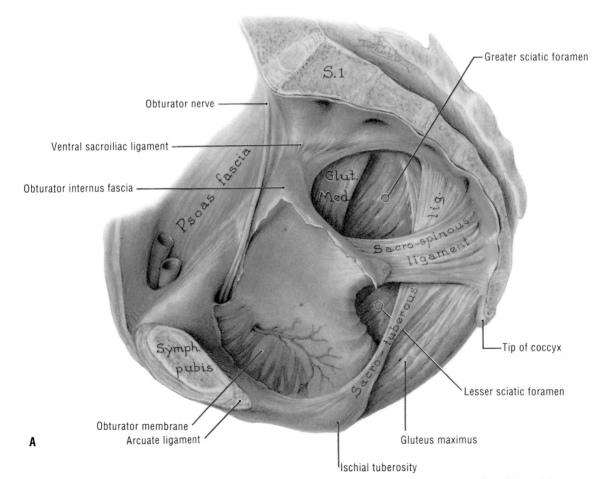

Figure 3-8. Dissections of the bony and ligamentous walls of the female pelvis. *A*, Right view of the pelvis minor. *B* and *C*, Anterior and posterior views, respectively, of the ligaments of the pelvis.

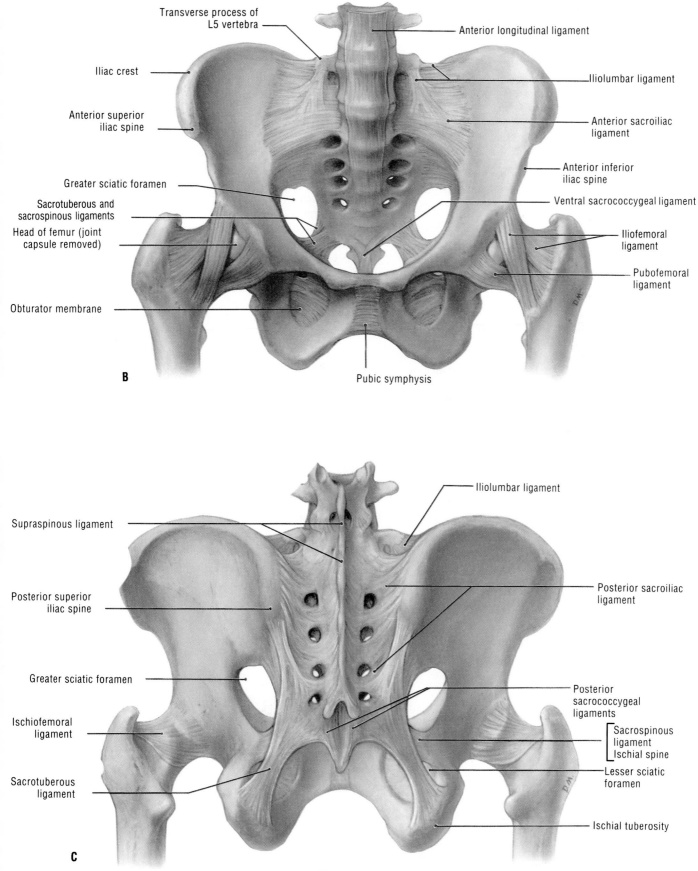

Transverse process of
L5 vertebra

Iliac crest

Anterior superior
iliac spine

Greater sciatic foramen

Sacrotuberous and
sacrospinous ligaments

Head of femur (joint
capsule removed)

Obturator membrane

Anterior longitudinal ligament

Iliolumbar ligament

Anterior sacroiliac
ligament

Anterior inferior
iliac spine

Ventral sacrococcygeal ligament

Iliofemoral
ligament

Pubofemoral
ligament

Pubic symphysis

B

Iliolumbar ligament

Supraspinous ligament

Posterior superior
iliac spine

Greater sciatic foramen

Ischiofemoral
ligament

Sacrotuberous
ligament

Posterior sacroiliac
ligament

Posterior
sacrococcygeal
ligaments

Sacrospinous
ligament
Ischial spine

Lesser sciatic
foramen

Ischial tuberosity

C

Figure 3-8*B–C*

lesser sciatic foramina by the sacrotuberous and sacrospinous ligaments (Fig. 3-8). These ligaments give the inferior pelvic aperture a diamond shape. Examine the *pubic arch* in Figs. 3-1, 3-4, and 3-8, noting that the *subpubic angle* is narrow in the male and wide in the female.

The **superior pelvic aperture** or pelvic inlet is routinely measured during a female pelvic examination for obstetrical reasons (Fig. 3-6). The *anteroposterior (AP) diameter* of the aperture is the measurement from the midpoint of the superior border of the pubic symphysis to the midpoint of the sacral promontory. The *transverse diameter* of the aperture is its greatest width. It is measured from the linea terminalis on one side to this line on the opposite side. The *oblique diameter* of the aperture is measured from one iliopubic eminence to the opposite sacroiliac joint (Figs. 3-1*A* and 3-2). The *midplane diameter* (interspinous diameter or distance between the ischial spines) of the aperture cannot be measured, but it may be estimated by palpating the sacrospinous ligament through the vagina (Fig. 3-6*B*). The length of this ligament is equal to about half the midplane diameter.

The **ischial spines** may be a barrier to the passage of the fetus during childbirth if they are closer than 9.5 cm. Pelvic measurements can also be made using radiographs such as that in Fig. 3-4. A good idea of the shape and position of the sacrum can be developed by palpating the sacral concavity during a pelvic examination. The ischial spines can also be palpated, and a general idea of their prominence can be obtained per vaginum. To pass through the **birth canal** (pelvic inlet, pelvis minor, cervix, vagina, and pelvic outlet, Fig. 3-34), the fetal head must make an almost 90 degree turn. The *gynecoid female pelvis* (Fig. 3-5*D*) has a wide, circular superior pelvic aperture with a *wide subpubic arch* and widely-spaced ischial spines. This female-type of pelvis is the most common and spacious obstetrically.

The bony pelvis is able to resist considerable trauma. Only violent injuries fracture it (*e.g.*, as occur in motor vehicle accidents). The lateral part of the hip bone is the strongest. One hears about elderly people falling and ''breaking their hips.'' Usually they do not fracture their hip bones; it is the necks of their femora that break (p. 379). *Weak areas of the pelvis* are: (1) the sacroiliac region, (2) the ala (L. wing) of the ilium, and (3) the pubic ramus. Fractures of the pelvis in the pubo-obturator area are relatively common and are often complicated owing to their relationship to the urinary bladder. Anteroposterior compression of the pelvis occurs during ''squeezing accidents'' (*e.g.*, when a heavy object falls on the pelvis). This type of trauma commonly causes *fractures of the pubic rami*. When the pelvis is compressed laterally, the acetabula and ilia are squeezed toward each other and may be broken. Some pelvic fractures result from the tearing away of bone by the strong posterior vertebropelvic ligaments associated with the sacroiliac joints (Figs. 3-8 and 3-9).

Falls on the feet or buttocks may produce the following injuries: (1) the pubic rami may be fractured, (2) the acetabula may be injured, and (3) the heads of the femora may be driven through the acetabula into the pelvis, injuring the pelvic organs. In persons under 17 years of age, the acetabula may fracture into their three developmental parts (see Fig. 5-3), or the acetabular margins may be torn away. Pelvic fractures are often complicated by injuries to the pelvic viscera (*e.g.*, rupture of the urinary bladder).

The Vertebropelvic Ligaments

The parts of the bony pelvis are bound together by dense ligaments (Figs. 3-8, 3-9, and 3-19). The ilium is united to L5 vertebra by the iliolumbar ligament and the sacrum is joined to the ischium by the sacrotuberous and sacrospinous ligaments. The sacrotuberous and sacrospinous ligaments bind the sacrum to the ischium and resist posterior rotation of the inferior end of the sacrum. They also hold the posterior part of the sacrum inferiorly, thereby preventing the body weight from depressing its anterior part at the sacroiliac joints. The sacrotuberous and sacrospinous ligaments permit some movement of the sacrum, giving resilience to this region when sudden weight increases are applied to the vertebral column (*e.g.*, when landing on the feet during a fall).

The Iliolumbar Ligament (Fig. 3-8*B* and *C*). This strong triangular ligament connects the tip of the transverse process of L5 vertebra (occasionally L4) to the iliac crest posteriorly. The inferior fibers of this ligament are attached to the lateral part of the sacrum; this band is called the *lateral lumbosacral ligament*. The iliolumbar ligaments are important because they limit rotation of L5 vertebra on the sacrum and assist the vertebral articular processes in preventing the anterior gliding of L5 on the sacrum.

The Sacrotuberous Ligament (Figs. 3-8 and 3-19). This ligament passes from the sacrum to the ischial tuberosity. It has a wide attachment to the dorsal surfaces of the sacrum and coccyx and the posterior superior iliac spine. Its fibers run inferolaterally to the superior medial impression on the ischial tuberosity and then extend along its medial margin.

The Sacrospinous Ligament (Figs. 3-6 and 3-8). This thin triangular ligament extends from the lateral margin of the sacrum and coccyx to the ischial spine. It is related anteriorly to the coccygeus muscle.

The Sacroiliac Ligaments (Figs. 3-1, 3-4, 3-8, and 3-9). The sacrum is wedged between the iliac bones and is held there by powerful interosseous and sacroiliac ligaments. These ligaments are discussed with the sacroiliac joints (p. 251).

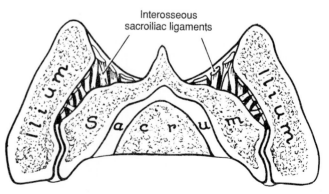

Figure 3-9. Transverse section of the sacroiliac joints illustrating the powerful interosseous sacroiliac ligaments. The sacrum is suspended between the iliac bones by these very strong interosseous ligaments and by the sacroiliac ligaments (see Fig. 3-8).

The vertebropelvic ligaments relax progressively during pregnancy, making movements freer between the inferior part of the vertebral column and the pelvis. Furthermore, the pubic symphysis relaxes owing to a hormone called *relaxin*. As a result, the distance between the pubic bones increases considerably. These changes facilitate passage of the fetus through the birth canal.

Joints of the Pelvis

The articulations or joints of the pelvis include the lumbosacral, sacrococcygeal, and sacroiliac joints and the pubic symphysis.

The Lumbosacral Joints

L5 and S1 vertebrae articulate with one another at an *anterior intervertebral joint* formed by the intervertebral disc between their bodies (Fig. 3-8) and at two *posterior synovial joints* between their articular processes. The L5/S1 intervertebral disc is wedge-shaped because it is thicker anteriorly. The *zygapophyseal (facet) joints* (see Fig. 4-27) are synovial joints between the inferior articular processes of L5 vertebra and the superior articular processes of S1 vertebra. The S1 facets face posteriorly and medially, thereby preventing L5 vertebra from sliding anteriorly. L5 vertebra is attached to the ilium and sacrum by the strong *iliolumbar ligaments* (Fig. 3-8*B* and *C*).

Large transverse processes on L5 vertebra are common (see Fig. 4-20), and they are more likely to strengthen the lumbosacral joint than to weaken it. A condition called **spondylolysis** is found in the inferior lumbar region in about 5% of white North American adults. It occurs more frequently in certain races (*e.g.*, Eskimos). In spondylolysis there is a *defect in the vertebral arch* between the superior and inferior facets in an area called the *pars interarticularis*. When bilateral, the defects result in the lumbar vertebra being divided into two pieces (see Fig. 4-38). If the two parts separate, the abnormality is called **spondylolisthesis**. The displacement of the anterior piece of L5 reduces the anteroposterior diameter of the superior pelvic aperture and may interfere with parturition (childbirth). Obstetricians routinely *test for spondylolisthesis* by running their digits down the lumbar spinous processes. If the process of L5 is prominent, it indicates that the anterior part of L5 vertebra and the vertebral column superior to it may have moved anteriorly (see Fig. 4-39). Radiographs are then taken to confirm the diagnosis and to measure the anteroposterior diameter of the superior pelvic aperture. Spondylolysis and spondylolisthesis are also discussed in the section on the back (see Chap. 4, p. 351).

The Sacrococcygeal Joint

The sacrococcygeal joint is a secondary cartilaginous joint in which fibrocartilage and ligaments join the articulating bones, the apex of the sacrum and the base of the coccyx. They are united by a thin, fibrocartilaginous intervertebral disc (Fig. 3-8*A*). The *sacrococcygeal ligaments* correspond to the anterior and posterior longitudinal ligaments of the other intervertebral joints (Fig. 3-8*B*; see also Fig. 4-30). The *sacral and coccygeal cornua* (L. horns; see Fig. 4-21*B*) are also united by *intercornual ligaments*. Until middle age, there is slight posterior movement of the coccyx at the sacrococcygeal joint during defecation, and there is considerable movement of the coccyx during childbirth. In some people the coccyx is freely movable and articulates with the sacrum by a synovial joint (Williams et al., 1989). In elderly persons the sacrococcygeal joint usually becomes ossified and is obliterated (see Fig. 4-1*A*).

The Sacroiliac Joints

The sacroiliac articulations are *strong synovial joints* between the articular surfaces of the sacrum and ilium (Figs. 3-1 to 3-4 and 3-8). These surfaces have irregular elevations and depressions, which result in a partial interlocking of the bones. The *strong articular capsule* is attached close to the articulating surfaces of the sacrum and ilium. The sacrum is suspended between the iliac bones, and the bones are firmly held together by the interosseous and posterior *sacroiliac ligaments* (Figs. 3-8 and 3-9). These are the strongest ligaments in the body.

The Interosseous Sacroiliac Ligaments (Figs. 3-8 and 3-9). These massive, *very strong ligaments* unite the iliac and sacral tuberosities. They consist of short, strong bundles of fibers that blend with and are supported by the thick firm *posterior sacroiliac ligaments*.

The Posterior (Dorsal) Sacroiliac Ligaments (Fig. 3-8; see also Fig. 4-21). These ligaments are composed of: (1) strong, short transverse fibers joining the ilium and the first and second tubercles of the lateral crest of the sacrum and (2) long vertical fibers uniting the third and fourth transverse tubercles of the sacrum to the posterior iliac spines. These ligaments blend with the *sacrotuberous ligaments*.

The Anterior (Ventral) Sacroiliac Ligaments (Fig. 3-8). This thin wide sheet of transverse fibers is located on the anterior and inferior aspects of the sacroiliac joint. It covers the abdominopelvic surface of this articulation. Replacement of most or all of these ligaments by bone often begins after 50 years of age.

The iliolumbar, sacrotuberous, and sacrospinous ligaments (Fig. 3-8), discussed on p. 250, are accessory ligaments of the sacroiliac joints. These articulations are covered posteriorly by the massive erector spine and gluteus maximus muscles (see Fig. 4-42). The skin dimples indicating the posterior superior iliac spines are located at the middle of the sacroiliac joints (see Fig. 2-86).

Functions and Movements of the Sacroiliac Joints. These joints are *strong weightbearing synovial joints* of an irregular plane type. They differ from most synovial joints in that they possess very little mobility. This provides for stability and is related to their responsibility for transmitting the weight of most of the body to the hip bones (see Fig. 4-3). Because the articular surfaces of the sacrum and ilium are irregular, they fit together securely and are not easily dislocated. This arrangement lessens the strain on the supporting ligaments of the joints. Movement of the sacroiliac joints is limited to a slight gliding and rotary movement, except when a considerable force is applied as occurs during a jump from a height. In this case the force is transmitted via the vertebral column to the superior end of the sacrum, which tends to rotate anteriorly. This rotation is counterbalanced by the interlocking articular surfaces and the strong supporting ligaments, especially the strong sacrotuberous and sacrospinous ligaments (Fig. 3-8). This allows the force to be transmitted to each ilium and lower limb.

Arterial Supply of the Sacroiliac Joints (Fig. 3-24). The articular branches to these joints are derived from the superior gluteal, iliolumbar, and lateral sacral arteries.

Innervation of the Sacroiliac Joints (Fig. 3-16). The articular branches to these joints are derived from the superior gluteal nerves, the sacral plexus, and the dorsal rami of S1 and S2 nerves.

When one falls from a height and lands on one's feet (*e.g.*, during parachuting), the very strong sacroiliac joints transfer most of the body weight to the hipbones. Flexion of the lower limbs at the hip and knee joints also helps to prevent injury to these joints and the vertebral column. The resilience of the sacrotuberous and sacrospinous ligaments also cushions the shock to the vertebral column.

The sacroiliac ligaments become softer and more yielding during the late stages of pregnancy, thereby increasing the range of movement of the sacroiliac joints. Combined with similar changes in the pubic symphysis and associated vertebropelvic ligaments (Fig. 3-8), passage of the fetus through the birth canal is facilitated. The sacroiliac joints often become partially ossified during old age, especially in men. Calcification in the anterior sacroiliac ligaments makes the joint cavities less visible on radiographs, even though they are still present.

The Pubic Symphysis

The pubic symphysis (symphysis pubis) is a median, secondary cartilaginous joint between the bodies of the two pubic bones (Figs. 3-1, 3-2, 3-4, and 3-8*B*). Each articular surface is covered by a thin layer of hyaline cartilage, which is connected to the cartilage of the other side by a thick fibrocartilaginous *interpubic disc* (Fig. 3-14). This disc is generally thicker in women than in men; it also contains a small cavity that is also larger in women and increases in size during pregnancy. These *sex differences in the interpubic disc* permit more mobility of the bones and increase the diameter of the pelvic cavity for passage of the fetus during childbirth (Fig. 3-6). The articular surface of each pubic bone is irregularly ridged and grooved (Fig. 3-2), but the irregularities of the two sides fit tightly together to form a strong joint.

The ligaments joining the pubic bones are thickened superiorly and inferiorly to form the *superior pubic ligament* and the arcuate pubic ligament, respectively (Figs. 3-8, 3-34, and 3-55). The superior pubic ligament connects the superior pubic rami along their superior surfaces. It extends as far laterally as the pubic tubercles. The *arcuate pubic ligament* is a thick arch of fibers that (1) connects the inferior borders of the joint, (2) rounds off the subpubic angle, and (3) forms the superior border of the pubic arch. The decussating tendinous fibers of the rectus ab-

dominis and external oblique muscles also strengthen the joint anteriorly (see Fig. 2-6).

Walls of the Pelvis

The walls of the pelvic cavity are composed of: (1) superficial muscles; (2) the hip bones, sacrum, and coccyx and their associated ligaments; and (3) deep muscles, blood vessels, nerves, lymphatics, and peritoneum. For descriptive purposes, the pelvis is subdivided into anterior, lateral, and posterior walls and a floor.

The Anterior Pelvic Wall

The anterior wall of the pelvis is formed on each side by (1) the bodies of the *pubic bones*, and its superior and inferior rami, which terminate laterally in the body and ramus of the *ischium*, and (2) the *obturator internus* muscle and its fascia (Figs. 3-1, 3-2, 3-8, and 3-10). The *pubic symphysis* also forms an important part of the anterior pelvic wall.

The Lateral Pelvic Walls

The **obturator internus muscles** cover most of these walls (Figs. 3-7 and 3-10 to 3-12). Their superior parts, near the pelvic brim, are covered by *pelvic fascia*. Medial to the obturator internus on each side are the *obturator nerve* and vessels and other branches of the internal iliac artery. The obturator internus passes from the pelvis through the *lesser sciatic foramen*, and its fibers converge to form a tendon that is attached to the *greater trochanter of the femur* (Figs. 3-8 and 3-12). The attachments, nerve supply, and main actions of the obturator internus are given in Table 3-2.

The Posterior Pelvic Wall

This wall is formed by the sacrum, adjacent parts of the ilium, and the sacroiliac joints and their ligaments. The **piriformis muscles** line the posterior wall laterally (Figs. 3-10 and 3-13 to 3-15). These pear-shaped muscles (L. *pirum*, pear + *forma*, form) occupy a key position in the gluteal region (buttock). Each piriformis muscle leaves the pelvis minor through the *greater sciatic foramen* (Fig. 3-8). Its attachments, nerve supply, and main actions are given in Table 3-2. Medial to the piriformis muscles are the *sacral plexus* and the *internal iliac vessels* and their subsidiaries. The piriformis muscles form a "muscular bed" for the sacral plexus (Figs. 3-14 and 3-16).

The Floor of the Pelvis

The pelvic floor is formed mainly by the funnel-shaped **pelvic diaphragm**, which is composed of the two *levator ani* and the two *coccygeus muscles* (Figs. 3-7, 3-11, and 3-14 to 3-18). This diaphragm closes the inferior pelvic aperture (pelvic outlet), except for a gap in the pelvic floor between the anterior edges of the levator ani muscles. This gap is filled with loose fascia around the prostate (or vagina) and is closed by the *urogenital diaphragm* and its superior fascia (Figs. 3-17, 3-18, and 3-58*B* and *C*). The

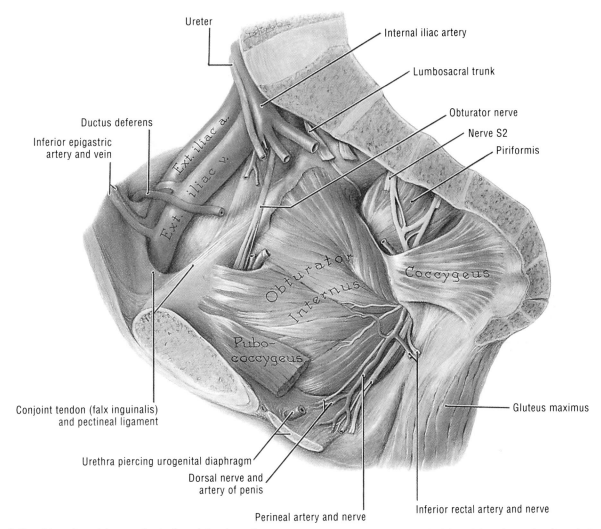

Figure 3-10. Dissection of the muscles in the pelvis minor of a man. Observe the obturator internus padding the lateral wall and escaping through the lesser sciatic foramen (see also Fig. 3-8). Note the piriformis padding the posterior wall of the pelvis and escaping through the greater sciatic foramen. Observe the pubococcygeus, the chief and strongest part of the levator ani muscle, springing from the body of the pubis.

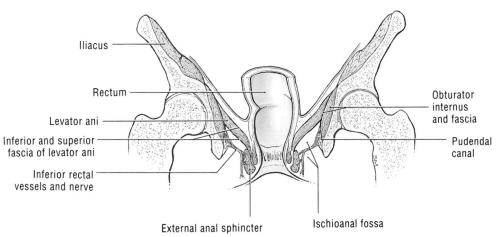

Figure 3-11. Coronal section primarily to show the lateral walls of the male pelvis. Observe that the intrapelvic surfaces of the muscles are covered with parietal pelvic fascia and that it is firmly attached to bone at the pelvic brim.

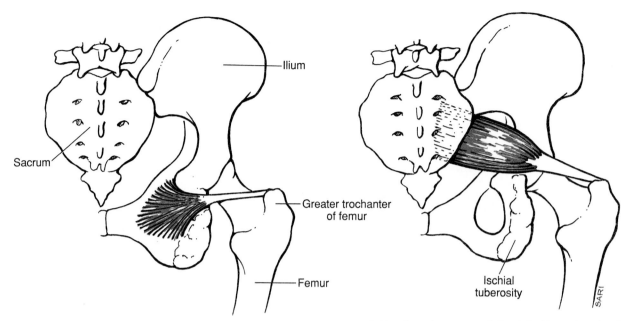

Figure 3-12. Posterior aspect of the right hip region showing the location and attachments of the obturator internus muscle.

Figure 3-13. Posterior aspect of the right hip region showing the location and attachments of the piriformis muscle.

pelvic diaphragm also separates the pelvic cavity from the perineum (see Fig. 2-29).

The musculofascial pelvic diaphragm supports the abdominopelvic viscera (*e,g*, sigmoid colon and rectum). It is slung somewhat like a funnel-shaped hammock between the pubis anteriorly and coccyx posteriorly (Figs. 3-11 and 3-18). Laterally it is attached to a thickening of the obturator fascia known as the *tendinous arch* (Fig. 3-14). The rectum and urethra (also the vagina in women) penetrate the pelvic diaphragm to reach the exterior (Fig. 3-56).

The Levator Ani Muscles (Figs. 3-7, 3-11, and 3-14 to 3-18). These are the largest and *most important muscles in the pelvic floor*. Posterior to them are the coccygeus muscles, which form the smaller part of the floor. The thin, broad levator ani muscle unites with its partner to form a hammocklike sheet of muscle between the pubis anteriorly and the coccyx posteriorly and from one lateral pelvic wall to the other. The levator ani muscles form most of the floor of the pelvic cavity which separates it from the wedge-shaped spaces known as the *ischioanal (ischiorectal) fossae*. For descriptive purposes *the levator ani muscle is divided into three principal parts*: puborectalis, pubococcygeus, and iliococcygeus (Figs. 3-14 and 3-35). These muscles form most of the pelvic floor.

The puborectalis muscle (Figs. 3-19 to 3-23) arises from the pubis and passes posteriorly, where it unites with its partner to form a *U-shaped muscular sling* around the anorectal junction. This sling maintains the *anorectal flexure*. Some fibers of the puborectalis sweep around the prostate in the male and the middle of the vagina in the female and are inserted into the *central perineal ligament* or tendon or **perineal body**, a fibromuscular mass anterior to the anus (Figs. 3-35, 3-55, and 3-60). These muscle fibers constitute the *levator prostatae* and *pubovaginalis*, respectively.

The pubococcygeus muscle (Figs. 3-14, 3-15*A*, 3-20*B*, 3-22, 3-34, 3-35, and 3-60) is *the main part of the levator ani*. It

arises from the pubis and runs posteromedially to insert into the coccyx and anococcygeal ligament. The *anococcygeal ligament* is the median fibrous intersection of the pubococcygeus muscles; it is located between the anal canal and the tip of the coccyx. As it courses inferiorly and medially in the female, the pubococcygeus muscle encircles the urethra, vagina, and anus and merges into the perineal body.

The iliococcygeus muscle (Figs. 3-14, 3-19, and 3-34) is the thin part of the levator ani muscle that arises on each side from the *tendinous arch of the obturator fascia* and the ischial spine (Fig. 3-14). Each muscle passes medially and posteriorly and attaches to the coccyx and anococcygeal ligament.

Innervation of the Levator Ani Muscles (Figs. 3-14, 3-15*A*, and 3-61). These important muscles are innervated by the perineal branches of the third and fourth sacral nerves, which enter its pelvic surface.

Actions of the Levator Ani Muscles. With the two coccygeus muscles, the levator ani muscles *form the pelvic diaphragm*, which constitutes the pelvic floor. This fibromuscular diaphragm supports the pelvic viscera and resists the inferior thrust that accompanies increases in intra-abdominal pressure (*e.g.*, that occur during forced expiration and coughing). *Acting together*, the levator ani muscles raise the pelvic floor, thereby assisting the anterior abdominal muscles in compressing the abdominal and pelvic contents. This action is an important part of forced expiration, coughing, vomiting, urinating, and fixation of the trunk during strong movements of the upper limbs (*e.g.*, when lifting a heavy object). The parts of the levator ani muscles that insert into the **perineal body** (tendinous center of the perineum) support the prostate (*levator prostatae*; Fig. 3-19). In the female these parts support the posterior wall of the vagina (*pubovaginalis*; Fig. 3-23). When the parts of the levator ani that insert into the wall of the anal canal and the perineal body contract (*puborectalis*), they raise the canal over a descending mass of feces, thereby aiding defecation. This part of the levator ani

Table 3-2.
The Muscles of the Pelvic Walls

Muscle	Proximal Attachment	Distal Attachment	Innervation	Main Actions
Obturator internus (covers most of lateral wall)	Pelvic surfaces of ilium and ischium; obturator membrane		Nerve to obturator internus (L5, S1, and S2)	Rotates thigh laterally; assists in holding the head of the femur in the acetabulum
		Greater trochanter of the femur		
Piriformis (posterior wall)	Pelvic surface of 2nd to 4th sacral segments; superior margin of greater sciatic notch, and sacrotuberous ligament		Ventral rami of S1 and S2	Rotates thigh laterally; abducts thigh; assists in holding the head of the femur in the acetabulum

holds the anorectal junction anteriorly (Fig. 3-20), thereby increasing the angle between the rectum and anal canal. This prevents passage of feces from the rectum into the anal canal when defecation is not desired or is inconvenient. The *anorectal angle* supports most of the weight of the fecal mass, thereby relieving much pressure from the *external anal sphincter* (Fig. 3-21). During parturition the levator ani muscles support the fetal head while the cervix of the uterus is dilating to permit delivery of the baby.

The Coccygeus Muscles (Figs. 3-14 and 3-15). This *triangular sheet of muscle* lies against the posterior part of the iliococcygeus muscle with which it is continuous. The coccygei, along with the levator ani muscles, form the pelvic diaphragm. The coccygei form the posterior and smaller part of the pelvic floor which is formed by the pelvic diaphragm.

Lateral Attachment (Figs. 3-14 and 3-15). Pelvic surface of the ischial spine and sacrospinous ligament.

Medial Attachment (Figs. 3-10 and 3-19). Lateral margin of the coccyx and S5 vertebra.

Innervation (Figs. 3-14 to 3-16). Branches of S4 and S5 nerves.

Actions. The coccygeus muscles assist the levator ani muscles in supporting the pelvic viscera. They also support the coccyx and pull it anteriorly, elevating the pelvic floor.

During childbirth the structures supporting the bladder, urethra, vagina, and rectum may be injured. The *pubococcygeus*, the main part of the levator ani, is the most likely muscle to be damaged. It is obstetrically important because it encircles and supports the urethra, vagina, and anus. To ease delivery and to avoid damaging the puborectalis, an *episiotomy* may be performed (Fig. 3-62 and p. 298). Injuries to the pelvic fascia and pubococcygeus muscle may result in *cystocele*, a herniation of the urinary bladder into the vagina (Fig. 3-15C). When the urethra is also involved, the condition is called *cystourethrocele* or urethrocystocele. Herniation of the rectum (*rectocele*) results from damage to the wall of the vagina and the pelvic diaphragm. *Urinary stress incontinence* may accompany weakening of the pelvic diaphragm. It is characterized by dribbling of urine whenever the intra-abdominal pressure is raised (*e.g.*, during coughing, sneezing, and lifting). This troublesome condition often results from *weakening of the pubovaginalis part of the levator ani* (Fig.

3-23), which may have stretched and been lacerated during parturition.

Nerves of the Pelvis

The pelvis is innervated mainly by the *sacral and coccygeal nerves* and the pelvic part of the autonomic nervous system (Fig. 3-16). Either the sympathetic trunk or its ganglia send gray rami communicantes to each sacral nerve and the coccygeal nerve. The piriformis muscle pads the posterior wall of the pelvis and forms a bed for the sacral and coccygeal nerve plexuses. The ventral rami of S2 and S3 nerves emerge between the digitations of the piriformis muscle.

The Lumbosacral Trunk

The descending part of L4 nerve unites with the ventral ramus of L5 nerve to form the thick, cordlike lumbosacral trunk (Fig. 3-16). It passes inferiorly, anterior to the ala of the sacrum, where it *joins the sacral plexus*. It descends obliquely over the sacroiliac joint and passes into the pelvis posterior to the pelvic fascia (see Fig. 2-99). It then *crosses the superior gluteal vessels* to join S1 nerve on the muscular bed formed by the piriformis.

The Sacral Plexus

This large plexus of nerves is located in the pelvis minor, where it is closely related to the anterior surface of the piriformis muscle (Fig. 3-16). *The sacral plexus is formed by the lumbosacral trunk and the ventral rami of S1 to S4 nerves.* The main nerves of this plexus lie external to the parietal pelvic fascia, and all its branches leave the pelvis through the *greater sciatic foramen*, except for the nerve to the piriformis muscle (S2), the perforating cutaneous nerves (S2 and S3), and those supplying the pelvic diaphragm. The two main nerves of the sacral plexus are the sciatic and pudendal nerves.

The Sciatic Nerve (Figs. 3-16 and 3-18; see also Fig. 5-73). This nerve to the lower limb is formed by the ventral rami of L4 to S3, which converge on the anterior surface of the piriformis muscle. The *largest nerve in the body* (about 2.0 cm in width), the sciatic passes through the *greater sciatic foramen* inferior to the piriformis muscle to enter the gluteal (buttock)

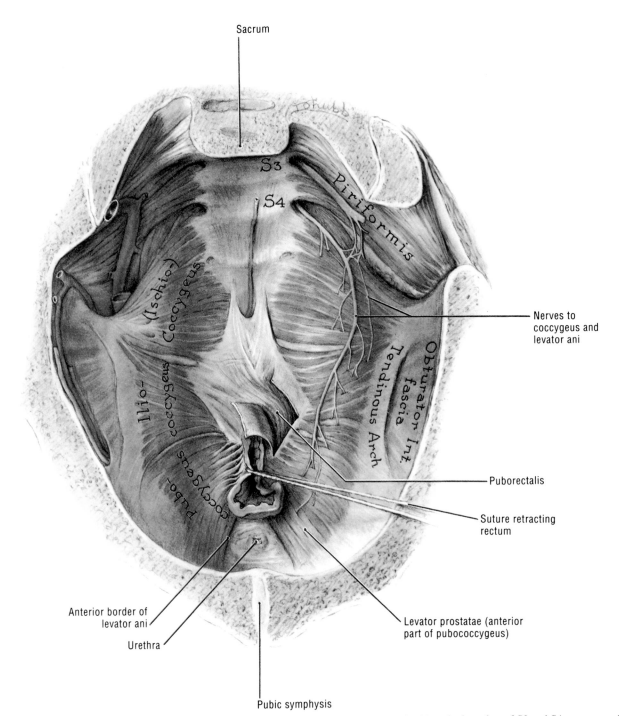

Sacrum

S3

S4

Piriformis

(Ischio-)Coccygeus

Ilio-coccygeus

Pubo-coccygeus

Obturator Int. fascia Tendinous Arch

Nerves to coccygeus and levator ani

Puborectalis

Suture retracting rectum

Levator prostatae (anterior part of pubococcygeus)

Anterior border of levator ani

Urethra

Pubic symphysis

Figure 3-14.   Superior view of a dissection of the floor of a man's pelvis. The pelvic viscera are removed and the bony pelvis is sawn through transversely. Note the branches of S3 and S4 nerves supplying the levator ani and coccygeus muscles, which form the pelvic diaphragm.

region. This very important nerve is discussed in the section on the lower limb (see Chap. 5, p. 413).

The Pudendal Nerve (Figs. 3-8, 3-16, and 3-18).   This nerve arises from the sacral plexus by separate branches of the ventral rami of S2, S3, and S4. It accompanies the internal pudendal artery and leaves the pelvis between the piriformis and coccygeus muscles. The pudendal nerve hooks around the sacrospinous ligament to enter the perineum through the *lesser sciatic foramen*. Here, it supplies the muscles of the perineum, including the external anal sphincter, and ends as the dorsal nerve of the penis or clitoris. The pudendal nerve is also sensory to the external genitalia.

The Superior Gluteal Nerve (Fig. 3-16).   This nerve arises from the dorsal divisions of L4, L5, and S1. It leaves the pelvis via the greater sciatic foramen, superior to the piriformis muscle. It supplies two of the gluteal muscles, medius and minimus, and

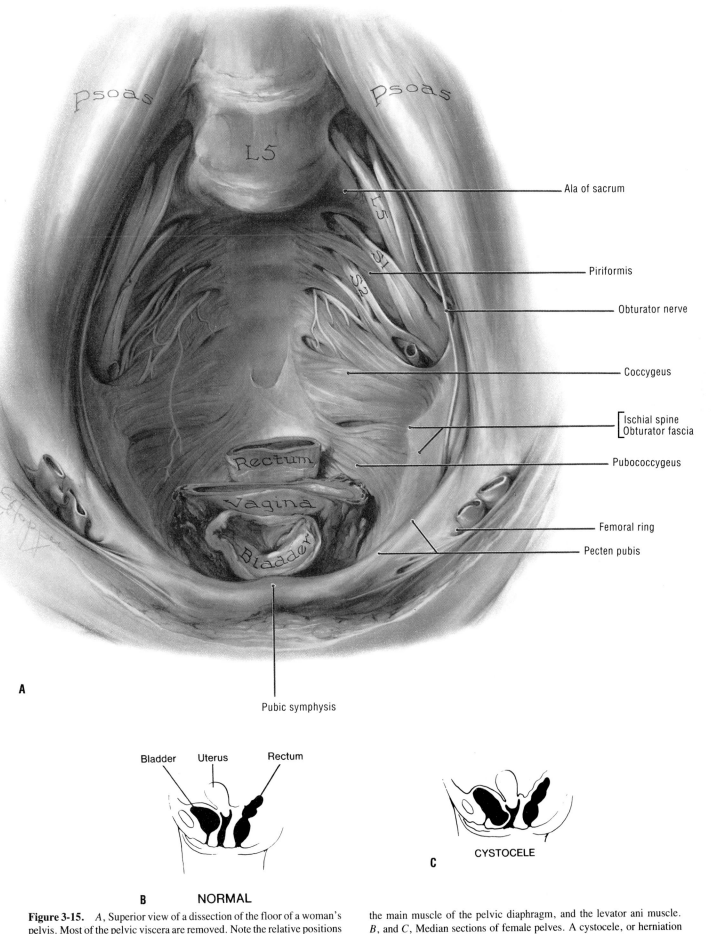

Psoas

Psoas

L5

—— Ala of sacrum

L5

S1

S2

—— Piriformis

—— Obturator nerve

—— Coccygeus

[Ischial spine
Obturator fascia

—— Pubococcygeus

Rectum

Vagina

Bladder

—— Femoral ring

—— Pecten pubis

A

Pubic symphysis

Bladder   Uterus   Rectum

B   **NORMAL**

C

CYSTOCELE

Figure 3-15. *A,* Superior view of a dissection of the floor of a woman's pelvis. Most of the pelvic viscera are removed. Note the relative positions of the bladder, vagina, and rectum. Observe the pubococcygeus muscle, the main muscle of the pelvic diaphragm, and the levator ani muscle. *B,* and *C,* Median sections of female pelves. A cystocele, or herniation of the bladder, results from weakening of the pelvic floor.

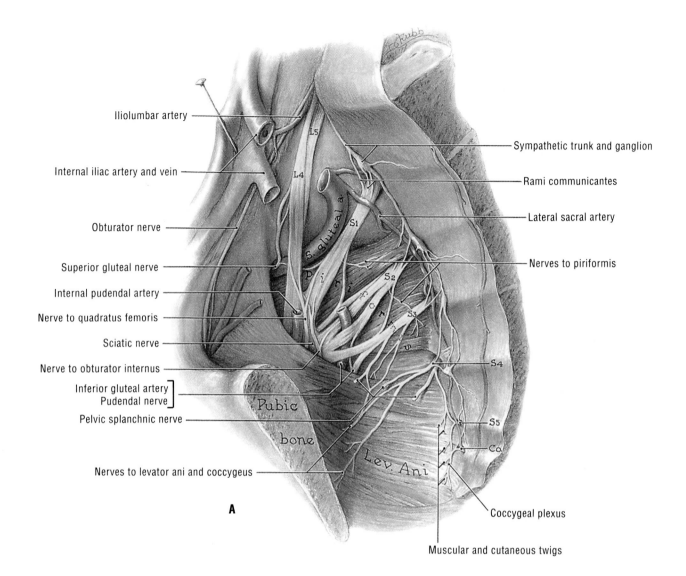

Iliolumbar artery

Internal iliac artery and vein

Obturator nerve

Superior gluteal nerve

Internal pudendal artery

Nerve to quadratus femoris

Sciatic nerve

Nerve to obturator internus

Inferior gluteal artery
Pudendal nerve

Pelvic splanchnic nerve

Nerves to levator ani and coccygeus

Sympathetic trunk and ganglion

Rami communicantes

Lateral sacral artery

Nerves to piriformis

Coccygeal plexus

Muscular and cutaneous twigs

A

Figure 3-16. *A*, The sacral and coccygeal nerve plexuses. Note the branch from L4 joining L5 to form the lumbosacral trunk. Note that the roots of S1 and S2 supply the piriformis muscle and that S3 and S4 supply the coccygeus and levator ani muscles. Note that S2, S3, and S4 each contribute a branch to the formation of the pelvic splanchnic nerve. *B*, The sacral plexus. Note that this network of nerves is pierced by the superior and inferior gluteal arteries.

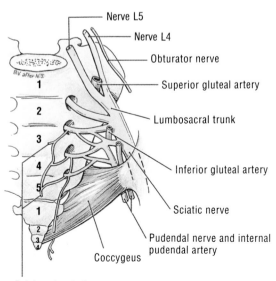

Nerve L5

Nerve L4

Obturator nerve

Superior gluteal artery

Lumbosacral trunk

Inferior gluteal artery

Sciatic nerve

Pudendal nerve and internal pudendal artery

Coccygeus

B   Pelvic splanchnic nerve

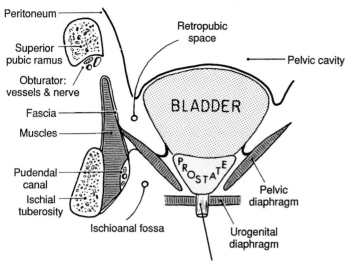

Peritoneum

Superior pubic ramus

Obturator: vessels & nerve

Fascia

Muscles

Pudendal canal

Ischial tuberosity

Ischioanal fossa

Retropubic space

Pelvic cavity

BLADDER

PROSTATE

Pelvic diaphragm

Urogenital diaphragm

Membranous urethra

Figure 3-17. Coronal section of the pelvis illustrating the funnel-shaped pelvic diaphragm, which forms the floor of the pelvic cavity. Observe that only the pelvic diaphragm (levator ani portion) intervenes between the ischioanal fossa and the retropubic space.

the tensor fasciae latae muscle. The superior gluteal nerve and these muscles are described in the section on the gluteal region (see Chap. 5, p. 407 and Tables 5-1 and 5-3).

Other components of the sacral plexus include (Fig. 3-16): (1) twigs to the piriformis muscle (S1 and S2); (2) twigs to the pelvic diaphragm (S3 and S4); (3) the nerve to the quadratus femoris muscle (L4, L5, and S1); and (4) the nerve to the obturator internus muscle (L5, S1, and S2).

The Obturator Nerve

This nerve is not derived from the sacral plexus; it *arises from the lumbar plexus* in the abdomen (L2, L3, and L4) and enters the pelvis minor (Figs. 3-10 and 3-16; see also Fig. 2-99). It runs along the lateral wall of the pelvis in the extraperitoneal fat to the obturator foramen, where it divides into anterior and posterior parts. It leaves the pelvis via the *obturator foramen* (Figs. 3-4 and 3-10) and supplies the medial thigh muscles (see Chap. 5, p. 393 and Table 5-2).

> Injuries to the sacral plexus are uncommon; however, the nerves arising from it may be compressed by pelvic tumors. This compression usually causes pain in the lower limb, and when associated with malignant pelvic tumors, the pain may be excruciating. The fetal head may also compress these nerves, producing pain in the mother's lower limbs. The obturator nerve supplies the obturator externus and the adductor muscles of the thigh (see Chap. 5, Table 5-2). It lies on the lateral wall of the pelvis (Fig. 3-16) where it is vulnerable to injury (*e.g.*, during removal of cancerous lymph nodes).

The Coccygeal Plexus (Fig. 3-16*A*). This small network of nerve fibers is formed by the ventral rami of S4 and S5 and the coccygeal nerves. It lies on the pelvic surface of the coccygeus muscle and supplies this muscle, part of the levator ani, and the sacrococcygeal joint. It then pierces the coccygeus muscle to supply a small area of skin over the coccyx.

The Pelvic Fascia

The pelvic viscera are surrounded by loose connective tissue called *visceral pelvic fascia*. The fascia lining the walls of the pelvic cavity is called *parietal pelvic fascia*.

The Visceral Pelvic Fascia[3] (Figs. 3-7, 3-11, 3-17, 3-18, and 3-24). This fascia binds the pelvic viscera together and to the parietal pelvic fascia. It envelopes the viscera, forming false capsules for them. The visceral pelvic fascia contains varying amounts of smooth muscle, which help to support the pelvic floor. In the female the visceral pelvic fascia is thickened laterally to form the *uterosacral and transverse cervical ligaments* (p. 285). These ligaments, passing from the sides of the uterine cervix to the posterior wall of the pelvis, are a major source of support for the uterus.

The Parietal Pelvic Fascia (Figs. 3-11, 3-17, 3-18, and 3-24). This fascia lines the pelvic cavity as far inferiorly as the ischiopubic rami. It is attached to the periosteum of the ilium just inferior to the pelvic brim and extends onto the superior surface of the pelvic diaphragm. It then passes to the pelvic viscera where it is continuous with the visceral pelvic fascia. The parietal fascia lining the abdominal and pelvic cavities is continuous and, as stated, is anchored to bone near the pelvic brim. It is separated from the parietal peritoneum by extraperitoneal fat. In addition to lining the pelvic cavity, the parietal pelvic fascia encloses its great vessels. *Superiorly the parietal pelvic fascia is continuous with the transversalis fascia* (see Fig. 2-11) and is anchored to the periosteum on the posterior aspect of the body of the pubis.

The parietal pelvic fascia covers the pelvic surfaces of the obturator internus, piriformis, coccygeus, sphincter urethrae, and levator ani muscles (Figs. 3-8, 3-11, 3-17, and 3-18). The fascia covering the obturator internus muscle (*obturator fascia*) is thicker than its other parts. The obturator fascia is separated superiorly from the *psoas fascia* (see also Figs. 2-82 and 2-102) by its attachment to the periosteum, just inferior to the pelvic brim (Fig. 3-11). As described on p. 254, the levator ani muscles attach to a thickening of the obturator fascia, known as the *tendinous arch*, which stretches between the body of the pubis and the ischial spine (Figs. 3-14 and 3-18). Superior to the tendinous arch the obturator fascia is thick and tough, whereas inferior to it the fascia is thin. This part lines the lateral wall of the *ischioanal fossa* and forms the medial wall of the *pudendal canal* (Figs. 3-11 and 3-17). The fascia of the pelvic diaphragm covers both surfaces of the levator ani muscles and forms part of the pelvic floor. It therefore helps to support the pelvic viscera (Figs. 3-11 and 3-17). The fascia lining its superior surface is called the *superior fascia of the pelvic diaphragm*. In females this fascia is attached to the posterior aspect of the body of the pubis, the neck of the urinary bladder, the vagina, and the rectum. *In males* the superior fascia of the pelvic diaphragm is attached to the prostate and rectum (Figs. 3-17 to 3-19).

The superior fascia of the pelvic diaphragm is thickened at the neck of the urinary bladder to form two cordlike bands called *pubovesical ligaments* in the female (Fig. 3-34) and *puboprostatic ligaments* in the male (Fig. 3-19). With one on each side

---

[3]This fascia is usually called *endopelvic fascia* by gynecologists.

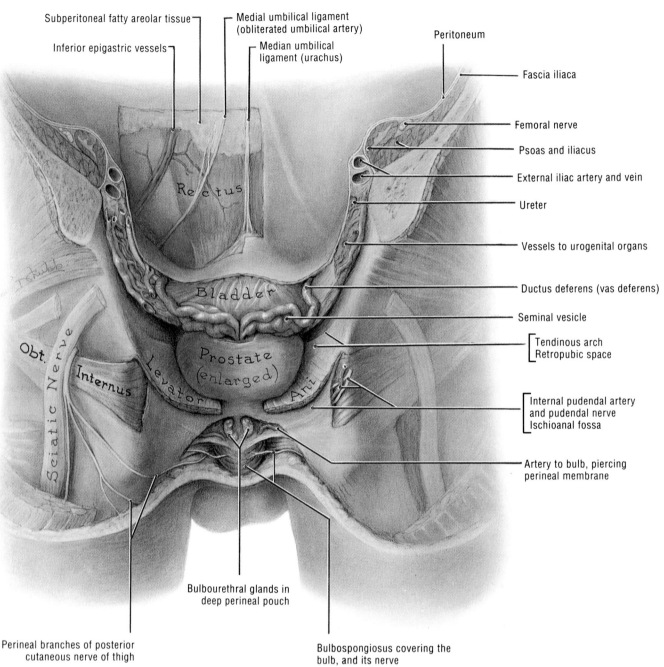

Subperitoneal fatty areolar tissue

Inferior epigastric vessels

Medial umbilical ligament
(obliterated umbilical artery)

Median umbilical
ligament (urachus)

Peritoneum

Fascia iliaca

Femoral nerve

Psoas and iliacus

External iliac artery and vein

Ureter

Vessels to urogenital organs

Ductus deferens (vas deferens)

Seminal vesicle

Tendinous arch
Retropubic space

Internal pudendal artery
and pudendal nerve
Ischioanal fossa

Artery to bulb, piercing
perineal membrane

Rectus

Bladder

Prostate
(enlarged)

Obt. Nerve

Internus

Levator

Ani

Sciatic Nerve

Perineal branches of posterior
cutaneous nerve of thigh

Bulbourethral glands in
deep perineal pouch

Bulbospongiosus covering the
bulb, and its nerve

Figure 3-18.   Dissection of a coronal section of the male pelvis just anterior to the rectum. This is a posterior view of the anterior portion. Note that the medial umbilical ligament (obliterated umbilical artery) and the median umbilical ligament (remnant of the urachus), like the urinary bladder, are in the subperitoneal fatty-areolar tissue. Examine the levator ani and its fascial coverings separating the retropubic space from the ischioanal fossa (see also Fig. 3-17). Observe that the free anterior borders of the levator ani muscles are about 1 cm apart.

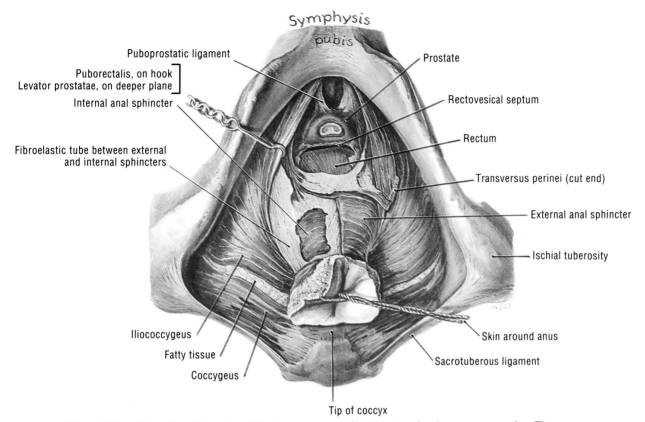

Symphysis pubis

Puboprostatic ligament

Puborectalis, on hook
Levator prostatae, on deeper plane

Internal anal sphincter

Fibroelastic tube between external
and internal sphincters

Prostate

Rectovesical septum

Rectum

Transversus perinei (cut end)

External anal sphincter

Ischial tuberosity

Iliococcygeus

Fatty tissue

Coccygeus

Skin around anus

Sacrotuberous ligament

Tip of coccyx

Figure 3-19.   Dissection of the male pelvic floor composed of the levator ani and coccygeus muscles. The urogenital diaphragm and its fasciae have been removed (see Fig. 3-17 for orientation).

of the median plane, these ligaments *anchor the neck of the urinary bladder to the pubis.* The pubovesical ligaments also attach to the wall of the vagina.

The inferior fascia of the pelvic diaphragm covers the inferior surface of the levator ani muscles and forms the medial walls of the ischioanal fossae (Figs. 3-11, 3-17, and 3-18). The inferior fascia of the pelvic diaphragm is also continuous with the fascia on the medial surface of the inferior half of the obturator internus muscle and on the inferior surface of the external anal sphincter.

The Retropubic Space (Figs. 3-17, 3-18, 3-33, 3-34, 3-35*B*, and 3-37).   This is the space between the parietal pelvic fascia and the anterior surface of the urinary bladder. It contains extraperitoneal fat, loose connective tissue, blood vessels, and nerves. The fat and connective tissue accommodate the expansion of the urinary bladder as urine accumulates.

Arteries of the Pelvis

Of the arteries that enter the pelvis minor, the median sacral and superior rectal arteries are unpaired, the internal iliac and ovarian arteries are paired.

The Internal Iliac Artery

This large vessel is one of the two terminal branches of the common iliac artery (Figs. 3-24 to 3-28; see also Fig. 2-95).

The internal iliac artery *supplies most of the blood to the pelvic viscera,* as well as supplying the musculoskeletal part of the pelvis and the gluteal region of the lower limb (p. 420). The internal iliac artery begins at the level of the intervertebral disc between L5 and S1, where it is crossed by the ureter. It is separated from the sacroiliac joint by the internal iliac vein and the lumbosacral trunk (Fig. 3-16*A*). It passes posteromedially into the pelvis minor, medial to the external iliac vein and obturator nerve, and lateral to the peritoneum (Fig. 3-24). It ends at the superior edge of the greater sciatic foramen by dividing into anterior and posterior divisions. The branches of the anterior division of the internal iliac artery are mainly visceral (*i.e.*, to the bladder, rectum, and reproductive organs). It also has two parietal branches, which pass to the buttock and thigh. The arrangement of the visceral branches is variable. The following *branches of the internal iliac artery* are listed in the order in which they commonly arise.

The Umbilical Artery (Fig. 3-24).   This vessel runs anteroinferiorly between the urinary bladder and the lateral wall of the pelvis. It gives off the *superior vesical artery,* which supplies numerous branches to the superior part of the urinary bladder.

Before birth the umbilical arteries carry blood to the placenta for reoxygenation (Moore, 1988). Postnatally their distal parts atrophy and become fibrous cords, called the *medial umbilical ligaments.* They raise folds of peritoneum called the *medial umbilical folds,* which run on the deep surface of the anterior abdominal wall (Figs. 3-18, 3-24, and 3-28; see also Fig. 2-12).

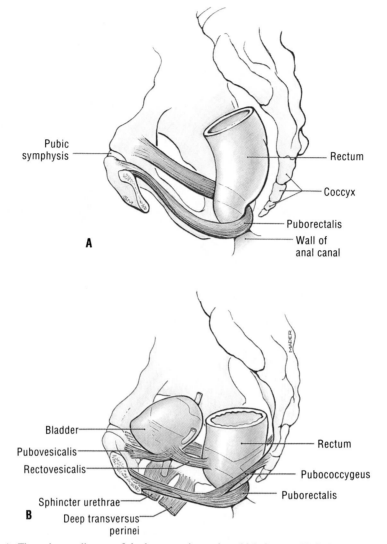

Figure 3-20.   *A*, The puborectalis part of the levator ani muscle, which forms a U-shaped sling around the anal canal. *B*, The pubovesicalis and rectovesicalis muscles, which support the urinary bladder.

The Obturator Artery (Fig. 3-24; see also Fig. 5-29).   The origin of this vessel is variable; usually it arises close to the umbilical artery, where it is crossed by the ureter. It then runs anteroinferiorly on the obturator fascia on the lateral wall of the pelvis and passes between the obturator nerve and vein. It then leaves the pelvis through the *obturator foramen* and supplies muscles of the thigh (p. 406). Within the pelvis, the obturator artery gives off muscular branches, a nutrient artery to the ilium, and a pubic branch. *The pubic branch* arises just before the obturator artery leaves the pelvis. It ascends on the pelvic surface of the ilium to anastomose with its fellow of the opposite side and the pubic branch of the *inferior epigastric artery*, which is a branch of the external iliac (see Fig. 2-7). The extrapelvic distribution of the obturator artery is described in the section on the lower limb (Chap. 5, p. 405).

The Inferior Vesical Artery (Fig. 3-24*A* and *B*; see also Fig. 2-19).   This vessel occurs only in the male. The inferior vesical artery passes to the fundus (base) of the urinary bladder, where it supplies the seminal vesicles, prostate, and the posteroinferior part of the bladder. It also gives rise to the *artery of the ductus deferens* and the *prostatic artery*.

The Vaginal Artery (Fig. 3-24*C* to 3-27).   This vessel is *homologous with the inferior vesical artery in the male*. It runs anteriorly and then passes along the side of the vagina, where it divides into numerous branches that supply the anterior and posterior surfaces of the vagina, the posteroinferior parts of the urinary bladder, and the pelvic part of the urethra. It anastomoses with the vaginal branch of the uterine artery. The vaginal artery, often double or triple, is usually a branch of the uterine artery, but it may arise independently from the internal iliac artery.

The Uterine Artery (Figs. 3-24*C* to 3-27).   This vessel usually arises separately from the internal iliac artery, but it may arise from the umbilical artery. It descends on the lateral wall of the pelvis, anterior to the internal iliac artery, and *enters the root of the broad ligament* where it passes superior to the lateral portion of the fornix of the vagina to reach the lateral margin of

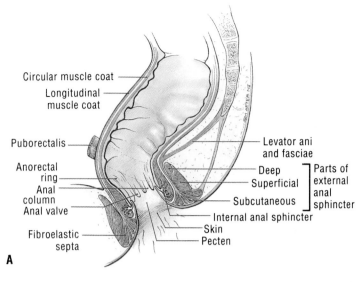

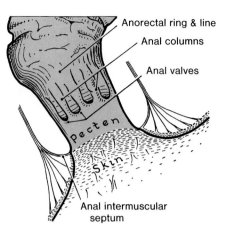

A

B

Figure 3-21. Median sections of the inferior part of the rectum and anal canal. In *A*, examine the large voluntary external anal sphincter forming a broad band on each side of the anal canal. In *B*, observe the anal columns, noting that they are vertical folds of the mucosa and that the inferior ends of these anal columns are joined by crescentic folds called anal valves. The concavities of the valves (*black*) are called *anal sinuses* (crypts). The comb-shaped inferior limit of the anal valves forms the clinically important *pectinate line*.

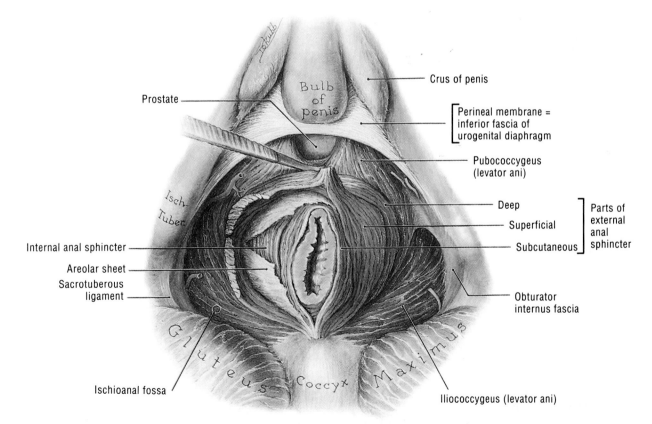

Figure 3-22. Dissection of the external anal sphincter. On the left, the superficial and deep parts of the sphincter are reflected and the underlying sheet—consisting of areolar tissue, levator ani fibers, and an outer longitudinal muscular coat of the anal canal—is cut. This was done to reveal the inner circular muscular coat of the anal canal, which is thickened to form the involuntary internal anal sphincter.

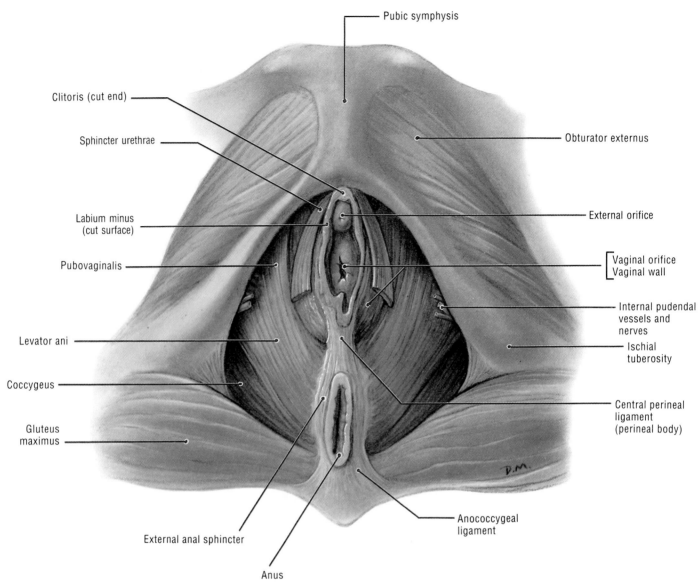

Pubic symphysis

Clitoris (cut end)

Sphincter urethrae

Obturator externus

Labium minus (cut surface)

External orifice

Pubovaginalis

Vaginal orifice
Vaginal wall

Internal pudendal vessels and nerves

Levator ani

Ischial tuberosity

Coccygeus

Central perineal ligament (perineal body)

Gluteus maximus

External anal sphincter

Anococcygeal ligament

Anus

Figure 3-23. Dissection of the female perineum. Examine the sphincter urethrae muscle arising from the inferior pubic ramus. Its fibers run medially and some of them cross each other between the urethra and the uterus. *The uterine artery passes anterior and superior to the ureter near the lateral portion of the fornix.* On reaching the vaginal orifice and attach to the wall of the vagina (Fig. 3-58C). Observe the clinically important central perineal ligament, the tendinous center of the perineum, which is commonly called the perineal body.

the uterus. *The uterine artery passes anterior and superior to the ureter near the lateral portion of the fornix.* On reaching the side of the cervix, the uterine artery divides into a large superior branch, which supplies the body and fundus of the uterus, and a smaller vaginal branch, which supplies the cervix and vagina. The uterine artery pursues a tortuous course along the lateral margin of the uterus and ends when its ovarian branch anastomoses with the ovarian artery between the layers of the broad ligament.

The fact that the uterine artery crosses anterior and superior to the ureter near the lateral portion of the fornix of the vagina is clinically important. The ureter is in danger of being inadvertently clamped or severed during a **hysterectomy** (G. *hystera*, uterus + *ektome*, excision) when the uterine artery is tied off. The point of crossing of the uterine artery and

ureter lies about 2 cm superior to the ischial spine. *The left ureter is particularly vulnerable* because it is very close to the lateral aspect of the cervix (Fig. 3-25). The ureter is also vulnerable to injury when the ovarian vessels are being tied off (*e.g.*, during an *ovariectomy*) because these structures lie very close to each other where they cross the pelvic brim (Fig. 3-28).

The Middle Rectal Artery (Figs. 3-24 and 3-50). This small vessel runs medially to the rectum. It also sends branches to the prostate and seminal vesicles in males and to the vagina in females.

The Internal Pudendal Artery (Figs. 3-8, 3-11, 3-16 to 3-18, and 3-24). This vessel, larger in the male than the female, passes inferolaterally, anterior to the piriformis muscle and sacral plexus. It leaves the pelvis between the piriformis and coccygeus muscles by passing through the inferior part of the *greater sciatic foramen*. The internal pudendal artery passes around the posterior aspect of the ischial spine or the sacrospinous ligament and enters the *ischioanal fossa* through the lesser sciatic foramen. The internal pudendal artery, along with the internal pudendal veins and branches of the pudendal nerve, *passes through the pudendal canal* in the lateral wall of the ischioanal fossa. Just before it reaches the pubic symphysis, the internal pudendal artery divides into its terminal branches: the deep and dorsal *arteries of the penis* or *clitoris* (Figs. 3-40, 3-59, and 3-71).

The Inferior Gluteal Artery (Figs. 3-16 and 3-24). This vessel passes posteriorly between the sacral nerves (usually S2 and S3), and leaves the pelvis through the inferior part of the *greater sciatic foramen*, inferior to the piriformis muscle. It supplies the muscles and skin of the buttock and the posterior surface of the thigh.

The three branches of the posterior division of the internal iliac artery (superior gluteal, iliolumbar, and lateral sacral) are mainly muscular.

The Superior Gluteal Artery (Figs. 3-16 and 3-24). This large artery arises from the posterior division of the internal iliac artery, passes posteriorly, and runs between the lumbosacral trunk and the ventral ramus of S1 nerve. It leaves the pelvis through the superior part of the *greater sciatic foramen*, superior to the piriformis muscle, to supply the gluteal muscles in the buttock.

The Iliolumbar Artery (Figs. 3-16 and 3-24*B* and *C*). This vessel arises from the posterior division of the internal iliac artery and runs superolaterally to the iliac fossa. It then passes anterior to the sacroiliac joint and posterior to the psoas major muscle. Here, it *separates the obturator nerve from the lumbosacral trunk*. Within the iliac fossa, the iliolumbar artery divides into: (1) an *iliac branch* that supplies the iliacus muscle and the *ilium* and (2) a *lumbar branch* that supplies the psoas major and quadratus lumborum muscles.

The Lateral Sacral Arteries (Figs. 3-16 and 3-24*B* and *C*). These vessels, usually superior and inferior ones on each side, arise from the posterior division of the internal iliac artery. They may arise from a common trunk. They pass medially and descend anterior to the sacral ventral rami, giving off spinal branches that pass through the pelvic sacral foramina and *supply the spinal meninges* enclosing the spinal cord and the roots of the *sacral nerves*. Some branches of these arteries pass from the sacral canal through the dorsal sacral foramina and supply the muscles and skin overlying the sacrum.

The Median Sacral Artery

This small unpaired artery usually arises from the posterior surface of the abdominal aorta, just superior to its bifurcation, but it may arise from its anterior surface (see Fig. 2-95). This small vessel runs anterior to the bodies of the last one or two lumbar vertebrae, the sacrum, and the coccyx and ends in a series of anastomotic loops. Before the median sacral artery enters the pelvis minor, it sometimes gives rise to a pair of *fifth lumbar arteries* (see Fig. 2-39). As it descends over the sacrum, the median sacral artery gives off small parietal branches, which anastomose with the lateral sacral arteries. It also gives rise to

small visceral branches to the posterior part of the rectum. These arteries anastomose with the superior and middle rectal arteries.

The median sacral artery represents the *caudal end of the embryonic dorsal aorta* which became reduced in size as the tail of the embryo disappeared (Moore, 1988).

The Superior Rectal Artery

This vessel is the *direct continuation of the inferior mesenteric artery* (Fig. 3-50; see also Fig. 2-81). It crosses the left common iliac vessels and descends in the sigmoid mesocolon to the pelvis minor. At the level of S3 vertebra, the superior rectal artery divides into two branches, which descend on each side of the rectum and supply it as far inferiorly as the internal anal sphincter (Fig. 3-21). The superior rectal artery anastomoses with branches of the middle rectal artery, a branch of the internal iliac artery, and with the inferior rectal artery, a branch of the internal pudendal artery.

The Ovarian Artery

The ovarian artery arises from the abdominal aorta inferior to the renal artery, but considerably superior to the inferior mesenteric artery (Figs. 3-25, 3-26, and 3-28; see also Fig. 2-95). As it passes inferiorly, it adheres to the parietal peritoneum and *runs anterior to the ureter* on the posterior abdominal wall. As it enters the pelvis minor, it crosses the proximal ends of the external iliac vessels. It then runs medially in the *suspensory ligament of the ovary* and enters the superolateral part of the broad ligament to supply the ovary and uterine tube. The ovarian artery anastomoses with the uterine artery (Fig. 3-26*B*).

Venous Drainage of the Pelvis

The pelvis is drained mainly by the *internal iliac veins* and their tributaries (Fig. 3-24*A*; see also 2-105), but there is some drainage through the superior rectal, median sacral, and ovarian veins. Some blood from the pelvis also passes to the *internal vertebral venous plexuses* (see Fig. 4-53).

The Internal Iliac Vein

This vein joins the external iliac vein to form the *common iliac vein*, which unites with its partner to form the *inferior vena cava* at the level of L5 vertebra (see Figs. 2-82, 2-83, and 2-105). The internal iliac vein lies posteroinferior to the internal iliac artery (Figs. 3-16 and 3-24), and its tributaries are similar to the branches of this artery, except for the *umbilical vein* that usually obliterates to form the *ligamentum teres* (see Fig. 2-67). This vessel, which is sometimes patent, drains into the left branch of the portal vein (see Fig. 2-83). The *iliolumbar vein* usually drains into the common iliac vein (see Fig. 2-105). The *superior gluteal veins*, the venae comitantes of the superior gluteal arteries (Fig. 3-24), are the *largest tributaries of the internal iliac veins*, except during pregnancy when the uterine veins become larger.

Pelvic venous plexuses are formed by the veins in the pelvis (Fig. 3-29). These intercommunicating *networks of veins* are

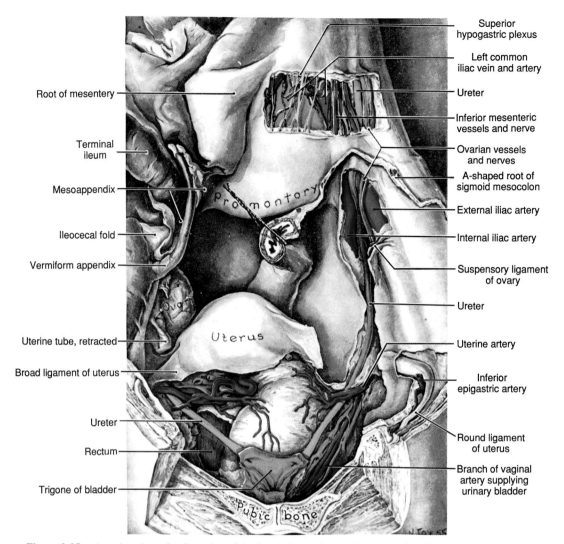

Figure 3-25.   Anterior view of a dissection of the female internal genital organs, primarily to show the clinically important pelvic course of the ureter.

Innervation of the Ureters (Figs. 3-36 and 3-47; see also Fig. 2-64). The nerve supply of the ureter is derived from adjacent *autonomic plexuses* (renal, testicular or ovarian, and inferior hypogastric), which contain pain fibers. Afferent fibers from the ureters reach the spinal cord through the dorsal roots of T11, T12, and L1 nerves.

The ureters are expansile muscular tubes that dilate if obstructed (Fig. 3-27). An acute obstruction results from a urinary or **ureteric calculus** (L. pebble). Although passage of small calculi or *kidney stones* usually causes little or no pain, larger ones produce severe pain. The symptoms and severity depend on the location, type, and size of the stones and on whether they are smooth or spiky. The pain caused by a calculus is referred to as colic or *colicky pain*. Colic results from hyperperistalsis in the ureter, superior to the level of the obstruction. Usually colic is accompanied or followed by a dull, more constant pain owing to distention of the ureter and renal pelvis. Ureteric calculi may cause complete or intermittent *obstruction of urinary flow*. The obstruction may occur anywhere along the ureter (Figs. 3-24*A*, 3-27, and 3-

28), but it occurs most often (1) where the ureter crosses the external iliac artery and the brim of the pelvis and (2) where it passes obliquely through the wall of the urinary bladder. The presence of calculi can often be confirmed by abdominal radiographs or by an intravenous urogram (see Fig. 2-89).

Ureteric calculi or stones may be removed in three ways: open surgery, endourology or lithotripsy (Farrow, 1989). Open surgery is performed uncommonly because the stones can usually be removed by endourology or lithotripsy. In *endourology*, a cystoscope is passed through the urethra into the bladder (Fig. 3-32). The *cystoscope* has a light, an observing lens, and various attachments (*e.g.*, for grasping a stone). Another instrument called a *ureteroscope* can be inserted into the cystoscope and passed up the ureter to grasp the stone. Larger stones may be broken up with an *ultrasonic probe* that is inserted through the ureteroscope. *Lithotripsy* uses shock waves to break up a stone (p. 221).

The Urinary Bladder (Figs. 3-18, 3-24, and 3-27 to 3-35). The bladder (L. *vesica*) is a hollow muscular vesicle for storing urine. *In adults* the empty bladder lies in the pelvis minor.

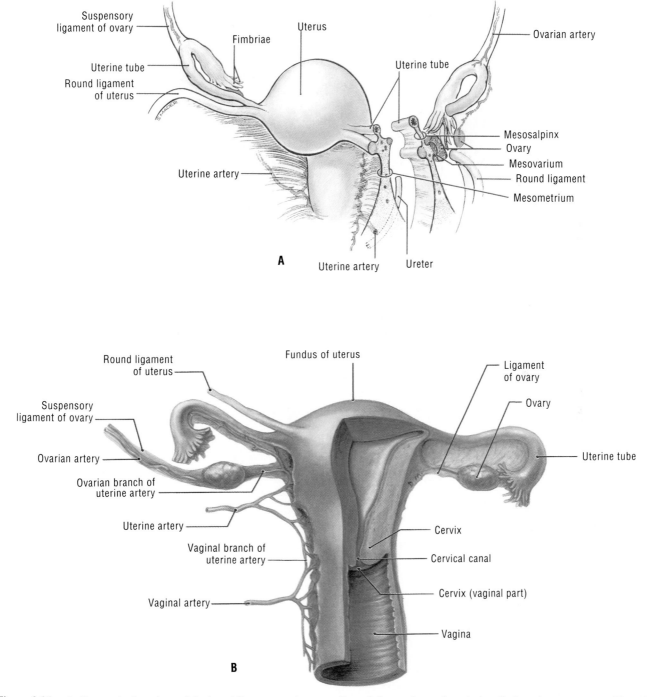

Figure 3-26. *A,* Two sagittal sections of the broad ligament to show the mesentery of the uterus. The mesentery of the uterine tube (mesosalpinx) and ovary (mesovarium) are also illustrated. *B,* Anterior view of the female internal genital organs after removal of the broad ligament. Part of the uterine and vaginal walls have been cut away. Note the anastomosis between the ovarian and uterine arteries and between the vaginal branch of the uterine artery and the vaginal artery. These communications occur between the layers of the broad ligament.

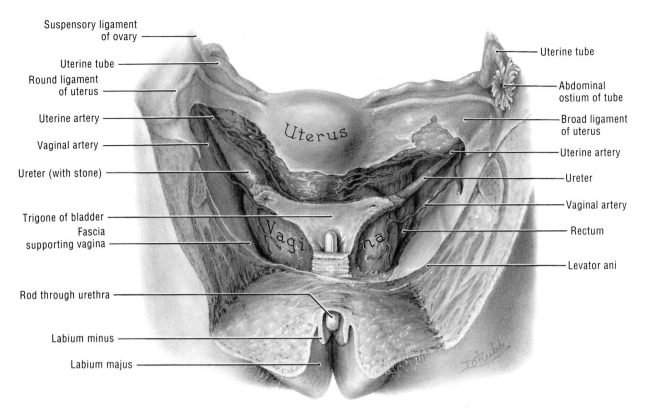

Figure 3-27. Anterior view of the uterus and uterine tubes. The pubic bones and urinary bladder, except for the trigone, are removed.

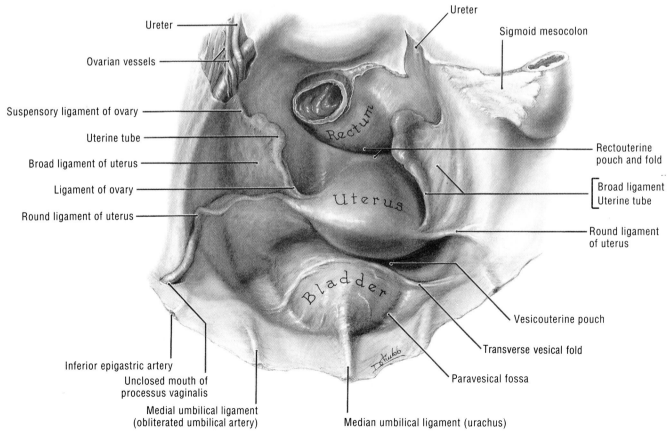

Figure 3-28. Superior view of the female pelvic organs. Observe the pear-shaped uterus that, as usual, is asymmetrically placed; here, it leans to the left. Note that the ovarian vessels cross the external iliac vessels very close to the ureter.

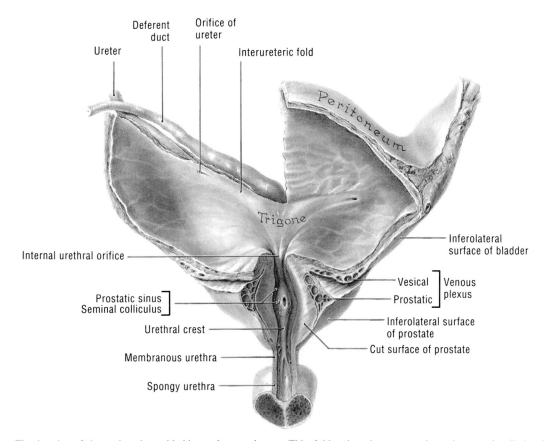

Figure 3-29. The interior of the male urinary bladder and prostatic urethra. The anterior parts of the bladder, prostate, and urethra are cut away. The knife was then carried through the posterior wall of the bladder at the superior border of the right ureter and interureteric fold. This fold unites the ureters along the superior limit of the trigone. Observe the slight fullness posterior to the internal urethral orifice, which, when exaggerated, becomes the uvula vesicae. This small projection is caused by the median lobe of the prostate.

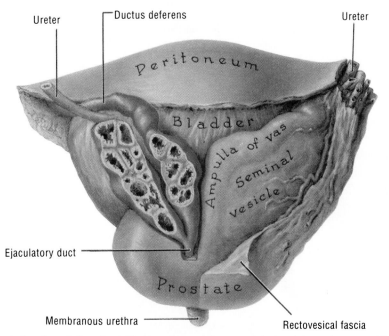

Figure 3-30. Posterior view of the urinary bladder, ductus deferens, seminal vesicle, and prostate. The left seminal vesicle and ampulla of the ductus deferens are dissected free and sliced open.

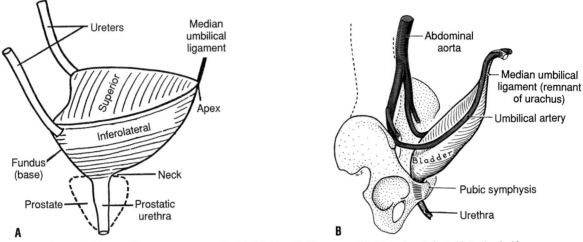

Figure 3-31.   *A*, The empty contracted adult bladder. *B*, The empty bladder of an infant. Note the fusiform shape in *B* and that the bladder is located in the abdominal cavity.

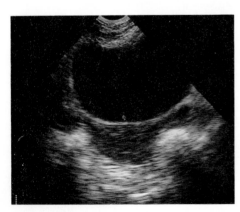

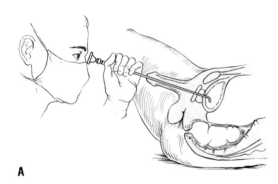

Figure 3-32.   *A*, Drawing illustrating the use of a cystoscope to examine the interior of the male urinary bladder. (Illustrated by Mrs. D. M. Hutchinson. Reprinted with permission from Laurenson RD: *An Introduction to Clinical Anatomy by Dissection of the Human Body*. Philadelphia, WB Saunders, 1968.). *B*, Transverse ultrasound scan of a female pelvis. *C*, Diagram of the ultrasound scan shown in *B*. (Courtesy of Dr. A. M. Arenson, Assistant Professor of Radiology, University of Toronto, Toronto, Ontario, Canada.)

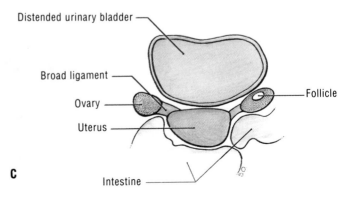

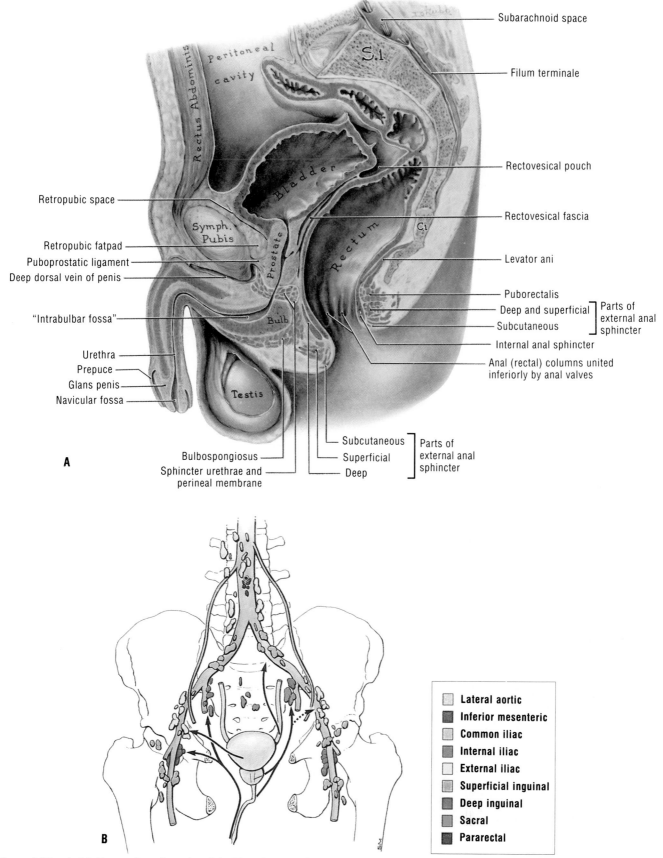

Figure 3-33. *A,* Median section of a male pelvis. Note the prostatic urethra descending vertically through a somewhat elongated prostate.

B, Lymphatic drainage of the ureters, urinary bladder, prostate, ductus deferens, and urethra.

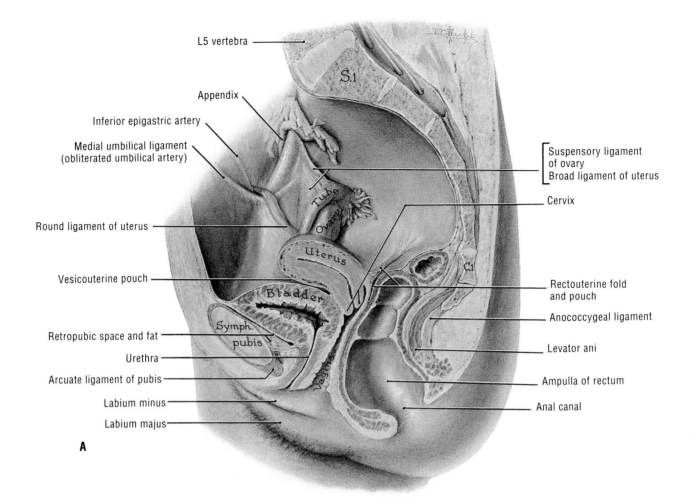

L5 vertebra

S.1

Appendix

Inferior epigastric artery

Medial umbilical ligament
(obliterated umbilical artery)

Suspensory ligament
of ovary
Broad ligament of uterus

Tube

Ovary

Cervix

Round ligament of uterus

Uterus

Vesicouterine pouch

Bladder

C.1

Rectouterine fold
and pouch

Anococcygeal ligament

Symph.
pubis

Levator ani

Retropubic space and fat

Urethra

Arcuate ligament of pubis

Vagina

Ampulla of rectum

Labium minus

Anal canal

Labium majus

A

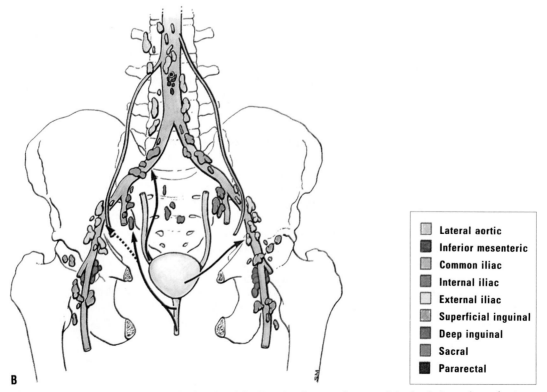

B

| Lateral aortic |
| Inferior mesenteric |
| Common iliac |
| Internal iliac |
| External iliac |
| Superficial inguinal |
| Deep inguinal |
| Sacral |
| Pararectal |

Figure 3-34. *A*, Median section of a female pelvis. Examine the posterior part of the fornix located anterior to the rectouterine pouch. *B*, Lymphatic drainage of the ureters, urinary bladder, and urethra.

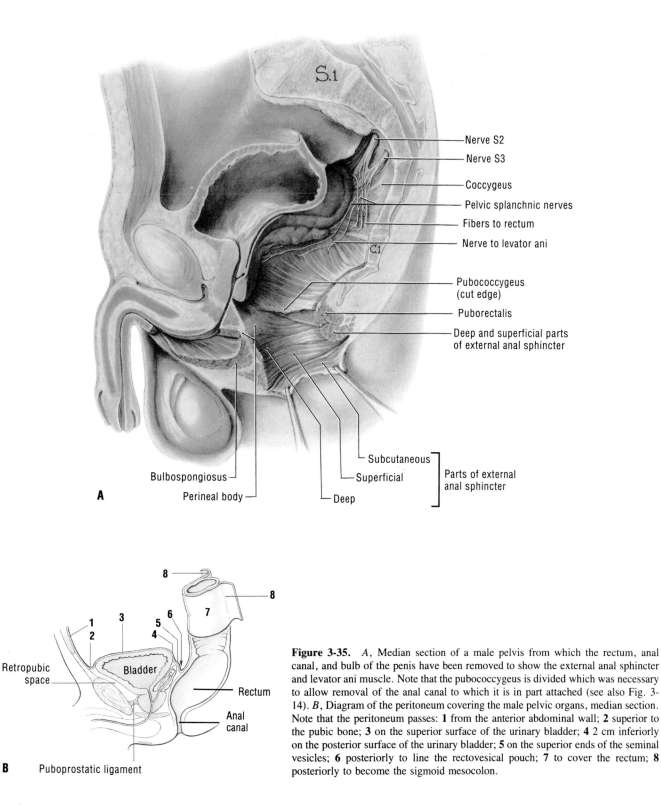

Figure 3-35. *A*, Median section of a male pelvis from which the rectum, anal canal, and bulb of the penis have been removed to show the external anal sphincter and levator ani muscle. Note that the pubococcygeus is divided which was necessary to allow removal of the anal canal to which it is in part attached (see also Fig. 3-14). *B*, Diagram of the peritoneum covering the male pelvic organs, median section. Note that the peritoneum passes: **1** from the anterior abdominal wall; **2** superior to the pubic bone; **3** on the superior surface of the urinary bladder; **4** 2 cm inferiorly on the posterior surface of the urinary bladder; **5** on the superior ends of the seminal vesicles; **6** posteriorly to line the rectovesical pouch; **7** to cover the rectum; **8** posteriorly to become the sigmoid mesocolon.

In adults the empty bladder lies posterior and slightly superior to the pubic bones. It is separated from the pubic bones by the **retropubic space** (Fig. 3-35*B*). *In infants* the bladder is in the abdomen even when empty (Fig. 3-31*B*). The bladder enters the pelvis major at about 6 years of age, but it is not entirely within the pelvic minor until after puberty. An empty bladder in an adult lies almost entirely in the pelvic minor, superior to the pelvic floor and posterior to the pubic symphysis. As it fills, it ascends into the pelvic major; a very full bladder may ascend to the level of the umbilicus.

The bladder is a hollow viscus with strong muscular walls; *it is characterized by its distensibility*. Its shape, size, position, and relations vary with the amount of urine it contains and with the age of the person. The mucous membrane of the bladder is loosely connected to its muscular wall, except in a triangular area in its fundus (base), called the *trigone of the bladder* (Figs. 3-25, 3-27, and 3-29). The mucous membrane in the empty bladder is thrown into numerous folds or rugae, except in the trigone where the mucous membrane is always smooth because it is firmly attached here to the muscular wall. Posterior to the internal urethral orifice in the male is a small elevation known as the *uvula vesicae*. It is produced by the middle lobe of the prostate.

In the female the peritoneum is reflected from the superior surface of the bladder near its posterior border and onto the anterior wall of the uterus at the junction of its body and cervix (Figs. 3-34 and 3-45). The *vesicouterine pouch* of peritoneum extends between the bladder and the uterus. This pouch is empty except when the uterus is retroverted (inclined posteriorly); in this case a loop of bowel may enter it.

In the male the peritoneum is reflected from the bladder over the superior surfaces of the deferent ducts (ductus deferentes) and seminal vesicles (Figs. 3-18, 3-24, 3-30, and 3-35). The bladder is relatively free within the loose extraperitoneal fatty tissue, except for its neck, which is held firmly by the *puboprostatic ligaments* (Fig. 3-19). As the bladder fills, it can easily expand superiorly into the extraperitoneal fatty tissue of the anterior abdominal wall (Figs. 3-18 and 3-33 to 3-35). This movement lifts the peritoneum from the transversalis fascia of the wall.

In cadaveric specimens the empty bladder has the form of a *triangular pyramid* (Fig. 3-31A), whereas *in living persons* the bladder always contains some urine and is usually *more or less rounded*. When filled, the bladder contains about 500 cc of urine. The empty pyramid-shaped bladder in a cadaver has four surfaces (Fig. 3-31A): a *superior surface*; *two inferolateral surfaces* facing inferiorly, laterally, and anteriorly; and a *posterior surface*. The inferolateral surfaces are in contact with the fascia covering the levator ani muscles (Fig. 3-17). The posteroinferior surface is referred to as the *fundus of the bladder* (base). In the female the fundus is closely related to the anterior wall of the vagina; in the male it is related to the rectum. The anterior end known as the *apex of the bladder* (Fig. 3-31A), points anteriorly toward the superior edge of the pubic symphysis (Figs. 3-18 and 3-34). The inferior part of the bladder, where the fundus and the inferolateral surfaces converge, is called the *neck of the bladder*. In the male this is where the lumen of the bladder opens into the prostatic urethra (Figs. 3-29 and 3-31A). In the male the neck of the bladder rests on the prostate (Fig. 3-33). From the apex of the bladder, the *median umbilical fold* of peritoneum passes superiorly to the umbilicus. This fold is raised by the *median umbilical ligament* (Figs. 3-18 and 3-28), a remnant of the embryonic *urachus* (Moore, 1988).

The Bladder Bed (Figs. 3-17, 3-18, 3-24, 3-27, 3-29, 3-30, 3-33, and 3-34).   The shape of the bladder is largely determined by the structures that are closely related to it. The entire organ is enveloped by loose connective tissue, called *vesical fascia*, in which is located the *vesical venous plexus*. The bladder bed is formed on each side by the pubic bones and the obturator internus and levator ani muscles and posteriorly by the rectum. *In the female* the fundus of the bladder is separated from the rectum by the cervix and the superior part of the vagina. Its neck lies

directly on the pelvic fascia surrounding the short urethra. *In the male* the fundus of the bladder is separated from the rectum by the ampullae of the deferent ducts and seminal vesicles. Its neck fuses with the prostate.

Structure of the Urinary Bladder (Figs. 3-27 to 3-31).   The wall of the bladder is composed chiefly of smooth muscle, called the *detrusor muscle* (L. *detrudere*, to thrust out). Its three layers run in many directions; there are external and internal layers of longitudinal fibers and a middle layer of circular fibers. Toward the neck of the bladder, these muscle fibers form the involuntary *internal sphincter*. Some of these fibers run radially and assist in opening the *internal urethral orifice*. The muscle fibers in the neck region *in the male* are continuous with the connective tissue stroma of the prostate, whereas *in the female* they are continuous with the muscle fibers in the wall of the urethra.

The mucous membrane of the bladder, lined with transitional epithelium, can undergo considerable stretching. The *ureteric orifices* (openings of the ureters) and the *internal urethral orifice* are located at the angles of the trigone (Figs. 3-27 and 3-29). The ureters pass obliquely through the bladder wall in an inferomedial direction, which helps to prevent urine from backing up into the ureters. An increase in bladder pressure presses the walls of the ureters together, preventing the pressure in the bladder from forcing urine up the ureters and damaging the kidneys. The small, slitlike orifices of the ureters are connected by a narrow *interureteric fold* (Fig. 3-29), which forms the superior margin of the trigone.

Arterial Supply of the Urinary Bladder (Figs. 3-24 and 3-25).   The main arteries supplying the bladder are branches of the internal iliac arteries. The *superior vesical arteries*, branches of the umbilical arteries, supply the anterosuperior parts of the bladder. In males the *inferior vesical arteries*, branches of the internal iliac arteries, supply the fundus of the bladder. In females the *vaginal arteries* replace the inferior vesical arteries and send small branches to the posteroinferior parts of the bladder. The obturator and inferior gluteal arteries also supply small branches to the bladder.

Venous Drainage of the Urinary Bladder (Figs. 3-24, 3-25, and 3-29).   The veins of the bladder correspond to the arteries and are tributaries of the internal iliac veins. **In the male** the *vesical venous plexus*, which is combined with the *prostatic plexus*, envelops the base of the bladder and prostate, the seminal vesicles, ductus deferentes, and inferior ends of the ureters. The *vesical venous plexus* mainly drains through the inferior vesical veins into the internal iliac veins, but it may drain via the sacral veins into the *vertebral venous plexuses* (see Fig. 4-53). **In the female** the vesical venous plexus envelops the pelvic part of the urethra and the neck of the bladder. It receives blood from the dorsal vein of the clitoris and communicates with the vaginal venous plexus.

Lymphatic Drainage of the Urinary Bladder (Figs. 3-33B, and 3-34B).   Lymph vessels from the superior part of the bladder pass to the *external iliac lymph nodes*, whereas those from the inferior part of the bladder pass to the *internal iliac lymph nodes*. Some lymph vessels from the neck region of the bladder drain into the sacral or common iliac lymph nodes.

Innervation of the Urinary Bladder (Figs. 3-35, 3-36, and 3-47).   Parasympathetic fibers to the bladder are from the *pelvic*

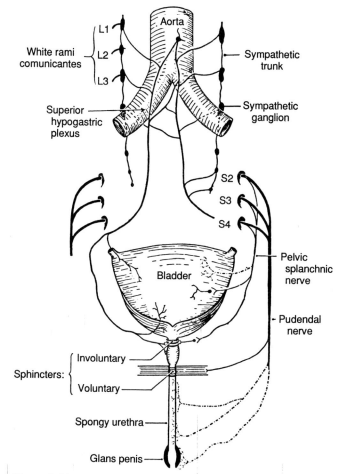

Figure 3-36. Nerve supply of urogenital structures in the male including the sphincter urethrae. The broken lines indicate afferent fibers. Parasympathetic fibers are in the pelvic splanchnic nerves (S2, S3, and S4) and are the motor nerves to the bladder (Fig. 3-35). They are also the sensory nerves of the bladder. The sympathetic fibers through the superior hypogastric plexus (presacral nerve; lower thoracic; L1, L2, and L3) are motor to a continuous muscle sheet comprising the ureteric musculature, the trigonal muscle, and the muscle of the urethral crest. They also supply the muscle of the epididymis, ductus deferens, seminal vesicle, and prostate. The pudendal nerve is motor to the sphincter urethrae muscle and sensory to the glans penis and spongy urethra.

splanchnic nerves (S2, S3, and S4). They are motor to the detrusor muscle and inhibitory to the internal sphincter. Hence, when these fibers are stimulated by stretching, the bladder contracts, the internal sphincter relaxes, and urine flows into the urethra. **Sympathetic fibers** to the bladder are derived from T11, T12, L1, and L2 nerves. These fibers are probably inhibitory to the bladder. The nerves supplying the bladder form the *vesical nerve plexus*, which consists of both sympathetic and parasympathetic fibers. This plexus is continuous with the *inferior hypogastric plexus* (Fig. 3-47; see also Fig. 2-64). Sensory fibers from the bladder are visceral and transmit pain sensations (*e.g.*, resulting from overdistention).

As the bladder expands, it rises from the pelvis within the extraperitoneal fat. When excessively distended, it rises to the level of the umbilicus. In so doing, it lifts several centimeters of parietal peritoneum from the suprapubic part of the anterior abdominal wall. The bladder then lies adjacent to this wall without the intervention of peritoneum. Consequently the distended bladder may be punctured (*suprapubic cystostomy*) or approached surgically superior to the pubic symphysis for the introduction of instruments without traversing the peritoneum and entering the peritoneal cavity. *Urinary calculi* (kidney stones), foreign bodies, and small tumors may also be removed from the bladder through a suprapubic extraperitoneal incision. Because of the superior position of the greatly distended bladder, it may be ruptured by injuries to the inferior part of the anterior abdominal wall or by fractures of the pelvis. The rupture may result in the escape of urine extraperitoneally or intraperitoneally. Rupture of the superior part of the bladder frequently tears the peritoneum, resulting in *extravasation of urine into the peritoneal cavity*. Posterior rupture of the bladder usually results in passage of urine extraperitoneally into the perineum (p. 313).

The interior of the bladder and its three orifices can be examined with a *cystoscope* inserted through the urethra and by various imaging techniques, *e.g.*, ultrasound (Fig. 3-32). The bladder can also be examined radiographically after it has been filled with radiopaque material via a catheter (see Fig. 2-89). A *cystogram* will reveal extravasation of dye in the peritoneal cavity or spread of it into the retroperitoneal space.

Cancers (CA) of the bladder develop in its lining epithelium; they tend to be superficial and multiple. In advanced stages they invade its muscular coat. The urologist can examine the tumors with the cystoscope and can take biopsies of them for microscopic study. Often these tumors can be removed with an instrument called a *resectoscope* that is passed through the urethra into the bladder. Using high-frequency electrical current, the tumors can be removed. For an account of the surgical treatment of CA of the bladder involving an *ileal-conduit urinary diversion technique*, see Farrow (1989).

The Male Urethra (Figs. 3-31, 3-33, 3-35, 3-38, and 3-67). The male urethra is a long muscular tube (15 to 20 cm long) and conveys urine from the urinary bladder to the exterior through the *external urethral orifice*, located at the tip of the glans penis. The urethra also provides an *exit for semen* (seminal fluid). Semen contains sperms (spermatozoa) and auxiliary sex gland secretions (*e.g.*, from the prostate). For descriptive purposes *the urethra is divided into three parts*: prostatic, membranous, and spongy. The membranous and spongy parts are in the perineum and are described in the section on this region (pp. 299 and 312, respectively).

The Prostatic Urethra (Figs. 3-29, 3-31A, and 3-33A). This first part which is about 3 cm long, begins at *the internal urethral orifice* at the apex of the trigone of the bladder. It descends through the prostate, forming a gentle curve that is concave anteriorly. It ends by piercing the superior fascia of the urogenital diaphragm (Figs. 3-11 and 3-17). The lumen of the prostatic urethra is narrower superiorly and inferiorly than in the middle, but it is contracted except when fluid is passing through it. The prostatic part is the *widest and most dilatable part of the urethra*, although it is within the solid prostate. The posterior wall of the prostatic urethra has notable features (Fig. 3-29). The most prominent feature is the median longitudinal ridge called the *urethral*

and 3-8). The interior of the vagina and the vaginal part of the uterine cervix can be examined with a *vaginal speculum* (Fig. 3-43). The cervix can also be palpated with the digits in the vagina or rectum. Pulsations of the uterine arteries may be felt via the lateral parts of the fornix (Fig. 3-41). An instrument called a *colposcope* (which has a magnifying lens) can be inserted into the vagina to examine the vagina and cervix. Owing to the anatomical relationships of the vagina, an instrument directed posteriorly into the vagina may be passed through the posterior wall of the vagina into the rectouterine pouch and the peritoneal cavity (Figs. 3-34 and 3-43). This could result in *peritonitis*. This has occurred when attempts have been made to induce an abortion with a hatpin, coathanger, or some other sharp object. A *peritoneoscope* or *culdoscope* may be inserted through the posterior part of the vaginal fornix to examine the ovaries or uterine tubes (*e.g.,* for a tubal pregnancy). This procedure is called *posterior colpotomy.*

Oocytes have also been obtained from hyperstimulated ovaries for *in vitro fertilization* (IVF) by transvaginal, sonographically controlled ovarian follicle puncture (Dellenbach et al., 1988a). The needle is introduced through the posterior part of the fornix into the *rectouterine pouch* close to the ovary (Figs. 3-43 and 3-45). Using ultrasound, the mature ovarian follicles may be located (Fig. 3-32*B*) and the oocytes can be removed for IVF. Using a similar route, sperms have been deposited in the rectouterine pouch where they can be picked up by the uterine tube at the same time that the oocyte is collected from the ovary. This technique is called *intraperitoneal fertilization* (Dellenback et al., 1988b).

Surgical operations on the vagina are usually performed via the perineal approach (Figs. 3-23 and 3-70). These procedures are commonly done to correct abnormal relaxation of the vaginal walls when childbearing has resulted in weakening of the pelvic diaphragm (p. 295). This may result in bulging of the bladder into the anterior wall of the vagina (*cystocele,* Fig. 3-15*C*) or bulging of the anterior wall of the rectum into the posterior wall of the vagina (*rectocele).*

The Uterus (Figs. 3-25 to 3-28, 3-34, and 3-40 to 3-46).   The uterus (L. womb) of a nonpregnant woman is a hollow, thick-walled, pear-shaped muscular organ located between the bladder and the rectum. It is 7 to 8 cm long, 5 to 7 cm wide, and 2 to 3 cm thick. The uterus normally projects superoanteriorly over the urinary bladder. During pregnancy the uterus enlarges greatly to accommodate the embryo and later the fetus. *The uterus consists of two major parts* (Figs. 3-41 and 3-44): (1) the expanded superior two-thirds is known as the **body**, and (2) the cylindrical inferior third is called the **cervix** (L. neck). A slight constriction called the **isthmus** marks the junction between the body and cervix (Fig. 3-41). The *fundus of the uterus* is the rounded superior part of the body, which is located superior to the line joining the points of entrance of the uterine tubes. The regions of the body where the uterine tubes enter are called the *cornua* (L. *horns*). Because it projects into the vagina, the cervix is divided into *vaginal* and *supravaginal parts* for descriptive purposes. The rounded vaginal part communicates with the vagina via the *external ostium* of the uterus. The Latin word *ostium* means door, entrance, or mouth. The ostium is bounded by *anterior* and *posterior lips* formed by the cervix.

The isthmus of the uterus, about 1 cm long, is the narrow transitional zone between the body and cervix (Fig. 3-41). This slight constriction is most obvious in *nulliparous women* (those who have not borne children). The uterus is usually bent anteriorly (anteflexed) between the cervix and body (Figs. 3-28 and 3-34*A*), and the entire uterus is normally bent or inclined anteriorly (anteverted). The uterus is frequently retroverted (inclined posteriorly) in older women.

The wall of the uterus consists of three layers (Fig. 3-34):

1. The outer serous coat called the *perimetrium*, consists of peritoneum supported by a thin layer of connective tissue;
2. The middle muscular coat called the *myometrium* consists of 12 to 15 mm of smooth muscle. The myometrium increases greatly during pregnancy. The main branches of the blood vessels and nerves of the uterus are located in this layer,
3. The inner mucous coat called *endometrium* is firmly adherent to the underlying myometrium.

The endometrium is partly sloughed off each month during menstruation. It lines only the body of the uterus. The cervix is lined by simple columnar, mucus-secreting epithelium. The middle coat of the cervix consists mostly of fibrous tissue;

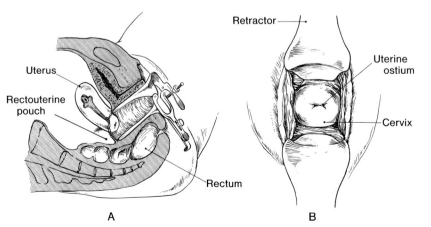

A   B

Figure 3-43.   Examination of the cervix. *A,* The patient is in the lithotomy position with a speculum in the vagina, separating the anterior and posterior walls of the vagina. *B,* Close-up view of the cervix.

(Illustrated by Mrs. D. M. Hutchinson. Reprinted with permission from Laurenson RD: *An Introduction to Clinical Anatomy by Dissection of the Human Body.* Philadelphia, WB Saunders, 1968.)

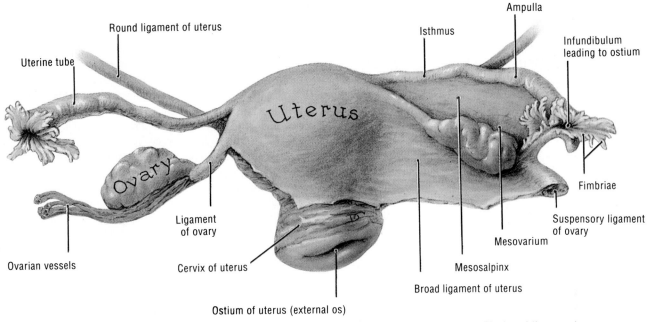

Figure 3-44.   Posterior view of the uterus, ovaries, uterine tubes, and related structures. The broad ligament is removed on the left side.

there are only small amounts of smooth muscle. Owing to its structure, the cervix is more firm and rigid than the body of the uterus.

The uterus has an anteroinferior or *vesical surface* related to the urinary bladder and a posterosuperior or *intestinal surface* related to the intestine (Fig. 3-45). These convex surfaces are separated by right and left borders. Each *uterine tube* enters the lateral border of the body of the uterus near its superior end. The tube opens at one end into the peritoneal cavity near the ovary and at the other end into the uterine cavity (Figs. 3-27, 3-40, 3-41, and 3-44). The *ligaments of the ovaries* are attached to the uterus, posteroinferior to the uterotubal junctions (Fig. 3-44). The *round ligaments of the uterus* are attached anteroinferiorly to these junctions.

Both of these ligaments are continuous within the wall of the uterus and are derivatives of the *gubernaculum* in the embryo (Moore, 1988). Each round ligament runs between the layers of the broad ligament across the pelvic wall to the deep inguinal ring. After traversing the inguinal canal, the round ligament merges with the subcutaneous tissue of the labium majus (see Fig. 2-18).

The body of the uterus is enclosed between the layers of the **broad ligament** and is freely movable (Figs. 3-27, 3-28 and 3-44). Hence, as the bladder fills, the uterus rises, and when the bladder is fully distended, the uterus is inclined posteriorly (*retroverted*) and lies in line with the vagina. As the bladder empties, it slowly moves to its normal anteverted position (Fig. 3-34A). The cervix of the uterus is not very mobile because it is held in position by several ligaments, which are condensations of pelvic fascia containing smooth muscle. The *transverse cervical ligaments* extend from the cervix and lateral parts of the vaginal fornix to the lateral walls of the pelvis. The *uterosacral ligaments*

pass superiorly and slightly posteriorly from the sides of the cervix to the middle of the sacrum. They are deep to the peritoneum and superior to the levator ani muscles. These pelvic ligaments can be palpated through the rectum as they pass posteriorly at the sides of the rectum. The uterosacral ligaments tend to hold the cervix in its normal relationship to the sacrum.

The principal support of the uterus is the pelvic floor (formed by the pelvic diaphragm). The pelvic viscera surrounding the uterus and the visceral fascia (endopelvic fascia) bind the pelvic viscera together (Figs. 3-15 and 3-34A). The two levator ani muscles, the two coccygeus muscles, and the muscles of the urogenital diaphragm are particularly important in supporting the uterus (Figs. 3-23 and 3-34A). *Peritoneum covers the uterus anteriorly and superiorly*, except for the vaginal part of the cervix (Fig. 3-45). It is reflected anteriorly from the uterus onto the bladder and posteriorly over the posterior part of the fornix of the vagina onto the rectum.

The Broad Ligament (Figs. 3-25 to 3-27, 3-28, and 3-43).   This is a fold of peritoneum with mesothelium on its anterior and posterior surfaces. This double-layered sheet extends from the sides of the uterus to the lateral walls and floor of the pelvis. The broad ligament holds the uterus in its normal position. The two layers of the broad ligament are continuous with each other at a free edge, which is directed anteriorly and superiorly to surround the uterine tube. Laterally the broad ligament is prolonged superiorly over the ovarian vessels as the *suspensory ligament of the ovary*. Enclosed in the free edge of each side of the broad ligament is a *uterine tube*. The ovarian ligament lies posterosuperiorly and the *round ligament* of the uterus lies anteroinferiorly within the broad ligament. The broad ligament contains extraperitoneal tissue (connective tissue and smooth muscle) called *parametrium*. The broad ligament gives attachment to the ovary through the *mesovarium* (Figs. 3-26A and 3-

44); this short peritoneal fold connects the anterior border of the ovary with the posterior layer of the broad ligament. The *mesosalpinx* is the part of the broad ligament between the ligament of the ovary, the ovary, and the uterine tube.

The Relationships of the Uterus (Figs. 3-25, 3-28, 3-34, 3-40, and 3-45).   **Anteriorly** the body of the uterus is separated from the urinary bladder by the *vesicouterine pouch*. Here, the peritoneum is reflected from the uterus onto the posterior margin of the superior surface of the bladder. The vesicouterine pouch is empty when the uterus is in its normal position, but it usually contains a loop of intestine when the uterus is retroverted. **Posteriorly** the body of the uterus and the supravaginal part of the cervix are separated from the *sigmoid colon* by a layer of peritoneum and the peritoneal cavity. The uterus is separated from the rectum by the *rectouterine pouch* (of Douglas). The inferior part of this pouch is closely related to the posterior part of the fornix of the vagina. **Laterally** the relationship of the ureter to the uterine artery is very important. The ureter is crossed superiorly by the uterine artery at the side of the cervix.

Arterial Supply of the Uterus (Figs. 3-24*C* to 3-28).   The blood supply of the uterus is derived mainly from the *uterine arteries*, which are branches of the internal iliac arteries. They enter the broad ligaments beside the lateral parts of the fornix of the vagina, superior to the ureters. At the isthmus of the uterus, the uterine artery divides into a large ascending branch that supplies the body of the uterus and a small descending branch that supplies the cervix and vagina. The uterus is also supplied by the *ovarian arteries*, which are branches of the aorta (Fig. 3-26*B*; see also Fig. 2-95). The uterine arteries pass along the sides of the uterus within the broad ligament and then turn laterally at the entrance to the uterine tubes, where they anastomose with the ovarian arteries.

Venous Drainage of the Uterus (Fig. 3-25; see also Fig. 2-83).   The uterine veins enter the broad ligaments with the uterine arteries. They form a *uterine venous plexus* on each side of the cervix and its tributaries drain into the *internal iliac veins*. The uterine venous plexus is connected with the superior rectal vein, forming a *portal-systemic anastomosis*.

Lymphatic Drainage of the Uterus (Fig. 3-42).   The lymph vessels of the uterus follow three main routes: (1) Most lymph vessels from the fundus pass with the ovarian vessels to the *aortic lymph nodes*, but some lymph vessels pass to the *external iliac lymph nodes* or run along the round ligament of the uterus to the *superficial inguinal lymph nodes*. (2) Lymph vessels from the body pass through the broad ligament to the *external iliac lymph nodes*. (3) Lymph vessels from the cervix pass to the *internal iliac and sacral lymph nodes*.

Innervation of the Uterus (Fig. 3-47).   The nerves of the uterus arise from the *inferior hypogastric plexus*, largely from the anterior and intermediate part known as the *uterovaginal plexus*, which lies in the broad ligament on each side of the cervix. Parasympathetic fibers are from the *pelvic splanchnic nerves* (S2, S3, and S4), and sympathetic fibers are from the above plexus. The nerves to the cervix form a plexus in which are located small paracervical ganglia, one of which is often large and is called the *uterine cervical ganglion*. The autonomic fibers of the uterovaginal plexus are mainly vasomotor. Most of the afferent fibers ascend through the inferior hypogastric plexus and enter the spinal cord via T10 to T12 and L1 spinal nerves.

The uterus is mainly an abdominal organ during infancy and the cervix is relatively large. During puberty the uterus grows rapidly. At *menopause* (46 to 52 years), the uterus becomes inactive and decreases in size (undergoes *atrophy*). Incomplete fusion of the embryonic paramesonephric ducts result in a variety of congenital malformations, *e.g.*, **double uterus** and double vagina (Moore, 1988).

The cervix and body of the uterus may be examined by bimanual palpation. Two digits of the right hand are passed high in the vagina while the other hand is pressed inferiorly and posteriorly on the pubic region of the anterior abdominal wall, just superior to the pubic symphysis. The size and other characteristics of the uterus can be determined in this way (*e.g.*, whether the uterus is in its normal anteverted position, Fig. 3-34). When *softening of the isthmus* of the uterus occurs (*Hegar's sign*), the cervix feels as though it were separate from the body. Softening of the uterine isthmus is an early sign of pregnancy. The part of the mesonephric duct, which forms the ductus deferens and ejaculatory duct in the male, may persist in the female as a *duct of Gartner*. It lies between the layers of the broad ligament along the lateral wall of the uterus or vagina. Vestigial remnants of the mesonephric ducts may also give rise to *Gartner's duct cysts* (Moore, 1988).

The Uterine Tubes (Figs. 3-26 to 3-28, 3-34, 3-40, 3-41, 3-44, and 3-46).   The uterine tubes (fallopian tubes), 10 to 12 cm long and 1 cm in diameter, extend laterally from the *cornua* of the uterus. The uterine tubes carry oocytes from the ovaries and sperms from the uterus to the fertilization site in the *ampulla of the uterine tube*. The uterine tube also conveys the dividing *zygote* to the uterine cavity. Each tube opens at its proximal end into the cornu or horn of the uterus and, at its distal end, into the peritoneal cavity near the ovary. Consequently, the uterine tubes allow communication between the peritoneal cavity and

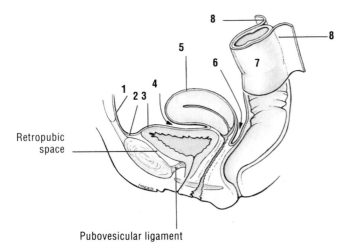

Figure 3-45.   The peritoneum covering the female pelvic organs. It passes: **1** from the anterior abdominal wall; **2** superior to the pubic bone; **3** on the superior surface of the urinary bladder; **4** from the bladder to the uterus (vesicouterine pouch); **5** on the fundus and body of the uterus, the posterior fornix, and the wall of the vagina; **6** between the rectum and uterus (rectouterine pouch); **7** on the anterior and lateral sides of the rectum; **8** posteriorly to become the sigmoid mesocolon.

the exterior of the body. For descriptive purposes, *the uterine tube is divided into four parts*: infundibulum, ampulla, isthmus, and uterine part.

The infundibulum (L. funnel) is the funnel-shaped lateral or distal end of the uterine tube, which is closely related to the ovary. Its opening into the peritoneal cavity is called the *abdominal ostium* (Figs. 3-27 and 3-46). About 2 mm in diameter, the ostium lies at the bottom of the infundibulum. Its margins have 20 to 30 *fimbriae* (L. fringes). These fingerlike processes spread over the surface of the ovary, and a large one, the *ovarian fimbria*, is attached to the ovary (Fig. 3-44). During ovulation the fimbriae trap the oocyte and sweep it through the abdominal ostium into the ampulla.

The ampulla of the uterine tube begins at the medial end of the infundibulum. It is in this tortuous part that fertilization of the oocyte by a sperm usually occurs. The ampulla is the widest and longest part of the uterine tube, making up over half of its length. *The isthmus* of the uterine tube is the short (about 2.5 cm), narrow, thick-walled part of the uterine tube that enters the cornu of the uterus. *The uterine (intramural portion) part* of the tube is the short segment that passes through the thick myometrium of the uterus and opens via the *uterine ostium* into the uterine cavity. This opening is smaller than the *abdominal ostium*.

The uterine tubes lie in the free edges of the broad ligaments of the uterus (Figs. 3-26A, 3-27, and 3-44). The part of the broad ligament attached to the uterine tube is called the mesentery of the tube or *mesosalpinx* (G. *salpinx*, a tube). The uterine tubes extend posterolaterally to the lateral walls of the pelvis, where they ascend and arch over the ovaries. Except for their uterine parts, the uterine tubes are clothed in peritoneum.

Arterial Supply of the Uterine Tubes (Figs. 3-24C and 3-25 to 3-28).   The arteries to the tubes are derived from the *uterine* and *ovarian arteries*. The tubal branches pass to the tube between the layers of the mesosalpinx.

Venous Drainage of the Uterine Tubes (Figs. 3-25 and 3-28).   The veins of the tubes are arranged similarly to the arteries and drain into the uterine and ovarian veins.

Lymphatic Drainage of the Uterine Tubes (Fig. 3-42).   The lymph vessels of the uterine tubes follow those of the fundus of the uterus and ovary and ascend with ovarian veins to the *aortic lymph nodes* in the lumbar region.

Innervation of the Uterine Tubes (Figs. 3-25 and 3-47).   The nerve supply of the uterine tubes comes partly from the *ovarian plexus* of nerves and partly from the *uterine plexus*. Afferent fibers from the tubes are contained in T11 and T12 and L1 nerves.

Usually the oocyte is fertilized in the ampulla of the uterine tube. As the dividing zygote passes slowly along it, it becomes a *morula*, which passes into the uterus (Moore, 1988). Fertilization of an oocyte cannot occur when both tubes are blocked because the sperms cannot reach the oocyte. *One of the major causes of infertility in women is blockage of the tubes* resulting from infection. Patency of a uterine tube may be determined by injecting a radiopaque material into the uterus, a radiographic procedure called *hysterosalpingogra-*

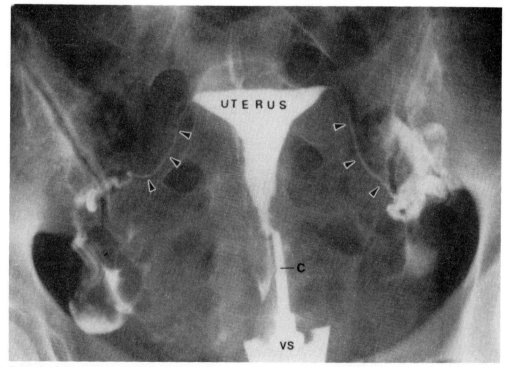

Figure 3-46.   Radiograph of the uterus and uterine tubes (*hysterosalpingogram*) after injection of radiopaque material into the uterus through the uterine ostium. The contrast medium has traveled through the triangular uterine cavity and uterine tubes (arrowheads), and has passed into the peritoneal cavity (lateral to the arrowheads). *C*, Indicates the catheter in the cervical canal. This radiograph illustrates that the female genital tract is in direct communication with the peritoneal cavity and is therefore a potential pathway for the spread of an infection from the vagina.

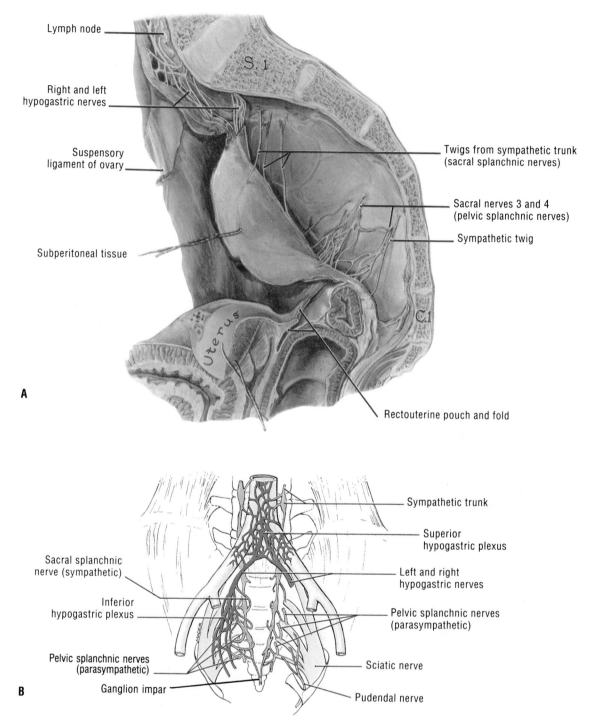

Figure 3-47. *A*, Dissection of a median section of the female pelvis showing the autonomic nerves. The rectum has been pulled anteriorly. *In the male*, the inferior hypogastric plexus is located on the side of the rectum, seminal vesicle, prostate, and posterior part of the bladder. *B*, The superior and inferior hypogastric plexuses. Observe the pudendal nerve, a branch of the sacral plexus, arising from the ventral primary rami of S2, S3, and S4. It is part of the somatic nervous system and innervates the perineal region (Figs. 3-63 and 3-74).

phy (Fig. 3-46). *Ligation of the uterine tubes* is one method of birth control. Oocytes discharged from the ovaries of these patients die in the uterine tube and soon disappear. Another technique is to block the tubes with removable plastic plugs. Over 90% of surgical sterilization is done by either abdominal tubal ligation or laparoscopic tubal ligation (Hannah, 1989). **Abdominal tubal ligation** is performed through a short pubic

incision made just at the pubic hairline (see Fig. 2-14*C*). **Laparoscopic tubal ligation** is done with an instrument known as a *laparoscope*. It is similar to a small telescope with a powerful light and is inserted through a small incision, usually near the umbilicus. A pubic incision is also made. For more information about these techniques, see Hannah (1989).

Because the female genital tract is in direct communication

with the peritoneal cavity via the abdominal ostia of the tubes (Fig. 3-46), infections of the vagina, uterus, and tubes may result in **peritonitis**. Conversely, inflammation of the uterine tube (*salpingitis*) may result from infections that spread from the peritoneal cavity. In some cases collections of pus may develop in the uterine tube (*pyosalpinx*) and the tube may be partly occluded by *adhesions*. In these cases the zygote may not be able to pass to the uterus and the blastocyst may implant in the mucosa of the uterine tube, producing an **ectopic tubal pregnancy**. Although implantation may occur in any part of the tube, the common site is in the ampulla. *Tubal pregnancy is the most common type of ectopic gestation*; it occurs about once in every 250 pregnancies in North America. Tubal pregnancy has increased in the last decade owing to an increase in sexually transmitted infections. If not diagnosed early (2 to 3 weeks after conception), ectopic tubal pregnancies may result in rupture of the uterine tube and hemorrhage into the abdominopelvic cavity during the first 8 weeks of gestation (Moore, 1988). Tubal rupture and the associated severe hemorrhage constitute *a threat to the mother's life*, and results in *death of the embryo*.

Occasionally the mesosalpinx between the uterine tube and the ovary contains embryonic remnants. The *epoöphoron* and *paraöphoron* are remnants of the cranial and caudal mesonephric tubules of the temporary embryonic kidney known as the *mesonephros* (Moore, 1988). There may also be a *duct of the epoöphoron*, which is a remnant of the mesonephric duct that forms the ductus deferens in the male. A *vesicular appendage* is sometimes attached to the infundibulum of the uterine tube. It represents the remains of the cranial end of the mesonephric duct that, in this region of the male, forms the *ductus epididymis*. Although these vestigial structures are mostly of embryological and morphological interest, they occasionally swell with fluid and form *cysts*, often becoming quite large.

The Ovaries (Figs. 3-25, 3-26, 3-34*A*, 3-40, and 3-44). In nulliparae (women who have not borne children), the ovaries are oval, almond-shaped, pinkish-white glands about 3 cm long, 1.5 cm wide, and 1 cm thick. Before puberty the surface of the ovaries is smooth. Thereafter it becomes progressively scarred and distorted owing to repeated ovulations, unless the woman has been taking *birth control pills*, which inhibit ovulation. The ovaries are located, one on each side, close to the lateral wall of the pelvis minor in a recess called the *ovarian fossa*. This fossa is bounded anteriorly by the medial umbilical ligament and posteriorly by the ureter and internal iliac artery. The anterior border of the ovary is attached to the posterior border of the broad ligament by a peritoneal fold called the *mesovarium*. The ampulla of the uterine tube curves over the lateral end of the ovary, and the infundibulum engulfs the ovary so that it can trap the oocyte at ovulation.

The superior (tubal) end of the ovary is connected to the lateral wall of the pelvis by the *suspensory ligament of the ovary* (Figs. 3-27 and 3-44). This ligament is a fold of the posterior layer of the broad ligament. The suspensory ligament *contains the ovarian vessels and nerves*, which pass into the mesovarium and the *hilum of the ovary* (Figs. 3-28 and 3-44). Each ovary is also attached to the uterus by a band of fibrous tissue, the *ligament of the ovary*, which runs in the mesovarium of the broad ligament. It connects the inferior (uterine) end of the ovary to the lateral angle of the uterus. This ligament is a remnant of the *guber-*

naculum in the embryo (Moore, 1988). *The surface of the ovary is not covered by peritoneum*; hence during ovulation the oocyte is expelled into the peritoneal cavity. However, its intraperitoneal life is short because it is trapped by the fimbriae of the tube and carried to the ampulla (Fig. 3-44). The surface of the ovary in young women is covered by cuboidal epithelium, which is continuous with the flattened mesothelium of the peritoneum forming the mesovarium.

Arterial Supply of the Ovaries (Figs. 3-24*C* and 3-26 to 3-28). The *ovarian arteries* arise from the abdominal aorta around the level of L2 vertebra and descend along the posterior abdominal wall. On reaching the pelvic brim, the ovarian arteries *cross over the external iliac vessels* and enter the suspensory ligaments. At the level of the ovary, the ovarian artery sends branches through the mesovarium to the ovary and continues medially in the broad ligament to supply the uterine tube. It anastomoses with the uterine artery.

Venous Drainage of the Ovaries (Figs. 3-25 and 3-28). The *ovarian veins* leave the hilum of the ovary and form a vinelike network of vessels, called the *pampiniform plexus* (L. *pampinus*, tendril + *forma*, form), in the broad ligament near the ovary and uterine tube. This plexus of veins communicates with the *uterine plexus of veins*. Each ovarian vein arises from the pampiniform plexus and leaves the pelvis minor with the ovarian artery. The right ovarian vein ascends to the inferior vena cava, whereas the left ovarian vein drains into the left renal vein.

Lymphatic Drainage of the Ovaries (Fig. 3-42). The lymph vessels follow the ovarian blood vessels and join those from the uterine tubes and the fundus of the uterus as they ascend to the *aortic lymph nodes* in the lumbar region.

Innervation of the Ovaries (Fig. 3-47). The nerves of the ovary descend along the ovarian vessels from the *ovarian plexus*. It is formed from the *aortic*, *renal*, and superior and inferior *hypogastric plexuses*. Nerves from the ovarian plexus supply the ovaries, broad ligaments, and uterine tubes. The parasympathetic fibers in the ovarian plexus are derived from the vagus nerves.

At ovulation, some women experience paraumbilical pain, called *mittelschmerz* (Ger. *mittel*, middle and *schmerz*, pain). As afferent impulses from the ovary reach the central nervous system through the dorsal root of T10 nerve, the pain is referred to its dermatome in the periumbilical area (see Figs. 1-24 and 2-13*A*). *The position of the ovaries varies* considerably in multiparae (women who have borne two or more children). During pregnancy the broad ligaments and ovaries are carried superiorly with the enlarging uterus. After childbirth the ovaries descend as the uterus contracts, but they may not return to their original locations. On the right side the vermiform appendix often lies close to the ovary and uterine tube (Fig. 3-25). This explains why a **ruptured tubal pregnancy** may be misdiagnosed as *acute appendicitis*. After menopause the formation of ovarian follicles, corpora lutea, and corpora albicantia ceases and the ovaries gradually atrophy. They become small and shriveled, as they appear in most female cadavers.

The Rectum

General Description. (Figs. 3-21, 3-33 to 3-35, 3-45, and 3-48) The rectum is the fixed terminal part of the large intestine. Continuous superiorly with the sigmoid colon, *the rectum begins*

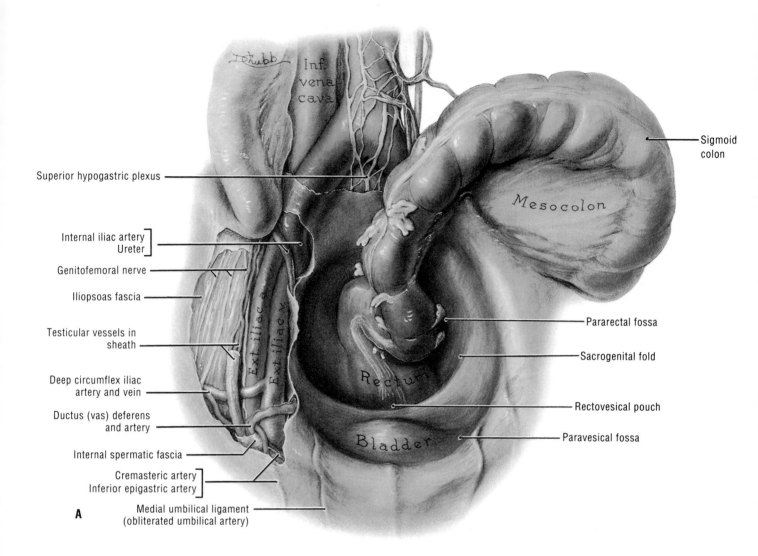

Inf. vena cava

Superior hypogastric plexus

Internal iliac artery
Ureter

Genitofemoral nerve

Iliopsoas fascia

Testicular vessels in
sheath

Deep circumflex iliac
artery and vein

Ductus (vas) deferens
and artery

Internal spermatic fascia

Cremasteric artery
Inferior epigastric artery

Medial umbilical ligament
(obliterated umbilical artery)

A

Ext iliaca
Ext iliaca v

Rectum

Bladder

Sigmoid
colon

Mesocolon

Pararectal fossa

Sacrogenital fold

Rectovesical pouch

Paravesical fossa

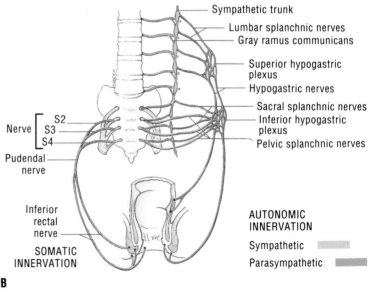

Sympathetic trunk

Lumbar splanchnic nerves
Gray ramus communicans

Superior hypogastric
plexus

Hypogastric nerves

Sacral splanchnic nerves

Inferior hypogastric
plexus

Pelvic splanchnic nerves

Nerve
S2
S3
S4

Pudendal
nerve

Inferior
rectal
nerve

SOMATIC
INNERVATION

AUTONOMIC
INNERVATION

Sympathetic

Parasympathetic

B

Figure 3-48. *A*, Anterosuperior view of a dissection of the
male pelvis. Observe the superior hypogastric plexus lying in
the bifurcation of the abdominal aorta. Note that the right ureter
adheres to the peritoneum, crosses the external iliac vessels,
and descends anterior to the internal iliac artery. *B*, Schematic
drawing illustrating the innervation of the rectum.

anterior to the level of the third sacral vertebra. The rectum (12 to 15 cm long) follows the curve of the sacrum and coccyx and terminates 3 to 4 cm anteroinferior to the tip of the coccyx. It ends by turning posteroinferiorly to become the anal canal. The *puborectalis muscle* (p. 254) forms a sling at the junction of the rectum and anal canal, producing the 90 degree *anorectal angle*. This U-shaped *puborectal sling* maintains this angle. The terminal part of the rectum has an anterior dilation known as the **rectal ampulla**, which is very distensible. The feces (L. *faex*, dregs) is held in the ampulla just before it is to be expelled during *defecation* (L. *defaecare*, to deprive of dregs). Inferiorly the rectum lies immediately *posterior to the prostate* in the male (Fig. 3-33A) and to *the vagina* in the female (Figs. 3-34 and 3-45). The termination of the rectum lies posterior to the *perineal body* in both sexes, and to the apex of the prostate in the male.

Peritoneal Relations of the Rectum (Figs. 3-14, 3-28, 3-33A, 3-34A, 3-45, and 3-48). Peritoneum covers the superior third of the rectum on its anterior and lateral surfaces. The middle third has peritoneum on its anterior surface only, and the inferior third has no peritoneal covering. *In the male* the peritoneum is reflected from the anterior surface of the rectum to the posterior wall of the bladder, where it forms the floor of the **rectovesical pouch**. In male children, at which time the bladder is in the abdomen, the peritoneum extends inferiorly as far as the base of the prostate. As the bladder moves into the pelvis during puberty, adult peritoneal relations are attained. *In the female* the peritoneum is reflected from the rectum to the posterior part of the fornix of the vagina, where it forms the floor of the **recto-uterine pouch** (of Douglas). *In both sexes*, lateral reflections of peritoneum from the rectum form *pararectal fossae* on each side of the rectum in its superior one-third (Fig. 3-48). These fossae permit the rectum to distend as it fills with feces. *The levator ani muscles support the ampulla of the rectum*, which stores the feces. These important muscles of the pelvic floor meet each other posteriorly in the *anococcygeal ligament* (body), a median musculotendinous structure that lies in the concavity of the anorectal flexure (Fig. 3-34A).

Posterior Relations of the Rectum (Figs. 3-14, 3-33, 3-34, 3-47, 3-48, and 3-50). Posteriorly the rectum rests on the inferior three sacral vertebrae, the coccyx, the anococcygeal ligament, the median sacral vessels, branches of the *superior rectal artery*, and the inferior ends of the *sympathetic trunks* and *sacral plexuses*. The rectum is surrounded by a fascial sheath and is loosely attached to the anterior surface of the sacrum.

Anterior Relations of the Rectum (Figs. 3-12, 3-33A, 3-34A, 3-39, 3-40, and 3-48). *In the male*, the rectum is related to the fundus of the urinary bladder, terminal parts of the ureters, deferent ducts (ductus deferentes), seminal vesicles, and the prostate. The two layers of the *rectovesical septum* lie in the median plane between the bladder and rectum and are closely associated with the seminal vesicles and the prostate. The rectovesical septum represents a potential cleavage plane between the rectum and prostate. *In the female* the rectum is related anteriorly to the vagina. It is separated from the posterior part of the fornix of the vagina and the cervix by the *rectouterine pouch*. Inferior to this pouch, the weak *rectovaginal septum* separates the vagina and rectum.

The Shape and Flexures of the Rectum (Figs. 3-20, 3-33, 3-34, and 3-48 to 3-50). The rectum (L. *rectus*, straight),

despite the origin of its name, is not straight.[5] Usually it has *three sharp flexures* or curves as it follows the sacrococcygeal curve. Its terminal part bends sharply in a posterior direction, where it joins the anal canal. The bend is the *anorectal flexure*. Although generally smaller in caliber than the sigmoid colon, the rectum increases in diameter as it passes inferiorly. Its terminal part, the rectal ampulla, which temporarily stores feces, is very distensible. *The rectum is S-shaped in the coronal plane*. At each of the three concavities, which are formed by the three flexures, there are infoldings of the mucous and submucous coats and most of the circular muscle layer, called **transverse rectal folds**. These folds partly close the lumen of the rectum. The largest of the three folds is located about 8 cm from the anus (Fig. 3-49A). Their form is maintained by prolongations of the *teniae coli* (longitudinal muscular bands) in the anterior and posterior walls of the rectum.

Arterial Supply of the Rectum (Figs. 3-24 and 3-50). The rectal arteries anastomose freely with each other. The continuation of the *inferior mesenteric artery*, called the **superior rectal artery**, supplies the terminal part of the sigmoid colon and the superior part of the rectum. Posterior to the superior end of the rectum, around the level of S3 vertebra, the superior rectal artery divides into two branches, which descend on each side of the rectum. Its right and left branches cross the left common iliac vessels and descend into the pelvis minor within the *sigmoid mesocolon* (Fig. 3-48; see also Fig. 2-81). The two **middle rectal arteries**, branches of the *internal iliac arteries*, supply the middle and inferior parts of the rectum. Branches of the *internal pudendal arteries*, the two **inferior rectal arteries**, originate in the ischioanal fossae. They supply the inferior part of the rectum. The rectum usually receives small branches posteriorly from the *median sacral artery* (see Fig. 2-95).

Venous Drainage of the Rectum (Figs. 3-24, 3-28, 3-29, and 3-51). The rectum is drained via superior, middle, and inferior rectal veins. There are many anastomoses between these vessels. The submucosal **rectal venous plexus** surrounds the rectum and communicates with the *vesical venous plexus* in the male and with the *uterovaginal venous plexus* in the female. The rectal venous plexus consists of two parts: (1) an *internal rectal venous plexus* that is just deep to the epithelium of the rectum and (2) an *external rectal venous plexus* that is external to the muscular coats of the wall of the rectum. The internal rectal plexus drains into the *superior rectal vein* but communicates freely with the external rectal venous plexus. The superior part of the external rectal venous plexus also drains into the *superior rectal vein*, the beginning of the inferior mesenteric vein. The inferior part of the external rectal venous plexus drains into the *internal pudendal vein*, and the middle part of the external rectal venous plexus drains into the *middle rectal vein* and then into the *internal iliac vein*. The superior rectal veins drain into the **portal system** (see Fig. 2-40), and the inferior rectal veins drain into systemic veins.

Lymphatic Drainage of the Rectum (Fig. 3-49B). Lymph vessels from the *superior half* or more of the rectum ascend along the superior rectal vessels to the *pararectal lymph nodes*, then pass to lymph nodes in the inferior part of the mesentery

[5]The rectum was originally named in monkeys in which it is straight (McMinn, 1990).

of the sigmoid colon, and from them to the *inferior mesenteric lymph nodes* and the **aortic lymph nodes**. The lymph vessels from the *inferior half* of the rectum pass superiorly with the middle rectal arteries and drain into the *internal iliac lymph nodes*.

Innervation of the Rectum (Figs. 3-16, 3-24*A*, 3-35, 3-47, and 3-48). The nerve supply to the rectum is derived from the sympathetic and parasympathetic systems. The *middle rectal plexus* is an offshoot from the **inferior hypogastric plexus**. Four

to eight nerves pass directly from this plexus to the rectum. The *parasympathetic nerves* are derived from S2, S3, and S4 nerves and run with the *pelvic splanchnic nerves* to join the inferior hypogastric plexus. The sensory nerves follow the path of the parasympathetic nerves; their fibers are stimulated by distention of the rectum.

The *rectovesical septum* in males (fascia of Denonvilliers) is important surgically (Fig. 3-19, 3-30, and 3-33). When resecting the rectum (*e.g.*, owing to cancer), the plane of this septum is located so that the prostate and urethra can be separated from the rectum. In this way they will not be damaged during excision of the rectum. The anastomosis of the arteries in the wall of the rectum is so extensive that the middle and inferior rectal arteries can supply the entire rectum if the inferior mesenteric artery (and thereby the superior rectal artery) has to be clamped (*e.g.*, during surgery for **colon cancer**). Because the superior rectal vein drains into the *portal venous system* (see Fig. 2-83) and the middle and inferior veins drain into the *systemic system*, this is an important area of *portacaval anastomosis* (Chap. 2, p. 209).

After anorectal surgery, men may be unable to ejaculate if there has been *damage to the pelvic splanchnic nerves* (Figs. 3-24*A*, 3-35, 3-36, and 3-47*B*). These nerves give off branches that pass anteriorly on the rectum to supply the bulbospongiosus muscles (Fig. 3-33*A*). **Ejaculation** consists of two phases: emission and ejaculation. *Emission of sperms* into the urethra follows reflex peristalsis in the deferent ducts and seminal vesicles and contraction of the smooth muscle in the prostate. These are sympathetic responses. *Ejaculation* (expulsion of semen) follows parasympathetic stimulation and is accompanied by clonic spasm of the bulbospongiosus muscles (p. 312).

Many of the structures related to the anteroinferior part of the rectum may be palpated through its walls (*e.g.*, the

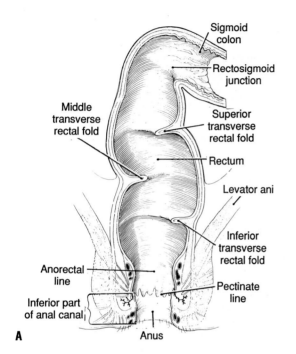

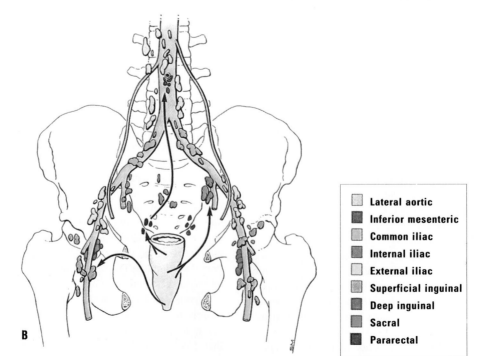

▨	**Lateral aortic**
▨	**Inferior mesenteric**
▨	**Common iliac**
▨	**Internal iliac**
▨	**External iliac**
▨	**Superficial inguinal**
▨	**Deep inguinal**
▨	**Sacral**
▨	**Pararectal**

Figure 3-49. *A*, Coronal section of the rectum and anal canal. Observe the acute flexion of the intestine at the rectosigmoid junction. The pectinate line is a clinically important vascular, lymphatic, and nerve boundary. *B*, Lymphatic drainage of the rectum and anal canal.

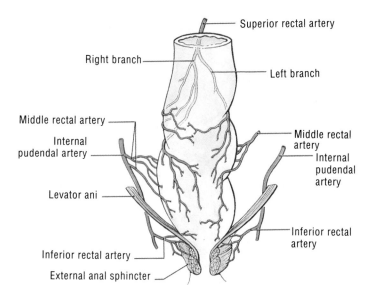

Figure 3-50. Anterior view of the arteries of the rectum and anal canal. In this specimen there are two right middle rectal arteries. Note that the inferior rectal arteries, which are branches of the internal pudendal arteries, mainly supply the inferior part of the anal canal. Observe the three sharp lateral flexures of the rectum, which produce the transverse rectal folds shown on the internal surface in Fig. 3-49. These flexures and folds help to support the weight of the feces.

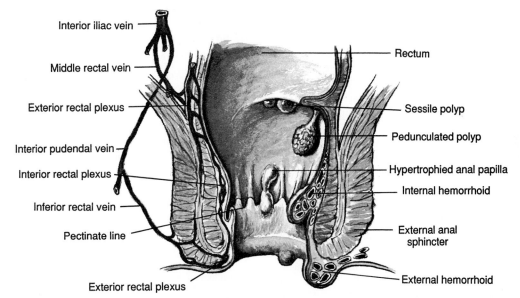

Figure 3-51. Coronal section of the rectum and anal canal, illustrating their venous drainage and various anorectal problems. Polyps are protruding growths from the mucous membrane. Internal hemorrhoids are covered with mucosa, whereas external hemorrhoids are covered with modified anal skin. (Reprinted with permission from Healey JE, Jr, Hodger J: *A Synopsis of Clinical Anatomy*, ed 2. Philadelphia, BC Decker, 1990.)

prostate and seminal vesicles in males [Fig. 3-39] and the cervix in females [Fig. 3-45]). *In both sexes* the pelvic surfaces of the sacrum and coccyx may be felt. The ischial spines and tuberosities may also be palpated (Figs. 3-1 and 3-6). Enlarged internal iliac lymph nodes, pathological thickening of the ureters, swellings in the ischioanal fossae (*e.g.*, ischioanal abscesses, Fig. 3-52), and abnormal contents in the *rectovesical pouch* in the male (Fig. 3-33A) or the *rectouterine pouch* in the female (Fig. 3-45) may be detected. Tenderness of an inflamed vermiform appendix can also be detected rectally if it hangs in the pelvis (see Fig. 2-79). The rectum can also be examined with a *proctoscope* (G. *proktos*, anus + *skopeo*, to view), and biopsies of lesions may be taken through

it. During insertion of a *sigmoidoscope* (p. 209), the curvatures of the rectum and the acute flexion at the rectosigmoid junction (Figs. 3-49A and 3-50) have to be kept in mind so that the patient will not undergo unnecessary discomfort. The operator must also know that the *transverse rectal folds* may temporarily impede passage of either of the above instruments.

Lymphatic Drainage of the Pelvis

The lymphatic drainage of the pelvis has been described in the sections on each of the pelvic organs (Figs. 3-33B, 3-34B,

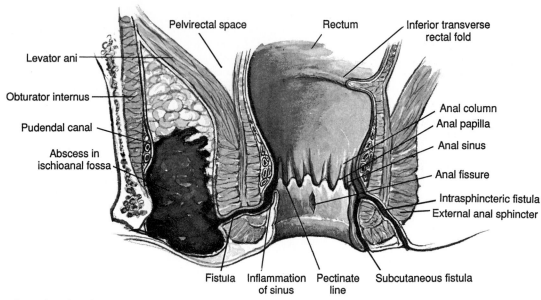

Figure 3-52.   Coronal section of the rectum and anal canal, illustrating an abscess in the right ischioanal fossa. (Reprinted with permission from Healey JE, Jr, Hodge J: *A Synopsis of Clinical Anatomy*, ed 2. Philadelphia, BC Decker, 1990).

3-43*B*, and 3-48*B*; see also Figs. 2-60 and 2-92). In general the pelvic organs drain through the *external and internal iliac lymph nodes* and the *sacral lymph nodes*. In addition, there are small lymph nodes between the layers of the broad ligament in females and in the fascial sheaths of the bladder and rectum in both sexes. Lymph drains from all these nodes to the **common iliac and lumbar aortic lymph nodes**.

The external iliac lymph nodes (8 to 10 in number) lie on the corresponding external iliac vessels and drain lymph from the lower limb, abdominal wall, bladder, and prostate (or uterus and vagina).

The internal iliac lymph nodes surround the internal iliac vessels and their branches. They receive lymph from all the pelvic viscera, deep parts of the perineum, and the gluteal and thigh regions.

The sacral lymph nodes lie along the median and lateral sacral arteries. They receive lymph from the posterior pelvic wall, rectum, neck of the bladder, and prostate (or cervix). Efferent vessels from these nodes pass to the medial nodes of the common iliac nodes.

The common iliac lymph nodes form two groups: (1) a lateral group lies along the common iliac vessels, and (2) a median group is located in the angle between these vessels (Figs. 3-42 and 3-49). The *lateral group of common iliac lymph nodes* receives lymph from the lower limb and pelvis via the external and internal iliac lymph nodes, whereas the *median group of common iliac lymph nodes* receives lymph directly from the pelvic viscera and indirectly through the internal iliac and sacral lymph nodes.

The lumbar aortic lymph nodes lie along the abdominal aorta and the inferior vena cava and receive lymph from the common iliac lymph nodes. The efferent vessels from the lumbar lymph nodes form right and left lumbar trunks, which drain into the *cisterna chyli* (see Fig. 1-70), a lymph sac about 5 cm long and 6 mm wide that lies on the first two lumbar vertebrae. From it lymph is carried via the *thoracic duct* through the thorax to the root of the neck, where it empties into the venous system (see Fig. 1-39).

The Pelvic Autonomic Nerves

The sacral sympathetic trunks are continuous with the lumbar sympathetic trunks posterior to the common iliac vessels (Figs. 3-16*A*; see also Figs. 2-82 and 2-99). The sacral trunks are smaller than the lumbar trunks and each of them has four ganglia. The trunks descend on the pelvic surface of the sacrum, just medial to the pelvic sacral foramina, and converge to form the small median *ganglion impar* (L. unequal, *i.e.*, unpaired) on the coccyx (Fig 2-99). The sympathetic trunks run in the presacral fascia, external to the peritoneum (Figs. 3-47 and 3-48).

Branches of the Sympathetic Trunks

The sympathetic trunks send gray rami communicantes to each of the ventral rami of the sacral and coccygeal nerves (Figs. 3-16, 3-36, and 3-47). They also send small branches to the median sacral artery and to the inferior hypogastric plexuses.

The Hypogastric Plexuses

The **superior hypogastric plexus** descends into the pelvis and lies just inferior to the bifurcation of the aorta (Figs. 3-36 and 3-47). The superior hypogastric plexus is the inferior prolongation of the *intermesenteric plexus* (see Fig. 2-64), which is joined by L3 and L4 splanchnic nerves. Branches from the superior hypogastric plexus enter the pelvis and descend anterior to the sacrum as the right and left *hypogastric nerves*. These nerves descend on the lateral walls of the pelvis where they mingle with the *pelvic splanchnic nerves* to form the right and left **inferior hypogastric plexuses** (Figs. 3-24*A*, 3-35, and 3-47).

The pelvic splanchnic nerves are parasympathetic and are derived from S2, S3, and S4 (Figs. 3-24*A*, 3-36, and 3-47). Hence, the inferior hypogastric plexuses contain both sympathetic and parasympathetic fibers. Each inferior hypogastric plexus surrounds the corresponding internal iliac artery. There are small ganglia within these plexuses and each plexus receives small branches from the *superior sacral ganglia* of the sympathetic trunks. Sympathetic and parasympathetic fibers contained in branches from the hypogastric plexuses are distributed to the pelvic viscera along the branches of the *internal iliac artery* (Fig. 3-24). The visceral plexuses are extensions of the inferior hypogastric plexuses in the walls of the pelvic viscera.

The middle rectal plexus arises from the superior part of the inferior hypogastric plexus and extends inferiorly as far as the internal anal sphincter. Branches from the middle rectal plexus pass directly to the rectum or along the middle rectal artery. Parasympathetic fibers also pass from this plexus to the sigmoid and descending parts of the colon.

The vesical plexus arises from the anterior part of the inferior hypogastric plexus and branches from it pass to the *urinary bladder* along the vesical arteries (Fig. 3-24). Branches from the vesical plexus also pass to the *seminal vesicles, deferent ducts,* and *prostate* (Fig. 3-36).

The prostatic plexus arises from the inferior part of the hypogastric plexus and is composed of rather large nerves, which enter the fundus and sides of the prostate. These nerves are also distributed to the seminal vesicles, ejaculatory ducts, urethra, bulbourethral glands, and penis (Fig. 3-36). The nerves supplying the corpora cavernosa of the penis, called *cavernous nerves,* arise from the anterior part of the prostatic plexus and join with branches from the pudendal nerve (Figs. 3-36 and 3-61). These nerves pass along the membranous urethra to the penis.

The uterovaginal plexus arises mainly from the part of the inferior hypogastric plexus that lies in the base of the broad ligament (Fig. 3-47). Some nerves from the uterovaginal plexus pass inferiorly to the vagina and cervix with the vaginal arteries; other nerves pass directly to the cervix or superiorly with the uterine arteries to the body of the uterus (Fig. 3-24*C*). Some of these nerves also supply medial parts of the uterine tube. *The uterovaginal plexus is homologous with the prostatic plexus.* Vaginal nerves from the uterovaginal plexus also supply the urethra, bulbs of the vestibule, greater vestibular glands, and clitoris (Figs. 3-40 and 3-60).

The Perineum

The perineum is the region of the trunk inferior to the pelvic diaphragm (Fig. 3-17; see also Fig. 2-29), which overlies the inferior pelvic aperture or pelvic outlet (Fig. 3-53*A*). Its posterior part, called the **anal triangle** or region, contains the anal canal with an ischioanal fossa on each side. Its anterior part, called the **urogenital triangle** or region, contains the external genitalia and terminal parts of the urogenital passages (*e.g.*, the urethra). In the anatomical position the perineum is a narrow area between the thighs, but when they are abducted, the perineum is a diamond-shaped area extending from the pubic symphysis to the tip of the coccyx[6] (Figs. 3-53 and 3-54). The term *perineum* is derived from the Greek word *perineos,* which means ''the space between the anus and scrotum.'' The perineum is of the highest importance to certain specialists: *obstetricians and gynecologists* who care for women during pregnancy and treat diseases of the female genital tract; *proctologists* who deal with the anorectal region and its diseases, and *urologists* who treat disorders of the urogenital organs.

Boundaries of the Perineum

The boundaries of the perineum, illustrated in Figs. 3-53 to 3-55, are: (1) the *pubic symphysis*; (2) the *inferior pubic rami*; (3) the *ischial rami*; (4) the *ischial tuberosities*; (5) the *sacrotuberous ligaments*; and (6) the *coccyx.* For descriptive purposes the perineum is divided into two unequal triangles by an imaginary traverse line joining the anterior ends of the ischial tuberosities (Figs. 3-53 and 3-54). The midpoint of this line is the central point of the perineum, which overlies the **perineal body** (central tendon of the perineum). The *anal triangle,* containing the anus, is posterior to this line and the *urogenital triangle,* containing the root of the scrotum and penis (or the external genitalia in the female), is anterior to this line.

The inferior pelvic aperture or pelvic outlet (Figs. 3-53 to 3-56) is closed, except where it transmits the urethra and anal canal; *in the female* it also transmits the vagina. The anterior part of the inferior pelvic aperture is closed by the *urogenital diaphragm* (Figs. 3-17 and 3-57 to 3-59*A*), and the posterior part is closed by the *pelvic diaphragm* (formed by the levator ani and coccygeus muscles [Figs. 3-14 to 3-18]). These two muscles form most of the *pelvic floor* (p. 252).

The Pelvic Diaphragm

The two levator ani muscles and two coccygeus muscles (pp. 254 and 255), which form the pelvic diaphragm (Fig. 3-58*A*), close the inferior pelvic aperture or pelvic outlet somewhat like a large funnel would if it were placed in the pelvic cavity (Figs. 3-7 and 3-17). The pelvic diaphragm divides the pelvic cavity into two parts: (1) a superior part containing the *pelvic viscera* (p. 267) and (2) an inferior part containing mainly fat, called the *ischioanal fossae.* The pelvic diaphragm forms the V-shaped floor of the pelvic cavity and the inverted V-shaped roof of each ischioanal fossa (Fig. 3-17). The fibromuscular *pelvic floor* and the muscles of the *pelvic diaphragm* are described on pp. 252 to 255.

> At one time the anal canal was considered to be part of the rectum. Therefore the wedge-shaped space on each side of it was called an *ischiorectal fossa.* Now that the anal canal is recognized as a separate part of the large intestine, it is more accurate to refer to these spaces as the *ischioanal fossae* because they are related to the anal canal. This is now their official name in *Nomina Anatomica* (Warwick, 1989).

---

[6]Obstetricians apply the term *perineum* to a more restricted region, the area between the vagina and anus.

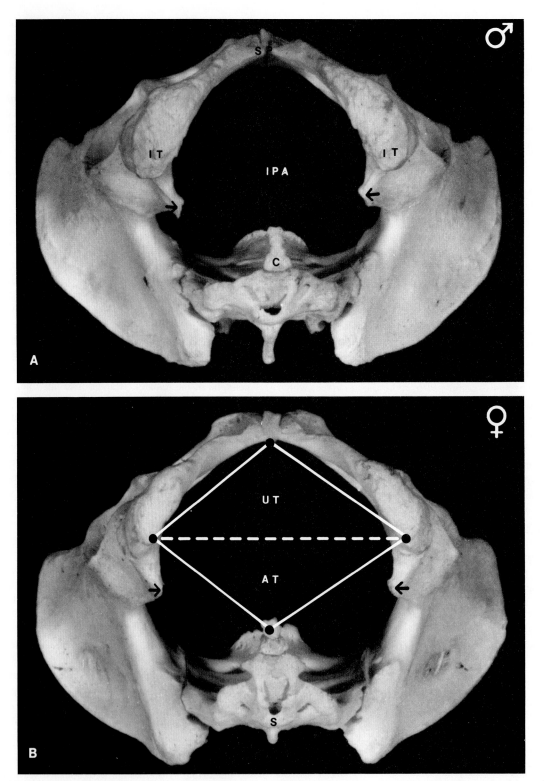

Figure 3-53.   The inferior pelvic apertures (pelvic outlets) of male and female pelves. Observe the sex difference in the size of the inferior pelvic aperture (*IPA*). The view of the female pelvis shown in *B* is the one that an obstetrician visualizes in his/her "mind's eye" when the patient is on the examining table. At the angles of the IPA are the symphysis pubis (*SP*), coccyx (*C*), and ischial tuberosities (*IT*). Note that the broken transverse white line between the right and left ischial tuberosities (*IT*) divides the diamond-shaped perineum into two triangles, the urogenital triangle (*UT*) and the anal triangle (*AT*). The arrows indicate the ischial spines. Note that the sacrum (*S*) is wedged between the iliac bones.

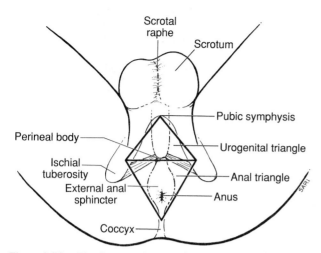

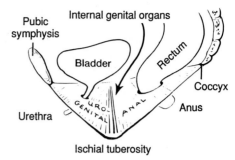

Figure 3-56. Median section of the male pelvis showing the urethra passing through the urogenital triangle and the anal canal traversing the anal triangle.

Figure 3-54. The diamond-shaped perineum, extending from the pubic symphysis to the coccyx. Note that a transverse line joining the anterior ends of the ischial tuberosities divides the perineum into two unequal triangular areas, the urogenital triangle anteriorly and the anal triangle posteriorly. The midpoint of the transverse line indicates the site of the perineal body (central perineal tendon).

stetricians, and gynecologists also consider the *urogenital diaphragm* to be part of the pelvic floor because when it is injured during childbirth, the pelvic floor is weakened and various types of herniation may occur (*e.g.*, cystocele; Fig. 3-15*C*). A compromise to these different views is to consider the pelvic diaphragm as the main pelvic floor and the urogenital as the subflooring.

Sphincter Urethrae Muscle

This muscle is attached to the medial surface of inferior pubic ramus, and its fibers pass medially toward the urethra, where they meet the fibers from the opposite side (Figs. 3-23, 3-33, and 3-58*C*). Some fibers encircle the membranous urethra in the male and form a true *voluntary sphincter that compresses the urethra*. It also extends to the base of the bladder and invests the prostate anteriorly and anterolaterally (Oelrich, 1980). *In the female* the inferior half of the sphincter urethrae blends with the anterolateral walls of the vagina, forming a *urethrovaginal sphincter* that compresses the urethra and vagina (Oelrich, 1983).

Innervation (Figs. 3-61 and 3-74*A*). *Perineal nerve*, which is a branch of the pudendal nerve (S2, S3, and S4).

Actions. This voluntary sphincter of the urethra constricts the membranous urethra *in the male* and compresses the urethra and vagina *in the female*.

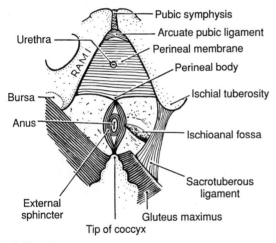

Figure 3-55. The boundaries of the perineum. Observe that the angles of the diamond-shaped region are at the arcuate pubic ligament, the tip of coccyx, and the ischial tuberosities.

Deep Transverse Perineal (Transversus Perinei) Muscle

This narrow slip of muscle is attached to the medial surface of the *ischial ramus* (Figs. 3-58*C*, 3-59*A*, and 3-60) and runs transversely to insert into the *perineal body* (central perineal tendon). In the female some fibers also insert into the vaginal wall.

Innervation (Figs. 3-61 and 3-74*A*). *Perineal nerve*, which is a branch of the pudendal nerve (S2, S3, and S4).

Action. This muscle steadies the perineal body, thereby contributing to the general supportive role of the urogenital diaphragm for the pelvic floor and viscera.

The Urogenital Diaphragm

This diaphragm is a thin sheet of striated muscle stretching between the two sides of the pubic arch (Figs. 3-17, 3-57, and 3-58*C*). It covers the anterior part of the inferior pelvic aperture or pelvic outlet. The most anterior and posterior fibers of the urogenital diaphragm (*deep transverse perineal muscles*) run transversely, whereas its middle fibers (*sphincter urethrae muscle*) surround the membranous urethra (Figs. 3-58*C*, 3-60, and 3-63).

Traditionally the *pelvic floor* has been considered to be the *pelvic diaphragm* formed by the two levatori ani and two coccygeus muscles and their fasciae. Some anatomists, ob-

The Perineal Body

The *tendinous center of the perineum* or perineal body is a small wedge-shaped mass of fibrous tissue located at the center

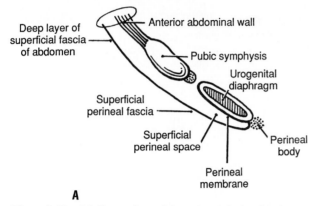

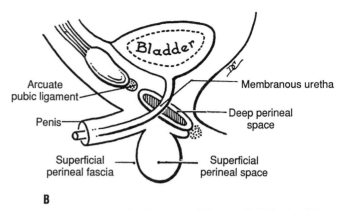

Figure 3-57. Median sections of the male pelvis showing the urogenital diaphragm and perineal spaces. Note that the superficial perineal fascia is a continuation of the deep layer of the superficial fascia of the abdomen.

of the perineum (Figs. 3-23, 3-35, 3-55, and 3-58*F*). The perineal body is the **landmark of the perineum** where several muscles converge (transverse perineal, bulbospongiosus, levator ani, and some fibers of the external anal sphincter).

> The perineal body is an especially important structure in women, particularly in those who bear children. Tearing or stretching of this tendinous center of the perineum during childbirth removes support from the inferior part of the posterior wall of the vagina (Figs. 3-34, 3-58*F*, and 3-60). As a result, *prolapse of the vaginal wall* through the vaginal orifice may occur. When tearing of the perineum, including the perineal body, appears inevitable during childbirth, an incision is often made in the perineum to enlarge the vaginal orifice (L. *introitus*). The reasoning is that a clean surgical incision is preferable to a jagged tear. This relaxing or appeasing incision is called an *episiotomy* (Fig. 3-62). In a *median episiotomy*, the incision starts posteriorly at the *frenulum of the labia minora* (Figs. 3-62*A* and 3-70) and extends through the skin, vaginal mucosa, perineal body, and superficial perineal muscle. The incision does not normally reach the external anal sphincter. Because there is a possibility of the incision tearing posteriorly and involving this sphincter, a *mediolateral episiotomy* is more common (Fig. 3-62*B*). The incision encounters the skin, vaginal wall, and bulbospongiosus muscle (Fig. 3-58*F*).

The Perineal Fascia

The urogenital diaphragm, like other muscles, is surrounded by deep fascia (Figs. 3-58 to 3-60, 3-63, and 3-64). The perineal fascia consists of two sheets, the *inferior and superior fasciae of the urogenital diaphragm*.

> The current concept of a urogenital diaphragm consisting of superior and inferior fascial layers enclosing the sphincter urethrae and deep transverse perineal muscles has been questioned (Oelrich, 1980; Woodburne and Burkel, 1988). It was decided to follow the traditional concept in this book until there is general agreement concerning this point.

The inferior fascia of the urogenital diaphragm, usually referred to as the **perineal membrane**, is continuous with the superior fascia of the urogenital diaphragm. It is also attached laterally to the pubic rami (Fig. 3-59*A*). The *superficial perineal fascia* (Figs. 3-57, 3-58, and 3-63), the membranous layer of the subcutaneous connective tissue of the perineum (Colles' fascia), is continuous with the membranous layer of the subcutaneous connective tissue of the inferior part of the anterior abdominal wall (p. 131). The *attachments of the superficial perineal fascia* are: (1) the *fascia lata* enveloping the thigh muscles (Figs. 3-64 and 3-71); (2) the *pubic arch*; and (3) the posterior edge of the *perineal membrane*. Anteriorly the superficial perineal fascia is prolonged over the penis and scrotum (Fig. 3-63), thereby forming a membranous covering for the testes and spermatic cords. *In the female* this fascia is prolonged over the clitoris and labia majora (Fig. 3-72).

The Superficial Perineal Space

This is the space between the superficial perineal fascia and the perineal membrane (Figs. 3-57 and 3-58). The superficial perineal fascia attaches medially to the superior border of the pubic symphysis and laterally to the body of the pubis.

Contents of the Superficial Perineal Space (Figs. 3-57 and 3-58).   *In the male* this space contains the root of the penis and the muscles associated with it, the contents of the scrotum, the proximal part of the spongy urethra, the superficial perineal muscles, branches of the internal pudendal vessels, and the pudendal nerves. *In the female* the contents of the superficial perineal space are: the root of the clitoris and the bulbs of the vestibule, the superficial perineal muscles, the related vessels and nerves, and the greater vestibular glands (p. 316).

> If there is **rupture of the male urethra** into the superficial perineal space, the attachments of the perineal fascia determine the direction of flow of the extravasated urine. Hence, it may pass into the loose connective tissue in the scrotum, around the penis, and superiorly into the anterior abdominal wall (Fig. 3-57). The urine cannot pass far into the thighs because the superficial perineal fascia blends with the *fascia lata* enveloping the thigh muscles (Fig. 3-71*A*), just distal to the inguinal ligament. In addition, urine cannot pass posteriorly into the anal triangle because the two layers of perineal fascia are continuous with each other around the superficial perineal muscles.

The Deep Perineal Space

This is the *fascial space enclosed by the superior and inferior fasciae of the urogenital diaphragm* (Figs. 3-18, 3-57, 3-58C, and 3-59A). The two layers of fascia are attached laterally to the pubic arch and blend with each other anteriorly at the apex and posteriorly at the base of the urogenital diaphragm. The deep dorsal vein of the penis enters the pelvis between its anterior edges (*transverse perineal ligament*) and the *arcuate pubic ligament* (Figs. 3-55, 3-57, and 3-59).

Contents of the Deep Perineal Space (Figs. 3-18, 3-58 C, and 3-61). *In the male* the deep perineal space contains the membranous urethra, sphincter urethrae muscle, bulbourethral glands, deep transverse perineal muscles, and related vessels and nerves (*e.g.*, the dorsal nerve of the penis). *In the female* the deep perineal space contains part of the urethra, the sphincter urethrae muscle, the deep transverse perineal muscle, and related vessels and nerves (*e.g.*, the dorsal nerve of the clitoris).

The Anal Triangle

This region of the perineum is posterior and faces inferoposteriorly (Figs. 3-17, 3-19, 3-53 to 3-57A, 3-58, 3-63, and 3-64). The *boundaries of the anal triangle* are: posteriorly the *tip of the coccyx* and anteriorly the imaginary line joining the anterior ends of the *ischial tuberosities*. It is also related anteriorly to the posterior border of the urogenital diaphragm and posterolaterally to the sacrotuberous ligaments. Overlying the anal triangle are the gluteus maximus muscles. The anal triangle contains the *anal canal*, *external anal sphincter*, and *ischioanal fossae*. The anal canal passes through the floor of the pelvis and opens on the surface of the perineum as the anus. The *perianal skin* contains large sebaceous and sweat glands and is pigmented (Fig. 3-58G). This dark skin is thrown into radiating folds, giving it a characteristic puckered appearance. The folds are produced by the pull of the underlying fibroelastic septa (Fig. 3-21A).

The Anal Canal

This canal, about 4 cm long in adults, is the terminal and most inferior part of the large intestine (Figs. 3-7, 3-20, 3-21, 3-33, 3-34, 3-49, and 3-63). It begins where the rectal ampulla narrows abruptly at the level of the *U-shaped sling* formed by the puborectalis muscle and ends at the **anus**, the external outlet of the GI tract (Figs. 3-60 and 3-64). The anus is contracted and forms an anteroposterior slit, except during defecation. The anal canal, surrounded by internal and external anal sphincters, descends posteroinferiorly between the *anococcygeal ligament* and the *perineal body* (Fig. 3-23). It is also surrounded by the *levator ani muscles*, which form the main part of the pelvic diaphragm.

> Because the anal canal slopes inferoposteriorly, the examining finger or an instrument introduced into the anal canal (*e.g.*, a sigmoidoscope) should be directed toward the umbilicus (Figs. 3-21, 3-33, and 3-39).

The Interior of the Anal Canal (Figs. 3-21, 3-33A, 3-34A, 3-49A, and 3-52). The superior half of the mucous membrane of the anal canal is characterized by a series of longitudinal ridges or folds called *anal columns* (rectal columns). These columns contain the terminal branches of the superior rectal artery and vein. It is here that the superior rectal veins of the portal system anastomose with the middle and inferior rectal veins of the caval system (see Fig. 2-83).

The anorectal line, indicated by the superior ends of the anal columns (Figs. 3-21B, 3-33A, and 3-49A), is where the rectum joins the anal canal. The inferior ends of these columns are joined to each other by *anal valves*, which are semilunar folds of epithelium. Superior to the valves are small recesses called *anal sinuses* (Fig. 3-52). When compressed by feces, the mucous-containing anal sinuses exude mucus, which aids in evacuation of the anal canal. The inferior comb-shaped limit of the anal valves forms an irregular line known as the pectinate line.

The pectinate line (Fig. 3-49) indicates the junction of the superior part of the anal canal (derived from the embryonic hindgut) and the inferior part (derived from the embryonic anal pit or proctodeum). The pectinate line (dentate line, mucocutaneous line) also approximates the line of junction of the columnar epithelium of the superior part of the anal canal and the stratified squamous epithelium of the inferior part. At the anus the moist, hairless mucosa of the anal canal becomes dry, hairy skin (Fig. 3-21B). The anal canal superior to the pectinate line differs from the part inferior to the pectinate line in its arterial supply, innervation, and venous and lymphatic drainage. These differences result from their different embryological origins (p. 302).

Arterial Supply of the Anal Canal (Figs. 3-10, 3-24, and 3-50; see also Fig. 2-81). The *superior rectal artery* supplies the part of the anal canal superior to the pectinate line. This artery is the pelvic continuation of the inferior mesenteric artery. The terminal branches of the superior rectal artery run distally in the anal columns and form anastomotic loops in the anal valves. The two *inferior rectal arteries* supply the inferior part of the anal canal (*i.e.*, inferior to the pectinate line), as well as the surrounding muscles and perianal skin. The *middle rectal arteries* assist with the blood supply to the anal canal by forming anastomoses with the superior and inferior rectal arteries.

Venous Drainage of the Anal Canal (Figs. 3-24, 3-49, and 3-51; see also Fig. 2-40). The *internal rectal venous plexus* drains in both directions from the level of the pectinate line. **Superior to the pectinate line**, the internal rectal plexus drains chiefly into the *superior rectal vein*, a tributary of the inferior mesenteric vein. **Inferior to the pectinate line**, the internal rectal plexus drains into *inferior rectal veins* around the margin of the external anal sphincter. The *middle rectal veins*, tributaries of the internal iliac veins, mainly drain the muscularis externa of the ampulla of the rectum. They form anastomoses with the superior and inferior rectal veins.

Lymphatic Drainage of the Anal Canal (Fig. 3-49B). *Superior to the pectinate line*, the lymph vessels drain into the *internal iliac lymph nodes* and through them into the common iliac and lumbar lymph nodes. *Inferior to the pectinate line*, the lymph vessels drain into the *superficial inguinal lymph nodes*.

Innervation of the Anal Canal (Figs. 3-24A, 3-47, 3-48, and 3-61). The nerve supply *superior to the pectinate line* is from the *inferior hypogastric plexus* (sympathetic) and the pelvic

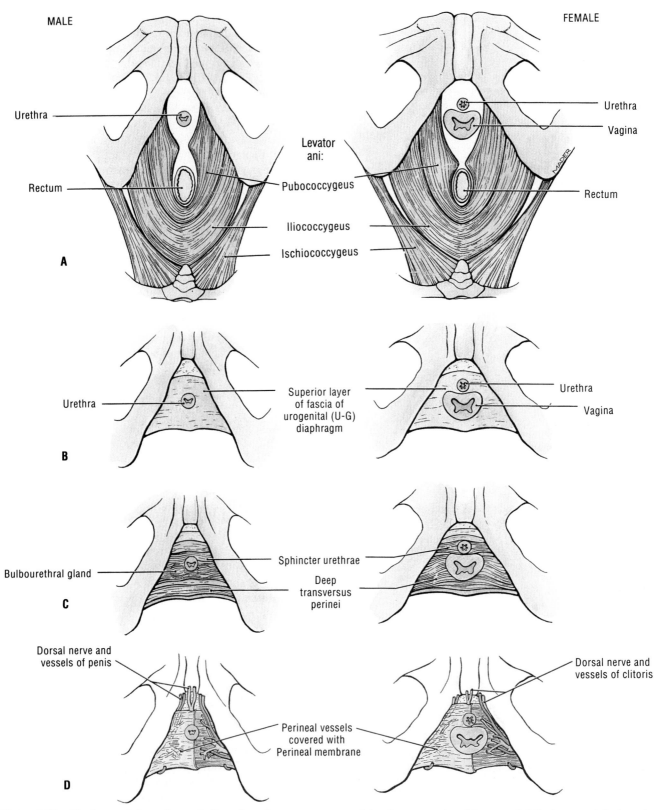

Figure 3-58. The layers of the perineum, built up from deep to superficial. In *A*, the angle between the two ischiopubic rami is almost filled by the muscles of the pelvic diaphragm. The urethra (and vagina) pass through anteriorly, the rectum posteriorly. A superior layer of fascia *B* and an inferior layer of fascia *D* enclose a deep perineal space *C* containing the two muscles of the urogenital diaphragm and, *in the male*, the bulbourethral glands. The superficial perineal space contains the structures shown in *E* and *F*. *G*, shows an overview of the urogenital diaphragm and related structures.

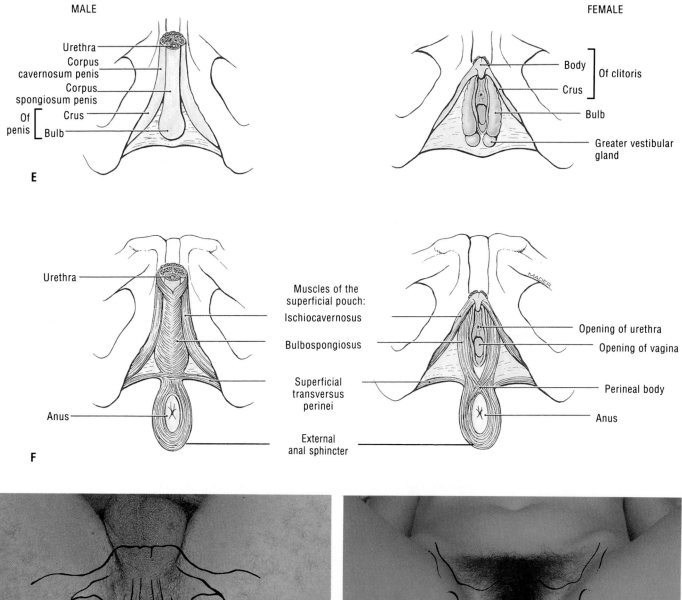

MALE

Urethra
Corpus cavernosum penis
Corpus spongiosum penis
Of penis { Crus
Of penis { Bulb

E

FEMALE

Body } Of clitoris
Crus } Of clitoris
Bulb
Greater vestibular gland

Urethra

Muscles of the superficial pouch:

Ischiocavernosus

Bulbospongiosus

Superficial transversus perinei

External anal sphincter

Anus

F

Opening of urethra
Opening of vagina

Perineal body

Anus

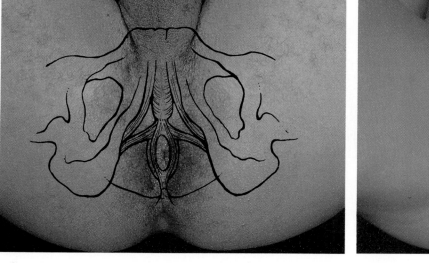

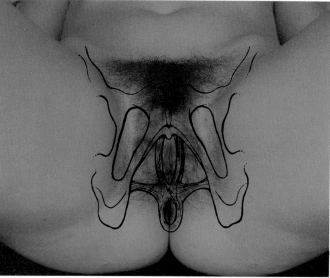

G

Figure 3-58E–G

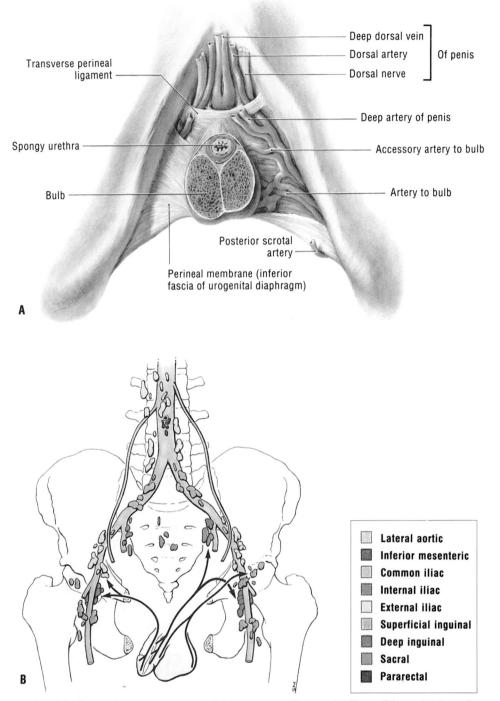

Figure 3-59. *A*, Dissection of the deep perineal space in a male pelvis. The crura of the penis have been removed. On the left side the perineal membrane is partly removed to show the structures in the deep perineal space. Observe the fibers of the perineal membrane converging on the bulb of the penis and mooring it to the pubic arch. *B*, Lymphatic drainage of the penis, scrotum, and spongy urethra.

splanchnic nerves (parasympathetic). The sympathetic nerves pass mainly along the inferior mesenteric and superior rectal arteries, whereas the parasympathetic nerves (from S2 to S4) run in the pelvic splanchnic nerves to join the *inferior hypogastric plexus*. The superior part of the anal canal is sensitive only to stretching. The nerve supply of the anal canal **inferior to the pectinate line** is derived from the *inferior rectal nerves*, branches of the *pudendal nerve*. This part of the anal canal is sensitive to pain, touch, and temperature.

> The anal canal superior to the pectinate line develops from the hindgut (*endoderm*), as does the rectum (Moore, 1988), whereas the anal canal inferior to the pectinate line is derived from the proctodeum (*ectoderm*). The pectinate line, indicated

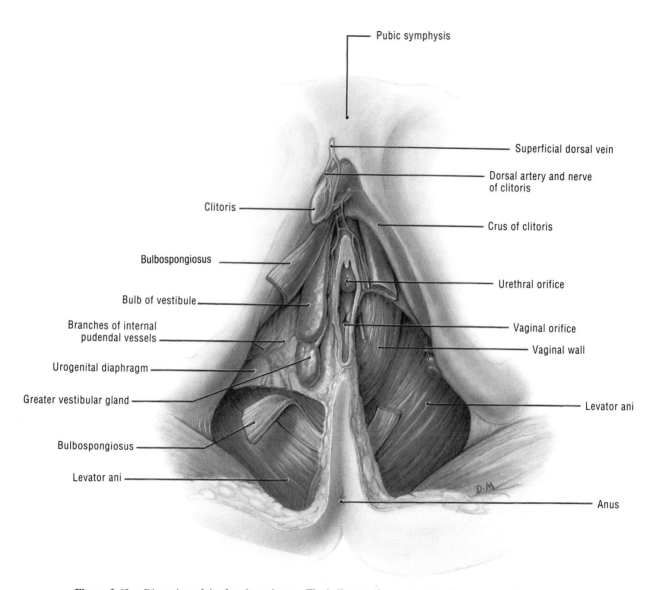

Pubic symphysis

Superficial dorsal vein

Dorsal artery and nerve of clitoris

Clitoris

Crus of clitoris

Bulbospongiosus

Urethral orifice

Bulb of vestibule

Branches of internal pudendal vessels

Vaginal orifice

Vaginal wall

Urogenital diaphragm

Greater vestibular gland

Levator ani

Bulbospongiosus

Levator ani

Anus

Figure 3-60.   Dissection of the female perineum. The bulbospongiosus muscle is divided and reflected on the right side, and is largely excised on the left side.

by the inferior limit of the anal valves, is located at the approximate site of the *anal membrane* in the embryo. Because of its hindgut origin, the superior part of the anal canal is supplied by the *superior rectal artery*, the terminal branch of the inferior mesenteric artery (*hindgut artery*), whereas the inferior part of the canal derived from the proctodeum is supplied by the *inferior rectal arteries*, branches of the *internal pudendal artery*. The venous and lymphatic drainage and the nerve supply of these regions also differ because of the different embryological origins of the superior and inferior parts of the anal canal. The change between the columnar or cuboidal epithelium in the superior part of the anal canal and the stratified epithelium in the inferior part occurs at or close to the pectinate line. Carcinomas (CA) developing from the two types of epithelium differ. This is related to the two different types of germ layer (endoderm in the superior part and surface ectoderm in the inferior part).

Internal hemorrhoids (''piles'') are varicosities of the

tributaries of the superior rectal veins and are *covered by mucous membrane* (Fig. 3-51). *External hemorrhoids* (G. very likely to bleed) are varicosities of the tributaries of the inferior rectal veins and are *covered by skin*. As there are multiple anastomoses between the venous plexuses of the rectal veins, these communicating veins may also be dilated. Hemorrhoids that prolapse through the external anal sphincter are often compressed, impeding blood flow. As a result they tend to ulcerate and strangulate. *Thrombus (clot) formation* is more common in external than in internal hemorrhoids.

The anastomoses between the superior, middle, and inferior rectal veins form clinically important *communications between the portal and systemic systems* (see Fig. 2-83). The superior rectal vein drains into the inferior mesenteric vein, whereas the middle and inferior rectal veins drain through the systemic system into the inferior vena cava. Any abnormal increase in pressure in the valveless portal system may cause enlargement of the superior rectal veins, resulting in *internal hemorrhoids*. In portal hypertension, as in *hepatic cirrhosis*, the anastomotic veins in the anal canal and elsewhere become varicose and may rupture (Fig. 3-51; see also Fig. 2-83).

In *chronically constipated persons*, the anal valves and mucosa may be torn by hard feces. A slitlike lesion, known as an *anal fissure* (Fig. 3-52), is usually inferior to the anal valves and is very painful because this region is supplied by sensory fibers of the inferior rectal nerves (Fig. 3-61). *Perianal abscesses* (collections of pus) may follow infection of anal fissures and the infection may spread to the ischioanal fossae or into the pelvis, forming *ischioanal and pelvirectal abscesses*, respectively (Fig. 3-52). An *anal fistula* may develop owing to the spread of an infection. One end of this abnormal canal opens into the *anal canal* and the other opens into an abscess in the *ischioanal fossa* or into the perianal skin.

As the anal canal superior to the pectinate line is supplied by *autonomic nerves*, an incision or a needle insertion in this region is painless. However, the anal canal inferior to the pectinate line is very sensitive (*e.g.*, to the prick of a hypodermic needle) because it is supplied by the *inferior rectal nerves*, which contain sensory fibers.

The External Anal Sphincter (Figs. 3-19, 3-21*A*, 3-22, 3-33*A* to 3-35, 3-58*F*, and 3-63).   This large voluntary sphincter is located in the perineum where it surrounds the inferior two-thirds of the anal canal. It forms a broad band (2 to 3 cm wide) on each side of the anal canal, which consists of *subcutaneous, superficial*, and *deep parts*. Many branches of the inferior rectal nerve and vessels pass between the superficial and deep parts of the muscle. Fibers of the external anal sphincter run from the perineal body to the coccyx and anococcygeal ligament. *The anal canal has two sphincters*, external and internal; both must relax before defecation can occur. The superior two-thirds is surrounded by the involuntary *internal anal sphincter* and is supported by the levator ani muscles (Figs. 3-49 and 3-50). The inferior two-thirds is surrounded by the voluntary *external anal sphincter*, which blends superiorly with the puborectalis muscle, part of the levator ani (Figs. 3-20*B*, 3-21*A*, 3-23, and 3-33). The external anal sphincter partly overlaps the inferior part of the internal anal sphincter.

Although the external anal sphincter is commonly considered to consist of three parts, they ae not distinct. For this reason some anatomists regard it as one muscular mass (Ayoub, 1979), whereas others describe only superficial and deep portions (Woodburne and Burkel, 1988). It was decided to follow the usual description in this book until there is general agreement concerning this point.

Parts of the External Anal Sphincter (Figs. 3-19 to 3-21*A*, 3-33, 3-35, and 3-63).   The *subcutaneous part*, its most inferior part, is slender and surrounds the anus. Its anular fibers, which cross anterior and posterior to the anus, have no bony attachments. The *superficial part* is elliptical or oval in shape. Its fibers extend anteriorly from the tip of the coccyx and anococcygeal ligament around the anus to the perineal body, mooring the anus in the median plane. The *deep part*, also anular, surrounds the anal canal like a collar. Some of its fibers cross to join the opposite superficial transverse perineal muscle. The deep

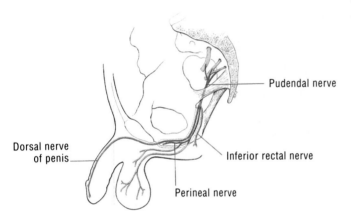

Figure 3-61.   The pudendal nerve, showing in color the five regions in which it runs. It supplies the skin, organs, and muscles of the perineum. It is therefore concerned with micturition, defecation, erection, ejaculation and in the female, parturition. Although the pudendal nerve is shown here in the male, its distribution is similar in the female because the parts of the female perineum are homologous.

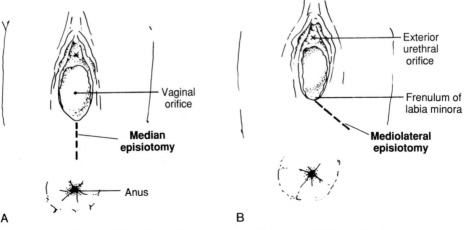

Figure 3-62.   Types of episiotomy. *A*, Median. *B*, Mediolateral.

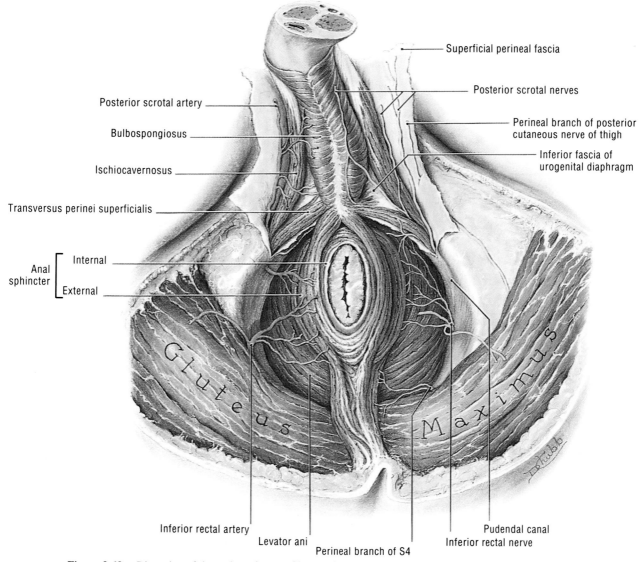

Figure 3-63.   Dissection of the male perineum. Observe the anus surrounded by the external anal sphincter with an ischioanal fossa on each side.

part arises from the perineal body and *blends with the puborectalis muscle*, part of the levator ani.

Innervation (Figs. 3-10 and 3-63).   Inferior rectal nerve and perineal branch of S4.

Actions.   The external anal sphincter assists in closing the anal canal and anus. It draws the canal anteriorly, thereby increasing the anorectal angle. The deep part of the sphincter is assisted in this action by the puborectalis muscle, part of the levator ani (Fig. 3-20*B*).

The Internal Anal Sphincter (Figs. 3-19, 3-21*A*, 3-24, 3-33, and 3-47).   This *involuntary sphincter*, surrounding the superior two-thirds of the anal canal, is formed by a thickening of the circular muscle layer of the intestine. It is innervated by the *pelvic splanchnic nerves* (parasympathetic). This sphincter reacts to the pressure of feces in the ampulla of the rectum.

The Ischioanal Fossae

On each side of the anal canal is a large fascia-lined, wedge-shaped space called the ischioanal fossa (ischiorectal fossa). It is located between the skin of the anal region and the pelvic diaphragm (Figs. 3-17, 3-18, 3-55, 3-63, 3-64, and 3-71). The apex of each ischioanal fossa lies superiorly, at the point where the levator ani muscle arises from the obturator fascia. Because the two levator ani are shaped like a funnel, the ischioanal fossae are wide inferiorly and narrow superiorly. The apex of each fossa is located about 6 cm superior to the ischial tuberosity. The base of each fossa is formed by perianal skin.

Anteriorly, the ischioanal fossae continue superior to the urogenital diaphragm where they form anterior recesses of the ischioanal fossae. These spaces are filled with loose connective

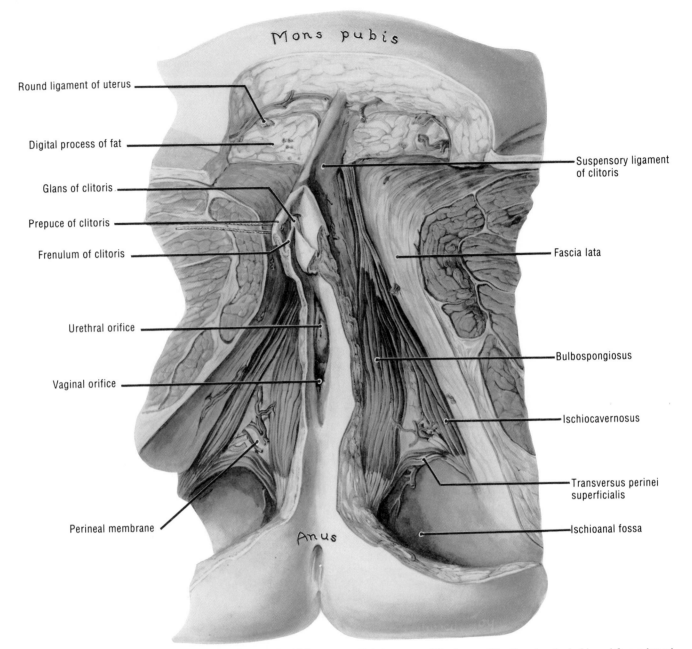

Mons pubis

Round ligament of uterus

Digital process of fat

Glans of clitoris

Prepuce of clitoris

Frenulum of clitoris

Urethral orifice

Vaginal orifice

Perineal membrane

Anus

Suspensory ligament of clitoris

Fascia lata

Bulbospongiosus

Ischiocavernosus

Transversus perinei superficialis

Ischioanal fossa

Figure 3-64. Dissection of the female perineum. Observe the thickness of the superficial fatty tissue in the mons pubis and the encapsulated tissue. There are also posterior recesses where the gluteus maximus muscles overhang the ischioanal fossae. *The two ischioanal fossae communicate* over the anococcygeal ligament (Figs. 3-23 and 3-34A). Posteriorly each fossa is also continuous with the lesser sciatic foramen, superior to the *sacrotuberous ligament* (Figs. 3-8A and 3-55).

digital process of fat deep to this. Examine the ischioanal fossae lateral to the anal canal.

Boundaries of the Ischioanal Fossae (Figs. 3-19, 3-55, and 3-63). Each fossa is bounded: *laterally* by the ischium and the inferior part of the obturator internus muscle; *medially* by the anal canal to which the levator ani and external anal sphincter are applied; *posteriorly* by the sacrotuberous ligament and gluteus maximus muscle; and *anteriorly* by the base of the urogenital diaphragm and its fasciae.

Contents of the Ischioanal Fossae (Figs. 3-17, 3-21, 3-61, and 3-63). These wedge-shaped fascial spaces, filled with soft fat called the *ischioanal pads of fat*, are traversed by many tough, fibrous bands and septa. These fibrofatty pads support the anal canal, but they can be readily displaced to permit the anal canal to expand when feces is present. The ischioanal fossae also contain the *internal pudendal artery and vein* and the *pudendal nerve*. These structures run on the lateral walls of the fossae in fibrous canals called the *pudendal canals*. Posteriorly these vessels and the pudendal nerve give off the inferior rectal vessels and nerves, which pass through the ischioanal fossae. These structures become superficial as they pass toward the surface to supply the external anal sphincter and the perianal skin.

Two other cutaneous nerves, the perforating branch of S2 and S3 nerves and the perineal branch of S4 nerve, also pass through the ischioanal fossae.

> The ischioanal fossae are occasionally the sites of infection, which may result in the formation of *ischioanal abscesses*. These collections of pus are annoying and painful (Fig. 3-52). Infections may reach the ischioanal fossae in the following ways: (1) following *cryptitis* or inflammation of the anal sinuses; (2) from extension of a *pelvirectal abscess*; (3) following a tear in the anal mucous membrane; or (4) from a penetrating wound in the anal region. *Diagnostic signs of an ischioanal abscess* are fullness and tenderness between the anus and the ischial tuberosity. An ischioanal abscess may spontaneously open into the anal canal, rectum, or perianal skin. Because the ischioanal fossae communicate, an abscess in one fossa may spread to the other one and form a semicircular abscess around the posterior aspect of the anus.

The Pudendal Canal

This fibrous tunnel is formed by splitting of the obturator fascia and is found on the lateral wall of the ischioanal fossa. The internal pudendal artery and vein and the pudendal nerve enter this canal at the lesser sciatic notch, inferior to the ischial spine (Figs. 3-17, 3-61, and 3-63). The pudendal canal begins at the posterior border of the ischioanal fossa and runs from the lesser sciatic notch adjacent to the ischial spine to the posterior edge of the urogenital diaphragm. There are three structures at the posterior end of the pudendal canal: the *internal pudendal artery*, the *internal pudendal vein*, and the *pudendal nerve* (Figs. 3-10 and 3-61).

The pudendal nerve supplies most of the innervation to the perineum. Toward the distal end of the pudendal canal, the pudendal nerve splits to form the *dorsal nerve of the penis* (or clitoris) and the *perineal nerve* (Figs. 3-59 to 3-61). These nerves run anteriorly on each side of the internal pudendal artery. The perineal nerve gives off scrotal (or labial) branches and continues to supply the muscles of the urogenital diaphragm (Fig. 3-10). The *dorsal nerve of the penis* (or clitoris), which is a sensory nerve, runs through the deep perineal space (Figs. 3-57*B* and 3-61) to reach its area of supply. The *inferior rectal vein* passes from the inferior end of the anal canal to empty into the *internal pudendal vein* in the pudendal canal. This vein anastomoses with the superior rectal veins of the *portal venous system* (see Fig. 2-83).

The Male Perineum

The male perineum includes the anal canal (p. 299), the membranous and spongy parts of the urethra, and the root of the scrotum and penis.

The Membranous Urethra (Figs. 3-10, 3-17, 3-29, 3-30, 3-33, 3-37, and 3-38). The prostatic urethra, the first part, is described in the section on the pelvis (p. 277). The membranous urethra, the second part, is the shortest (1 to 2 cm long), thinnest, and (except for the external urethral orifice) narrowest portion of the urethra. It begins at the apex of the prostate and ends at the *bulb of the penis* where it joins the spongy urethra. The

membranous urethra traverses the deep perineal space where it is surrounded by the sphincter urethrae muscle and the perineal membrane (inferior fascia of the urogenital diaphragm). Posterolateral to the membranous urethra, on each side, is a small *bulbourethral gland* and its slender duct (Fig. 3-18).

> The narrowness of the membranous urethra results from contraction of the sphincter urethrae muscle. This circular investment of muscle also makes the membranous urethra the least distensible part. Owing to its thin wall, the inferior part of it is vulnerable to penetration by a urethral catheter or to rupture during an accident.

The Scrotum

The contents of the scrotum, the testes, and their coverings (Fig. 3-33) are described in the section on the abdomen (see Chap. 2, p. 149). As stated there, the scrotum develops from an outpouching of the skin of the anterior abdominal wall (see Figs. 2-12 and 2-15; Table 2-2, p. 147). The scrotum is a loose **cutaneous fibromuscular sac** that is situated posteroinferior to the penis and inferior to the pubic symphysis (Figs. 3-33, 3-54, 3-56, and 3-65). Its bilateral formation is indicated by the midline *scrotal raphe* (Fig. 3-54), which continues on the ventral surface of the penis as the *penile raphe* and posteriorly along the median line of the perineum to the anus as the *perineal raphe*. The scrotum is composed of skin and dartos muscle (see Fig. 2-12). The *dartos muscle*, firmly attached to the skin, consists largely of smooth muscle fibers that contract under the influence of cold, exercise, and sexual stimulation. Under these conditions the wall of the scrotum becomes contracted and firm, and its skin becomes rugose.

> Contraction of the dartos and cremasteric muscles (p. 147) causes the testes to be drawn up against the body. In hot weather the scrotum relaxes and allows the testes to hang freely away from the body. This provides a larger skin surface for the dissipation of heat. These reflexes of the scrotum in response to temperature help to maintain a stable temperature, an important function because spermatogenesis (formation of sperms) is impaired by extremes of heat or cold. In older men the dartos muscle loses its tone and the scrotum tends to be smoother and to hang down further.

Arterial Supply of the Scrotum (Figs. 3-24*A*, 3-59*A*, 3-63, and 3-66). The *external pudendal arteries* supply the anterior aspect of the scrotum, and the *internal pudendal arteries* supply its posterior aspect. It also receives branches from the *testicular* and *cremasteric arteries*.

Venous Drainage of the Scrotum (Fig. 3-66). The scrotal veins accompany the arteries and join the external pudendal veins.

Innervation of the Scrotum (Figs. 3-18, 3-61, and 3-63). Its anterior part is supplied by the *ilioinguinal nerve*, and its posterior part by the medial and lateral scrotal branches of the *perineal nerve* and the perineal branch of the *posterior femoral cutaneous nerve*.

Lymphatic Drainage of the Scrotum (Figs. 3-59*B* and 3-73; see also Fig. 2-22). The lymph vessels from the scrotum drain into the *superficial inguinal lymph nodes*.

The Penis

The penis (L. tail) is the male organ of copulation and the common outlet for urine and semen (Figs. 3-33, 3-38, 3-59, and 3-65 to 3-69). It is composed of three cylindrical bodies (L. *corpora*) of erectile cavernous tissue that are enclosed by a dense white fibrous capsule, the *tunicia albuginea*. Superficial to this layer is the *deep fascia of the penis* which forms a common covering for the two corpora cavernosa and the corpus spongiosum. The skin of the penis is very thin, dark in color, and loose. Two of the three erectile bodies, the *corpora cavernosa penis,* are arranged side by side in the dorsal part of the penis. The *corpus spongiosum penis* (corpus cavernosum urethrae), which lies ventrally, contains the *spongy urethra.* The corpora cavernosa are fused with each other in the median plane, except posteriorly, where they separate to form two *crura* (L. legs). The crura are attached on each side to the conjoint rami of the pubis and ischium. They support the corpus spongiosum penis lying between and inferior to the conjoint rami. The penis consists of a *root* and a *body* (shaft). The dorsal (posterior) surface of the penis faces posteriorly when the penis is erect, and anteriorly when it is in the flaccid (nonerect) state. The dorsum of the penis is continuous with the anterior abdominal wall. The other aspect is referred to as the ventral (urethral) surface.

The root of the penis (radix), its attached portion (Figs. 3-38, 3-57B, 3-63, 3-65, and 3-66), is located in the superficial perineal space between the perineal membrane superiorly and the superficial perineal fascia inferiorly. The root consists of the crura, bulb, and the muscles associated with them (bulbospongiosus). The *bulb of the penis* is located between the crura in the superficial perineal space. The enlarged posterior part of the bulb is penetrated superiorly by the spongy urethra.

The body of the penis (Figs. 3-33, 3-65, and 3-66) is the free part which is pendulous in the flaccid condition. Except for a

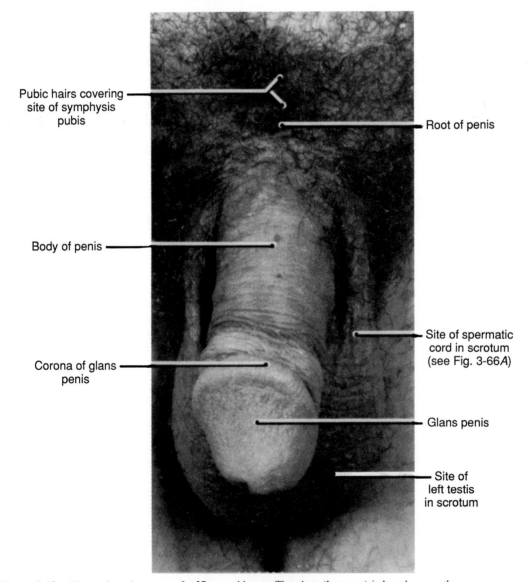

Figure 3-65.   The penis and scrotum of a 27-year-old man. The glans (L. acorn) is bare because the prepuce was removed by circumcision. The arrow indicates the external urethral orifice.

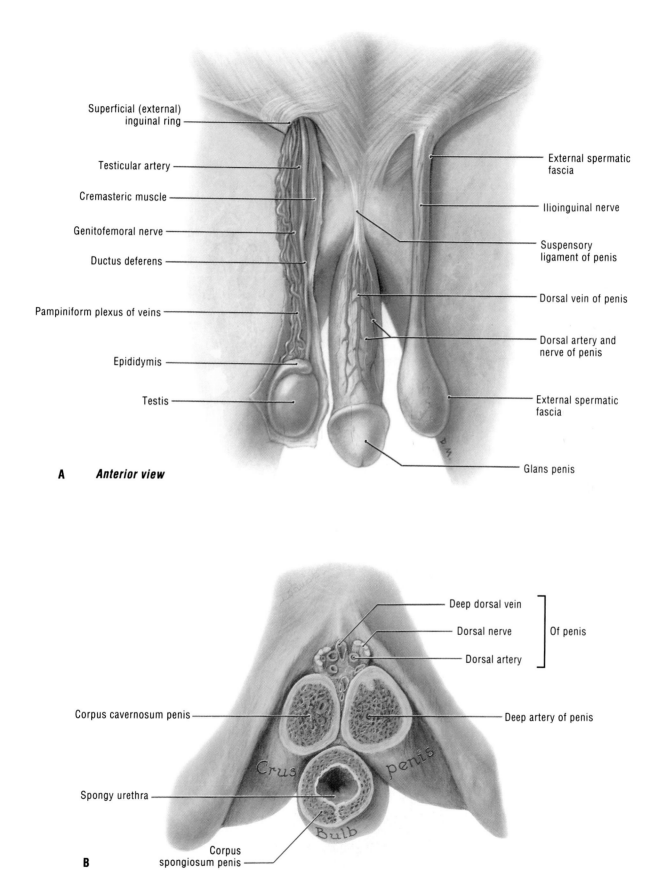

Superficial (external) inguinal ring

Testicular artery

Cremasteric muscle

Genitofemoral nerve

Ductus deferens

Pampiniform plexus of veins

Epididymis

Testis

External spermatic fascia

Ilioinguinal nerve

Suspensory ligament of penis

Dorsal vein of penis

Dorsal artery and nerve of penis

External spermatic fascia

Glans penis

A   *Anterior view*

Deep dorsal vein

Dorsal nerve

Dorsal artery

Of penis

Corpus cavernosum penis

Deep artery of penis

Crus penis

Spongy urethra

Bulb

Corpus spongiosum penis

B

Figure 3-66. *A*, Dissection of the penis, spermatic cord, and scrotum showing their vessels and nerves. *B*, Transverse section through the dissected root of a penis. The skin and fibrous tissue that bind the corpora together have been removed. Observe that the spongy urethra is dilated in the bulb of the penis to form the intrabulbar fossa (see also Fig. 3-33).

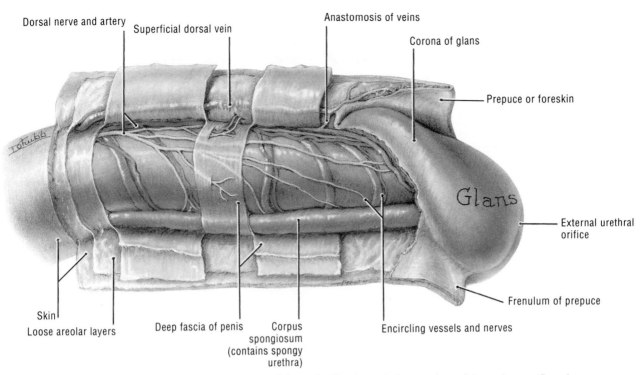

Dorsal nerve and artery

Superficial dorsal vein

Anastomosis of veins

Corona of glans

Prepuce or foreskin

Glans

External urethral orifice

Frenulum of prepuce

Encircling vessels and nerves

Skin

Loose areolar layers

Deep fascia of penis

Corpus spongiosum (contains spongy urethra)

Figure 3-67.   Dissection of the body of the penis. The three tubular coverings of the penis are reflected.

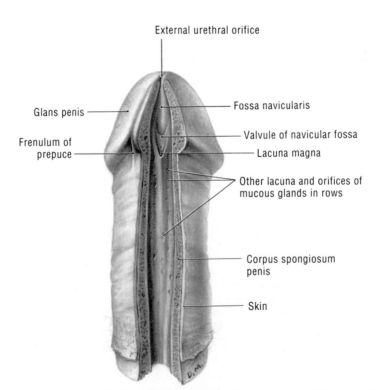

External urethral orifice

Glans penis

Frenulum of prepuce

Fossa navicularis

Valvule of navicular fossa

Lacuna magna

Other lacuna and orifices of mucous glands in rows

Corpus spongiosum penis

Skin

Figure 3-68.   Part of the body of a penis in which a longitudinal incision has been made on its ventral (urethral) surface. Hence the view is of the dorsal surface of the interior of the spongy urethra. Note the large recess (lacuna magna), which is clinically important because it may impede the passage of a catheter.

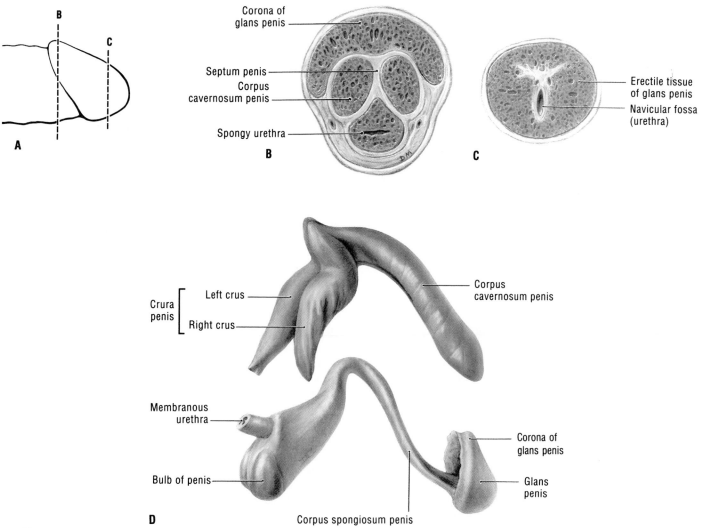

Figure 3-69. *A*, The levels of the transverse sections of the glans penis illustrated in *B* and *C*. The parts of a dissected penis are shown in *D*. Observe that the corpus spongiosum contains the spongy urethra. The glans penis fits like a cap on the blunt ends of the corpora cavernosa penis as shown in Fig. 3-38.

few fibers of the bulbospongiosus near the root, the body of *the penis has no muscles*. It consists of the corpora cavernosa and corpus spongiosum and is covered by skin. At the distal end of the body is the *glans penis* (head). It consists entirely of corpus spongiosum. *The spongy urethra* is within the corpus spongiosum and runs through the root and body of the penis. Hence, it is the longest part of the urethra. Distally the corpus spongiosum penis expands to form the conical *glans penis*, the concavity of which covers the free blunt ends of the corpora cavernosa (Figs. 3-38 and 3-69). The glans has a higher concentration of sensory nerve endings than the rest of the body of the penis, and is thus particularly sensitive to physical stimulation (Masters et al., 1988). The prominent margin of the glans penis, called the *corona* (L. crown), projects beyond the ends of the corpora cavernosa penis. It overhangs an obliquely grooved constriction called the neck of the penis. The slitlike opening of the spongy urethra, called the *external urethral orifice* (meatus), is near the tip of the glans penis (Figs. 3-65, 3-67, and 3-68). The skin and fasciae of the penis are prolonged as a free fold or double layer of skin, called

the *prepuce* (L. *praeputium*, foreskin), which covers the glans penis to a variable extent (Figs. 3-33, 3-35, 3-67, and 3-68). A median fold, called the *frenulum of the prepuce*, passes from the deep layer of the prepuce to a point just inferior to the external urethra orifice.

The root, body and glans of the penis feel spongy because they are composed of *cavernous erectile tissue*, which consists of interlacing and intercommunicating spaces (Fig. 3-69*B* and *C*). In the three corpora of the penis these spaces are separated by fibrous trabeculae and are filled with blood in the erect penis. The weight of the body of the penis is supported by two ligaments that are continuous with the fasciae of the penis. The *fundiform ligament* (see Fig. 2-9) arises from the membranous layer of subcutaneous tissue of the lower abdomen (Scarpa's fascia). The *suspensory ligament* (Fig. 3-66) is a condensation of deep fascia in the form of a thick, triangular fibroelastic band. It arises from the anterior surface of the pubic symphysis and passes inferiorly. It splits to form a sling, which is attached to the deep fascia of the penis at the junction of its fixed and mobile parts (*ie.*, where it bends in the flaccid state; Figs. 3-65 and 3-66).

Arterial Supply of the Penis (Figs. 3-10, 3-18, 3-63, 3-66, and 3-67). The arteries to the penis are: (1) the *dorsal arteries*, which run in the interval between the corpora cavernosa on each side of the deep dorsal vein; (2) the *deep arteries*, which pierce the crura and run within the corpora cavernosa; and (3) the *artery to the bulb*, which enters on each side. The dorsal and deep arteries are branches of the *internal pudendal arteries*, which arise in the pelvis from the internal iliac arteries. The deep arteries are the principal vessels that supply the cavernous spaces (erectile tissue) in the three corpora and therefore are involved in erection of the penis. They give off numerous branches that open directly into these spaces. When the penis is flaccid, these arterial branches are coiled; hence, they are called *helicine arteries* (G. *helix*, a coil).

Venous Drainage of the Penis (Figs. 3-29, 3-55, 3-66, and 3-67). Blood from the cavernous spaces is drained by a *venous plexus* that joins the *deep dorsal vein* located in the deep fascia. This vein passes deep to the arcuate pubic ligament and joins the *prostatic venous plexus*. Blood from the superficial coverings of the penis drain into the *superficial dorsal vein* and then into the *superficial external pudendal vein*.

Lymphatic Drainage of the Penis (Fig. 3-59*B*). The lymph vessels from most of the penis drain into the *superficial inguinal lymph nodes*. Vessels from the glans penis drain into the *deep inguinal lymph nodes*.

Innervation of the Penis (Figs. 3-10, 3-36, 3-61, 3-63, 3-66, and 3-67). The *dorsal nerve of the penis* is one of the two terminal branches of the *pudendal nerve*; the perineal nerve is the other. The dorsal nerve arises in the pudendal canal and passes anteriorly into the deep perineal space. It then runs to the dorsum of the penis where it passes lateral to the arteries. It supplies both the skin and glans penis. The penis is richly provided with a great variety of sensory nerve endings, especially the glans; thus it is very sensitive. The *cavernous nerves* from the inferior hypogastric plexus pass through the urogenital diaphragm to reach the penis. The skin covering the root of the penis is supplied by the *ilioinguinal nerve*, the perineal branch of the *posterior cutaneous nerve of the thigh*, and the posterior scrotal branches of the perineal nerve.

> **Ejaculation of semen** consists of two phases. First there is *emission of sperms* and genital duct secretions into the prostatic and membranous parts of the urethra. This results from peristalsis in the deferent ducts and seminal vesicles and contraction of the bulbourethral glands and the smooth muscle in the prostate. *Emission of semen is a sympathetic response.* Expulsion of the semen from the spongy urethra (*ejaculation*) is accompanied by clonic spasm of the bulbospongiosus muscle in response to innervation via the perineal branch of the pudendal nerve (Fig. 3-61).
>
> **The Anatomical Basis of Erection** (Figs. 3-63, 3-66, and 3-69). When a male is stimulated erotically, the smooth muscle in the fibrous trabeculae and helicine arteries relaxes owing to *parasympathetic stimulation*. As a result, the arteries straighten and their lumina enlarge, allowing blood to flow into the cavernous spaces. Blood engorges and dilates these

> spaces; the bulbospongiosus and ischiocavernosus muscles compress the venous plexuses at the periphery of the corpora cavernosa and impede the return of venous blood. As a result, the three corpora become enlarged, rigid, and the penis erects. Following ejaculation and orgasm (climax of the sexual act or passage of erotic thoughts), the penis gradually returns to its flaccid state, a subsiding process called *detumescence* (resolution). This results from sympathetic stimulation that causes constriction of the smooth muscle in the helicine arteries. The bulbospongiosus and ischiocavernosus muscles relax, allowing more blood to flow into the veins. Blood is slowly drained from the cavernous spaces into the deep dorsal vein.

The Spongy Urethra (Figs. 3-33, 3-35, and 3-66*B* to 3-69). This is the longest part of the urethra (15 to 16 cm). It passes through the bulb of the penis and the corpus spongiosum and ends at the *external urethral orifice* (meatus), the narrowest part of the urethra. The lumen of the spongy urethra is about 5 mm in diameter, but it is expanded in the bulb of the penis to form the *intrabulbar fossa* and in the glans penis to form the *navicular fossa*. On each side the slender *ducts of the bulbourethral glands* open into the proximal part of the spongy urethra, about 3 cm distal to the perineal membrane. The orifices of these ducts are very small. There are also minute openings into the urethra of the ducts of the mucus-secreting *urethral glands*. These are most numerous on the dorsal surface of the spongy urethra.

Arterial Supply and Venous Drainage of the Male Urethra (Figs. 3-18, 3-24*A*, 3-59*A*, and 3-66). The arteries are derived from the structures that the urethra crosses. Hence, branches of the arteries to the prostate supply the urethra as it passes through this gland (p. 279); the artery to the bulb and the urethral artery supply the urethra as it passes distally. The latter vessels are branches of the *internal pudendal artery*. Veins accompany these arteries and have similar names.

Innervation of the Male Urethra (Figs. 3-24*A*, 3-36, and 3-61). The nerves of the male urethra are branches of the *pudendal nerve*. Most afferent fibers from the urethra run in the *pelvic splanchnic nerves*. Nerves from the *prostatic plexus*, which arise from the inferior hypogastric plexus, are distributed to all parts of the urethra.

> Although the size of the flaccid penis varies widely in different males (average 9.5 cm), this variation is less apparent in the erect state. Hence, men with penes (penises) that are smaller when flaccid usually have larger volume increases during erection than men who have large flaccid penes (Masters et al., 1988). Very uncommonly, boys are born with a *micropenis* (less than 2 cm); sometimes this condition results from a treatable deficiency of testosterone.
>
> The spongy urethra, about 5 mm in diameter, will expand enough to permit passage of an instrument about 8 mm in diameter. The external urethral orifice is its narrowest and least distensible part. Hence, an instrument that passes through this opening should pass through all other parts of the urethra. *Urethral stricture* may result from external trauma of the penis or from infection of the urethra. Instruments called *sounds* are used to dilate the urethra (*e.g.*, for the insertion of a *cystoscope*, Fig. 3-32). The interior of the urethra may be observed by an electrically lit instrument called an *urethroscope*.

Urethral catheterization is done to remove urine from a patient who is unable to micturate. It is also performed to irrigate the bladder and to obtain an uncontaminated sample of urine. When inserting catheters and sounds, the curves in the urethra must be considered. The membranous part runs inferiorly and anteriorly as it passes through the *urogenital diaphragm* (Figs. 3-17, 3-33, and 3-38). The prostatic part takes a slight curve that is concave anteriorly as it traverses the prostate. Just inferior to the perineal membrane, the spongy urethra is well covered inferiorly and posteriorly by erectile tissue of the bulb of the penis (Figs. 3-33 and 3-66B), but a short segment of the membranous urethra is unprotected. Because the urethral wall is thin and distensible here, it is vulnerable to injury during instrumentation.

Rupture of the spongy urethra within the bulb of the penis (Fig. 3-66B) is fairly common in "straddle injuries." The urethra is torn when it is caught between a hard object (*e.g.*, a steel beam or the crossbar of a bicycle) and the person's equally hard pubic arch (Fig. 3-1). Urine escapes into the *superficial perineal space* (Fig. 3-57) and passes from there inferiorly into the scrotum and superiorly into the anterior abdominal wall (p. 277).

A common congenital malformation of the penis and urethra (1 in 500 newborns) is **hypospadias**. In the simplest form, *glandular hypospadias*, the external urethral orifice is on the ventral aspect of the glans penis (Behrman and Vaughan, 1992). In other infant males the defect is in the skin and ventral wall of the spongy urethra. Hence, the external urethral orifice is on the ventral surface of the penis. Hypospadias results from failure of fusion of the urogenital folds during early fetal development (Moore, 1988).

The prepuce (*foreskin*) of the penis is usually sufficiently elastic to permit it to be retracted over the glans penis; however, in some males it fits tightly over the glans and cannot be retracted easily, if at all. This condition is called **phimosis** (F. *phimos*, a muzzle). As there are modified sebaceous glands in the prepuce, the oily secretions of cheesy consistency from them, called *smegma*, accumulate in the preputial sac and cause irritation. The preputial sac is located between the glans and the prepuce (Fig. 3-67). In some persons there is a narrow *preputial opening* and retraction of the prepuce over the glans penis constricts the neck of the penis so much that there is interference with the drainage of blood and tissue fluid from the glans. In patients with this condition, called *paraphimosis*, the glans may enlarge so much that the prepuce cannot be drawn over it. Circumcision is commonly performed in such cases. **Circumcision** (L. *circumcido*, to cut around) is the surgical removal of the prepuce (Fig. 3-65). This is the most commonly performed minor surgical operation on male infants in North America. In adults, circumcision is usually performed when there is phimosis or paraphimosis. Although it is a religious practice in Islam and Judaism, it is often done routinely for nonreligious reasons in North America (*e.g.*, mostly related to hygiene).

The Superficial Perineal Muscles (Figs. 3-18, 3-22, 3-58F, and 3-63). There are three of these muscles: (1) *superficial transverse perineal*; (2) *bulbospongiosus*; and (3) *ischiocavernosus*. The last two are muscles of the penis. The superficial perineal muscles are in the superficial perineal space, which is bounded inferiorly by the superficial perineal fascia and supe-

riorly by the perineal membrane. The **perineal nerve** supplies all three superficial perineal muscles.

The Superficial Transverse Perineal Muscles. These slender, narrow strips of muscle pass transversely, anterior to the anus. Each muscle extends from the ischial tuberosity to the perineal body; they probably help to fix this wedge-shaped tendon.

The Bulbospongiosus Muscles (Figs. 3-33A and 3-63). These muscles of the bulb of the penis lie in the median plane of the perineum, anterior to the anus. The two symmetrical parts are united by a median tendinous raphe inferior to the bulb. These muscles arise from this median raphe and the perineal body. The bulbospongiosus muscles form a *sphincter*, which compresses the bulb of the penis and the corpus spongiosum, thereby emptying the spongy urethra of residual urine and/or semen. The anterior fibers of the *bulbospongiosus also assist erection* by increasing the pressure on the erectile tissue. They also compress the deep dorsal vein of the penis, which impedes venous drainage of the cavernous spaces and helps to promote enlargement and turgidity of the penis.

The Ischiocavernosus Muscles (Figs. 3-58F, 3-63, and 3-69). These muscles surround the crura in the root of the penis. Each muscle arises from the internal surface of the *ischial tuberosity* and *ischial ramus* and passes anteriorly on the crus of the penis, where it is inserted into the sides and ventral surface of the crus. The ischiocavernosus muscles force blood from the cavernous spaces in the crura into the distal parts of the corpora cavernosa; this increases the turgidity of the penis. Contraction of the ischiocavernosus muscles also compresses the deep dorsal vein of the penis as it leaves the crus of the penis, thereby cutting off the venous return from the crus of the penis and helping to maintain the erection.

The Female Perineum

The muscles, nerves, and vessels of the female perineum are almost identical with those of the male (Figs. 3-23, 3-40, 3-58, and 3-70). The external genital organs (genitalia) also have similarities to those of males because they develop from the same *primordia* (Moore, 1988). The main differences in the female perineum are: (1) the vagina pierces the urogenital diaphragm; (2) the urethra is in the anterior wall of the vagina; (3) the clitoris does not contain part of the urethra; and (4) the embryonic labioscrotal and urogenital folds remain unfused and form the labia majora and minora, respectively. The female external genital organs are known collectively as the **vulva** (L. covering) or *pudendum* (Figs. 3-40, 3-60, 3-64, and 3-70 to 3-72). The vulva consists of: the mons pubis, labia majora and minora, vestibule of the vagina, clitoris, bulb of the vestibule, and greater vestibular glands.

The Superficial Perineal Fascia (Figs. 3-71 and 3-72; see also Fig. 2-18). As in the male, there is a fatty layer of superficial fascia and a membranous layer of subcutaneous connective tissue. These layers are continuous over the labia majora. The membranous fascia attaches medially to the pubic symphysis and laterally to the body of the pubis.

The Superficial Perineal Muscles (Figs. 3-58F and 3-64). The *superficial transverse perineal*, a slender muscle, passes

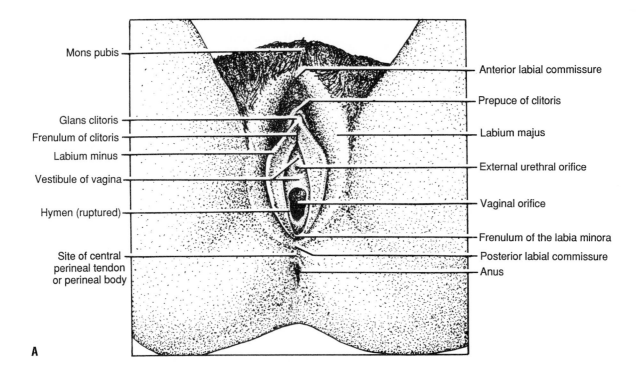

Mons pubis

Glans clitoris
Frenulum of clitoris
Labium minus
Vestibule of vagina

Hymen (ruptured)

Site of central
perineal tendon
or perineal body

Anterior labial commissure

Prepuce of clitoris

Labium majus

External urethral orifice

Vaginal orifice

Frenulum of the labia minora

Posterior labial commissure

Anus

A

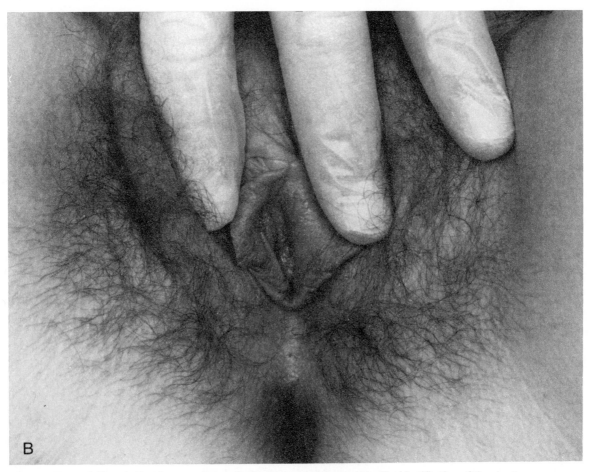

B

Figure 3-70.   *A*, Female external genitalia. *B*, Close-up of the genitals. For identification of the structures, see *A*.

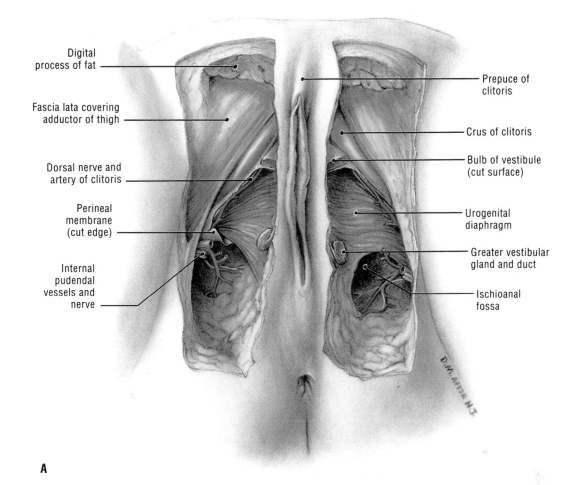

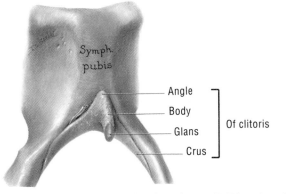

Figure 3-71. *A*, Dissection of the female perineum. *B*, Dissection of the clitoris.

in the base of the superficial perineal space from the ischial ramus to the perineal body. The *ischiocavernosus*, another slender muscle, also attaches to the ischial ramus but inserts into the crus of the clitoris. The *bulbospongiosus*, which is a thin, wide muscle, is separated from its contralateral part by the vagina. It arises from the perineal body, passes around the vagina, and inserts into the clitoris.

The Female External Genital Organs

The Mons Pubis (Figs. 3-64 and 3-70*A*).   The mons (L. mountain) pubis is a rounded fatty elevation (eminence) located anterior to the pubic symphysis and lower pubic region. It consists mainly of a pad of fatty connective tissue deep to the skin. This fat is a specialized part of the fatty layer of subcutaneous connective tissue of the anterior abdominal wall. The amount of fat increases during puberty and decreases after menopause. The mons pubis becomes covered with coarse *pubic hairs* during puberty, which also decrease after menopause. The typical female distribution of pubic hair has a horizontal superior limit across the pubic region.

The Labia Majora (Figs. 3-27, 3-34, and 3-70 to 3-72).   The labia (L. large lips) are two symmetrical folds of skin which provide protection for the urethral and vaginal orifices that open into the *vestibule of the vagina*. Each labium majus, largely filled with subcutaneous fat, passes posteriorly from the mons pubis to about 2.5 cm from the anus. They are situated on each side of the *pudendal cleft*, which is the slit between the labia majora into which the vestibule of the vagina opens. The labia majora meet anteriorly at the *anterior labial commissure*. They do not join posteriorly but a transverse bridge of skin called the *posterior labial commissure* passes between them.

Embryologically the labia majora are homologous to the male scrotum. They develop from the unfused *labioscrotal*

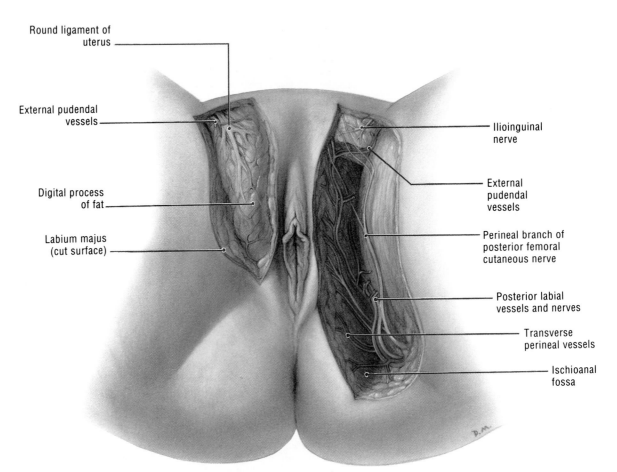

Round ligament of uterus

External pudendal vessels

Digital process of fat

Labium majus (cut surface)

Ilioinguinal nerve

External pudendal vessels

Perineal branch of posterior femoral cutaneous nerve

Posterior labial vessels and nerves

Transverse perineal vessels

Ischioanal fossa

Figure 3-72.   Dissection of the labia majora demonstrating their vessels and nerves. *On the right side*, the lobulated digital process of fat has been opened to show the anastomotic vessels, which unite the external and internal pudendal vessels. *On the left side*, the digital process of fat is largely removed to show that the posterior labial vessels and nerves (S2 and S3) are joined by the perineal branch of the posterior femoral cutaneous nerve (S1, S2, and S3).

folds that fuse in males to form the scrotum (Moore, 1988). The *round ligaments of the uterus* (remnants of the embryonic gubernacula) pass through the inguinal canals and enter the labia majora (Figs. 3-64 and 3-72; see also Fig. 2-18), where they end as branching bands of fascia that are attached to the skin.

The Labia Minora (Figs. 3-23, 3-27, 3-34*A*, 3-40, 3-62, 3-64, and 3-70 to 7-72).   The labia minora (L. small lips) are thin, delicate folds of fat-free hairless skin, which are located between the labia majora. They contain a core of spongy tissue with many small blood vessels but no fat. Although the internal surface of each labium minus consists of thin skin, it has the typical pink color of a mucous membrane and contains many sensory nerve endings. In young females the labia minora are usually concealed by the labia majora, but in multiparae (women who have borne several children), the labia minora may protrude through the pudendal cleft. Sebaceous and sweat glands open on both of their surfaces. The labia minora enclose the vestibule of the vagina and lie on each side of the orifices of the urethra and vagina. They meet just superior to the clitoris to form a fold of skin called the *prepuce* (clitoral hood). In young females the labia minora are usually united posteriorly by a small fold of the skin, the *frenulum of the labia minora*. Some obstetricians refer to the frenulum (L. small bridle) as the *fourchette* (F. fork). The frenulum is frequently torn during childbirth or incised during an *episiotomy*.

The Vestibule of the Vagina (Figs. 3-34*A*, 3-40, 3-60, 3-64, and 3-70 to 3-72).   The vestibule (L. *vestibulum*, antechamber) is the space between the labia minora. The urethra, vagina, and ducts of the greater vestibular glands open into the vestibule. The inferior half of the *urethra* is about 2 cm long. Like the superior half, it lies in close contact with the anterior wall of the vagina.

The External Urethral Orifice (Figs. 3-23, 3-34*A*, 3-64, and 3-70*A*).   This median aperture is located 2 to 3 cm posterior to the clitoris and immediately *anterior to the vaginal orifice*. On each side of this orifice are the openings of the ducts of the *paraurethral glands* (Skene's glands); these glands are homologous to the prostate in the male.

The Vaginal Orifice (Figs. 3-23, 3-40, 3-64, and 3-70*A*).   This large opening is located inferior and posterior to the much smaller external urethral orifice. The size and appearance of the vaginal orifice (hymenal opening) varies with the condition of the *hymen* (G. membrane), a thin fold of mucous membrane that surrounds the vaginal orifice.

The Greater Vestibular Glands (Figs. 3-60 and 3-

71*A*). These glands, about 0.5 cm in diameter, are located on each side of the vestibule of the vagina, posterolateral to the vaginal orifice. They are round or oval in shape and the *bulbs of the vestibule* partly overlap them posteriorly. From the anterior parts of the glands, slender ducts pass deep to the bulbs of the vestibule and open into the vestibule of the vagina on each side of the vaginal orifice. These glands secrete a small amount of lubricating mucus into the vestibule of the vagina during sexual arousal. The greater vestibular glands (Bartholin's glands) are homologous with the *bulbourethral glands* in the male (Fig. 3-18).

The Lesser Vestibular Glands. These are small glands on each side of the vestibule of the vagina that open into it between the urethral and vaginal orifices. These glands also secrete mucus into the vestibule, which moistens the labia and the vestibule.

The Clitoris (Figs. 3-40, 3-60, 3-64, 3-67, and 3-70 to 3-72). The clitoris, 2 to 3 cm in length, is homologous with the penis and, like it, is an *erectile organ*. Unlike the penis, the clitoris is not traversed by the urethra; therefore it has no corpus spongiosum. The clitoris is located posterior to the anterior labial commissure, where the labia majora meet. It is usually hidden by the labia when it is flaccid (nonerect). The clitoris consists of a *root* and a *body* which are composed of two crura, two corpora cavernosa, and a *glans*. Like the penis, it is suspended by a *suspensory ligament*. The parts of the labia minora passing anterior to the clitoris form the *prepuce of the clitoris* (homologous with the male prepuce). The parts of the labia passing posterior to the clitoris form the *frenulum of the clitoris*, which is homologous with the frenulum of the penile prepuce. *The clitoris is a small sexual organ*, which is composed of erectile

tissue. Like the penis, it will enlarge upon tactile stimulation, but it does not lengthen significantly. It is *highly sensitive* and very important in the sexual arousal of a female. Because it has many nerve endings (Fig. 3-60), it is highly sensitive to touch, pressure, and temperature.

The Bulbs of the Vestibule (Figs. 3-58*E*, 3-60, and 3-71). These two large, elongated masses of erectile tissue are about 3 cm in length. They lie along the sides of the vaginal orifice, deep to the bulbospongiosus muscles and are homologous with the bulb of the penis, but unlike the penis, the bulbs are separated from the clitoris and are separated by the vestibule of the vagina.

Arterial Supply of the Female External Genitalia (Figs. 3-60, 3-71, and 3-72). The rich arterial supply to the vulva is from two *external pudendal arteries* and one *internal pudendal artery* on each side. The internal pudendal artery supplies the skin, sex organs, and the perineal muscles. The *labial arteries* are branches of the internal pudendal artery, as are the dorsal and deep arteries of the clitoris.

Venous Drainage of the Female External Genitalia (Figs. 3-60, 3-71, and 3-72). The labial veins are tributaries of the *internal pudendal veins* and venae comitantes of the internal pudendal artery.

Lymph Drainage of the Female External Genitalia (Fig. 3-73). The vulva contains an exceedingly rich network of lymphatic channels. Most lymph vessels pass to the *superficial inguinal lymph nodes* and deep inguinal nodes.

Innervation of the Female External Genitalia (Figs. 3-61, 3-71, 3-72, and 3-74). The nerves to the vulva are branches of the *ilioinguinal nerve*, the genital branch of the *genitofemoral*

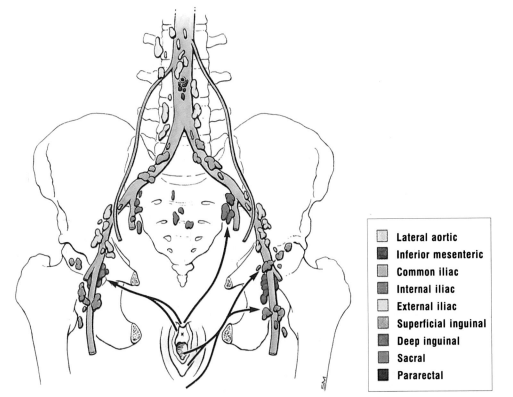

	Lateral aortic
	Inferior mesenteric
	Common iliac
	Internal iliac
	External iliac
	Superficial inguinal
	Deep inguinal
	Sacral
	Pararectal

Figure 3-73. Lymphatic drainage of the female external genitalia.

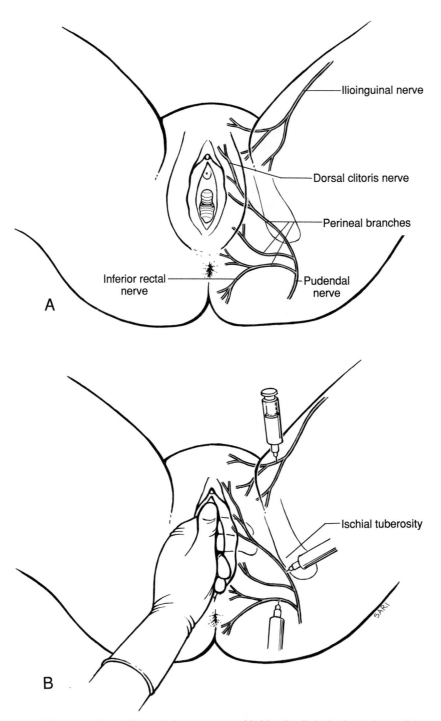

Figure 3-74. *A*, Distribution of the pudendal and ilioinguinal nerves to the female perineum. *B*, The pudendal nerves may be blocked with anesthetic. The needle is inserted toward the ischial tuberosity, where the pudendal nerve emerges from the pudendal canal. The needle is guided by the digits in the vagina until its tip is posterior and inferior to the ischial spine where the pudendal nerve lies. The injection from the upper needle blocks (anesthetizes) the ilioinguinal nerve and its labial branches, which supply the vulva (see Fig. 3-72).

nerve, the perineal branch of the *femoral cutaneous nerve of thigh*, and the *perineal nerve*.

The Anatomical Basis of Erection. The clitoris is sometimes said to be a miniature penis. This is an incorrect and sexist statement (Masters et al., 1988). Although it is an important sexual organ, it has no reproductive or urinary function. Because it is very sensitive, rough stroking of it can be irritating and even painful. The clitoris erects—although it does not lengthen—by the same mechanism described in the male (p. 312) because its parts are similar to those of the penis. In the female, parasympathetic stimulation leads to increased secretions from the *paraurethral glands* (homologous with the prostate) and the *greater vestibular* glands (homologous with the bulbourethral glands). This stimulation also causes erection of the clitoris and enlargement of the bulbs of the vestibule.

The female urethra is easily infected because it is open to the exterior via the vestibule of the vagina (Figs. 3-64 and 3-70A). *Urethritis* (inflammation of the urethra) may result from pathological organisms, *e.g.*, caused by a gonococcal infection. The gonococci may enter the ducts of the urethral glands and, although the condition may be asymptomatic, may enter and infect the male urethra during sexual intercourse. The short female urethra is very distensible because it contains considerable elastic tissue, as well as smooth muscle. It can easily be dilated to 1 cm without injuring it; consequently, the passage of catheters or cystoscopes is much easier in females than in males. Urine may be readily removed from a distended female bladder by passing a catheter through the urethra into the bladder.

The hymen was the inferior closed end of the *uterovaginal canal* in the embryo (Moore, 1988). During infancy and early childhood the vaginal orifice is usually closed by the *hymen*, but it normally ruptures before puberty to allow the menstrual fluid to escape. In virgins the vaginal orifice may be only a few mm in diameter. Usually the opening will admit the tip of one digit. In some cases dilation of the vaginal orifice may be necessary to prevent painful tearing of the hymen during intercourse. Even after dilation the hymen may tear, usually posteriorly or posterolaterally. Some bleeding may occur. After childbirth very little of the hymen is left, except a few fleshy tags called *hymenal caruncles* (Fig. 3-23). Uncommonly the hymen fails to rupture before puberty, a condition known as *imperforate hymen*. In these cases menstrual blood distends the vagina, a condition called *hematocolpos*. The hymen is incised to allow the menstrual fluid to escape.

Historically it has been considered important for women to have an "intact hymen" at the time of marriage as proof of their virginity. It is, however, very uncommon for women to have an imperforate hymen. Usually it stretches across some but not all of the vaginal orifice. It normally becomes torn or stretched at an early age by various activities (*e.g.*, physical exercises). *Clitoral circumcision* (surgical removal of the prepuce of the clitoris) is sometimes performed, supposedly to enhance sexual arousal (Masters et al., 1988). In some tribes in Africa and South America, surgical removal of the clitoris (*clitoridectomy*) is also performed as a rite of puberty. This disfiguring procedure inhibits sexual arousal significantly; however, it is fortunate other areas (*e.g.*, the labia minora) are also sensitive sexual areas.

The greater vestibular glands are usually not palpable, but they become readily so when infected. *Bartholinitis*, inflammation of the greater vestibular glands (Bartholin's glands), may result from a number of pathogenic organisms. Infected glands may enlarge to a diameter of 4 to 5 cm and impinge on the wall of the rectum.

Because there is a such rich arterial supply to the labia majora and minora (Fig. 3-72), hemorrhage resulting from injuries may be severe. During parturition painful labor often occurs, and most anguish is usually felt when the fetal head passes through the vulva owing to stretching of its parts. To relieve the pain associated with childbirth, *pudendal block anesthesia* may be performed by injecting a local anesthetic agent into the tissues surrounding the **pudendal nerve**. The injection may be made where the pudendal nerve crosses the lateral aspect of the sacrospinous ligament (Figs. 3-8A, 3-72, and 3-74), near its attachment to the ischial spine. Sometimes a long needle is inserted through the vaginal wall and guided by a digit in the vagina to the ischial spine (Ellis, 1983). To ease delivery of a fetus and to avoid laceration of the perineum, an *episiotomy* or relaxing incision is frequently made in the perineum (Fig. 3-62). The two common types of episiotomy are described on p. 298.

PATIENT ORIENTED PROBLEMS

Case 3-1

A 23-year-old primigravida had been in labor for nearly 24 hours. The crown of the fetal head was visible through the vaginal orifice. The obstetrician, fearing that her perineum might be torn, decided to perform a mediolateral episiotomy to enlarge the inferior opening of the birth canal.

> **Problems.** What perineal structures would probably be cut during this surgical procedure, known as episiotomy? What structures might have been injured if the perineum had been allowed to tear in an uncontrolled fashion? In severe perineal lacerations, what muscles may be torn? These problems are discussed on p. 320.

Case 3-2

A 31-year-old construction worker was walking along a steel beam when he fell, straddling it. He was in severe pain owing to trauma to his testes and perineum. Later he observed swelling and discoloration of his scrotum and when he attempted to urinate, only a few drops of bloody urine appeared. He went to a hospital emergency department. After examining the patient, the physician consulted a urologist who ordered radiographic studies of the patient's urethra and bladder.

Radiologist's Report. The radiographic studies revealed a rupture of the spongy urethra just inferior to the inferior fascia of the urogenital diaphragm. The urethrograms showed passage of the contrast material out of the urethra into the surrounding tissue of the perineum (*extravasation of urine*).

Diagnosis. Rupture of the proximal part of the spongy urethra with extravasation of urine into the surrounding tissues.

> **Problems.** When the patient tried to urinate, practically no urine came from his external urethral orifice. Where did it go? Explain why extravasated urine cannot pass posteriorly, laterally, or into the pelvis minor. These problems are discussed on p. 320.

Case 3-3

A 49-year-old man complained of tenderness and pain on the right side of his anus. The pain was aggravated by defecation and sitting. Because he had a history of hemorrhoids, he suspected that he might be having a recurrence of this problem. After he had explained his symptoms and history, his physician examined his rectum and anal canal. When he asked the patient to strain as if to defecate, prolapsing internal hemorrhoids came into view. During careful digital examination of the anal canal and rectum, the doctor detected some swelling in the patient's right ischioanal fossa. The swelling produced severe pain when it was compressed.

Diagnosis. Ischioanal abscess.

Treatment. The abscess was drained through an incision in the skin between the anus and the ischial tuberosity.

> **Problems.** Differentiate between internal and external hemorrhoids. What is an ischioanal abscess? What nerve is vulnerable to injury during surgical treatment of an ischioanal abscess? If this nerve were severed, what structure(s) would be partly denervated? These problems are discussed on p. 321.

Case 3-4

A 40-year-old unconscious woman was rushed to the hospital because she had sustained multiple injuries during an automobile accident. Priority was given to securing a patent airway by inserting an endotracheal tube. Next, emergency care was directed toward controlling bleeding and treating the shock. When the patient's general condition had stabilized, radiographs were taken of the injured regions of her body. Because she had not urinated since being admitted, she was catheterized. The presence of blood in her urine (*hematuria*) suggested rupture of her urinary bladder. Therefore a sterile dilute contrast solution was injected into her bladder through a catheter and radiographs of the pelvis and abdomen were taken.

Radiologist's Report. There are fractures of the pubic rami on both sides. The cystogram showed extravasation of contrast material from the superior surface of the bladder.

Diagnosis. Fractured pelvis and ruptured urinary bladder.

> **Problems.** Where would the extravasated urine from the bladder go? What covers the superior surface of the urinary bladder? Thinking anatomically, what route do you think the surgeon would take when repairing the ruptured bladder? These problems are discussed on p. 321.

Case 3-5

A 28-year-old woman was experiencing pregnancy for the first time (*primigravida*). Toward the end of the gestational period, she suffered painful uterine contractions at night which subsided toward morning (*false pains*). When she called her physician, she told her that her labor was imminent. In a few days she observed a discharge of mucus and some blood. When she reported that her "pains" (*uterine contractions*) were occurring every 10 minutes, her obstetrician asked her to go to the hospital. Following admission, the physician palpated her cervix and informed the intern that the uterine ostium was open about one fingertip and that the patient was still in the *first stage of labor* (period of dilation of the uterine ostium). Later a large volume of fluid was expelled (rupture of the fetal membranes). When the patient entered the *second stage of labor* (period of expulsive effort beginning with complete dilation of the cervix and ending with delivery of the baby), she began to experience considerable pain. Although she had wanted to have a *natural*

birth without the use of anesthetics, she was unable to bear the pain. Medication for pain relief was administered as ordered by her physician.

When it was determined that her contractions were 2 minutes apart and lasting 40 to 60 seconds, she was moved to the case room and placed on a delivery table. As the fetal head dilated the birth canal, it was obvious that the woman was suffering intense pain. The obstetrician decided to do a *median episiotomy* when it appeared possible that a tear might occur in her perineum. She administered an intradermal injection of an anesthetic agent into the patient's perineum. Although the local anesthetic enabled the incision to be made without pain, it did not alleviate the severe pain of her labor. The obstetrician decided to perform bilateral pudendal nerve blocks. Thereafter the patient completed the second stage and proceeded through the *third stage of labor* (beginning after delivery of the child and ending with expulsion of the placenta and fetal membranes).

> **Problems.** What structures are usually incised during a median episiotomy? Name the structures supplied by the pudendal nerve. Based on your knowledge of the anatomy of this nerve, where do you think the obstetrician would inject the anesthetic agent to perform a pudendal nerve block? When complete perineal anesthesia is required, branches of what other nerves would have to be blocked? These problems are discussed on p. 321.

DISCUSSION OF PATIENTS' PROBLEMS

Case 3-1

During a mediolateral episiotomy the following structures are usually cut: perineal skin, posterior wall of the vagina, and the bulbospongiosus muscle. An episiotomy is performed to ease delivery of a fetus and/or when a perineal laceration seems inevitable. If a tear is allowed to occur spontaneously in whatever direction it may, the central perineal tendon or *perineal body*, the external anal sphincter, and the wall of the rectum may be torn. Episiotomy thus makes a clean cut away from important structures. If the perineum had been allowed to tear in an uncontrolled fashion, the levator ani muscles, which form the pelvic diaphragm, might have been torn. This results in poor perineal support for the pelvic organs, which may cause sagging of the pelvic floor in later life. This could lead to difficulty in bladder control and could be the basis of subsequent prolapse of the urinary bladder (cystocele).

Case 3-2

Traumatic rupture of the man's spongy urethra in the bulb of his penis resulted in superficial or subcutaneous *extravasation of urine* when he attempted to urinate. Urine from the torn urethra would pass into the perineum, superficial to the inferior fascia of the urogenital diaphragm but deep to the *superficial perineal fascia* (Fig. 3-57). The urine in the superficial perineal space passes inferiorly into the loose connective tissue of the scrotum, anteriorly into the penis, and superiorly into the anterior wall of the abdomen.

The inferior fascia of the urogenital diaphragm and the superficial fascia of the perineum are firmly attached to the ischiopubic rami. Therefore the urine cannot pass posteriorly because the two layers are continuous with each other around the superficial transverse perineal muscles. The urine does not extend laterally because these two layers are connected to the rami of the pubis and ischium. It cannot extend into the pelvis minor because the opening into this cavity is closed by the inferior fascia of the urogenital diaphragm. Urine cannot pass into the thighs because the membranous layer of the superficial fascia of the anterior abdominal wall blends with the *fascia lata*, just distal to the inguinal ligament. The fascia lata is the strong fascia enveloping the muscles of the thigh (Fig. 3-71).

Case 3-3

Hemorrhoids are varicosities of one or more of the veins draining the anal canal. *Internal hemorrhoids* are varicosities of the tributaries of the *superior rectal vein*. This vein is a tributary of the inferior mesenteric vein and belongs to the portal system of veins. The tributaries of the superior rectal vein arise in the *internal rectal plexus*, which lies in the anal columns. Here, they frequently become varicose (dilated and tortuous). Internal hemorrhoids are covered by mucous membrane. At first they are contained in the anal canal, but as they enlarge they may protrude through the anal canal on straining during defecation. Bleeding from internal hemorrhoids is common.

External hemorrhoids are varicosities of the tributaries of the inferior rectal vein arising in the *external rectal plexus*, which drains the inferior part of the anal canal. External hemorrhoids are covered by anal skin and are usually not painful.

Perianal abscesses often result from injury to the anal mucosa by hardened fecal material. Inflammation of the anal sinuses may result, producing a condition called *cryptitis*. The infection may spread through a small crack or lesion in the anal mucosa and pass through the anal wall into the ischioanal fossa, producing an *ischioanal abscess*. The ischioanal fossa is a wedge-shaped space lateral to the anus and levator ani muscle. The main component of the ischioanal fossae is fat. The branches of the nerves and vessels (pudendal nerve, internal pudendal vessels, and the nerve to the obturator internus muscle) enter the ischioanal fossa through the lesser sciatic foramen. The pudendal nerve and internal pudendal vessels pass in the pudendal canal lying in the lateral wall of the ischioanal fossa.

The *inferior rectal nerve* leaves the pudendal canal and runs anteromedially and superficially across the ischioanal fossa. It passes to the external anal sphincter and supplies it. It is vulnerable during surgery in the ischioanal fossa. Damage to the inferior rectal nerve results in impaired action of this voluntary anal sphincter.

Case 3-4

The urine which escaped from the superior surface of the patient's ruptured urinary bladder would pass into the peritoneal cavity. Although pelvic fractures are sometimes complicated by bladder rupture, the radiographs indicated that the rupture was not likely caused by a sharp bone fragment. Probably the bladder was ruptured by the same compressive blow to the region of the pubic symphysis that fractured the pelvis. A full bladder is especially liable to rupture at its superior surface following a non-penetrating blow.

The superior surface of the bladder is almost completely covered with peritoneum. In the female the peritoneum is reflected onto the uterus at the junction of its body and cervix, forming the *vesicouterine pouch*. In a patient with an intraperitoneal bladder rupture, signs and symptoms of peritoneal irritation are likely to develop. *Septic peritonitis* (G. *sepsis*, putrefaction) may develop if there are pathogenic organisms in the urine. As the urine accumulates in the peritoneal cavity, dullness will be detected over the paricolic gutters during percussion of the abdomen. This dullness will disappear from the left side when the patient is rolled onto the right side and vice versa. This finding indicates free fluid in the peritoneal cavity from ruptured viscus (the bladder in this case).

Access to the urinary bladder for surgical repair of its ruptured superior wall would most likely be via the suprapubic route. The bladder is separated from the pubic bones by a thin layer of loose connective tissue, which may contain fat. When the bladder is full, its anteroinferior surface is in contact with the anterior abdominal wall, without the interposition of peritoneum.

Case 3-5

During a median episiotomy the incision starts at the frenulum of the labia minora (Fig. 3-62A) and extends through the skin, vaginal mucosa, perineal body, and superficial perineal muscle. The *pudendal nerve* arising from the sacral plexus (S2, S3, and S4), is the main nerve of the perineum. It is both motor and sensory to this region and also carries some postganglionic sympathetic fibers to the perineum. In the female the pudendal nerve divides into the perineal nerve and the dorsal nerve of the clitoris. The perineal nerve gives off two posterior labial nerves and divides into small terminal muscular branches. They enter the superficial and deep perineal spaces to supply the muscles in them and the bulb of the vestibule. The dorsal nerve of the clitoris supplies the prepuce and glans of the clitoris and the associated skin.

When the pudendal nerve is blocked via the perineal route, the chief bony landmark is the *ischial tuberosity* (Fig. 3-74). With the patient in the lithotomy position, the ischial tuberosity is palpated. Here, the pudendal nerve emerges from the pudendal canal to distribute itself over the perineum. The needle is inserted superomedially for about 2.5 cm before the injection is made. When complete perineal anesthesia is required, genital branches of the genitofemoral and ilioinguinal nerves and the perineal branch of the posterior cutaneous nerve of the thigh must also be anesthetized by making an injection along the lateral margin of the labia majora.

SUGGESTED ADDITIONAL READING

Ayoub SF: Anatomy of the external anal sphincter in man, Acta Anat 105:25, 1979.

Behrman RE: *Nelson Textbook of Pediatrics*, ed 14. Philadelphia, WB Saunders, 1992.

Dellenbach P, et al.: The transvaginal method for oocyte retrieval. An update on our experience (1984–1987), In Jones HW, Jr, Schrader C (Eds): *In Vitro Fertilization and Other Assisted Reproduction.* New York, Annals of the New York Academy of Sciences, vol. 541, p.111, 1988a.

Dellenbach P, Forrler A, Moreau L, Rouard M, Badoc E: Direct intraperitoneal insemination. New treatment for cervical and unexplained infertility, In Jones HW, Jr, Schrader C (Eds): *In Vitro Fertilization and Other Assisted Reproduction*. New York, Annals of the New York Academy of Sciences, vol. 541, p.761, 1988b.

Dilts PV, Jr, Greene JW, Jr, Roddick JW, Jr: *Core Studies in obstetrics and Gynecology*, ed 3. Baltimore, Williams & Wilkins, 1981.

Ellis H: *Clinical Anatomy. A Revision and Applied Anatomy for Clinical Students*, ed 7. Oxford, Blackwell Scientific Publications, 1983.

Farrow GA: Urology. *In* Gross A, Gross P, Langer B (Eds): *Surgery. A Complete Guide for Patients and their Families*, Toronto, Harper & Collins, 1989.

Healey JE, Jr, Hodge J: *Surgical Anatomy*, ed 2. Toronto, BC Decker, 1990.

Hannah WJ: Obstetrics and Gynecology, In Gross A, Gross P, Langer B (Eds): *Surgery. A Complete Guide for Patients and their Families*, Toronto, Harper & Collins, 1989.

Masters WH, Masters VE, Kolodny RC: *Human Sexuality*, ed 3. Glenview Illinois, Scott, Foresman, 1988.

Moore KL: *The Developing Human*: *Clinically Oriented Embryology*, ed 4. Philadelphia, WB Saunders, 1988.

Oelrich TM: The urethral sphincter muscle in the male. Am J Anat 158:229, 1980.

Oelrich TM: The striated urogenital sphincter muscle in the female, Anat Rec 205:223, 1983.

Warwick R (Ed): *Nomina Anatomica*, ed 6. Edinburgh, Churchill Livingstone, 1989.

Williams PL, Warwick R, Dyson M, Bannister LH: *Gray's Anatomy*, ed 37. New York, Churchill Livingstone, 1989.

Woodburne RT, Burkel WE: *Essentials of Human Anatomy*, ed 8. New York, Oxford University Press, 1988.

The back, or *posterior aspect of the trunk*, is the main part of the body to which the head, neck, and limbs are attached. It consists of skin, superficial fascia containing fatty tissue, deep fascia, muscles, vertebrae, intervertebral discs, ribs (in the thoracic region), vessels, and nerves (see Fig. 2-96). **Low back pain** is a commonly encountered complaint. To understand the anatomical basis of back problems that cause disabling pain, an understanding of the structure and function of the back is required. *The common sites of aches and pains are the cervical (neck) and lumbar (lower back) regions*, mainly because they are the most mobile parts of the vertebral column.

The Vertebral Column

The vertebral column (spine, spinal column, backbone) *forms the skeleton of the back* and the main part of the **axial skeleton**[1] (Figs. 4-1 and 4-2). It consists of 33 bones called *vertebrae* which articulate with each other at anterior and posterior *intervertebral joints*. The vertebral column forms a strong but flexible support for the trunk. It extends from the base of the skull through the neck and trunk. The vertebrae are stabilized by ligaments which limit the movements that are produced by the trunk muscles. The **spinal cord**, spinal *nerve roots*, and their coverings, called *meninges*, are located within the *vertebral canal* (Fig. 4-1B), which is formed by the *vertebral foramina* in successive vertebrae. The **spinal nerves** and their branches are located outside the vertebral canal, except for the *meningeal nerves*, which return through the *intervertebral foramina* to innervate the **spinal meninges** (G. membranes).

The vertebral column provides a partly rigid and partly flexible axis for the body and a pivot for the head. Consequently, it has important roles in: posture, support of body weight, locomotion, and protection of the spinal cord and nerve roots. When sitting, the vertebral column transmits the weight of the body across the *sacroiliac joints* to the ilia (Fig. 3-1 on p. 244), and then to the *ischial tuberosities* (Fig. 4-3). When standing, body weight is transferred from the sacroiliac joints to the *acetabula* (hip sockets) and then to the *femora* or thigh bones (see Figs. 3-4 and 5-2). *The vertebral column usually consists of 33 vertebrae* (backbones), arranged in five regions, but only 24 of them (7 cervical, 12 thoracic, and 5 lumbar) are movable (Fig. 4-1). In adults the five sacral vertebrae are fused to form the *sacrum* and the four coccygeal vertebrae are more or less fused together to form the *coccyx*. The abbreviations C, T, L, S, and Co are used for the regions of the vertebral column. The *24*

movable vertebrae give the vertebral column considerable flexibility (Figs. 4-8 and 4-36). The stability of the vertebral column is provided by the shape and strength of the vertebrae and by the intervertebral discs, ligaments, and muscles. The movable vertebrae are connected by resilient *intervertebral discs* (Figs. 4-4 and 4-7), which play an important role in movements between the vertebrae, and in absorbing shocks transmitted up or down the vertebral column. The movable vertebrae are also connected to each other by paired, posterior *zygapophyseal joints* (facet joints) between the articular processes (Figs. 4-26 to 4-28), and by strong anterior and posterior *longitudinal ligaments*. These ligaments, extending the length of the vertebral column, are attached to the intervertebral discs and vertebral bodies. The intervertebral ligaments and joints generally prevent excessive flexion and extension of the vertebral column. Movements beyond the normal limits, *i.e.*, *hyperextension and/or hyperflexion of the neck* ("whiplash injury") usually cause damage to the joints and ligaments and the associated muscles, nerves, and vessels (p. 347).

The bodies of the vertebrae contribute about three-fourths of the length of the movable part of the vertebral column, and the intervertebral discs contribute the other one-fourth (Fig. 4-1). *When counting the vertebrae*, it is important to *begin at the base of the neck* (Fig. 4-8), because what may appear to be an extra lumbar vertebra in a radiograph may be an extra thoracic or sacral vertebra. The vertebral bodies gradually become larger as the sacrum is approached and then become progressively smaller toward the coccyx. These structural differences are related to the fact that the lumbosacral region carries more weight than the cervical and thoracic regions.

Normal Curvatures of the Vertebral Column

In the articulated vertebral column and in various images used clinically, *e.g.*, *MRI*s (magnetic resonance images), four curvatures are normally visible in the adult. The thoracic and sacral curvatures are concave anteriorly, whereas the cervical and lumbar curvatures are concave posteriorly (Figs. 4-1 and 4-6A). The thoracic and sacral curvatures are called **primary curvatures** because they develop during the fetal period (Fig. 4-5). The cervical and lumbar curvatures begin to appear in the cervical and lumbar regions before birth, but they are not very obvious until after birth. They are called **secondary curvatures**. The *cervical curvature* (Figs. 4-1 and 4-16) is accentuated when an infant begins to hold its head erect, and the *lumbar curvature* becomes obvious when a child begins to walk. The cervical curvature can be obliterated by flexing the neck. *The thoracic curvature is permanent* and is formed by the 12 articulated thoracic vertebral bodies (Figs. 4-1 and 4-6A). The lumbar curvature, generally more pronounced in women, ends at the lumbosacral angle. *The sacral curvature* is also permanent and dif-

---

[1]The bones of the cranium, vertebral column, ribs, and sternum.

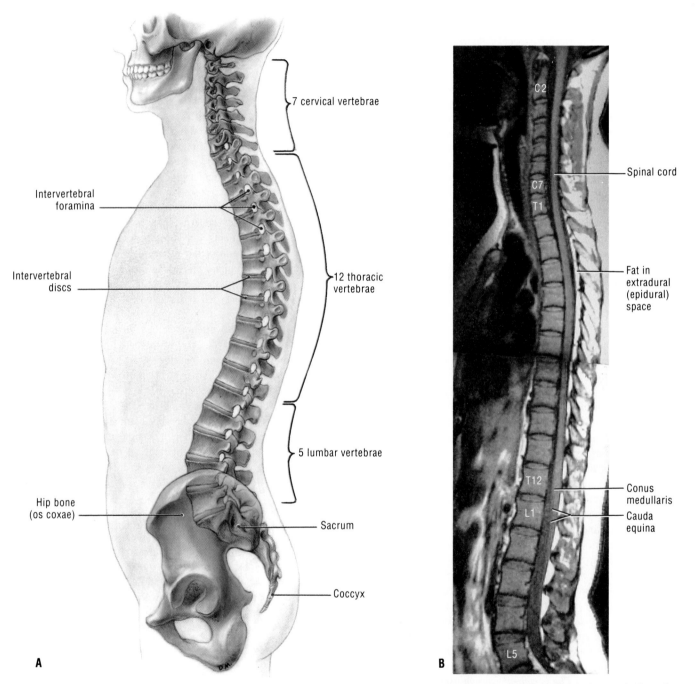

A

B

Figure 4-1. *A*, Lateral view of the vertebral column showing its normal curvatures (cervical, thoracic, lumbar, and sacrococcygeal) and its relationship to the skull and hip bone. *B*, Sagittal MRI (magnetic resonance image) showing the spinal cord and cauda equina in the vertebral canal. (Courtesy of Dr. W. Kucharczyk, Clinical Director of Tri-Hospital Resonance Centre, Toronto, Ontario, Canada.)

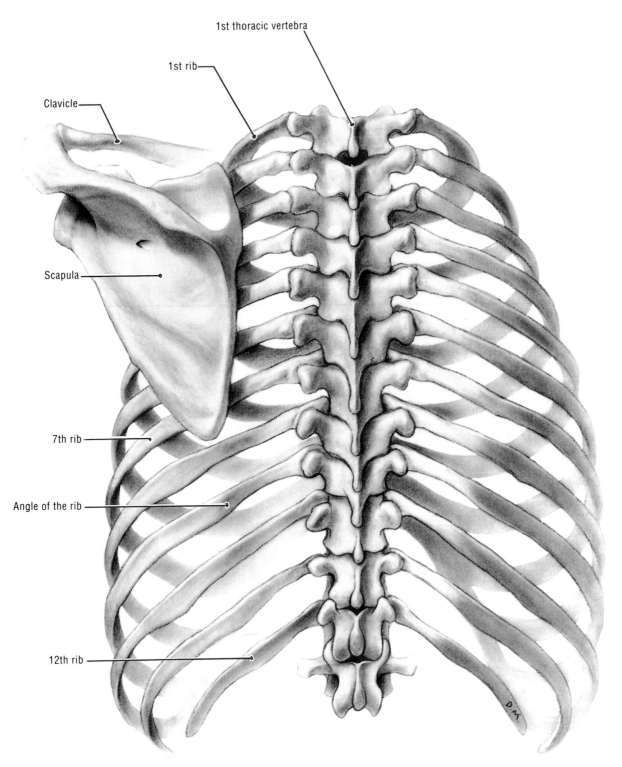

1st thoracic vertebra

1st rib

Clavicle

Scapula

7th rib

Angle of the rib

12th rib

Figure 4-2. Posterior aspect of the osteocartilaginous thoracic cage and the left pectoral girdle (clavicle and scapula). Its anterior aspect is shown in Figure 1-1. Note that the scapula crosses T2 to T7 ribs and that the seventh intercostal space is located just inferior to the inferior angle of the scapula.

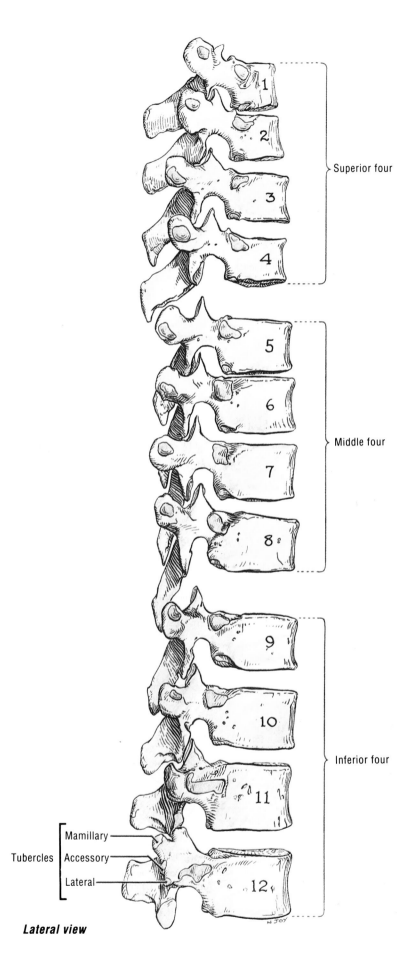

Superior four

Middle four

Inferior four

Tubercles
- Mamillary
- Accessory
- Lateral

Lateral view

Figure 4-19. Inferior four thoracic vertebrae. They are atypical in that they have tubercles typical of lumbar vertebrae (Fig. 4-20).

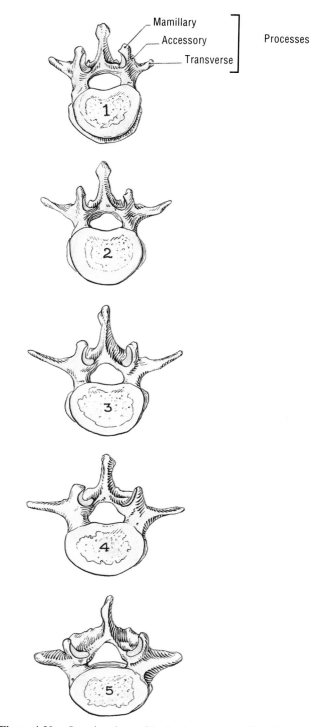

Mamillary
Accessory ⎤
Transverse ⎦ Processes

Figure 4-20. Superior views of the lumbar vertebrae. Note their massive bodies and sturdy vertebral arches.

also be injected into the sacral canal through the posterior sacral foramina (Fig. 4-21B).

In about 5% of people, the L5 vertebra is partly or completely incorporated into the sacrum, conditions known as *hemisacralization and sacralization of L5 vertebra*, respectively (Fig. 4-22). In other people, S1 vertebra is more or less separated from the sacrum and is partly or completely fused with L5 vertebra. This is known as *lumbarization of S1 vertebra*. In some cases of lumbarization of S1 (Fig. 4-23), or of sacralization of L5 (Fig. 4-22), it is the first "normal articulation" (intervertebral disc plus two zygapophyseal joints) that bears the body weight (*i.e.*, takes the strain). Consequently this articulation may degenerate prematurely. For example, when L5 is sacralized (Fig. 4-22), the L5/S1 level is strong and the L4/L5 level degenerates, often producing painful symptoms (*e.g.*, backache).

The Coccyx (Figs. 4-1, 4-21, and 4-24B).   The coccyx (tailbone) is the remnant of the tail which human embryos have until the beginning of the 8th week. Usually four rudimentary vertebrae are present, but there may be one less or one more. The first three vertebrae consist of bodies only (*i.e.*, they have no vertebral arches with their processes). The Co1 vertebra is the largest and broadest of all the coccygeal bones and its rudimentary articular processes form *coccygeal cornua* (horns), which articulate with the *sacral cornua*. The last three coccygeal vertebrae often fuse during middle life, forming a beak-like bone; this accounts for its name, *coccyx* (G. cuckoo). The curvature of the coccyx is more anterior in males than females where the inferior pelvic aperture (pelvic outlet) is larger because it is part of the birth canal (Chap. 3, p. 243). During old age, the Co1 vertebra often fuses with the sacrum (Fig. 4-1). The coccyx does not participate with the vertebral column in the support of the body weight, but it provides attachments for parts of the gluteus maximus and coccygeus muscles and the *anococcygeal ligament* (Fig. 3-34).

The posterior surface and tip of the coccyx can be palpated in the natal cleft (Fig. 4-9). The anterior surface of the coccyx can be palpated with the gloved index finger in the rectum and the thumb in the natal cleft. The coccyx articulates with the sacrum at the *sacrococcygeal joint* (see Fig. 3-34 and p. 251), which permits flexion and extension movements that are important during childbirth. In unusual cases an injury may separate the coccyx from the sacrum, producing *coccydynia* (pain in the coccygeal region, most noticeable during sitting). Coccydynia usually results from a fall directly on the buttocks which injures the coccyx (*e.g.*, during roller skating).

Ossification of Vertebrae

Typical vertebrae begin to ossify toward the end of the embryonic period (7 to 8 weeks). *Three primary ossification centers* develop in each cartilaginous vertebra, one in the centrum and one in each half of the vertebral arch (Moore, 1988). At birth the inferior sacral vertebrae and all of the coccygeal vertebrae are cartilaginous. They begin to ossify during infancy. *At birth each typical vertebra consists of three bony parts*, united by hyaline cartilage (Fig. 4-25A).

arachnoid mater and enters the cerebrospinal fluid in the subarachnoid space (Fig. 4-49). Here it acts directly on the spinal nerves in the *cauda equina* (Figs. 4-48 and 4-51). Sensation is lost inferior to the **epidural block**. As the sacral hiatus is located between the sacral cornua, these horns are important bony landmarks for locating the hiatus. Anesthetic agents can

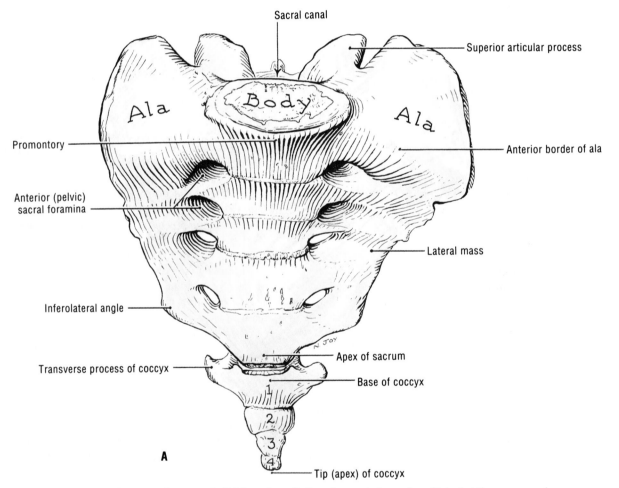

Figure 4-21.   Sacrum and coccyx. *A*, Pelvic surface. *B*, Dorsal or posterior surface. Note that there are no spinous processes or laminae on S4 and S5; this resulted in a large sacral hiatus.

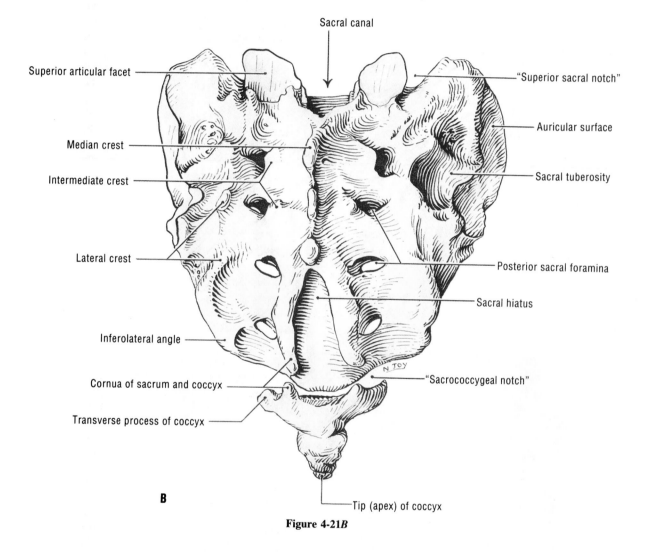

Figure 4-21*B*

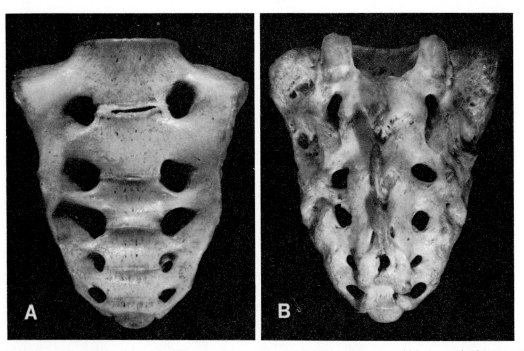

Figure 4-22. Sacralization of the fifth lumbar vertebra. Note that there are five anterior and posterior sacral foramina. *A*, Pelvic surface. *B*, Dorsal surface. As the L5 vertebra is fused with the sacrum, only four typical lumbar vertebrae would be recognized in a radiograph. In *A*, note the remnant of the L5/S1 intervertebral disc (transverse black space).

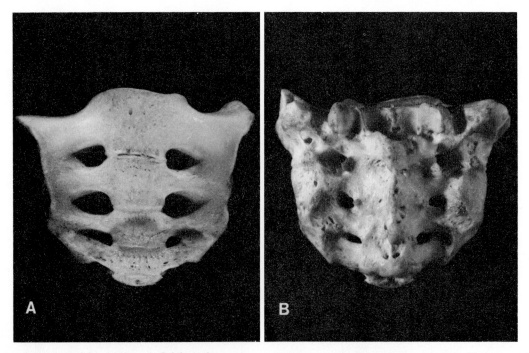

Figure 4-23. Lumbarization of S1 vertebra. *A*, Pelvic surface; note that there are only three anterior sacral foramina. *B*, Dorsal surface. Note how small this sacrum is compared with the normal one shown in Figure 4-21.

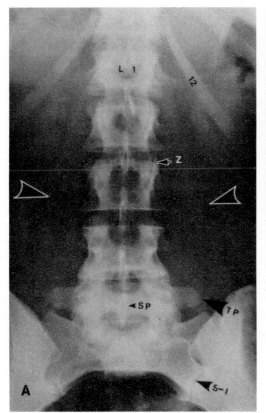
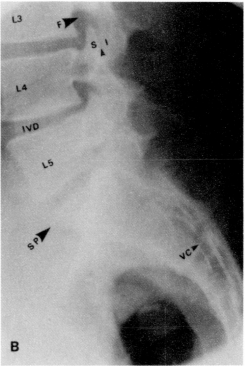

Figure 4-24. Radiographs of the lumbosacral region. *A*, Anteroposterior (*AP*) view. Observe the five lumbar vertebrae. The spinous process (*SP*) and transverse process (*TP*) of L5 vertebra are labeled. The left zygapophyseal joint (*Z*) between L2/L3 vertebrae and the sacroiliac joint (S-I) are also indicated. The *large arrows* point to the lateral margins of the psoas major muscles. *B*, Lateral view. Observe the last three lumbar vertebrae and the spaces between them representing the inter-vertebral discs (*IVD*). Note the angulation at the lumbosacral junction, producing the sacral promontory (*SP*). A *small arrow* points to the zygapophyseal joint between the superior articular process of L4 (*S*) and the inferior articular process of L3 (*I*). A *small arrow* points to the anterior margin of the vertebral canal (*VC*) and the *large arrow* at the top points to the intervertebral foramen (*F*).

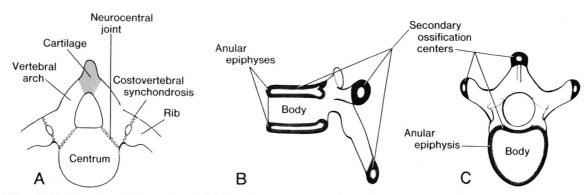

Figure 4-25. Ossification of a typical vertebra. *A*, Midthoracic vertebra at birth consisting of three bony parts. *B* and *C*, Lateral and superior views of a midthoracic vertebra, respectively, showing the location of the secondary centers of ossification. The anular epiphyses form the smooth rounded rims on the circumferences of the superior and inferior surfaces of the vertebral body (Figs. 4-11 and 4-13). (Modified from Moore KL: *The Developing Human: Clinically Oriented Embryology*, ed 4. Philadelphia, WB Saunders, 1988.)

The halves of the vertebral arch begin to fuse in the cervical region during the 1st year, and fusion is usually complete in the lumbar region by the 6th year. At birth the halves of the vertebral arch articulate with the centrum at *neurocentral joints*, which are primary cartilaginous joints (Fig. 4-25A). The vertebral arch fuses with the centrum during childhood (5 to 8 years). *Five secondary ossification centers* develop during puberty (12 to 15 years) in each typical vertebra: one at the tip of the spinous process; one at the tip of each transverse process; and *two anular epiphyses* (ring epiphyses), one on the superior and one on the inferior edge of the centrum (Fig. 4-25B and C). The body of the vertebra forms mainly from growth of the embryonic *centrum*, the ossification center of the central mass of the vertebral body.

All secondary ossification centers are usually united with the vertebra by the 25th year, but the times of their union are variable. Caution must be exercised so that a persistent epiphysis will not be mistaken in a radiograph for a fracture. Exceptions to the typical ossification of vertebrae occur in C1, 2, and 7, the lumbar vertebrae, the sacrum, and coccyx. For details about the ossification of vertebrae, see O'Rahilly (1986) or Williams et al. (1989).

The common congenital malformation of vertebrae is *spina bifida occulta*. This defect, usually in the vertebral arch of L5 and/or S1 vertebrae, *occurs in about 10% of people*. The bifid vertebral arch is concealed by skin, but its location is often marked by a tuft of hair. Spina bifida occulta results from failure of fusion of the laminae of the vertebral arch during the embryonic period (Moore, 1988). Although most persons with spina bifida occulta have no back problems, some people with this bony defect may have low back pain. Clinically significant types of spina bifida associated with neurological symptoms are described on p. 368.

Joints of the Vertebral Column

The vertebrae from C2 to S1 articulate with one another at joints between their bodies and between their articular processes.

Joints of the Vertebral Bodies

The anterior intervertebral joints are *secondary cartilaginous joints* (symphyses) which are designed for strength and weight bearing. The articulating surfaces of the adjacent vertebrae are covered with hyaline cartilage and are connected by a fibrocartilaginous intervertebral disc and ligaments (Figs. 4-4, 4-7, and 4-26 to 4-29). The **intervertebral discs** provide the strongest attachment between the bodies of the vertebrae. In addition to these discs, the bodies are united by strong anterior and posterior longitudinal ligaments. The discs vary in size and thickness in the different regions of the vertebral column. For example, the discs are thinnest in the thoracic region and thickest in the lumbar region where they constitute a third or more of its length. In the cervical and lumbar regions the discs are thicker anteriorly, making them wedge-shaped (Figs. 4-16, 4-24, and 4-29). This structure of the discs is related to the normal curvatures in these regions (Fig. 4-1A). The muscles producing movements of the joints of the vertebral bodies are listed in Tables 4-2 and 4-3.

The anterior longitudinal ligament (Figs. 4-26 and 4-30) is a strong, broad fibrous band which covers and connects the anterior aspects of the bodies of the vertebrae and intervertebral discs. This ligament, which is thickest opposite the discs, extends from the pelvic surface of the sacrum to the anterior tubercle of C1 (atlas) and the occipital bone of the skull, anterior to the foramen magnum. The fibers of the ligament are firmly fixed to the intervertebral discs and the periosteum of the vertebral bodies. This strong ligament maintains the stability of the joints between the vertebral bodies and helps prevent *hyperextension of the vertebral column* (*i.e.*, extension beyond the normal limit of the person concerned).

The posterior longitudinal ligament (Figs. 4-26, 4-30, 4-35, and 4-45) is a narrower, somewhat weaker band than the anterior longitudinal ligament. It runs along the posterior aspect of the vertebral bodies *within the vertebral canal*. It is broadest superiorly where it is continuous with *tectorial membrane*, which is attached to the occipital bone on the internal aspect of the foramen magnum. The posterior longitudinal ligament is also broad in the thoracic and lumbar regions where it is related to the intervertebral discs. It is attached to the intervertebral discs and the posterior edges of the vertebral bodies from the axis (C2) to the sacrum. The posterior longitudinal ligament helps to prevent *hyperflexion of the vertebral column* and posterior protrusion of the nucleus pulposus of the disc.

The Intervertebral Discs (Figs. 4-7, 4-16, 4-24, 4-26, and 4-28). These *plates of fibrocartilage* correspond in shape to the articular surfaces of the vertebral bodies. *The discs play a leading role in weight bearing* and a lesser role in movement. Each disc is composed of an external **anulus fibrosus** which surrounds the internal gelatinous **nucleus pulposus**. The anuli fibrosi insert into the smooth, rounded rims on the articular surfaces of the vertebral bodies. The nuclei pulposi contact the *hyaline cartilage plates*, which are attached to the rough articular surfaces of the vertebral bodies. There are poorly developed intervertebral discs between the bodies of the sacral and coccygeal vertebrae in young persons, but they usually ossify with advancing age (Fig. 4-21A). The most superior intervertebral disc is between C2 and 3. *There is no disc between C1 (atlas) and C2 (axis)*. The most inferior functional disc is between L5 and S1.

The Anulus Fibrosus (Figs. 4-26 to 4-30). This fibrous rim of the intervertebral disc is composed of *concentric lamellae of fibrocartilage*, which run obliquely from one vertebra to another. Some fibers in one lamella are at right angles to those in adjacent ones. This arrangement, while allowing some movement between adjacent vertebrae, provides a very strong bond between them. The lamellae are thinner and less numerous posteriorly than they are anteriorly or laterally.

The Nucleus Pulposus (Figs. 4-7, 4-28, and 4-29). This central core of the intervertebral disc is *more cartilaginous than fibrous* and is normally *highly elastic*. It is located more posteriorly than centrally and has a high water content until old age. The nucleus pulposus (L. *pulpa*, fleshy) acts like a shock absorber for axial forces and like a semifluid ball bearing during flexion, extension, rotation, and lateral flexion of the vertebral column (Fig. 4-36). It becomes broader when compressed (Fig. 4-7B). *The nucleus pulposus is avascular*. It receives its nourishment by diffusion from blood vessels at the periphery of the

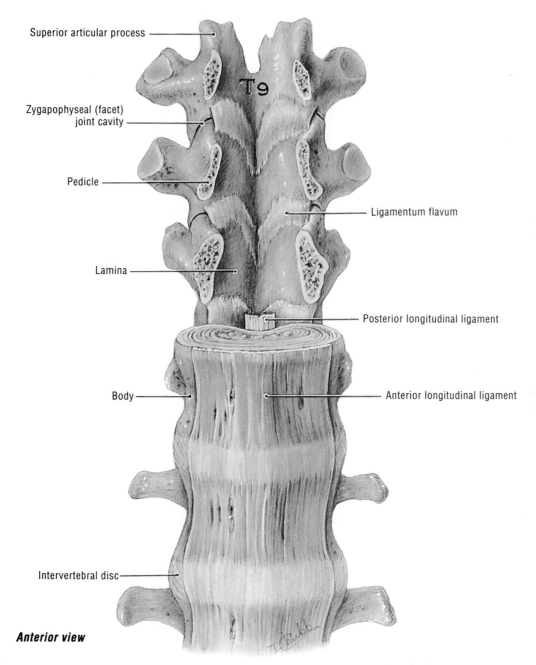

Superior articular process

Zygapophyseal (facet) joint cavity

Pedicle

Lamina

Body

Intervertebral disc

T9

Ligamentum flavum

Posterior longitudinal ligament

Anterior longitudinal ligament

Anterior view

Figure 4-26. Parts of the thoracic and lumbar regions of the vertebral column showing their joints and ligaments. The pedicles of T9 to T11 vertebrae have been sawn through and their bodies have been removed. Note the broad anterior longitudinal ligament and the narrow posterior longitudinal ligament.

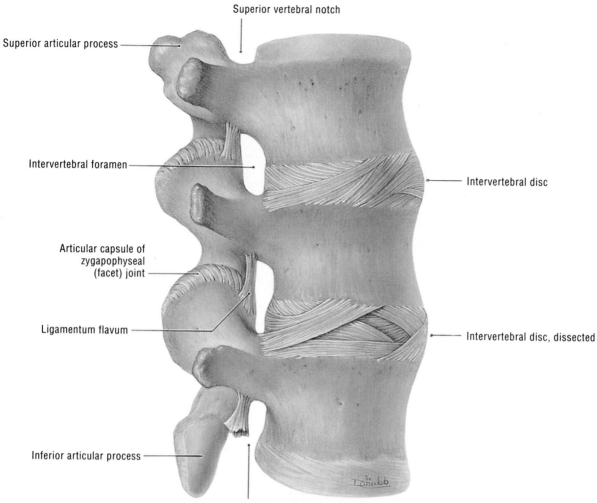

Figure 4-27.   Superior lumbar region of the vertebral column, lateral view, primarily to show the structure of the anuli fibrosi of the intervertebral discs.

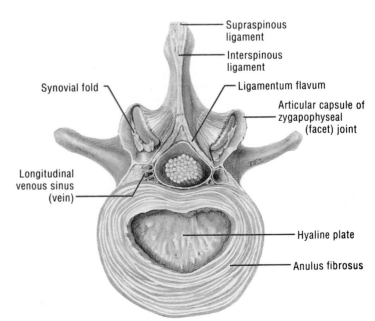

Figure 4-28.   Transverse section of an intervertebral disc and the associated intervertebral ligaments. The nucleus pulposus has been removed to show the hyaline cartilage plate covering the superior surface of the body of this L3 vertebra. Examine the zygapophyseal (facet) joints, the articulations of the vertebral arches.

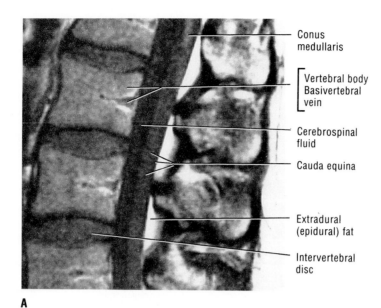

Conus medullaris

Vertebral body
Basivertebral vein

Cerebrospinal fluid

Cauda equina

Extradural (epidural) fat

Intervertebral disc

A

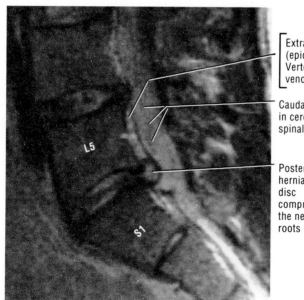

Extradural (epidural fat)
Vertebral venous plexus

Cauda equina in cerebro- spinal fluid

Posterolateral herniation of disc compressing the nerve roots

L5

S1

B

Figure 4-29.   Sagittal MRIs (magnetic resonance images) of the vertebral column and intervertebral discs. *A*, Normal intervertebral disc. *B*, Herniated (ruptured) intervertebral disc. Protrusion of the nucleus pulposus through the anulus fibrosus usually occurs in the lumbosacral region. (Courtesy of Dr. W. Kucharczyk, Clinical Director of the Tri-Hospital Magnetic Resonance Centre, Toronto, Ontario, Canada.)

Table 4-2.
Principal Muscles Producing Movements of the Thoracic and Lumbar Intervertebral Joints

Flexion	Extension	Lateral Flexion	Rotation
Bilateral action of: 　Rectus abdominis (Table 2.1) 　Psoas major (Table 2.5) 　Gravity	Bilateral action of: 　Erector spinae 　Multifidus	Unilateral action of: 　Iliocostalis thoracis and 　　lumborum 　Longissimus thoracis 　Longus thoracis 　Multifidus 　External and internal oblique 　　(Table 2.1) 　Quadratus lumborum (Table 　　2.5)	Unilateral action of: 　Rotatores 　Multifidus 　External oblique acting 　　synchronously with the 　　opposite internal oblique 　　(Table 2.1)

Table 4-3.
Principal Muscles Producing Movements of the Cervical Intervertebral Joints

Flexion	Extension	Lateral Flexion	Rotation
Bilateral action of: 　Longus colli 　Scalene mm. (Fig. 8-11) 　Sternocleidomastoid (Fig. 8-7)	Bilateral action of: 　Splenius capitis (Fig. 4-41) 　Semispinalis capitis and 　　cervicis (Fig. 4-43)	Unilateral action of: 　Iliocostalis cervicis 　Longissimus capitis and 　　cervicis (Fig. 4-42) 　Splenius capitis and cervicis 　　(Fig. 4-41)	Unilateral action of: 　Rotatores (Fig. 4-45) 　Semispinalis capitis and 　　cervicis (Fig. 4-43) 　Multifidus (Fig. 4-44) 　Splenius cervicis (Fig. 4-41)

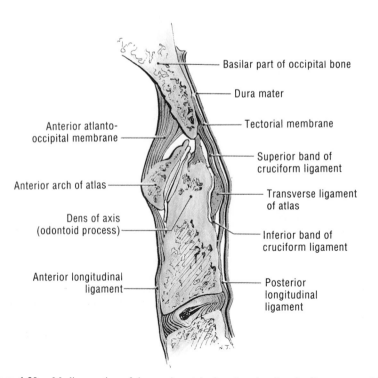

Figure 4-30.   Median section of the craniovertebral region showing the ligaments and joints.

anulus fibrosus and from the adjacent surfaces of the vertebral bodies.

The anterior longitudinal ligament is severely stretched and may be torn during severe *hyperextension of the neck*, e.g., the so-called "whiplash injury" (Fig. 4-31). There may also be a *hyperflexion injury* to the cervical region of the vertebral column as the head snaps back onto the thorax. "Facet jumping" or locking of the cervical vetebrae may occur owing to dislocation of the vertebral arches. The association of rear end automobile collision and these injuries is well known, especially to litigation lawyers. The intervertebral discs in young persons are very strong and the water content of their nuclei pulposi is high (about 88%). This gives them greater turgor (fullness). In young adults, the intervertebral discs are usually so strong that the vertebrae often break before the discs rupture during a fall. However, *violent hyperflexion of the vertebral column*, may rupture an intervertebral disc and fracture the adjacent vertebral bodies.

As people get older, their nuclei pulposi lose their turgor and become thinner owing to dehydration and degeneration. These age changes account in part for the slight loss in height that occurs during old age. During flexion of the vertebral column the nucleus pulposus is pushed further posteriorly. If wearing of the anulus fibrosus has occurred, either posteriorly or laterally, the nucleus pulposus may rupture and protrude or prolapse (Fig. 4-32). It then presses on the spinal cord, cauda equina, or the emerging spinal nerves. A herniated or *protruding intervertebral disc* is often inappropriately called a "slipped disc" by lay people. Sports announcers often call the injury a "ruptured disc." **Protrusions of the nucleus pulposus** usually occur posterolaterally, where the anulus fibrosus is weak and poorly supported by the posterior longitudinal ligament (Figs. 4-26 and 4-32). The protruding part of the nucleus pulposus may compress an adjacent spinal nerve root, causing severe low back pain and/or leg pain.

Acute mid and low back pain (lumbago) that radiates down the posterolateral aspect of the thigh and leg, is often caused by a *posterolateral protrusion of a lumbar intervertebral disc* at the L5/S1 level (Fig. 4-29*B*). The clinical picture varies considerably, but pain of acute onset in the lower back is a common presenting symptom. Because there is muscle spasm associated with low back pain, the lumbar region of the vertebral column becomes rigid and movement is painful. With treatment, this type of back pain usually begins to fade after a few days, but it may be gradually replaced by **sciatica** (pain resulting from irritation of the sciatic nerve). *About 95% of lumbar disc protrusions occur at the L4/5 or L5/S1 levels.* The remaining protrusions are at the L3/L4 level. Any maneuver that stretches the sciatic nerve, such as flexing the thigh with the leg extended, will produce or exacerbate the pain caused by a disc protrusion. Intervertebral discs may also be damaged by violent rotation or flexing of the vertebral column. Posterior herniations of the nucleus pulposus, although less common than posterolateral ones, may also exert pressure on the spinal cord in cervical or thoracic levels, and on the cauda equina at inferior levels (Figs. 4-28 and 4-32).

Symptom-producing disc protrusions occur in the cervical region almost as often as in the lumbar region. A forcible hyperflexion of the cervical region (*e.g.*, during a head-on collision) may rupture the disc posteriorly without fracturing the vertebral body. The cervical discs most commonly ruptured are those between C5/C6 and C6/C7, compressing spinal nerve roots C6 and C7, respectively. The general rule is that when a disc protrudes, it may compress the nerve roots numbered one inferior to the disc, *i.e.*, S1 by L5 disc, L5 nerve by L4 disc, and C7 nerve by C6 disc. Disc protrusions result in pain in the neck, shoulder, arm, and hand. Rupture of an intervertebral disc may be caused by a sudden injury or result from chronic stress, such as occurs with constant lifting or prolonged obesity. Any sport in which movement causes downward or twisting pressure on the neck or lower back may produce a herniated nucleus pulposus (Griffith, 1986). The most common sports involved are bowling, tennis, jogging, football, hockey, weight-lifting, and gymnastics.

Joints of the Vertebral Arches

These synovial joints are between the inferior articular processes of a superior vertebrae and the superior articular processes (zygapophyses) of an inferior vertebrae (Figs. 4-26 to 4-28). These plane joints are known as **zygapophyseal joints**.[3] The flat surfaces of the articular facets are covered with hyaline cartilage. Each joint is surrounded by a thin, loose articular capsule which is attached to the articular margins of the processes (Figs. 4-27 and 4-28). The fibrous capsules are longer and looser in the cervical region than in the thoracic and lumbar regions. Consequently, flexion is most extensive in the cervical region. The fibrous capsule of each joint is lined with a synovial membrane. *The zygapophyseal (facet) joints permit gliding movements* between the vertebrae. The muscles producing these movements are listed in Tables 4-2 and 4-3. In the cervical and lumbar regions, these joints bear some weight, sharing this function with the intervertebral discs. These joints help to control flexion, extension, and rotation of adjacent cervical and lumbar vertebrae (Fig. 4-36). Most of the movement of the vertebral column takes place in these two regions.

Innervation of the Zygapophyseal (Facet) Joints (Figs. 4-15, 4-33, and 4-45). These articulations are innervated by nerves that arise from medial branches of the dorsal primary rami of spinal nerves. As these nerves pass posteroinferiorly, they lie

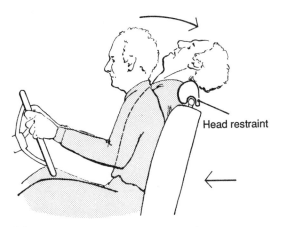

Figure 4-31. Hyperextension of the neck during a rear end motor vehicle collision. Hyperflexion of the neck may also occur later when the head is thrown anteriorly. Note that the head restraint is not raised to a position where it would likely have prevented a neck injury.

Head restraint

---

[3]For brevity they are usually called *facet joints*.

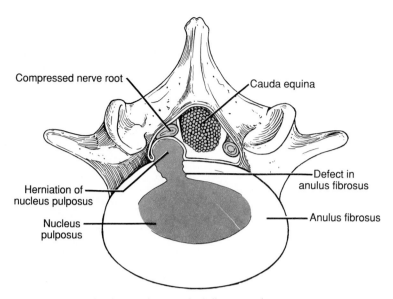

Figure 4-32. An illustration showing how an intervertebral disc protrusion may exert pressure on a spinal nerve root and/or the cauda equina.

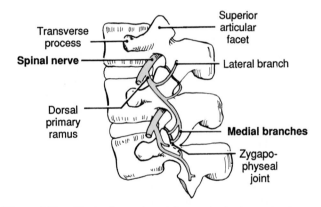

Figure 4-33. The lumbar region of the vertebral column showing innervation of the zygapophyseal joints of the vertebral arches. Observe that the dorsal primary ramus arises from the spinal nerve outside the intervertebral foramen, where it divides into medial and lateral branches.

in grooves on the posterior surfaces of the medial parts of the transverse processes. Each articular branch supplies the joint nearby, and it may send twigs to the subjacent joint as well.

The zygapophyseal (facet) joints are of clinical interest because they are close to the intervertebral foramina through which the spinal nerves emerge from the vertebral canal (Figs. 4-15, 4-27, and 4-33). When these joints are injured or diseased (*e.g.*, owing to *osteoarthritis*), the related spinal nerves are often affected. This causes pain along the distribution patterns of the *dermatomes* (see Fig. 1-24), and spasm in the muscles derived from the associated myotomes. *Denervation of lumbar zygapophyseal joints* is one procedure used for treatment of back pain which is thought to be caused by disease of these joints. In some cases the nerves are sectioned near these joints; in other cases the nerves are destroyed by radiofrequency *percutaneous rhizolysis* (G. *rhiza*, root, + *lysis*, dissolution). In each procedure the destructive process

is directed at the medial branches of the dorsal primary rami of the spinal nerves which supply these joints (Figs. 4-15 and 4-33).

Accessory Ligaments of the Intervertebral Joints (Figs. 4-26 to 4-28). The laminae of adjacent vertebral arches are joined by broad, elastic bands called *ligamenta flava* (yellow ligaments), which extend almost vertically from the lamina above to the lamina below. The ligamentum flavum was given its name because its fibers consist mainly of yellow elastic tissue (L. *flavus, yellow*). The ligaments are attached superiorly to the anterior surfaces of the inferior borders of a pair of laminae, and inferiorly to the posterior surfaces of the superior border of the next succeeding pair. Some of their fibers extend to the articular capsules of the zygapophyseal (facet) joints and contribute to the posterior boundaries of the intervertebral foramina (Fig. 4-27). The ligamenta flava help to preserve the normal curvature of the vertebral column and to straighten the column after it has been flexed.

Adjacent spinous processes are joined by weak *interspinous ligaments* and a strong cord-like *supraspinous ligament* (Figs. 4-28 and 4-55B). These ligaments are represented superiorly by the *ligamentum nuchae*, a triangular median septum between the muscles on each side of the posterior aspect of the neck (Fig. 4-41). The *intertransverse ligaments*, connecting adjacent transverse processes, consist of a few scattered fibers, except in the lumbar region where they are membranous and more substantial.

The Craniovertebral Joints

The suboccipital joints are between the skull and C1 (atlas), and between C1 and 2 (atlas and axis). They are called the atlanto-occipital and atlanto-axial joints (*atlanto* refers to the atlas). The main differences between these joints and others in the vertebral column are: (1) they are synovial only. There are *no intervertebral discs*; and (2) there are *no zygapophyseal (facet) joints*.

The Atlanto-occipital Joints (Figs. 4-16, 4-17, 4-30, 4-34, and 4-35). These articulations between C1 (atlas) and the occipital condyles *permit nodding of the head (i.e.*, the flexion and extension of the neck that occurs when indicating approval). The muscles producing movements of these joints are listed in Table 4-4. The atlanto-occipital joints, one on each side, are between the superior articular facets on the lateral masses of C1 and the occipital condyles. They are *synovial joints of the condyloid type* (p. 19). They have thin, loose articular capsules which are lined

with synovial membrane. The skull and C1 are also connected by anterior and posterior *atlanto-occipital membranes*, which extend from the anterior and posterior arches of C1 to the anterior and posterior margins of the foramen magnum. They prevent excessive movement of the atlanto-occipital joints.

The transverse ligament of the atlas (Figs. 4-14, 4-17, 4-30, 4-34, and 4-35) is a strong band extending between the tubercles on the lateral masses of C1 vertebrae. It holds the dens of C2 (axis) against the anterior arch of C2 (atlas). There is a synovial joint between them. Vertically oriented superior and inferior bands pass from the transverse ligament to the occipital bone superiorly and to the body of C2 inferiorly. They form the *cruciform ligament* (L. *crux*, cross) which was given this name because of its resemblance to a cross.

The alar ligaments extend from the sides of the dens to the lateral margins of the foramen magnum (Figs. 4-34 and 4-35).

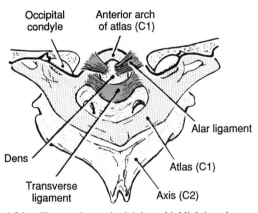

Figure 4-34. The craniovertebral joints, highlighting the transverse ligament of the atlas and the alar ligament. The cranium and atlas (C1) rotate as a unit on the axis (C2). Excessive rotation is prevented by the alar ligaments, which act as "check ligaments."

Table 4-4.
Principal Muscles Producing Movements of the Atlanto-occipital Joints

Flexion	Extension	Lateral Flexion
Longus capitis	Rectus capitis	Sternocleidomastoid
Rectus capitis	posterior major	(acting
anterior	and minor	unilaterally)
Anterior fibers of	Obliquus capitis	Obliquus capitis
sternocleidomastoid	superior	superior and
(Fig. 8-7)	Semispinalis capitis	inferior
	Splenius capitis	Rectus capitis
	Longissimus capitis	lateralis
	Trapezius (acting	Longissimus capitis
	bilaterally, Fig.	Splenius capitis
	8-8)	

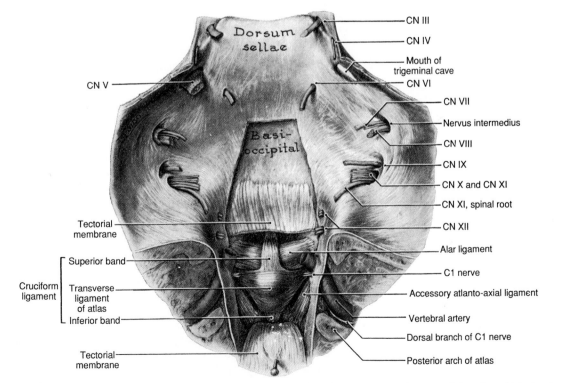

Figure 4-35. Superior view of the craniovertebral region showing the ligaments of the atlanto-axial and atlanto-occipital joints. Observe the bow-shaped transverse ligament of the atlas which, by the addition of superior and inferior bands, becomes a cross-shaped cruciform ligament.

These short, strong, rounded cords attach the skull to C2 vertebra. They *check rotation and side-to-side movements* of the head.

The tectorial membrane is the superior continuation of the posterior longitudinal ligament (Fig. 4-30). It runs from the body of C1 to the internal surface of the occipital bone and covers the alar and transverse ligaments (Fig. 4-35).

The Atlanto-axial Joints (Fig. 4-30).   These are synovial joints between C1 and C2 vertebrae. There are two lateral joints and one median joint. *The movement at these joints is rotation,* which permits the head to be turned from side to side (*e.g.*, when rotating the head to indicate disapproval.) During this movement, the skull and C1 rotate as a unit on C2. The muscles producing movement of these joints are described on p. 358 and are listed in Table 4-5. Excessive rotation of these joints is prevented by the *alar ligaments* (Fig. 4-34). Movements of the joints between the skull and C1 and between C1 and C2 vertebrae are augmented by the flexibility of the neck owing to movements of the joints of the vertebral column in the middle and inferior cervical regions. During rotation of the head, the dens of C2 is held in a collar formed by the anterior arch of the atlas and the transverse ligament of the atlas (Figs. 4-17 and 4-34). The articulation of the dens with C1 is a *pivot joint* (p. 19).

If the transverse ligament of the atlas ruptures or is destroyed by disease (*e.g.*, *rheumatoid arthritis*), the dens is set free and may be driven into the cervical region of the spinal cord, causing *quadriplegia*, or into the medulla (inferior end of the brain stem), causing sudden death. The latter injury usually occurs when people hang themselves. *All neck injuries are potentially serious* because of the possibility of fracturing vertebrae and injuring the spinal cord (Figs. 4-1, 4-15, and 4-37). This is the reason first aid people are trained not to move a person with a neck or back injury, unless his/her life is in danger. *The cervical region is especially vulnerable to injury.* Injuries to the spinal cord in the cervical region, associated with compression or transection of the spinal cord, may result in loss of all sensation and voluntary movement inferior to the lesion; or in sudden death, depending on the level of injury. Compression of any part of the central nervous system rendering it ischemic for 3 to 5 minutes results in death of nervous tissue, particularly nerve cells.

Movements of the Vertebral Column

The range of movement of the vertebral column varies considerably according to the individual. It is extraordinary in some people (*e.g.*, acrobats) who begin to train during early childhood.

Table 4-5.
Principal Muscles Producing Rotation at the Atlanto-axial Joints[1]

Ipsilateral[2]	Contralateral
Obliquus capitis inferior	Sternocleidomastoid
Rectus capitis posterior, major and minor	(Fig. 8-7)
Longissimus capitis	
Splenius capitis	

[1]Rotation is the specialized movement at these joints. Movement of one joint involves the other.
[2]The same side to which the head is rotated.

The extreme extension exhibited by acrobats who can put their heads between their lower limbs would be *hyperextension* in almost everyone. Forced extension to this degree in a nonacrobat would not be possible without straining or rupturing the anterior longitudinal ligament and dislocating the intervertebral joints (Fig. 4-26). The normal range of movement is limited by: (1) the thickness and compressibility of the intervertebral discs, (2) the resistance of the muscles and ligaments of the back, and (3) the tension of the articular capsules of the zygapophyseal (facet) joints (Fig. 4-27). Movements between adjacent vertebrae take place on the resilient nuclei pulposi of the intervertebral discs (Figs. 4-4 and 4-7), and at the zygapophyseal (facet) joints. Although movements between adjacent vertebrae are relatively small, especially in the thoracic region, the summation of all of the small movements produces a considerable range of movement of the vertebral column as a whole (Figs. 4-8 and 4-36).

Movements of the vertebral column are freer in the cervical and lumbar regions than elsewhere. As stated previously, these regions are also the most frequent sites of aches, pain, and serious injuries. The movements of the vertebral column are *flexion*, *extension*, *lateral flexion* (bending), and *rotation*. Rocking, rotation, and gliding occur at the joints of the vertebral bodies, and gliding movements occur at the zygapophyseal (facet) joints (Figs. 4-24 and 4-27). *The thoracic region of the vertebral column is relatively stable* owing to its connection to the sternum via the ribs and costal cartilages. In addition, the intervertebral discs are slightly thinner and their spinous processes overlap (Figs. 4-1 and 4-2). *Extension is most marked in the lumbar region* and generally it is more extensive than flexion. *Flexion is greatest in the cervical region* and is almost nonexistent in the thoracic region (Fig. 4-36C).

During flexion of the lumbar region the *nucleus pulposus* moves posteriorly, putting tension on the posterior part of the *anulus fibrosus*. This is the major reason that posterolateral herniations of the nucleus pulposus through the anulus fibrosus are most common in the lower lumbar and lumbosacral regions (Figs. 4-29B and 4-32). *Lateral flexion is greatest in the lumbar region*; it is restricted in the thoracic region by the ribs. Because of the greater rotation and gliding movements between its vertebrae, *rotation is most marked in the thoracic region* (Fig. 4-36C).

Fractures, dislocations, and fracture-dislocations of the vertebral column usually result from *sudden forceful flexion*, as may occur in a car accident, or from a violent blow on the back of the head. The common fracture is a *compression fracture of the body of one or more vertebrae*. In severe flexion injuries the posterior longitudinal and interspinous ligaments are torn, and the vertebral arches may be dislocated and/or fractured along with fractures of the vertebral bodies. *With most severe flexion injuries, there are injuries to the spinal cord.* When a person falls from a height and lands on the crown of the head, the violence is transmitted along the axis of the vertebral column (Fig. 4-1). A hard fall may force the occipital condyles into the atlas, splitting it into two or more fragments. In other cases, the thin bone around the occipital condyles fractures. Falling on the feet or the buttocks from a height produces a similar axial force.

Sudden forceful extensions can also injure the vertebral column. Extension fractures and/or dislocations vary from one

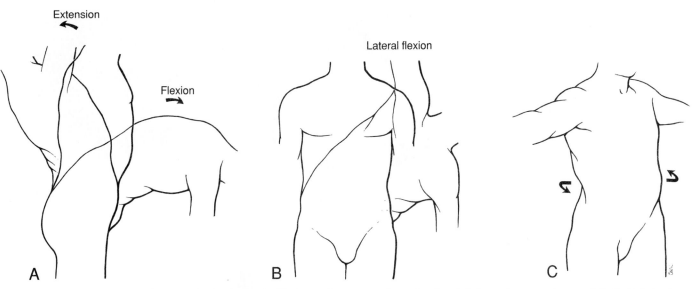

Figure 4-36. Movements of the vertebral column. *A*, Flexion and extension; extension beyond the anatomical position is called hyper-extension only when it is beyond a person's normal limit. *B*, Lateral flexion. *C*, Rotation.

vertebral region to another, but posterior portions of the vertebral column are most likely to be injured. *Severe hyperextension of the neck* (Fig. 4-31) may pinch the posterior arch of C1 between the occiput and C2. In these cases C1 usually breaks at one or both grooves for the vertebral arteries (Figs. 4-14 and 4-17). If the extension force is severe, the anterior longitudinal ligament and adjacent anulus fibrosus of the C2/C3 intervertebral disc may rupture. In these cases the skull, C1, and C2 are separated from the rest of the axial skeleton and the spinal cord is usually severed. Persons with this injury seldom survive more than 5 minutes because the injury to the spinal cord is superior to the *phrenic outflow*, the origin of the phrenic nerves (C3, C4, and C5). As these nerves innervate the diaphragm (see Fig. 2-101), respiration is severely affected.

Sometimes when the neck is severely hyperextended while the head is turned to the side, the spinal ganglion of C2 nerve on the opposite side is compressed between the posterior arches of C1 and C2 (Figs. 4-46 and 4-47). This may be followed by prolonged headaches in the occipital region, so severe that they produce suicidal tendencies. *Dislocation of vertebrae without fractures is uncommon*, except in the cervical region, because of the interlocking of the thoracic and lumbar articular processes. The vertebral canal in the cervical region is usually somewhat larger than the spinal cord (Fig. 4-1*B*), therefore there can be some displacement of the vertebrae without causing serious damage to the spinal cord (Fig. 4-37*B*).

Displacement of the Vertebral Column (Figs. 4-38 and 4-39). In a few people, the L5 vertebra consists of two parts. The posterior fragment remains in normal relation to the arch of the sacrum, but the anterior fragment and the superimposed vertebra may move anteriorly. This anterior displacement of most of the vertebral column is called *spondylolisthesis*. If the anterior part of the bone does not move anteriorly, the condition is called *spondylolysis*. Spondylolisthesis at L5 may result in pressure on the spinal nerves as they pass into the superior part of the sacrum, causing back pain.

Muscles of the Back

There are *three groups of muscles* in the back: *superficial*, *intermediate*, and *deep* (Fig. 4-40). The superficial and intermediate groups are **extrinsic back muscles** that are concerned with limb movements and respiration, respectively. The deep group constitutes the **intrinsic back muscles** that are concerned with movements of the vertebral column (Fig. 4-36). The extrinsic muscles are superficial to the intrinsic muscles.

The Extrinsic Back Muscles

The superficial muscles of the back (*e.g.*, trapezius and latissimus dorsi, Fig. 4-40) connect the upper limbs to the trunk and are concerned with movements of these limbs. They are described in Chap. 6 (p. 530 and Table 6-4). *The intermediate muscles of the back* (serratus posterior) are superficial respiratory muscles (Fig. 4-40). They are described with the thorax (Chap. 1, p. 50).

The Intrinsic Back Muscles

The intrinsic or **deep muscles of the back** (*e.g.*, the erector spinae, Figs. 4-40 and 4-41) are concerned with the maintenance of posture and movements of the vertebral column and head. The muscles are named according to their relationship to the surface: (1) *a superficial layer*, (2) *an intermediate layer*, and (3) *a deep layer*.

Superficial Layer of Intrinsic or Deep Back Muscles

The Splenius Muscles (Fig. 4-41). These bandage-like muscles (G. *splenion*, bandage) are applied to the sides and

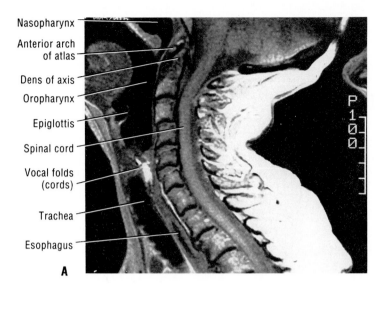

Nasopharynx

Anterior arch of atlas

Dens of axis

Oropharynx

Epiglottis

Spinal cord

Vocal folds (cords)

Trachea

Esophagus

A

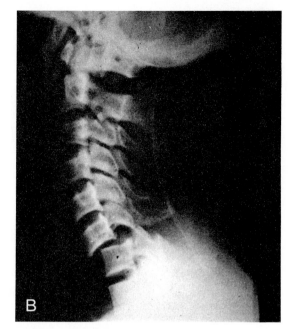

B

Figure 4-37.   *A*, Sagittal MRI (magnetic resonance image) of the neck. (Courtesy of Dr. W. Kucharczyk, Clinical Director of Tri-Hospital Resonance Centre, Toronto, Ontario, Canada.) *B*, Lateral radiograph of the neck showing the cervical region of the vertebral column. Note the dislocation of C6 vertebra on C7.

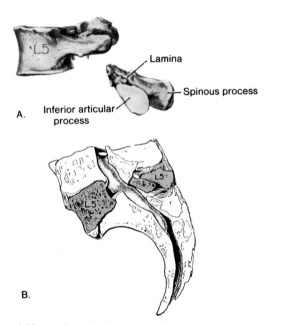

Lamina

Spinous process

A.   Inferior articular process

B.

Figure 4-38.   *A*, Bipartite L5 vertebra. Note that one piece, consisting of the spinous process, laminae, and inferior articular processes, is separated from the body, pedicles, and superior articular processes. *B*, When standing, the body and superimposed vertebrae tend to slide anteriorly, a condition known as spondylolisthesis.

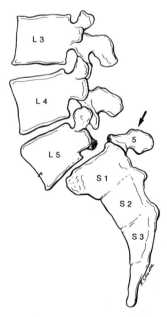

Figure 4-39.   Drawing based on a lateral radiograph of a patient with spondylolisthesis of L5 vertebra. Note the anterior movement of the body of this vertebra.

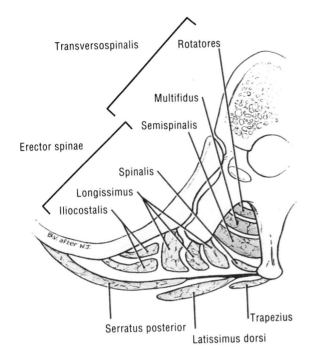

Figure 4-40. Transverse section of the muscles of the back. Note that the erector spinae muscles are in three columns and the transversospinalis muscle is in three layers.

back of the neck, somewhat like spiral bandages. They ascend from the median plane of the neck and the transverse processes of the superior cervical vertebrae to the base of the skull. Each muscle is divided into a cranial portion, *splenius capitis* (L. *caput*, head), and a cervical portion, *splenius cervicis* (L. *cervix*, neck).

Origin. Inferior half of the ligamentum nuchae and the spinous processes of T1 to T6 vertebrae.

Insertion. *Splenius capitis* inserts into the lateral aspect of the mastoid process and the lateral third of the superior nuchal line of the occipital bone (deep to the sternocleidomastoid muscle). *Splenius cervicis* inserts into the posterior tubercles of the transverse processes of C1 to C4 vertebrae (posterior to the levator scapulae muscle).

Actions (Table 4-3). Acting alone, the splenius muscles laterally flex and rotate the head and neck to the same side. Acting together, they extend the head and neck.

Innervation. Dorsal rami of inferior cervical nerves.

Intermediate Layer of Intrinsic or Deep Back Muscles

The Erector Spinae Muscle (Figs. 4-9 and 4-40 to 4-42). This *massive muscle* forms a prominent bulge on each side of the vertebral column. It lies within a fascial compartment between the posterior and anterior layers of the **thoracolumbar fascia** (Fig. 4-44; see also Fig. 2-102). The erector spinae is arranged in three vertical columns: *iliocostalis* (lateral column); *longissimus* (intermediate column); and *spinalis* (medial column). The *common origin* of the three columns is through a broad tendon which is attached inferiorly to the posterior part of the *iliac crest*, the posterior aspect of the *sacrum*, the *sacroiliac ligaments*, and the *sacral* and *inferior lumbar spinous processes*.

Details of their superior attachments, being relatively unimportant, are set in intermediate type.

The Iliocostalis Muscle (Figs. 4-40 to 4-42). This lateral column of the erector spinae arises from the common origin and inserts into the angles of the ribs (L. *costae*). It may be divided into three parts according to the region involved.

The iliocostalis is often divided into: (1) the *iliocostalis lumborum*, which is attached to the inferior six ribs; (2) the *iliocostalis thoracis*, which is attached to all of the ribs; and (3) the *iliocostalis cervicis*, which is attached to the superior six ribs and the posterior tubercles of C4 to C6 vertebrae.

The Longissimus Muscle (Figs. 4-40 to 4-42). This intermediate column of the erector spinae arises from the common origin and is attached to the transverse processes of the thoracic and cervical vertebrae, and the mastoid process of the temporal bone of the skull. This gives the muscle a herring-bone appearance. The longissimus may also be divided into three parts according to the regions it traverses.

The *longissimus thoracis* inserts into the tips of the transverse processes of all of the thoracic vertebrae, and into the tubercles of the inferior nine to 10 ribs. The *longissimus cervicis* extends from the superior thoracic transverse processes to the cervical transverse processes. The *longissimus capitis* arises in common with the cervical part and attaches to the mastoid process of the temporal bone.

The Spinalis Muscle (Figs. 4-40 to 4-42). This narrow medial column of the erector spinae is relatively insignificant. It arises from the common origin and extends from the spinous processes in the superior lumbar and inferior thoracic regions to the spinous processes in the superior thoracic region. It may also be divided into three parts (spinalis thoracis, spinalis cervicis, and spinalis capitis).

Actions of the Erector Spinae Muscles (Figs. 4-4 and 4-36; Tables 4-2 and 4-3). Acting bilaterally, all three columns of the erector spinae extend the head and part or all of the vertebral column. Acting unilaterally, the erector spinae laterally flexes the head or the vertebral column. In addition, the longissimus capitis muscle rotates the head so that it is turned to the same side. *The erector spinae muscles are the chief extensors of the vertebral column.* They straighten the flexed column and can bend it posteriorly. They also "pay out" (release) during its flexion so that the movement is slow and controlled.

> *Back strain* is a common back problem in persons who participate in sports. It results from extreme movements of the vertebral column, such as excessive extension or rotation. The term "strain" is used to indicate some degree of stretching or microscopic tearing of the muscle fibers and/or ligaments of the back. The muscles usually involved are those producing movements of the lumbar intervertebral joints, especially the *parts of the erector spinae* (Table 4-2).

Deep Layer of Intrinsic or Deep Back Muscles

When the massive erector spinae muscles are removed, several short muscles (semispinalis, multifidus, and rotatores) are visible in the groove between the transverse and spinous processes of the vertebrae (Fig. 4-40). Collectively this obliquely

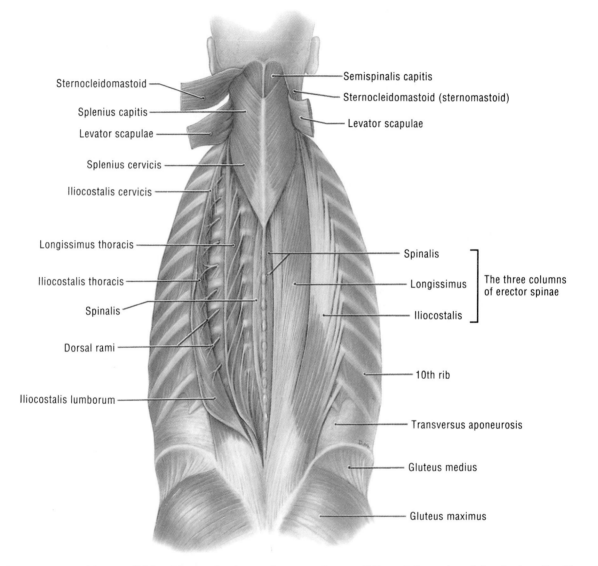

Sternocleidomastoid

Splenius capitis

Levator scapulae

Splenius cervicis

Iliocostalis cervicis

Longissimus thoracis

Iliocostalis thoracis

Spinalis

Dorsal rami

Iliocostalis lumborum

Semispinalis capitis

Sternocleidomastoid (sternomastoid)

Levator scapulae

Spinalis

Longissimus — The three columns of erector spinae

Iliocostalis

10th rib

Transversus aponeurosis

Gluteus medius

Gluteus maximus

Figure 4-41.   Dissection of the superficial and intermediate layers of deep muscles in the back. Part of the superficial layer (splenius capitis) is reflected on the left side. The intermediate layer (erector spinae) is intact on the right side, lying between the spinous processes of the vertebrae medially and the angles of the ribs laterally. Most of the ligamentum nuchae was removed with the trapezius muscles (see also Fig. 6-40). This nuchal (neck) ligament is a sheet of mixed white fibrous tissue that is interposed between the cervical muscles of the two sides.

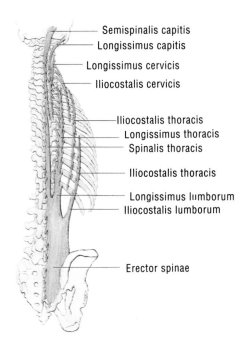

Figure 4-42. The erector spinae and semispinalis muscles. In Fig. 4-40 and here, observe that the erector spinae muscle is arranged in three columns that have a common origin.

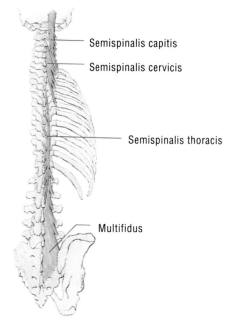

Figure 4-43. The back muscles which lie deep to the erector spinae muscles. As most of them originate from transverse processes and insert into spinous processes, they are known collectively as the transversospinal muscles (Fig. 4-40).

disposed group is known as the *transversospinal muscle* because their fibers run from the transverse processes to the spinous processes (spines) of the vertebrae.

The Semispinalis Muscle (Figs. 4-40 and 4-43 to 4-47; Table 4-3). As its name indicates, this muscle arises from about half of the vertebral column (spine). It can be divided into three parts according to their superior attachments: *semispinalis thoracis*, *semispinalis cervicis*, and *semispinalis capitis*. Semispinalis thoracis and cervicis pass superomedially from the transverse processes to the thoracic and cervical spinous processes superiorly. *Semispinalis capitis* arises from the transverse processes of T1 to T6 vertebrae and inserts into the medial half of the area between the superior and inferior nuchal lines on the occipital bone. It forms the largest muscle mass in the posterior aspect of the neck.

Actions (Fig. 4-36 and Table 4-3). Bilaterally, the semispinalis thoracis and cervicis extend the cervical and thoracic regions of the vertebral column. Unilaterally, they rotate these regions toward the opposite side. Bilaterally the semispinalis capitis muscles extend the head.

Innervation (Fig. 4-46). These muscles are supplied by the dorsal rami of the cervical spinal nerves.

The Multifidus Muscle (Figs. 4-40, 4-43, and 4-44; Tables 4-2 and 4-3). The name multifidus (L. *multus*, many, + *findo*, to cleave) indicates that this muscle is divided into several bundles. It covers the laminae of S4 to C2 vertebrae. Its fibers pass superomedially from the vertebral arches to the spinous processes, spanning one to three vertebrae.

Actions. Acting unilaterally, the multifidus muscle flexes the trunk laterally and rotates it to the opposite side. Acting bilaterally, they extend the trunk and stabilize the vertebral column.

Innervation (Fig. 4-46). These muscles are supplied by the dorsal rami of the cervical spinal nerves.

The Rotatores Muscles (Figs. 4-40 and 4-45; Table 4-3). These short muscles, the deepest ones in the groove between the spinous and transverse processes, run the entire length of the vertebral column. They are easiest to observe in the thoracic region where rotation of the vertebral column is greatest. These rotators arise from the transverse process of one vertebra and insert into the base of the spinous process of the vertebra superior to it. They *rotate the superior vertebra to the opposite side; they also stabilize it.* They are innervated by the dorsal rami of the spinal nerves.

The Interspinales and Intertransversarii Muscles (Fig. 4-44; see also Fig. 2-102). These small muscles unite the spinous and transverse processes of consecutive vertebrae. They are well developed in the cervical region. *The interspinales help to extend the vertebral column.* The intertransversarius can produce lateral flexion of the superior vertebra, and acting bilaterally, they help to extend the vertebral column. The interspinales are innervated by the dorsal rami of the cervical spinal nerves; the intertransversarii are supplied mainly by ventral rami of the cervical spinal nerves, but some are supplied by the dorsal rami.

The Levator Costarum Muscles (Fig. 4-45; see also Fig. 1-21). In the thoracic region, these muscles represent the posterior intertransversarius muscles of the neck. There are 12 on each side. They extend inferolaterally from the tip of the transverse process to the rib just inferior to it (between the tubercle and angle; see Chap. 1, p. 36). They *raise the ribs during inspiration* and are supplied by the lateral division of the dorsal rami of spinal nerves (Fig. 1-18).

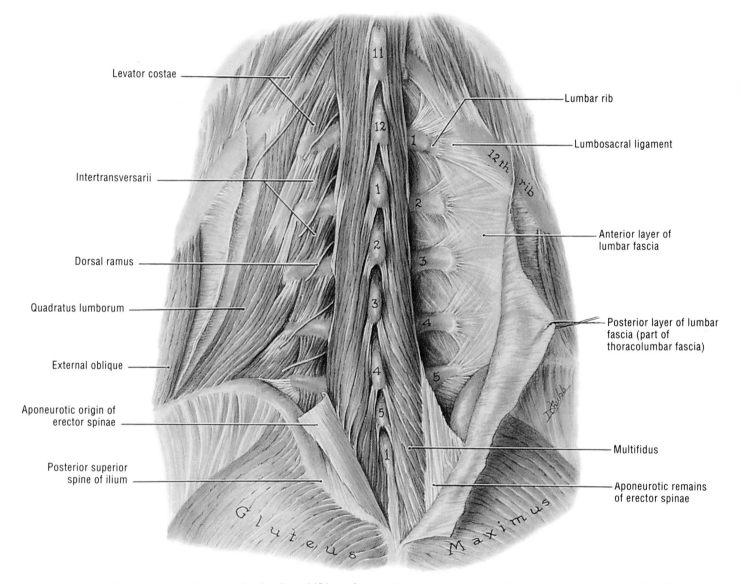

Figure 4-44. Deep dissection of the back showing the multifidus and other deep back muscles. Note that a short lumbar rib is present; this common malformation is usually of no clinical significance. Awareness of its possible presence will avoid confusion during the identification of vertebral levels in radiographs and other diagnostic images (p. 38).

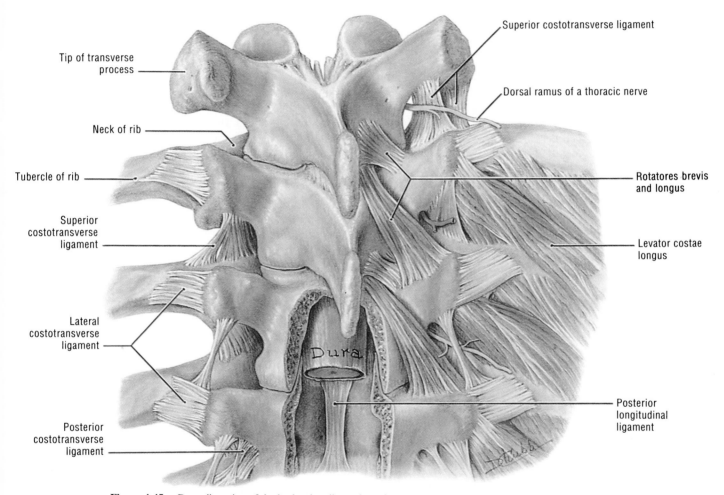

Tip of transverse process

Neck of rib

Tubercle of rib

Superior costotransverse ligament

Lateral costotransverse ligament

Posterior costotransverse ligament

Superior costotransverse ligament

Dorsal ramus of a thoracic nerve

Rotatores brevis and longus

Levator costae longus

Posterior longitudinal ligament

Dura

Figure 4-45. Deep dissection of the back primarily to show the rotatores muscles and the costotransverse ligaments.

The Suboccipital Region

This is the triangular area around the articulation between the skull and the superior end of the vertebral column (Figs. 4-46 and 4-47). It is located between the occipital bone of the skull and the posterior aspects of C1 (atlas) and C2 (axis), deep to the trapezius and semispinalis capitis muscles (p. 355). This region contains two joints: atlanto-occipital (p. 349) and atlanto-axial (p. 350).

The Suboccipital Muscles

There are *four small muscles in this region*; they lie deep to the semispinalis capitis muscle (Fig. 4-46). They are *mainly postural muscles*, but they help to move the head (Tables 4-4 and 4-5). They are all innervated by the dorsal ramus of C1 nerve.

The Rectus Capitis Posterior Muscles (Figs. 4-14, 4-46, and 4-47; Tables 4-4 and 4-5). The *rectus capitis posterior major* is a small triangular muscle that arises from the posterior edge of the spinous process of C2 (axis), whereas the *rectus capitis posterior minor* arises from the posterior tubercle on the posterior arch of C1. These muscles insert, side by side, into the occipital bone inferior to the inferior nuchal line. They are *mainly postural muscles*, but they help to rotate the head to the same side. Acting bilaterally, they help to extend the head at the atlanto-occipital joints (p. 350).

The Obliquus Capitis Muscles (Figs. 4-46 and 4-47; Table 4-5). *The obliquus capitis inferior is a small rectangular mus-

cle that arises from the lateral surface of the spinous process of C2 vertebra and runs obliquely and anteriorly to insert on the inferior surface of the transverse process of the C1 vertebra. Although it is not attached to the skull, the obliquus capitis inferior muscle helps to turn the head by pulling on the atlas (C1) which supports the head. This is why ''capitis'' is a part of its name, even though it is not attached to the cranium. *The obliquus capitis superior* is also a small triangular muscle. It arises from the superior surface of the transverse process of C1 and inserts into the smaller lateral impression between the superior and inferior nuchal lines on the posterior aspect of the occipital bone (see Fig. 7-3). It is also *mainly a postural muscle*, but it helps to extend the head and to laterally flex it. Obliquus capitis superior and inferior are innervated by the dorsal ramus of C1 nerve.

The Suboccipital Triangle

Boundaries of the Suboccipital Triangle (Figs. 4-35, 4-46, and 4-47). Three muscles form the boundaries: *rectus capitis posterior major*, superiorly and medially; *obliquus capitis superior*, superiorly and laterally; and *obliquus capitis inferior*, inferiorly and laterally. The floor of the suboccipital triangle is formed by the posterior atlanto-occipital membrane and the posterior arch of C1 vertebra, and *the roof of the suboccipital triangle is formed by the semispinalis capitis muscle*.

Contents of the Suboccipital Triangle (Figs. 4-14, 4-35, 4-46, and 4-47). This triangle contains the *vertebral artery* and the *suboccipital nerve* (dorsal ramus of C1). These structures lie

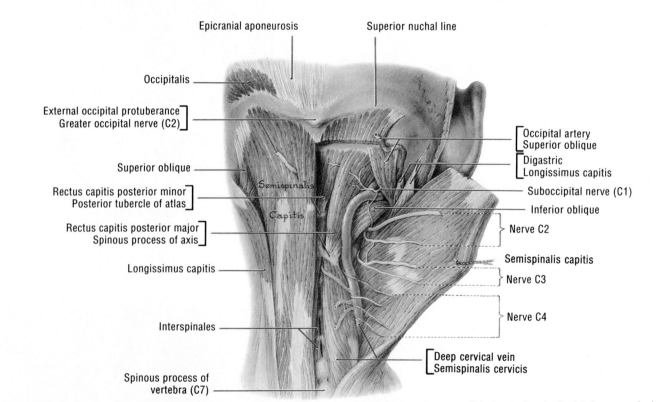

Epicranial aponeurosis

Superior nuchal line

Occipitalis

External occipital protuberance
Greater occipital nerve (C2)

Occipital artery
Superior oblique

Digastric
Longissimus capitis

Superior oblique

Semispinalis

Capitis

Suboccipital nerve (C1)

Rectus capitis posterior minor
Posterior tubercle of atlas

Inferior oblique

Nerve C2

Rectus capitis posterior major
Spinous process of axis

Semispinalis capitis

Nerve C3

Longissimus capitis

Nerve C4

Interspinales

Deep cervical vein
Semispinalis cervicis

Spinous process of
vertebra (C7)

Figure 4-46. Dissection of the suboccipital region. The trapezius, sternocleidomastoid, and splenius muscles have been removed. The semispinalis capitis is the largest muscle in the posterior aspect of the neck and is largely responsible for the longitudinal bulge on each side of the median plane.

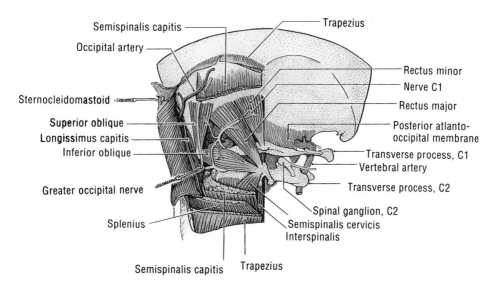

Semispinalis capitis

Occipital artery

Trapezius

Sternocleidomastoid

Superior oblique

Longissimus capitis

Inferior oblique

Greater occipital nerve

Splenius

Semispinalis capitis

Trapezius

Rectus minor

Nerve C1

Rectus major

Posterior atlanto-occipital membrane

Transverse process, C1

Vertebral artery

Transverse process, C2

Spinal ganglion, C2

Semispinalis cervicis

Interspinalis

Figure 4-47. Dissection of the suboccipital region. The suboccipital triangle is bounded by three muscles: obliquus capitis inferior, obliquus capitis superior, and rectus capitis posterior major. Observe the vertebral artery winding posterior to the superior articular process of the atlas to enter the foramen magnum of the skull.

in a groove on the superior surface of the posterior arch of the atlas. The **vertebral arteries** wind their way from the vertebral column, posterior to the superior articular process of C1 (atlas), to enter the foramen magnum of the skull. They have spinal and cranial branches.

> The winding course of the vertebral arteries becomes clinically significant when blood flow through them is reduced (*e.g.*, owing to *arteriosclerosis*). Under these conditions, prolonged turning of the head (as may occur when backing up a car) may cause dizziness and other symptoms owing to interference with the blood supply to the brain stem.

Spinal Cord and Meninges

The Spinal Cord

The spinal cord, part of the central nervous system (p. 27), is located in the *vertebral canal* (neural canal) formed by successive vertebral foramina (Figs. 4-1, 4-15, 4-48, and 4-49*B*). In addition to the spinal cord, the vertebral canal contains its protective membranes, called *spinal meninges*, as well as the associated vessels embedded in loose connective and fatty tissue. The spinal cord is a cylindrical structure that is slightly flattened anteriorly and posteriorly. It is protected by the vertebrae, their associated ligaments and muscles, the spinal meninges, and CSF (cerebrospinal fluid).

The spinal cord begins as a continuation of the medulla (oblongata), the inferior part of the brain stem. It extends from the *foramen magnum* in the occipital bone (see Fig. 7-7) to the level of L2 vertebra (Fig. 4-48). It ranges from 42 to 45 cm in length. *In adults* the spinal cord usually ends opposite the intervertebral

disc between L1 and L2 vertebrae, but it may terminate at T12 or L3 (Fig. 4-55*B*). *The spinal cord occupies only the superior two-thirds of the vertebral canal* (Fig. 4-48). It is enlarged in two regions for innervation of the limbs (Fig. 4-52). The **cervical enlargement** extends from C4 to T1 segments of the spinal cord, and most of the ventral rami of the spinal nerves arising from it form the *brachial plexus of nerves* for innervation of the upper limb (see Fig. 6-17). The **lumbosacral enlargement** extends from T11 to L1 segments of the spinal cord, and the corresponding nerves make up the *lumbar* and *sacral plexuses* for innervation of the lower limbs (see Figs. 2-99 and 3-16).

The spinal cord segments do not correspond with the vertebral levels. For example, the lumbosacral enlargement (L2 to S3 segments of the spinal cord) extends from the body of T11 to the level of the body of the L1 vertebra. In Fig. 4-48, note that *the thoracic region of the spinal cord is the longest part* and that the sacral region of the spinal cord is considerably superior to the sacrum. Also note that *the sacral region is the shortest part* of the spinal cord (Figs. 4-1*B*, 4-48, and 4-51).

Age Differences in the Position of the Spinal Cord. During the embryonic period the spinal cord extends the entire length of the vertebral canal, and the spinal nerves form just outside the intervertebral foramina at their levels of origin. Because the vertebral column grows more rapidly than the spinal cord, this relationship does not persist (*i.e.*, *the vertebral column outgrows the spinal cord*). The inferior end of the spinal cord comes to lie at relatively higher levels. *In the newborn infant* the spinal cord terminates at L2 or L3, and *in the adult* it usually ends at the inferior border of L1 vertebra (Fig. 4-48). However, in about 1% of people the spinal cord extends inferior to L2 vertebra and in unusual cases it may reach the level of L3 vertebra (Barr and Kiernan, 1988).

Structure of the Spinal Nerves (Figs. 4-15 and 4-48 to 4-51). There are *31 pairs of spinal nerves* attached to the spinal

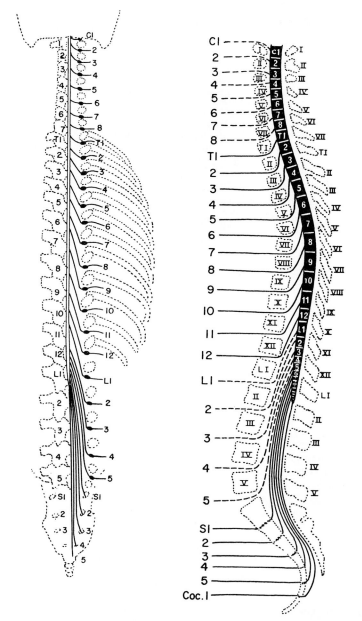

Figure 4-48. Diagrams illustrating the relation of spinal cord segments to the adult vertebral column. *Left*, Posterior view. *Right*, Schematic lateral view.

cord by dorsal and ventral roots. The *ventral roots* leaving the cord contain efferent or motor fibers which are distributed to muscles and glands. The *dorsal roots* entering the cord contain afferent or sensory fibers which convey sensation from sensory nerve endings (*e.g.*, in the skin). The cell bodies of axons making up the ventral roots are in the *ventral gray horn* of the spinal cord, whereas the cell bodies of axons making up the dorsal roots are outside the spinal cord in the *spinal ganglia* (dorsal root ganglia). These ganglia are located in the *intervertebral foramina*, where they rest on the pedicles of the vertebral arches. Distal to the spinal ganglia, just outside the intervertebral foramina, the dorsal and ventral nerve roots unite to form a *spinal*

nerve. Each spinal nerve divides almost immediately into a *ventral primary ramus* (L. branch) and a *dorsal primary ramus*.

Owing to the inequality in length of the adult spinal cord and vertebral column, there is a progressive obliquity of the dorsal and ventral nerve roots (Fig. 4-48). Consequently, the length and obliquity of these roots increase progressively as the inferior end of the vertebral column is approached because of the increasing distance between the spinal cord segments and the corresponding vertebrae. Thus, the lumbar and sacral spinal nerve rootlets (L1 to L5, S1 to S5, and Co1) are the longest. They must descend until they reach their intervertebral foramina of exit (*i.e.*, in lumbar and sacral regions). This collection of root-

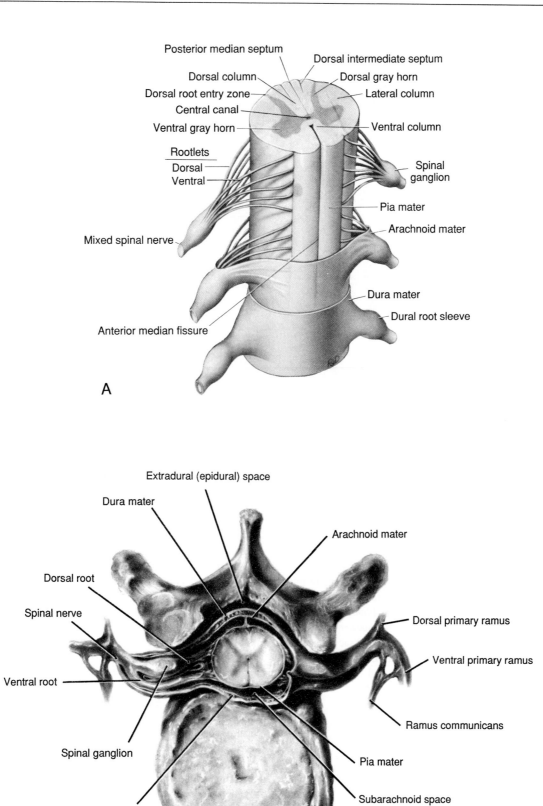

Figure 4-49. *A*, Spinal cord, nerve roots, and spinal meninges. Observe that each nerve root emerges from the spinal cord as a series of rootlets and that each spinal nerve is formed by the union of dorsal and ventral spinal nerve rootlets. *B*, Diagrammatic drawing of a transverse section of the upper lumbar region of the vertebral column at the level of the intervertebral foramina. Note the relation of the spinal meninges to the spinal cord and nerve roots (From Cormack DH: *Ham's Histology*, ed 9. Philadelphia, JB Lippincott, 1987.)

dotomy, the use of high frequency electricity, has almost completely replaced open chordotomy. Because this procedure can be performed on a conscious patient, the position and size of the lesion in the spinal cord may be controlled by asking the patient what she/he feels while the electrode is in place. The electrode is usually inserted between C1 and C2 vertebrae to destroy all ascending pain fibers in the spinal cord on one side.

CSF functions as a shock absorber, decreasing the possibility of spinal cord injury when the spinal cord is forced against the wall of the vertebral canal during sudden traumatic acceleration or deceleration (*e.g.*, during professional automobile races). It also helps to protect the cord when there is a heavy blow to the back (*e.g.*, when a small tree falls on an amateur logger).

CSF can be obtained from the lumbar cistern (Fig. 4-55). As the spinal cord usually ends at the inferior border of L1 vertebra, or the superior border of L2, there is no danger of injuring the spinal cord when a *lumbar puncture needle* is inserted into the lumbar cistern between the spinous processes of L3/L4 or L4/L5 vertebrae to obtain a CSF sample. During **lumbar puncture** there is seldom any damage to the spinal nerve roots because, being suspended in CSF, they tend to move away from the needle. It is more likely to touch a spinal nerve root if it is not inserted exactly in the median plane. Touching a nerve root with the point of the needle may result in sharp pain in the *dermatome* or area of skin supplied by the nerve root concerned (see Fig. 1-24). It is worth remembering that an imaginary line joining the iliac crests passes through L4 vertebra (see Fig. 2-86), because it serves as a guide to the L3/L4 or L4/L5 levels. Another practical anatomical fact is that the vertebral interspinous spaces widen when the lumbar region is fully flexed (Figs. 4-8 and 4-55*A*). This provides a good entry for the lumbar puncture needle.

Lumbar punctures are performed to obtain CSF for diagnostic procedures (*e.g.*, for determination of alterations in the concentrations of chemicals). Spinal anesthetic agents can also be injected by lumbar puncture. In a **spinal block**, the local anesthetic agent is injected directly into the CSF (Hew, 1989). The effect on the nerve roots usually occurs within 1 minute, whereas an epidural block often takes 10 to 20 minutes. The amount of anesthetic agent used for a spinal block is about 10% of that required for an epidural block because the anesthetic does not have to diffuse through the meninges to reach the CSF (Hew, 1989). Spinal anesthesia blocks the nerve roots in the dural sac. The number of segmental levels of the spinal cord that are blocked is controlled by the amount of anesthetic injected and through positioning of the patient. During this procedure care is taken not to let the operating table be tilted so that the patient's head is lower than the feet. Otherwise the anesthetic agent will flow toward the head and affect the nerves that control vital functions (*e.g.*, respiration).

Spina bifida is used to describe a wide range of developmental defects of the vertebral column. In its most simple form, *spina bifida occulta* (p. 342), the laminae of the vertebral arch fail to develop fully and fuse. *Spina bifida cystica* is a more serious congenital malformation in which there is herniation of the meninges and/or spinal cord through the defect in the vertebral arch (Moore, 1988). When the meninges alone are herniated, the condition is known as *spina bifida with meningocele*, whereas when the meninges and spinal cord and/or nerve roots herniate, the malformation is known as *spina bifida with meningomyelocele*. Patients with this severe malformation often exhibit spinal cord and/or spinal nerve root malfunction (*e.g.*, paralysis of the limbs and incontinence of urine and feces).

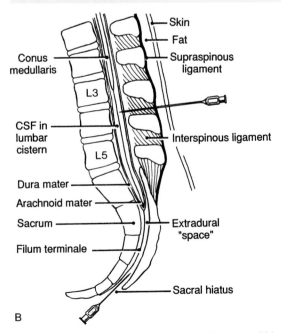

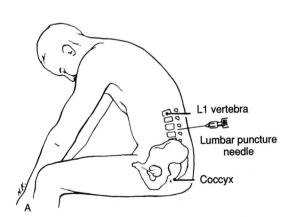

Figure 4-55. *A*, Lumbar puncture ("spinal tap"). Note that the person's back is flexed to open the spaces between the spinous processes and laminae of the vertebrae. *B*, Median section of the inferior end of the vertebral column containing the spinal cord and its membranes. A lumbar puncture needle has been inserted between L3/L4 vertebrae for withdrawal of CSF. A needle is also shown in the sacral hiatus, the site sometimes used for epidural anesthesia. Note that the spinal cord in this person ends at the middle of the body of L3 vertebra, an unusually low position. This shows why a lumbar puncture should never be inserted between L2/L3 vertebrae in adults.

PATIENT ORIENTED PROBLEMS

Case 4-1

During a swarming attack by several skinheads, a 16-year-old boy was stabbed in the posterior aspect of the neck with a knife. As he ducked to avoid his attacker, he flexed his neck. Much to the surprise of his assailants, the boy fell to the ground and was completely immobilized from the neck down.

> **Problems.** How did this serious injury probably occur? Using your anatomical knowledge of the vertebral column and its contents, explain the basis of the injury. How might this knowledge be utilized in the diagnosis and treatment of diseases of the nervous system and in the administration of anesthetic agents? These problems are discussed on p. 370.

Case 4-2

A 51-year-old man was waiting for a traffic light to turn green when his car was "rear-ended." His body was pushed forward and his head was thrown violently backward. He suffered a slight concussion and felt shaky. When he talked to the man who had hit his car and to the traffic officer, he informed them that he was not badly hurt. The officer noted that the head restraint in his car was not raised to the level that would prevent hyperextension of his neck. The next morning the man's neck was stiff and painful, and he felt pain on the left side of his neck and in his left arm. The *neck pain* was aggravated by movement of his head. He decided to visit his physician.

The physician observed that the man held his head rigidly and tilted it to the right. She also observed that his chin was pointed to the left and that his neck was slightly flexed. Her palpation of the posterior aspect of his neck revealed some tenderness over the spinous processes of his inferior cervical vertebrae. His biceps reflex was also weak on the left side. She ordered a radiographic study of his cervical region.

Radiologist's Report. The intervertebral discs between C5 and C6, and C6 and C7 are thin and there were small fringes of bone on the opposing edges of the bodies of C5, C6, and C7.

Diagnosis. Hyperextension injury of the neck.

> **Problems.** What is the anatomical basis for the patient's concussion, stiff neck, and pain in the neck and arm? What spinal nerve root was probably compressed? What muscles were probably injured? What probably caused the thinning of the patient's intervertebral discs and the formation of the bony fringes on the edges of his cervical vertebral bodies? These problems are discussed on p. 370.

Case 4-3

While helping you carry a heavy box of books, your father suddenly experienced severe pain in his lower back. Later he developed a dull ache in the posterolateral aspects of his left thigh that extended along the calf of his leg into his foot. Lateral deviation (tilt) of the lumbar region of his vertebral column was also observed. He limped when he walked because he did not fully extend his thigh. After consulting his family physician, who recommended bed rest, he was referred to a back specialist.

As you were studying to be a physiotherapist, you were per-mitted to observe the examination. The orthopaedist told you that your father's back muscles were in spasm. When asked to indicate the site of the most severe pain, your father pointed to his lower lumbar region. During the examination you noted that your father had no ankle reflex on the left side and that he experienced increased pain when the physician raised his extended lower limb on that side. He arranged for a radiographic study of your father's back.

Radiologist's Report. The radiographs showed a slight narrowing of the space between the vertebral bodies of L5 and S1. MRI revealed that the nucleus pulposus of one of your father's intervertebral disc was protruding.

Diagnosis. Posterolateral protrusion of the nucleus pulposus of the intervertebral disc between L5 and S1.

> **Problems.** What is the anatomical basis for protrusion (prolapse) of an intervertebral disc and the resulting low back pain? What produced the lumbar deviation? Why did the patient experience pain in the posterolateral aspect of his thigh and leg? Why did the pain increase when the orthopaedist raised the patient's extended lower limb? These problems are discussed on p. 371.

Case 4-4

An 18-year-old woman was thrown from a horse and she sustained a *spinal cord injury* as the result of severe hyperextension of her neck. Although she was rushed to hospital, she died in about 5 minutes. An autopsy was performed.

Diagnosis. Transection of the superior end of the spinal cord resulting from multiple fractures of the cervical vertebrae.

> **Problems.** What vertebrae were most likely fractured and dislocated? What associated structures of the vertebral column were probably also ruptured? Although one would expect the patient to be quadriplegic following a cervical spinal cord transection, what probably caused her death? These problems are discussed on p. 371.

Case 4-5

A 62-year old man, who was a heavy drinker and smoker, consulted his physician about feeling a *strong pulse in his abdomen*. He said it felt like a second heart. He also complained about pain in his abdomen, back, and groin. The physician arranged for radiographic studies, including CT scans.

Radiologist's Report. The plain radiographs showed calcium deposits in the wall of the abdominal aorta and an apparent aneurysm. The CT scans revealed an *abdominal aortic aneurysm* that was 11 cm in diameter. Before he could be admitted to hospital for repair of his abdominal aortic aneurysm, the patient passed out on his way home and was involved in a minor car accident. He was rushed to hospital and admitted to surgery for repair of the ruptured aneurysm. During surgery for repair of the *ruptured aortic aneurysm*, there was extensive mobilization of the aorta and several segmental arteries were ligated. Although the aorta was successfully repaired using a Dacron graft, the patient was paraplegic, impotent, and his bladder and bowel functions were no longer under voluntary control.

Diagnosis. Paraplegia and other neurological deficits resulting in sphincter paralysis in the urinary bladder and anal canal.

> **Problems.** What arteries supply the spinal cord? What is the most likely anatomical basis for the patient's paraplegia, sphincter dysfunctions, and impotence? What arteries were probably ligated during surgery? Name the important artery supplying the spinal cord that was likely deprived of blood. Why is its supply to the spinal cord so important? What "bad habits" are known to be associated with the development of aneurysms? These problems are discussed on p. 371.

Case 4-6

A 21-year-old man was involved in a head-on collision. When removed from his sports car, he complained of loss of sensation and voluntary movements in his lower limbs. There was also impaired ability of upper limb movements, particularly in his hands. The patient was kept warm and immobilized until the ambulance arrived. After examination at the hospital, radiographs of his vertebral columns were taken.

Radiologist's Report. Radiographs showed dislocation of C6 vertebra on C7 and a chip fracture of the anterosuperior corner of the body of C7.

Surgical Treatment. Open reduction was carried out and the spinous processes of C6 and C7 were wired together to hold these vertebrae in normal relation to each other. The reduction was maintained by immobilization of the neck in a plastic collar, thereby allowing the patient to exercise his upper limbs and to sit up within a day or so after the injury.

> **Problems.** What joints of the cervical region of the vertebral column were dislocated? What ligaments binding the vertebrae together were probably strained and/or torn? What was the most likely cause of the patient's paralysis? What other physiological functions would no longer be under voluntary control? These problems are discussed on p. 371.

DISCUSSION OF PATIENTS' PROBLEMS

Case 4-1

The spinous processes and laminae of the vertebral arches usually protect the spinal cord during injuries to the posterior aspect of the neck, even those resulting from stab wounds. However, when the neck is flexed, the spaces between the vertebral arches increase. This could permit a knife to pass between them, enter the vertebral canal, and sever the cervical region of the spinal cord. Verify this movement of the cervical vertebrae during flexion by placing your hand on the back of your neck and then flexing it as much as possible. Note that the space between the external occipital protuberance and the spinous process of the axis (C1) widens, as do the spaces between the spinous processes of C2 to C7 vertebrae. Had the knife severed the spinal cord superior to C3, the lesion would have stopped the patient's breathing. At this site the injury would have interfered with the patient's *phrenic outflow* (C3, **C4**, and C5), the nerve supply to the diaphragm. As a result, the patient may have died in a few minutes.

Complete transection of the spinal cord results in loss of all sensation and voluntary movement inferior to the lesion. The patient is quadriplegic when the lesion is superior to C5 segment because the nerves supplying the upper limb are derived from C5 to T1 segments of the spinal cord. Had the young man been stabbed in the same place when his head was erect, he probably would not have been severely injured. Very likely the knife would have struck the spinous processes or laminae of the cervical vertebrae and glanced off without damaging the spinal cord.

Similar gaps exist between the lumbar spinous processes when the back is flexed. Their presence is important because they enable clinicians to insert a lumbar puncture needle. In adults the needle is usually inserted between the spinous processes of L3 and L4 vertebrae into the subarachnoid space (Fig. 4-55), inferior to the termination of the spinal cord. This procedure, known as a *lumbar puncture*, is performed to obtain a sample of cerebral spinal fluid (CSF). These punctures are performed during the investigation of some diseases of the nervous system (*e.g.*, meningitis). Anesthetic solutions may also be injected into the extradural or epidural space. The extradural space is not a real space because it is filled with areolar tissue, fat, and veins. For *epidural anesthesia*, called an **epidural block**, the needle may also be inserted through the sacral hiatus at the inferior end of the sacrum. This is inferior to the level of the dural sac (Fig. 4-55B).

Case 4-2

The association of rear end automobile collisions and *hyperextension injuries* of the soft tissues of the cervical region of the vertebral column is well known. Head restraints ("headrests") and bucket seats have been designed to minimize these injuries. However, a head restraint is useless if it is not raised so that the head will hit it if a rear end collision occurs. The mechanism of injury in the present case was primarily one of *rapid hyperextension of the neck*. Because the head restraint was not in the correct position, there was nothing to restrict posterior movement of the head and neck. There also was probably a *hyperflexion injury* of the neck when the head was flung back onto the chest. This would likely have occurred because the muscles of the patient's neck, the chief stabilizers of the cervical region of the vertebral column, were relatively relaxed when the patient was caught off guard.

During severe hyperextension of the neck, the anterior longitudinal ligament and neck muscles would be severely stretched, and some of its fibers were probably torn, leading to small hemorrhages. The resulting muscle spasms would account for the patient's stiff and painful neck. The concussion experienced by the patient probably resulted from the sudden impact of the frontal and sphenoid bones against the frontal and temporal poles of his brain. The pain in his left shoulder and the weakness of his biceps reflex on the left very likely resulted from *compression of the left C6 nerve root*, probably by a posterolateral protrusion of the intervertebral disc between the C5 and C6 vertebrae. The musculocutaneous nerve (C5 and C6) supplies the biceps brachii muscle (see Table 6-7, p. 549), and the *biceps reflex* is also mediated through C5 and C6. A hyperextension injury of the neck is popularly called a "*whiplash injury*." Many doctors consider this term an unacceptable medical designation because there is no well defined clinical syndrome or fixed pathological condition associated with the injury.

The thinning of the intervertebral discs in the cervical region probably resulted from desiccation of the nuclei pulposi of the intervertebral discs. **Degenerative disc disease** often occurs with advancing age and results in bulging of the anuli fibrosi of the intervertebral discs (Fig. 4-32). Formation of fringes of subperiosteal new bone on the edges of the vertebral bodies also occurs in older persons.

Case 4-3

The patient's *low back pain and muscle spasm*, sometimes called "**lumbago**," was probably caused by rupture of the posterolateral part of the anulus fibrosus and protrusion of the nucleus pulposus of the intervertebral disc between L5 and S1 vertebrae. The lumbar deviation of the patient's vertebral column was produced by spasm of the intrinsic back muscles. Muscle spasm has a protective, splinting effect on the vertebral column. As your father lifted the heavy box of books, the strain on his intervertebral discs was so severe that the anulus fibrosus of one of them tore, resulting in *herniation of the nucleus pulposus of the disc.*

The disc protrusion exerted pressure on a nerve root or roots of the **sciatic nerve**. Disc protrusions most commonly occur posterolaterally where the anulus is thin. As the dorsal and ventral nerve roots cross the posterolateral region (Fig. 4-15), the protruding nucleus pulposus often affects one or more spinal nerve roots. Some hemorrhage, muscle spasm, and edema would be present at the site of the rupture. This probably caused the patient's initial back pain. In the present case, pressure appears to have affected the *S1 component of the sciatic nerve* as it passes inferiorly, posterior to the intervertebral disc between L5 and S1. As a result, the patient experienced pain over the posterolateral region of his thigh and leg. When the physician raised the patient's extended lower limb, the sciatic nerve was stretched. As its S1 component was compressed by the protruding disc, the lower limb pain increased because of stretching of the compressed fibers in that root.

Sciatica is the name given to pain in the area of distribution of the sciatic nerve (L4 to S3; see Fig. 5-40*B*). Pain is felt in one or more of the following areas: the buttock, especially the region of the greater sciatic notch, the posterior aspect of the thigh; the posterior and lateral aspects of the leg; and usually parts of the lateral aspect of the ankle and foot (see Fig. 5-82). The variation in the location of the pain results from the fact that a posterolateral protrusion of a single lumbar disc presses on only one nerve root. However, the sciatic nerve is composed of several inferior lumbar and superior sacral roots. The *paravertebral muscle spasm and pain* results from the muscles being in continuous tonic contraction to prevent the vertebrae from moving and causing more severe pain. The narrowing or thinning of the space between the vertebral bodies noted in the radiograms is caused by the reduction of the disc material between the adjacent vertebral bodies that normally occurs with advancing age.

Case 4-4

Severe hyperextension of the neck resulting from a fall on the head, usually causes a fracture of C1 vertebra (atlas) at one or both grooves for the vertebral arteries (Figs. 4-14 and 4-17). The vertebral arch of C1 may break at the isthmus between the lateral mass and the inferior articular process. Probably the patient's anterior longitudinal ligament and the anterior part of the intervertebral disc between C2 and C3 were also ruptured. As the patient hit the ground, hyperextending her neck, her skull and C1 and C2 vertebrae were probably separated from the rest of her vertebral column. As a result, her spinal cord was probably torn in the superior cervical region. Patients with this severe injury rarely survive more than a few minutes because the injury to the spinal cord is superior to the *phrenic outflow* (origin of the phrenic nerves). As these nerves are the sole motor supply to the diaphragm, respiration is severely affected; in addition, the actions of the intercostal muscles are lost.

Case 4-5

During certain surgical procedures in the abdomen, such as resection of an *aortic aneurysm* (see Fig. 2-104), it is necessary to ligate some aortic segmental branches (*e.g.*, a lumbar artery). If the *large radicular artery* (arteria radicularis magna) arises from one of the intercostal or lumbar arteries that has been ligated, the blood supply to the *lumbosacral enlargement of the spinal cord* may be severely impaired (Fig. 4-52). As a result, spinal cord infarction, paraplegia, impotence, and loss of sensation inferior to the lesion may follow. Arising more frequently on the left from an inferior intercostal (T6 to T12) or lumbar (L1 to L3) artery, the large radicular artery enters the vertebral canal through an intervertebral foramen. It supplies blood mainly to the inferior two-thirds of the spinal cord (Figs. 4-50 and 4-52); therefore it is understandable why function is lost in the lower limbs, bladder, and intestine when this artery and part of the spinal cord are deprived of blood.

The development of an aneurysm is accelerated by smoking; aneurysms occur three times more frequently in smokers than non-smokers. For a description of the diagnosis and surgical treatment of abdominal aortic aneurysm, including illustrations of aneurysm repair, see Ameli (1989).

Case 4-6

The patient is paraplegic and the condition is known as **paraplegia**. Both the intervertebral disc and the *zygapophyseal (facet) joints* between the bodies and vertebral arches of C6 and C7, respectively, were dislocated in this case (Fig. 4-37*B*). Probably the posterior longitudinal and interspinous ligaments, as well as the anulus fibrosus, ligamenta flava, and articular capsules of the zygapophyseal joints were severely injured and some of them may have been torn.

The cervical region, being the most mobile part of the vertebral column, is the most vulnerable to injuries such as dislocations and fracture-dislocations. Most injuries occur when a person's head moves forward suddenly and violently, as in the present case, or when the back of the head is struck by a hard blow. In *hyperflexion injuries of the neck*, the anterior longitudinal ligament is usually not torn, and when the patient's neck is placed in a position of extension, this ligament tightens and, along with the plastic cervical collar, tends to hold the vertebrae together.

During the surgical treatment of the injury, the spinous processes of C6 and C7 were wired together to help stabilize the vertebral column during the initial part of the rehabilitation program, and to promote healing of the strained and/or torn ligaments. Normally the vertebral bodies are bound together by the longitudinal ligaments and anuli fibrosi of the intervertebral discs. The *posterior longitudinal ligament*, a narrower and weaker band than the anterior longitudinal ligament, is attached to the intervertebral discs and to the edges of the vertebral bodies. It lies inside the vertebral canal and tends to prevent excessive flexion of the vertebral column. As dislocation occurred in this case, the posterior longitudinal ligament and the ligamenta flava were severely stretched and were probably torn.

As the anulus fibrosus of the intervertebral disc attaches to the compact bony rims on the articular surfaces of the vertebral discs, its posterior part would also have been stretched and torn at the C6 and C7 level. It is possible that protrusion of the nucleus pulposus of the intervertebral disc between these vertebrae also occurred because these nuclei are semifluid in young adults. Because the vertebral canal in the cervical region is usually larger than the spinal cord, there can be some displacement of the vertebrae without causing damage to the spinal cord. In view of the patient's paraplegia, it is likely that the spinal cord was severely stretched and/or torn. At the moment of impact, the displacement of C6 on C7 was undoubtedly greater than shown in the radiograph (Fig. 4-37). There is an initial period of **spinal shock** in these cases, lasting from a few days to several weeks, during which all somatic and visceral activity is abolished. On return of reflex activity, there is spasticity of muscles and exaggerated tendon reflexes inferior to the level of the lesion. In addition, bladder and bowel functions are no longer under voluntary control.

SUGGESTED READINGS

Ameli FW: Vascular Surgery. *In* Gross A, Gross P, Langer B (Eds): *Surgery. A Complete Guide for Patients and their Families*, Toronto, Harper & Collins, 1989.

Armstrong JR: *Lumbar Disc Lesions. Pathogenesis and Treatment of Low Back Pain and Sciatica*, ed. 3. Edinburgh, E & S Livingstone, 1965.

Barr ML, Kiernan JA: *The Human Nervous System: An Anatomical Viewpoint*, ed. 5. Hagerstown, Harper & Row Publishers, 1988.

Bates B: *A Guide To Physical Examination*, ed 2. Philadelphia, JB Lippincott, 1979.

Behrman RE: *Nelson Textbook of Pediatrics*, ed. 14. Philadelphia, WB Saunders, 1992.

Bergman RA, Thompson SA, Afifi AK, Saadeh FA: *Compendium of Human Anatomic Variation. Text, Atlas, and World Literature*, Baltimore, Urban & Schwarzenberg, 1988.

Bertram EG, Moore KL: *An Atlas of the Human Brain and Spinal Cord*, Baltimore, Williams & Wilkins, 1982.

Griffith HW: *Complete Guide To Sports Injuries*, Los Angeles, Price Stern Sloan, 1986.

Hew E: Anesthesia. *In* Gross A, Gross P, Langer B (Eds): *Surgery. A Complete Guide for Patients and their Families*, Toronto, Harper & Collins, 1989.

Moore KL: *The Developing Human: Clinically Oriented Embryology*, ed 4. Philadelphia, WB Saunders, 1988.

O'Rahilly R: *Gardner-Gray-O'Rahilly Anatomy. A Regional Study of Human Structure*, ed 5. Philadelphia, WB Saunders, 1986.

Rothman RH, Simeone FA (Eds): *The Spine*, Philadelphia, WB Saunders, vol. 1, 1975.

Salter R: Orthopedic Surgery: Pediatric. *In* Gross A, Gross P, Langer B (Eds): *Surgery. A Complete Guide for Patients and their Families*, Toronto, Harper & Collins, 1989.

Tobias PV, Arnold M, Allan JC: *Man's Anatomy: A Study In Dissection*, ed. 4. Johannesburg, Witwatersrand University Press, vol 2, 1988.

Weinstein PR, Ehi G, Wilson LB: *Lumbar Spondylosis*, Chicago, Year Book Medical Publishers, 1977.

Williams PL, Warwick R, Dyson M, Bannister LH (Eds): *Grays Anatomy*, ed 37. London, Churchill Livingstone, 1989.

The lower limb (extremity) is specialized for (1) *locomotion*, (2) *bearing weight*, and (3) *maintaining equilibrium*. It consists of four major parts (Figs. 5-1 and 5-2): (1) the **hip**, containing the *hip bone* (os coxae, innominate bone), which connects the skeleton of the lower limb to the vertebral column; (2) the **thigh**, containing the *femur* (thigh bone) and connecting the hip and knee; (3) the **leg** containing the *tibia* (medial leg bone or "shin bone") and *fibula* (lateral leg bone or "splint bone"), which connect the knee and ankle; and (4) the **foot** containing the *tarsus* (bones of posterior and middle parts of the foot), *metatarsus* (bones of anterior part of the foot), and *phalanges* (bones of digits or toes). The Latin word for the foot, *pes*, is the basis of the words *pedialgia* (neuralgic pain in the foot) and *pedal*. The Greek word for the foot, *podos*, appears in the terms *podiatry* (treatment of the foot) and *podalgia* (foot pain).

The parts of the lower limb are comparable to those of the upper limb (*e.g.*, foot and hand; knee and elbow). This is understandable when you recall that the limbs rotate in opposite directions during embryonic development. The upper limbs rotate laterally through a 90 degree angle on their longitudinal axes, bringing the thumbs to the lateral side, whereas the lower limbs rotate medially through an almost 90 degree angle bringing the great toes to the medial side. Hence the knee faces anteriorly when one stands in the anatomical position and the extensor muscles lie on the anterior aspect of the lower limb. Therefore extension occurs in opposite directions in the upper and lower limbs (see Fig. 5C and *D* in the Overview of Anatomy, p. 7).

The hip bones articulate posteriorly with the sacrum (see Fig. 3-53) and meet anteriorly at the *pubic symphysis* (Fig. 5-2). The **pelvic girdle** (which is formed by the two hip bones), together with the sacrum and coccyx, form the skeleton of the *bony pelvis*[1] (see Figs. 3-1 and 3-4). Some of the muscles that act on the lower limbs arise from the pelvic girdle and the inferior part of the vertebral column. Consequently, it is customary when describing the lower limb to include regions that are transitional between it and the trunk, *e.g.*, the buttock or gluteal region (Figs. 5-5 and 5-6).

The Hip and Thigh Areas

The hip and thigh include the area from the iliac crest to the knee (Figs. 5-2 and 5-5). The hip is the region between the iliac crest and the greater trochanter of the femur, lateral to and in-cluding the hip joint. The thigh is the region between the greater trochanter and the knee, *i.e.*, between the hip and knee.

The Bone and Surface Anatomy of the Hip

The Hip Bone

This large, irregularly shaped bone is formed by three bones: *ilium*, *ischium*, and *pubis*. Before puberty, these bones are separated by cartilage (Fig. 5-1). They begin to fuse at the acetabulum at 15 to 17 years of age to form one hip bone. Fusion is usually complete by age 23 years; hence these bones are indistinguishably joined in the adult (Fig. 5-4). The hip bone has a cup-shaped socket, the **acetabulum**, on its lateral aspect for articulation with the head of the femur. It was given its name because of its resemblance to a shallow Roman vinegar cup (L. *acetabulum*).

The Ilium (Figs. 5-1 to 5-6; see also Fig. 3-2). This bone is fanshaped; its *ala* (L. wing) resembles the spread of a fan and its *body* represents the handle. The *iliac fossa* is a concavity in the ala of the ilium and forms part of the posterior abdominal wall (p. 229). The ilium forms the superior two-thirds of the hip bone and the superior two-fifths of the cup-shaped *acetabulum*. When you put your hand on your "hip," it rests on the superior margin of the ilium, which is called the **iliac crest**. It has *internal and external lips* and its posterior part is thicker than other parts. This crest is easily palpated because it extends through the inferior margin of the flank or side of the trunk. Its highest point, as palpated posteriorly, is at the level of the fourth lumbar vertebra (Fig. 5-6; see also Fig. 2-86). Clinically, this level is commonly used as a surface marking when performing *lumbar punctures* (see Fig. 4-55). The iliac crest ends anteriorly in a rounded **anterior superior iliac spine**, which is easily felt and may be visible. The iliac crest ends posteriorly in a sharp **posterior superior iliac spine**, which is difficult to palpate in most people, but its position can be located because it lies at the bottom of a *skin dimple*, about 4 cm lateral to the median plane (see Figs. 2-86 and 4-9). These dimples exist because the skin and underlying fasciae are attached to the posterior superior iliac spine. A line connecting the right and left skin dimples is at the level of the S2 vertebra and the middle of the sacroiliac joints (see Figs. 3-1 and 3-4).

Another palpable bony landmark, the *tubercle of the iliac crest* (Figs. 5-2 and 5-3), is located on the external lip about 5 cm posterior to the anterior superior iliac spine. The *anterior inferior iliac spines* and the *posterior inferior iliac spines* are often difficult to identify by palpation, except in very thin people. The posterior part of the internal surface of the ilium articulates with the sacrum at the sacroiliac joint (see Figs. 3-1 and 3-4). Just inferior to this joint is the large **greater sciatic notch** (Fig. 5-3) through which pass the sciatic nerve and other important structures (Fig. 5-37).

---

[1]The bony pelvis is described on p. 243. The osteology of the sacrum and coccyx is studied with the vertebral column (Chap. 4, pp. 332 and 337).

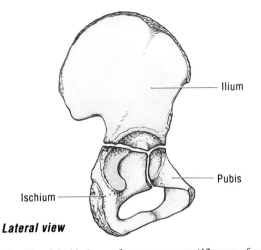

Lateral view

Figure 5-1. The right hip bone of a young person (13 years of age). Observe that it is composed of three bones (ilium, ischium, and pubis), which meet in the cup-shaped acetabulum, the socket for the head of the femur. Note that the bones are not fused at this stage of development and are united by cartilage along a Y-shaped line. Fusion is usually complete by age 23.

The skin dimples indicating the location of the posterior superior iliac spines are useful landmarks for **obtaining bone marrow** from the ilium. The needle is usually inserted 1 cm inferolateral to the dimple into the bone, and bone marrow is aspirated for examination. *Study of bone marrow* provides valuable information for evaluating many hematologic diseases (Behrman, 1992). Another common site for obtaining bone marrow is the sternum (p. 42). The iliac crest is also a common site for obtaining pieces of bone for **bone grafts**, *e.g.*, for treating fractures of the tibia (Gross, 1989).

The Ischium (Figs. 5-1 to 5-6). This bone forms the posteroinferior third of the hip bone and the posterior two-fifths of the acetabulum. The ischium (G. hip) is the roughly L-shaped part of the hip bone which passes inferiorly from the acetabulum and then turns anteriorly to join the pubis. The ischium consists of two parts, a body and a ramus.

The **body of the ischium**, its superior thick portion, is fused with the ilium and the pubis at the acetabulum. The inferior end of the body has a rugged, blunt projection called the **ischial tuberosity**. This large prominence is covered by the gluteus maximus muscle when the thigh is extended, but it is uncovered when the thigh is flexed. It bears the weight of the body when one sits (see Fig. 4-3) and can be felt through the distal part of the gluteus maximus, just superior to the medial portion of the *gluteal fold*, a prominent skin fold delimiting the buttock inferiorly (Fig. 5-6). The gluteal fold coincides with the inferior border of the gluteus maximus muscle (Fig. 5-35A), but this muscle does not produce the fold. Inferior to the gluteal fold formed by fat is the *gluteal sulcus*, a skin crease that separates the buttock from the posterior aspect of the thigh. The **ischial spine** projects medially and separates the *greater sciatic notch* superiorly from the lesser sciatic notch inferiorly (Figs. 5-3 and 5-4). The *lesser sciatic notch* is located between the ischial spine and the ischial tuberosity. The *sacrospinous ligament* spans the

greater sciatic notch converting it into the *greater sciatic foramen* (see Fig. 3-8) through which pass the piriformis muscle and the vessels and nerves on their way to the gluteal region and the thigh (Fig. 5-37). The sacrotuberous and sacrospinous ligaments convert the lesser sciatic notch into the *lesser sciatic foramen* through which pass the tendon and nerve of the obturator internus muscle, the pudendal nerve, and the internal pudendal vessels (see Figs. 3-8 and 3-18).

The **ramus of the ischium** is an inferior, thinner bar of bone than the body of the ischium. The ramus extends medially from the body and joins the inferior ramus of the pubis to form the *ischiopubic ramus*, which completes the *obturator foramen* (Figs. 5-2 and 5-4).

Alteration in the degree of prominence of the gluteal fold occurs in certain abnormal conditions, *e.g.*, *wasting of the gluteus maximus* muscle resulting from *spinal poliomyelitis*. The muscular atrophy results from denervation of the muscle plus disuse owing to the paralysis.

The Pubis (Figs. 5-1, 5-2, and 5-4). This L-shaped bone forms the *inferoanterior part of the hip bone* and the anteromedial one-fifth of the acetabulum. The pubis consists of three parts: a **body** and two **rami**. Its flattened body lies medially. The *superior ramus of the pubis* passes superolaterally to the acetabulum, where it is fused with the ilium and ischium. The *inferior ramus of the pubis* passes posteriorly, inferiorly, and laterally to join the ramus of the ischium and form half of the pubic arch. The body of the pubis joins the body of the opposite pubis in the median plane at a fibrocartilaginous joint, called the **pubic symphysis**. The superior border of the body of the pubis is thickened to form a **pubic crest**. At its lateral end there is an anterior-projecting prominence, known as the pubic tubercle, which provides the main pubic attachment for the inguinal ligament (see Fig. 2-3). The **pubic tubercle**, which can be palpated about 2.5 cm from the median plane (Fig. 5-14), is a very important bony landmark when an inguinal hernia is present (p. 147). From the pubic tubercle two ridges diverge laterally into the superior ramus (Fig. 5-2). The superior ridge, called the *pecten pubis* (pectineal line) is sharp and forms part of the pelvic brim (p. 243). The inferior ridge, called the *obturator crest* is more rounded (Fig. 5-89).

Orientation of the Hip Bone (Figs. 5-2 and 5-4; see also Fig. 3-3). To place the hip bone in the *anatomical position*, move it until the anterior superior iliac spine and the pubic symphysis are in the same coronal plane. Observe that in the anatomical position the pubic tubercle and the anterior superior iliac spine are in the same vertical plane. Also observe that the ischial spine and the superior end of the pubic symphysis are in the same horizontal plane. In the anatomical position the medial aspect of the body of the pubis faces almost directly superiorly.

The Obturator Foramen (Figs. 5-2 and 5-4; see also Fig. 3-8). This large, round (male) or oval (female) aperture is surrounded by the bodies and rami of the pubis and ischium. It lies inferomedial to the acetabulum and is nearly closed by the fibrous *obturator membrane*, which is attached to its margin.

Fractures of the hip bone are common in serious vehicular accidents. Anteroposterior compression of the hip bones com-

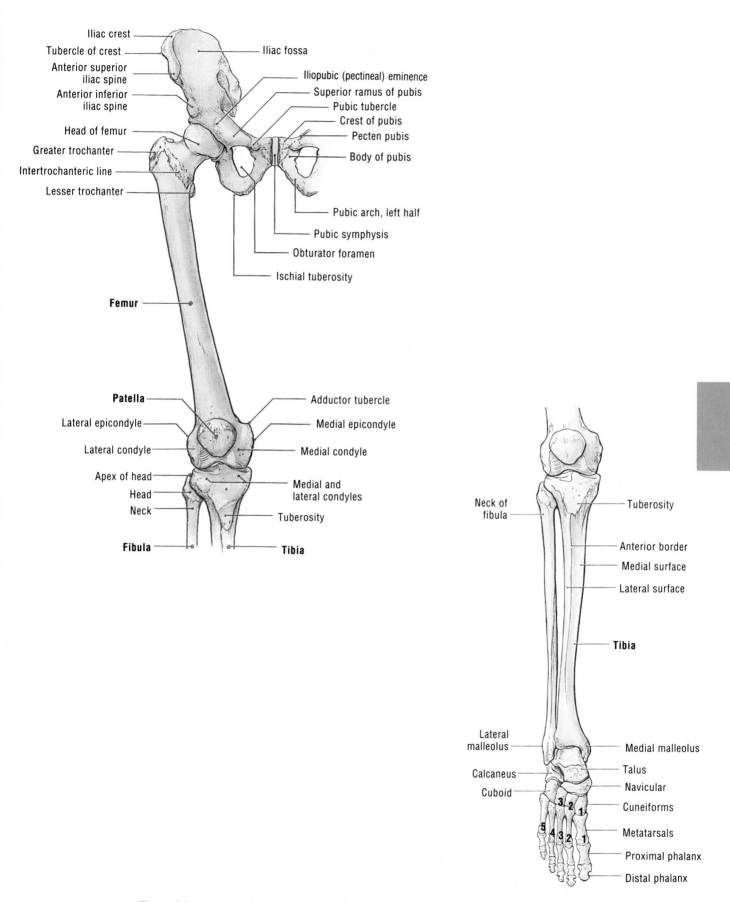

Iliac crest
Tubercle of crest
Anterior superior iliac spine
Anterior inferior iliac spine
Head of femur
Greater trochanter
Intertrochanteric line
Lesser trochanter
Femur

Iliac fossa
Iliopubic (pectineal) eminence
Superior ramus of pubis
Pubic tubercle
Crest of pubis
Pecten pubis
Body of pubis
Pubic arch, left half
Pubic symphysis
Obturator foramen
Ischial tuberosity

Patella
Lateral epicondyle
Lateral condyle
Apex of head
Head
Neck
Fibula

Adductor tubercle
Medial epicondyle
Medial condyle
Medial and lateral condyles
Tuberosity
Tibia

Neck of fibula
Lateral malleolus
Calcaneus
Cuboid

Tuberosity
Anterior border
Medial surface
Lateral surface
Tibia
Medial malleolus
Talus
Navicular
Cuneiforms
Metatarsals
Proximal phalanx
Distal phalanx

Figure 5-2.   Anterior view of the bones of the lower limb. Articular cartilages are colored blue.

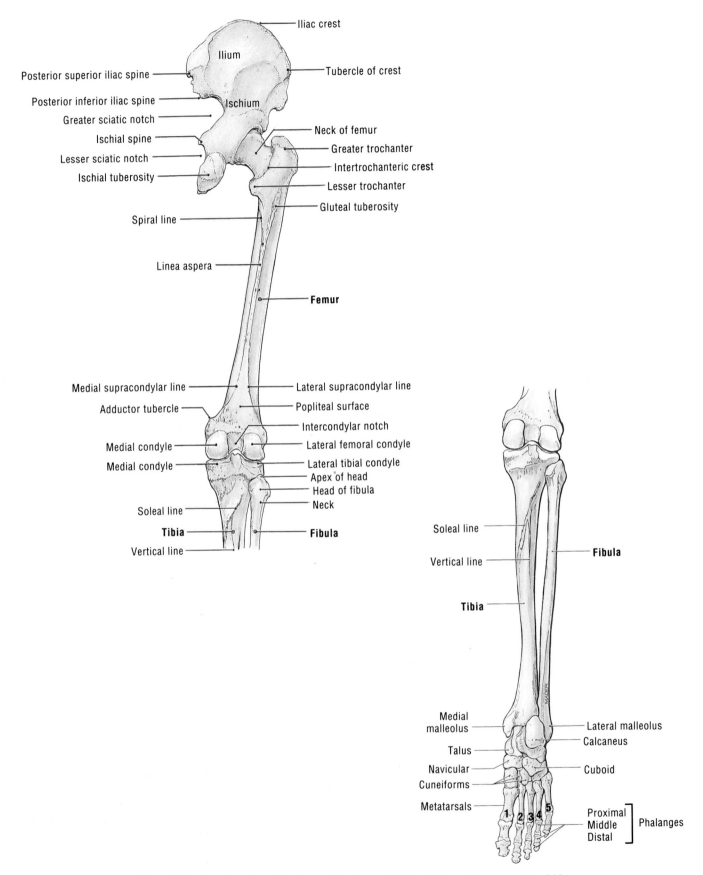

Figure 5-3. Posterior view of the bones of the lower limb. Articular cartilages are colored blue.

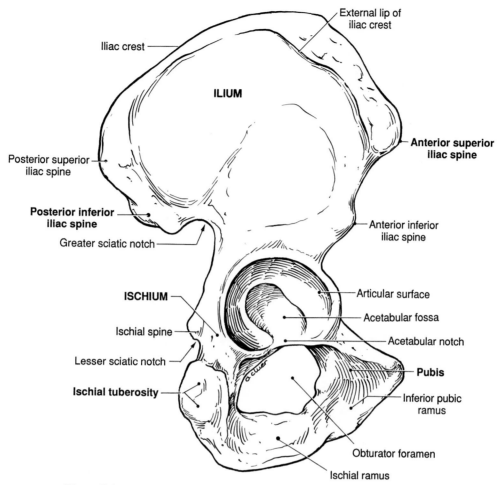

Figure 5-4.   Lateral aspect of an adult right hip bone in the anatomical position.

monly fractures the pubic rami. Lateral compression of the pelvis may fracture the acetabulum, as may "falls on the feet" (*e.g.*, from a roof). See pages 250 and 499 for a description of pelvic fractures.

The Bone and Surface Anatomy of the Thigh

The Femur

The femur (thigh bone) is the longest, strongest, and heaviest bone in the body (Figs. 5-2 and 5-3). A person's height is roughly four times the length of his/her femur. It extends from the hip joint where its rounded head articulates with the acetabulum, to the knee joint, where its condyles articulate with the tibia. The femur consists of a *body* (shaft) and two ends (extremities). The proximal end consists of a head, neck, and greater and lesser trochanters. The distal end is broadened by medial and lateral *condyles* where it articulates with the tibia and patella to form the knee joint. The femur is so covered with muscles that it is palpable only near its ends.

The head of the femur (Figs. 5-2 and 5-3) is smooth and forms about two-thirds of a sphere. It is directed medially, superiorly,

and slightly anteriorly to fit into the acetabulum of the hip bone. A little inferior and posterior to its center is a *fovea* (L. pit) where the ligament of the head is attached (Fig. 5-89). The head can sometimes be palpated, particularly in thin males, when the thigh is rotated laterally.

The neck of the femur connects the head to the body. It runs obliquely in an inferolateral direction to meet the body (shaft) at an angle of about 125 degrees. The neck is limited laterally by the greater trochanter and is narrowest in diameter at its middle. A broad, rough *intertrochanteric line* runs inferomedially from the greater trochanter. This lines passes inferior to the lesser trochanter and becomes continuous with the *spiral line* on the posterior aspect of the femur (Fig. 5-3). The intertrochanteric line is produced by the attachment of the massive *iliofemoral ligament* (Fig. 5-88). The intertrochanteric line separates the anterior surface of the neck from the body of the femur. A prominent ridge, the *intertrochanteric crest*, unites the two trochanters posteriorly. The neck has several prominent pits, especially on its posterior aspect, for the entrance of blood vessels.

The greater trochanter of the femur (Figs. 5-2 and 5-3) is a large, somewhat rectangular projection from the junction of the neck and body. It provides an attachment for several muscles of the gluteal region (Fig. 5-36). The greater trochanter lies later-

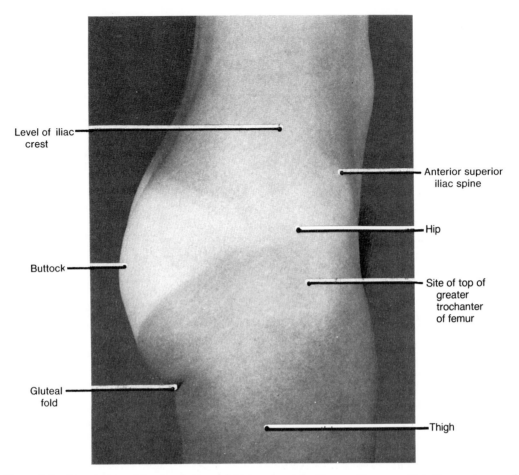

Figure 5-5.   Lateral view of the gluteal, hip, and thigh regions of a 27-year-old woman. Note the prominence of the buttock formed mainly by fat covering the gluteus maximus muscle (see Fig. 5-42).

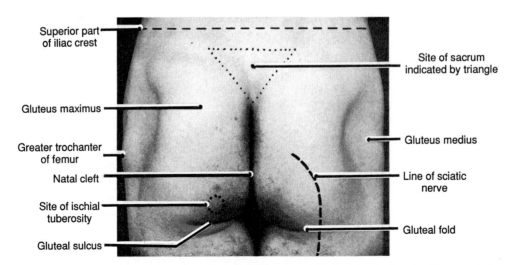

Figure 5-6.   The gluteal region of a 27-year-old man showing the principal surface landmarks. He was asked to stand in the anatomical position and to press his heels together firmly.

ally, close to the skin, and can be easily palpated on the lateral side of the thigh (Fig. 5-6). Because it is the most lateral point of the hip region, the greater trochanter causes you discomfort when you lie on your side on a hard surface. In the anatomical position, a line joining the tips of the greater trochanters normally passes through the center of the femoral heads and the pubic tubercles (Fig. 5-2). The degree of prominence of the greater trochanter increases when the gluteal muscles atrophy (*e.g.*, waste away owing to injury to the gluteal nerves, p. 420).

The lesser trochanter of the femur (Figs. 5-2 and 5-3) projects from the posteromedial surface of the femur at the inferior end of the intertrochanteric crest. It is located in the angle between the neck and body of the femur.

The body (shaft) of the femur is slightly bowed anteriorly and is narrowest at its midpoint. Its middle two-quarters are approximately circular in transverse section. Inferior to the neck, the body is smooth and featureless except for a rough ridge of bone, called the *linea aspera* (L. rough line), in the middle of its posterior surface. The linea aspera has medial and lateral lips, which diverge inferiorly to form medial and lateral *supracondylar lines* (Fig. 5-3). The body of the femur is not usually palpable because it is so well covered with large muscles. The *pectineal line* runs from the lesser trochanter to the medial lip of the linea aspera. The tendon of the pectineus muscle is attached to it (Figs. 5-18 and 5-22A).

The distal end (extremity) of the femur is broadened for articulation with the tibia. Two large, oblong *condyles* (G. knuckles) project posteriorly and are separated by a deep U-shaped *intercondylar notch* (Fig. 5-3). The medial and lateral condyles blend with each other anteriorly and with the body of the femur superiorly (Fig. 5-2). Although the articular surfaces are confluent anteriorly, each condyle is separated from the patellar surface by a slight groove. The patellar surface is where the patella (kneecap) slides during flexion and extension of the leg at the knee joint. The lateral and medial margins of the patellar surface can be palpated when the leg is flexed. The *adductor tubercle*, a small prominence of bone, may be felt at the superior part of the medial femoral condyle (Fig. 5-2). The medial and lateral **condyles of the femur** are subcutaneous and easily palpable when the knee is flexed and extended. Superior to each condyle is a prominent *epicondyle* to which the tibial and fibular collateral ligaments of the knee joint are attached (Fig. 5-91). The medial and lateral epicondyles are easily palpable.

The neck of the femur is frequently fractured in elderly people. As stated, it has many pits for the entrance of blood vessels (Fig. 5-29*B*). Consequently these vessels are vulnerable to injury during fractures of the femoral neck. The femur is large and strong, particularly its body, but a violent direct injury may fracture it, and it may take up to 20 weeks for firm union of the fragments to occur. Fractures of the neck of the femur and between the greater and lesser trochanters (*intertrochanteric fractures*), or through the trochanters (*pertrochanteric fractures*), are common in persons over 60 years of age.

Fractures of the neck of the femur usually result from indirect violence and often result from tripping over something (*e.g.*, a rug). These fractures are more common in older women than in men because their bones become markedly

weakened owing to *postmenopausal osteoporosis*. In this condition, bone resorption is greater than bone formation. When one hears that an old person has a "broken hip", the usual injury is a *fracture of the femoral neck* (see Case 5-1, p. 497). These fractures usually result from the combined effects of osteoporosis and a fall.

The distal end of the femur undergoes ossification just before birth. The visibility of this center of ossification in radiographs is commonly used as medicolegal evidence that a newborn infant found dead was near full term and viable.

The Fasciae of the Thigh

The fasciae consist of superficial and deep layers. The superficial fascia (tela subcutanea) underlies the skin and contains cutaneous nerves and vessels. The deep fascia is a dense layer of connective tissue between the superficial fascia and the muscles.

The Superficial Fascia of the Thigh

The subcutaneous connective tissue, including that over certain bones and prominences, constitutes the superficial fascia of the thigh. It lies deep to the dermis of the skin and consists of loose connective tissue, containing a variable amount of fat. Over the ischial tuberosities, which bear the sitting weight, the fat is within much fibrous tissue. In certain regions the superficial fascia splits into two layers, between which run the superficial vessels and nerves (Fig. 5-7). These layers are thick in the inguinal region where the superficial layer is continuous with the superficial fascia of the abdomen (see Figs. 2-6 and 2-18). The *superficial inguinal lymph nodes* and the great saphenous vein are also located between the layers of the superficial fascia (Fig. 5-31*B*). The superficial veins of the lower limb terminate in the small and great *saphenous veins* (Fig. 5-7), which drain most of the blood from the superficial fascia. There is only one saphenous vein in the thigh, the great one, because the small saphenous vein passes from the foot to the posterior aspect of the knee, where it ends in the popliteal vein (Fig. 5-46).

The Great Saphenous Vein (Figs. 5-7 to 5-11).   This large vein, the longest in the body, ascends from the foot to the groin in the subcutaneous connective tissue. The great (long) saphenous vein begins at the medial end of the dorsal venous arch of the foot and passes *anterior to the medial malleolus* of the tibia, where it is accompanied by the saphenous nerve. It then ascends obliquely across the inferior third of the tibia to the medial aspect of the knee. Here it lies superficial to the medial epicondyle, about a handsbreadth or 10 cm posterior to the medial border of the patella (Fig. 5-8*A*). From here, it ascends superolaterally to the *saphenous opening* in the deep fascia and enters the *femoral vein* (p. 385).

In Fig. 5-7, observe that the great saphenous vein anastomoses freely with the small saphenous vein. Examine the clinically important **perforating (anastomotic) veins** that connect the superficial veins to the deep veins. The main perforating veins from the great saphenous vein (indicated by arrows) are arranged in three sets: one related to the *adductor canal* (Fig. 5-15), one related to the calf muscles in the posterior part of the leg, and one just proximal to the ankle joint. The great saphenous vein

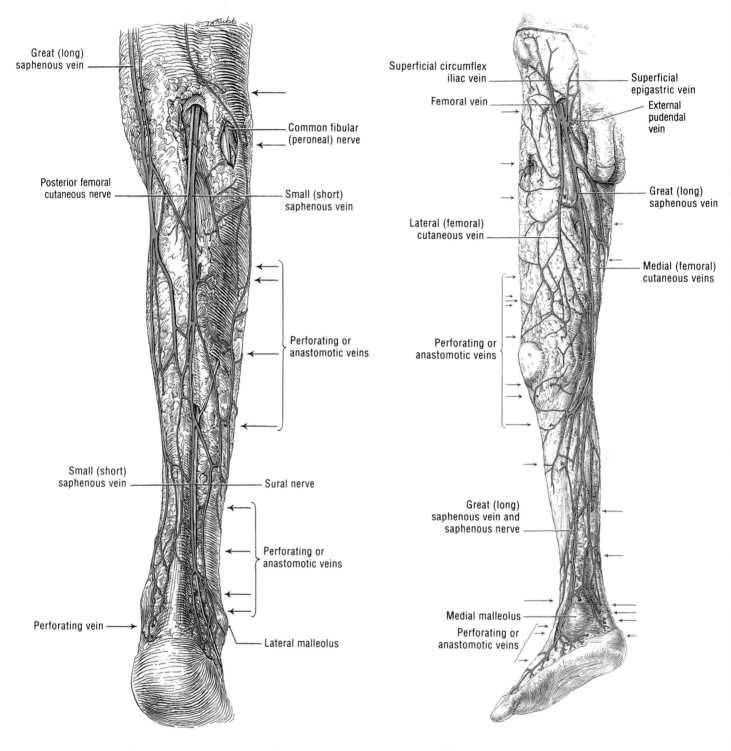

Great (long)
saphenous vein

Common fibular
(peroneal) nerve

Posterior femoral
cutaneous nerve

Small (short)
saphenous vein

Perforating or
anastomotic veins

Small (short)
saphenous vein

Sural nerve

Perforating or
anastomotic veins

Perforating vein

Lateral malleolus

Superficial circumflex
iliac vein

Superficial
epigastric vein

Femoral vein

External
pudendal
vein

Great (long)
saphenous vein

Lateral (femoral)
cutaneous vein

Medial (femoral)
cutaneous veins

Perforating or
anastomotic veins

Great (long)
saphenous vein and
saphenous nerve

Medial malleolus

Perforating or
anastomotic veins

A, *Posterior view*

B, *Anteromedial view*

Figure 5-7.   Superficial veins of the lower limb; some nerves are also
shown. The arrows indicate where perforating veins pierce the deep

fascia, bringing the superficial and deep veins into communication with
each other.

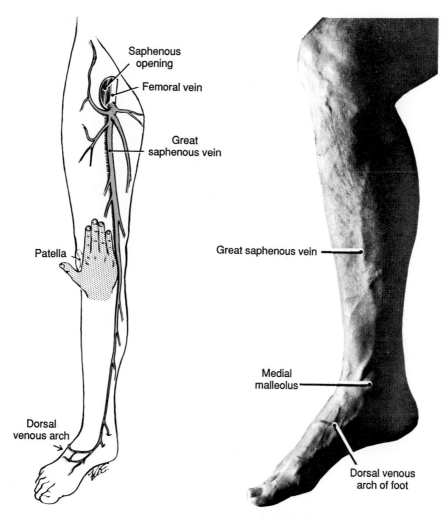

Figure 5-8. *A*, Drawing of the right lower limb showing how to locate the great saphenous vein at the knee. Observe that it is about 10 cm or a handsbreadth posterior to the medial border of the patella. *B*, The right leg of a 75-year-old man. Note that the great saphenous vein begins at the medial end of the dorsal venous arch of the foot and passes anterior to the medial malleolus.

may be duplicated, especially distal to the knee (Fig. 5-7*A*). It has 10 to 20 valves, which are more numerous in the leg than in the thigh; the perforating veins also have valves. These *venous valves* are usually located just inferior to the perforating veins. Usually there are three valves in its leg part and three in its thigh part. The valves are cup-like flaps of endothelium that fill from above. When they are full, they occlude the lumen of the great saphenous vein, thereby preventing reflux of the blood distally. This valvular mechanism enables the blood in the saphenous vein to overcome the force of gravity as it passes to the heart.

Tributaries of the Great Saphenous Vein (Figs. 5-7 and 5-8). This large vein drains the dorsum of the foot through the *dorsal venous arch* and the *medial dorsal vein* of the great toe. It also drains the sole of the foot through *medial marginal veins*. As it ascends in the leg and thigh it receives numerous tributaries and usually communicates in several locations with the small saphenous vein. Tributaries from the medial and posterior aspects of the thigh frequently unite to form an *accessory saphenous vein*. There are also fairly large vessels called the *lateral and anterior cutaneous veins*. They arise from networks of veins in the inferior part of the thigh and

enter the great saphenous vein superiorly, just before it enters the femoral vein. Near its termination, the great saphenous vein also receives the *superficial circumflex iliac*, *superficial epigastric*, and external *pudendal veins*.

The Small Saphenous Vein (Figs. 5-7*A* and 5-8). This lateral superficial vein *begins posterior to the lateral malleolus*. It is formed by the union of veins arising from the lateral part of the dorsal venous arch, the dorsum of fifth digit (little toe), and the lateral edge of the foot and sole. The small (short) saphenous vein passes along the lateral side of the foot with the *sural nerve* and ascends along the lateral side of the *tendo calcaneus* (Achilles tendon[2]). The small saphenous vein passes on the deep fascia between the two heads of the gastrocnemius muscle to the *popliteal fossa* (the diamond-shaped area posterior to the knee) where it perforates the deep popliteal fascia and

---

[2]This common tendon for the gastrocnemius and soleus muscles is often referred to as Achilles tendon after the mythical Greek warrior who was vulnerable only in the heel. Another common term is the Achilles reflex (ankle jerk), a contraction of the calf muscles that occurs when the tendo calcaneus or Achilles tendon (Fig. 5-66) is struck sharply.

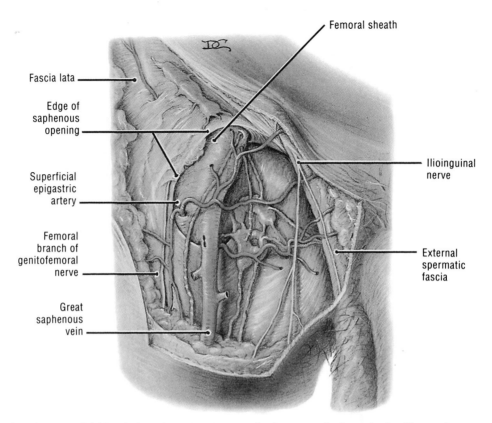

Fascia lata

Edge of
saphenous
opening

Superficial
epigastric
artery

Femoral
branch of
genitofemoral
nerve

Great
saphenous
vein

Femoral sheath

Ilioinguinal
nerve

External
spermatic
fascia

Figure 5-9. Dissection of the superficial inguinal arteries, veins, lymphatics, and nerves. Observe the saphenous opening in the fascia lata through which the great saphenous vein passes through the cribriform fascia to enter the femoral vein. Observe the arrangement of the superficial inguinal lymph nodes (green).

usually ends in the *popliteal vein* (Fig. 5-46). Sometimes it ends in the great saphenous vein or in one of the superior gluteal veins. The small saphenous vein has several communications with the great saphenous vein on the medial side of the leg. Before piercing the popliteal fascia just inferior to the knee flexion crease, the small saphenous vein frequently gives off a branch that unites with another vein to form the *accessory saphenous vein*. When present, this vein becomes the main communication between the great and small saphenous veins. The small saphenous vein has several valves.

When the valves of the perforating lower limb veins become incompetent (*i.e.*, dilated so that their cusps do not close the veins), contractions of the calf muscles, which normally propel the blood superiorly, cause a reverse flow through the perforating veins (*i.e.*, from deep to superficial veins). As a result, the perforating and superficial veins become tortuous and dilated (**varicose veins**).

Vein grafts using the great saphenous vein have been used to bypass obstructions in blood vessels (*e.g.*, atheromatous occlusion of the coronary arteries, p. 99). When a portion of the great saphenous vein is removed and used as a bypass, the vein is reversed so that the valves do not obstruct blood flow. Removal of the great saphenous veins rarely produces any significant problem in the lower limb because there are so many other veins. Following removal of the great saphenous vein, most blood from superficial parts of the lower limb passes the deep veins via the perforating veins (Fig. 5-7).

It is clinically important to know that *the great saphenous vein lies anterior to the medial malleolus* (Figs. 5-7 and 5-8). Even when it may not be visible in infants and obese persons or in patients in shock whose veins are collapsed, the great saphenous vein can always be located by making a skin incision anterior to the medial malleolus. This procedure, called a "*saphenous cutdown*," is used to insert a cannula for prolonged administration of blood, plasma expanders, electrolytes, or drugs. In Fig. 5-7*B*, observe that the *saphenous nerve* accompanies the great saphenous vein anterior to the medial malleolus. Should this nerve be caught by a ligature during a saphenous cutdown, the patient is likely to complain of pain along the medial border of the foot (Fig. 5-82). In the standing position, venous return from the lower limbs depends almost completely on muscular activity, especially of the calf muscles. The action of this "*calf pump*" is assisted by the tight sleeve of deep fascia surrounding these muscles (p. 441).

Varicose veins are common in posterior and medial parts of the lower limb, particularly in older persons. They cause considerable discomfort. Varicose saphenous veins have a caliber greater than normal and their cusps are incompetent. Consequently they allow blood to pass from the deep veins into the superficial veins. A localized dilation of the terminal part of the great saphenous vein, known as a *saphenous varix*,

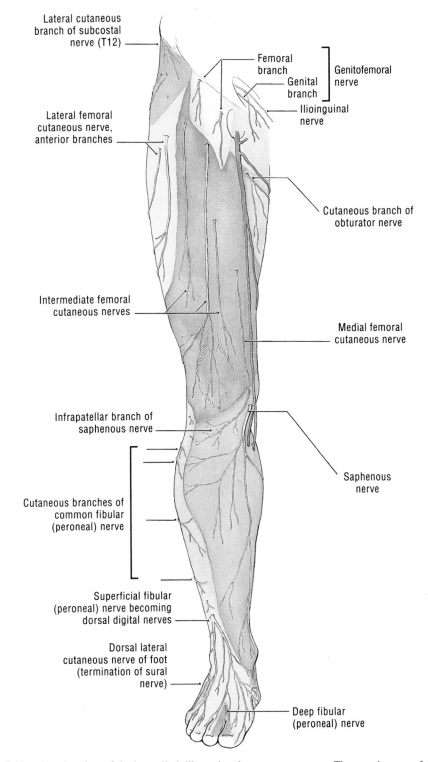

Figure 5-10. Anterior view of the lower limb illustrating the cutaneous nerves. The superior part of the great saphenous vein is also shown.

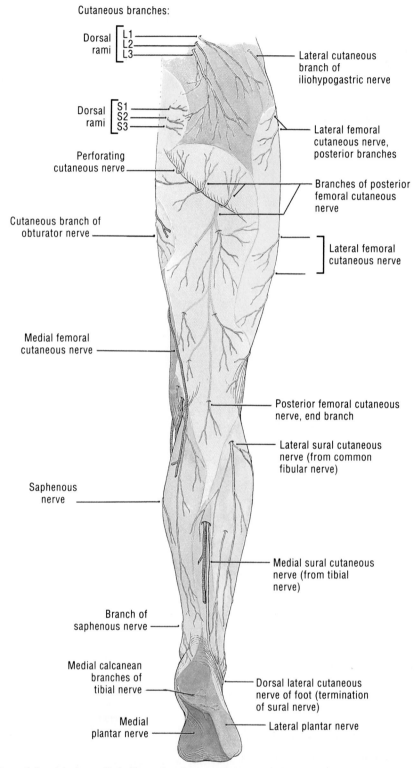

Cutaneous branches:

Dorsal rami [L1, L2, L3]

Dorsal rami [S1, S2, S3]

Perforating cutaneous nerve

Cutaneous branch of obturator nerve

Medial femoral cutaneous nerve

Saphenous nerve

Branch of saphenous nerve

Medial calcanean branches of tibial nerve

Medial plantar nerve

Lateral cutaneous branch of iliohypogastric nerve

Lateral femoral cutaneous nerve, posterior branches

Branches of posterior femoral cutaneous nerve

Lateral femoral cutaneous nerve

Posterior femoral cutaneous nerve, end branch

Lateral sural cutaneous nerve (from common fibular nerve)

Medial sural cutaneous nerve (from tibial nerve)

Dorsal lateral cutaneous nerve of foot (termination of sural nerve)

Lateral plantar nerve

Figure 5-11. Posterior view of the right lower limb illustrating the cutaneous nerves. *Sural* is the Latin term for the calf of the leg. Observe that the chief nerve supply to the gluteal region (buttock) is from the dorsal primary rami of lumbar and sacral nerves.

causes a swelling in the femoral triangle, located just inferior to the inguinal ligament (Fig. 5-14). A saphenous varix may be confused with other groin swellings (*e.g.*, a femoral hernia). Varicose veins have various causes, *e.g.*, when the normal venous return is impeded in constipated persons, pregnant females, and persons with large abdominal tumors. Varicose veins often develop in members of the same family, which may indicate a genetic weakness in the walls of the veins. Varicose veins and hemorrhoids (p. 303) are often associated.

Thrombophlebitis (inflammation of a vein with secondary thrombus or clot formation) of the deep veins destroys their valves, resulting in much of the blood from the leg being returned to the femoral vein via the superficial veins. This causes them to become tortuous and dilate (*i.e.*, varicose). If a thrombus breaks loose, it will be carried through the right side of the heart to the lung. Here, it will be stopped as the branches of the pulmonary artery become progressively smaller. If the pulmonary embolus is small, it may produce few or no symptoms, but if it is very large, it may result in sudden death. *Pulmonary thromboembolism* (PTE) is discussed on p. 76.

The Cutaneous Nerves of the Thigh (Figs. 5-9 to 5-11). Several cutaneous nerves in the superficial fascia supply the skin on the anterior, medial, and lateral aspects of the thigh. The *ilioinguinal nerve* (L1) is distributed though its anterior scrotal (or anterior labial) branch to the skin of the superomedial area of the thigh, adjacent to the scrotum or labium majus. Femoral branches of the *genitofemoral nerve* (L1 and L2) supply the skin just inferior to the middle part of the inguinal ligament. The *lateral femoral cutaneous nerve*[3] (L2 and L3) is a direct branch of the *lumbar plexus* (see Fig. 2-99). It enters the thigh deep to the lateral end of the inguinal ligament, near the anterior superior iliac spine. It supplies the skin on the anterior and lateral aspects of the thigh. The lateral cutaneous branches of the *subcostal nerve* (T12) descend across the crest of the ilium into the superior part of the thigh, at a point several centimeters posterior to the anterior superior iliac spine. It supplies the skin and subcutaneous tissue of the thigh anterior to the greater trochanter of the femur.

The many anterior cutaneous branches of the **femoral nerve** supply the skin on the anterior and medial aspects of the thigh (Fig. 5-10). The *posterior femoral cutaneous nerve*[4] is a branch of the *sacral plexus* (see Fig. 5-37). It supplies branches to the skin on the posterior aspect of the thigh (Fig. 5-11) and over the popliteal fossa (posterior to the knee).

The Deep Fascia of the Thigh

This fascia, known as the **fascia lata** (L. broad), is a strong, dense, broad layer that *invests the muscles of the thigh like an elastic stocking*. It provides a dense tubular sheath for the thigh muscles, which prevents them from bulging excessively when they contract. It therefore improves their effectiveness. The fascia lata is attached superiorly to the proximal part (root) of the lower limb as follows: the inguinal ligament, the external lip of the iliac crest, the posterior surface of the sacrum and coccyx,

the sacrotuberous ligament, the ischial tuberosity, the margin of the pubic arch, the body of the pubis, and the pubic tubercle.

The fascia lata is extremely strong laterally where it runs from the tubercle of the iliac crest to the tibia (Figs. 5-2 and 5-19*A*). This part of the fascia lata, known as the **iliotibial tract** (Fig. 5-20), receives tendinous reinforcements from the tensor fasciae latae and gluteus maximus muscles (Figs. 5-19*A* and 5-35). The distal end of the straplike iliotibial tract is attached to the lateral condyle of the tibia. It forms a prominent band when the thigh is flexed and the leg is extended (Fig. 5-26). Just posterior and inferior to the anterior superior iliac spine, the *fascia lata encloses the tensor fasciae latae muscle* (Figs. 5-15, 5-19*A*, and 5-20). This muscle pulls on the iliotibial tract, thereby steadying the trunk on the thigh and preventing posterior displacement of the iliotibial tract by the gluteus maximus muscle, three-quarters of which inserts into the iliotibial tract (Fig. 5-35).

The Saphenous Opening in the Fascia Lata of the Thigh (Figs. 5-8 and 5-9). Just inferior to the inguinal ligament, there is a gap or deficiency in the fascia lata, known as the saphenous opening. The great saphenous vein passes through it to join the femoral vein. The center of the saphenous opening is located about 3 cm inferolateral to the pubic tubercle (Fig. 5-2). The opening is about 4 cm long and 1 to 2 cm wide. Its medial margin is smooth, but its superior, lateral, and inferior margins form a sharp crescentic edge, called the *falciform margin* (L. *falx*, sickle + *forma*, form). This sickle-shaped margin of the saphenous opening is joined to its medial margin by fibrofatty tissue known as the *cribriform fascia*. This thin part of the deep fascia spreads over the saphenous opening (Fig. 5-9). This fascia is pierced by the great saphenous vein and the efferent lymph vessels from the superficial inguinal lymph nodes (Figs. 5-7*B* and 5-31*B*). This gives the fascia a sieve-like appearance from which it derives its name (L. *cribrum*, sieve + *forma*, form).

The Thigh Muscles

These large powerful muscles are organized into *three main groups* (anterior, medial, and posterior) on the basis of their location, actions, and nerve supply (Tables 5-1, 5-2, and 5-4). These muscle groups are separated by three fascial **intermuscular septa** (Fig. 5-16), which arise from the deep aspect of the fascia lata and are attached to the linea aspera of the femur (Fig. 5-3). The lateral intermuscular septum is strong; the other two are relatively weak.

The Anterior Thigh Muscles

This group of muscles consists of the iliopsoas, tensor fasciae latae, sartorius, and quadriceps femoris muscles (Figs. 5-12 to 5-15; Table 5-1).

The Iliopsoas Muscle (Fig. 5-12).   This powerful muscle is composed of the psoas major and iliacus muscles.

The Psoas Major Muscle (Figs. 5-12, 5-15, 5-19, and 5-32; see also Fig. 2-99).   This long, thick, powerful muscle passes from the abdomen to the thigh deep to the inguinal ligament. Consequently, it is also discussed with the abdomen (see

---

[3]Formerly this nerve was known as the lateral cutaneous nerve of the thigh.

[4]Formerly this nerve was known as the posterior cutaneous nerve of the thigh.

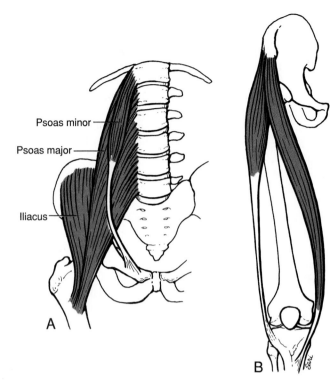

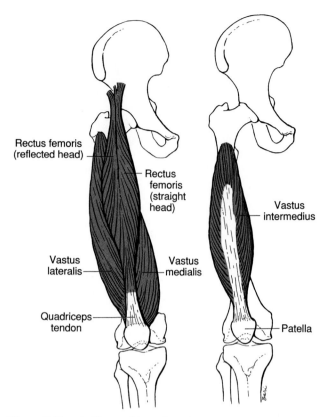

Figure 5-12. *A*, Psoas and iliacus muscles. Combined, the iliacus and psoas major muscles are called the iliopsoas muscle. *B*, Sartorius muscle and tensor fasciae latae.

Figure 5-13. *A*, The quadriceps femoris muscle. *B*, The vastus intermedius is visible when the other three parts of the muscle are removed.

p. 229). Its attachments, nerve supply, and main actions are given in Table 5-1.

The Iliacus Muscle (Figs. 5-12, 5-14, 5-15, and 5-19). This large triangular or fan-shaped muscle lies along the lateral side of the psoas major in the pelvis. Its attachments, nerve supply, and main actions are given in Table 5-1.

Actions of the Iliopsoas Muscle. This composite muscle is the *strongest flexor of the thigh* at the hip joint (Table 5-10). The iliopsoas muscles are also important antigravity *postural muscles*, which help to maintain erect posture at the hip joints. Acting inferiorly, the iliopsoas muscles *flex the trunk*, as in raising the trunk from the supine position to the sitting position.

The Tensor Fasciae Latae Muscle (Figs. 5-12*B*, 5-19 to 5-21, and 5-26). This fusiform muscle lies on the lateral side of the thigh, enclosed between two layers of the fascia lata, which form its sheath. The deep layer of its sheath is fused with the anterior surface of the capsule of the hip joint. Its attachments, nerve supply, and main actions are given in Table 5-1. As its name implies, it **tightens the fascia lata**, thereby enabling the thigh muscles to act with increased power. *It also tightens the iliotibial tract* (p. 414), enabling the gluteus maximus muscle (p. 407) to keep the knee joint in the extended position. In addition, when in the standing position, it steadies the trunk on the thigh and counteracts the posterior pull of the gluteus maximus on the iliotibial tract.

The Sartorius Muscle (Figs. 5-12*B*, 5-14, 5-15, 5-18, and 5-27). This narrow, elongated, straplike muscle is the longest one in the body and is the *most superficial muscle in the anterior part of the thigh*. It acts across two joints. It was given its name

because it is used to cross the legs in the tailor's squatting position[5] (L. *sartor*, a tailor). Throughout much of its course, the sartorius covers the femoral artery as it runs in the adductor (subsartorial) canal. Its attachments, nerve supply, and main actions are given in Table 5-1. During squatting it *flexes*, *abducts*, and *laterally rotates the thigh* at the hip joint, and it *flexes the leg* at the knee joint. Acting inferiorly, it also flexes the trunk on the thigh and rotates the trunk to the opposite side.

The Quadriceps Femoris Muscle (Figs. 5-13 to 5-19, 5-21, 5-23, 5-25, and 5-26). This *great extensor muscle of the leg*, the biggest muscle in the body, covers almost all of the anterior surface and sides of the femur. It is divided into **four parts**: rectus femoris, vastus lateralis, vastus medialis, and vastus intermedius. *Quadriceps* means four (L. *quadri*) heads (L. cipital or -ceps from L. caput, head). *Vastus* is a Latin term meaning large. The names of the parts of the quadriceps muscle indicate their form or location: the *rectus femoris* (L. *rectus*, straight) has deep fibers that run straight down the thigh (Fig. 5-15); the *vastus lateralis* lies on the lateral side of the thigh; the *vastus medialis* covers the medial aspect of the thigh; and the *vastus intermedius* is located between the vastus medialis and vastus lateralis parts. The attachments, nerve supply, and main actions of this large muscle are given in Table 5-1. The tendons of all four parts of the muscle unite to form the *quadriceps tendon*. This broad tendon attaches to and surrounds the patella (p. 388) and then continues as the *patellar ligament* (ligamentum patel-

---

[5]Jewelers in Japan and other eastern countries also cross their legs in this manner.

Table 5-1.
The Anterior Thigh Muscles

Muscle	Proximal Attachment	Distal Attachment	Innervation	Main Actions
Iliopsoas				
Psoas major (Fig. 5-12*A*)	Sides of T12 to L5 vertebrae and intervertebral discs between them	Lesser trochanter of femur	Ventral rami of lumbar nerves (**L1, L2,** and L3)[1]	Act conjointly in flexing thigh at hip joint and in stabilizing this joint
Iliacus (Fig. 5-12*A*)	Iliac crest, iliac fossa, ala of sacrum, and anterior sacroiliac ligaments	Tendon of psoas major and body of femur, inferior to lesser trochanter	Femoral nerve (**L2** and L3)	
Tensor fasciae latae (Figs. 5-19 and 5-20)	Anterior superior iliac spine and anterior part of external lip of iliac crest	Iliotibial tract that attaches to lateral condyle of tibia	Superior gluteal (L4 and L5)	Abducts, medially rotates, and flexes thigh; helps to keep knee extended; steadies trunk on thigh
Sartorius (Figs. 5-12*B* and 5-19*A*)	Anterior superior iliac spine and superior part of notch inferior to it	Superior part of medial surface of tibia	Femoral nerve (L2 and L3)	Flexes, abducts, and laterally rotates thigh at hip joint
Quadriceps femoris Rectus femoris	Anterior inferior iliac spine and groove superior to acetabulum	Base of patella and via patellar ligament to tibial tuberosity	Femoral nerve (L2, **L3,** and **L4,** posterior divisions)	Extend leg at knee joint; rectus femoris also steadies hip joint and helps iliopsoas to flex thigh
Vastus lateralis	Great trochanter and lateral lip of linea aspera of femur			
Vastus medialis	Intertrochanteric line and medial lip of linea aspera of femur			
Vastus intermedius	Anterior and lateral surfaces of body of femur			

[1]In this and subsequent tables, the numbers indicate the spinal cord segmental innervation of the nerves. For example, **L1, L2,** and L3 indicate that the nerves supplying the psoas major muscle are derived from the first three lumbar segments of the spinal cord; the bold face type (**L1, L2**) indicates the main segmental innervation. Damage to one or more of these spinal cord segments or to the motor nerve roots arising from them results in paralysis of the muscles concerned.

lae), which is attached to the tuberosity of the tibia. Expansions of the aponeuroses (sheet-like tendons) of the vasti muscles, called the medial and lateral *retinacula of the patella*, insert into the condyles of the tibia. **All parts** of the quadriceps, acting through the patellar ligament, **extend the leg** at the knee joint, and through the actions of the rectus femoris, they *flex the hip joint*. All four parts of this muscle are used during climbing, running, jumping, and rising from a chair.

A small flat ribbon of muscle, called the *articularis genu* (Fig. 5-19*C*), may be blended with the vastus intermedius. It arises from the inferior part of the anterior aspect of the femur and inserts into the synovial capsule of the knee joint and the walls of the *suprapatellar bursa* (Fig. 5-90). This small muscle pulls the synovial capsule superiorly during extension of the leg so it will not get caught in the knee joint.

The quadriceps muscle is tested with the patient lying on his/her back with the knee partly flexed. The patient is asked to extend his/her knee against resistance. If the muscle is functioning normally, the quadriceps can be seen and felt easily. If the quadriceps femoris muscle is paralyzed, the leg

cannot be extended, but the person can stand erect because the body weight tends to overextend the knee joint. Such a person can also walk with short steps if the pelvis is rotated to prevent extension of the hip far enough to flex the knee. Patients with *paralysis of the quadriceps femoris* muscle often press on the distal end of their thigh during walking to prevent flexion of the knee joint.

During football broadcasts, you often hear about a **thigh injury** known as a "hip pointer," which is a contusion or *bruise of the bone of the iliac crest*, usually its anterior part. Contusions cause bleeding from ruptured small capillaries, which allow blood to infiltrate muscles, tendons, or other soft tissues. The term "hip pointer" is also used to refer to avulsion (tearing off) of muscle attachments to the iliac crest, *e.g.*, iliacus, rectus femoris, sartorius, and tensor fasciae latae (Figs. 5-12 and 5-13*A*). Another common term used by sports broadcasters is "charley horse." This refers to contusion and tearing of muscle fibers sufficient enough to result in the formation of a **thigh hematoma** (a local collection of blood that has escaped into the muscles and surrounding tissues from damaged vessels). The most common site of a "charley horse" is in the quadriceps muscle. It is associated with localized pain and/or muscle stiffness and commonly follows

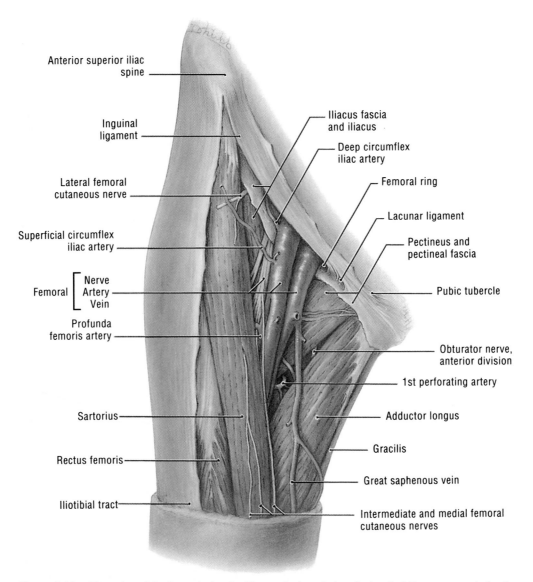

Anterior superior iliac spine

Inguinal ligament

Lateral femoral cutaneous nerve

Superficial circumflex iliac artery

Femoral { Nerve / Artery / Vein

Profunda femoris artery

Sartorius

Rectus femoris

Iliotibial tract

Iliacus fascia and iliacus

Deep circumflex iliac artery

Femoral ring

Lacunar ligament

Pectineus and pectineal fascia

Pubic tubercle

Obturator nerve, anterior division

1st perforating artery

Adductor longus

Gracilis

Great saphenous vein

Intermediate and medial femoral cutaneous nerves

Figure 5-14.   Dissection of the femoral triangle. Observe its boundaries: the inguinal ligament superiorly; the medial border of the adductor longus medially; and the medial border of the sartorius laterally.

direct trauma (*e.g.*, from a stick slash in hockey or a tackle in football). A common term in countries where cricket is frequently played is "cricket thigh." This injury refers to rupture of some fibers of the rectus femoris muscle. Sometimes the quadriceps tendon is also torn.

The Patella (Figs. 5-13 and 5-19*A*).   The patella (kneecap) is a triangular *sesamoid bone* (p. 13) with its apex pointing inferiorly. It is embedded in the quadriceps femoris tendon. The *patellar ligament* (ligamentum patellae), which attaches the patella (L. little plate) to the tibial tuberosity, is a continuation of the tendon of the quadriceps muscle. The apex of the patella indicates the level of the knee joint when the patellar ligament is taut. The patella is subcutaneous and can be easily palpated. It lies anterior to the distal end of the femur; hence it articulates posteriorly with the condyles of the femur (Fig. 5-2). The patella

is thought to increase the power of the already strong quadriceps femoris muscle by increasing its leverage. Palpate your patella as you flex your leg, noting that it is pulled down. Stand in a relaxed position and note that you can move your patella from side to side because the quadriceps femoris muscle is relaxed. Kneel on the floor and verify that it is the tibial tuberosity and the patellar ligament that bear most of the weight, not the patella.

A common knee problem for runners is *chondromalacia patellae* ("runner's knee"). This soreness and aching around or deep to the patella is caused by quadriceps muscle imbalance or compression of the knee, which pulls the patella sideways. This condition may be caused by a blow to the patella or from extreme flexing of the knee joint, as in squatting and kneeling (Griffith, 1986). Such overstressing of the

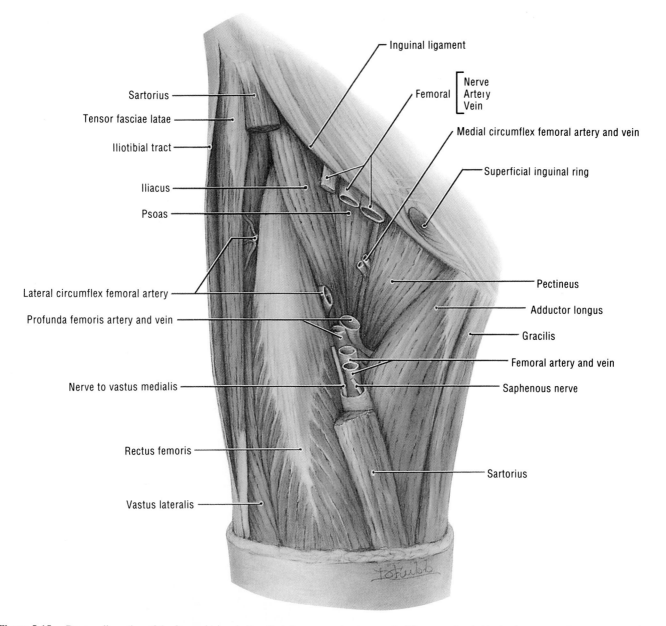

Inguinal ligament

Sartorius

Tensor fasciae latae

Iliotibial tract

Femoral ⎡ Nerve
        | Artery
        ⎣ Vein

Medial circumflex femoral artery and vein

Iliacus

Psoas

Superficial inguinal ring

Lateral circumflex femoral artery

Profunda femoris artery and vein

Pectineus

Adductor longus

Gracilis

Femoral artery and vein

Nerve to vastus medialis

Saphenous nerve

Rectus femoris

Sartorius

Vastus lateralis

Figure 5-15. Deeper dissection of the femoral triangle than that shown in Fig. 5-14. Sections are removed from the sartorius muscle and the femoral vessels and nerve. Observe that the floor of the femoral triangle is composed of four muscles (adductor longus, pectineus, psoas major, and iliacus). At the apex of the femoral triangle (see also Fig. 5-28), observe four vessels and two nerves in the adductor canal (p. 407).

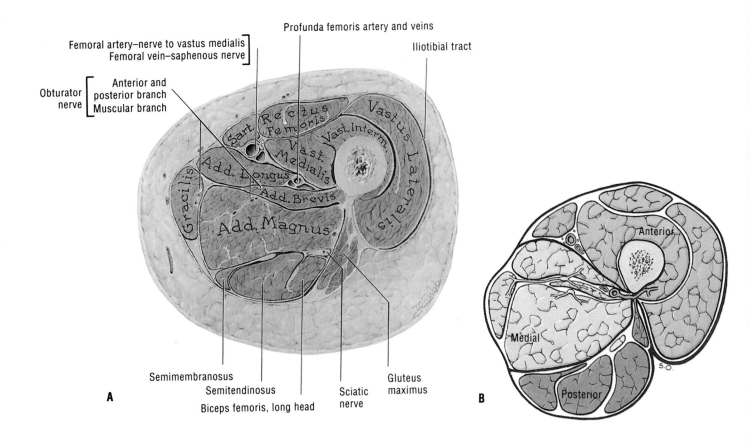

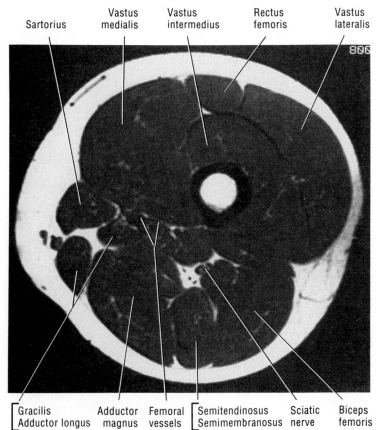

Figure 5-16. *A*, Transverse section through the thigh of a woman, 10 to 15 cm inferior to the inguinal ligament. *B*, Diagram showing the three groups of thigh muscles, each with its own nerve supply and primary function: *anterior*, femoral nerve—extensors of the knee; *medial*, obturator nerve—adductors of the hip; and *posterior*, sciatic nerve—flexors of the knee. *C*, Transverse MRI (magnetic resonance image) of the thigh. (Courtesy of Dr. W. Kucharczyk, Clinical Director of the Tri-Hospital Resonance Centre, Toronto, Ontario, Canada.)

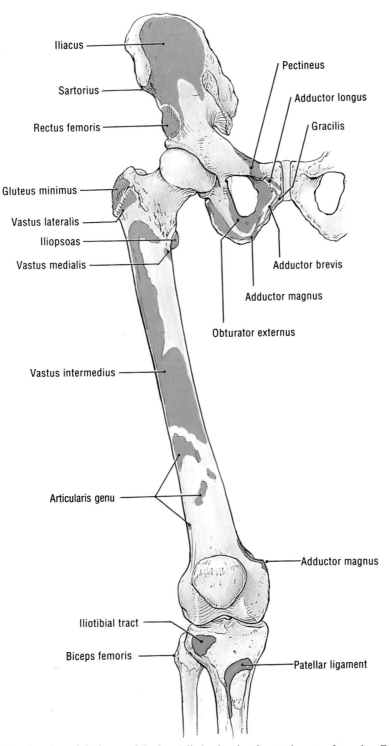

Figure 5-17.   Anterior view of the bones of the lower limb, showing the attachments of muscles. For distal parts of the leg bones, see Fig. 5-49. Salmon color indicates proximal attachments; blue, distal attachments.

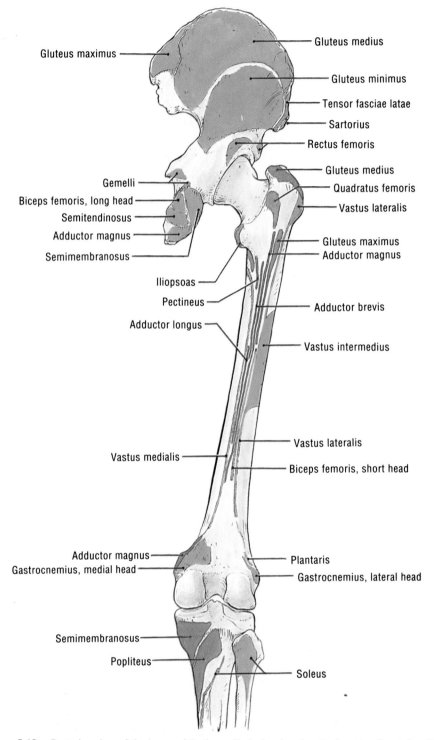

Figure 5-18.   Posterior view of the bones of the lower limb showing the attachments of muscles. For distal parts of the leg bones see Fig. 5-48.

knee can occur in any running sport, such as jogging, basketball, or soccer.

The patella is cartilaginous at birth and becomes ossified during the third to sixth year of life, frequently from more than one center. Although these centers usually coalesce, forming a single bone, they may remain separate on one or both sides, giving rise to a bipartite or tripartite patella. An unwary observer might interpret this condition on a radiograph as a fractured patella. A direct blow on the patella may fracture it in two or more fragments. The patella may also be fractured transversely by sudden contraction of the quadriceps (*e.g.*, when one slips and attempts to prevent a backward fall). In these cases the proximal fragment is pulled superiorly with the quadriceps tendon, and the distal fragment remains with the patellar ligament. This condition tends to occur more frequently in persons with unfused or poorly fused ossification centers in this bone.

Tapping the patellar ligament with a percussion hammer normally elicits the *quadriceps reflex* (patellar reflex or knee jerk; see Fig. 19 and p. 21). This reflex is routinely tested during a physical examination. Tapping the patellar ligament activates muscle spindles in the quadriceps femoris muscle. Afferent impulses from these spindles travel in the femoral nerve to the spinal cord (L2, L3, and L4 segments). From here, efferent impulses are transmitted via motor fibers in the femoral nerve to the quadriceps femoris, resulting in a jerklike contraction of the muscle and extension of the leg at the knee joint. Diminution or *absence of the quadriceps reflex* may result from any lesion that interrupts the reflex arc just described, (*e.g.*, peripheral nerve disease).

The Medial Thigh Muscles

The main action of the medial group of muscles is **adduction of the thigh** (*e.g.*, when squeezing your knees against a horse while riding). Hence they form the *adductor group of muscles*, including the pectineus, gracilis and adductors magnus, brevis, and longus muscles (Fig. 5-22; Table 5-2). All adductors of the thigh, except the pectineus, are supplied by the obturator nerve (L2, L3, and L4). The pectineus is supplied by the femoral nerve (**L2** and L3). The "hamstring" part of the adductor magnus is supplied by the sciatic nerve (L4).

The Pectineus Muscle[6] (Figs. 5-14, 5-15, 5-19, 5-22*A*, and 5-25). This short, flat quadrangular muscle forms part of the floor of the femoral triangle. Its attachments, nerve supply, and main actions are given in Table 5-2.

The Adductor Longus Muscle (Figs. 5-14 to 5-16, 5-19, 5-21, and 5-22*B*). This *long adductor* is a triangular muscle and is the most anterior muscle in the adductor group. Its attachments, nerve supply, and main actions are given in Table 5-2.

The Adductor Brevis Muscle (Figs. 5-16, 5-19*B*, and 5-22*C*). This *short adductor* lies deep to the pectineus and adductor longus muscles and anterior to the adductor magnus. Its attachments, nerve supply, and main actions are given in Table 5-2.

The Adductor Magnus Muscle (Figs. 5-16, 5-19*B*, 5-22*D*, 5-23, 5-25, and 5-27). This *large adductor* is the largest in the muscle adductor group. It is a composite, triangular muscle and *comprises adductor and hamstring parts*. The two portions differ in their attachments, nerve supply, and main actions (Table 5-2). There is a hiatus (L. aperture) in the aponeurotic attachment of the adductor magnus to the supracondylar line, called the *adductor hiatus*. It is located at the junction of the adductor canal (p. 407) and the popliteal fossa (p. 423), about a handsbreadth superior to the adductor tubercle of the femur (Figs. 5-3 and 5-28). This opening enables the femoral vessels to pass into the popliteal fossa (Fig. 5-46).

The Gracilis Muscle (Figs. 5-14, 5-15, 5-19, 5-22*E*, 5-23, and 5-27). This long, straplike fusiform muscle lies along the medial side of the thigh and knee. The gracilis (L. slender) is the most superficial of the adductor group of muscles and is the weakest member. It is the only one of the group to cross the knee joint. Its attachments, nerve supply, and main actions are given in Table 5-2.

The Obturator Externus Muscle (Figs. 5-23 to 5-25). This flat, relatively small, fan-shaped muscle is deeply placed in the superomedial part of the thigh. Its tendon crosses the posterior aspect of the neck of the femur. Its attachments, nerve supply, and main action are given in Table 5-2.

Because the gracilis is a relatively weak member of the adductor group, it can be removed without noticeable loss of its actions on the leg. Hence surgeons often transplant the gracilis with its nerves and blood vessels to replace a damaged muscle (*e.g.*, in the hand). Once the muscle has been transplanted, it can provide digital flexion and extension (McKee et al., 1990).

Occasionally during sports broadcasts, reference is made to an injury called a "pulled groin" or a "groin strain." This means that there has been a strain, stretching, and probably some tearing away of the attachments of the anterior and medial muscles of the thigh. The injury usually involves the iliopsoas (Table 5-1) and/or the adductor group of muscles (Table 5-2). The superior attachments of these muscles are located in the inguinal region at the junction of the abdomen and thigh (L. *inguen*, groin). This injury usually occurs in sports that require quick starts, such as short-distance racing (*e.g.*, 50-meter dash), basketball, football, and soccer.

Ossification sometimes occurs in the tendons of the adductor longus muscles. You may hear these ossified tendons referred to as "riders' bones" because the condition is common in persons who ride horses and actively adduct their thighs.

The Femoral Triangle

This is a clinically important *triangular subfascial space* in the superomedial one-third part of the thigh. It appears as a depression inferior to the inguinal ligament when the thigh is actively flexed at the hip joint (Figs. 5-14 and 5-21). Its main contents are the femoral vessels and branches of the femoral nerve.

Boundaries of the Femoral Triangle (Figs. 5-14, 5-15, 5-21, and 5-28). The femoral triangle is bounded *superiorly*

---

[6]The name of this muscle is derived from the Latin word *pecten* meaning a comb and is related to its attachment to the pecten pubis (Fig. 5-2). This sharp ridge was thought to resemble a comb.

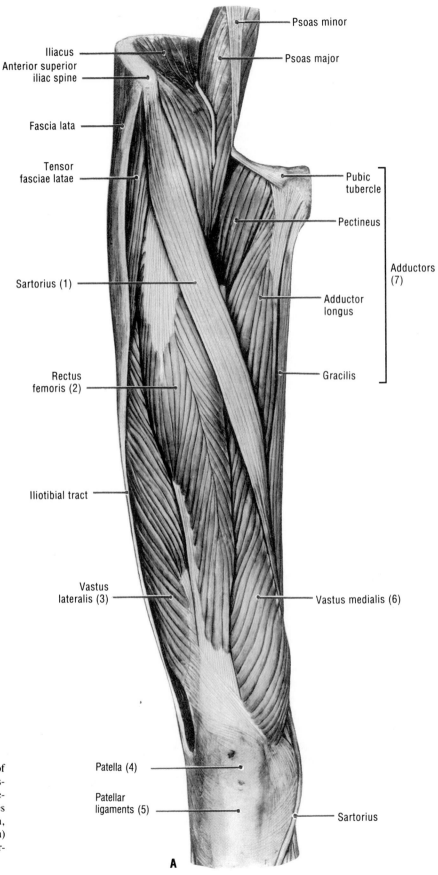

Figure 5-19. Muscles of the anterior aspect of the thigh. *A*, Superficial dissection. *B*, Deep dissection with sections of the sartorius, rectus femoris, pectineus, and adductor longus muscles excised. *C*, Deep dissection of the knee region, medial view, showing the articularis genus (genu) muscle, which often blends with the vastus intermedius muscle.

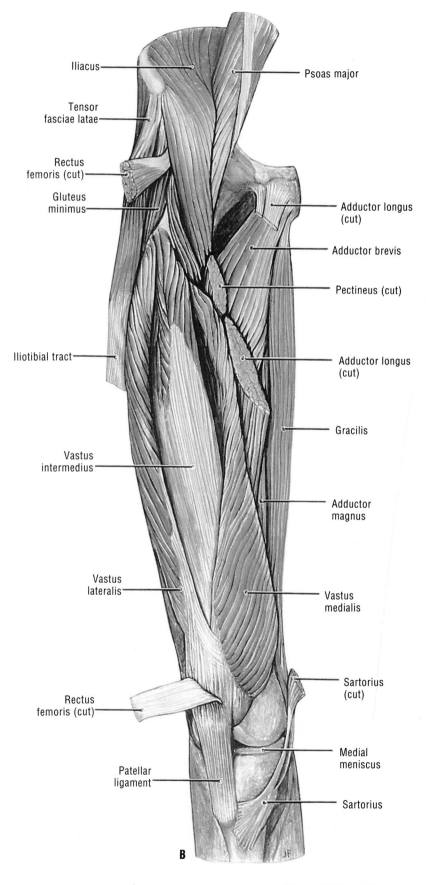

Iliacus

Psoas major

Tensor fasciae latae

Rectus femoris (cut)

Adductor longus (cut)

Gluteus minimus

Adductor brevis

Pectineus (cut)

Iliotibial tract

Adductor longus (cut)

Gracilis

Vastus intermedius

Adductor magnus

Vastus lateralis

Vastus medialis

Sartorius (cut)

Rectus femoris (cut)

Medial meniscus

Patellar ligament

Sartorius

B

Femur

Articularis genu

Vastus intermedius (cut)

Vastus lateralis (cut)

Retinaculum

Fibrous capsule of knee joint

Adductor magnus

Vastus medialis (reflected)

C

Figure 5-19*B* and *C*

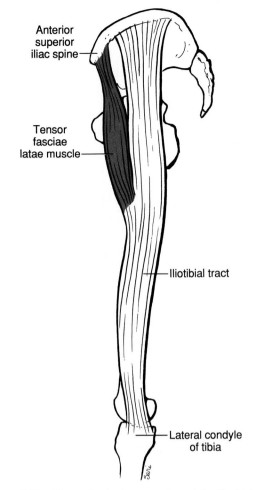

Anterior
superior
iliac spine

Tensor
fasciae
latae muscle

Iliotibial tract

Lateral condyle
of tibia

Figure 5-20. Tensor fasciae latae muscle and the iliotibial tract.

by the inguinal ligament, *medially* by the medial border of the adductor longus muscle,[7] and *laterally* by the medial border of the sartorius muscle. The **base of the femoral triangle** is formed by the inguinal ligament. The inferior boundary and *apex of the femoral triangle* are located where the medial borders of the sartorius and the adductor longus muscles meet. The muscular **floor of the femoral triangle** is not flat, but gutter-shaped. It is formed from medial to lateral by the *adductor longus, pectineus,* and *iliopsoas muscles.* It is the juxtaposition of the iliopsoas and pectineus muscles that forms the deep gutter in the muscular floor. The **roof of the femoral triangle** is formed by the *fascia lata,* which includes the cribriform fascia (p. 385). The skin covering the triangle is supplied by the *ilioinguinal nerve,* which emerges from the anterior abdominal wall through the superficial inguinal ring (see Fig. 2-16), and by the *femoral branch of the genitofemoral nerve,* which enters the thigh just lateral to the femoral artery (Fig. 5-10).

Surface Anatomy of the Femoral Triangle (Figs. 5-21 and

---

[7]Formerly, the medial border of the femoral triangle was given as the lateral border of the adductor longus. Some authors still use this former designation. The widely accepted new medial boundary is used in this text.

5-28). When the thigh is actively flexed at the hip joint, the femoral triangle appears as a triangular depression in its proximal third. You can easily palpate and usually observe its base, the *inguinal ligament.* Its lateral boundary, the medial border of the sartorius muscle, is obvious in most people, but its medial boundary (medial border of adductor longus) is not usually so easy to identify. The **femoral pulse** can easily be palpated in the femoral triangle, 2 to 3 cm inferior to the midpoint of the inguinal ligament. The head of the femur lies posterior to the femoral artery at this site, making compression of the vessel easy (Fig. 5-28).

Contents of the Femoral Triangle

This triangular space in the anterior aspect of the thigh contains the *femoral artery* and its branches, the *femoral vein* and its tributaries, the *femoral nerve* and its branches (*e.g.,* the saphenous nerve), the lateral femoral cutaneous nerve, the femoral branch of the genitofemoral nerve, lymphatic vessels, and some *inguinal lymph nodes.*

The Femoral Artery (Figs. 5-14 to 5-16, 5-28 to 5-30*A,* and 5-33; see also Fig. 2-11). This large vessel, *the continuation of the external iliac artery,* provides the chief arterial supply to the lower limb. It enters the femoral triangle deep to the midpoint of the inguinal ligament and *lateral to the femoral vein.* The femoral artery is located posterior to the deep fascia, whereas the great saphenous vein is in the superficial fascia (Fig. 5-7). These relationships are clinically important. The femoral artery descends on the psoas major, pectineus, and adductor longus muscles in the floor of the femoral triangle. It *bisects the femoral triangle* and at its apex the femoral artery runs deep to the sartorius muscle within the *adductor canal.* Its course in this canal is discussed on p. 407.

The Surface Marking of the Femoral Artery (Fig. 5-28). With the thigh slightly flexed and laterally rotated, the artery runs from the midpoint between the pubic symphysis and the anterior superior iliac spine and along the superior two-thirds of a line running towards the adductor tubercle.

The Profunda Femoris Artery (Figs. 5-14 to 5-16, 5-29, and 5-30*A*). This deep (L. *profundus*) vessel is the largest branch of the femoral artery and is the **chief artery to the thigh.** It arises from the lateral side of the femoral artery within the femoral triangle, about 4 cm inferior to the inguinal ligament. It runs lateral to the femoral artery and then passes posterior to it and the femoral vein. The profunda femoris artery leaves the femoral triangle between the pectineus and adductor longus muscles and descends posterior to the latter muscle, giving off perforating arteries that supply the adductor magnus and hamstring muscles. The medial and lateral **circumflex femoral arteries,** which are branches of the profunda femoris, supply the thigh muscles and the proximal end of the femur. *The medial circumflex femoral artery is clinically important because it supplies most of the blood to the head and neck of the femur.* It passes deeply between the iliopsoas and pectineus muscles to reach the posterior part of the thigh. The *lateral circumflex femoral artery* passes laterally, deep to the sartorius and rectus femoris muscles, and between the branches of the femoral nerve. Here, it divides into branches that supply the muscles on the lateral side of the thigh and the head of the femur.

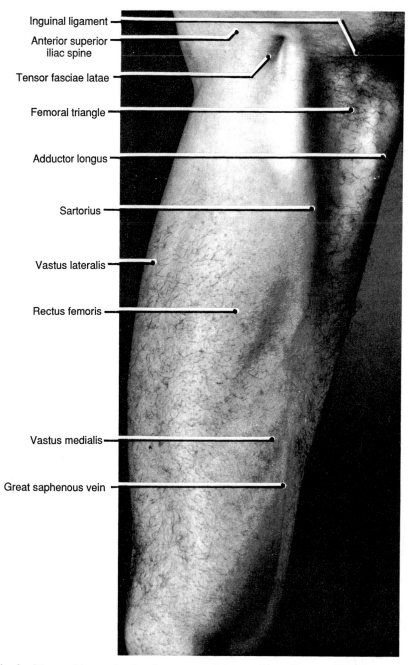

Inguinal ligament

Anterior superior
iliac spine

Tensor fasciae latae

Femoral triangle

Adductor longus

Sartorius

Vastus lateralis

Rectus femoris

Vastus medialis

Great saphenous vein

Figure 5-21.   The right thigh of a 27-year-old man showing the sartorius muscle in action. To display this muscle, he was asked to flex, abduct, and laterally rotate his thigh. This photograph was taken with his pelvis slightly rotated to the left.

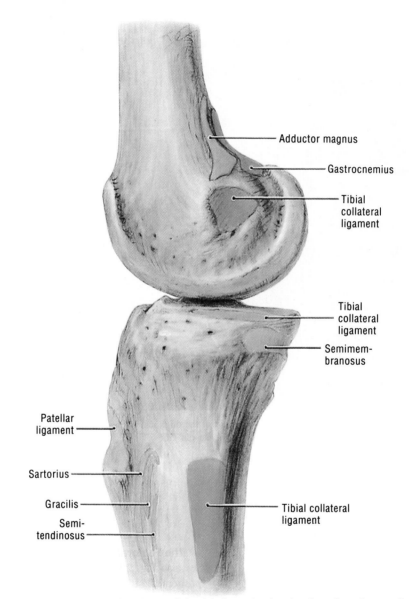

Figure 5-27.   Medial view of the bones of the right knee showing the sites of attachment of muscles and ligaments.

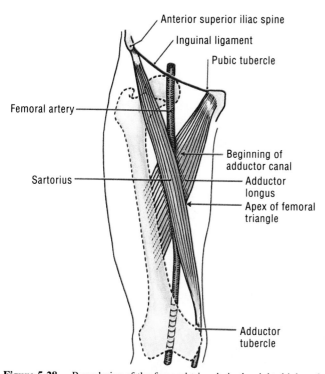

Anterior superior iliac spine
Inguinal ligament
Pubic tubercle
Femoral artery
Beginning of adductor canal
Sartorius
Adductor longus
Apex of femoral triangle
Adductor tubercle

Figure 5-28. Boundaries of the femoral triangle in the right thigh and the site of the adductor canal. Note that the head of the femur (yellow) lies posterior to the femoral artery, just inferior to the midpoint of the inguinal ligament.

femoral vessels. The saphenous nerve accompanies the femoral artery in the adductor canal and becomes superficial by passing between the sartorius and gracilis muscles. It passes antero-inferiorly to supply the skin and fascia of the anterior and medial aspects of the knee, leg, and foot.

The Femoral Sheath

This oval, funnel-shaped fascial tube *encloses the proximal parts of the femoral vessels*, which lie inferior to the inguinal ligament (Figs. 5-32 to 5-34). It also *surrounds the femoral canal*, but it **does not enclose the femoral nerve**. The femoral sheath is a diverticulum or inferior prolongation of the fasciae lining the abdomen (transversalis fascia anteriorly and iliac fascia posteriorly). It is covered by the fascia lata (Fig. 5-32). The femoral sheath ends about 4 cm inferior to the inguinal ligament by becoming continuous with the adventitia or external loose connective tissue covering of the femoral vessels. The medial wall of the femoral sheath is pierced by the great saphenous vein and lymphatic vessels (Fig. 5-32).

The presence of the femoral sheath allows the femoral artery and vein to glide in and out, deep to the inguinal ligament, during movements of the hip joint. The sheath does not project into the thigh when the thigh is fully flexed, but it is drawn further into the femoral triangle when the thigh is extended.

Compartments of the Femoral Sheath (Figs. 5-32 to 5-34). This sheath is subdivided by two vertical septa into three

compartments: (1) a *lateral compartment* for the femoral artery; (2) an *intermediate compartment* for the femoral vein; and (3) a *medial compartment* or space called the femoral canal.

The Femoral Canal (Figs. 5-33 and 5-34). This short, conical medial compartment of the femoral sheath lies between the medial edge of the sheath and the femoral vein. This space allows the femoral vein to expand during times of increased venous return from the lower limb. It contains a few lymph vessels, sometimes a deep inguinal lymph node, loose connective tissue, and fat. It is also the route by which the efferent lymph vessels from the deep inguinal lymph nodes pass to the external iliac lymph nodes. The canal is *widest at its abdominal end, the femoral ring*, and extends distally to the level of the proximal end of the saphenous opening (Fig. 5-9).

The femoral ring (Figs. 5-14, 5-33, and 5-34) is the small superior end or *mouth of the femoral canal*. It is about 1 cm wide and is closed by extraperitoneal fatty tissue called the **femoral septum**, which is pierced by the lymph vessels connecting the inguinal and external iliac lymph nodes. The four *boundaries of the femoral ring* are: *laterally*, the partition between the femoral ring and the femoral vein; *posteriorly*, the superior ramus of the pubis covered by the pectineus muscle and its fascia; *medially*, the lacunar ligament and conjoint tendon (Fig. 5-14; see also Fig. 2-11); and *anteriorly*, the medial part of the inguinal ligament.

The femoral ring is a weak area in the anterior abdominal wall that normally admits the tip of the fifth digit. This is important in understanding the mechanism of **femoral hernia**, a protrusion of abdominal viscera (often small intestine) through the femoral ring into the femoral canal. The hernia passes through the femoral ring into the femoral canal, pushing a covering of peritoneum before it. The hernial sac compresses the contents of the femoral canal (lymph vessels, connective tissue, and fat) and distends its wall. Usually a femoral hernia can be palpated easily just inferior to the inguinal ligament. Initially, it is relatively small because it is contained within the femoral canal, but it can enlarge by passing inferiorly through the saphenous opening into the loose connective tissue of the thigh (Figs. 5-8 and 5-9). Strangulation of a femoral hernia may occur owing to the sharp, rigid boundaries of the femoral ring, particularly the concave margin of the *lacunar ligament* (Fig. 5-14). Sometimes this ligament has to be incised to release a strangulated hernia. In these cases an *abnormal obturator artery*, present in about 20% of people, is vulnerable to injury (McMinn, 1990).

Strangulation of a femoral hernia interferes with the blood supply to the herniated intestine, and this vascular impairment may result in death of the tissues concerned. Because the distal end of the femoral canal reaches the proximal part of the saphenous opening in the fascia lata (Fig. 5-9), a femoral hernia may herniate through the cribriform fascia and bulge anteriorly under the skin over the saphenous opening. A femoral hernia presents as a mass inferolateral to the pubic tubercle and medial to the femoral vein. This type of hernia is more common in women than men, largely because the femoral ring is larger owing to the greater breadth of the female pelvis (p. 243).

The obturator artery (Fig. 5-29A), a branch of the internal iliac, passes through the obturator foramen to supply adjacent muscles. In about 20% of people, an enlarged pubic branch

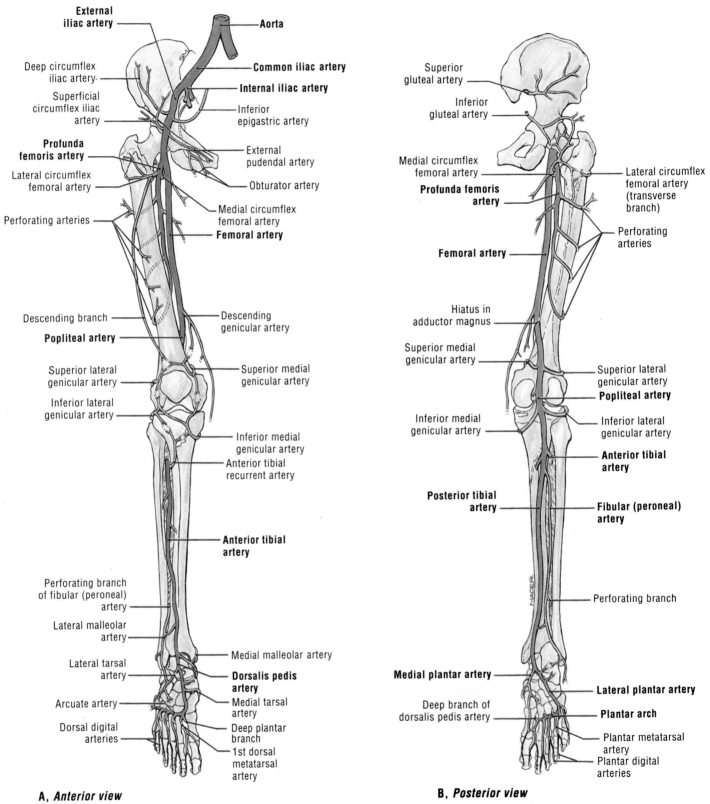

External iliac artery

Deep circumflex iliac artery·

Superficial circumflex iliac artery

Profunda femoris artery

Lateral circumflex femoral artery

Perforating arteries

Descending branch

Popliteal artery

Superior lateral genicular artery

Inferior lateral genicular artery

Perforating branch of fibular (peroneal) artery

Lateral malleolar artery

Lateral tarsal artery

Arcuate artery

Dorsal digital arteries

Aorta

Common iliac artery

Internal iliac artery

Inferior epigastric artery

External pudendal artery

Obturator artery

Medial circumflex femoral artery

Femoral artery

Descending genicular artery

Superior medial genicular artery

Inferior medial genicular artery

Anterior tibial recurrent artery

Anterior tibial artery

Medial malleolar artery

Dorsalis pedis artery

Medial tarsal artery

Deep plantar branch

1st dorsal metatarsal artery

A, *Anterior view*

Superior gluteal artery

Inferior gluteal artery

Medial circumflex femoral artery

Profunda femoris artery

Femoral artery

Hiatus in adductor magnus

Superior medial genicular artery

Inferior medial genicular artery

Posterior tibial artery

Medial plantar artery

Deep branch of dorsalis pedis artery

Lateral circumflex femoral artery (transverse branch)

Perforating arteries

Superior lateral genicular artery

Popliteal artery

Inferior lateral genicular artery

Anterior tibial artery

Fibular (peroneal) artery

Perforating branch

Lateral plantar artery

Plantar arch

Plantar metatarsal artery

Plantar digital arteries

B, *Posterior view*

Figure 5-29.   Arteries of the lower limb.

of the inferior epigastric artery takes the place of the obturator artery or forms an *accessory obturator artery*. This artery runs close to or across the femoral ring to reach the obturator foramen. Here, it is closely related to the free margin of the lacunar ligament and the neck of a femoral hernia. Consequently, this artery could be involved in a strangulated femoral hernia.

The Adductor Canal

The adductor canal (subsartorial canal or Hunter's canal) is about 15 cm in length and is a *narrow, fascial tunnel in the thigh* (Figs. 5-15 and 5-30*A*). Located deep to the middle third of the sartorius muscle, it provides an intermuscular passage through which the femoral vessels pass to reach the popliteal fossa, where they become the popliteal vessels (Figs. 5-28 and 5-46). The adductor canal begins about 15 cm inferior to the inguinal ligament, where the sartorius muscle crosses over the adductor longus muscle (Fig. 5-28). It ends at the *adductor hiatus* in the tendon of the adductor magnus muscle.

Boundaries of the Adductor Canal (Figs. 5-15, 5-16, and 5-28). This canal is bounded *laterally* by the vastus medialis, *posteromedially* by the adductor longus and adductor magnus muscles, and *anteriorly* by the sartorius muscle. The sartorius and subsartorial fascia form the *roof of the adductor canal*. About the middle third of the thigh, a *subsartorial plexus of nerves* lies on this fascia. It supplies the overlying skin.

Contents of the Adductor Canal (Figs. 5-15, 5-16, 5-28, and 5-30*A*). The femoral vessels enter the adductor canal where the sartorius muscle crosses over the adductor longus muscle, the vein lying posterior to the artery. The *femoral artery and vein* leave the adductor canal through the tendinous opening in the adductor magnus muscle, known as the **adductor hiatus**. As soon as the femoral artery enters the popliteal fossa, it is called the popliteal artery; similarly the femoral vein becomes the popliteal vein. In Fig. 5-15, observe that the *profunda femoris artery* and vein do not enter the adductor canal. The perforating branches of these deep vessels pierce the fibers of the adductor muscles to reach the posterior aspect of the thigh.

The **saphenous nerve**, a cutaneous branch of the femoral nerve, accompanies the femoral artery through the adductor canal (Fig. 5-30). It enters the adductor canal lateral to this artery, crosses it anteriorly, and lies medial to it at the distal end of the canal. The saphenous nerve does not leave the adductor canal via the adductor hiatus. It passes between the sartorius and gracilis muscles, pierces the deep fascia on the medial aspect of the knee (Fig. 5-10), and passes down the medial side of the leg with the great saphenous vein. The *nerve to the vastus medialis* muscle accompanies the femoral artery through the proximal part of the adductor canal (Figs. 5-15 and 5-30) and then divides into branches that supply this muscle and the knee joint.

The Gluteal Region

The gluteal (buttock) region lies posterior to the pelvis and is the prominence formed on each side by the gluteal muscles

(Figs. 5-5, 5-6, 5-35, and 5-40). It is *bounded superiorly by the iliac crest and inferiorly by the inferior border of the gluteus maximus muscle*. The gluteal sulcus (groove), located inferior to the gluteal fold (Fig. 5-6), indicates the inferior border of the gluteus maximus. It is important clinically to know the boundaries of the gluteal region. Some people wrongly consider the buttock to be only the prominence formed by the gluteus maximus. The gluteal region is largely formed by the gluteus maximus muscle, but the gluteus medius forms its superolateral part (Fig. 5-35).

Bony Landmarks and Surface Markings of the Gluteal Region

Palpate your anterior and posterior superior *iliac spines* and iliac crests. The *iliac crest* extends from the anterior superior iliac spine to the posterior superior iliac spine, which is indicated by a skin dimple (Fig. 4-9). Between the two posterior superior iliac spines is the posterior aspect of the sacrum. The spinous process of the S2 vertebra is at the level of the posterior superior iliac spines, and the summit of the iliac crest corresponds to the level of the spinous process of L4 vertebra. Inferomedially, the *ischial tuberosity* can be palpated deep to the gluteus maximus muscle.

Ligaments of the Gluteal Region. The parts of the bony pelvis are bound together by dense ligaments (p. 250). The region between the sacrum and the bony pelvis is bridged by two important accessory ligaments of the sacroiliac joint, called the *sacrotuberous and sacrospinous ligaments* (p. 249). They close the sciatic notches in the hip bones, producing the greater and lesser sciatic foramina (Fig. 5-24; see also Fig. 3-8). The *greater sciatic foramen* is a passageway for structures entering or leaving the pelvis (*e.g.*, the sciatic nerve, Fig. 5-35*B*), whereas the *lesser sciatic foramen* is a doorway for structures entering or leaving the perineum (*e.g.*, the pudendal nerve; see Fig. 3-61). The greater and lesser sciatic notches are separated by a beaklike process, the *ischial spine* (Figs. 5-3 and 5-4), which gives attachment to the sacrospinous ligament.

The Gluteal Muscles

The gluteal muscles consist of: (1) the large glutei (maximus, medius, and minimus), which are mainly extensors and abductors of the thigh at the hip joint (Table 5-3), and (2) a deeper group of smaller muscles (piriformis, obturator internus, gemelli, and quadratus femoris), which are mainly lateral rotators of the thigh at the hip joint.

The Gluteus Maximus Muscle (Figs. 5-6 and 5-35 to 5-38). This is the largest, heaviest, and most coarsely fibered muscle in the gluteal region and is one of the largest muscles in the body. *It forms a thick, quadrilateral pad over the ischial tuberosity* when the thigh is extended. The ischial tuberosity can be felt on deep palpation through the inferior portion of the gluteus maximus, just superior to the medial portion of the gluteal fold. When the thigh is flexed, the distal border of the gluteus maximus moves superiorly, leaving the ischial tuberosity subcutaneous. The attachments, nerve supply, and main actions of this muscle are given in Table 5-3. A thick flat sheet of muscle, the gluteus maximus slopes inferolaterally from the pelvis across

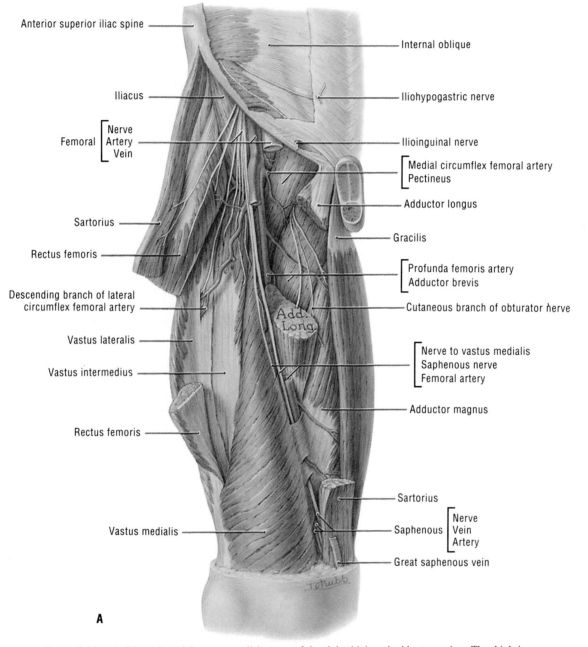

Anterior superior iliac spine

Internal oblique

Iliacus

Iliohypogastric nerve

Femoral ⎡ Nerve
        | Artery
        ⎣ Vein

Ilioinguinal nerve

Medial circumflex femoral artery
Pectineus

Adductor longus

Sartorius

Gracilis

Rectus femoris

Profunda femoris artery
Adductor brevis

Descending branch of lateral
circumflex femoral artery

Cutaneous branch of obturator nerve

Add.
Long.

Vastus lateralis

Nerve to vastus medialis
Saphenous nerve
Femoral artery

Vastus intermedius

Adductor magnus

Rectus femoris

Sartorius

Saphenous ⎡ Nerve
          | Vein
          ⎣ Artery

Vastus medialis

Great saphenous vein

A

Figure 5-30.   *A*, Dissection of the anteromedial aspect of the right thigh and adductor region. The thigh is rotated laterally. *B*, Scheme of the motor distribution of the femoral and obturator nerves.

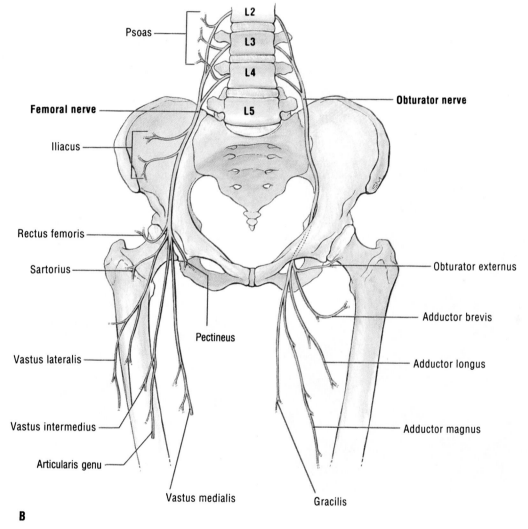

Psoas

L2

L3

L4

Femoral nerve

L5

Obturator nerve

Iliacus

Rectus femoris

Sartorius

Obturator externus

Adductor brevis

Pectineus

Vastus lateralis

Adductor longus

Vastus intermedius

Adductor magnus

Articularis genu

Vastus medialis

Gracilis

B

Figure 5-30*B*

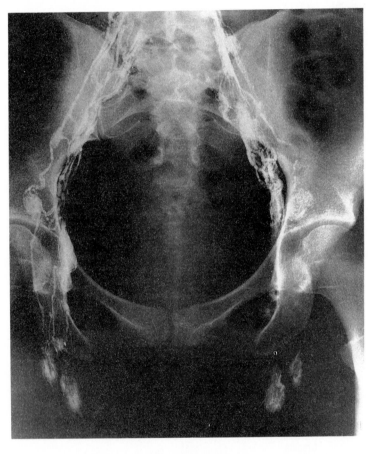

A

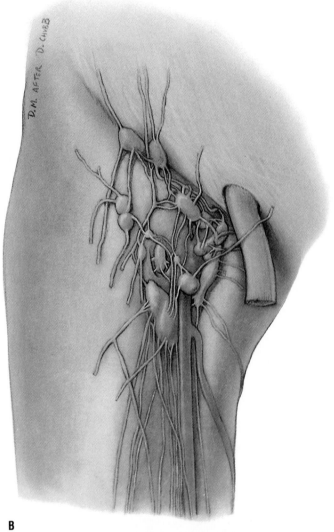

B

Figure 5-31.   Inguinal lymph nodes on the right side. *A,* Lymphangiogram taken after injection of lipiodol. (Courtesy of Dr. E. L. Lansdown, Professor of Radiology, University of Toronto, Toronto, Ontario, Canada.) *B,* Dissection of the inguinal lymph nodes. The *superficial nodes* are arranged in the form of a **T**. Most of these nodes are along the horizontal part of the T, parallel to and about 1 cm inferior to the inguinal ligament. A few larger nodes are located along the vertical part of the T which follows the termination of the great saphenous vein (see also Fig. 5-7). The *deep nodes* are on the medial side of the femoral vein. Often one of the three nodes is in the superior end of the femoral ring (p. 405 and Fig. 5-33).

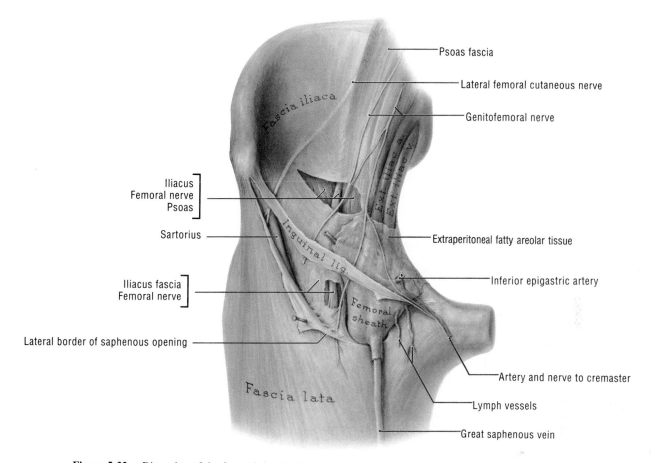

Figure 5-32.   Dissection of the femoral sheath. The falciform margin of the saphenous opening in the fascia lata is cut and reflected. Note that the femoral nerve is external and lateral to the femoral sheath.

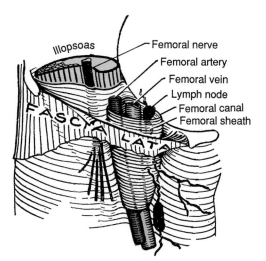

Figure 5-33.   Superior part of the right thigh showing the femoral sheath and its contents.

the buttock at a 45 degree angle (Fig. 5-35A). Its inferior edge lies just superior to the gluteal fold produced by bulging fat.

Actions of the Gluteus Maximus Muscle.   It extends the thigh and steadies it. The gluteus maximus is the **chief extensor of the thigh**. When acting with its distal attachment fixed, it is also a strong *extensor of the pelvis* (e.g., when rising from a

seated or stooped position). The gluteus maximus also assists with lateral rotation of the thigh. It is used very little during ordinary walking and is relaxed when one stands still. *It acts when force is required, e.g., in running,* climbing, and similar activities. As the thigh is being flexed, *e.g.,* during sitting, the gluteus maximus steadies the movement by relaxing gradually. When acting from its distal attachment, it tilts the superior part of the pelvis posteriorly. For instance, this action occurs when one stands at attention.

Bursae are closed sacs, which are lined with a synovial-like membrane and contain synovial fluid (see Fig. 6-46). They are usually located in areas that are subject to friction, *e.g.,* over an exposed or prominent part (*e.g.,* the ischial tuberosity in the buttock) or where a tendon passes over a bone. The bursae formed at such sites are simply *flattened, membranous sacs containing a capillary layer of synovial fluid.* They are supported by dense irregular connective tissue and are located in the superficial fascia between the skin and bone. The purpose of bursae is to reduce friction.

Bursae Associated with the Gluteus Maximus Muscle.   Usually there are three bursae separating this large muscle from underlying structures: (1) the *trochanteric bursa* separates the gluteus maximus muscle from the lateral side of the greater trochanter of the femur (Fig. 5-37); (2) the *gluteofemoral bursa*

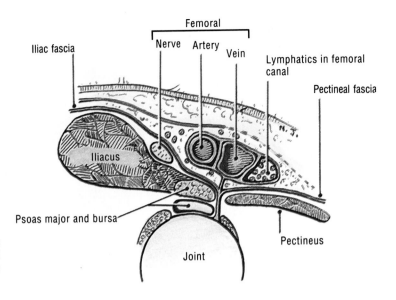

Figure 5-34. Transverse section of the superior part of the thigh showing the femoral sheath and the relationship of the femoral vessels and nerve to each other and the hip joint. Note that the femoral nerve lies deep to the iliac fascia.

separates this muscle from the superior part of the proximal attachment of the vastus lateralis muscle; and (3) the *ischial bursa* (sciatic bursa) separates it from the ischial tuberosity.

> The ischial bursa, superficial to the ischial tuberosity, may become inflamed as the result of excessive friction. This produces a friction bursitis known as *ischial bursitis* ("weaver's bottom"). Weavers first extend one lower limb and then the other. This repeated friction on the ischial bursae may cause inflammation of the walls of the bursae. The inflammatory reaction results in painful swelling of the bursae. As the ischial tuberosities bear the weight during sitting (see Fig. 4-3) and when supine, these pressure points may lead to *pressure sores* in debilitated patients, particularly paraplegic persons, if good nursing care is not exercised. *Trochanteric bursitis* causes diffuse deep pain in the gluteal and lateral thigh regions. This type of *pelvic girdle pain* is characterized by tenderness over the greater trochanters of the femora.

The Gluteus Medius Muscle (Figs. 5-35 to 5-38). Most of this thick, fan-shaped or triangular muscle lies deep to the gluteus maximus on the external surface of the ilium. Its attachments, nerve supply, and main actions are given in Table 5-3. The gluteus medius is a *powerful abductor of the hip joint* and plays an essential role during locomotion; it is largely responsible for the tilt of the pelvis. When the left muscle pulls the left side of the pelvis down, the right side is prevented from sagging as the right limb is raised during walking. Conversely, the right gluteus muscle permits the left foot to clear the ground during walking. It also helps to rotate the thigh medially.

The Gluteus Minimus Muscle (Figs. 5-36 to 5-38). This fan-shaped or triangular muscle, the smallest of the gluteal muscles, lies deep to the gluteus medius. Its attachments, nerve supply, and main actions are given in Table 5-3. During locomotion *its actions are similar to those of the gluteus medius.* Its anterior fibers help to rotate the thigh medially, more strongly than do those of the gluteus medius.

> If the gluteus medius and minimus muscles are weakened following surgery or are paralyzed (*e.g.*, owing to injury of the superior gluteal nerve or to a disease such as *poliomyelitis*), the supportive and steadying effect of these muscles on the pelvis is lost. Hence, when the foot is raised on the normal side, the pelvis falls on that side. Similarly, the person walks with a waddling gait known as a *gluteal gait* or a **gluteus medius limp**, characterized by a falling of the pelvis toward the unaffected side at each step. A similar gait occurs in persons with unilateral *posterior dislocation of the hip joint*. This condition prevents normal functioning of the gluteus medius and minimus muscles.

The Piriformis Muscle (Figs. 5-35 to 5-38; see also Fig. 3-16). This muscle is located partly on the posterior wall of the pelvis minor and partly posterior to the hip joint. It leaves the pelvis through the *greater sciatic foramen* to reach its distal attachment to the greater trochanter of the femur. Because of its position, this narrow, pear-shaped (L. *piriform*) muscle is **the landmark of the gluteal region**. It is the key to understanding relationships in this area because it determines the names of the blood vessels and nerves; *e.g.*, the superior gluteal vessels and nerve emerge superior to it and the inferior gluteal vessels and nerve emerge inferior to it. The surface marking of the superior border of the piriformis muscle is indicated by a line joining the skin dimple, formed by the posterior superior iliac spine (Figs. 5-3 and 5-4; see also Fig. 2-86), to the superior border of the greater trochanter of the femur. The attachments, nerve supply, and main actions of the piriformis are given in Table 5-3.

The Obturator Internus Muscle (Figs. 5-24 and 5-35 to 5-38). This thick fan-shaped muscle, like the piriformis, is located partly in the pelvis where it covers most of the lateral wall of the pelvis minor (see Fig. 3-10). It leaves the pelvis through the *lesser sciatic foramen* to reach its attachment to the greater trochanter of the femur. Its attachments, nerve supply, and main actions are given in Table 5-3. The *gemelli assist the* *obturator internus* in the performance of its actions. The two **gemelli muscles** are narrow, triangular extrapelvic parts of the obturator internus. The superior and inferior gemelli (Figs. 5-24 and 5-36*F*) arise from the ischial spine and ischial tuberosity, respectively. The obturator internus and gemelli muscles laterally rotate the extended thigh and abduct it when it is flexed. They also help to hold the head of the femur in the acetabulum; *i.e.*, they stabilize the hip joint.

The Quadratus Femoris Muscle (Figs. 5-35 to 5-38).   True to its name (L. *quadratus*, squared; four sided), this is a short, flat, *rectangular muscle*. It is located inferior to the obturator internus and gemelli muscles. Its attachments, nerve supply, and main actions are given in Table 5-3.

The Gluteal Nerves

There are many nerves in the gluteal region; the deep ones are the most important clinically.

The Superficial Gluteal Nerves

The skin of the gluteal region is richly innervated (Fig. 5-11). It receives cutaneous branches from several lumbar and sacral segments. These cutaneous nerves are called cluneal (clunial) nerves (L. *clunes*, buttocks). The *superior cluneal nerves* are lateral branches of the dorsal rami of the first three lumbar nerves. They emerge from the deep fascia just superior to the iliac crest and supply the skin on the superior two-thirds of the buttock. The middle cutaneous nerves or *middle cluneal nerves* are also lateral branches of the dorsal rami of the first three sacral nerves. They become cutaneous along a line connecting the posterior superior iliac spine and the tip of the coccyx. They supply the skin over the sacrum and adjacent areas of the buttock. The inferior cutaneous nerves or *inferior cluneal nerves* are gluteal branches of the posterior femoral cutaneous nerves and are larger than the other cutaneous nerves. As branches of the ventral rami of the first three sacral nerves, they become cutaneous and curve around the inferior border of the gluteus maximus muscle, just superior to the *gluteal fold* (Figs. 5-5 and 5-11), to supply the inferior one-third of the buttock. The iliohypogastric nerve arises from the ventral ramus of L1, with a small contribution from T12. This nerve supplies a lateral cutaneous branch to the lateral area of the hip over the iliac crest and the greater trochanter of the femur (Fig. 5-11).

The Deep Gluteal Nerves

There are **seven nerves** in the gluteal region; they are *branches of the sacral plexus* and leave the pelvis via the greater sciatic foramen (Figs. 5-37 and 5-73). Except for the superior gluteal nerve, they emerge inferior to the piriformis muscle. The site of exit of these nerves can be detected by deep pressure (usually painful), just superior to the midpoint of a line joining the posterior superior iliac spine and the ischial tuberosity.

The Superior Gluteal Nerve (Fig. 5-37).   This nerve arises from the posterior divisions of the ventral rami of L4, L5, and S1. It leaves the pelvis through the superior part of the greater sciatic foramen, superior to the piriformis muscle with the superior gluteal artery. It runs laterally between the gluteus medius and minimus muscles with the deep branch of the superior gluteal artery. The superior gluteal nerve divides into a *superior branch* that supplies the *gluteus medius* and an *inferior branch* that passes between *gluteus medius* and *gluteus minimus*. It supplies both of these muscles and the *tensor fasciae latae*.

The Inferior Gluteal Nerve (Fig. 5-37).   This nerve arises from the posterior divisions of the ventral rami of L5, S1, and S2. It leaves the pelvis through the inferior part of the greater sciatic foramen, inferior to the piriformis muscle and superficial to the sciatic nerve. It accompanies the inferior gluteal artery and breaks up into several branches that supply the overlying *gluteus maximus* muscle.

The Sciatic Nerve (Figs. 5-35, 5-37, 5-39, 5-40, and 5-73). This is the *main branch of the sacral plexus*. The largest nerve in the body, it is formed by the ventral rami of L4, L5, S1, S2, and S3, which converge at the inferior border of the piriformis muscle. The sciatic nerve leaves the pelvis as a thick, flattened band (about 2 cm wide) and travels through the inferior part of the *greater sciatic foramen*. It enters the gluteal region inferior to the piriformis muscle and is the most lateral of all the structures emerging inferior to the piriformis. Medial to it are the inferior gluteal nerve and vessels, the internal pudendal vessels, and the pudendal nerve. It runs inferolaterally deep to the gluteus maximus muscle, midway between the greater trochanter of the femur and the ischial tuberosity. The sciatic nerve rests on the ischium and then passes posterior to the obturator internus, quadratus femoris, and adductor magnus muscles. *The sciatic nerve usually supplies no structures in the gluteal region*. The sciatic is really two nerves, the tibial and common fibular (peroneal), which are bound together in the same connective tissue sheath (epineurium). The two nerves usually separate from each other about halfway or more down the thigh, but occasionally they are separate when they leave the pelvis (Fig. 5-39*B* and *C*). In these cases the tibial nerve passes inferior to the piriformis and the common fibular pierces this muscle or passes superior to it. The tibial nerve supplies flexor muscles and the common fibular nerve supplies extensor and abductor muscles.

The Posterior Femoral Cutaneous Nerve (Figs. 5-11, 5-37, and 5-40*B*).   This posterior cutaneous nerve of the thigh arises from the posterior divisions of the ventral rami of S1 and S2 and the anterior divisions of S2 and S3. *It supplies more skin than any other cutaneous nerve*. The posterior femoral cutaneous nerve leaves the pelvis with the inferior gluteal nerve and vessels and the sciatic nerve. Its fibers from the posterior divisions of S1 and S2 supply the skin of the inferior part of the buttock; those from the anterior divisions of S2 and S3 supply the skin of the perineum (see Figs. 3-63 and 3-72); other branches continue inferiorly where they supply the skin of the posterior thigh and proximal part of the leg.

The Nerve to the Quadratus Femoris Muscle.   This nerve arises from the anterior divisions of the ventral rami of L4, L5, and S1. It leaves the pelvis deep (anterior) to the sciatic nerve and the obturator internus muscle and passes over the posterior surface of the hip joint. It supplies an articular branch to this joint and innervates the quadratus femoris and inferior gemellus muscles.

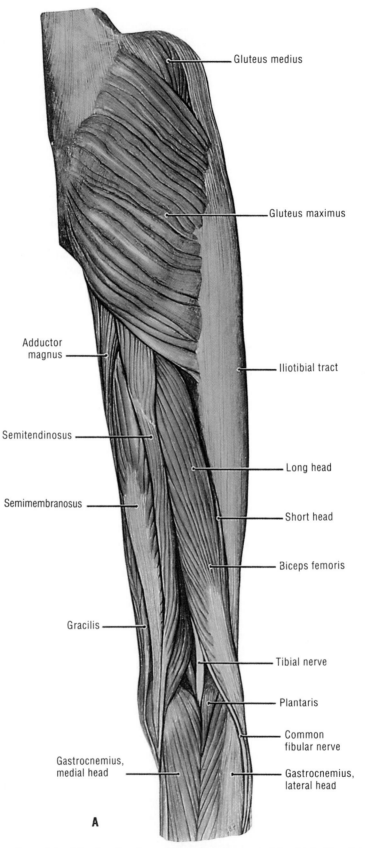

Figure 5-35. Dissections of the right gluteal region and the posterior aspect of the thigh. The gluteal fascia has been removed in *A* to expose the gluteal muscles. The gluteus maximus muscle has been reflected in *B* to show the large sciatic nerve (yellow).

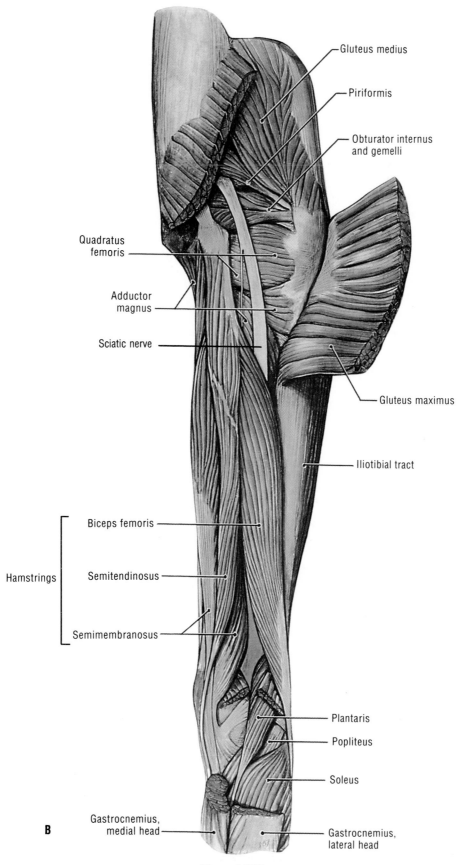

Gluteus medius

Piriformis

Obturator internus
and gemelli

Quadratus
femoris

Adductor
magnus

Sciatic nerve

Gluteus maximus

Iliotibial tract

Biceps femoris

Hamstrings

Semitendinosus

Semimembranosus

Plantaris

Popliteus

Soleus

B

Gastrocnemius,
medial head

Gastrocnemius,
lateral head

Figure 5-35*B*

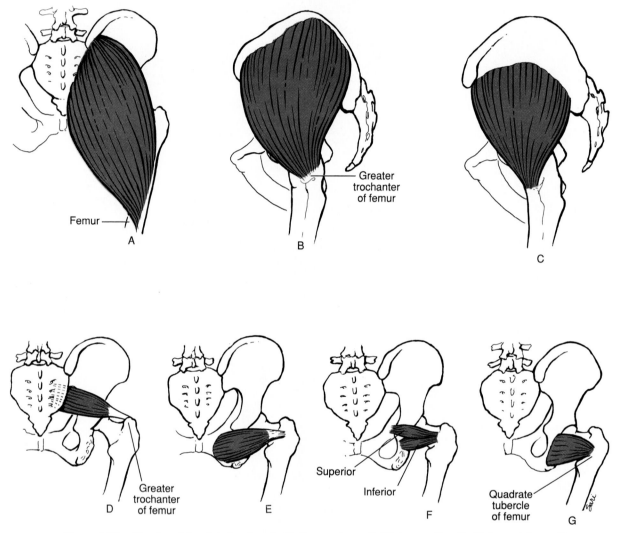

Figure 5-36.   Muscles of the gluteal region. *A*, Gluteus maximus. *B*, Gluteus medius. *C*, Gluteus minimus. *D*, Piriformis. *E*, Obturator internus. *F*, Gemelli. *G*, Quadratus femoris.

Table 5-3.
The Muscles of the Gluteal Region

Muscle	Proximal Attachment	Distal Attachment	Innervation	Main Actions
Gluteus maximus (Fig. 5-36A)	External surface of ala of ilium, including iliac crest, dorsal surface of sacrum and coccyx, and sacrotuberous ligament	Most fibers end in iliotibial tract which inserts into lateral condyle of tibia; some fibers insert on gluteal tuberosity of femur	Inferior gluteal nerve (L5, **S1**, and **S2**)	Extends thigh and assists in its lateral rotation; also steadies thigh and assists in raising trunk from flexed position
Gluteus medius (Figs. 5-36B and 5-18)	External surface of ilium between anterior and posterior gluteal lines	Lateral surface of greater trochanter of femur	Superior gluteal nerve (**L5** and S1)	Abduct and medially rotate thigh; steady pelvis
Gluteus minimus (Fig. 5-36C)	External surface of ilium between anterior and inferior gluteal lines	Anterior surface of greater trochanter of femur		
Piriformis (Fig. 5-36D)	Anterior surface of sacrum and sacrotuberous ligament	Superior border of greater trochanter of femur	Branches from ventral rami of **S1** and S2	
Obturator internus (Fig. 5-36E)	Pelvic surface of obturator membrane and surrounding bones	Medial surface of greater trochanter of femur[1]	Nerve to obturator internus (L5 and **S1**)	Laterally rotate extended thigh and abduct flexed thigh; steady femoral head in acetabulum
Gemelli, superior and inferior (Fig. 5-36F)	*Superior*, ischial spine; *Inferior*, ischial tubersity		*Superior gemellus*, same nerve supply as obturator internus *Inferior gemellus*, same nerve supply as quadratus femoris	
Quadratus femoris (Fig. 5-36G)	Lateral border of ischial tuberosity	Quadrate tubercle on intertrochanteric crest of femur and inferior to it	Nerve to quadratus femoris (L5 and S1)	Laterally rotates thigh[2] and steadies femoral head in acetabulum

[1]The gemelli muscles blend with the tendon of the obturator internus muscle as it attaches to the greater trochanter.
[2]There are six lateral rotators of the thigh; piriformis, obturator internus, gemelli (superior and inferior), quadratus femoris, and obturator externus (p. 393). These muscles also stabilize the hip joint.

The Nerve to the Obturator Internus (Fig. 5-37).   This nerve arises from the anterior divisions of the ventral rami of L5, S1, and S2. It leaves the pelvis through the greater sciatic foramen, inferior to the piriformis muscle and medial to the sciatic nerve. It winds around the base of the ischial spine to supply the superior gemellus muscle and then passes posterior to the ischial spine, reentering the pelvis via the lesser sciatic foramen. It lies on the lateral pelvic wall and supplies the obturator internus muscle.

The Pudendal Nerve (Fig. 5-37; see also Fig. 3-61).   This nerve arises from the anterior divisions of the ventral rami of S2, S3, and S4. It is the most medial structure to pass through the greater sciatic foramen inferior to the piriformis muscle. It passes lateral to the sacrospinous ligament, reentering the pelvis via the lesser sciatic foramen, to supply structures in the perineum (genitalia, sphincter urethrae, and anular fibers of the external anal sphincter; see Chap. 3, p. 256).

Injury to the gluteal nerves may occur in wounds of the buttock (*e.g.*, gunshot or stab wounds). With respect to the sciatic nerve, the buttock has a side of safety (its lateral side) and a side of danger (its medial side). Wounds or surgery on the medial side are liable to injure the sciatic nerve and its branches to the hamstring muscles on the posterior aspect of

the thigh (Fig. 5-35B). Paralysis of these muscles results in impairment of extension of the thigh and flexion of the leg (Table 5-4). "Driver's thigh" is a *sciatic neuralgia*[8] that is caused by pressure on the gluteal nerves resulting from the use of the accelerator in driving a car or truck. Cruise control was designed to prevent this condition. For a discussion of other types of *sciatica*, the term used to describe pain in the area of distribution of the sciatic nerve, see Chap. 4 (p. 347).

The gluteal region is a common site for the intramuscular injection of drugs. **Gluteal intramuscular injections** penetrate the skin, subcutaneous tissue, and muscles. The gluteal region is a favorite injection site because the gluteal muscles are thick and large (Fig. 5-35); consequently, they provide a large surface area for absorption of drugs. It is essential to know the extent of the gluteal region and the safe region for giving injections. Too many people restrict the area of the buttock to the "cheek" or most prominent part; this is a dangerous concept. The full extent of the buttock is shown in Fig. 5-40. As there are a number of important nerves and blood vessels in the gluteal region, *injections can be made safely only into the superolateral part of the buttock*, that is, superior to the tubercle of the iliac crest (Figs. 5-3 and

[8]Neuralgia (G. *neuron*, nerve + *algos*, pain + *ia*, condition) is paroxysmal (intensification) pain that extends along the course of one or more nerves.

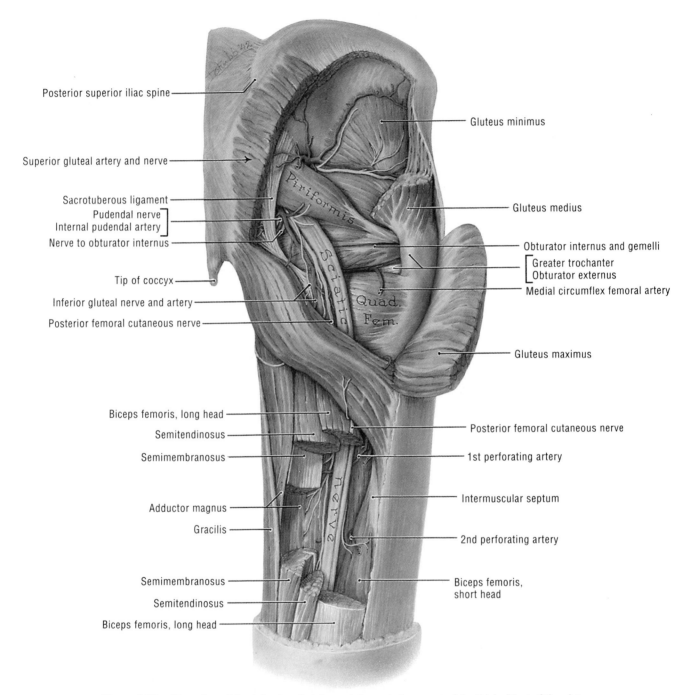

Posterior superior iliac spine

Superior gluteal artery and nerve

Sacrotuberous ligament

Pudendal nerve
Internal pudendal artery

Nerve to obturator internus

Tip of coccyx

Inferior gluteal nerve and artery

Posterior femoral cutaneous nerve

Piriformis

Sciatic

Quad. Fem.

Gluteus minimus

Gluteus medius

Obturator internus and gemelli

Greater trochanter
Obturator externus

Medial circumflex femoral artery

Gluteus maximus

Biceps femoris, long head

Semitendinosus

Semimembranosus

Adductor magnus

Gracilis

Semimembranosus

Semitendinosus

Biceps femoris, long head

nerve

Posterior femoral cutaneous nerve

1st perforating artery

Intermuscular septum

2nd perforating artery

Biceps femoris, short head

Figure 5-37.   Dissection of the right gluteal region and the posterior aspect of the thigh. Most of the gluteus maximus is reflected and parts of the gluteus medius and hamstring muscles are excised.

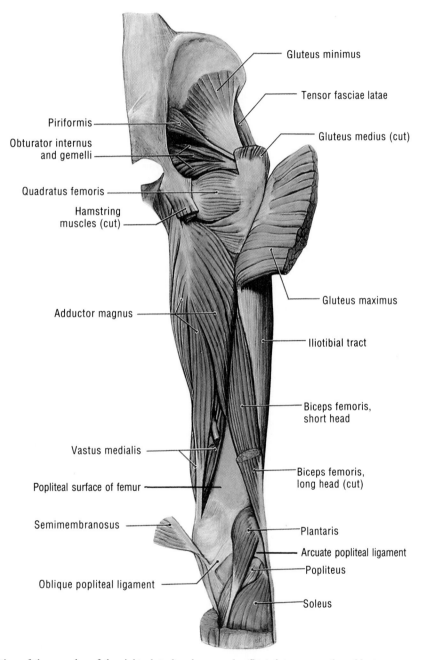

Gluteus minimus

Tensor fasciae latae

Piriformis

Gluteus medius (cut)

Obturator internus
and gemelli

Quadratus femoris

Hamstring
muscles (cut)

Gluteus maximus

Adductor magnus

Iliotibial tract

Biceps femoris,
short head

Vastus medialis

Biceps femoris,
long head (cut)

Popliteal surface of femur

Semimembranosus

Plantaris

Arcuate popliteal ligament

Popliteus

Oblique popliteal ligament

Soleus

Figure 5-38.   Deep dissection of the muscles of the right gluteal and posterior thigh regions. Most of the hamstring muscles have been cut and reflected to expose the adductor magnus muscle and the floor of the popliteal fossa.

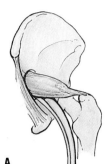

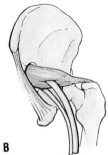

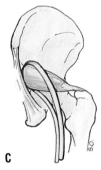

Figure 5-39. Relationship of the sciatic nerve to the piriformis muscle. Usually it passes inferior to the muscle (*A*). In 12.2% of 640 limbs studied, the sciatic nerve divided before it entered the gluteal region and the common fibular (peroneal) division (yellow) passed through the piriformis muscle (*B*). In 0.5% of cases the common fibular division passed superior to the muscle (*C*), where it is vulnerable to injury during gluteal intramuscular injections.

5-40). The needle should go deeply enough to pass through the thick subcutaneous tissue and enter the gluteus medius muscle, where it is not covered by the gluteus maximus muscle (Fig. 5-35). Injections into other areas could endanger and possibly injure the sciatic or other nerves and vessels.

The sciatic nerve sometimes divides within the pelvis into the tibial and common fibular (peroneal) nerves, in which case the common fibular nerve passes through the piriformis muscle or superior to it (Fig. 5-39). Hence, injection into the inferomedial part of the superolateral quadrant could injure the common fibular nerve and produce **foot-drop** (p. 429). Improper gluteal intermuscular injections may also injure gluteal branches of the posterior femoral cutaneous nerve resulting in pain and *dysesthesia* (loss of sensation) in the area

of skin supplied by its lateral cutaneous branches (Fig. 5-11). The hazards of injecting drugs into the gluteal region of small infants are well recognized. Because of the danger of injuring the gluteal nerves, drugs are commonly injected into the muscles of the anterolateral region of the thigh. To avoid injury to the femoral nerve, injections are administered inferolateral to the anterior superior iliac spine into the tensor fasciae latae and vastus lateralis muscles (Fig. 5-19). Some physicians also use this site for giving intramuscular injections to adults.

The Gluteal Arteries

The arteries supplying the gluteal region directly are branches of the *internal iliac arteries* (Fig. 5-29).

The Superior Gluteal Artery

This short vessel is the largest branch of the internal iliac artery (Figs. 5-29*B* and 5-37; see also Figs. 3-16 and 3-24). It passes posteriorly between the lumbosacral trunk and the first sacral ventral ramus. The superior gluteal artery leaves the pelvis through the greater sciatic foramen, superior to the piriformis muscle. It divides immediately into superficial and deep branches. The *superficial branch* supplies the gluteus maximus muscle and the skin over this muscle's proximal attachment (Table 5-3); the *deep branch* supplies the gluteus medius, gluteus minimus, and tensor fasciae latae muscles. The superior gluteal artery anastomoses with the inferior gluteal and medial circumflex femoral arteries (Fig. 5-29*B*).

The Inferior Gluteal Artery

This vessel arises from the internal iliac artery and passes posteriorly through the parietal pelvic fascia (p. 259), between the first and second (or second and third) sacral ventral rami (Figs. 5-29*B* and 5-37; see also Figs. 3-16 and 3-24). The inferior gluteal artery leaves the pelvis through the greater sciatic foramen, inferior to the piriformis muscle. It supplies the gluteus maximus, obturator internus, quadratus femoris, and superior parts of the hamstring muscles. It anastomoses with the superior gluteal artery and participates in the *cruciate anastomosis of the thigh*, involving the first perforating arteries of the profunda femoris and the medial and lateral circumflex femoral arteries (Fig. 5-29*B*).

The Internal Pudendal Artery

This vessel arises from the internal iliac artery and lies anterior to the inferior gluteal artery. It passes lateral to the pudendal nerve and leaves the pelvis via the greater sciatic foramen inferior to the piriformis muscle (Fig. 5-37; see also Figs. 3-16, 3-18, and 3-24). It then descends posterior to the ischial spine, reenters the pelvis via the lesser sciatic foramen, and enters the perineum with the pudendal nerve and lies on its lateral side. It crosses the ischial spine, against which it can be compressed to control hemorrhage in the perineum (p. 317). The internal pudendal artery supplies the external genitalia and muscles in the pelvic and gluteal regions (see Figs. 3-60, 3-71, and 3-72).

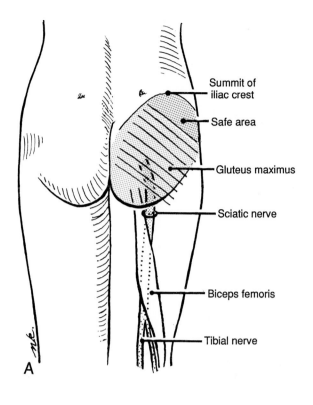

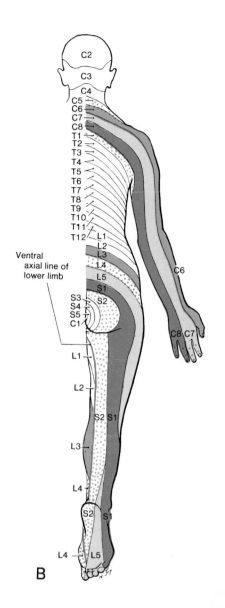

Figure 5-40. *A*, Diagram showing the extent of the gluteal region or buttock and the location of the sciatic nerve deep to the gluteus maximus muscle. The safe area for giving intramuscular injections is shown in green. The area to be avoided is shown in red. *B*, Diagram showing the dermatomes of the left lower limb, the striplike areas of skin supplied by the dorsal (sensory) roots of spinal nerves.

The Gluteal Veins

The veins supplying the gluteal region are tributaries of the *internal iliac veins* (see Fig. 3-24).

The Gluteal Veins.   The superior and inferior gluteal veins accompany the corresponding arteries through the greater sciatic foramen (Fig. 5-37). Usually these veins are double, *i.e.*, *venae comitantes* (L. accompanying veins). They communicate with tributaries of the femoral vein and thus can provide an alternate route for return of blood from the lower limb if the femoral vein is occluded or has to be ligated.

The Internal Pudendal Veins.   These veins accompany the corresponding arteries and join to form a single vein that enters the internal iliac vein. The internal pudendal veins drain blood from the external genitalia and perineal region (p. 317).

The Posterior Thigh Muscles

Three large muscles (semitendinosus, semimembranosus, and biceps femoris) in the posterior aspect of the thigh are commonly called the **hamstring muscles** (Figs. 5-35, 5-37, 5-38, and 5-41 to 5-44). They can be made to stand out by flexing the leg against resistance. The hamstrings have a common proximal attachment to the *ischial tuberosity* (Table 5-4) deep to the gluteus maximus, but one of them, the biceps femoris, has an additional attachment to the body of the femur (Fig. 5-41*B*). They also have a common nerve supply from the sciatic nerve. *The hamstring muscles span the hip and knee joints*; hence, they are extensors of the thigh and flexors of the leg. Both actions cannot be performed fully at the same time. The hamstrings descend in the posterior aspect of the thigh and their tendons are visible posterior to the knee (Figs. 5-26, 5-43, and 5-44).

The posterior thigh muscles became known as the "hamstrings" because their tendons posterior to the knee are used to hang up hams (hip and thigh regions) of animals such as pigs. Furthermore, in ancient times it was common for soldiers to slash their opponents' horses posterior to the knees in order to cut the tendons of their posterior thigh muscles. This would bring the horse and its rider down. Similarly,

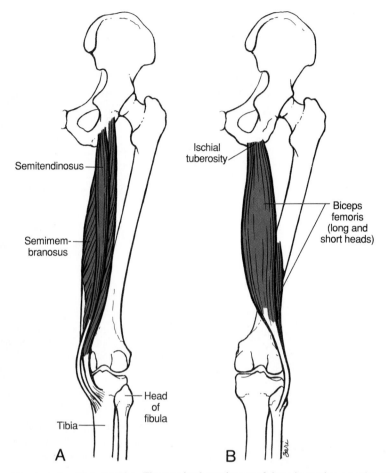

Figure 5-41. The posterior thigh muscles. The proximal attachment of these hamstring muscles (except for the short head of the biceps femoris) is the ischial tuberosity.

they cut these tendons of soldiers so they could not run; this was called "hamstringing" the enemy.

The Semitendinosus Muscle (Figs. 5-18, 5-23, 5-26, 5-27, 5-35, 5-37, 5-38, and 5-41 to 5-45). As its name indicates, this muscle is *half tendinous*. It is fusiform with a long, rounded, cordlike tendon, which begins about two-thirds of the way down the thigh. Its attachments, nerve supply, and main actions are given in Table 5-4. In Figs. 5-35*B* and 5-41*A*, observe that it has a common proximal attachment to the ischial tuberosity with the semimembranosus and the long head of the biceps femoris muscle.

The Semimembranosus Muscle (Figs. 5-18, 5-23, 5-27, 5-35, 5-37, 5-38, and 5-41 to 5-45). True to its name, this broad muscle is *half membranous*. It is located deep to the semitendinosus muscle. Its attachments, nerve supply, and main actions are given in Table 5-4.

Actions of the Semitendinosus and Semimembranous Muscles. In addition to the common actions of the hamstrings in extending the thigh and flexing the leg, these two muscles can medially rotate the tibia on the femur, particularly when the knee is semiflexed.

The Biceps Femoris Muscle (Figs. 5-18, 5-23, 5-35, 5-37, 5-38, and 5-41 to 5-45). As its name indicates, this fusiform muscle has *two heads*, (L. *ceps*), long and short. Its attachments, nerve supply, and main actions are given in Table 5-4. The rounded tendon of the biceps femoris can easily be seen and felt as it passes the knee to insert into the head of the fibula, especially when the knee is flexed against resistance. Like the other hamstrings, the long head of the biceps femoris extends the thigh at the hip joint. Both heads *flex the leg* at the knee joint and *laterally rotate the leg* when it is flexed.

Because the heads of the biceps femoris have a different nerve supply, *i.e.*, from different divisions of the sciatic nerve (Table 5-4), a wound in the thigh may sever a nerve, paralyzing one head and not the other. The length of the hamstrings varies. In some people they are not long enough to allow them to touch their toes when they flex their vertebral column and keep their knees straight. In other people the hamstrings are long and they can easily touch the floor with their palms or perform a high kick. Some athletes are unable to excel in certain sports (*e.g.*, gymnastics) despite rigorous exercises because of their short hamstrings.

Table 5-4.
The Posterior Thigh Muscles[1]

Muscle	Proximal Attachment	Distal Attachment	Innervation	Main Actions
Semitendinosus (Figs. 5-18 and 4-41A)	Ischial tuberosity	Medial surface of superior part of tibia	Tibial division of sciatic nerve (**L5, S1,** and S2)	Extend thigh; flex leg and rotate it medially; when thigh and leg are flexed, they can extend trunk
Seminimembranosus (Figs. 5-18 and 5-41A)		Posterior part of medial condyle of tibia		
Biceps femoris (Figs. 5-18 and 5-41B)	*Long head*: Ischial tuberosity *Short head*: Lateral lip of linea aspera and lateral supracondylar line	Lateral side of head of fibula; tendon is split at this site by fibular collateral ligament of knee joint	*Long head*: tibial division of sciatic nerve (L5, **S1**, and S2) *Short head*: common fibular (peroneal) division of sciatic nerve (L5, **S1**, and S2)	Flexes leg and rotates it laterally; long head extends thigh (*e.g.*, when starting to walk)

[1]Collectively these muscles are known as the *hamstrings*.

Hamstring injuries, often called *"pulled hamstrings,"* are common in persons who run very hard and/or kick (*e.g.*, in running, jumping, and quick-start sports such as baseball, football, and soccer). The violent muscular exertion required to excel in these sports tears or avulses part of the tendinous proximal attachments of the hamstrings from the ischial tuberosity. Usually there is also contusion and tearing of muscle fibers, resulting in rupture of some blood vessels supplying the muscles. The resultant *hematoma* (pool of blood) is contained by the dense fascia lata. The tearing of hamstring fibers is often so painful when the athlete moves or stretches the leg that the person will fall and writhe in pain. These injuries may result from an inadequate warming up before practice or competition.

The Popliteal Fossa

This diamond-shaped region is posterior to the knee, between the semitendinosus and biceps femoris tendons (Fig. 5-44). It lies posterior to the distal third of the femur, the knee joint, and the proximal part of the tibia. This fossa appears as a hollow when the knee joint is slightly flexed. The *roof of the popliteal fossa* (its posterior wall) is formed by skin and fasciae. The *superficial popliteal fascia* contains fat, the small saphenous vein, and three cutaneous nerves (Figs. 5-7A and 5-45). The roof is pierced proximally by the *posterior femoral cutaneous nerve* and distally by the *small saphenous vein*. This vein perforates the deep popliteal fascia and ends in the **popliteal vein** (Fig. 5-46). The *deep popliteal fascia* forms a strong, dense sheet of deep fascia, which affords a protective covering for the neurovascular structures passing from the thigh to the leg (Figs. 5-45 and 5-46). The deep fascia of the thigh is strengthened posterior to the knee by transverse fibers. When the leg is extended, the semimembranosus muscle moves laterally, offering further protection for these neurovascular structures.

The *floor of the popliteal fossa* (its anterior wall) is formed by the popliteal surface of the femur, the *oblique popliteal ligament*, an expansion of the semimembranosus tendon, and the popliteus fascia (Figs. 5-3, 5-38, and 5-46).

Because the deep popliteal fascia is strong and limits expansion, pain from an abscess or tumor in the popliteal fossa is usually severe. In addition, popliteal abscesses tend to spread superiorly and inferiorly owing to the toughness of this deep fascia.

Boundaries of the Popliteal Fossa (Figs. 5-44 to 5-46). The muscles surrounding the popliteal fossa delineate this diamond-shaped area. It is bounded *superolaterally* by the biceps femoris muscle, *superomedially* by the semimembranosus and semitendinosus muscles, and *inferolaterally* and *inferomedially* by the lateral and medial heads of the gastrocnemius muscle, respectively.

Contents of the Popliteal Fossa

When the muscles forming the boundaries of the fossa are pulled apart, especially the heads of the gastrocnemius, the popliteal fossa and its contents can be observed (Figs. 5-45 and 5-46). Although the fossa appears large when this is done, normally the muscles are packed closely together and the fossa is relatively small. The contents of the popliteal fossa are: fat; the *popliteal vessels* (artery, vein, and lymphatics); the *tibial and common fibular nerves*; the *small saphenous vein*; the end branch of the *posterior femoral cutaneous nerve* (Fig. 5-11); an articular branch of the *obturator nerve*; four to six *popliteal lymph nodes*; and the *popliteus bursa*.

The Popliteal Artery (Figs. 5-29B, 5-35A, and 5-46). This vessel begins as soon as the femoral artery passes through the adductor hiatus in the tendon of the adductor magnus muscle. The popliteal artery is the *continuation of the femoral artery*. From its origin at the adductor hiatus (p. 393), it passes inferolaterally through the fat of the popliteal fossa. It ends by dividing into the anterior and posterior tibial arteries at the inferior border of the popliteus muscle. *The popliteal artery is located deeply throughout its course.* Anteriorly, from proximal to distal, it lies against the fat on the posterior surface of the femur, the fibrous capsule of the knee joint, and the popliteus fascia. Posteriorly, from proximal to distal, it lies deep to the semimembranosus

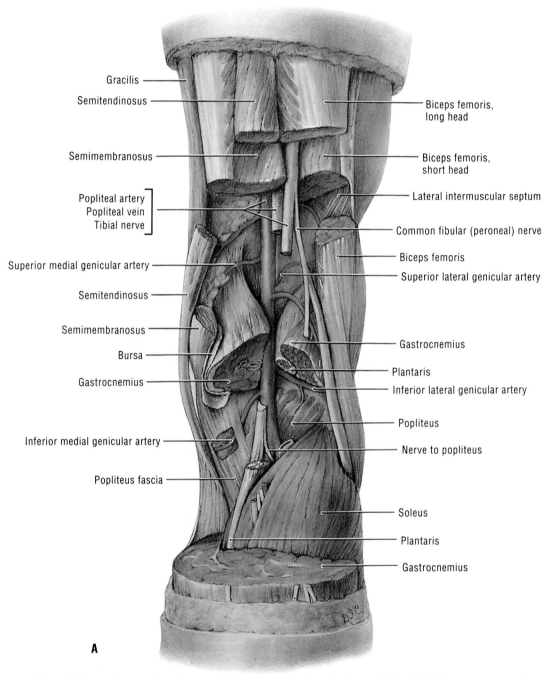

Gracilis

Semitendinosus

Semimembranosus

Popliteal artery
Popliteal vein
Tibial nerve

Superior medial genicular artery

Semitendinosus

Semimembranosus

Bursa

Gastrocnemius

Inferior medial genicular artery

Popliteus fascia

Biceps femoris,
long head

Biceps femoris,
short head

Lateral intermuscular septum

Common fibular (peroneal) nerve

Biceps femoris

Superior lateral genicular artery

Gastrocnemius

Plantaris

Inferior lateral genicular artery

Popliteus

Nerve to popliteus

Soleus

Plantaris

Gastrocnemius

A

Figure 5-46.   *A*, Deeper dissection the right popliteal fossa than that shown in Fig. 5-45. *B*, Anastomoses of
arteries around the knee, posterior view.

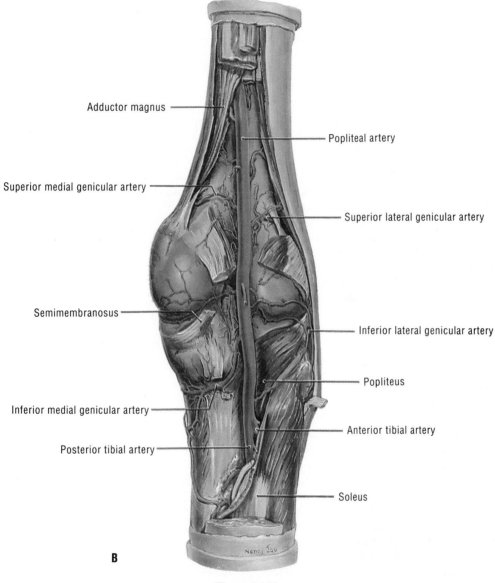

Adductor magnus

Superior medial genicular artery

Semimembranosus

Inferior medial genicular artery

Posterior tibial artery

Popliteal artery

Superior lateral genicular artery

Inferior lateral genicular artery

Popliteus

Anterior tibial artery

Soleus

B

Figure 5-46*B*

fibular (peroneal) nerves. *The tibial nerve is not commonly injured* because of its protected position in the popliteal fossa (Fig. 5-45). It may, however, be injured by deep lacerations. *Severance of the tibial nerve* results in paralysis of the flexor muscles in the leg and the intrinsic muscles of the sole of the foot (Tables 5-7 and 5-8). Persons with tibial nerve injury are unable to plantarflex their foot or flex their toes. There is also a loss of sensation on the sole of the foot (Fig. 5-82).

The Leg

The leg is the inferior part of the lower limb, between the knee and ankle joints. Although the lower limb is commonly called the "leg" by nonmedical people, anatomically this term refers only to the region between the knee and foot.

Bones of the Leg

The tibia and fibula are the bones of the leg (Fig. 5-47). The tibia ("shin bone") supports most of the weight. It articulates with the condyles of the femur superiorly and the talus inferiorly. The fibula ("calf bone") is mainly for the attachment of muscles (Figs. 5-48 and 5-49), but it also provides stability to the ankle joint. The bodies (shafts) of the tibia and fibula are connected by an *interosseous membrane* composed of strong oblique fibers (Fig. 5-50).

The Tibia

The tibia, the second largest bone of the skeleton, is located on the anteromedial side of the leg. The proximal end of the tibia is large because its medial and lateral condyles articulate with the large condyles of the femur (Fig. 5-2). The superior surface of the tibia is flat and consists of medial and lateral *tibial plateaus*. In Figs. 5-3 and 5-47, note that the *intercondylar eminence* of the tibia fits into the *intercondylar notch* between the condyles of the femur. In Fig. 5-49*B*, observe that the lateral condyle of the tibia has a facet inferiorly for the head of the fibula. Also note the prominent *tibial tuberosity* anteriorly into which the *patellar ligament* (ligamentum patellae) inserts (Figs. 5-19, 5-42, and 5-43). The distal end of the tibia is small and has facets for the fibula and talus (Fig. 5-49*B*). The distal end projects medially and inferiorly as the *medial malleolus*, which has a facet on its lateral surface for articulation with the talus (Fig. 5-47). The body (shaft) of the tibia is approximately triangular in transverse section (Fig. 5-50) and has medial, lateral, and posterior surfaces. In Figs. 5-49*B* and 5-50, observe that muscles attach to its lateral surface.

The lateral border of the tibia is sharp where it gives attachment to the *interosseous membrane*, uniting the two leg bones (Fig. 5-50). This border is referred to as the *interosseous border*. On the posterior surface of the proximal part of the body of the tibia, observe a rough diagonal ridge known as the *soleal line* (Fig. 5-47). It runs inferomedially to the medial border, about

a third of the way down the body of the tibia. The *nutrient foramen* of the tibia is the largest in the skeleton. It is located on the posterior surface of the superior third of the bone. The *nutrient canal* runs a long inferior course in the compact bone before it opens into the medullary (marrow) cavity.

Fracture of the tibia through the nutrient canal predisposes to nonunion of the fragments, owing to damage to the nutrient artery. The body of the tibia is narrowest at the junction of its middle and inferior thirds, which is the most frequent site of fracture and the region where rickets (a disease of growing bone) has its effect during infancy and childhood (p. 13). *March fractures* of the inferior third of the tibia are common in persons who take long walks when they are not used to this activity. The strain may fracture the anterior cortex of the tibia. Indirect violence may be applied to the tibia when the body turns during a fall with the foot fixed (*e.g.*, when tackled in a football game). In addition, severe torsion during skiing may produce a *spiral fracture of the tibia* at the junction of the middle and inferior thirds, as well as a fracture of the neck of the fibula. An anterior or posterior fall may produce a "boot-top fracture" owing to the rigidity of most ski boots.

Fractures of the tibia may also result from a direct blow, *e.g.*, when the bumper of a car strikes the leg. Because of this common cause, they have been called "bumper fractures." As the tibia lies subcutaneously, the blow often tears the skin, permitting the bone fragments to protrude (*compound fracture*). Because the body of the tibia is unprotected anteromedially throughout its course and is relatively slender at the junction of its inferior and middle thirds, it is not surprising that *the tibia is the most common long bone to be fractured and to suffer compound injury*. The tibia has a relatively poor blood supply, hence even undisplaced stable fractures may take up to six months to heal (Gross, 1989). Because of its extensive subcutaneous surface, the tibia is accessible for obtaining pieces of bone for grafting.

The Fibula

This long pinlike bone (L. *fibula*, pin) lies posterolateral to the tibia (Figs. 5-47 and 5-50). The fibula is the lateral bone of the leg. Its slender **body** (shaft) has little or no function in weight bearing, but its malleolus helps to hold the talus in its socket (Fig. 5-55). The fibula's main function is to provide sites for muscles to attach. It also acts as a brace and provides support for the tibia. The fibula enables the tibia to withstand some bending and twisting. Without fibular support, tibial fractures would occur more frequently. The slightly constricted part of the body near the head is called the **neck of the fibula** (Figs. 5-47 and 5-49*A*). The sharp interosseous border of the fibula provides the surface for attachment of the *interosseous membrane*. A small *nutrient foramen* is usually present in the middle third of the fibula, entering on the posterior surface (Fig. 5-47). The proximal end or **head of the fibula** is irregular in shape and knoblike. The head has a facet on its superior surface for articulation with the inferior surface of the lateral tibial condyle (Fig. 5-49*B*). The distal end of the fibula or *lateral malleolus* forms a knoblike subcutaneous prominence on the lateral surface of the ankle (Figs. 5-2, 5-47, 5-49, and 5-65). The medial surface of the fibula articulates with the lateral side of the tibia and talus.

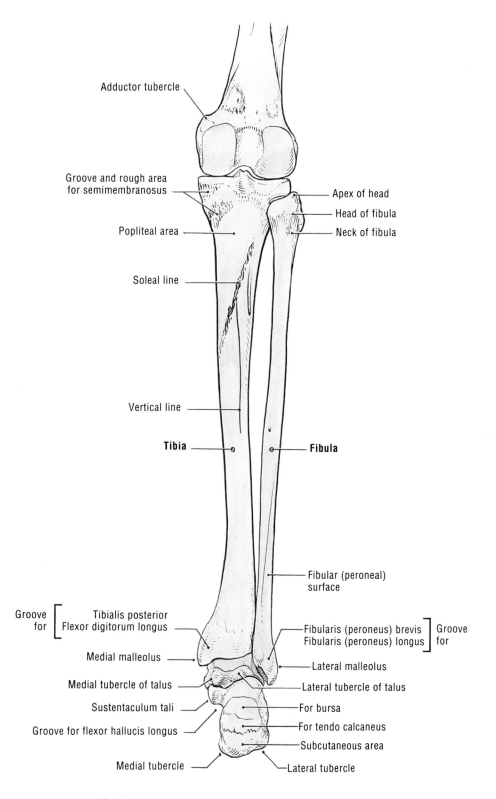

Adductor tubercle

Groove and rough area
for semimembranosus

Popliteal area

Soleal line

Vertical line

Tibia

Apex of head

Head of fibula

Neck of fibula

Fibula

Fibular (peroneal)
surface

Groove
for
| Tibialis posterior
| Flexor digitorum longus

Medial malleolus

Medial tubercle of talus

Sustentaculum tali

Groove for flexor hallucis longus

Medial tubercle

Fibularis (peroneus) brevis
Fibularis (peroneus) longus | Groove
for

Lateral malleolus

Lateral tubercle of talus

For bursa

For tendo calcaneus

Subcutaneous area

Lateral tubercle

Posterior view

Figure 5-47. Bones of the right lower limb. The proximal part of the femur is not illustrated (see Fig. 5-3). Observe that the lateral malleolus lies more inferior (1 to 2 cm) and posterior than does the medial malleolus. Also observe the large nutrient foramen in the tibia near the soleal line and lateral to the vertical line.

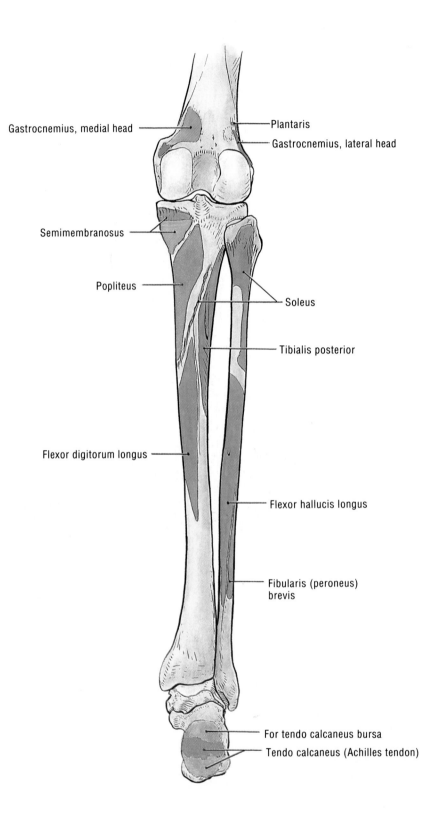

Gastrocnemius, medial head

Plantaris

Gastrocnemius, lateral head

Semimembranosus

Popliteus

Soleus

Tibialis posterior

Flexor digitorum longus

Flexor hallucis longus

Fibularis (peroneus) brevis

For tendo calcaneus bursa

Tendo calcaneus (Achilles tendon)

Posterior view

Figure 5-48. Bones of the right lower limb showing the sites of attachment of muscles. The proximal part of the femur is not illustrated (see Fig. 5-18). Proximal attachments are shown in salmon color and distal attachments in blue.

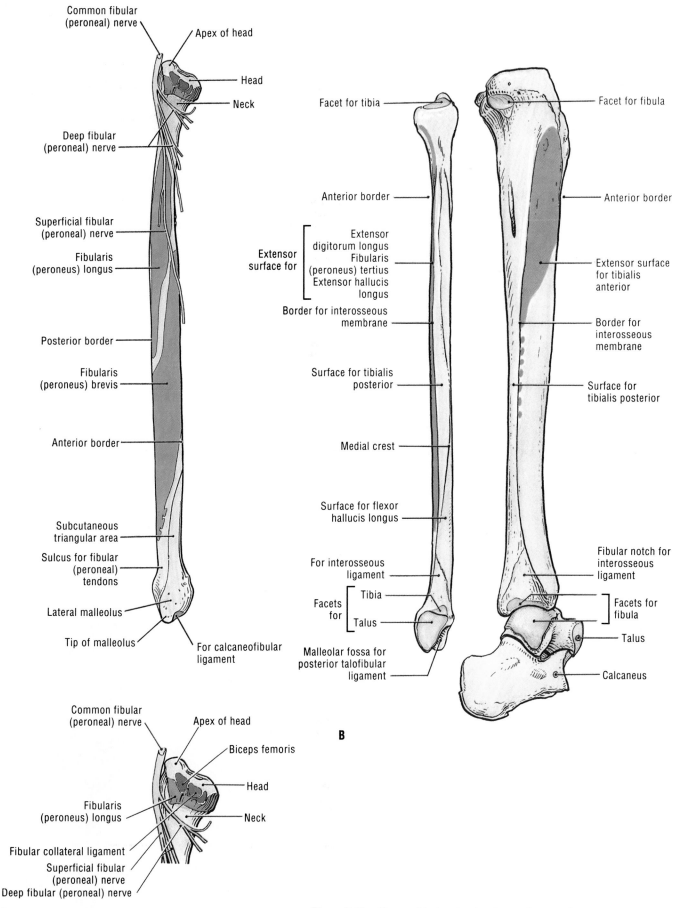

A

Common fibular
(peroneal) nerve

Apex of head

Head

Neck

Deep fibular
(peroneal) nerve

Superficial fibular
(peroneal) nerve

Fibularis
(peroneus) longus

Posterior border

Fibularis
(peroneus) brevis

Anterior border

Subcutaneous
triangular area

Sulcus for fibular
(peroneal)
tendons

Lateral malleolus

Tip of malleolus

For calcaneofibular
ligament

Common fibular
(peroneal) nerve

Apex of head

Biceps femoris

Head

Neck

Fibularis
(peroneus) longus

Fibular collateral ligament

Superficial fibular
(peroneal) nerve

Deep fibular (peroneal) nerve

B

Facet for tibia

Facet for fibula

Anterior border

Anterior border

Extensor
surface for

Extensor
digitorum longus
Fibularis
(peroneus) tertius
Extensor hallucis
longus

Extensor surface
for tibialis
anterior

Border for interosseous
membrane

Border for
interosseous
membrane

Surface for tibialis
posterior

Surface for
tibialis posterior

Medial crest

Surface for flexor
hallucis longus

For interosseous
ligament

Fibular notch for
interosseous
ligament

Facets for

Tibia

Talus

Facets for
fibula

Talus

Malleolar fossa for
posterior talofibular
ligament

Calcaneus

Figure 5-49. Bones of the leg showing the attachments of muscles to the lateral surface of the fibula. *A*, Lateral surface of the fibula. *B*, Apposed aspects of the tibia and fibula. Two foot bones are also shown.

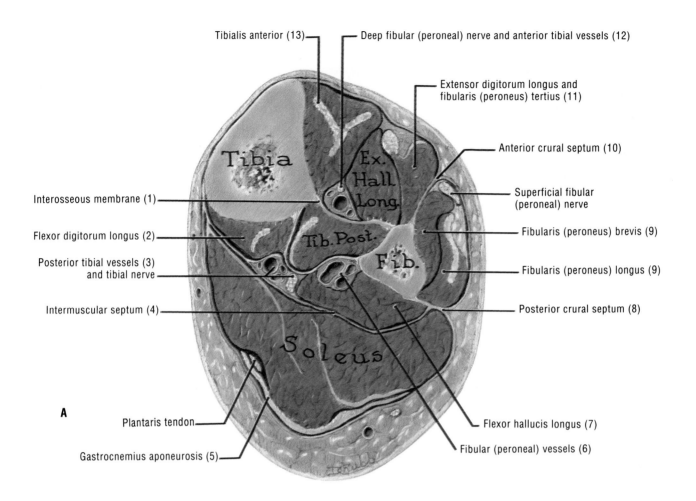

Tibialis anterior (13)

Deep fibular (peroneal) nerve and anterior tibial vessels (12)

Extensor digitorum longus and fibularis (peroneus) tertius (11)

Anterior crural septum (10)

Superficial fibular (peroneal) nerve

Interosseous membrane (1)

Fibularis (peroneus) brevis (9)

Flexor digitorum longus (2)

Posterior tibial vessels (3) and tibial nerve

Fibularis (peroneus) longus (9)

Intermuscular septum (4)

Posterior crural septum (8)

Plantaris tendon

Flexor hallucis longus (7)

Gastrocnemius aponeurosis (5)

Fibular (peroneal) vessels (6)

Tibia

Ex. Hall. Long.

Tib. Post.

Fib.

Soleus

A

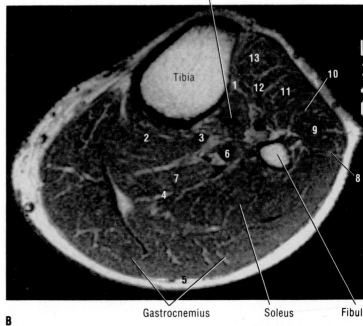

Tibialis posterior

Gastrocnemius

Soleus

Fibula

B

Figure 5-50. *A*, Transverse section through the right leg of an adult male showing its three compartments and their contents (p. 443). *B*, Transverse MRI (magnetic resonance image) of the leg. (Courtesy of Dr. W. Kucharczyk, Clinical Director of the Tri-Hospital Resonance Centre, Toronto, Ontario, Canada.)

Posteroinferior to the facet for the talus is a depression, called the *malleolar fossa* (Fig. 5-49*B*), for the attachment of the posterior talofibular ligament (p. 489). In Fig. 5-47, observe that the lateral malleolus lies more inferior and posterior than does the medial malleolus.

> **Fractures of the fibula** commonly occur 2 to 6 cm proximal to the distal end of the lateral malleolus and are often associated with fracture-dislocations of the ankle joint, *e.g.*, *Pott's fracture* (Fig. 5-99 and p. 490). When a person slips and the foot is forced into an excessively inverted position, the ankle ligaments tear and the talus is forcibly tilted against the lateral malleolus, shearing it off. Fracture of the lateral malleolus is relatively common in certain athletes (*e.g.*, soccer and basketball players). If a part of a major bone is destroyed by injury or disease, the limb becomes useless (McKee, 1989). Without a bone transplant, the affected part of the limb would have to be amputated.
>
> *The fibula is a common source of bone for grafting.* Even after a long piece of the fibula has been removed, walking, running, and jumping can be normal. Free vascularized fibulae have been used to restore skeletal integrity to upper and lower limbs in which congenital bone defects exist, or to replace segments of bone following trauma or excision of a malignant tumor. Removal of portions of the fibula does not affect leg or foot function. The missing piece of bone usually does not regenerate because the periosteum and nutrient artery are generally removed with the piece of bone so that the graft will remain alive when transplanted to another site. When the transplanted piece of fibula is secured in its new site, the blood supply of the bone is restored. Healing proceeds as if there were merely a fracture at each of its ends.
>
> Awareness of the location of the *nutrient foramen* in the fibula (Fig. 5-47) is important when performing free *vascularized fibular transfers*. As the nutrient foramen is located in the middle third of the fibula in most cases, this segment of the bone should be used for transplanting when it is desirable for the graft to include an endosteal, as well as a periosteal, blood supply (see Fig. 14, p. 15).

Surface Anatomy of the Leg

The anteromedial surface of the tibia is subcutaneous, smooth and flat (Fig. 5-50). The skin covering it is freely movable. The prominence at the ankle known as the *medial malleolus* is also subcutaneous and its inferior end is blunt. The ankle joint is about 1.5 cm proximal to the tip of the medial malleolus (Fig. 5-97). Feel the anterior border of your tibia, noting that it is sharp and subcutaneous (Fig. 5-50). The skin is very close to the bone here. No wonder it hurts so much and bruises so easily when you hit this surface of your leg (shin) on something hard. Move your digits proximally along the anterior aspect of the tibia until you feel the rounded elevation called the *tibial tuberosity* (Fig. 5-43). It is about 5 cm distal to the apex of the patella. Palpate your *patellar ligament*, which extends from the apex and margins of the patella to the tuberosity of the tibia (Fig. 5-42). It is most easily felt when your leg is extended. Flex your leg and feel the depression on each side of the patellar ligament. Usually some indentation is visible at these sites when the leg is extended. The capsule of the knee joint (p. 477) is very superficial in these depressions.

The tibial tuberosity is a useful bony landmark because it roughly indicates the level of the division of the popliteal artery into its terminal branches, the anterior and posterior tibial arteries, at the distal border of the popliteus muscle. The smooth superior part of the tibial tuberosity is at the level of the head of the fibula. The subcutaneous, rough inferior part of the tibial tuberosity, which bears the weight during kneeling, is at the level of the neck of the fibula. The *head of the fibula* can be easily palpated at the level of the superior part of the tibial tuberosity, because this knob of bone is subcutaneous at the posterolateral aspect of the knee (Fig. 5-43). Do not mistake the lateral condyle of the tibia for the head of the fibula (Fig. 5-2). A good guide to the location of the head of the fibula is the distal end of the tendon of the biceps femoris (Fig. 5-42). The neck of the fibula can be palpated just distal to the head. The common fibular nerve can be rolled under your digit here (Fig. 5-49*A*). This may cause a tingling sensation on the anterolateral aspect of the leg and the dorsal surfaces of the toes, the areas of skin supplied by branches of this nerve.

Only the distal portion of the body of the fibula is subcutaneous. Hence the fibula just proximal to the lateral malleolus is the part that is commonly fractured (Figs. 5-56 and 5-99). Palpate your lateral malleolus noting that it is subcutaneous and that its inferior end is sharp. Note that the tip of the lateral malleolus extends further distally (1 to 2 cm) and more posteriorly than does the tip of the medial malleolus (Fig. 5-47). This relationship is important in the diagnosis and treatment of injuries in the ankle region (p. 490).

Bones of the Foot

Because the distal attachments of the leg muscles are in the foot, the bones of the ankle and foot are described here. These bones comprise the tarsus, metatarsus, and phalanges (p. 439). Using an articulated skeleton of the foot (Fig. 5-51), observe that its medial border is almost straight. Note also that the line joining the midpoints of the medial and lateral borders of the foot is oblique and that the metatarsal bones and phalanges are located anterior to this line and the tarsal bones are posterior to it. The **tarsus** (G. *tarsos*, flat) consists of seven *tarsal bones* (Figs. 5-51 to 5-55): talus, calcaneus, cuboid, navicular, and three cuneiforms. Only one of them, the talus, articulates with the leg bones.

The Talus

The talus (L. ankle bone) has a body, neck, and head. It looks somewhat like a saddle when viewed from its dorsal aspect (Fig. 5-51). The talus rests on the anterior two-thirds of the calcaneus (Figs. 5-54 and 5-55.) It also articulates with the tibia, fibula, and navicular bone. The saddle-shaped superior surface of the talus bears the weight of the body transmitted via the tibia (Figs. 5-49*B* and 5-55).

> The *body of the talus* is cuboidal in shape. Its pulley-shaped superior surface, often called the *trochlea* (L. pulley), articulates with the inferior surface of the tibia as part of the ankle joint (p. 488). The body of the talus has three continuous facets for articulation (Fig. 5-47): one for the facet on the

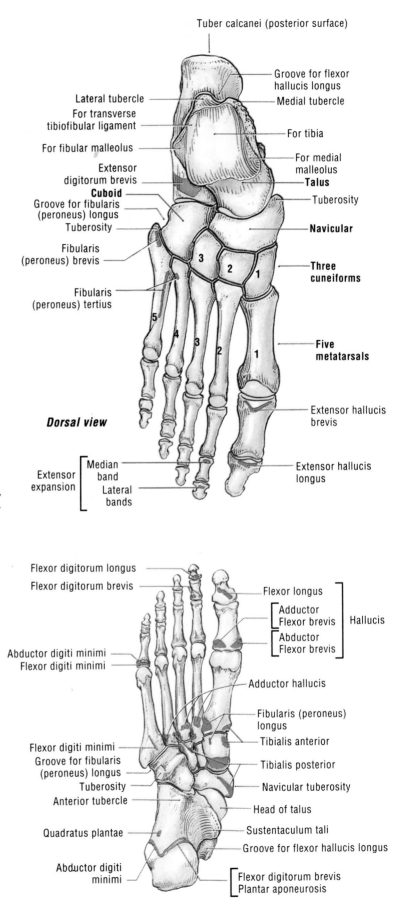

Figure 5-51. Dorsal aspect of the bones of the right foot, showing the muscle attachments and articular cartilages (blue).

Figure 5-52. Plantar aspect of the bones of the right foot showing the attachments of muscles. Proximal attachments are shown in salmon color and distal attachments in blue.

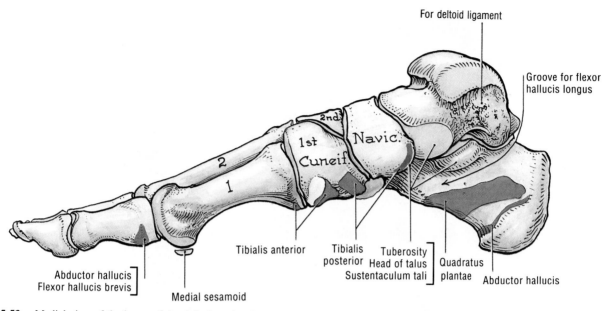

Figure 5-53. Medial view of the bones of the right foot showing the muscle attachments. Proximal attachments are shown in red and distal attachments in blue. Observe that the medial part of the longitudinal arch of the foot is formed by the calcaneus, talus, navicular, cuneiforms, and the first, second, and third metatarsals.

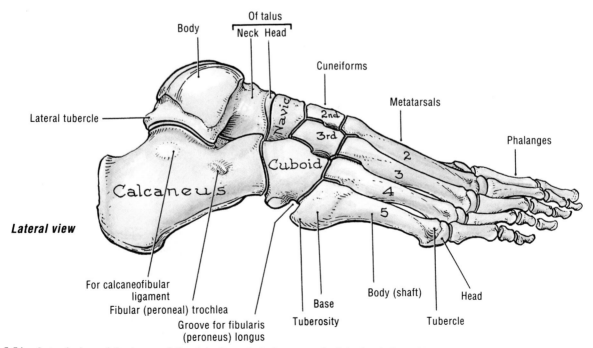

Figure 5-54. Lateral view of the bones of the right foot. Articular cartilages are shown in blue. Note that the lateral part of the longitudinal arch of the foot is formed by the calcaneus, cuboid, and fourth and fifth metatarsals.

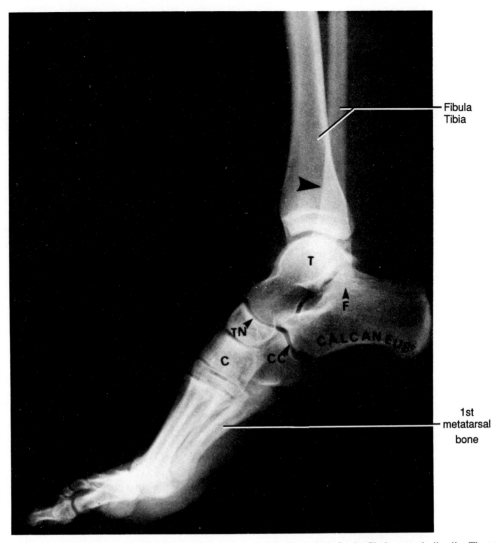

Figure 5-55. Lateral radiograph of the bones of the right leg (distal part) and foot. This radiograph was taken with the foot raised as in walking. The large arrow points to the edge of the triangular area where the tibia and fibula are superimposed on each other. The small arrow (F) indicates how far the fibula extends distally. The talus (T) participates in the talonavicular joint (TN) and the calcaneus in the calcaneocuboid (CC) joint. The cuneiforms (C) and the proximal ends of the metatarsals are superimposed upon each other.

inferior surface of the tibia; one for the facet on the lateral surface of the medial malleolus; and one for the facet on the medial surface of the lateral malleolus. The inferior surface of the body of the talus has an oval, deeply concave area for articulation with the calcaneus. The body also has a posterior process that has medial and lateral tubercles (Fig. 5-47). There is a groove between these tubercles for the tendon of the flexor hallucis longus muscle (Figs. 5-47 and 5-53). The *neck of the talus* is the slightly constricted part between the head and body (Fig. 5-54). Inferiorly there is a deep groove called the *sulcus tali* for the ligaments connecting the talus and calcaneus. The *head of the talus* is its rounded anterior end; it is directed anteromedially (Figs. 5-51 to 5-54). It has a large facet for articulation with the navicular bone and one for articulation with the shelflike projection of the calcaneus, known as the *sustentaculum tali* (Figs. 5-47 and 5-52), as well as a small facet for articulation with the plantar calcaneonavicular ligament.

Occasionally during ossification, the lateral tubercle of the posterior process (Fig. 5-47) fails to unite with the body of the talus. This results in an extra bone, known as the *os trigonum*, which could be misinterpreted as a fracture by an inexperienced viewer of radiographs. *Fractures of the neck of the talus* occur during severe dorsiflexion of the ankle (*e.g.*, when a person is pressing extremely hard on the brake pedal of a car during a head-on collision). In some injuries the body of the talus is dislocated posteriorly.

The Calcaneus

The calcaneus is a rectangular block of bone (Fig. 5-54). It is *the largest and strongest bone of the foot*. It is also the first one to ossify. It articulates with the talus superiorly and the cuboid anteriorly. The calcaneus (L. heel) lies inferior to the talus; thus its superior surface has articular facets for this bone. The posterior facet is demarcated anteriorly by a groove, the *sulcus calcanei*. Anterior to this sulcus is the *sustentaculum tali* (Figs. 5-47, 5-52, and 5-53), a shelf that projects from the superior border of the medial surface of the calcaneus. It helps to support the talus. The lateral surface has an oblique ridge called the *fibular (peroneal) trochlea* (Fig. 5-54). The tendon of the fibularis (peroneus) longus muscle passes inferior to this ridge (Fig. 5-56).

As described by Woodburne and Burkel (1988), "the calcaneus is shaped somewhat like a pistol grip, the thumb gliding naturally into the hollow under the sustentaculum tali." The calcaneus projects posteriorly and forms the prominence of the heel (Figs. 5-55 and 5-65). At the posterior end of its inferior surface, a tuber calcanei projects inferiorly. It is deep to the fibrous tissue and fat of the heel pad and it transmits the weight of the body to the floor or ground. The *tuber calcanei* or calcanean tuberosity forms the *projection of the heel*; it has medial and lateral processes for the attachment of muscles. On the medial surface of the calcaneus is a groove on the inferior surface of the sustentaculum tali for the attachment of the flexor hallucis longus tendon (Fig. 5-53), and on its lateral surface is a tubercle, the *fibular (peroneal) trochlea* (Fig. 5-54).

Persons who fall on their heels (*e.g.*, from a ladder) often fracture their calcanei, usually breaking them into several fragments. A *calcanean fracture* is usually very disabling because it disrupts the subtalar joint (p. 490).

The Navicular

The navicular (L. little ship) is a flattened, oval, boat-shaped bone. Located between the head of the talus and the three cuneiform bones (Fig. 5-51), it has facets for articulation with each of them. The navicular also has an occasional facet for articulation with the cuboid bone. Medially and inferiorly, there is a rough *navicular tuberosity* to which the tendon of the tibialis posterior muscle attaches (Figs. 5-52 and 5-53).

The Cuboid

This rather wedge-shaped bone, approximately cubical in shape, is the most lateral bone in the distal row of the tarsus (Figs. 5-51 and 5-54). Posteriorly it presents an articular surface for the calcaneus and anteriorly two facets for the fourth and fifth metatarsal bones. On its medial surface are facets for the lateral cuneiform and navicular bones. Anterior to the tuberosity of the cuboid, on the lateral and inferior surfaces of the bone, there is a groove for the tendon of the fibularis (peroneus) longus muscle (Fig. 5-54).

The Cuneiform Bones

The name of these three bones is derived from a Latin word meaning "wedge-shaped." They are referred to as the medial (first), intermediate (second), and lateral (third) cuneiforms (Figs. 5-51 to 5-55). The medial cuneiform is the largest bone and the intermediate cuneiform the smallest. Each cuneiform articulates with the navicular bone posteriorly, and with the base of its appropriate metatarsal, anteriorly. In addition, the lateral cuneiform articulates with the cuboid bone.

The Metatarsus

The metatarsus *consists of five metatarsal bones* (Figs. 5-51 to 5-55). These miniature long bones are numbered from the medial side of the foot and each bone consists of a base (proximally), a body or shaft, and a head (distally). The bases of the metatarsals articulate with the cuneiform and cuboid bones, and their heads articulate with the proximal phalanges. The second metatarsal bone is wedged between the medial and lateral cuneiforms and between the first and third metatarsals. The second metatarsal bone is the longest bone. On the plantar surface of the head of the first metatarsal bone, there are prominent medial and lateral *sesamoid bones* (Figs. 5-53 and 5-81). The heads of the metatarsals bear some of the weight of the body (Fig. 5-104). The base of the fifth metatarsal has a large tuberosity that projects over the lateral margin of the cuboid bone (Fig. 5-54). The tuberosity of the fifth metatarsal bone provides attachment on its dorsal surface for the fibularis (peroneus) brevis tendon (Fig. 5-56).

The *sesamoid bones* of the first digit or great toe *bear the weight of the body*, especially during the latter part of the stance phase of walking (Figs. 5-55 and 5-104). Occasionally, an accessory bone called the *os vesalianum pedis* (Vesalius' bone), appears near the base of the fifth metatarsal. When examining radiographs, it is important to know of its possible presence so it will not be diagnosed as a fracture of this tuberosity. When this accessory bone is large, the tuberosity of the fifth metatarsal is small. *Fractures of the metatarsals* usually occur when a heavy object falls on the foot or the foot is run over by a heavy metal wheel. When the foot is suddenly and violently inverted, the tuberosity of the fifth metatarsal may be avulsed (pulled off) by the tendon of the fibularis (peroneus) brevis muscle.

The Phalanges

There are 14 phalanges (Fig. 5-54): the first digit or great toe (L. *hallux*) has two strong phalanges (proximal and distal); the other four digits have three each (proximal, middle, and distal). Each phalanx consists of a base (proximally), a body or shaft, and a head (distally).

The Crural Fascia

The *deep fascia of the leg* (L. *crus*) or crural fascia is continuous with the *fascia lata* (deep fascia of the thigh). It forms an incomplete covering (elastic stocking) for the muscles of the leg (Figs. 5-50 and 5-56). The crural fascia is attached to the anterior and medial borders of the tibia, where it is continuous with its periosteum. Deep fascia is absent over the subcutaneous part of the medial surface of the tibia and over the triangular subcutaneous surface of the inferior quarter of the fibula. Here, it is attached to the borders of the fibula. The crural fascia is very thick in the proximal part of the anterior aspect of the leg

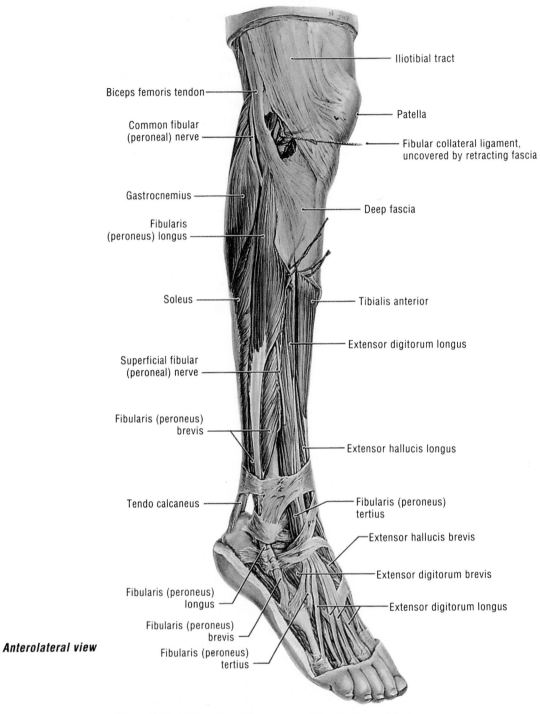

Iliotibial tract

Biceps femoris tendon

Patella

Common fibular
(peroneal) nerve

Fibular collateral ligament,
uncovered by retracting fascia

Gastrocnemius

Deep fascia

Fibularis
(peroneus) longus

Soleus

Tibialis anterior

Extensor digitorum longus

Superficial fibular
(peroneal) nerve

Fibularis (peroneus)
brevis

Extensor hallucis longus

Tendo calcaneus

Fibularis (peroneus)
tertius

Extensor hallucis brevis

Extensor digitorum brevis

Fibularis (peroneus)
longus

Extensor digitorum longus

Fibularis (peroneus)
brevis

Anterolateral view

Fibularis (peroneus)
tertius

Figure 5-56.   Dissection of the muscles of the right leg and foot.

(Fig. 5-50) where it forms part of the proximal attachments of the underlying muscles (*e.g.*, tibialis anterior; Fig. 5-56). Although thin in the distal part of the leg, the fascia is thickened where it forms the superior and inferior extensor retinacula (Fig. 5-57). From the deep surface of the crural fascia, anterior and posterior *intermuscular septa* pass deeply to attach to the corresponding margins of the fibula (Fig. 5-50). These septa form three *muscular compartments* of the leg.

The Superior Extensor Retinaculum (Figs. 5-56 to 5-58). This strong, *broad band of deep fascia* passes from the fibula to the tibia, proximal to the malleoli. It binds down the tendons of muscles in the anterior crural compartment (Table 5-5), preventing them from bowstringing anteriorly during dorsiflexion of the ankle joint.

The Inferior Extensor Retinaculum (Figs. 5-56 and 5-57). This *Y-shaped band of deep fascia* is attached laterally to the anterosuperior surface of the calcaneus. It forms a strong loop around the tendons of the fibularis (peroneus) tertius and the extensor digitorum longus. The proximal limb of the inferior extensor retinaculum is attached to the medial malleolus. During its superomedial course it passes over the tendons of the tibialis anterior and extensor hallucis longus muscles, the dorsalis pedis vessels, and the deep fibular (peroneal) nerve. The distal limb of the inferior extensor retinaculum is attached medially to the deep fascia of the medial margin of the foot and the *plantar aponeurosis* (Fig. 5-76). During its inferomedial course, it also passes over the tendons of the tibialis anterior and extensor hallucis longus muscles, the dorsalis pedis vessels, and the deep fibular nerve.

Compartments of the Leg

The tibia and fibula, the interosseous membrane, and the crural intermuscular septa divide the leg into *three crural compartments* (Fig. 5-50): **anterior** (extensor), **lateral** (fibular or peroneal), and **posterior** (flexor). The fascial septa, called *crural*

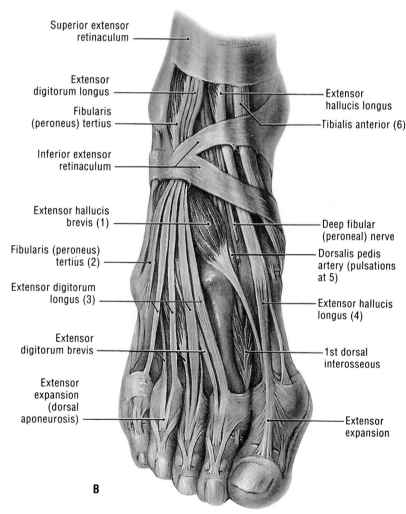

Figure 5-57. Dorsum of the foot. *A*, Surface anatomy. *B*, Dissection. Observe the medial (*M*) and lateral (*L*) malleoli. Observe the dorsalis pedis artery. Its pulsations can be felt at the site indicated by the number 5.

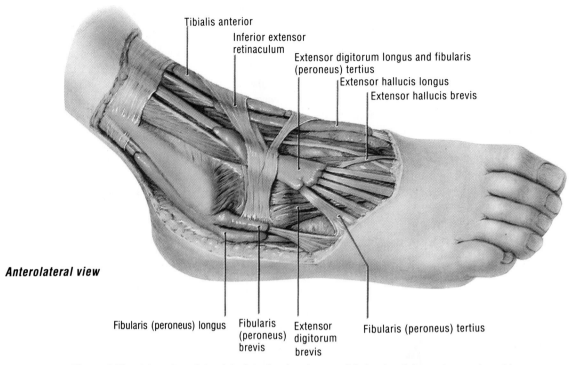

Tibialis anterior

Inferior extensor retinaculum

Extensor digitorum longus and fibularis (peroneus) tertius

Extensor hallucis longus

Extensor hallucis brevis

Anterolateral view

Fibularis (peroneus) longus

Fibularis (peroneus) brevis

Extensor digitorum brevis

Fibularis (peroneus) tertius

Figure 5-58.   Dissection of the right foot showing the synovial sheaths of the tendons at the ankle.

intermuscular septa, are attached superficially to the ensheathing deep fascia and the fibula. The anterolateral part of the leg contains the anterior and lateral crural compartments, which are separated by the *anterior crural intermuscular septum* (Fig. 5-50). The muscles in the lateral crural compartment are separated from muscles in the posterior crural compartment by the *posterior crural intermuscular septum*. The much larger posterior compartment of the leg, often called the calf, is subdivided by a broad intermuscular septum (formed by the deep transverse fascia of the leg) into superficial and deep posterior crural compartments, containing the superficial and deep crural muscles, respectively (Fig. 5-50; Tables 5-7 and 5-8). **In summary**, the leg muscles are arranged in three compartments (Fig. 5-50): the *anterior compartment*, between the tibia and the anterior crural septum; the *posterior compartment*, between the tibia and the posterior crural septum; and the *lateral compartment*, between the anterior and posterior crural septa. The muscles in a compartment share the *same general function*, the *same nerve supply*, and the *same blood supply*.

Anterior Compartment of the Leg

This **extensor compartment**, located anterior to the interosseous membrane, is located between the lateral surface of the tibia and the anterior crural intermuscular septum (Fig. 5-50). The **four muscles in the anterior compartment** are extensor (dorsiflexor) muscles: tibialis anterior, extensor hallucis longus, extensor digitorum longus, and fibularis (peroneus) tertius muscles (Table 5-5). These muscles are mainly concerned with dorsiflexion of the ankle joint and extension of the toes (Fig.

5-61A). They are supplied by the *deep fibular (peroneal) nerve*, which is derived from the sciatic (Fig. 5-73) via the common fibular (peroneal), and by the *anterior tibial artery* (Figs. 5-50 and 5-60). The muscles of this group are *true extensors of the foot*, although their action is usually referred to as dorsiflexion.

The Tibialis Anterior Muscle (Figs. 5-50, 5-56 to 5-59A, 5-71B, and 5-75).   This long, thick muscle lies against the lateral surface of the tibia, where it is easy to palpate. Its attachments, nerve supply, and main actions are given in Table 5-5.

When the tibialis anterior is paralyzed owing to injury of the common fibular (peroneal) nerve or to its branch, the deep fibular (peroneal) nerve, the foot drops (*i.e.*, it falls into plantarflexion when it is raised from the ground). This condition, known as *foot-drop*, was discussed earlier (p. 429). "Shin splints" is a lay person's expression for a painful condition of the anterior compartment of the leg that follows vigorous and/or lengthy exercise. Often persons who lead sedentary lives develop pains in the anterior part of their legs when they take long walks. Their anterior tibial muscles swell from sudden overuse and the swollen muscles in the anterior compartment reduce the blood flow to the muscles. Cramps may develop if use of the muscles is continued. The swollen muscles are painful and tender to pressure. "Shin splints" may also occur in trained athletes who do not warm up adequately or warm down sufficiently.

The Extensor Hallucis Longus Muscle (Figs. 5-50 and 5-56 to 5-59).   This thin muscle lies between and partly deep to the tibialis anterior and extensor digitorum longus muscles. Its attachments, nerve supply, and main actions are given in Table 5-5.

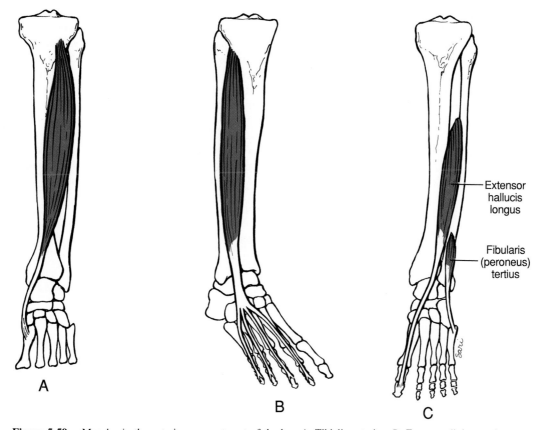

Figure 5-59.   Muscles in the anterior compartment of the leg. *A*, Tibialis anterior. *B*, Extensor digitorum longus. *C*, Extensor hallucis longus and fibularis (peroneus) tertius.

Table 5-5.
The Muscles in the Anterior Compartment of the Leg

Muscle	Proximal Attachments	Distal Attachments	Innervation	Main Actions
Tibialis anterior (Figs. 5-49*B* and 5-59*A*)	Lateral condyle and superior half of lateral surface of tibia	Medial and inferior surfaces of medial cuneiform bone and base of first metatarsal bone	Deep fibular (peroneal) nerve (**L4** and L5)	Dorsiflexes and inverts foot (Fig. 5-61*A* and *D*)
Extensor hallucis longus (Fig. 5-59*C*)	Middle part of anterior surface of fibula and interosseous membrane	Dorsal aspect of base of distal phalanx of great toe (hallux)		Extends great toe and dorsiflexes foot (Fig. 5-61*A*)
Extensor digitorum longus (Fig. 5-59*B*)	Lateral condyle of tibia, superior three-fourths of anterior surface of fibula, and interosseous membrane	Middle and distal phalanges of lateral four digits	Deep fibular (peroneal) nerve (L5 and S1)	Extends lateral four digits and dorsiflexes foot (Fig. 5-61*A*)
Fibularis (peroneus) tertius (Fig. 5-59*C*)	Inferior third of anterior surface of fibula and interosseous membrane	Dorsum of base of fifth metatarsal bone		Dorsiflexes foot and aids in eversion of it (Fig. 5-51*A* and *D*)

The Extensor Digitorum Longus Muscle (Figs. 5-50, 5-56 to 5-58, and 5-59B). This pennate (featherlike) muscle lies lateral to the tibialis anterior and can be easily palpated. Its tendons may be seen and felt when the toes are dorsiflexed (Fig. 5-57). Its attachments, nerve supply, and actions are given in Table 5-5. *A common synovial sheath surrounds its four tendons*, which diverge on the dorsum of the foot as they pass to their distal attachments. Each tendon forms a membranous extensor expansion over the dorsum of the proximal phalanx, which divides into two lateral slips and one central slip. The central slip inserts into the base of the middle phalanx and the lateral slips converge to insert into the base of the distal phalanx.

The Fibularis (Peroneus) Tertius Muscle (Figs. 5-50 and 5-56 to 5-59). This small variable muscle is a partially *separated part of the extensor digitorum longus* muscle. The two muscles are fused at their proximal attachments, but distally the tendon of the fibularis tertius does not attach to a digit. The attachments, nerve supply, and main actions of this muscle are given in Table 5-5.

The Deep Fibular (Peroneal) Nerve (Figs. 5-49A, 5-50, 5-57, and 5-60). This *nerve of the anterior crural compartment* is one of the two terminal branches of the common fibular (peroneal) nerve. It begins between the fibula and the superior part of the fibularis (peroneus) longus muscle. It then runs infero-

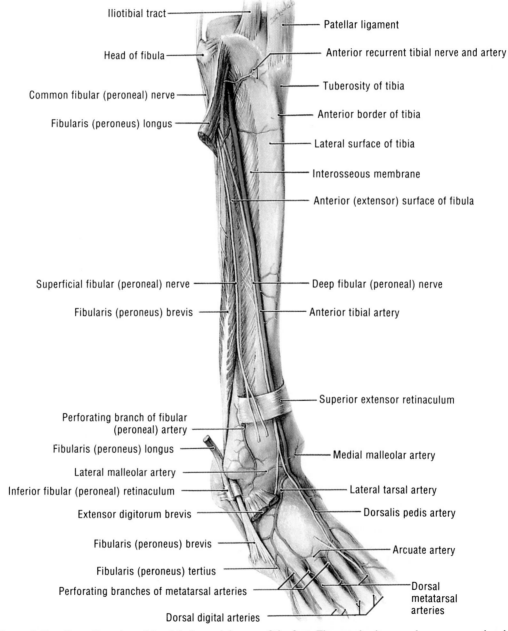

Figure 5-60. Deep dissection of the right leg and dorsum of the foot. The anterior leg muscles are removed and the fibularis (peroneus) longus is excised.

medially on the fibula, deep to the extensor digitorum longus. After piercing the anterior crural intermuscular septum and the extensor digitorum longus, the deep fibular nerve descends anterior to the interosseous membrane in the anterior crural compartment. It accompanies the anterior tibial artery between the extensor hallucis longus and tibialis anterior muscles. The deep fibular nerve passes deep to the extensor retinacula with the anterior tibial artery, where it ends by dividing into medial and lateral branches. In addition to supplying the muscles in the anterior compartment, the deep fibular nerve gives branches to the posterior tibial and fibular (peroneal) arteries. It also sends articular branches to the ankle joint and other articulations that it crosses, and supplies the skin between the first and second digits (Fig. 5-82).

The Anterior Tibial Artery (Figs. 5-50, 5-57, 5-60, 5-77, and 5-83). Structures in the anterior crural department are supplied by the anterior tibial artery and its branches. The smaller of the terminal branches of the popliteal artery, the anterior tibial begins opposite the inferior border of the popliteus muscle. It ends at the ankle joint, midway between the malleoli where it **becomes the dorsalis pedis artery**. From its origin, the anterior tibial artery passes anteriorly through the interosseous mem-

brane. It then descends on the anterior surface of this membrane between the extensor hallucis longus and tibialis anterior muscles with the deep fibular nerve. In the distal part of the leg, the anterior tibial artery lies on the tibia. In addition to supplying muscles in the anterior crural compartment, the anterior tibial artery has several other named branches. The anterior and posterior *tibial recurrent arteries* join the anastomoses around the knee (Fig. 5-46), and the medial and lateral anterior *malleolar arteries* ramify over the medial and lateral malleoli, respectively (Fig. 5-60), contributing to the arterial networks around the ankle.

Lateral Compartment of the Leg

This compartment is bounded by the lateral surface of the fibula, the anterior and posterior crural intermuscular septa, and the crural fascia (Fig. 5-50). The lateral (fibular or peroneal) compartment contains the fibularis (peroneus) longus and brevis muscles, which plantarflex and evert the foot (Figs. 5-61 and 5-62; Table 5-6). They are supplied by the superficial fibular (peroneal) nerve, a branch of the common fibular (peroneal) nerve (Fig. 5-73).

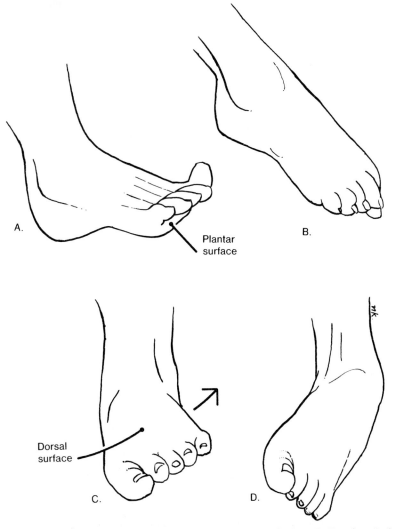

A.

Plantar surface

B.

Dorsal surface

C.

D.

Figure 5-61. Movements of the foot. *A*, Dorsiflexion. *B*, Plantarflexion. *C*, Eversion. *D*, Inversion.

In the current *Nomina Anatomica* (Warwick, 1989), terms derived from the Greek word *peronē* for the fibula (peroneus and peroneal) are replaced by ''fibularis'' and ''fibular'' to indicate more clearly the relationship of muscles, arteries, and nerves to the fibula, a Latin word. Parentheses are used to indicate familiar and still valid terms, *e.g.*, fibularis (peroneus) longus muscle, fibular (peroneal) artery, and common fibular (peroneal) nerve. This terminology change was also made because some people have difficulty differentiating between ''peroneal'' and ''perineal'' structures. Because the terms *peroneal* and *peroneus* are used by so many people, they are used as alternatives as in this text.

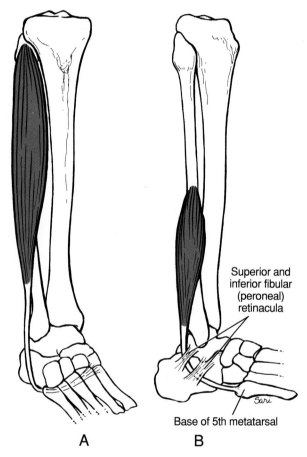

Figure 5-62.   Muscles in the lateral compartment of the leg. *A*, Fibularis (peroneus) longus. *B*, Fibularis (peroneus) brevis.

The Fibularis (Peroneus) Longus Muscle (Figs. 5-49, 5-50, 5-56, 5-58, 5-60, and 5-62*A*).   This is the longer and more superficial of the two fibularis (peroneal) muscles and it arises much more superiorly on the body of the fibula. The fibularis longus is a narrow muscle that extends from the head of the fibula to the sole of the foot. Its tendon can be palpated and observed proximal and posterior to the lateral malleolus. Its attachments, nerve supply, and main actions are given in Table 5-6. When one stands on one foot, the fibularis longus also helps to steady the leg on the foot. *The fibularis longus is enclosed in a common synovial sheath with the fibularis (peroneus) brevis muscle*. It passes inferior to the fibular (peroneal) trochlea on the calcaneus to enter a groove on the anteroinferior aspect of the cuboid bone. It then crosses the sole of the foot, running obliquely and distally to reach its distal attachment to the first metatarsal and medial cuneiform bones (Fig. 5-102).

The Fibularis (Peroneus) Brevis Muscle (Figs. 5-49, 5-50, 5-56, 5-58, 5-60, and 5-62*B*).   This fusiform muscle lies deep to the fibularis (peroneus) longus and, as its name indicates, is shorter than its partner in the lateral compartment of the leg. Its attachments, nerve supply, and main actions are given in Table 5-6. Its tendon grooves the posterior aspect of the lateral malleolus and can be felt inferior to the lateral malleolus, where it lies in a common tendon sheath with the fibularis (peroneus) longus. The tendon of the fibularis brevis can be easily traced to its distal attachment to the base of the fifth metatarsal bone.

The tendon of a small slip from the fibularis brevis muscle often attaches to the long extensor tendon of the small toe (extensor digitorum longus) or continues to attach to the proximal phalanx of this digit. This small muscle is often referred to as the fibularis (peroneus) tertius (Fig. 5-60).

The tuberosity of the fifth metatarsal bone may be avulsed (torn away), by the fibularis (peroneus) brevis tendon during *violent inversion of the foot*. This kind of fracture is associated with a severely sprained ankle. Injury to the associated superficial fibular (peroneal) nerve (Fig. 5-56) results in an inverted foot owing to paralysis of the fibular muscles, which evert the foot (Table 5-6).

In some children and adolescents, there may be a secondary ossification center for the lateral surface of the tuberosity of the fifth metatarsal bone. This results in the formation of a chiplike piece of bone, which should not be mistaken for a ''flake'' fracture of the tuberosity of the fifth metatarsal

Table 5-6.
The Muscles in the Lateral Compartment of the Leg[1]

Muscle	Proximal	Distal	Innervation	Main Actions
Fibularis (peroneus) longus (Figs. 5-49*A* and 5-62*A*)	Head and superior two-thirds of lateral surface of fibula	Base of first metatarsal bone and medial cuneiform bone	Superficial fibular (peroneal) nerve (**L5, S1**, and **S2**)	Evert foot and weakly plantarflex it
Fibularis (peroneus) brevis (Figs. 5-49*A* and 5-62*B*)	Inferior two-thirds of lateral surface of fibula	Dorsal surface of tuberosity on lateral side of base of fifth metatarsal bone		

[1]The fibularis (peroneus) longus and brevis were named because their proximal attachment is to the fibula. *Peroneus* is the Greek word for the Latin term *fibula* and was formerly used to describe these muscles.

bone. The presence of similar secondary centers in both feet usually indicates that a fracture is not present. These centers are not usually observed in adults because, by this age they have fused with the tuberosity.

The Superficial Fibular (Peroneal) Nerve (Figs. 5-50, 5-56, and 5-60). This *nerve of the lateral compartment* of the leg is one of the two terminal branches of the common fibular (peroneal) nerve. The superficial fibular nerve begins between the fibularis (peroneus) longus muscle and the fibula and descends posterolateral to the anterior crural intermuscular septum. It lies anterolateral to the fibula between the fibular muscles and the extensor digitorum longus. The superficial fibular nerve supplies the fibular (peroneal) muscles and then pierces the deep fascia to become superficial in the distal third of the leg. It passes in the superficial fascia to supply the skin on the distal part of the anterior surface of the leg, nearly all the dorsum of the foot, and most of the digits (Fig. 5-82).

There are no arteries in the lateral crural compartment, except for the muscular branches to the fibular (peroneal) muscles, which arise from the fibular (peroneal) artery (Fig. 5-60), a branch of the posterior tibial artery.

Posterior Compartment of the Leg

From medial to lateral, this compartment lies posterior to the tibia, interosseous membrane, fibula, and the posterior crural intermuscular septum (Fig. 5-50). The *calf muscles* in this compartment are divided into superficial and deep groups by the *transverse crural intermuscular septum* formed by the deep transverse fascia of the leg. The superficial group of muscles forms a powerful mass in the calf of the leg, which plantarflexes the foot (Fig. 5-61B; Table 5-7). The large size of these muscles is a human characteristic, which is directly related to our upright stance. They are strong and heavy because they support and move the weight of the body. In Figs. 5-50, 5-67, and 5-69, observe that *the tibial nerve and posterior tibial vessels supply both divisions of the posterior compartment*. They run between the superficial and deep groups of muscle. Also observe that the tibial nerve and posterior tibial vessels are deep to the transverse crural intermuscular septum, and that the superficial muscles are much larger than the deep ones.

Three muscles comprise the superficial group (Fig. 5-64; Table 5-7): *gastrocnemius, soleus,* and *plantaris*. The gastrocnemius and soleus form a tripartite muscle that is referred to as the *triceps surae*, which forms the prominence of the calf (Figs. 5-63 and 5-65). These muscles act together in plantarflexing the foot at the ankle joint (Fig. 5-61B). They raise the heel against the weight of the body (*e.g.*, in walking, dancing, and standing on the toes).

It is the gastrocnemius that produces the rapid movements occurring during running and jumping. As Tobias et al. (1988) have said, "One strolls along with the soleus, but one wins the long jump with the gastrocnemius."

The Gastrocnemius Muscle (Figs. 5-45 and 5-63 to 5-65). This is the most superficial of the muscles in the pos-

terior compartment; it *forms most of the prominence of the calf*. The gastrocnemius is a fusiform, two-headed, two-joint muscle. Its medial head is slightly larger and extends a little more distally than does its lateral head. The heads come together at the inferior margin of the popliteal fossa where they form the inferolateral and inferomedial boundaries of this fossa (p. 423). The attachments, nerve supply, and main actions of the gastrocnemius are given in Table 5-7. As its fibers are mainly vertical, *contraction of this muscle produces rapid movement during running and jumping*. These muscles help to steady the legs, consequently they are active during standing, even when at ease. When standing on the toes (Fig. 5-65) or when high heels are worn, these muscles are especially active. Although *the gastrocnemius acts on both the knee and ankle joints*, it is unable to exert its full power on both joints at the same time. The lateral head of the gastrocnemius often contains a *sesamoid bone*, called a *fabella* (L. bean), which is close to its proximal attachment. It is often visible on lateral radiographs of the knee. The distal attachment of the gastrocnemius to calcaneus via the *tendo calcaneus* or *calcaneal tendon* (Achilles tendon) is shared with the soleus muscle. The inferior expanded end of the tendo calcaneus attaches to the middle of the posterior surface of the calcaneus (Fig. 5-48). A *tendo calcaneus bursa* separates the tendo calcaneus from the superior part of this bony surface.

The Soleus Muscle (Figs. 5-50, 5-63 to 5-67, and 5-69). This broad, flat, fleshy, multipennate muscle was named because of its resemblance to sole, a flat fish. It lies deep to the gastrocnemius (Fig. 5-66) and can be palpated on each side of it, inferior to the midcalf, when a person is standing on his/her tiptoes (Fig. 5-65). Its attachments, nerve supply, and main actions are given in Table 5-7. The soleus has a horseshoe-shaped proximal attachment to the tibia and fibula. It acts with the gastrocnemius in plantarflexing the ankle (*e.g.*, in walking and dancing), but *it does not act on the knee joint*. It is also concerned with the maintenance of posture by steadying the leg on the foot, *e.g.*, it prevents the body from falling anteriorly when standing. Because it is broad and multipennate (p. 22), it is a very powerful muscle, but its contractions are slower than those of the gastrocnemius.

The Plantaris Muscle (Figs. 5-45 and 5-64B). This small muscle is variable in size and extent. It may be absent; sometimes it is double. It has a small, fusiform fleshy belly and a long slender tendon, which runs obliquely between the gastrocnemius and soleus muscles. This feeble muscle, the rudiment of a large muscle, is of no practical importance, but it is clinically significant (see column 1, p. 452). Its attachments, nerve supply and actions are given in Table 5-7.

Inflammation, strain, or rupture of the tendo calcaneus or calcaneal tendon is common. Although this painful injury often occurs during running and games such as squash, which require quick starts, it may occur when a person stumbles or is startled, causing him/her to jump or to suddenly start to run (*e.g.*, when crossing a street). This tendon is often strained in young persons, frequently at the start of a 100-meter dash when the tendon is severely stressed during take-off. Complete rupture of the tendo calcaneus usually results in abrupt pain in the posterior aspect of the leg. The person is unable

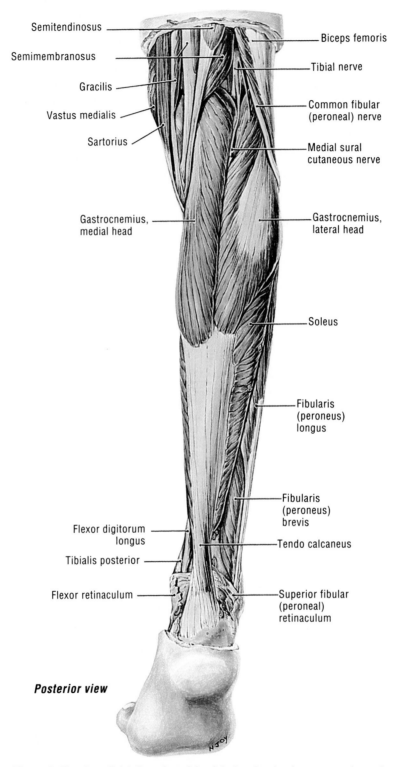

Semitendinosus

Semimembranosus

Gracilis

Vastus medialis

Sartorius

Gastrocnemius, medial head

Flexor digitorum longus

Tibialis posterior

Flexor retinaculum

Biceps femoris

Tibial nerve

Common fibular (peroneal) nerve

Medial sural cutaneous nerve

Gastrocnemius, lateral head

Soleus

Fibularis (peroneus) longus

Fibularis (peroneus) brevis

Tendo calcaneus

Superior fibular (peroneal) retinaculum

Posterior view

Figure 5-63.   Superficial dissection of the right leg showing its nerves and muscles.

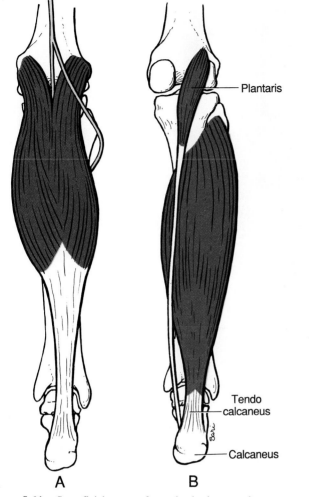

to use the limb and a lump or increase in the prominence of the calf occurs, owing to shortening of the triceps surae muscle (gastrocnemius and soleus). Following rupture of the tendo calcaneus, the foot can be dorsiflexed to a greater extent than normal and the patient is unable to plantarflex his/her foot against resistance.

The ankle reflex (ankle jerk) is the twitch of the triceps surae muscle, which is induced by striking the tendo calcaneus with a reflex hammer. The reflex center for the ankle reflex is the S1 and S2 segments of the spinal cord. "Tennis leg" is a painful calf injury resulting from partial tearing of the medial belly of the gastrocnemius at or near its musculotendinous junction. It is caused by overstretching the muscle by concomitant full extension of the knee and dorsiflexion of the ankle joint. It usually occurs when a middle-aged tennis player is serving the ball or stretches for a difficult shot. The gastrocnemius is one of a few muscles with only one source of blood supply, (*i.e.*, the sural arteries, which are branches of the popliteal). They are virtually end arteries, *i.e.*, there are no anastomoses except by capillaries. If one branch is blocked, the part supplied by it dies. Although the soleus muscle is mainly supplied by the sural arteries, it also receives branches from the fibular (peroneal) artery (Fig. 5-69). Inflammation and swelling of the tendo calcaneus or calcaneal bursa, called *calcaneal bursitis*, is fairly common in long distance runners, owing to excessive friction on the bursa as the tendo calcaneus continuously slides over it.

When standing, *the venous return of the leg depends largely on muscular activity*, especially of the triceps surae muscle. The efficiency of this "**calf pump**" is improved by the tight stocking of deep fascia covering these muscles (Fig. 5-50 and p. 441). When the calf muscles contract, blood is pumped superiorly in the deep veins. Normally blood is prevented from flowing into the superficial veins by the valves in the perforating veins (Fig. 5-7). If these valves become incompetent, blood is forced into the superficial veins during contraction of the triceps surae muscles and by hydrostatic pressure

Figure 5-64. Superficial group of muscles in the posterior compartment of the leg. *A*, Gastrocnemius. *B*, Soleus and plantaris.

Table 5-7.
The Superficial Muscles in the Posterior Compartment of the Leg

Muscle	Proximal Attachment	Distal Attachment	Innervation	Main Actions
Gastrocnemius (Figs. 5-48 and 5-64*A*)	*Lateral head:* Lateral aspect of lateral condyle of femur *Medial head:* Popliteal surface of femur, superior to medial condyle	Posterior surface of calcaneus via tendo calcaneus	Tibial nerve (S1 and S2)	Plantarflexes foot, raises heel during walking, and flexes knee joint
Soleus (Figs. 5-48 and 5-64*B*)	Posterior aspect of head of fibula, superior fourth of posterior surface of fibula, soleal line, and medial border of tibia			Plantarflexes foot and steadies leg on foot
Plantaris (Figs. 5-48 and 5-68*B*)	Inferior end of lateral supracondylar line of femur and oblique popliteateal ligament			Weakly assists gastrocnemius in plantarflexing foot and flexing knee joint

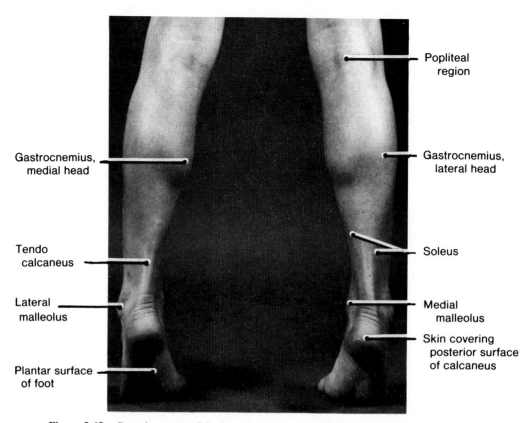

Figure 5-65. Posterior aspect of the legs of a 12-year-old girl who is standing on her tiptoes.

when straining or standing. The distended perforating and superficial veins are called **varicose veins**.

The clinical importance of the plantaris muscle lies in the possibility of its rupture during violent ankle movements. Sudden dorsiflexion of the ankle joint may rupture the long slender tendon of this feeble muscle. In most cases of apparent rupture of the plantaris tendon, muscle fibers of the triceps surae are also torn. This injury is common in basketball players, sprinters, and ballet dancers. Surprisingly, the pain following *rupture of the plantaris tendon* may be so severe that the person is unable to bear weight on the foot. The long tendon of the plantaris muscle is commonly used in reconstructive surgery of the tendons of the hand. It can be removed completely without causing any disability of knee or ankle movements. Although the plantaris acts with the gastrocnemius (Table 5-7), its role is minor.

Four muscles comprise the deep group in the posterior compartment of the leg (Fig. 5-50; Table 5-8): popliteus, flexor digitorum longus, flexor hallucis longus, and tibialis posterior. The popliteus acts on the knee joint, whereas the other muscles act on the ankle and foot joints.

The Popliteus Muscle (Figs. 5-45, 5-68*A*, and 5-69).   This thin, flat triangular muscle forms the floor of the inferior part of the popliteal fossa. *The proximal attachment of the popliteus tendon is inside the fibrous capsule of the knee joint*, deep to the fibular collateral ligament; thus its deep surface is covered by synovial membrane. Its attachments, nerve supply, and main actions are given in Table 5-8. The popliteus weakly flexes the

knee, but *its important action is in unlocking the knee*. It unlocks the extended leg by rotating the femur laterally on the fixed tibia (*e.g.*, when the foot is set on the ground). As the lateral condyle of the femur moves posteriorly, the lateral meniscus of the knee joint (p. 485) is also drawn posteriorly so it will not be injured. When the femur is fixed, the popliteus muscle unlocks the locked knee by rotating the tibia medially on the femur.

The tendons of three deep muscles (flexor hallucis longus, flexor digitorum longus, and tibialis posterior) pass deep to the *flexor retinaculum of the ankle* (Figs. 5-67 and 5-69), where they steady the leg on the foot when one is standing. The flexor retinaculum is a thickening of the deep fascia of the leg, which passes from the medial side of the calcaneus to the medial malleolus.

The Flexor Hallucis Longus Muscle (Figs. 5-66 to 5-70). This long, powerful *flexor of the great toe* is the largest of the four deep muscles. It lies laterally and is closely attached to the fibula. Its attachments, nerve supply, and main actions are given in Table 5-8. Its tendon passes posterior to the distal end of the tibia and deep to the flexor retinaculum. The tendon occupies a shallow groove on the posterior surface of the *sustentaculum tali* (Fig. 5-52). This tendon then crosses deep to the tendon of the flexor digitorum longus in the sole of the foot, giving a tendinous slip to its tendon. As it passes to the great toe, the tendon runs between two *sesamoid bones* in the tendons of the flexor hallucis brevis (Fig. 5-8). These bones protect the tendon from the pressure of the head of the first metatarsal bone. The flexor hallucis longus is *the powerful "push-off" muscle*

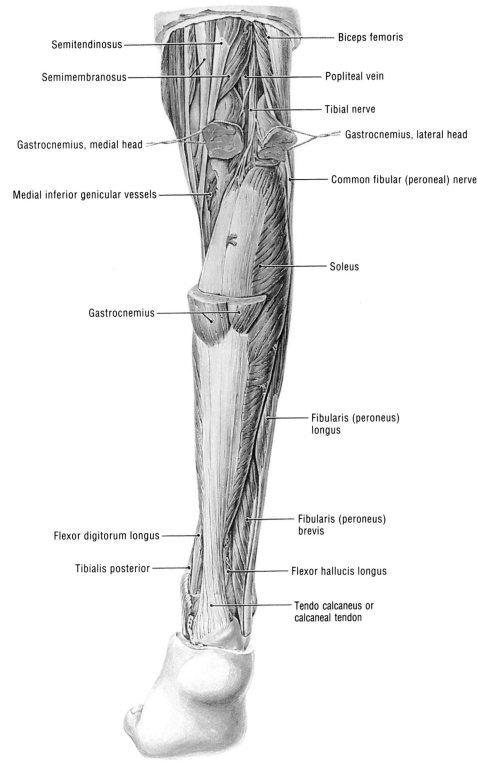

Semitendinosus

Semimembranosus

Gastrocnemius, medial head

Medial inferior genicular vessels

Gastrocnemius

Flexor digitorum longus

Tibialis posterior

Biceps femoris

Popliteal vein

Tibial nerve

Gastrocnemius, lateral head

Common fibular (peroneal) nerve

Soleus

Fibularis (peroneus) longus

Fibularis (peroneus) brevis

Flexor hallucis longus

Tendo calcaneus or calcaneal tendon

Figure 5-66. Dissection of the superficial muscles in the posterior compartment of the leg. The fleshy bellies of the gastrocnemius muscle are largely excised, exposing the proximal attachment of the soleus muscle.

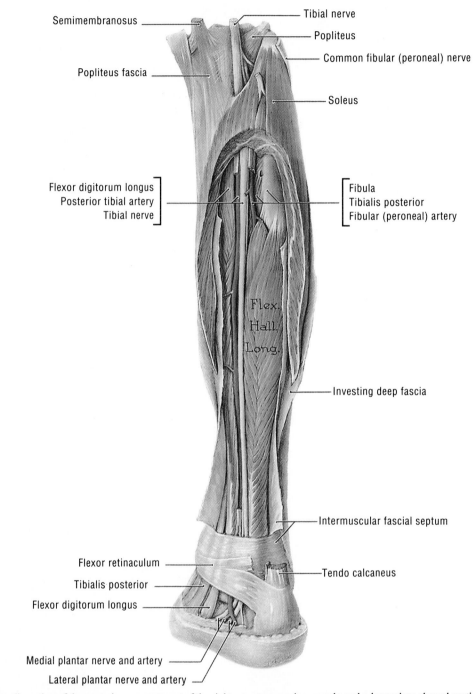

Semimembranosus

Popliteus fascia

Flexor digitorum longus
Posterior tibial artery
Tibial nerve

Tibial nerve

Popliteus

Common fibular (peroneal) nerve

Soleus

Fibula
Tibialis posterior
Fibular (peroneal) artery

Flex.
Hall.
Long.

Investing deep fascia

Intermuscular fascial septum

Flexor retinaculum

Tibialis posterior

Flexor digitorum longus

Tendo calcaneus

Medial plantar nerve and artery

Lateral plantar nerve and artery

Figure 5-67.   Deep dissection of the posterior compartment of the right leg. The tendo calcaneus or calcaneal tendon is divided and the gas-trocnemius muscle and a horseshoe-shaped section of the soleus muscle are removed to show the vessels and nerves.

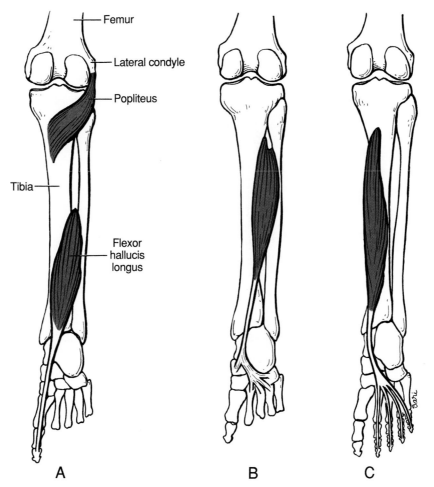

Figure 5-68.   Deep group of muscles in the posterior compartment of the leg. *A*, Popliteus and flexor hallucis longus. *B*, Tibialis posterior. *C*, Flexor digitorum longus.

Table 5-8.
The Deep Muscles in the Posterior Compartment of the Leg

Muscle	Proximal Attachment	Distal Attachment	Innervation	Main Actions
Popliteus (Figs. 5-48, 5-68*A* and 5-90)	Lateral surface of lateral condyle of femur and lateral meniscus	Posterior surface of tibia, superior to soleal line	Tibial nerve (**L4, L5,** and S1)	Weakly flexes knee and unlocks it (discussed on p. 452)
Flexor hallucis longus (Figs. 5-48 and 5-68*A*)	Inferior two-thirds of posterior surface of fibula and inferior part of interosseous membrane	Based of distal phalanx of great toe (hallux)	Tibial nerve (**S2** and S3)	Flexes great toe at all joints and plantarflexes foot; supports longitudinal arch of foot
Flexor digitorum longus (Figs. 5-48 and 5-68*C*)	Medial part of posterior surface of tibia, inferior to soleal line, and by a broad aponeurosis to fibula	Bases of distal phalanges of lateral four digits		Flexes lateral four digits and plantarflexes foot; supports longitudinal arch of foot
Tibialis posterior (Figs. 5-47, 5-48, and 5-68*B*)	Interosseous membrane, posterior surface of tibia inferior to soleal line, and posterior surface of fibula	Tuberosity of navicular, cuneiform, and cuboid bones, and bases of second, third, and fourth metatarsal bones	Tibial nerve (L4 and L5)	Plantarflexes and inverts foot

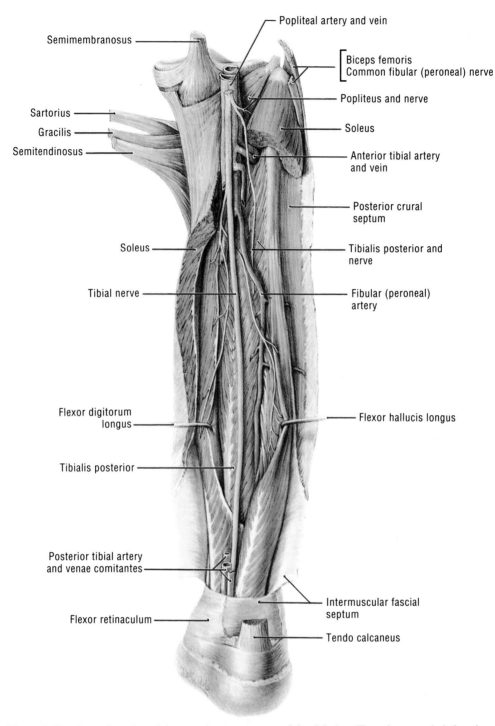

Figure 5-69. Deep dissection of the posterior compartment of the right leg. The soleus muscle is largely cut away, the two long digital flexors are pulled apart, and most of the posterior tibial artery is excised.

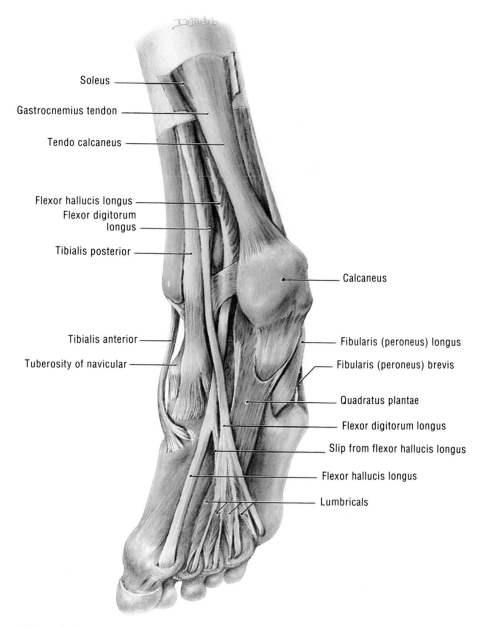

Figure 5-70. Dissection of the distal part of the right leg and foot, displaying the second layer of plantar muscles.

during walking, running, and jumping. It provides much of the spring to the step. Aware of this, successful sprinters prepare for a quick start by bracing their great toe. The flexor hallucis longus is also important in holding the leg in the normal position on the foot.

The Flexor Digitorum Longus Muscle (Figs. 5-66 to 5-70). This long flexor of the lateral four toes lies medially and is closely attached to the tibia. It is smaller than the flexor hallucis longus, even though it moves four digits. Its attachments, nerve supply, and main actions are given in Table 5-8. Its tendon runs inferiorly, passing posterior to the tibialis posterior tendon and the medial malleolus. It then passes diagonally in the sole of the foot, superficial to the tendon of the flexor hallucis longus. As the tendon reaches the middle of the sole,

it divides into four tendons, which pass to the distal phalanges of the lateral four digits.

The Tibialis Posterior Muscle (Figs. 5-67 to 5-70 and 5-75). This large, fusiform muscle is the deepest one in the posterior crural compartment. It lies between the flexor digitorum longus and the flexor hallucis longus in the same plane as the tibia and fibula. Its tendon can be seen and felt posterior to the medial malleolus, especially when the foot is inverted against resistance. Its attachments, nerve supply, and main actions are given in Table 5-8.

The Tibial Nerve (Figs. 5-45, 5-50, 5-67, 5-69, 5-72, and 5-73). This nerve *supplies all muscles in the posterior compartment* of the leg (Tables 5-7 and 5-8). As the larger terminal branch of the sciatic nerve, the tibial nerve arises from the an-

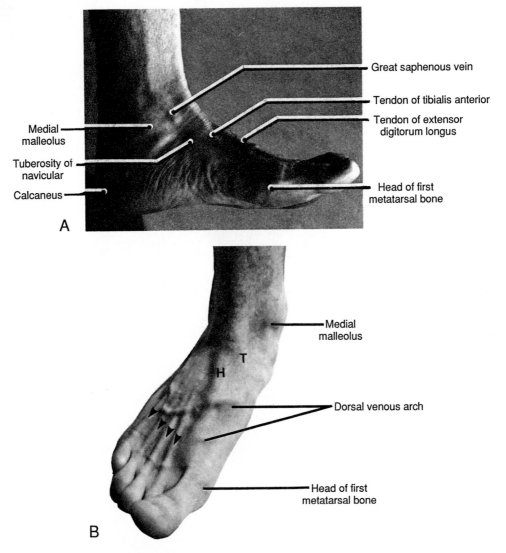

Figure 5-71. *A*, Medial aspect of the left ankle and foot of a 45-year-old man. His foot is dorsiflexed. *B*, Dorsal aspect of the right foot. The arrows show the individual tendons of the extensor digitorum longus muscle. Observe the tendon of the extensor hallucis longus (*H*) passing to the base of the distal phalanx of the great toe. The tendon of the tibialis anterior muscle (*T*) disappears as it approaches its distal attachments to the medial cuneiform bone and the base of the first metatarsal bone.

terior branches of the ventral rami of L4 to S3. The tibial nerve descends through the middle of the popliteal fossa, posterior to the popliteal vein and artery. At the distal border of the popliteus muscle, the tibial nerve passes with the posterior tibial vessels deep to the tendinous arch of the soleus muscle. It then descends straight down the median plane of the calf, deep to the soleus. It runs inferiorly on the tibialis posterior muscle, in company with the posterior tibial vessels. The tibial nerve leaves the posterior compartment of the leg by passing deep to the flexor retinaculum in the interval between the medial malleolus and calcaneus. The tibial nerve lies between the posterior tibial vessels and the tendon of the flexor hallucis longus muscle. Posteroinferior to the medial malleolus, *the tibial nerve divides into the medial and lateral plantar nerves* (Figs. 5-72 and 5-73*B*).

The tibial nerve gives branches to all muscles in the posterior compartment of the leg. A cutaneous branch of the tibial, the *medial sural cutaneous nerve*, usually unites with the commu-

nicating branch of the common fibular (peroneal) nerve to form the *sural nerve* (Fig. 5-45). This nerve supplies the skin of the lateral and posterior part of the inferior third of the leg and the lateral side of the foot (Figs. 5-7 and 5-82). Articular branches of the tibial nerve supply the knee joint and medial calcaneal branches supply the skin of the heel, including the weightbearing surface (Figs. 5-72 and 5-76).

Because the tibial nerve is deep and well protected, it is not commonly injured. However, lacerations in the popliteal fossa and posterior dislocations of the knee joint may damage this nerve, producing paralysis of all muscles in the posterior compartment of the leg and the intrinsic muscles in the sole of the foot (Tables 5-8 and 5-9). When the plantarflexors of the foot are paralyzed, the patient is unable to curl the toes or stand on them. In addition, there is loss of sensation in the sole of the foot, making it vulnerable to the development of pressure sores.

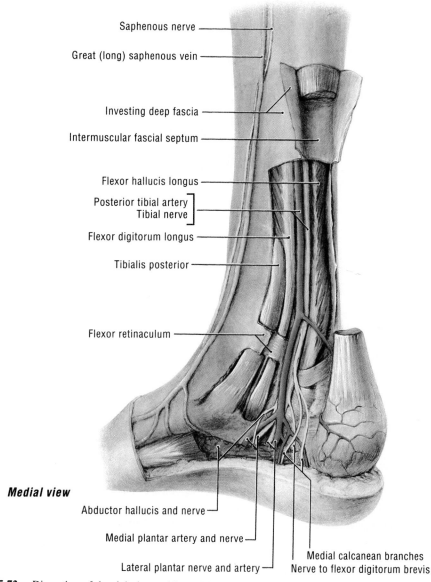

Saphenous nerve

Great (long) saphenous vein

Investing deep fascia

Intermuscular fascial septum

Flexor hallucis longus

Posterior tibial artery
Tibial nerve

Flexor digitorum longus

Tibialis posterior

Flexor retinaculum

Medial view

Abductor hallucis and nerve

Medial plantar artery and nerve

Lateral plantar nerve and artery

Medial calcanean branches
Nerve to flexor digitorum brevis

Figure 5-72.   Dissection of the right leg, ankle, and heel. The posterior part of the abductor hallucis muscle is excised, as is most of the tendo calcaneus or calcaneal tendon.

The Posterior Tibial Artery (Figs. 5-50, 5-67, 5-69, 5-72, and 5-74).   This vessel, the larger *terminal branch of the popliteal artery*, begins at the distal border of the popliteus muscle. The posterior tibial artery passes deep to the origin of the soleus muscle, and after giving off the fibular (peroneal) artery, its largest branch, it passes inferomedially on the posterior surface of the tibialis posterior muscle. During its descent, it is accompanied by the tibial nerve and two venae comitantes, deep to the transverse crural intermuscular septum. At the ankle the posterior tibial artery runs posterior to the medial malleolus, from which it is separated by the tendons of the tibialis posterior and flexor digitorum longus muscles. Inferior to the medial malleolus, it

runs between the tendons of the flexor hallucis longus and flexor digitorum longus muscles. Deep to flexor retinaculum and the origin of the abductor hallucis muscle, the posterior tibial artery divides into medial and lateral plantar arteries.

Branches of Posterior Tibial Artery (Figs. 5-67, 5-69, and 5-74).   Most of the muscular branches of the posterior tibial artery are unnamed, but its most important one is the fibular artery.

The **fibular (peroneal) artery**, the largest and most important branch of the posterior tibial artery, begins inferior to the distal border of the popliteus muscle and the tendinous arch of the soleus (Fig. 5-69). It descends obliquely toward the fibula and

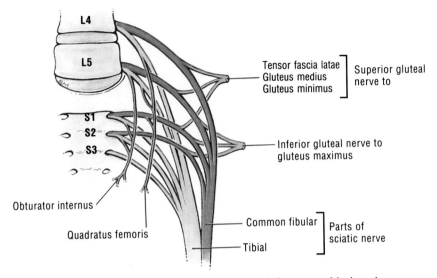

Figure 5-73.   *A*, Formation of the sciatic nerve in the pelvis. *B*, The sciatic nerve and its branches.

passes along its medial side within the flexor hallucis longus muscle, or between it and the intermuscular septum and tibialis posterior muscle. The fibular artery gives off muscular branches to the popliteus and other muscles in the posterior and lateral compartments of the leg. It also supplies a *nutrient artery to the fibula* and a communicating branch, which joins that of the posterior tibial artery (Fig. 5-74). The fibular artery usually pierces the interosseous membrane and passes to the dorsum of the foot, where it anastomoses with the *arcuate artery* (Fig. 5-60).

The *circumflex fibular artery*, a branch of the posterior tibial, arises from the posterior tibial at the knee and passes laterally over the neck of the fibula to the anastomoses around the knee (Fig. 5-46).

The **nutrient artery of the tibia**, the largest nutrient artery in the body, arises from the posterior tibial artery near its origin (Figs. 5-29*B* and 5-74). The nutrient foramen through which it passes is just distal to the soleal line on the posterior surface of the tibia (Fig. 5-47). Other branches of the posterior tibial artery are the *calcanean arteries*, which supply the tissues of the heel (Figs. 5-72 and 5-74). They pass medial and posterior to the tendo calcaneus and anastomose with branches of the fibular artery. A *malleolar branch* joins the network of vessels on the medial malleolus (Fig. 5-29).

> Absence of the posterior tibial artery with compensatory enlargement of the fibular (peroneal) artery occurs in about 5% of people. The *pulse of the posterior tibial artery* can usually be palpated about halfway between the posterior surface of the medial malleolus and the medial border of the tendo calcaneus (Figs. 5-72 and 5-75). Knowing how to palpate this pulse is essential for examining patients with a condition called *intermittent claudication*, which is caused by ischemia of the leg muscles owing to arteriosclerotic stenosis or *occlusion of the leg arteries*. Intermittent claudication is characterized by leg cramps that develop during walking and disappear soon after rest.

The Foot

This is the part of the lower limb distal to the ankle joint and is *concerned mainly with support and locomotion* of the body. The bones of the foot are described and illustrated on pages 437 to 441. The clinical importance of the foot is indicated by an estimate that the average orthopaedic surgeon devotes about 20% of his/her practice to foot problems, and the entire practice of *podiatry* is concerned with the diagnosis and treatment of diseases, injuries, and abnormalities of the foot.

Surface Anatomy of the Foot Bones

The Talus (Figs. 5-51 to 5-55 and 5-71).   Its head is often visible and is *palpable in two places*: (1) anteromedial to the proximal part of the lateral malleolus on inversion of the foot and (2) anterior to the medial malleolus on eversion of the foot. The head occupies the space between the sustentaculum tali and the navicular tuberosity. When the foot is plantarflexed (Fig. 5-61*B*), the superior surface of the body of the talus can be palpated on the anterior aspect of the ankle, anterior to the inferior end of the tibia.

The Calcaneus (Figs. 5-47, 5-49, 5-53, 5-71, 5-75, and 5-76).   The posterior, medial, and lateral surfaces of this large *heel bone* can be easily palpated, but its inferior surface is not easily felt owing to the overlying plantar aponeurosis and pad of fat. The *sustentaculum tali* can be felt as a small prominence distal to the tip of the medial malleolus. The *fibular (peroneal) trochlea* may be detectable as a small tubercle on the lateral aspect of the calcaneus. It lies anteroinferior to the tip of the

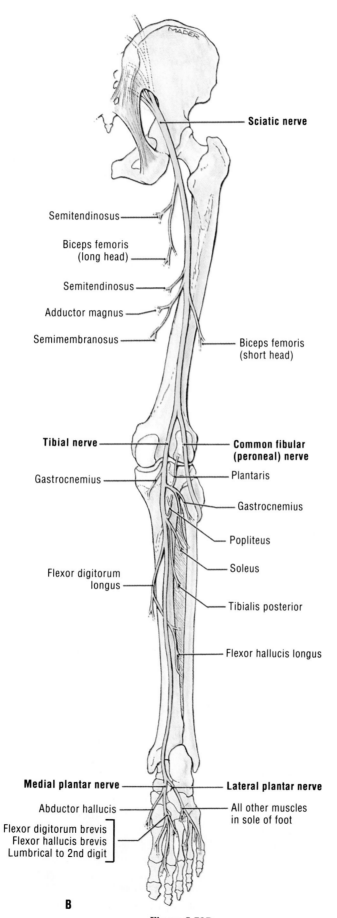

Sciatic nerve

Semitendinosus

Biceps femoris
(long head)

Semitendinosus

Adductor magnus

Semimembranosus

Biceps femoris
(short head)

Tibial nerve

**Common fibular
(peroneal) nerve**

Gastrocnemius

Plantaris

Gastrocnemius

Popliteus

Soleus

Flexor digitorum
longus

Tibialis posterior

Flexor hallucis longus

Medial plantar nerve

Lateral plantar nerve

Abductor hallucis

All other muscles
in sole of foot

Flexor digitorum brevis
Flexor hallucis brevis
Lumbrical to 2nd digit

B

Figure 5-73*B*

lateral malleolus. Evert your foot and palpate the tendons of the fibularis (peroneus) longus and brevis, which are separated by this small tubercle.

The Navicular (Figs. 5-53 and 5-71). The tuberosity of the navicular bone is easily seen and palpated on the medial aspect of the foot, inferoanterior to the tip of the medial malleolus. It is a prominent and important bony landmark in the foot. Actively

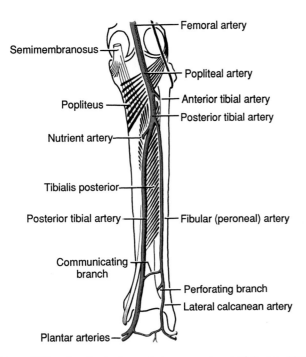

Figure 5-74. Arteries in the posterior compartment of the leg. Note that the posterior tibial artery, the larger of the two terminal branches of the popliteal artery, begins at the inferior border of the popliteus muscle.

invert your foot and palpate the tendon of the tibialis posterior muscle (Fig. 5-75), which attaches to this tuberosity.

The Cuboid and Cuneiforms (Figs. 5-51 and 5-55). These bones are difficult for most people to identify individually by palpation. The cuboid can be felt somewhat indistinctly on the lateral aspect of the foot, posterior to the base of the fifth metatarsal. The medial cuneiform can be indistinctly palpated between the tuberosity of the navicular and the base of the first metatarsal bone.

The Metatarsal Bones (Figs. 5-54, 5-55, 5-71, and 5-75). The head of the first metatarsal bone forms a prominence on the medial aspect of the foot. The medial and lateral sesamoids inferior to the head of this metatarsal can be felt to slide when the great toe is moved passively. The base of the fifth metatarsal forms a prominent landmark on the lateral aspect of the foot and the large tuberosity of the fifth metatarsal can easily be palpated at the midpoint of the lateral border of the foot. In some people it even produces a prominence in their shoe. The bodies of the metatarsals can be felt indistinctly on the dorsum of the foot between the extensor tendons.

The Phalanges (Fig. 5-54). The dorsal surfaces of these bones can be felt indistinctly through the extensor tendons. *Fractures of the phalanges* most often result from heavy objects falling on them or from stubbing the bare toes.

Skin of the Foot

The skin on the dorsal surface or dorsum of the foot is thin and mobile; hair is sparse and there is relatively little subcutaneous fat. Because of the thinness of the skin and superficial fascia on the dorsum of the foot, the tendons are usually visible, especially during dorsiflexion (Fig. 5-71). The skin of the dorsum of the foot is supplied mainly by the superficial fibular (peroneal) nerve (Fig. 5-10).

The skin on the plantar surface or sole of the foot is thin on the toes and instep, but it is thick over the heel and ball of the

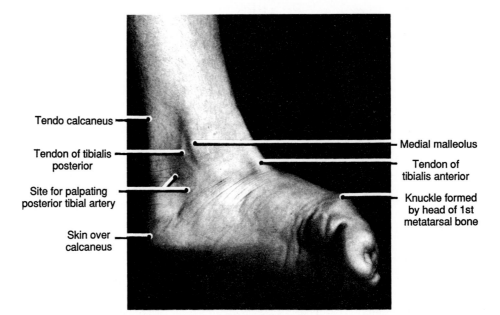

Figure 5-75. Medial aspect of the left ankle of a 12-year-old girl whose foot is inverted.

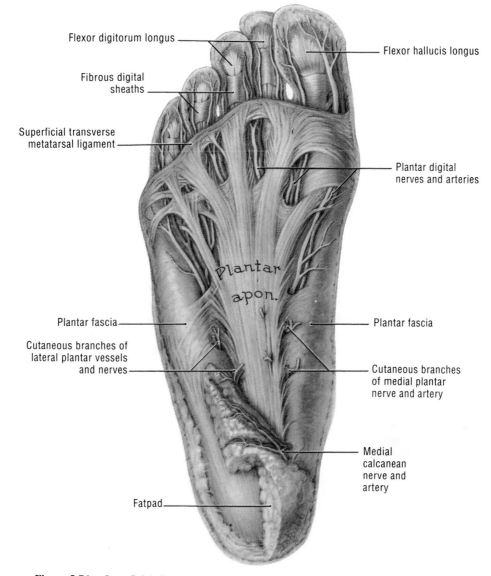

Flexor digitorum longus

Flexor hallucis longus

Fibrous digital sheaths

Superficial transverse metatarsal ligament

Plantar digital nerves and arteries

Plantar fascia

Plantar fascia

Cutaneous branches of lateral plantar vessels and nerves

Cutaneous branches of medial plantar nerve and artery

Medial calcanean nerve and artery

Fatpad

Figure 5-76.   Superficial dissection of the plantar aspect or sole (L. *planta*) of the right foot.

foot (head of great toe). The plantar skin contains many sweat glands and much fat in the subcutaneous tissue, especially over the heel (Figs. 5-75 and 5-76), which is firmly bound down to underlying structures by fibrous connective tissue. The sole of the foot is designed for weightbearing and protection of the underlying nerves and vessels.

Deep Fascia of the Foot

This fascia is continuous with that of the ankle. It is thin on the dorsum of the foot (Fig. 5-72), where it is continuous with the inferior extensor retinaculum (Fig. 5-58). Over the lateral and posterior aspects of the foot, the deep fascia is continuous with the plantar fascia or deep fascia of the sole.

The Plantar Aponeurosis (Fig. 5-76).   The central part of the plantar fascia is greatly thickened to form the plantar apo-

neurosis. It consists of a strong, *thick central part* and weaker and thinner medial and lateral portions. The plantar aponeurosis, which covers the whole length of the sole, consists of *longitudinally arranged bands of dense fibrous connective tissue*. It helps to support the longitudinal arches of the foot and to hold the parts of the foot together. It arises posteriorly from the tuber calcanei (Fig. 5-51) and fans out over the sole, where it becomes broader and somewhat thinner. The plantar aponeurosis *divides into five bands* that split to enclose the digital tendons. They are attached to the margins of the fibrous digital sheaths and to the sesamoids of the great toe (Fig. 5-81). From the margins of the central part of the plantar aponeurosis, vertical septa extend deeply to form **three compartments of the sole** of the foot: a *medial compartment*, a *lateral compartment*, and a *central compartment*. The muscles, nerves, and vessels in the sole may be described according to these compartments, but the muscles are usually dissected and described by layers (Figs. 5-77 and 5-79; Table 5-9).

Muscles of the Foot

Muscles on the Dorsum of the Foot

There are two closely connected muscles on the dorsum of the foot, the extensor digitorum brevis and extensor hallucis brevis. The latter muscle is part of the extensor digitorum brevis.

The Extensor Digitorum Brevis and Extensor Hallucis Brevis Muscles (Figs. 5-56 to 5-60). These broad thin muscles form a fleshy mass on the lateral part of the dorsum of the foot, anterior to the lateral malleolus, which can be seen in most feet and felt in all of them.

Proximal Attachments (Figs. 5-51 and 5-58). The two muscles are attached to the anterior part of the dorsal surface of the calcaneus, anteromedial to the lateral malleolus. They are also attached to the inferior extensor retinaculum.

Distal Attachments (Fig. 5-57). These muscles attach by four tendons. The most medial tendon is the extensor hallucis brevis. It is attached to the base of the proximal phalanx of the great toe. The other three tendons are those of the extensor digitorum brevis, which attach to the lateral edge of the corresponding tendons of the extensor digitorum longus to the second to fourth digits.

Innervation (Figs. 5-57 and 5-60). Deep fibular (peroneal) nerve (S1 and S2).

Actions. The extensor digitorum brevis extends the second to fourth digits at the metatarsophalangeal joints, and the extensor hallucis brevis extends the first digit or great toe at the metatarsophalangeal joint. These muscles help the long extensor muscles extend the toes (Tables 5-5 and 5-14).

> Functionally the extensor digitorum brevis and extensor hallucis brevis muscles are relatively unimportant. Probably the only clinical reason for knowing about their presence is that contusion and tearing of their fibers result in a *hematoma*, which produces a swelling anteromedial to the lateral malleolus. Most people who have not seen these swollen muscles before think they indicate a badly sprained ankle, but state that the ankle joint is not sore.

Muscles in the Sole of the Foot

There are **four muscular layers** in the sole of the foot (Figs. 5-77 and 5-79; Table 5-9). They are specialized to help maintain the arches of the foot and to enable one to stand on uneven ground. Consequently, these muscles have gross functions rather than delicate individual functions like those in the hand. As a result, several muscles in the foot have names implying functions that they rarely perform or are unable to perform (Table 5-9). The muscles in the sole of the foot are of little importance individually because the fine control of the individual toes is not important to most people. There are **two neurovascular planes** in the foot (Figs. 5-76 to 5-78 and 5-80): a *superficial one* between the first and second muscular layers and a *deep one* between the third and fourth muscular layers.

The First Layer of Plantar Muscles (Figs. 5-77 to 5-79; Table 5-9). This superficial layer *contains three short muscles*, all of which extend from the posterior part of the calcaneus to the phalanges. It contains the abductor of the great and small digits and the short flexor of the lateral four digits. These muscles comprise a functional group that acts as an elastic spring for supporting the arches of the foot and maintaining the concavity of the foot (p. 494).

The Abductor Hallucis Muscle (Figs. 5-52, 5-77, 5-78, and 5-79A). This abductor of the great toe (L. *hallux*) lies superficially along the medial border of the foot. Its attachments, nerve supply, and main actions are given in Table 5-9. When the foot is bearing weight, it supports the medial longitudinal arch of the foot.

The Flexor Digitorum Brevis Muscle (Figs. 5-52 and 5-77 to 5-79B). This short flexor of the toes lies between the abductor hallucis and abductor digiti minimi muscles. Its attachments, nerve supply, and main action are given in Table 5-9. When the foot is bearing weight, it supports the medial and lateral longitudinal arches of the foot.

The Abductor Digiti Minimi Muscle (Figs. 5-52 and 5-78 to 5-80). This abductor of the small digit is the most lateral of the three muscles in the first layer. Its attachments, nerve supply, and main actions are given in Table 5-9. When the foot is bearing weight, it supports the lateral longitudinal arch of the foot.

The Second Layer of Plantar Muscles (Figs. 5-70, 5-77, 5-79; Table 5-9). This layer, located deep to the first layer, consists of the quadratus plantae and lumbrical muscles. The long tendons of two leg muscles (the flexor hallucis longus and flexor digitorum longus) are also located in this layer. The tendon of the flexor hallucis longus muscle crosses deep to the tendon of the flexor digitorum longus as it passes to the first digit or great toe.

The Quadratus Plantae Muscle (Figs. 5-52, 5-53, 5-70, 5-79C, and 5-80). This small, flat muscle, also known as *flexus accessorius*, joins the tendon of the flexor digitorum longus to

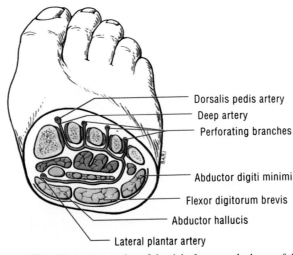

Figure 5-77. Transverse section of the right foot near the bases of the metatarsal bones (see also Fig. 5-54), showing the plantar arteries and the four layers of muscles in the sole of the foot. (Also see Fig. 5-79; Table 5-9).

Dorsalis pedis artery
Deep artery
Perforating branches
Abductor digiti minimi
Flexor digitorum brevis
Abductor hallucis
Lateral plantar artery

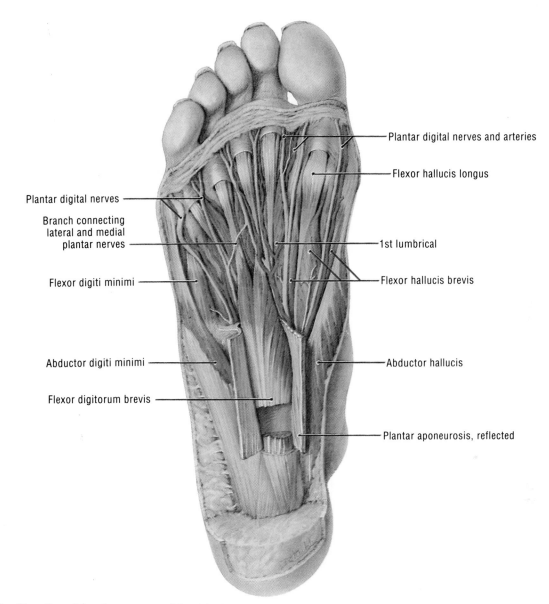

Plantar digital nerves and arteries

Flexor hallucis longus

Plantar digital nerves

Branch connecting lateral and medial plantar nerves

1st lumbrical

Flexor digiti minimi

Flexor hallucis brevis

Abductor digiti minimi

Abductor hallucis

Flexor digitorum brevis

Plantar aponeurosis, reflected

Figure 5-78. Dissection of the plantar aspect of the right foot demonstrating the *first layer of plantar muscles* and the digital nerves and arteries. The plantar aponeurosis and fascia are reflected or removed, and a section is removed from the flexor digitorum brevis.

the calcaneus and forms a fleshy sheet of muscle in the posterior half of the foot. Its two heads of attachment embrace the calcaneus. Its attachments, nerve supply, and main actions are given in Table 5-9. This muscle assists the flexor digitorum longus muscle in flexing the lateral four digits by adjusting the pull of the flexor digitorum longus (*i.e.*, it brings its tendons more directly in line with the long axes of the digits).

The Lumbrical Muscles (Figs. 5-70, 5-78, and 5-79*C*). There are four of these small, wormlike muscles called lumbricals (L. *lumbricus*, earthworm). Their attachments, nerve supply, and main actions are given in Table 5-9.

The Third Layer of Plantar Muscles (Figs. 5-79 to 5-81; Table 5-9). This layer of three muscles consists of the short muscles of the great and small digits, which lie in the anterior half of the sole of the foot. Two act on the great digit, one on the little digit.

The Flexor Hallucis Brevis Muscle (Figs. 5-79*D* to 5-81). This fleshy muscle has two heads that cover the plantar surface of the first metatarsal bone. Its attachments, nerve supply, and main actions are given in Table 5-9. A *sesamoid bone* adheres to each of its tendons. These bones protect the tendons from pressure from the head of the first metatarsal bone during standing and walking. This muscle also prevents excessive extension at the metatarsophalangeal of the great digit.

The Adductor Hallucis Muscle (Figs. 5-79*D* to 5-81). This flat, triangular adductor of the great toe has two heads. Its attachments, nerve supply, and main actions are given in Table 5-9. It adducts the great digit (*i.e.*, moves it toward the second digit) and assists in flexing the metatarsophalangeal joint. It also helps to maintain the transverse arch of the foot.

The Flexor Digiti Minimi Brevis Muscle (Figs. 5-79*D*, 5-80, and 5-81). This slender, relatively insignificant muscle

Table 5-9.
The Muscles in the Sole of the Foot

Muscle	Proximal Attachment	Distal Attachment	Innervation	Main Actions
First Layer				
Abductor hallucis (Fig. 5-79A)	Medial process of tuber calcanei, flexor retinaculum, and plantar aponeurosis	Medial side of base of proximal phalanx of great toe (hallux)		Abducts and flexes great toe
Flexor digitorum brevis (Fig. 5-79B)	Medial process of tuber calcanei, plantar aponeurosis, and intermuscular septa	Both sides of middle phalanges of lateral four digits	Medial plantar nerve (S2 and **S3**)	Flexes lateral four digits (toes)
Abductor digiti minimi (Fig. 5-79A)	Medial and lateral processes of tuber calcanei, plantar aponeurosis and intermuscular septa	Lateral side of base of proximal phalanx of fifth digit (little toe)	Lateral plantar nerve (S2 and **S3**)	Abducts and flexes fifth digit
Second Layer				
Quadratus plantae (Fig. 5-79C)	Medial surface and lateral margin of plantar surface of calcaneus	Posterolateral margin of tendon of flexor digitorum longus	Lateral plantar nerve (S2 and **S3**)	Assists flexor digitorum longus in flexing lateral four digits
Lumbricals (Fig. 5-79C)	Tendons of flexor digitorum longus	Medial sides of bases of proximal phalanges of lateral four digits and extensor expansions of tendons of extensor digitorum longus	*Medial one:* medial plantar nerve (S2 and **S3**) *Lateral three:* lateral plantar nerve (S2 and **S3**)	Flex proximal phalanges and extend middle and distal phalanges of lateral four digits
Third Layer				
Flexor hallucis brevis (Fig. 5-79D)	Plantar surfaces of cuboid and lateral cuneiform bones	Both sides of base of proximal phalanx of great toe	Medial plantar nerve (S1 and **S2**)	Flexes proximal phalanx of great toe (hallux)
Adductor hallucis (Fig. 5-79D)	*Oblique head:* Bases of metatarsals 2–4 *Transverse head:* Plantar ligaments of metatarsophalangeal joints	*Tendons of both heads* attached to lateral side of base of proximal phalanx of great toe (hallux)	Deep branch of lateral plantar nerve (S2 and **S3**)	Adducts great toe; assists in maintaining transverse arch of foot
Flexor digiti minimi brevis (Fig. 5-79D)	Base of fifth metatarsal bone	Base of proximal phalanx of fifth digit	Superficial branch of lateral plantar nerve (S2 and **S3**)	Flexes proximal phalanx of fifth digit, thereby assisting with its flexion
Fourth Layer				
Plantar interossei (3 muscles) (Fig. 5-79E)	Bases and medial sides of metatarsal bone 3–5	Medial sides of bases of proximal phalanges of digits 3–5		Adduct digits (2–4) and flex metatarsophalangeal joints
Dorsal interossei (4 muscles) (Fig. 5-79F)	Adjacent sides of metatarsal bones 1–5	*1st:* medial side of proximal phalanx of second digit *2nd–4th:* lateral sides of digits 2–4	Lateral plantar nerve (S2 and **S3**)	Abduct digits (2–4) and flex metatarsophalangeal joints

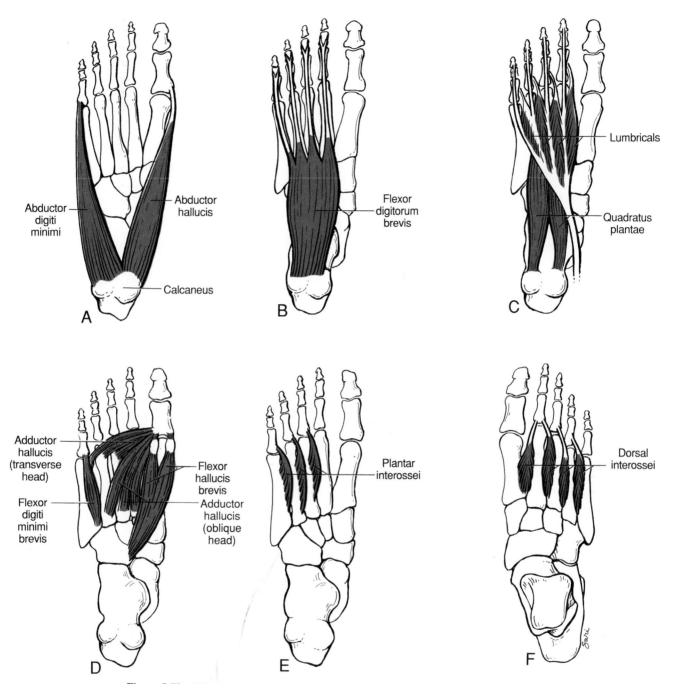

Figure 5-79. Muscle layers in the sole of the foot. *A* and *B*, First layer. *C*, Second layer.
D, Third layer. *E* and *F*, Fourth layer.

is only a fleshy slip. Its attachments, nerve supply, and action are given in Table 5-9.

The Fourth Layer of Plantar Muscles (Figs. 5-77, 5-79*E* and *F*, and 5-81). This layer consists of the interosseous muscles or *interossei* and the tendons of the fibularis (peroneus) longus and tibialis posterior muscles, which cross the sole of the foot to reach their distal attachments.

The Interosseous Muscles (Figs. 5-57, 5-79*E* and 5-81). There are three plantar and four dorsal interossei, and as their name indicates, these slender muscles *occupy the intermetatarsal spaces* **between the metatarsal bones**. The dorsal interossei are

larger than the plantar interossei and are attached proximally by two heads. Their attachments, nerve supply, and main actions are given in Table 5-9. The plantar interossei adduct the digits (*PAD* is the key, *Plantar ADduct*), and the dorsal interossei abduct the digits (*DAB* is the key, *Dorsal ABduct*). Adduction is moving the digits toward the second digit and abduction is moving the digits away from the second digit. These actions are not important to most people; however, the interossei maintain the integrity of the forefoot by approximating the bones during weightbearing. The interossei also flex the toes at the metatarsophalangeal joints.

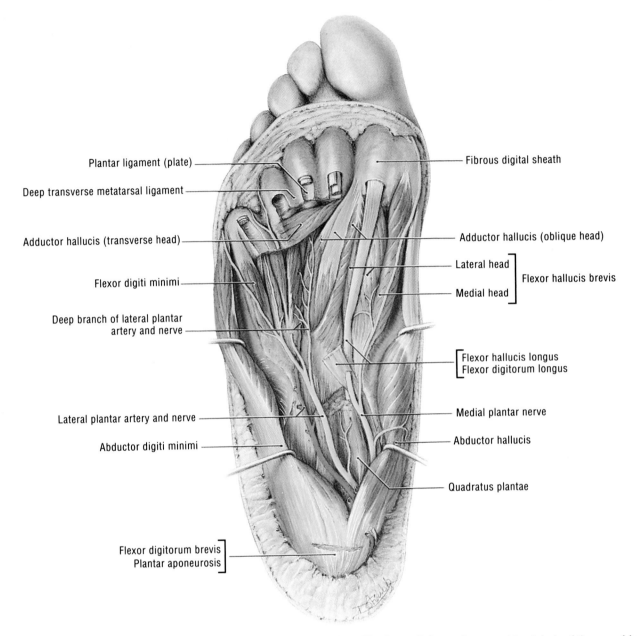

Plantar ligament (plate)

Deep transverse metatarsal ligament

Adductor hallucis (transverse head)

Flexor digiti minimi

Deep branch of lateral plantar artery and nerve

Lateral plantar artery and nerve

Abductor digiti minimi

Flexor digitorum brevis
Plantar aponeurosis

Fibrous digital sheath

Adductor hallucis (oblique head)

Lateral head
Medial head } Flexor hallucis brevis

Flexor hallucis longus
Flexor digitorum longus

Medial plantar nerve

Abductor hallucis

Quadratus plantae

Figure 5-80. Dissection of the *third layer of plantar muscles* in the right foot. The abductor digiti minimi and the abductor hallucis muscles of the first layer are pulled aside and the flexor digitorum brevis is cut short. The flexor digitorum longus and lumbricals of the second layer are excised and the quadratus plantae is cut.

Nerves of the Foot

The tibial nerve divides posterior to the medial malleolus into medial and lateral *plantar nerves* (Figs. 5-11, 5-67, 5-72, 5-73*B*, 5-78, 5-80, and 5-82). They supply the intrinsic muscles of the foot, except for the extensor digitorum brevis, which is supplied by the deep fibular (peroneal) nerve. These nerves also supply the skin of the foot.

The Medial Plantar Nerve (Figs. 5-11, 5-67, 5-72, 5-73*B*, 5-76, 5-78, 5-80, and 5-82). This is the larger of the two terminal branches of the **tibial nerve**. It passes deep to the abductor hallucis muscle and runs anteriorly between this muscle and the flexor digitorum brevis on the lateral side of the medial plantar artery. The medial plantar nerve terminates near the bases of the metatarsal bones by dividing into three sensory branches (which supply cutaneous branches to the medial three and a half digits), and motor branches to the abductor hallucis, flexor digitorum brevis, flexor hallucis brevis muscles, and the most medial lumbrical muscle.

The Lateral Plantar Nerve (Figs. 5-11, 5-67, 5-72, 5-73*B*, 5-76, 5-78, 5-80, and 5-82). This is the smaller of the two terminal branches of the **tibial nerve**. It begins deep to the flexor retinaculum and the abductor hallucis muscle and runs antero-laterally, medial to the lateral plantar artery and between the first and second layers of plantar muscles. The lateral plantar nerve terminates by dividing into superficial and deep branches. The superficial branch divides into two digital nerves which send cutaneous branches to the lateral one and a half digits. The superficial and deep branches of the lateral plantar nerve supply motor branches to muscles of the sole that are not supplied by the medial plantar nerve (Table 5-9).

The Sural Nerve (Figs. 5-45 and 5-82).   This nerve usually forms in the popliteal fossa and descends between the two heads of the gastrocnemius muscle. It pierces the deep fascia around the middle of the posterior aspect of the leg, where it is joined by the fibular (peroneal) communicating branch of the common fibular (peroneal) nerve. The sural nerve supplies the skin on the lateral and posterior part of the inferior one-third of the leg. It enters the foot posterior to the lateral malleolus and supplies the skin along the lateral margin of the foot and the lateral side of the fifth digit.

The Saphenous Nerve (Figs. 5-11, 5-30, and 5-82).   This is the largest cutaneous branch of the femoral nerve. In addition to supplying skin and fascia on the anterior and medial sides of the leg, the saphenous nerve passes to the dorsum of the foot, anterior to the medial malleolus. It supplies the skin along the medial side of the foot as far anteriorly as the head of the first metatarsal bone.

Arteries of the Foot

The arteries of the foot are the terminal branches of the anterior and posterior tibial arteries.

Arteries of the Dorsum of the Foot

The Dorsalis Pedis Artery (Figs. 5-57, 5-60, 5-77, 5-83, and 5-84).   This vessel is the direct continuation of the anterior tibial artery distal to the ankle joint. It begins midway between the malleoli and runs anteromedially, deep to the inferior extensor retinaculum, to the posterior end of the first interosseous space. Here it divides into a deep plantar artery, which passes

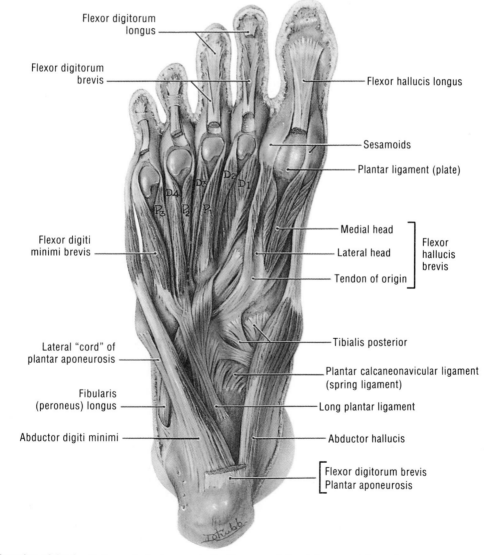

Figure 5-81.   Dissection of the *fourth layer of plantar muscles* in the right foot. The abductor and flexor brevis of the fifth digit (small toe) and the abductor and flexor brevis of the first digit (great toe) of the first and third layers of muscles remain for orientation.

Figure 5-82.   The cutaneous distribution of the nerves of the distal part of the leg and foot (also see Fig. 5-10). *A*, Leg and dorsum of the right foot. *B*, Plantar surface of the left foot.

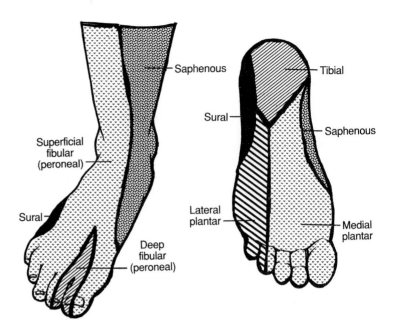

Figure 5-83.   Arteries of the dorsum of the right foot.

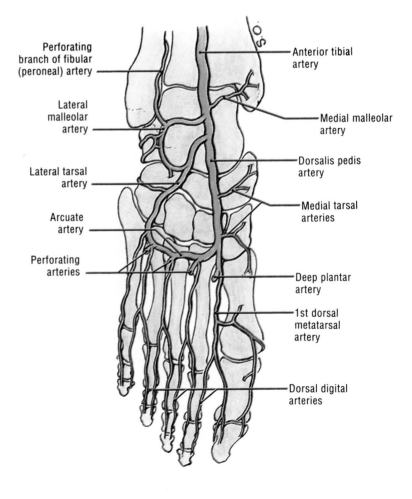

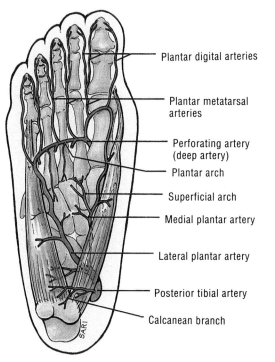

Figure 5-84. Plantar arteries in the right foot.

Labels on figure:
Plantar digital arteries
Plantar metatarsal arteries
Perforating artery (deep artery)
Plantar arch
Superficial arch
Medial plantar artery
Lateral plantar artery
Posterior tibial artery
Calcanean branch

to the sole of the foot, and an arcuate artery. The *deep plantar artery* passes deeply through the first interosseous space to join the lateral plantar artery and form the *deep plantar arch*. The *arcuate artery* runs laterally across the bases of the metatarsal bones, deep to the extensor tendons, where it gives off the second, third, and fourth dorsal metatarsal arteries. These vessels run to the clefts of the toes where each of them divides into two dorsal digital arteries for the sides of adjoining toes (Fig. 5-83).

> Palpation of the **dorsalis pedis pulse** is essential, particularly in suspected cases of *intermittent claudication* (cramps in the calf brought on by exercise and relieved by rest). The dorsalis pedis pulse can usually be felt on the dorsum of the foot, where the artery passes over the navicular and cuneiform bones just lateral to the extensor hallucis longus tendon (Figs. 5-57 and 5-60). It may also be felt distal to this at the proximal end of the first interosseous space (Fig. 5-83). A diminished or absent dorsalis pedis pulse suggests *arterial insufficiency*. In 14% of people the dorsalis pedis artery is absent or is too small to palpate, or it may not be in its usual position. Consequently, failure to detect a dorsalis pedis pulse does not always indicate the presence of arteriosclerotic disease.

Arteries in the Sole of the Foot

These arteries are derived from the posterior tibial artery. It divides deep to the abductor hallucis muscle to form the medial and lateral plantar arteries, which run parallel to the similarly named nerves.

The Medial Plantar Artery (Figs. 5-29*B*, 5-67, 5-72, 5-78, and 5-84). This vessel, the smaller of the two terminal branches of the posterior tibial artery, arises deep to the flexor retinaculum, midway between the medial malleolus and the prominence of the heel. It passes distally on the medial side of the foot between

the abductor hallucis and flexor digitorum brevis muscles. The medial plantar artery supplies branches to the medial side of the great digit and gives off muscular, cutaneous, and articular branches during its course.

The Lateral Plantar Artery (Figs. 5-29*B*, 5-77, 5-80, and 5-84). This is the larger of the two terminal branches of the posterior tibial artery. It arises deep to the flexor retinaculum and runs obliquely across the sole of the foot on the lateral side of the lateral plantar nerve, between the flexor digitorum brevis and quadratus plantae muscles. The lateral plantar artery gives off calcaneal, cutaneous, muscular, and articular branches. When it reaches the base of the fifth metatarsal bone, it curves medially between the third and fourth muscular layers. It terminates at the point where it joins the deep plantar branch of the dorsalis pedis artery to form the plantar arterial arch.

The Plantar Arterial Arch (Figs. 5-29*B*, 5-77, 5-78, and 5-84). This arch begins at the base of the fifth metatarsal bone as the continuation of the lateral plantar artery. It is completed medially by union with the deep plantar artery, a branch of the dorsalis pedis artery. As it crosses the foot, the plantar arch gives off four *plantar metatarsal arteries*, three perforating arteries, and branches to the tarsal joints and the muscles in the sole of the foot. These arteries join with the superficial branches of the medial and lateral plantar arteries to form the *plantar digital arteries* (Fig. 5-84).

> Wounds of the foot involving the plantar arterial arch result in severe bleeding. Ligature of the arch is difficult owing to its depth and the structures that are related to it.

Veins of the Foot

The *dorsal digital veins* run along the dorsal margins of each toe and unite in their webs to form common dorsal digital veins (Figs. 5-7, 5-8, 5-71*B*, 5-72, and 5-85). These veins join to form a **dorsal venous arch** on the dorsum of the foot. Veins leave the dorsal venous arch and converge medially to form the *great saphenous vein* and laterally to form the *small saphenous vein*. The superficial veins of the sole unite to form a *plantar venous arch* from which efferents pass to medial and lateral marginal veins that join the great and small saphenous veins. The *deep veins of the sole* begin as plantar digital veins on the plantar aspects of the digits. They communicate with the dorsal digital veins via *perforating veins*. Most blood returns from the foot via the deep veins, which are connected with the superficial veins by perforating veins.

Joints of the Lower Limb

The joints of the lower limb include those of the *pelvic girdle*. It consists of the two hip bones, which connect the lower limbs to the trunk. The joints of the pelvis are described with the pelvis in Chap. 3 (p. 251).

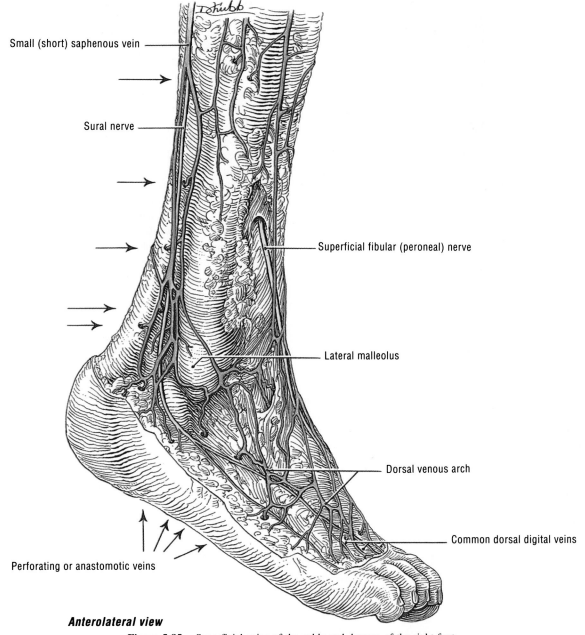

Small (short) saphenous vein

Sural nerve

Superficial fibular (peroneal) nerve

Lateral malleolus

Dorsal venous arch

Common dorsal digital veins

Perforating or anastomotic veins

Anterolateral view

Figure 5-85.   Superficial veins of the ankle and dorsum of the right foot.

The Hip Joint

This is a *multiaxial ball and socket type of synovial joint* (p. 19) between the head of the femur and the acetabulum of the hip bone.

Articular Surfaces of the Hip Joint (Figs. 5-2 and 5-86 to 5-89).   The globular head of the femur articulates with the cuplike acetabulum of the hip bone. The wide superior part of the articular surface is the weightbearing area. Thus it is the ilium that bears the weight. The rim of the acetabulum is defective inferiorly at the *acetabular notch*, which is bridged by the *transverse acetabular ligament*. The head of the femur forms about

two-thirds of a sphere and is covered with hyaline cartilage, except over the roughened *fovea* or pit, to which the ligament of the head of the femur is attached. More than half of the femoral head is contained within the acetabulum. The articular or *lunate surface of the acetabulum* is horseshoe-shaped. The acetabulum has a centrally located nonarticular acetabular fossa, which is occupied by a fatpad that is covered with synovial membrane. This nonarticular bone is paper thin and translucent.

The Acetabular Labrum (Figs. 5-86, 5-87, and 5-89).   The depth of the acetabulum is increased by this fibrocartilaginous labrum (L. lip). It is attached to the bony rim of the acetabulum and to the transverse acetabular ligament. The labrum deepens

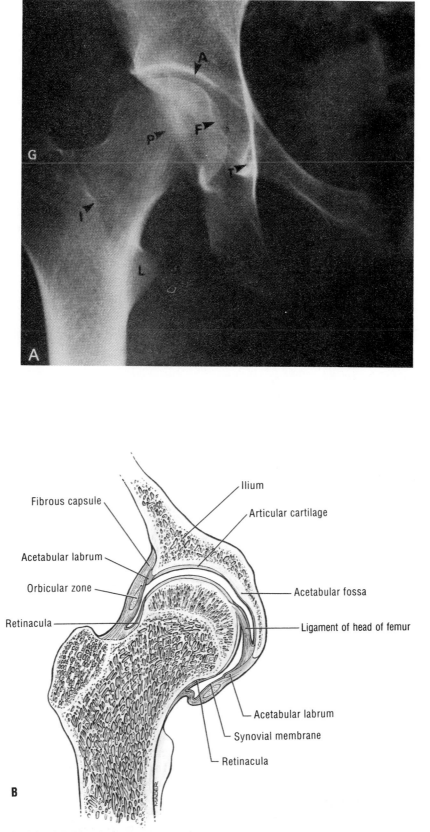

Figure 5-86. *A*, Radiograph of the right hip joint (AP projection). On the femur, observe the greater (*G*) and lesser (*L*) trochanters, the intertrochanteric crest (*I*), and the fovea (*F*) for the ligament of the head. On the pelvis, observe the roof (*A*) and posterior rim (*P*) of the acetabulum and the "teardrop" appearance (*T*) caused by the superimposition of structures at the inferior margin of the acetabulum. *B*, Coronal section of the hip joint.

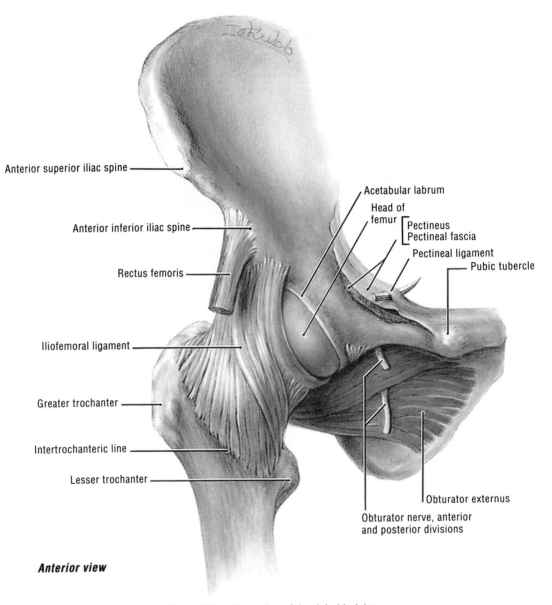

Anterior superior iliac spine

Anterior inferior iliac spine

Rectus femoris

Iliofemoral ligament

Greater trochanter

Intertrochanteric line

Lesser trochanter

Acetabular labrum

Head of femur

Pectineus
Pectineal fascia

Pectineal ligament

Pubic tubercle

Obturator externus

Obturator nerve, anterior and posterior divisions

Anterior view

Figure 5-87.   Dissection of the right hip joint.

the socket for the femoral head and its free thin edge clasps the head beyond its widest diameter. This helps to hold it firmly in the acetabulum (*i.e.*, preventing its dislocation).

Movements of the Hip Joint (Table 5-10).   The range of movement of the hip joint is decreased somewhat to provide stability and strength. Its range of mobility results from the femur having a neck that is much narrower than the diameter of the head. The movements of the thigh at the hip joint are: flexion-extension, abduction-adduction, medial and lateral rotation, and circumduction (Figs. 5, 6, and 7; pp. 7–9).

There is disagreement about the rotatory action of the adductor muscles in the medial side of the thigh (p. 393). If they have any rotatory action it must be weak because their line of pull is close to the axis about which the femur rotates (Tobias et al., 1988).

The Articular Capsule of the Hip Joint (Figs. 5-87 and 5-88).   The fibrous capsule is strong and dense. Proximally, it is attached to the edge of the acetabulum, just distal to the acetabular labrum, and to the transverse acetabular ligament. Distally, the fibrous capsule is attached to the neck of the femur as follows: anteriorly to the *intertrochanteric line* and the root of the greater trochanter and posteriorly to the neck proximal to the *intertrochanteric crest*. The **fibrous capsule** forms a cylindrical sleeve that encloses the hip joint and most of the neck of the femur. Most of its fibers take a spiral course from the hip bone to the lateral portion of the intertrochanteric line of the femur, but some deep fibers form an *orbicular zone* (zona orbicularis) and pass circularly around the neck of the femur. These fibers form a collar around the neck of the femur, which constricts the capsule and helps to hold the femoral head in the acetabulum. Some deep longitudinal fibers of the fibrous capsule form *retinacula* (Fig. 5-86*B*), which are reflected superiorly along the neck of the femur as longitudinal bands that blend with the periosteum. The *retinacula contain blood vessels* that supply the head and neck of the femur. Four main groups of longitudinal capsular fibers or intrinsic ligaments are given names according to the region of the hip bone which they attach to the femur. These **intrinsic ligaments** are thickened parts of the fibrous capsule that strengthen the hip joint.

The Iliofemoral Ligament (Figs. 3-8*B*, 5-87, and 5-88). This ligament is a very strong band that covers the anterior aspect of the hip joint. It is Y-shaped and attached proximally to the anterior inferior iliac spine and the acetabular rim. The iliofemoral ligament is attached distally to the *intertrochanteric line* of the femur. The capsule of the hip joint is taut and the iliofemoral ligament is tense in full extension of the joint. This strong ligament has an important role in preventing overextension of the hip joint during standing (*i.e.*, it helps to maintain the erect posture). *It screws the head of the femur into the acetabulum* and thereby maintains the integrity of the joint.

The Pubofemoral Ligament (Figs. 3-8*B*). This large ligament arises from the pubic part of the acetabular rim and the iliopubic eminence and blends with the medial part of the iliofemoral ligament. It strengthens the inferior and anterior parts of the fibrous capsule of the hip joint. The pubofemoral ligament tightens during extension of the hip joint and becomes tense during abduction. Although it is relatively weak, this ligament tends to prevent overabduction of the thigh at the hip joint.

The Ischiofemoral Ligament (Figs. 5-8*C* and 5-88). This ligament reinforces the fibrous capsule of the hip joint posteriorly. It arises from the ischial portion of the acetabular rim and spirals superolaterally to the neck of the femur, medial to the base of the greater trochanter. Its anatomical construction tends to screw the femoral head medially into the acetabulum during extension of the thigh at the hip joint, thereby preventing hyperextension of it.

The Ligament of the Head of the Femur (Figs. 5-86*B* and 5-89). This intracapsular ligament (ligamentum capitis femoris), about 3.5 cm long, is weak and appears to be of little importance in strengthening the hip joint. Its wide end is attached

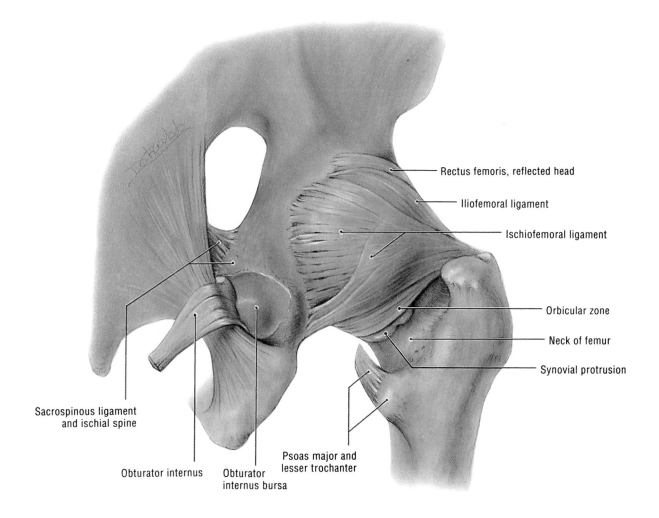

Rectus femoris, reflected head

Iliofemoral ligament

Ischiofemoral ligament

Orbicular zone

Neck of femur

Synovial protrusion

Sacrospinous ligament and ischial spine

Obturator internus

Obturator internus bursa

Psoas major and lesser trochanter

Posterior view

Figure 5-88.   Dissection of the right hip joint. Observe that the synovial membrane protrudes inferior to the fibrous capsule, forming a bursa for the tendon of the obturator externus muscle.

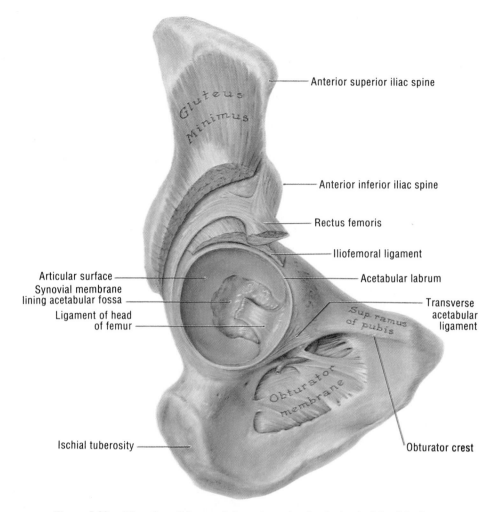

Figure 5-89. Dissection of the acetabulum, the socket for the head of the right femur.

Table 5-10.
The Main Muscles Producing Movements of the Hip Joint[1]

Flexion	Extension	Abduction
Iliopsoas	**Gluteus maximus**	**Gluteus medius**
Iliacus	**Semitendinosus**	**Gluteus minimus**
Psoas major	**Semimembranosus**	Tensor fasciae latae
Tensor fasciae	**Biceps femoris**	Sartorius
latae	(long head)	Piriformis
Rectus femoris	Adductor magnus	Obturator externus
Pectineus	(ischial fibers)	
Sartorius		
Adductor muscles,		
especially		
adductor longus		

Adduction	Medial Rotation	Lateral Rotation
Adductor magnus	**Tensor fasciae**	**Obturator internus**
Adductor longus	**latae**	**and gemelli**
Adductor brevis	**Gluteus medius**	**Obturator externus**
Pectineus	**Gluteus minimus**	**Quadratus femoris**
Gracilis		Piriformis
		Gluteus maximus
		Sartorius

[1]**Bold face** indicates the muscles that are chiefly responsible for the movement; the other muscles assist them.

to the margins of the acetabular notch and to the transverse acetabular ligament, and its narrow end is attached to the fovea or pit in the femur (Fig. 5-86). Usually it contains a small *artery to the head of the femur*, which is a branch of the obturator artery. The ligament of the head of the femur (LHF) is stretched when the flexed thigh is adducted or laterally rotated. It is *located inside the fibrous capsule of the hip joint* and is surrounded by a sleeve of synovial membrane.

The LHF varies in size and strength; sometimes it is absent. Its function, other than as a pathway for the artery to the femoral head, is unclear. It appears to be of limited value in strengthening the hip joint. Crelin (1976) found the LHF to be very important in stabilizing the hip joint in fetuses and neonates.

The synovial capsule of the hip joint lines the internal surface of the fibrous capsule and is reflected from it on the neck of the femur (Figs. 5-86*B* and 5-88). The synovial capsule forms a sleeve for the LHF, which is attached to the margins of the fovea. It also lines the acetabular fossa, covers the fatty pad in the acetabular notch, and is attached to the edges of the acetabular fossa and to the transverse acetabular ligament. The synovial capsule protrudes inferior to the fibrous capsule posteriorly (Fig. 5-88) and forms the *obturator externus bursa*, which protects the tendon of the obturator externus muscle.

Stability of the Hip Joint (Figs. 5-87 to 5-89).   The hip joint is a very strong and stable articulation. It is surrounded by powerful muscles and the articulating bones are united by a dense fibrous capsule, which is strengthened by strong intrinsic ligaments, particularly the iliofemoral ligament. Its stability is largely the result of the adaptation of the articulating surfaces of the acetabulum and the femoral head to each other.

Blood Supply of the Hip Joint (Figs. 5-29 and 5-37).   The articular arteries are branches of the medial and lateral circumflex femoral arteries, the deep division of the superior gluteal artery, and the inferior gluteal artery. The artery to the head of the femur is a branch of the posterior division of the obturator artery.

Nerve Supply of the Hip Joint (Figs. 5-30, 5-37, and 5-73).   The articular nerves are derived from the *femoral nerve* via the nerve to the rectus femoris muscle; the *obturator nerve* via its anterior division; the *sciatic nerve* via the nerve to the quadratus femoris muscle; and the *superior gluteal nerve*.

Fractures of the femoral neck close to the head often disrupt the blood supply to the head of the femur. In some cases the blood supplied via the artery in the ligament of the head may be the only blood received by the proximal fragment of the femoral head. If the ligament is ruptured, the fragment of bone may receive no blood and undergo *aseptic necrosis* (death in the absence of infection). Because the femoral, sciatic, and obturator nerves also supply the knee joint, hip disease may cause referred pain to the knee.

Congenital dislocation of the hip joint is common, occurring in about 1.5 per 1000 live births. This abnormality is bilateral in about half of the cases. Despite its name, this condition is not usually obvious at birth and may not be noticed for several months. Females are affected much more often than males (8:1). Some studies have shown that the articular capsule of the hip joint is loose at birth and that there is hypoplasia of the acetabulum and femoral head. A characteristic clinical sign of congenital dislocation of the hip joint is inability to abduct the thigh. In addition, the affected limb seems to be shorter because the dislocated femoral head is more superior than on the normal side.

Acquired dislocation of the hip joint is uncommon because this articulation is so strong and stable. Nevertheless, dislocation may occur during an automobile accident when the hip joint is flexed, adducted, and medially rotated. When a person's knee strikes a dashboard with the thigh in this position, the force transmitted superiorly along the femur drives the femoral head out of the acetabulum. In this position the femoral head is covered posteriorly by capsule rather than bone. As a result, the fibrous capsule ruptures inferiorly and posteriorly, which allows the head to pass through the tear in the capsule and over the posterior margin of the acetabulum. Often the acetabular margin fractures, producing a *fracture-dislocation of the hip joint*. When the femoral head dislocates, it usually carries the acetabular bone fragment and the acetabular labrum with it.

Because of the close relationship of the *sciatic nerve* to the hip joint (Figs. 5-37 and 5-73B), it may be injured (stretched and/or compressed) during posterior dislocations or fracture-

dislocations of the hip joint. This may result in paralysis of the hamstring muscles and those muscles distal to the knee supplied by the sciatic nerve (Fig. 5-73B). Sensory changes may also occur in the skin over the posterior and lateral aspects of the leg and over much of the foot (Figs. 5-11 and 5-82).

The Knee Joint

This is a *hinge type of synovial joint* (p. 19) that permits some rotation. Its structure is complicated because it **consists of three articulations**: an intermediate one between the patella and femur and lateral and medial ones between the femoral and tibial condyles.

Articular Surfaces of the Knee Joint (Figs. 5-27, 5-42, and 5-90 to 5-93).   The bones involved are the femur, tibia, and patella. The articular surfaces are the large curved condyles of the femur, the flattened condyles of the tibia, and the facets of the patella. When you stand in the anatomical position, your knees are in contact, but your femora are set obliquely because their heads are separated by the width of the pelvis (Fig. 5-2). This produces an open angle at the lateral side of the knee, toward which the patella tends to be displaced when the quadriceps femoris muscle contracts. The knee joint is relatively weak mechanically because of the configurations of its articular surfaces. It relies on the ligaments that bind the femur to the tibia for strength. On the superior surface of each tibial condyle, there is an articular area for the corresponding femoral condyle (Fig. 5-93). These areas, commonly referred to as the medial and lateral *tibial plateaus*, are separated from each other by a narrow, nonarticular area, which widens anteriorly and posteriorly into anterior and posterior *intercondylar areas*, respectively.

Surface Anatomy of the Knee Joint (Figs. 5-26, 5-42B, 5-43, and 5-44).   This joint may be felt as a slight gap on each side between the corresponding femoral and tibial condyles. When the leg is flexed or extended, a depression appears on each side of the patellar ligament. The articular capsule of the knee joint is very superficial in these depressions. The knee joint lies deep to the apex of the patella.

Movements of the Knee Joint (Table 5-11).   The principal movements occurring at this joint are *flexion and extension of the leg*, but some rotation also occurs in the flexed position. Flexion and extension of the knee joint are very free movements. Flexion normally stops when the calf contacts the thigh. Extension of the leg is stopped by the ligaments of the knee. When the knee is fully extended, as when sitting on a chair with the heel of the foot resting on another chair, the skin anterior to the patella is loose and can easily be picked up. This laxity of the skin helps flexion to occur. Note that the slackness disappears as the leg is flexed. When the leg is fully extended, the knee "locks" owing to medial rotation of the femur on the tibia. This makes the lower limb a solid column and more adapted for weightbearing. To "unlock" the knee the popliteus muscle contracts, thereby rotating the femur laterally so that flexion of the knee can occur (also see p. 452).

The Articular Capsule of the Knee Joint (Figs. 5-38, 5-90, and 5-91).   The *fibrous capsule* is strong, especially where local thickenings of it form ligaments. Those continuous with

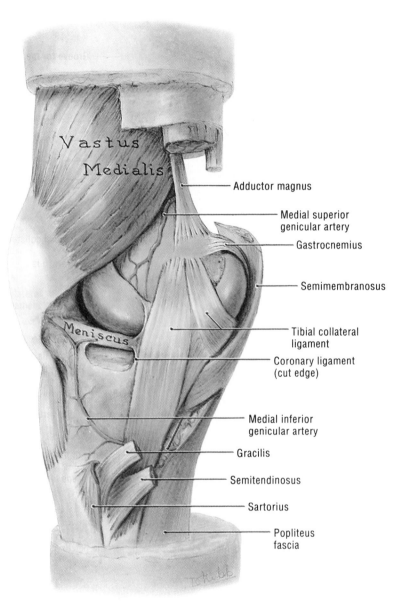

Vastus Medialis

Meniscus

- Adductor magnus
- Medial superior genicular artery
- Gastrocnemius
- Semimembranosus
- Tibial collateral ligament
- Coronary ligament (cut edge)
- Medial inferior genicular artery
- Gracilis
- Semitendinosus
- Sartorius
- Popliteus fascia

Figure 5-92. Dissection of the medial aspect of the right knee joint. Observe the bandlike part of the tibial collateral ligament, which is attached to the medial epicondyle, almost in line with the adductor magnus tendon and crossing the insertion of the semimembranosus muscle.

of the medial surface of the tibia. It is a thickening of the fibrous capsule of the knee joint and is partly continuous with the tendon of the adductor magnus muscle. The inferior end of the ligament is separated from the tibia by the medial inferior genicular vessels and nerve. *The deep fibers of the tibial collateral ligament are firmly attached to the medial meniscus* and the fibrous capsule of the knee joint.

The tibial and fibular collateral ligaments normally prevent disruption of the sides of the knee joint. They are tightly stretched when the leg is extended and they prevent rotation of the tibia laterally or the femur medially. As the collateral ligaments are slack during flexion of the leg, they permit some rotation of the tibia on the femur in this position.

The fibular collateral ligament is not commonly torn be-

cause it is very strong. Furthermore, severe blows to the medial side of the knee that might force it laterally and tear it are uncommon. However, lesions (*e.g.*, sprains or tears) of the fibular collateral ligament can have serious consequences. Usually it is the distal end of the ligament that tears, and sometimes the head of the fibula is pulled off because the ligament is stronger than bone. Complete tears of the fibular collateral ligament are often associated with *stretching of the common fibular (peroneal) nerve* (Fig. 5-56). This affects the muscles of the anterior and lateral compartments of the leg (Tables 5-5 and 5-6) and may produce *foot-drop* owing to paralysis of the dorsiflexor and eversion muscles of the foot (p. 429).

The firm attachment of the tibial collateral ligament to the medial meniscus is of considerable clinical significance be-

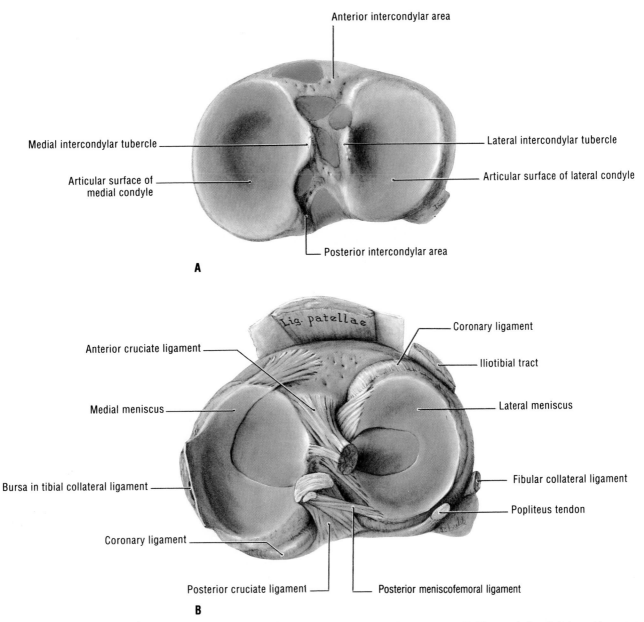

Figure 5-93. Dissections of the right knee joint. *A*, Superior aspect of the proximal end of the tibia showing the medial and lateral plateaus (articular surfaces). The sites of attachment of the cruciate ligaments are colored green; those of the medial meniscus, purple; and those of the lateral meniscus, orange. *B*, The menisci and their attachments to the intercondylar area of the tibia. The tibial attachments of the cruciate ligaments are also shown.

Table 5-11.
The Main Muscles Producing Movements of the Knee Joint[1]

Flexion	Extension	Medial Rotation of Tibia on Femur	Lateral Rotation of Tibia on Femur
Hamstrings	**Quadriceps femoris**	**Popliteus**	**Biceps femoris**
Semimembranosus	**Rectus femoris**	**Semimembranosus**	
Semitendinosus	**Vastus lateralis**	**Semitendinosus**	
Biceps femoris	**Vastus intermedius**	Sartorius	
Gracilis	**Vastus medialis**	Gracilis	
Sartorius	Tensor fasciae latae		
Popliteus			

[1]**Bold face** indicates the muscles that are chiefly responsible for the movement; the other muscles assist them.

cause *injury to the tibial collateral ligament frequently results in concomitant injury to the medial meniscus.* Rupture of the tibial collateral ligament, often associated with tearing of the medial meniscus and the anterior cruciate ligament, is a common type of football injury. The damage is frequently caused by a blow to the lateral side of the knee. When considering soft tissue injuries of the knee, always think of *the three Cs* which indicate those structures that may be damaged: *Co*llateral ligaments, *Cr*uciate ligaments, and *Ca*rtilages (menisci). Sprains of the tibial collateral ligament result in tenderness over the femoral or tibial attachments of this ligament, owing to tearing of these parts.

The Oblique Popliteal Ligament (Fig. 5-38). This broad band is an expansion of the tendon of the semimembranosus muscle. The oblique popliteal ligament strengthens the fibrous capsule of the knee joint posteriorly. It arises posterior to the medial condyle of the tibia and passes superolaterally to attach to the central part of the posterior aspect of the fibrous capsule of the knee joint.

The Arcuate Popliteal Ligament (Fig. 5-38). This Y-shaped band of fibers also strengthens the fibrous capsule posteriorly. The stem of the ligament arises from the posterior aspect of the head of the fibula. As it passes superomedially over the tendon of the popliteus muscle, the arcuate popliteal ligament spreads out over the posterior surface of the knee joint. It inserts into the intercondylar area of the tibia and the posterior aspect of the lateral epicondyle of the femur.

The synovial capsule of the knee joint is extensive (Fig. 5-90). It lines the inner aspect of the fibrous capsule and reflects onto the articulating bones as far as the edges of their articular cartilages. This capsule is more extensive than that of any other joint; thus the synovial cavity is the largest joint space in the body. The synovial capsule is also attached to the periphery of the patella. It is separated from the patellar ligament by the infrapatellar fatpad.

Injections may be made into the synovial cavity of the knee joint for diagnostic and/or therapeutic purposes. They are generally made into the lateral side of the knee joint, with the patient sitting on the side of a table with the knee flexed and the leg hanging. When the knee joint is inflamed (*e.g.,* owing to infection or arthritis), the amount of synovial fluid may increase. Aspiration of this fluid may be necessary to relieve pressure in the joint or to obtain a sample of it for diagnostic studies. In other cases it may be necessary to evacuate blood that has entered the joint (*e.g.,* after a tibial fracture in which the fracture line extends into the joint cavity).

In *pneumoarthrography* (air contrast study of a joint), air is injected into the joint cavity of the knee to facilitate study of the structures in it. Being less opaque than these structures (*e.g.,* the menisci) it appears black on radiographs and outlines the soft tissues which appear as gray images. Sometimes the cavity of the knee joint is examined for injury or disease using an *arthroscope*; this procedure is called *arthroscopy.* The joint cavity can also be irrigated or washed out using saline.

Bursae Around the Knee (Figs. 5-90 and 5-91). Several bursae are present around the knee because most tendons around the knee joint run parallel to the bones and pull lengthwise across the joint. Four bursae communicate with the synovial cavity of the knee joint; they lie deep to the tendons of the quadriceps femoris, the popliteus, and the medial head of the gastrocnemius muscle.

The Suprapatellar (Quadriceps) Bursa (Fig. 5-90). This large saccular extension of the synovial capsule passes superiorly between the femur and the tendon of the quadriceps femoris muscle. The clinically important suprapatellar bursa extends about 8 cm superior to the base of the patella. The suprapatellar bursa permits free movement of the quadriceps tendon over the distal end of the femur and facilitates full extension and flexion of the knee joint. The bursa is held in position by the part of the vastus intermedius muscle called the *articularis genus muscle* (Fig. 5-19C).

Because the suprapatellar bursa communicates freely with the synovial cavity of the knee joint, it is regarded as a part of it. Hence, stab or puncture wounds superior to the patella may infect the knee joint via the suprapatellar bursa. It may also be involved in fractures of the distal end of the femur, resulting in *hemarthrosis* of the knee joint (blood in the joint cavity).

The Popliteus Bursa (Fig. 5-90). This extension of the synovial cavity lies between the tendon of the popliteus muscle and the lateral condyle of the tibia. The bursa opens into the lateral part of the synovial cavity, inferior to the lateral meniscus. Sometimes the bursa is also continuous with the synovial cavity of the proximal tibiofibular joint (p. 486) as a result of perforation of the partition between the popliteus bursa and its joint cavity.

The Anserine Bursa (Bursa anserina). This complicated bursa has several diverticula and separates the tendons of the sartorius, gracilis, and semitendinosus muscles from the proximal part of the medial surface of the tibia and from the tibial collateral ligament.

The area where the tendons of the sartorius, gracilis, and semitendinosus muscles attach distally (Fig. 5-27) has been called the *pes anserinus* (L. foot of a goose). The anserine bursa is associated with these tendons.

The Gastrocnemius Bursa. This extension of the synovial cavity of the knee joint lies deep to the proximal attachment of the tendon of the medial head of the gastrocnemius muscle. As it separates the tendon from the femur, it is often called the subtendinous bursa of the gastrocnemius.

The Semimembranosus Bursa (Fig. 5-46A). This bursa is related to the distal attachment of the semimembranosus muscle and is located between the medial head of the gastrocnemius and the semimembranosus tendon. Frequently it is a prolongation of the gastrocnemius bursa and communicates with the knee joint cavity.

Occasionally synovial fluid escapes from the knee joint (*synovial effusion*) and accumulates in the popliteal fossa, forming a **popliteal cyst** (Baker's cyst). Such cysts are common in children, and although often quite large, they seldom produce symptoms. In adults, popliteal cysts usually communicate with the synovial cavity of the knee joint by a narrow stalk, which passes through the fibrous capsule of the joint. In cases in which synovial fluid escapes owing to either rheumatoid or degenerative joint disease, the cyst becomes greatly distended and may extend inferiorly as far as the middle of the calf. In such cases the cyst usually interferes with movements of the knee joint.

The Subcutaneous Prepatellar Bursa (Fig. 5-90). This bursa lies between the skin and the anterior surface of the patella. It allows free movement of the skin over the patella during flexion and extension of the leg. Because of its superficial and exposed position, this bursa may become inflamed after prolonged periods of weightbearing on the hands and knees (*e.g.*, when scrubbing a hard floor).

> **Prepatellar bursitis** is a friction bursitis caused by friction between the skin and the patella. If the inflammation is chronic, the bursa becomes distended with fluid and forms a soft, fluctuant swelling anterior to the knee (Fig. 5-94). This condition is commonly called "housemaid's knee," but other people who work on their knees without using kneepads, also develop prepatellar bursitis.

The Subcutaneous Infrapatellar Bursa. This bursa is located between the skin and the tibial tuberosity. It allows the skin to glide over the tibial tuberosity and to withstand pressure when kneeling with the trunk upright (*e.g.*, when one kneels or genuflects during praying).

The Deep Infrapatellar Bursa. This small bursa lies between the patellar ligament and the anterior surface of the tibia, superior to the tibial tuberosity. It is separated from the knee joint by the infrapatellar fatpad, a mass of fatty tissue between the superior part of the patellar ligament and the synovial capsule.

> *Subcutaneous infrapatellar bursitis* results from excessive friction between the skin and the tibial tuberosity. The swelling occurs over the proximal end of the tibia. This condition has been called "clergyman's knee," but it occurs more often in roofers, floor tilers, and carpet layers who do not wear kneepads.
>
> *Deep infrapatellar bursitis* results in a swelling between the patellar ligament and the tibia, superior to the tibial tuberosity. The swelling is less pronounced than that associated with subcutaneous prepatellar bursitis. Enlargement of this

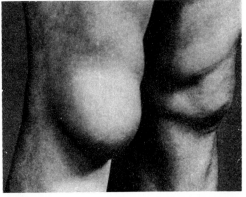

Figure 5-94. The knees of a patient with prepatellar bursitis ("housemaid's knee"), resulting from inflammation of the subcutaneous prepatellar bursae (Fig. 5-90).

bursa obliterates the dimples on each side of the patellar ligament when the leg is extended. Obliteration of these dimples may also result from synovial effusion (p. 482).

The Cruciate Ligaments of the Knee Joint (Figs. 5-91, 5-93, and 5-95). These very strong ligaments are within the capsule of the joint but outside the synovial cavity. *Joining the femur and tibia*, they are located between the medial and lateral condyles and are separated from the joint cavity by the synovial membrane. The synovial capsule lines the fibrous capsule, except posteriorly where it is reflected anteriorly around the cruciate ligaments. The cruciate (L. resembling a cross) ligaments are strong, rounded bands that cross each other obliquely in a manner similar to the limbs of St. Andrew's cross or an X. They are named anterior and posterior according to their site of attachment to the tibia, *i.e*, the anterior cruciate ligament attaches to the tibia anteriorly and the posterior cruciate ligament attaches to it posteriorly. These ligaments are essential to the anteroposterior stability of the knee joint, especially when it is flexed.

The Anterior Cruciate Ligament (Figs. 5-91, 5-93, and 5-95). The weaker of the two ligaments, the anterior cruciate ligament arises from the anterior part of the intercondylar area of the tibia, just posterior to the attachment of the medial meniscus. It extends superiorly, posteriorly, and laterally to attach to the posterior part of the medial side of the lateral condyle of the femur. The anterior cruciate ligament, which is slack when the knee is flexed and taut when it is fully extended, *prevents posterior displacement of the femur on the tibia* and hyperextension of the knee joint. When the joint is flexed at a right angle, the tibia cannot be pulled anteriorly because it is held by the anterior cruciate ligament.

The Posterior Cruciate Ligament (Figs. 5-91, 5-93, and 5-95). This is the stronger of the two cruciate ligaments. It arises from the posterior part of the intercondylar area of the tibia and passes superiorly and anteriorly on the medial side of the anterior cruciate ligament to attach to the anterior part of the lateral surface of the medial condyle of the femur. The posterior cruciate ligament is the first structure observed when the knee joint is surgically opened posteriorly. The posterior cruciate ligament, which tightens during flexion of the knee joint, *prevents anterior displacement of the femur on the tibia* or posterior displacement of the tibia. It also helps to prevent hyperflexion of the knee joint. In the weightbearing flexed knee, it is the main stabilizing factor for the femur, *e.g.*, when walking downhill or downstairs.

> The relatively weak anterior cruciate ligament may be torn when the tibial collateral ligament ruptures after the knee is hit hard from the lateral side while the foot is fixed (*e.g.*, in the ground). First the tibial collateral ligament ruptures, opening the joint on the medial side. This may tear the medial meniscus and the anterior cruciate ligament. The anterior cruciate ligament may also be torn when: (1) the tibia is driven anteriorly on the femur; (2) the femur is driven posteriorly on the tibia; or (3) the knee joint is severely hyperextended. The knee joint becomes very unstable when the anterior cruciate ligament is torn. To test its stability, the tibia is pulled in an anterior direction. If there is anterior movement (*anterior*

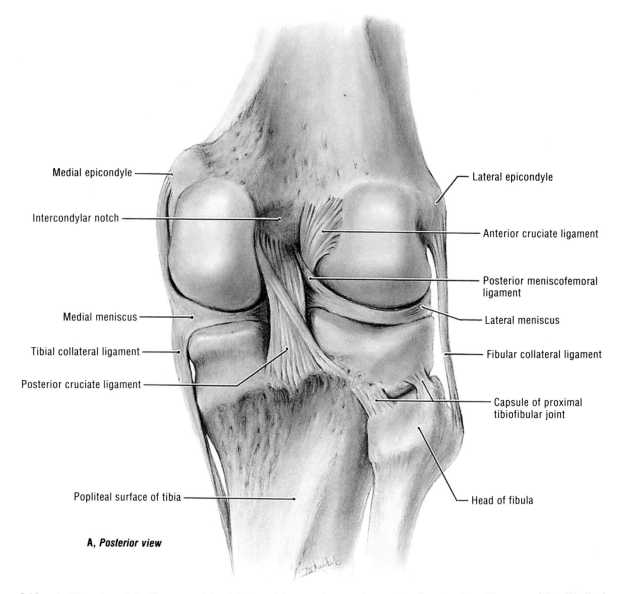

Medial epicondyle

Intercondylar notch

Medial meniscus

Tibial collateral ligament

Posterior cruciate ligament

Popliteal surface of tibia

Lateral epicondyle

Anterior cruciate ligament

Posterior meniscofemoral ligament

Lateral meniscus

Fibular collateral ligament

Capsule of proximal tibiofibular joint

Head of fibula

A, *Posterior view*

Figure 5-95. *A,* Dissection of the ligaments of the right knee joint. Observe that the bandlike tibial collateral ligament is attached to the medial meniscus. *B,* Coronal MRI (magnetic resonance image) of the knee and an orientation drawing. (Courtesy of Dr. W. Kucharczyk, Clinical Director of Tri-Hospital of Magnetic Resonance Centre, Toronto, Ontario, Canada.)

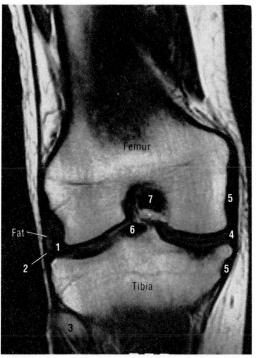

B

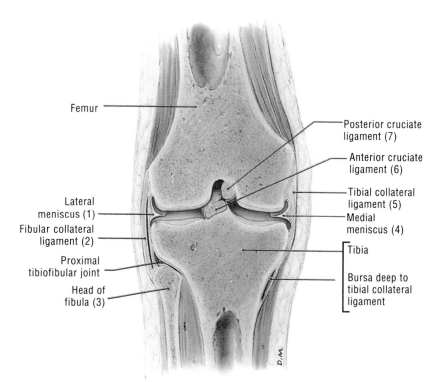

Figure 5.95*B*

drawer sign), a tear of the anterior cruciate ligament is indicated.

The posterior cruciate ligament may be injured when the superior part of the tibia is struck with the knee flexed. This kind of injury may occur when a passenger's leg is driven against the dashboard. If the tibia is driven posteriorly on the femur, if the femur is driven anteriorly on the tibia, or if the knee joint is severely hyperflexed, the posterior cruciate ligament may be torn. The flexed knee is unstable when the posterior cruciate ligament is torn. To test its stability, the tibia is forced in a posterior direction. If there is posterior movement (*posterior drawer sign*), a tear of the posterior cruciate ligament is indicated.

The Menisci of the Knee Joint (Figs. 5-90 to 5-93 and 5-95). The medial and lateral menisci (G. crescents) are crescentic plates of fibrocartilage on the articular surface of the tibia. They act like *shock absorbers*. Because they are basically C-shaped, they were formerly called semilunar cartilages. Wedge-shaped in transverse section, the menisci are firmly attached at their ends to the intercondylar area of the tibia. The menisci deepen the articular surfaces of the tibia where they articulate with the femoral condyles. Their superior surfaces are slightly concave for reception of these condyles, whereas their inferior surfaces that rest on the tibial condyles are flatter. The menisci are thick at their peripheral attached margins and thin at their internal unattached edges. Being smooth and slightly movable, *the menisci fill the gaps between the femur and tibia* that would otherwise be present during movements of the knee joint. Their external margins are attached to the fibrous capsule of the knee joint and through it to the edges of the articular surfaces of the tibia. The capsular fibers that attach the thick, convex margins

of the menisci to the tibial condyles are called *coronary ligaments*. A slender fibrous band, called the *transverse ligament of the knee*, joins the anterior edges of the two menisci. This connection allows them to move together during movements of the femur on the tibia. The thickness of this ligament varies in different people; sometimes it is absent. The thick peripheral margins of the menisci are vascularized by genicular branches of the popliteal artery (Fig. 5-46), but their thin unattached edges in the interior of the joint are avascular.

The Medial Meniscus (Figs. 5-91 to 5-93 and 5-95). This *C-shaped cartilage* is broader posteriorly than anteriorly. Its anterior end or horn (L. cornu) is attached to the anterior intercondylar area of the tibia, anterior to the attachment of the anterior cruciate ligament. Its posterior end or horn is attached to the posterior intercondylar area, anterior to the attachment of the posterior cruciate ligament, and between the attachments of the lateral meniscus and the posterior cruciate ligament. *The medial meniscus is firmly adherent to the deep surface of the tibial collateral ligament.*

The Lateral Meniscus (Figs. 5-90, 5-91, 5-93, and 5-95). This C-shaped cartilage is nearly circular and conforms to the rather circular lateral tibial condyle. The lateral meniscus is smaller and more freely movable than the medial meniscus, but it covers a larger area of articular surface than does the medial meniscus. The tendon of the popliteus muscle separates the lateral meniscus from the fibular collateral ligament. The anterior and posterior ends or horns of the lateral meniscus are attached close together in the anterior and posterior intercondylar areas. A strong tendinous slip, called the *posterior meniscofemoral ligament*, joins the lateral meniscus to the posterior cruciate ligament and the medial femoral condyle. The popliteus tendon and bursa pass between the lateral meniscus and the fibular collateral ligament.

Blood Supply of the Knee Joint (Figs. 5-29, 5-46, and 5-92).   The articular arteries are branches of the vessels that form the *genicular anastomoses* around the knee. The middle genicular artery, a branch of the popliteal artery, penetrates the fibrous capsule and supplies the cruciate ligaments, synovial capsule, and the peripheral margins of the menisci.

Nerve Supply of the Knee Joint (Figs. 5-30, 5-45, and 5-73).   The articular nerves are branches of the obturator, femoral, tibial, and common fibular (peroneal) nerves.

Knee joint injuries are common because it is a major weightbearing joint and its stability depends almost entirely on its associated ligaments and muscles. Ligamentous injuries of the knee joint may result from any blow that forces it to move in an abnormal plane. A blow on the lateral side of the knee when a person is bearing weight on the leg stresses the tibial collateral ligament. If the blow is relatively minor, the fibers are stretched and some of them may be torn; this is called a *sprained tibial collateral ligament*. When the blow is severe, all the fibers may be torn partially or completely. This is called a *torn tibial collateral ligament*; the tear usually occurs near its attachment to the medial epicondyle of the femur.

Meniscus Injury (Knee-Cartilage Injury).   Localized tenderness and pain in the flexed knee on the medial side of the patellar ligament, just proximal to the medial tibial plateau, suggests injury to the medial meniscus. Injury to this cartilage is about 20 times more common than injury to the lateral meniscus (McMinn, 1990). *Injury to the medial meniscus* results from a twisting strain that is applied to the knee joint when it is flexed. Because the medial meniscus is firmly adherent to the tibial collateral ligament, twisting strains of this ligament may tear and/or detach the medial meniscus from the fibrous capsule. Part of the torn cartilage may become displaced toward the center of the joint and become lodged between the tibial and femoral condyles. This ''locks the knee'' in the flexed position, preventing the patient from fully extending the knee. When weight is borne by the flexed knee joint, a sudden twist of the knee may also rupture the medial meniscus, usually splitting it longitudinally. This injury is common in athletes who twist their flexed knees while running (*e.g.*, in football and basketball). It also occurs in coal miners and other persons who can topple over when they are working in a crouched or squatting position. Because the internal edges of the menisci are poorly supplied with blood, tears in them heal poorly. Tears near the peripheral margin, which are vascularized by genicular branches of the popliteal artery, usually heal well. The menisci can be observed during **arthroscopy** (p. 482) and when air and/or dense contrast material is injected into the synovial cavity of the knee joint before radiographs are taken. *Pneumarthrograms* or double contrast arthrograms are helpful in demonstrating soft tissue lesions of the knee joint. Because air is less opaque than the menisci, it appears black in the radiograph and outlines the soft tissues (*e.g.*, the menisci). When dense contrast materials are used, the articular cartilages and menisci appear as radiolucent images within the dense contrast medium. Good images of the ligaments and menisci of the knee are also produced by magnetic resonance imaging (Fig. 5-95*B*).

Stability of the Knee Joint.   Although the knee joint is well constructed and one of the strongest joints in the body,

particularly when extended, its function is commonly impaired (*e.g.*, in body contact sports). The stability of the knee joint depends upon the strength of the surrounding muscles and ligaments. Of these supports, the muscles are most important. Thus many sports injuries are preventable through appropriate conditioning and training. The most important muscle in stabilizing the knee joint is the **quadriceps femoris**, particularly the inferior fibers of the vastus medialis and vastus lateralis (Fig. 5-19). The evidence for this is that the knee joint will function surprisingly well following a strain of the ligaments if the quadriceps femoris is well developed.

The quadriceps femoris muscle undergoes considerable atrophy during periods of disuse (*e.g.*, while the limb is in a cast). Thus it is exercised to prevent disuse atrophy and instability of the knee. Without good muscular support, repaired knee ligaments may be more easily sprained or retorn, either partially or completely. Residual instability of the knee joint is one of the most troublesome complications of ligamentous injuries of the knee. Traumatic *dislocation of the knee joint* is not common; however, this injury may occur, *e.g.*, during an automobile accident. In addition to disruption of the knee ligaments, the popliteal artery and tibial nerve may be injured, resulting in partial or complete paralysis of the muscles supplied by this nerve (Fig. 5-73*B*).

The Tibiofibular Joints

The tibia and fibula articulate at their proximal and distal ends. Movement at the proximal tibiofibular joint is impossible without movement at the distal one.

The Proximal (Superior) Tibiofibular Joint.   This is a plane type of synovial joint (p. 19) between the head of the fibula and the lateral condyle of the tibia.

Articular Surfaces of the Proximal Tibiofibular Joint (Figs. 5-49*B* and 5-95).   The flat, oval-to-circular facet on the head of the fibula articulates with a similar facet located posterolaterally on the inferior aspect of the lateral condyle of the tibia.

Movements of the Proximal Tibiofibular Joint.   Slight movement occurs at the superior tibiofibular joint during dorsiflexion of the foot at the ankle joint. This presses the lateral malleolus laterally and causes movement of the body and head of the fibula. Some movement of the joint also occurs during plantarflexion of the foot.

The Articular Capsule of the Proximal Tibiofibular Joint (Figs. 5-90 and 5-95).   The *fibrous capsule* surrounds the joint and is attached to the margins of the articular facets on the fibula and tibia. It is strengthened by the anterior and posterior ligaments of the head of the fibula. The fibers of these ligaments run superomedially from the fibula to the tibia. The tendon of the popliteus muscle is intimately related to the posterosuperior aspect of the proximal tibiofibular joint. The **synovial membrane** lines the fibrous capsule. The pouch of synovial membrane passing under the tendon of the popliteus muscle, known as the *popliteus bursa*, sometimes communicates with the synovial cavity of the proximal tibiofibular joint through an opening in the superior part of the synovial capsule. Consequently, the proximal tibiofibular joint may be indirectly in communication with the synovial cavity of the knee joint.

Blood Supply of the Proximal Tibiofibular Joint (Figs. 5-29, 5-46, and 5-92). The articular arteries are derived from the inferior lateral genicular and anterior tibial recurrent arteries.

Nerve Supply of the Proximal Tibiofibular Joint (Figs. 5-45, 5-60, and 5-73). The articular nerves are derived from the common fibular (peroneal) nerve and the nerve to the popliteus muscle.

The Distal (Inferior) Tibiofibular Joint. This is a fibrous joint of the syndesmosis type (p. 17). It is located between the inferior ends of the tibia and fibula.

Articular Surfaces of the Distal Tibiofibular Joint (Figs. 5-49*B*, 5-60, 5-96, and 5-97). The rough, convex, triangular articular area on the medial surface of the inferior end of the fibula articulates with a facet on the inferior end of the tibia. A small superior projection of the synovial capsule of the ankle joint extends into the inferior part of the distal tibiofibular joint. A strong *interosseous ligament* continuous superiorly with the interosseous membrane (Figs. 5-50 and 5-60), forms the principal connection between the tibia and fibula at this joint. It consists of strong bands that extend from the fibular notch of the tibia to the medial surface of the distal end of the fibula. The distal tibiofibular joint is also strengthened anteriorly and posteriorly by the strong *anterior and posterior tibiofibular ligaments*. They extend from the borders of the fibular notch of the tibia (Fig. 5-49*B*) to the anterior and posterior surfaces of the lateral malleolus, respectively (Fig. 5-96). The inferior, deep part of the posterior tibiofibular ligament is called the *transverse tibiofibular ligament*. This strong band closes the posterior angle between the tibia and fibula.

Stability of the Distal Tibiofibular Joint. This articulation forms a strong union between the distal ends of the tibia and fibula; much of the strength of the ankle joint is dependent on this union.

Movement of the Distal Tibiofibular Joint. Slight movement of the distal tibiofibular joint occurs to accommodate the talus during dorsiflexion of the foot at the ankle joint.

Blood Supply of the Distal Tibiofibular Joint (Figs. 5-60 and 5-74). The articular arteries are derived from the perforating branch of the fibular (peroneal) artery and the medial malleolar branches of the anterior and posterior tibial arteries.

Nerve Supply of the Distal Tibiofibular Joint (Figs. 5-60 and 5-72). The articular nerves are derived from the deep fibular (peroneal), tibial, and saphenous nerves.

> The posterior tibiofibular ligament is much stronger than the anterior tibiofibular ligament. In severe ankle injuries, the posterior ligament may avulse the posteroinferior part of the tibia. In these cases the fracture enters the ankle joint. If, in addition, the medial and lateral malleoli are fractured, the injury is referred to as a "trimalleolar fracture" (*i.e.*, a fracture of both the malleoli and the posterior part of the inferior border of the tibia).

The Ankle Joint

The ankle joint or *talocrural joint* is a hinge type of synovial joint (p. 19). It is located between the inferior ends of the tibia and fibula and the superior part of the talus. The ankle joint can

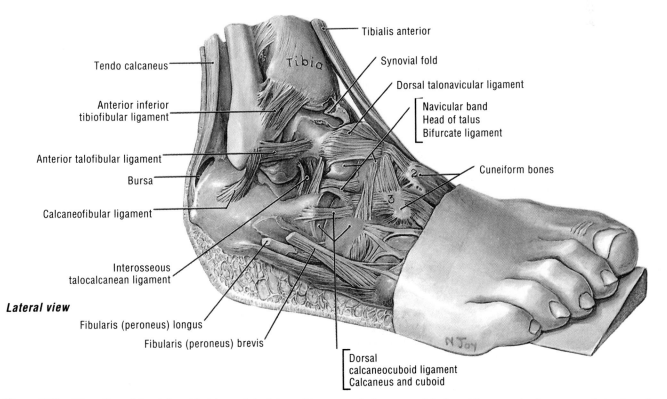

Tibialis anterior
Synovial fold
Dorsal talonavicular ligament
Navicular band
Head of talus
Bifurcate ligament
Cuneiform bones

Tendo calcaneus
Anterior inferior tibiofibular ligament
Anterior talofibular ligament
Bursa
Calcaneofibular ligament
Interosseous talocalcanean ligament

Lateral view

Fibularis (peroneus) longus
Fibularis (peroneus) brevis

Dorsal calcaneocuboid ligament
Calcaneus and cuboid

Figure 5-96. Dissection of the right ankle joint and the joints of inversion and eversion. The foot has been inverted to demonstrate the articular areas and the lateral ligaments that become taut during inversion of the foot.

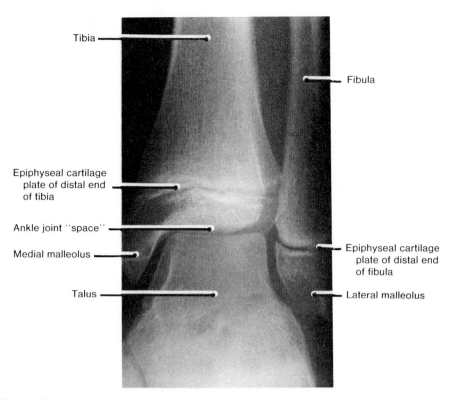

Tibia

Fibula

Epiphyseal cartilage
plate of distal end
of tibia

Ankle joint "space"

Medial malleolus

Epiphyseal cartilage
plate of distal end
of fibula

Talus

Lateral malleolus

Figure 5-97.   Slightly oblique frontal radiograph of the ankle joint of a 14-year-old boy, showing how the trochlea of the body of the talus fits into the mortise formed by the medial and lateral malleoli.

be felt between the tendons on the anterior surface of the ankle as a slight depression, about 1 cm proximal to the tip of the medial malleolus.

Articular Surfaces of the Ankle Joint (Figs. 5-49*B*, 5-51, and 5-96 to 5-98).   The inferior ends of the tibia and fibula form a deep socket or boxlike **mortise** into which the pulley-shaped *trochlea of the talus* fits (p. 437). The three-sided mortise (socket) is formed by the two malleoli and the inferior end of the tibia (Fig. 5-97). The fibula has an articular facet on its lateral malleolus which faces medially and articulates with the facet on the lateral surface of the talus. The tibia articulates with the talus in two places: (1) its inferior surface forms the roof of the mortise, which is wider anteriorly than posteriorly and slightly concave from anterior to posterior; and (2) the lateral surface of its medial malleolus articulates with the talus. The talus has three articular facets, which articulate with the inferior surface of the tibia and the malleoli. The superior articular surface of the talus is often called the trochlea (L. pulley) because of its pulleylike shape. It is wider anteriorly than posteriorly, convex from anterior to posterior, and slightly concave from side to side.

Movements of the Ankle Joint (Figs. 5-61*A* and *B* and 5-97; Table 5-12).   The ankle joint is uniaxial; its main movements are dorsiflexion and plantarflexion. When the foot is plantarflexed, some rotation, abduction, and adduction of the ankle joint are possible. During dorsiflexion the trochlea of the talus rocks posteriorly in the three-sided mortise, and the malleoli are forced apart because the superior articular surface of the talus is wider anteriorly than posteriorly. The separation of the malleoli requires some movement of the proximal tibiofibular joint. The

range of plantarflexion is greater than that of dorsiflexion, but there is considerable variation in these movements.

The Articular Capsule of the Ankle Joint (Figs. 5-96 and 5-98).   The *fibrous capsule* is thin anteriorly and posteriorly, but it is supported on each side by strong collateral ligaments. It is attached superiorly to the borders of the articular surfaces of the tibia and the malleoli. It is attached inferiorly to the talus close to the superior articular surface, except anteroinferiorly, where it is attached to the dorsum of the neck of the talus. The fibrous capsule is strengthened medially and laterally by two strong collateral ligaments (the deltoid and lateral ligaments).

The Medial or Deltoid Ligament[10] (Fig. 5-98).   This strong ligament attaches the medial malleolus to the tarsus (tarsal bones). The apex of the ligament is attached to the margins and tip of the medial malleolus. Its broad base fans out and attaches to three tarsal bones (talus, navicular, and calcaneus). *The deltoid ligament consists of four parts*, which are named according to their bony attachments: (1) *tibionavicular*, (2) and (3) anterior and posterior *tibiotalar*, and (4) *tibiocalcanean ligaments*. They strengthen the joint and hold the calcaneus and navicular bones against the talus. In addition, they help to maintain the medial side of the foot and the medial longitudinal arch (Fig. 5-105).

The Lateral Ligaments of the Ankle (Fig. 5-96).   On the lateral side of the ankle there are three ligaments that attach the

---

[10]Only the superficial part of this ligament is triangular like the Greek letter *delta*. The complete ligament is rectangular, hence *Nomina Anatomica* (Warwick, 1989) recommends that the term medial ligament be used, but the name *deltoid ligament* is given as an officially recognized alternative.

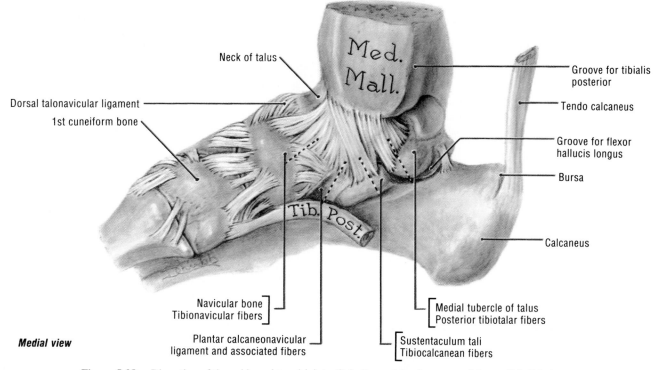

Med. Mall.

Neck of talus

Dorsal talonavicular ligament

1st cuneiform bone

Groove for tibialis posterior

Tendo calcaneus

Groove for flexor hallucis longus

Bursa

Calcaneus

Tib. Post.

Navicular bone
Tibionavicular fibers

Medial tubercle of talus
Posterior tibiotalar fibers

Medial view

Plantar calcaneonavicular ligament and associated fibers

Sustentaculum tali
Tibiocalcanean fibers

Figure 5-98.   Dissection of the ankle and tarsal joints. Only three of the four parts of the medial (deltoid) ligament are visible from this aspect.

Table 5-12.
The Main Muscles Producing Movements of the Ankle Joint[1]

Dorsiflexion	Plantarflexion[2]	
Tibialis anterior	**Gastrocnemius**	} Triceps surae
Extensor digitorum longus	**Soleus**	
Extensor hallucis longus	Plantaris	
Fibularis (peroneus) tertius	Tibialis posterior	
	Flexor hallucis longus	
	Flexor digitorum longus	

[1]**Bold face** indicates the muscles that are chiefly responsible for the movement; the other muscles assist them.
[2]During plantarflexion, slight side-to-side movements are possible, as is some abduction, adduction, and rotation.

lateral malleolus to the talus and calcaneus. They are not nearly so strong as the medial ligament. The "lateral ligament"[11] consists of three distinct parts (anterior and posterior talofibular ligaments and the calcaneofibular ligaments). The *anterior talofibular ligament* is a flat band that extends anteromedially from the lateral malleolus to the neck of the talus. It is not very strong. The *posterior talofibular ligament* is thick and fairly strong. It runs horizontally medially and slightly posteriorly from the malleolar fossa to the lateral tubercle of the posterior process of the talus. The *calcaneofibular ligament* is a round cord that passes posteroinferiorly from the tip of the lateral malleolus to the lateral surface of the calcaneus. It is crossed superficially by the tendons of the fibularis (peroneus) longus and brevis muscles.

[11]Collectively the three lateral ligaments are often referred to as the lateral ligament, but this term has not been officially recognized by *Nomina Anatomica* (Warwick, 1989). Clinicians prefer to use *lateral ligament* because of its brevity (Healey and Hodge, 1990).

The synovial capsule of the ankle joint lines the fibrous capsule and occasionally projects superiorly for a short distance into the inferior tibiofibular ligament between the tibia and fibula. The synovial cavity of the ankle joint is somewhat superficial on each side of the tendo calcaneus. Hence, when the ankle joint is inflamed, the synovial fluid may increase, causing swelling in these locations.

Stability of the Ankle Joint (Figs. 5-96 to 5-98).   This joint is *very strong during dorsiflexion* because (1) it is supported by powerful ligaments and (2) it is crossed by several tendons that are tightly bound down by thickenings of the deep fascia called *retinacula*. Its stability is also greatest in dorsiflexion because in this position the trochlea of the talus fills the mortise formed by the malleoli. The malleoli grip the talus tightly as it rocks anteriorly and posteriorly during movements of the ankle joint. The grip of the malleoli on the trochlea of the talus is strongest during dorsiflexion of the foot because this movement forces the anterior part of the trochlea posteriorly, spreading the tibia and fibula slightly apart. This spreading is limited by the strong interosseous ligament and by the anterior and posterior tibiofibular ligaments that unite the leg bones.

The ankle joint is relatively unstable during plantarflexion. During this movement the trochlea of the talus moves anteriorly in the mortise, causing the malleoli to come together. However, the grip of the malleoli on the trochlea is not so strong as during dorsiflexion. Some side movement can be demonstrated in full plantarflexion of the foot. The wedge-shaped form of the trochlea assists the ankle ligaments in preventing posterior displacement of the foot during sudden jumps and stops. Although the ankle joint is unstable in plantarflexion, appropriate training and conditioning strengthens the joint in this position (*e.g.*, in ballet dancers and persons who always wear shoes with high heels).

A bony outgrowth often develops on the anterior aspect of the distal end of the tibia and on the superior surface of the neck of the talus in persons who repeatedly kick a football or soccer ball. The plantarflexion associated with kicking pulls on the attachments of the anterior tibiotalar ligament, inducing a characteristic bony outgrowth. The condition has been called "footballer's ankle." Medial dislocations of the ankle joint are not common owing to the strength of the medial (deltoid) ligament. In fracture-dislocations of the ankle, the distal end of the tibia and/or the fibula is usually fractured. The common *Pott's fracture* occurs when the foot is forcibly everted (Fig. 5-99). This pulls on the extremely strong medial (deltoid) ligament, often tearing off the medial malleolus. The talus then moves laterally, shearing off the lateral malleolus or, more commonly, breaking the fibula superior to the inferior tibiofibular joint. If the tibia is carried anteriorly, the posterior margin of the distal end of the tibia is also sheared off by the talus.

The ankle joint is the most frequently injured major articulation in the body. The lateral ligaments of the joint are the ones most frequently injured. One or more of its four parts may be stretched and/or torn (Figs. 5-96 and 5-100). A sprained ankle results from twisting of the weightbearing foot and is nearly always an inversion injury (*i.e.*, the foot is forcefully inverted). A typical history follows: (1) The person steps on an uneven surface and falls; (2) this stretches most of the fibers of the lateral ligaments and tears some of them; (3) the ankle soon becomes painful and localized swelling and tenderness appear anteroinferior to the tip of the lateral malleolus. In severe ankle sprains many fibers of the lateral

ligaments are torn, either partially or completely, resulting in *instability of the ankle joint.* As inversion of the foot tends to occur with active plantarflexion, *inversion injuries of the ankle* tend to occur more often than eversion injuries when the foot is in plantarflexion. When one or more of the lateral ligaments are stretched or torn, the ankle becomes very unstable.

Fracture-dislocation of the ankle joint may occur in severe injuries in which the tip of the lateral malleolus is avulsed (Fig. 5-100). This injury often occurs when the foot is fixed against some object (*e.g.*, a large stone) and is thrown into an inverted position. The body weight is then violently transferred to the lateral ligaments of the ankle. Usually the calcaneofibular ligament tears, partially or completely, often along with the anterior talofibular ligament. As the latter ligament is fused with the fibrous capsule of the ankle joint, this part of the capsule may also be torn. The fibularis (peroneus) brevis muscle (Figs. 5-56 and 5-62B) tends to prevent overinversion of the foot, thereby aiding the lateral ligaments in preventing severe inversion injuries of the ankle joint. Violent inversion of the foot may result in avulsion of the tuberosity of the fifth metatarsal bone, into which the tendon of fibularis brevis muscle inserts. For this reason, radiologists routinely ensure that this bone is visible in radiographs of persons with ankle injuries so they may look for avulsion of the tuberosity of the fifth metatarsal.

Blood Supply of the Ankle Joint (Figs. 5-29, 5-60, and 5-83). The articular arteries are derived from the malleolar branches of the fibular (peroneal) and anterior and posterior tibial arteries.

Nerve Supply of the Ankle Joint (Figs. 5-49 and 5-72). The articular nerves are derived from the tibial nerve and the deep fibular (peroneal) nerve, a division of the common fibular (peroneal) nerve.

Joints of the Foot

There are many joints in the foot involving the tarsal and metatarsal bones and the phalanges (Figs. 5-2, 5-3, and 5-101). The important **intertarsal joints** are the subtalar, talocalcaneonavicular, and calcaneocuboid joints. The other joints are relatively small and are so tightly joined by ligaments that only slight movement occurs between them. All the foot bones are united by dorsal and plantar ligaments (Figs. 5-96 and 5-102).

The Subtalar Joint[12]

The subtalar (talocalcanean) joint is distal to the ankle joint, where the talus rests on and articulates with the calcaneus (Figs. 5-96 and 5-101). The subtalar joint is a synovial joint between the inferior surface of the body of the talus and the superior surface of the calcaneus. It is surrounded by an **articular capsule**, which is attached near the margins of the articular facets. The fibrous capsule is weak but it is supported by medial, lateral,

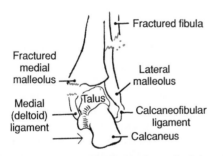

Figure 5-99. Fracture-dislocation (Pott's fracture) of the ankle joint caused by forced eversion of the right foot. Note that the strong medial (deltoid) ligament has not ruptured, but has avulsed the medial malleolus. Observe the associated fracture of the fibula.

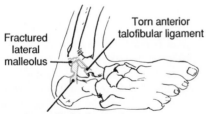

Figure 5-100. Torn calcaneofibular and talofibular ligaments and a fracture of the lateral malleolus, resulting from forced inversion of the right foot.

---

[12]Anatomically the subtalar joint is the commonly used term for the talocalcaneal joint. However, many orthopedic surgeons use the designation subtalar joint as a functional term to include both the talocalcaneal and the talocalcaneal part of the talocalcaneonavicular joint, *i.e.*, they use the term subtalar joint as a composite term for the two joints inferior to the talus.

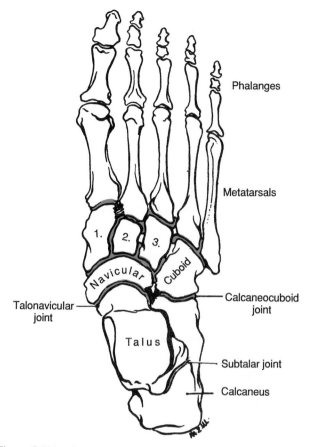

Phalanges

Metatarsals

Talonavicular joint

Navicular

Cuboid

Calcaneocuboid joint

Talus

Subtalar joint

Calcaneus

Figure 5-101. Dorsal aspect of the bones of the right foot, showing the six separate joint cavities (red). The transverse tarsal joint is the articular plane that extends from side-to-side across the foot. It is composed of the talocalcaneonavicular joint medially and the calcaneocuboid joint laterally. Although anatomically separate, these joints act together during movements of the foot.

and posterior *talocalcaneal ligaments*. In addition, it is supported anteriorly by the interosseous talocalcanean ligament. The fibrous capsule is lined with synovial membrane.

Movements of the Subtalar Joint (Fig. 5-61). Inversion and eversion of the foot are the main movements that occur at this joint. The joint permits slight gliding and rotation that assist with inversion and eversion of the posterior part of the foot. The muscles producing these movements are listed in Table 5-13. Movements of the subtalar joint are closely associated with those at the talocalcaneonavicular and calcaneocuboid joints (parts of the transverse tarsal joint).

The Talocalcaneonavicular Joint

This joint is located where the head of the talus articulates with the socket formed by the posterior surface of the navicular bone, the superior surface of the plantar calcaneonavicular ligament, the sustentaculum tali, and the articular surface of the calcaneus (Figs. 5-96, 5-98 and 5-101). The talocalcaneonavicular articulation is a synovial joint of the ball and socket type. The ball is the head of the talus and the socket comprises two bones and two ligaments. The talocalcaneonavicular joint is *part*

of the transverse tarsal joint. The head of the talus has three facets, one for the navicular and two for the calcaneus. All these articular surfaces are surrounded by a single articular capsule that blends with the interosseous talocalcaneal ligament posteriorly. The talocalcaneonavicular joint is reinforced dorsally by the dorsal talonavicular ligament, a broad band connecting the neck of the talus and the dorsal surface of the navicular bone. The *plantar calcaneonavicular ligament* (''**spring ligament**''), is a triangular band that extends from the sustentaculum tali to the posteroinferior surface of the navicular bone. This ligament blends with the deltoid ligament medially and forms part of the socket for the head of the talus. It plays an important role in maintaining the longitudinal arch of the foot (Figs. 5-102 and 5-105).

The Calcaneocuboid Joint

This articulation is a synovial joint between the anterior surface of the calcaneus and the posterior surface of the cuboid (Figs. 5-55, 5-96, and 5-101). It is *part of the transverse tarsal joint.* The capsule of the calcaneocuboid joint is strengthened by the dorsal calcaneocuboid ligament and plantar calcaneocuboid ligaments (Fig. 5-98). This joint is also supported by the long plantar ligament.

The *long plantar ligament* (Fig. 5-102) covers the plantar surface of the calcaneus. It passes from the plantar surface of the calcaneus, including its anterior and posterior tubercles, to both lips of the groove on the cuboid bone. Some of its fibers extend to the bases of the second, third, and fourth metatarsal bones, thereby forming a tunnel for the tendon of the fibularis (peroneus) longus muscle. This tendon passes through the groove in the cuboid bone to insert into the base of the first metatarsal and the adjoining part of the medial cuneiform bone. The long plantar ligament is important in maintaining the arches of the foot (Fig. 5-105). The *plantar calcaneocuboid ligament* (short plantar ligament) is deep to the long plantar ligament (Fig. 5-102). It extends from the anterior aspect of the inferior surface of the calcaneus to the inferior surface of the cuboid.

The Transverse Tarsal Joint

The talocalcaneonavicular and calcaneocuboid joints are separated from each other, but together they constitute the transverse tarsal joint or midtarsal joint (Figs. 5-55, 5-96, and 5-101). They extend across the tarsus and lie in almost the same transverse plane.

Movements of the Transverse Tarsal Joint (Fig. 5-61C and D and Table 5-13). Movements occurring at this joint produce *inversion and eversion of the foot.* The muscles producing these movements are listed in Table 5-13. During inversion, the foot is adducted and directed so that its medial border is raised and its lateral border is depressed. In other words, inversion directs the sole of the foot toward the median plane of the body (Figs. 5-61D and 5-75). During eversion, the foot is abducted and directed so that the lateral border is raised and the medial border lowered. In other words, eversion directs the sole of the foot away from the median plane of the body. *Inversion and eversion of the feet* commonly occur during walking on cobblestones or rough ground as the feet adjust to the stones or

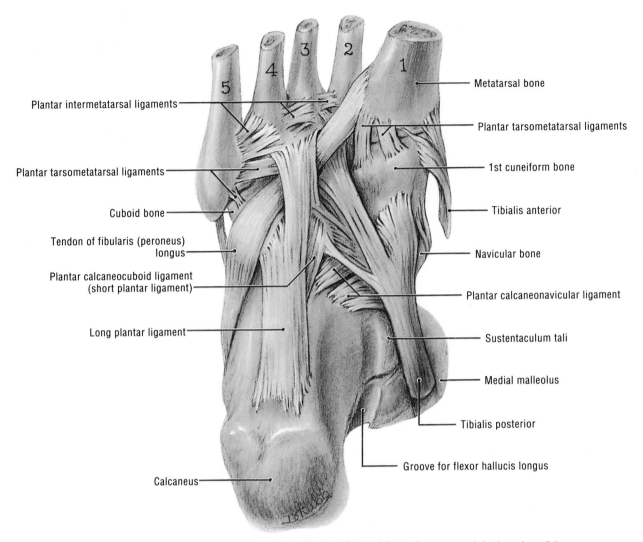

Figure 5-102.   Dissection of the sole of the right foot showing the plantar ligaments and the insertion of three long tendons: the fibularis (peroneus) longus, tibialis anterior, and tibialis posterior.

Table 5-13.
The Main Muscles Producing Movements of the Intertarsal Joints[1]

Inversion[2]	Eversion[2]	Plantarflexion
Tibialis anterior	**Fibularis (peroneus) longus**	**Tibialis posterior**
Tibialis posterior	**Fibularis (peroneus) brevis**	**Abductor hallucis**
	Fibularis (peroneus) tertius	**Abductor digiti minimi**
		Flexor digitorum brevis
		Fibularis (peroneus) longus
		Fibularis (peroneus) brevis

[1]**Bold face** indicates the muscles that are chiefly responsible for the movement; the other muscles assist them.
[2]Inversion and eversion occur mainly at the subtalar and transverse tarsal joints (Fig. 5-101).

depressions. The strong medial (deltoid) ligament tends to prevent overeversion of the foot; the weaker lateral ligaments (with the assistance of the fibularis [peroneus] longus and brevis muscles) tend to prevent overinversion of the foot.

The Tarsometatarsal Joints

These articulations are of the plane type of synovial joint that permit only gliding or sliding movements (Figs. 5-96, 5-101, and 5-102). The four anterior tarsal bones articulate with the bases of the metatarsal bones. The metatarsal bones are firmly attached to the tarsal bones by dorsal, plantar, and interosseous ligaments. There are three separate tarsometatarsal joint cavities.

The medial tarsometatarsal joint occurs between the medial cuneiform bone and the base of the first metatarsal bone. The medial tarsometatarsal joint has more range of movement than the other two joints.

The intermediate tarsometatarsal joint is between all three cuneiform bones and the second and third metatarsal bones. The base of the second metatarsal bone fits into a socket formed by the cuneiform bones; thus the intermediate tarsometatarsal joint is the strongest of the three tarsometatarsal joints. As the second metatarsal bone is firmly attached to the cuneiform bones, it has little independent movement. The cavity of the intermediate tarsometatarsal joint is continuous with that between the two medial cuneiform bones and through it with the synovial cavity of the cuneonavicular joint.

The lateral tarsometatarsal joint is between the cuboid and the fourth and fifth metatarsal bones (Figs. 5-101 and 5-102). There is more movement permitted at this joint than at the intermediate tarsometatarsal joint, but not so much as at the medial tarsometatarsal joint.

> Because the second metatarsal bone has little movement at the second tarsometatarsal joint, it is particularly liable to fracture when sudden, unaccustomed stresses are applied to the distal part of the foot. For example, when a person who is "out of condition" begins to participate in a strenuous exercise program, long walks, or track and field activities, a *stress fracture* may occur in one of the weightbearing bones. Stress fractures of the metatarsals are sometimes referred to as "march fractures" or "fatigue fractures."

The Intermetatarsal Joints

These articulations between the bases of the metatarsal bones are the plane type of synovial joint that permits a slight gliding movement (Figs. 5-101 and 5-102). Their joint cavities are extensions of the tarsometatarsal joints. The bases of the second to fifth metatarsal bones are very firmly bound together by dorsal, plantar, and interosseous ligaments. The *deep transverse metatarsal ligament* connects the heads of the metatarsal bones. This ligament, along with the interosseous ligaments, helps to maintain the transverse arch of the foot. Owing to the tight binding of the bases of the metatarsal bones together, little individual movement of the metatarsal bones is possible.

The Metatarsophalangeal Joints

These articulations between the heads of the metatarsals and the bases of the proximal phalanges are the knucklelike condyloid type of synovial joint (Fig. 5-101 and p. 19). They permit flexion, extension, and some abduction, adduction, and circumduction (Table 5-14). The first metatarsophalangeal joint is by far the largest articulation because of the size of the head of the first metatarsal bone and the presence of the *sesamoid bones* in the two tendons of the flexor hallucis brevis muscle (Fig. 5-81). An *articular capsule* surrounds each joint and is attached near the margins of the articular surfaces of the involved bones. The articular surfaces pass well onto the dorsal and plantar surfaces of the metatarsal bones (Figs. 5-51 and 5-52). The articular surfaces of the first metatarsophalangeal joint are particularly large and are related to dorsiflexion of the great toe during walking. The *fibrous capsules* of the metatarsophalangeal joints are strengthened on each side by thick collateral ligaments. The plantar part of the capsule is greatly thickened to form the *plantar ligament* (Fig. 5-81). This fibrocartilaginous plate is firmly attached to the proximal border of the phalanx and forms part of the socket for the head of the first metatarsal. The margins of the plantar ligament give attachment to the fibrous flexor sheath, to slips of the plantar aponeurosis, and to the deep transverse metatarsal ligaments (Fig. 5-76).

> The first metatarsophalangeal joint may become enlarged and deformed with permanent lateral displacement of the great toe (L. hallux). This condition, known as *hallux valgus* (Fig. 5-103), is common in persons who wear pointed shoes. They are unable to move their great toe away from their second toe because the sesamoids under the head of the first metatarsal bone (Fig. 5-81) are usually displaced and lie in the space between the heads of the first and second metatarsal bones. **Gout**, a metabolic disorder, is characterized by urate deposits in connective tissue, including cartilage and bone. Gout commonly affects the first metatarsophalangeal joint,

Table 5-14.
The Main Muscles Producing Movements of the Metatarsophalangeal Joints[1]

Flexion	Extension	Abduction	Adduction
Flexor digitorum brevis	**Extensor hallucis longus**	**Abductor hallucis**	**Adductor hallucis**
Lumbricals	**Extensor digitorum longus**	**Abductor digiti minimi**	**Plantar interossei**
Interossei	**Extensor digitorum brevis**	**Dorsal interossei**	
Flexor hallucis brevis			
Flexor hallucis longus			
Flexor digiti minimi brevis			
Flexor digitorum longus			

[1]**Bold face** indicates the muscles that are chiefly responsible for the movement; the other muscles assist them.

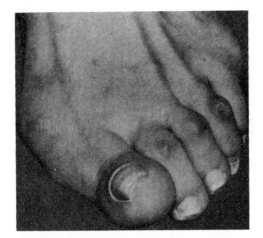

Figure 5-103. The left foot of a woman with hallux valgus and a localized swelling (bunion) on the medial aspect of her first metatarsophalangeal joint. She also has hard corns over the proximal interphalangeal joints of her lateral four digits.

which becomes swollen and painful, a condition referred to as *gouty arthritis*. Degenerative joint disease (*osteoarthritis*) is also common in the metatarsophalangeal joint of the great toe. Often it is the first joint to be affected.

The Interphalangeal Joints

The interphalangeal joints are between the head of one phalanx and the base of the one distal to it (Fig. 5-101). They are the hinge type of synovial joint, permitting only flexion and extension (Fig. 5-15). In most people the lateral four digits are, to a varying degree, partially flexed at the interphalangeal joints at all times.

Hammer toe is a common deformity in which the proximal phalanx is permanently dorsiflexed at the metatarsophalangeal joint and the middle phalanx is plantarflexed at the interphalangeal joint. The distal phalanx is also flexed or extended, giving the digit (usually the second) a hammerlike appearance. This deformity may result from weakness of the lumbrical and interosseus muscles, which flex the metatarsophalangeal joints and extend the interphalangeal joints (Fig. 5-79; Table 5-9).

The Arches of the Foot

The bones of the foot are arranged in longitudinal and transverse arches, which are designed as shock absorbers for supporting the weight of the body and for propelling it during movement. The constituent bones of the arches can be seen in the articulated skeleton of the foot (Figs. 5-53 to 5-55 and 5-105). The design of the foot makes it adaptable to surface and weight changes. The resilient arches of the foot provide it with this adaptability. The weight of the body is transmitted to the talus from the tibia and fibula (Fig. 5-97). Then it is transferred posteroinferiorly to the calcaneus and anteroinferiorly to the heads of the metatarsal bones (Figs. 5-104 and 5-105). Body weight is divided about equally between the calcaneus and the heads of

Table 5-15.
The Main Muscles Producing Movements of the Interphalangeal Joints[1]

Flexion	Extension
Flexor hallucis longus	**Extensor hallucis longus**
Flexor digitorum longus	**Extensor digitorum longus**
Flexor digitorum brevis	**Extensor digitorum brevis**
Quadratus plantae	

[1]**Bold face** indicates the muscles that are chiefly responsible for the movement; the other muscles assist them.

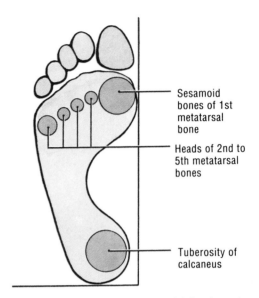

Figure 5-104. A footprint illustrating the weightbearing points of the right foot. Body weight is divided about equally between the calcaneus and the heads of the metatarsal bones. The anterior part of the foot has six points of contact with the ground: two with the sesamoid bones associated with the head of the first metatarsal bone and four with the heads of the lateral four metatarsals. Hence, the first metatarsal bone supports a double load.

the metatarsal bones. Between these weightbearing points are the arches of the foot, formed by the tarsal and metatarsal bones. The relatively elastic arches, convex superiorly and disposed both longitudinally and transversely, become slightly flattened by the body weight during standing, but they normally resume their curvature when the body weight is removed (*e.g.*, during sitting). There are two arches, longitudinal and transverse.

The Longitudinal Arch of the Foot

For purposes of description, the longitudinal arch is regarded as being composed of medial and lateral parts (Figs. 5-53, 5-54, and 5-105). The medial longitudinal arch is higher and more important. Functionally, both parts of the longitudinal arch act as a unit with the transverse arch, spreading the weight in all directions. The *medial part of the longitudinal arch* is obvious when the normal living foot is examined from the medial aspect. A normal *footprint* also shows the longitudinal arch (Fig. 5-104) and reveals that the weightbearing parts of the foot are the heel, the lateral border of the sole, and the *ball of the foot*, which is formed by the heads of the metatarsal bones. The medial longitudinal arch is composed of the calcaneus, talus, navicular, three cuneiforms, and three metatarsal bones (Figs. 5-53 to 5-55 and 5-105). *The head of the talus is the "keystone."* It is located at the summit of the medial longitudinal arch and receives the weight of the body. At the articular surfaces between the talus and navicular, and between the navicular and the three cuneiforms, the medial longitudinal arch yields slightly when weight is put on it and recoils when weight is removed. The *lateral part of the longitudinal arch* is much flatter than the medial part of the arch and rests on the ground during standing (Fig. 5-104). It is composed of the calcaneus, cuboid, and lateral two metatarsal bones (Figs. 5-54 and 5-105).

The Transverse Arch of the Foot

This arch runs from side to side. It is formed by the cuboid, the three cuneiforms, and the bases of the metatarsal bones. The cuneiform bones are wedge-shaped and are therefore constructed to maintain the transverse arch. The medial and lateral parts of the longitudinal arch serve as pillars for the transverse arch. The tendon of the fibularis (peroneus) longus muscle, crossing the sole of the foot obliquely, helps to maintain the curvature of the transverse arch (Fig. 5-102). The plantar calcaneonavicular (spring) ligament, a strong fibrocartilaginous band, acts as a tie between the calcaneus and navicular bones, preventing collapse of the medial longitudinal arch.

Maintenance of the Arches of the Foot (Figs. 5-52, 5-76, and 5-102). The integrity of the bony arches of the foot is maintained by: (1) the shape of the interlocking bones; (2) the strength of the plantar ligaments and the plantar aponeurosis; and (3) the action of muscles through the bracing action of their tendons. Of these three factors, the plantar ligaments and the plantar aponeurosis bear the greatest stress and are most important in maintaining the arches of the foot. *Electromyographic studies* indicate that the muscles are relatively inactive until walking begins. The inversion and eversion muscles of the foot (Table

5-13) appear to control weight distribution in the foot (*e.g.*, when walking on rough ground). The maintenance of the arches of the foot is also dependent on the intertarsal, tarsometatarsal, and intermetatarsal joints, because the bones of these articulations are bound together as parts of the arches. The plantar ligaments of these joints are the strongest and they are supported by robust interosseous ligaments. The following fibrous structures, *listed in order of importance*, are essential for maintaining the arches of the foot.

The plantar calcaneonavicular ("spring") ligament is the most important ligament because it is the main supporter of the medial longitudinal arch of the foot (Figs. 5-98 and 5-102). Its principal attachments are the sustentaculum tali of the calcaneus and the tuberosity of the navicular bone (Fig. 5-53). The plantar calcaneonavicular ligament is a strong fibrocartilaginous band that acts as a tie between the calcaneus and navicular bones, preventing collapse of the medial longitudinal arch.

The *long plantar ligament* is the next most important ligament for supporting the arches of the foot (Fig. 5-102). Its principal attachments are the tubercle of the calcaneus and the plantar surface of the cuboid bone. The long plantar ligament is longer and more superficial than the plantar calcaneonavicular ligament (Fig. 5-102). It stretches like a tie beam under nearly the whole length of the lateral longitudinal arch, thus the long plantar ligament provides the main support for the lateral longitudinal arch of the foot.

The *plantar aponeurosis* (Fig. 5-76 and p. 463) acts as a strong tie-beam for the maintenance of the longitudinal arch. One part of it, a dense fibrous band known as the *calcaneometatarsal ligament*, extends from the lateral process of the tuberosity of the calcaneus to the tuberosity of the fifth metatarsal bone. It is particularly important in helping the long plantar ligament maintain the lateral longitudinal arch. The medial part of the plantar aponeurosis, through its attachment to the sesamoid bones of the flexor hallucis brevis (Fig. 5-81), is important in strengthening the medial longitudinal arch when standing on the toes.

The *plantar calcaneocuboid (short plantar) ligament* is short and deep to the long plantar ligament (Fig. 5-102). It is attached to the anterior end of the calcaneus and the proximal edge of the cuboid. This ligament aids the plantar calcaneonavicular ligament and the long plantar ligament in supporting the longitudinal arch. The fibularis (peroneus) longus tendon forms a sling beneath the lateral part of the longitudinal arch and acts as a tie beam for the transverse arch (Fig. 5-102).

In infants the flat appearance of the feet is normal and results from the subcutaneous fatpads in the soles of their feet. The arches of the foot are present at birth, but they do not become visible until after the infant has walked for a few months. **Flatfeet** are common and are not painful in some people. Flatfeet in adolescents and adults are caused by "fallen arches," usually the medial parts of the longitudinal arches. During standing the plantar ligaments and plantar aponeurosis, important in maintaining the arches of the foot, stretch somewhat under the body weight. If these ligaments become abnormally stretched during long periods of standing, the plantar calcaneonavicular (spring) ligament can no longer adequately support the head of the talus. As a result, some

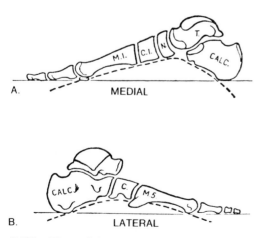

Figure 5-105. The medial and lateral longitudinal arches of the right foot. Observe that the foot is arched longitudinally and that the posterior pillar of the medial and lateral parts of the arch is the calcaneus. See Figs. 5-53, 5-54, and 5-105 for the bones forming these arches. Calc. indicates the calcaneus; T, the talus; N, navicular; C.1., the first cuneiform; M.1. and M5, the first and fifth metacarpals.

flattening of the medial part of the longitudinal arch occurs and there is concomitant lateral deviation of the forefoot.

In the common type of flatfoot, the foot resumes its arched form when the weight is removed from it. *Flatfeet are common in older persons*, particularly if they undertake much unaccustomed standing or gain weight rapidly. This results from the added stress on the muscles and the increased strain on the ligaments supporting the arches. Fallen arches cause pain owing to stretching of the plantar muscles and straining of the plantar ligaments. Flattening of the transverse arch of the foot may also occur. Frequently callus formation occurs on the plantar surfaces of the heads of the lateral four metatarsal bones (Fig. 5-104). This thickening of the skin develops as a protective measure where abnormal pressure is exerted.

PATIENT ORIENTED PROBLEMS

Case 5-1

Your elderly grandmother slipped on the polished floor in the front hall. As you approached her, she was lying on her back in severe pain. Her right lower limb immediately attracted your attention because it was laterally rotated and noticeably shorter than her left limb. She was unable to get up or lift her limb off the floor, and when she attempted to do so, she experienced excruciating pain.

> **Problems.** In view of the abnormal position of her leg, what bone was probably fractured? Name the common fracture site of this bone in elderly people. Why is this bone so fragile? Explain anatomically why her injured limb was shorter than the other one. What are the anatomical reasons for the complications (nonunion and avascular necrosis) commonly associated with these fractures? These problems are discussed on p. 497.

Case 5-2

While playing in an old-timer's hockey game, a 45-year-old man was accidentally kicked with a skate on the lateral surface of his right leg just inferior to the knee. The superficial wound was treated by the trainer, but the man was unable to continue playing because of pain in the region of the laceration and loss of power in his leg and foot. He also experienced numbness and tingling on the lateral surface of his leg and the dorsum of his foot. When he removed his skate, he found that he was unable to move his right foot or his toes superiorly (*i.e.*, dorsiflex them). He was advised to see his physician immediately. As he walked into the examining room, the physician observed that he had an *abnormal gait*—he raised his right foot higher than usual and brought it down suddenly, making a flapping noise. During the physical examination, the physician detected tenderness over the head and neck regions of the patient's fibula and a sensory deficit on the lateral side of the distal part of his leg, including the dorsum of his foot. Radiographs were taken.

Diagnosis. Fracture of the neck of the fibula and a peripheral nerve injury.

> **Problems.** What is the anatomical basis of the loss of sensation and impaired function in the patient's foot? What

nerve appears to have been injured? What is its relationship to the neck of the fibula? If the skate blade had not severed the nerve, what probably would have injured it if he had continued to play? What is the name given to the foot condition exhibited by the patient when he walked? These problems are discussed on p. 498.

Case 5-3

While a 26-year-old worker was loading a heavy crate, it fell on his knee. He suffered severe pain and was unable to get up. The first aid team carried him to the physician's office on a stretcher. Following a physical examination, the physician requested radiographs of the man's knee.

Diagnosis. Comminuted fractures of the proximal end of the tibia and a fracture of the neck of the fibula.

> **Problems.** What artery or arteries might have been torn by the bone fragments? Using your anatomical knowledge, where would you check the patient's pulse to determine whether there has been damage to these arteries? What nerve may have been injured by the fracture of the fibula? These problems are discussed on p. 498.

Case 5-4

A 32-year-old man slipped on a patch of ice and fell. After he was helped up, he was unable to bear weight on his right foot. When he noticed that his ankle was beginning to swell, he hailed a cab and went to the hospital for treatment of what he thought was a badly "sprained ankle." On examination it was found that the patient could barely move his ankle because of pain. Maximum tenderness was located over the lateral malleolus, about 2 cm proximal to its tip. Radiographs of the ankle were taken.

Diagnosis. Transverse fracture of the lateral malleolus at the level of the superior articular surface of the talus.

> **Problems.** What excessive movement usually results in a sprained ankle? Discuss what is meant by the term "sprain." Explain anatomically how this fracture probably occurred. What structures were probably torn or ruptured? Does the patient have what is usually referred to as a *Pott's fracture*? These problems are discussed on p. 499.

Case 5-5

A 22-year-old woman was a front seat passenger in a car that was involved in a head-on collision. Although she sustained head injuries, her chief complaint was a sore right hip that prevented her from standing up. Believing that she might have broken her hip, she was rushed to the nearest hospital. A physical examination revealed that her lower limb was slightly flexed, adducted, medially rotated, and appeared shorter than the other limb (Fig. 5-106). Radiographs were taken.

Diagnosis. Posterior dislocation of the right hip joint with a fracture of the posterior margin of the acetabulum.

> **Problems.** Explain anatomically how this injury probably occurred. What nerve may have been injured? When

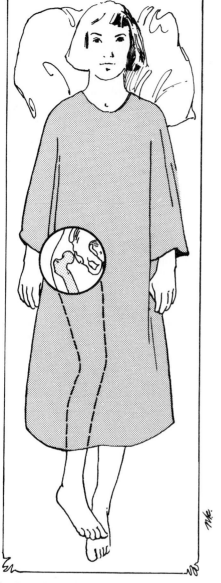

Figure 5-106. Posterior dislocation of the right hip. Observe that the women's lower limb is flexed, adducted, medially rotated, and is shorter than her left limb.

paralysis of this nerve is complete, what muscles are paralyzed. Where may cutaneous sensation be lost? These problems are discussed on p. 499.

Case 5-6

A 62-year-old man presented with an aching pain in his left buttock that extended along the posterior aspect of his left thigh. During the examination the patient pointed to the area where he felt most pain, which was in the region of the greater sciatic notch. Tenderness was also elicited by pressure along a line beginning from a point midway between the top of the greater trochanter of his femur and the ischial tuberosity to a point in the midline of the thigh about halfway to the knee. When seated, the patient was unable to extend his left leg fully because of

severe pain. With the patient in the supine position, the physician grasped the patient's left ankle and placed his other hand on the anterior aspect of the knee to keep the leg straight. He then slowly raised the left lower limb; when it reached about a 75 degree angle the man grimaced with pain. Even more pain was elicited when the patient's foot was dorsiflexed.

> **Problems.** What nerve is involved in this case? From which segments of the spinal cord does it arise? Why does the *straight leg-raising test* elicit pain? Why did the pain increase when his foot was dorsiflexed? What back lesion probably produced the pain in the buttock and posterior thigh region? Thinking anatomically, what other lesions (*e.g.*, resulting from disease or injury) do you think might have caused the patient's symptoms? These problems are discussed on p. 499.

Case 5-7

A football player was clipped (blocked from the rear) as he was about to tackle the ball carrier. The lineman's hip hit the runner's knee from the side. It was obvious on the slow motion videotape replay that the tackler's knee was slightly flexed and his foot firmly planted on the turf when he was hit. As he lay on the ground clutching his knee, it was obvious from his face that he was in severe pain. While he was being helped to the sidelines, you said to your friend, "I'm afraid he has torn knee ligaments." Not knowing much about the functioning of the knee joint, your friend said, "Which knee ligaments are probably torn?"

> **Problems.** How would you explain this injury to your friend, assuming that he has little knowledge of the anatomy of the knee joint? What ligament was probably ruptured? What ligament may have been torn? Would the menisci be injured? These problems are discussed on p. 499.

DISCUSSION OF PATIENTS' PROBLEMS

Case 5-1

Very likely your grandmother fractured the neck of her femur, a common injury in elderly women (Fig. 5-107). This fracture is often wrongly referred to as a "fractured hip," implying that the hip bone is broken. As your grandmother stumbled and tried to "catch herself," she probably exerted a torsional force on one hip, producing a fracture of the femoral neck, the most fragile part of the femur. She fell when the bone fractured; hence, the fracture was probably the cause of her fall rather than the result of it. Lateral rotation and shortening of the injured limb are characteristic clinical features following fractures of the femoral neck. The rotation results from the change in the axis of the limb owing to the separation of the body and head of the femur. The shortening of the lower limb results from the superior pull of the muscles connecting the femur to the hip bone (Fig. 5-22; Table 5-2). Spasm of the muscles (sudden involuntary muscular contractions) causes the pull.

The body's total bone mass becomes progressively diminished

often not patent in elderly patients because they commonly have arteriosclerosis. Sometimes other blood vessels supplying the femoral head are torn when the femoral neck fractures. Generally, the more proximal the fracture, the greater are the chances of interrupting the vascular supply.

A poor blood supply may result in nonunion and *avascular necrosis* of the femoral head (death and collapse of the proximal bone fragment owing to poor blood supply). *Intracapsular fractures* (high in the neck) almost always present healing problems because they usually interfere with the blood supply to the proximal bone fragment. The importance of preserving the blood supply to the proximal part of the femur is one reason why patients with this type of injury are handled with extreme care; another is that this injury is very painful.

Case 5-2

The close relationship of the *common fibular (peroneal) nerve* to the neck of the fibula makes it vulnerable to injury when this region of the bone is fractured. As the nerve lies on the lateral aspect of the neck of the fibula, it can be easily injured by superficial lacerations. The present patient's signs and symptoms make it obvious that the common fibular nerve was injured. Superficial wounds, prolonged pressure by hard objects (*e.g.*, the sharp edge of a bed during sleep), or compression by a tight plaster cast may present similar clinical features.

Injury to the common fibular nerve (Fig. 5-73*B*) affects muscles (1) in the lateral compartment of the leg (the fibularis [peroneus] longus and brevis muscles supplied by the superficial fibular [peroneal] nerve) and (2) in the anterior compartment of the leg (muscles in the anterior part of the leg and the extensor digitorum brevis supplied by the deep fibular [peroneal] nerve). Consequently, eversion and dorsiflexion of the foot and extension of the toes are impaired (Tables 5-5 and 5-6). This patient showed a characteristic *foot-drop* (plantarflexion and slight inversion) and stepping gait. As the patient walked, his toes dragged and his foot slapped the floor. In an attempt to prevent this from happening, he raised his foot higher than usual.

The *dysesthesia* (impairment of sensation) on the patient's leg and foot resulted from injury to sensory fibers in the cutaneous branches of the common fibular nerve (Figs. 5-10 and 5-82). The injury to the nerve resulted from the skate grazing the nerve or the nerve being compressed or torn by the bone fragments. Although the fibula is not a weightbearing bone, fractures of its proximal end cause pain on walking because the pull of muscles attached to it causes the fragments to move, which is painful.

Case 5-3

Because the *popliteal artery* lies deep in the popliteal fossa against the fibrous capsule of the knee joint, it could have been torn by fragments from the comminuted fractures of the proximal ends of the tibia and fibula. The popliteal artery divides into its terminal branches (*anterior and posterior tibial arteries*) at the inferior end of the popliteal fossa, therefore these vessels may also have been torn when the bones fractured. Undoubtedly one or more of the five *genicular arteries*, branches of the popliteal artery, were also torn. They supply the articular capsule and ligaments of the knee joint (Figs. 5-46 and 5-91). Pulsations of the posterior tibial artery may be felt halfway between the medial

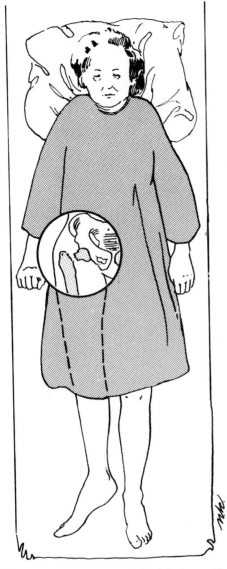

Figure 5-107. Fracture of the neck of the right femur. Observe that there is shortening and lateral rotation of the injured lower limb.

with advancing age, a skeletal disorder called *osteoporosis*.[13] As a result, the femoral neck becomes weaker. Fractures of the proximal part of the femur can result from little or no trauma. In this bone disorder of postmenopausal women and elderly men, absorption of bone is greater than bone formation. The blood vessels to the proximal part of the femur are derived mostly from the medial and lateral *circumflex femoral arteries* (Fig. 5-29). Branches of these arteries run in the retinacula of the fibrous capsule of the hip joint. A variable amount of blood may be supplied to the femoral head through a branch of the obturator artery that runs in the ligament of the head of the femur, called the *artery of the ligament of the head*. This ligament may be ruptured during fractures of the femoral neck. This vessel is

---

[13]The problem is not failure of new bone to become adequately calcified. The disorder results primarily because the amount of new bone matrix produced is insufficient to keep pace with bone resorption. As a result the body's total bone mass becomes progressively diminished (Cormack, 1987).

malleolus and the heel. The *dorsalis pedis artery*, the continuation of the anterior tibial artery, can also be palpated where it passes over the navicular and cuneiform bones in the foot (Fig. 5-57). These are good places to take the pulse of the arteries because they are superficial here and can be compressed against bones. Loss of a pulse in these arteries in the present case would have suggested a torn popliteal and/or tibial artery.

There is also a chance that the *tibial nerve* was injured in this patient, as it is the most superficial of the three main central structures in the popliteal fossa. Severance of the tibial nerve results in paralysis of the popliteus and muscles of the calf (gastrocnemius, soleus, flexor hallucis longus, and tibialis posterior), together with those in the sole of the foot (Fig. 5-73). Probably some of the genicular branches (articular nerves) to the knee joint might also be severed. The close relationship of the common fibular nerve to the neck of the fibula (Fig. 5-49A) makes it vulnerable to injury when this part of the bone is fractured. For the signs and symptoms resulting from severance of this nerve, see the discussion of Case 5-2 (p. 498).

Case 5-4

The usual sprained ankle results from excessive inversion of the weightbearing foot, which causes rupture of the anterolateral portion of the fibrous capsule of the ankle joint and the calcaneofibular and talofibular ligaments (Fig. 5-100). The term *sprain* is used to indicate some degree of tearing of the ligaments without fracture or dislocation. In severe *sprains* many fibers of the ligaments are completely torn, and often, considerable instability of the ankle joint results. In the present case the severe sprain and fracture occurred when the patient slipped in such a way that his foot was forced into an excessively inverted position. His body weight then caused a forceful inversion of the ankle joint. Probably the calcaneofibular and anterior talofibular ligaments were partly or completely torn. Normally the deep mortise formed by the distal end of the tibia and malleoli holds the talus firmly in position (Fig. 5-97). When the ankle ligaments tear, the talus is forcibly tilted against the lateral malleolus, shearing it off (Fig. 5-100).

Had the man's ankle been forced in the opposite direction (*i.e.*, in an extremely everted position), the strong *medial (deltoid) ligament* may have avulsed the medial malleolus (Fig. 5-99). As the force continued, it would have tilted the talus, moving it and the lateral malleolus laterally. Because the interosseous tibiofibular ligament acts as a pivot, the fibula breaks proximal to the superior tibiofibular joint.

Probably this patient's injury was not *Pott's fracture of the ankle joint*, because it is not the kind of fracture-dislocation sustained and described by Dr. Percival Pott, an English surgeon.

Case 5-5

Dislocation of the hip joint is uncommon owing to the stability of this articulation. The head of the femur is deeply seated in the acetabulum and is held there by an exceedingly strong fibrous capsule. Traumatic dislocations of the hip joint may occur during automobile accidents when the hip joint is flexed, and the thigh is adducted and medially rotated. Very likely, the patient's knee struck the dashboard when her right lower limb was in the po-

sition just described. Consequently the force was transmitted along the femur, driving its head and the posterior margin of the acetabulum posteriorly. As the head in this position is covered posteriorly by capsule rather than bone, the articular capsule probably ruptured inferiorly and posteriorly. This permitted the head to dislocate posteriorly and carry the fractured posterior margin of the acetabulum and acetabular labrum with it. As a result, the head of the femur came to lie on the gluteal surface of the ilium.

The close relationship of the *sciatic nerve* (L4 to S3) to the posterior surface of the hip joint makes it vulnerable to injury in posterior dislocations. If the paralysis is complete, which is rarely the case, the hamstring muscles and those distal to the knee would all be paralyzed (Fig. 5-73B). In addition, there probably would be anesthesia in the lower leg and foot, except for the skin on the medial side, which is supplied by the *saphenous nerve* (L3 and L4), a terminal branch of the femoral nerve.

Case 5-6

The site of the patient's pain and its course down the posterior aspect of the thigh clearly indicates damage to the *sciatic nerve*, the largest branch of the sacral plexus. It arises from spinal cord segments L4 to S3. The sciatic nerve leaves the pelvis through the inferior part of the greater sciatic notch and extends from the inferior border of the piriformis muscle to the distal third of the thigh, along the course clearly indicated by the patient's pain. The *straight leg-raising test* elicits pain because the sciatic nerve is stretched when the limb is raised. Dorsiflexion of the foot further increases the pull on the sciatic nerve and its roots.

A posterolateral *protrusion of an intervertebral disc* (see Figs. 4-29B and 4-32) is a common cause of sciatica, most often affecting the first sacral nerve roots (p. 347). A protruding L5/S1 disc exerts pressure on the dorsal and ventral roots, producing *sciatica*, which may be accompanied by low back pain. Sciatic pain can also result from pressure (*e.g.*, a tumor) on the sciatic nerve or its components in the pelvis, in the gluteal region, or in the thigh. Pain can also be produced by (1) irritation of the sciatic nerve resulting from inflammation of the nerve (neuritis) or its sheath or (2) compression of the nerve (see ''driver's thigh,'' p. 417).

Case 5-7

The knee joint is one of the most secure joints in the body, particularly when it is extended. Although these bones are bound together by strong ligaments, the knee is subject to a wide range of injuries because of the severe strain that is placed on its attachments, especially in contact sports like hockey and football. The blow on the lateral side of the tackler's knee by the lineman's hip occurred while he was running and his foot was fixed on the ground. As he was bearing weight on his leg, the hard blow bent the runner's knee medially relative to the fixed tibia. This severely stressed the *tibial collateral ligament*. Some fibers of this ligament may have ruptured and he may only have a sprain; however, because of the severity of the blow, probably the entire ligament ruptured near its attachment to the medial femoral epicondyle. Because the *medial meniscus* is attached to this ligament, it may tear or become detached from it.

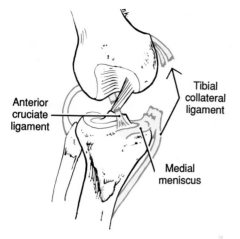

Figure 5-108.   Anterior view of the knee joint showing rupture of the tibial collateral ligament, medial meniscus, and anterior cruciate ligament.

This may have been the full extent of the athlete's injury, but he may also have ruptured his *anterior cruciate ligament* (Fig. 5-108). This ligament, which prevents posterior displacement of the femur and hyperextension of the knee joint, is sometimes torn when the knee is hit hard from the lateral side. In summary, the forced abduction and lateral rotation of the runner's leg by the block probably resulted in the simultaneous rupture of three structures: the tibial collateral ligament, the medial meniscus, and the anterior cruciate ligament.

SUGGESTED READINGS

Behrman RE: *Nelson Textbook of Pediatrics*, ed 14. Philadelphia, WB Saunders, 1992.

Cormack DH: *Ham's Histology*, ed 9. Philadelphia, JB Lippincott, 1987.

Crelin ES: An experimental study of hip stability in human newborn cadavers. Yale J Biol Med 49: 109–221, 1976.

Devinsky O, Feldman E: *Examination of Cranial and Peripheral Nerves*, New York, Churchill Livingstone, 1988.

Griffith HW: *Complete Guide To Sports Injuries*, Los Angeles, Price Stern Sloan, 1986.

Gross A: Orphopedic Surgery Adult. In Gross A, Gross P, Langer B (Eds): *Surgery: A Complete Guide for Patients and their Families*, Toronto, Harper & Collins, 1989.

Healey JE, Jr, Hodge J: *Surgical Anatomy*, ed 2. Toronto, BC Decker, 1990.

Jenkins, DB: *Functional Anatomy of the Limbs and Back*, ed 6. Philadelphia, WB Saunders, 1990.

McKee NH, Fish JS, Manktelow RT, McAvoy GV, Young S, Zuker RM: Gracilis muscle anatomy as related to function of a free functioning muscle transplant. Clin Anat 3:87–92, 1990.

McMinn RMH: *Last's Anatomy. Regional and Applied*, ed 8. Churchill Livingstone, 1990.

Salter, RB: *Textbook of Disorders and Injuries of the Musculoskeletal System*, ed 2. Baltimore, Williams & Wilkins, 1983.

Tobias PV, Arnold M, Allan JC: *Man's Anatomy*, ed 4. vol III. Johannesburg, Witwatersrand University Press, 1988.

Warwick R: *Nomina Anatomica*, ed 6. Edinburgh, Churchill Livingstone, 1989.

Williams PH, Warwick M, Dyson M, Bannister LH: *Gray's Anatomy*, ed 37. New York, Churchill Livingstone, 1989.

Woodburne RT, Burkel WE: *Essentials of Human Anatomy*, ed 8. New York, Oxford University Press, 1988.

The upper limb (extremity) is the *organ of manual activity*. It is freely movable, especially the hand, which is adapted for grasping and manipulating. For purposes of description, the upper limb is divided into the following: the **shoulder** (junction of the arm and trunk); **arm** (brachium) between the shoulder and elbow; the **forearm** (antebrachium) between the elbow and wrist; the **wrist** (carpus) between the forearm and hand; and the **hand** (manus). Although some people refer to the upper limb as the arm, note that it is only one part of the upper limb.

The upper limb is not usually involved in weightbearing; as a result, *its stability has been sacrificed to gain mobility*. The digits (fingers) are the most mobile, but other parts are still more mobile than comparable parts of the lower limb (Chap. 5). Because the disabling effect of an injury to the upper limb, particularly the hand, is far out of proportion to the extent of the injury, a sound understanding of its structure and function is of the highest clinical importance. Furthermore, knowledge of its structure without an understanding of its functions is almost useless clinically, because the aim of treating an injured limb is to preserve or restore its functions.

Bones of the Upper Limb

These bones form the superior part of the *appendicular skeleton* (Fig. 6-1). They comprise the *clavicle* (collar bone) and *scapula* (shoulder blade) in the pectoral girdle; the *humerus* in the arm; the *radius and ulna* in the forearm; the *carpal bones* in the wrist; the *metacarpal bones* in the hand; and the *phalanges* in the digits (fingers).

The Pectoral Girdle

This girdle, formed by the clavicles and scapulae (Fig. 6-1), connects the upper limb to the *axial skeleton*, comprised of the skull, vertebral column, ribs and their cartilages, and the sternum. The *pectoral (shoulder) girdle* articulates with the sternum at the sternoclavicular joint and with the upper limb at the shoulder joint. The girdle is incomplete posteriorly, but it is completed anteriorly by the manubrium of the sternum; however, the sternum is not part of the girdle. Although very mobile, in keeping with the need for mobility of the upper limb, the pectoral girdle is supported and stabilized by *pectoral muscles*, which are connected to the ribs and vertebrae. These muscles act on the upper limb (see Table 6-1).

Bones of the Pectoral Girdle

The Clavicle (Figs. 6-1 to 6-3). This bone extends laterally and almost horizontally across the root of the neck. It extends from the manubrium of the sternum to the acromion of the scapula. The clavicle (L. little key) *connects the upper limb to the axial skeleton* and the trunk.

The triangular-shaped medial (sternal) end of the clavicle articulates with the sternum at the *sternoclavicular joint*. The medial two-thirds of the *body* (shaft) of the clavicle are convex anteriorly, whereas the lateral one-third is flattened and concave anteriorly. Its curvatures increase its resilience. The broad lateral (acromial) end of the clavicle articulates with the acromion of the scapula at the *acromioclavicular joint*. **The clavicle has three functions:**

1. To *act as a strut* for holding the upper limb free from the trunk so it may have maximum freedom of action;
2. To *provide attachments for muscles*;
3. To *transmit forces from the upper limb* to the axial skeleton.

The Scapula (Figs. 6-1 and 6-3). This flattened, triangular bone lies on the posterolateral aspect of the thorax, covering parts of the 2nd to 7th ribs. The scapula connects the clavicle to the humerus. It is highly mobile and has a *head*, *neck*, and *body*. The body is thin and translucent. The scapula has a concave costal or anterior surface (*subscapular fossa*) and a convex posterior surface from which the *spine of the scapula* projects. The smaller part, which is superior to the spine, is called the *supraspinous fossa*, and the larger part, which is inferior to the spine, is called the *infraspinous fossa*. The spine continues laterally into a flattened process called the *acromion* (acromion process). It projects anteriorly and articulates with the clavicle. Superolaterally the scapula has a shallow *glenoid cavity* (fossa) for articulation with the head of the humerus. This part of the scapula, called the *head*, is connected to its bladelike *body* by a short *neck*. The *coracoid process*, like a bird's beak, arises from the superior border of the head and projects superoanteriorly. The *scapular notch* is in the superior border.

Surface Anatomy of the Pectoral Girdle

The Clavicle (Figs. 6-2 and 6-3). Located in the root of the neck at the thoracocervical junction, the clavicle can be palpated throughout its entire length. Its medial end can be easily palpated because it projects superior to the manubrium of the sternum. Between the medial elevations of the clavicles is the deep *jugular notch* (suprasternal notch). As the clavicle passes laterally in the horizontal plane, its medial part can be felt to be convex anteriorly. The large vessels and nerves to the upper limb pass posterior to this convexity (Fig. 6-18). The lateral end of the clavicle does not reach the "*point of the shoulder*," which

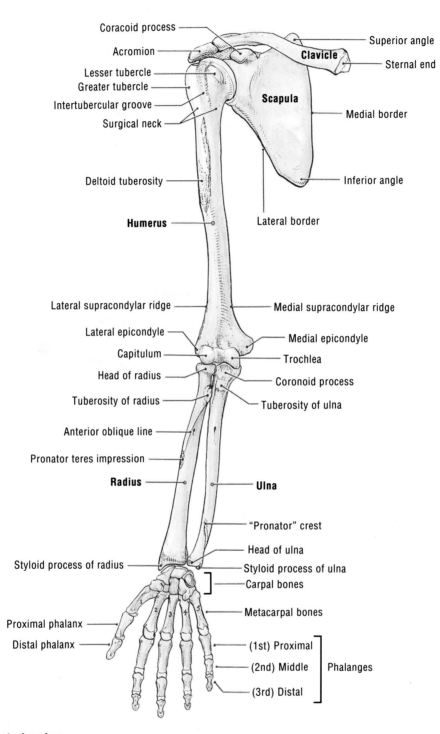

Coracoid process
Acromion
Lesser tubercle
Greater tubercle
Intertubercular groove
Surgical neck

Superior angle
Clavicle
Sternal end
Scapula
Medial border

Deltoid tuberosity

Humerus

Inferior angle

Lateral border

Lateral supracondylar ridge
Lateral epicondyle
Capitulum
Head of radius
Tuberosity of radius

Medial supracondylar ridge
Medial epicondyle
Trochlea
Coronoid process
Tuberosity of ulna

Anterior oblique line

Pronator teres impression

Radius

Ulna

"Pronator" crest
Head of ulna
Styloid process of radius
Styloid process of ulna
Carpal bones

Metacarpal bones

Proximal phalanx
Distal phalanx

(1st) Proximal
(2nd) Middle   Phalanges
(3rd) Distal

A, *Anterior view*

Figure 6-1. Bones of the upper limb. The ribs and thoracic vertebrae are also shown in *B*.

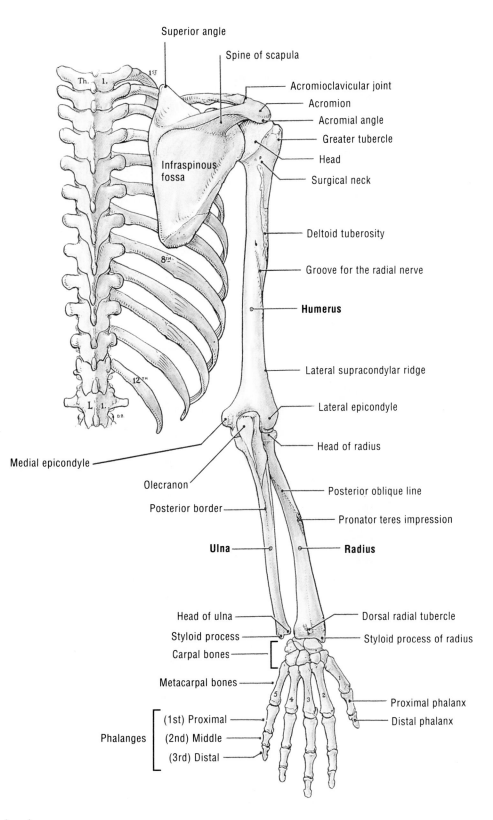

Superior angle

Spine of scapula

Acromioclavicular joint

Acromion

Acromial angle

Greater tubercle

Head

Surgical neck

Infraspinous fossa

Deltoid tuberosity

Groove for the radial nerve

Humerus

Lateral supracondylar ridge

Lateral epicondyle

Head of radius

Medial epicondyle

Olecranon

Posterior oblique line

Posterior border

Pronator teres impression

Ulna

Radius

Head of ulna

Dorsal radial tubercle

Styloid process

Styloid process of radius

Carpal bones

Metacarpal bones

Proximal phalanx

Distal phalanx

Phalanges

(1st) Proximal

(2nd) Middle

(3rd) Distal

B, *Posterior view*

Figure 6-1*B*

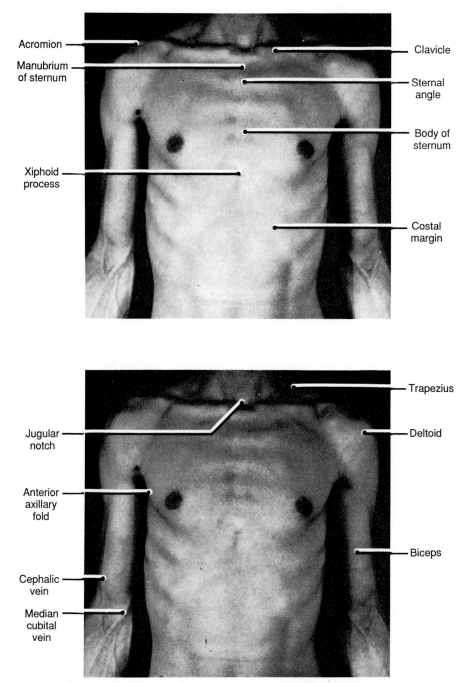

Figure 6-2. Surface anatomy of the pectoral region, arm, and proximal part of the forearm.

is formed by the lateral part of the acromion of the scapula (Fig. 6-3A). The small *acromioclavicular joint* (Fig. 6-117), where the clavicle and acromion articulate, can be palpated 2 to 3 cm medial to the lateral border of the acromion, particularly when the upper limb is swung slowly anteriorly and posteriorly.

The Scapula (Figs. 6-1 to 6-3 and 6-7).   The *crest of the spine* of the scapula is subcutaneous throughout and is easily felt. Its medial end, called the *root of the spine*, is opposite the spinous process of the third thoracic vertebra when the upper limb is in the anatomical position.

The *acromion*, projecting anteriorly from the lateral end of the spine of the scapula, is clearly visible in many people, especially those who are in good condition (Fig. 6-2A). The prominence produced by the acromion is often referred to as the "point of the shoulder." The acromion is important because it is *the proximal point from which clinicians measure the length of the upper limb*. Inferior to the acromion is the smooth, rounded curve of the shoulder formed by the deltoid muscle. The *superior angle* of the scapula lies at the level of T2 vertebra. Its *inferior angle* lies opposite the spinous process of T7 vertebra, near the inferior

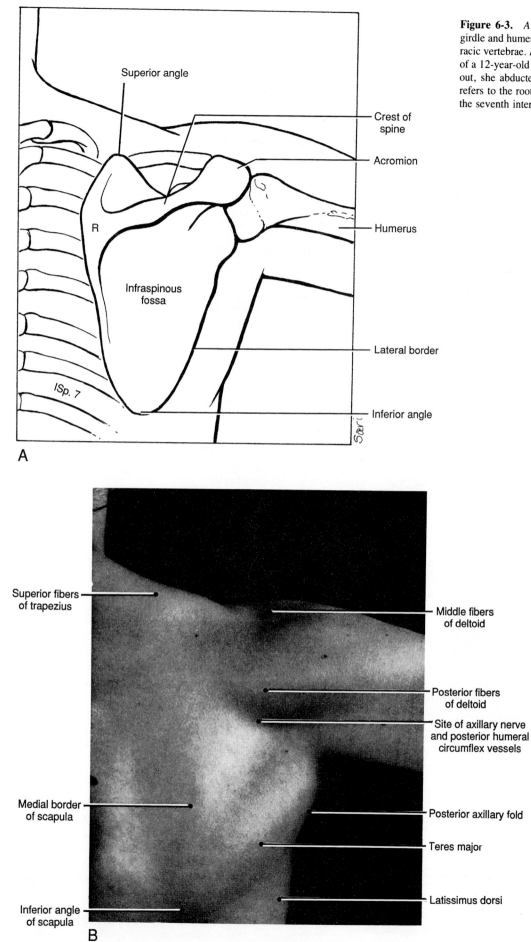

Figure 6-3. *A*, Posterior view of the pectoral girdle and humerus, including some ribs and thoracic vertebrae. *B*, Shoulder and scapular regions of a 12-year-old girl. To make her muscles stand out, she abducted her arm against resistance. *R* refers to the root of the scapular spine; *ISp.7*, to the seventh intercostal space.

A labels: Superior angle; Crest of spine; Acromion; Humerus; R; Infraspinous fossa; Lateral border; ISp. 7; Inferior angle

B labels: Superior fibers of trapezius; Middle fibers of deltoid; Posterior fibers of deltoid; Site of axillary nerve and posterior humeral circumflex vessels; Medial border of scapula; Posterior axillary fold; Teres major; Latissimus dorsi; Inferior angle of scapula

505

border of the seventh rib. The inferior angle is a good guide posteriorly to the *seventh intercostal space* when the arm is in the anatomical position (Fig. 6-1B). *The coracoid process* of the scapula (Fig. 6-1A) can be palpated toward the lateral side of the *deltopectoral triangle* (Fig. 6-5). The superior half of the *medial (vertebral) border of the scapula* is covered by the trapezius muscle (Fig. 6-37), but it can be easily palpated and observed from the superior to the inferior angles of the scapula. The medial border of the scapula crosses the second to seventh ribs (Fig. 6-3A). The *lateral (axillary) border of the scapula* is not easily palpated, except for its inferior part, owing to the presence of the teres major muscle (Figs. 6-3B and 6-40).

Because the clavicle is the area of bony contact with the axial skeleton, fractures of the clavicle are relatively common (Fig. 6-4). The weakest part of the clavicle is at the junction of its medial two-thirds and its lateral third. A main function of the clavicle is to transmit forces from the upper limb to the axial skeleton. Hence, in falls on the shoulder if the force is greater than the strength of the clavicle, a fracture results. **Fractures of the clavicle** that are medial to the attachment of the coracoclavicular ligament are common (Fig. 6-4), especially in children and young adults. The fracture is often incomplete in children, *i.e.*, it is a *green-stick fracture* in which one cortex of the bone breaks and the opposite one bends. Delayed union of the fracture is common owing to its

poor blood supply or to interruption of its blood supply (Griffith, 1986). After fracture of the clavicle, the sternocleidomastoid muscle elevates the medial fragment of bone. As the trapezius muscle is unable to hold up the lateral fragment, owing to the weight of the upper limb, it drops. Patients with a fractured clavicle frequently present with their upper limb in a sling or supporting the sagging limb with the other hand. In addition to being depressed, the lateral fragment of the clavicle is pulled medially by the adductors of the arm, principally the latissimus dorsi and pectoralis major muscles (see Tables 6-1 and 6-4). This overriding of the bone fragments shortens the clavicle.

Occasionally a communicating vein from the cephalic vein in the *deltopectoral triangle* (Fig. 6-5) passes anterior to the clavicle to join the external jugular vein (Fig. 6-33). This communicating vein may be torn when the clavicle fractures.

The clavicle is the first bone in the body to ossify. Intramembranous ossification begins in it during the seventh embryonic week. In unusual instances, the clavicle is incomplete or absent. This congenital abnormality is often associated with delayed ossification of the skull. The combined condition, known as *cleidocranial dysostosis*, is characterized by drooping and excessive mobility of the shoulder. Sometimes only the middle of the clavicle is absent, and the two ends are joined by a fibrous band.

The Pectoral Muscles

The pectoral region contains four muscles, all of which are attached to the pectoral girdle and are associated with movements of the pectoral girdle and upper limb (Table 6-1).

The Pectoralis Major Muscle (Figs. 6-5 to 6-11 and 6-15). This large, thick, fan-shaped muscle covers the superior part of the thorax. Its lateral border forms the *anterior axillary fold* and most of the anterior wall of the axilla. The fascial sheath enclosing the pectoralis major muscle is attached at its origin to the clavicle and sternum. It leaves the lateral border of this muscle to form the *axillary fascia* in the floor of the axilla.

The pectoralis major and deltoid muscles diverge slightly from each other superiorly (Figs. 6-5 and 6-8) and, along with the clavicle, form the *deltopectoral triangle* (infraclavicular fossa). The *cephalic vein*, one of the two major superficial veins of the upper limb, occupies the furrow between the deltoid and pectoralis major muscles before it enters the deltopectoral triangle on its way to the *axillary vein* (Fig. 6-33). The attachments, nerve supply, and main actions of the pectoralis major are given in Table 6-1. When both parts of the muscle act together, the pectoralis major adducts and medially rotates the humerus at the shoulder joint. Acting alone, *the clavicular head helps to flex the humerus* and from this position *the sternocostal head extends the humerus*. When the arm is flexed, the sternocostal head raises the ribs during forced inspiration.

Absence of part of the pectoralis major muscle, usually its sternocostal part, is uncommon and when it occurs there

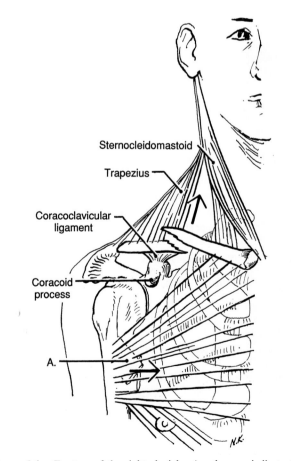

Figure 6-4. Fracture of the right clavicle. *A* and arrow indicate the pectoralis major muscle.

Sternocleidomastoid

Trapezius

Coracoclavicular ligament

Coracoid process

A.

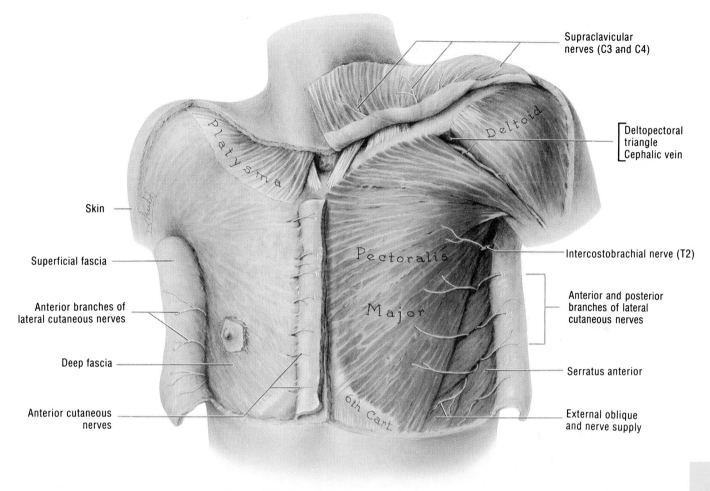

Figure 6-5.   Superficial dissection of the pectoral region.

is usually no disability. However, the anterior axillary fold is absent on the affected side and the nipple is more inferior than usual.

The Pectoralis Minor Muscle (Figs. 6-7*B*, 6-9 to 6-11, 6-16, and 6-17).   This triangular muscle lies in the anterior wall of the axilla, where it is largely covered by the much larger pectoralis major. *The pectoralis minor is the landmark of the axilla*, as illustrated in Fig. 6-19. Along with the coracoid process of the scapula, it forms an arch deep to which pass the vessels and nerves to the upper limb.

The pectoralis minor is surrounded by *clavipectoral fascia* (Figs. 6-11 and 6-15), a thin sheet of fibrous tissue that runs from the clavicle superiorly to the *axillary fascia* inferiorly. The connection of the clavipectoral fascia with the clavicle supports and suspends the floor of the axilla, composed of axillary fascia and skin. The attachments, nerve supply, and main actions of the pectoralis minor are given in Table 6-1. It stabilizes the scapula by drawing it inferiorly and anteriorly against the thoracic wall. It also rotates the scapula, thereby tilting its glenoid cavity inferiorly. When the scapula is fixed, it raises the third to fifth ribs in forced inspiration (Fig. 1-49).

The Subclavius Muscle (Figs. 6-7*C*, 6-16, and 6-17).   This small, rounded, triangular muscle lies inferior to the clavicle. Because of its location, it serves as a protective cushion between the clavicle and subclavian vessels when this bone is broken. Its attachments, nerve supply, and main actions are given in Table 6-1. The subclavius muscle draws the clavicle medially and pulls the point of the shoulder anteriorly. It also steadies the clavicle during shoulder movements and resists the tendency for the clavicle to dislocate at the sternoclavicular joint (*e.g.*, when pulling very hard in tug-of-war).

The Serratus Anterior Muscle (Figs. 6-5, 6-7 to 6-10, and 6-13).   This large, foliate muscle overlies the lateral portion of the thorax and the intercostal muscles. It was given its name (L. *serratus*, a saw) because of the saw-toothed appearance of the fleshy digitations at its proximal attachment. Its attachments, nerve supply, and main actions are given in Table 6-1. The serratus anterior protracts the scapula[1] and holds or fixes it against

---

[1]Protraction is the sliding superoanteriorly of the scapula over the thoracic cage. Retraction is the resumption of the anatomical position.

Table 6-1.
The Muscles in the Pectoral Region

Muscle	Proximal Attachments	Distal Attachments	Innervation	Main Actions
Pectoralis major (Figs. 6-5, 6-7A, 6-9, and 6-10)	*Clavicular head*: anterior surface of the medial half of clavicle *Sternocostal head*: anterior surface of sternum, superior six costal cartilages, and aponeurosis of external oblique muscle	Lateral lip of intertubercular groove of humerus	Lateral and medial pectoral nerves; clavicular head (C5 and **C6**)[1] Sternocostal head (**C7, C8,** and T1)	Adducts and medially rotates humerus *Acting alone*: Clavicular head flexes humerus and sternocostal head extends it
Pectoralis minor (Figs. 6-7B, 6-9, and 6-16)	Ribs 3 to 5 near their costal cartilages	Medial border and superior surface of coracoid process of scapula	Medial pectoral nerve (C8 and T1)	Stabilizes scapula by drawing it inferiorly and anteriorly against thoracic wall
Subclavius (Figs. 6-7C and 6-16)	Junction of rib 1 and its costal cartilage	Inferior surface of middle third of clavicle	Nerve to subclavius (**C5** and C6)	Anchors and depresses clavicle
Serratus anterior (Figs. 6-5, 6-7D, and 6-9)	External surfaces of lateral parts of ribs 1 to 8	Anterior surface of medial border of scapula	Long thoracic nerve (C5, **C6,** and **C7**)	Protracts scapula and holds it against thoracic wall; rotates scapula

[1]In this and subsequent tables, the numbers indicate the spinal cord segmental innervation of the nerves (*e.g.,* C5 and C6 indicate that the nerves supplying the clavicular head of the pectoralis major muscle are derived from the 5th and 6th cervical segments of the spinal cord). **Boldface** indicates the main segmental innervation. Damage to these segments of the spinal cord, or to the motor nerve roots arising from them, results in paralysis of the muscles concerned.

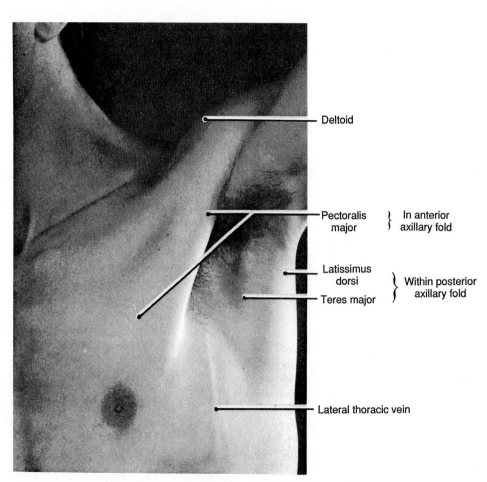

Figure 6-6.   Pectoral region and axilla of a 27-year-old man.

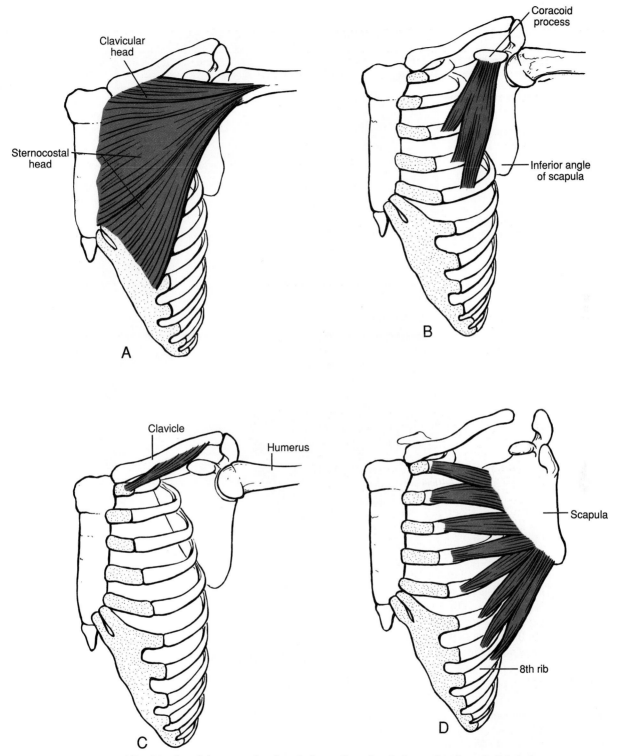

Figure 6-7.   Muscles of the pectoral region. *A*, Pectoralis major. *B*, Pectoralis minor. *C*, Subclavius. *D*, Serratus anterior.

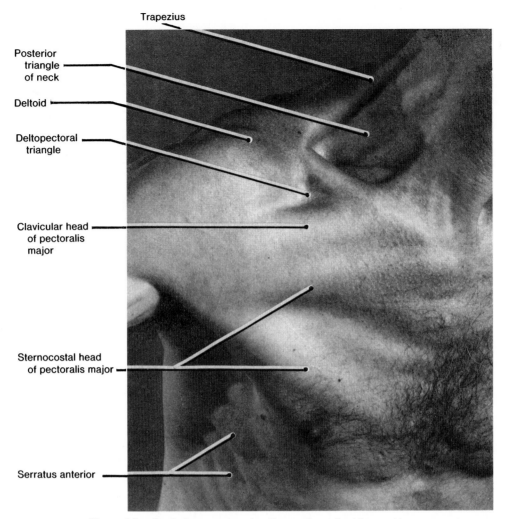

Trapezius

Posterior
triangle
of neck

Deltoid

Deltopectoral
triangle

Clavicular head
of pectoralis
major

Sternocostal head
of pectoralis major

Serratus anterior

Figure 6-8. Cervical, pectoral, and axillary regions of a 46-year-old man.

the thoracic wall. Because it is active during punching, it has been called "*the boxer's muscle*." By fixing the scapula to the thorax, it acts as an anchor for this bone and permits other muscles to use it as a fixed bone to produce movements of the humerus. Inferior fibers of the serratus anterior help to raise the glenoid cavity of scapula (*e.g.*, when the upper limb is raised superior to the head).

When a person's serratus anterior muscle is paralyzed owing to *injury of the long thoracic nerve*, the medial border of the scapula stands out, especially its inferior angle, giving it the appearance of a wing when the person presses anteriorly, *e.g.*, against a wall (Fig. 6-29). Consequently this condition is called a "winged scapula." When the arm is raised, the scapula is pulled away from the thoracic wall. In addition the arm cannot be abducted farther than the horizontal position because the serratus anterior is unable to rotate the scapula and raise the glenoid cavity. Consequently, a patient with a *paralyzed serratus anterior muscle* is unable to raise the upper limb fully or to push with it.

The Axilla

The axilla (armpit) is a roughly *pyramidal space at the junction of the arm and thorax* (Figs. 6-6 and 6-10 to 6-14). It provides a passageway for the large, important nerves and vessels to reach the upper limb (Fig. 6-16). The axilla has an apex, a base, and four walls.

The Apex of the Axilla. The apex is directed toward the root of the neck and is located at the medial side of the root of the coracoid process of the scapula. It is formed by the convergence of the bones in its three major walls: the clavicle in its anterior wall, the scapula in its posterior wall, and the first rib in its medial wall. The interval between these three bones is the *entrance to the axilla* through which all nerves and vessels pass to the upper limb.

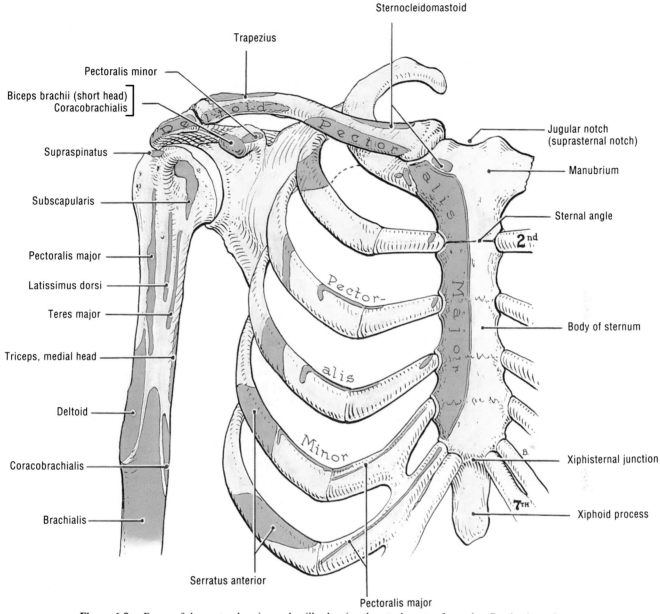

Figure 6-9.   Bones of the pectoral region and axilla showing the attachments of muscles. Proximal attachments are shown in red and distal attachments in blue.

The Base of the Axilla (Figs. 6-6 and 6-10).   The base, facing inferiorly, is formed by the fascia and skin of the concave axilla (*i.e.*, the armpit). The skin of the base is normally covered with hair in postpubertal persons. The boundaries of the axilla can be visualized best in a transverse section (Fig. 6-14).

The Anterior Wall of the Axilla (Figs. 6-5, 6-6, 6-11, and 6-15).   The clavicle and pectoral muscles form the anterior wall. The lateral border of the pectoralis major forms the *anterior axillary fold*. Posterior to the pectoralis major, the pectoralis minor and subclavius muscles form the deep layer of the anterior wall.

The Posterior Wall of the Axilla (Figs. 6-3*B*, 6-6, 6-10, and 6-12).   This wall is formed chiefly by the scapula and the subscapularis muscle. Inferior to the subscapularis is the teres major muscle, which combines with the latissimus dorsi to form the *posterior axillary fold*. The tendon of the latissimus dorsi muscle wraps around the lateral part of the teres major and forms part of the posterior wall.

The Medial Wall of the Axilla (Figs. 6-5, 6-10, and 6-13).   This wall is formed by the ribs and intercostal muscles, which are covered by the serratus anterior muscle. The latter muscle forms the main part of this wall.

The Lateral Wall of the Axilla (Figs. 6-1, 6-10, and 6-14).   This narrow wall is formed by the floor·of the *intertubercular groove* in the humerus, which lodges the tendon of the long head of the biceps brachii muscle.

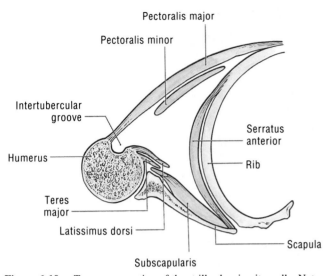

Figure 6-10. Transverse section of the axilla showing its walls. Note two pectoral muscles in the anterior wall; the scapula and subscapularis muscle in the posterior wall, a rib and the serratus anterior muscle in the medial wall, and the intertubercular groove in the humerus that forms the narrow lateral wall.

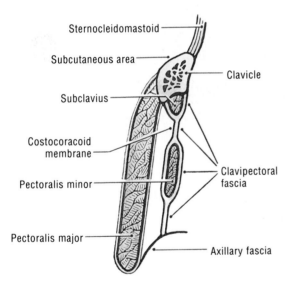

Figure 6-11. Sagittal section of the axilla illustrating its anterior wall formed by the clavicle and pectoral muscles. The clavipectoral fascia is a strong sheet of connective tissue that is attached to the clavicle and encloses the subclavius muscle superiorly and the pectoralis minor muscle inferiorly.

Contents of the Axilla (Figs. 6-14, 6-16, 6-18, and 6-35). The axilla contains large nerves that are branches of the *brachial plexus*. These nerves pass from the neck to the upper limb. The axilla also contains the *axillary vessels* (axillary artery and its branches, axillary vein and its tributaries, and axillary lymph vessels). There are also several groups of *axillary lymph*

nodes that are of clinical importance because of their frequent invasion by cancer cells from the breast (p. 47).

The Brachial Plexus

This *large network of nerves to the upper limb* extends from the neck into the axilla. It is a large and very important plexus that is situated partly in the neck and partly in the axilla. It is formed by rami, trunks, divisions, cords, and their branches (Fig. 6-18). The supraclavicular part of the brachial plexus (rami and trunks with their branches) is in the *posterior triangle of the neck* (Figs. 6-8 and 6-17), and its infraclavicular part (cords and their branches) is in the *axilla* (Figs. 6-16 to 6-19).

The brachial plexus is formed by the union of the ventral rami of nerves C5 to C8 and the greater part of the T1 ventral ramus (Fig. 6-18). The rami that form the plexus, sometimes inappropriately referred to as the "roots of the branchial plexus" (Fig. 6-19), lie between the scalenus anterior and scalenus medius muscles (Fig. 6-17).

The Typical Plan of the Brachial Plexus

As the ventral rami enter the posterior triangle of the neck (Figs. 6-16 to 6-19), those from C5 and C6 unite to form a *superior trunk* (upper trunk). The ventral ramus of C7 continues as a *middle trunk*, and the ventral rami of C8 and T1 unite at the neck of the first rib to form an *inferior trunk* (lower trunk). The inferior trunk lies on the first rib posterior to the subclavian artery. Each of the three trunks divides into *anterior and posterior divisions*, posterior to the clavicle. These divisions are of fundamental significance because the anterior divisions supply anterior (flexor) parts and the posterior divisions supply posterior (extensor) parts of the upper limb.

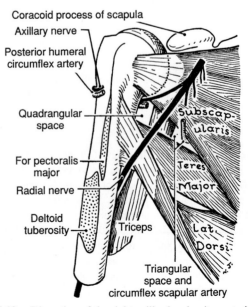

Figure 6-12. Dissection of the right axilla showing its posterior wall formed by the scapula and the subscapularis, teres major, and latissimus dorsi muscles.

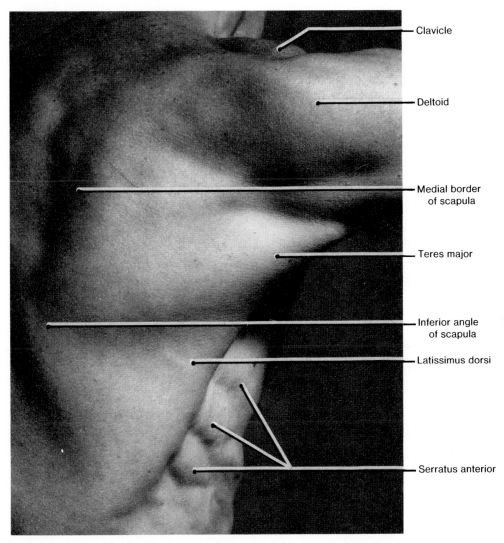

Clavicle

Deltoid

Medial border
of scapula

Teres major

Inferior angle
of scapula

Latissimus dorsi

Serratus anterior

Figure 6-13. Posterior aspect of the right shoulder and lateral thoracic wall of a 46-year-old man.

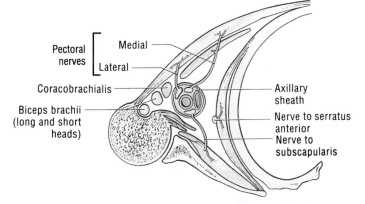

Pectoral
nerves

Medial

Lateral

Coracobrachialis

Biceps brachii
(long and short
heads)

Axillary
sheath

Nerve to serratus
anterior

Nerve to
subscapularis

Figure 6-14. Transverse section of the axilla showing its contents.

The three posterior divisions unite to form the *posterior cord*. The anterior divisions of the superior and middle trunks unite to form the *lateral cord*, and the anterior division of the inferior trunk continues as the *medial cord*. In Fig. 6-19, observe that the cords of the plexus bear the relationship to the second part of the axillary artery that is indicated by their names (*e.g.*, the lateral cord is lateral to the axillary artery).

Each cord of the brachial plexus divides into two terminal branches (Figs. 6-18 to 6-20). The lateral cord divides into the musculocutaneous nerve and the lateral root of the median nerve. The medial cord divides into the ulnar nerve and the medial root of the median nerve. The posterior cord divides into the axillary and radial nerves. In Fig. 6-19, note that three nerves (musculocutaneous, median, and ulnar) are arranged like the limbs of a capital **M**.

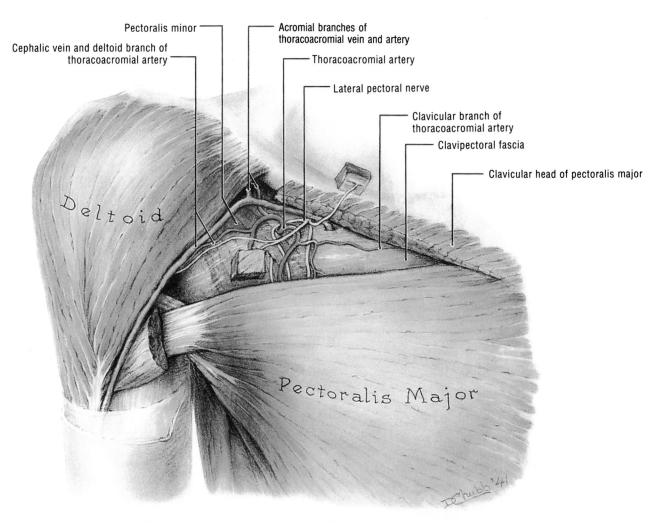

Cephalic vein and deltoid branch of thoracoacromial artery

Pectoralis minor

Acromial branches of thoracoacromial vein and artery

Thoracoacromial artery

Lateral pectoral nerve

Clavicular branch of thoracoacromial artery

Clavipectoral fascia

Clavicular head of pectoralis major

Deltoid

Pectoralis Major

Figure 6-15.   Dissection of the pectoral region. Most of the clavicular head of the pectoralis major is excised, but two cubes of it remain to identify its nerves.

Variations in the Plan of the Brachial Plexus.   In addition to the five ventral rami (C5 to C8 and T1) that unite to form the brachial plexus, small contributions may come from the ventral rami of C4 or T2. When the superior ramus (root) of the plexus is C4 and the inferior ramus (root) is C8, it is called a *prefixed plexus*. Alternately, when the superior ramus is C6 and the inferior one is T2, it is called a *postfixed plexus*. In the latter type of plexus, the inferior trunk may be compressed by the first rib, producing nervous and vascular symptoms in the upper limb. In some persons, trunk divisions or cord formation may be absent in one or other parts of the plexus. However, the makeup of the terminal branches is unchanged. In addition the lateral or medial cords may receive fibers from ventral rami inferior or superior to the usual levels, respectively.

Branches of the Brachial Plexus (Figs. 6-16 to 6-22).   These nerves are divided into supraclavicular and infraclavicular branches.

Supraclavicular Branches of the Brachial Plexus (Figs. 6-17 and 6-22; Table 6-2).   These nerves arise from the rami and

trunks of the brachial plexus and are approachable through the posterior triangle of the neck (Fig. 6-8).

The dorsal scapular nerve (nerve to the rhomboids) arises chiefly from the posterior aspect of the ventral ramus of C5 with a frequent contribution from C4. It pierces the scalenus medius muscle, runs deep to the levator scapulae muscle, which it helps to supply, and then enters the deep surface of the *rhomboid muscles*, which it supplies.

The long thoracic nerve arises from the posterior aspect of the ventral rami of C5, C6, and C7, and passes through the apex of the axilla posterior to the other components of the brachial plexus to supply the *serratus anterior muscle* (Fig. 6-23). The roots from C5 and C6 pierce the scalenus medius, and the root from C7 passes anterior to this muscle.

The nerve to the subclavius muscle, a slender nerve, arises from the anterior aspect of the superior trunk of the brachial plexus. It receives fibers chiefly from C5, with occasional additions from C4 and C6. It descends posterior to the clavicle and anterior to the brachial plexus to supply the subclavius muscle.

The suprascapular nerve arises from the posterior aspect of the superior trunk of the brachial plexus, receiving fibers from

C5 and C6, and often C4 (about 50% of people). The suprascapular nerve supplies the *supraspinatus and infraspinatus muscles* and the shoulder joint. To reach these muscles it passes laterally across the posterior triangle of the neck, superior to the brachial plexus, and passes through the scapular notch (Fig. 6-23).

Infraclavicular Branches of the Brachial Plexus (Figs. 6-16, 6-19, and 6-20; Table 6-3). These nerves arise from the cords of the plexus and are approachable through the axilla.

The lateral cord of the brachial plexus has three branches (Figs. 6-15 to 6-19; Table 6-3): one side branch, called the lateral pectoral nerve, and two terminal branches, called the musculocutaneous nerve and the lateral root of the median nerve.

The lateral pectoral nerve contains nerve fibers from the anterior divisions of C5 to C7. It pierces the clavipectoral fascia to supply the pectoralis major muscle and sends a branch to the medial pectoral nerve, which supplies the pectoralis minor mus-

cle. This nerve is called the lateral pectoral nerve because it *arises from the lateral cord of the brachial plexus*. Remembering this avoids confusion when you observe it running deep to the pectoralis major muscle, more medially than the medial pectoral nerve (Fig. 6-16).

The musculocutaneous nerve (C5 to C7), supplying the muscles of the anterior aspect of the arm, is one of the two terminal branches of the lateral cord of the brachial plexus (Figs. 6-19 and 6-21). It enters the deep surface of the coracobrachialis muscle, supplies it, and then continues in the arm to supply the biceps brachii and brachialis muscles. Just proximal to the elbow joint, the musculocutaneous nerve pierces the deep fascia and becomes superficial. From here it is called the *lateral antebrachial cutaneous nerve* (lateral cutaneous nerve of the forearm) and supplies skin on the lateral aspect of the forearm.

The lateral root of the median nerve (Figs. 6-16 and 6-18 to 6-20) is the continuation of the lateral cord of the brachial plexus;

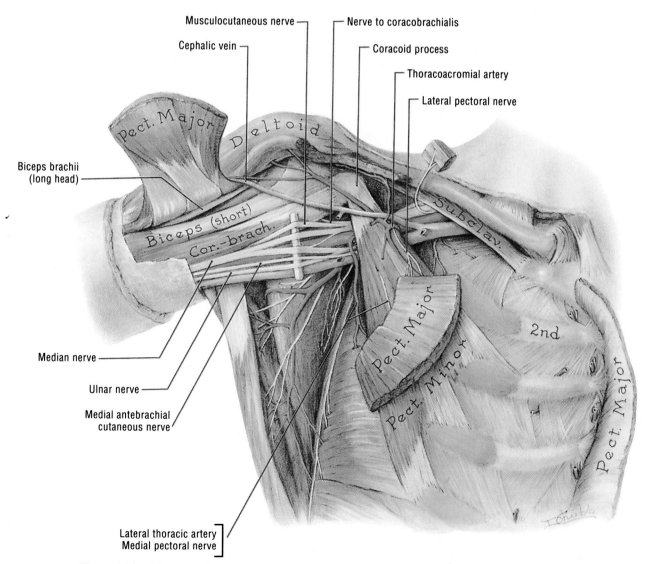

Figure 6-16. Dissection of the pectoral region and axilla. The pectoralis major is reflected and the clavipectoral fascia is removed (also see Fig. 6-11).

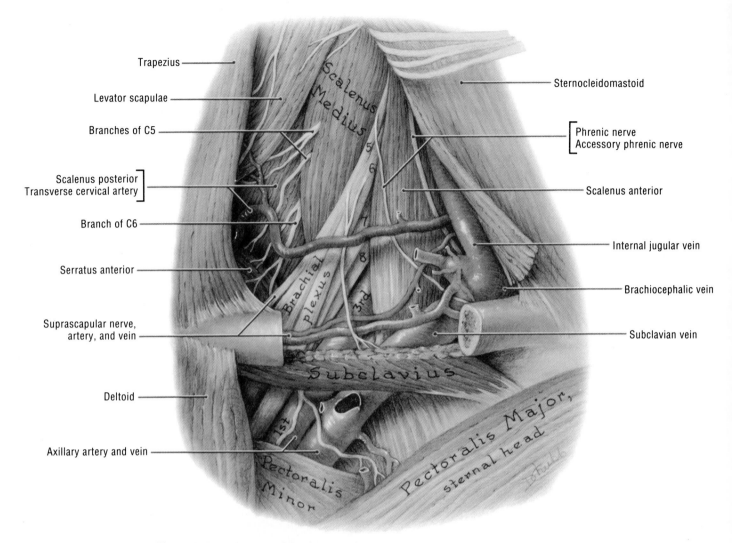

Labels on figure:
Trapezius
Levator scapulae
Branches of C5
Scalenus posterior
Transverse cervical artery
Branch of C6
Serratus anterior
Suprascapular nerve, artery, and vein
Deltoid
Axillary artery and vein
Scalenus Medius
Brachial plexus
Subclavius
Pectoralis Minor
Pectoralis Major, sternal head
Sternocleidomastoid
Phrenic nerve
Accessory phrenic nerve
Scalenus anterior
Internal jugular vein
Brachiocephalic vein
Subclavian vein

Figure 6-17.   Dissection of the right posterior triangle of the neck, pectoral region, and axilla.

i.e., it is the other terminal branch of the lateral cord. It is joined by the medial root of the median nerve, lateral to the axillary artery, to form the *median nerve.*

The medial cord of the brachial plexus has five branches (Table 6-3).

The medial pectoral nerve (C8 and T1) enters the deep surface of the pectoralis minor muscle, supplying it and part of the pectoralis major (Figs. 6-14, 6-16, and 6-22A). This nerve is called the medial pectoral nerve because it *arises from the medial cord of the brachial plexus.* In Fig. 6-16, note that it lies lateral to the lateral pectoral nerve.

The medial brachial cutaneous nerve (medial cutaneous nerve of the arm) contains fibers from C8 and T1. It is a slender nerve that supplies skin over the medial surface of the arm and the proximal part of the forearm (Fig. 6-21A). This nerve usually communicates with the *intercostobrachial nerve*, which supplies the skin on the floor of the axilla and adjacent regions of the arm (Fig. 6-21B).

The medial antebrachial cutaneous nerve (medial cutaneous nerve of the forearm) also contains fibers from C8 and T1. It

runs between the axillary artery and vein to supply skin over the medial surface of the forearm (Fig. 6-21A).

The ulnar nerve (C8, T1, and sometimes C7) is a terminal branch of the medial cord of the brachial plexus (Figs. 6-16 and 6-18 to 6-21). It passes through the arm into the forearm and hand, where it supplies one and one-half muscles in the forearm, most small muscles in the hand, and some skin.

The medial root of the median nerve is the other terminal branch of the medial cord of the brachial plexus (Figs. 6-18 to 6-20). It joins with the lateral root to form the *median nerve*, which supplies the flexor muscles in the forearm, except the flexor carpi ulnaris, and the skin on the hand.

The posterior cord of the brachial plexus has five branches (Figs. 6-18 to 6-22; Table 6-3). In general, these nerves supply muscles that extend the joints of the upper limb. These branches also supply cutaneous nerves to the extensor surface of the upper limb.

The upper subscapular nerve (C5 and C6) is a small nerve that supplies the subscapularis muscle (Fig. 6-22B).

The thoracodorsal nerve (C6, C7 and C8) or *nerve to the*

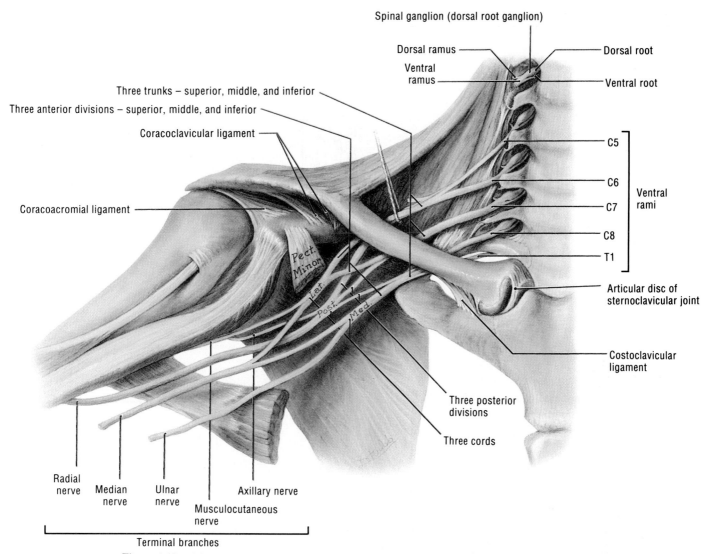

Figure 6-18. Dissection of the brachial plexus of nerves. The ligaments of the clavicle are also shown.

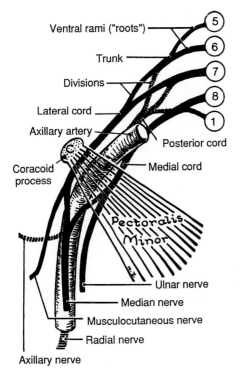

Figure 6-19. Diagrammatic representation of the relationship of the brachial plexus to the axillary artery. Note that this vessel is surrounded by the three cords of the plexus and that they bear the relation indicated by their names to the second part of the axillary artery, *i.e.*, the part posterior to the pectoralis minor.

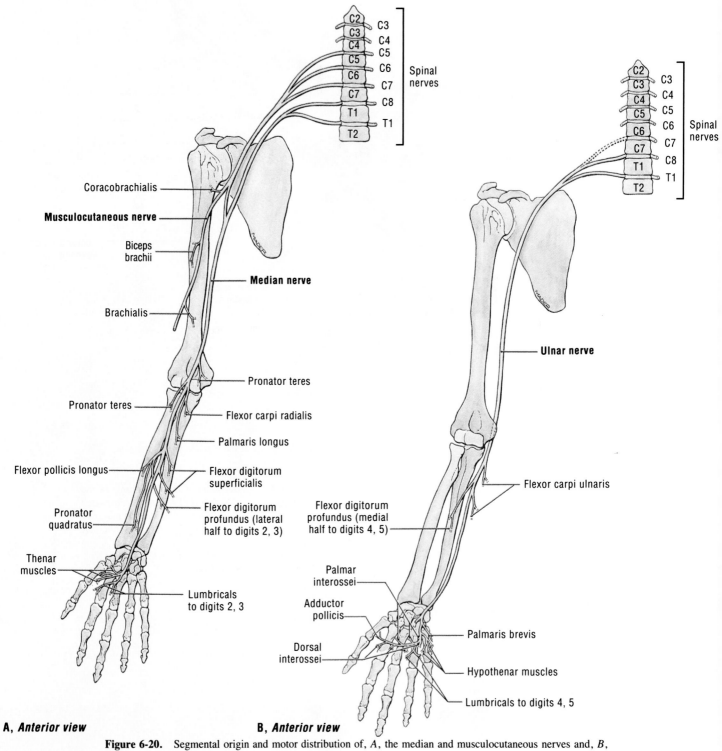

A, *Anterior view*

B, *Anterior view*

Figure 6-20.   Segmental origin and motor distribution of, *A*, the median and musculocutaneous nerves and, *B*, the ulnar nerve.

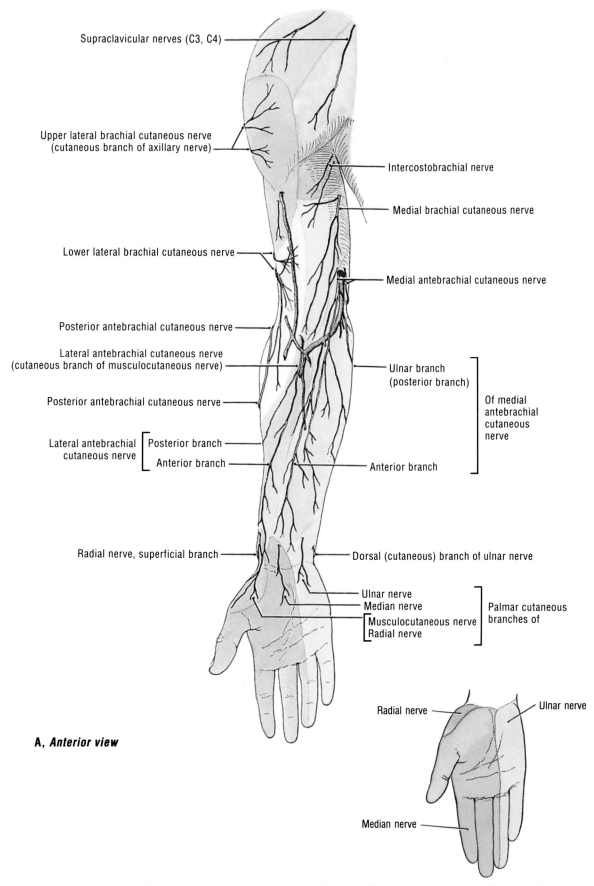

Supraclavicular nerves (C3, C4)

Upper lateral brachial cutaneous nerve
(cutaneous branch of axillary nerve)

Intercostobrachial nerve

Medial brachial cutaneous nerve

Lower lateral brachial cutaneous nerve

Medial antebrachial cutaneous nerve

Posterior antebrachial cutaneous nerve

Lateral antebrachial cutaneous nerve
(cutaneous branch of musculocutaneous nerve)

Ulnar branch
(posterior branch)

Posterior antebrachial cutaneous nerve

Of medial antebrachial cutaneous nerve

Lateral antebrachial cutaneous nerve

Posterior branch

Anterior branch

Anterior branch

Radial nerve, superficial branch

Dorsal (cutaneous) branch of ulnar nerve

Ulnar nerve

Median nerve

Palmar cutaneous branches of

Musculocutaneous nerve
Radial nerve

A, Anterior view

Radial nerve

Ulnar nerve

Median nerve

Figure 6-21. Distribution of the cutaneous nerves of the upper limb. *A*, Anterior view. *B*, Posterior view.

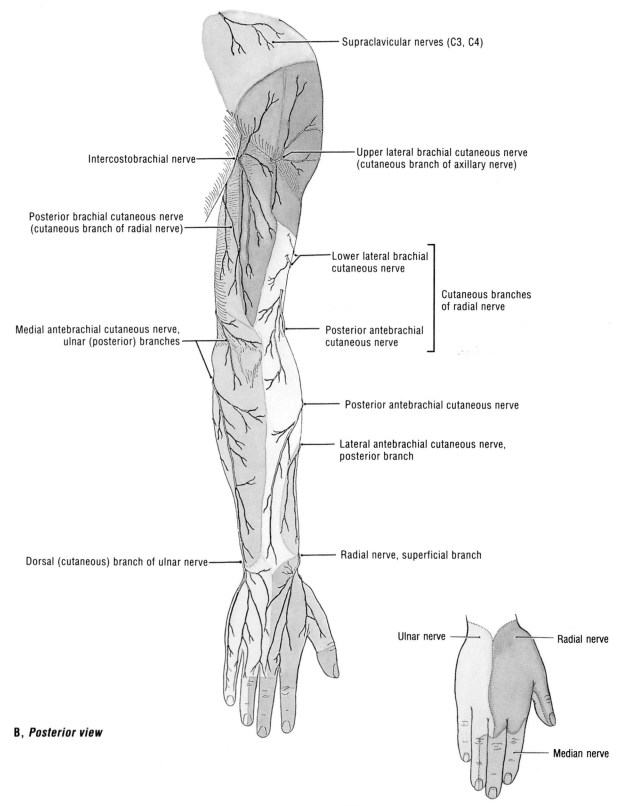

Supraclavicular nerves (C3, C4)

Upper lateral brachial cutaneous nerve
(cutaneous branch of axillary nerve)

Intercostobrachial nerve

Posterior brachial cutaneous nerve
(cutaneous branch of radial nerve)

Lower lateral brachial
cutaneous nerve

Cutaneous branches
of radial nerve

Posterior antebrachial
cutaneous nerve

Medial antebrachial cutaneous nerve,
ulnar (posterior) branches

Posterior antebrachial cutaneous nerve

Lateral antebrachial cutaneous nerve,
posterior branch

Radial nerve, superficial branch

Dorsal (cutaneous) branch of ulnar nerve

Ulnar nerve

Radial nerve

Median nerve

B, Posterior view

Figure 6-21B

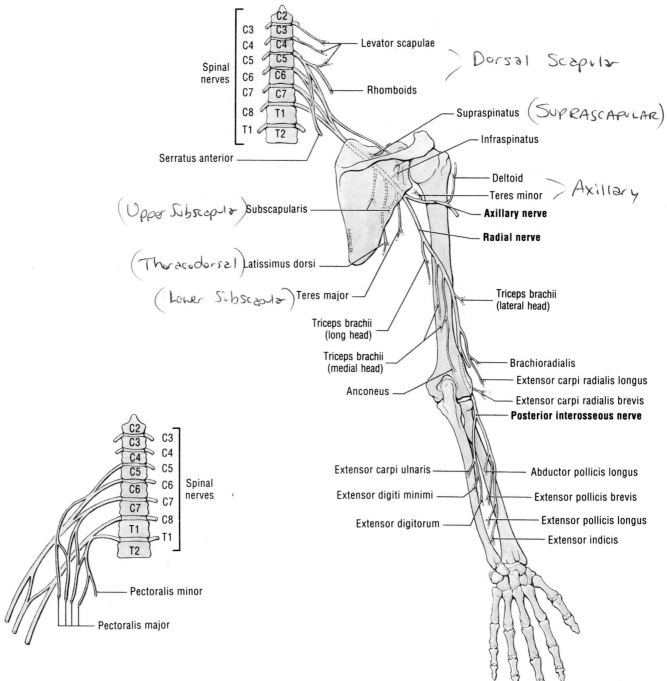

Dorsal Scapular

Suprascapular (SUPRASCAPULAR)

Axillary

(Upper Subscapular) Subscapularis

(Thoracodorsal) Latissimus dorsi

(Lower Subscapular) Teres major

A, Anterior view   **B, Posterior view**

Figure 6-22.   Segmental origin and motor distribution of, *A*, the medial and lateral pectoral nerves supplying the pectoralis muscles and, *B*, the radial nerve.

Table 6-2.
The Supraclavicular Branches of the Brachial Plexus

Origin	Nerves	Muscles Supplied	Segmental Innervation[1]
From the cervical ventral rami	Dorsal scapular	Rhomboids	**C4** and **C5**
		Levator scapulae	C5
	Long thoracic	Serratus anterior	C5, **C6**, and **C7**
From the trunks of brachial plexus	Nerve to subclavius	Subclavius	**C5** and C6
	Suprascapular	Supraspinatus	**C4, C5**, and C6
		Infraspinatus	**C5** and C6

[1]**Boldface** indicates the main spinal cord segmental innervation.

Table 6-3.
The Infraclavicular Branches of the Brachial Plexus

Origin	Nerves	Muscles	Segmental Innervation[1]
From lateral cord of brachial plexus	Lateral pectoral	Pectoralis major	C5, **C6**, and C7
	Musculocutaneous	Coracobrachialis	C5, **C6**, and C7
		Biceps brachii } Brachialis	C5 and **C6**
	Lateral root of median	Flexor muscles in forearm (except FCU[2] and ulnar half of FDP) and five hand muscles	(C5), C6, and C7
From medial cord of brachial plexus	Medial pectoral	Pectoralis major	**C8** and T1
		Pectoralis minor	C8 and T1
	Medial brachial cutaneous		C8 and T1
	Ulnar	1½ muscles of forearm and most small hand muscles	**C8** and T1[3]
	Medial root of median	Flexor muscles in forearm (except FCU and ulnar half of FDP) and five hand muscles	C8 and T1
From posterior cord of plexus	Upper subscapular	Subscapularis	C5 and **C6**
	Thoracodorsal	Latissimus dorsi	**C6, C7**, and C8
	Lower subscapular	Subscapularis	C5 and **C6**
		Teres major	**C6** and C7
	Axillary	Teres minor } Deltoid	**C5** and C6
	Radial	Triceps and anconeus	C5, **C6, C7**, C8, and T1
		Brachioradialis and extensor muscles of forearm	C5, **C6**, and C7

[1]**Boldface** indicates the main spinal cord segmental innervation.
[2]FCU, Flexor carpi ulnaris; FDP, Flexor digitorum profundus.
[3]Sometimes the ulnar nerve also receives fibers from C7.

latissimus dorsi arises between the upper and lower subscapular nerves and runs inferolaterally to supply the latissimus dorsi muscle (Figs. 6-22B and 6-23).

The lower subscapular nerve (C5 and C6) passes inferolaterally, deep to the subscapular artery and vein, gives a branch to the subscapularis muscle, and ends by supplying the teres major muscle (Figs. 6-22B and 6-23).

The axillary nerve (C5 and C6) is a large terminal branch of the posterior cord of the brachial plexus (Figs. 6-22B and 6-23). It passes to the posterior aspect of the arm through the *quadrangular space* (quadrilateral space) in company with the posterior circumflex humeral vessels. On emerging from the quadrangular space, **the axillary nerve winds around the surgical neck of the humerus** to supply the teres minor and deltoid muscles. The axillary nerve ends as the upper lateral brachial cutaneous nerve and supplies skin over the inferior half of the deltoid and adjacent areas of the arm (Fig. 6-21B).

The radial nerve (C5 to C8 and T1) is the other terminal branch of the posterior cord of the brachial plexus (Figs. 6-18 to 6-20). This nerve provides the major nerve supply to the extensor muscles of the upper limb (Fig. 6-22). It also supplies cutaneous sensation to the skin of the extensor region, including the hand (Fig. 6-21). As it leaves the axilla the radial nerve runs posteriorly, inferiorly, and laterally between the long and medial heads of the triceps muscle. *It enters the radial groove in the humerus* (Figs. 6-12, 6-23, and 6-49). The radial nerve gives branches to the triceps, anconeus, and brachioradialis muscles and to the extensor muscles of the forearm (Figs. 6-22B and 6-23).

Injuries to the brachial plexus and its branches are of great importance because they affect movements and cutaneous sensations in the upper limb. The plexus can be injured by disease, stretching, and wounds in the neck or axilla. The signs and symptoms of an injury depend on which part of the brachial plexus is involved. Injuries to the brachial plexus result in loss of muscular movement (*paralysis*) and often a loss of cutaneous sensation (G. *anesthesia*, absence of feeling). The degree of paralysis may be assessed by testing the patient's ability to perform movements. In *complete paralysis* no movement can be detected, whereas in *incomplete paralysis* movement can be performed, but it is weak compared with that on the normal side. The explanation for this is that in incomplete paralysis not all muscles concerned with the movement are paralyzed. The degree of anesthesia may be tested by determining the ability of the person to feel pain (*e.g.*, a pinprick).

Upper Brachial Plexus Injuries (Figs. 6-24 to 6-26). Injuries to superior (upper) parts of the plexus usually result from excessive separation of the neck and shoulder. This may occur during a football game when one tackler is pulling a person's arm as another one hits or pulls the person's head (*e.g.*, pulling on the face mask). These injuries can also occur when a person is thrown from a motorcycle or a horse and lands on his/her shoulder in a way that widely separates the neck and shoulder.

Most major upper brachial plexus injuries result from motorcycle accidents When thrown, the patient's shoulder often hits something (*e.g.*, a tree) and stops, but the head and trunk continue to move, which stretches or tears part of the

Carrying a heavy object on the [...]
a steel beam) may also compre[...]
vicle and the lateral part of th[...]
stretching, or cutting the long[...]
ralysis of the serratus anteri[...]
scapula (Fig. 6-29). The media[...]
the scapula become unusually p[...]
accentuated when the person pu[...]
hands. Instead of keeping the [...]
wall, as is normal, the paraly[...]
allows the scapula to move out[...]
experienced in flexing or abduct[...]
angle from the side of the body[...]
muscle normally protracts and [...]
1), so the glenoid cavity of the [...]
carrying out such a movement.[...]

Injury to the Axillary Ner[...]
winds around the neck of the [...]
injured during fracture of this [...]
be damaged during dislocation[...]
ing *severance of the axillary* [...]
paralyzed and undergoes *atrop*[...]
sation (anesthesia) may also oc[...]
proximal part of the arm (Fig. [...]
deltoid and teres minor muscle[...]
6-5). Injuries of other nerves an[...]
are described in the discussion [...]
and hand.

Injury to the Thoracodors[...]
This nerve is vulnerable to inju[...]
in the axilla. In its course infe[...]
wall, it slopes anteriorly to en[...]
latissimus dorsi muscle, just p[...]
(Fig. 6-23). Injury results in *par*[...]
(p. 533).

The Axillary Artery

This large vessel begins at th[...]
as the continuation of the subcl[...]
and 6-19). The axillary artery en[...]
teres major muscle, where it pas[...]
the brachial artery. During its [...]
axillary artery passes posterior t[...]
(Fig. 6-19). For purposes of des[...]
divided into three parts by this m[...]
in the branching pattern of the axi[...]
For this reason, it is important to [...]
axillary artery are named accord[...]
than by their point of origin.

The First Part of the Axillar[...]
6-31). This part is located betw[...]
first rib and the superior border[...]
first part of the axillary artery is [...]
along with the axillary vein and t[...]
It has only one branch, the *sup*[...]
or highest thoracic artery). This [...]
and second intercostal spaces an[...]
ratus anterior muscle.

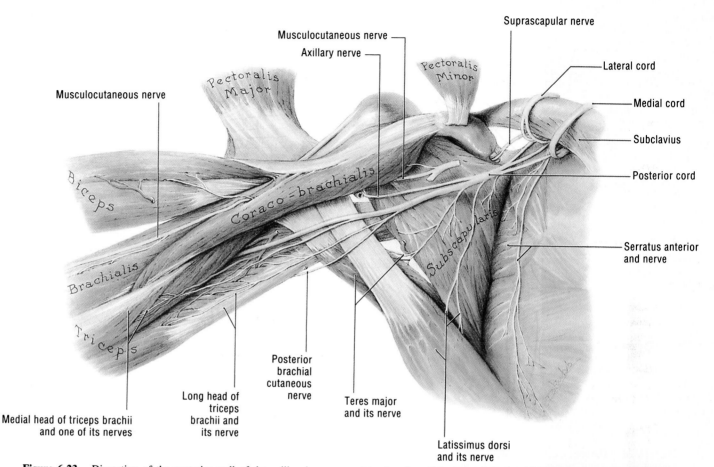

Figure 6-23. Dissection of the posterior wall of the axilla, demonstrating the posterior cord of the brachial plexus and its branches. The pectoralis major and pectoralis minor muscles are turned laterally; the lateral and medial cords of the brachial plexus are turned superiorly; and the arteries, veins, and median and ulnar nerves are removed. Note the large *radial nerve*, a terminal branch of the posterior cord.

plexus (Fig. 6-24). Similar damage to the plexus can result from violent stretching of an infant's neck during delivery (Fig. 6-25). During these injuries, the dorsal and ventral roots of the spinal nerves from C5 and C6 may be pulled out of the spinal cord (Fig. 6-18). In such cases there is paralysis of the scapular muscles (Table 6-3) and loss of sensation over the region of the back supplied by the posterior primary rami, in addition to paralysis of muscles and loss of sensation in the upper limb. If the lesion is confined to C5, usually no sensory changes can be detected because this segment is not responsible for the exclusive supply of any area of skin. However, when both C5 and C6 are involved, there is usually a detectable loss of sensation on the lateral aspect of the upper limb (see Fig. 5-40*B*).

In *stab and bullet wounds of the neck*, the superior trunk of the brachial plexus may be torn or severed where it emerges between the scalenus anterior and scalenus medius muscles (Fig. 6-17). These injuries result in loss of flexion, abduction, and lateral rotation of the shoulder joint, as well as loss of flexion of the elbow joint. Injury to the superior trunk of the brachial plexus may be recognized by the *characteristic position of the limb* (Fig. 6-26). It hangs by the side in medial rotation, a position referred to as the "waiter's tip" position because it is the way modest waiters indicate their desire for a tip. The following muscles that receive nerve fibers from C5 and C6 are most severely affected when there is an upper

Figure 6-24. The neck and shoulder may be violently separated during a fall on the shoulder, producing an upper brachial plexus injury.

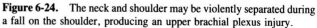

brachial plexus injury: deltoid, biceps brachii, brachialis, brachioradialis, supraspinatus, infraspinatus, and teres minor (Fig. 6-20; Table 6-3).

Poorly fitting crutches (*e.g.*, ones that are too long) may injure the posterior cord of the brachial plexus. Often only the radial nerve is affected; as a result, the triceps, anconeus,

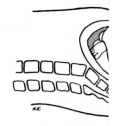

Figure 6-25. The superior pa
by violent stretching of the nec

Figure 6-26. The appearance
brachial plexus injury. Note tha
rotation, the characteristic wai
usually anesthetic is shown in

and extensor muscles of
The person is unable to
This type of paralysis pr
tend the wrist joint and

Lower Brachial Ple
Injuries to inferior part:
they may occur when th
periorly, *e.g.*, a forcefu
birth. It may also occur
break a fall. These acci
brachial plexus (C8 an
ventral roots of the spin
resulting paralysis and a
and skin supplied by t
disabilities are in wrist a
ment of wrist flexion an
of the hand (Fig. 6-20
reduced sensation along
and hand (Fig. 6-21A).

A **cervical rib** (Gra
sure on the inferior tru
symptoms of nerve cor
pressure on the inferior
ularly when the upper
carrying a heavy suitca

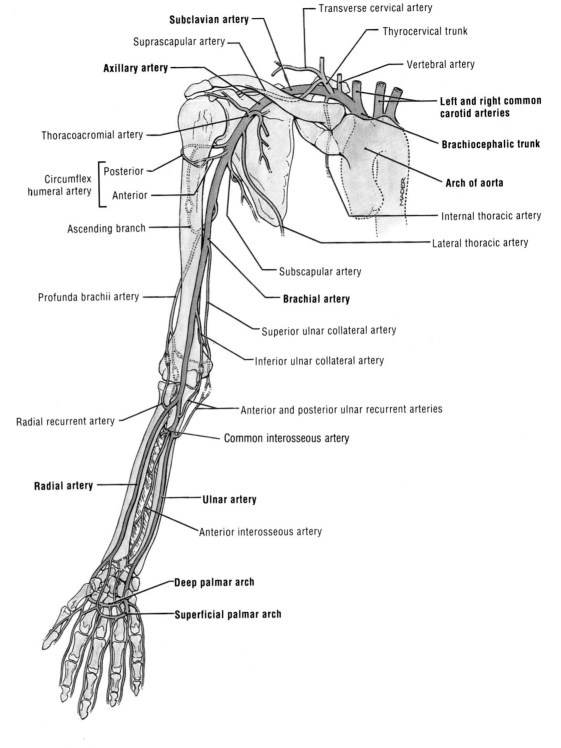

Anterior view

Figure 6-31. Arteries of the upper limb. The subclavian artery becomes the axillary artery at the lateral border of the first rib; the axillary
artery becomes the brachial artery at the inferior border of the teres
major muscle (Fig. 6-32); and the brachial artery ends by dividing into
the radial and ulnar arteries, usually opposite the neck of the radius.

It has **three branches** (subscapular, anterior circumflex humeral, and posterior circumflex humeral).

The subscapular artery (Fig. 6-31), the largest branch of the axillary artery, descends along the lateral border of the subscapularis muscle and divides into the circumflex scapular and thoracodorsal arteries. The *circumflex scapular artery* passes around the lateral border of the scapula to supply muscles on the dorsum of the scapula (Fig. 6-32). The *thoracodorsal artery* continues the general course of the subscapular artery to supply adjacent muscles, principally the latissimus dorsi.

The circumflex humeral arteries pass around the surgical neck of the humerus and anastomose with each other (Fig. 6-31). The *anterior circumflex humeral artery* passes laterally, deep to the coracobrachialis and biceps brachii muscles. It gives off an ascending branch that supplies the shoulder, but the main artery winds around the surgical neck of the humerus. The larger *posterior circumflex humeral artery* passes through the posterior wall of the axilla through the *quadrangular space* with the axillary nerve (Fig. 6-23) to supply the surrounding muscles (*e.g.*, the deltoid and triceps brachii).

The axillary artery can be palpated in the lateral wall of the inferior part of the axilla. Compression of this artery may be necessary (*e.g.*, in injuries of the axilla). This can be done in the inferior part of its course by pressing the artery against the humerus (Fig. 6-31).

There are many arterial anastomoses around the scapula (Figs. 6-31 and 6-32). Several vessels join to form networks on both of its surfaces: dorsal scapular, suprascapular, and subscapular. The clinical importance of the *collateral circulation* that is possible owing to these anastomoses becomes apparent during ligation of an injured axillary or subclavian artery. For example, the axillary artery may be ligated between the thy-

rocervical trunk and the subscapular artery. In this case, the direction of blood flow in the subscapular artery is reversed and blood reaches the distal portion of the axillary artery first. Note that the subscapular artery receives its blood through several anastomoses with the suprascapular artery, transverse cervical artery, and some intercostal arteries (Figs. 6-31 and 6-32). *Ligation of the axillary artery* distal to the subscapular artery cuts off the blood supply to the arm.

Because of the thinness of the axillary sheath (Figs. 6-14 and 6-34), an *aneurysm of the axillary artery* commonly enlarges rapidly and compresses the nerves of the brachial plexus. This causes pain and subsequently anesthesia in the areas of the upper limb supplied by the nerves concerned.

The Axillary Vein

This large vessel lies on the medial side of the axillary artery (Figs. 6-17 and 6-33). It completely overlaps the artery anteriorly when the arm is abducted. The axillary vein, *the continuation of the basilic vein*, begins at the inferior border of the teres major muscle. It ends at the lateral border of the first rib, where it *becomes the subclavian vein*. The axillary vein receives tributaries that correspond to the branches of the axillary artery and at the inferior margin of the subscapularis muscle, the axillary vein receives the venae comitantes of the brachial artery. Superior to the pectoralis minor, the axillary vein is *joined by the cephalic vein* (Figs. 6-15, 6-16, and 6-33).

Wounds in the axilla often involve the axillary vein owing to its large size and exposed position (Fig. 6-33). A wound in the superior part of the vein, where it is largest, is particularly dangerous not only because of the profuse hemorrhage, but also owing to the risk of air entering the vessel.

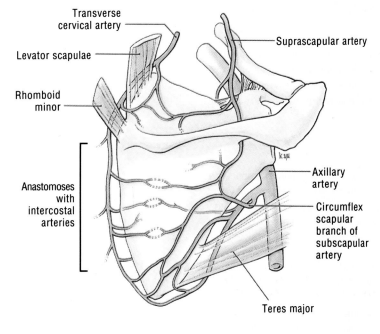

Posterior view

Figure 6-32. Anastomoses around the scapula. Note that the various branches of the axillary artery communicate with other branches of this artery and with branches of other arteries.

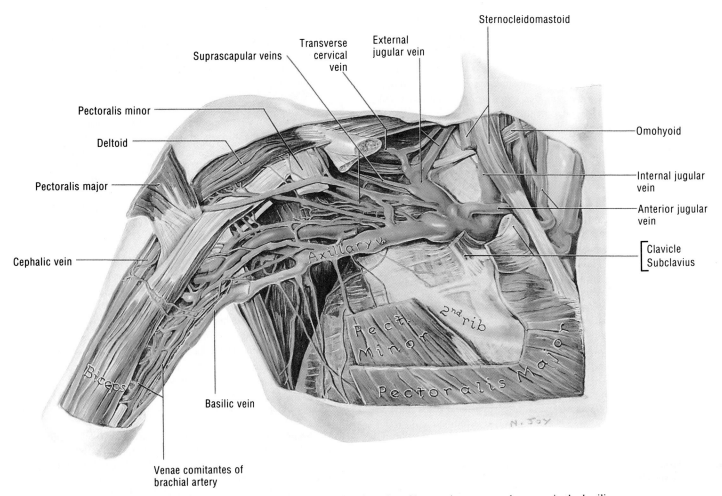

Figure 6-33.   Dissection of the axilla, primarily to show its veins. Observe the venous valves: one in the basilic, three in the axillary, and one in the subclavian veins.

The Axillary Sheath (Figs. 6-14 and 6-34).   The axillary artery and vein and the cords of the brachial plexus are enclosed in this thin fascial sheath. Anterior to the subclavian artery the prevertebral layer of cervical fascia is prolonged laterally, where it forms the axillary sheath.

The Axillary Lymph Nodes

There are 20 to 30 lymph nodes in the fibrofatty connective tissue of the axilla, which are *the main lymph nodes of the upper limb* (Fig. 6-35; see Fig. 1-12). These lymph nodes are arranged in five principal groups, four of which lie inferior to the pectoralis minor tendon and one, the apical group, superior to it.

The pectoral group of axillary lymph nodes consists of three to five lymph nodes that lie along the medial wall of the axilla, around the lateral thoracic artery and the inferior border of the pectoralis major (Fig. 6-35). The pectoral group of nodes receives lymph mainly from the anterior thoracic wall including the breast. The efferent lymph vessels from these nodes pass to the central and apical groups of axillary lymph nodes.

The lateral group of axillary lymph nodes (Fig. 6-35) consists of four to six lymph nodes that lie along the lateral wall of the axilla, medial and posterior to the axillary vein. These lymph nodes receive lymph from most of the upper limb.

The subscapular group of axillary lymph nodes (Fig. 6-35) consists of six or seven lymph nodes situated along the posterior axillary fold and the subscapular blood vessels. This group of lymph nodes receives lymph from the posterior aspect of the thoracic wall and scapular region. Efferent vessels pass from them to the central group of axillary lymph nodes.

The central group of axillary lymph nodes consists of three or four large lymph nodes situated deep to the pectoralis minor near the base of the axilla, in association with the axillary artery. As its name indicates, the central group receives lymph from other groups of axillary lymph nodes. Efferent vessels from the central group pass to the apical group of axillary lymph nodes (Fig. 6-35).

The apical group of axillary lymph nodes consists of lymph nodes situated in the apex of the axilla, along the medial side of the axillary vein and the first part of the axillary artery. This group of lymph nodes *receives lymph from all other axillary lymph nodes*. The efferent vessels from the apical group of axillary lymph nodes unite to form the *subclavian lymphatic trunk*, which joins the jugular and bronchomediastinal trunks to form

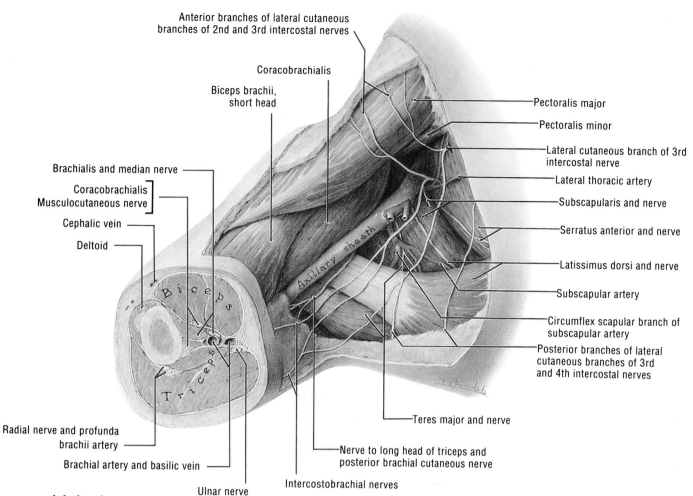

Anterior branches of lateral cutaneous
branches of 2nd and 3rd intercostal nerves

Coracobrachialis

Biceps brachii,
short head

Brachialis and median nerve

Coracobrachialis
Musculocutaneous nerve

Cephalic vein

Deltoid

Radial nerve and profunda
brachii artery

Brachial artery and basilic vein

Ulnar nerve

Inferior view

Pectoralis major

Pectoralis minor

Lateral cutaneous branch of 3rd
intercostal nerve

Lateral thoracic artery

Subscapularis and nerve

Serratus anterior and nerve

Latissimus dorsi and nerve

Subscapular artery

Circumflex scapular branch of
subscapular artery

Posterior branches of lateral
cutaneous branches of 3rd
and 4th intercostal nerves

Teres major and nerve

Nerve to long head of triceps and
posterior brachial cutaneous nerve

Intercostobrachial nerves

Figure 6-34. Dissection of the axilla and transverse section of the arm. Note that the axillary sheath forms a fascial tube for the axillary artery and vein. It also encloses the three cords of the brachial plexus (also see Fig. 6-14).

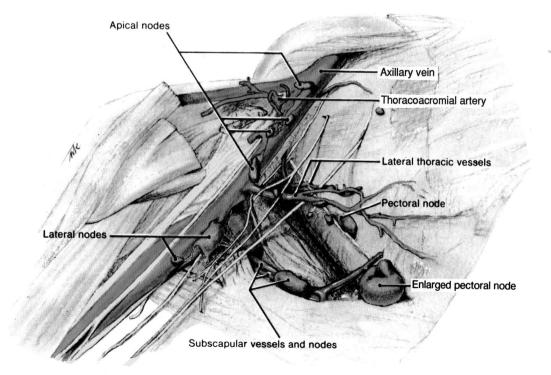

Apical nodes

Axillary vein

Thoracoacromial artery

Lateral thoracic vessels

Pectoral node

Lateral nodes

Enlarged pectoral node

Subscapular vessels and nodes

Figure 6-35. Dissection of the axilla, primarily to show its lymph nodes.

the *right lymphatic duct* (see Fig. 1-39). On the left side, the subclavian lymphatic trunk joins the *thoracic duct*.

The lateral group of axillary lymph nodes is the first one to be involved in *lymphangitis* (inflammation of lymphatic vessels), *e.g.*, resulting from a hand infection. Lymphangitis is characterized by the development of red, warm, tender streaks in the skin. The axillary nodes often become enlarged and tender when there are infections of the upper limb. Infections in the pectoral region and breast, including the superior part of the abdomen, can also produce enlargement of these nodes. In *breast cancer* the axillary lymph nodes may be involved in any stage of the disease (p. 47). In carcinoma of the apical group of axillary nodes, the lymph nodes often adhere to the axillary vein, which may necessitate excision of part of this vessel. Enlargement of the apical group of lymph nodes sometimes obstructs the cephalic vein superior to the pectoralis minor muscle.

The Back and Shoulder Region

The back is described in Chapter 4, but the superficial and intermediate groups of muscles (*extrinsic back muscles*), which attach the upper limb to the axial skeleton, are also described in this chapter (Figs. 6-36 to 6-40).

The muscles of the shoulder are divided into three groups: (1) the superficial extrinsic muscles (trapezius and latissimus dorsi); (2) the deep extrinsic muscles (levator scapulae, rhomboids, and serratus anterior); and (3) the intrinsic muscles (deltoid, supraspinatus, infraspinatus, teres minor, teres major, and subscapularis).

Muscles Connecting the Upper Limb to the Vertebral Column

These muscles (*trapezius*, *latissimus dorsi*, *levator scapulae*, and *rhomboids*) are **extrinsic muscles of the back** (Table 6-4). They are supplied by the ventral rami of cervical nerves, not by dorsal rami as one would expect. The explanation for this is that the superficial back muscles develop in the embryo as a ventrolateral sheet that migrates posteriorly to gain attachment to the vertebral column.

The Trapezius Muscle (Figs. 6-36 to 6-40). This large, flat, triangular muscle covers the posterior aspect of the neck and superior half of the trunk. It was given its name because the muscles of the two sides form a *trapezion* (G. irregular four-sided figure). The trapezius muscle attaches the pectoral girdle to the skull and the vertebral column and assists in suspending it. Its attachments, nerve supply, and main actions are given in Table 6-4. The superior fibers of the trapezius *elevate the scapula* (*e.g.*, when squaring the shoulders); its middle fibers *retract the scapula* (*i.e.*, pull it posteriorly toward the median plane), and its inferior fibers *depress the scapula and lower the shoulder*. The superior and inferior fibers act together in the superior rotation of the scapula (Table 6-4). The trapezius muscles brace the shoulders by pulling the scapulae posteriorly, hence weakness of these muscles results in drooping of the shoulders.

The Latissimus Dorsi Muscle (Figs. 6-12, 6-13 and 6-36 to 6-38). The Latin name of this muscle, meaning "widest of the back," is a good one because it covers the inferior half of the back (T6 vertebra to the iliac crest). This large, wide, *fan-shaped muscle* passes between the trunk and the humerus and acts on the shoulder joint and indirectly on the pectoral girdle. Its attachments, nerve supply, and main actions are given in Table 6-4. The latissimus dorsi *extends, adducts, and medially rotates the humerus* at the shoulder joint. These movements are

Table 6-4.
The Muscles Connecting the Upper Limb to the Vertebral Column

Muscle	Medial Attachments	Lateral Attachments	Innervation[1]	Main Actions
Trapezius (Fig. 6-36*A*)	Medial third of superior nuchal line; external occipital protuberance, ligamentum nuchae, spinous processes of C7 to T12 vertebrae, and lumbar and sacral spinous processes	Lateral third of clavicle, acromion and spine of scapula	Spinal root of accessory n. (CN XI) and cervical nn. (C3 and C4)	Elevates, retracts, and rotates scapula; *superior fibers* elevate, *middle fibers* retract, and *inferior fibers* depress scapula; superior and inferior fibers act together in superior rotation of scapula
Latissimus dorsi (Figs. 6-10 and 6-36*B*)	Spinous processes of the inferior six thoracic vertebrae, thoracolumbar fascia, iliac crest, and inferior 3 or 4 ribs	Floor of intertubercular groove of humerus	Thoracodorsal n. (**C6, C7,** and **C8**)	Extends, adducts, and medially rotates humerus; raises body toward arms during climbing
Levator scapulae (Figs. 6-36*C* and 6-39)	Posterior tubercles of transverse processes of C1 to C4 vertebrae	Superior part of medial border of scapula	Dorsal scapular (C5) and cervical (C3 and C4) nn.	Elevates scapula and tilts its glenoid cavity inferiorly by rotating scapula
Rhomboid minor and major (Fig. 6-36*D*)	*Minor*: Ligamentum nuchae and spinous processes of C7 and T1 vertebrae *Major*: spinous processes of T2 to T5 vertebrae	Medial border of scapula from level of spine to inferior angle	Dorsal scapular n. (C4 and **C5**)	Retracts scapula and rotates it to depress glenoid cavity; fixes scapula to thoracic wall

[1]**Boldface** indicates the main spinal cord segmental innervation.

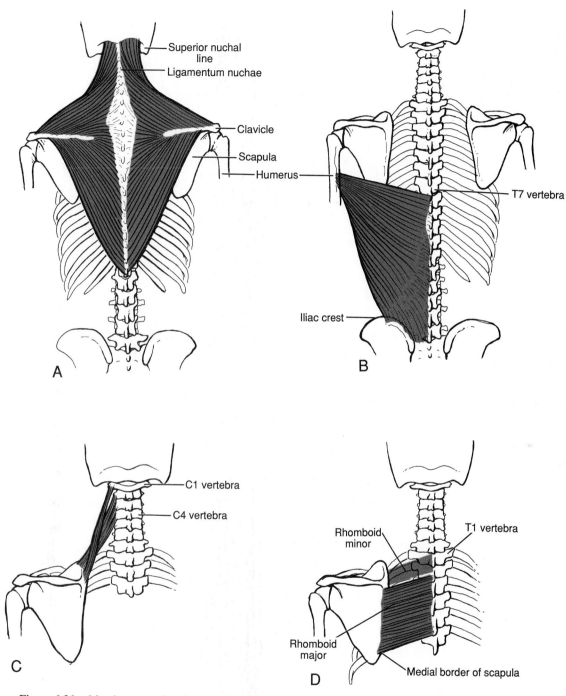

Figure 6-36.   Muscles connecting the upper limb to the vertebral column. *A*, Trapezius. *B*, Latissimus dorsi. *C*, Levator scapulae. *D*, Rhomboid minor and major.

used when chopping wood, climbing, paddling a canoe, and swimming (particularly during the crawl stroke). When climbing, these muscles raise the trunk. In conjunction with the pectoralis major muscle, the latissimus dorsi muscle raises the trunk to the arm, which occurs when performing chin-ups (i.e., hoisting oneself up on an overhead bar).

In Fig. 6-37, observe that the superior border of the latissimus dorsi and a part of the rhomboid major are overlapped

by the trapezius. The triangle formed by the borders of these three muscles is called the *triangle of auscultation* (L. to listen). When the scapula is drawn anteriorly by folding the arms across the chest and the trunk is flexed, the auscultatory triangle enlarges and its lateral side is formed by the medial border of the scapula. Parts of the sixth and seventh ribs and the sixth intercostal space become subcutaneous; consequently respiratory sounds may be heard clearly with a stethoscope in this location.

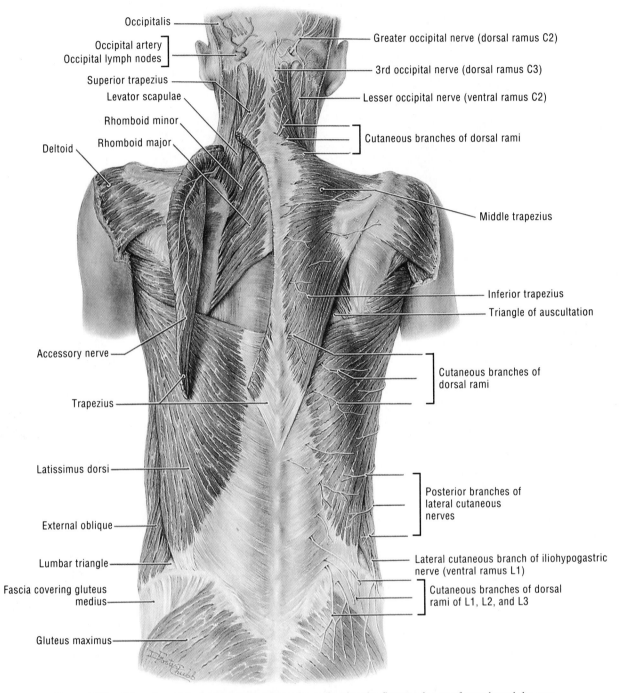

Occipitalis

Occipital artery
Occipital lymph nodes

Superior trapezius

Levator scapulae

Rhomboid minor

Rhomboid major

Deltoid

Accessory nerve

Trapezius

Latissimus dorsi

External oblique

Lumbar triangle

Fascia covering gluteus
medius

Gluteus maximus

Greater occipital nerve (dorsal ramus C2)

3rd occipital nerve (dorsal ramus C3)

Lesser occipital nerve (ventral ramus C2)

Cutaneous branches of dorsal rami

Middle trapezius

Inferior trapezius

Triangle of auscultation

Cutaneous branches of
dorsal rami

Posterior branches of
lateral cutaneous
nerves

Lateral cutaneous branch of iliohypogastric
nerve (ventral ramus L1)

Cutaneous branches of dorsal
rami of L1, L2, and L3

Figure 6-37. Dissection of the back and shoulder regions, showing the first two layers of muscle and the cutaneous nerves. The trapezius muscle is severed and reflected on the left side.

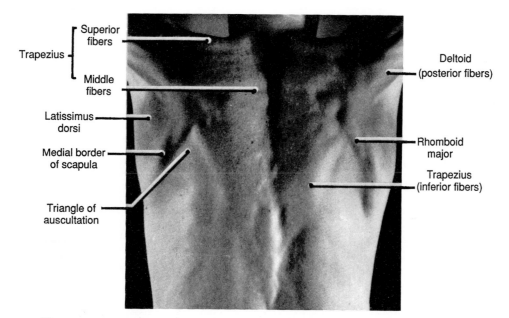

Figure 6-38. Back and shoulder regions of a 46-year-old man with his upper limbs elevated.

The latissimus dorsi and the inferior portion of the pectoralis major form an anteroposterior sling from the trunk to arm, but the latissimus dorsi forms the more powerful part of the sling. When there is *paralysis of the latissimus dorsi*, the patient is unable to raise the trunk as occurs during climbing or "chinning" one's self. Furthermore, one cannot use a crutch because the shoulder is pushed superiorly by it.

The Levator Scapulae Muscle (Figs. 6-36 to 6-40). The superior third of this straplike muscle lies deep to the sternocleidomastoid muscle; the inferior third is deep to the trapezius. Its attachments, nerve supply, and main actions are given in Table 6-4. The levator scapulae *elevates the scapula* and helps to tilt its glenoid cavity inferiorly by rotating the scapula. It also helps to retract the scapula and fix it against the trunk and to flex the neck laterally.

The Rhomboid Muscles (Figs. 6-36 to 6-40). These two muscles lie deep to the trapezius and are not always distinct from each other. The *rhomboid major* is about two times wider than the *rhomboid minor*. They appear as parallel bands that pass inferolaterally from the vertebrae to the scapulae. They have a rhomboid appearance, *i.e.*, they form an oblique parallelogram. Their attachments, nerve supply, and main actions are given in Table 6-4. The rhomboids *retract the scapula and rotate it* to depress its glenoid cavity. They also help the serratus anterior to hold the scapula against the thoracic wall and to fix the scapula during movements of the upper limb. The rhomboid muscles are used when forcibly lowering the raised upper limbs, *e.g.*, when driving a stake with a sledge hammer.

The Scapular Muscles

Six short muscles (*deltoid, supraspinatus, infraspinatus, subscapularis, teres major,* and *teres minor*) pass from the scapula to the humerus and act on the shoulder joint (Table 6-6).

The Deltoid Muscle (Figs. 6-3*B*, 6-5, 6-8, 6-13, 6-15, 6-37, and 6-41 to 6-44*A*). This thick, powerful *shoulder muscle* covers the shoulder joint and forms the rounded contour of the shoulder. As its name "deltoid" indicates, it is triangular in outline; *i.e.*, it is *shaped like an inverted Greek letter delta*. Its attachments, nerve supply, and main actions are given in Table 6-6. For descriptive purposes, the deltoid may be divided into three parts: anterior, middle, and posterior. It is capable of acting in part or as a whole. All three parts are active in movements of the arm.

The **actions of the deltoid** are as follows: (1) the *anterior part* is a strong *flexor and medial rotator* of the humerus; (2) the *middle part* is the chief *abductor* of the humerus; and (3) the *posterior part* is a strong *extensor and lateral rotator* of the humerus. In performing these movements, the deltoid works with other muscles, *e.g.*, the anterior part acts with the pectoralis major and coracobrachialis muscles in flexing the arm, whereas the middle part acts with the supraspinatus in abducting the arm. The deltoids are used every day when swinging the upper limbs during walking. The anterior parts flex the arms and the posterior parts extend them. The deltoid muscle tends to *stabilize the shoulder joint* and helps to hold the head of the humerus in the glenoid cavity of the scapula during arm movements.

The deltoid muscle atrophies when the axillary nerve is severely damaged (*e.g.*, during a fracture of the surgical neck of the humerus). As the deltoid atrophies, the rounded contour of the shoulder disappears; this gives it a flattened appearance. To test the strength of the deltoid clinically, the patient's arm is abducted, and then the patient is asked to hold it in that position against resistance. Inability to do this indicates injury to the axillary nerve.

The Teres Major Muscle (Figs. 6-3*B*, 6-34, 6-40, 6-42, and 6-44*E*). The teres (L. round) major is a somewhat flattened

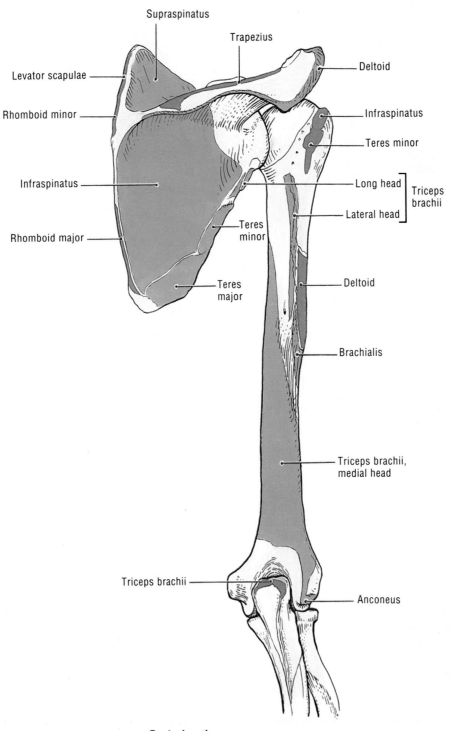

Posterior view

Figure 6-39. Bones of the upper limb (right side) showing the sites of attachment of muscles to them. Proximal attachments are shown in red and distal attachments in blue.

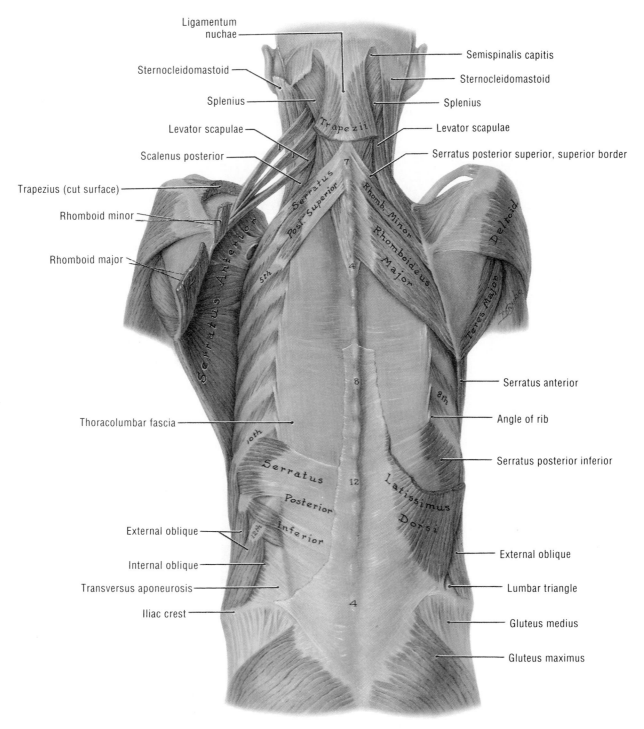

Ligamentum nuchae

Sternocleidomastoid

Splenius

Levator scapulae

Scalenus posterior

Trapezius (cut surface)

Rhomboid minor

Rhomboid major

Thoracolumbar fascia

External oblique

Internal oblique

Transversus aponeurosis

Iliac crest

Semispinalis capitis

Sternocleidomastoid

Splenius

Levator scapulae

Serratus posterior superior, superior border

Serratus anterior

Angle of rib

Serratus posterior inferior

External oblique

Lumbar triangle

Gluteus medius

Gluteus maximus

Figure 6-40. Dissection showing the intermediate back muscles. The trapezius and latissimus dorsi are largely cut away. On the left side, the rhomboids are severed, allowing the vertebral border of the scapula to separate from the thoracic wall. Note the three digitations (slips) of the levator scapulae muscle.

Table 6-5.
The Muscles Producing Movements of the Scapula[1]

Elevation	Depression	Protraction	Retraction	Superior Rotation[2]	Inferior Rotation
Trapezius-superior fibers	**Pectoralis minor**	**Serratus anterior**	**Trapezius**	**Trapezius**-superior and inferior fibers	**Levator scapulae**
Levator scapulae	**Latissimus dorsi**	**Pectoralis minor**	**Rhomboids**	**Serratus anterior**-	**Pectoralis minor**
Serratus anterior- superior fibers	Pectoralis major	Levator scapulae	Latissimus dorsi	inferior fibers	**Rhomboids**
					Latissimus dorsi
					Pectoralis major

[1]The principal muscles producing these movements appear in **boldface**.
[2]Superior rotation results in the glenoid cavity being elevated so that it faces superiorly; inferior rotation of the scapula is the reverse movement.

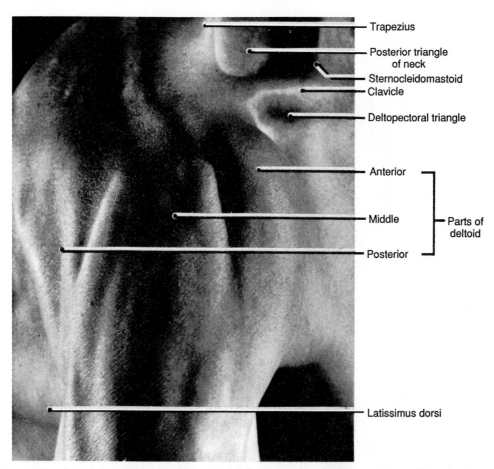

Figure 6-41.   Lateral view of the root of the neck and shoulder region of a 46-year-old man. To make the parts of his deltoid muscle stand out, he abducted his arm against resistance.

rectangular muscle that forms a raised oval area on the dorsum of the scapula, beginning at the inferior angle. The inferior border of this rounded muscle forms the inferior border of the posterior wall of the axilla. The teres major and the tendon of the latissimus dorsi form the *posterior axillary fold*. Its attachments, nerve supply, and main actions are given in Table 6-6. The teres major muscle *adducts and medially rotates the arm*. It can also help to extend it from the flexed position. It is also an important stabilizer of the head of the humerus in the glenoid cavity during abduction of the arm; *i.e.*, it steadies the head of the humerus in its socket.

The Rotator Cuff Muscles

Four scapular muscles (supraspinatus, infraspinatus, teres minor, and subscapularis) are called rotator cuff muscles of the shoulder joint (Figs. 6-44 and 6-45) because they form a musculotendinous *rotator cuff for the shoulder joint* (see Fig. 6-115). The tendons of these four muscles blend with the articular capsule of the shoulder joint. Consequently, *the rotator cuff is a mass of tendons fused with the lateral part of the capsule of the shoulder joint*. All the rotator cuff muscles, except the supraspinatus, are **rotators of the humerus**; this explains why they are called rotator cuff muscles. *The musculotendinous rotator cuff protects the shoulder joint and gives it stability* by holding the head of the humerus in the glenoid cavity of the scapula (p. 612).

There are *bursae around the shoulder joint* (Figs. 6-46 and 6-47), which are located between the tendons of the rotator cuff muscles and the fibrous capsule of the shoulder joint. These bursae reduce the friction on tendons passing over bones or other areas of resistance.

The Supraspinatus Muscle (Figs. 6-44*B* and 6-47). This rounded, conical muscle lies in the supraspinous fossa located superior to the spine of the scapula, deep to the trapezius muscle and the *coracoacromial arch.*[2] Its tendon is covered by the deltoid muscle. Its attachments, nerve supply, and main actions are given in Table 6-6. The supraspinatus, a rotator cuff muscle, *helps to abduct the arm*, and it helps the other rotator cuff muscles hold the head of the humerus in the glenoid cavity of the scapula during all movements of the shoulder joint (*i.e.*, it helps to stabilize this joint). The supraspinatus acts with the deltoid during abduction of the arm and *acts strongly when a heavy weight is carried with the upper limb adducted* (*e.g.*, when carrying a heavy suitcase). If the deltoid is paralyzed, the supraspinatus can partially abduct the humerus.

> The tendon of the supraspinatus is separated from the coracoacromial ligament, the acromion, and the deltoid muscle by the *subacromial bursa* (Fig. 6-47). When this bursa is inflamed (**subacromial bursitis**), abduction of the arm is painful. The supraspinatus tendon is the most commonly torn part of the rotator cuff (p. 536).

---

[2]A protective arch for the shoulder joint formed by the coracid process, coracoacromial ligament, and the acromion (p. 611).

The Subscapularis Muscle (Figs. 6-10, 6-12, and 6-44*F*). This thick triangular muscle lies on the costal surface of the scapula and forms part of the posterior wall of the axilla. It crosses the anterior aspect of the shoulder joint on its way to the humerus. Its attachments, nerve supply, and main actions are given in Table 6-6. The subscapularis *medially rotates the arm and adducts it*. It also helps the supraspinatus, infraspinatus, and teres minor to hold the head of the humerus in the glenoid cavity during all movements of the shoulder joint (*i.e.*, it helps to stabilize this joint).

The Teres Minor Muscle (Figs. 6-44*D* and 6-45). This elongated, tapering muscle is often inseparable from the infraspinatus muscle, which lies along its superior border. Its attachments, nerve supply, and main actions are given in Table 6-6. The teres minor laterally rotates the arm and assists in its adduction. It also helps the subscapularis, supraspinatus, and infraspinatus to hold the head of the humerus in the glenoid cavity during movements of the shoulder joint (*i.e.*, it helps to stabilize this joint).

The Infraspinatus Muscle (Figs. 6-44*C* and 6-45). This triangular muscle occupies most of the infraspinous fossa. Its attachments, nerve supply, and main actions are given in Table 6-6. The infraspinatus *laterally rotates the arm* and helps the subscapularis, supraspinatus, and teres minor to hold the head of the humerus in the glenoid cavity during movements of the shoulder joint (*i.e.*, it helps to stabilize this joint).

> The rotator cuff holds the head of the humerus in the glenoid cavity of the scapula. It may be damaged by injury or disease, resulting in instability of the shoulder joint. Trauma may tear or rupture one or more of the tendons of the rotator cuff muscles. As has been mentioned, the supraspinatus ten-

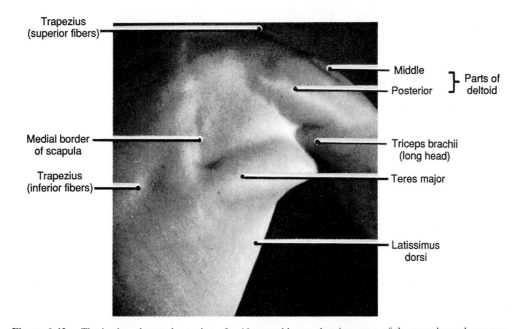

Figure 6-42.   The back and scapular region of a 46-year-old man showing some of the scapular and arm muscles. The teres major muscle was made to stand out by asking him to adduct his arm against resistance.

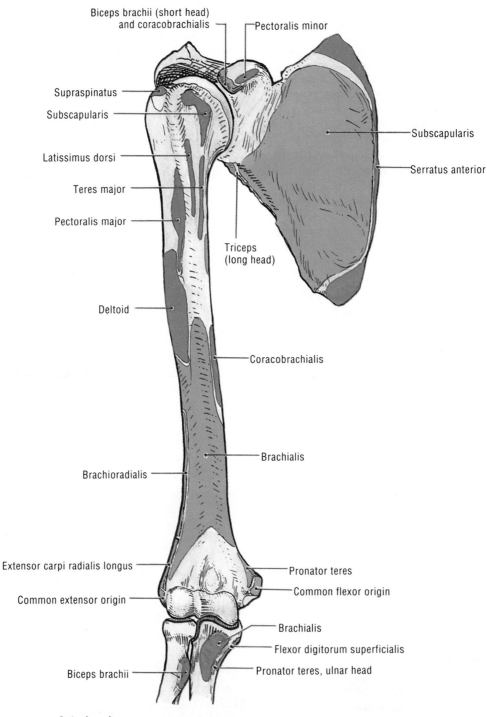

Biceps brachii (short head)
and coracobrachialis

Pectoralis minor

Supraspinatus

Subscapularis

Latissimus dorsi

Teres major

Pectoralis major

Triceps
(long head)

Deltoid

Subscapularis

Serratus anterior

Coracobrachialis

Brachialis

Brachioradialis

Extensor carpi radialis longus

Pronator teres

Common flexor origin

Common extensor origin

Brachialis

Flexor digitorum superficialis

Biceps brachii

Pronator teres, ulnar head

Anterior view

Figure 6-43.   Scapula and some bones of the upper limb (right side) showing the attachment of muscles to them. Proximal attachments are shown in deep salmon and distal attachments in blue.

don is the most commonly torn part of the rotator cuff. This injury is common in baseball pitchers.

Degenerative tendonitis of the musculotendinous rotator cuff is a common disease, especially in older people. Calcium deposits may be demonstrated on radiographs in the supraspinatus tendon and in other tendons of the cuff. Tendonitis and inflammation of the subacromial bursa (*subacromial bursitis*) result in shoulder pain that is intensified by attempts to abduct the arm. The supraspinatus tendon does not rupture very often in young people because their tendons are usually so strong that they will tear away (avulse) the tip of the greater tubercle of the humerus rather than rupture the tendon.

The main stability of the shoulder joint is provided by the musculotendinous rotator cuff, which is formed by the fusion of the tendons of the rotator cuff muscles with the fibrous capsule of the joint. It strengthens the shoulder joint everywhere, except inferiorly. Consequently, if a person falls when the humerus is abducted, the head of this bone may be levered out of the glenoid cavity of the scapula, producing *dislocation of the shoulder joint* (p. 615).

The Arm

The arm (L. *brachium*) extends from the shoulder to the elbow. The rounded prominence on the anterior surface of the arm is formed by the biceps brachii muscle (Figs. 6-48 and 6-50), which is commonly referred to as the *biceps*. The triceps brachii muscle, usually called the *triceps*, occupies the posterior part of the arm (Figs. 6-50, 6-51*D*, and 6-53).

The Humerus

The humerus is the largest bone in the upper limb (Fig. 6-1). Its smooth, ball-like head articulates with the glenoid cavity of the scapula. Close to the head are the *greater and lesser tubercles* (Fig. 6-49) for the insertion of the muscles that surround and move the shoulder joint. The lesser tubercle is separated from the greater tubercle by the *intertubercular groove* (sulcus), in which lies the tendon of the long head of the biceps brachii muscle. The *anatomical neck* separates the head and the tubercles. Distal to the anatomical neck is the surgical neck (Fig. 6-49). It is located where the bone narrows to become the *body* (shaft). This region is called the *surgical neck* because it is the most frequent fracture site of the proximal end of the humerus.

The body, or shaft, of the humerus is easy to palpate, as are its medial and lateral epicondyles (Fig. 6-1). Its superior half is cylindrical. Anterolaterally there is a roughness known as the *deltoid tuberosity* for the insertion of the deltoid muscle (Figs. 6-43, 6-44*A*, and 6-49). There is a shallow, oblique *radial groove* (spiral groove) for the radial nerve that extends inferolaterally on the posterior aspect of the body. The distal end of the humerus is expanded from side to side. The *trochlea* (L. pulley) fits into the *trochlear notch of the ulna* (Figs. 6-49 and 6-61*A*), which swings on this pulley when the elbow is flexed. Just proximal

to the trochlea are the *coronoid fossa* and the *olecranon fossa*, which accommodate corresponding parts of the ulna. Adjoining the lateral part of the trochlea is a rounded ball of bone called the *capitulum* (L. little head). A prominent process, the *medial epicondyle*, projects from the trochlea, and the *lateral epicondyle* projects from the capitulum. The epicondyles, being subcutaneous, are easily felt. The medial epicondyle is more prominent (Fig. 6-48). From each epicondyle, a bony ridge runs proximally; these are known as the medial and lateral *supracondylar ridges*, respectively (Fig. 6-49*A*).

Fractures of the proximal end of the humerus may be through its anatomical or surgical necks (Fig. 6-1). *Fractures of the surgical neck are common in elderly persons* and usually result from falls on the elbow when the arm is abducted. The fracture line occurs superior to the insertion of the pectoralis major, teres major, and latissimus dorsi muscles (Fig. 6-43). Because nerves are in contact with the humerus (surgical neck-axillary nerve; radial groove-radial nerve; and medial epicondyle-ulnar nerve), *the axillary, radial, and ulnar nerves may be injured in fractures of the humerus*.

Traumatic separation of the proximal epiphysis of the humerus (Fig. 6-52) can occur in young persons because this epiphysis does not fuse with the body of the humerus until about 18 years of age in females and 20 years of age in males. *Fracture-separation of the proximal humeral epiphysis* occurs in children because the articular capsule of the shoulder joint is stronger than the epiphyseal cartilaginous plate.

The Brachial Fascia and Intermuscular Septa

The arm is enclosed in a strong sheath of deep fascia known as the *brachial fascia* (Figs. 6-50 and 6-56). It is continuous superiorly with the pectoral and axillary fasciae. The brachial fascia is attached inferiorly to the epicondyles of the humerus and to the olecranon of the ulna. It also is continuous with the deep fascia of the forearm (Fig. 6-56). Two *intermuscular septa* extend from the brachial fascia and are attached to the medial and lateral supracondylar ridges of the humerus (Figs. 6-49, 6-50, and 6-54). The medial and lateral intermuscular septa divide the arm into anterior and posterior *fascial compartments*, each of which contains muscles, nerves, and blood vessels. The *anterior fascial compartment of the arm* (flexor compartment) contains three muscles (biceps brachii, brachialis, and coracobrachialis) and their nerves and vessels. The *posterior fascial compartment of the arm* (extensor compartment) contains one muscle (triceps) and its nerve and vessels.

Muscles of the Arm

There are *four muscles* in the arm: *three flexors* in the anterior fascial compartment supplied by the musculocutaneous nerve and *one extensor* in the posterior fascial compartment supplied by the radial nerve (Table 6-7). The anconeus muscle is located chiefly in the forearm, but it is described with the brachial muscles because it is morphologically and functionally related to the triceps brachii muscle.

The Biceps Brachii Muscle (Figs. 6-48, 6-50, 6-51, and 6-54). As its name "biceps" indicates, the proximal attach-

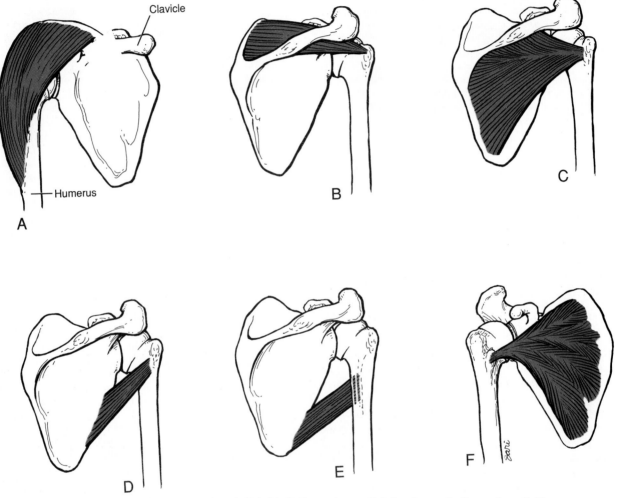

Figure 6-44.   The scapular muscles. *A*, Deltoid. *B*, Supraspinatus. *C*, Infraspinatus. *D*, Teres minor. *E*, Teres major. *F*, Subscapularis.

Table 6-6.
The Scapular Muscles

Muscle	Proximal/Medial Attachments	Distal/Lateral Attachments	Innervation[1]	Main Actions
Deltoid (Fig. 6-44A)	Lateral third of clavicle, acromion and spine of scapula	Deltoid tuberosity of humerus	Axillary n. (**C5** and **C6**)	*Anterior part:* flexes and medially rotates arm *Middle part:* abducts arm *Posterior part:* extends and laterally rotates arm
Supraspinatus[2] (Fig. 6-44B)	Supraspinous fossa of scapula	Superior facet on greater tubercle of humerus	Suprascapular n. C4, **C5**, and C6	Helps deltoid to abduct arm and acts with rotator cuff muscles[2]
Infraspinatus[2] (Fig. 6-44C)	Infraspinous fossa of scapula	Middle facet on greater tubercle of humerus	Suprascapular n. (**C5** and C6)	Laterally rotate arm; Help to hold humeral head in glenoid cavity of scapula
Teres minor[2] (Fig. 6-44D)	Superior part of lateral border of scapula	Inferior facet on greater tubercle of humerus	Axillary n. (**C5** and **C6**)	
Teres major (Fig. 6-44E)	Dorsal surface of inferior angle of scapula	Medial lip of intertubercular groove of humerus	Lower subscapular n. (**C6** and C7)	Adducts and medially rotates arm
Subscapularis[2] (Fig. 6-44F)	Subscapular fossa	Lesser tubercle of humerus	Upper and lower subscapular nn. (C5, **C6**, and C7)	Medially rotates arm and adducts it; helps to hold humeral head in glenoid cavity

[1]**Boldface** indicates the main spinal cord segmental innervation.
[2]Collectively, the supraspinatus, infraspinatus, teres minor, and subscapularis muscles are referred to as the **rotator cuff muscles**. Their prime function during all movements of the shoulder joint is to hold the head of the humerus in the glenoid cavity of the scapula (Fig. 6-115).

ment of this long fusiform muscle has *two heads* (*bi*, two + *cipital*, from L. *caput*, head). The two bellies of the muscle unite just distal to the middle of the arm. The biceps is located in the anterior aspect of the arm in the anterior fascial compartment. Its attachments, nerve supply, and main actions are given in Table 6-7. When the elbow is extended, the biceps is a simple *flexor of the forearm*. It is also a *powerful supinator* when the forearm is flexed and when more power is needed against resistance (e.g., when right-handed persons drive a screw into hard wood). The biceps barely operates at all during flexion of the prone forearm. It is used when inserting a corkscrew and pulling out the cork of a wine bottle.

The tendon of the long head of the biceps crosses the head of the humerus within the capsule of the shoulder joint and descends in the *intertubercular groove* of the humerus (Figs. 6-10, 6-49A, and 6-114A). Distally its tendon attaches to the tuberosity of radius. The *bicipitoradial bursa* separates this tendon from the anterior part of the tuerosity. The biceps brachii is also attached through the **bicipital aponeurosis** (Figs. 6-54 and 6-56), a triangular membranous band that runs from the biceps tendon across the cubital fossa and merges with the deep fascia over the flexor muscles in the medial side of the forearm. The proximal part of the bicipital aponeurosis can be easily felt where it passes obliquely over the brachial artery and median nerve (Figs. 6-48, 6-54, and 6-56). This aponeurosis affords protection for these and other structures in the cubital fossa. It also helps to lessen the pressure of the biceps tendon on the radial tuberosity during pronation and supination of the forearm.

Sometimes the tendon of the long head of the biceps brachii is dislocated from the intertubercular groove in the humerus. This may occur during traumatic separation of the proximal epiphysis of the humerus (Fig. 6-52). When this occurs, the arm is fixed in the abducted position and the head of the humerus can be felt in its normal position (Fig. 6-114).

The Brachialis Muscle (Figs. 6-50, 6-51B, 6-53, 6-54, 6-56, and 6-70). This flattened, fusiform muscle lies posterior to the biceps brachii. Its attachments, nerve supply, and main actions are given in Table 6-7. The strong brachialis is the **main flexor of the forearm**. It flexes the forearm in all positions and in slow or quick movements. When the forearm is being extended slowly, the brachialis steadies the movement by slowly relaxing. It always contracts during flexion of the elbow joint and is primarily responsible for maintaining flexion. Because of this, it is the *workhorse among the flexor muscles* of the elbow joint.

The Coracobrachialis Muscle (Figs. 6-16, 6-23, 6-34, 6-51C, and 6-54). This elongated, narrow muscle in the superomedial part of the arm is important mainly as a landmark (*e.g.*, the musculocutaneous nerve pierces it). Its attachments, nerve supply, and main actions are given in Table 6-7. The coracobrachialis *helps to flex and adduct the arm* at the shoulder joint. It also helps to stabilize this joint.

The Triceps Brachii Muscle (Figs. 6-45, 6-50, 6-51D, 6-53, and 6-54). This large, fusiform muscle is located in the posterior fascial compartment of the arm. It is *associated with the small anconeus muscle* at the elbow. As its name "triceps" indicates, it has three heads of proximal attachment: long, lateral, and medial. Its attachments, nerve supply, and main actions are given in Table 6-7. Just proximal to its distal attachment is a subtendinous *olecranon bursa* between the triceps tendon and the olecranon. The triceps is the **main extensor of the forearm**. Because the long head of the triceps crosses the shoulder joint, it also *aids in extension and adduction of the arm*. Its long head also steadies the head of the abducted humerus.

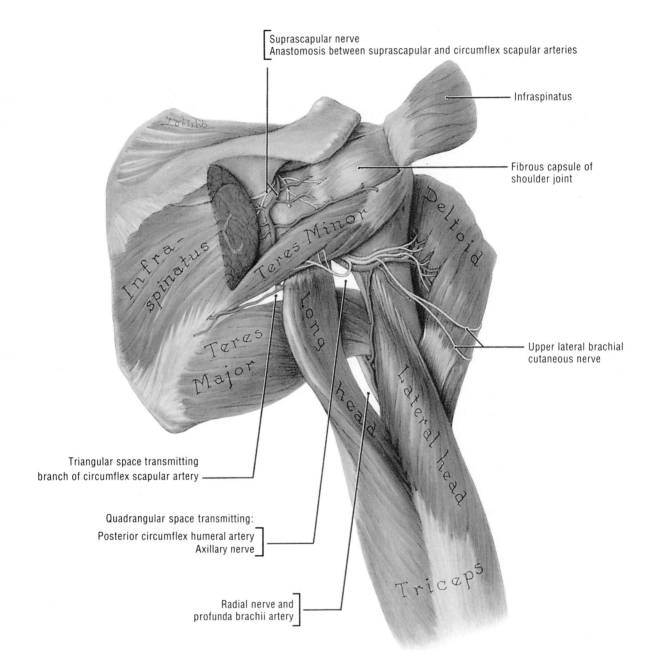

Suprascapular nerve
Anastomosis between suprascapular and circumflex scapular arteries

Infraspinatus

Fibrous capsule of
shoulder joint

Upper lateral brachial
cutaneous nerve

Triangular space transmitting
branch of circumflex scapular artery

Quadrangular space transmitting:
Posterior circumflex humeral artery
Axillary nerve

Radial nerve and
profunda brachii artery

Figure 6-45. Dissection of the posterior scapular and subdeltoid regions.

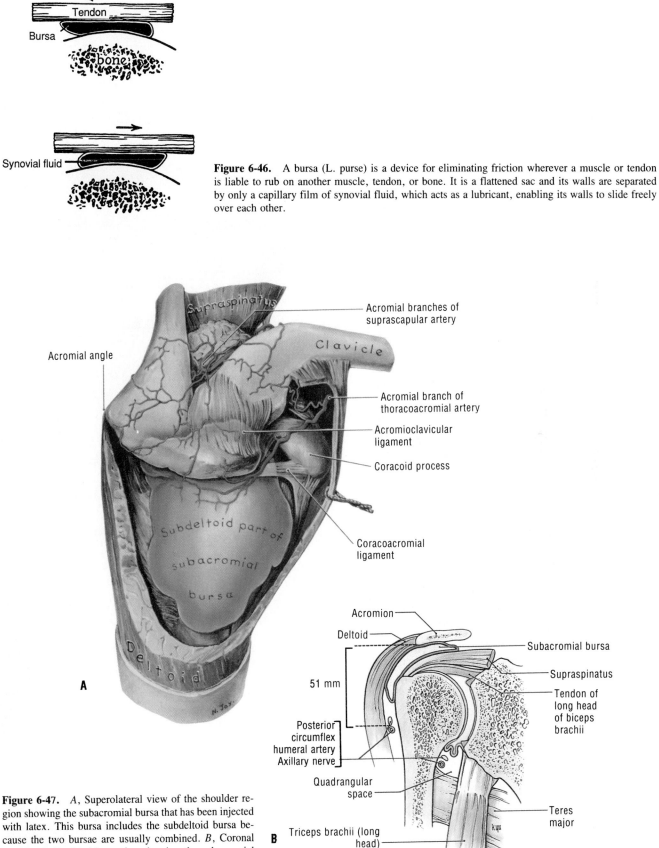

Figure 6-46.   A bursa (L. purse) is a device for eliminating friction wherever a muscle or tendon is liable to rub on another muscle, tendon, or bone. It is a flattened sac and its walls are separated by only a capillary film of synovial fluid, which acts as a lubricant, enabling its walls to slide freely over each other.

Figure 6-47.   *A,* Superolateral view of the shoulder region showing the subacromial bursa that has been injected with latex. This bursa includes the subdeltoid bursa because the two bursae are usually combined. *B,* Coronal section of the shoulder region showing the subacromial bursa.

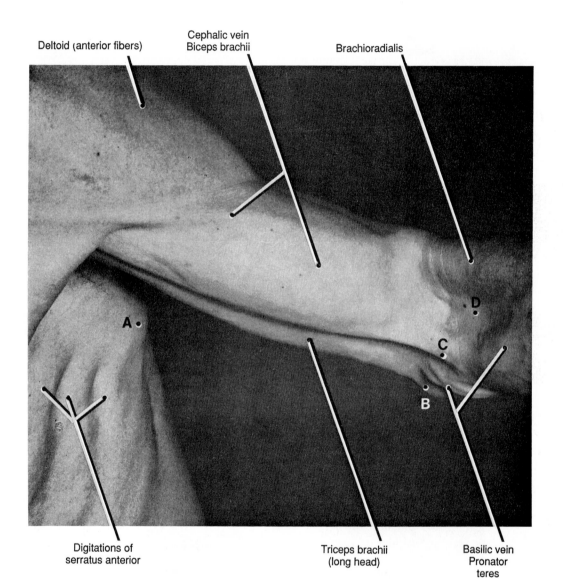

Figure 6-48.   Anterior surface of the left shoulder, arm, elbow, and lateral thoracic regions of a 46-year-old man. *A* indicates inferior angle of scapula; *B*, medial epicondyle of the humerus; *C*, bicipital aponeurosis; *D*, cubital fossa.

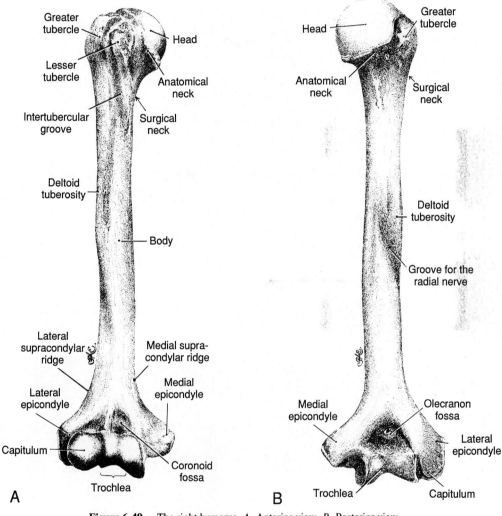

Figure 6-49. The right humerus. *A*, Anterior view. *B*, Posterior view.

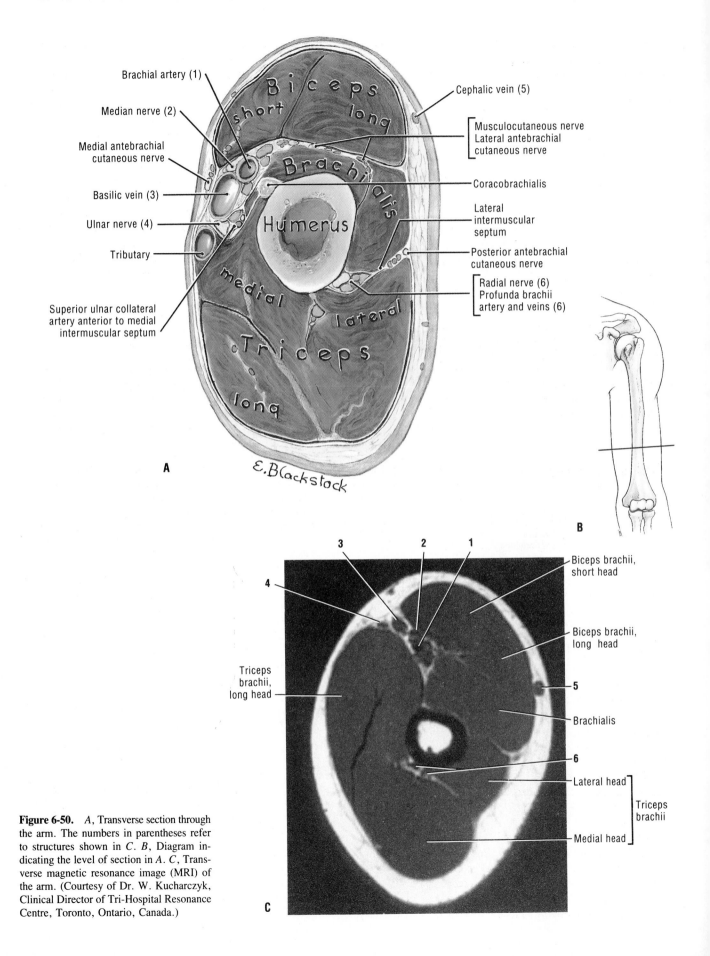

Figure 6-50. *A,* Transverse section through the arm. The numbers in parentheses refer to structures shown in *C. B,* Diagram indicating the level of section in *A. C,* Transverse magnetic resonance image (MRI) of the arm. (Courtesy of Dr. W. Kucharczyk, Clinical Director of Tri-Hospital Resonance Centre, Toronto, Ontario, Canada.)

The Anconeus Muscle (Figs. 6-51 and 6-63). This small, triangular muscle is on the lateral part of the posterior aspect of the elbow. It is usually partially blended with the triceps and should be considered as part of the medial head of this muscle. Its attachments, nerve supply, and actions are given in Table 6-7. It helps the triceps to extend the forearm (p. 544). It also *abducts the ulna during pronation of forearm* (p. 9) and contracts whenever this joint needs to be stabilized against flexion.

The Cubital Fossa

This triangular, *hollow area on the anterior surface of the elbow* (Figs. 6-48D and 6-60) is bound superiorly by an imaginary line connecting the epicondyles of the humerus, medially by the pronator teres muscle, and laterally by the brachioradialis muscle. The *floor of the cubital fossa* is formed by the brachialis and supinator muscles of the arm and forearm, respectively (Fig. 6-62). The *roof of the cubital fossa* is formed by deep fascia and includes the bicipital aponeurosis. The roof is covered by superficial fascia and skin (Figs. 6-57 and 6-58).

Contents of the Cubital Fossa (Figs. 6-54 to 6-58 and 6-62). This fossa is an important clinical area because it contains the biceps tendon, the brachial artery and its terminal branches (radial and ulnar arteries), the brachial veins, and parts of the median and radial nerves.

Arteries of the Arm

The Brachial Artery (Figs. 6-31, 6-34, 6-50, 6-54 to 6-56, and 6-62). This main artery provides the principal arterial supply to the arm. It begins at the inferior border of the teres major as the *continuation of the axillary artery*. It runs inferiorly and slightly laterally on the medial side of the biceps brachii muscle to the cubital fossa, where it ends opposite the neck of the radius. Under cover of the bicipital aponeurosis, *the brachial artery divides into the radial and ulnar arteries*. Its course through the arm is represented by a line connecting the midpoint of the clavicle with the midpoint of the cubital fossa.

The brachial artery is superficial and palpable throughout its course. At first it lies medial to the humerus and then anterior to it. It lies anterior to the triceps and brachialis muscles and is overlapped by the coracobrachialis and biceps muscles (Fig. 6-54). As the brachial artery passes inferiorly and slightly laterally, it *accompanies the median nerve*, which crosses anterior to the artery in the middle of the arm. In the cubital fossa *the bicipital aponeurosis covers and protects the median nerve and brachial artery* (Fig. 6-56), and it separates them from the median cubital vein (Fig. 6-57). During its course through the arm, the brachial artery gives rise to many unnamed muscular branches, mainly from its lateral side. The named branches of the brachial artery are (Figs. 6-54 and 6-55): the *profunda brachii artery* (deep brachial artery), the *nutrient humeral artery*, and the *ulnar collateral arteries* (superior and inferior).

The Profunda Brachii Artery (Figs. 6-31, 6-34, 6-50, 6-55, and 6-59). This is the largest branch of the brachial artery, and it has the most superior origin. It *accompanies the radial nerve in its posterior course in the radial groove*. Posterior to the humerus, the profunda brachii, or **deep brachial, artery** divides into anterior and posterior descending branches, which help to form the arterial anastomoses of the elbow region.

The Nutrient Humeral Artery. This vessel arises from the brachial artery around the middle of the arm and enters the *nutrient canal* on the anteromedial surface of the humerus. This artery and canal run toward the elbow.

The Superior Ulnar Collateral Artery (Figs. 6-54 and 6-55). This vessel arises from the brachial artery near the middle of the arm and accompanies the ulnar nerve posterior to the medial epicondyle of the humerus. Here, it anastomoses with the posterior ulnar recurrent branch of the ulnar artery and the inferior ulnar collateral artery, a branch of the brachial artery.

The Inferior Ulnar Collateral Artery (Figs. 6-54 and 6-55). This vessel arises from the brachial artery about 5 cm proximal to the elbow crease. It then passes inferomedially, anterior to the medial epicondyle of the humerus, where it joins the anastomoses of the elbow region.

The brachial artery is doubled in 20% of cases during all or part of its course. In these people one of the arteries lies superficial to the median nerve and is called the *superficial brachial artery*. The arterial *anastomoses around the elbow* provide a functionally and surgically important collateral circulation. The brachial artery may be clamped or even ligated distal to the inferior ulnar collateral artery without producing tissue damage. The anatomical basis for this is that the ulnar and radial arteries still receive sufficient blood through the anastomoses around the elbow (Fig. 6-55).

Arterial blood pressure levels are routinely measured using a *sphygmometer*, consisting of an inflatable cuff and a mercury manometer. The cuff is placed around the arm and inflated with air until it compresses the brachial artery against the humerus and occludes it. A stethoscope is placed over the artery in the cubital fossa, just medial to the biceps tendon (Fig. 6-56). As the pressure in the cuff is reduced, blood begins to spurt through the artery. The first audible spurt indicates the *systolic blood pressure*.

Compression of the brachial artery may be produced in almost its entire course (*e.g.*, to control hemorrhage). The best place to compress the brachial artery is near the middle of the arm, where it lies on the tendon of the coracobrachialis muscle medial to the humerus (Figs. 6-50, 6-54, and 6-55). In elderly people the brachial artery is often tortuous and subcutaneous here, and its pulsations are often visible in very thin people. To compress the brachial artery in the inferior part of the arm, pressure has to be directed posteriorly because the artery lies anterior to the humerus in this region (Fig. 6-31).

Occlusion or *laceration of the brachial artery* represents a surgical emergency because, within a few hours, the *paralysis of muscles* that results from the associated ischemia of the deep flexors of the forearm (flexor pollicis longus and flexor digitorum profundus) may be irreversible (Table 6-9). Necrotic muscle is replaced by fibrous scar tissue that causes the involved muscles to become permanently shortened. This produces a flexion deformity called *Volkmann's ischemic contracture*. There is contraction of the digits and sometimes of the wrist, with loss of hand power. Extended and improper use of a tourniquet can also produce this injury, but it results more often from brachial artery injury (*e.g.*, tearing during a fracture of the elbow).

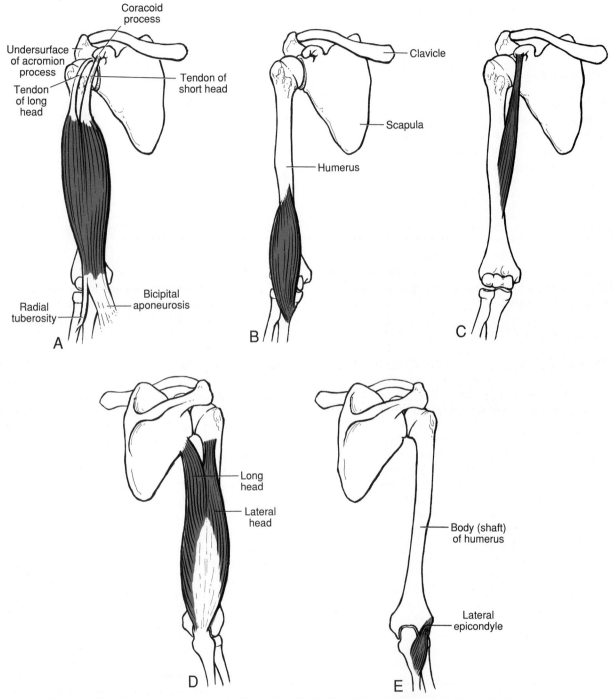

Figure 6-51.   Muscles of the arm. *A*, Biceps brachii, *B*, Brachialis, *C*, Coracobrachialis, *D*, Triceps brachii. The medial head is not shown. It is attached to the deep surface of the triceps tendon. *E*, Anconcus.

Table 6-7.
Muscles of the Arm

Muscle	Proximal Attachments	Distal Attachments	Innervation[1]	Main Actions
Biceps brachii (Fig. 6-51A)	*Short head:* tip of coracoid process of scapula *Long head:* supraglenoid tubercle of scapula	Tuberosity of radius and fascia of forearm via bicipital aponeurosis	Musculocutaneous n. (C5 and **C6**)	Supinates forearm and, when it is supine, flexes forearm
Brachialis (Fig. 6-51B)	Distal half of anterior surface of humerus	Coronoid process and tuberosity of ulna		Flexes forearm in all positions
Coracobrachialis (Fig. 6-51C)	Tip of coracoid process of scapula	Middle third of medial surface of humerus	Musculocutaneous n. (C5, **C6**, and C7)	Helps to flex and adduct arm
Triceps brachii (Fig. 6-51D)	*Long head:* infraglenoid tubercle of scapula *Lateral head:* posterior surface of humerus, superior to radial groove *Medial head:* posterior surface of humerus, inferior to radial groove	Proximal end of olecranon ulna and fascia of forearm	Radial n. (C6, **C7**, and **C8**)	Extends the forearm; it is *chief extensor of forearm;* long head steadies head of abducted humerus
Anconeus (Fig. 6-51E)	Lateral epicondyle of humerus	Lateral surface of olecranon and superior part of posterior surface of ulna	Radial n. (C7, C8, and T1)	Assists triceps in extending forearm; stabilizes elbow joint; abducts ulna during pronation (p. 9)

[1] **Boldface** indicates the main spinal cord segmental innervation.

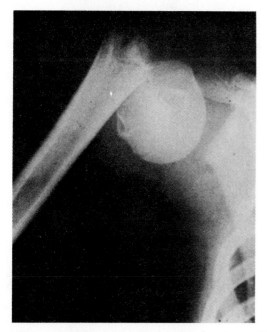

Figure 6-52. Radiograph of the right shoulder of a 14-year-old boy with a separation of the proximal epiphysis of the humerus. Although the humeral head has rotated, it has otherwise retained a fairly normal relationship with the glenoid cavity of the scapula.

Veins of the Arm and Cubital Fossa

Two deep **brachial veins** accompany the brachial artery (Figs. 6-33, 6-50, and 6-56). These veins and their connections encompass the artery in an anastomotic network. The pulsations of the brachial artery help to move the blood through this venous network. The brachial veins begin at the elbow by union of the venae comitantes of the ulnar and radial arteries, and they end in the axillary vein. The brachial veins contain valves and are connected at intervals by short transverse branches. Not uncommonly, the deep veins join to form one brachial vein during part of their course. The two main superficial veins of the arm are the cephalic and basilic veins.

The Cephalic Vein (Figs. 6-5, 6-15, 6-16, 6-33, 6-34, 6-50, and 6-56 to 6-58). This vein is located in the superficial fascia along the anterolateral surface of the biceps and is often visible through the skin. Superiorly the cephalic vein passes between the deltoid and pectoralis major muscles and through the deltopectoral triangle, where it *empties into the axillary vein.*

The Basilic Vein (Figs. 6-33, 6-35, 6-50, 6-56, and 6-57). This vein is also located in the superficial fascia and passes on the medial side of the inferior part of the arm. Often, it is visible through the skin. Near the junction of the middle and inferior thirds of the arm, the basilic vein passes deep to the brachial fascia and runs superiorly into the axilla, where it *becomes the axillary vein.*

The Median Cubital Vein (Figs. 6-57 and 6-58). This vein forms the communication between the basilic and cephalic veins in the cubital fossa, where it *lies anterior to the bicipital aponeurosis.* The basilic veins receive some of the blood carried by the cephalic vein through this communication.

Because of the prominence and accessibility of the veins in the cubital fossa, they are commonly used for *venipuncture* (taking blood). Considerable variation occurs in the connection of the basilic and cephalic veins in this fossa. If the median cubital vein is very large, most blood from the cephalic vein is transferred to the basilic vein. In these cases the proximal part of the cephalic vein may be absent or much diminished.

Nerves of the Arm

The four nerves of the arm (*median, ulnar, musculocutaneous,* and *radial*) are **terminal branches of the brachial plexus**

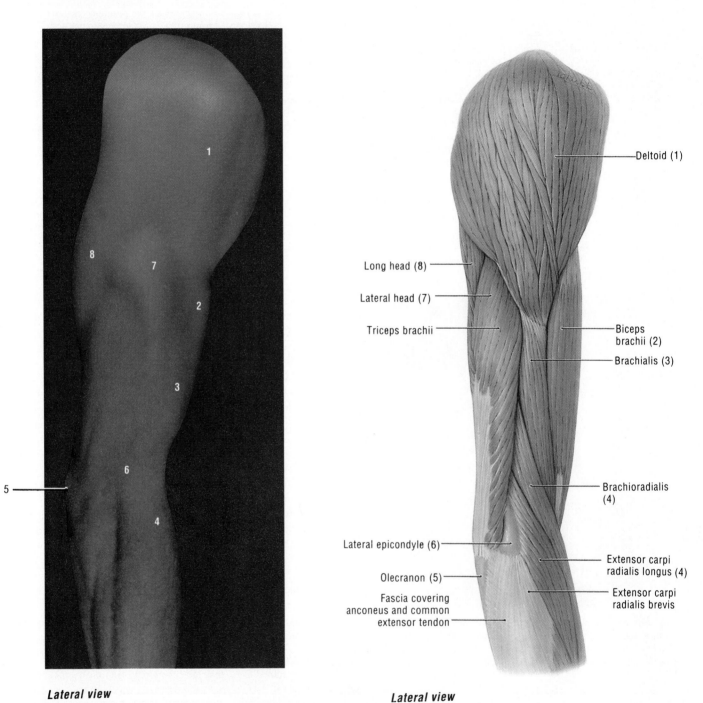

Lateral view

Lateral view

Figure 6-53.   Lateral views of the arm. *A*, Surface anatomy. *B*, Dissection. The numbers in parentheses refer to the muscles shown in *A*.

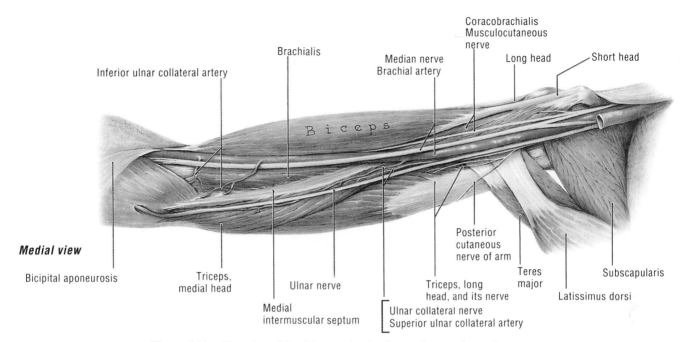

Figure 6-54. Dissection of the right arm showing its muscles, arteries, and nerves.

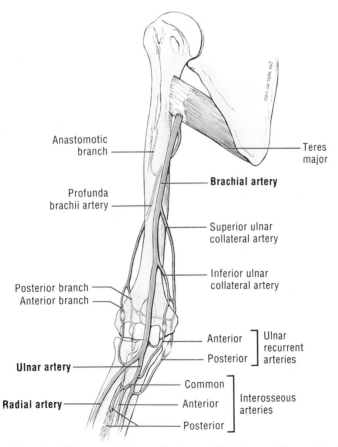

Figure 6-55. Arterial supply of the arm and the proximal part of the forearm. Observe the clinically important arterial anastomoses around the elbow.

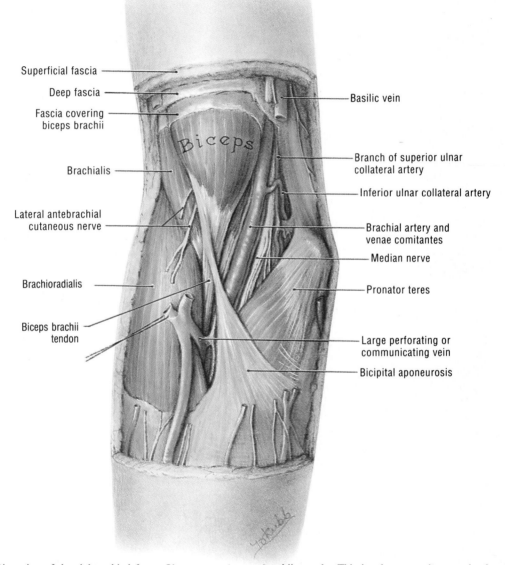

Superficial fascia

Deep fascia

Fascia covering
biceps brachii

Biceps

Brachialis

Lateral antebrachial
cutaneous nerve

Brachioradialis

Biceps brachii
tendon

Basilic vein

Branch of superior ulnar
collateral artery

Inferior ulnar collateral artery

Brachial artery and
venae comitantes

Median nerve

Pronator teres

Large perforating or
communicating vein

Bicipital aponeurosis

Figure 6-56. Dissection of the right cubital fossa. Observe that the distal end of the brachial artery lies medial to the tendon of the biceps brachii muscle. This is where a stethoscope is placed for listening to pulsations of this artery when measuring blood pressure.

(Figs. 6-16 to 6-19 and 6-23), as is the axillary nerve that supplies the skin of the arm over the inferior half of the deltoid and adjacent regions of the arm (Fig. 6-30). Two brachial nerves (median and ulnar) supply no brachial muscles, but supply the elbow joint and muscles in the anterior aspect of the forearm (Fig. 6-20).

The Median Nerve (Figs. 6-16, 6-18 to 6-21, 6-34, 6-50, 6-54, and 6-56; Table 6-3).   This major nerve of the arm is formed in the axilla by the union of a lateral root from the lateral cord and a medial root from the medial cord of the brachial plexus. The median nerve runs distally in the arm on the lateral side of the brachial artery until it reaches the middle of the arm, where it crosses to its medial side and contacts the brachialis muscle. The median nerve descends into the cubital fossa where it lies deep to the bicipital aponeurosis and the median cubital vein.

The median nerve has no branches in the axilla or the arm (Fig. 6-20). It passes deeply into the forearm to supply all but one and one-half of the muscles in the anterior part of the forearm (Table 6-3). It also supplies articular branches to the elbow joint.

Injury to the median nerve proximal to the elbow results in a loss of sensation on the lateral portion of the palm, the palmar surface of the thumb, and the lateral two and one-half digits (Fig. 6-21). As the median nerve supplies no muscles in the arm, they are not affected. However, pronation of the forearm (p. 9), flexion of the wrist and digits, and important movements of the thumb are lost or are severely affected because the muscles producing these movements are supplied by the median nerve after it enters the forearm (Table 6-3).

The Ulnar Nerve (Figs. 6-16, 6-18 to 6-21, 6-34, 6-50, 6-54, and 6-63; Table 6-3).   This is the larger of the two terminal branches of the medial cord of the brachial plexus. It passes distally, anterior to the triceps muscle, on the medial side of the brachial artery. Around the middle of the arm, it pierces the medial intermuscular septum and descends between it and the medial head of the triceps muscle. *The ulnar nerve passes between the medial epicondyle of the humerus and the olecranon to enter the forearm.* Posterior to the medial epicondyle, the ulnar nerve is superficial and easily palpable. *The ulnar nerve has no branches in the arm,* but it supplies one and one-half muscles in the forearm (Fig. 6-20; Table 6-3). It also supplies articular branches to the elbow joint.

Injury to the ulnar nerve in the arm results in impaired flexion and adduction of the wrist and impaired movement of the thumb, ring, and little fingers, resulting in a poor grasp. The characteristic clinical sign of ulnar nerve damage is the *inability to adduct or abduct the medial four digits* owing to loss of power of the interosseous muscles (p. 600). Ulnar nerve injuries are discussed in more detail with the forearm and hand.

The Musculocutaneous Nerve (Figs. 6-16, 6-18 to 6-20, 6-23, 6-34, 6-50, and 6-54; Table 6-3).   This nerve, one of the terminal branches of the lateral cord of the brachial plexus, *begins opposite the inferior border of the pectoralis minor* muscle. It pierces the coracobrachialis muscle and then continues distally between the biceps and brachialis muscles. It supplies all three

of these muscles. At the lateral border of the tendon of the biceps, the musculocutaneous nerve becomes the lateral antebrachial cutaneous nerve (lateral cutaneous nerve of the forearm), which supplies the skin of the forearm (Fig. 6-21).

Injury to the musculocutaneous nerve in the axilla before it innervates any muscles results in paralysis of the coracobrachialis, biceps, and brachialis muscles (Tables 6-3 and 6-7). As a result, flexion of the elbow joint and supination of the forearm are greatly weakened. There may also be loss of sensation on the lateral surface of the forearm supplied by the lateral antebrachial cutaneous nerve.

The Radial Nerve (Figs. 6-18, 6-19, 6-21 to 6-23, 6-50, 6-59, and 6-62; Table 6-3).   This nerve is the direct continuation of the posterior cord of the brachial plexus. The *largest branch of the brachial plexus,* the radial nerve enters the arm posterior to the brachial artery, medial to the humerus, and anterior to the long head of the triceps muscle. It passes inferolaterally with the profunda brachii artery around the body of the humerus in the *radial groove.* When it reaches the lateral border of this bone, the radial nerve pierces the lateral intermuscular septum and then continues inferiorly between the brachialis and brachioradialis muscles to the level of the lateral epicondyle of the humerus, where it divides into deep and superficial branches. *The deep branch of the radial nerve* is entirely muscular and articular in its distribution (Fig. 6-22; Table 6-3). *The superficial branch of the radial nerve* supplies sensory fibers to the dorsum of the hand and digits (Figs. 6-21 and 6-109).

Injury to the radial nerve proximal to the origin of the triceps (Fig. 6-22) results in paralysis of the triceps, brachioradialis, supinator, and extensors of the wrist and digits (Table 6-3). There is also loss of sensation in the areas of skin supplied by this nerve (Fig. 6-21). When the radial nerve is injured in the radial groove, the triceps is not completely paralyzed, but there is paralysis of other muscles supplied by it (Fig. 6-22). The characteristic *clinical sign of radial nerve injury* is **wrist-drop** (inability to extend the wrist and digits).

The Forearm

The forearm (antebrachium) extends from the elbow to the wrist and contains two bones, the *radius* and *ulna,* which are parallel in the anatomical position (Figs. 6-1, 6-61, and 6-64). *An interosseous membrane joins these bones.* Although thin, this membrane is a very strong fibrous sheet. In addition to tying the forearm bones together, it provides attachment for some deep forearm muscles (Fig. 6-65; Table 6-9). The *ulna* is more firmly connected to the humerus than is the *radius.* The latter bone is broadened distally to be more fully in contact with the wrist bones. The head of the ulna is at its distal end, whereas the head of the radius is at its proximal end.

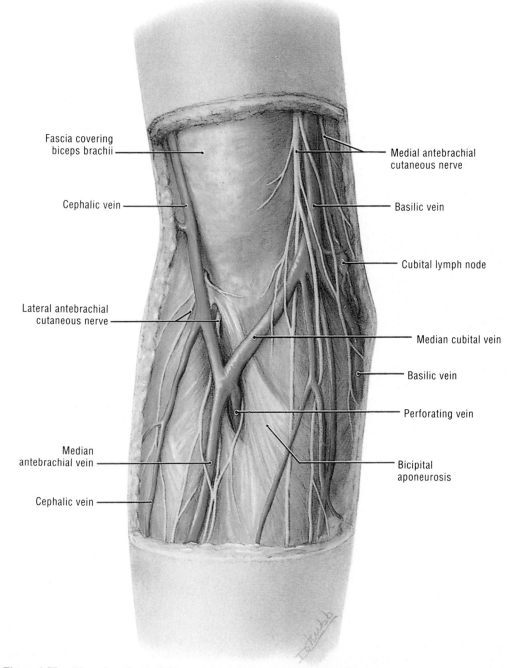

Fascia covering
biceps brachii

Cephalic vein

Lateral antebrachial
cutaneous nerve

Median
antebrachial vein

Cephalic vein

Medial antebrachial
cutaneous nerve

Basilic vein

Cubital lymph node

Median cubital vein

Basilic vein

Perforating vein

Bicipital
aponeurosis

Figure 6-57.   Dissection of superficial structures in the anterior aspect of the right elbow. Note that the cephalic
and basilic veins lie on each side of the biceps brachii muscle.

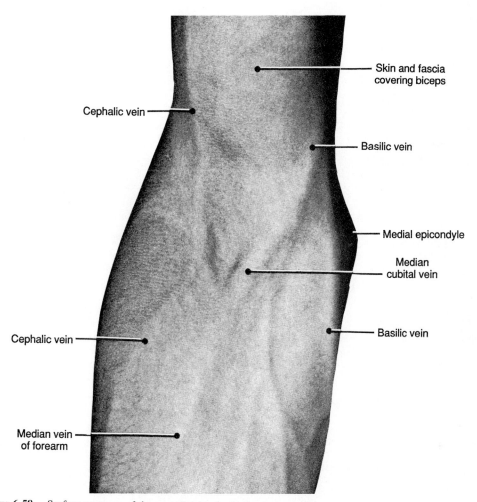

Cephalic vein

Skin and fascia
covering biceps

Basilic vein

Medial epicondyle

Median
cubital vein

Basilic vein

Cephalic vein

Median vein
of forearm

Figure 6-58.   Surface anatomy of the superficial veins in the anterior aspect of the elbow of a 27-year-old man.
The largest vein, usually the median cubital, is selected for blood sampling.

Bones of the Forearm

The Radius (Figs. 6-1, 6-60, 6-61, 6-64, and 6-65).   This is the shorter of the two forearm bones. It was given its name because of its resemblance to the spoke of a wheel, which is what the translated Latin word *radius* means. The proximal end of the radius has a disc-shaped **head**, a smooth cylindrical **neck**, and an oval prominence or *tuberosity*, distal to the neck. The **body** (shaft) of the radius increases in size from its proximal to its distal end; it has a slight lateral convexity or bowing. The body is concave anteriorly in its proximal three-fourths and flattened in its distal one-fourth. The *anterior oblique line* of the radius runs obliquely across the body from the region of the radial tuberosity to the area of greatest bowing. The medial aspect of the body has a sharp *interosseous border* for attachment of the *interosseous membrane* (Fig. 6-65). Its lateral border is rounded.

The distal end of the radius has a median *ulnar notch* into which the head of the ulna fits, forming the *distal radioulnar joint* (Fig. 6-61B). Laterally the distal end of the radius tapers abruptly into a prominent pyramidal **styloid process**. The inferior surface of the distal end of the radius is smooth and concave

where it articulates with the wrist or carpal bones. Posteriorly there is a prominent *dorsal tubercle* on the distal end of the radius (Fig. 6-61B).

A fall on the outstretched hand may result in a *fracture of the distal end of the radius* (p. 591). Sometimes there is also a fracture of the styloid process of the ulna. In the common **Colles' fracture**, the distal fragment of the radius is displaced posteriorly (Fig. 6-94). This results in the radial and ulnar styloid processes being at approximately the same horizontal level, which is an abnormal condition.

The Ulna (Figs. 6-1 and 6-59 to 6-61).   The ulna (L. elbow) is the longer bone of the forearm. This prismatic bone looks somewhat like a pipe wrench, with the *olecranon* resembling the upper jaw, the *coronoid process* the lower jaw, and the *trochlear notch* the mouth. The olecranon and coronoid processes clasp the trochlea of the humerus, somewhat like a pipe wrench clasps a pipe. The proximal "wrenchlike" end of the ulna is larger than the small, rounded distal end, called the **head**. The lateral side of the *coronoid process* has a small, shallow *radial notch* for the disc-shaped head of the radius. Inferior to the *radial notch* is a triangular *supinator fossa*, which provides an attachment for

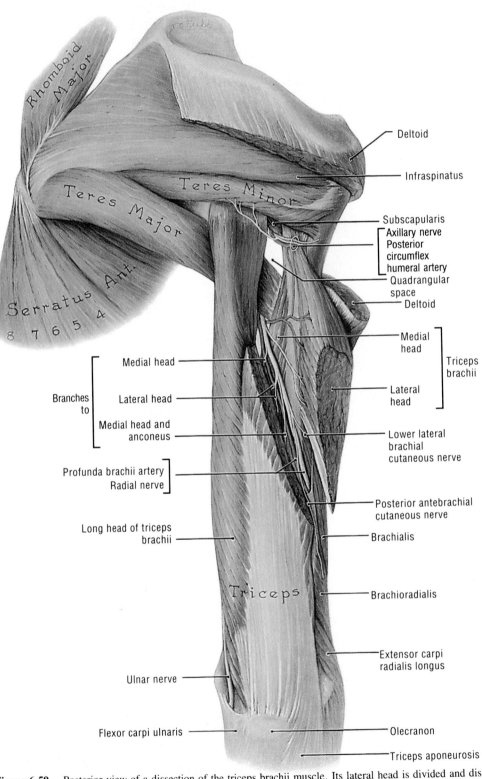

Figure 6-59. Posterior view of a dissection of the triceps brachii muscle. Its lateral head is divided and displaced to show its three related nerves and the profunda brachii artery.

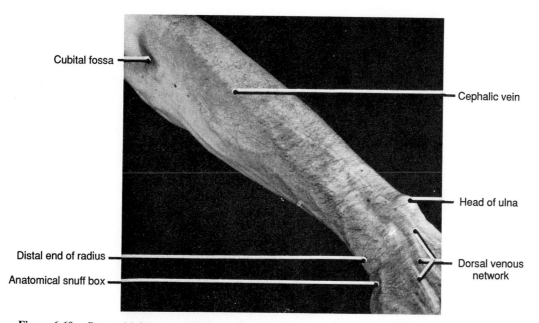

Cubital fossa

Cephalic vein

Head of ulna

Distal end of radius

Dorsal venous network

Anatomical snuff box

Figure 6-60.   Pronated left forearm and hand of a 46-year-old man showing the principal surface landmarks. (For a close-up of his anatomical snuff box, see Fig. 6-87.)

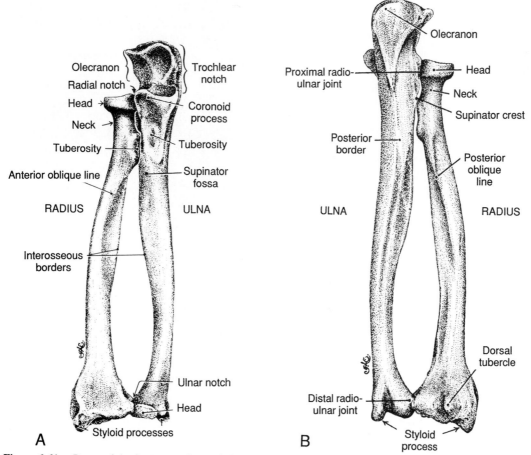

Olecranon

Radial notch

Head

Neck

Tuberosity

Anterior oblique line

RADIUS

Trochlear notch

Coronoid process

Tuberosity

Supinator fossa

ULNA

Interosseous borders

Ulnar notch

Head

Styloid processes

A

Olecranon

Head

Neck

Supinator crest

Proximal radio-ulnar joint

Posterior border

ULNA

Posterior oblique line

RADIUS

Dorsal tubercle

Distal radio-ulnar joint

Styloid process

B

Figure 6-61.   Bones of the forearm: radius and ulna. *A*, Anterior view. *B*, Posterior view. Note that the head of the radius is at the elbow, whereas the head of the ulna is at the wrist (see also Fig. 6-60).

the supinator muscle (Figs. 6-61, 6-62, and 6-67). This fossa is bounded posteriorly by a distinct *supinator crest*. The irregular anterior surface of the *coronoid process* is rough and ends distally in a *tuberosity* onto which the brachialis, the chief flexor muscle of the forearm, inserts (Fig. 6-67).

The **body** (shaft) of the ulna is thick proximally. Its prominent lateral edge, the *interosseous border* (Fig. 6-61), is where the interosseous membrane is attached (Fig. 6-65). The small, slender distal end of the ulna has a rounded *head* and a conical styloid process. The **styloid process** projects distally, about 1 cm proximal to the styloid process of the radius. The distal end of the ulna has a convex articular surface on its lateral side for articulation with the ulnar notch of the radius (Fig. 6-61A).

Surface Anatomy of the Forearm

The *head of the radius* can be palpated and felt to rotate in the depression on the posterolateral aspect of the extended elbow joint, just distal to the lateral epicondyle of the humerus (Fig. 6-64). The *radial styloid process* can be easily palpated on the lateral side of the wrist, particularly when the tendons that cover it are relaxed (Fig. 6-60). It is *located about 1 cm more distal than the ulnar styloid process* (Fig. 6-61). The relationship of these processes of the radius and ulna is important in the diagnosis of certain injuries in the wrist region (Fig. 6-94). Proximal to the radial styloid process, the anterior, lateral, and posterior surfaces of the radius are palpable for a few centimeters. The lateral surface of the distal half of the radius is easy to palpate.

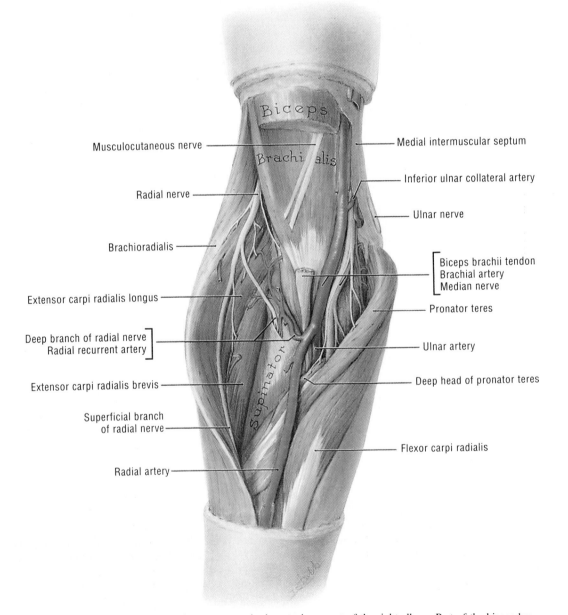

Figure 6-62.   Dissection of the deep structures in the anterior aspect of the right elbow. Part of the biceps brachii muscle is excised and the cubital fossa is opened widely.

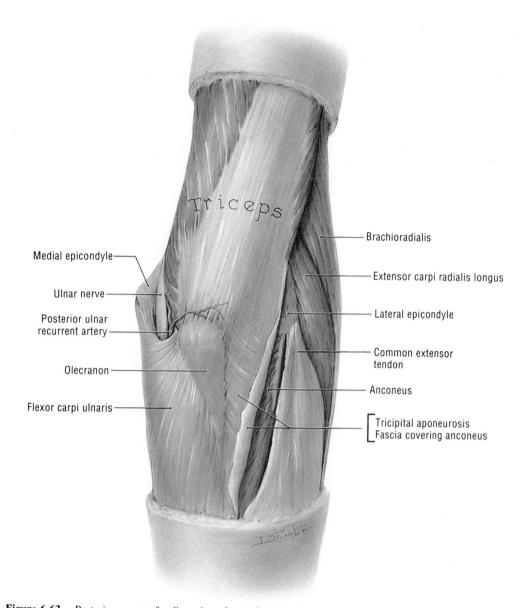

Medial epicondyle

Ulnar nerve

Posterior ulnar
recurrent artery

Olecranon

Flexor carpi ulnaris

Triceps

Brachioradialis

Extensor carpi radialis longus

Lateral epicondyle

Common extensor
tendon

Anconeus

Tricipital aponeurosis
Fascia covering anconeus

Figure 6-63.   Posterior aspect of a dissection of superficial structures in the right elbow. The fascia covering the anconeus muscle is cut and lifted.

The olecranon of the ulna can be easily palpated (Fig. 6-64). The loose skin covering it is often rough because the elbow frequently rests on it. The subcutaneous posterior border of the ulna (Fig. 6-65) can be palpated along its entire length. When palpated, the *ulnar nerve* resembles a rounded cord where it lies posterior to the medial epicondyle of the humerus (Figs. 6-63 and 6-64). The *head of the ulna* forms a large, rounded subcutaneous prominence that can be easily seen and felt on the medial part of the dorsal aspect of the wrist (Figs. 6-60 and 6-64). The *ulnar styloid process*, also subcutaneous, may be felt slightly distal to its head when the hand is supinated.

Muscles in the Cubital Region

The pronator teres muscle lies medially in the cubital fossa and the brachioradialis muscle is located laterally (Fig. 6-62).

The floor of the cubital fossa is formed by the brachialis and supinator muscles. The brachialis muscle was described with the muscles of the arm (p. 541).

The Supinator Muscle (Figs. 6-62, 6-67, 6-70, and 6-83). This muscle lies deep in the cubital fossa and, along with the brachialis, forms its floor. The humeral and ulnar heads of attachment of the supinator muscle envelop the neck and proximal part of the body of the radius. Its attachments, nerve supply, and main action are given in Table 6-11 (p. 588). The supinator muscle supinates the forearm (p. 8) by rotating the radius (Fig. 6-126A). *It is the prime mover in supination of the forearm and hand.* The biceps brachii assists the supinator in rapid and forceful supination, particularly when resistance is required and the forearm is flexed (*e.g.*, when a right-handed person turns a screw).

The Brachioradialis Muscle (Figs. 6-48, 6-53, 6-56, 6-62,

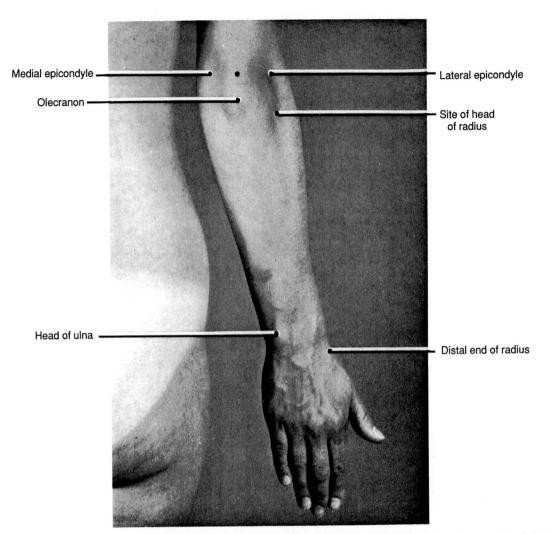

Medial epicondyle

Olecranon

Lateral epicondyle

Site of head
of radius

Head of ulna

Distal end of radius

Figure 6-64. Posterior aspect of the forearm and hand of a 27-year-old woman showing the principal surface landmarks of the bones of the arm and forearm (also see Fig. 6-1*B*). In full extension of the elbow joint, as here, the superior part of the olecranon (*black dot*) and the two humeral epicondyles are in a straight line.

6-63, 6-65, 6-66, and 6-70). This fusiform muscle forms the lateral boundary of the cubital fossa and is the most superficial muscle on the radial side of the forearm. Its attachments, nerve supply, and main actions are given in Table 6-10 (p. 587). The *brachioradialis flexes the forearm, especially when quick movement is required* or when a weight is lifted during slow flexion of the forearm. It is used to give power and speed and acts to its best advantage when the forearm is in the midprone position. It is therefore capable of initiating both pronation and supination.

The Pronator Teres Muscle (Figs. 6-48, 6-62, 6-65, 6-66, 6-73, and 6-69*A*). This fusiform muscle forms the medial boundary of the cubital fossa. It has two heads of proximal attachment. Its attachments, nerve supply, and main actions are given in Table 6-8. The pronator teres *pronates the forearm and flexes it*. It also assists the pronator quadratus (p. 571) during pronation and is a weak flexor of the forearm.

Bones of the Wrist and Hand

To facilitate understanding of distal attachments of forearm muscles, the carpal and hand bones are described.

The Carpus

The eight small bones of the wrist, called *carpal bones*, are referred to collectively as the *carpus* (L. wrist). They are arranged in proximal and distal rows, each containing four bones (Figs. 6-1, 6-67, 6-68, 6-71, and 6-72).

The proximal row of carpal bones (lateral to medial) consists of the scaphoid (navicular), lunate, triquetrum, and pisiform. The boat-shaped *scaphoid* is the largest bone of the proximal row and was given its name because of its resemblance to a rowboat (G. *scaphe*). The lunate is moon-shaped. The pea-shaped pisiform (L. *pisum*, pea) is included in the proximal row, even though it is a sesamoid bone in the tendon of the flexor carpi ulnaris muscle (Figs. 6-73 and 6-74). The **pisiform bone** is a clinically important landmark that is easily palpable.

The distal row of carpal bones (lateral to medial) consists of the trapezium, trapezoid, capitate, and hamate. The *hamate* can be identified by its prominent process, the *hook of the hamate*, which projects anteriorly (Fig. 6-71*A*). The *capitate* has a rounded head (L. *caput*). The carpal bones articulate with each other at synovial *intercarpal joints* and are bound together with ligaments

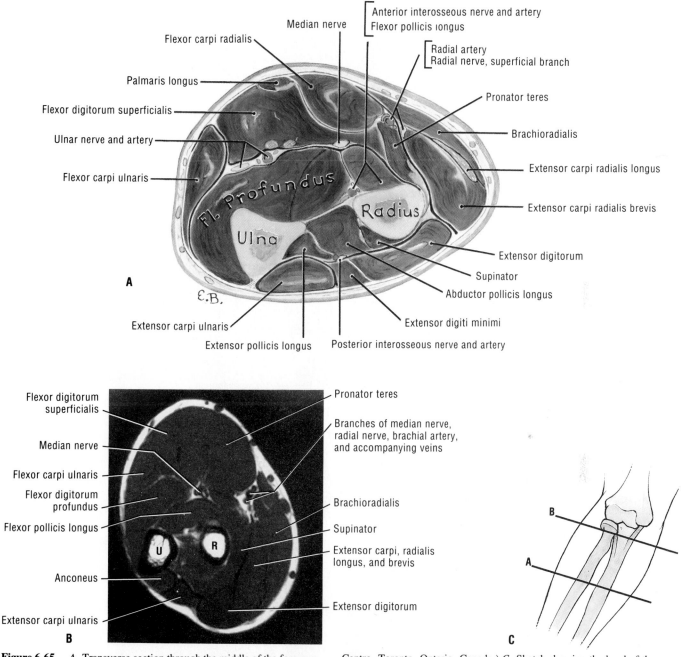

Extensor carpi radialis longus

Extensor carpi radialis brevis

Extensor digitorum

Supinator

Abductor pollicis longus

Extensor digiti minimi

Posterior interosseous nerve and artery

Extensor pollicis longus

Extensor carpi ulnaris

A

E.B.

Flexor carpi ulnaris

Flexor carpi radialis

Palmaris longus

Flexor digitorum superficialis

Ulnar nerve and artery

Median nerve

Anterior interosseous nerve and artery

Flexor pollicis longus

Radial artery

Radial nerve, superficial branch

Pronator teres

Brachioradialis

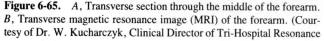

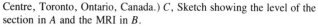

B

C

Figure 6-65. *A*, Transverse section through the middle of the forearm. *B*, Transverse magnetic resonance image (MRI) of the forearm. (Courtesy of Dr. W. Kucharczyk, Clinical Director of Tri-Hospital Resonance Centre, Toronto, Ontario, Canada.) *C*, Sketch showing the level of the section in *A* and the MRI in *B*.

to form a compact mass (p. 626). The carpus has an anterior concavity known as the *carpal groove* (sulcus). The groove is converted into an osseofibrous **carpal tunnel** (canal) by the *flexor retinaculum*, which is attached to the scaphoid and trapezium laterally and to the pisiform and the hook of the hamate bone medially (Fig. 6-72). The carpal tunnel is filled with tendons and the median nerve (Figs. 6-70, 6-75, and 6-76). Compression of the median nerve in the carpal tunnel produces the *carpal tunnel syndrome* (p. 603).

The Metacarpus

The *five metacarpal bones*[3] are miniature long bones (Figs. 6-1 and 6-71). They extend from the carpus (wrist) to the digits (thumb and fingers) and are numbered from the lateral side. In Fig. 6-71, note that the first metacarpal is shorter than the others. Although covered with tendons, the metacarpals can be easily

[3]In common usage, the metacarpal bones are referred to as *metacarpals*.

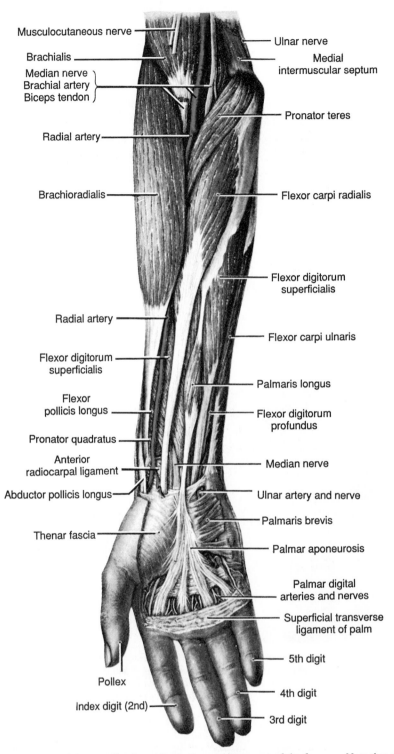

Figure 6-66.   Dissection of the superficial muscles in the anterior aspect of the forearm. Note that the tendon of the palmaris longus muscle is continuous with the palmar aponeurosis.

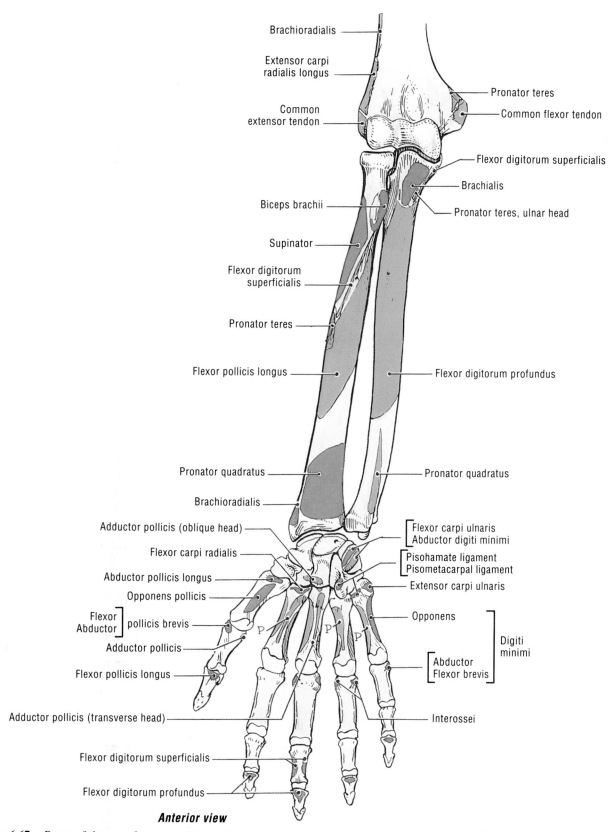

Brachioradialis

Extensor carpi
radialis longus

Common
extensor tendon

Biceps brachii

Supinator

Flexor digitorum
superficialis

Pronator teres

Flexor pollicis longus

Pronator quadratus

Brachioradialis

Adductor pollicis (oblique head)

Flexor carpi radialis

Abductor pollicis longus

Opponens pollicis

Flexor ⎱ pollicis brevis
Abductor ⎰

Adductor pollicis

Flexor pollicis longus

Adductor pollicis (transverse head)

Flexor digitorum superficialis

Flexor digitorum profundus

Pronator teres

Common flexor tendon

Flexor digitorum superficialis

Brachialis

Pronator teres, ulnar head

Flexor digitorum profundus

Pronator quadratus

Flexor carpi ulnaris
Abductor digiti minimi

Pisohamate ligament
Pisometacarpal ligament

Extensor carpi ulnaris

Opponens

Digiti
minimi

Abductor
Flexor brevis

Interossei

P   P   P

Anterior view

Figure 6-67. Bones of the arm, forearm, and hand showing the attachments of muscles. Proximal attachments are shown in deep salmon; distal attachments in blue. Observe that the proximal attachments of the three palmar interossei are indicated by the letters P,P,P; those of the four dorsal interossei by color only; and the proximal attachments of the three thenar and two of the hypothenar muscles are omitted.

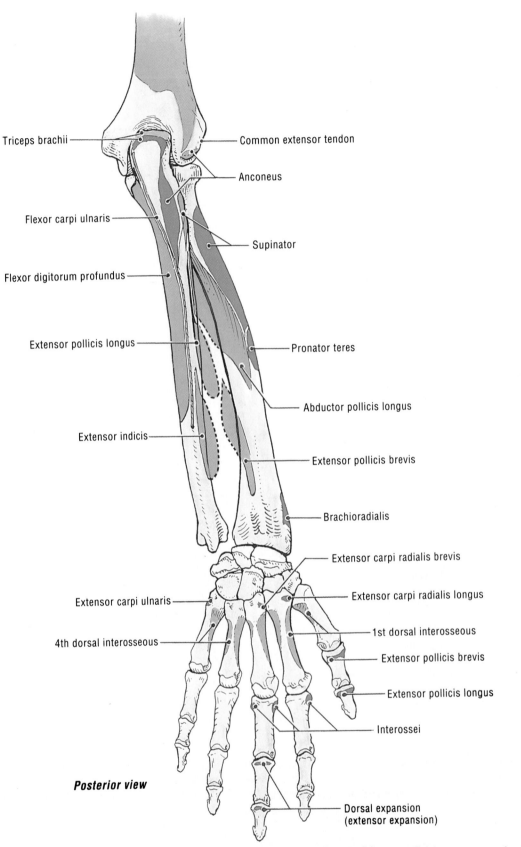

Triceps brachii

Common extensor tendon

Anconeus

Flexor carpi ulnaris

Supinator

Flexor digitorum profundus

Extensor pollicis longus

Pronator teres

Abductor pollicis longus

Extensor indicis

Extensor pollicis brevis

Brachioradialis

Extensor carpi radialis brevis

Extensor carpi radialis longus

Extensor carpi ulnaris

1st dorsal interosseous

4th dorsal interosseous

Extensor pollicis brevis

Extensor pollicis longus

Interossei

Posterior view

Dorsal expansion
(extensor expansion)

Figure 6-68. Bones of the right arm, forearm, and hand showing the attachments of muscles. Note the site of attachment of the common extensor tendon from the lateral epicondyle of the humerus. This is the proximal attachment of four superficial extensor muscles (see Table 6-10).

palpated throughout their whole length on the dorsum of the hand. The *heads of the metacarpals* are at their distal ends (Fig. 6-71A), where they articulate with the *phalanges* (bones of the digits). They form the *knuckles of the hand* that become visible when the fist is clenched. On the dorsal surface of each head is a small tubercle on each side for attachment of collateral ligaments and joint capsules. The *bodies (shafts) of the metacarpals* are slightly concave on their medial and lateral sides, where the dorsal interosseous muscles attach (Fig. 6-106C). The *bases of the metacarpals* are arranged in a fan-shaped manner from the distal row of carpal bones (Fig. 6-71B).

The Phalanges

Each phalanx (bone of a digit) is a miniature long bone, which consists of a *body* (shaft), a larger proximal end or *base*, and a smaller distal end or *head* (Fig. 6-71). The thumb (first digit) has two phalanges (proximal and distal) and each finger (second to fifth digits) has three phalanges (proximal, middle, and distal). The phalanges in the first digit are shorter and broader than those in the other digits. The proximal phalanges are the longest and the distal ones the shortest.

Muscles of the Forearm

These muscles act on the elbow and wrist joints and on those of the digits. In the proximal part of the forearm, the muscles form fleshy masses inferior to the medial and lateral epicondyles of the humerus (Fig. 6-66). The tendons of these muscles pass through the distal part of the forearm and continue into the hand (Figs. 6-74 and 6-75). These muscles can be divided into flexor-pronator and extensor-supinator groups. *The flexor-pronator group* arises by a common flexor tendon from the *medial epicondyle* of the humerus (Fig. 6-67); this is referred to as the **common flexor attachment** (origin). *The extensor-supinator group* arises by a common extensor tendon from the *lateral epicondyle* (Figs. 6-67 and 6-68); this is referred to as the **common extensor attachment** (origin).

The deeper flexor and extensor muscles originate from the anterior and posterior aspects of the bodies of the ulna and radius, respectively (Figs. 6-67 and 6-68). The dividing line on the posterior aspect of the forearm between the extensor and flexor groups is the posterior border of the ulna, which is palpable from the olecranon to the wrist (Figs. 6-64 and 6-65). All the flexor tendons are located on the anterior surface of the wrist, and most of them are held in place by the **flexor retinaculum**, a thickening of the deep fascia of the forearm (Figs. 6-70, 6-72, and 6-78). *The eight muscles in the anterior aspect of the forearm are flexor-pronator muscles* (Figs. 6-65 and 6-66; Tables 6-8 and 6-9), and they can be organized into **three functional muscle groups**: (1) *muscles that rotate the radius* on the ulna, *e.g.*, those that pronate the forearm and hand (pronator teres and pronator quadratus); (2) *muscles that flex the hand* (flexor carpi radialis, flexor carpi ulnaris, and palmaris longus); and (3) *muscles that flex the digits* (flexor digitorum superficialis, flexor digitorum profundus, and flexor pollicis longus).

The anterior forearm muscles are divided into **three muscular layers** as follows (Tables 6-8 and 6-9): (1) a *superficial layer* (pronator teres, flexor carpi radialis, palmaris longus, and flexor carpi ulnaris); (2) an *intermediate layer* (flexor digitorum superficialis); and (3) a *deep layer* (flexor digitorum profundus, flexor pollicis longus, and pronator quadratus). A septum of deep fascia separates the deep layer of flexor muscles from the superficial and intermediate layers (Fig. 6-65).

The Superficial and Intermediate Muscles on the Anterior Surface of the Forearm

The Pronator Teres Muscle (Figs. 6-62, 6-65, 6-66, and 6-69A). This elongated, narrow muscle is a *pronator of the forearm and a flexor of the elbow* joint. It was described with the muscles of the cubital region (p. 560). It is prominent when the forearm is strongly flexed and pronated, and its lateral border forms the medial boundary of the cubital fossa. Its attachments, nerve supply, and main actions are given in Table 6-8.

The Flexor Carpi Radialis Muscle (Figs. 6-65, 6-66, 6-69B, 6-70, and 6-73 to 6-75). This long, fusiform muscle *is located medial to the pronator teres*. In the middle of the forearm its fleshy belly is replaced by a long, flattened tendon that becomes cordlike as it approaches the wrist, where it is readily palpable. Its attachments, nerve supply, and main actions are given in Table 6-8. To reach its insertion, its long tendon passes through a canal in the lateral part of the flexor retinaculum (Fig. 6-76) and through the vertical groove in the trapezium (Figs. 6-71 and 6-72). The tendon of the flexor carpi radialis is a good *guide to the radial artery*, which lies just lateral to it (Figs. 6-66, 6-74, and 6-75). The flexor carpi radialis *flexes the hand and abducts it*. It is tested with the posterior aspect of the forearm and hand flat on a table. The patient is asked to flex the wrist against resistance while the examiner palpates the tendon of the muscle.

The Palmaris Longus Muscle (Figs. 6-65, 6-66, 6-69C, and 6-73 to 6-76). This small fusiform muscle is absent on one or both sides (usually the left) in about 14% of people, but its actions are not missed. It may have two bellies and two tendons. When present, the palmaris longus tendon is readily visible and palpable. Its long, thin tendon passes superficial to the flexor retinaculum (Fig. 6-66). *The palmaris longus tendon is a useful guide to the median nerve*, which passes lateral to it at the wrist (Figs. 6-66 and 6-75). The attachments, nerve supply, and actions of the palmaris longus are given in Table 6-8.

> When treating lacerations of the wrist and during practical examinations, the tendon of the palmaris longus can be mistaken for the median nerve. In Figs. 6-66, 6-75, and 6-76, note that the median nerve is much larger than the palmaris longus tendon and lies deep to it.

The Flexor Carpi Ulnaris Muscle (Figs. 6-65, 6-66, 6-69D, 6-70, and 6-73 to 6-76). This is the most medial of the superficial flexor muscles in the forearm. It *has two heads* of proximal attachment, between which the ulnar nerve passes distally in the forearm. Its tendon is readily visible and palpable. The tendon of the flexor carpi ulnaris is a good *guide to the ulnar nerve* and artery, which are on its lateral side at the wrist (Figs. 6-75 and 6-76). The attachments, nerve supply, and main actions of the flexor carpi ulnaris are given in Table 6-8. It *flexes*

Table 6-8.
Superficial and Intermediate Layers of Muscles on the Anterior Surface of the Forearm[1]

Muscle	Proximal Attachments	Distal Attachments	Innervation[2]	Main Actions
Pronator teres (Fig. 6-69A)	Medial epicondyle of humerus and coronoid process of ulna	Middle of lateral surface of radius	Median n. (C6 and **C7**)	Pronates forearm and flexes it
Flexor carpi radialis (Fig. 6-69B)	Medial epicondyle of humerus	Base of 2nd metacarpal bone		Flexes hand and abducts it
Palmaris longus (Fig. 6-69C)	Medial epicondyle of humerus	Distal half of flexor retinaculum and palmar aponeurosis	Median n. (C7 and C8)	Flexes hand and tightens palmar aponeurosis
Flexor carpi ulnaris[2] (Fig. 6-69D)	*Humeral head:* medial epicondyle of humerus *Ulnar head:* olecranon and posterior border of ulna	Pisiform bone, hook of hamate bone and 5th metacarpal bone	Ulnar n. (C7 and **C8**)	Flexes hand and adducts it
Flexor digitorum superficialis[3] (Fig. 6-69E)	*Humeroulnar head:* medial epicondyle of humerus, ulnar collateral ligament, and coronoid process of ulna *Radial head:* superior half of anterior border of radius	Bodies of the middle phalanges of medial four digits	Median n. (C7, **C8**, and T1)	Flexes middle phalanges of medial four digits; acting more strongly, it flexes proximal phalanges and hand

[1]The superficial muscles of the *flexor-pronator group* are attached to the anterior surface of the medial epicondyle by a *common flexor tendon* (Fig. 6-67). **Boldface** indicates the main spinal cord segmental innervation.
[2]In contrast to the other superficial flexor muscles, the flexor carpi ulnaris is supplied by the ulnar nerve.
[3]This muscle comprises the *intermediate muscle layer* in the anterior part of the forearm. In some clinical texts, *e.g.*, Healey (1990), this muscle is referred to by its old name, "flexor digitorum sublimis."

and adducts the hand. It also fixes the pisiform bone when the hypothenar muscles act (p. 599). The flexor carpi ulnaris is tested with the posterior aspect of the forearm and hand on a flat table. The patient is asked to flex the wrist against resistance while the examiner palpates the tendon of the muscle.

The Flexor Digitorum Superficialis Muscle (Figs. 6-65, 6-66, 6-69E, and 6-74 to 6-76). This is the largest superficial muscle in the forearm; *it has two heads.* It **forms an intermediate layer between the superficial and deep groups of forearm muscles.** The median nerve and the ulnar artery pass deeply between the heads of this muscle. Near the wrist, the flexor digitorum superficialis *gives rise to four tendons,* which pass deep to the flexor retinaculum (Figs. 6-75 and 6-76), where they are surrounded by a *common flexor synovial sheath* (Fig. 6-78). The superficial pair of tendons passes to the third (middle) and fourth (ring) digits (fingers) within **synovial sheaths** in osseofibrous tunnels in the digits (Figs. 6-78 and 6-79). The attachments, nerve supply, and main actions of the flexor digitorum superficialis are given in Table 6-8. It *flexes the middle phalanges of the medial four digits* by flexing the proximal interphalangeal joints. In continued action, it *also flexes the proximal phalanges* (*i.e.*, the metacarpophalangeal joints) and the wrist joint.

The Deep Muscles on the Anterior Surface of the Forearm

The deep group of flexor muscles in the forearm (Table 6-9) is composed of the flexor digitorum profundus, flexor pollicis longus, and pronator quadratus (Fig. 6-70). None of these muscles is attached proximally to the humerus; they are all attached to the radius or ulna (Figs. 6-65 and 6-67).

The Flexor Digitorum Profundus Muscle (Figs. 6-65, 6-66, 6-70, 6-76, and 6-80A). This long, thick (L. *profundus*, deep) muscle is the only one that can flex the distal interphalangeal joints of the digits (*i.e.*, it flexes all these joints). It has an extensive proximal attachment to the ulna and interosseous membrane, and "clothes" the anterior aspect of the ulna. Its attachments, nerve supply, and main actions are given in Table 6-9. The flexor digitorum profundus *flexes the distal phalanges of the second to fifth digits* after the flexor digitorum superficialis muscle has flexed the middle phalanges (*i.e.*, it "rolls up" the digits and hand). Each tendon flexes two interphalangeal joints, a metacarpophalangeal joint, and the wrist joint. *The flexor digitorum profundus divides into four parts that end in four tendons.* They pass posterior to the tendons of the flexor digitorum superficialis and the flexor retinaculum (Figs. 6-66, 6-70, 6-76, and 6-78). The portion of the muscle going to the second digit (index finger) usually separates from the rest of the muscle for some distance in the distal part of the forearm. Each tendon enters the fibrous sheath of its digit, posterior to the tendon of the flexor digitorum superficialis.

The Flexor Pollicis Longus Muscle (Figs. 6-65, 6-66, 6-70, 6-73 to 6-76, and 6-78 to 6-80). This long flexor of the thumb (L. *pollex*) lies lateral to the flexor digitorum profundus, where it "clothes" the anterior aspect of the radius. Its flat tendon passes deep to the flexor retinaculum, enveloped in its own synovial sheath on the lateral side of the common flexor synovial sheath. The attachments, nerve supply, and main actions of this muscle are given in Table 6-9. *It flexes the distal phalanx of the first digit or thumb* and, secondarily, the proximal phalanx and the first metacarpal bone. *The flexor pollicis longus is the only muscle that flexes the interphalangeal joint of the thumb.* It also

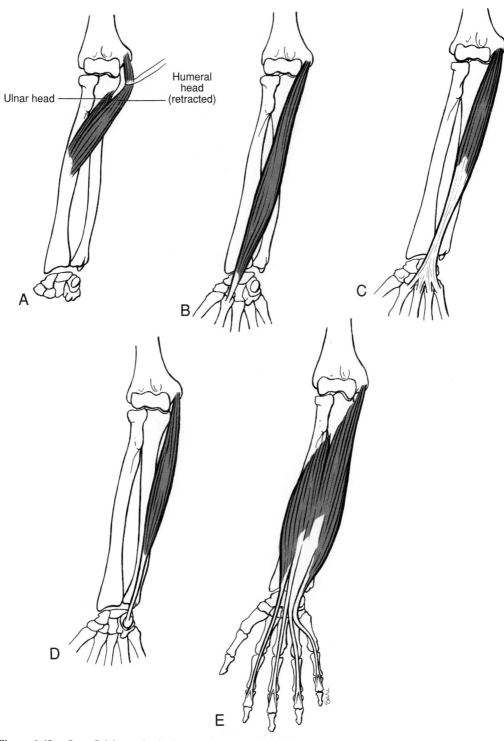

Ulnar head

Humeral
head
(retracted)

A

B

C

D

E

Figure 6-69.   Superficial muscles in the anterior aspect of the forearm. *A*, Pronator teres. *B*, Flexor carpi radialis. *C*, Palmaris longus. *D*, Flexor carpi ulnaris. *E*, Flexor digitorum superficialis.

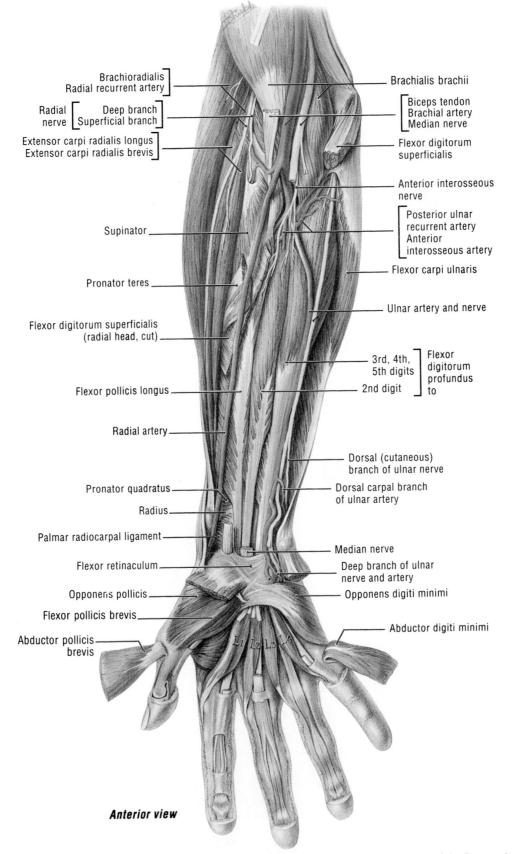

Brachioradialis
Radial recurrent artery

Radial nerve — Deep branch / Superficial branch

Extensor carpi radialis longus
Extensor carpi radialis brevis

Supinator

Pronator teres

Flexor digitorum superficialis (radial head, cut)

Flexor pollicis longus

Radial artery

Pronator quadratus

Radius

Palmar radiocarpal ligament

Flexor retinaculum

Opponens pollicis

Flexor pollicis brevis

Abductor pollicis brevis

Brachialis brachii

Biceps tendon / Brachial artery / Median nerve

Flexor digitorum superficialis

Anterior interosseous nerve

Posterior ulnar recurrent artery / Anterior interosseous artery

Flexor carpi ulnaris

Ulnar artery and nerve

3rd, 4th, 5th digits / 2nd digit — Flexor digitorum profundus to

Dorsal (cutaneous) branch of ulnar nerve

Dorsal carpal branch of ulnar artery

Median nerve

Deep branch of ulnar nerve and artery

Opponens digiti minimi

Abductor digiti minimi

L1 L2 L3 L4

Anterior view

Figure 6-70. Dissection of the deep flexor muscles of the digits and related structures. Part of the flexor retinaculum has been removed. L1 to L4 indicate the lumbrical muscles.

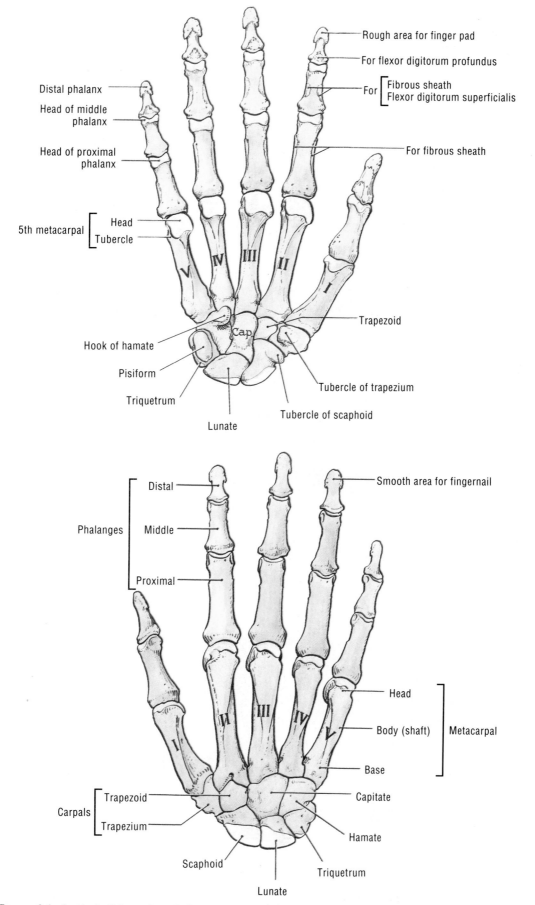

Figure 6-71. Bones of the hand. *A*, Palmar view. *B*, Dorsal view. Observe that the skeleton consists of three segments: (1) the carpal bones of the wrist; (2) the metacarpal bones of the hand; and (3) the phalanges of the digits. The heads of the metacarpals form the knuckles of the hand and the heads of the phalanges form the knuckles of the digits.

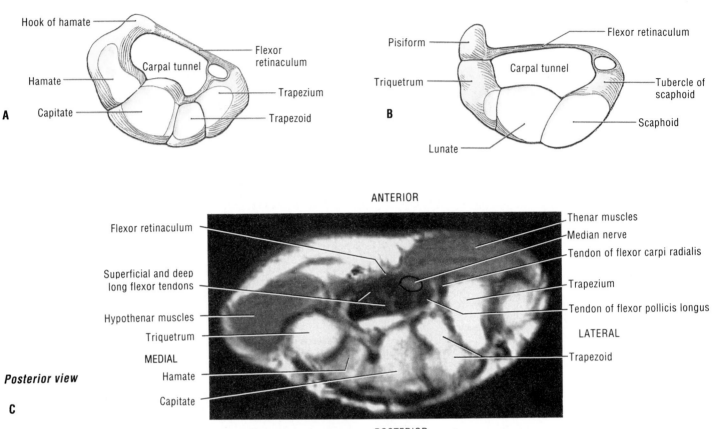

Figure 6-72. Distal (*A*) and proximal (*B*) rows of carpal bones (see Fig. 6-71), showing how the flexor retinaculum stretches between the ends of these bones to form an osseofibrous carpal tunnel. *C*, Transverse magnetic resonance image (MRI) of the wrist showing the carpal tunnel. (Courtesy of Dr. W. Kucharczyk, Clinical Director of Tri-Hospital Resonance Centre, Toronto, Ontario, Canada.)

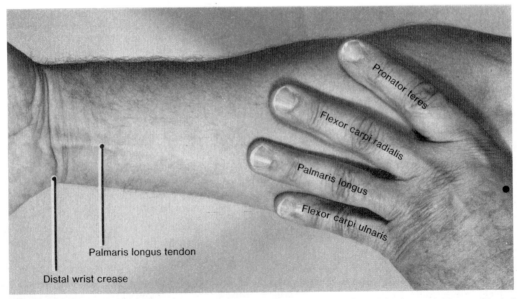

Figure 6-73. Anterior surface of the right forearm and wrist, showing how to locate the position of the four superficial flexor muscles. The thumb of the person's left hand is placed posterior to the elbow, around the medial epicondyle of the humerus (deep to *black dot*), to which the common flexor tendon of these muscles is attached (see Fig. 6-67).

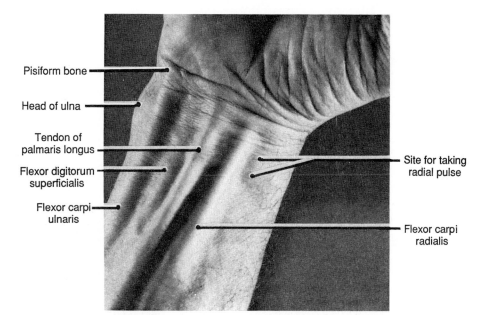

Pisiform bone

Head of ulna

Tendon of palmaris longus

Flexor digitorum superficialis

Flexor carpi ulnaris

Site for taking radial pulse

Flexor carpi radialis

Figure 6-74. Anterior aspect of the forearm and wrist of a 46-year-old man, showing the principal surface markings and the usual site for taking the pulse of the radial artery. The cordlike tendon of the flexor carpi radialis serves as a guide to the radial artery, which lies lateral to it (see Fig. 6-75).

flexes the metacarpophalangeal and carpometacarpal joints of the thumb, and it may assist in flexion of the wrist joint.

The Pronator Quadratus Muscle (Figs. 6-66, 6-70, 6-76, and 6-80*C*).   As its name states, this small muscle is quadrangular (*i.e.*, it has four angles and four sides) and pronates the forearm (p. 9). It cannot be palpated or observed, except in dissections, because it is the deepest muscle in the anterior aspect of the forearm. It "clothes" the distal fourth of the radius and ulna and the interosseous membrane between them. The pronator quadratus is the only muscle that is attached proximally only to the ulna and is attached distally only to the radius. Its attachments, nerve supply, and main actions are given in Table 6-9. It pronates the forearm at the superior and inferior radioulnar joints and is the **prime mover in pronation of the forearm**. The pronator quadratus initiates pronation and is *assisted by the pronator teres* when more speed and power are needed. The pronator quadratus also helps the interosseous membrane to hold the radius and ulna together, particularly when upward thrusts are transmitted through the wrist (*e.g.*, during a fall on the outstretched hand).

Summary of the Muscles in the Anterior Part of the Forearm. *All muscles are supplied by the median and ulnar nerves*, except the brachioradialis (Figs. 6-20 and 6-22; Tables 6-8 and 6-9). The brachioradialis muscle forms the lateral boundary of the cubital fossa (Fig. 6-77). Although belonging to the extensor group of muscles (Table 6-10), it lies in the lateral part of the anterior aspect of the forearm. Although functionally a flexor of the forearm, *the brachioradialis is supplied by the radial nerve* (see Fig. 6-22). Hence, this muscle is the one major exception to the rule that the radial nerve supplies only extensor muscles. The long flexors of the digits (flexor digitorum superficialis and flexor digitorum profundus) also flex the metacarpophalangeal and wrist joints (Table 6-18). The flexor digitorum profundus flexes the digits in slow action, but this activity is reinforced by the flexor digitorum superficialis when speed and flexion against resistance are required. When the wrist, metacarpophalangeal, and interphalangeal joints are flexed, the flexor muscles are shortened and their action is consequently weakened. In addition, some weakening results from the ligamentous action of the extensor muscles. Verify this by flexing your wrist and gripping a pencil. Now extend your wrist and note that your grip is much firmer.

> **Golfer's elbow** (elbow tendinitis) is a painful musculoskeletal condition that may follow repetitive use of the superficial muscles of the anterior aspect of the forearm, such as occurs during golfing. Pain is experienced on the medial side of the elbow. *Repeated forceful movements strain the common flexor tendon* of the flexor muscles (Table 6-8) and produce inflammation of the medial epicondyle (Fig. 6-67).

Nerves of the Forearm

The nerves of the forearm are the *median, ulnar, and radial*. The median nerve is the principal nerve of the anterior fascial compartment (Fig. 6-65). Although the radial nerve appears in the cubital region (Fig. 6-62), it soon enters the posterior fascial compartment. Aside from cutaneous branches, the nerves of the anterior aspect of the forearm are only two in number: the median and ulnar.

The Median Nerve

The median nerve enters the forearm with the brachial artery (Figs. 6-62, 6-65, 6-66, 6-70, and 6-74 to 6-76). It lies on the brachialis muscle and passes between the two heads of the pronator teres muscle, giving branches to them. It then descends deep to the flexor digitorum superficialis, to which it is closely at-

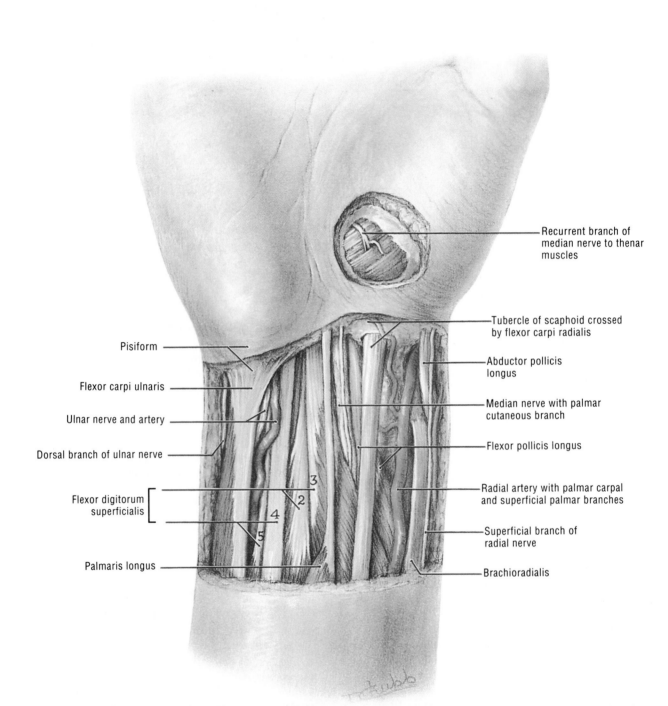

Recurrent branch of
median nerve to thenar
muscles

Tubercle of scaphoid crossed
by flexor carpi radialis

Abductor pollicis
longus

Median nerve with palmar
cutaneous branch

Flexor pollicis longus

Radial artery with palmar carpal
and superficial palmar branches

Superficial branch of
radial nerve

Brachioradialis

Pisiform

Flexor carpi ulnaris

Ulnar nerve and artery

Dorsal branch of ulnar nerve

Flexor digitorum
superficialis

Palmaris longus

3
2
4
5

Figure 6-75.   Dissection of the structures in the anterior aspect of the forearm and wrist. The distal skin incision was made along the transverse wrist crease (see Fig. 6-73). This crease crosses the pisiform bone to which the tendon of the flexor carpi ulnaris is attached.

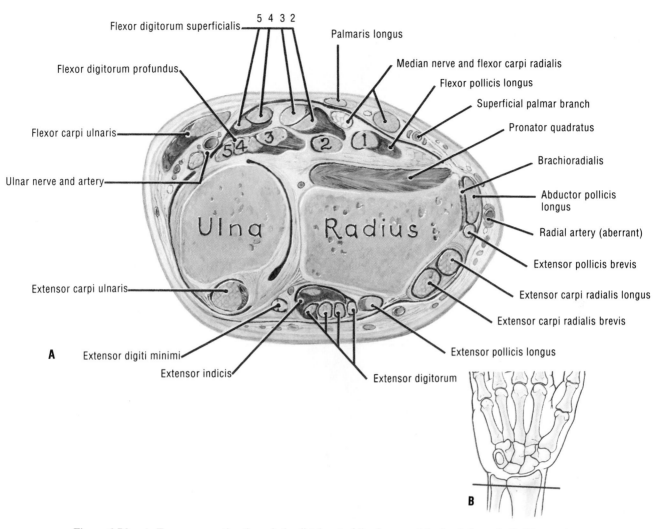

Figure 6-76.   *A*, Transverse section through the distal part of the forearm at the level shown in *B*. Observe the synovial cavity of the distal radioulnar joint (black slit-like space between the ulna and radius in *A*).

tached by the muscle's fascial sheath. It continues distally between this muscle and the flexor digitorum profundus. Near the wrist, the median nerve becomes superficial by passing between the tendons of the flexor digitorum superficialis and the flexor carpi radialis, deep to the tendon of the palmaris longus.

Branches of the Median Nerve (Fig. 6-20).   The median nerve has no branches in the arm. They arise in the forearm and hand as follows.

Articular branches pass to the elbow joint as the median nerve passes it (Fig. 6-70).

Muscular branches supply the pronator teres, pronator quadratus, and all the flexor muscles except the flexor carpi ulnaris and the medial half of the flexor digitorum profundus, which are supplied by the ulnar nerve (Table 6-8).

The anterior interosseous nerve (Fig. 6-70) arises from the median nerve in the distal part of the cubital fossa. It passes inferiorly on the interosseous membrane with the anterior interosseous branch of the ulnar artery. It runs between the flexor

digitorum profundus and flexor pollicis longus muscles to reach the pronator quadratus muscle. It supplies all three of these muscles, although the ulnar nerve supplies half of the flexor digitorum profundus (Fig. 6-20). The anterior interosseous nerve then passes deep to the pronator quadratus and ends by sending articular branches to the wrist joint.

The palmar cutaneous branch (Fig. 6-75) arises from the median nerve just proximal to the flexor retinaculum and becomes cutaneous between the tendons of the palmaris longus and flexor carpi radialis muscles. It passes superficial to the flexor retinaculum to supply the skin of the lateral part of the palm (Fig. 6-21).

> Occasionally there are *communications between the median and ulnar nerves* in the forearm. These branches are usually represented by several slender nerves, but these communications are important clinically because, even with a complete lesion of the median nerve, some muscles may not

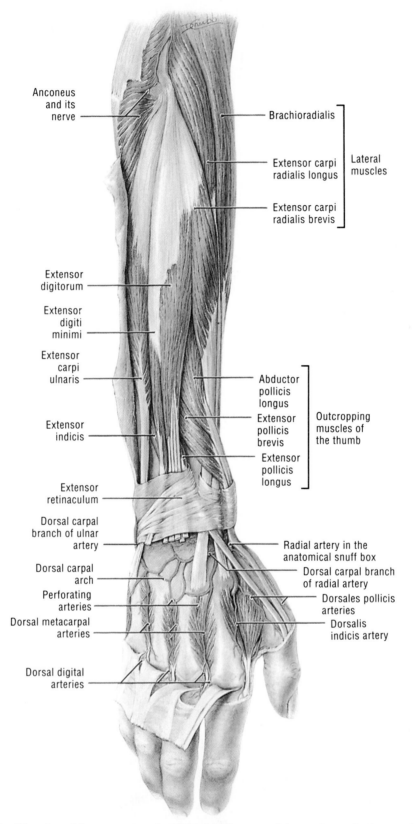

Anconeus and its nerve

Brachioradialis

Extensor carpi radialis longus

Extensor carpi radialis brevis

Lateral muscles

Extensor digitorum

Extensor digiti minimi

Extensor carpi ulnaris

Extensor indicis

Abductor pollicis longus

Extensor pollicis brevis

Extensor pollicis longus

Outcropping muscles of the thumb

Extensor retinaculum

Dorsal carpal branch of ulnar artery

Dorsal carpal arch

Perforating arteries

Dorsal metacarpal arteries

Dorsal digital arteries

Radial artery in the anatomical snuff box

Dorsal carpal branch of radial artery

Dorsales pollicis arteries

Dorsalis indicis artery

Figure 6-77.   Dissection of the extensor muscles of the right forearm and the arteries on the dorsum of the hand. The digital extensor tendons have been reflected.

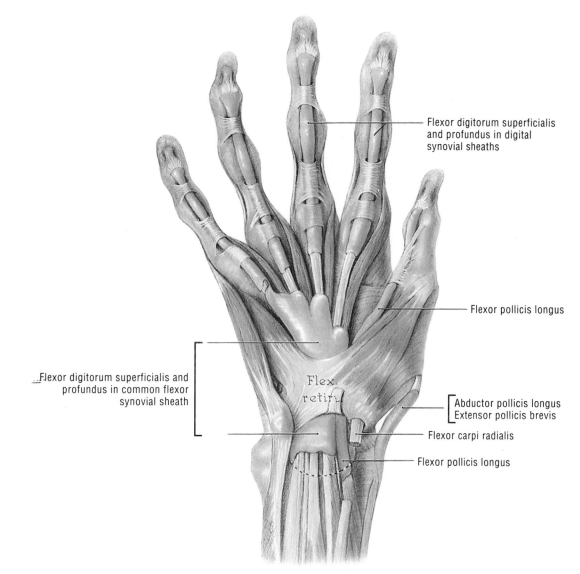

Flexor digitorum superficialis
and profundus in digital
synovial sheaths

Flexor pollicis longus

Flexor digitorum superficialis and
profundus in common flexor
synovial sheath

Flex.
retin.

Abductor pollicis longus
Extensor pollicis brevis

Flexor carpi radialis

Flexor pollicis longus

Figure 6-78. Dissection of the anterior aspect of the wrist and hand showing the synovial sheaths of the long flexor tendons of the digits. Observe the two sets: (1) proximal or carpal, posterior to the flexor retinaculum and (2) distal or digital, posterior to the fibrous sheaths of the digital flexors.

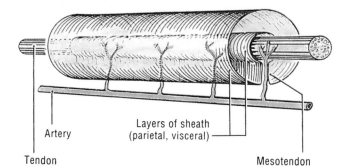

Artery

Tendon

Layers of sheath
(parietal, visceral)

Mesotendon

Figure 6-79. Synovial sheath of a long flexor tendon. The tubular sheath is a lubricating device (bursa) that envelops the long digital tendons where they pass through the osseofibrous tunnels in the digits (see Fig. 6-78). The layers of the synovial sheath are separated by a capillary layer of synovial fluid. Note that the mesotendon conveys small blood vessels to the tendons.

be paralyzed. This may lead to an erroneous conclusion that the median nerve has not been damaged.

Median Nerve Injury. When the median nerve is severed in the elbow region, there is loss of flexion of the proximal interphalangeal joints of all the digits. There is also loss of flexion of the distal interphalangeal joints of the second and third digits (Table 6-20). Flexion of the distal interphalangeal joints of the fourth and fifth digits is not affected because the medial part of the flexor digitorum profundus, which produces these movements, is supplied by the ulnar nerve (Fig. 6-20; Table 6-20). The ability to flex the metacarpophalangeal joints of the second and third digits will be affected because the digital branches of the median nerve supply the first and second lumbrical muscles (Fig. 6-20; Table 6-19). The median nerve is commonly injured just

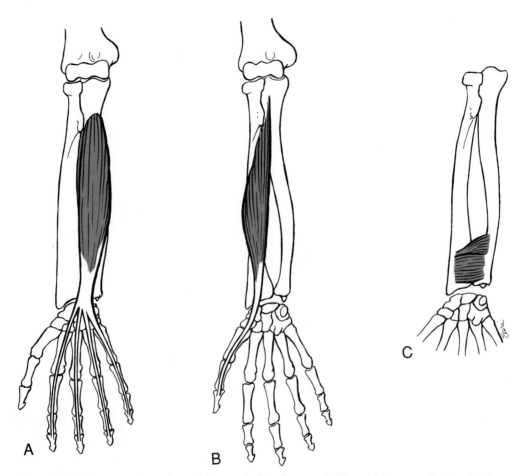

Figure 6-80. Deep muscles on the anterior aspect of the forearm. *A*, Flexor digitorum profundus. *B*, Flexor pollicis longus. *C*, Pronator quadratus.

proximal to the flexor retinaculum (Figs. 6-70 and 6-75), owing to the frequency of *wrist slashing in attempted suicides*. Although severance of the palmaris longus tendon is also common in these cases because it is superficial to the median nerve (Fig. 6-76), the loss of function of this muscle is not missed.

The Pronator Syndrome.   This nerve entrapment syndrome is caused by *compression of the median nerve near the elbow* (Devinsky and Feldmann, 1988). The nerve may be compressed between the two heads of the pronator teres owing to trauma, muscular hypertrophy, or fibrous bands. Patients are first seen clinically with pain and tenderness in the proximal aspect of the anterior forearm. Symptoms often follow activities that involve repeated elbow movements.

The Ulnar Nerve

After passing posterior to the medial epicondyle of the humerus, the ulnar nerve enters the forearm by passing between the two heads of the flexor carpi ulnaris muscle (Figs. 6-54, 6-63, 6-65, and 6-66). It then descends deep to this muscle on the flexor digitorum profundus, where it accompanies the ulnar artery near the middle of the forearm. Next it passes on the medial

side of this artery and the lateral side of the tendon of the flexor carpi ulnaris. In the distal part of the forearm, the ulnar nerve becomes relatively superficial, covered only by fascia and skin (Figs. 6-75 and 6-76). It pierces the deep fascia and passes superficial to the flexor retinaculum with the ulnar artery, lateral to the pisiform, between this bone and the hook of the hamate. This passage for the ulnar and artery, covered by a slip of the flexor retinaculum, is referred to clinically as the *canal of Guyon*.

Branches of the Ulnar Nerve (Fig. 6-20).   The ulnar nerve has no branches in the arm. They arise in the forearm and hand as follows:

Articular branches pass to the elbow joint, where the nerve is in the groove between the olecranon and medial epicondyle (Figs. 6-54 and 6-63).

Muscular branches supply the flexor carpi ulnaris and the medial half of the flexor digitorum profundus muscles (Figs. 6-20 and 6-70).

The palmar cutaneous branch arises from the ulnar nerve near the middle of the forearm and pierces the deep fascia in its distal third to supply skin on the medial part of the palm (Fig. 6-21).

The dorsal cutaneous branch arises from the ulnar nerve in the distal half of the forearm (Fig. 6-70) and passes posteroinferiorly between the ulna and the flexor carpi ulnaris. It supplies the posterior surface of the medial part of the hand (Fig. 6-21).

Table 6-9.
Deep Layer of Muscles on the Anterior Surface of the Forearm

Muscle	Proximal Attachments	Distal Attachments	Innervation[1]	Main Actions
Flexor digitorum profundus (Fig. 6-80*A*)	Proximal three-fourths of medial and anterior surfaces of ulna and interosseous membrane	Bases of distal phalanges of medial four digits	*Medial part:* Ulnar n. (**C8** and T1) *Lateral part:* Median n. (**C8** and T1)	Flexes distal phalanges of medial four digits (fingers)
Flexor pollicis longus (Fig. 6-80*B*)	Anterior surface of the radius and adjacent interosseous membrane	Base of distal phalanx of thumb	Anterior interosseous n. from median (**C8** and T1)	Flexes phalanges of first digit (thumb)
Pronator quadratus (Fig. 6-80*C*)	Distal fourth of anterior surface of ulna	Distal fourth of anterior surface of radius		Pronates forearm; deep fibers bind radius and ulna together

[1]**Boldface** indicates the main spinal cord segmental innervation.

Ulnar nerve injury commonly occurs where the nerve passes posterior to the medial epicondyle of the humerus (Fig. 6-63). Often the damage occurs when the elbow hits a hard surface and the epicondyle is fractured. *Ulnar nerve injury may result in extensive motor and sensory loss to the hand.* There is impaired power of adduction, and when an attempt is made to flex the wrist joint, the hand is drawn to the radial side by the flexor carpi radialis. Following ulnar nerve injury, patients are likely to have difficulty in making a fist because they cannot flex their fourth and fifth digits at the distal interphalangeal joints (Fig. 6-81). This characteristic appearance of the hand is known as *clawhand* when an effort is made to straighten the fingers.

Ulnar Nerve Entrapment. Compression of the ulnar nerve at the elbow is common (Fig. 6-63). It usually produces numbness and tingling of the medial part of the palm and the fourth and fifth digits.

The Radial Nerve

The radial nerve descends in the arm between the brachialis and brachioradialis muscles and crosses the anterior aspect of the lateral epicondyle of the humerus (Figs. 6-22, 6-62, 6-70, 6-75, and 6-76). Soon after it enters the forearm, it divides into the superficial and deep branches.

The superficial branch of the radial nerve (Fig. 6-82), the smaller of the two terminal branches, is the direct continuation of the radial nerve. It passes distally, anterior to the pronator teres muscle and under cover of the brachioradialis muscle. In the distal one-third of the forearm, the superficial branch passes posteriorly, deep to the tendon of the brachioradialis, and enters the posterior fascial compartment of the forearm. It pierces the deep fascia 3 to 4 cm proximal to the wrist and supplies skin on the dorsum of the wrist, hand, thumb, and lateral one (or two) and one-half digits (Figs. 6-21 and 6-82).

The deep branch of the radial nerve (Fig. 6-62), the larger of the two terminal branches, is entirely muscular and articular in its distribution. As it passes posteroinferiorly, it gives branches to the extensor carpi radialis brevis and supinator muscles (Figs. 6-22 and 6-62). It then *pierces the supinator muscle,* giving additional branches to it, and curving around the lateral side of the radius to enter the posterior fascial compartment of the forearm (Fig. 6-65). On reaching the posterior aspect of the forearm,

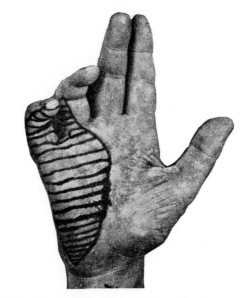

Figure 6-81. Clawhand (main en griffe) resulting from severance of the ulnar nerve at the wrist. (Reprinted with permission from Haymaker W, Woodhall B: *Peripheral Nerve Injuries,* ed 2. Philadelphia, WB Saunders, 1953.)

the deep branch of the radial nerve gives many branches to the extensor muscles (Fig. 6-90; Table 6-11). One of these branches, the *posterior interosseous nerve,* accompanies the posterior interosseous artery and supplies the deep extensor muscles (Fig. 6-83).

Radial nerve injury may occur in deep wounds of the forearm. Severance of the deep branch of the radial nerve results in an inability to extend the thumb and the metacarpophalangeal joints of the other digits. There is no loss of sensation because the deep branch of the radial is entirely muscular and articular in distribution. See Fig. 6-22 and Tables 6-10 and 6-11 to determine the muscles that are paralyzed when the deep branch of the radial nerve is severed.

Severance of the superficial branch of the radial nerve results in loss of sensation on the posterior surface of the forearm, hand, and proximal phalanges of the lateral three and one-half digits (Fig. 6-21). However, absence of cuta-

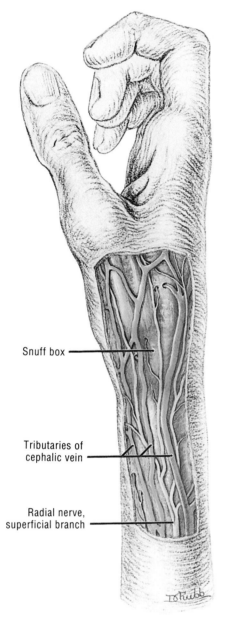

Snuff box

Tributaries of
cephalic vein

Radial nerve,
superficial branch

Figure 6-82. Dissection of the lateral aspect of the distal part of the forearm, wrist, and hand. Observe the superficial veins and nerves crossing the anatomical snuff box.

neous anesthesia is not necessarily indicative of an intact radial nerve because there is so much overlap between the cutaneous nerves of the hand (Fig. 6-109).

Radial Nerve Entrapment. The most common site of compression of this nerve is in the middle of the arm. This results in *wrist-drop* (p. 524), but there is normal strength of the triceps muscle. Sensory loss is variable and, if present, is usually limited to a patch on the dorsum of the hand between the first and second digits. *Compression of the radial nerve may result from improper positioning of the upper limb during sleeping* (Devinsky and Feldmann, 1988), especially in intoxicated persons ("Saturday night palsy"). Radial nerve compression can also result from improper use of a tourniquet.

Arteries of the Forearm

The two main arteries in the forearm are the *radial and ulnar arteries*. Each of these vessels has several branches. The brachial artery ends opposite the neck of the radius in the inferior part of the cubital fossa by dividing into its two terminal branches, the radial and ulnar arteries.

The Radial Artery. The radial artery *begins in the cubital fossa opposite the neck of the radius*, just medial to the biceps tendon (Figs. 6-55, 6-62, 6-66, and 6-70). The radial artery is the smaller of the two terminal branches of the brachial and continues the direct line of this vessel. The course of the radial artery in the forearm can be represented by a line connecting the midpoint of the cubital fossa to a point just medial to the tip of the styloid process of the radius (Fig. 6-55). The proximal part of the radial artery is overlapped by the fleshy belly of the brachioradialis muscle (Fig. 6-66), which, when pulled laterally, reveals the entire length of the artery in the forearm. The radial artery lies on muscle until it comes into contact with the distal end of the radius, where it is covered only by superficial and deep fasciae and skin. This is the common site for measuring the **pulse rate** (Figs. 6-74 and 6-75).

The radial artery leaves the forearm by winding around the lateral aspect of the radius and passing posteriorly between the lateral collateral ligament of the wrist joint and the tendons of the abductor pollicis longus and extensor pollicis brevis (Fig. 6-83). *The radial artery crosses the floor of the anatomical snuff box*, formed by the scaphoid and trapezium bones (Figs. 6-71 and 6-77). It ends by completing the *deep palmar arterial arch* in conjunction with the ulnar artery (Figs. 6-84 and 6-109).

Branches of the Radial Artery in the Forearm and at the Wrist. *Muscular branches* of the radial artery supply muscles on the lateral side of the forearm (Fig. 6-70).

The radial recurrent artery arises from the lateral side of the radial artery, just distal to its origin, and ascends between the brachioradialis and brachialis muscles (Fig. 6-62). It supplies these muscles and the elbow joint and anastomoses with the radial collateral artery, a branch of the profunda brachii. Thereby it participates in the arterial anastomosis around the elbow (Fig. 6-55).

The superficial palmar branch of the radial artery arises at the distal end of the radius, just proximal to the wrist (Fig. 6-84). It passes through and sometimes over the muscles of the thenar eminence, which it supplies. It usually anastomoses with the terminal part of the ulnar artery to form a *superficial palmar arterial arch* (Figs. 6-84B and 6-103).

The palmar carpal branch of the radial artery is a small vessel that arises near the distal border of the pronator quadratus (Figs. 6-75 and 6-76). It then runs across the wrist deep to the flexor tendons, where it anastomoses with the carpal branch of the ulnar artery and recurrent branches of the deep palmar arch to form the *palmar carpal arch* or network (Fig. 6-84).

The dorsal carpal branch of the radial artery (Fig. 6-77) runs medially across the dorsal surface of the wrist, deep to the extensor tendons, where it anastomoses with the dorsal carpal branch of the ulnar artery and with the terminations of the anterior and posterior interosseus arteries to form the *dorsal carpal arch*.

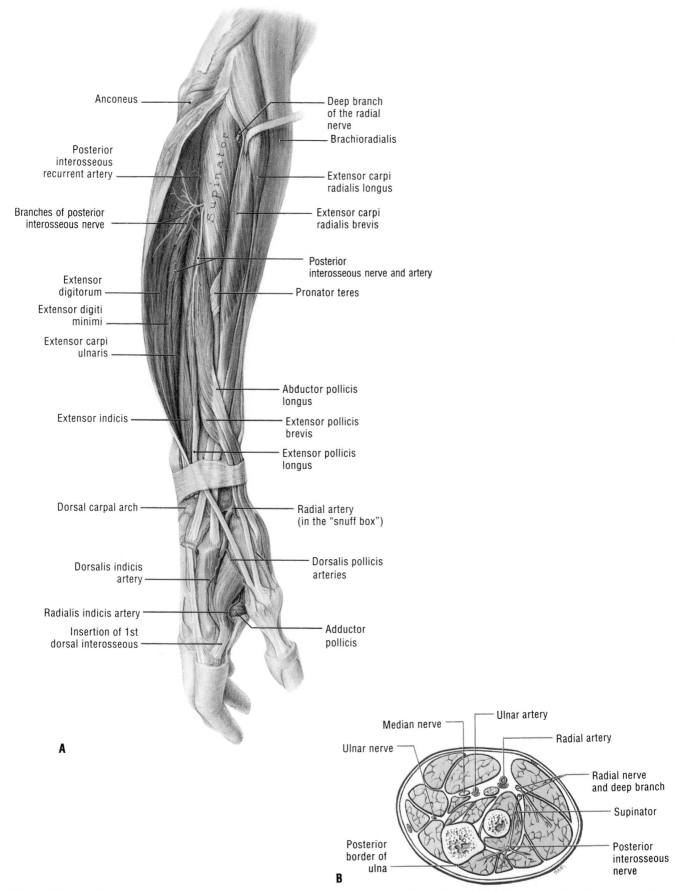

Anconeus

Posterior interosseous recurrent artery

Branches of posterior interosseous nerve

Extensor digitorum

Extensor digiti minimi

Extensor carpi ulnaris

Extensor indicis

Dorsal carpal arch

Dorsalis indicis artery

Radialis indicis artery

Insertion of 1st dorsal interosseous

Supinator

Deep branch of the radial nerve

Brachioradialis

Extensor carpi radialis longus

Extensor carpi radialis brevis

Posterior interosseous nerve and artery

Pronator teres

Abductor pollicis longus

Extensor pollicis brevis

Extensor pollicis longus

Radial artery (in the "snuff box")

Dorsalis pollicis arteries

Adductor pollicis

A

Median nerve

Ulnar nerve

Posterior border of ulna

B

Ulnar artery

Radial artery

Radial nerve and deep branch

Supinator

Posterior interosseous nerve

Figure 6-83. *A*, Posterolateral view of a dissection of deep structures in the posterior part of the right forearm and hand. *B*, Transverse section of the forearm illustrating its nerve supply. The flexor territory (*orange*), supplied by the ulnar and median nerves, is separated from the extensor territory (*red*), supplied by the radial nerve, by the radial artery laterally, and by the posterior, sharp, palpable border of the ulna posteromedially. No motor nerve crosses either line.

579

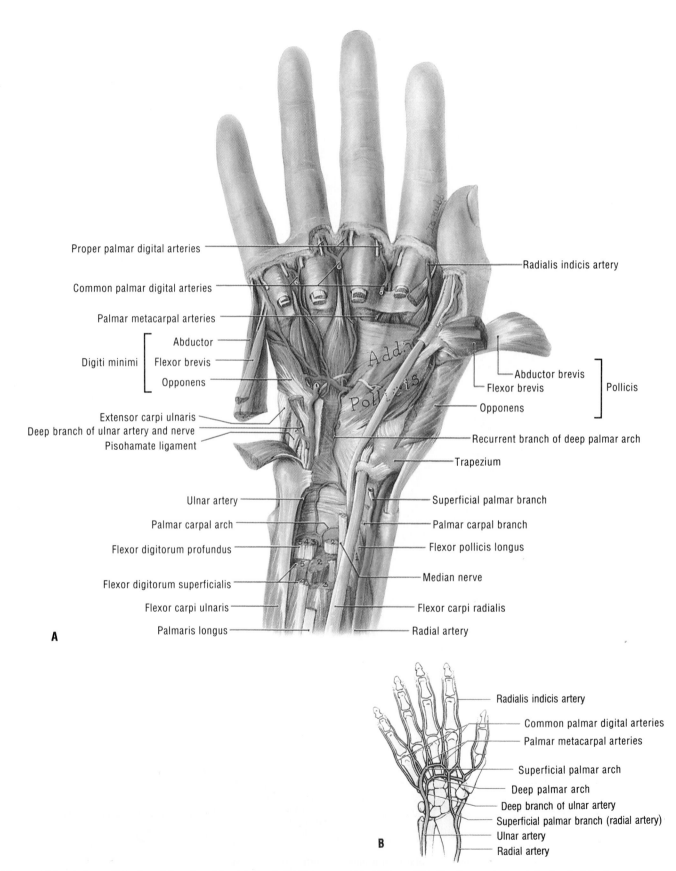

Figure 6-84. *A*, Deep dissection of the palm of the right hand showing its muscles and arteries. Observe the deep branch of the artery joining the radial artery to form the deep palmar arch. *B*, The palmar arterial arches. Observe that the deep arch lies at the level of the bases of the metacarpal bones and that the superficial arch is located more distally.

The most *common place for measuring the pulse rate* is where the radial artery lies on the anterior surface of the distal end of the radius, lateral to the tendon of the flexor carpi radialis muscle (Figs. 6-74, 6-75, and 6-84). Here, it is covered only by deep and superficial fasciae and skin. About 4 cm of this artery can be compressed against the distal end of the radius, where it lies between the tendons of the flexor carpi radialis and abductor pollicis longus muscles. When measuring the **radial pulse rate**, the pulp of the thumb should not be used because it has its own pulse, which could be interpreted as the patient's pulse. If a pulse cannot be felt, try the other wrist because an *aberrant radial artery* on one side may make the pulse difficult to palpate. A radial pulse may also be felt in the *anatomical snuff box* (Fig. 6-77).

The Ulnar Artery

The ulnar artery *begins near the neck of the radius*, just medial to the biceps tendon (Fig. 6-62). It is the larger of the two terminal branches of the brachial artery. It makes a gentle curve as it passes from the cubital fossa to the medial side of the forearm (Figs. 6-55 and 6-70). The ulnar artery passes inferomedially deep to the pronator teres muscle. In company with the median nerve, *the ulnar artery passes between the ulnar and radial heads of the flexor digitorum superficialis* (Fig. 6-70). About midway between the elbow and wrist, the ulnar artery crosses posterior to the median nerve to reach the medial side of the forearm, where it lies on the flexor digitorum profundus (Figs. 6-65 and 6-70).

In the distal two-thirds of the forearm, the ulnar artery lies lateral to the ulnar nerve (Fig. 6-70). It leaves the forearm by passing superficial to the flexor retinaculum on the lateral side of the pisiform bone (Figs. 6-75 and 6-76). At the wrist, the ulnar artery and nerve lie lateral to the tendon on the flexor carpi ulnaris, where they are covered only by fascia and skin. The ulnar nerve and artery are usually covered by the superficial part of the flexor retinaculum (Fig. 6-66). The *pulsations of the ulnar artery* can sometimes be felt where it passes anterior to the head of the ulna (Figs. 6-74 and 6-75).

Branches of the Ulnar Artery in the Forearm (Figs. 6-55, 6-70, 6-75, and 6-77).   These branches supply medial muscles in the forearm and hand, the common flexor synovial sheath, and the ulnar nerve.

The anterior ulnar recurrent artery (Fig. 6-55) arises from the ulnar just inferior to the elbow joint and runs superiorly between the brachialis and pronator teres muscles. It supplies these muscles and *anastomoses with the inferior ulnar collateral artery*, a branch of the brachial, thereby participating in the arterial anastomoses around the elbow.

The posterior ulnar recurrent artery (Figs. 6-55 and 6-70) arises from the ulnar distal to the anterior ulnar recurrent artery. It passes superiorly, posterior to the medial epicondyle, where it lies deep to the tendon of the flexor carpi ulnaris. It supplies adjacent muscles and then takes part in the *arterial anastomosis around the elbow.*

The common interosseous artery, a short branch of the ulnar, arises in the distal part of the cubital fossa (Fig. 6-55) and divides into anterior and posterior interosseous arteries.

The anterior interosseous artery (Figs. 6-70 and 6-77) passes distally on the interosseous membrane to the proximal border of the pronator quadratus muscle. Here it pierces this membrane and continues distally to join the dorsal carpal arch.

The posterior interosseous artery (Figs. 6-55 and 6-83) passes posteriorly between the bones of the forearm, just proximal to the interosseous membrane. It supplies adjacent muscles and then gives off the *posterior interosseous recurrent artery*, which passes superiorly, posterior to the lateral epicondyle, and participates in the arterial anastomosis around the elbow.

The muscular branches of the ulnar artery supply muscles on the medial side of the forearm, mainly the flexor-pronator group of muscles (Fig. 6-70).

The palmar carpal branch of the ulnar artery is a small branch that runs across the anterior aspect of the wrist, deep to the tendons of the flexor digitorum profundus. This branch anastomoses with the palmar carpal branch of the radial artery to form the *palmar carpal arch* (Fig. 6-84).

The dorsal carpal branch of the ulnar artery (Fig. 6-77) arises just proximal to the pisiform bone. It passes across the dorsal surface of the wrist, deep to the extensor tendons, where it anastomoses with the dorsal carpal branch of the radial artery to form the *dorsal carpal arch* (Figs. 6-77 and 6-83).

The superficial branch of the ulnar artery continues into the palm as the *superficial palmar arch* (p. 607).

The deep palmar branch of the ulnar artery passes deeply in the hand, where it anastomoses with the radial artery and completes the *deep palmar arch* (p. 607).

Sometimes the brachial artery divides at a more proximal level than usual. In other cases the ulnar artery passes superficial to the flexor muscles within the superficial fascia. These variations must be kept in mind when performing *venesections* (incisions into a vein) at the elbow (*e.g.*, for inserting a cannula or catheter into a vein for intravenous injection of fluids, blood, or medication). If an aberrant artery is mistaken for a vein and certain drugs are injected into it, the result can be disastrous (*e.g.*, gangrene resulting in partial or total loss of the hand).

The Extensor Muscles of the Forearm

The eleven muscles in the posterior part of the forearm are extensors (Tables 6-10 and 6-11). They can be organized into three functional groups as follows: (1) *muscles that extend the hand* at the wrist (extensor carpi radialis longus, extensor carpi radialis brevis, and extensor carpi ulnaris); (2) *muscles that extend the medial four digits* (extensor digitorum, extensor indicis, and extensor digiti minimi); and (3) *muscles that extend the first digit or thumb* (abductor pollicis longus, extensor pollicis brevis, and extensor pollicis longus). For purposes of description, the muscles of the forearm are usually divided into *superficial and deep extensor groups.*

Four of the muscles in the superficial group of extensor muscles (extensor carpi radialis brevis, extensor digitorum, extensor digiti minimi, and extensor carpi ulnaris) are attached by a flattened *common extensor tendon* (Figs. 6-63, 6-67, and 6-68). It is attached to the *lateral epicondyle of the humerus*, the adjacent fascia, and the lateral supracondylar ridge of the humerus (Figs. 6-49A and 6-68). This is known as the *common extensor attachment* (origin) of the common extensor tendon. The brachioradialis and extensor carpi radialis longus muscles are attached superiorly to the superior and inferior parts of the lateral supracondylar ridge, respectively (Fig. 6-67). The *brachioradialis muscle*, a flexor of the forearm at the elbow joint (Figs. 6-70 and 6-88A), is included with the extensor muscles because it is *supplied by the radial nerve*. It is also described with the muscles of the cubital region because it forms the lateral boundary of the cubital fossa (Figs. 6-60 and 6-62).

The tendons of these extensor muscles occupy the lateral side, as well as the dorsum of the wrist, where they lie within *extensor synovial sheaths* located in osseofibrous canals (Fig. 6-85). These sheaths reduce friction between the extensor tendons and the walls of these canals. The extensor tendons are held in place by a strong fibrous band, called the **extensor retinaculum** (Figs. 6-77 and 6-85), which is attached laterally to the distal end of the radius and medially to the styloid process of the ulna and the triquetral and pisiform bones (Fig. 6-71). *The extensor retinaculum prevents bowstringing of the long extensor tendons* when the hand is hyperextended at the wrist joint.

Sometimes a cystic, usually nontender swelling of a tendon sheath appears on the dorsum of the wrist or hand. Usually the swelling is the size of a grape, but it can be as large as a plum. Flexion of the wrist makes the swelling enlarge and extension of it tends to make it smaller. Clinically, this type of swelling is called a *ganglion* (G. swelling or knot); anatomically, this is a misnomer because a ganglion refers to a collection of nerve cells (*e.g.*, a spinal ganglion). These cystic swellings of tendons ("ganglia") are close to and often communicate with the extensor synovial sheaths (Fig. 6-85). The distal attachment of the extensor carpi radialis brevis tendon into the base of the third metacarpal bone is a common site for such a cystic swelling.

The Superficial Muscles on the Posterior or Extensor Surface of the Forearm

The Brachioradialis Muscle (Figs. 6-48, 6-53, 6-56, 6-62, 6-63, 6-66, and 6-88A).   This fusiform muscle is located superficially on the anterolateral surface of the forearm. Its attachments, nerve supply, and main actions are given in Table 6-10 (also see pp. 559 and 560).

The Extensor Carpi Radialis Longus Muscle (Figs. 6-62, 6-65, 6-70, 6-77, and 6-88B).   This fusiform muscle is partly overlapped by the brachioradialis with which it is often blended. Its attachments, nerve supply, and main actions are given in Table 6-10.

The Extensor Carpi Radialis Brevis Muscle (Figs. 6-62, 6-65, 6-70, 6-77, 6-83, 6-86, 6-87, and 6-88C).   As its name indicates, this fusiform muscle is shorter than the extensor carpi radialis longus, which covers it. Its attachments, nerve supply, and main actions are given in Table 6-10. It *extends and abducts the hand* at the wrist joint. This muscle and the extensor carpi radialis longus act together to steady the wrist during flexion of the medial four digits.

The Extensor Digitorum Muscle (Figs. 6-65, 6-77, 6-84, 6-85, 6-88D, and 6-89).   This *principal extensor of the medial four digits* occupies much of the posterior surface of the forearm. It divides into four tendons proximal to the wrist, which pass through a *common synovial sheath*, deep to the extensor retinaculum, with the tendon of the extensor indicis muscle. Its attachments, nerve supply, and main actions are given in Table 6-10. It *extends the proximal phalanges* and, through its collateral reinforcements, the middle and distal phalanges as well. It also helps to extend the hand at the wrist joint after exerting its traction primarily on the digits.

The Extensor Digiti Minimi Muscle (Figs. 6-77, 6-83, 6-85, 6-88E, and 6-91).   This fusiform slip of muscle is a partially detached part of the extensor digitorum. Its tendon runs through a separate compartment in the extensor retinaculum and then divides into two slips; the lateral one is joined to the tendon of the extensor digitorum. Its attachments, nerve supply, and main actions are given in Table 6-10. It *extends the proximal phalanx of the fifth digit* (little finger) at the metacarpophalangeal joint and assists with extension at its interphalangeal joints. It also assists with extension of the hand after exerting its traction, primarily on the fifth digit.

The Extensor Carpi Ulnaris Muscle (Figs. 6-65, 6-66, 6-76, 6-77, 6-83, 6-85, and 6-88F).   This long fusiform muscle, located on the medial border of the forearm, has two heads. Its tendon runs in a groove between the head and the styloid process of the ulna, within a special compartment of the extensor retinaculum. Its attachments, nerve supply, and main actions are given in Table 6-10. It *extends and adducts the hand* at the wrist joint. Acting with the extensor carpi radialis, it extends the hand; acting with the flexor carpi ulnaris, it adducts the hand.

The Deep Muscles on the Posterior or Extensor Surface of the Forearm

The deep extensors of the forearm consist of three muscles that act on the first digit or thumb (abductor pollicis longus, extensor pollicis brevis, and extensor pollicis longus) and the extensor indicis, which helps to extend the second digit or index finger (Figs. 6-77, 6-83, 6-85, 6-86, and 6-87). The abductor pollicis longus, extensor pollicis brevis, and extensor pollicis longus are attached deep to the superficial extensors and appear from concealment (*i.e.*, they crop out) along a furrow that divides the extensor muscles into lateral and medial groups. Because of this characteristic, these three muscles are referred to as the *outcropping (thumb) muscles* (Figs. 6-77 and 6-83).

Boundaries of the Anatomical Snuff Box (Figs. 6-77, 6-82, 6-83, and 6-85 to 6-87).   The tendons of the abductor pollicis longus and extensor pollicis brevis bound the anatomical snuff box anteriorly, and the tendon of the extensor pollicis longus bounds it posteriorly.

The scaphoid and trapezium lie in the floor of the anatomical snuff box (Figs. 6-71 and 6-76). *The scaphoid is the most frequently fractured carpal bone* (p. 591). Injury to this bone results in localized tenderness in the snuff box. Initial radiographs may not reveal a fracture of the scaphoid. However, repeat radiographs taken two to three weeks after an

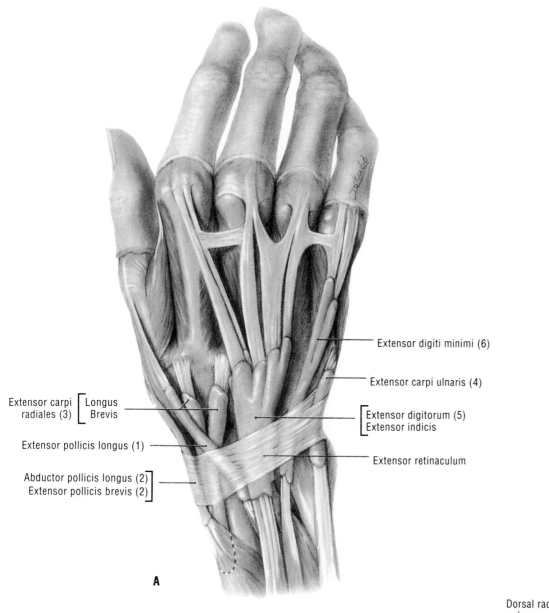

Extensor digiti minimi (6)

Extensor carpi ulnaris (4)

Extensor carpi radiales (3) [Longus, Brevis]

Extensor digitorum (5)
Extensor indicis

Extensor pollicis longus (1)

Extensor retinaculum

Abductor pollicis longus (2)
Extensor pollicis brevis (2)

A

Figure 6-85. *A*, The synovial sheaths (*blue*) on the dorsum of the right wrist. Observe that the six sheaths occupy six osseofibrous tunnels deep to the extensor retinaculum and contain nine tendons: three for the first digit or thumb in sheaths (*1* and *2*); three for the extensors of the wrist in two sheaths (*3* and *4*); and three for the extensors of the digits in two sheaths (*5* and *6*). *B*, Transverse section of the wrist showing the tendons and their synovial sheaths on the dorsum of the distal ends of the radius and ulna. The numbers refer to structures named in *A*.

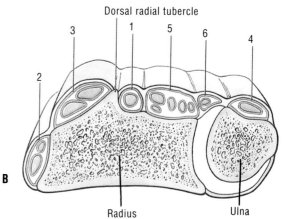

Dorsal radial tubercle

3   1   5   6   4

2

B

Radius   Ulna

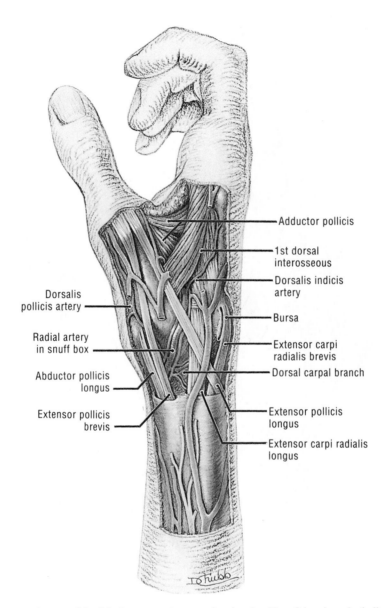

Adductor pollicis

1st dorsal interosseous

Dorsalis indicis artery

Bursa

Extensor carpi radialis brevis

Dorsal carpal branch

Extensor pollicis longus

Extensor carpi radialis longus

Dorsalis pollicis artery

Radial artery in snuff box

Abductor pollicis longus

Extensor pollicis brevis

Figure 6-86.   Dissection of the lateral aspect of the right forearm, wrist, and hand. Observe the three long tendons of the first digit (thumb) forming the sides of the triangular hollow known as the anatomical snuff box.

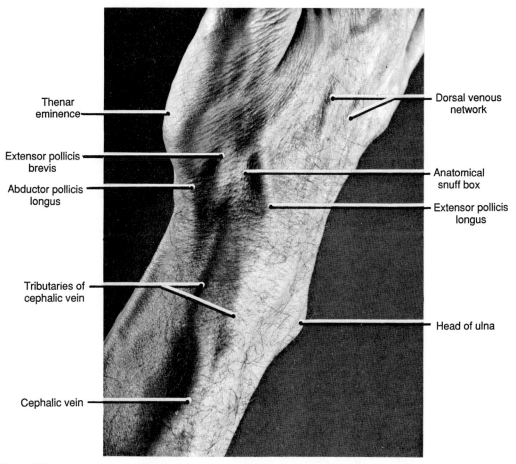

Thenar eminence

Dorsal venous network

Extensor pollicis brevis

Abductor pollicis longus

Anatomical snuff box

Extensor pollicis longus

Tributaries of cephalic vein

Head of ulna

Cephalic vein

Figure 6-87. Lateral aspect of the distal part of the right forearm, wrist, and hand of a 46-year-old man showing the surface landmarks of the anatomical snuff box.

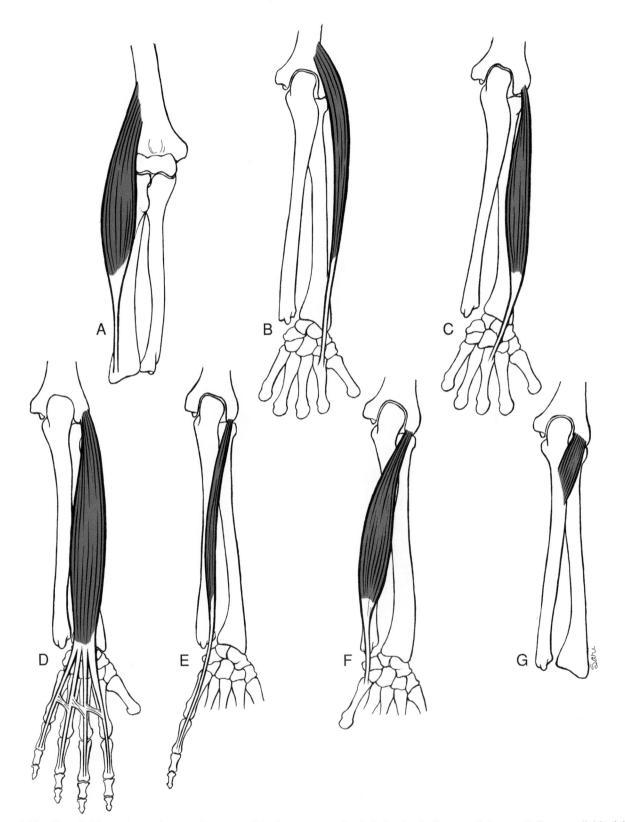

Figure 6-88. Superficial muscles on the posterior aspect of the forearm. *A*, Brachioradialis. *B*, Extensor carpi radialis longus. *C*, Extensor carpi radialis brevis. *D*, Extensor digitorum. *E*, Extensor digiti minimi. *F*, Extensor carpi ulnaris. *G*, Anconeus.

Table 6-10.
Superficial Muscles on the Posterior or Extensor Surface of the Forearm

Muscle	Proximal Attachments	Distal Attachments	Innervation[1]	Main Actions
Brachioradialis[2] (Figs. 6-67 and 6-88A)	Proximal two-thirds of Lateral supracondylar ridge of humerus	Lateral surface of distal end of radius	Radial n. (C5,**C6**, and C7)	Flexes forearm
Extensor carpi radialis longus (Figs. 6-68 and 6-88B)	Lateral supracondylar ridge of humerus	Base of 2nd metacarpal bone	Radial n. (C6 and C7)	Extend and abduct hand at wrist joint
Extensor carpi radialis brevis (Fig. 6-88C)	Lateral epicondyle of humerus	Base of 3rd metacarpal bone	Deep branch of radial n. (**C7** and C8)	
Extensor digitorum (Figs. 6-85 and 6-88D)	Lateral epicondyle of humerus	Extensor expansions of medial four digits		Extends medial four digits at metacarpophalangeal joints; extends hand at wrist joint
Extensor digiti minimi (Fig. 6-88E)	Lateral epicondyle of humerus	Extensor expansion of fifth digit	Posterior interosseous n. (**C7** and C8), a branch of the radial n.	Extends digit 5 at metacarpophalangeal and interphalangeal joints
Extensor carpi ulnaris (Fig. 6-88F)	Lateral epicondyle of humerus and posterior border of ulna	Base of 5th metacarpal bone		Extends and adducts hand at wrist joint
Anconeus[3] (Fig. 6-88G)	Lateral epicondyle of humerus	Lateral surface of olecranon and superior part of posterior surface of ulna	Radial n. (C7, C8 and T1)	Assists triceps in extending elbow joint; stabilizes elbow joint; abducts ulna during pronation

[1]**Boldface** indicates the main spinal cord segmental innervation.

[2]The brachioradialis muscle is described with the muscles of the cubital fossa (p. 559) because it forms the lateral boundary of this space in the anterior aspect of the elbow (Fig. 6-62).

[3]This small triangular muscle is part of the triceps (Fig. 6-63 and p. 547). It does not belong to the posterior fascial compartment of the forearm, but for convenience and as is the custom, it is listed with it.

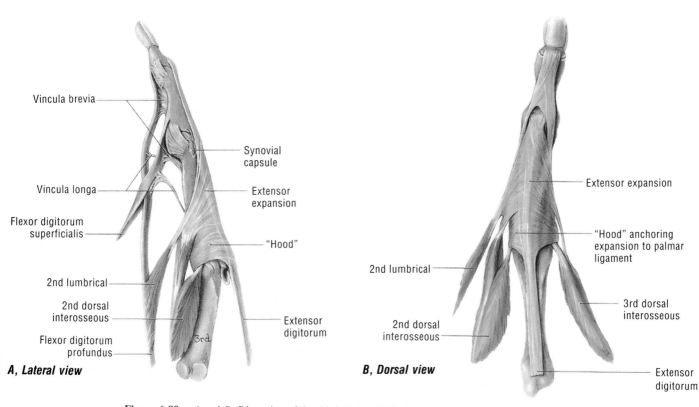

A, Lateral view

Vincula brevia
Vincula longa
Flexor digitorum superficialis
2nd lumbrical
2nd dorsal interosseous
Flexor digitorum profundus
3rd
Synovial capsule
Extensor expansion
"Hood"
Extensor digitorum

B, Dorsal view

2nd lumbrical
2nd dorsal interosseous
Extensor expansion
"Hood" anchoring expansion to palmar ligament
3rd dorsal interosseous
Extensor digitorum

Figure 6-89. *A* and *B*, Dissection of the third digit (middle finger) showing its extensor expansion and interosseous muscles.

injury to the wrist may reveal a fracture owing to resorption of bone at the fracture site.

The Abductor Pollicis Longus Muscle (Figs. 6-77, 6-82, 6-83, to 6-87, and 6-90B).   This long fusiform abductor of the thumb (L. *pollex*) lies just distal to the supinator muscle and is closely related to the extensor pollicis brevis muscle. Its attachments, nerve supply, and main actions are given in Table 6-11. It *abducts and extends the thumb* at the carpometacarpal joint. This muscle acts with the abductor pollicis brevis during abduction of the thumb and with the extensor pollicis brevis during extension of this digit. Although deeply situated, the abductor pollicis longus emerges at the wrist as *one of the outcropping muscles* (p. 582). Its tendon passes deep to the extensor retinaculum in a common synovial sheath with the tendon of the extensor pollicis brevis.

The Extensor Pollicis Brevis Muscle (Figs. 6-77, 6-82, 6-83, 6-85 to 6-87, 6-90C, and 6-91).   This short fusiform extensor of the thumb lies distal to the long abductor of the thumb (abductor pollicis longus) and is partly covered by it. Its attachments, nerve supply, and main actions are given in Table 6-11. It *extends the proximal phalanx of the thumb* at the metacarpophalangeal joint. In continued action, it helps to extend the metacarpal bone of the thumb. It also helps to extend and abduct the hand.

The Extensor Pollicis Longus Muscle (Figs. 6-77, 6-83 to 6-87, 6-90D, and 6-91).   This long extensor of the thumb is larger and its tendon is longer than that of the extensor pollicis brevis. Its attachments, nerve supply, and main actions are given in Table 6-11. It *extends the distal phalanx of the thumb*, and in continued action, it extends the metacarpophalangeal and interphalangeal joints of the thumb. It also adducts the extended thumb and rotates it laterally. It can also abduct the hand.

The Extensor Indicis Muscle (Figs. 6-77, 6-83, 6-85, 6-90E, 6-91, and 6-92).   This narrow, elongated muscle lies medial to and alongside the extensor pollicis longus. Its attachments, nerve supply, and main action are given in Table 6-11. Acting with the extensor digitorum (Table 6-10), it *extends the index finger* at the proximal interphalangeal joint, as in pointing. It also assists in extending the hand.

> **Elbow tendonitis** (*tennis elbow*, lateral epicondylitis) is a painful musculoskeletal condition that may follow repetitive forceful pronation-supination of the forearm. Elbow tendonitis is characterized by pain and point tenderness at or just distal to the lateral epicondyle of the humerus and appears to result from *premature degeneration of the common extensor attachment* of the superficial extensor muscles of the forearm, *i.e.*, the attachment of the common extensor tendon (Figs. 6-63, 6-67, and 6-68). The pain is aggravated by activities that put tension on the common extensor tendon (*e.g.*, grasping something such as a tennis racquet). Elbow tendonitis is *common in persons who play tennis* because of the repeated strenuous contraction of the extensor muscles, especially during the backhand stroke. These movements strain the common extensor tendon of these muscles and produce inflammation of the lateral epicondyle. *Elbow tendonitis is not confined to those who play tennis.* It may develop following an injury to the elbow or result from any continuous activity that involves extensive use of the superficial extensor muscles of the forearm (*e.g.*, using a screwdriver, golfing, or shoveling snow).

The Posterior Nerves of the Forearm

Just superior to the elbow, the radial nerve divides into deep and superficial branches (p. 577).

The deep branch of the radial nerve supplies the extensor carpi radialis brevis and supinator muscles and then enters the

Table 6-11.
Deep Muscles on the Posterior or Extensor Surface of the Forearm

Muscle	Proximal Attachments	Distal Attachments	Innervation[1]	Main Actions
Supinator[2] (Fig. 6-90A)	Lateral epicondyle of humerus, radial collateral and anular ligaments, supinator fossa and crest of ulna	Lateral, posterior, and anterior surfaces of proximal third of radius	Deep branch of radial n. (C5 and **C6**)	Supinates forearm, *i.e.*, rotates radius to turn palm anteriorly (p. 8)
Abductor pollicis longus (Fig. 6-90B)	Posterior surfaces of ulna and radius and interosseous membrane	Base of 1st metacarpal bone		Abducts thumb and extends it at carpometacarpal joint
Extensor pollicis brevis (Fig. 6-90C)	Posterior surface of radius and interosseous membrane	Base of proximal phalanx of thumb	Posterior interosseous n. (C7 and **C8**)[3]	Extends proximal phalanx of thumb at metacarpophalangeal joint
Extensor pollicis longus (Fig. 6-90D)	Posterior surface of middle third of ulna and interosseous membrane	Base of distal phalanx of thumb		Extends distal phalanx of thumb at metacarpophalangeal and interphalangeal joints
Extensor indicis (Fig. 6-90E)	Posterior surface of ulna and interosseous membrane	Extensor expansion of second digit (index finger)		Extends digit 2 and helps to extend hand

[1]**Boldface** indicates the main spinal cord segmental innervation.
[2]The supinator is also described with the muscles of the cubital fossa (p. 559) because it forms the floor of this space in the anterior aspect of the elbow (Fig. 6-62).
[3]The terminal branch of the deep branch of the radial nerve.

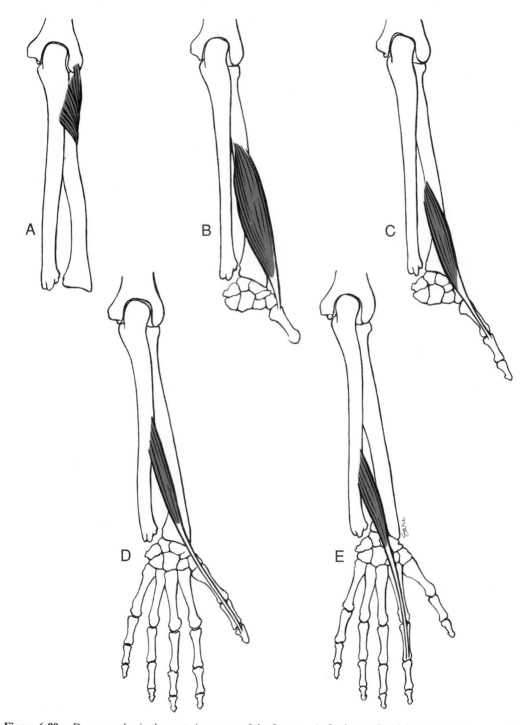

Figure 6-90. Deep muscles in the posterior aspect of the forearm. *A*, Supinator. *B*, Abductor pollicis longus. *C*, Extensor pollicis brevis. *D*, Extensor pollicis longus. *E*, Extensor indicis.

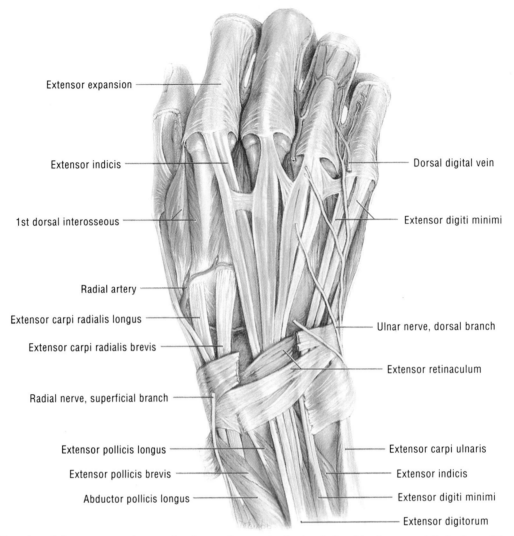

Extensor expansion

Extensor indicis

1st dorsal interosseous

Radial artery

Extensor carpi radialis longus

Extensor carpi radialis brevis

Radial nerve, superficial branch

Extensor pollicis longus

Extensor pollicis brevis

Abductor pollicis longus

Dorsal digital vein

Extensor digiti minimi

Ulnar nerve, dorsal branch

Extensor retinaculum

Extensor carpi ulnaris

Extensor indicis

Extensor digiti minimi

Extensor digitorum

Figure 6-91.   Dissection of the extensor tendons passing deep to the extensor retinaculum on the dorsum of the right hand. Note the bands proximal to the knuckles that connect the tendons of the digital extensors and thereby restrict the independent action of the digits.

latter muscle (Fig. 6-70). After it emerges from the supinator and enters the posterior compartment of the forearm, the deep branch is referred to as the *posterior interosseous nerve* (Fig. 6-83). It passes deep to the extensor pollicis longus muscle and lies on the interosseous membrane, where it is accompanied by the posterior interosseous artery. The posterior interosseous nerve terminates on the dorsum of the wrist, where it sends articular branches to the distal radioulnar joint and the carpal joints. **The posterior interosseous nerve**, *the terminal branch of the deep radial*, supplies the extensor digitorum, the extensor digiti minimi, the extensor carpi ulnaris, the extensor indicis, and the three outcropping thumb muscles (Fig. 6-22; Tables 6-10 and 6-11).

The superficial branch of the radial nerve is the continuation of the radial nerve (Fig. 6-70). It passes distally, deep to the brachioradialis muscle in company with the radial artery. It perforates the antebrachial fascia along the lateral border of the forearm and divides into two branches. At the wrist, it *divides into four or five digital nerves* (Fig. 6-21). The distribution of the superficial branch is cutaneous and articular. It supplies the lateral two-thirds of the posterior surface of the hand and the posterior surface of the lateral two and one-half digits over the proximal phalanx (Figs. 6-21 and 6-109). For a discussion of radial nerve injury, see p. 606.

The Posterior Arteries of the Forearm

The posterior aspect of the forearm and hand is supplied by the *posterior interosseous artery* (Fig. 6-83), which arises from the common interosseous branch of the ulnar artery (Fig. 6-70). The posterior interosseous artery passes posteriorly between the radius and ulna, just proximal to the interosseous membrane, and appears in the posterior part of the forearm between the supinator and abductor pollicis longus muscles (Fig. 6-83). It then descends between the superficial and deep muscles, supplying muscles in the posterior part of the forearm. As it reaches the posterior aspect of the wrist, the posterior interosseous artery becomes very small and ends by anastomosing with the termi-

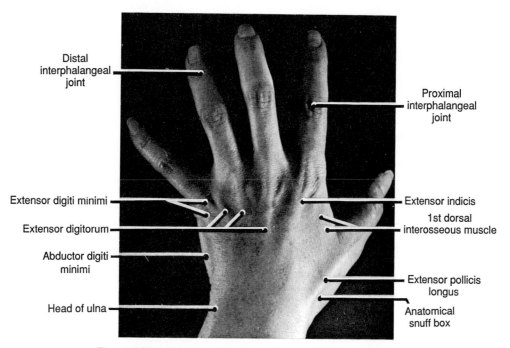

Distal interphalangeal joint

Proximal interphalangeal joint

Extensor digiti minimi

Extensor indicis

1st dorsal interosseous muscle

Extensor digitorum

Abductor digiti minimi

Extensor pollicis longus

Head of ulna

Anatomical snuff box

Figure 6-92.   Dorsum of the wrist and hand of a 36-year-old woman.

nation of the anterior interosseous artery and the *dorsal carpal arch* (Figs. 6-77 and 6-83).

The Wrist and Hand

The wrist (carpal region) is between the forearm and hand. Despite its name, a "wrist" watch is usually not worn around the wrist; commonly the band encircles the distal end of the forearm, just proximal to the head of the ulna (Fig. 6-60). *Movements of the hand occur primarily at the wrist joint.* The wrist or carpal bones were described previously (p. 560) and are illustrated in Figs. 6-1 and 6-71. The *antebrachial fascia* (fascia of the forearm) is thickened posteriorly at the wrist to form a transverse band known as the *extensor retinaculum* (Figs. 6-77 and 6-83). This fibrous band retains the extensor tendons in position, thereby increasing their efficiency. The deep fascia is also thickened anteriorly at the wrist to form the *flexor retinaculum* (Fig. 6-70), a fibrous band that converts the anterior concavity of the carpus into a *carpal tunnel* (Fig. 6-72) through which the flexor tendons pass (Figs. 6-70, 6-76, and 6-78). The *distal wrist crease* indicates the proximal border of the flexor retinaculum (Figs. 6-70 and 6-72) and the level of the wrist joint (Fig. 6-93).

The Hand

The hand (L. *manus*) forms the distal part of the upper limb; it contains the metacarpal bones and phalanges. Because of the importance of manual dexterity in occupational and recreational

activities, a good understanding of the structure and function of the hand is essential for all who are involved in maintaining or restoring its activities: free motion, power grip, precision handling, and pinching. The skeleton of the hand is illustrated in Figs. 6-1, 6-71, and 6-95.

We pray with our hands and often communicate with them. We use them to eat, work, and make love. We employ them as marvelously sophisticated instruments of flexibility and strength, and when they are damaged, we anguish.

When a person falls on the outstretched hand with the forearm pronated, the main force of the fall is transmitted through the carpus to the distal ends of the forearm bones, particularly the radius, and then proximally to the humerus, scapula, and clavicle. During such falls, fractures may occur in the wrist, forearm, or clavicle (Figs. 6-4 and 6-94).

Wrist Injuries.   The radius tends to break proximal to the wrist joint (*Colles' fracture*). In this injury the distal fragment of the radius is often comminuted (broken into pieces), and the fragments are usually displaced posteriorly and superiorly, producing shortening of the radius (Fig. 6-94). When there is a single fragment, it may be impacted (*i.e.*, the jagged ends of bone are driven into each other). Displacement of the distal part of the radius often breaks off the ulnar styloid process, owing to the violent pull of the articular disc connecting the radius and ulna. *Fracture of the scaphoid* bone is also a common injury, especially when the person falls on the palm with the hand abducted (p. 635). Commonly the bone fractures at its narrow part or "waist," producing two fragments. Because the scaphoid lies in the floor of the anatomical snuff box, the clinical sign of fracture of the scaphoid is *tenderness in the snuff box* or on the anterior aspect of the wrist over the tubercle of the scaphoid (Fig. 6-71*A*).

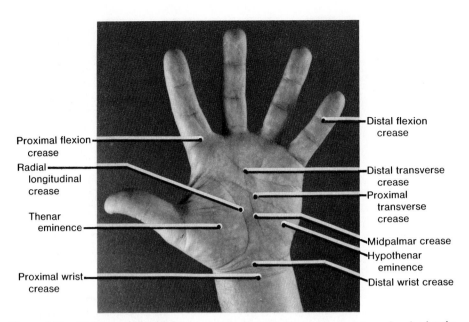

Figure 6-93. Wrist and palm of the left hand of a 53-year-old man showing its surface landmarks.

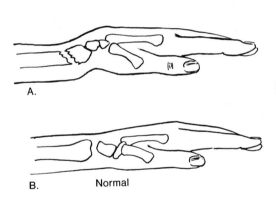

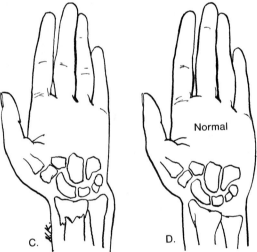

Figure 6-94. *A*, Colles' fracture of the wrist showing a bend that is known as the "dinner fork deformity." *B*, Normal position of the radius and carpal bones. *C*, Anterior view of the wrist before reduction of the fracture. Note that the styloid processes of the radius and ulna are at the same level. *D*, During reduction of the radial fracture, the shortening was corrected by placing the displaced fragment in its correct position. Observe that the radial styloid is now distal to the ulnar styloid, which is its normal position (also see Fig. 6-1).

> *Fractures of the hand* are common, and disability can result if normal relationships are not restored. Street fighters commonly fracture the distal end of their fifth metacarpal bone (Fig. 6-95). In this type of fracture, the head of the metacarpal is bent toward the palm (Figs. 6-71A and 6-95).

Surface Anatomy of the Hand

Dorsum of the Hand (Figs. 6-87, 6-92, and 6-110). The skin covering this region is thin and loose when the hand is relaxed. Hair is present on the dorsum of the hand and on the proximal parts of the digits, especially in males. If the dorsum of the hand is examined with the wrist extended and the digits abducted, the extensor tendons of the fingers usually stand out clearly, particularly in thin persons. These tendons are not visible far beyond the knuckles because they flatten here to form the extensor expansions (Fig. 6-91). The knuckles that become visible when a fist is made are produced by the heads of the metacarpal bones (Fig. 6-71A). Under the loose subcutaneous tissue and extensor tendons on the dorsum of the hand, the metacarpal bones can be palpated. A prominent feature of the dorsum of the hand is the *dorsal venous network* (Fig. 6-110).

Palm of the Hand (Fig. 6-93). The skin on the palm is thick because it is required to withstand the wear and tear of work and play. It is richly supplied with sweat glands, but it

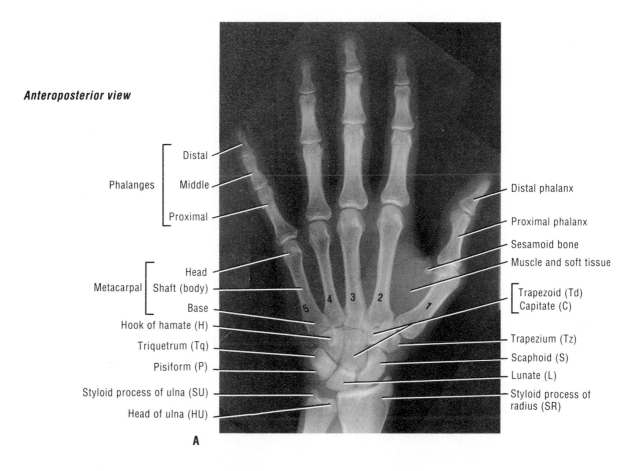

Anteroposterior view

Phalanges
- Distal
- Middle
- Proximal

Metacarpal
- Head
- Shaft (body)
- Base

Hook of hamate (H)
Triquetrum (Tq)
Pisiform (P)
Styloid process of ulna (SU)
Head of ulna (HU)

Distal phalanx
Proximal phalanx
Sesamoid bone
Muscle and soft tissue
Trapezoid (Td)
Capitate (C)
Trapezium (Tz)
Scaphoid (S)
Lunate (L)
Styloid process of radius (SR)

A

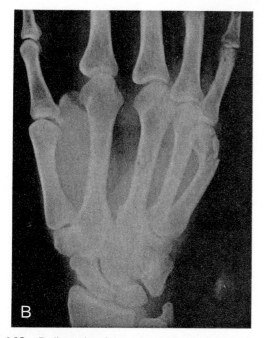

B

Figure 6-95. Radiographs of the wrist and hand. *A*, Normal appearance. *AP* (anteroposterior projection). (Courtesy of Dr. E. L. Lansdown, Professor of Radiology, University of Toronto, Toronto, Ontario, Canada.) *B*, Slightly oblique view showing an angulated fracture of the distal end of the body (shaft) of the fifth metacarpal bone (also see Fig. 6-71).

contains no hair or sebaceous glands. The skin presents several more or less constant longitudinal and transverse *flexion creases*, where the skin is firmly bound to the deep fascia (Fig. 6-96). Usually the four major palmar creases form an M-shaped pattern (Fig. 6-93). The transverse creases indicate where folding of the skin occurs during flexion of the hand. The longitudinal creases deepen when the thumb is opposed (Fig. 6-97), and the transverse creases deepen when the metacarpophalangeal joints are flexed.

The Palmar Flexion Creases (Fig. 6-93). Some creases are useful surface landmarks. The radial longitudinal crease partially encircles the *thenar eminence* (ball of thumb), formed by the short muscles of the first digit. The midpalmar crease indicates the *hypothenar eminencě* (ball of little finger), formed by the short muscles of the fifth digit.

The *proximal transverse palmar crease* commences on the lateral border of the palm, in common with the *radial longitudinal crease*, and superficial to the head of the second metacarpal bone. It extends medially and slightly proximally across the palm, superficial to the bodies of the third to fifth metacarpal bones.

The *distal transverse palmar crease* begins at or near the cleft between the index and middle fingers and crosses the palm with a slight convexity, superficial to the heads of the second to fourth metacarpal bones.

The Digital Flexion Creases (Fig. 6-93). Each of the medial four digits usually has three *transverse flexion creases*. The proximal flexion crease is located at the root of the digit,

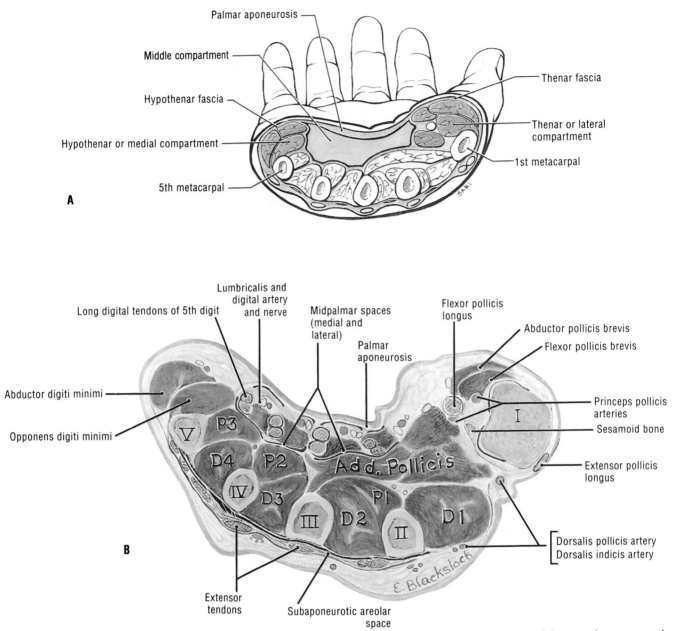

Palmar aponeurosis

Middle compartment

Hypothenar fascia

Hypothenar or medial compartment

5th metacarpal

Thenar fascia

Thenar or lateral compartment

1st metacarpal

A

Lumbricalis and digital artery and nerve

Long digital tendons of 5th digit

Midpalmar spaces (medial and lateral)

Palmar aponeurosis

Flexor pollicis longus

Abductor pollicis brevis

Flexor pollicis brevis

Abductor digiti minimi

Opponens digiti minimi

Princeps pollicis arteries

Sesamoid bone

Extensor pollicis longus

Add. Pollicis

Dorsalis pollicis artery
Dorsalis indicis artery

B

Extensor tendons

Subaponeurotic areolar space

E. Blackstock

Figure 6-96. Transverse sections through the middle of the palm of the right hand. *A,* Fascial compartments of the hand. *B,* Contents of the compartments. Observe (1) the thenar or lateral compartment containing vessels, nerves, and the thenar muscles; (2) a central compartment containing vessels, nerves, and the flexor tendons and their sheaths; (3) the hypothenar or medial compartment containing vessels, nerves, and hypothenar muscles; and (4) the adductor compartment containing the adductor pollicis muscle. Also observe that the four dorsal interosseous muscles (abductors) fill the spaces between the five metacarpal bones.

about 2 cm distal to the metacarpophalangeal joint (Figs. 6-93 and 6-131). There are two *middle flexion creases:* the proximal one lies over the proximal interphalangeal joint and the distal one lies proximal to the distal interphalangeal joint. The thumb, having two phalanges, has only two flexion creases. Like other digital creases, they deepen when the thumb is flexed. The proximal flexion crease crosses the thumb obliquely, proximal to the first metacarpophalangeal joint. The distal flexion crease on the thumb lies proximal to the interphalangeal joint. The skin ridges on the ventral ends of the digits, known as *fingerprints,* are used for identification (*e.g.*, in

criminal investigations) because of their unique patterns. Their anatomical function is to reduce slippage when grasping objects.

The science of studying the configurations of the dermal ridge patterns of the palm is known as dermatoglyphics. The most important landmarks are the patterns on the distal phalanges. *Dermatoglyphics* can be a valuable extension of the conventional physical examination of patients with certain congenital malformations and genetic diseases. For example,

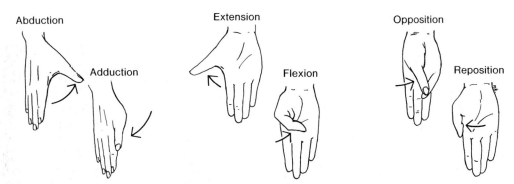

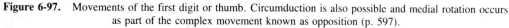

Figure 6-97. Movements of the first digit or thumb. Circumduction is also possible and medial rotation occurs as part of the complex movement known as opposition (p. 597).

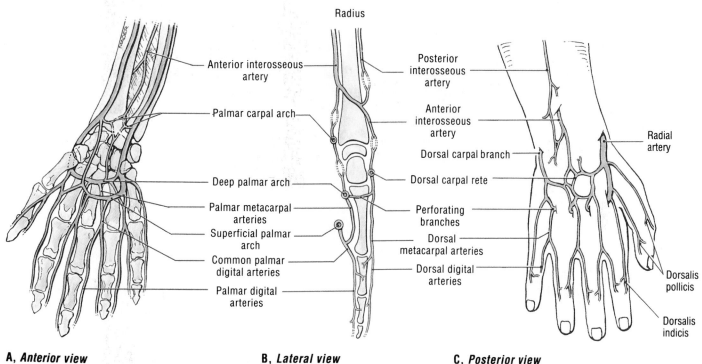

A, *Anterior view*   **B, *Lateral view***   **C, *Posterior view***

Figure 6-98. *A* to *C*, Arteries in the right forearm and hand. Observe the anastomoses in the hand. Note the radial artery emerging from the anatomical snuff box in *C* (also see Fig. 6-86).

persons with trisomy 21 or *Down syndrome*, often have only one transverse palmar crease, usually referred to as a *simian crease*. However it is important to know that about 1% of the general population has this crease with no other clinical features of the syndrome (Moore, 1988).

Fascia of the Palm

The deep fascia of the palm is continuous proximally with the antebrachial fascia (fascia of the forearm) and at the borders of the palm with the fascia on the dorsum of the hand (Figs. 6-66, 6-96, and 6-99). The fascia is thin over the thenar and hypothenar eminences (thenar and hypothenar fasciae), but it is thick in the palm where it forms the *palmar aponeurosis* (Fig. 6-66) and in the digits where it forms the fibrous digital sheaths (Fig. 6-100).

The Palmar Aponeurosis (Figs. 6-66, 6-96, 6-99, and 6-103). This strong, well-defined *triangular part of the deep fascia of the hand* covers the soft tissues and overlies the long flexor tendons of the palm. The proximal end of the palmar aponeurosis is continuous with the flexor retinaculum and the tendon of the palmaris longus muscle. The distal end of the aponeurosis divides at the roots of the digits into four longitudinal bands. Each band is attached to the base of the proximal phalanx and is fused with the fibrous digital sheath (Fig. 6-100*A*).

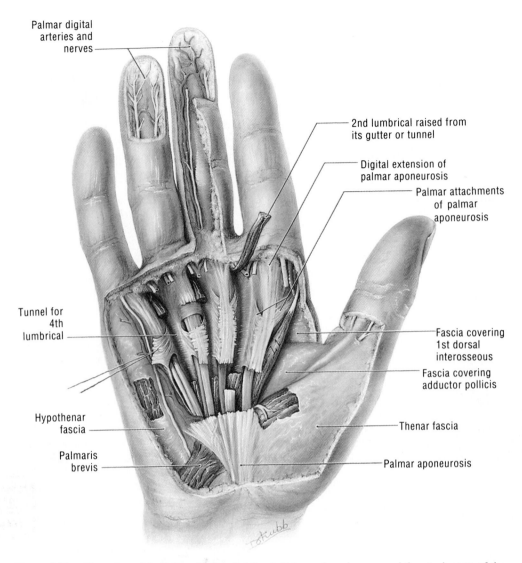

Palmar digital arteries and nerves

2nd lumbrical raised from its gutter or tunnel

Digital extension of palmar aponeurosis

Palmar attachments of palmar aponeurosis

Tunnel for 4th lumbrical

Fascia covering 1st dorsal interosseous

Fascia covering adductor pollicis

Hypothenar fascia

Thenar fascia

Palmaris brevis

Palmar aponeurosis

Figure 6-99.   Dissection of the right palm showing the digital vessels and nerves and the attachments of the palmar aponeurosis (also see Fig. 6-66).

Dupuytren's contracture is a progressive fibrosis (increase in fibrous tissue) of the palmar aponeurosis, resulting in shortening and thickening of the fibrous bands that extend from the aponeurosis to the bases of the phalanges. These fibrotic bands pull the digits into such marked flexion at the metacarpophalangeal joints that they cannot be straightened (Fig. 6-101).

The Fascial Compartments and Potential Spaces of the Palm (Figs. 6-96 and 6-102).   Between the palmar aponeurosis and the deep muscles of the palm are two potential spaces that are surgically important. They lie between the flexor tendons and the fascia covering the deep muscles in the floor of the palm. The spaces are bounded medially and laterally by fibrous septa passing from the edges of the palmar aponeurosis to the metacarpal bones. A fibrous *medial septum* extends deeply from the medial border of the palmar aponeurosis to the fifth metacarpal

bone. Medial to this septum is the medial or *hypothenar compartment*, containing the three *hypothenar muscles* and concerned with movements of the little finger (Table 6-13). Similarly, a fibrous *lateral septum* extends deeply from the lateral border of the palmar aponeurosis to the first metacarpal bone. Lateral to this septum is the lateral or *thenar compartment* containing the *thenar muscles* concerned with movements of the thumb (Table 6-12). Between the thenar and hypothenar compartments is the intermediate or *central compartment* containing the *flexor tendons and their sheaths*, the superficial palmar arch, and branches of the median and ulnar nerves. From the lateral border of the palmar aponeurosis, another fibrous septum passes obliquely and posteriorly to the third metacarpal bone. This creates potential medial and lateral **midpalmar spaces**. The *adductor compartment* is the deepest muscular plane of the palm. It contains the *adductor pollicis muscle*. Between the adductor pollicis and 1st dorsal interosseous muscle, there is a potential **retroadductor space**.

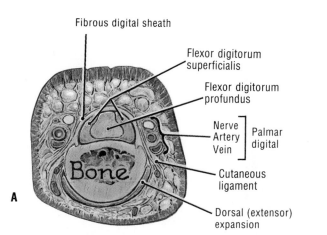
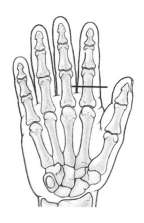

Figure 6-100. *A,* Transverse section through the proximal phalanx of the second digit. Note the osseofibrous tunnel containing the tendons of its flexor muscles (superficialis and profundus). Observe that the palmar digital nerves and vessels are applied to the fibrous digital sheath, not to the bone. Note that the skin is thickest on the palmar surface. *B,* Sketch showing the level of the section shown in *A.*

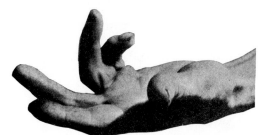

Figure 6-101. Left hand of a 56-year-old man with Dupuytren's contracture. Note the marked flexion of the fourth (ring) and fifth (little) digits at the metacarpophalangeal and proximal interphalangeal joints. This resulted from fibrosis and shortening of the digital slips of the palmar aponeurosis that pass to these digits (see Fig. 6-66).

The potential fascial spaces of the palm are clinically important because they may become infected (*e.g.*, following a puncture wound of the hand). The fascial spaces determine the extent and direction of the spread of pus formed by these infections. Depending on the site of infection, pus will accumulate in the thenar, hypothenar, or adductor compartments. Owing to the widespread use of antibiotics, infection rarely spreads from one of these fascial compartments, but an untreated infection can spread proximally from them through the carpal tunnel into the forearm, anterior to the pronator quadratus muscle and its fascia (Fig. 6-66).

Muscles of the Hand

The intrinsic muscles of the hand are on the palmar aspect and are innervated by branches of the ulnar or median nerves. They can be divided into three groups (Figs. 6-96 and 6-103): (1) the thumb or *thenar muscles* in the thenar compartment; (2) the little finger or *hypothenar muscles* in the hypothenar compartment; and (3) the *lumbrical muscles* in the central compartment and the interosseous muscles between the metacarpal bones.

The long flexor tendons of the *extrinsic muscles of the hand* arise in the forearm and pass to the digits. They are located in the central compartment of the palm with the lumbrical muscles.

The three short thenar muscles (abductor pollicis brevis, flexor pollicis brevis, and opponens pollicis) produce the *thenar eminence* and are chiefly responsible for the movement known as *opposition of the thumb* (Fig. 6-104). The pressure that the opposed thumb can exert on a fingertip is increased by the reinforcing action of the adductor pollicis and flexor pollicis longus muscles. The thenar muscles are located in the thenar compartment of the palm (Fig. 6-96) and are supplied by the recurrent branch of the median nerve (Figs. 6-75 and 6-103).

The Abductor Pollicis Brevis Muscle (Figs. 6-96, 6-97, 6-102, 6-103, 6-104*A,* and 6-108). This thin, short, relatively broad muscle forms the anterolateral part of the thenar eminence. Its attachments, nerve supply, and main action are shown in Table 6-12. It abducts the thumb at the carpometacarpal joint and assists the opponens pollicis muscle during the early stages of opposition of the thumb by rotating its proximal phalanx slightly medially.

The Flexor Pollicis Brevis Muscle (Figs. 6-97, 6-102, 6-103, 6-104*B,* and 6-108). This short, wide fusiform muscle is located medial to the abductor pollicis brevis. Its attachments, nerve supply, and main action are given in Table 6-12. It *flexes the thumb* at the carpometacarpal and metacarpophalangeal joints and aids in opposition of the thumb.

The Opponens Pollicis Muscle (Figs. 6-70, 6-97, 6-102, and 6-104*C*). This quadrangular muscle lies deep to the abductor pollicis brevis and lateral to the flexor pollicis brevis. Its attachments, nerve supply, and main actions are given in Table 6-12. It **opposes the thumb**, *i.e.*, it flexes and rotates it medially as during grasping. In Fig. 6-97, note that the tip of the thumb is brought into contact with the palmar surface of the fifth digit (little finger). The thumb can also be opposed to the other digits. *Opposition is the most important movement of the thumb.* Opposition involves extension initially, then abduction, flexion, medial rotation, and usually adduction. Several muscles are involved in this complex movement (see Table 6-21, p. 632).

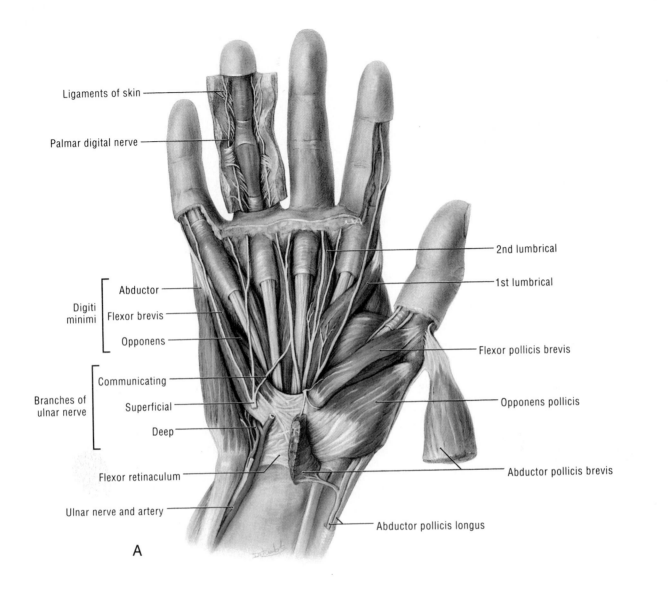

Ligaments of skin

Palmar digital nerve

Digiti minimi
- Abductor
- Flexor brevis
- Opponens

Branches of ulnar nerve
- Communicating
- Superficial
- Deep

Flexor retinaculum

Ulnar nerve and artery

2nd lumbrical

1st lumbrical

Flexor pollicis brevis

Opponens pollicis

Abductor pollicis brevis

Abductor pollicis longus

A

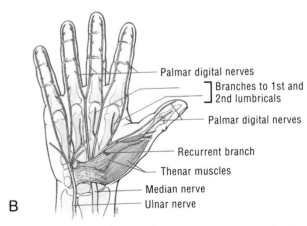

Palmar digital nerves

Branches to 1st and 2nd lumbricals

Palmar digital nerves

Recurrent branch

Thenar muscles

Median nerve

Ulnar nerve

B

Figure 6-102. *A*, Superficial dissection of the right palm showing the three thenar and three hypothenar muscles attached to the flexor retinaculum and the four marginal carpal bones united by it. Observe the four lumbrical muscles arising from the lateral sides of the four profundus tendons and inserting into the lateral sides of the extensor expansions of the corresponding digits (also see Fig. 6-89A). *B*, Nerve supply of the hand. Observe that the recurrent branch (motor branch) of the median nerve arises from the lateral side of the nerve at the distal border of the flexor retinaculum. It lies superficially and can easily be severed (see Fig. 6-75).

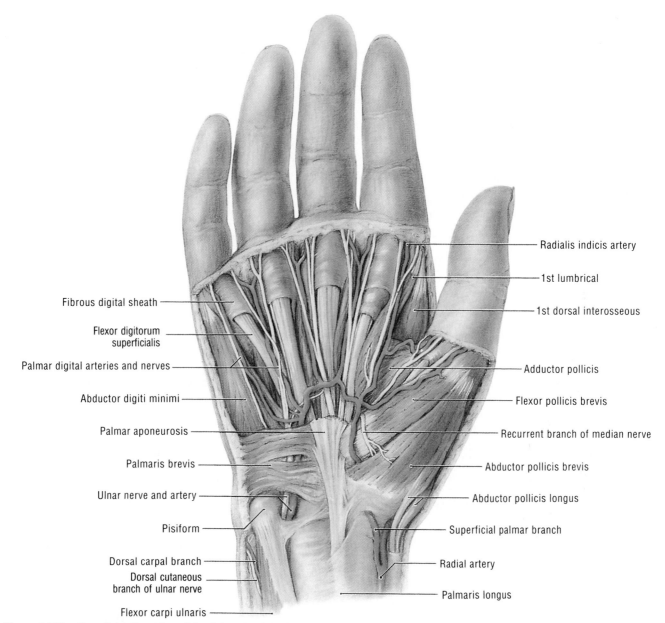

Radialis indicis artery

1st lumbrical

1st dorsal interosseous

Fibrous digital sheath

Flexor digitorum superficialis

Palmar digital arteries and nerves

Adductor pollicis

Flexor pollicis brevis

Abductor digiti minimi

Palmar aponeurosis

Recurrent branch of median nerve

Palmaris brevis

Abductor pollicis brevis

Ulnar nerve and artery

Abductor pollicis longus

Pisiform

Superficial palmar branch

Dorsal carpal branch

Radial artery

Dorsal cutaneous branch of ulnar nerve

Palmaris longus

Flexor carpi ulnaris

Figure 6-103. Superficial dissection of the right palm. The skin and superficial fascia have been removed, as have the palmar aponeurosis and the thenar and hypothenar fasciae. Observe the superficial palmar arterial arch (also see Fig. 6-84B). Note that the pisiform bone protects the ulnar nerve and artery as they pass into the palm.

Because of the complexity of the movement, **opposition of the thumb may be affected by most nerve injuries in the upper limb**. Obviously injuries to the nerves supplying the intrinsic muscles of the hand, especially the median nerve (Fig. 6-20), have the most severe effects on this movement. If the median is severed in the forearm or at the wrist, the thumb cannot be opposed. However, the intact abductor pollicis longus and adductor pollicis muscles, supplied by the posterior interosseous and ulnar nerves, respectively, may imitate opposition. The recurrent branch of the median nerve that supplies the thenar muscles lies superficially (Figs. 6-75 and 6-103) and may be severed by relatively minor lacerations of the palm involving the thenar eminence. If this nerve is severed, the thenar muscles are paralyzed and the thumb loses much of its usefulness.

The Adductor Pollicis Muscle (Figs. 6-83, 6-84, 6-86, 6-96, 6-97, 6-103, 6-104D, and 6-108). This fan-shaped muscle is located in the adductor compartment of the hand. It has two heads that are separated by a gap through which the radial artery passes. Its attachments, nerve supply, and main action are given in Table 6-12. It *adducts the thumb* and gives power to the grasp.

The three short hypothenar muscles are concerned with movements of the fifth digit. They lie in the hypothenar compartment of the palm with the fifth metacarpal bone (Fig. 6-96). *The hypothenar muscles produce the hypothenar eminence*, or

Table 6-12.
The Short Muscles of the Thumb or Thenar Muscles[1]

Muscle	Proximal Attachments	Distal Attachments	Innervation[2]	Main Actions
Abductor pollicis brevis (Fig. 6-104*A*)	Flexor retinaculum and tubercles of scaphoid and trapezium bones	Lateral side of base of proximal phalanx of thumb	Recurrent branch of median n. (**C8** and T1)	Abducts thumb and helps oppose it
Flexor pollicis brevis (Fig. 6-104*B*)				Flexes thumb (Fig. 6-97)
Opponens pollicis (Fig. 6-104*C*)	Flexor retinaculum and tubercle of trapezium bone	Lateral side of 1st metacarpal bone		Opposes thumb toward center of palm and rotates it medially
Adductor pollicis (Fig. 6-104*D*)	*Oblique head*: bases of 2nd and 3rd metacarpals, capitate, and adjacent carpal bones	Medial side of base of proximal phalanx of thumb	Deep branch of ulnar n. (C8 and **T1**)	Adducts thumb toward middle digit (Fig. 6-108)
	Transverse head: anterior surface of body of 3rd metacarpal bone			

[1]The thenar muscles (abductor pollicis brevis, flexor pollicis brevis, and opponens pollicis) form the thenar eminence or ball of the thumb (Fig. 6-93). The actions of the three muscles are indicated by their names to some extent, but they are all involved in opposition (Fig. 6-97), the pincer-like grip between the thumb and index finger that is an indispensable movement to most people, *e.g.*, when they pinch. The thumb can also be opposed to the other digits.
[2]**Boldface** indicates the main spinal cord segmental innervation.

ball of the fifth digit (Fig. 6-93). They are all supplied by the deep branch of the ulnar nerve (Table 6-13).

The Abductor Digiti Minimi Muscle (Figs. 6-102, 6-103, and 6-105*A*).   This short, wide fusiform muscle is the most superficial of the three muscles forming the hypothenar eminence. Its attachments, nerve supply, and main action are given in Table 6-13. It *abducts the fifth digit* and helps to flex its proximal phalanx.

The Flexor Digiti Minimi Brevis Muscle (Figs. 6-102 and 6-105*B*).   This short, wide, fusiform muscle is variable in size and lies lateral to the abductor digiti minimi. Its attachments, nerve supply, and main action are given in Table 6-13. It *flexes the proximal phalanx of the fifth digit* at the metacarpophalangeal joint.

The Opponens Digiti Minimi Muscle (Figs. 6-70, 6-102, and 6-105*C*).   This quadrangular muscle lies deep to the abductor and flexor muscles of the fifth digit. Its attachments, nerve supply, and main actions are given in Table 6-13. It *draws the fifth metacarpal bone anteriorly and rotates it laterally*, thereby deepening the hollow of the palm and bringing the fifth digit into opposition with the thumb.

The Palmaris Brevis Muscle (Figs. 6-66, 6-99, and 6-103).   This small, thin, quadrilateral muscle lies in the fascia, deep to the skin of the hypothenar eminence. It is a *relatively unimportant muscle*, except that it covers and protects the ulnar nerve and artery.

Proximal Attachment.   Flexor retinaculum and palmar aponeurosis.

Medial Attachment.   Skin on the medial side of the palm.

Innervation (Fig. 6-20).   Superficial branch of ulnar nerve (C8 and **T1**).

Actions.   It wrinkles the skin on the medial side of the palm and deepens the hollow of the palm, as in cupping the hand, thereby aiding the grip.

There are 11 short muscles in the hand (four lumbrical and seven interosseous muscles). The lumbricals act only on the medial four digits; the interossei act on all five digits.

The Lumbrical Muscles (Figs. 6-89*B*, 6-99, 6-102, 6-103, and 6-106*A*).   The four slender lumbricals (L. *lumbricus*, earthworm), one for each digit, were named because of their elongated, wormlike form. Their attachments, nerve supply, and actions are given in Table 6-14. *They flex the digits at the metacarpophalangeal joints and extend the interphalangeal joints*; *i.e.*, they place the digits in the writing position.

The Interosseous Muscles (Figs. 6-89*B*, 6-96, 6-99, 6-103, 6-106*B* and *C*, 6-107, and 6-108).   Seven interossei are located between the metacarpal bones. They are arranged in two layers: three palmar and four dorsal muscles. As their name indicates, they are located between bones (*i.e.*, the metacarpals). Their attachments, nerve supply, and main actions are given in Table 6-14. The dorsal interossei abduct the digits (**DAB** is the key, *i.e.*, **D**orsal **AB**duct), and the palmar interossei adduct the digits (**PAD** is the key, *i.e.*, **P**almar **AD**uct). They also assist the lumbrical muscles with flexion of the metacarpophalangeal joints and extension of the interphalangeal joints. These are important movements in typing, writing, and playing the piano.

The Long Flexor Tendons of the Extrinsic Muscles of the Hand (Figs. 6-66, 6-78, 6-99, 6-102, 6-103, and 6-107).   The tendons of the flexor digitorum superficialis and flexor digitorum profundus *enter a common synovial sheath* deep to the flexor retinaculum. They then pass deep to the palmar aponeurosis and *enter osseofibrous digital tunnels*. There are two tendons in each tunnel. To enable these tendons to slide freely over each other during movements of the digits, each of them is covered with

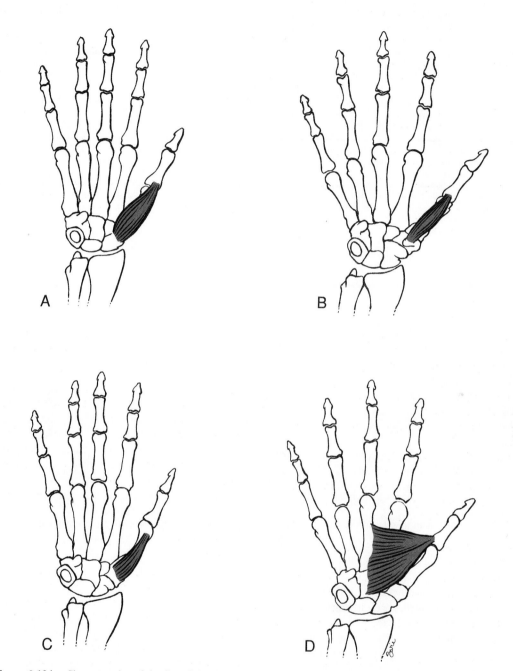

Figure 6-104. Short muscles of the first digit (thumb). *A*, Abductor pollicis brevis. *B*, Flexor pollicis brevis. *C*, Opponens pollicis. *D*, Adductor pollicis.

Table 6-13.
The Short Muscles of the Fifth Digit or Hypothenar Muscles[1]

Muscle	Proximal Attachments	Distal Attachments	Innervation[2]	Main Actions
Abductor digiti minimi (Fig. 6-105*A*)	Pisiform bone	Medial side of base of proximal phalanx of digit 5 (little finger)	Deep branch of ulnar n. (C8 and **T1**)	Abducts digit 5 (little finger)
Flexor digiti minimi brevis (Fig. 105*B*)	Hook of hamate bone and flexor retinaculum			Flexes proximal phalanx of digit 5
Opponens digiti minimi (Fig. 6-105*C*)		Medial border of 5th metacarpal bone		Draws 5th metacarpal bone anteriorly and rotates it, bringing digit 5 into opposition with thumb

[1]The hypothenar muscles form the hypothenar eminence or ball of the fifth digit (Fig. 6-93).
[2]**Boldface** indicates the main spinal cord segmental innervation.

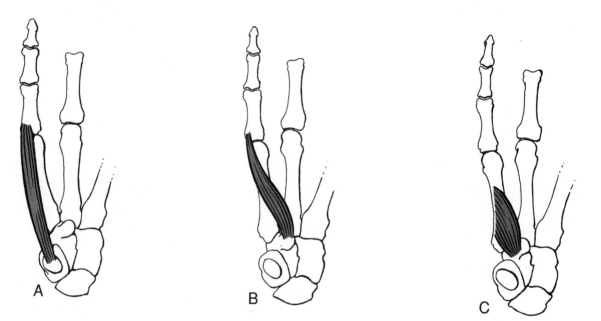

Figure 6-105.   Short muscles of the fifth digit (little finger). *A*, Abductor digiti minimi. *B*, Flexor digiti minimi brevis. *C*, Opponens digiti minimi.

synovial membrane (Figs. 6-78 and 6-79). Near the base of the proximal phalanx, the tendon of the flexor digitorum superficialis splits and surrounds the tendon of the flexor digitorum profundus (Figs. 6-89 and 6-107). The halves of the tendon of the flexor digitorum superficialis are attached to the margins of the middle phalanx (Fig. 6-67). The tendon of the flexor digitorum profundus, after passing through the split in the tendon of the flexor digitorum superficialis, passes distally to attach to the base of the distal phalanx (Figs. 6-67, 6-89, and 6-107).

The long flexor tendons are supplied with blood by small blood vessels, which pass from the periosteum of the phalanges within special folds of connective tissue called *vincula tendinum* (Fig. 6-79). There are two kinds of vincula (L. fetter or chain): vincula brevia and vincula longa (Fig. 6-89*B*). These folds connect the tendons of the flexor digitorum superficialis and flexor digitorum profundus muscles to the fibrous digital sheaths in the digits (Figs. 6-89*B* and 6-100). *The tendon of the flexor pollicis*

longus passes to the thumb, deep to the flexor retinaculum, within its own synovial sheath (Fig. 6-78), and enters the osseofibrous digital tunnel in the thumb. At the head of the metacarpal, the tendon runs between two *sesamoid bones*, one in the combined tendon of the flexor pollicis brevis and abductor pollicis brevis and the other in the tendon of the adductor pollicis (Fig. 6-103).

The synovial sheaths of the flexor tendons of the hand may become infected (*e.g.*, by entry of a foreign object into a digit, such as a rusty nail). When *tenosynovitis* (inflammation of the tendon and digital synovial sheath) occurs, the digit swells and movement of it becomes painful. Because the tendons of the second, third, and fourth digits nearly always have separate digital synovial sheaths (Fig. 6-78), the infection is usually confined to the digit concerned. In neglected infections, however, the proximal ends of these sheaths may rupture and infection may spread to the midpalmar fascial spaces (Fig. 6-96).

Because the synovial sheaths of the thumb and fifth digit are often continuous with the common flexor synovial sheath (Fig. 6-78), tenosynovitis in these digits may spread to the common synovial sheath. As there are variations in the connections between the common synovial sheath and the digital sheaths, the degree of spreading of infections from the digits depends on whether or not there are connections between them. Infections of the second, third, and fourth digits are likely to remain localized because their digital synovial sheaths are connected with the common flexor synovial sheath in only about 10% of cases.

Nerves of the Hand

The median, ulnar, and radial nerves supply the hand (Figs. 6-20, 6-21, 6-102, 6-107, and 6-109).

The Median Nerve (Figs. 6-20, 6-21, 6-72, 6-75, and 6-109). This nerve enters the hand through the *carpal tunnel*, deep to the flexor retinaculum, between the tendons of the flexor digitorum superficialis and the tendon of the flexor carpi radialis. The median nerve *supplies motor fibers to the three thenar muscles and the first and second lumbrical muscles*. It also sends cutaneous sensory fibers to the lateral palmar surface, the sides of the first three digits, the lateral half of the fourth digit, and the dorsum of the distal halves of these digits.

Carpal Tunnel Syndrome (Figs. 6-70, 6-72, 6-75, and 6-76). Any lesion that significantly reduces the size of the carpal tunnel (*e.g.*, inflammation of the flexor retinaculum, anterior dislocation of the lunate bone, arthritic changes, or tenosynovitis of the tendon sheaths) may cause **compression of the median nerve**. Because this nerve has two terminal branches (lateral and medial) that supply skin of the hand (Fig. 6-21), there is often tingling (*paresthesia*), absence of

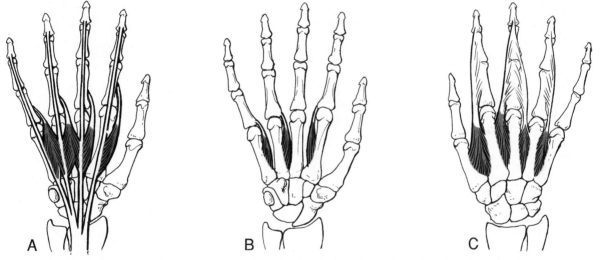

Figure 6-106. Short muscles of the hand. *A*, Lumbricals. *B*, Palmar interossei. *C*, Dorsal interossei.

Table 6-14.
The Short Muscles of the Hand

Muscles	Proximal Attachments	Distal Attachments	Innervation[1]	Main Actions
Lumbricals 1 and 2 (Fig. 6-106*A*)	Lateral two tendons of flexor digitorum profundus	Lateral sides of extensor expansions of digits 2 to 5	*Lumbricals 1 and 2,* median n. (C8 and **T1**)	Flex digits at metacarpophalangeal joints and extend interphalangeal joints
Lumbricals 3 and 4 (Fig. 6-106*A*)	Medial three tendons of flexor digitorum profundus		*Lumbricals 3 and 4,* deep branch of ulnar n. (C8 and **T1**)	
Dorsal interossei 1 to 4 (Fig. 6-106*C*)	Adjacent sides of two metacarpal bones	Extensor expansions and bases of proximal phalanges of digits 2 to 4	Deep branch of ulnar n. (C8 and **T1**)	Abduct digits and assist lumbricals
Palmar interossei 1 to 3 (Fig. 6-106*B*)	Palmar surfaces of 2nd, 4th, and 5th metacarpal bones	Extensor expansions of digits and bases of proximal phalanges of digits 2, 4, and 5		Adduct digits and assist lumbricals

[1]**Boldface** indicates the main spinal cord segmental innervation.

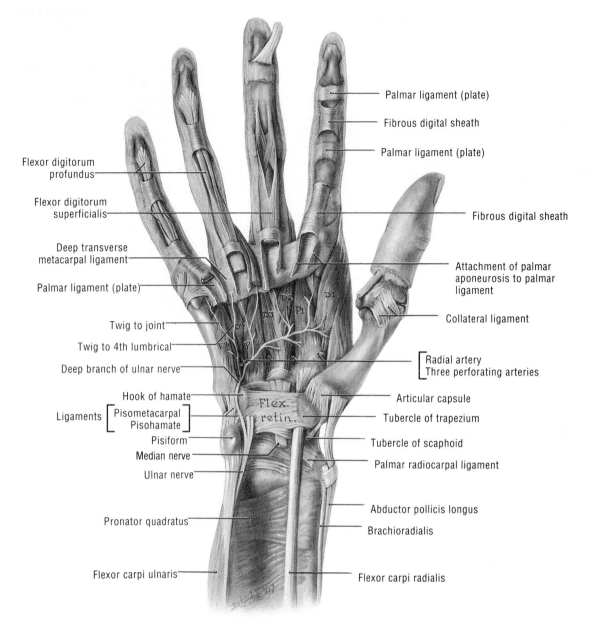

Flexor digitorum profundus

Flexor digitorum superficialis

Deep transverse metacarpal ligament

Palmar ligament (plate)

Twig to joint

Twig to 4th lumbrical

Deep branch of ulnar nerve

Hook of hamate

Ligaments [Pisometacarpal / Pisohamate]

Pisiform

Median nerve

Ulnar nerve

Pronator quadratus

Flexor carpi ulnaris

Palmar ligament (plate)

Fibrous digital sheath

Palmar ligament (plate)

Fibrous digital sheath

Attachment of palmar aponeurosis to palmar ligament

Collateral ligament

Radial artery
Three perforating arteries

Articular capsule

Tubercle of trapezium

Tubercle of scaphoid

Palmar radiocarpal ligament

Abductor pollicis longus

Brachioradialis

Flexor carpi radialis

D1 D2 D3 P1 P2 P3 P4

Flex. retin.

Figure 6-107. Deep dissection of the anterior surface of the right forearm, palm, digits, and ulnar nerve. The first palmar interosseous muscle, together with the small muscles of the thumb, has been removed.

Figure 6-108. Abduction and adduction of the digits. Observe that the axial line (median plane) of the hand passes through the third (middle) digit. Note that abduction of the digits is movement away from the axial line and that adduction of them is movement toward it. Note also that the third digit can be abducted to either side of the axial line. Adduction restores it to the axial line. Observe that in abduction of the thumb it points anteriorly, *i.e.*, away from the palm at a right angle, and that adduction closes the first digit or thumb on the second digit (index finger).

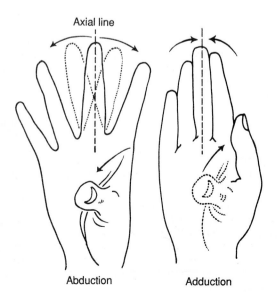

Axial line

Abduction

Adduction

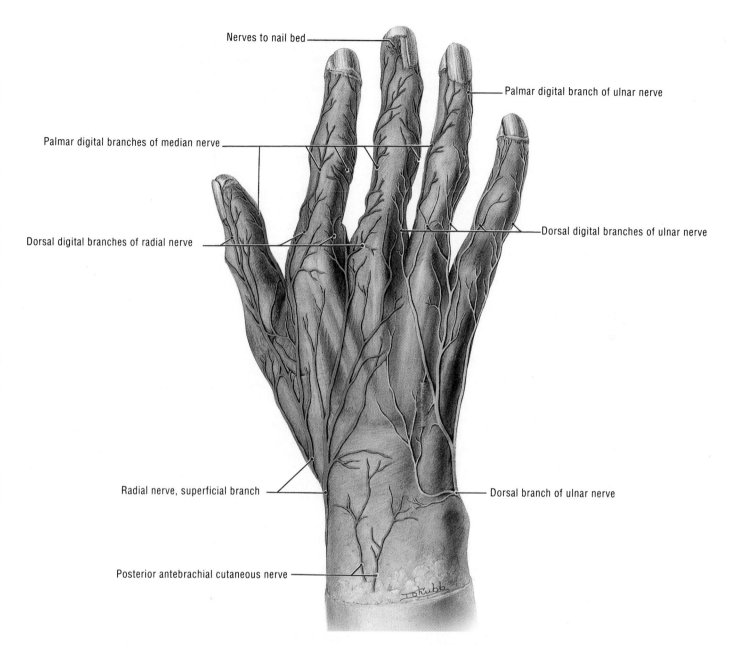

Nerves to nail bed

Palmar digital branch of ulnar nerve

Palmar digital branches of median nerve

Dorsal digital branches of ulnar nerve

Dorsal digital branches of radial nerve

Radial nerve, superficial branch

Dorsal branch of ulnar nerve

Posterior antebrachial cutaneous nerve

Figure 6-109.   Cutaneous nerves on the dorsum of the hand. Observe that branches of the radial nerve and the dorsal branch of the ulnar nerve are distributed nearly equally and symmetrically on the hand and digits. The radial nerve supplies the lateral half of the hand and the lateral 2½ digits; the ulnar nerve has a similar distribution on the medial side of the hand. Observe that palmar digital branches of the median and ulnar nerves supply the distal halves of the three middle digits, including the nail beds. Note that there are numerous communications between adjacent nerves (see also Fig. 6-21).

tactile sensation (*anesthesia*), or diminished sensation (*hypoesthesia*) in the digits. Because the median nerve sends a palmar cutaneous branch superficial to the flexor retinaculum to supply most of the palm, there is often no sensory impairment of this area. Most cases (about 60%) are seen in patients who are between 40 and 60 years of age and the syndrome is more common in women.

Often there is a progressive loss of coordination and strength in the thumb, owing to *weakness of the abductor pollicis brevis and opponens pollicis muscles* (Table 6-12), if the

cause of the median nerve compression is not alleviated. This results in difficulty in performing fine movements of the thumb. As the thenar muscles and the lateral two lumbrical muscles of the other digits are also supplied by the median nerve (Fig. 6-20), the usefulness of the first to third digits may be diminished. In severe cases of compression of the median nerve, there may be wasting or atrophy of the thenar muscles. To relieve symptoms of the carpal tunnel syndrome, partial or complete *division of the flexor retinaculum* may be necessary. This operation is called a *carpal tunnel release*.

Median nerve injury frequently occurs just proximal to the flexor retinaculum, the common site of wrist slashing in *suicide attempts* (Fig. 6-75). The median nerve can be palpated here just before it passes deep to the flexor retinaculum (Fig. 6-66) and enters the carpal tunnel (Figs. 6-75 and 6-76). *When the median nerve is severed at the wrist*, the patient is unable to oppose his/her thumbs owing to paralysis of the thenar muscles (Tables 6-12 and 6-21). The first and second lumbrical muscles will also be paralyzed, resulting in weakening of flexion of the metacarpophalangeal joints and extension of the interphalangeal joints of the second and third digits (Tables 6-19 and 6-20). The fine control of movements of these digits will also be lost. There will also be a loss of cutaneous sensation in the lateral portion of the palm, the palmar surface of the thumb, and the lateral two and one-half digits, including the nail beds of these digits (Fig. 6-21). The areas of anesthesia may not be sharply defined, owing to variations in the distribution of the nerves in the hand.

The Ulnar Nerve (Figs. 6-20, 6-21, 6-66, 6-70, 6-75, 6-102, 6-103, 6-107, and 6-109).   This nerve leaves the forearm by emerging from deep to the tendon of the flexor carpi ulnaris muscle. It passes distally on the flexor retinaculum alongside the lateral border of the pisiform bone; the ulnar artery is on its lateral side. The ulnar nerve and artery are bridged over by a slender band of connective tissue, which forms a small tunnel or canal (of Guyon). Just proximal to the wrist, the ulnar nerve gives off a *palmar cutaneous branch*, which passes superficial to the flexor retinaculum and to the palmar aponeurosis and supplies the skin of the medial side of the palm. It also gives off a *dorsal cutaneous branch*, which supplies the medial half of the dorsum of the hand, the fifth digit, and the medial half of the fourth digit. At the distal border of the flexor retinaculum, the ulnar nerve ends by dividing into a superficial and a deep branch.

The *superficial branch of the ulnar nerve* supplies cutaneous fibers to the anterior surfaces of the medial one and one-half digits (Figs. 6-21 and 6-102). The *deep branch of the ulnar nerve* supplies motor fibers to the hypothenar muscles, the medial two lumbrical muscles, the adductor pollicis muscle, and all the interosseous muscles (Fig. 6-20). The deep branch of the ulnar nerve also supplies several joints (wrist, intercarpal, carpometacarpal, and intermetacarpal).

The ulnar nerve is referred to as the nerve of fine movements because it innervates muscles that are concerned with fine movements of the hand (Tables 6-13 and 6-14).

Ulnar nerve injury is common because this nerve lies superficially in the distal part of the forearm. It may be injured by lacerations at these sites, producing sensory alteration in the medial part of the hand, the fifth digit, and the medial half of the fourth digit (Figs. 6-21 and 6-109). There is also impaired power of adduction and abduction of the fingers owing to paralysis of the interossei (Table 6-14). Some adduction of the digits may be possible on flexion of the digits by the long flexor muscles. Adduction of the thumb is lost owing to paralysis of the adductor pollicis (Fig. 6-20; Table 6-12), but other movements of the thumb are normal. After an ulnar nerve injury, the fourth and fifth digits are hyperextended at the metacarpophalangeal joints and somewhat flexed at the interphalangeal joints. This occurs because the medial two lumbrical muscles are paralyzed. This deformity, called *clawhand* (Fig. 6-81), does not become obvious until a considerable time after the injury to the nerve.

Compression of the ulnar nerve may occur as it passes through a connective tissue canal (of Guyon), that is located superficial to the flexor retinaculum. *Ulnar nerve entrapment* in this canal usually causes pure motor neuropathy, which affects the hand muscles that are supplied by the ulnar nerve (Tables 6-12 and 6-13).

The Radial Nerve (Figs. 6-21, 6-22, 6-70, 6-82, and 6-109).   This nerve *supplies no hand muscles*. Its terminal branches, superficial and deep, arise in the cubital fossa. The *deep branch of the radial* is muscular and articular in its distribution (*e.g.*, the elbow joint). The *superficial branch of the radial* is the direct continuation of the radial nerve along the anterolateral side of the forearm and is entirely sensory. It pierces the deep fascia near the dorsum of the wrist and supplies skin and fascia over the lateral two-thirds of the dorsum of the hand, the dorsum of the thumb, and proximal parts of the lateral one and one-half digits.

Although the radial nerve supplies no muscles in the hand, **radial nerve injury** in the arm or forearm produces serious disability of the hand. The characteristic handicap is the inability to extend the wrist owing to *paralysis of the extensor muscles of the forearm* (Tables 6-10 and 6-11). The hand is flexed at the wrist and lies flaccid, a condition known as **wrist-drop**. The digits are also flexed at the metacarpophalangeal joints. The interphalangeal joints can be extended weakly through the action of the intact lumbrical and interosseous muscles, which are supplied by the median and ulnar nerves. The radial nerve has only a small area of exclusive cutaneous supply on the hand. The extent of anesthesia is minimal, even in serious radial nerve injuries, and is usually confined to a small area on the lateral part of the dorsum of the hand.

Arteries of the Hand

The radial and ulnar arteries and their branches provide all the blood to the hand.

The Radial Artery (Figs. 6-55, 6-62, 6-77, 6-83, 6-86, 6-98, 6-99, and 6-103).   This is the smaller of the two terminal branches of the brachial artery; it continues the direct line of this vessel. Just before passing from the anterior to the posterior surface of the wrist, the radial artery gives off a *superficial palmar branch*. This artery passes through the thenar muscles and runs superficial to the long flexor tendons, where it joins the *superficial palmar arterial arch*, the continuation of the ulnar artery (Figs. 6-98 and 6-103). As the radial artery curves dorsally over the wrist, it passes deep to the tendons of the abductor pollicis longus and extensor pollicis brevis muscles. It then *crosses the floor of the anatomical snuff box* and enters the palm of the hand by passing between the heads of the first dorsal interosseous muscle.

The radial artery gives off the *princeps pollicis artery* and the *radialis indicis artery* (Figs. 6-77, 6-83, and 6-98), the digital arteries to the first digit (thumb) and the lateral side of the second digit (index finger), respectively. The radial artery then passes between the two heads of the adductor pollicis muscle and joins the deep branch of the ulnar artery to form the *deep palmar arterial arch* (Fig. 6-98). This arch is located between the long flexor tendons and the metacarpal bones. Three *palmar metacarpal arteries* arise from the deep palmar arch and run distally, where they join the *common palmar digital arteries*, which arise from the *superficial palmar arterial arch*.

The Ulnar Artery (Figs. 6-65, 6-70, 6-84, 6-98, and 6-103). This is the larger of the two terminal branches of the brachial artery. It enters the palm on the lateral side of the ulnar nerve superficial to the flexor retinaculum. The ulnar artery *passes lateral to the pisiform* bone and then gives off a *deep palmar branch* before continuing across the palm as the *superficial palmar arterial arch*. The deep palmar branch passes through the hypothenar muscles and anastomoses with the radial artery, thereby completing the deep palmar arterial arch.

The Superficial Palmar Arterial Arch (Figs. 6-84*B*, 6-98, and 6-103). This arch is located distal to the deep palmar arterial arch. *Formed mainly by the ulnar artery*, it is convex toward the digits and the middle of its convexity lies deep to the center of the proximal transverse crease of the palm. This arch gives rise to *three common palmar digital arteries* that anastomose with the palmar metacarpal arteries from the deep palmar arch. Each common palmar digital artery divides into a pair of *proper palmar digital arteries*, which run along the sides of the second to fourth digits.

The Deep Palmar Arterial Arch (Figs. 6-84 and 6-98). This arch lies across the metacarpal bones, just distal to their bases. *Formed mainly by the radial artery*, it is about a fingerbreadth closer to the wrist than the superficial palmar arterial arch. The deep arch gives rise to three *palmar metacarpal arteries* that run distally and join the common palmar digital arteries from the superficial palmar arterial arch.

> The ulnar artery descends superficial to the flexor muscles in about 3% of limbs. This must be kept in mind when performing intravenous injections in the cubital area. An accidental intraarterial injection could have serious consequences.
>
> The palmar arterial arches shown in Fig. 6-98 are typical, but several variations occur. For example, the superficial palmar arterial arch may be formed by the ulnar artery alone. Because of the large number of arteries in the hand and digits, bleeding is usually profuse when these structures are lacerated. Often both ends of the bleeding artery must be tied to stop the hemorrhage because there are four transversely placed arterial arches that communicate with each other. In lacerations of the palmar arterial arches, it may be useless to ligate only one of the forearm arteries because they usually have numerous communications in the forearm and hand. Even simultaneous clamping of the ulnar and radial arteries proximal to the wrist sometimes fails to stop all bleeding. To obtain a bloodless surgical operating field in the hand for treating complicated injuries, it may be necessary to compress the *brachial artery* and its branches proximal to the elbow (*e.g.*, using a pneumatic tourniquet). This prevents blood from reaching the arteries of the forearm and hand through the anastomosis around the elbow (Fig. 6-55*B*).

Veins of the Hand

The superficial and deep palmar arterial arches are accompanied by venae comitantes, known as the superficial and deep *venous arches*, respectively. The *dorsal digital veins* (Figs. 6-82, 6-83, and 6-91) drain into three *dorsal metacarpal veins*, which unite to form a **dorsal venous network** (Fig. 6-110). Located superficial to the metacarpus, this network is prolonged proximally as the *cephalic vein*, which winds superiorly around the lateral border of the forearm to its anterior surface (Figs. 6-86 and 6-110).

Joints of the Upper Limb

The *pectoral girdle* (clavicle and scapula) connects the upper limb to the trunk (Fig. 6-111); therefore, its articulations are included with those of the upper limb.

The Sternoclavicular Joint

This is the *saddle type of synovial joint* and is the only bony articulation between the upper limb and the axial skeleton (Fig. 6-111). An important function of the clavicle is to hold the upper limb away from the trunk (p. 501), *i.e.*, it acts as a strut for keeping the shoulder away from the chest to give the upper limb the maximum freedom of motion. The sternoclavicular joint can be readily palpated because the medial end of the clavicle lies superior to the manubrium of the sternum.

The Articular Surfaces of the Sternoclavicular Joint (Figs. 6-18, 6-111, and 6-112; see Fig. 1-6). The enlarged medial end of the clavicle articulates in a shallow socket formed by the superolateral part of the manubrium of the sternum and the medial part of the first costal cartilage. Unlike most articular surfaces, the articular cartilage is mainly fibrocartilaginous. The articular surfaces are separated by a strong, thick, densely fibrous or fibrocartilaginous **articular disc**. It is located inside the joint and divides it into two synovial cavities. The disc is attached superiorly to the medial end of the clavicle and inferiorly to the junction of the sternum and the first costal cartilage. The articular disc is continuous with the anterior and posterior *sternoclavicular ligaments*, which are thickenings of the fibrous capsule. This disc prevents medial displacement of the clavicle; it is also an important shock absorber of forces transmitted along the clavicle.

The Articular Capsule of the Sternoclavicular Joint (Fig. 6-112). The *fibrous capsule* surrounds the entire joint, including the epiphysis at the medial end of the clavicle. Although thin inferiorly, other parts of the fibrous capsule are strong because they are reinforced by the anterior and posterior *sternoclavicular ligaments* and superiorly by the *interclavicular ligament*. The latter ligament extends across the jugular notch of the sternum. The sternoclavicular and interclavicular ligaments are thickenings of the fibrous capsule of this joint.

The costoclavicular ligament is a strong, extracapsular ligament. It ascends from the first rib and its costal cartilage to the

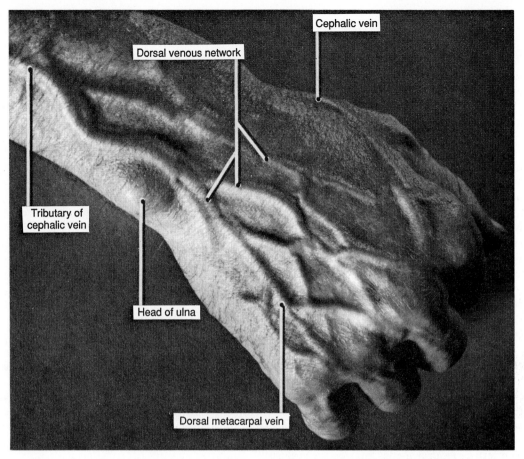

Figure 6-110. Dorsum of the distal end of the forearm, wrist, and hand of a 46-year-old man showing the superficial veins.

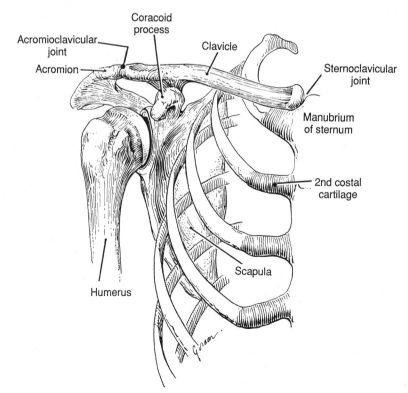

Figure 6-111. The bones of the right pectoral and shoulder regions. Note that the clavicle is the only bone uniting the upper limb to the axial skeleton, represented here by the sternum.

inferior margin of the medial end of the clavicle. This ligament reinforces the sternoclavicular joint laterally and limits elevation of the medial end of the clavicle. Any movement of the clavicle, except inferior movement, tightens its fibers and is resisted.

The *synovial capsule* lines the fibrous capsule and both surfaces of the articular disc. Because this disc divides the joint into two cavities, there are two synovial membranes. The lateral one reflects from the articular margin of the medial end of the clavicle to the margins of the articular disc. The medial one lines the capsule between its sternal attachments and the disc.

Movements of the Sternoclavicular Joint. Despite the saddlelike form of its articular surfaces, this joint *moves in many directions like a ball and socket joint*. Verify this by elevating and depressing your shoulder and protracting and retracting it. Elevate your upper limb as far as possible and verify by palpation that your clavicle is raised to about a 60-degree angle from its anatomical position.

Stability of the Sternoclavicular Joint. Because the bony surfaces involved are incongruent and the surrounding muscles offer little support, the sternoclavicular joint depends on its ligaments and articular disc for stability. The strong disc is largely responsible for preventing medial displacement of the clavicle (*e.g.*, when carrying a heavy suitcase). The disc prevents the medial end of the clavicle from being pushed out of its socket and superior to the manubrium. The sternoclavicular ligaments also help to prevent displacement of the medial end of the clavicle. Excessive protraction and elevation of the clavicle are also restrained by these ligaments.

Blood Supply of the Sternoclavicular Joint (Figs. 6-31 and 6-32). The articular arteries are branches of the internal thoracic and suprascapular arteries.

Nerve Supply of the Sternoclavicular Joint (Figs. 6-5 and 6-17). The articular nerves are branches of the medial supraclavicular nerve and the nerve to the subclavius muscle.

> The rarity of dislocation of the sternoclavicular joint attests to its strength. When a blow is received to the acromion of the scapula or when a force is transmitted to the pectoral girdle during a fall on the outstretched hand, the force of the blow is usually transmitted along the long axis of the clavicle. The clavicle may break near the junction of its middle and lateral thirds (Fig. 6-4 and p. 506), but it is very uncommon for the sternoclavicular joint to dislocate.

The Acromioclavicular Joint

This is the *plane type of synovial joint* that is located between the lateral end of the clavicle and the acromion of the scapula (Figs. 6-111, 6-113, 6-114, and 6-117). It is located 2 to 3 cm medial to the acromion, which projects anteriorly from the lateral end of the spine of the scapula. The acromion forms a palpable and sometimes visible prominence (Fig. 6-2), known as the "point of the shoulder."

The Articular Surfaces of the Acromioclavicular Joint (Figs. 6-1, 6-111, 6-113, 6-114, and 6-117). The small oval articular facet on the lateral end of the clavicle articulates with a similar facet on the anterior part of the medial surface of the medial end of the acromion. Both articular surfaces are covered

with fibrocartilage and slope inferomedially so that the clavicle tends to override the acromion and project over it. A wedge-shaped, incomplete, fibrocartilaginous **articular disc** projects into the joint from the superior part of the articular capsule. It is located in the superior part of the joint and partially divides the joint cavity into two parts.

The Articular Capsule of the Acromioclavicular Joint (Figs. 6-113 and 6-114). The *fibrous capsule* enclosing the joint is attached to the margins of its articular surfaces. Although weak, it is strengthened superiorly by the *acromioclavicular ligament* and by fibers from the trapezius muscle (Fig. 6-17). This ligament extends from the superior part of the lateral end of the clavicle to the superior surface of the acromion. A synovial capsule lines the fibrous capsule.

The Coracoclavicular Ligament (Figs. 6-111 and 6-114). This ligament anchors the lateral part of the clavicle to the coracoid process of the scapula. It is the strongest of the ligaments that bind the clavicle to the scapula. It consists of two parts, the *conoid and trapezoid ligaments*, which are directed in such a way that they enable the clavicle to hold the scapula and upper limb laterally.

Movements of the Acromioclavicular Joint. This articulation allows the acromion to rotate on the clavicle and to move anteriorly and posteriorly. These movements are associated with

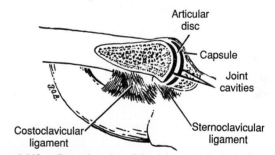

Figure 6-112. Coronal section of the right sternoclavicular joint. Note that the fibrocartilaginous articular disc completely divides the joint.

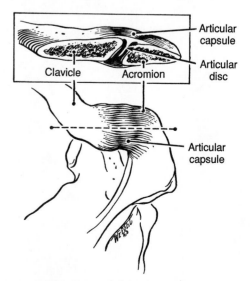

Figure 6-113. Superior view of the right acromioclavicular joint. Inset is a coronal section of the joint showing its articular disc.

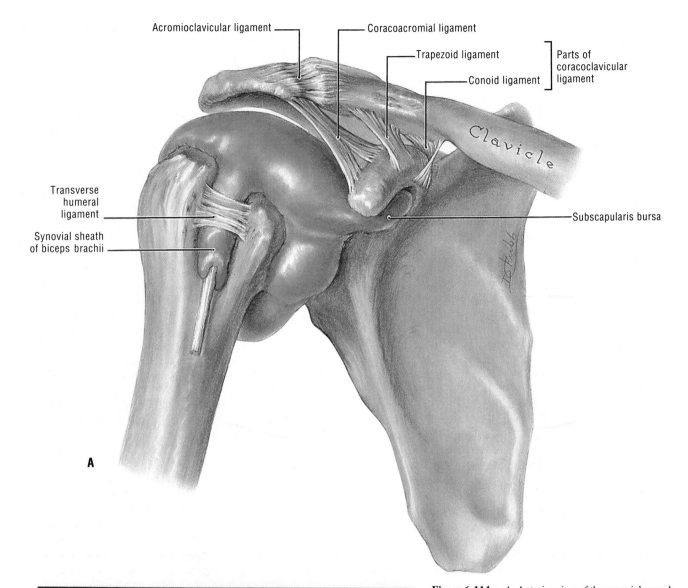

Acromioclavicular ligament —⌐ ⌐— Coracoacromial ligament

⌐— Trapezoid ligament

⌐— Conoid ligament ⌐ Parts of coracoclavicular ligament

Clavicle

Transverse humeral ligament —

Synovial sheath of biceps brachii —

— Subscapularis bursa

A

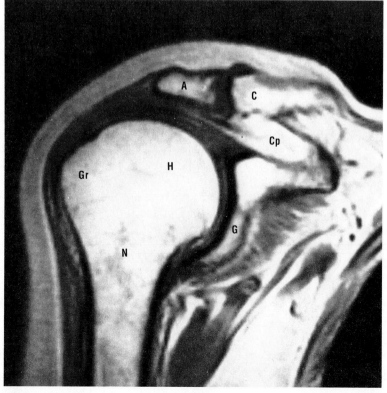

A

C

Cp

Gr

H

N

G

B

Figure 6-114. *A*, Anterior view of the synovial capsule of the right shoulder joint and the ligaments at the lateral end of the clavicle. Note that the synovial capsule has two prolongations: (1) where it forms a synovial sheath for the tendon of the long head of the biceps brachii muscle and (2) inferior to the coracoid process where it forms the subscapular bursa between the subscapularis tendon and the margin of the glenoid cavity. *B*, Coronal magnetic resonance image (MRI) of the right shoulder joint. *A*, indicates acromion; *C*, clavicle; *Cp*, coracoid process; *Gr*, greater tubercle; *H*, head of humerus; *G*, glenoid cavity; *N*, surgical neck of humerus. (Courtesy of Dr. W. Kucharczyk, Clinical Director of Tri-Hospital Resonance Centre, Toronto, Ontario, Canada.)

movement of the scapula and with those at the sternoclavicular joint.

Stability of the Acromioclavicular Joint (Fig. 6-114). The coracoclavicular ligament, an extrinsic ligament of the acromioclavicular joint, is principally responsible for providing stability to the articulation. It prevents the clavicle from losing contact with the acromion of the scapula.

Blood Supply of the Acromioclavicular Joint (Figs. 6-31, 6-32, 6-34, and 6-47). The articular arteries are branches of the suprascapular and thoracoacromial arteries.

Nerve Supply of the Acromioclavicular Joint (Figs. 6-5, 6-15, and 6-22). The articular nerves are branches of the supraclavicular, lateral pectoral, and axillary nerves.

> In contact sports such as football, soccer, and hockey, it is not uncommon for *dislocation of the acromioclavicular joint* to result from a hard fall on the shoulder or to occur when a hockey player is driven viciously into the boards. This injury, often inaccurately called a "shoulder separation," is serious when both the acromioclavicular and coracoclavicular ligaments are torn. When the coracoclavicular ligament ruptures, the shoulder falls away from the clavicle owing to the weight of the upper limb. The fibrous capsule of the acromioclavicular joint also ruptures, resulting in the acromion passing inferior to the lateral end of the clavicle. Dislocation of this joint makes the acromion more obvious.

The Shoulder Joint

This is a multiaxial *ball and socket type of synovial joint* that permits a wide range of movement. However, mobility is gained at the expense of stability.

The Articular Surfaces of the Shoulder Joint (Figs. 6-1, 6-111, and 6-114 to 6-117). The spheroidal head of the humerus (the ball) articulates with the shallow glenoid cavity of the scapula (the socket). Both articular surfaces are covered with hyaline cartilage. The shallow glenoid cavity accepts little more than a third of the large humeral head, but the glenoid cavity is deepened slightly and enlarged by a fibrocartilaginous rim called the *glenoid labrum* (L. lip). The superior portion of the labrum blends with the tendon of the long head of the biceps brachii muscle (Fig. 6-117A).

The Articular Capsule of the Shoulder Joint (Figs. 6-114 to 6-116). The *fibrous capsule* enclosing the shoulder joint is thin and loose; thus it allows a wide range of movement. The capsule is attached medially to the glenoid cavity, beyond the glenoid labrum. Superiorly it encroaches on the root of the coracoid process so that the fibrous capsule encloses the attachment of the long head of the biceps muscle within the joint. Laterally the fibrous capsule is attached to the anatomical neck of the humerus. *The inferior part of the capsule is its weakest area.* The capsule is lax and lies in folds when the arm is adducted, but it becomes taut when the arm is abducted. There are two *apertures in the articular capsule* of the shoulder joint. The opening between the tubercles of the humerus is for passage of the tendon of the long head of the biceps brachii muscle. The other opening is situated anteriorly, inferior to the coracoid process. It allows communication between the subscapular bursa and the synovial cavity of the joint.

The *synovial membrane* lines the fibrous capsule and is reflected from it onto the glenoid labrum and the neck of the humerus, as far as the articular margin of the head. The synovial capsule forms a tubular sheath for the tendon of the long head of the biceps brachii muscle (Fig. 6-114), where it passes into the joint cavity and lies in the *intertubercular groove*, extending as far as the surgical neck of the humerus.

Intrinsic Ligaments of the Capsule of the Shoulder Joint (Figs. 6-114 to 6-116). These ligaments are thickenings of the fibrous capsule, which strengthen the shoulder joint.

The glenohumeral ligaments (Fig. 6-116) are thickenings of the anterior part of the fibrous capsule. The superior, middle, and inferior glenohumeral ligaments run from the supraglenoid tubercle of the scapula to the lesser tubercle and the anatomical neck of the humerus. These ligaments are frequently indistinct or absent.

The transverse humeral ligament (Fig. 6-114) is a broad band of transverse fibers passing from the greater to the lesser tubercles of the humerus. It forms a bridge over the superior end of the intertubercular groove, converting it into a canal that holds the synovial sheath and tendon of the long head of the biceps as they emerge from the capsule of the shoulder joint.

The coracohumeral ligament (Fig. 6-117A) is a strong, broad band that strengthens the superior part of the capsule of the shoulder joint. It passes from the lateral side of the base of the coracoid process of the scapula to the anatomical neck of the humerus, adjacent to the greater tubercle.

The coracoacromial arch (Fig. 6-117A) is formed by the coracoid process, coracoacromial ligament, and acromion. When force is transmitted superiorly along the humerus (*e.g.*, when standing at a desk and partly supporting the body with the outstretched limbs), the head of the humerus is pressed against this protective arch. *The coracoacromial arch prevents displacement of the humeral head* superiorly from the glenoid cavity of the scapula. The supraspinatus muscle passes under this arch and lies between the deltoid muscle and the capsule of the shoulder joint. The *supraspinatus tendon*, passing to the greater tubercle of the humerus, is separated from the arch by the *subacromial bursa* (Figs. 6-47 and 6-115A).

The coracoacromial ligament (Figs. 6-47 and 6-117A) is a strong triangular ligament, the base of which is attached to the lateral border of the coracoid process. Its apex is inserted into the edge of the acromion. Superiorly the coracoacromial ligament is covered by the deltoid muscle.

Movements of the Shoulder Joint (Table 6-15). This joint has more freedom of movement than any other joint in the body. This freedom results from the laxity of the joint's articular capsule and the large size of the humeral head compared with the small size of the glenoid cavity. The shoulder articulation is a *multiaxial ball-and-socket joint* that allows movements around three axes and permits flexion-extension, abduction-adduction, circumduction, and rotation. In circumduction, the distal end of the humerus describes the base of a cone, the apex of which is at the head of the humerus.

Stability of the Shoulder Joint. The free movement of this joint leads to instability. The shallowness of the glenoid cavity and the laxity of the fibrous capsule also result in a considerable loss of stability. The strength of the joint results mainly from the muscles that surround it (Figs. 6-44, 6-45, and 6-115), particularly the *rotator cuff muscles* (supraspinatus, infraspina-

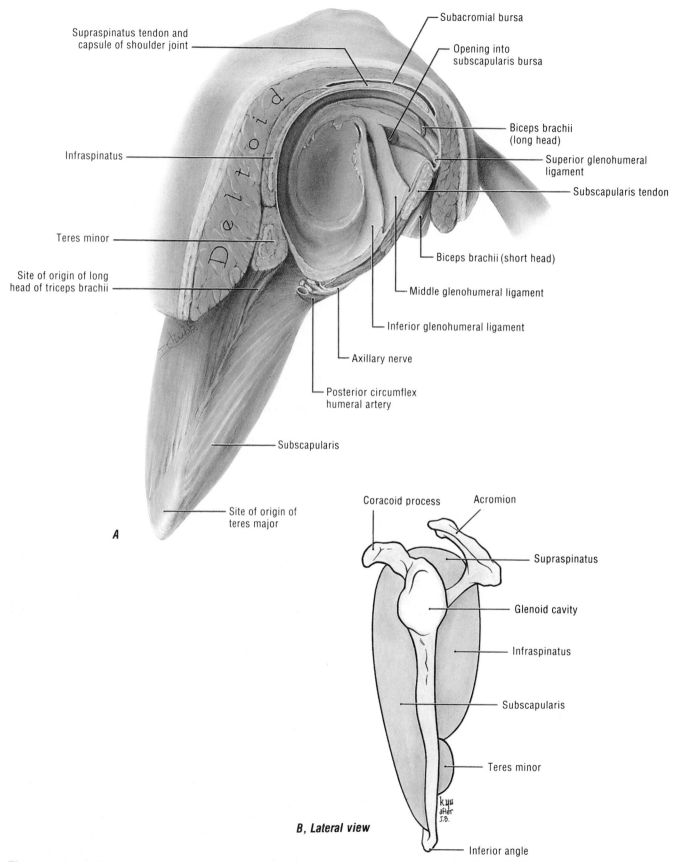

Subacromial bursa

Opening into subscapularis bursa

Supraspinatus tendon and capsule of shoulder joint

Biceps brachii (long head)

Infraspinatus

Superior glenohumeral ligament

Subscapularis tendon

Teres minor

Biceps brachii (short head)

Site of origin of long head of triceps brachii

Middle glenohumeral ligament

Inferior glenohumeral ligament

Axillary nerve

Posterior circumflex humeral artery

Subscapularis

Site of origin of teres major

A

Coracoid process   Acromion

Supraspinatus

Glenoid cavity

Infraspinatus

Subscapularis

Teres minor

Inferior angle

B, Lateral view

Figure 6-115. *A*, Dissection of the right glenoid cavity of the scapula, as viewed from the anterolateral aspect. Note the four short rotator cuff muscles (teres minor, infraspinatus, supraspinatus, and subscapularis) crossing the joint and blending with the capsule. *B*, Diagram of the rotator cuff. The prime function of these muscles is to hold the head of the humerus in the glenoid cavity of the scapula.

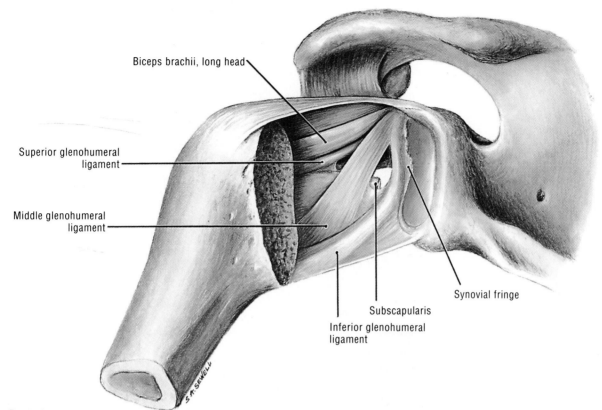

Biceps brachii, long head

Superior glenohumeral
ligament

Middle glenohumeral
ligament

Synovial fringe

Subscapularis

Inferior glenohumeral
ligament

S.A. SEWELL

Posterior view

Figure 6-116. Dissection of the interior of the right shoulder joint, exposed posteriorly by cutting away the posterior part of the articular capsule and sawing off the head of the humerus. Observe the three thickenings of the anterior part of the fibrous capsule, called the superior, middle, and inferior glenohumeral ligaments.

tus, teres minor, and subscapularis). These four scapular muscles, joining the scapula to the humerus (Table 6-5), are attached near the articular areas of the articulation and are closely related to the fibrous capsule of the joint.

Although they have separate functions, *the rotator cuff muscles work as a group in holding the head of the humerus in the glenoid cavity* (Table 6-5). They give stability to the shoulder joint in several positions, especially when the arm is abducted. The supraspinatus muscle and the coracoacromial arch guard the shoulder joint superiorly; the infraspinatus and teres minor muscles stabilize it posteriorly (Fig. 6-45); and the subscapularis muscle protects it anteriorly (Figs. 6-23 and 6-44*F*). In Figure 6-115, observe that no tendons support the shoulder joint inferiorly; consequently, this is where it usually dislocates.

Bursae Around the Shoulder Joint (Figs. 6-47, 6-114, and 6-115). There are several bursae containing capillary films of synovial fluid in the vicinity of this joint. Bursae are located where tendons rub against bone, ligaments, or other tendons and where skin moves over a bony prominence. The bursae around the shoulder joint are of special clinical importance. Some of them communicate with the joint cavity (*e.g.*, the subscapular bursa); hence, to open a bursa may mean entering this cavity.

The Subscapular Bursa (Figs. 6-114 and 6-115). This bursa is located between the tendon of the subscapularis muscle and the neck of the scapula. The bursa protects this tendon where it passes inferior to the root of the coracoid process and over the

neck of the scapula. It usually communicates with the cavity of the shoulder joint through an opening in its fibrous capsule; thus it is really an extension of the cavity of the shoulder joint.

The Subacromial Bursa (Figs. 6-47 and 6-115). This large bursa lies between the deltoid muscle, the supraspinatus tendon, and the fibrous capsule of the shoulder joint. Its size varies, but it does not normally communicate with the cavity of the shoulder joint. The subacromial bursa is located inferior to the acromion and the coracoacromial ligament, between them and the supraspinatus muscle. This bursa facilitates movement of the deltoid muscle over the fibrous capsule of the shoulder joint and the supraspinatus tendon.

Blood Supply of the Shoulder Joint (Figs. 6-31, 6-32, 6-45, 6-47, and 6-115). The articular arteries to the shoulder joint are branches of the anterior and posterior *circumflex humeral arteries* from the axillary and the suprascapular artery from the subclavian.

Nerve Supply of the Shoulder Joint (Figs. 6-23, 6-45, 6-59, and 6-115). The articular nerves are branches of the suprascapular, axillary, and lateral pectoral nerves.

Calcific Supraspinatus Tendonitis (Tendinitis). In these cases there is inflammation and calcification of the subacromial bursa. This results in pain, tenderness, and limitation of movement of the shoulder joint. The condition is also called *calcific scapulohumeral bursitis.* Deposition of calcium in the

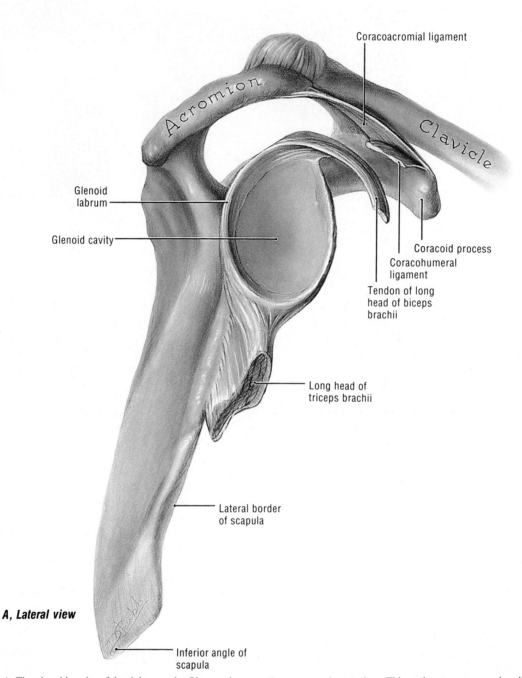

A, Lateral view

Figure 6-117. *A*, The glenoid cavity of the right scapula. Observe that it is deepened by the glenoid labrum, a dense fibrocartilaginous lip (L. *labrum*) that is attached to the rim of the glenoid cavity. Observe the coracoacromial arch formed by the coracoid process, coracoacromial ligament, and acromion. This arch prevents superior displacement of the head of the humerus. *B*, Radiograph of the right shoulder. (Courtesy of Dr. E. L. Lansdown, Professor of Radiology, University of Toronto, Toronto, Ontario, Canada.)

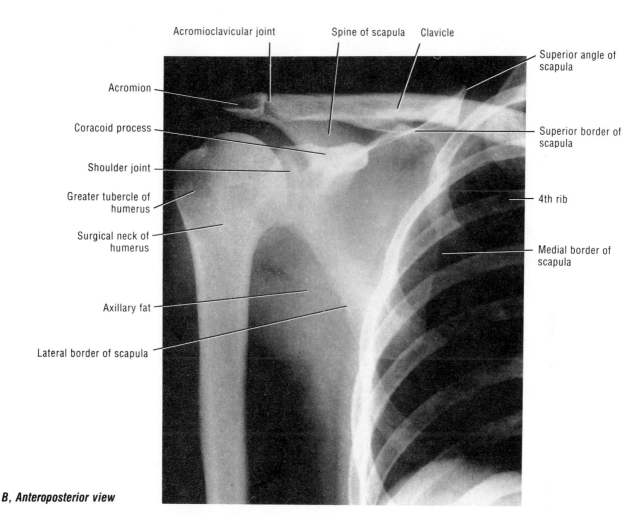

Acromioclavicular joint   Spine of scapula   Clavicle

Acromion

Coracoid process

Shoulder joint

Greater tubercle of humerus

Surgical neck of humerus

Axillary fat

Lateral border of scapula

Superior angle of scapula

Superior border of scapula

4th rib

Medial border of scapula

B, Anteroposterior view

Figure 6-117*B*

supraspinatus portion of the musculotendinous rotator cuff of the shoulder joint is common. This condition causes increased local pressure, which often causes pain during abduction of the arm. The calcium deposit may irritate the overlying subacromial bursa, producing an inflammatory reaction known as *subacromial bursitis*. So long as the shoulder joint is adducted, there is usually no pain because in this position the painful lesion is away from the acromion. In most patients, pain occurs during 50 to 130 degrees of abduction because, during this arc, the supraspinatus tendon is in intimate contact with the inferior surface of the acromion.

Rupture of the Rotator Cuff of the Shoulder Joint.   When an older person strains to lift something (*e.g.*, a window that is stuck), a previously degenerated musculotendinous rotator cuff may rupture. Often the capsule of the shoulder joint also tears. As a result the joint cavity communicates with the subacromial bursa. Tears of the rotator cuff also occur in athletes owing to strain on the shoulder joint (*e.g.*, base ball pitchers). This causes pain in the shoulder region when the arm is moved (*e.g.*, to throw a ball).

Dislocation of the Shoulder Joint.   Because of its freedom of movement and instability, this joint is dislocated more

often than any other joint in adults. The dislocation may result from direct or indirect injury. Anterior dislocation of the shoulder joint occurs most often in young adults, particularly athletes. It is usually caused by excessive extension and lateral rotation of the humerus. The head of the humerus is driven anteriorly, and usually the fibrous capsule and glenoid labrum are stripped from the anterior aspect of the glenoid cavity. A hard blow to the humerus when the shoulder joint is fully abducted (*e.g.*, when a quarterback is about to release a football), tilts the head of the humerus inferiorly onto the inferior weak part of the articular capsule. This may tear the capsule and dislocate the shoulder so that the humeral head comes to lie inferior to the glenoid cavity. The strong flexor and abductor muscles of the shoulder joint (Table 6-15) usually pull the humeral head anterosuperiorly into a subcoracoid position. Unable to use the arm, the patient commonly supports it with the other hand.

The axillary nerve may be injured when the shoulder is dislocated because of its close relation to the inferior part of the articular capsule of this joint (Fig. 6-115). The subglenoid displacement of the head of the humerus into the quadrangular space damages the axillary nerve (Figs. 6-45 and 6-59). This nerve injury is indicated by *paralysis of the deltoid muscle* and loss of skin sensation in the shoulder region (Fig. 6-30 and p. 525).

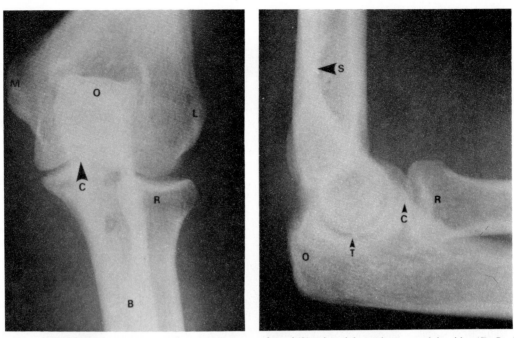

Figure 6-118.   Radiographs of the elbow region: an AP projection on the left with the elbow extended and a lateral projection on the right with the elbow flexed. On the humerus, observe the medial (*M*) and lateral (*L*) epicondyles and supracondylar ridge (*S*). On the ulna, observe the olecranon (*O*), coronoid process (*C*), and trochlear notch (*T*). Observe the head (*R*) and tuberosity (*B*) of the radius.

The Elbow Joint

This is a *hinge type of synovial joint* that is formed where the distal end of the humerus articulates with the proximal ends of the radius and ulna (Fig. 6-118). The elbow is a *uniaxial joint*; its movements consist of flexion and extension.

The Articular Surfaces of the Elbow Joint (Figs. 6-1, 6-49, and 6-118 to 6-122).   The trochlea and capitulum of the humerus articulate with the trochlear notch of the ulna and the head of the radius, respectively. The articular surfaces, covered with hyaline cartilage, are most fully in contact when the forearm is in a position midway between pronation and supination and is flexed to a right angle.

The elbow joint includes three articulations:

1. *The humeroulnar articulation* is between the trochlea of the humerus and the trochlear notch of the ulna. They form a *uniaxial hinge joint*, permitting movement in one axis: flexion and extension.
2. *The humeroradial articulation* is between the capitulum of the humerus and the head of the radius. The capitulum fits into the slightly cupped surface of the head.
3. *The proximal radioulnar joint* is between the head of the radius and the radial notch of the ulna. This is a pivot joint, permitting rotation of the radius about the ulna.

The Articular Capsule of the Elbow Joint (Figs. 6-119 to 6-122 and 6-124*B*).   The fibrous capsule completely encloses the joint. Its anterior and posterior parts are thin and weak, but its sides are strengthened by collateral ligaments. The fibrous capsule is attached to the proximal margins of the coronoid and radial fossae anteriorly, but not quite to the superior limit of the olecranon fossa posteriorly. Distally the fibrous capsule is at-tached to the margins of the trochlear notch, the anterior border of the coronoid process, and the anular ligament.

The Collateral Ligaments (Figs. 6-121 and 6-122).   These strong triangular bands are medial and lateral thickenings of the fibrous capsule; hence, they are intrinsic ligaments.

The radial collateral ligament is a strong triangular band. Its apex is attached proximally to the lateral epicondyle of the humerus and its base blends with the anular ligament of the radius.

The ulnar collateral ligament is also triangular in shape. It is composed of anterior and posterior bands (parts), which are connected by a thinner, relatively weak oblique band. Its apex is attached to the medial epicondyle of the humerus. The strong, cordlike anterior part is attached to the tubercle on the coronoid process of the ulna and the weaker, fanlike posterior part is attached to the medial edge of the olecranon. The *ulnar nerve* passes posterior to the medial epicondyle and is closely applied to the ulnar collateral ligament (Figs. 6-63 and 6-120). It enters the forearm between the heads of the flexor carpi ulnaris muscle.

The synovial membrane of the elbow joint (Figs. 6-119 to 6-121) lines the fibrous capsule and is reflected onto the humerus, lining the coronoid and radial fossae anteriorly and the olecranon fossa posteriorly. The synovial capsule is continued into the proximal radioulnar joint. A redundant fold of the synovial capsule, called the *sacciform recess*, emerges distal to the anular ligament and facilitates rotation of the head of the radius, *e.g.*, during pronation and supination of the forearm (p. 8).

Movements of the Elbow Joint (Tables 6-15 and 6-16). This joint can be flexed or extended. Flexion is produced by the brachialis and brachioradialis muscles, but the *main flexor muscle is the brachialis*. When the forearm is supinated, the biceps brachii muscle also flexes this joint; when it is pronated, the pronator teres does. Flexion is limited by apposition of the an-

Table 6-15.
The Muscles Producing Movements of the Shoulder Joint[1]

Movements	Muscles	Reference Tables	Chief Innervation[2]
Flexion	**Pectoralis major**, clavicular head	6-1	Lateral and medial pectoral nn. (C5 and **C6**)
	Deltoid, anterior fibers	6-6	Axillary n. (C5 and **C6**)
	Coracobrachialis	6-7	Musculocutaneous n. (C5, **C6** and C7)
	Biceps brachii (short head)	6-7	Musculocutaneous n. (C5 and **C6**)
Extension	**Pectoralis major**, sternocostal head	6-1	Lateral and medial pectoral nn. (**C7, C8** and T1)
	Deltoid, posterior fibers	6-6	Axillary n. (**C5** and C6)
	Latissimus dorsi	6-4	Thoracodorsal n. (**C6, C7** and C8)
	Teres major	6-6	Lower subscapular n. (**C6** and C7)
	Triceps brachii, long head	6-7	Radial n. (C5, **C6**, and C7)
Abduction	**Deltoid**	} 6-6	Axillary n. (**C5** and C6)
	Supraspinatus		Suprascapular n. (C4, **C5**, and C6)
Adduction	**Pectoralis major**, both heads	6-1	Lateral and medial pectoral nn. (**C7, C8**, and T1)
	Latissimus dorsi	6-4	Thoracodorsal n. (**C6, C7**, and C8)
	Teres major	6-6	Lower subscapular n. (**C6** and C7)
	Coracobrachialis	} 6-7	Musculocutaneous n. (C5, **C6**, and C7)
	Triceps brachii, long head		Radial n. (C5, **C6**, and C7)
Medial Rotation	**Pectoralis major**, both heads	6-1	Lateral and medial pectoral nn. (**C7, C8**, and T1)
	Latissimus dorsi	6-4	Thoracodorsal n. (**C6, C7**, and C8)
	Deltoid, anterior fibers	6-6	Axillary n. (**C5** and C6)
	Subscapularis	} 6-6	Upper and lower subscapular nn. (C5, **C6**, and C7)
	Teres major		Lower subscapular n. (**C6** and C7)
Lateral Rotation	**Deltoid**, posterior fibers	} 6-6 {	Axillary (**C5** and C6)
	Teres minor **Infraspinatus**		Suprascapular (**C5** and C6)

[1]The principal muscles producing these movements are printed in **boldface**.
[2]**Boldface** indicates the main spinal cord segmental innervation.

terior surfaces of the forearm and arm, by tension of the posterior arm muscles, and by the radial and ulnar collateral ligaments (Figs. 6-121 and 6-122). The flexion of the forearm, or twitch of the biceps brachii that occurs following tapping of the bicipital aponeurosis without movement, is known as the *biceps jerk*. The reflex center is in C5 and C6 segments of the spinal cord.

The main extensor of the elbow joint is the triceps brachii muscle (Table 6-16). Gravity and the anconeus muscle assist with this movement. Extension is limited by impingement of the olecranon of the ulna on the olecranon fossa of the humerus (Fig. 6-118) and by tension of the anterior arm muscles and collateral ligaments (Figs. 6-120 to 6-122). The anconeus muscle stabilizes the elbow joint and may assist in its extension. The extension of the forearm, or twitch of the triceps without movement that occurs following tapping of the triceps tendon, is known as the *triceps jerk*. The reflex center is in C6, **C7**, and **C8** segments of the spinal cord.

When the forearm is fully extended and supinated in the anatomical position, the arm and forearm are not in the same line. The explanation for this is that the articular surfaces of the distal end of the humerus are not set at a right angle to the body (shaft). Normally the forearm is directed laterally, forming a *carrying angle* of about 165 degrees. This angle permits the extended forearm to clear the side of the hip in swinging movements during walking, which is important when carrying heavy loads. The angle is diminished when the forearm is pronated or flexed.

The escape of synovial fluid from the joint cavity (*effusion*) generally occurs posteriorly, as do dislocations of the elbow joint (Fig. 6-123). The elbow joint is also most easily approached surgically from its posterior aspect (Fig. 6-120).

Knowledge of the normal *carrying angle of the forearm* is essential for aligning the bones of the arm and forearm during reduction of a fracture or dislocation of the elbow joint. An increase in the normal carrying angle at the elbow is known as *cubitus valgus*. This is one of the clinical manifestations of patients with *Turner syndrome* (45, XO chromosome constitution). Other characteristics of this syndrome are webbing of the neck, broad chest, and short stature (Moore, 1988).

Stability of the Elbow Joint. In adults this joint is quite stable because of the hingelike arrangement formed by the jaw-like trochlear notch of the ulna into which the spool-shaped trochlea of the humerus fits. In addition, the joint is strengthened by very strong ulnar and radial collateral ligaments (Figs. 6-121 and 6-122).

The elbow joint of children is not so stable owing to the late fusion of the epiphyses of the ends of the bones involved in the articulation (humerus, radius, and ulna). For example, the proximal part of the olecranon fuses with the body of the ulna between 16 and 19 years of age, and the head of the radius fuses with its body at 15 to 17 years of age. As a consequence, separation of the epiphyses can occur during a fall on the elbow because the epiphyseal cartilaginous plate is weaker than the surrounding bone.

Bursae Around the Elbow Joint (Fig. 6-124). Only some of the many bursae around the elbow are clinically important. There are two olecranon bursae. The *subcutaneous olecranon bursa* is located in the subcutaneous connective tissue over the olecranon, whereas the *subtendinous olecranon bursa* is located between the tendon of the triceps muscle and the olecranon, just proximal to its insertion into the olecranon. The *radioulnar bursa* lies between the extensor digitorum, the radiohumeral joint, and

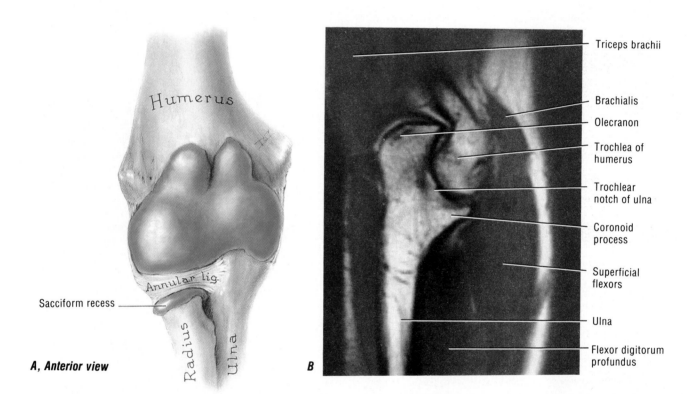

A, Anterior view

Humerus

Annular lig.

Sacciform recess

Radius

Ulna

B

Triceps brachii

Brachialis

Olecranon

Trochlea of
humerus

Trochlear
notch of ulna

Coronoid
process

Superficial
flexors

Ulna

Flexor digitorum
profundus

Figure 6-119. *A*, Dissection of the elbow joint showing the synovial membrane or capsule (*purple*), which lines the fibrous capsule. To display the synovial capsule, the joint cavity was filled with wax and then the fibrous capsule was removed. *B*, Sagittal magnetic resonance image (MRI) of the elbow. (Courtesy of Dr. W. Kucharczyk, Clinical Director of Tri-Hospital Resonance Centre, Toronto, Ontario, Canada.)

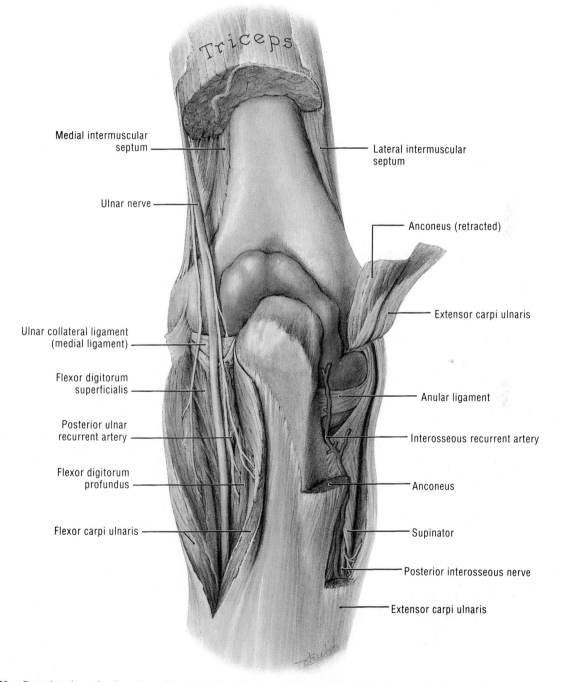

Triceps

Medial intermuscular septum

Lateral intermuscular septum

Ulnar nerve

Anconeus (retracted)

Extensor carpi ulnaris

Ulnar collateral ligament (medial ligament)

Flexor digitorum superficialis

Anular ligament

Posterior ulnar recurrent artery

Interosseous recurrent artery

Flexor digitorum profundus

Anconeus

Flexor carpi ulnaris

Supinator

Posterior interosseous nerve

Extensor carpi ulnaris

Figure 6-120. Posterior view of a dissection of the right elbow from which the distal portion of the triceps muscle has been removed. Observe the synovial membrane (*purple*) protruding between the head of the radius and the anular ligament.

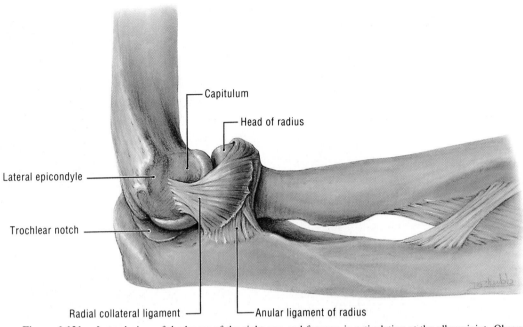

Capitulum

Head of radius

Lateral epicondyle

Trochlear notch

Radial collateral ligament

Anular ligament of radius

Figure 6-121. Lateral view of the bones of the right arm and forearm in articulation at the elbow joint. Observe that the fan-shaped radial collateral ligament is attached to the anular ligament of the radius.

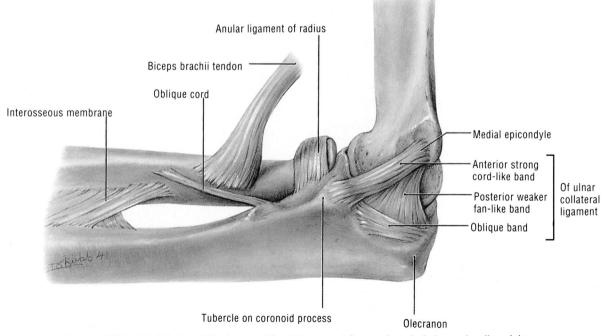

Anular ligament of radius

Biceps brachii tendon

Oblique cord

Interosseous membrane

Medial epicondyle

Anterior strong cord-like band

Posterior weaker fan-like band

Of ulnar collateral ligament

Oblique band

Tubercle on coronoid process

Olecranon

Figure 6-122. Medial view of the bones of the right arm and forearm in articulation at the elbow joint.

Table 6-16.
The Muscles Producing Movements of the Elbow Joint[1]

Movements	Muscles	Reference Tables	Chief Innervation[2]
Flexion	**Brachialis**[3] **Biceps brachii**	6-7	Musculocutaneous n. (C5 and **C6**)
	Brachioradialis	6-10	Radial n. (C5, **C6**, and C7)
	Extensor carpi radialis longus[4]	6-10	Radial n. (C6 and C7)
	Pronator teres[4] Flexor carpi radialis[4]	6-8	Median n. (C6 and **C7**)
Extension	**Triceps brachii**	6-7	Radial n. (C5, **C6**, and **C7**)
	Anconeus	6-10	Radial (C7, C8, and T1)
	Brachioradialis[5]	6-10	Radial (C5, **C6**, and C7)

[1]The principal muscles producing these movements are printed in **boldface**.
[2]**Boldface** indicates the main spinal cord segmental innervation.
[3]The brachialis is the main flexor of the forearm.
[4]Although not normally used to flex the forearm, these muscles can assist with the movement if necessary, *e.g.*, the pronator teres acts as a flexor when flexion is resisted.
[5]During active extension the brachioradialis assists with this movement of the forearm.

the supinator muscle. The bursa lies posterior to the supinator muscle, lateral to the tendon of the biceps muscle, and medial to the ulna. The *bicipitoradial bursa* (biceps bursa) lies between the biceps tendon and the anterior part of the tuberosity of the radius.

The *subcutaneous olecranon bursa* is exposed to injury during falls on the elbow and to infection from abrasions of the skin covering the olecranon (Fig. 6-64). Repeated excessive friction may cause this bursa to become inflamed, producing a friction bursitis, *e.g.*, "student's elbow" (Fig. 6-125). Inflammation of the subtendinous olecranon bursa is much less common. This type of bursitis results from excessive friction between the triceps tendon and the olecranon, *e.g.*, owing to repeated flexion-extension of the forearm as occurs during certain assembly line jobs. The pain is most severe during flexion of the forearm because of pressure exerted on the inflamed subtendinous olecranon bursa by the triceps tendon.

Radioulnar bursitis may result from repeated or violent extension of the wrist with the forearm pronated, as occurs during the backhand stroke in tennis. It may be associated with elbow tendonitis ("tennis elbow," p. 588). Pain occurs on elbow extension when the forearm is pronated.

Bicipitoradial bursitis (biceps bursitis) results in pain when the forearm is pronated because this action compresses the bicipitoradial bursa against the anterior half of the tuberosity of the radius.

Blood Supply of the Elbow Joint (Figs. 6-55, 6-62, 6-63, and 6-70). The articular arteries are derived from the anastomosis around the elbow, which are formed by collateral branches of the brachial and recurrent branches of the ulnar and radial arteries.

Nerve Supply of the Elbow Joint (Figs. 6-56, 6-62, and 6-70). The articular nerves are derived mainly from the mus-

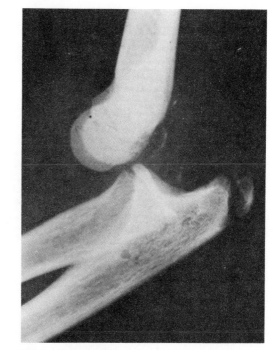

Figure 6-123. Radiograph of the elbow region of a child showing a posterior dislocation of the elbow joint.

culocutaneous and radial nerves, but the ulnar, median, and anterior interosseous nerves may also supply articular branches.

Fracture-separation of the proximal radial epiphysis can result when a young person falls and exerts compression and abduction forces on the elbow joint. The anatomical basis of the injury is the late fusion (at 15 to 17 years of age) of the proximal epiphysis of the head with the body of the radius.

Avulsion of the medial epicondyle of the humerus in children can also result from a fall that causes abduction of the extended elbow joint, an abnormal movement of this articulation. The resulting traction on the ulnar collateral ligament pulls the medial epicondyle distally. The anatomical basis of avulsion of the medial epicondyle is that the epiphysis for the medial epicondyle may not fuse with the distal end of the humerus until up to 20 years of age. Usually fusion is complete radiographically at 14 years of age in females and 16 years of age in males.

Traction injury of the ulnar nerve is a frequent complication of the abduction type of avulsion of the medial epicondyle. The anatomical basis for this stretching of the ulnar nerve is that it passes posterior to the medial epicondyle before entering the forearm (Fig. 6-63).

Posterior dislocation of the elbow joint (Fig. 6-123) may occur when children fall on their hands with their elbows flexed. The distal end of the humerus is driven through the weak anterior portion of the fibrous capsule of the elbow joint as the radius and ulna dislocate posteriorly.

The Radioulnar Joints

The radius and ulna articulate with each other at their proximal and distal ends at synovial joints, called the proximal and distal

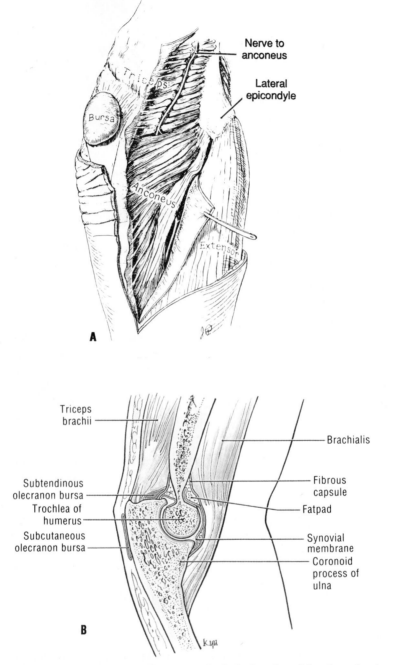

Nerve to
anconeus

Lateral
epicondyle

Triceps

Bursa

Anconeus

Extensor

A

Triceps
brachii

Brachialis

Subtendinous
olecranon bursa

Fibrous
capsule

Trochlea of
humerus

Fatpad

Subcutaneous
olecranon bursa

Synovial
membrane

Coronoid
process of
ulna

B

Figure 6-124. Bursae around the elbow. *A*, Dissection of the poster-olateral aspect of the right elbow. Observe the subcutaneous olecranon bursa lying on the tendinous expansion of the triceps brachii muscle. *B*, Sagittal section of the elbow showing the subcutaneous and subtendinous olecranon bursae.

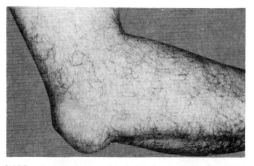

Figure 6-125. Lateral view of a student's right elbow exhibiting enlargement of the subcutaneous olecranon bursa (also see Fig. 6-124).

radioulnar joints. These articulations are the *pivot type of synovial joint* that produce pronation and supination of the forearm. The interosseous borders of the radius and ulna are connected by an *interosseous membrane* (Fig. 6-126). They are also joined by an oblique cord (Fig. 6-122). The interosseous membrane is a strong, broad fibrous sheet that stretches between the interosseous borders of the radius and the ulna, commencing 2 to 3 cm distal to the tuberosity of the radius. In addition to providing a flexible and strong attachment between the forearm bones, it provides the proximal attachments for the deep muscles of the forearm (Table 6-11). A thin, fibrous layer, called the *quadrate ligament*, extends between the radial notch of the ulna and the medial surface of the neck of the radius. This ligament covers the synovial membrane and probably supports it.

The oblique cord (Fig. 6-122) is a fibrous band that extends inferolaterally from the lateral border of the tuberosity of the ulna to the radius, just distal to its tuberosity. Its fibers are at right angles to those in the interosseous membrane. This cord is not always present and is not known to be of much functional significance.

The Proximal Radioulnar Joint (Figs. 6-119 to 6-121 and 6-126). This is the *pivot type of synovial joint* that allows movement of the radius on the ulna.

The Articular Surfaces of the Proximal Radioulnar Joint (Figs. 6-61, 6-119, 6-120, and 6-126). The radial head articulates with the radial notch of the ulna. The head of the radius is held in position by the strong *anular* (annular) *ligament*, a U-shaped fibrous collar which is attached to the anterior and posterior margins of the radial notch.

The Articular Capsule of the Proximal Radioulnar Joint (Figs. 6-119 to 6-122). The fibrous capsule enclosing the joint is continuous with the fibrous capsule of the elbow joint. The synovial capsule, which lines the fibrous capsule, is an inferior prolongation of the synovial capsule of the elbow joint. The deep surface of the anular ligament is lined with synovial membrane, which continues distally as a *sacciform recess* on the neck of the radius. This arrangement allows the radius to rotate within the anular ligament without tearing the synovial capsule. The synovial cavities of the elbow and proximal radioulnar joints are in free communication with each other.

Preschool children, especially 1- to 3-year-old children, are particularly vulnerable to an injury usually known as *"pulled*

elbow" (Fig. 6-127). Synonyms for this minor injury, known clinically as *subluxation (incomplete dislocation) of the head of the radius* are: "slipped elbow" and "nursemaid's elbow." The last term is a particularly inappropriate one because it implies that it is the nursemaid that is injured. The history of these cases is typical. The child is suddenly lifted by the upper limb when the forearm is pronated (*e.g.*, when lifting children into a bus or pulling them away from danger). The child cries out and refuses to use the limb, which he/she protects by holding it with the elbow flexed and the forearm pronated. *The sudden pulling of the upper limb tears the distal attachment of the anular ligament*, where it is loosely attached to the neck of the radius. The radial head is pulled distally, partially out of the torn anular ligament. The proximal part of the ligament may become trapped between the head of the radius and the capitulum of the humerus.

The Distal Radioulnar Joint (Figs. 6-61, 6-76, 6-126, and 6-128 to 6-130). This is also a pivot type of synovial joint. The radius moves around the relatively fixed inferior end of the ulna.

The Articular Surfaces of the Distal Radioulnar Joint (Figs. 6-61, 6-126, 6-128, and 6-129). The rounded side of the head of the ulna articulates with the ulnar notch in the distal end of the radius. A fibrocartilaginous *articular disc* binds the ends of the ulna and radius together and is the main uniting structure of the joint. The base of the articular disc is attached to the medial edge of the ulnar notch of the radius, and the apex of the disc is attached to the lateral side of the base of the styloid process of the ulna. The proximal surface of this triangular disc articulates with the distal aspect of the head of the ulna. Hence, the joint cavity is L-shaped in a coronal section. The articular disc separates the cavity of the distal radioulnar joint from the cavity of the wrist joint.

The Articular Capsule of the Distal Radioulnar Joint (Figs. 6-128 to 6-130). The fibrous capsule encloses the joint. It is formed by relatively weak transverse bands that extend from the radius to the ulna across the anterior and posterior surfaces of the joint. The *synovial membrane* lines the fibrous capsule and the proximal surface of the articular disc. The synovial capsule extends proximally a short distance between the radius and ulna as the *sacciform recess*. This redundancy of the synovial capsule accommodates the twisting of the capsule that occurs when the distal end of the radius travels around the relatively fixed distal end of the ulna during pronation of the forearm.

Movements of the Radioulnar Joints (Fig. 6-126A; Table 6-17). Movements at these joints make pronation and supination of the forearm and hand possible. *Pronation* is rotation that turns the palm posteriorly or inferiorly when the forearm is flexed. *Supination* carries the palm anteriorly or superiorly when the forearm is flexed. The axis for these movements passes proximally through the center of the head of the radius and distally through the site of attachment of the apex of the articular disc to the head of the ulna. During pronation and supination, it is mainly the radius that rotates. Its head rotates within the cup-shaped ring formed by the anular ligament and the radial notch on the ulna. Distally the end of the radius rotates around the head of the ulna.

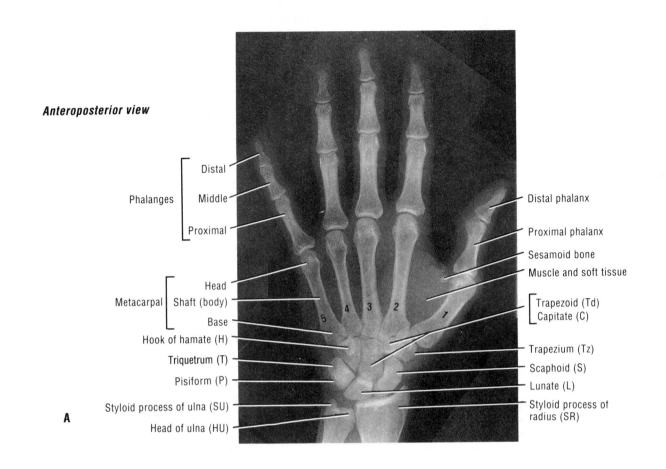

Anteroposterior view

Phalanges
- Distal
- Middle
- Proximal

Metacarpal
- Head
- Shaft (body)
- Base

Hook of hamate (H)
Triquetrum (T)
Pisiform (P)
Styloid process of ulna (SU)
Head of ulna (HU)

Distal phalanx
Proximal phalanx
Sesamoid bone
Muscle and soft tissue
Trapezoid (Td)
Capitate (C)
Trapezium (Tz)
Scaphoid (S)
Lunate (L)
Styloid process of radius (SR)

A

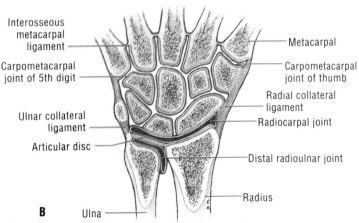

Interosseous metacarpal ligament
Carpometacarpal joint of 5th digit
Ulnar collateral ligament
Articular disc

Metacarpal
Carpometacarpal joint of thumb
Radial collateral ligament
Radiocarpal joint
Distal radioulnar joint
Radius

B Ulna

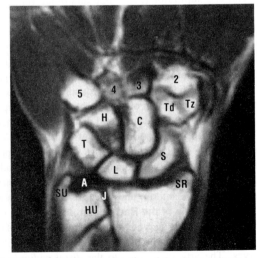

C

Figure 6-128. *A*, Radiograph of the wrist and hand (Courtesy of Dr. E. L. Lansdown, Professor of Radiology, University of Toronto, Toronto, Ontario, Canada). *B*, Coronal section of the right wrist and hand. Observe the distal radioulnar, wrist, intercarpal, carpometacarpal, and intermetacarpal joints. Note that the cavities of the distal radioulnar and wrist joints are separated by the articular disc of the distal radioulnar joint. *C*, Coronal magnetic resonance image (MRI) of the wrist and hand. (Courtesy of Dr. W. Kucharczyk, Clinical Director of Tri-Hospital Magnetic Resonance Centre, Toronto, Ontario, Canada.)

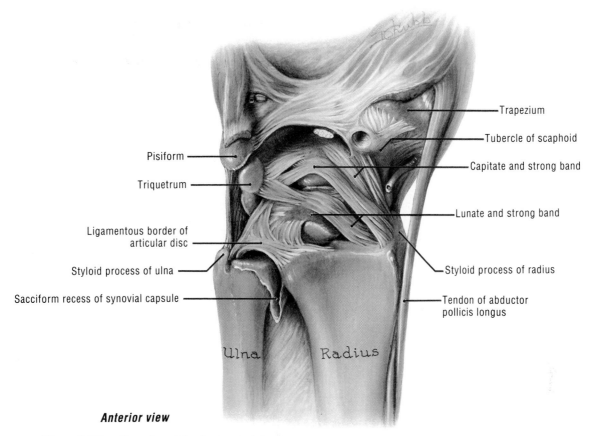

Pisiform

Triquetrum

Ligamentous border of
articular disc

Styloid process of ulna

Sacciform recess of synovial capsule

Trapezium

Tubercle of scaphoid

Capitate and strong band

Lunate and strong band

Styloid process of radius

Tendon of abductor
pollicis longus

Ulna   Radius

Anterior view

Figure 6-129.   Dissection of the distal end of the right forearm and wrist with the hand forcibly extended. Observe the ligaments of the distal radioulnar, radiocarpal, and intercarpal joints.

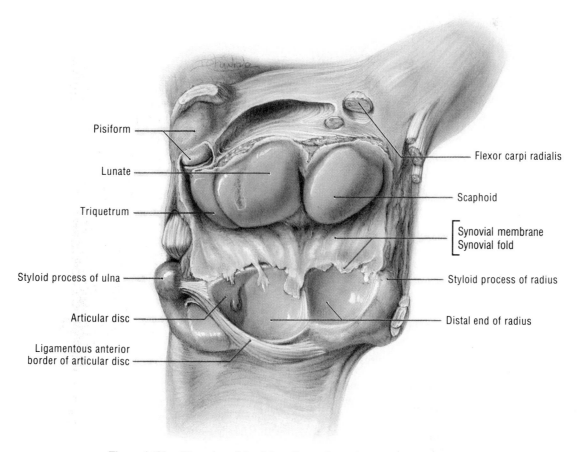

Pisiform

Lunate

Triquetrum

Styloid process of ulna

Articular disc

Ligamentous anterior
border of articular disc

Flexor carpi radialis

Scaphoid

Synovial membrane
Synovial fold

Styloid process of radius

Distal end of radius

Figure 6-130.   Dissection of the right radiocarpal or wrist joint, opened anteriorly.

Table 6-17.
The Muscles Producing Movements of the Radioulnar Joints[1]

Movements	Muscles	Reference Tables	Chief Innervation[2]
Supination	**Supinator**[3]	6-11	Radial n. (C5 and **C6**)
	Biceps brachii	6-7	Musculocutaneous n. (C5 and **C6**)
Pronation	**Pronator quadratus**[4]	6-9	Ant. interosseous n. (**C8** and T1)
	Pronator teres	6-8	Median n. (C6 and **C7**)
	Flexor carpi radialis		
	Anconeus	6-10	Radial n. (C7, C8, and T1)

[1]The principal muscles producing these movements are printed in **boldface**.
[2]**Boldface** indicates the main spinal cord segmented innervation.
[3]The supinator is the chief supinator of the forearm. It is aided by the biceps during fast or resisted movement.
[4]The pronator quadratus is the prime mover in pronation of the forearm. It is assisted by the pronator teres during rapid and/or forceful movements.

the triquetral bone, forming a separate, small synovial joint with it, called the *pisotriquetral joint*.

The Articular Capsule of the Intercarpal Joints (Figs. 6-128 and 6-129).   The *fibrous capsule* encloses these joints and helps to unite the articulating carpal bones. The bones of each row are connected to each other by dorsal, palmar, and interosseous ligaments. The two rows are connected by dorsal, palmar, ulnar, and radial ligaments. The *synovial membrane* lines the fibrous capsule and is attached to the margins of the articular surfaces of the carpal bones. In Figure 6-128, observe that the cavity of the midcarpal joint is part of the general joint cavity, extending between the bones of each row. The common joint space includes the carpometacarpal and intermetacarpal joints.

Movements of the Intercarpal Joints (Table 6-18).   Movements of these joints increase the range of movement at the wrist joint. Movement of the head of the capitate in its socket and the gliding movement of the bones on each side of it result in considerable flexion of the hand. The midcarpal joint is largely concerned with flexion and abduction of the hand. It also increases the range of abduction of the hand. Extension of the wrist joint and flexion of the intercarpal joints improve the grasp of the hand.

Blood Supply of the Intercarpal Joints (Figs. 6-70, 6-84, and 6-98).   The articular arteries are derived from the palmar and dorsal carpal arches.

Nerve Supply of the Intercarpal Joints (Figs. 6-83 and 6-102).   The articular nerves are derived from the anterior interosseous nerve of the median, the posterior interosseous nerve of the radial, and the dorsal and deep branches of the ulnar nerve.

The Carpometacarpal and Intermetacarpal Joints

These are *plane synovial joints* that permit a gliding movement. They share a common joint cavity with the intercarpal joints (Fig. 6-128). The articulating bones are united by dorsal, palmar, and interosseous ligaments.

The Carpometacarpal Joint of the Thumb (Fig. 6-128; Table 6-21).   This articulation is a *saddle type of synovial joint.*

Table 6-18.
The Muscles Producing Movements of the Wrist and Midcarpal Joints[1]

Movements	Muscles	Reference Tables	Chief Innervation[2]
Flexion	**Flexor carpi radialis**	6-8	Median n. (C6 and **C7**)
	Flexor carpi ulnaris		Ulnar n. (C7 and **C8**)
	Palmaris longus		Median n. (C7 and C8)
	Abductor pollicis longus	6-11	Posterior interosseous n. (C7 and **C8**)
	Flexor digitorum profundus	6-9	*Medial part,* ulnar n. *Lateral part,* median n. (**C8** and T1)
	Flexor digitorum superficialis	6-8	Median n. (C7, **C8** and T1)
	Flexor pollicis longus	6-9	Anterior interosseous n. (**C8** and T1)
Extension	**Extensor carpi radialis longus**	6-10	Radial n. (C6 and C7)
	Extensor carpi radialis brevis		Radial n. (**C6** and C7)
	Extensor carpi ulnaris	6-10	Posterior interosseous n. (C7 and C8)
	Extensor digitorum	6-11	
	Extensor pollicis longus		
	Extensor indicis		
	Extensor digiti minimi	6-10	Posterior interosseous n. (**C7** and C8)
Abduction	**Extensor carpi radialis longus**	6-10	Radial n. (C6 and C7)
	Extensor carpi radialis brevis	6-10	Radial n. (**C7** and C8)
	Flexor carpi radialis	6-8	Median n. (C6 and **C7**)
	Abductor pollicis longus		Posterior interosseous n. (C7 and **C8**)
	Extensor pollicis brevis	6-11	Posterior interosseous n. (C7 and C8)
	Extensor pollicis longus		
Adduction	**Flexor carpi ulnaris**	6-8	Ulnar n. (C7 and **C8**)
	Extensor carpi ulnaris	6-10	Posterior interosseous n. (C7 and **C8**)

[1]The principal muscles producing these movements are printed in **boldface**. The wrist joint is mainly concerned with extension and adduction of the hand, whereas the midcarpal joint is largely concerned with flexion and abduction of the hand. However movements at the wrist and midcarpal joints are best considered together because these articulations form parts of the same mechanism and are acted on by the same muscle groups.
[2]**Boldface** indicates the main spinal cord segmental innervation.

Articular Surfaces of the Carpometacarpal Joint of the Thumb (Figs. 6-71*B* and 6-128).   The trapezium articulates with the saddle-shaped base of the first metacarpal bone.

The Articular Capsule of the Carpometacarpal Joint of the Thumb (Fig. 6-128).   The fibrous capsule encloses the joint and is attached to the margins of the articular surfaces. The looseness of its capsule facilitates its movements. The synovial membrane lines the fibrous capsule and forms a separate joint cavity from the rest of the carpus.

Movements of the Carpometacarpal Joint of the Thumb (Figs. 6-97 and 6-108; Table 6-21). This joint permits angular movements in any plane and a restricted amount of axial rotation. Only ball and socket joints are more mobile. The following thumb movements are possible: flexion, extension, abduction, adduction, and opposition. The functional importance of the thumb lies in its ability to be opposed to the other digits.

Blood Supply of the Carpometacarpal Joint of the Thumb (Figs. 6-77, 6-84, and 6-98). The articular arteries are derived from the dorsal and palmar *metacarpal arteries* and from the *dorsal carpal and deep palmar arterial arches*. These vessels are branches of the ulnar and radial arteries.

Nerve Supply of the Carpometacarpal Joint of the Thumb (Figs. 6-83 and 6-102). The articular nerves are derived from the anterior interosseous nerve of the median, the posterior interosseous nerve of the radial, and the dorsal and deep branches of the ulnar nerve.

The Metacarpophalangeal Joints

These articulations are *condyloid (knucklelike) synovial joints* that allow movement in two directions.

Articular Surface of the Metacarpophalangeal Joints (Figs. 6-128 and 6-131). The heads of the metacarpal bones articulate with the bases of the proximal phalanges. The unique feature of their bony surfaces is that they both have oval articular surfaces.

The Articular Capsule of the Metacarpophalangeal Joints (Figs. 6-128 and 6-131). A fibrous capsule encloses each joint. They are strengthened on each side by a triangular *collateral ligament*. It extends from the sides of the head of the proximal bone to the sides of the base of the distal bone. The *palmar ligaments* are strong thick plates that are firmly attached to the phalanx and loosely attached to the metacarpal. The palmar ligaments of the second to fifth joints are united by deep transverse metacarpal ligaments that hold the heads of the metacarpals together. The *synovial capsule* lines the fibrous capsule of each joint and is attached to the margins of the articular surfaces.

Movements of the Metacarpophalangeal Joints (Fig. 6-131; Tables 6-19 and 6-21). The following movements occur at these articulations: flexion, extension, abduction, adduction, and circumduction.

Blood Supply of the Metacarpophalangeal Joints (Figs. 6-98, 6-99, and 6-103). The articular arteries are branches of the *digital arteries* that arise from the superficial palmar arterial arch.

Nerve Supply of the Metacarpophalangeal Joints (Figs. 6-99, 6-102, and 6-103). The articular nerves are derived from the *digital nerves* that arise from the ulnar and median nerves.

The Interphalangeal Joints

These articulations are *uniaxial hinge joints*, which permit only flexion and extension (Fig. 6-131; Table 6-20). They join the head of one phalanx with the base of the more distal one. They are structurally similar to the metacarpophalangeal joints and are reinforced dorsally by the extensor expansions of the digits (Fig. 6-89). The articular arteries and nerves are derived from the adjacent digital arteries and nerves (Figs. 6-98, 6-99, 6-102, and 6-103).

> Sudden tension on a long extensor tendon may avulse part of its attachment to a phalanx (Fig. 6-89). The most common injury is called a *mallet finger* (also known as a *cricket or baseball finger*). This condition results from the distal interphalangeal joints suddenly being forced into extreme flexion (*i.e.*, hyperflexion). This avulses the attachment of the terminal tendon into the base of the distal phalanx (Fig. 6-89B). As a result, the patient is unable to extend the distal interphalangeal joint (Fig. 6-132).

PATIENT ORIENTED PROBLEMS

Case 6-1

A 20-year-old man complained that he was unable to raise his right upper limb. He held it limply at his side with the forearm and hand pronated. During questioning he stated that he had been thrown from his motorcycle about 2 weeks previously and that he had hit his shoulder against a tree. He also recalled that his neck felt sore shortly after the accident. On examination it was found that he was unable to flex, abduct, or laterally rotate his arm. In addition, there was loss of flexion of the elbow joint. A lack of sensation was detected on the lateral surface of his arm and forearm.

> **Problems.** Using your anatomical knowledge of the nerve supply to the upper limb, discuss the probable cause of this patient's loss of motor and sensory functions. What muscles were probably paralyzed? Is he likely to recover full use of his paralyzed limb? These problems are discussed on p. 633.

Case 6-2

One of your friends injured his shoulder during a hockey game when he was driven heavily into the boards, hitting his shoulder. As you assisted him to the dressing room, you noted that his injury was very painful. When his sweater and shoulder pads were removed, you observed that the lateral end of his clavicle produced an abnormal prominence. At first you thought he may have what sportswriters call a ''*shoulder pointer*.'' Later the team physician informed you that your friend had a *dislocation of his acromioclavicular joint* (''shoulder separation'') and would be out of the lineup for several weeks.

> **Problems.** Explain what sportswriters mean by the terms ''shoulder pointer'' and ''shoulder separation.'' How would you explain the structure of the shoulder joint to a nonmedical student? What ligaments would be torn? What made the patient's shoulder fall? These problems are discussed on p. 633.

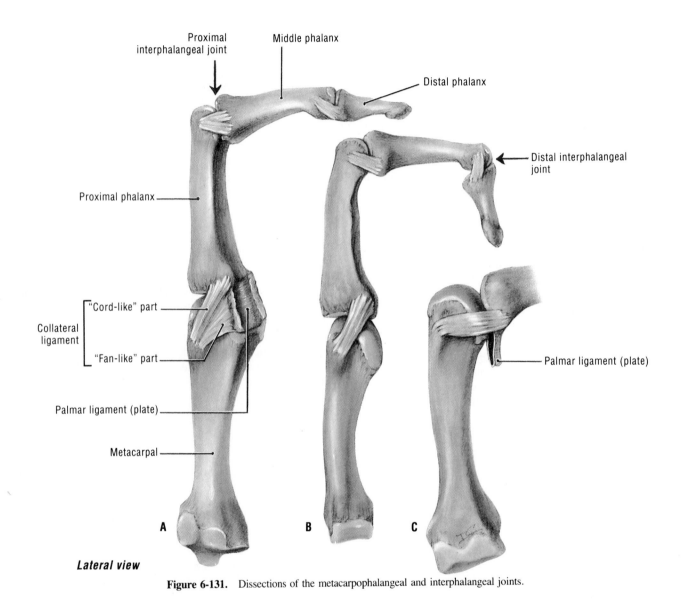

Proximal interphalangeal joint

Middle phalanx

Distal phalanx

Distal interphalangeal joint

Proximal phalanx

Collateral ligament
  "Cord-like" part
  "Fan-like" part

Palmar ligament (plate)

Metacarpal

Palmar ligament (plate)

A   B   C

Lateral view

Figure 6-131.   Dissections of the metacarpophalangeal and interphalangeal joints.

Table 6-19.
The Muscles Producing Movements of the Medial Four Digits at the Metacarpophalangeal Joints[1]

Movements	Muscles	Reference Tables	Chief Innervation[2]
Flexion	**Flexor digitorum profundus**	6-9	*Medial part*, ulnar n. (**C8** and T1) *Lateral part*, median n. (**C8** and T1)
	Flexor digitorum superficialis	6-8	Median n. (C7, **C8** and T1)
	Flexor digiti minimi Abductor digiti minimi	6-13	Ulnar n. (C8 and **T1**)
	Interossei Lumbricals	6-14	Ulnar n. (C8 and T1) Median n., 1st and 2nd (C8 and **T1**) Ulnar n., 3rd and 4th (C8 and **T1**)
Extension	**Extensor digitorum** (3rd and 4th digits)	6-10	Posterior interosseous n. (**C7** and C8)
	Extensor digiti minimi		Posterior interosseous n.
	Extensor indicis	6-11	(**C7** and C8)
Abduction	**Dorsal interossei**[3]	6-14	Ulnar n. (C8 and T1)
	Extensor digitorum **Extensor digiti minimi**	6-10	Posterior interosseous n. (**C7** and C8)
	Extensor indicis	6-11	Posterior interosseous n. (**C7** and C8)
	Abductor digiti minimi	6-13	Ulnar n. (C8 and T1)
Adduction	**Palmar interossei**[3]	6-14	Ulnar n. (C8 and T1)
	Flexor digitorum superficialis	6-8	Median n. (C7, **C8** and T1)
	Flexor digitorum profundus	6-9	*Medial part*, ulnar n. (**C8** and T1) *Lateral part*, median n. (**C8** and T1)

[1]The principal muscles producing these movements are printed in **boldface**.
[2]**Boldface** indicates the main spinal cord segmental innervation.
[3]Of the extended digits.

Table 6-20.
The Muscles Producing Movements of the Interphalangeal Joints of the Digits

Movements	Muscles	Reference Tables	Chief Innervation[1]
Flexion	Flexor digitorum profundus (proximal and distal joints)	6-9	*Medial part*, ulnar n. (C7 and **C8**) *Lateral part*, median n. (**C8** and T1)
	Flexor digitorum superficialis (proximal joints only)	6-8	Median (C7, **C8** and T1)
Extension	Extensor digitorum Extensor digiti minimi	6-10	Posterior interosseous n. (**C7** and C8)
	Extensor indicis	6-11	Posterior interosseous n. (C7 and **C8**)
	Lumbricals	6-14	Median n. (1st and 2nd) Ulnar n. (3rd and 4th)
	Interossei	6-14	Ulnar n. (C8 and T1)

[1]**Boldface** indicates the main spinal cord segmental innervation.

and displacement of the epiphysis of the medial epicondyle of the right humerus.

> **Problems.** Explain the numbness of the boy's fifth digit and his inability to hold a piece of paper between his digits. Drawing on your knowledge of degeneration and regeneration of peripheral nerves, make an attempt to forecast the probable degree of recovery of the boy's motor and sensory functions that may occur. These problems are discussed on p. 633.

Case 6-4

A young man was hit very hard by a hockey stick in the midhumeral region of his left arm. He presented with signs of tenderness, swelling, deformity, and abnormal movements of his left upper limb. The physical examination revealed a *wrist-drop*, an inability to extend the digits at the metacarpophalangeal joints, and loss of sensation on a small area of skin on the dorsum of the hand proximal to the first two digits. There was also weakness of extension of the interphalangeal joints. Measurement of the limb indicated that there was some shortening. Radiographs of his arm were ordered.

Radiologist's Report. Radiographs showed the presence of a fracture of the humerus just distal to the attachment of the deltoid muscle. The proximal fragment of bone was abducted and the distal fragment was displaced proximally.

> **Problems.** Using your anatomical knowledge, determine what peripheral nerve has been damaged and what artery may have been torn. Would elbow flexion be weakened? Explain the observed effects of this peripheral nerve injury. Why are the fragments of humerus displaced in the manner described? These problems are discussed on p. 634.

Case 6-3

A 12-year-old boy fell off his skateboard, hitting his right elbow on the sidewalk. Because he was suffering considerable elbow pain and numbness on the side of his hand, his mother took him to a pediatrician. The boy told the physician: "I fell on my funny bone and right away my little finger began to tingle." The physician noted that the boy showed no response to pinprick over his right fifth digit and the medial border of the palm. He was unable to grip a piece of paper placed between his digits. Suspecting a fracture of the elbow and peripheral nerve damage, the physician arranged to have the boy's elbow radiographed.

Radiologist's Report. The radiographs showed separation

Table 6-21.
The Muscles Producing Movements of the First Digit or Thumb[1]

Movements	Muscles	Reference Tables	Chief Innervation[2]
1. *At the Carpometacarpal Joint*			
Flexion	**Flexor pollicis brevis**	6-12	Median n. (C8 and **T1**)
	Flexor pollicis longus	6-9	Anterior interosseous n. (**C8** and T1)
	Opponens pollicis	6-12	Median n. (**C8** and T1)
Extension	**Extensor pollicis longus** **Extensor pollicis brevis** **Abductor pollicis longus**	6-11	Posterior interosseous n. (C7 and **C8**)
Abduction	**Abductor pollicis longus**	6-11	Posterior interosseous n. (C7 and **C8**)
	Abductor pollicis brevis	6-12	Median n. (C8 and **T1**)
Adduction	**Adductor pollicis**	6-12	Ulnar n. (C8 and **T1**)
	Dorsal interosseous, 1st	6-14	Ulnar n. (C8 and **T1**)
	Extensor pollicis longus	6-11	Posterior interosseous n. (C7 and **C8**)
	Flexor pollicis longus	6-9	Anterior interosseous n. (**C8** and T1)
Opposition	**Opponens pollicis** **Abductor pollicis brevis** **Flexor pollicis brevis**	6-12	Median n. (C8 and **T1**)
	Flexor pollicis longus	6-9	Anterior interosseous n. (**C8** and T1)
	Adductor pollicis	6-12	Ulnar n. (C8 and **T1**)
2. *At the Metacarpophalangeal Joint*			
Flexion	**Flexor pollicis longus**	6-9	Anterior interosseous n. (**C8** and T1)
	Flexor pollicis brevis	6-12	Median n. (**C8** and T1)
	Palmar interosseous, 1st	6-14	Ulnar n. (C8 and **T1**)
	Abductor pollicis brevis	6-12	Median n. (C8 and **T1**)
Extension	**Extensor pollicis longus** **Extensor pollicis brevis**	6-11	Posterior interosseous n. (C7 and **C8**)
Abduction	**Abductor pollicis brevis**	6-12	Median n. (C8 and **T1**)
Adduction	**Adductor pollicis**	6-12	Ulnar n. (C8 and **T1**)
	Palmar interosseous, 1st	6-14	Ulnar n. (C8 and **T1**)
3. *At the Interphalangeal Joint*			
Flexion	**Flexor pollicis longus**	6-9	Anterior interosseous n. (**C8** and T1)

Table 6-21.
Continued

Movements	Muscles	Reference Tables	Chief Innervation[2]
Extension	**Extensor pollicis longus**	6-11	Posterior interosseous n. (C7 and **C8**)
	Abductor pollicis brevis	6-12	Median n. (C8 and T1)
	Adductor pollicis		Ulnar n. (C8 and **T1**)

[1]The principal muscles producing these movements are printed in **boldface**.
[2]**Boldface** indicates the main spinal cord segmental innervation.

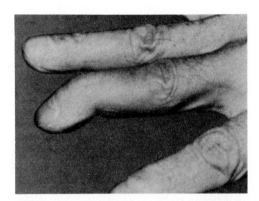

Figure 6-132. A mallet or baseball finger resulting from avulsion of the attachment of the long tendon of the extensor digitorum muscle from the base of the distal phalanx of the third digit (middle finger).

Case 6-5

While you were playing touch football, you fell on your open hand with your wrist hyperextended and laterally deviated. You told your friends that you had just sprained your wrist and did not pay much attention to the injury for about 2 weeks. You later sought medical advice because the wrist pain was still present. When the physician deeply palpated your anatomical snuff box, there was localized tenderness. You experienced most pain on the lateral side of your wrist, particularly when she asked you to extend it as far as you could. Suspecting a fracture, she ordered radiographs of your wrist.

Radiologist's Report. The radiographs revealed a small hairline fracture of one of the lateral carpal bones.

Problems. Which carpal bones lie in the floor of the anatomical snuff box? The distal end of which forearm bone is also in the floor of this depression? Which carpal bone was most likely fractured? These problems are discussed on p. 634.

Case 6-6

A 15-year-old girl who had slashed her wrists with a razor blade was rushed to the emergency department of a hospital. The moderate bleeding from the left wrist was soon stopped with slight pressure. The small spurts of blood coming from the lateral side of the right wrist were more difficult to stop. Examination of her left hand and wrist revealed that her hand movements were normal and that there was no loss of sensation. The following observations were made on her right wrist and hand: two superficial tendons and a large nerve were cut; she could adduct her thumb but was unable to oppose it; she had lost some fine control of the movements of her second and third digits; and there was anesthesia over the lateral half of her palm and digits.

> **Problems.** Which tendon was almost certainly severed? What large nerve was undoubtedly cut? Which tendon may have been severed? What superficial artery appears to have been lacerated? Would flexion of her wrist be affected? These problems are discussed on p. 635.

Case 6-7

An elderly lady slipped on a patch of ice and fell on the palm of her outstretched hand. She told the intern that she heard her wrist crack and that it was very sore. The dorsal aspect of her wrist was unduly prominent and resembled a dinner fork. It was obvious that a forearm bone was fractured.

> **Problems.** What bone in the forearm was probably fractured? What carpal bone may have been fractured? What do you call this kind of fracture? Explain the cause of the dinner fork appearance of the patient's wrist. These problems are discussed on p. 635.

DISCUSSION OF PATIENTS' PROBLEMS

Case 6-1

When the patient was thrown from his motorcycle and hit a tree, his right shoulder was pulled violently away from his head (Fig. 6-24). This pulled on the superior trunk of the brachial plexus (Fig. 6-18), stretching or tearing the ventral primary rami of C5 and C6. As a result, the nerves arising from these rami and the superior trunk are affected and the muscles supplied by them are paralyzed. The muscles involved would be the deltoid, biceps brachii, brachialis, brachioradialis, supraspinatus, infraspinatus, teres minor, and supinator (Figs. 6-20 and 6-22).

The patient's arm was medially rotated because the infraspinatus and teres minor muscles (lateral rotators of the shoulder) were paralyzed. His forearm was pronated because the supinators were paralyzed, including the biceps muscle. Flexion of his elbow was weak because of paralysis of the brachialis and biceps muscles. The inability of the patient to flex his humerus resulted from paralysis of the deltoid and coracobrachialis muscles and probably the clavicular head of the pectoralis major muscle. Loss of abduction of the humerus resulted from paralysis of the supraspinatus and deltoid muscles. The paralysis would be permanent if the rootlets making up the rami (C5 and C6) have been pulled out of the spinal cord. As these rootlets cannot be sutured back into the cord, the axons of the nerves would not regenerate and the muscles supplied by them would soon undergo atrophy.

Movements of the shoulder and elbow would be greatly affected, *e.g.*, the patient will always have difficulty lifting a glass to his mouth with his right arm. The loss of sensation in his arm resulted from damage to sensory fibers of C5 and C6 that are conveyed

in the upper lateral brachial cutaneous nerve (from the axillary), the lower lateral brachial cutaneous nerve (from the radial), and the lateral antebrachial cutaneous nerve (from the musculocutaneous nerve).

Case 6-2

A "shoulder pointer" is a sportswriter's term for contusion over the "point of the shoulder," formed by the acromion of the scapula. To explain a "shoulder separation," first you should make a simple diagram of the scapula and clavicle, showing the ligaments attached to them (Fig. 6-114). You should emphasize that it is the coracoclavicular ligament that provides most stability to the acromioclavicular joint. You should explain that the scapula and clavicle are parts of the upper limb and make up what is called the pectoral girdle and that the clavicle articulates laterally with the acromion to form the acromioclavicular joint. Also explain that the scapula and clavicle are held together by the acromioclavicular and coracoclavicular ligaments.

When your friend hit the boards with the "point of his shoulder" (acromion), the acromioclavicular and coracoclavicular ligaments were torn (Fig. 6-133). As a result, the shoulder fell under the weight of the upper limb and the acromion was pulled inferiorly, relative to the clavicle. Also, the lateral end of the clavicle was displaced superiorly, relative to the acromion, and produced an obvious prominence. Stress that the expression "separation of the shoulder" is a misnomer. Explain that it is the acromioclavicular joint that is separated, not the shoulder joint. Rupture of the acromioclavicular ligament alone is not a serious injury, but when combined with rupture of the coracoclavicular ligament, the dislocation is complicated because the scapula and clavicle are separated and the scapula and upper limb are displaced inferiorly.

Case 6-3

The medial epicondyle of the humerus does not completely fuse with the side of the diaphysis until 16 years of age in males (Fig. 6-134). Although an *epiphyseal separation* is sometimes called "an epiphyseal fracture," or a fracture-dislocation, it is

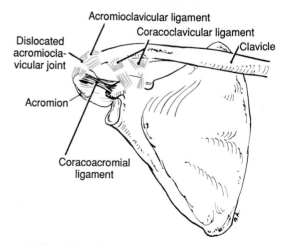

Figure 6-133. Dislocation of the acromioclavicular joint ("separation of the shoulder"), following tearing of the acromioclavicular and coracoclavicular ligaments (also see Fig. 6-114).

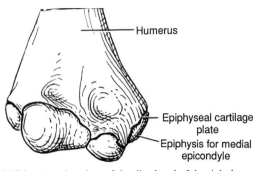

Figure 6-134. Anterior view of the distal end of the right humerus of a child, showing the epiphysis for the medial epicondyle.

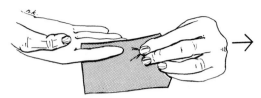

Figure 6-135. The method for testing the interosseous muscles. The examiner is attempting to pull the paper away from the patient in the direction of the arrow. Inability to adduct the fingers is a sign of paralysis of the palmar interosseous muscles and ulnar nerve injury.

best to refer to this injury as a *separation of the epiphysis of the medial epicondyle*. Had this accident occurred in a person over age 16 years, a fracture of the medial epicondyle might have occurred. Because the epiphyseal cartilaginous plate is weaker than the surrounding bone in children, a direct blow that causes a fracture in adolescents and adults is likely to cause an *epiphyseal cartilaginous plate injury* in children. Because the ulnar nerve passes posterior to the medial epicondyle (Fig. 6-63), between it and the olecranon, it is vulnerable to injuries at the elbow. In the present case, it is likely that the ulnar nerve was compressed and that the axons were damaged at the site of the injury. This kind of injury causes paralysis of muscles and some loss of sensation in the area of skin supplied by the ulnar nerve (Fig. 6-21).

Appreciation of light touch is usually lost over the medial one and one-half digits and response to pinprick is lost over the fifth digit and the medial border of the palm. Knowing that the interosseus muscles are supplied by the ulnar nerve, the physician tested them for weakness by placing a piece of paper between the boy's fully extended digits and asking him to grip it as tightly as possible while he pulled on it (Fig. 6-135). Inability to adduct the digits is a classic sign of paralysis of the palmar interosseous muscles and ulnar nerve injury.

Loss of other muscle movements would likely have occurred: inability to abduct the digits (*paralysis of the dorsal interossei*), loss of adduction of the thumb (*paralysis of the adductor pollicis*), weakness of flexion of the fourth and fifth digits at the metacarpophalangeal joints (*paralysis of the medial two lumbricals*), impaired flexion and adduction of the wrist (*paralysis of the flexor carpi ulnaris*), a poor grasp in the fourth and fifth digits, and inability to flex the distal interphalangeal joints of the fourth and fifth digits (*paralysis of the lumbricals, interossei, and part of the flexor digitorum profundus*).

Because all but five of the intrinsic muscles of the hand are supplied by the ulnar nerve, injury of it at the elbow has its primary effect in the hand. Because the nerve was only crushed, the nerve does not require suturing because new axons can grow into the part of the nerve distal to the injury within the original endoneurial sheaths and neurolemmal sheaths and reinnervate the paralyzed muscles. Hence, after a crush injury, as in this case, restoration of function should occur in a few months' time with appropriate physiotherapy.

Case 6-4

The inability of the patient to extend his hand at the wrist indicates *injury to the radial nerve*. Because the fracture is in the middle of the humerus, it is likely that the radial nerve was damaged where it passes diagonally across the humerus in the radial groove (Figs. 6-12 and 6-23). The nerve is particularly susceptible to injury in this location because of its close relationship to the humerus. *Section of the radial nerve* totally paralyzes the extensor muscles of the forearm and hand (Fig. 6-22). As a result, extension of the wrist is impossible and the hand assumes the flexed position referred to clinically as **wrist-drop**. The radial nerve supplies no muscles in the hand, but it supplies muscles whose tendons pass into the hand; hence the patient is unable to extend his metacarpophalangeal joints. Because the lumbrical muscles (supplied by the median and ulnar nerves) and the interossei (supplied by the ulnar nerve) are intact, the patient is able to flex his metacarpophalangeal joints and extend his interphalangeal joints. However, he would not have normal power of extension of his digits.

Elbow flexion would be very painful and would be weakened when the forearm is in the position midway between pronation and supination. Recall that the radial nerve innervates the brachioradialis muscle, a strong flexor of the elbow in this position. The area of sensory loss is often minimal following radial nerve injury because its area of exclusive supply is very small. The degree of sensory loss varies from patient to patient, depending on the extent to which the territory is overlapped by adjacent nerves. Sometimes there is no detectable loss of sensation. The shortening of the patient's arm occurred because the broken fragments of bone were pulled apart. Contraction of the deltoid muscle abducts the proximal part of the humerus. The proximal contraction of the triceps, biceps, and coracobrachialis muscles pulls the distal fragment superiorly.

Although the profunda brachii artery accompanies the radial nerve through the radial groove (Fig. 6-59) and may be severed by bone fragments, the muscles and structures supplied by this artery (*e.g.*, the humerus) are not likely to show ischemia because the radial recurrent artery anastomoses with the profunda brachii artery (Fig. 6-55). This communication should provide sufficient blood for the structures supplied by the damaged artery.

Case 6-5

The lateral bones of the carpus, the scaphoid and trapezium, lie in the floor of the anatomical snuff box. This depression at the base of the thumb is limited proximally by the styloid process of the radius and distally by the base of the first metacarpal bone.

Fracture of the scaphoid is the most common type of carpal injury and usually results from a fall on the hand. Because of the position of the scaphoid and its small size, it is a difficult bone to immobilize. Continued movement of the wrist often results in nonunion of the fragments of bone. Usually there is displacement and tearing of ligaments that may interfere with the blood supply to one of the fragments. *Ischemic necrosis* of part of the scaphoid bone may result. Usually the bone is supplied by two nutrient arteries, one to the proximal and one to the distal half. Occasionally both vessels supply the distal half; the separated proximal half receives no blood. The resulting ischemia may result in delay or lack of union of the fragments.

Case 6-6

Obviously the patient had not cut her wrist deeply on the left side; the slight bleeding was probably from severed superficial veins (Fig. 6-82). On the right side, she would have certainly cut the tendon of her palmaris longus (Figs. 6-73 and 6-75). She probably also cut the tendon of her flexor carpi radialis muscle. In view of the clinical findings, it is obvious that *her median nerve was severed.* At the wrist this nerve lies deep to and lateral to the tendon of the palmaris longus muscle (Fig. 6-66). The slight spurting of blood in her right wrist suggests that she probably cut the superficial palmar branch of her radial artery. This artery arises from the radial just proximal to the wrist (Fig. 6-75). Had she severed her radial artery, the bleeding would have been severe.

Cutting the median nerve at her wrist resulted in paralysis of her thenar muscles and first two lumbricals (Tables 6-13 and 6-14). Paralysis of the thenar muscles explains her inability to oppose her thumb. Because the posterior interosseus nerve (a branch of the radial) was unaffected, she could abduct her thumb with her abductor pollicis longus, but there would be some impairment of this movement owing to paralysis of the abductor pollicis brevis, supplied by the recurrent branch of the median nerve. The patient could extend her thumb normally using her extensor pollicis longus and brevis muscles. Because the nerve supply to her adductor pollicis muscle by deep branch of ulnar is intact, she could also adduct her thumb. Owing to paralysis of her first two lumbrical muscles and the loss of sensation over the thumb and adjacent two and one-half digits and the radial

two-thirds of her palm (Fig. 6-21), fine control of movements of her second and third digits is lacking. Thus, *cutting the median nerve produces a serious disability of the hand.* In a few weeks there will be atrophy of the thenar muscles.

The cutting of the tendons of the palmaris longus and flexor carpi radialis muscles would weaken flexion of her wrist. In addition, if she attempted to flex her wrist, her hand would be pulled to the ulnar side by the flexor carpi ulnaris, which is unaffected because it is supplied by the ulnar nerve.

Case 6-7

The common injury of the wrist in persons over 50 years of age, particularly women, is fracture of the distal end of the radius, known as a *Colles' fracture* (Fig. 6-94). The distal fragment of the radius tilts posteriorly, producing the typical "*dinner fork deformity*" of the wrist (Fig. 6-94A). The styloid processes of the ulna and radius are at the same level (Fig. 6-94C), instead of the radial styloid being more distal than the ulnar styloid as is normal. There is also subluxation or partial dislocation of the distal radioulnar joint.

SUGGESTED READINGS

Behrman RE: *Nelson's Textbook of Pediatrics*, ed 14. Philadelphia, WB Saunders, 1992.

Dravinsky O, Feldmann E: *Examination of the Cranial and Peripheral Nerves*, New York, Churchill Livingstone, 1988.

Ellis H: *Clinical Anatomy: A Revision and Applied Anatomy for Clinical Students*, ed 7. Oxford, Blackwell Scientific Publications, 1983.

Griffith HW: *Complete Guide to Sports Injuries*, Los Angeles, Price Stern Sloan, 1986.

Healey JE, Jr., Hodge J: *Surgical Anatomy*, ed 2. Toronto, BC Decker, 1990.

McMinn RM: *Last's Anatomy: Regional and Applied*, ed 8. Edinburgh, Churchill Livingstone, 1990.

McVay CB: *Anson and McVay Surgical Anatomy*, ed 6. Philadelphia, WB Saunders, 1984.

Moore KL: *The Developing Human: Clinically Oriented Embryology*, ed 4. Philadelphia, WB Saunders, 1988.

Salter RB: *Textbook of Disorders and Injuries of the Musculoskeletal System*, ed 2. Baltimore, Williams & Wilkins, 1983.

Tobias PV, Arnold M, Allan JC: *Man's Anatomy: A Study in Dissection*, ed 4. Johannesburg, Witwatersrand University Press, vol. 3, 1988.

Williams PL, Warwick R, Dyson M, Bannister LH (Eds): *Gray's Anatomy*, ed 37. Edinburgh, Churchill Livingstone, 1989.

Few complaints are more common than headache and facial pain. These terms are used to describe diffuse painful sensations in the head, whereas localized aches are given specific names such as *earache* (otalgia) and *toothache* (odontalgia). Headache often accompanies fever, tension, and fatigue, but sometimes it indicates a serious intracranial problem (*e.g.*, a **brain tumor**, subarachnoid hemorrhage, or *meningitis*). Consequently, all health care professionals require a sound knowledge of the anatomy of the head to understand the anatomical basis of headaches and facial pains.

The head contains a number of important structures, the diseases of which form the basis of several medical and surgical specialties: *neurology* (study of the nervous system and its disorders); *neuroradiology* (study of the skull and nervous system using imaging techniques); *neuropsychiatry* (study of organic and functional diseases of the brain); *neurosurgery* (surgery of the nervous system); *ophthalmology* (study of the eye and its disorders); *otology* (study of diseases of the ear and related structures); *rhinology* (study of the nose and its diseases); *maxillofacial surgery* (surgery of the face and jaws), *oral surgery*, (surgery of the mouth), and *dentistry* (study and treatment of the oral-facial complex, especially the teeth).

The Skull

The skull, the *skeleton of the head,* is the most complex bony structure in the body because it: (1) *encloses the brain,* which is irregular in shape; (2) *houses the organs of special senses* for seeing, hearing, tasting, and smelling; and (3) *surrounds the openings into the digestive and respiratory tracts.* In the anatomical position, the skull is oriented so that the inferior margin of the orbit (eye socket) and the superior margin of the external acoustic meatus (auditory canal) are horizontal. This is called the *orbitomenial plane* (Frankfort plane).

The term *cranium* (L. skull) is sometimes used when referring to *the skull without the mandible* (lower jaw), but the cranium is often used when referring to the part of the skull containing the brain. Its superior part is a boxlike structure called the **calvaria** (cranial vault, brain case); the remainder of the cranium, including the maxilla (upper jaw), orbits (eyeball sockets) and nasal cavities, forms the *facial skeleton*. The term *skullcap* (calotte) refers to the superior part of the calvaria (calvarium is incorrect), which is removed during autopsies and dissections. The inferior aspect of the cranium is called the *cranial base.*

Anterior Aspect of the Skull (Figs. 7-1 and 7-2). This aspect comprises the *anterior part of the calvaria* superiorly and

the *skeleton of the face* inferiorly. Notable features are: the *forehead* formed by the frontal bone; the *orbits* (eyeball sockets); the *prominences of the cheek* formed by the zygomatic bones; the *anterior nasal apertures* (piriform apertures), which open into the nasal cavities; the paired *maxillae* (forming the upper jaw) containing the maxillary (upper) teeth; and the *mandible* (lower jaw bone) containing the mandibular (lower) teeth. The anterior aspect of the skull may be divided into five areas: frontal, orbital, maxillary, nasal, and mandibular.

Posterior Aspect of the Skull (Fig. 7-3). This aspect, round or ovoid in outline, is formed mainly by the paired *parietal bones* and the *occipital bone*. They meet the mastoid parts of the temporal bones laterally. The most prominent feature of this aspect of the skull is the rounded posterior pole, called the *occiput*. Hence, this region is often referred to as the occipital area. The **external occipital protuberance** is a median projection, that is easily palpable. It is identifiable in living persons at the superior end of the median furrow (groove) in the posterior aspect of the neck. The center of the external occipital protuberance and its most prominent projection is called the *inion* (Fig. 7-8). Curved *superior nuchal lines* run laterally from the external occipital protuberance toward the *mastoid processes* of the temporal bones. These lines represent the superior limit of the posterior aspect of the neck. The posterior part of the *sagittal suture* and the *lambdoid sutures* are visible posteriorly. The *lambda*, where the lambdoid and sagittal sutures intersect (Fig. 7-8), is used when measuring the skull. The posterior aspect of the *mandible* is a prominent feature of this aspect of the skull. It is described on pp. 651, 653, and 654.

The adjective *nuchal* relates to the *nucha* (Fr. *nuque*), the nape or posterior aspect of the neck. The nuchal lines indicate where certain neck muscles attach, *e.g.*, semispinalis capitis (see Fig. 4-46).

Superior Aspect of the Skull (Figs. 7-4, 7-5, and 7-8). This aspect, also round or ovoid in outline, is broadened posteriorly by projections called *parietal eminences* (Fig. 7-9). The *superciliary arches* of the frontal bone (Fig. 7-1) form the anterior limit of this aspect of the skull. The zygomatic arches are visible laterally (Fig. 7-2). Four bones are united by **interlocking sutures**. The two parietal bones are joined at the *sagittal suture*; the frontal and parietal bones are united by the *coronal suture*; and the parietal bones are united to the occipital bone by the *lambdoid suture*. The intersection of the sagittal and coronal sutures is called the **bregma**. This landmark is used when measuring the skull. The *vertex*, the most superior part of the skull, is located near the center of the sagittal suture (Fig. 7-3). A *parietal foramen* is located in the parietal bone on each side of the sagittal suture (Fig. 7-3). These foramina transmit emissary veins, which connect the intracranial dural venous sinuses with the veins covering the skull (Fig. 7-30).

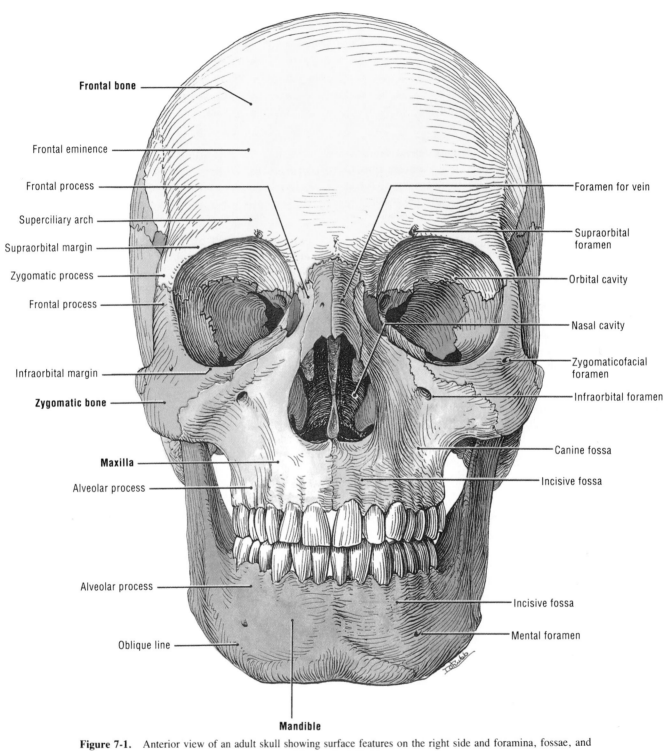

Frontal bone

Frontal eminence

Frontal process

Superciliary arch

Supraorbital margin

Zygomatic process

Frontal process

Infraorbital margin

Zygomatic bone

Maxilla

Alveolar process

Alveolar process

Oblique line

Mandible

Foramen for vein

Supraorbital foramen

Orbital cavity

Nasal cavity

Zygomaticofacial foramen

Infraorbital foramen

Canine fossa

Incisive fossa

Incisive fossa

Mental foramen

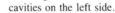

Figure 7-1. Anterior view of an adult skull showing surface features on the right side and foramina, fossae, and cavities on the left side.

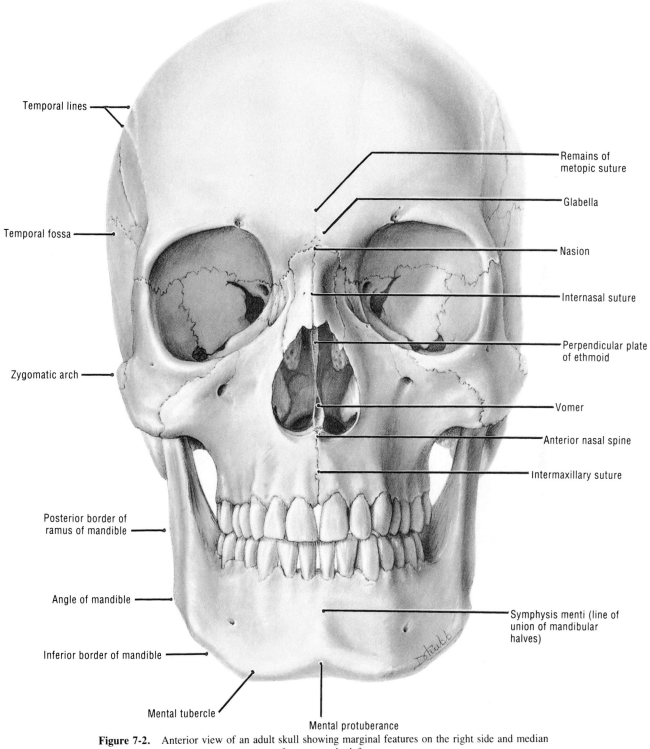

Figure 7-2. Anterior view of an adult skull showing marginal features on the right side and median features on the left.

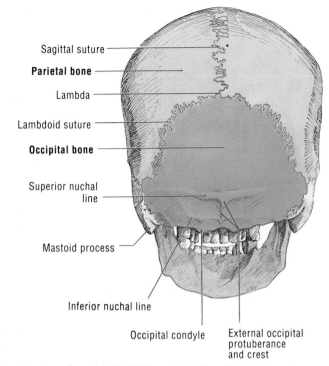

Sagittal suture

Parietal bone

Lambda

Lambdoid suture

Occipital bone

Superior nuchal line

Mastoid process

Inferior nuchal line

Occipital condyle

External occipital protuberance and crest

Figure 7-3. Posterior view of an adult skull. The occipital bone (*blue*) forms a large part of this aspect.

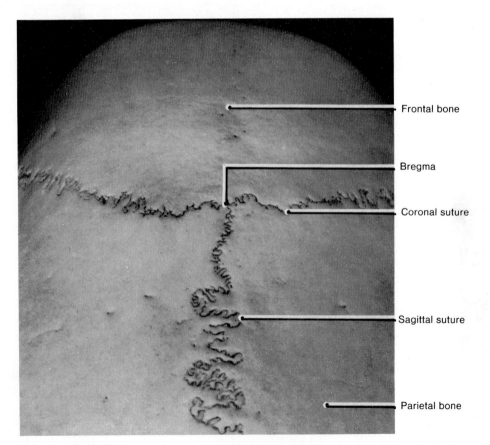

Frontal bone

Bregma

Coronal suture

Sagittal suture

Parietal bone

Figure 7-4. Anterior half of the superior aspect of an adult skull, showing a close-up of the sagittal and coronal sutures. Note the variation in the form and degree of interlocking of the cranial bones.

Inferior Aspect of the Skull (Figs. 7-6 to 7-9).   The external surface of the *cranial base* shows the inferior surface of the *maxillae* (upper jaw), *bony palate*, and *maxillary teeth*. The inferior surfaces of the *zygomatic arches*, which curve posteriorly on each side, are also visible. Each zygomatic arch is formed by the union of the *temporal process of the zygomatic bone* and the *zygomatic process of the temporal bone* (zygoma). Centrally, the inferior surface of the skull or **cranial base** is irregular owing to the many foramina, processes, and articulations. Laterally the cranial base exhibits the *temporal bones* with their prominent *mastoid and styloid processes.* The **foramen magnum** is one of the most conspicuous features of the base of the skull. It is bordered anterolaterally by the *occipital condyles*, which articulate with the atlas (C1 vertebra; see Fig. 4-17). The medulla oblongata of the brain stem (Fig. 7-55) passes through the foramen magnum where it is continuous with the spinal cord. Because it has many foramina (L. openings) and thin areas of bone, *the cranial base is fragile;* hence fractures of it are common.

Lateral Aspect of the Skull (Figs. 7-8, 7-9, and 7-14).   This aspect includes the parietal, frontal, and parts of the temporal and sphenoid bones. The division of the skull into the large, ovoid *calvaria* (cranial vault) and the smaller, uneven *facial skeleton* is clearly demonstrated. The calvaria is formed by the frontal bone anteriorly, the sphenoid and parietal bones laterally, and the occipital bone posteriorly. The **pterion** is an *important clinical landmark* on the lateral aspect of the skull. Observe that it is the area at which four bones articulate (Fig. 7-8): frontal, parietal, temporal, and sphenoid. The group of sutures joining these bones usually form an H-shape, which somewhat resembles wings (G. *pterion,* wing). The pterion is located in the *temporal fossa,* a region that is commonly known as the *temple* (Fig 7-2). The pterion is two fingerbreadths superior to the zygomatic arch and a thumbbreadth posterior to the zygomatic process of the frontal bone.

Other features of the lateral aspect of the skull are: the *external acoustic meatus* (external auditory meatus); *zygomatic bone*; **zygomatic arch**; mastoid process; and mandible. In Fig. 7-9 observe the body, ramus, coronoid process, head, and neck of the mandible. The **mastoid process** projects anteroinferiorly, medial to the lobule (ear lobe) of the auricle (external ear), where it is readily palpable (Fig. 7-8). The size of this process varies with the age and muscularity of the person. It is *not present at birth* and it is small during childhood. After puberty the mastoid processes enlarge. They form part of the superior attachment of the sternocleidomastoid muscles (see Figs. 8-4 and 8-7). The *mental protuberance* is an obvious feature in lateral views of the skull in most people (Fig. 7-8).

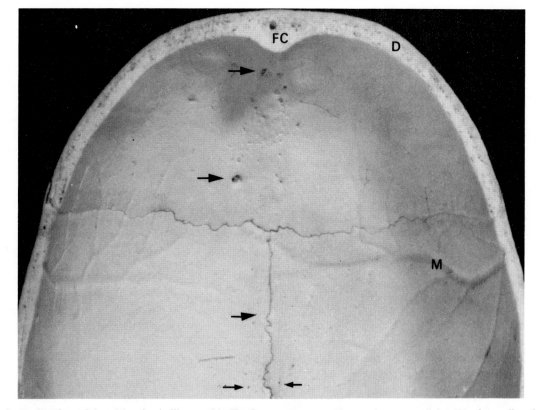

Figure 7-5.   Internal surface of the adult calvaria illustrated in Fig. 7-4. Observe the pits in the frontal bone (*large arrows*) produced by arachnoid granulations (Fig. 7-44A). On each side of the sagittal suture, note the parietal foramina (*small arrows*), through which emissary veins pass to connect the superior sagittal sinus with veins in the diploë (D) and scalp. Observe the spongy diploë (*D*) of cancellous bone that contained red marrow in life. Note also the sinuous groove (*M*) formed by the frontal branch of the middle meningeal artery (Fig. 7-44A). Note the frontal crest (*FC*) to which the falx cerebri was attached (Fig. 7-45).

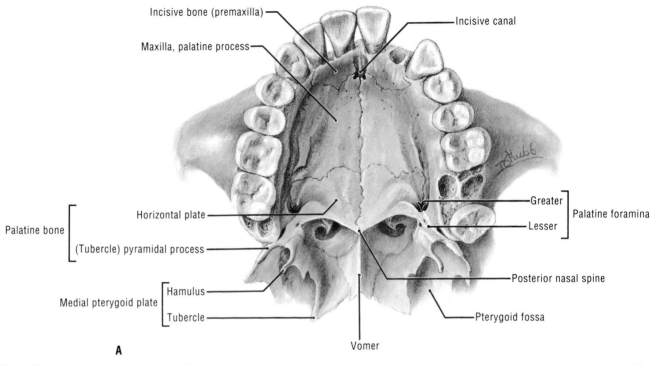

Incisive bone (premaxilla)

Maxilla, palatine process

Incisive canal

Palatine bone

Horizontal plate

(Tubercle) pyramidal process

Greater

Lesser

Palatine foramina

Medial pterygoid plate

Hamulus

Tubercle

Posterior nasal spine

Pterygoid fossa

A

Vomer

Figure 7-6. *A*, Inferior view of the anterior part of the skull, showing the bones of the hard palate and the maxillary (upper) teeth. Observe that the palate is formed by the palatine processes of the maxillae and the horizontal plates of the palatine bones. *B*, Similar view of another skull. Observe the large choanae (posterior nasal apertures) at either side of the posterior border of the nasal septum. They communicate with the nasal cavities.

The **pterion** is very important clinically because the anterior (frontal) branch of the *middle meningeal artery*, which is embedded between the layers of the dura mater (Fig. 7-44A), lies in a groove on the internal aspect of the lateral wall of the calvaria (Figs. 7-5 and 7-38). Here, it is vulnerable to tearing when there are *fractures of the bones forming the pterion*. The groove for this artery and its accompanying vein is usually very deep; sometimes it is roofed over with bone. In these cases it more liable to tear when the calvaria fractures. The resulting collection of blood (*extradural hematoma*; Fig. 7-126), may exert pressure on the underlying cerebral cortex. The pterion is also a useful landmark for locating certain parts of the brain. For example, an oblique line drawn from the frontozygomatic suture to the pterion is level with the inferior surface of the frontal lobe (Fig. 7-36).

Internal Aspect of the Skull (Figs. 7-5 and 7-10 to 7-12). To expose the interior of the skull, the calvaria has to be sawn circumferentially and removed. The bones that can be seen in the internal aspect of the base of the skull are the frontal, ethmoid, sphenoid, temporal, and occipital bones. The internal surface of the calvaria is fairly smooth and concave, particularly from side to side. The striking features of the internal aspect of the calvaria are the grooves in the parietal bones made by the anterior branches of the middle meningeal artery and its accompanying vein (Fig. 7-5). The sutures of the skull are more distinct on the external surface (Fig. 7-4) than on the internal surface (Fig. 7-5) because bony fusion begins on the inside of the calvaria between the ages of 20 and 30 years (10 years earlier than on

the external surface). *Arachnoid granulations* (Fig. 7-44A) project into venous sinuses, particularly into the lacunae at the side of the superior sagittal sinus. They protrude sufficiently to indent the bone of the calvaria, forming small pits along the groove formed by the superior sagittal sinus (Fig. 7-5). These pits increase in size and number with age. The internal aspect of the cranial base presents three distinct tiered areas: the anterior, middle, and posterior *cranial fossae* (Fig. 7-12B). These fossae are described subsequently (p. 675).

Walls of the Cranial Cavity (Figs. 7-9 to 7-11). These walls vary in thickness in different regions and in different persons. The cranium is usually thinner in females than in males and is thinner in children and elderly people. The bone tends to be thinnest in areas that are well covered with muscles (Fig. 7-23), *e.g.*, the squamous part of the temporal bone and the posteroinferior part of the skull, posterior to the foramen magnum. You can observe these thin areas of bone if you hold a cranium up to a light. Most bones of the calvaria consist of inner and outer tables of compact bone, separated by spongy diploë (Figs. 7-10 and 7-34A). The *diploë* is cancellous bone containing *red bone marrow* during life, through which run the canals formed by the *diploic veins*.

Examine the diploë in the calvaria from a laboratory specimen. It is not red in a dried skull because the protein was removed during preparation of the specimen. Also observe that the inner table of bone is thinner than the outer table and that in some areas, there is a thin plate of compact bone with no diploë.

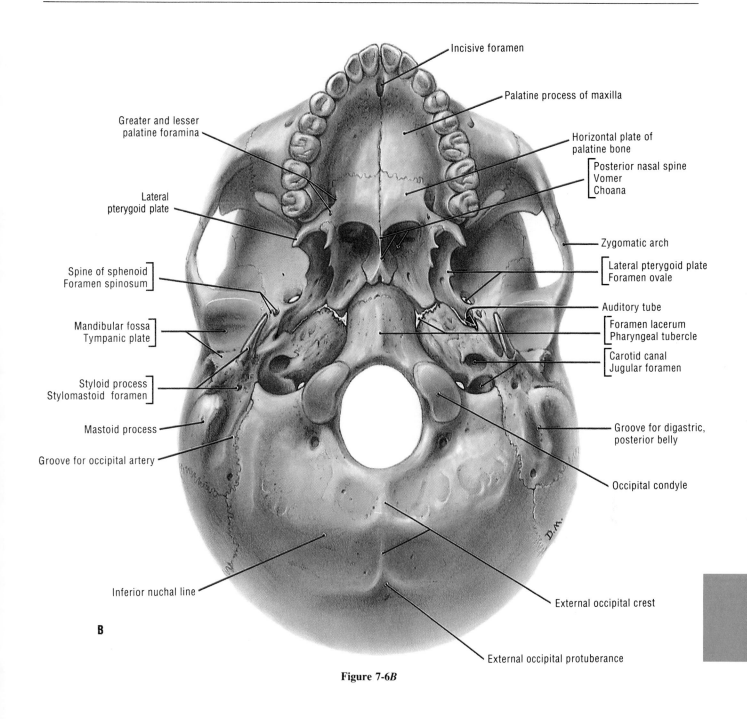

Incisive foramen

Palatine process of maxilla

Greater and lesser
palatine foramina

Horizontal plate of
palatine bone

Posterior nasal spine
Vomer
Choana

Lateral
pterygoid plate

Zygomatic arch

Lateral pterygoid plate
Foramen ovale

Spine of sphenoid
Foramen spinosum

Auditory tube

Foramen lacerum
Pharyngeal tubercle

Mandibular fossa
Tympanic plate

Carotid canal
Jugular foramen

Styloid process
Stylomastoid foramen

Mastoid process

Groove for digastric,
posterior belly

Groove for occipital artery

Occipital condyle

Inferior nuchal line

External occipital crest

B

External occipital protuberance

Figure 7-6B

Bones of the Skull

In radiographs of the skull the *diploic canals* containing the diploic veins may be mistaken for fractures of the cranium by inexperienced observers, especially in the parietal area. The convexity of the calvaria distributes and thereby minimizes the effects of a blow to it. However, hard blows to the head in areas where the calvaria is thin, *e.g.*, in the temporal fossa in the region of the pterion (Figs. 7-2 and 7-8) are likely to produce fractures. In depressed fractures of the skull, the inner table of the calvaria is often more extensively fractured than the outer table.

The *skeleton of the head*, or skull, is composed of many bones that are closely fitted together. As described previously, a series of flat bones are united by interlocking sutures to form the calvaria, and a group of irregular bones form the face and the cranial base. *The skull as a whole is of greater importance to most health professionals than are its constituent bones*, but it is important for everyone to understand how the skull is constructed. Except for the mandible and the auditory ossicles (ear bones) of the middle ear, the bones of the adult skull are joined by rigid

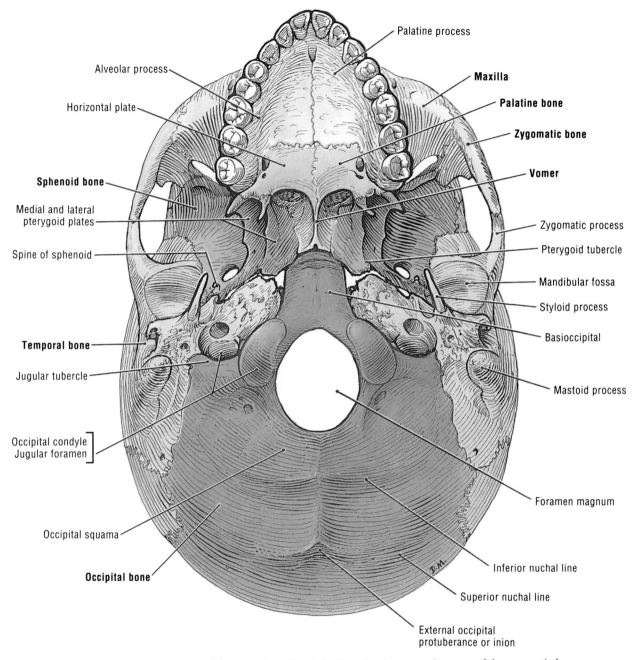

Figure 7-7.   External surface of the base of an adult skull. Note that the temporal process of the zygomatic bone unites with the zygomatic process of the temporal bone to form the zygomatic arch (Fig. 7-9).

sutures (p. 16). The cranium is essentially a single complex bone. Although the adult skull is rigid, the bones forming it in infants and children grow as individual bones and undergo remodeling. Furthermore, relationships among the various bones are constantly changing during these developmental periods.

Bones of the Calvaria

The Frontal Bone (Figs. 7-1, 7-2, 7-4, 7-5, and 7-8 to 7-14).   The forehead (L. *frons*) is formed by the smooth, broad, convex plate of bone called the *frontal squama*. In fetal skulls

(p. 17), the halves of the frontal squama are divided by a *metopic suture* (G. *metopon*, forehead). In most people the halves of the frontal bone begin to fuse during infancy and the suture between them is usually not visible after 6 years of age. The frontal bone forms the thin roof of the orbits (eye sockets). Just superior to and parallel with each supraorbital margin is a bony ridge, the *superciliary arch* (Fig. 7-1), which overlies the *frontal sinus* (Figs. 7-13 and 7-14). This arch is more pronounced in males. Between these arches there is a gently, rounded, median elevation called the *glabella* (Fig. 7-2); this term derives from the Latin word *glabellus* meaning smooth and hairless. In most people the skin over the glabella is hairless. The slight prominences of the forehead on each side, superior to the superciliary arches (Figs. 7-1 and 7-9), are called *frontal eminences* (tubers). The *supraorbital foramen* (occasionally a notch), which transmits the supraorbital vessels and nerve (Figs. 7-23 and 7-24), is located in the medial part of the supraorbital margin.

The frontal bone articulates with the two parietal bones at the *coronal suture* (Figs. 7-4 and 7-9). It also articulates with the nasal bones at the *frontonasal suture*. At the point where this suture crosses the *internasal suture* in the median plane, there is an anthropological landmark called the *nasion* (L. *nasus*, nose). This depression is located at the root of the nose, where it joins the cranium (Fig. 7-2). The frontal bone also articulates with the zygomatic, lacrimal, ethmoid, and sphenoid bones.

> In about 8% of adult skulls, a remnant of the inferior part of the metopic (interfrontal) suture is visible (Fig. 7-2). It may be mistaken in radiographs for a fracture line by inexperienced observers. As the superciliary arches are relatively sharp ridges of bone (Fig. 7-1), a blow to them may lacerate the skin and cause bleeding (*e.g.*, a head butt during a boxing match). Bruising of the skin over a superciliary arch causes tissue fluid and blood to accumulate in the surrounding connective tissue, which gravitates into the upper eyelid and around the eye. This results in swelling and a "black eye."
>
> *Compression of the supraorbital nerve* as it emerges from its foramen (Fig. 7-21) causes considerable pain, a fact that may be used by anesthesiologists and *anesthetists*[1] to determine the depth of anesthesia and by physicians attempting to arouse a moribund (L. dying) patient.

The Parietal Bones (Figs. 7-3 to 7-5 and 7-8 to 7-11). The two parietal bones (L. *paries*, wall) form large parts of the walls of the calvaria. On the outside of these smooth convex bones, there are slight elevations near the center called *parietal eminences*. The middle of the lateral surfaces of the parietal bones is crossed by two curved lines, the superior and inferior *temporal lines*. The superior temporal line indicates an attachment of the *temporal fascia* (Fig. 7-75); the inferior temporal line marks the superior limit of the *temporalis muscle*. The parietal bones articulate with each other in the median plane at the *sagittal suture* (Figs. 7-3 to 7-5). The median plane of the body passes through the sagittal suture (p. 3). The inverted V-shaped suture between the parietal bones and the occipital bone is called the *lambdoid suture* (Figs. 7-3 and 7-9) because of its resemblance to the letter **lambda** in the Greek alphabet. The point where the parietal and occipital bones join is a useful reference point called the lambda. It can be felt as a depression in some people. In addition to articulating with each other and the frontal and occipital bones, the parietal bones articulate with the temporal bones and the greater wings of sphenoid bone.

> In fetal and infant skulls, the bones of the calvaria are separated by dense connective tissue membranes at fibrous joints called *sutures*. The large fibrous areas where several sutures meet are called fonticuli or *fontanelles* (p. 16). The softness of the bones and the looseness of their connections

at these sutures enable the calvaria to undergo changes of shape during birth called *molding*. Within a day or so after birth, the shape of the infant's calvaria returns to normal. The loose construction of the newborn calvaria also allows the skull to enlarge and undergo remodeling during infancy and childhood. Relationships between the various bones are constantly changing during the active growth period. The increase in the size of the cranium is greatest during the first 2 years of life, the period of most rapid postnatal growth of the brain. The cranium normally increases in capacity until about 15 or 16 years of age; thereafter the cranium usually increases only slightly in size as its bones thicken for 3 to 4 years.

The Temporal Bones (Figs. 7-1 to 7-3, 7-8, and 7-9). The sides and base of the skull are formed partly by these bones. Each bone consists of four morphologically distinct parts that fuse during development (squamous, petromastoid, and tympanic parts and the styloid process). The flat *squamous part* is external to the lateral surface of the temporal lobe of the brain (Fig. 7-36). The *petromastoid part* encloses the internal ear and *mastoid cells* and forms part of the base of the skull. The *tympanic part* contains the bony passage from the auricle (external ear), called the *external acoustic meatus*. The petromastoid part also forms a portion of the bony wall of the tympanic cavity (middle ear). The meatus and tympanic cavity are concerned with the transmission of sound waves (p. 763).

The slender, pointed *styloid process of the temporal bone* gives attachment to certain ligaments and muscles (*e.g.*, the stylohyoid muscle that elevates the hyoid bone). The temporal bones articulate at sutures with the parietal, occipital, sphenoid, and zygomatic bones. The *zygomatic process*[2] of the temporal bone unites with the temporal process of the zygomatic bone to form the **zygomatic arch**. The zygomatic arches form the widest parts of the face. The *head of the mandible* articulates with the *mandibular fossa* on the inferior surface of the zygomatic process of the temporal bone (Fig. 7-7). Anterior to the mandibular fossa is the *articular tubercle*.

> Because the zygomatic arches are the widest parts of the face and are such prominent facial features, they are commonly fractured and depressed. A fracture of the temporal process of the zygomatic bone would likely involve the lateral wall of the orbit and could injure the eye.

The Sphenoid Bone (Figs. 7-7 to 7-12). This wedge-shaped bone (G. *sphen*, wedge) is located anterior to the temporal bones. *It is a key bone in the cranium because it articulates with eight bones* (frontal, parietal, temporal, occipital, vomer, zygomatic, palatine, and ethmoid). Its main parts are the **body** and the greater and lesser **wings**, which spread laterally from the body. The superior surface of its body is shaped like a Turkish saddle (L. *sella*, a saddle); hence its name *sella turcica* (Fig. 7-12A). It forms the hypophyseal fossa (Fig. 7-14) which contains the hypophysis cerebri or *pituitary gland* (Fig. 7-110). The sella turcica is bounded posteriorly by the *dorsum sellae*, a square plate of

---

[1]Anesthesiologists are physicians who administer anesthetics and are distinct from anesthetists who may not be physicians.

[2]The zygomatic process of the temporal bone, which forms the posterior root of the zygomatic arch, is sometimes referred to as the *zygoma*. This term is used most often as an abbreviated term for the *zygomatic arch* (Figs. 7-6*B* and 7-23*A*).

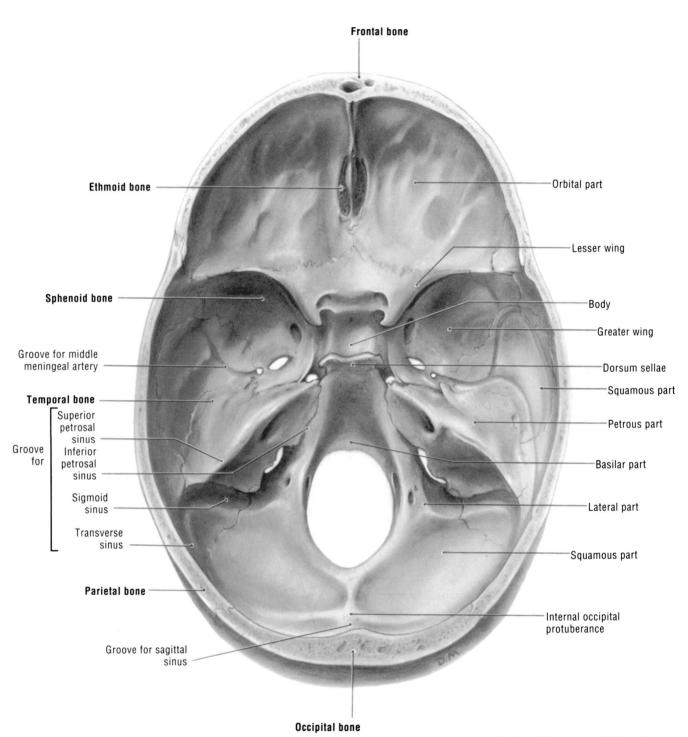

Frontal bone

Ethmoid bone

Sphenoid bone

Groove for middle
meningeal artery

Temporal bone

Groove
for
— Superior
petrosal
sinus
Inferior
petrosal
sinus

Sigmoid
sinus

Transverse
sinus

Parietal bone

Groove for sagittal
sinus

Occipital bone

Orbital part

Lesser wing

Body

Greater wing

Dorsum sellae

Squamous part

Petrous part

Basilar part

Lateral part

Squamous part

Internal occipital
protuberance

Figure 7-10. Interior of the base of an adult skull. Observe that the sphenoid bone resembles a bat with its wings outstretched; however, it was given its name because of its wedge shape (G. *sphen*, wedge). Note that the orbital parts or plates of the frontal bone show shallow, sinuous convolutional depressions (brain markings). They were formed by the convolutions of the frontal lobe gyri (Fig. 7-36).

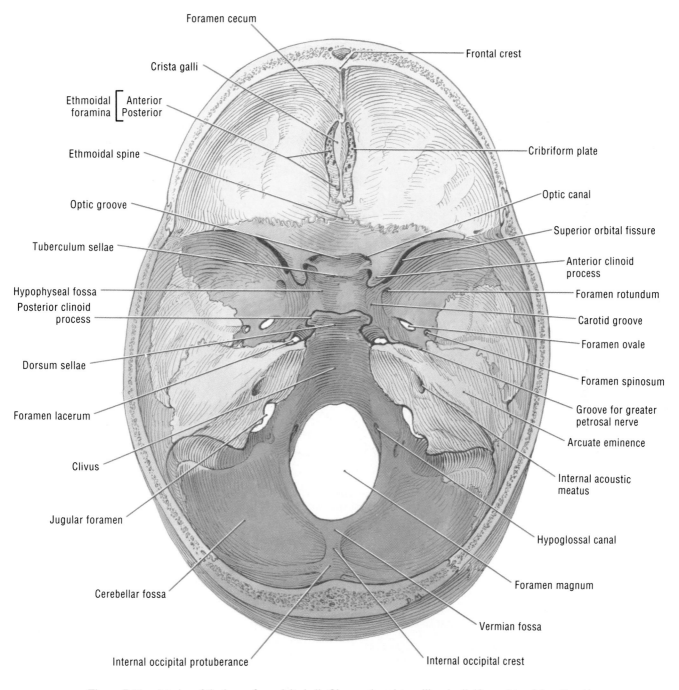

Foramen cecum

Frontal crest

Crista galli

Ethmoidal foramina [Anterior / Posterior]

Cribriform plate

Ethmoidal spine

Optic canal

Optic groove

Superior orbital fissure

Tuberculum sellae

Anterior clinoid process

Hypophyseal fossa

Foramen rotundum

Posterior clinoid process

Carotid groove

Foramen ovale

Dorsum sellae

Foramen spinosum

Foramen lacerum

Groove for greater petrosal nerve

Clivus

Arcuate eminence

Internal acoustic meatus

Jugular foramen

Hypoglossal canal

Cerebellar fossa

Foramen magnum

Internal occipital protuberance

Vermian fossa

Internal occipital crest

Figure 7-11. Interior of the base of an adult skull. Observe the crista galli and cribriform plate of the ethmoid bone in the anterior cranial fossa.

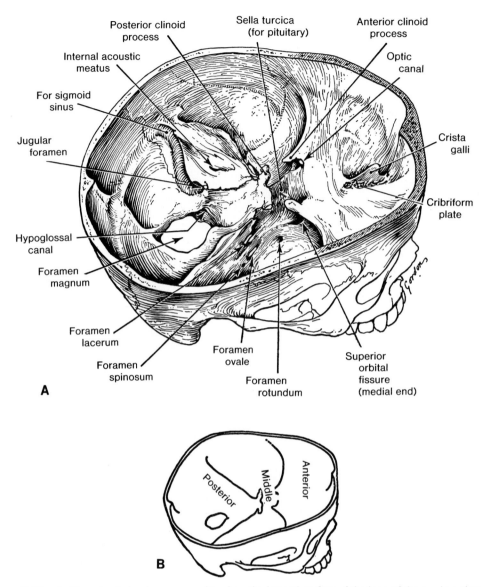

Figure 7-12. *A*, The calvaria has been removed to show the internal surface of the base of the cranium showing three tiered areas: the anterior, middle, and posterior cranial fossae. *B*, Orientation drawing.

bone that projects superiorly and has a *posterior clinoid process* on each side. Inside the body of the sphenoid bone, there are right and left *sphenoidal sinuses* (Fig. 7-14). The floor of the sella turcica forms the roof of these paranasal sinuses.

> Studies of the sella turcica and hypophyseal fossa in radiographs or by other imaging techniques (Fig. 7-14) are important because they may reflect pathological changes such as a *pituitary tumor* or an aneurysm of the internal carotid artery. Decalcification of the dorsum sellae is one of the signs of a generalized increase in intracranial pressure.

The Occipital Bone (Figs. 7-3 and 7-7 to 7-12).   This bone forms much of the base and posterior aspect of the skull. It has a large oval opening called the *foramen magnum*, through which

the cranial cavity communicates with the vertebral canal. It is also where the spinal cord becomes continuous with the medulla (oblongata) of the brain stem (Fig. 7-37). The occipital bone is saucer-shaped and can be divided into four parts: a *squamous part* (squama), a *basilar part* (basioccipital part), and two *lateral parts* (condylar parts). These four parts develop separately around the foramen magnum and unite at about the age of 6 years to form one bone. On the inferior surfaces of the lateral parts of the occipital bone are **occipital condyles**, where the skull articulates with C1 vertebra (the atlas) at the atlanto-occipital joints (see Fig. 4-34). The internal aspect of the squamous part of the occipital bone is divided into four fossae: the superior two for the occipital poles of the cerebral hemispheres, and the inferior two, called *cerebellar fossae*, for the cerebellar hemispheres (Figs. 7-11 and 7-36).

Bones of the Face

Most of the facial skeleton is formed by nine bones: four paired (nasal, zygomatic, maxilla, and palatine) and one unpaired (mandible).

The calvaria of the newborn infant is large compared with the relatively small facial skeleton. This results from the small size of the jaws and the almost complete absence of the maxillary and other *paranasal sinuses* in the newborn skull. These sinuses form large spaces in the adult facial skeleton (Figs. 7-13 and 7-14). As the teeth and sinuses develop during infancy and childhood, the facial bones enlarge. The growth of the maxillae between the ages of 6 and 12 years accounts for the vertical elongation of the child's face.

The Nasal Bones (Figs. 7-1, 7-2, 7-8, and 7-9).   These bones may be felt easily because they form the *bridge of the nose*. The right and left nasal bones articulate with each other at the *internasal suture*. They also articulate with the frontal bones, the maxillae, and the ethmoid bones.

The mobility of the anteroinferior portion of the nose, supported only by cartilages, serves as a partial protection against injury (*e.g.*, a punch in the nose). However, a hard blow to the anterosuperior bony portion of the nose (*e.g.*, with a hockey stick or a baseball), may fracture the nasal bones (*broken nose*). Often the bones are displaced sideways and/or posteriorly (Fig. 7-13). The outer table of bone forming the cranium of a living person is somewhat resilient, unlike that in a dried laboratory skull from which the proteins have been removed. This elasticity, especially in infants and children, tends to prevent many blows to the head from producing skull fractures. Furthermore, in places where the bone is very thin, *e.g.*, the squamous part of the temporal bone (Fig. 7-9), the overlying muscles afford some assistance in cushioning blows (Fig. 7-75).

Fractures of the skull are caused by external violence. **Linear skull fractures** are the most frequent type. The fracture usually occurs at the point of impact, but fracture lines may radiate away from it in two or more directions (*e.g.*, the way a window breaks when hit by a stone). The directions of the radiating fracture lines are determined by the thick and thin areas of the skull. Sometimes there is no fracture at the point of impact, but one occurs at the opposite side of the skull. This is called a *contrecoup fracture* (from the French word *contrecoup* meaning counterblow).

The Maxillae (Figs. 7-1, 7-2, 7-6 to 7-9, 7-13, and 7-14).   The skeleton of the face between the mouth and eyes is formed by the two maxillae. They surround the *anterior nasal apertures* and are united in the median plane at the *intermaxillary suture* to form the maxilla (upper jaw). This suture is also visible in the *hard palate*, where the palatine processes of the maxillae unite. Each adult maxilla consists of: a hollow *body* that contains a large **maxillary sinus**; a *zygomatic process* that articulates with the zygomatic bone; a *frontal process* that articulates with the frontal and nasal bones; a *palatine process* that articulates with its mate on the other side to form most of the hard palate; and *alveolar processes* that form sockets for the maxillary (upper) teeth. The maxillae also articulate with the vomer, lacrimal, sphenoid, and palatine bones.

The **body of the maxilla** has a *nasal surface* that contributes to the lateral wall of the nasal cavity; an *orbital surface* that forms most of the floor of the orbit; an *infratemporal surface* that forms the anterior wall of the infratemporal fossa (Fig. 7-76); and an *anterior surface* that faces partly anteriorly and partly anterolaterally and is covered by facial muscles. The relatively large *infraorbital foramen*, which faces inferomedially, is located about 1 cm inferior to the infraorbital margin; it transmits the *infraorbital nerve and vessels* (Fig. 7-21). The *incisive fossa* is a shallow concavity overlying the roots of the incisor teeth, just inferior to the nasal cavity (Fig. 7-1). This fossa is the injection site for anesthesia of the maxillary incisor teeth (Liebgott, 1986).

If infected maxillary teeth are removed, the bone of the alveolar processes of the maxillae begins to resorb. As a result, the maxilla becomes smaller (*i.e.*, decreased in height) and the shape of the face changes. Owing to absorption of the alveolar processes, there is a marked reduction in the height of the lower face, which produces deep creases in the facial skin that pass posteriorly from the corners of the mouth.

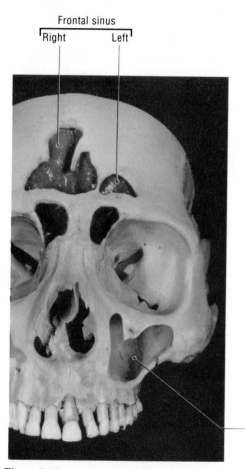

Frontal sinus
⌈Right   Left⌉

—— Left maxillary sinus

Figure 7-13.   Anterior view of an adult skull that has been specially prepared to demonstrate the paranasal sinuses. Note the crooked bony nasal septum. (For its normal appearance, see Fig. 7-2.) This deformity may have resulted from a facial injury involving the nose.

The Mandible (Figs. 7-1 to 7-3, 7-8, 7-9, and 7-14 to 7-16).   This U-shaped bone forms the skeleton of the lower jaw and the inferior part of the face. It is the *largest and strongest*

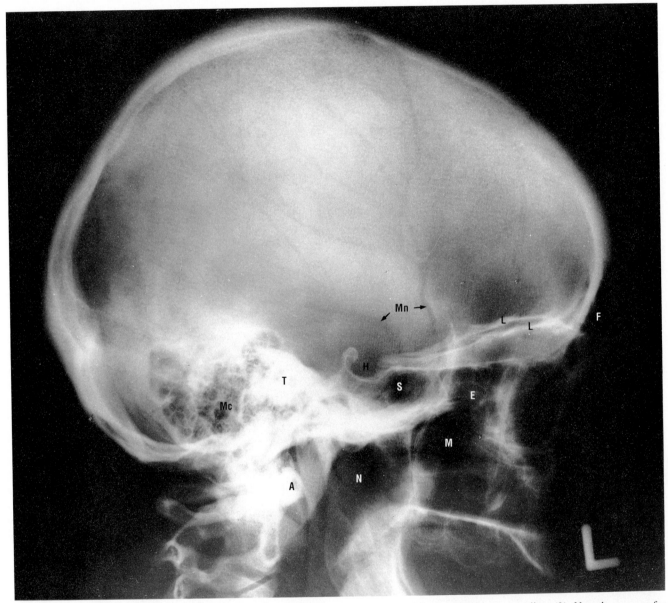

Figure 7-14. Lateral radiograph of the skull showing the paranasal sinuses: frontal (*F*), ethmoidal (*E*), sphenoidal (*S*), maxillary (*M*). Also observe the hypophyseal fossa (*H*); the density of the petrous part of the temporal bone (*T*); and the mastoid (air) cells (*Mc*). The orbital parts or plates of the frontal bone (*F*) are not superimposed, thus the floor of the anterior cranial fossa appears as two lines (*L*). Note the grooves for the meningeal vessels (*Mn*), the arch of the atlas or C1 vertebra (*A*), and the nasopharynx (*N*). (Courtesy of Dr. E. Becker, Associate Professor of Radiology, University of Toronto, Toronto, Ontario, Canada.)

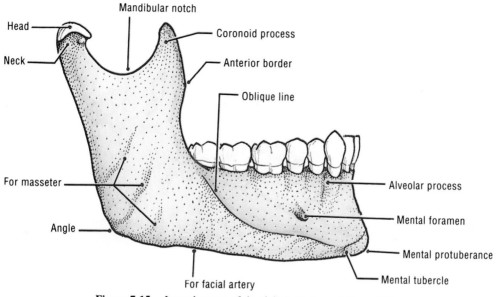

Figure 7-15.   Lateral aspect of the right half of an adult mandible.

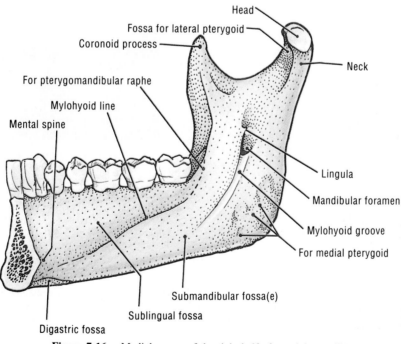

Figure 7-16.   Medial aspect of the right half of an adult mandible.

facial bone. The mandibular (lower) teeth project superiorly from their sockets in the alveolar processes of the mandible. The mandible (L. *mandere,* to masticate) consists of two parts: a horizontal part called the *body,* and two vertical oblong parts, called *rami.* Each **ramus** ascends almost vertically from the posterior aspect of the body. The superior part of the ramus has two processes: a posterior *condylar process* with a head or **condyle** and a neck, and a sharp anterior *coronoid process.* The condylar process is separated from the coronoid process by the *mandibular notch,* which forms the concave superior border of

the mandible. Viewed from the superior aspect, the mandible is horseshoe-shaped (Fig. 7-89), whereas each half is L-shaped when viewed laterally. The rami and body meet posteriorly at the *angle of the mandible.* Inferior to the second premolar tooth on each side of the mandible is a *mental foramen* (L. *mentum,* chin) for transmission of the mental vessels and the mental nerve (Fig. 7-21).

In the anatomical position, *the rami of the mandible are almost vertical* (Fig. 7-15), except in infants and in edentulous (toothless) adults (Fig. 7-17). On the internal aspect of the ramus,

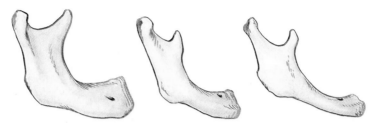

Figure 7-17. Lateral views of the right side of edentulous mandibles, showing how the position of the mental foramen varies with the extent of absorption of the alveolar processes (tooth sockets). *A,* Several months after removal of the permanent mandibular teeth. *B,* Several years after loss of the teeth. Observe that much of the alveolar bone has been resorbed and that the angle of the mandible has increased. *C,* Many years after loss of the teeth showing that the alveolar processes are entirely resorbed. The mental foramen now lies at the superior border of the mandible. Note that the angle between the body and the ramus has further increased and that the neck of the condylar process is bent posteriorly.

there is a large **mandibular foramen** (Fig. 7-16). It is the oblong entrance to the *mandibular canal* that transmits the inferior alveolar vessels and nerve to the roots of the mandibular teeth (Fig. 7-25). Branches of these vessels and the mental nerve emerge from the mandibular canal at the *mental foramen* (Figs. 7-15 and 7-21).

Anterior to the mandibular foramen is a thin, tonguelike projection of bone, called the *lingula of the mandible* (L. *lingua,* tongue). It somewhat overlaps and guards the superoanterior border of the foramen like a tongue or shield (Fig. 7-16). Anesthetic agents are injected into the mandibular foramen when performing an inferior *alveolar nerve block* (p. 731). Sometimes the lingula interferes with these injections. When it is necessary to *anesthetize the mental nerve* and the incisive branch of the inferior alveolar nerve (*e.g.,* to suture a laceration in the lower lip and chin), a *mental and incisive nerve block* is performed by injecting anesthetic fluid into the mouth of the mental foramen. As this foramen opens superiorly and slightly posteriorly, access to it is somewhat difficult. During injection the mental artery and vein must be avoided so that the anesthetic agent is not injected into them.

Running inferiorly and slightly anteriorly on the internal surface of the mandible from the mandibular foramen is a small *mylohyoid groove* (sulcus), which indicates the course taken by the *mylohyoid nerve* and vessels (Figs. 7-16 and 7-25). These structures arise from the inferior alveolar nerve and vessels, just before they enter the mandibular foramen. The internal surface of the mandible is divided into two areas by the *mylohyoid line,* which commences posterior to the third molar tooth. Just superior to the anterior end of the mylohyoid line are two small, sharp *mental spines,* which serve as attachments for the genioglossus muscles (Fig. 7-94).

The size and shape of the mandible and the number of teeth it bears change with age. In the newborn the mandible consists of two halves united in the median plane by fibrous tissue joint called the *symphysis menti* or mandibular symphysis (see Fig. 15, p. 17). The former site of this articulation is evident in the adult skull as a bony ridge (Fig. 7-2). The two halves of the mandible begin to fuse during the first year and are fused by the end of the second year of life. The body of the newborn's mandible is a mere shell, each half enclosing five primary teeth. The teeth usually begin to erupt in infants of about 6 months (p. 740). The body of the mandible elongates, particularly posterior to the mental foramen, to accom-

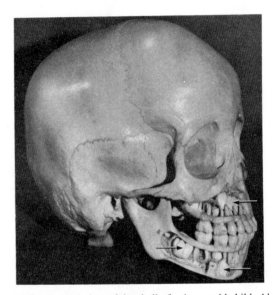

Figure 7-18. Lateral view of the skull of a 4-year-old child. Alveolar bone has been ground away and the jaws have been dissected to show the relations of the developing permanent teeth (*arrows*) to the deciduous teeth. Between the sixth and twelfth years, the 20 primary teeth are shed as the permanent teeth erupt. (Reprinted with permission from Moore KL: *The Developing Human: Clinically Oriented Embryology,* ed 4. Philadelphia, WB Saunders, 1988).

modate the eight secondary teeth (Fig. 7-18), which begin to erupt during the sixth year of life. Eruption of the permanent teeth is not complete until early adulthood.

If infected mandibular teeth are removed, the alveolar processes of the mandible begin to resorb. Gradually the mental and mandibular foramina come to lie near the superior border of the body of the mandible (Fig. 7-17). In extreme cases the mental foramen and part of the mandibular canal may disappear, exposing the mental and inferior alveolar nerves to injury. *Pressure of a dental prosthesis* (*e.g.,* a denture resting on an exposed nerve in an edentulous jaw) may produce pain during eating. Edentulousness results in a decrease in the vertical facial dimensions and *mandibular prognathism* (overclosure or protrusion).

The common sites of **fractures of the mandible** are explained and illustrated in Fig. 7-19. Usually there are two fractures and they frequently occur on opposite sides of the mandible; thus if one fracture is observed, a search should be made for another. For example, a hard blow to the jaw often fractures the neck of the mandible and its body in the region of the opposite canine tooth.

The Zygomatic Bones (Figs. 7-1, 7-8, 7-9, and 7-14). The prominences of the cheeks (L. *mala*), the anterolateral rims and much of the infraorbital margins of the orbits, are formed by the zygomatic bones (malar bones, cheek bones). They articulate with the frontal, maxilla, sphenoid, and temporal bones. The frontal process of the zygomatic bone passes superiorly, where it forms the lateral border of the orbit (eye socket) and articulates with the frontal bone at the lateral edge of the supraorbital margin.

The zygomatic bones articulate medially with the greater wings of the sphenoid bone. The site of their articulation may be observed on the lateral wall of the orbit (Figs. 7-7 and 7-9). On the anterolateral aspect of the zygomatic bone near the infraorbital margin is a small *zygomaticofacial foramen* for the nerve and vessels of the same name (Figs. 7-1 and 7-21). The posterior surface of the zygomatic bone near the base of its frontal process is pierced by a small *zygomaticotemporal foramen* for the nerve of the same name. The zygomaticofacial and zygomaticotemporal nerves, leaving the orbit through the previously mentioned foramina, enter the zygomatic bone through small *zygomatico-orbital foramina* that pierce its orbital surface.

The temporal process of the zygomatic bone unites with the zygomatic process of the temporal bone to form the *zygomatic arch* (Figs. 7-2 and 7-7 to 7-9). This arch can be easily palpated on the side of the head, posterior to the *zygomatic prominence*

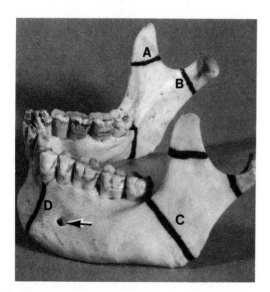

Figure 7-19.   Lateral aspect of an adult mandible from the left side, indicating the common sites of fracture. Usually there are two fractures, frequently on opposite sides. *A*, Fractures of the coronoid process are usually single. *B*, Fractures of the neck are often transverse and may be associated with dislocation of the temporomandibular joint on the same side. *C*, Fractures of the angle of the mandible are usually oblique and may involve the socket of the third molar tooth. *D*, Fractures of the body of the mandible frequently pass through the socket of the canine tooth. The arrow indicates the mental foramen.

(malar eminence) at the inferior boundary of the *temporal fossa* (temple). The zygomatic arches form one of the useful landmarks for determining the location of the *pterion* (Fig. 7-8; p. 641). These arches are especially prominent in emaciated persons. A horizontal plane passing medially from the zygomatic arch separates the temporal fossa superiorly from the infratemporal fossa inferiorly (Fig. 7-2).

You may hear the term *malar flush*; this redness of the skin over the zygomatic prominence (malar eminence) occurs with certain diseases (*e.g.*, tuberculosis). The common variants of *fractures of the maxilla and zygomatic bones* were classified by Le Fort, a Paris surgeon. The three types of fracture are remarkably constant; they are described and illustrated in Figure 7-20.

The Face

The face is the part of the head that is visible in a frontal or anterior view, (*i.e.*, all that is anterior to the external ears and between the hairline [or where it was] and the tip of the chin). The term *facies* (L. face) refers to the appearance of the face. Some diseases produce a typical facies, *e.g.*, the masklike, expressionless facies of patients with *Parkinson's disease* (Parkinson's syndrome), a disturbance of the function of the facial muscles (Barr and Kiernan, 1988).

At birth the face is small compared with the rest of the head because the jaws are not developed fully and most of the paranasal sinuses in the facial bones have not started to form. As these structures develop during infancy and childhood, the face becomes longer, the zygomatic prominences more prominent, and the cheeks lose some of their fat. An infant's cheeks appear full because of the *buccal fatpad* between the buccinator muscle and the superficial facial muscles (Fig. 7-21). These fatpads in the cheeks are present in adults, but they are much smaller than in infants.

Muscles of the Face

The *muscles of facial expression* lie in the subcutaneous tissue and are attached to the skin of the face (Figs. 7-21 to 7-23). They enable us to move our skin and change our facial expression to convey mood during communication. Most facial muscles are attached to bone or fascia. They produce their different effects by pulling on the skin; they do not move the facial skeleton. All facial muscles receive their motor innervation from the *facial nerve* (cranial nerve [CN] VII). These muscles surround the facial orifices (mouth, eyes, nose, and ears) and act as sphincters and dilators, *i.e.*, they open and close these orifices.

The facial muscles develop from the second branchial or pharyngeal arch as part of the subcutaneous muscle sheet in the head and neck which forms the *platysma* (Figs. 7-21 and 7-22; see Fig. 6-5). It spreads like a sheet over the neck and

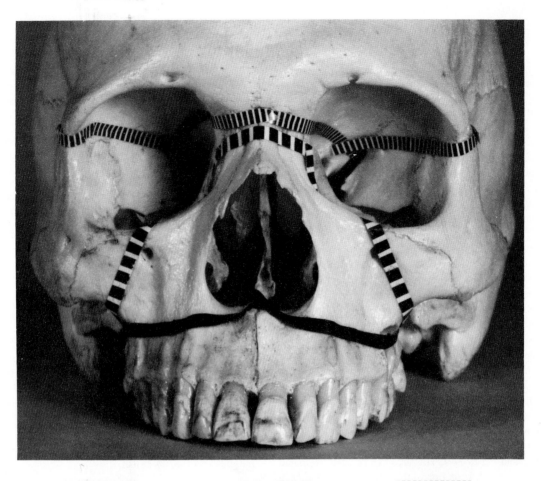

Le Fort I

Le Fort II

Le Fort III

Figure 7-20. A skull showing the common fractures of the maxillae and concurrent fracturing of other bones. *A*, Le Fort I. A horizontal fracture of the maxillae, located just superior to the alveolar processes, crossing the bony nasal septum and the pterygoid plates of the sphenoid bone. *B*, Le Fort II. A fracture that passes from the posterolateral parts of the maxillary sinuses, superomedially through the infraorbital notch, the lacrimals, or the ethmoid to the bridge of the nose. As a result, the entire central part of the face, including the hard palate and the alveolar processes, is separated from the rest of the skull. *C*, Le Fort III. A horizontal fracture that passes through the superior orbital fissures, the ethmoid and nasal bones, and extends laterally through the greater wings of the sphenoid bone and the frontozygomatic sutures. As there is concurrent fracturing of the zygomatic arches, the maxillae and zygomatic bones are separated from the rest of the skull.

face during embryonic development, bringing branches of the facial nerve with it (Moore, 1988). Because of their common origin, the sheetlike platysma and facial muscles are often fused and their fibers, which are attached to the skin, are frequently intermingled (Fig. 7-21).

> Because the face has no distinct deep fascia and the superficial fascia between the cutaneous attachments of the facial muscles is loose, *facial lacerations* tend to gape (part widely). Consequently the skin has to be sutured with great care to prevent scarring. The looseness of the superficial fascia also enables fluid and blood to accumulate in the loose connective tissue following bruising of the face (*e.g.*, a "black eye"). Similarly, facial inflammation causes considerable swelling (*e.g.*, the swelling resulting from a bee sting on the bridge of the nose can close both eyes). *As a person ages, the skin loses its resiliency.* As a result, ridges and wrinkles occur in the skin perpendicular to the direction of the facial

muscle fibers. Incisions along these ridges heal with minimal scarring.

Muscle of the Forehead

The **frontalis muscle** is part of the scalp muscle called the *occipitofrontalis* (Figs. 7-21, 7-22*B*, and 7-23). The frontalis elevates the eyebrows, giving the face a surprised look, and produces transverse wrinkles in the forehead when one frowns.

Muscles Around the Mouth

Several muscles alter the shape of the mouth and lips (*e.g.*, during singing, speaking, and mimicry). The *sphincter of the mouth* (used when whistling) is the **orbicularis oris** (Fig. 7-21). The *dilator muscles* radiate outward from the lips like the spokes of a wheel. The individual muscles (described in the following

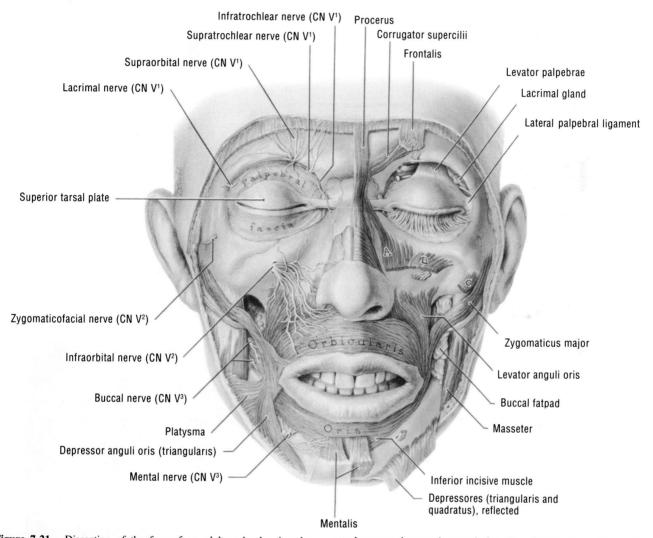

Infratrochlear nerve (CN V¹)
Supratrochlear nerve (CN V¹)
Supraorbital nerve (CN V¹)
Lacrimal nerve (CN V¹)
Procerus
Corrugator supercilii
Frontalis
Levator palpebrae
Lacrimal gland
Lateral palpebral ligament
Superior tarsal plate
Zygomaticofacial nerve (CN V²)
Infraorbital nerve (CN V²)
Buccal nerve (CN V³)
Platysma
Depressor anguli oris (triangularis)
Mental nerve (CN V³)
Zygomaticus major
Levator anguli oris
Buccal fatpad
Masseter
Inferior incisive muscle
Depressores (triangularis and quadratus), reflected
Mentalis

Figure 7-21. Dissection of the face of an adult male showing the cutaneous branches of the trigeminal nerve (*yellow*) and the muscles of facial expression. A pin has been inserted posterior to the aponeurosis of the levator palpebrae superioris muscle to demonstrate its fan-shaped attachment to the superior tarsal plate (Fig. 7-57A). Some fibers of its tendon insert into the skin of the upper eyelid. *A*, Levator labii superioris alaeque nasi. *B*, Levator labii superioris. *C*, Zygomaticus major.

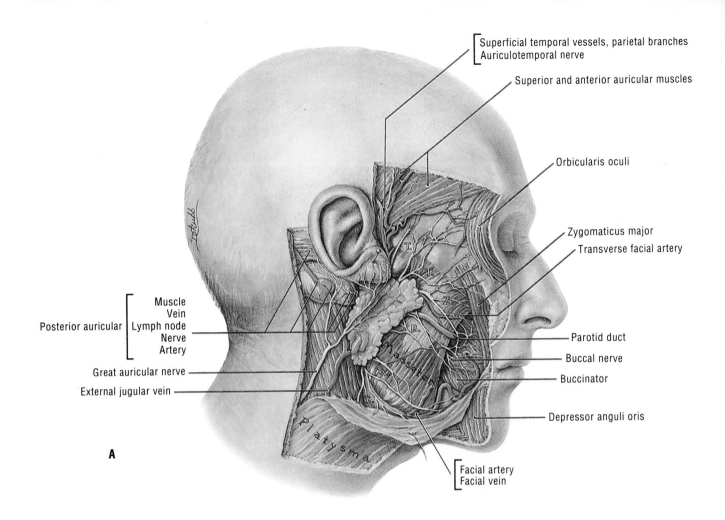

Superficial temporal vessels, parietal branches
Auriculotemporal nerve

Superior and anterior auricular muscles

Orbicularis oculi

Zygomaticus major
Transverse facial artery

Posterior auricular {
Muscle
Vein
Lymph node
Nerve
Artery
}

Great auricular nerve

External jugular vein

Parotid duct

Buccal nerve

Buccinator

Depressor anguli oris

Facial artery
Facial vein

A

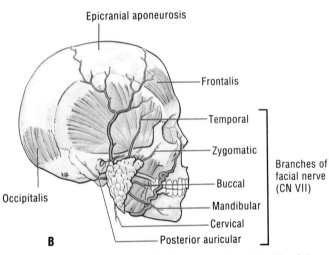

Epicranial aponeurosis

Frontalis

Temporal

Zygomatic

Buccal

Mandibular

Cervical

Posterior auricular

Occipitalis

Branches of
facial nerve
(CN VII)

B

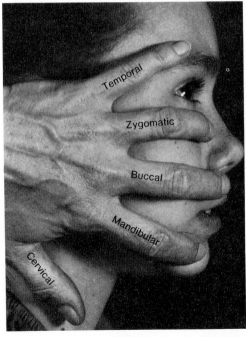

Temporal

Zygomatic

Buccal

Mandibular

Cervical

C

Figure 7-22. *A*, Lateral view of a dissection of the right side of the head, showing the great auricular nerve (C2 and C3), and terminal branches of the facial nerve (CN VII): *T*, temporal, *Z*, zygomatic, *B*, buccal, *M*, mandibular and *C*, cervical. Note that the parotid duct turns medially at the anterior border of the masseter muscle to pierce the buccinator muscle. It enters the oral cavity opposite the crown of the second maxillary molar tooth. *B*, Branches of the facial nerve. The occipitalis and frontalis muscles and the epicranial aponeurosis are also shown. *C*, Lateral view of the face of a 12-year-old girl, illustrating a simple method for remembering the general course of the five main branches of the facial nerve (CN VII).

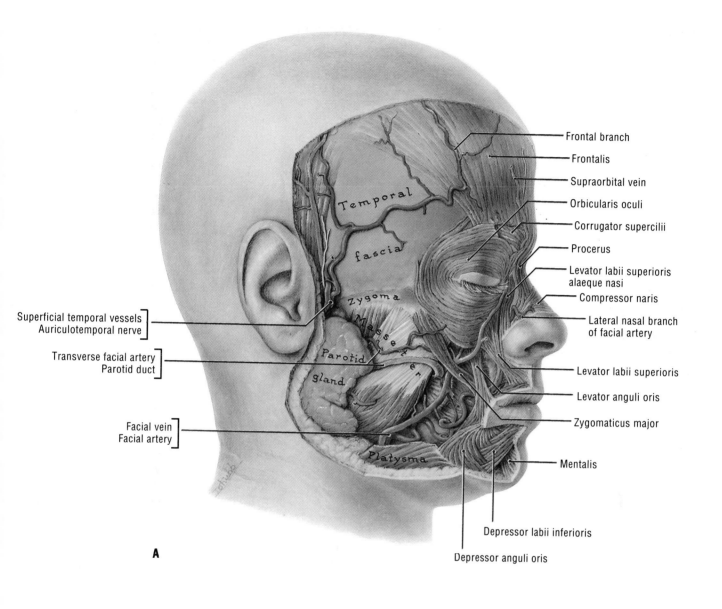

Frontal branch
Frontalis
Supraorbital vein
Orbicularis oculi
Corrugator supercilii
Procerus
Levator labii superioris
alaeque nasi
Compressor naris
Lateral nasal branch
of facial artery
Levator labii superioris
Levator anguli oris
Zygomaticus major
Mentalis
Depressor labii inferioris
Depressor anguli oris

Superficial temporal vessels
Auriculotemporal nerve

Transverse facial artery
Parotid duct

Facial vein
Facial artery

Temporal fascia
Zygoma
Masseter
Parotid gland
Platysma

A

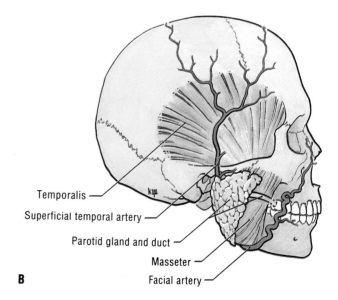

Temporalis
Superficial temporal artery
Parotid gland and duct
Masseter
Facial artery

B

Figure 7-23. *A,* Lateral view of a dissection of the face of a young man, exposing the muscles of facial expression and the arteries of the face. The masseter, one of the muscles of mastication and a powerful closer of the mandible, is also shown. *B,* The parotid gland and the facial and superficial temporal arteries.

section) are of little practical importance to most health professionals. Therefore only their main features and functions are mentioned.[3]

The *levator labii superioris alaeque nasi* muscle (Fig. 7-23) is attached superiorly to the maxilla. It divides into two slips which attach to the alar cartilage of the nose and the upper lip. It elevates both of these structures.

The *depressor anguli oris* muscle (Figs. 7-21 to 7-23), as its name indicates, depresses the corner of the mouth. Posterior fibers of the platysma muscle assist with this movement.

The *levator anguli oris* muscle (Fig. 7-23) is attached superiorly to the infraorbital margin and inferiorly to the angle of the mouth. It elevates the corner of the mouth.

The *zygomaticus major* muscle (Figs. 7-21 to 7-23), extending from the zygomatic bone to the angle of the mouth, draws the angle of the mouth superolaterally as during smiling or laughing.

The *zygomaticus minor*, a narrow slip of muscle, passes obliquely from the zygomatic bone to the orbicularis oris muscle. It helps to raise the upper lip when showing contempt or to deepen the nasolabial sulcus when showing sadness.

The *levator labii superioris* muscle (Fig. 7-23), descending from the infraorbital margin to the upper lip (L. *labium*, lip), raises and everts the upper lip. It also helps the zygomaticus minor to deepen the nasolabial sulcus when showing sadness.

The *mentalis* (Figs. 7-21 and 7-23) is a small muscle that is attached to the mandible and descends to attach to the skin of the chin. It raises the skin of the chin during the expression of doubt.

The *depressor labii inferioris* is located lateral to the mentalis muscle (Fig. 7-23). It is attached inferiorly to the mandible and merges superiorly with its partner and the orbicularis oris muscle. It draws the lip inferiorly and slightly laterally, as when showing impatience.

The *risorius* is a variable muscle that is closely related to the platysma. It is attached to the fascia covering the parotid gland and to the angle of the mouth. It draws the corner of the mouth laterally when grinning.

The Orbicularis Oris Muscle (Figs. 7-21 and 7-22). This muscle encircles the mouth and is the *sphincter of the oral aperture*. Its fibers, located within the upper and lower lips, are derived from the buccinator and other facial muscles. The orbicularis oris closes the lips, (*i.e.*, purses them, as during whistling and sucking), protrudes them, and compresses them against the teeth. It plays an important role in articulation and mastication (chewing). In association with the buccinator muscle, it helps to hold food between the teeth during mastication.

The Buccinator Muscle (Figs. 7-16 and 7-22). This thin, flat, rectangular muscle is attached laterally to the alveolar processes of the maxilla and mandible, opposite the molar teeth, and to the *pterygomandibular raphe*. Medially its fibers mingle with those of the orbicularis oris muscle. The buccinator aids mastication and swallowing by pressing the cheeks against the molar teeth during chewing. This pushes the food against the occlusal surfaces of the teeth. The buccinator is also used during whistling and sucking by forcing the cheeks against the teeth.

The buccinator muscle was given its name because it compresses the cheeks (L. *buccae*) during blowing, *e.g.*, when a musician plays a wind instrument such as a trumpet (L. *buccinator*, trumpeter).

Some trumpeters have stretched their buccinator and other buccal muscles so much that their cheeks balloon out when they blow forcibly on their instruments.

Muscles Around the Eyelids

The function of the eyelids (L. *palpebrae*) is to protect the eye from injury and excessive light. They also keep the cornea moist (p. 712).

The Orbicularis Oculi Muscle (Figs. 7-22 and 7-23). This is the *sphincter muscle of the eye*. Its fibers sweep in concentric circles around the orbital margin and eyelids. Contraction of its fibers narrows the orbital (eye) opening and encourages the flow of tears by helping to empty the lacrimal (tear) sac (Fig. 7-64). The orbicularis oculi muscle consists of three parts: (1) a thick *orbital part* for closing the eyes to protect against the glare of light and dust in the air; (2) a thin *palpebral part* for closing the eyelids lightly to keep the cornea from drying; and (3) a *lacrimal part* for drawing the eyelids and *lacrimal puncta* medially (Figs. 7-61 and 7-64). This part lies deep to the palpebral part and is often considered to be a portion of it. It aids the spreading of tears by holding the eyelids close to the eyeballs. When all three parts of the orbicularis oris contract, the eyes are firmly closed and the adjacent skin becomes wrinkled. Similar wrinkling occurs when a person scrutinizes something. It is supplied by a *zygomatic branch of the facial nerve* (CN VII).

The Levator Palpebrae Superioris Muscle (Fig. 7-21). As its name indicates (L. *levare*, to raise), this muscle raises the upper eyelid to open the *palpebral fissure*. It is supplied by the oculomotor nerve (CN III). This important muscle is also discussed with the orbit (p. 715).

Muscles Around the Nose

All muscles around the nose are supplied by the facial nerve (CN VII). Those that are relatively unimportant to most health professionals appear in intermediate type.

The Nasalis Muscle (Fig. 7-23). This is the main muscle of the nose. It consists of transverse (compressor naris) and alar (dilator naris) parts. The *compressor naris* part arises from the superior part of the canine ridge of the maxilla, superior to the incisor teeth, and passes superomedially to the dorsum of the nose. It compresses the anterior nasal aperture (nostril). The *dilator naris* arises from the maxilla superior to the compressor naris and attaches to the alar cartilages of the nose (Fig. 7-107). It widens the anterior nasal aperture, *i.e.*, it "flares the nostrils." The dilator naris also draws the nostril down as occurs during fright and anger. Each part of the nasalis muscle is supplied by a buccal branch of the facial nerve.

The actions of the nasalis muscle are insignificant; however, observant clinicians study its action because it may be of diagnostic value. For example, true *nasal breathers* can

---

[3]Persons interested in more information should consult Williams et al. (1989).

distinctly "flare their nostrils." Habitual mouth breathing, *e.g.*, caused by chronic nasal obstruction, diminishes and sometimes eliminates the ability to "flare the nostrils." Children who are chronic *mouth breathers* often develop *dental malocclusions* (improper bite).

The procerus muscle (Figs. 7-21 and 7-22) is a small slip of muscle that is continuous with the occipitofrontalis muscle. It passes from the forehead over the bridge of the nose, where it is attached to the skin over the glabella. The procerus draws the medial part of the eyebrow inferiorly, producing transverse wrinkles over the bridge of the nose. This action probably reduces the glare of bright sunlight; it is also used when frowning. It is supplied by a buccal branch of the facial nerve (CN VII).

The depressor septi muscle, often regarded as part of the dilator naris portion of the nasalis muscle, arises from the maxilla superior to the central incisor tooth and inserts into the mobile part of the nasal septum. It assists the dilator naris to widen the nasal aperture during deep inspiration. It is supplied by a buccal branch of the facial nerve (CN VII).

Injury to the facial nerve (CN VII) or some of its branches (Figs. 7-22 and 7-27) produces *paresis* (weakness) or *paralysis* (loss of voluntary movement) of all or some of the facial muscles on the affected side. Paralysis of the facial nerve for no obvious reason, known as *Bell's palsy* (paralysis), may occur after exposure to a cold draft. The most common cause of Bell's palsy is *inflammation of the facial nerve* near the stylomastoid foramen (Fig. 7-27). This causes edema and swelling of the nerve and compression of its fibers in the facial canal or stylomastoid foramen (Figs. 7-27 and 7-117). Patients with facial paralysis are unable to close their lips and eyelids on the affected side. The eye on the affected side is not lubricated, and these people are unable to whistle, to blow a wind instrument, or to chew effectively. Because the buccinator muscle is also weakened or paralyzed, food and saliva dribbles out that side of the mouth or collects in its vestibule. Expressive lines in the facial skin produced by the facial muscles are obliterated. Facial distortion occurs owing to contractions of unopposed contralateral facial muscles. Facial paralysis is discussed further on p. 665.

Nerves of the Face

Innervation of the skin of the face is largely through the three branches of CN V, the **trigeminal nerve** (Figs. 7-21 and 7-24). Some skin over the angle of the mandible and anterior and posterior to the auricle is supplied by the *great auricular nerve* from the cervical plexus (Fig. 7-22A). Some cutaneous fibers of the auricular branch of the facial nerve also supply skin on both sides of the auricle (Fig. 7-27).

The Trigeminal Nerve[4]

The *fifth cranial nerve* (CN V) is the largest of the 12 cranial nerves (Figs. 7-24 and 7-25). It is the principal *general sensory*

[4]This nerve is described in more detail in Chapter 9.

nerve to the head, particularly the face, and is the *motor nerve to the muscles of mastication* (masseter, temporalis, and several others).

The Ophthalmic Nerve (Figs. 7-24 and 7-25). This is the *superior division of the trigeminal nerve* and is the smallest of its three branches. It is wholly sensory and supplies the area of skin derived from the embryonic *frontonasal prominence* (Moore, 1988). The ophthalmic nerve (designated as CN V[1]) divides into three branches: *nasociliary*, *frontal*, and *lacrimal*, just before entering the orbit through the superior orbital fissure. The five branches of these nerves participate in the sensory supply to the skin of the forehead, upper eyelid, and nose (Fig. 7-24C).

The **nasociliary nerve** supplies the tip of the nose through the external nasal branch of the anterior ethmoidal nerve and the root of the nose through the infratrochlear nerve. The **frontal nerve**, the direct continuation of CN V[1], divides into two branches: supratrochlear and supraorbital. The *supratrochlear nerve* supplies the middle part of the forehead, and the *supraorbital nerve* supplies the lateral part and the front of the scalp. The **lacrimal nerve**, the smallest of the main ophthalmic branches, emerges over the superolateral orbital margin to supply the lacrimal gland and the lateral part of the upper eyelid.

The Maxillary Nerve (Figs. 7-21, 7-24, and 7-25). This nerve (designated as CN V[2]) is the *intermediate division of the trigeminal nerve*. It has three cutaneous branches that supply the area of skin derived from the embryonic *maxillary prominence* (Moore, 1988). The **infraorbital nerve**, the large terminal branch of CN V[2], passes through the infraorbital foramen and breaks up into branches that convey sensation from skin on the lateral aspect of the nose, upper lip, and lower eyelid. The *zygomaticofacial nerve*, a small branch of the maxillary, emerges from the zygomatic bone through a small foramen with the same name. It supplies the skin of the face over the zygomatic bone (*i.e.*, the zygomatic prominence). The *zygomaticotemporal nerve* emerges from the zygomatic bone through a foramen of the same name and supplies the skin over the temporal region.

For local anesthesia of the inferior part of the face, the infraorbital nerve is often infiltrated with an anesthetic agent at the infraorbital foramen or in the infraorbital canal (*e.g.*, for treatment of wounds of the upper lip and cheek or for repairing the maxillary [upper] incisor teeth). The site of emergence of this nerve can easily be determined by exerting pressure on the maxilla in the region of the infraorbital foramen and nerve. Pressure on the nerve causes considerable pain. Care is exercised when performing an **infraorbital nerve block** because companion infraorbital vessels leave the infraorbital foramen with the nerve. Careful aspiration of the syringe during injection prevents inadvertent injection of the anesthetic fluid into a blood vessel. In Fig. 7-1, note that the orbit is located just superior to the injection site. A careless injection could result in the passage of anesthetic fluid into the orbit, causing temporary *paralysis of the extraocular muscles* (Fig. 7-65).

The Mandibular Nerve (Figs. 7-21, 7-24, and 7-25). This nerve (designated as CN V[3]) is the *inferior division of the trigeminal*. CN V[3] has three sensory branches that supply the area of skin derived from the embryonic *mandibular prominence* (Moore, 1988). It also supplies motor fibers to the muscles of

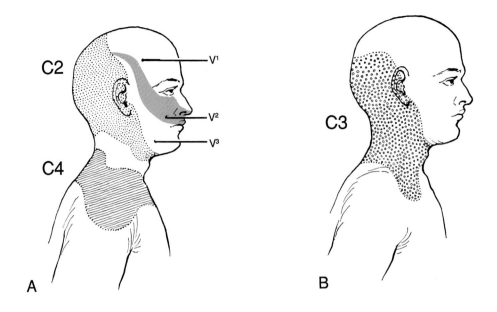

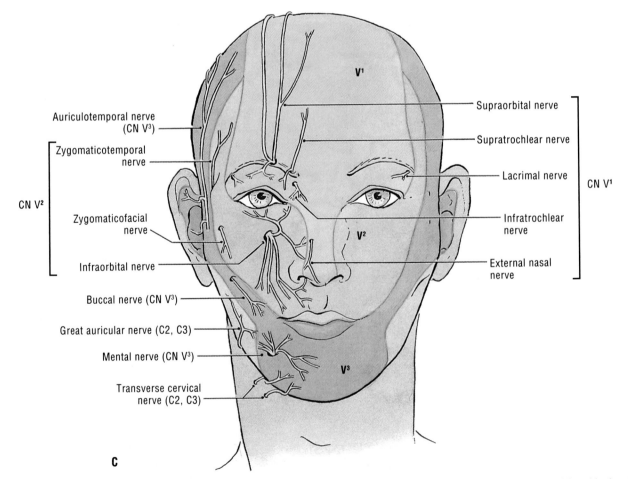

Figure 7-24. *A* and *B*, Distribution of the cutaneous areas (dermatomes) supplied by spinal nerves (C2, C3 and C4). In *A*, observe that the trigeminal nerve (CN V) is responsible for general sensation from the skin of the face and forehead. *C*, The distribution of the three divisions of the trigeminal nerve correspond roughly with the three embryological regions of the face (see Moore, 1988): V^1, frontonasal prominence; V^2, maxillary prominence; and V^3, mandibular prominence.

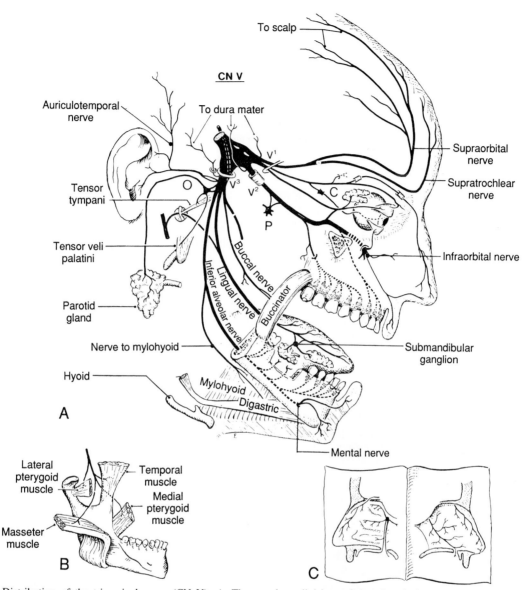

Figure 7-25. Distribution of the trigeminal nerve (CN V). *A*, The three divisions of CN V (V¹, V², and V³) arise from the large trigeminal ganglion (black). Note that the maxillary nerve (CN V²) gives off two pterygopalatine nerves, which suspend the small pterygopalatine ganglion (*P*). Note also the ciliary (*C*) and otic (*O*) ganglia. Each of the three divisions of the trigeminal nerve is connected with a parasympathetic ganglion: CN V¹ with the ciliary, CN V² with the pterygopalatine, and CN V³ with the submandibular and otic ganglia. *B*, The motor fibers passing to the muscles of mastication. *C*, The sensory and secretory fibers to the nasal mucosa and palate.

mastication. Of the three divisions of the trigeminal, *CN V³ is the only division of CN V that carries motor fibers*. The main sensory branches of the mandibular nerve are the buccal, auriculotemporal, inferior alveolar, and lingual nerves. The **buccal nerve** (Figs. 7-21, 7-22, and 7-25) is a small branch of CN V³ that emerges from deep to the ramus of the mandible to supply the skin of the cheek over the buccinator muscle. It also supplies the mucous membrane lining the cheek and the posterior part of the buccal surface of the gingiva (gum). The **auriculotemporal nerve** (Figs. 7-22 to 7-25) passes medial to the neck of the mandible and then turns superiorly, posterior to its head and anterior to the auricle. It then crosses over the root of the zygomatic process of the temporal bone, deep to the superficial temporal artery (Fig. 7-28). As its name suggests, it supplies

parts of the auricle, external acoustic meatus, tympanic membrane (eardrum), and skin in the temporal region.

The **inferior alveolar nerve** (Fig. 7-25) is the large terminal branch of the posterior division of CN V³; the lingual nerve is the other terminal branch. It enters the mandibular canal through the mandibular foramen (Fig. 7-16). In the canal it gives off branches that supply the mandibular (lower) teeth. Opposite the mental foramen, the inferior alveolar nerve divides into its terminal incisive and mental branches. The *incisive nerve* supplies the incisor teeth, the adjacent gingiva, and the mucosa of the lower lip. The *mental nerve* (Figs. 7-21, 7-24, and 7-25) emerges from the mental foramen and divides into three branches, which supply the skin of the chin and the skin and mucous membrane of the lower lip and gingiva (gum).

The **lingual nerve** (Figs. 7-25 and 7-96), the smaller terminal branch of the posterior division of the CN V^3, supplies general sensory fibers to the anterior two-thirds of the tongue, the floor of the mouth, and the gingivae (gums) of the mandibular teeth.

Dentists anesthetize the inferior alveolar nerve before repairing or removing the mandibular teeth. Because the mental and incisive nerves are its terminal branches, it is understandable why one's chin and lower lip on the affected side also lose sensation. The mental nerve can be blocked by injecting the anesthetic fluid around the nerve as it emerges from the mental foramen (Figs. 7-21, 7-24, and 7-25).

Trigeminal neuralgia (tic douloureux) is a condition characterized by sudden attacks of excruciating pain that are initiated by a mere touch in the area of distribution of one of the divisions of the trigeminal nerve, usually CN V^2. The cause(s) of the neuralgia (G. nerve pain) is unknown. In some cases the symptoms are removed if a small aberrant artery is moved away from the sensory root of CN V (Barr and Kiernan, 1988).

The *infraorbital nerve* (Figs. 7-21, 7-24C, and 7-25) is commonly injured in fractures of the maxilla because such fractures often pass through or close to the infraorbital foramen through which the nerve exits the skull (Figs. 7-20 and 7-21). The *inferior alveolar nerve* may be damaged by a fracture of the ramus of the mandible (Figs. 7-19 and 7-25A).

A lesion of the entire trigeminal nerve (Figs. 7-24C, 7-25, and 7-26) causes widespread anesthesia involving: (1) the corresponding anterior half of the scalp; (2) the face, except for an area around the angle of the mandible; (3) the cornea and conjunctiva; and (4) the mucous membranes of the nose, mouth, and tongue (anterior two-thirds). Paralysis and *atrophy of the muscles of mastication* also occur so that when the mouth is opened, the mandible moves to the paralyzed side.

Herpes zoster ophthalmicus (shingles) is an infection of the face that involves the region supplied by the ophthalmic nerve; hence the cornea is often involved (Fig. 7-25). In some cases there is partial paralysis or paresis of the ocular muscles indicating that the infection has also involved CN III, CN IV, CN VI, or a combination of these nerves which supply the muscles that move the eye (Fig. 7-65).

The Facial Nerve[5]

The seventh cranial nerve (CN VII) supplies the superficial muscle of the neck (platysma), the muscles of facial expression, the auricular muscles, the scalp muscles (Figs. 7-21 to 7-23 and 7-27), and certain other muscles derived from the mesoderm in the embryonic second pharyngeal or branchial arch (Moore, 1988). *CN VII is the sole motor supply to the muscles of facial expression* and is sensory to the taste buds in the anterior two-thirds of the tongue (Fig. 7-96). It also conveys general sensation from a small area around the external acoustic meatus and is secretomotor to the submandibular, sublingual, and intralingual salivary glands (Fig. 7-27).

The facial nerve emerges from the skull through the *stylomastoid foramen* (Figs. 7-7 and 7-27) between the mastoid and styloid processes of the temporal bone. Almost immediately, it enters the parotid gland (Figs. 7-22 and 7-27). It runs superficially within this gland before giving rise to its five terminal branches: temporal, zygomatic, buccal, mandibular, and cervical. These nerves emerge from the superior, anterior, and inferior margins of the gland and spread out like the abducted digits of the hand to supply the muscles of facial expression (Figs. 7-22 and 7-24). The motor component of the facial nerve, supplying the muscles of facial expression and certain other muscles, is a clinically important part of the nerve. Tests for the facial nerve concern the facial muscles; hence health professionals should know where the five major branches of the facial nerve go. The names of the nerves indicate the regions supplied by them.

The *temporal branches of CN VII* cross the zygomatic arch to supply all the superficial facial muscles superior to it, including the orbital and forehead muscles (Figs. 7-22 and 7-23). The *zygomatic branches of CN VII* pass transversely over the zygomatic bone to supply muscles in the zygomatic, orbital, and infraorbital regions. The *buccal branches of CN VII* (Figs. 7-21, 7-22, and 7-24) pass horizontally, external to the masseter muscle, to supply the buccinator and the muscles in the upper lip. The *mandibular branch of CN VII* (Figs. 7-22 and 7-24)

---

[5]This nerve is described in more detail in Chapter 9 (p. 866).

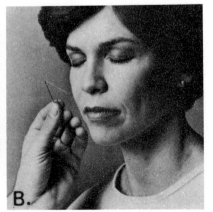

Figure 7-26. Two methods for testing the trigeminal nerve (CN V). Note that these tests are conducted with the patient's eyes closed. *A*, Test tubes filled with warm and cold fluid are pressed alternately against the cheek. Differences in response on opposite sides of the face indicate increased or decreased sensitivity to temperature. *B*, Differences in response to pinpricks on opposite sides of the face indicate increased or decreased sensitivity to pain. (Used with permission from Smith Kline Corporation: *Essentials of Neurological Examination*, 1978.)

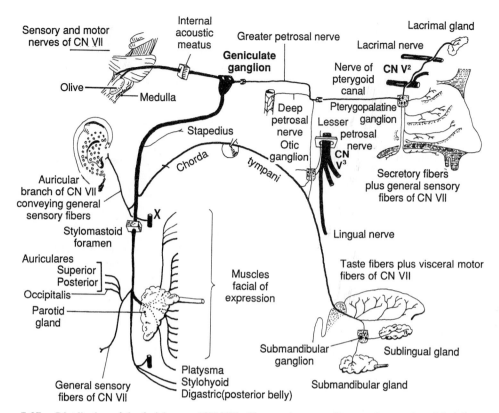

Figure 7-27.   Distribution of the facial nerve (CN VII). Observe the motor fibers to the muscles of facial expression and the special sense (taste) fibers to the tongue.

supplies the muscles in the lower lip and chin. The *cervical branch of CN VII* (Figs. 7-22A and 7-27) supplies the platysma, the superficial muscle of the neck.

> Peripheral **facial paralysis** may be caused by chilling of the face, middle ear infections, tumors, fractures, and other disorders. About 75% of all facial nerve lesions are of this type. The signs and symptoms depend on the location of the lesion. (See discussion of Bell's palsy on p. 661.) As CN VII runs superficially within the parotid gland (Figs. 7-22A and 7-27), it may be infiltrated by malignant cells (*e.g.*, from a carcinoma of the parotid gland). This commonly results in paresis (incomplete paralysis) of the muscles of facial expression. Benign tumors usually do not infiltrate and cause facial nerve paralysis, but care must be taken to preserve the facial nerve and its branches during excision of these tumors. On rare occasions, *parotid inflammation* (*e.g.*, owing to mumps) may temporarily affect this nerve.
>
> *The mastoid process of the temporal bone is not present at birth* and does not develop until the second year of life. Consequently, the stylomastoid foramen opens beneath the skin and the facial nerve is vulnerable to injury. As a result, the facial nerve may be injured by forceps during delivery of a baby. Similarly, because the mastoid process is not well developed in childhood, CN VII may be more easily injured at its site of emergence from the stylomastoid foramen in children than it is in adults.

Arteries of the Face

The face is richly supplied by arteries, the terminal branches of which anastomose freely (Figs. 7-22, 7-23, and 7-28). The superficial arteries of the face are derived from the *external carotid arteries*.

The Facial Artery

This is *the chief artery of the face*. It arises from the external carotid[6] and winds its way to the inferior border of the mandible, just anterior to the masseter muscle (Figs. 7-22A and 7-28). It hooks around the inferior border of the mandible, usually grooving the bone (Fig. 7-15). The artery lies superficially here, immediately beneath the platysma muscle, where its pulsations can be easily felt. In its course over the face to the medial angle of the eye, the facial artery crosses the mandible, buccinator muscle, and maxilla. It lies deep to the zygomaticus major and levator labii superioris muscles (Fig. 7-23). Near the termination of its sinuous course, the facial artery passes about a fingerbreadth lateral to the angle of the mouth. Here, it can be felt in living persons when the cheek is grasped between the first and second

---

[6]Sometimes the facial and lingual arteries arise from the external carotid by a common trunk (see Fig. 8-17).

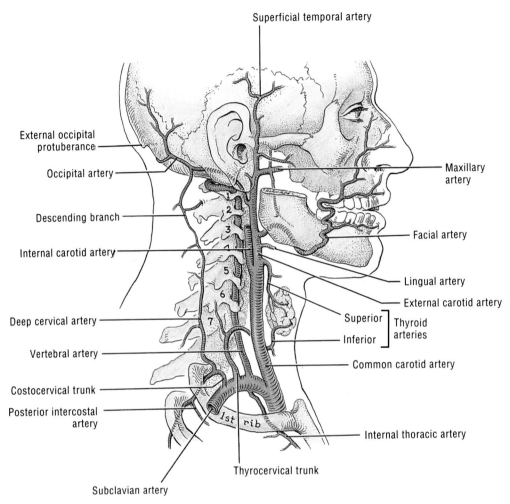

Figure 7-28.   Arteries of the right side of the head and neck.

digits. The facial artery ends by sending branches to the lip and the side of the nose (Figs. 7-23 and 7-28). The part of the artery that runs along the nose to the inner angle of the eye to supply the eyelids is called the *angular artery*.

> The facial artery can be occluded by pressure against the mandible where the vessel crosses it (Fig. 7-28). Because there are numerous anastomoses between the branches of the facial artery and other arteries of the face, compressing the facial artery against the mandible on one side does not stop all bleeding from a lacerated facial artery or one of its branches. In lacerations of the lip, pressure must be applied on both sides of the cut to stop the bleeding. In general, facial wounds bleed freely and heal quickly.

The Superficial Temporal Artery

This artery is the smaller of the two terminal branches of the external carotid artery (Fig. 7-28); the other terminal branch is the maxillary artery. The superficial temporal artery begins deep to the parotid gland, posterior to the neck of the mandible, and ascends superficial to the posterior part of the zygomatic process of the temporal bone. It then enters the *temporal fossa* (Figs. 7-

2 and 7-28). The superficial temporal artery ends in the scalp by dividing into frontal and parietal branches (Fig. 7-23). *Pulsations of this artery can be felt* by compressing it against the root of the zygomatic process of the temporal bone (Fig. 7-28). This artery is often visible, particularly in thin elderly persons.

The Transverse Facial Artery (Figs. 7-22A and 7-23).   This small artery of the face arises from the superficial temporal artery before it emerges from the parotid gland. It crosses the face superficial to the masseter muscle, about a fingerbreadth inferior to the zygomatic arch, in company with one or two branches of the facial nerve. It divides into numerous branches that supply the parotid gland and duct, the masseter muscle, and the skin of the face. It anastomoses with branches of the facial artery.

> The pulses of the superficial temporal and facial arteries are often measured when it is inconvenient to measure the pulse of other arteries. For example, anesthesiologists at the head of the operating table often measure the temporal pulse just anterior to the auricle of the external ear, where the superficial temporal artery crosses the root of the zygomatic process of the temporal bone (Fig. 7-28). They palpate the facial pulse where the facial artery winds around the inferior border of the mandible.

Veins of the Face

The external facial veins anastomose freely and are drained by veins that accompany the arteries of the face. As with most superficial veins, they are subject to many variations; a common pattern is shown in Fig. 7-29.

The Supratrochlear Vein (Fig. 7-29). This vessel begins on the forehead from a network of veins connected to the frontal tributaries of the superficial temporal vein. It descends near the median plane with its fellow of the other side. These veins diverge near the orbits, each joining a supraorbital vein to form the facial vein near the medial canthus or angle of the eye.

The Supraorbital Vein (Figs. 7-23, 7-29, and 7-30). This vessel begins near the zygomatic process of the temporal bone, where it joins the tributaries of the superficial and middle temporal veins. It passes medially and joins the supratrochlear vein to form the facial vein near the medial canthus.

The Facial Vein (Figs. 7-22, 7-23, 7-29, and 7-30). *This vein provides the major venous drainage of the face.* It begins at the medial canthus of the eye by the union of the supraorbital and supratrochlear veins. It runs inferoposteriorly through the face, posterior to the facial artery, but it takes a straighter and more superficial course than the artery. Inferior to the margin of the mandible, the facial vein is joined by the anterior branch of the *retromandibular vein*. The facial vein ends by draining into the *internal jugular vein* (Fig. 7-43). The superior part of the facial vein near the confluence of the supratrochlear and supraorbital veins is often called the *angular vein* because of its relationship to the **medial canthus** (G. corner of the eye).

The facial vein makes clinically important connections with (1) the *cavernous sinus* (Fig. 7-30)—a venous sinus of the dura mater covering the brain—through the superior ophthalmic vein and (2) with the *pterygoid plexus*—a network of very small veins within the lateral pterygoid muscle—through the deep facial vein. Blood from the medial canthus of the eye, nose, and lips usually drains inferiorly through the facial vein, especially when the patient is erect. However, because *the facial vein has no valves*, blood may run through it in the opposite direction. Consequently, blood from the face may enter the *cavernous sinus*. In patients with *thrombophlebitis of the facial vein* (inflammation of the vein with secondary thrombus or clot formation), pieces of an infected clot may extend into the intracranial venous system. Here, it may produce *thrombophlebitis of the cavernous sinuses or the cortical veins*. Infection of the facial veins spreading to the dural venous sinuses may result from lacerations of the nose, or be initiated by squeezing pustules on the side of the nose and upper lip. Consequently this area is often called the *danger triangle of the face* (Fig. 7-31).

The Superficial Temporal Vein (Figs. 7-22, 7-23, and 7-29). This vein drains the forehead and scalp and receives tributaries from the veins of the temple and face. In the region of the temporomandibular joint, this vein enters the parotid gland.

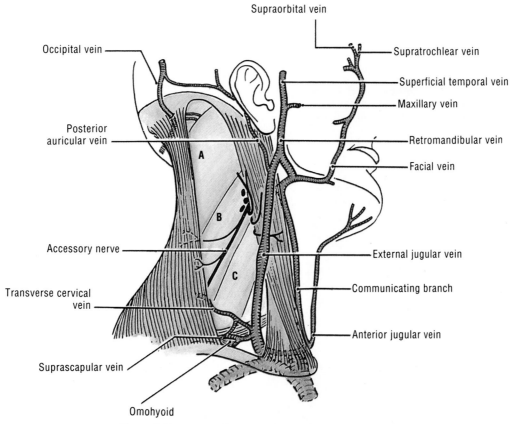

Figure 7-29.   Veins of the right side of the head and neck.

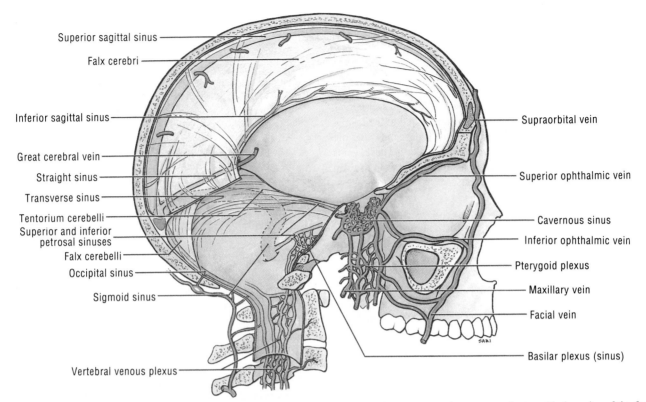

Figure 7-30.   The facial vein and venous sinuses of the dura mater. The cavernous sinuses are situated on each side of the sphenoid bone. Note that the cavernous sinus communicates with the veins of the face via the ophthalmic veins and the pterygoid plexus.

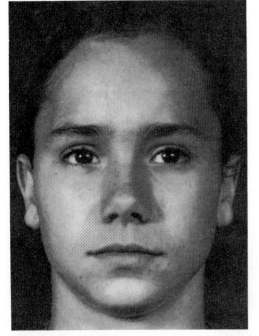

Figure 7-31.   The danger triangle of the face (*red*). Veins in this area communicate with the superior and inferior ophthalmic veins, which drain into the cavernous venous sinuses of the dura mater (Fig. 7-30).

The Retromandibular Vein (Fig. 7-29). This vessel is formed by the union of the superficial temporal and maxillary veins, posterior to the neck of the mandible. It descends within the parotid gland, superficial to the external carotid artery but deep to the facial nerve. It divides into an anterior branch that unites with the facial vein and a posterior branch that joins the *posterior auricular vein* to form the external jugular vein.

Lymphatic Drainage of the Face

The lymphatic vessels in the forehead and anterior part of the face accompany the other facial vessels and drain into the *sub-mandibular lymph nodes*, located along the inferior border of the mandible (Fig. 7-32). The lymph vessels from the lateral part of the face, including the eyelids, drain inferiorly toward the *superficial parotid lymph nodes* (Figs. 7-22A and 7-32). These nodes drain into the *deep parotid nodes*, which in turn drain into the deep **cervical lymph nodes**. Lymphatics in the upper lip and in lateral parts of the lower lip drain into the *submandibular lymph nodes*, whereas lymphatics in the central part of the lower lip and in the chin drain into the *submental lymph nodes*, from which lymph may drain directly into the *jugulo-omohyoid lymph nodes*.

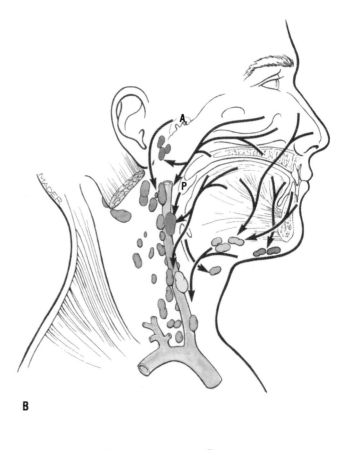

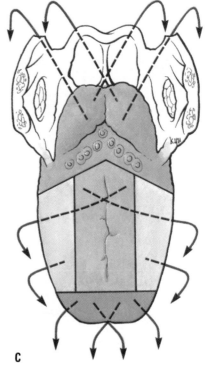

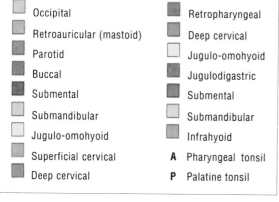

Occipital	Retropharyngeal
Retroauricular (mastoid)	Deep cervical
Parotid	Jugulo-omohyoid
Buccal	Jugulodigastric
Submental	Submental
Submandibular	Submandibular
Jugulo-omohyoid	Infrahyoid
Superficial cervical	**A** Pharyngeal tonsil
Deep cervical	**P** Palatine tonsil

Figure 7-32. Lymphatic drainage of the head and neck. *A*, Superficial. *B*, Deep. *C*, Tongue.

The Parotid Gland

This large salivary gland is located near the auricle of the external ear (G. *para*, near + *otis*, ear), where it is wedged between the ramus of the mandible and the mastoid process (Figs. 7-22, 7-23, and 7-33). The parotid gland is wrapped with a fibrous capsule (parotid fascia) that is continuous with the deep investing fascia of the neck. It occupies the side of the face anterior and inferior to the auricle. In living persons the parotid gland is an irregular, lobulated, yellowish mass, that is closely related to the ramus of the mandible. It is irregular in shape because it grew between the mandible and mastoid process into the cervical fascia during development. During this growth it enclosed structures in the area it invaded (*e.g.*, the facial nerve). There is a constriction in the parotid gland, sometimes called the *isthmus*, between ramus of the mandible and the masseter muscle anteriorly and the posterior belly of the digastric muscle posteriorly.

The parotid gland is the largest of the paired salivary glands and, like the other two, develops as an outgrowth from the primitive mouth (Moore, 1988). As viewed superficially, the parotid gland is somewhat triangular in shape (Fig. 7-22A). Its apex is posterior to the angle of the mandible and its base is along the zygomatic arch (Fig. 7-33). The parotid gland overlaps the posterior part of the masseter muscle (Fig. 7-23). The **parotid duct** (Stensen's duct), which is about 5 cm long and 5 mm in diameter, passes horizontally from the anterior edge of the gland. At the anterior border of the masseter muscle, the parotid duct turns medially and pierces the buccinator muscle (Figs. 7-22A, 7-23, and 7-33). It enters the oral cavity opposite the crown of the second maxillary molar tooth.

To locate the approximate course of the parotid duct, place your second digit (index finger) along the inferior border of the zygomatic arch and point it toward the upper lip. The parotid duct courses along the inferior border of your digit and then curves medially. If you tense your masseter muscle by clenching your teeth, you may be able to palpate the thick-walled parotid duct in your cheek, about a fingerbreadth inferior to your zygomatic arch, as you roll it with your digit over the anterior border of the masseter muscle. Look into someone's mouth with a flashlight and observe the opening of the parotid duct opposite the second maxillary molar tooth. If the person is asked to suck a lemon slice, you may be able to see saliva flowing out of the duct. There is a small papilla where the duct opens into the mouth.

Blood Vessels of the Parotid Gland (Figs. 7-22A, 7-23, 7-28, and 7-29). This gland is supplied by branches of the *external carotid artery*. The veins from the parotid gland drain into the *retromandibular vein*, which enters the external jugular vein.

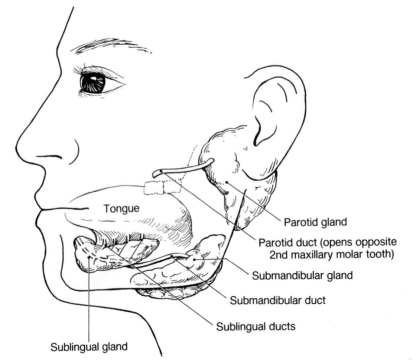

Tongue

Parotid gland

Parotid duct (opens opposite 2nd maxillary molar tooth)

Submandibular gland

Submandibular duct

Sublingual ducts

Sublingual gland

Figure 7-33. The three salivary glands. Note that the parotid gland, the largest of them, is wrapped around the neck of the mandible.

Lymphatic Drainage of the Parotid Gland (Figs. 7-22*A* and 7-32).   The lymph vessels from this gland end in the superficial and deep *cervical lymph nodes*. There are two or three lymph nodes on the surface of the parotid gland and within the substance of the gland.

Nerves of the Parotid Gland (Figs. 7-22 to 7-25, 7-27, and 7-74).   These nerves are derived from the *auriculotemporal nerve* and from the sympathetic and parasympathetic systems. The parasympathetic fibers are derived from the *glossopharyngeal nerve* (CN IX). Stimulation of these fibers causes a thin watery saliva to flow from the parotid duct. The sympathetic fibers are derived from the *cervical ganglia* through a plexus of sympathetic nerve fibers on the external carotid artery (*external carotid plexus*). Stimulation of the sympathetic fibers produces a thick mucous saliva.

Although the facial nerve (CN VII) passes through the parotid gland (Figs. 7-22*A* and 7-27) and carries parasympathetic fibers, it does not supply the parotid gland. Because CN VII passes through the substance of the parotid gland and divides into numerous branches, it is in jeopardy during surgery (p. 665).

The parotid gland may become infected through the blood stream, *e.g.*, as occurs in **mumps** (*epidemic parotiditis*), an acute communicable viral disease that affects the salivary glands, particularly the parotid. Infection of the parotid gland causes inflammation (*parotiditis*). Severe pain soon occurs because the gland's capsule, derived from the cervical fascia, limits swelling. The pain results from stretching of the parotid capsule. Often the pain is worse during chewing because the gland is wrapped around the posterior border of the ramus of the mandible and is compressed against the mastoid process when the mouth is opened.

The mumps virus may also cause *inflammation of the parotid duct*, producing redness of its papilla. Because mumps (from an old English word meaning lumps) may be confused with a toothache, redness of the papilla of the parotid duct is often an early sign indicating disease involving the parotid gland and not the teeth. Sucking a lemon is also used as a test for parotiditis because lemon juice, being acid, stimulates the secretion of saliva. This increases the swelling of the gland and stimulates the sensory nerve fibers in its capsule. *Parotid gland disease* often causes pain in the auricle, external acoustic meatus, temple, and temporomandibular joint because the auriculotemporal nerve, from which the parotid gland receives sensory fibers, also supplies sensory fibers to the skin of the auricle and over the temporal fossa (Fig. 7-25*A*).

A radiopaque material can be injected into the duct system of the parotid gland through a cannula inserted through the oral cavity into the parotid duct. This technique, followed by radiography, is called **sialography**. *Parotid sialograms* (G. *sialon*, saliva + *grapho*, to write) demonstrate parts of the duct system that have been displaced or dilated by disease. The parotid duct may become blocked by a calcified deposit called a *sialolith* or calculus (L. a pebble). The resulting pain in the parotid gland is made worse by eating. Again, sucking a lemon slice is painful because of the buildup of saliva in the proximal part of the duct. Sometimes there is an *accessory parotid gland*. It usually lies on the masseter muscle between the parotid duct and the zygomatic arch. Several ducts open from this accessory gland into the parotid duct.

The Scalp

The scalp *consists of five layers* of soft tissue that cover the calvaria (Fig. 7-34). It extends from the superior nuchal line on the posterior aspect of the skull to the supraorbital margins. Laterally the scalp extends into the temporal fossae to the level of the zygomatic arches (Figs. 7-1 to 7-3).

Although the scalp consists of five layers, the first three layers, called the *scalp proper*, are often clinically regarded as a single layer because they remain together when a scalp flap is made during a *craniotomy* (surgical opening of the calvaria) or when the scalp is torn off during accidents. It is difficult to separate the skin and the subcutaneous tissues of the scalp from the dense connective tissue of the *epicranial aponeurosis* or galea aponeurotica (L. *galea*, helmet).

Layers of the Scalp

The scalp proper is composed of three fused layers (Fig. 7-34*A*). It is separated from the pericranium (periosteum of the calvaria) by loose connective tissue. Because of this potential areolar cleavage plane, the scalp is fairly mobile. As you move your scalp note in a mirror that your eyebrows and the skin over the root of your nose rise and that your forehead wrinkles. This occurs because your scalp muscle (frontalis, p. 656), a muscle of facial expression, is also used to express surprise and horror. The looseness of the scalp proper explains why people can be scalped so easily (*e.g.*, during automobile and industrial accidents). Each letter of the word **S C A L P** serves as a memory key for one of the layers of the scalp (Fig. 7-34*C*).

Layer 1: Skin

Hair covers the scalp in most people. The skin of the scalp is thin, especially in elderly people, except in the occipital region. The skin contains many sweat and sebaceous glands and hair follicles. The skin of the scalp has an abundant arterial supply and good venous and lymphatic drainage systems (Figs. 7-32 and 7-35).

Scalp lacerations are the most common type of head injury requiring surgical care. These wounds bleed profusely because its communicating arteries enter around the periphery of the scalp (Fig. 7-35), and they do not retract when lacerated because the scalp is tough. Hence, unconscious patients may bleed to death from scalp lacerations if the bleeding is not controlled. If these wounds are not treated appropriately, a scalp infection may develop and spread into the underlying bones of the calvaria, causing *osteomyelitis*.[7] The infection can also spread to the cranial cavity, producing an *extradural (epidural) abscess*[8], *meningitis*[9], or both.

---

[7]Inflammation of the bone marrow (diploë in this location) and adjacent bone.
[8]Collection of pus outside the dura mater.
[9]Inflammation of the membranes covering the brain (Fig. 7-34*C*).

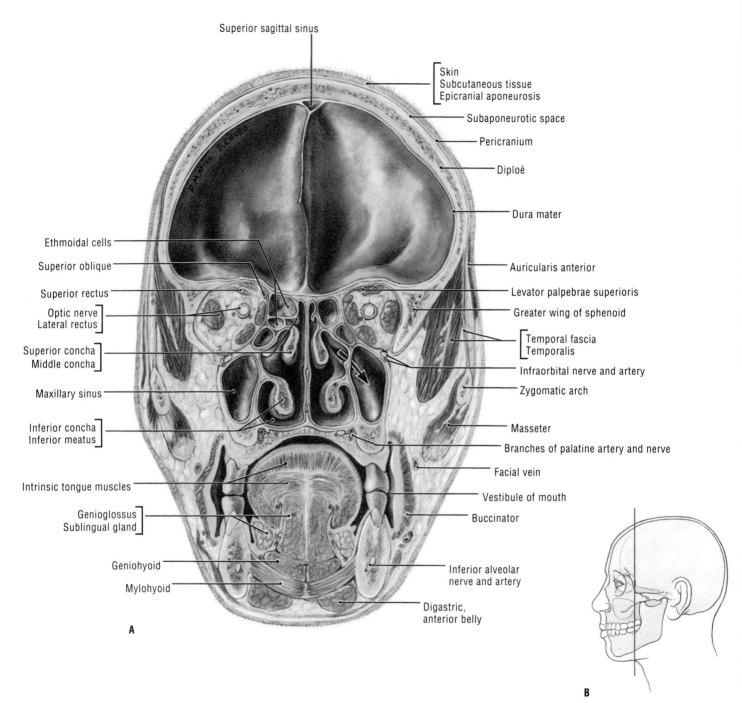

Superior sagittal sinus

Skin
Subcutaneous tissue
Epicranial aponeurosis

Subaponeurotic space

Pericranium

Diploë

Dura mater

Ethmoidal cells

Superior oblique

Superior rectus

Optic nerve
Lateral rectus

Auricularis anterior

Levator palpebrae superioris

Greater wing of sphenoid

Temporal fascia
Temporalis

Superior concha
Middle concha

Infraorbital nerve and artery

Maxillary sinus

Zygomatic arch

Inferior concha
Inferior meatus

Masseter

Branches of palatine artery and nerve

Intrinsic tongue muscles

Facial vein

Genioglossus
Sublingual gland

Vestibule of mouth

Buccinator

Geniohyoid

Mylohyoid

Inferior alveolar
nerve and artery

Digastric,
anterior belly

A

B

Figure 7-34. *A*, Coronal section of the head. *B*, Orientation drawing of *A*. *C*, Section of the scalp, calvaria and meninges, giving the key to remembering the five layers of the scalp: **S**, skin; **C**, connective tissue; **A**, aponeurosis epicranialis; **L**, loose connective tissue; **P**, pericranium.

The ducts of the sebaceous glands that are associated with hair follicles in the scalp may become obstructed, resulting in the retention of secretions and the formation of **sebaceous cysts** (wens). Because they are in the skin of the scalp proper, *sebaceous cysts move with the scalp*. Hair follicles in the scalp go through alternate growing and resting phases, *i.e.*, hairs grow and eventually drop out of their follicles (*e.g.*, during combing). After a while new hairs begin to grow in the same follicles.

Layer 2: **Connective Tissue**

This thick, subcutaneous layer of connective tissue is richly vascularized and well supplied with nerves (Figs. 7-34 and 7-35). Its collagenous and elastic fibers criss-cross in all directions, attaching the skin to the third layer of the scalp (aponeurosis epicranialis). *Fat is enclosed in lobules* between the connective fibers. The amount of subcutaneous fat in the scalp is relatively constant, varying little in emaciation or obesity, but decreases with advancing age. As a result, the scalp is thinner in elderly people.

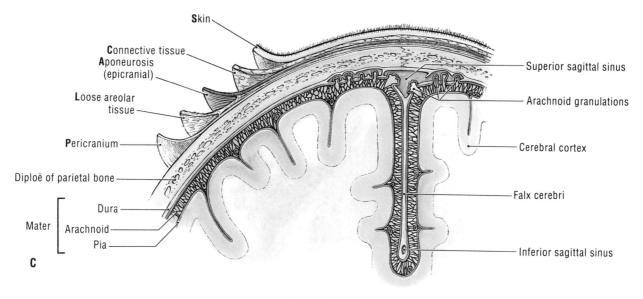

Figure 7-34C

Layer 3: Aponeurosis Epicranialis[10]

The *epicranial aponeurosis* is a strong *membranous sheet* that covers the superior aspect of the calvaria (Fig. 7-22B). The term galea (L. *helmet*) aponeurotica is sometimes used to indicate the helmetlike nature of the epicranial aponeurosis. This aponeurosis is the membranous tendon of the fleshy bellies of the *epicranius muscle* (Figs. 7-22B, 7-23A, and 7-34).

During embryonic development, the epicranius was part of a broad sheet of muscle that developed from the second branchial or pharyngeal arch (Moore, 1988). The muscles of facial expression and the platysma muscle also developed from this muscular sheet (Fig. 7-21). Hence, the epicranius or occipitofrontalis muscle is also supplied by CN VII, the nerve that innervates this arch in the embryo.

The **epicranius muscle** consists of the *occipitofrontalis*, a broad musculoaponeurotic layer that covers the superior aspect of the calvaria from the highest nuchal line posteriorly to the supraorbital margins anteriorly (Figs. 7-1, 7-3, and 7-22B). *The epicranius muscle consists of four parts*: two occipital bellies (occipitalis) and two frontal bellies (frontalis) that are *connected by the epicranial aponeurosis*. The **frontal bellies** pull the scalp anteriorly and wrinkle the forehead transversely, whereas the **occipital bellies** pull the scalp posteriorly and wrinkle the skin on the posterior aspect of the neck. Each occipital belly arises from the lateral two-thirds of the highest nuchal line and the mastoid part of the temporal bone and ends in the epicranial aponeurosis (Figs. 7-3, 7-22B, and 7-34B). Each frontal belly arises from the epicranial aponeurosis and inserts into the skin and dense subcutaneous connective tissue near the eyebrows. *The frontalis muscle has no bony attachments.* All four parts of the occipitofrontalis muscles are supplied by the facial nerve (CN VII)—the occipitalis by posterior auricular branches and the frontalis by temporal branches (Fig. 7-22). The epicranial aponeurosis is continuous laterally with the temporal fascia covering the temporalis muscle. This fascia is attached to the zygomatic arch or zygoma (Figs. 7-23A and 7-75).

The epicranial aponeurosis is a clinically important layer of the scalp. Owing to its strength, a superficial laceration in the skin does not gape because its margins are held together by this aponeurosis. When suturing a superficial wound, deep sutures are not necessary because the epicranial aponeurosis does not allow wide separation of the scalp proper. *Deep scalp wounds gape widely* when the epicranial aponeurosis is split or lacerated in the coronal plane, owing to the pull of the frontal and occipital parts of the epicranius muscle in different directions (anteriorly and posteriorly, respectively). For this reason, deep coronal lacerations gape most widely. As mentioned (p. 671), bleeding from scalp wounds is severe because the arteries cannot retract owing to the density of the connective tissue in the second layer of the scalp.

Layer 4: Loose Areolar Tissue

This subaponeurotic layer of areolar or *loose connective tissue* is somewhat like a sponge because it contains innumerable potential spaces that are capable of being distended with fluid. It is this loose connective tissue layer that allows free movement of the scalp proper, composed of layers 1-3 (Fig. 7-34A).

The loose areolar connective tissue layer is the dangerous area of the scalp because pus or blood in it can spread easily (Fig. 7-34C). Infection in this layer can also be transmitted to the cranial cavity through *emissary veins* that pass from this layer through apertures in the cranial bones, *e.g.*, the parietal foramina (Fig. 7-5). The emissary veins connect with the intracranial venous sinuses, *e.g.*, the *superior sagittal sinus* (Figs. 7-30, 7-34A, and 7-47A). Infections in the loose connective tissue layer may produce inflammatory processes in the emissary veins, leading to *thrombophlebitis* of the intracranial venous sinuses and the cortical veins (p. 692).

---

[10]Latin terminology is used here so that it will agree with the memory key for the scalp.

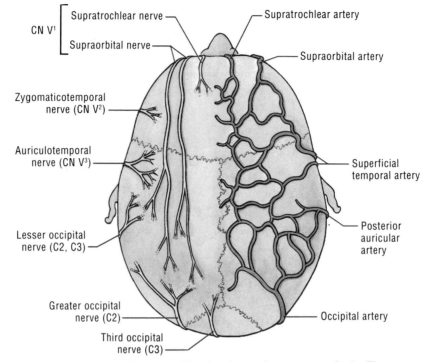

Figure 7-35.   Arteries and nerves of the scalp. Note that the arteries anastomose freely. The nerves appearing in sequence are: CN V^1, CN V^2, CN V^3, ventral rami of C2 and C3, and dorsal rami of C2 and C3.

Awareness of the limits of the loose connective tissue layer (also known as the *subaponeurotic space*) is important so that the possible spread of an infection can be anticipated. An infection is unable to spread into the neck because the occipitalis muscle is attached posteriorly to the highest nuchal line of the occipital bone and posterolaterally to the mastoid parts of the temporal bones (Fig. 7-3). An infection is unable to spread laterally beyond the zygomatic arches because the epicranial aponeurosis is continuous with the temporal fascia, where it is attached to these arches (Figs. 7-23 and 7-75). An infection or fluid (*e.g.*, blood or pus) can enter the eyelids and the root of the nose because the frontalis muscle is inserted into the skin and dense subcutaneous tissue and is not attached to the frontal bone (Fig. 7-34*B*). Because of the free movement permitted by the loose connective tissue layer, it is through it that the scalp proper separates during accidents (*e.g.*, when the hair is caught in machinery).

Layer 5: Pericranium

The pericranium is a *dense layer of specialized connective tissue* that has relatively poor osteogenic properties in adults. The periosteum or pericranium of the calvaria is firmly attached to the bones by connective tissue fibers known as *Sharpey's fibers*. They penetrate the bones and firmly anchor the pericranium to the calvaria; however, it can be stripped fairly easily from the cranial bones of living persons, except where it is continuous with the fibrous tissue in the cranial sutures. Here, the pericranium passes inwardly and is continuous with the endocranium on the internal surface of the calvaria (Fig. 7-34*A*).

During birth, bleeding sometimes occurs between the pericranium and calvaria, usually over one parietal bone. The bleeding results from rupture of multiple, minute periosteal arteries that enter and nourish the bones of the calvaria. The resulting swelling, which develops several hours after birth, is called a *cephalohematoma*. Because the adult pericranium has poor osteogenic properties, there is very little regeneration if bone loss occurs (*e.g.*, when pieces of bone have to be removed following a fracture). Surgically produced *bone flaps* are usually put back into place and wired to the other parts of the calvaria. Large defects in the adult calvaria (*e.g.*, owing to severe trauma) usually do not "fill in," necessitating the insertion of a metal or plastic plate to protect the area of the brain that is related to the defect.

Nerves of the Scalp

The sensory innervation of the scalp anterior to the auricles (external ears) is through nerves that are branches of *all three divisions of CN V, the trigeminal nerve* (Figs. 7-24*C* and 7-35). Posterior to the auricles, the nerve supply of the scalp is from the spinal cutaneous nerves (C2 and C3) of the neck from the *cervical plexus* (Fig. 7-24*A* and *B*). The areas of distribution of the branches of the trigeminal and cervical nerves are usually about equal.

Arteries of the Scalp

The blood supply of the scalp is from the *external carotid arteries*—through the occipital, posterior auricular, and superficial temporal arteries—and from the *internal carotid arteries*—through the supratrochlear and supraorbital arteries (Figs. 7-28 and 7-35). All these arteries anastomose freely with each other in *layer 2*, the subcutaneous connective tissue layer of the scalp (Fig. 7-34). Very few branches of these arteries cross *layer 4*, the loose connective tissue layer, to supply the calvaria. Its bones are supplied mainly by the *middle meningeal arteries*, branches of the maxillary arteries (Figs. 7-44*A* and 7-79), but these bones receive some blood from pericranial vessels that enter the cranium through *Volkmann's canals* (vascular canals in bones).

> The arteries of the scalp supply very little blood to the bones of the calvaria; as stated, they are supplied by the middle meningeal artery (p. 687). Hence, scalping does not produces necrosis of the cranial bones.

Veins of the Scalp

The vena comitantes accompany the arteries of the scalp and have the same names. The *supraorbital and supratrochlear veins* unite at the medial angle or canthus of the eye to form the **facial vein** (Figs. 7-23 and 7-29). Here, it communicates with the *superior ophthalmic vein*, thereby making a link that may allow facial infections to reach the cavernous sinus (Fig. 7-30 and p. 667). The *superficial temporal vein* joins the maxillary vein to form the *retromandibular vein* posterior to the neck of the mandible (Figs. 7-23 and 7-29). The *posterior auricular vein* drains the scalp posterior to the auricle and often receives a *mastoid emissary vein* from the sigmoid sinus, an intracranial venous sinus (Fig. 7-30).

Lymphatic Drainage of the Scalp

Most lymph from the scalp, including the forehead, drains into the *superficial ring of lymph nodes* that is located at the junction of the head and neck (Fig. 7-32). However, some vessels drain directly into the *deep cervical lymph nodes*. The superficial ring of lymph nodes form several groups that are named according to their position: submental, submandibular, parotid (preauricular), retroauricular (mastoid), and occipital (suboccipital). These lymph nodes receive lymph from the scalp and lymph vessels from them drain into the cervical lymph nodes. *The deep cervical lymph nodes, located along the internal jugular vein, are the most important ones* (Figs. 7-32*B* and 7-43). Lymph from them passes through the jugular trunk into the *thoracic duct* or the right lymphatic duct (see Fig. 1-39). The *superficial cervical lymph nodes* lie along the external jugular vein (Figs. 7-29 and 7-32*A*). Lymph from the forehead drains into the *submandibular lymph nodes*, and lymph from the occipital region of the scalp drains into the *occipital lymph nodes*. Lymph from the temporoparietal region drains into the *retroauricular lymph nodes*, and lymph from the frontoparietal region enters the superficial parotid lymph nodes. Lymph from the entire scalp eventually drains into the deep cervical group of lymph nodes.

> The nerves and vessels of the scalp enter inferiorly and ascend through the connective tissue (layer 2) of the scalp

into the skin; consequently, **surgical pedicle flaps**[11] of the scalp are made so that they remain attached inferiorly to preserve the nerves and vessels and to promote good healing. Reflection of the scalp and removal of the superior part of the calvaria (skullcap) exposes the meninges (G. membranes) enveloping the brain (Fig. 7-44*A*). This is done during dissections and autopsies by pulling the anterior half of the scalp over the face and the posterior half over the back of the neck. The calvaria is sawed horizontally, about 5 cm superior to the external acoustic meatus (Fig. 7-8). Following sectioning of the superficial attachment of the cranial nerves at the cranial base (Fig. 7-45) and separation of the attachments of the cranial meninges from the calvaria, the brain and meninges can be removed intact by severing the spinal cord just inferior to the foramen magnum (Figs. 7-37 and 7-51).

The Cranial Fossae

When the superior part of the calvaria (skullcap) and the brain and meninges (Fig. 7-44*A*) are removed, the internal aspect of the *cranial base* is exposed (Figs. 7-10 to 7-12). This bowl-shaped area that supports the brain has three levels (like the steps of a staircase). They are called the anterior, middle, and posterior cranial fossae (L. depressions). On a dried skull note that each fossa is at a slightly more inferior level than the one rostral to it. The posterior cranial fossa is the largest and deepest of the three fossae. The floors of the fossae are irregular owing to projections of bones from the base of the skull.

The Anterior Cranial Fossa

The inferior part and anterior extremities of the frontal lobes of the cerebral hemispheres, known as the **frontal poles** (Fig. 7-36), occupy the anterior cranial fossa, *the shallowest of the three fossae*. The anterior cranial fossa is formed mainly by the frontal bone (Figs. 7-10 to 7-12). Most of its floor is composed of the convex *orbital parts* (plates) of this bone, which constitute the bony roofs of the *orbits* (eye sockets). The orbital parts of the frontal bone show sinuous, shallow depressions called convolutional impressions, or *brain markings*. They are formed by the *frontal gyri* (convolutions) of the frontal lobes of the brain (Figs. 7-10, 7-36, and 7-51*A*). Between these impressions of the brain there are low sinuous ridges. They are produced by the sulci (L. furrows) between the gyri on the inferior surface of the frontal lobe. The **crista galli** (L. *crista*, crest + *gallus*, cock) is a median process or crest resembling a cock's comb that projects superiorly from the ethmoid bone (Figs. 7-11 and 7-12*A*). On both sides of the crista galli is a narrow, perforated **cribriform plate** (L. *cribrum*, sieve). Its numerous foramina transmit the olfactory nerves (CN I) that enter the *olfactory bulbs*

---

[11]The term *pedicle flap* is used to describe a detached mass of tissues cut away from underlying parts but attached at the edge to the pedicle, which contains its vessels and nerves.

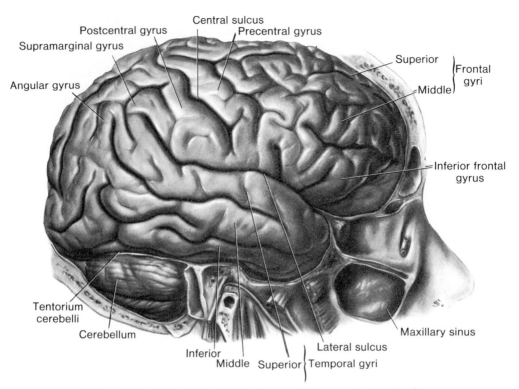

Figure 7-36. Lateral view of the brain exposed by removing the calvaria.

of the brain (Figs. 7-51 and 7-108). The crista galli and *frontal crest* give attachment to a median septum or *fold of dura mater*, called the *falx cerebri* (Figs. 7-11 and 7-37). It is located in the longitudinal cerebral fissure between the cerebral hemispheres of the brain (Fig. 7-51*A*).

In addition to the orbital parts (plates) of the frontal bone, the *lesser wings of the sphenoid bone* and the part of this bone that joins them, called the *jugum*, form the posterior part of the floor of the anterior cranial fossa (Fig. 7-10). The lesser wings present sharp posterior margins, called *sphenoidal ridges*, that overhang the anterior part of the middle cranial fossa and project into the anterior part of the *lateral sulci or fissures* of the cerebral hemispheres (Fig. 7-36). Each lesser wing ends medially in an *anterior clinoid process* (Figs. 7-11 and 7-12*A*), which gives attachment to a dural septum called the *tentorium cerebelli* (Fig. 7-37). The clinoid processes (G. *kline*, bed + *eidos*, resemblance) reminded Greek anatomists of bedposts.

The Middle Cranial Fossa

The rounded anterior extremities of the temporal lobes of the cerebral hemispheres, known as the *temporal poles* (Fig. 7-51), and about half of the inferior surface of the temporal lobe fit into the middle cranial fossa (Fig. 7-36). This fossa is located posterior and inferior to the anterior cranial fossa and is marked off from the posterior cranial fossa by a median, rectangular bony projection called the **dorsum sellae** (Figs. 7-10 to 7-12). More laterally the middle cranial fossa is separated from the posterior cranial fossa by crests or prominences formed by the superior borders of the petrous parts of the temporal bones (Figs. 7-10 and 7-38). The saddlelike part of the sphenoid bone, be-

tween the anterior and posterior clinoid processes, is known as the **sella turcica** (L. Turkish saddle). It is composed of three parts (Figs. 7-10 to 7-12): (1) anteriorly an olive-shaped swelling, called the *tuberculum sellae* (the knoblike protuberance at the anterior end of the sella turcica); (2) a seatlike depression, called the *hypophyseal fossa*, for the hypophysis cerebri or *pituitary gland* (Figs. 7-11 and 7-110); and (3) the posterior part of the "saddle" or sella turcica, known as the *dorsum sellae*.

The Posterior Cranial Fossa

This is the largest and deepest of the three cranial fossae. It lodges the cerebellum, pons, and medulla (Figs. 7-11, 7-36, and 7-37). It is formed largely by the inferior and anterior parts of the occipital bone, but the body of the sphenoid and the petrous and mastoid parts of the temporal bones also contribute to its formation (Figs. 7-10 to 7-12). The *occipital lobes* of the cerebral hemispheres lie on the *tentorium cerebelli* (Figs. 7-36 and 7-37), superior to the posterior cranial fossa. In Fig. 7-10, observe the broad bony grooves formed by the *transverse sinuses* (Fig. 7-47). They lie between diverging folds of the peripheral attachments of the tentorium cerebelli (Figs. 7-30 and 7-36). The groove for the right sinus is usually larger because the superior sagittal sinus commonly enters the transverse sinus on this side. At the center of the posterior cranial fossa is the **foramen magnum**.

The tentorium (L. tent) of the cerebellum is a tent-shaped dural septum that roofs over most of the posterior cranial fossa, intervening between the occipital lobes of the cerebral hemispheres and the cerebellum (Figs. 7-36 and 7-37). Between the anteromedial parts of the right and left leaves of the tentorium

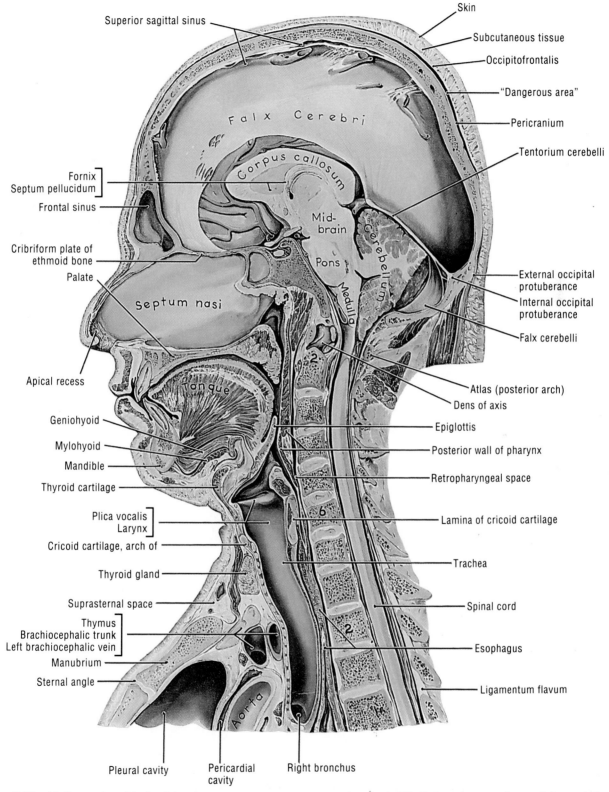

Superior sagittal sinus

Skin

Subcutaneous tissue

Occipitofrontalis

"Dangerous area"

Pericranium

Falx Cerebri

Tentorium cerebelli

Corpus callosum

Fornix
Septum pellucidum

Frontal sinus

Mid-brain

Cerebellum

Pons

Medulla

External occipital
protuberance

Internal occipital
protuberance

Falx cerebelli

Cribriform plate of
ethmoid bone

Palate

Septum nasi

Apical recess

Tongue

Atlas (posterior arch)

Dens of axis

Geniohyoid

Mylohyoid

Mandible

Thyroid cartilage

Epiglottis

Posterior wall of pharynx

Retropharyngeal space

Plica vocalis
Larynx

Cricoid cartilage, arch of

Thyroid gland

Suprasternal space

Lamina of cricoid cartilage

Trachea

Spinal cord

Thymus
Brachiocephalic trunk
Left brachiocephalic vein

Manubrium

Sternal angle

Esophagus

Ligamentum flavum

Aorta

Pleural cavity

Pericardial
cavity

Right bronchus

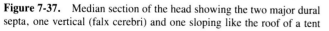

Figure 7-37.   Median section of the head showing the two major dural septa, one vertical (falx cerebri) and one sloping like the roof of a tent (tentorium cerebelli). Observe that a small part of the cerebellum has herniated through the foramen magnum in this cadaveric specimen.

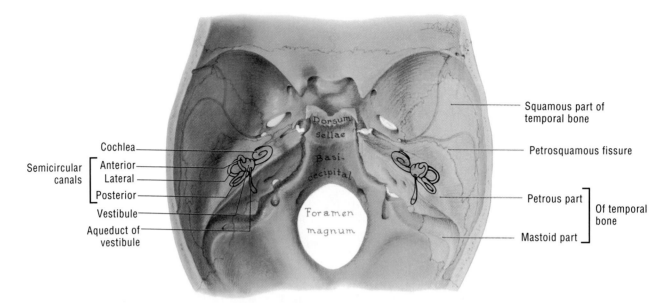

Figure 7-38. Interior of the base of the skull, primarily to show the middle cranial fossa. The parts of the temporal bone are colored. Observe the bone markings formed by the branches of the middle meningeal artery; usually a corresponding vein runs in the groove with the artery (see also Fig. 7-44A). The location of the bony labyrinth of the internal ear (p. 775) is also shown.

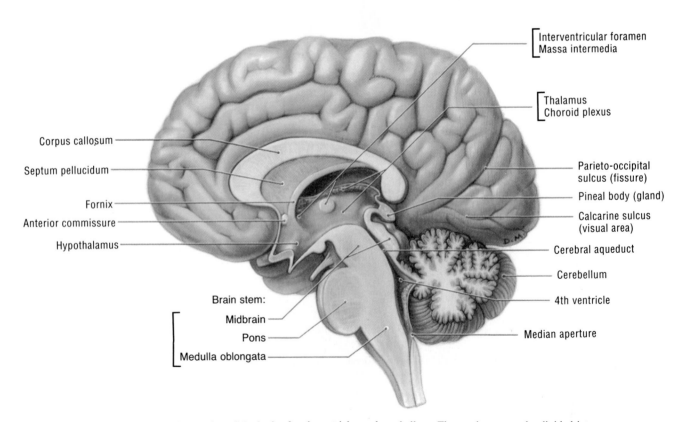

Figure 7-39. Median section of the brain, fourth ventricle, and cerebellum. The cerebrum may be divided into lobes roughly in relation to the overlying cranial bones: frontal, parietal, occipital, and temporal (Fig. 7-51B).

is an oval opening, called the *tentorial incisure* (notch), for the brain stem (midbrain, pons, and medulla). This defect in the tentorium permits the brain stem to pass from the middle to the posterior cranial fossa (Fig. 7-37). In the median plane, superior and posterior to the edge of the foramen magnum, there is a prominent bony ridge called the *internal occipital crest* (Fig. 7-11). It partly divides the posterior cranial fossa into two **cerebellar fossae** that lodge the cerebellar hemispheres. This crest ends superiorly and posteriorly in an irregular elevation called the *internal occipital protuberance*, which more or less matches the external occipital protuberance (Figs. 7-3, 7-7, 7-8, and 7-10 to 7-12).

The *vermis* (L. worm), the region in and near the median portion of the cerebellum, lies in the *vermian fossa* (Fig. 7-11). Directly anterior to this fossa is the largest opening in the base of the skull, the **foramen magnum** (Figs. 7-10 to 7-12), through which the spinal cord passes to join the medulla (Figs. 7-37 and 7-39). Rostral to the foramen magnum, the basilar part of the occipital bone rises to meet the body of the sphenoid bone (basisphenoid). This inclined bony surface, called the **clivus** (L. slope), is located anterior to the pons and medulla of the brain stem (Figs. 7-11, 7-37, and 7-40).

The Cranial Foramina

Many foramina perforate the cranial fossae (Figs. 7-10 to 7-12 and 7-41; Table 7-1). Most of these openings are for transmission of cranial nerves, but some are for the passage of blood vessels. The presence of so many foramina and thin areas of bone in the cranial base makes it fragile and vulnerable to fractures, e.g., when the base of the head is injured.

Foramina in the Anterior Cranial Fossa (Figs. 7-11 and 7-12; Table 7-1). The sievelike appearance of the *cribriform plate* of the ethmoid bone is caused by the numerous small foramina in them. Axons of olfactory cells in the olfactory epithelium in the roof of each nasal cavity converge to form about 20 small *olfactory nerves*. These bundles of axons pass through the tiny **foramina in the cribriform plate** and enter the olfactory bulbs of the brain (Figs. 7-106 and 7-108). The foramina for the *anterior ethmoidal nerve and artery* to the nasal cavity are located lateral to the anterior end of a slitlike fissure in the anterior end of the cribriform plate (Fig. 7-11). There are also posterior ethmoidal foramina for the posterior ethmoidal nerves and arteries. They are small and relatively unimportant clinically.

Between the frontal crest and crista galli there is often a small *foramen cecum*[12] (Fig. 7-11). Usually it is a blind pit, but in about 1% of people a canal forms in the frontal bone through which a *nasal emissary vein* passes from the nasal cavity to the beginning of the *superior sagittal sinus* (Fig. 7-30). It is present in children and a few adults, but it is relatively unimportant clinically.

Foramina in the Middle Cranial Fossa (Figs. 7-10 to 7-12*A*, 7-38, 7-41, and 7-42; Table 7-1). In the anteromedial part of this fossa, the **optic canals** establish communication with the orbits. These canals are traversed by the *optic nerves* (CN II) and the *ophthalmic arteries*. On each side of the base of the body of the sphenoid bone is an important **crescent of foramina** in the greater wing of the sphenoid bone. Only the four constant

---

[12]Do not confuse the foramen cecum of the cranium with the foramen cecum of the tongue (Fig. 7-94). They are unrelated.

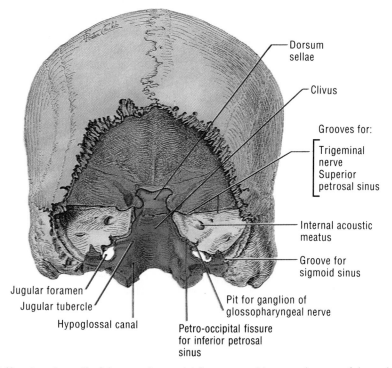

Dorsum sellae
Clivus
Grooves for:
Trigeminal nerve
Superior petrosal sinus
Internal acoustic meatus
Groove for sigmoid sinus
Jugular foramen
Jugular tubercle
Hypoglossal canal
Pit for ganglion of glossopharyngeal nerve
Petro-occipital fissure for inferior petrosal sinus

Figure 7-40. Anterior wall of the posterior cranial fossa exposed by removing part of the occipital bone.

Table 7-1.
Summary of the Foramina and Other Apertures in the Cranial Fossae and Their Contents

Foramina/Openings	Contents
Anterior Cranial Fossa	
Foramen cecum (Fig. 7-11)	Nasal emissary vein
Foramina in cribriform plates (Figs. 7-11 and 7-12)	Axons of olfactory cells in olfactory epithelium that form olfactory nn.
Anterior and posterior ethmoidal foramina (Fig. 7-11)	Vessels and nn. with same names
Middle Cranial Fossa	
Optic canals (Figs. 7-11 and 7-42)	Optic nn. (CN II) and the ophthalmic arteries
Superior orbital fissures[1] (Figs. 7-11 and 7-41)	Ophthalmic veins, CN V[1] (ophthalmic nerve), CN III, IV, and VI, and sympathetic fibers
Foramen rotundum[1] (Fig. 7-41)	CN V[2], the maxillary n.
Foramen ovale[1] (Fig. 7-41)	CN V[3], the mandibular n. and the accessory meningeal artery
Foramen spinosum[1] (Figs. 7-11 and 7-41)	Middle meningeal artery and vein and meningeal branch of CN V[3]
Foramen lacerum[1] (Fig. 7-11)	Internal carotid artery and its accompanying sympathetic and venous plexuses
Hiatus of greater petrosal nerve (Fig. 7-11)	Greater petrosal n. and petrosal branch of middle meningeal artery
Posterior Cranial Fossa	
Foramen magnum (Figs. 7-7 and 7-11; see also Fig. 4-1B)	Medulla and meninges, vertebral arteries, spinal roots of CN XI, dural veins, anterior and posterior spinal arteries
Jugular foramen (Figs. 7-7 and 7-11)	CNs IX, X, and XI, superior bulb of internal jugular vein, inferior petrosal and sigmoid sinuses, and meningeal branches of ascending pharyngeal and occipital arteries
Hypoglossal canal (Figs. 7-11 and 7-40)	CN XII, the hypoglossal n.
Condylar canal	Emissary vein that passes from sigmoid sinus to vertebral veins in neck
Mastoid foramen	Mastoid emissary vein from sigmoid sinus and meningeal branch of occipital artery

[1]Part of the crescent of foramina (Fig. 7-41).

foramina and the nerves and vessels traversing them warrant special note.

The **superior orbital fissure** is an elongated slit between the greater and lesser wings of the sphenoid bone that permits communication between the middle cranial fossa and the orbit. It is *traversed by four cranial nerves* (CN III, CN IV, CN VI, and branches of CN V[1]) and the *ophthalmic vein.*

The **foramen rotundum** is located in the greater wing of the sphenoid bone (Figs. 7-11, 7-12, and 7-41), immediately inferior and a little posterior to the medial end of the superior orbital fissure. It transmits CN V[2], the **maxillary nerve** (Fig. 7-42), the second division of the trigeminal nerve. Although the name of this foramen (L. *rotundus*, round) indicates that it is circular, it is frequently oval rather than round and is often a short canal rather than a foramen.

The **foramen ovale** is the next largest aperture in the crescent of foramina (Figs. 7-41 and 7-42). This oval foramen passes through the greater wing of the sphenoid bone, posterior and slightly lateral to the foramen rotundum, before it opens into the infratemporal fossa (Fig. 7-76). The foramen ovale transmits CN V[3], the **mandibular nerve**, the third division of the trigeminal nerve (Fig. 7-42) and the accessory meningeal artery (Fig. 7-79).

The **foramen spinosum**, the smallest of the important foramina in the crescent of foramina (Figs. 7-11, 7-12A, 7-38, and 7-41), was so named because it appears in the skull close to the spine of the sphenoid bone (Fig. 7-7). It is located posterolateral to the foramen ovale and transmits the clinically important **middle meningeal artery** (Fig. 7-42), the largest of the meningeal arteries. It also transmits the middle meningeal vein and the meningeal branch of CN V[3].

The **foramen lacerum** (Fig. 7-11) is a ragged foramen (really a canal). It is not part of the crescent of foramina (Fig. 7-41). It is located between the body of the sphenoid bone and the apex of the petrous part of the temporal bone (Fig. 7-38), posteromedial to the foramen ovale. The carotid and pterygoid canals open into the foramen lacerum, which is filled with fibrocartilage, except in dried skulls. The superior end of the foramen lacerum contains the *internal carotid artery* (Figs. 7-28 and 7-42) and its accompanying sympathetic and venous plexuses. The internal carotid artery enters the skull through the carotid canal, which opens into the foramen lacerum.

Hiatuses of the Canals for the Petrosal Nerves (Fig. 7-11). Extending posteriorly and laterally from the foramen lacerum, there is a narrow groove for the greater petrosal nerve on the anterior surface on the petrous part of the temporal bone. At the lateral end of this sulcus, there is a small hiatus (slit), called the *greater petrosal hiatus* (foramen). It transmits the petrosal branch of the middle meningeal artery and **greater petrosal nerve** as it leaves the geniculate ganglion of the facial nerve (Fig. 7-27). Sometimes there is also a groove for the *lesser petrosal nerve*, but its hiatus is so small that it usually cannot be identified.

Foramina in the Posterior Cranial Fossa (Figs. 7-7, 7-10 to 7-12, and 7-40; Table 7-1). There are several clinically important foramina and canals in the posterior cranial fossa.

The **foramen magnum** (Figs. 7-7, 7-10 to 7-12, and 7-38) is a unique opening for three reasons: (1) it is the largest foramen in the skull; (2) the junction of the medulla and spinal cord occurs within it (Fig. 7-37); and (3) it is unpaired. This large foramen, usually oval, lies at the most inferior part of the posterior cranial fossa, midway between the mastoid processes. Its long diameter is anteroposterior. The posterior cranial fossa communicates with the vertebral canal (p. 329) through the foramen magnum.

Structures passing through the foramen magnum are: (1) the

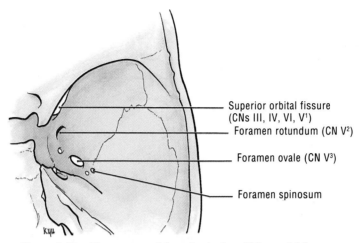

Superior orbital fissure
(CNs III, IV, VI, V¹)
Foramen rotundum (CN V²)

Foramen ovale (CN V³)

Foramen spinosum

Figure 7-41. The crescent of foramina in the middle cranial fossa.

medulla (Fig. 7-37); (2) the spinal roots of the *accessory nerves* (CN XI) (see Fig. 8-13); (3) the meningeal branches of the superior *cervical nerves* (C1 to C3); (4) the *meninges* (coverings of the brain and spinal cord); (5) the *vertebral arteries*, ascending to supply parts of the brain (Fig. 7-28); and (6) the anterior and posterior *spinal arteries*, descending to supply the superior part of the spinal cord (see Fig. 4-52).

The **jugular foramen** (Figs. 7-11 and 7-40) is an interosseous foramen that is located between the occipital bone and the petrous part of the temporal bone. It lies at the posterior end of the *petrooccipital suture*. The posterior part of the jugular foramen contains the superior bulb of the internal jugular vein (Fig. 7-43) into which the *sigmoid sinus* enters (Fig. 7-30). The **tympanic body** (jugular glomus) is a small ovoid body consisting of chemoreceptor tissue. It has a structure similar to the carotid body (p. 801) and presumably has a similar function. It is enclosed in the adventitia of the jugular bulb of the internal jugular vein. Anteromedial to this vein, several structures descend through the jugular foramen: (1) the glossopharyngeal nerve (CN IX); (2) the vagus nerve (*CN X*); (3) the accessory nerve (*CN XI*); and (4) the *inferior petrosal sinus* on its way to the superior end of the internal jugular vein (Fig. 7-30).

The **hypoglossal canal** (Figs. 7-11 and 7-40) transmits the *hypoglossal nerve* (CN XII). Its internal opening lies between the jugular foramen and the occipital condyle. The hypoglossal canal is frequently divided into two canals by a bony septum that separates the two bundles of rootlets of the hypoglossal nerve.

The **condylar canal**, located at the inferior end of the groove for the sigmoid sinus (Fig. 7-10), is inconstant. It transmits an emissary vein from the sigmoid sinus to the vertebral veins in the neck (Fig. 7-30). The condylar canal passes inferoposteriorly to emerge just posterior to the occipital condyle (Fig. 7-7).

The **internal acoustic meatus** (Figs. 7-11 and 7-40) lies superior to the anterior part of the jugular foramen in the *petrous part of the temporal bone*. Running to it in a transverse direction from the brain stem are the facial (CN VII) and vestibulocochlear (CN VIII) nerves, the nervous intermedius, and the labyrinthine vessels. The internal acoustic meatus is closed laterally by a perforated plate of bone, which separates it from the internal ear

(Fig. 7-123). As the sigmoid groove reaches the mastoid portion of the temporal bone, the *mastoid foramen* opens into it. It is close to the occipital suture. The mastoid foramen transmits an emissary vein from the sigmoid sinus and a meningeal branch of the occipital artery.

Cranial Meninges and Cerebrospinal Fluid

A clear understanding of the membranes (L. *meninges*) covering the brain and their relationship to the *cerebrospinal fluid* (*CSF*) is an essential basis for understanding intracranial disease and head injuries. The soft fresh brain tends to flatten when it is removed from the skull and placed on a table. In the living body, this vital organ is enveloped by three membranes (Figs. 7-34C and 7-44): (1) an external, thick, tough **dura mater** (L. *dura*, hard, tough + *mater*, mother); (2) an intermediate, thin, cobweblike **arachnoid mater** (G. *arachne*, spider + *eidos*, resemblance); and (3) an internal, delicate, vascular **pia mater** (L. *pius*, tender). These three membranes covering the brain, known collectively as the *cranial meninges*, are continuous with the spinal meninges covering the spinal cord. *The cranial meninges and CSF provide support and protection for the brain*, in addition to that afforded by the calvaria and scalp.

The dura mater is occasionally called the pachymeninx (G. *pachys*, thick + *menix*, membrane); hence, you may hear the term *pachymeningitis*, which refers to inflammation of the dura mater. Since the advent of antibiotics, this condition is uncommon. The pia and arachnoid together are referred to as the *leptomeninges* (G. *leptos*, slender + *meninges*, membranes). When clinicians refer to *piarachnitis*, *leptomeningitis*, or *meningitis*, you should realize that all these terms mean essentially the same thing (*i.e.*, inflammation of the pia-arachnoid or leptomeninges).

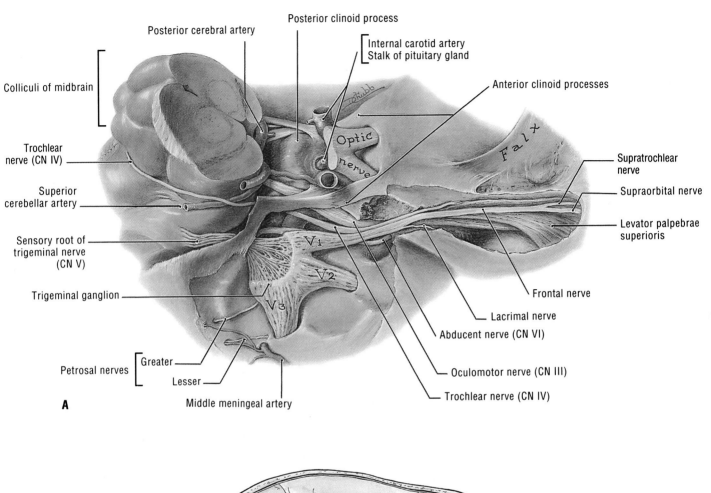

Posterior cerebral artery

Posterior clinoid process

Internal carotid artery
Stalk of pituitary gland

Anterior clinoid processes

Colliculi of midbrain

Optic nerve

Falx

Supratrochlear nerve

Trochlear nerve (CN IV)

Superior cerebellar artery

Supraorbital nerve

Levator palpebrae superioris

Sensory root of trigeminal nerve (CN V)

V1

V2

Trigeminal ganglion

V3

Frontal nerve

Lacrimal nerve

Abducent nerve (CN VI)

Petrosal nerves { Greater
Lesser

Oculomotor nerve (CN III)

Trochlear nerve (CN IV)

Middle meningeal artery

A

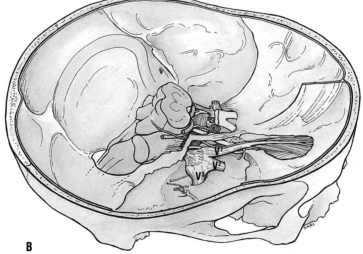

B

Figure 7-42. *A*, Dissection of the nerves in the middle cranial fossa. The roof of the orbit has been partly removed, and the tentorium cerebelli (Figs. 7-37 and 7-45) has been removed on the right side to show the trochlear nerve (CN IV) and the sensory root of the trigeminal (CN V). *B*, Orientation drawing.

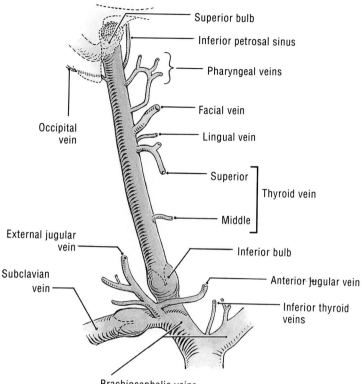

Figure 7-43.   Anterior view of the right internal jugular vein and its tributaries (see also Fig. 7-47).

The Dura Mater

This is the outermost and toughest of the membranes covering the central nervous system. It consists of collagenous connective tissue. When the superior part of the calvaria is removed during dissection or at autopsy to expose the brain, you will find that the bones can be pried loose from the dura without difficulty. When you do this, you also strip the internal periosteum or endocranium from the internal surface of these bones (Fig. 7-44A). Although the cranial dura mater[13] consists of one layer, similar to the spinal dura (p. 366), it is described as a two-layered membrane because the dura adheres so closely to the internal periosteum or endocranium, except where there are dural venous sinuses (*e.g.*, the superior sagittal sinus) and *venous lacunae* (Figs. 7-34A, 7-44A, and 7-45). The dura also extends medially to form dural folds or septa, *e.g.*, the *falx cerebri* (Fig. 7-37).

It is important to understand that the "external layer of the dura" is really the internal periosteum or endocranium of the calvaria (Fig. 7-34A). It is attached to the bones of the calvaria by *Sharpey's fibers*, which penetrate and attach it to the cranial bones. This periosteal layer ("outer layer of the dura") is also continuous with the external periosteum or pericranium of the cranial bones at the margins of the foramen magnum and at the smaller foramina for nerves and vessels.

The internal meningeal layer of cranial dura is continuous with the spinal dura at the foramen magnum. It also provides tubular sheaths for the cranial nerves as they pass through the foramina in the floors of the cranial fossae (Fig. 7-45). Outside the skull, the dural sheaths fuse with the epineurium of the cranial nerves. The **dural sheaths of the cranial nerves** extend approximately to the *cranial ganglia*; *e.g.*, the trigeminal ganglion of the fifth cranial nerve (CN V) is surrounded by an extension of the cranial meninges. The dura-enclosed space at the *trigeminal impression* in the petrous part of the temporal bone, called the *trigeminal cave*, is occupied by the trigeminal ganglion (Figs. 7-46 and 7-81). The dural sheath of the optic nerve is also continuous with the internal meningeal layer of the cranial dura and the sclera of the eye (Fig. 7-62A). These relationships are clinically important.

> The attachment of the dura to the bones in the floors of the cranial fossae is firmer than it is to the calvaria (Fig. 7-45). Thus, a blow to the head can detach the dura from the calvaria without fracturing the bones, whereas a basal fracture usually tears the dura and results in leakage of CSF into the soft tissues of the neck, nose, ear, or nasopharynx.

Dural Septa or Reflections (Figs. 7-36, 7-37, 7-44, and 7-45).   During development of the brain, the dura is duplicated or reflected to form four inwardly projecting dural septa (folds).

---

[13]For brevity during clinical discussions of this layer, the term *dura* is used. This usage is also followed here. Similary after the opening statements, the arachnoid mater will be referred to as the *arachnoid* and the pia mater as the *pia*.

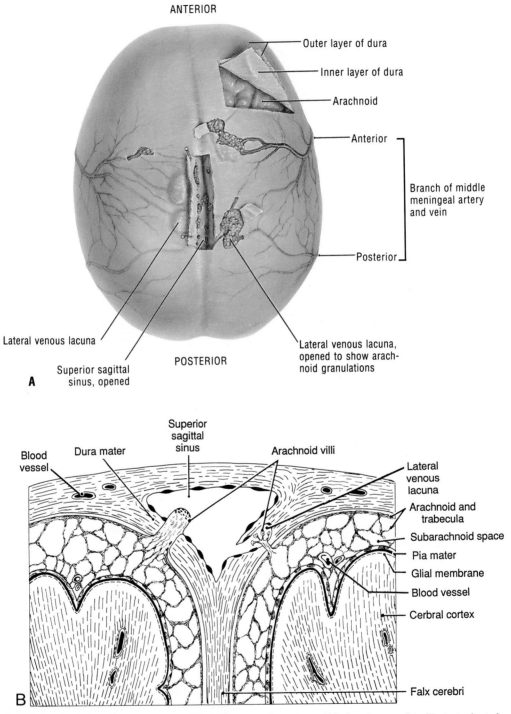

Figure 7-44. *A*, The external surface of the dura mater has been reflected to show the arachnoid mater covering the brain. *B*, Meningeal-cortical relationship. The subarachnoid space (containing CSF) is shown in greater width than is normal to illustrate the trabeculae that cross it and connect the pia mater and arachnoid mater. Note that the pia is firmly anchored to the cerebral cortex.

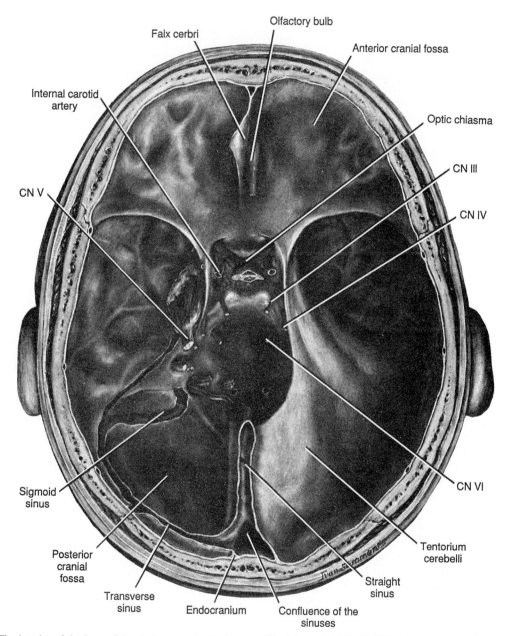

Figure 7-45. The interior of the base of the skull covered with dura mater. Observe that the meningeal layer of dura and the endocranium are separated by venous sinuses (*e.g.*, at the confluence of the sinuses).

The falx cerebri (Fig. 7-37) has been cut close to its anterior attachment and most of it has been removed. The tentorium cerebelli has been cut away on the left to expose the posterior cranial fossa.

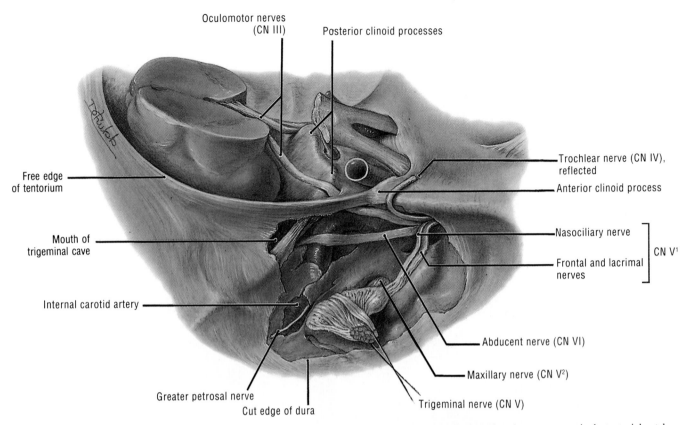

Oculomotor nerves (CN III)

Posterior clinoid processes

Trochlear nerve (CN IV), reflected

Anterior clinoid process

Free edge of tentorium

Nasociliary nerve

CN V¹

Mouth of trigeminal cave

Frontal and lacrimal nerves

Internal carotid artery

Abducent nerve (CN VI)

Maxillary nerve (CN V²)

Greater petrosal nerve

Trigeminal nerve (CN V)

Cut edge of dura

Figure 7-46.   Nerves in the middle cranial fossa. The trigeminal nerve is divided and turned anterolaterally from the mouth of the trigeminal cave. Observe the midbrain and oculomotor nerves in the tentorial notch. Note the sinuous course of the internal carotid artery.

These septa divide the cranial cavity into three intercommunicating compartments, one subtentorial and two supratentorial (*i.e.*, inferior and superior to the tentorium cerebelli). The dural septa provide support for parts of the brain, particularly the cerebral hemispheres (Fig. 7-37).

The Falx Cerebri (L. *falx*, a sickle).   This is a large, sickle-shaped, vertical partition in the longitudinal fissure between the two cerebral hemispheres (Figs. 7-37, 7-44B, 7-45, and 7-51). This thick, tough dural fold is attached in the median plane to the internal surface of the calvaria from the frontal crest of the frontal bone and the *crista galli* of the ethmoid bone anteriorly to the *internal occipital protuberance* posteriorly (Fig. 7-11). The falx cerebri is also attached to the midline of the *tentorium cerebelli*, another dural fold that lies between the occipital lobes of the cerebral hemispheres and the cerebellum. At the superior convex border of the falx cerebri, its two layers separate to enclose the *superior sagittal sinus* (Figs. 7-34A and 7-44). The *inferior sagittal sinus* (Fig. 7-30) is enclosed within the free inferior edge of the falx cerebri. Throughout life the falx cerebri forms a rigid partition between the cerebral hemispheres that reduces side-to-side movement of them.

The Tentorium Cerebelli (L. *tentorium*, tent).   This is a wide crescentic, arched fold of dura that separates the occipital lobes of the cerebral hemispheres from the cerebellum (Figs. 7-36, 7-37, and 7-45). The attachment of the falx cerebri to the median portion of the tentorium cerebelli holds the latter fold up, similar to the ridgepole of a tent. The tentorium cerebelli is attached anterolaterally to the superior edges of the petromastoid parts of the temporal bones and to the anterior and posterior **clinoid processes** (Figs. 7-12A and 7-46). Posteriorly, the tentorium is attached to the occipital bone along the grooves for the transverse sinuses, which it encloses (Fig. 7-45). Its concave anteromedial border is free, and between it and the dorsum sellae of the sphenoid bone is an opening called the *tentorial incisure* or notch (Fig. 7-45). This oval opening surrounds the midbrain as it passes from the middle to the posterior cranial fossa (Figs. 7-37 and 7-46). **CSF** from the *lateral ventricles* of the brain passes inferiorly through the tentorial incisure via the third ventricle and *cerebral aqueduct* (Figs. 7-52 and 7-53A) into the *fourth ventricle*. After leaving the ventricular system, CSF passes superiorly through the tentorial incisure into the subarachnoid space surrounding the midbrain.

There are two chances of obstructing the flow of ventricular fluid, CSF, or both at the tentorial incisure, and each produces a different type of *hydrocephalus*.[14] The tentorium cerebelli prevents inferior displacement of the occipital lobes of the brain when a person is erect (Fig. 7-36). The tentorial incisure is slightly larger than is necessary to accommodate the midbrain. Hence, when intracranial pressure superior to the tentorium is considerably higher than that inferior to it,

---

[14]Excessive accumulation of ventricular fluid, CSF, or both, resulting in enlargement of the ventricles and the head (p. 700).

part of the adjacent temporal lobe may herniate through the tentorial incisure. *During tentorial herniation the temporal lobe may be lacerated* by the tough, taut tentorium, and CN III, the *oculomotor nerve, may be stretched*, compressed, or both (Figs. 7-45 and 7-46).

The Falx Cerebelli (Fig. 7-37).   This is a small, sickle-shaped, median dural fold in the posterior part of the posterior cranial fossa. It extends almost vertically, inferior to the inferior surface of the tentorium cerebelli. Its free edge projects slightly between the cerebellar hemispheres. The *occipital venous sinus* is located in the base of the falx cerebelli (Fig. 7-30).

The Diaphragma Sellae (Figs. 7-12, 7-37, and 7-42).   This is a small, circular, horizontal sheet of dura that forms a roof for the *hypophyseal fossa* in the sella turcica. This fold is formed by the dura surrounding the *hypophysis cerebri* (pituitary gland) and encircling the stalk of this gland. The diaphragma sellae covers the hypophysis cerebri, which lies in the hypophyseal fossa. The diaphragma sellae has a central aperture for passage of the hypophyseal veins and the hypophyseal stalk or *infundibulum*, which connects the hypothalamus and the hypophysis.

> Tumors of the hypophysis cerebri (**pituitary tumors**) may extend superiorly through the aperture in the diaphragma sellae and/or cause bulging of it. These tumors often expand the diaphragma sellae and may produce endocrine symptoms early or late (*i.e.*, before or after enlargement of the diaphragma sellae). *Superior extension of a tumor causes visual symptoms owing to pressure on the optic chiasma* (Figs. 7-42 and 7-45), the place where the optic nerve fibers cross. The Greek term *chiasma* is derived from the letter *chi* (χ), which has two crossed lines.

Arteries of the Dura Mater (Fig. 7-44).   There are many *meningeal arteries* in the periosteum that are inappropriately named because they supply more blood to the bones of the calvaria than to the dura. Only very fine branches of these arteries are distributed to the dura. The **middle meningeal artery** is the largest of the meningeal arteries (Figs. 7-42, 7-44A, and 7-78). It is a *branch of the maxillary artery* and is embedded in the external layer of the dura. It is clinically important largely because it is often torn when the skull is fractured. The middle meningeal artery enters the cranial cavity through the *foramen spinosum* in the floor of the middle cranial fossa (Fig. 7-42). It then runs laterally on the floor of this fossa and turns superoanteriorly on the greater wing of the sphenoid bone, where it divides into anterior (frontal) and posterior (parietal) branches. The *anterior branch* runs superiorly to the pterion and then curves posteriorly to ascend toward the vertex of the skull. In the region of the pterion, the artery is often in a bony tunnel for 1 cm. The *posterior branch* runs horizontally in a posterior direction in a groove in the squamous part of the temporal bone. It then ramifies over the posterior aspect of the skull. The middle meningeal artery and all its branches are accompanied by meningeal veins (Fig. 7-44A); they lie between them and the bones.

> A fracture in the temporal region of the skull (*e.g.*, at the *pterion*) may tear the anterior branches of the middle meningeal vessels, producing a profuse hemorrhage between the calvaria and the dura. This results in an **extradural hema-**

toma that compresses the *motor area of the brain* (Fig. 7-51B and C). Trephination (removal of a circular piece of the calvaria) may be required for removal of the hematoma (mass of clotted blood). Knowledge of the surface markings of the middle meningeal artery are required for *trephining for an extradural hemorrhage*. It is approached surgically through a bur hole at the pterion that lies 3 cm superior to the midpoint of the zygomatic arch, posterosuperior to the frontozygomatic suture (Fig. 7-8).

Veins of the Dura Mater (Figs. 7-11, 7-30, and 7-44). Meningeal veins accompany the meningeal arteries and lie between them and the bones. Hence, they may also be torn in fractures of the calvaria and contribute to the extradural bleeding. The *middle meningeal veins* leave the skull through the foramen spinosum and foramen ovale to join the *pterygoid plexus*.

Nerve Supply to the Dura Mater (Figs. 7-24 and 7-100).   The rich sensory supply to the dura is largely through the three divisions of the *trigeminal nerve* (CN V). For example, most of the supratentorial part of the dura is supplied by CN V¹. Sensory branches are also received from the vagus nerve (CN X) and the superior three cervical nerves accompanying the hypoglossal nerve (CN XII). The sensory endings are more numerous in the dura along each side of the superior sagittal sinus and in the tentorium cerebelli than they are in the floor of the cranium. Pain fibers are also numerous where arteries and veins pierce the dura.

> The dura is sensitive to pain, especially around the venous sinuses of the dura (Figs. 7-30, 7-45, and 7-47). Consequently, pulling on arteries at the base of the skull or veins in the vertex of the skull or cranial base where they pierce the dura, causes pain. Although there are many *causes of headache*, distention of the scalp, meningeal vessels, or both, is believed to be one cause. Many headaches appear to be dural in origin; *e.g.*, the headache that occurs after a *lumbar puncture* (p. 368) for the removal of CSF (see Fig. 4-55) is thought to result from stimulation of sensory nerve endings in the dura. When CSF is removed the brain sags slightly, pulling on the dura. This may cause pain and headache. It is for this reason that patients are asked to keep their heads down after a lumbar puncture to minimize or prevent headaches (hence, use of a pillow is not advisable).

The Arachnoid Mater

This delicate, *transparent membrane* is composed of weblike tissue (Fig. 7-44). The arachnoid forms the intermediate covering of the brain and is separated from the dura by what has been clinically described as a film of fluid in a potential *subdural space*.[15] The arachnoid does not form a close investment of the brain but passes over the sulci and fissures without dipping into them. The term *arachnoid* is derived from the Greek *arachne + -oid*, meaning "resembling a spider's web." The arachnoid is partly separated from the pia by the *subarachnoid space* containing CSF (Fig. 7-44B). Numerous trabeculae, which are del-

---

[15]Recent evidence indicates that the so-called subdural space is not a potential space and that the creation of a cleft in this area of the meninges is the result of tissue damage. For a discussion on this new concept, see Haines (1991).

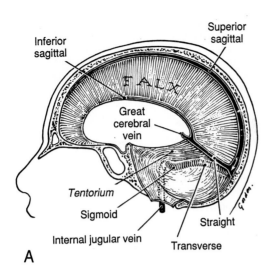

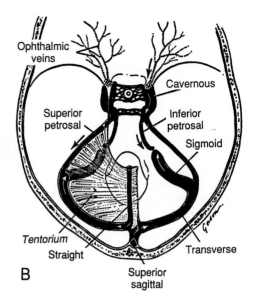

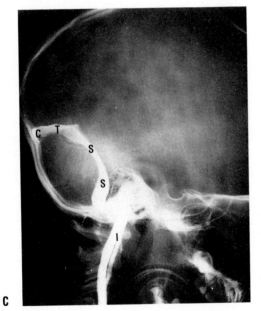

Figure 7-47. Venous sinuses of the dura mater. *A*, Sinuses in the falx cerebri and tentorium cerebelli. *B*, The sinuses are opened superiorly. The arrows show the direction of blood flow. *C*, Lateral view of a venogram of the sinuses. *C* indicates the confluence of sinuses; *T*, transverse sinus; *S*, sigmoid sinus; and *I*, internal jugular vein. (Courtesy of Dr. D. Armstrong, Associate Professor of Radiology, University of Toronto, Toronto, Ontario, Canada.)

icate strands of connective tissue, pass from the arachnoid to the pia, giving it a weblike structure.

The Pia Mater

This membrane is very thin (L. *pius*, tender), but it is thicker than the arachnoid. The pia is the innermost of the three layers of meninges and is a *highly vascularized*, loose connective tissue membrane that adheres closely to the surface of the brain. It dips into all sulci and fissures and carries small blood vessels with it (Fig. 7-44*B*). The cerebral veins run on the pia within the subarachnoid space. When branches of cerebral vessels penetrate the brain, the pia follows them for a short distance, forming a sleeve of pia. Hence, the *perivascular spaces* (Virchow-Robin spaces) are continuous with the subarachnoid space. They extend in an increasingly attenuated form as far as the arterioles and venules in the brain.

The central artery and vein of the retina cross the subarachnoid space to become enclosed in the space around the posterior part of the optic nerve (Fig. 7-62). Consequently, an increase in CSF pressure around the optic nerve slows the return of venous blood and results in *edema of the retina*. This is most apparent as a swelling of the optic papilla (*papilledema*). Thus inspection of the ocular fundus is an important part of neurological examinations.

In cases of ruptured *aneurysm of the cerebral arterial circle* (Fig. 7-54), there is hemorrhage into the subarachnoid space. *Blood in the CSF* results in an increase in intracranial pressure, which is transmitted to the subarachnoid spaces around the optic nerves, compressing the retinal vein where it runs in the subarachnoid space around the optic nerve. This results in increased pressure in the retinal capillaries and *subhyaloid hemorrhages* between the retina and vitreous body (Fig. 7-62*A*).

The Meningeal Spaces

The *extradural (epidural) space* is superficial to the dura, *i.e.*, between the bone and the internal periosteum (Fig. 7-44). Because the dura is intimately attached to the periosteum of the calvaria, the extradural space is only a potential space. It becomes a real space when blood accumulates in it from torn meningeal vessels (*extradural hemorrhage*).

The so-called *subdural space* is in the deepest part of the dura. When a tissue space is created in this area, it appears to be the result of tissue damage (Haines, 1991). It represents, in most instances, a cleaving open of the dural border cell layer. In these cases the extracellular spaces in this layer are enlarged, creating what has classically been called the subdural space.

Head injuries may be associated with various types of *intracranial bleeding* (hemorrhage).

Extradural (Epidural) Hemorrhage. Bleeding between the dura and the internal periosteum of the calvaria (*i.e.*, superficial to the dura) may follow a blow to the head. Typically, there is a brief concussion (loss of consciousness), followed by a lucid interval of some hours. This is succeeded by drowsiness and **coma** (profound unconsciousness). When extradural hemorrhage occurs, there is tearing of meningeal vessels. *The artery most commonly injured is the middle meningeal artery* (p. 687) and/or one of its branches, often the anterior one (Fig. 7-44*A*). Similarly, the superior sagittal sinus may be torn in fractures of the superior part of the calvaria.

Although veins and venous sinuses may be torn, most bleeding is usually from torn arteries. Because the internal periosteum is firmly attached to calvaria, there is a slow, localized accumulation of blood; this forms an *extradural hematoma* (Fig. 7-126). As the blood mass increases in size, compression of the brain occurs. This necessitates evacuation of the fluid and occlusion of the bleeding vessels. The formation of an extradural hematoma, superior to the tentorium cerebelli (Figs. 7-36, 7-37, and 7-45) causes a rise in *supratentorial pressure*. If the pressure rises very high, part of the cerebellum (*cerebellar tonsils*) may be forced through the foramen magnum (Fig. 7-37), compressing the medulla. This may cause fatal effects owing to interference with the respiratory and cardiovascular centers.

Subdural (Intradural) Hemorrhage. Bleeding within the dura may follow a blow to the head that jerks the brain inside the skull. Such displacement of the brain is greatest in elderly people in whom some shrinkage of the brain has occurred. This type of hemorrhage commonly results from tearing of a superior cerebral vein as it enters the superior sagittal sinus (Fig. 7-37).

Subarachnoid Hemorrhage. Bleeding into the subarachnoid space usually follows the rupture of an *aneurysm of an intracranial artery*. These thin-walled outpouchings or evaginations occur chiefly at bifurcations of the arteries at the base of the brain (Fig. 7-54). Subarachnoid hemorrhages are also associated with skull fractures and cerebral lacerations. These hemorrhages result in meningeal irritation, which produces a severe headache, stiff neck, and often loss of consciousness.

Intracerebral Hemorrhage. Bleeding into the brain, often from one of the branches of the *middle cerebral artery*, is frequent in persons with hypertension. One of these arteries was named the *artery of cerebral hemorrhage* by Charcot because he observed that certain hemorrhages produced a **paralytic stroke**. Paralysis occurs because of the interruption of motor pathways from the motor cortex to the brain stem and spinal cord.

Venous Sinuses of the Dura Mater

These sinuses are venous channels located between the dura and the internal periosteum lining the cranium, usually along the lines of attachment of the dural septa (Figs. 7-34*A*, 7-37, 7-44, and 7-45). *The venous sinuses drain all the blood from the brain.* The sinuses are lined with endothelium that is continuous with that of the cerebral veins entering them. These sinuses have no valves and no muscle in their walls. Several of the sinuses are triangular in transverse section (Fig. 7-44*B*) because their bases are on bone and their side walls are formed by the origins of the dural folds.

The Superior Sagittal Sinus (Figs. 7-11, 7-30, 7-34*A*, 7-37, and 7-44). This venous sinus lies in the median plane, along the attached border of the falx cerebri. It begins at the *crista galli*, runs the entire length of the superior attached portion of the falx cerebri, and ends at the *internal occipital protuberance*. In about 60% of cases the superior sagittal sinus ends by becoming the right *transverse sinus* (Figs. 7-45 and 7-47). In other cases it turns to the left and ends in the left transverse sinus or in both. At the termination of the superior sagittal sinus is a dilation, known as the *confluence of the sinuses*. Occasionally this dilation is referred to as the sinus confluens or the *torcular Herophili* (L. winepress of Herophilus, an early anatomist and surgeon). Five sinuses often communicate at the confluence; however, considerable variation in venous pathways occurs.

The superior sagittal sinus is triangular in transverse section (Figs. 7-34*A* and 7-44), the superior wall being formed by the endocranium (internal periosteum) lining the calvaria and the lateral walls by the dura. The superior sagittal sinus becomes larger as it passes posteriorly and receives more and more superior cerebral veins. These veins enter the sinus in an anterior direction, resulting from their fixation to the superior sagittal sinus in early fetal life. During later development, their more distal parts are carried posteriorly by the growth of the brain. The superior sagittal sinus communicates on each side through slitlike openings with venous spaces in the dura, called *lateral venous lacunae*, into which some arachnoid villi project (Fig. 7-44). Usually there are three **venous lacunae** on each side of the superior sagittal sinus. In very old people, these lacunae often coalesce to form one long sinus on each side.

The main site of passage of CSF into venous blood is through the **arachnoid villi**, especially those projecting into the superior sagittal sinus and the adjacent lateral venous lacunae (Fig. 7-44). *Hypertrophied aggregations of arachnoid villi*, associated with advancing age, are called *arachnoid granulations* (Pacchionian bodies). These granulations often produce erosion or pitting of the internal surface of the superior portions of the frontal and parietal bones (Fig. 7-5). The superior sagittal sinus also communicates with veins in the scalp through emissary veins that pass through the *parietal foramina*.

The Inferior Sagittal Sinus (Figs. 7-30 and 7-47*A*). This venous sinus is much smaller than the superior sagittal sinus. It occupies the posterior two-thirds of the free inferior edge of the falx cerebri. It ends by joining the *great cerebral vein* (of Galen) to form the *straight sinus*. The inferior sagittal receives cortical veins from the medial aspects of the cerebral hemispheres.

The Straight Sinus (Figs. 7-30, 7-45, and 7-47). This venous sinus is formed by the union of the inferior sagittal sinus with the great cerebral vein. It runs inferoposteriorly along the line of attachment of the falx cerebri to the tentorium cerebelli, where it becomes continuous with one of the transverse sinuses, usually the left.

The Transverse Sinuses (Figs. 7-11, 7-30, 7-45, and 7-47*B*). These venous sinuses pass laterally from the confluence of the sinuses in the attached border of the tentorium cerebelli. They groove the occipital bones and the posteroinferior angles of the parietal bones. They then leave the tentorium and become the sigmoid sinuses.

The Sigmoid Sinuses (Figs. 7-11, 7-30, 7-40, 7-45, and 7-47). These venous sinuses (shaped like the Greek letter *sigma* [S]), follow S-shaped courses in the posterior cranial fossa, forming deep grooves in the inner surface of the posterior part of the mastoid parts of the temporal bones, and in the lateral surfaces of the jugular tubercles of the occipital bone. The sigmoid sinuses then turn anteriorly and enter venous enlargements called the *superior bulbs of the internal jugular veins* (Fig. 7-43), which occupy the jugular foramina. These large bulbs receive the inferior petrosal sinuses and continue as the internal jugular veins. *All the venous sinuses of the dura eventually deliver most of their blood to the sigmoid sinuses* and from them to the internal jugular veins, except for the inferior petrosal sinuses, which enter these veins directly (Figs. 7-43 and 7-49).

The Occipital Sinus (Fig. 7-30). This is the smallest of the dural venous sinuses. It begins near the posterior margin of the foramen magnum as two or more venous channels around the edges of this large aperture. These channels usually join to form an unpaired sinus that lies in the attached border of the falx cerebelli. The occipital sinus communicates inferiorly with the *internal vertebral plexus* (Fig. 7-30; see also Fig. 4-53) and ends superiorly in the confluence of sinuses (Fig. 7-45).

The Cavernous Sinuses (L. *caverna*, cave). These large venous sinuses are about 2 cm long and 1 cm wide (Figs. 7-30 and 7-47). They are *located on each side of the sella turcica and the body of the sphenoid* bone (Figs. 7-10 and 7-12). They were named cavernous sinuses because of their cavelike appearance, which results from the many blood channels formed by numerous trabeculae. Each cavernous sinus extends from the superior orbital fissure anteriorly to the apex of the petrous part of the temporal bone posteriorly. Here, it joins the posterior *intercavernous sinus* and the superior and inferior petrosal sinuses (Figs. 7-30 and 7-47*B*). Each cavernous sinus receives blood from the superior and inferior ophthalmic veins, the superficial middle cerebral vein in the lateral fissure of the cerebral hemisphere, and the *sphenoparietal sinus*. The latter sinus lies inferior to the edge of the lesser wing of the sphenoid bone. The cavernous sinuses communicate with each other through *intercavernous sinuses*, which pass anterior and posterior to the hypophyseal stalk or stalk of the pituitary gland (Figs. 7-42 and 7-47*B*). The cavernous sinuses drain posteriorly and inferiorly through the superior and inferior petrosal sinuses and the *pterygoid plexuses* (Figs. 7-30, 7-47*B*, and 7-49). *The lateral wall of each cavernous sinus contains*, from superior to inferior, the following structures (Fig. 7-48): (1) the oculomotor nerve (*CN III*); the trochlear nerve (*CN IV*); and the ophthalmic and maxillary divisions of the trigeminal nerve (*CN V¹ and CN V²*). Inside each cavernous sinus is the *internal carotid artery* with its sympathetic plexus and the abducent nerve (*CN VI*).

The Superior Petrosal Sinuses (Figs. 7-30, 7-47*B*, and 7-49). These venous sinuses are small channels that *drain the cavernous sinuses*. They run from the posterior ends of the cavernous sinuses to the transverse sinuses, at the points where they curve inferiorly to form the sigmoid sinuses. Each superior petrosal sinus lies in the attached margin of the tentorium cerebelli, running in a small groove on the superior margin of the petrous part of the temporal bone.

The Inferior Petrosal Sinuses (Figs. 7-30, 7-43, 7-47*B*, and 7-49). These small venous sinuses drain the cavernous sinuses directly into the *internal jugular veins*, just inferior to the skull. Commencing at the posterior end of the cavernous sinus, each sinus runs posteriorly, laterally, and inferiorly in a groove between the petrous part of the temporal bone and the basilar part of the occipital bone. The inferior petrosal sinus enters the jugular foramen and *joins the superior bulb of the internal jugular vein*, usually inferior to the skull. It receives cerebellar and labyrinthine (internal auditory) veins.

The Basilar Sinus (Figs. 7-30 and 7-40; see also Fig. 4-53). This sinus consists of several interconnecting venous channels on the *clivus*, the posterior surfaces of the basilar part of the occipital bone and the basisphenoid. It connects the two inferior petrosal sinuses and *communicates inferiorly with the internal vertebral venous plexus*.

Emissary Veins. These vessels connect the intracranial venous sinuses with veins outside the cranium. Although they are valveless and blood may flow in both directions, the flow in the emissary veins is usually away from the brain. The size and number of emissary veins vary. A *frontal emissary vein* is present in children and some adults. It passes through the foramen cecum of the skull (Fig. 7-11), connecting the superior sagittal sinus with the veins of the frontal sinus and nasal cavities or both. The *parietal emissary veins*, one on each side, pass through the parietal foramina in the calvaria (Fig. 7-5) and connect the superior sagittal sinus with the veins external to it, particularly those in the scalp (Fig. 7-37). A *mastoid emissary vein* connects each sigmoid sinus through the mastoid foramen with the occipital or posterior auricular vein (Fig. 7-29). A *posterior condylar emissary vein* may also be present and pass through the condylar canal (Fig. 7-40), connecting the sigmoid sinus with the suboccipital plexus of veins.

> The basilar and occipital sinuses communicate through the foramen magnum with the internal vertebral plexuses (Fig. 7-30; see also Fig. 4-53). Consequently, blood may pass to or from the *vertebral venous system*. As these venous channels are valveless, compression of the thorax, abdomen, or pelvis—as occurs during heavy coughing and straining—may force venous blood from these regions into the vertebral ven-

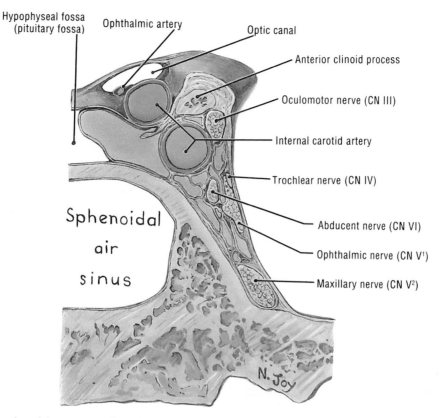

Figure 7-48. *A*, Coronal section of the cavernous sinus. Observe that this dural venous sinus is situated beside the sphenoidal sinus and the hypophyseal fossa. Note that cranial nerves III, IV, V^1, and V^2 are in a sheath in the lateral wall of the cavernous sinus. Observe that the internal carotid artery and the abducent nerve pass through the cavernous sinus. They are vulnerable when there is thrombosis of the cavernous sinus. Note that the internal carotid artery, having made a hairpin bend, is cut twice, once in the cavernous sinus and again superior to the cavernous sinus in the subarachnoid space.

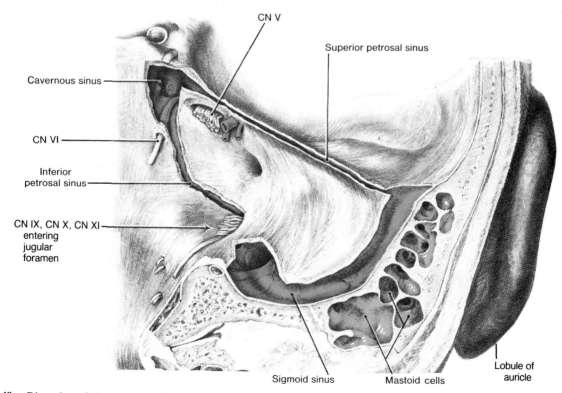

Figure 7-49. Dissection of the anterior wall of the right posterior cranial fossa, as viewed posteriorly. Some venous sinuses of the dura mater are illustrated. Note that the posterior surface of the petrous part of the temporal bone is encircled by three sinuses: sigmoid, superior petrosal, and inferior petrosal. The two petrosal sinuses drain the cavernous sinus (see also Fig. 7-30). Observe the mastoid (air) cells lined with mucus membrane (*pink*) in the mastoid process of the temporal bone.

ous system and from it into the venous sinuses of the dura. As a result, tumor cells and pus in abscesses in the head, thorax, abdomen, or pelvis may spread to the vertebrae and the brain. These connections of the vertebral plexuses with the dural venous sinuses also explain how a blood clot or thrombus (*e.g.*, from the pelvic veins after childbirth) can reach the dural venous sinuses without passing through the heart and lungs. Such a thrombus may block a dural venous sinus and produce a *venous infarction of the cerebral cortex*.

Infections of the scalp may pass through emissary veins into the venous sinuses of the dura. A scalp infection can be transmitted to the superior sagittal sinus through the parietal emissary veins. The bacteria may produce inflammatory processes in the walls of the emissary veins that could cause *thrombophlebitis of the dural venous sinuses* and later of the cortical veins (p. 667). The facial veins connect with the cavernous sinuses through communications with the ophthalmic veins or with their tributaries, the supraorbital veins (Figs. 7-30 and 7-47*B*). Consequently, *thrombophlebitis of the facial vein* may extend into the dural venous sinuses. Thrombophlebitis in one cavernous sinus commonly spreads to the other one because they are connected by intercavernous sinuses (Fig. 7-47*B*). Furthermore, thrombosis could spread from the cavernous sinus into the superior and inferior petrosal sinuses, the transverse sinus, the sigmoid sinus, and the internal jugular vein. In cases of *mastoiditis* (uncommon since the advent of antibiotics), thrombosis of the sigmoid sinus may occur owing to its proximity to the mastoid (air) cells in the mastoid process (Fig. 7-49).

Excision of an internal jugular vein is sometimes necessary, *e.g.*, during a radical neck dissection to remove deep cervical lymph nodes (Fig. 7-32). These nodes are commonly involved with malignant tumors of the head and neck. In most cases circulatory complications are minimal because the connections between the dural venous sinuses and veins outside the skull (*e.g.*, the vertebral venous plexus and the facial veins) enlarge enough to carry all the blood from intracranial structures.

The fact that the cavernous sinus envelops the internal carotid artery and several cranial nerves is of clinical significance (Figs. 7-48 and 7-50). In fractures of the base of the skull, the internal carotid artery may tear within the cavernous sinus, producing an *arteriovenous fistula*. In such cases arterial blood rushes into the cavernous sinus, enlarging it and forcing blood into the connecting veins (Fig. 7-30), especially the ophthalmic veins that normally drain the orbital cavity (Fig. 7-47). As a result, the eye protrudes (*exophthalmos*) and the conjunctiva becomes engorged (*chemosis*) on the side of the torn artery. In these circumstances, the bulging eye pulsates in synchrony with the radial pulse (see Fig. 6-74). Owing to this phenomenon, the condition is called *pulsating exophthalmos*. Because CNs III, IV, V^1, V^2, and VI lie in or close to the lateral wall of the cavernous sinus (Fig. 7-48), they may also be affected when injuries or infections of this sinus occur.

The Brain

Because neuroanatomy may be studied after gross anatomy, some features of the brain are described here to assist with the understanding of the relationship of the brain to the cranial nerves,

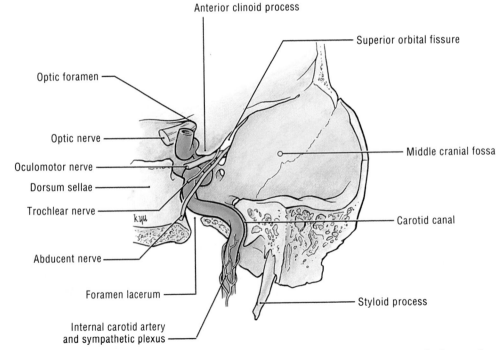

Anterior clinoid process

Superior orbital fissure

Optic foramen

Optic nerve

Oculomotor nerve

Dorsum sellae

Trochlear nerve

Abducent nerve

Foramen lacerum

Internal carotid artery and sympathetic plexus

Middle cranial fossa

Carotid canal

Styloid process

Figure 7-50. Coronal section through the carotid canal showing the course of the internal carotid artery in this canal. Note that it takes an inverted L-shaped course from the inferior surface of the petrous part of the temporal bone to its apex. At the superior end of the foramen lacerum, the internal carotid artery enters the cavernous sinus, turns anteriorly, makes a hairpin turn, and enters the subarachnoid space.

skull, CSF, and meninges. This material is set in smaller type to indicate that it is unlikely to be examined in most gross anatomy courses. The *cerebrum* consists of two **cerebral hemispheres** (Figs. 7-36 and 7-51), which are incompletely separated by the *longitudinal cerebral fissure*. The term *cerebral cortex* refers to the gray matter covering the surface of these hemispheres (Fig. 7-44*B*).

When the calvaria and dura are removed (Figs. 7-36 and 7-51), convolutions (gyri) and depressions or grooves (sulci or fissures) of the cerebral cortex are visible through the delicate layers of arachnoid and pia (Fig. 7-44*A*). The complicated folding of the surface of the cerebral hemispheres greatly increases the surface area of the brain. Over 60% of the total cortical area is buried in the various sulci and fissures.

Sulci and Fissures of the Cerebral Cortex.[16]   These elongated depressions are distinctive landmarks that are used to subdivide the cerebral hemispheres into smaller areas for descriptive purposes. There are six main sulci and fissures: (1) the longitudinal cerebral fissure; (2) the transverse cerebral fissure; (3) the lateral sulcus (fissure); (4) the central sulcus; (5) the parieto-occipital sulcus (fissure); and (6) the calcarine sulcus.

The Longitudinal Cerebral Fissure (Figs. 7-37 and 7-51*A*).   This long, deep, median cleft incompletely separates the two cerebral hemispheres. In situ it contains the falx cerebri (p. 686).

The Transverse Cerebral Fissure (Figs. 7-36 and 7-37).   This cleft separates the cerebral hemispheres superiorly from the cerebellum, midbrain, and diencephalon inferiorly. The tentorium cerebelli (p. 686) lies in the posterior part of this fissure.

The Lateral Sulcus (Fissure) (Figs. 7-36 and 7-51).   This deep cleft begins inferiorly on the inferior surface of the cerebral hemisphere as a deep furrow and extends posteriorly, separating the frontal and temporal lobes. Posteriorly, it separates parts of the parietal and temporal lobes.

The Central Sulcus (Figs. 7-36 and 7-51*B* and *C*).   This prominent groove runs inferoanteriorly from about the middle of the superior margin of the cerebral hemisphere, stopping just short of the lateral sulcus. *The central sulcus is an important landmark of the cerebral cortex* because the **motor area** (precentral gyrus) lies anterior to it and the general **sensory cortex** (postcentral gyrus) posterior to it.

The Parieto-occipital Sulcus (Fissure) (Fig. 7-39).   This deep fissure, as its name indicates, separates the parietal and occipital lobes of the brain on the medial aspect of the brain. It extends from the calcarine sulcus to the superior border of the cerebral hemisphere and continues for a short distance on the dorsolateral surface.

The Calcarine Sulcus (Fig. 7-39).   This groove on the medial surface of the cerebral hemisphere commences inferior to the posterior end of the corpus callosum and follows an arched course to the occipital pole.

Gyri of the Cerebral Cortex (Figs. 7-36, 7-39, 7-44, and 7-51).   A gyrus or convolution is an area of cerebral cortex that lies between two sulci or fissures. Most of the cortex of a gyrus is not visible on the surface of the brain.

The important gyri are: the *precentral gyrus*, between the central and precentral fissures; the *postcentral gyrus*, between the central and postcentral fissures; the *supramarginal gyrus*, around the end of the lateral sulcus; the *angular gyrus*, around the end of the superior temporal fissures; the *superior temporal gyrus*, between the lateral sulcus and the superior temporal fissure; and the *cingulate gyrus* on the medial aspect of the hemisphere, between the cingulate gyrus and the corpus callosum.

The Main Lobes of the Cerebral Hemispheres (Figs. 7-36, 7-39, and 7-51).   There are four main lobes of the cerebral hemispheres: frontal, parietal, temporal, and occipital.

The Frontal Lobes (Figs. 7-36, 7-39, and 7-51*B*).   These are the largest of all the lobes, and they form the anterior parts of the cerebral hemispheres. They are located anterior to the central sulci and superior to the lateral sulci. Their lateral and superior surfaces extend posterior to the coronal suture of the skull. The basal surfaces of the frontal lobes rest on the orbital parts of the frontal bone in the anterior cranial fossa (Figs. 7-10 and 7-11). The related gyri form impressions in the orbital parts (plates) of the frontal bone.

The Parietal Lobes (Figs. 7-36 and 7-51*B*).   These lobes are related to the internal aspects of the posterior and superior parts of the parietal bones. Each lobe is bounded anteriorly by the central sulcus and posteriorly by the superior part of the line joining the parieto-occipital sulcus and the preoccipital notch. The inferior boundary of each lobe is indicated by an imaginary line extending from the posterior ramus of the lateral sulcus to the inferior end of the posterior boundary of the lobe.

The Temporal Lobes (Figs. 7-36 and 7-51*B*).   These lobes lie inferior to the lateral sulci. Their convex anterior ends, called *temporal poles*, fit into the anterior and lateral parts of the middle cranial fossa. Their posterior parts lie against the middle one-third of the inferior part of the parietal bone.

The Occipital Lobes (Figs. 7-36, 7-39, and 7-51*B*).   These lobes are relatively small and are located posterior to the parieto-occipital sulci. They rest on the tentorium cerebelli, superior to the posterior cranial fossa. Although they are relatively small, these lobes are very important because they *contain the visual area*.

The Main Parts of the Brain

To understand the origin and course of the cranial nerves (Chap. 9), the circulation of CSF, and the structure of the ventricular system (Figs. 7-39 and 7-51 to 7-53), the major divisions of the brain must be understood.

The Cerebral Hemispheres

The cerebral hemispheres *form the largest part of the brain*. They occupy the anterior and middle cranial fossae and extend posteriorly over the tentorium cerebelli and the cerebellum to the internal occipital protuberance (Figs. 7-12, 7-36, 7-39, and 7-51). They comprise the cerebral cortex, the basal nuclei or ''ganglia,'' their fiber connections, and the lateral ventricles (Fig. 7-53). The cavity in each cerebral hemisphere, known as a *lateral ventricle*, is part of the ventricular system of the brain (Figs. 7-52 and 7-53).

---

[16]The terminology for the sulci and fissures of the cerebral cortex varies in current gross anatomy and neuroanatomy texts. The terms used in this book are in accordance with the latest edition of *Nomina Anatomica* (Warwick, 1989). Parentheses are used to indicate unofficial but widely used alternatives, *e.g.*, lateral sulcus (fissure).

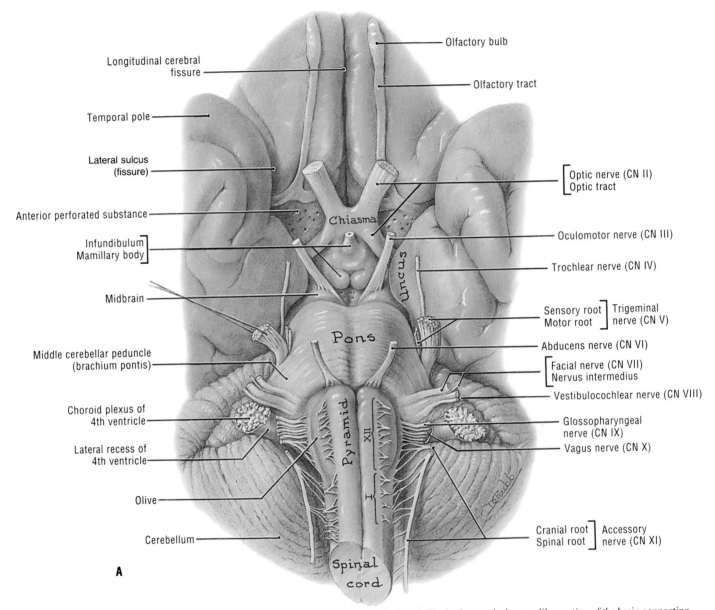

Labels (clockwise from top):
Olfactory bulb
Olfactory tract
Optic nerve (CN II) / Optic tract
Oculomotor nerve (CN III)
Trochlear nerve (CN IV)
Sensory root / Motor root — Trigeminal nerve (CN V)
Abducens nerve (CN VI)
Facial nerve (CN VII) / Nervus intermedius
Vestibulocochlear nerve (CN VIII)
Glossopharyngeal nerve (CN IX)
Vagus nerve (CN X)
Cranial root / Spinal root — Accessory nerve (CN XI)

Longitudinal cerebral fissure
Temporal pole
Lateral sulcus (fissure)
Anterior perforated substance
Infundibulum / Mamillary body
Midbrain
Middle cerebellar peduncle (brachium pontis)
Choroid plexus of 4th ventricle
Lateral recess of 4th ventricle
Olive
Cerebellum

Chiasma
Uncus
Pons
Pyramid
XII
I
Spinal cord

A

Figure 7-51. *A*, Inferior surface of the brain showing the superficial origin of the 12 cranial nerves. The olfactory bulb, into which the olfactory nerves (CN I) end, are related to the cribriform plate (Figs. 7-11 and 7-108). The infundibulum and mamillary bodies are in the middle cranial fossa, as are the temporal lobes lateral to them (Fig. 7-12). The pons, medulla, and cerebellum are in the posterior cranial fossa. *I* indicates nerve rootlets from the first cervical segment of the spinal cord. The brain stem is the stemlike portion of the brain connecting the cerebral hemispheres to the spinal cord. *B*, Lateral surface of the brain. *C*, Lobes of the brain: frontal (*purple*); parietal (*orange*); occipital (*green*), and temporal (*blue*). The primary **motor area** is located in the precentral gyrus, including the anterior wall of the central sulcus. *D*, Blood supply of the cerebral hemispheres. Anterior cerebral artery (*green*), middle cerebral artery (*purple*), and posterior cerebral artery (*yellow*).

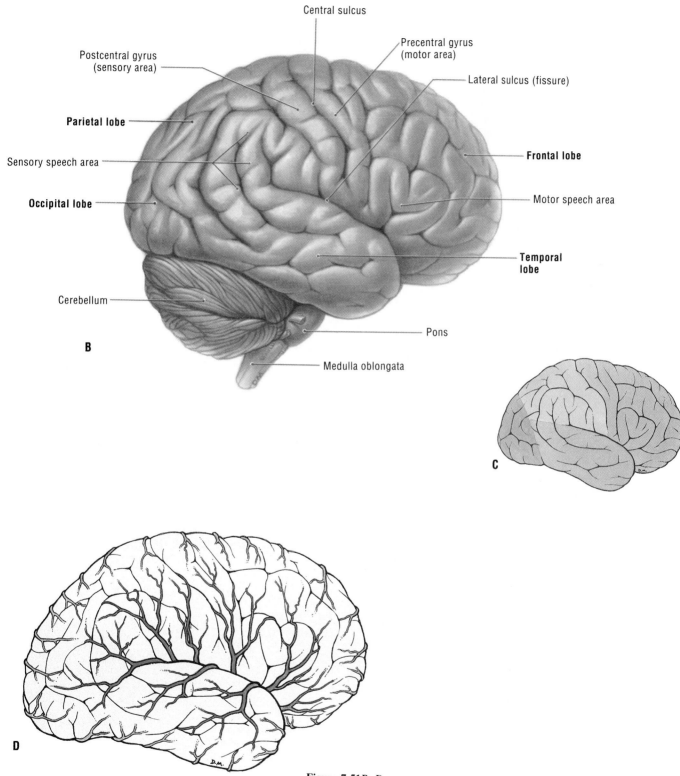

Central sulcus

Precentral gyrus
(motor area)

Postcentral gyrus
(sensory area)

Lateral sulcus (fissure)

Parietal lobe

Frontal lobe

Sensory speech area

Motor speech area

Occipital lobe

**Temporal
lobe**

Cerebellum

Pons

B

Medulla oblongata

C

D

Figure 7-51*B–D*

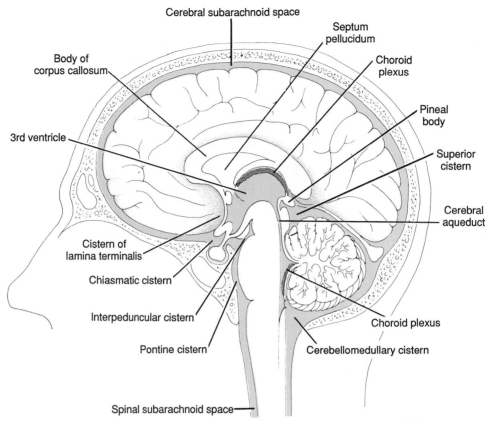

Figure 7-52.   Median section of the brain showing the subarachnoid spaces and cisterns.

The Diencephalon

This part of the brain is composed of the thalamus, hypothalamus, epithalamus, and subthalamus. It *surrounds the third ventricle* of the brain (Figs. 7-51 and 7-53) and forms the central core of the brain. It is surrounded by the cerebral hemispheres. The two **thalami** make up four-fifths of the diencephalon (Fig. 7-53*B*). The cavity of the diencephalon is the narrow *third ventricle* lying between the right and left thalami (Figs. 7-52 and 7-53).

The Midbrain

This is the smallest part of the brain (Figs. 7-37, 7-39, 7-46, and 7-51 to 7-53). It is located at the junction of the middle and posterior cranial fossae, lying partly in each. The cavity of the midbrain is represented by a narrow canal, called the *cerebral aqueduct*. It conducts CSF from the lateral and third ventricles to the fourth ventricle.

The Pons

The pons lies in the anterior part of the posterior cranial fossa, posterior to the superior part of the clivus and the posterior surface of the dorsum sellae (Figs. 7-12, 7-37, 7-39, 7-40, and 7-51). The Latin term *pons* means a bridge, which is an appropriate name for this part of the brain because the part of it that is visible from the inferior surface is a wide, bridgelike transverse band of nerve fibers (Fig. 7-51*A*). Al-

though it appears to be a bridge between the cerebellar hemispheres, it is not. The pontine fibers connect one cerebral hemisphere with the opposite cerebellar hemisphere. The cavity in the pons forms part of the fourth ventricle.

The Medulla Oblongata

This is the most caudal part of the **brain stem**, composed of the midbrain, pons, and medulla.[17] It is located in the posterior cranial fossa with its ventral aspect facing the clivus (Fig. 7-40). The medulla is continuous with the spinal cord at the foramen magnum (Figs. 7-37 and 7-39), but the transition is gradual. The distinctive characteristics of its ventral surface are the elongated **pyramids**, which contain the corticospinal tracts from the cerebral cortex (Fig. 7-51*A*). The medulla *contains the cardiovascular and respiratory centers* for the automatic control of heartbeat and respiration, respectively. The cavity of the medulla forms the inferior part of the fourth ventricle (Figs. 7-39, 7-52, and 7-53*A*).

The Cerebellum

This "little brain" (L. *cerebellum*) overlies the posterior aspect of the pons and medulla and extends laterally beneath the tentorium cerebelli (Figs. 7-36, 7-37, 7-39, and 7-51 to

---

[17]For brevity during clinical discussions, the medulla oblongata is referred to as the *medulla*. This usage is also followed here.

7-53). It occupies most of the posterior cranial fossa (Figs. 7-11 and 7-12). The cerebellum consists of a midline portion, the *vermis* (L. worm), and two lateral lobes or hemispheres. These hemispheres lie posterior to the petrous parts of the temporal bones and rest in the concavities of the posterior cranial fossae (Figs. 7-11 and 7-45). The cerebellum is mainly *concerned with motor functions that regulate posture, muscle tone, and muscular coordination.*

Brain injuries are commonly associated with severe head injuries. The main purpose of the calvaria and CSF is to protect the brain. Helmets are used by motorcycle riders, construction workers, soldiers, some athletes, and others for additional protection. Various terms are used to describe injuries to the brain resulting from trauma.

Cerebral Concussion. This is an abrupt but *transient loss of consciousness* immediately following a blow to the head. A concussion can also follow the sudden stopping of the moving head, as occurs during rear end collisions. These accidents result in the brain hitting the stationary skull. When a person is hit (*e.g.*, by a club or baseball), they may be stunned and fall to the ground. The concussion appears to result from a mechanical disturbance of nerve cells because there is no visible bruising of the brain on gross or microscopic examination. However, repeated cerebral concussions, as may occur in professional fighters, lead to *cerebral atrophy*; thus some submicroscopic damage appears to occur with each concussion. The *punch-drunk syndrome*, caused by repeated cerebral concussions, is characterized by weakness in the lower limbs, unsteady gait, slowness of muscular movements, tremors of the hands, hesitancy of speech, and slow cerebration (use of one's brain).

Cerebral Contusion (L. *contusio*, a bruising). There is visible bruising of the brain owing to trauma and blood leaking from small vessels. The pia is stripped from the brain in the injured area and may be torn, allowing blood to enter the

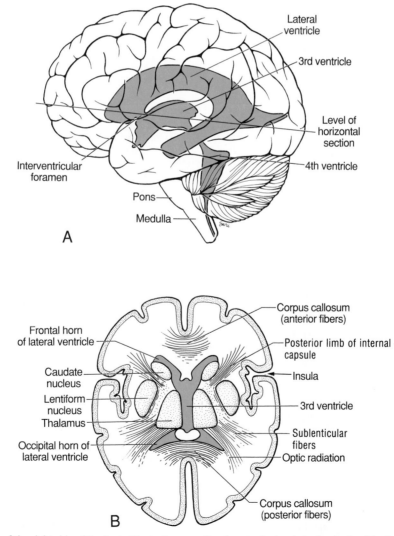

Figure 7-53. *A*, Lateral view of the right side of the brain illustrating the ventricular system. *B*, Horizontal section of the brain. Between the right and left thalami is the third ventricle. The internal capsule separates the thalamus and the caudate nucleus from the lentiform nucleus, which lies deep to the insula in the depth of the lateral sulcus (Fig. 7-36). The corpus callosum consists of nerve fibers connecting the cerebral hemispheres.

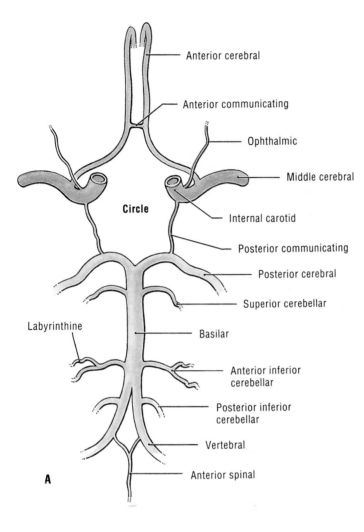

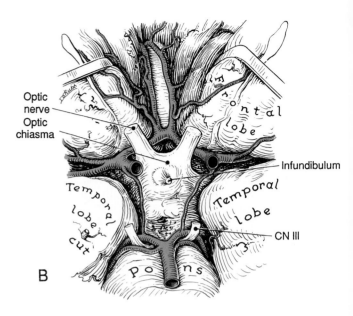

Figure 7-54. *A*, The arteries that supply the brain. Note that the middle cerebral artery is not a part of the cerebral arterial circle (of Willis) and that it is the continuation of the internal carotid artery. *B*, Anastomosis of major arteries supplying the brain. They join together at the base of the brain to form the cerebral arterial circle.

subarachnoid space. Contusions are frequently found at the frontal, temporal, and occipital poles of the brain. The bruising results either from the sudden impact of the moving brain against the stationary skull or the skull against the brain when there is a blow to it. The living calvaria is somewhat flexible and may sometimes bend without breaking. As a consequence, bruising of the brain may occur when there is no fracture. A contusion usually results in an *extended loss of consciousness* (several minutes to many hours). Blows to the forehead tend to produce contusions of the frontal poles of the brain and perhaps the temporal poles, whereas blows to the occiput (posterior part of the head) generally produce contusions of the occipital poles. Contusion is soon followed by swelling (*edema of the brain*), which contributes to a rise in intracranial pressure. This is a serious complication that may result in a life-threatening situation, as discussed with cerebral compression.

Cerebral Laceration (L. *lacero*, to tear). Tearing of the brain is often associated with a depressed skull fracture or a gunshot wound. These lacerations result in rupture of large blood vessels with bleeding into the brain and subarachnoid space. This leads to the formation of a *cerebral hematoma*, edema, and an increase in local and general intracranial pressure.

Cerebral Compression. Pressure on the brain may be produced by: (1) intracranial collections of blood (hematomas) or other fluids; (2) obstruction of the circulation or absorption of CSF; (3) intracranial tumors or abscesses; and (4) edema of the brain (*e.g.*, owing to contusions). Cerebral compression is associated with increased intracranial pressure. A marked increase in intracranial pressure results in *herniation of parts of the brain*. Often the uncus of the temporal lobe (Fig. 7-51*A*) herniates through the tentorial incisure (Figs. 7-37 and 7-45), causing pressure on the midbrain and kinking of the oculomotor nerves on the edge of the tentorium. This results in **third nerve palsy**, usually commencing with progressive enlargement of the pupil and fixation of it to light. In some cases the cerebellar tonsils herniate through the foramen magnum (Fig. 7-37), compressing the medulla and disturbing the vital cardiovascular and respiratory centers.

Ventricular System and CSF

The ventricular system in the brain consists mainly of four cavities called ventricles (Figs. 7-39, 7-52, 7-53, and 7-55). The first and second ventricles, called **lateral ventricles** are the largest components of the system. They occupy a considerable part of the cerebral hemispheres. Each lateral ventricle opens in the third ventricle through an *interventricular foramen*. The **third ventricle** is a narrow, slitlike cavity between the two thalami. It is continuous posteriorly with the cerebral aqueduct in the midbrain, which connects the third and fourth ventricles. The **fourth ventricle** is diamond-shaped when viewed superiorly and tent-shaped when observed laterally. It is located in the pons, anterior to the cerebellum, and extends posteriorly into the central canal of the spinal cord. CSF escapes from the fourth ventricle into the subarachnoid space through the *median and lateral apertures*. CSF is formed by **choroid plexuses** in the ventricles and circulates through the ventricular system and the subarachnoid space before being reabsorbed into the dural venous sinuses.

Formation of CSF (Figs. 7-52 and 7-53). The main source of CSF is the choroid plexuses of the lateral, third, and fourth ventricles of the brain. **Choroid plexuses** are located in the roofs of the third and fourth ventricles and on the floors of the bodies and inferior horns of the lateral ventricles. *The choroid plexuses in the lateral ventricles are the largest and most important producers of CSF*. They are continuous with the choroid plexus in the roof of the third ventricle through the interventricular foramina. The lateral apertures

of the fourth ventricle are also partially occupied by parts of the choroid plexuses, which protrude through them from the fourth ventricle and secrete CSF into the subarachnoid space. Each choroid plexus (G. *choroid*, delicate membrane + *eidos*, resemblance) is composed of highly vascular pia called *tela choroidea* (L. *web* + G. *chorioeides*, like a membrane), covered by a simple cuboid or low-columnar epithelium.

The Subarachnoid Cisterns[18] (Fig. 7-52). At certain places, mainly at the base of the brain, the arachnoid is widely separated from the pia and forms *large pools of CSF*, called subarachnoid cisterns. They communicate freely with each other and with the subarachnoid space of the spinal cord and cover all surfaces of the cerebral hemispheres.

The Cerebellomedullary Cistern (Cisterna Magna) (Fig. 7-52). This cistern is located in the space between the cerebellum and the inferior part of the medulla. It receives CSF from the median aperture of the fourth ventricle and is continuous with the large subarachnoid space around the brain and spinal cord.

---

[18]A cistern is a closed space that serves as reservoir for fluid such as CSF. It is an enlarged part of the subarachnoid space (Fig. 7-52).

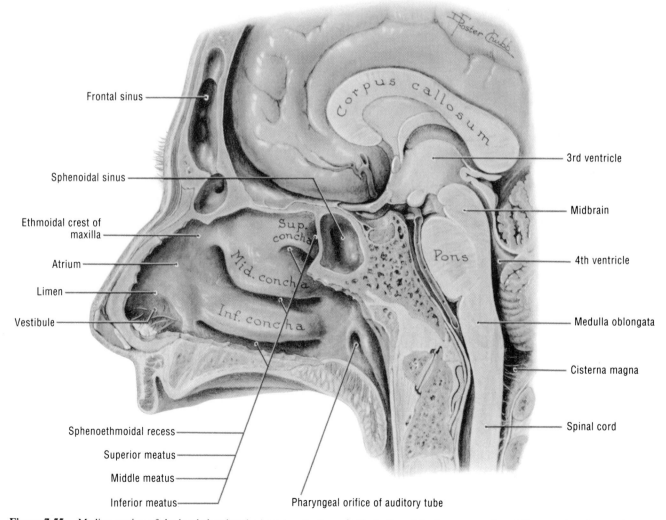

Figure 7-55. Median section of the head showing the lateral wall of the nasal cavity, a medial view of the brain, and the spinal cord. Collectively the midbrain, pons, and medulla are referred to as the brain stem (see also Fig. 7-39).

The Pontine Cistern (Figs. 7-52 and 7-55). This is an extensive space along the ventral and lateral surfaces of the pons, containing the *basilar artery* and some of the cranial nerves. It is continuous inferiorly with the cerebellomedullary cistern and superiorly with the interpeduncular cistern.

The Interpeduncular Cistern (Figs. 7-37, 7-51, 7-52, and 7-54). This cistern is located between the cerebral peduncles and contains the posterior part of the *cerebral arterial circle* (of Willis). This vascular circle is formed by the major arteries supplying the cerebrum. The interpeduncular cistern is continuous anterosuperiorly with the *chiasmatic cistern*, which continues as the *cistern of the lamina terminalis*. The latter cistern continues into the shallow *cistern of the corpus callosum*, which is located anterosuperior to this large commissure.

The Cistern of the Lateral Sulcus (Fissure) (Fig. 7-52). This cistern contains the *middle cerebral artery* and is located anterior to each temporal lobe where the arachnoid covers the lateral sulcus.

The Superior Cistern (Cistern of the Great Cerebral Vein). This cistern lies between the splenium of the corpus callosum and the superior surface of the cerebellum (Fig. 7-52). It contains the great cerebral vein (Figs. 7-30 and 7-47A) and the pineal body. Because it lies dorsal to the colliculi of the midbrain (corpora quadrigemina), this cistern is often referred to clinically as the *quadrigeminal cistern*.

The **pineal body** (Fig. 7-52) serves as a useful neuroradiological and neurosurgical landmark. It becomes calcified in adolescence and is visible on skull radiographs. Charts showing the usual position of the pineal body (gland) are used to diagnose displacement of it in patients with suspected expanding intracranial lesions.

CSF may be obtained from the cerebellomedullary cistern, a procedure known as *cisternal puncture*. Samples of CSF may also be obtained from the *anterior fontanelle* in infants or through bur holes drilled in the adult calvaria. The subarachnoid space or the ventricular system may also be entered for measuring or monitoring CSF pressure, injecting antibiotics, or administering contrast media for radiography (*ventriculography*; see column 2).

Circulation of CSF (Figs. 7-52 and 7-53). CSF produced by the *choroid plexuses* in the lateral and third ventricles passes through the cerebral aqueduct into the fourth ventricle, where more CSF is produced by the choroid plexus in its roof. CSF leaves the fourth ventricle through its median and lateral apertures and passes into the *subarachnoid space*, where it collects in the cerebellomedullary and pontine cisterns. From these cisterns, some CSF passes inferiorly into the spinal subarachnoid space around the spinal cord and posterosuperiorly over the cerebellum. However, most CSF flows superiorly through the tentorial incisure into the subarachnoid space around the midbrain (interpeduncular cistern and superior cistern). CSF from the various cisterns spreads superiorly through the sulci and fissures on the medial and superolateral surfaces of the cerebral hemispheres. This flow of CSF is aided by pulsations of the cerebral arteries and movements of the cerebral hemispheres. CSF also passes into the extensions of the subarachnoid space around the cranial nerves, the most important of which are those surrounding the optic nerves (Fig. 7-57A).

Absorption of CSF (Fig. 7-44). The main site of absorption or passage of CSF into venous system is through the *arachnoid villi*, which are protrusions of the arachnoid into the dural venous sinuses, especially the superior sagittal sinus and its adjacent lateral lacunae. The rate of CSF absorption is pressure dependent, and the arachnoid villi appear to act as one-way ''valves.'' When CSF pressure is greater than venous pressure, the valves open and CSF passes into the blood in the venous sinuses of the dura (Fig. 7-47). However, when venous pressure is higher than CSF pressure, the valves close, preventing blood from entering the CSF. Some CSF appears to be absorbed by the ependymal lining of the ventricles (ependyma) in the spinal subarachnoid space and through the walls of capillaries in the pia.

By replacing some CSF by contrast media of various types, certain parts of the brain can be outlined by radiography. Air may be used to replace CSF to demonstrate the ventricular system and the subarachnoid spaces and to detect or exclude displacement or deformities of their walls. Radiographic study of the brain using techniques known as **ventriculography** and *encephalography* enables radiologists to locate atrophic or space-occupying lesions (*e.g.*, a brain tumor). These radiological methods may also be used to determine the site of an obstruction in the circulation of CSF. Because these techniques have certain disadvantages, they have been almost completely replaced by computerized tomographic (CT) scans and magnetic resonance images (MRIs). These scans show the presence of hemorrhage, infarction, tumors, cysts, and other lesions.

Hydrocephalus. Overproduction of CSF, obstruction of its flow, or interference with its absorption results in an excess of CSF in the head, a condition known as hydrocephalus (G. *hydor*, water + *kephale*, head). The condition is marked by an excessive accumulation of CSF that dilates the ventricles of the brain and causes a separation of the bones of the calvaria in infants (Moore, 1988). *Internal hydrocephalus is an accumulation of fluid in the ventricles.* A blockage of the median aperture of the fourth ventricle results in the thin wall of the fourth ventricle herniating through the foramen magnum into the superior part of the vertebral canal. *External hydrocephalus is an accumulation of fluid in the subarachnoid space.* There may also be an accumulation of fluid in the subdural space owing to a communication between the subarachnoid and subdural spaces. The subdural space is a potential space between the dura and the external surface of the arachnoid. See footnote on page 687.

Blockage of CSF circulation results in dilation of the ventricles superior to the point of obstruction and in pressure on the cerebral hemispheres. This squeezes the brain between the ventricular fluid and the bones of the calvaria. In infants the internal pressure results in expansion of the brain and calvaria because the sutures and fontanels are still open. It is possible to form an artificial drainage system (*ventriculoatrial shunt*) to allow the CSF to escape, thereby lessening the damage to the brain.

Functions of CSF. Along with the meninges and calvaria, *CSF protects the brain* by providing a cushion against blows to the head. Because the brain is slightly heavier than CSF, the gyri on the basal surface of the brain are in contact with the cranial fossae in the floor of the cranial cavity when a person is erect. In many places at the base of the brain, only the cranial meninges intervene between the brain and the cranial bones. In this position the CSF is in the subarachnoid cisterns and in the sulci on the superior and lateral parts of the brain. Hence, CSF normally separates the superior part of the brain from the calvaria.

There are small, rapidly recurring changes in intracranial pressure owing to the heartbeat, as well as slow recurring changes resulting from unknown causes. In addition, momentarily large changes in intracranial pressure occur during coughing and straining. Any change in the volume of the intracranial contents (*e.g.*, a brain tumor, ventricular fluid as a result of blockage of the cerebral aqueduct, and blood owing to hemorrhage) will be reflected by a change in intracranial pressure. This is called the *Monro-Kellie doctrine*, which states that the cranial cavity is a closed rigid box and that a change in the quantity of intracranial blood can occur only through the displacement or replacement of CSF.

Contusions of the brain resulting from sudden acceleration or deceleration of the head are frequent at the base of the brain and at the frontal, temporal, and occipital poles, partly because of the paucity of CSF in these areas when a person is in the erect position. This situation allows the brain to strike the cranial bones with only the meninges intervening.

Fractures of the floor of the middle cranial fossa may result in leakage of CSF from the ear (*CSF otorrhea*), if the meninges superior to the middle ear and mastoid antrum are torn and the tympanic membrane (eardrum) is also ruptured. Fractures of the floor of the anterior cranial fossa may involve the cribriform plate of the ethmoid bone (Fig. 7-11), resulting in leakage of CSF through the nose (*CSF rhinorrhea*). In these cases the subarachnoid space is in communication with the outside through the nose, as the result of a tear in the meninges. *CSF otorrhea and CSF rhinorrhea present a risk of meningitis* because an infection may spread from the ear or the nose, especially if the nose is blown hard.

Blood Supply of the Brain

The brain is supplied through an extensive system of branches from two pairs of vessels, the *internal carotid arteries* and the *vertebral arteries*.

The Internal Carotid Arteries (Figs. 7-28, 7-42, 7-45, 7-46, 7-50, 7-51*D*, and 7-54). Each artery arises in the neck from the common carotid artery opposite the superior border of the thyroid cartilage. The *cervical part* of the artery ascends almost vertically to the base of the skull, where it turns and enters the carotid canal in the petrous part of the temporal bone. The *petrous part* of the artery enters the middle cranial fossa through the superior part of the foramen lacerum and then runs anteriorly in the *cavernous sinus* (Figs. 7-48 and 7-50). The *cavernous part* of the internal carotid artery is covered by the endothelium of this sinus. At the anterior end of the cavernous sinus, the internal carotid artery makes a hairpin turn and leaves the sinus to enter the subarachnoid space. *The cerebral part* of the internal carotid artery (supracavernous or supraclinoid part) immediately gives off the important *ophthalmic artery*, which supplies the eye. The internal carotid artery then passes inferior to the optic nerve. Finally it turns obliquely superiorly, lateral to the optic chiasma (Figs. 7-42 and 7-46) for a variable distance, before branching into the anterior and middle cerebral arteries at the medial

end of the lateral sulcus (Figs. 7-51*D* and 7-54). The sinuous course taken by the cavernous and cerebral parts of the internal carotid artery forms a U-shaped bend (Fig. 7-50), often called the "*carotid siphon*" (G. a bent tube). Within the cranial cavity, the internal carotid artery and its branches supply the hypophysis cerebri (pituitary gland), the orbit, and much of the supratentorial part of the brain.

The Vertebral Arteries (Figs. 7-28 and 7-54; see also Fig. 4-35). Each artery begins in the root of the neck as a branch of the first part of the subclavian artery. It ascends vertically through the transverse foramina of the cervical vertebrae and then inclines laterally in the transverse foramen of C2 vertebra. Superior to this foramen, the vertebral artery ascends vertically into the transverse foramen of C1. It then bends posteriorly at right angles and winds around the superior part of the lateral mass of the atlas. The vertebral artery pierces the posterior atlanto-occipital membrane (see Figs. 4-30 and 4-35), the dura, and the arachnoid. It enters the subarachnoid space of the cerebellomedullary cistern at the level of the foramen magnum (Fig. 7-52). The vertebral artery runs anteriorly on the anterolateral surface of the medulla and unites with its fellow of the opposite side at the caudal border of the pons to form the basilar artery.

The Basilar Artery (Figs. 7-52, 7-54, and 7-55). This artery is formed by the union of the two vertebral arteries. It runs through the pontine cistern to the superior border of the pons, where it ends by dividing into the two posterior cerebral arteries.

To investigate the vertebral and basilar arteries using angiography, injections of contrast material are made into them by various routes, *e.g.*, through a femoral catheter advanced up the aorta into the inferior part of the vertebral artery or by retrograde injections into the brachial artery in the cubital fossa (see Fig. 6-62).

Occlusion or stenosis (narrowing) of one vertebral artery results in the brain stem being dependent mainly on the other vertebral artery for its blood supply. If that supply is reduced by injury, brain stem symptoms may occur such as dizziness, fainting, spots before the eyes, and transient *diplopia* (double vision).

The Cerebral Arterial Circle (Figs. 7-42 and 7-54). This circle (of Willis) is an important anastomosis between the four arteries that supply the brain (the two vertebral and the two internal carotid arteries). It is formed by the posterior cerebral, posterior communicating, internal carotid, anterior cerebral, and anterior communicating arteries. The circle is located at the base of the brain, principally in the interpeduncular fossa. It extends from the superior border of the pons to the longitudinal fissure between the cerebral hemispheres. The cerebral arterial circle *encircles the optic chiasma*, the infundibulum, and the mamillary bodies (Fig. 7-51*A*). Two types of branches, central and cortical, arise from the cerebral arterial circle and the main cerebral arteries. *Central arteries* penetrate the substance of the brain and supply deep structures, *e.g.*, the basal nuclei or "ganglia." *Cortical branches* pass in the pia and supply the more superficial parts of the brain. In general each of the cerebral arteries, (anterior, middle, and posterior) supplies a surface and a pole of the brain as follows (Fig. 7-51*D*): (1) the *anterior cerebral artery* supplies most of the medial and superior surfaces and the frontal pole; (2) the *middle cerebral artery* supplies the lateral

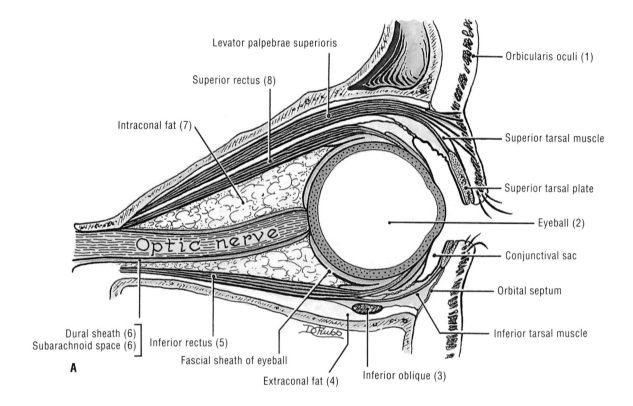

Levator palpebrae superioris

Superior rectus (8)

Intraconal fat (7)

Orbicularis oculi (1)

Superior tarsal muscle

Superior tarsal plate

Eyeball (2)

Conjunctival sac

Orbital septum

Inferior tarsal muscle

Optic nerve

Dural sheath (6)
Subarachnoid space (6)   Inferior rectus (5)

A

Fascial sheath of eyeball

Extraconal fat (4)   Inferior oblique (3)

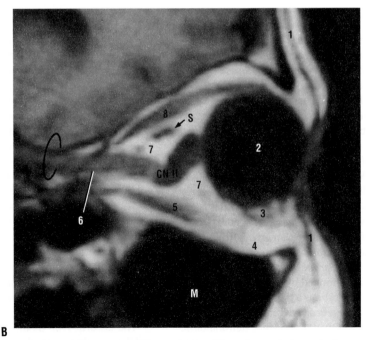

B

Figure 7-57. *A*, Sagittal section of the orbit and its contents. The numbers in parentheses refer to structures in *B*. Observe the subarachnoid space around the optic nerve. *B*, Sagittal magnetic resonance image (MRI) through the optic nerve and eyeball. *S*, superior ophthalmic vein; *M*, maxillary sinus; *circle*, optic foramen. (Courtesy of Dr. W. Kucharczyk, Clinical Director of Tri-Hospital Resonance Centre, Toronto, Ontario, Canada.) *C*, Lateral view of the eye of a 46-year-old man.

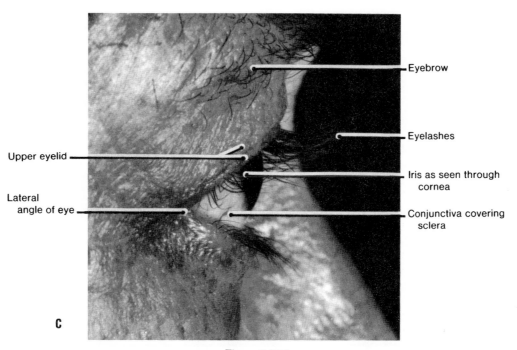

Upper eyelid

Lateral angle of eye

Eyebrow

Eyelashes

Iris as seen through cornea

Conjunctiva covering sclera

C

Figure 7-57C

lies between the orbit and the temporal fossa. The lateral wall and the roof of the orbit are partially separated posteriorly by the *superior orbital fissure*. This cleft communicates with the middle cranial fossa and transmits the *oculomotor* (CN III), *trochlear* (CN IV), and *abducent* (CN VI) nerves and the terminal branches of the *ophthalmic nerve* (CN V^1), as well as the superior ophthalmic vein. The **apex of the orbit** is at the medial ends of the superior and inferior orbital fissures (Fig. 7-56). These fissures form a V-shaped area that encloses the orbital part of the greater wing of the sphenoid bone.

Owing to the thinness of the medial and inferior walls of the orbital cavity, a blow to the eye may produce a *blow-out fracture of the orbit* as the result of the sudden increase in intraorbital pressure. Because the medial wall of the orbit is so delicate, surgery of the ethmoidal sinuses must be performed with extreme care. Although the superior wall (roof) of the orbit is stronger than the medial and inferior walls, it is thin enough to be translucent and may be readily penetrated. Thus a sharp object, even a pencil, may pass through it into the frontal lobe of the brain. One type of prefrontal *lobotomy* or *leukotomy* (G. *lekos*, white + *tome*, a cutting), that was used in the past in attempts to modify the behavior of severely psychotic patients made use of this anatomical fact by driving a sterile instrument (*leukotome*) through the roof of the orbit and sectioning nerve fiber connections in the anterior part of the frontal lobe.

An object accidentally pushed through the anteromedial part of the roof of the orbit may penetrate the *frontal sinus* (Figs. 7-13, 7-55, and 7-68), whereas an object pushed through its floor would enter the *maxillary sinus*. Similarly, a foreign object (*e.g.*, a piece of wire or a bullet) penetrating the eye in an anteroposterior direction could traverse the superior orbital fissure and enter the middle cranial fossa and the

temporal lobe of the brain (Fig. 7-36). Because the petrous part of the temporal bone is very hard, it might stop the bullet so that it may remain in the middle cranial fossa.

Owing to the closeness of the optic nerve to the sphenoidal and posterior ethmoidal sinuses (Fig. 7-58), a malignant tumor in these sinuses may erode the thin bony walls and compress the optic nerve and orbital contents. Tumors in the orbit produce bulging of the eyeball (*proptosis* or *exophthalmos*). The easiest entrance to the orbital cavity for a tumor in the middle cranial fossa is through the superior orbital fissure, whereas tumors in the temporal or infratemporal fossa gain access to this cavity through the inferior orbital fissure. Although the lateral wall of the orbit is nearly as long as the medial wall, it does not reach so far anteriorly; thus nearly 2.5 cm of the eyeball is exposed when the pupil is turned medially as far as possible. This is why the lateral side affords a good approach for operations on the eyeball.

The Orbital Contents

The most important contents of the orbit are the **eyeball** and **optic nerve** (Figs. 7-57 to 7-62). The orbit also contains the muscles of the eyeball, their nerves and vessels, the lacrimal gland, and other structures.

The Living Eye

Examine your eyes in a mirror. Note that, from the anterior aspect (Fig. 7-61), most of the eyeball appears to be in the orbit, but from the side much of it protrudes between the eyelids, *i.e.*, through the palpebral fissure (Fig. 7-57C). The "white of the eye" or anterior aspect of the **sclera** (G. *scleros*, hard) is continuous with the dural sheath of the optic nerve and the dura

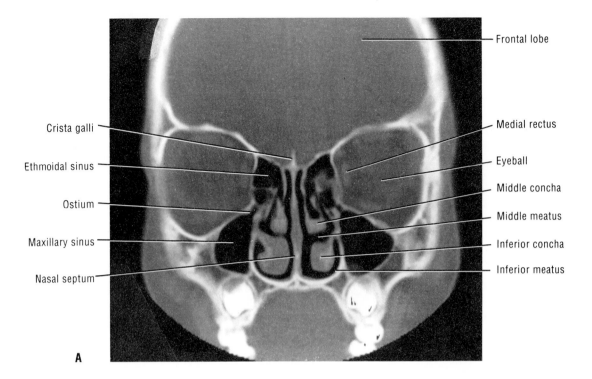

Frontal lobe

Crista galli

Ethmoidal sinus

Ostium

Maxillary sinus

Nasal septum

Medial rectus

Eyeball

Middle concha

Middle meatus

Inferior concha

Inferior meatus

A

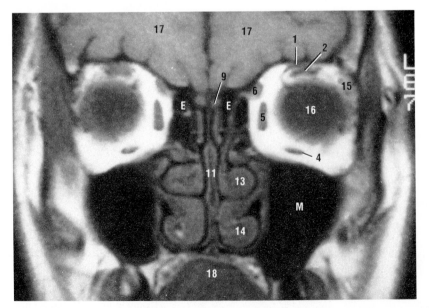

B

Figure 7-58. *A*, Coronal computed tomographic (CT) scan of the head, showing the relations of the orbital cavity. Note that the ostium (aperture) of the maxillary sinus is high on its medial wall. Observe the conchae in the lateral walls of the nasal cavities. (Courtesy of Dr. D. Armstrong, Associate Professor of Radiology, University of Toronto, Toronto, Ontario, Canada.) *B*, Coronal magnetic resonance image (MRI) through the orbit. *M* indicates maxillary sinus; *E*, ethmoidal sinus; *1*, levator palpabrae superioris; *2*, superior rectus; *4*, inferior rectus; *5*, medial rectus; *6*, superior oblique; *9*, olfactory bulb; *11*, nasal septum; *13*, middle concha; *14*, inferior concha; *17*, frontal lobe; and *18*, tongue. (Courtesy of Dr. W. Kucharczyk, Clinical Director of Tri-Hospital Resonance Centre, Toronto, Ontario, Canada.)

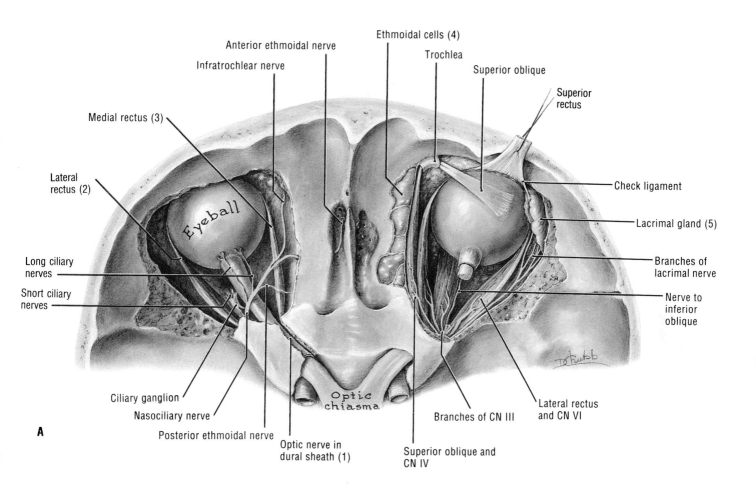

Anterior ethmoidal nerve

Infratrochlear nerve

Medial rectus (3)

Lateral rectus (2)

Ethmoidal cells (4)

Trochlea

Superior oblique

Superior rectus

Check ligament

Lacrimal gland (5)

Branches of lacrimal nerve

Nerve to inferior oblique

Long ciliary nerves

Snort ciliary nerves

Eyeball

Ciliary ganglion

Nasociliary nerve

Posterior ethmoidal nerve

Optic nerve in dural sheath (1)

Optic chiasma

Superior oblique and CN IV

Branches of CN III

Lateral rectus and CN VI

A

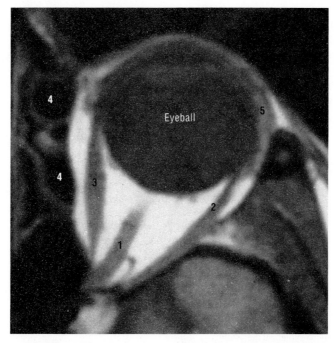

Eyeball

B

Figure 7-59. *A,* Dissection of the orbital cavities (superior view) showing the nasociliary nerve. Observe the optic nerve in its dural sheath, which is continuous with the dura mater of the brain. The numbers in parentheses refer to structures in *B.* Observe the nerves to the six ocular muscles. *B,* Transverse magnetic resonance image (MRI) of the orbit showing its contents. (Courtesy of Dr. W. Kucharczyk, Clinical Director of Tri-Hospital Resonance Centre, Toronto, Ontario, Canada.)

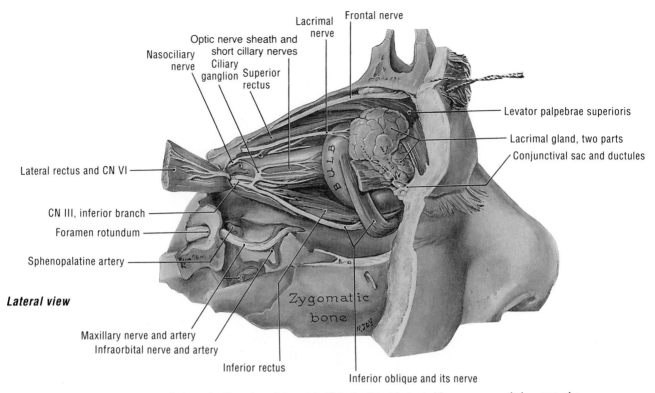

Nasociliary nerve

Ciliary ganglion

Optic nerve sheath and short cillary nerves

Superior rectus

Lacrimal nerve

Frontal nerve

Levator palpebrae superioris

Lacrimal gland, two parts

Conjunctival sac and ductules

Lateral rectus and CN VI

CN III, inferior branch

Foramen rotundum

Sphenopalatine artery

Lateral view

BULB

Zygomatic bone

Maxillary nerve and artery

Infraorbital nerve and artery

Inferior rectus

Inferior oblique and its nerve

Figure 7-60.   Lateral view of a dissection of the orbit. Note the 8 to 10 short ciliary nerves on their way to the eyeball. Observe the orbital muscles and lacrimal gland.

covering the brain (Fig. 7-62). The tough, opaque **sclera** appears slightly blue in infants and children and has a yellow hue in many older people. The anterior transparent part of the eye is the **cornea**; at the margins, *it is continuous with the sclera* (Figs. 7-61 and 7-62). The dark circular aperture that you see through the cornea is the **pupil**. This opening is surrounded by a circular, pigmented diaphragm known as the **iris**. The white sclera that you see in a mirror or when looking at someone else's eye is covered by a thin, moist, mucous membrane called the *bulbar conjunctiva* (Fig. 7-61). Because it is transparent, the white sclera is seen easily through it.

The bulbar conjunctiva is reflected off the sclera on to the deep surface of the eyelids (L. *palpebrae*), lining them to their margins. This *palpebral conjunctiva* becomes continuous with the skin of the eyelid (Figs. 7-61 and 7-63). As the palpebral **conjunctiva** is normally red and very vascular, it is commonly examined in cases of suspected *anemia*, a blood condition that is commonly manifested by pallor of the mucous membranes. As the bulbar conjunctiva is continuous with the anterior epithelium of the cornea and with the palpebral conjunctiva, it forms a *conjunctival sac* (Figs. 7-57A and 7-60). The opening between the eyelids, called the *palpebral fissure*, is the mouth of the conjunctival sac; hence, when the eyelids are closed, the bulbar and palpebral conjunctivae form a closed sac. Sensory innervation of the conjunctiva is from the trigeminal nerve through the infratrochlear, maxillary, and lacrimal nerves (Fig. 7-24C).

The bulbar conjunctiva is colorless, except when its vessels are dilated and congested (*e.g.*, "bloodshot eyes"). This

hyperemia of the conjunctiva is caused by local irritations (*e.g.*, dust, chlorine, and smoke) and infections (*e.g.*, "pinkeye"). *Subconjunctival hemorrhages* are common and are manifested by bright or dark red patches deep to and in the bulbar conjunctiva. These hemorrhages may result from injury or inflammation. The conjunctiva is frequently inflamed owing to infection, a condition called *conjunctivitis*.

The Eyelids

The eyelids protect the eyes from injury and excessive light and keep the cornea moist. Look into a mirror and slowly move your digit toward your eye. Note that your eyelids close when your digit gets close to your eye. Verify that the upper eyelid is larger and more movable than the lower one, and that the upper eyelid partly covers the iris (Figs. 7-57C and 7-61), whereas the entire inferior half of the eye is normally uncovered. The eyelids are essentially movable folds that are covered externally by thin skin and internally by the highly vascular palpebral conjunctiva (Figs. 7-61 and 6-63). The **palpebral conjunctiva** is reflected onto the eyeball, where it is *continuous with the bulbar conjunctiva*. The palpebral conjunctiva forms deep recesses known as the superior and inferior *conjunctival fornices*. Each eyelid is strengthened by a dense connective tissue band, about 2.5 cm wide, called the *tarsal plate* or tarsus (Figs. 7-21, 7-57A, and 7-63). In the connective tissue between this plate and the skin of the eyelid are fibers of the *orbicularis oculi* muscle.

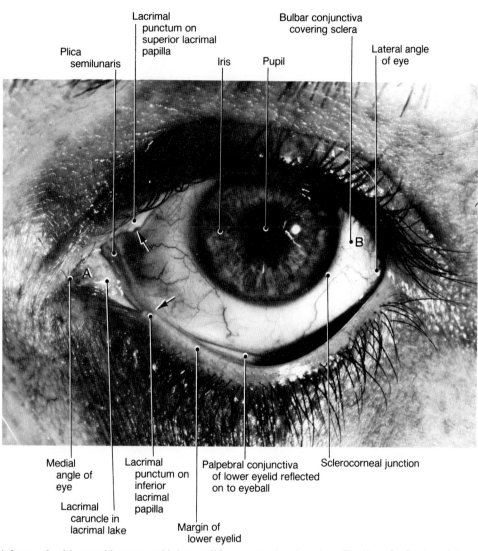

Plica
semilunaris

Lacrimal
punctum on
superior lacrimal
papilla

Iris

Pupil

Bulbar conjunctiva
covering sclera

Lateral angle
of eye

A

B

Medial
angle of
eye

Lacrimal
punctum on
inferior
lacrimal
papilla

Palpebral conjunctiva
of lower eyelid reflected
on to eyeball

Sclerocorneal junction

Lacrimal
caruncle in
lacrimal lake

Margin of
lower eyelid

Figure 7-61.   The left eye of a 36-year-old woman with her eyelids slightly everted to show the structures at the medial angle or canthus. Note that the pupil, an aperture in the iris, appears black because you are looking through the pupil toward the pigmented posterior aspect of the eye. You are looking through the transparent cornea to see the iris and pupil. Observe the fine vascular network of the bulbar conjunctiva covering the sclera. The line of reflection of the palpebral conjunctiva from the eyelids on the eyeball is called the conjunctival fornix; hence, there are superior and inferior fornices (Fig. 7-57A). The white spot on the iris is a highlight from the photographer's lamp. The arrows indicate the lacrimal puncta on the upper and lower lacrimal papillae; *A*, lacrimal caruncle in the lacrimal lake; *B*, ocular conjunctiva covering the sclera.

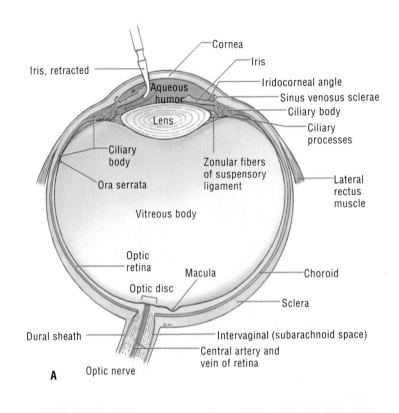

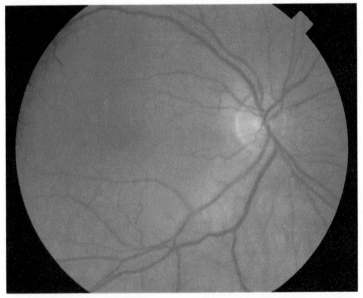

Figure 7-62. *A,* Horizontal section of the eyeball. Observe its three coats: (1) external or fibrous coat (sclera and cornea); (2) middle or vascular coat (choroid, ciliary body, and iris); and (3) internal or retinal layer. The arrow on the left indicates the flow of aqueous humor from the posterior chamber to the anterior chamber. *B,* The retina as seen through an ophthalmoscope. Observe the pale, oval optic disc with retinal vessels radiating from its center. Note that the retinal veins are wider than the arteries. Near the disc the central artery is medial to the vein. The round dark area lateral to the optic disc is the macula. Its center, a depressed spot called the fovea centralis, is avascular and is the area of most acute vision. (Courtesy of Dr. R. Bunzic, Professor of Ophthalmology, University of Toronto, Toronto, Ontario, Canada.)

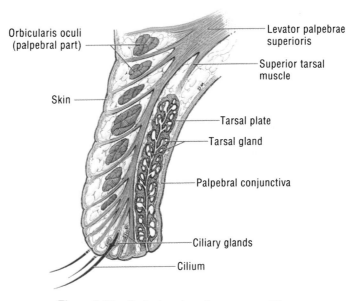

Orbicularis oculi (palpebral part)

Skin

Levator palpebrae superioris

Superior tarsal muscle

Tarsal plate

Tarsal gland

Palpebral conjunctiva

Ciliary glands

Cilium

Figure 7-63.   Sagittal section of an upper eyelid.

Embedded in the tarsal plates of the upper and lower eyelids are a number of *tarsal glands*, the fatty secretion of which lubricates the edges of the eyelids. This prevents them from sticking together when the eyelids are closed. The *eyelashes* (cilia) project from the margins of the eyelids (Figs. 7-57, 7-60, and 7-63) and are arranged in two or three irregular rows. The large sebaceous glands associated with the eyelashes are known as *ciliary glands*. Between the hair follicles there are modified sweat glands. The place where the eyelids meet is called the angle or **canthus** (G. *kanthos*, corner of eye). Thus each eye has medial and lateral canthi (Figs. 7-57A and 7-61). When the eyelids are closed the **palpebral fissure** is nearly in the horizontal plane, except in certain races (*e.g.*, Asian). In these people there is a slight superior slant of the palpebral fissure toward the nose because the medial ends of the upper eyelids project superomedially. Furthermore, their medial canthi are covered by an extra skin fold called the *epicanthal fold* that varies in size.

> Slanted palpebral fissures and inner epicanthal folds are also present in persons with the *Down syndrome* (the trisomy 21 chromosomal abnormality) and other syndromes, *e.g.*, the *cri du chat syndrome*, resulting from a terminal deletion of chromosome number 5 (Moore, 1988).

In the medial canthus (angle) of your eye, observe the reddish area known as the *lacrimal lake* (Fig. 7-61), within which there is a small mound of moist, unattractive skin called the *lacrimal caruncle* (L. small fleshy mass). Lateral to the caruncle is a vertical curved fold of conjunctiva, called the *semilunar fold* (plica semilunaris). This fold slightly overlaps the eyeball and is a remnant of the nictitating membrane present in some animals (*e.g.*, amphibians). Evert the edge of your lower eyelid, as shown in Fig. 7-61, and locate a small black pit at its medial end, the *lacrimal punctum* (L. a point), on the summit of a small elevation called the **lacrimal papilla** (Figs. 7-61 and 7-64). There is a similar punctum and papilla on the upper eyelid. The lacrimal punctum is the opening of a slender canal called the *lacrimal canaliculus* (L., canal), which carries the tears to the *lacrimal*

sac. From here, they are conducted to the nose through the *nasolacrimal duct*. The **lacrimal sac** has some fibers of the orbicularis oculi muscle posterior to it, which insert into the crest of the lacrimal bone, called the *lacrimal crest* (Fig. 7-56). When these muscle fibers contract, the lacrimal sac is squeezed, forcing the tears into the nasolacrimal duct, which opens into the nasal cavity (Fig. 7-64).

Observe that the lacrimal puncta face posteriorly, enabling them to "suck up" the tears. Carefully press your fingertip between your nose and the medial canthus of your eye. You should feel a horizontal cord, the *medial palpebral ligament* (Fig. 7-64). It connects the eyelids, including their muscles, to the medial margin of the orbit. If you pull your eyelid laterally, this ligament may raise a small skin fold. A similar *lateral palpebral ligament* attaches the eyelids to the lateral margin of the orbit. The medial and lateral palpebral ligaments are connected by *tarsal plates* (Figs. 7-63 and 7-64). These dense connective tissue plates enable the eyelids to be everted. It is easy to evert the lower eyelid, but difficult to evert the upper one because its tarsal plate is rigid. Hence when the upper eyelid is everted (*e.g.*, over a wooden match stick), it tends to stay that way until it is turned inferiorly.

The *palpebral fascia* is a thin fibrous membrane that connects the tarsal plates to the margins of the orbit and, with them, forms an *orbital septum*. This septum passes posterior to the lacrimal sac and is pierced by the levator palpebrae superioris muscle (Fig. 7-64). On the deep surface of your eyelids you may be able to see the *tarsal glands* (Fig. 7-63) because they appear as yellowish streaks through the palpebral conjunctiva. The ducts of these glands open on the free flat margin of the eyelid near its posterior edge.

Tears (lacrimal fluid) are produced by a small, almond-shaped *lacrimal gland*, located in the superolateral part of the orbit (Figs. 7-60 and 7-64). Its three to nine excretory ducts open into the *superior fornix of the conjunctival sac* (Figs. 7-57A and 7-60). In addition to the main gland, there are *accessory lacrimal glands*. As the lacrimal fluid drains into the lacrimal sac, it moistens the corneal surface. When the cornea becomes dry, the eye blinks and the eyelids carry a film of lacrimal fluid over the cornea, somewhat like the wipers on a car wash the windshield. In this way, foreign material (*e.g.*, dust) is carried to the medial canthus of the eye where you can remove it, *e.g.*, with a tissue. When you cry, the excess tears overflow the lacrimal lakes and roll down your cheeks.

In *third nerve palsy*, the upper eyelid droops (**ptosis**) and cannot be voluntarily raised. This results from damage to the superior division of the *oculomotor nerve* (CN III), which supplies the *levator palpebrae superioris* muscle (Figs. 7-21, 7-24, 7-60, and 7-64). You can tell by its name that this muscle normally elevates the upper lid. *When the facial nerve (CN VII) is damaged, the eyelids cannot be closed* owing to **paralysis of the orbicularis oculi muscle**, which closes the eyelids (p. 660). *When the facial nerve is paralyzed, the eyelids cannot be closed* and protective blinking of the eye is lost. As a result, tears cannot be washed across the cornea. Irritation of the unprotected eyeball results in excessive *lacrimation* (tear formation). Excessive tearing also occurs when the lacrimal drainage apparatus is obstructed or when the lower eyelid is lax and everted, thereby preventing the tears from reaching the inferior lacrimal punctum (Figs. 7-61 and 7-64).

Any of the glands in the eyelid may become inflamed and swollen owing to infection or obstruction of their ducts (Fig. 7-63). If the ducts of the ciliary glands become obstructed or inflamed, a painful red swelling known as a **sty** develops on the eyelid. *Cysts of the sebaceous glands* of the eyelid, called *chalazia*, may also form. An obstruction of a tarsal gland produces an inflammation, called a *tarsal chalazion*, that protrudes toward the eyeball and rubs against it as the eyelids blink. Usually chalazia are more painful than sties.

Structure of the Eyeball

The eyeball (about 2.5 cm long) is like a miniature camera suspended in the anterior half of the orbital cavity in such a way that the six ocular muscles can move it in all directions (Figs. 7-60 and 7-65). Do not be misled by the position of the eyeballs in cadavers; they are sunken owing to dehydration and post-mortem atrophy of the fat and muscles in the orbit. Sunken eyes are also characteristic of emaciated or dehydrated living people, owing to the scantiness of fat and fluid in their orbital cavities.

The eyeball has three concentric coats:

1. The External or Fibrous Coat (Fig. 7-62). This supporting coat of the eye consists of a white, opaque posterior five-sixths, called the *sclera*, and a transparent anterior one-sixth, called the *cornea*.

2. The Middle or Vascular Coat (Fig. 7-62). From posterior to anterior, this heavily pigmented vascular layer consists of the choroid, ciliary body, and iris.

The **choroid** is a dark brown membrane that is *located between the sclera and retina*. It forms the largest part of the middle coat of the eye and lines most of the sclera. It terminates anteriorly in the ciliary body. The choroid is firmly attached to the retina, but it can easily be stripped from the sclera. The choroid contains venous plexuses and layers of capillaries that are responsible for nutrition of the adjacent layers of the retina.

Figure 7-64. Parts of the lacrimal apparatus. Tears are secreted from the lacrimal gland into the superolateral angle of the conjunctival sac. After passing over the cornea, the tears drain into the lacrimal puncta in the upper and lower eyelids (see also Fig. 7-61). The puncta open into the lacrimal canaliculi (little canals) through which the lacrimal fluid is transported to the lacrimal sac. Medially, the superior and inferior canaliculi usually meet to form a common canaliculus, which carries the tears to the lacrimal sac (*upper right drawing*). This sac drains into the nasolacrimal duct, which empties into the inferior meatus of the nose (Fig. 7-55).

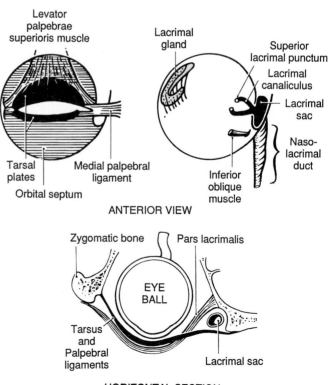

ANTERIOR VIEW

HORIZONTAL SECTION

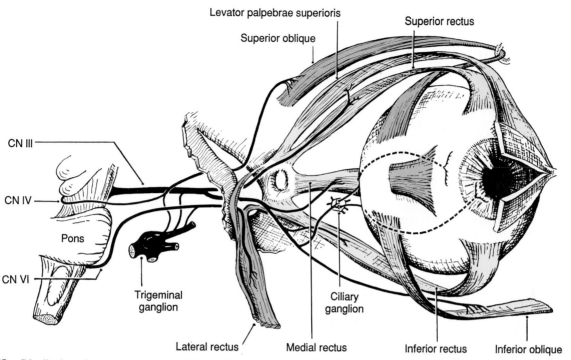

Levator palpebrae superioris

Superior oblique

Superior rectus

CN III

CN IV

Pons

CN VI

Trigeminal
ganglion

Lateral rectus

Medial rectus

Ciliary
ganglion

Inferior rectus

Inferior oblique

Figure 7-65.   Distribution of the oculomotor (CN III), trochlear (CN IV), and abducens (CN VI) nerves to the muscles of the eyeball. They enter the orbit through the superior orbital fissure. Note that CN IV supplies the superior oblique, CN VI the lateral rectus, and CN III the remaining five muscles.

The ciliary body connects the choroid with the circumference of the iris; it *is continuous posteriorly with the choroid.* This body has protrusions or folds on its internal surface, called *ciliary processes* (Fig. 7-62A), which secrete *aqueous humor,* a watery fluid that fills the anterior and posterior **chambers of the eye**. These chambers are the fluid-filled spaces anterior and posterior to the iris. The direction of flow of the aqueous humor is shown by the arrow hooking around the retracted iris in Figure 7-62. Externally *the ciliary body contains ciliary muscle,* which on contraction permits the lens to bulge by relaxing its suspensory ligament.

The iris is anterior to the lens (Fig. 7-62A). This *contractile diaphragm* has a central, circular aperture for transmitting light, called the *pupil* (Fig. 7-61). The iris is between the cornea and the lens (Fig. 7-62). When awake, the size of the pupils continually varies to regulate the amount of light entering the eye through the lens. *Eye color* depends on the amount and the distribution of pigment-containing cells (chromatophores) in the iris. The pupil appears black because one looks through the pupil toward the posterior aspect of the eye at the *optic fundus,* which is heavily pigmented. In persons with blue eyes, the pigment is limited to the posterior surface of the iris, whereas in persons with dark brown eyes the pigment is scattered throughout the loose connective tissue of the iris.

3. Internal or Retinal Coat (Fig. 7-62).   The retina is a very thin, delicate membrane. It is covered externally by the choroid and internally by the vitreous body. The **retina** is composed of two layers: an outer pigment cell layer and an internal neural layer. The light-sensitive neural layer ends at the posterior edge of the ciliary body in a wavy border, called the *ora serrata retinae.* A thin, insensitive layer continues anteriorly onto the ciliary body and iris. In the posterior portion of the interior of the eye, called the *optic fundus,* there is a circular depressed, white to pink area in the retina known as the optic papilla or *optic disc.* This is where the optic nerve enters the eyeball and its fibers spread out in the neural layer of the retina. Because it contains nerve fibers and no photoreceptor cells, *the optic disc is insensitive to light.* For this reason it is sometimes referred to as the *blind spot.* Just lateral to the optic disc is a small, oval, yellowish area called the *macula lutea* (L. yellow spot). The central depressed part of the macula, known as the **fovea** centralis, is the *area of most acute vision.* The retina is more adherent to the choroid at the optic disc and ora serrata than elsewhere.

The retina is supplied by the *central artery of the retina,* a branch of the ophthalmic artery, which enters the eyeball with the optic nerve (Figs. 7-62 and 7-69A). A corresponding system of retinal veins unites to form the *central vein of the retina* (Fig. 7-69B). The retinal arteries and veins usually accompany each other and sometimes cross one another.

Pulsation is usually visible in the retinal arteries through an *ophthalmoscope* (Fig. 7-62B). The retina and the optic nerve develop from an outgrowth of the embryonic forebrain, known as the *optic vesicle,* which takes its covering of meninges with it (Moore, 1988). Hence the meningeal layers and subarachnoid space extend around the optic nerve up to its attachment to the eyeball (Fig. 7-62A). The central artery and vein of the retina, which are branches of the ophthalmic artery and vein, run within the anterior part of the optic nerve and cross the extension of subarachnoid space around it. Consequently an increase in CSF pressure slows venous return from the retina, causing edema as the result of fluid accumulation. **Edema of the retina** is obvious during *ophthal-*

moscopy as a swelling of the optic disc or optic papilla (*i.e.,* *papilledema*). For this reason, inspection of the optic fundus (*funduscopy*) is an essential part of a neurological examination.

The pigment epithelium of the retina develops from the outer layer of the optic cup, which is a derivative of the embryonic optic vesicle, and the neural layer develops from the inner layer of the optic cup. As they are first separate in the embryo and then fuse during the early fetal period, these layers are separated by a *potential intraretinal space*. Although the pigment layer becomes firmly fixed to the choroid, its attachment to the neural layer is not so firm. **Detachment of the retina**, which may follow a blow to the eye, is a separation of the pigment layer from the neural layer, a condition that is normally present in the early embryo.

Refractive Media of the Eye

On their way to the retina, light waves pass through structures with different densities: the cornea, aqueous humor, lens, and vitreous body (Fig. 7-62*A*). These structures are the refractive media of the eye.

The Cornea (Figs. 7-57, 7-61, and 7-62*A*). This is the clear, circular area of the anterior part of the outer fibrous coat of the eyeball. *It is transparent and avascular, i.e.,* it has no blood vessels. It is largely responsible for refraction of the light that enters the eye. The cornea is continuous peripherally with the sclera at the *sclerocorneal junction* and consists chiefly of a special kind of dense connective tissue. The degree of curvature of the cornea varies in different people and is greater in young than in old persons.

Homologous *corneal transplants* can be performed surgically for patients with scarred or opaque corneas. The surface epithelium is regenerated by the host and covers the transplant in a few days. *Corneal implants* of nonreactive plastic material are also used. As the central part of the cornea receives its oxygen from the air, soft plastic lenses worn for long periods must be gas permeable.

The Aqueous Humor (Fig. 7-62*A*). This clear watery fluid in the anterior and posterior chambers of the eye is continuously *produced by the ciliary processes*. The aqueous humor provides nutrients for the avascular cornea and lens. After passing through the pupil from the posterior chamber into the anterior chamber, the aqueous humor is drained off through spaces at the *iridocorneal angle* (filtration angle). These spaces, visible through an instrument known as a *slitlamp*, open into a circular venous canal called the *sinus venosus sclerae* (canal of Schlemm). This sinus drains through aqueous veins into the scleral venous plexuses.

The Lens (Fig. 7-62*A*). This transparent, flexible, biconvex structure is enclosed in a transparent capsule. Encircled by the ciliary processes, it is located posterior to the iris and anterior to the vitreous humor. Like the cornea, *the lens is transparent and avascular*. About 1 cm in diameter, the highly elastic lens is held in position by a series of radially arranged zonular fibers, known collectively as the *suspensory ligament of the lens*. The fibers are attached to the capsule of the lens and laterally to the ciliary processes. The curvatures of the surfaces of the lens, particularly the anterior surface, constantly vary to focus near or distant objects on the retina.

As one gets older, the lens becomes harder and more flattened. During old age the lens gradually acquires a yellow tint. These changes gradually reduce the person's focusing power, a condition known as *presbyopia*. In some elderly people, there is also a loss of transparency of the lens. An area of opaqueness is known as a *cataract*. Cataracts may result from disease (*e.g.,* diabetes) or injury to the eye. They may occur in children, sometimes as a congenital defect. *Cataract extraction is the most common of all eye operations.* For information on how this operation is performed, see Mortimer and Kraft (1989).

The Vitreous Body (Fig. 7-62*A*). This material fills the eyeball posterior to the lens. It consists of a jellylike substance called *vitreous humor*, in which there is a meshwork of fine collagenic fibrils. A colorless, transparent gel, the vitreous body, occupies the space, called the *vitreous chamber*, between the lens and the retina. The vitreous body consists of about 99% water and forms about four-fifths of the eyeball. In addition to transmitting light, it holds the retina in place and provides support for the lens. In contrast to the aqueous humor, the vitreous body is not continuously replaced. It forms during the embryonic period and is not exchanged. At the periphery of the vitreous body, there is a *vitreous membrane*. It is formed by condensation of the vitreous humor.

Fascial Sheath of the Eyeball

This thin cuplike sheath surrounds the eyeball, except for its corneal part, and separates it from the fat and other contents in the orbit (Fig. 7-57*A*). The fascial sheath (bulbar fascia, Tenon's capsule) is attached posteriorly to the sclera close to the optic nerve and anteriorly, just posterior to the cornea (*sclerocorneal junction*). The tendons of the muscles that rotate the eyeball pierce the fascial sheath on their way to their attachments. This sheath also blends with the fascial sheaths of the ocular muscles. A potential space between the eyeball and the fascial sheath allows the eyeball to move inside this cup-shaped structure. There are triangular expansions from the sheaths of the medial and lateral rectus muscles (Figs. 7-62 and 7-65), which are attached to the lacrimal and zygomatic bones, respectively. Because they check the actions (prevent excessive movement) of these muscles, they are called the medial and lateral *check ligaments*. There is also a check ligament associated with the levator palpebrae superioris muscle (Fig. 7-60). The fascial sheath is perforated posteriorly by the ciliary nerves and vessels, and it fuses with the dural sheath of the optic nerve (Figs. 7-59*A*, 7-60, and 7-62). A thickening of the inferior part of the fascial sheath of the eyeball, called the *suspensory ligament of the eyeball*, is attached to the anterior parts of the medial and lateral walls of the orbit. Because it is a slinglike hammock inferior to the eyeball, it supports this organ.

The cuplike fascial sheath of the eyeball helps to form a socket for an artificial eye when the eyeball is removed (*enucleation*). After this operation, the eye muscles cannot retract far because their fascial sheaths are attached to the fascial sheath of the eyeball. Because the suspensory ligament in the orbit supports the eyeball, it is preserved when surgical removal of the bony floor of the orbit is carried out (*e.g.*, during the removal of a tumor).

Muscles of the Orbit

Levator Palpebrae Superioris Muscle (Figs. 7-21, 7-57*A*, 7-60, and 7-63 to 7-65).   This thin, triangular muscle elevates the upper eyelid.

Origin.   Roof of the orbit, anterior to the optic canal.

Insertion.   This thick muscle fans out into a wide aponeurosis that inserts into the skin of the upper eyelid. The inferior part of the aponeurosis contains some smooth muscle fibers that form the *superior tarsal muscle* (Figs. 7-57*A* and 7-63). These involuntary muscle fibers insert into the *tarsal plate*.

Action.   Elevates the upper eyelid. It is continuously active except during sleeping and when the eyelid is closing.

Innervation (Fig. 7-65).   Superior fibers are innervated by the *oculomotor nerve* (*CN III*), whereas the superior tarsal muscle is innervated by fibers from the cervical sympathetic trunk and the internal carotid plexus.

> In third (cranial) nerve palsy, the levator palpebrae superioris muscle is affected and the upper eyelid cannot be raised voluntarily. If the cervical sympathetic trunk is interrupted, the superior tarsal muscle is paralyzed, which causes drooping of the eyelid (*ptosis*). This is a sign of *cervical sympathetic trunk injury* and is part of the *Horner syndrome* (p. 815).

The Rectus Muscles

The four rectus muscles (L. *rectus*, straight) arise from a tough tendinous cuff, called the *common tendinous ring* (common ring tendon), which surrounds the optic canal and the junction of the superior and inferior orbital fissures (Figs. 7-60 and 7-65 to 7-67). From their common origin, these muscles run anteriorly, close to the walls of the orbit, and attach to the eyeball just posterior to the sclerocorneal junction. Each muscle (superior, inferior, medial, and lateral) passes anteriorly in the position implied by its name. Note that the lateral and medial rectus muscles lie in the same horizontal plane, whereas the superior and inferior rectus muscles lie in the same vertical plane.

The Oblique Muscles

The **superior oblique muscle** (Figs. 7-65 to 7-67) is fusiform and arises from the body of the sphenoid bone, superomedial to the common tendinous ring, and passes anteriorly, superior and medial to the superior and medial rectus muscles. It ends in a round tendon, which runs through a pulleylike loop, called the *trochlea* (Figs. 7-67 and 7-68), which is attached to the superomedial angle of the orbital wall. After passing through the trochlea (L. pulley), the tendon of the superior oblique turns posterolaterally and inserts into the sclera at the posterosuperior aspect of the lateral side of the orbit.

The **inferior oblique muscle** is a thin, narrow muscle that arises from the maxilla in the floor of the orbit (Figs. 7-65 to 7-67). It passes laterally and posteriorly, inferior to the inferior rectus muscle, and inserts into the sclera at the posteroinferior aspect of the lateral side of the orbit.

Nerve Supply of the Muscles of the Orbit (Figs. 7-65 and 7-66; Table 7-2).   All three cranial nerves supplying the muscles of the eyeball (oculomotor, *CN III*; trochlear, *CN IV*; and abducent, *CN VI*) *enter the orbit through the superior orbital fissure*. Cranial nerves IV and VI each supply one muscle, whereas CN III supplies the remaining five muscles. CN IV supplies the superior oblique (SO); CN VI supplies the lateral rectus (LR); and CN III supplies the levator palpebrae superioris, superior rectus (SR), medial rectus (MR), inferior rectus (IR), and inferior

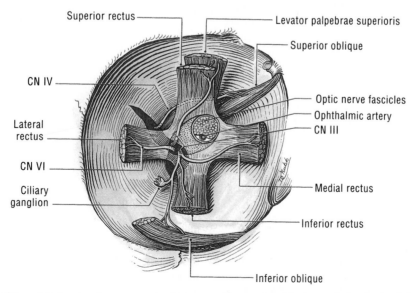

Superior rectus — Levator palpebrae superioris
Superior oblique
CN IV
Optic nerve fascicles
Ophthalmic artery
CN III
Lateral rectus
CN VI
Medial rectus
Ciliary ganglion
Inferior rectus
Inferior oblique

Figure 7-66.   Dissection of the orbit showing the common tendinous ring and the motor nerves of the orbit. Note the four rectus muscles arising from the fibrous cuff, known as the common tendinous ring.   Observe that this ring encircles the dural sheath of the optic nerve (CN II), the abducent nerve (CN VI), and the superior and inferior divisions of the oculomotor nerve (CN III).

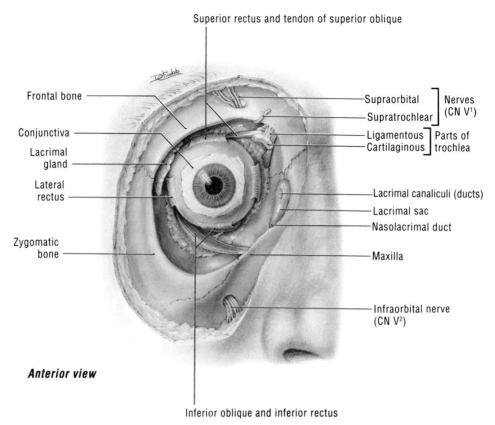

Superior rectus and tendon of superior oblique

Frontal bone

Conjunctiva

Lacrimal gland

Lateral rectus

Zygomatic bone

Supraorbital — Supratrochlear | Nerves (CN V¹)

Ligamentous — Cartilaginous | Parts of trochlea

Lacrimal canaliculi (ducts)

Lacrimal sac

Nasolacrimal duct

Maxilla

Infraorbital nerve (CN V²)

Anterior view

Inferior oblique and inferior rectus

Figure 7-67.   Dissection of the orbital cavity (anterior view). The eyelids, orbital septum, levator palpebrae superioris muscle, and some fat have been removed.

oblique (IO). Hence, all three nerves carry fibers that are motor to extraocular muscles (muscles of the eyeball).

In summary, all orbital muscles are supplied by CN III except the superior oblique and the lateral rectus, which are supplied by CN IV and VI, respectively (*i.e.*, SO IV, LR VI, all others III). The following ''formula'' is a *useful memory key*: $SO_4(LR_6)_3$.

Actions of the Extraocular Muscles (Table 7-2).   The six muscles rotate the eyeball in the orbit around three axes (sagittal, horizontal, and vertical). The four rectus muscles are arranged around the orbital axis; thus the medial, superior, and inferior recti are adductors.

Paralysis of one or more extraocular muscles owing to injury of the nerves supplying them, results in **diplopia** (*double vision*). Paralysis of a muscle of the eyeball is noted by the limitation of eye movement in the field of action of the paralyzed muscle, and by the production of two images when an attempt is made to use the paralyzed muscle. When the abducent nerve is paralyzed, the patient is unable to abduct the eye on the affected side. Usually there is also double vision (diplopia). During clinical testing, each muscle is examined in its position of greatest efficiency, *i.e.*, when its action is at a right angle to the axis around which it is moving the eyeball. The patient is asked to look: superolaterally (upward and outward) to test the superior rectus; inferolaterally (downward and outward) to test the inferior rectus; superomedially (upward and inward) to test the inferior oblique; inferomedially (downward and inward) to test the superior

oblique; medially (inward) to test the medial rectus; and laterally (outward) to test the lateral rectus.

The Orbital Vessels

The orbital contents are supplied chiefly by the *ophthalmic artery* (Figs. 7-69A and 7-70). The *infraorbital artery*, the continuation of the maxillary, also contributes blood to this region. Venous drainage is through the *ophthalmic veins*, superior and inferior (Fig. 7-69B), which pass through the superior orbital fissure to enter the cavernous sinus.

The Ophthalmic Artery (Figs. 7-48, 7-69A, and 7-70).   This artery arises from the internal carotid artery as it emerges from the cavernous sinus. It passes through the optic foramen within the dural sheath of the optic nerve and runs anteriorly, close to the superomedial wall of the orbit. It gives off branches to structures in the orbit and to the ethmoid bone.

The Central Artery of the Retina (Figs. 7-62A and 7-69A).   This is one of the smallest but most important branches of the ophthalmic artery. It arises inferior to the optic nerve. It runs within the dural sheath of the optic nerve until it approaches the eyeball. It then pierces the optic nerve and runs within it to emerge through the optic disc. The central artery of the retina spreads over the internal surface of the retina and supplies it. Twigs of this artery anastomose with the ciliary arteries, but the

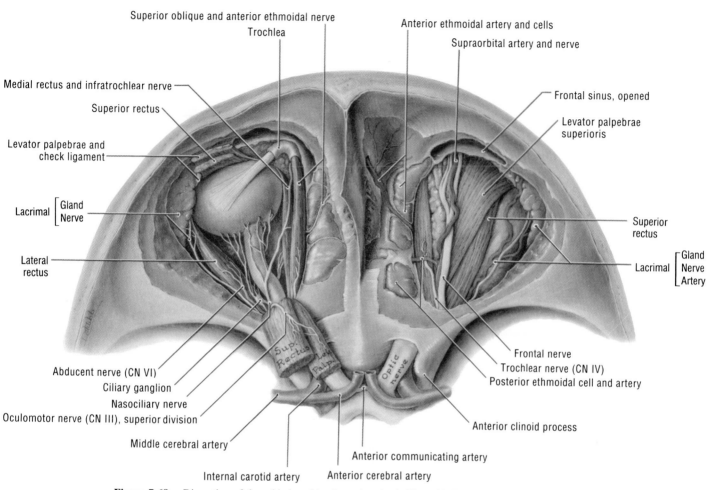

Figure 7-68.   Dissection of the orbital cavities (superior view). The orbital part or plate of the frontal bone is removed on the right side.

Table 7-2.
The Actions and Nerve Supply of the Ocular Muscles

Muscle	Action(s) on the Eyeball	Nerve Supply[1]
Medial rectus[2]	Adducts	CN III
Lateral rectus[2]	Abducts	CN VI[3]
Superior rectus	Elevates, adducts, and rotates medially	CN III
Inferior rectus	Depresses, adducts, and rotates laterally	CN III
Superior oblique[4-5]	Depresses medially rotated eye, abducts, and rotates medially	CN IV[3]
Inferior oblique[4-5]	Elevates medially rotated eye, abducts, and rotates laterally	CN III

[1]These cranial nerves are described in Chapter 9.
[2]The medial and lateral rectus muscles move the eyeball in one axis only, whereas each of the other four muscles moves it in all three axes.
[3]CN IV and VI each supply one muscle, whereas CN III supplies the other four muscles.
[4]The two oblique muscles protrude the eyeball, whereas the four rectus muscles retract it.
[5]The superior and inferior oblique muscles are used with the medial rectus muscle in turning both eyes medially for near vision. This movement, accompanied by pupillary constriction, is known as accommodation.

terminal branches of the central artery of the retina are essentially end arteries.

Because the retinal artery is essentially an end artery, obstruction of it by an embolus or thrombosis leads to instant and total blindness in the eye concerned.

The Ciliary Arteries (Fig. 7-69A).   The ciliary arteries, branches of the ophthalmic, supply the sclera, choroid, ciliary body, and iris. Two long posterior ciliary arteries pierce the sclera and supply the ciliary body and iris. Several short posterior ciliary arteries pierce the sclera and supply the choroid.

The Lacrimal Artery (Fig. 7-69A).   This branch of the ophthalmic artery supplies the lacrimal gland, conjunctiva, and eyelids. A recurrent meningeal branch anastomoses with the middle meningeal artery; hence, there is an anastomosis between branches of the internal and external carotid arteries.

Muscular Branches of the Ophthalmic Artery.   These branches to the eye muscles frequently arise from a common trunk and accompany the branches of the oculomotor nerve. Muscular branches also give rise to *anterior ciliary arteries*, which give branches to the conjunctiva and then pierce the sclera to supply the iris.

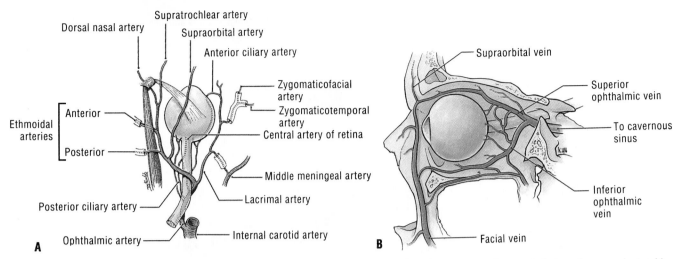

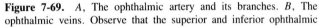

Figure 7-69. *A*, The ophthalmic artery and its branches. *B*, The ophthalmic veins. Observe that the superior and inferior ophthalmic veins empty into the cavernous sinus posteriorly and communicate with the facial and superior orbital veins anteriorly (Fig. 7-30).

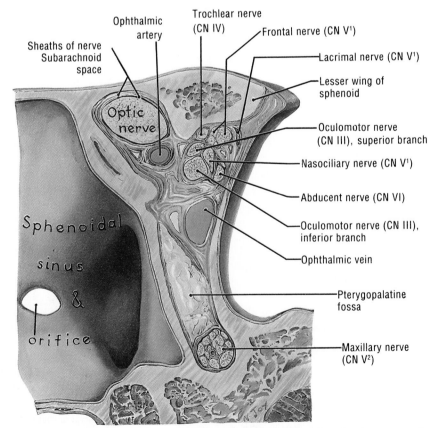

Figure 7-70. Coronal section of the apex of the orbital cavity. The optic nerve is surrounded by its sheaths, which are extensions of the cranial meninges. Note that the nerves to the orbit are crowded together as they pass through the medial end of the superior orbital fissure.

Other Branches of the Ophthalmic Artery (Figs. 7-35 and 7-69*A*).   Five branches leave the orbit with correspondingly named nerves and anastomose with branches of the external carotid artery. They are the supraorbital, supratrochlear, and dorsal nasal arteries, which supply the forehead or face. The anterior and posterior ethmoidal arteries enter the skull but end in the nasal mucosa.

The Ophthalmic Veins (Figs. 7-30, 7-69*B*, and 7-70).   The *superior ophthalmic vein* anastomoses with the facial vein; because it has no valves, blood can flow in either direction. It crosses superior to the optic nerve, passes through the superior orbital fissure, and ends in the cavernous sinus. The *inferior ophthalmic vein* begins as a plexus on the floor of the orbit. It communicates through the inferior orbital fissure with the *pterygoid plexus*, crosses inferior to the optic nerve, and ends in either the superior ophthalmic vein or in the cavernous sinus. The *central vein of the retina* (Fig. 7-62) usually enters the cavernous sinus directly, but it may join one of the ophthalmic veins, usually the superior ophthalmic.

> As the cavernous sinuses of the dura communicate with the veins of the face through the valveless superior and inferior ophthalmic veins (Fig. 7-69*B*), *thrombophlebitis of a facial vein*, resulting from an infection of the face in the area drained by these veins (Figs. 7-29 and 7-33), may spread to the cavernous sinus (Figs. 7-30 and 7-69*B*). As the central vein of the retina enters either the cavernous sinus or the superior ophthalmic vein, thrombosis may sometimes extend along the central vein of the retina and produce thromboses in the small retinal veins.

The Optic Nerve[19]

This is the second cranial nerve (CN II) and the *nerve of sight* (Figs. 7-45, 7-46, 7-51, 7-57*A*, 7-59, 7-62, 7-66, 7-68, and 7-70). It is about 5 cm in length and extends from the **optic chiasma** (G. a crossing) to the eyeball. The optic nerve is slightly longer than the distance it travels; this permits free movement of the eyeball. Most fibers of the optic nerve are afferent and arise from *ganglion cells* in the retina. The optic nerve passes posteromedially within a cone formed by the extraocular muscles and leaves the orbit through the **optic canal** to enter the optic chiasma slightly superior and anterior to the tuberculum sellae (Figs. 7-11 and 7-42).

The optic nerve is a special sensory nerve. Posterior to the optic chiasma, the optic nerves are continued as the **optic tracts**. Within the optic chiasma is a partial decussation of the optic nerve fibers. Fibers from the nasal half of each retina cross to the opposite side, whereas those from the temporal half of each retina are uncrossed. Thus fibers from the right half of the retina of both eyes form the right optic tract and those from the left halves form the left tract. This crossing of nerve fibers results in the right optic tract conveying impulses from the left visual field and vice versa.

> *Injury to any part of the optic pathway results in visual defects*, the nature of which depends on the location and extent of the injury (Table 9-3, p. 857). Severe degenerative disease or a complete lesion of the optic nerve (*e.g.*, sectioning) causes *total blindness* in the corresponding eye. A lesion at the lateral border of the optic chiasma results in loss of the nasal half of the visual field of the eye on the same side as the lesion. A localized dilation or *aneurysm of the internal carotid artery*, superior to the cavernous sinus, could exert this kind of pressure on the optic chiasma.

The Parotid Region

This region is mainly the space between the mastoid process of the temporal bone and the neck and ramus of the mandible. It contains the parotid gland, as well as the structures related to it.

The Parotid Gland

The parotid gland and parotid duct are described with the face (p. 670). The parotid gland, enclosed within fascia called the *parotid sheath*, occupies all the space available to it in the area around the posterior margin of the mandible.

The parotid bed, occupied by the gland, is a small space between the mastoid process posteriorly and the ramus of the mandible anteriorly (Figs. 7-71 to 7-73). *Superiorly* the parotid bed is bounded by the floor of the external acoustic meatus and the zygomatic process of the temporal bone. *Medially* there is the styloid process of the temporal bone and its associated muscles. *Laterally* the parotid bed is bounded by the superficial layer of the parotid fascia and skin (Fig. 7-23). The posterior wall of the parotid bed extends between the mastoid and styloid processes; thus the muscles attached to them are closely related to the parotid gland. The anterior wall of the parotid bed is formed by the ramus of the mandible and the muscles attached to it (masseter and medial pterygoid).

Structures Within the Parotid Gland (Figs. 7-22, 7-71, and 7-73).   From superficial to deep, structures traversing the parotid gland are the facial nerve, the retromandibular vein, and the external carotid artery. There are also parotid lymph nodes on the parotid fascia and within the gland. The **facial nerve (CN VII)** is unique in traversing the parotid gland, an occurrence of considerable clinical significance. *The facial nerve emerges from the stylomastoid foramen* of the skull and can be exposed in the notch between the mastoid process and the external acoustic meatus (Fig. 7-71). Beyond this, the nerve enters the posterior part of the parotid gland and divides almost at once into superior and inferior divisions. These divisions give rise to temporal, zygomatic, and buccal branches and to mandibular and cervical branches, respectively (Fig. 7-22). The branches of the facial nerve emerge on the anterior aspect of the periphery of the parotid gland and lie on the lateral surface of the masseter muscle (Fig. 7-73). From here, they pass to the muscles of facial expression, which they supply (Figs. 7-22 and 7-27).

---

[19]This nerve is described in more detail in Chapter 9 (p. 854).

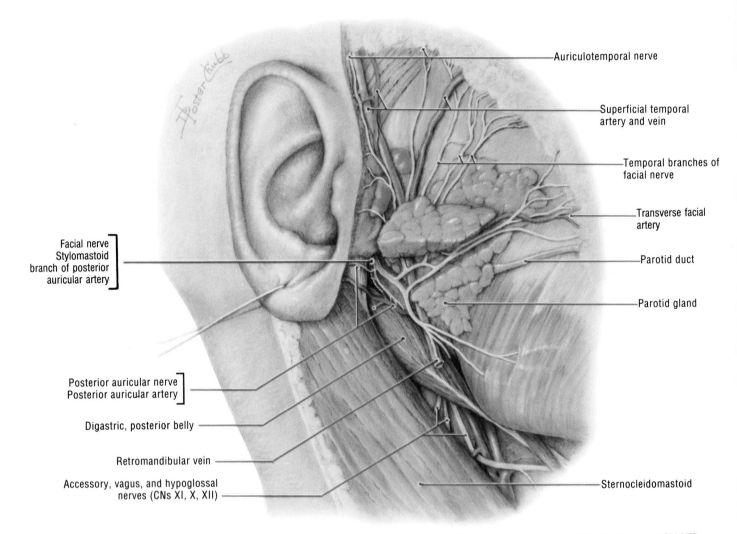

Auriculotemporal nerve

Superficial temporal artery and vein

Temporal branches of facial nerve

Transverse facial artery

Parotid duct

Parotid gland

Sternocleidomastoid

Facial nerve
Stylomastoid branch of posterior auricular artery

Posterior auricular nerve
Posterior auricular artery

Digastric, posterior belly

Retromandibular vein

Accessory, vagus, and hypoglossal nerves (CNs XI, X, XII)

Figure 7-71.   Dissection of the parotid region. Part of the parotid gland has been removed to expose the branches of the facial nerve (CN VII).

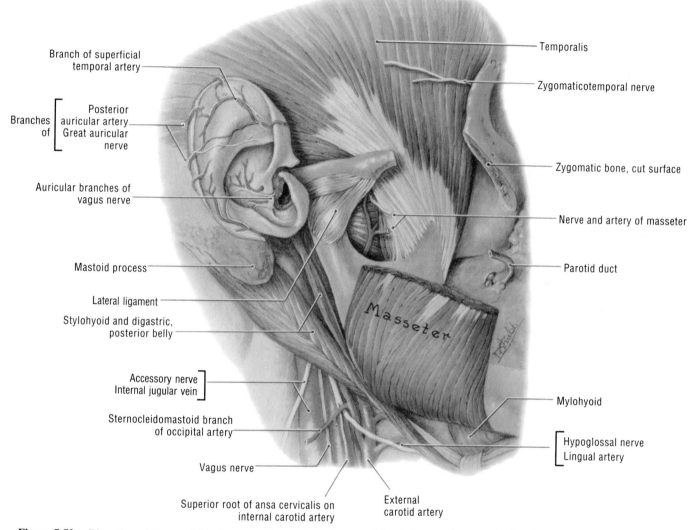

Branch of superficial
temporal artery

Branches { Posterior
of { auricular artery
Great auricular
nerve

Auricular branches of
vagus nerve

Mastoid process

Lateral ligament

Stylohyoid and digastric,
posterior belly

Accessory nerve
Internal jugular vein

Sternocleidomastoid branch
of occipital artery

Vagus nerve

Superior root of ansa cervicalis on
internal carotid artery

External
carotid artery

Temporalis

Zygomaticotemporal nerve

Zygomatic bone, cut surface

Nerve and artery of masseter

Parotid duct

Masseter

Mylohyoid

Hypoglossal nerve
Lingual artery

Figure 7-72. Dissection of the parotid bed, temporalis muscle, and auricular vessels and nerves. The sternocleidomastoid, splenius capitus, and longissimus capitus muscles have been removed from the mastoid process.

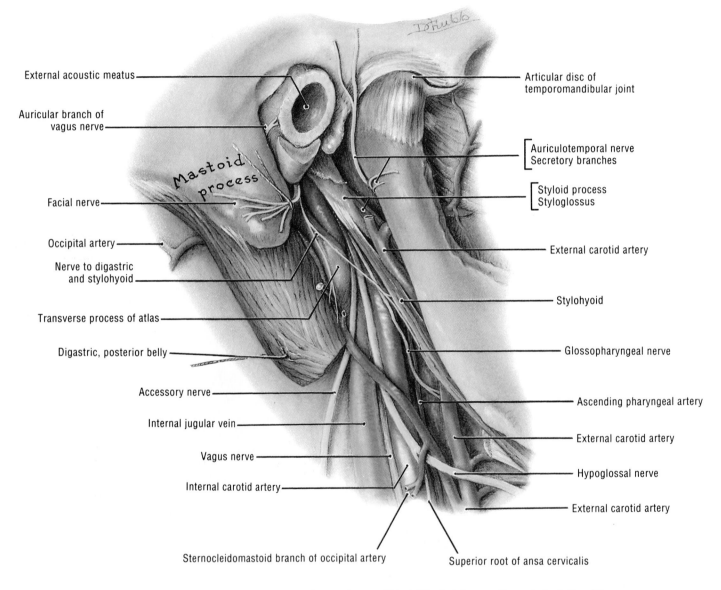

External acoustic meatus

Auricular branch of vagus nerve

Facial nerve

Occipital artery

Nerve to digastric and stylohyoid

Transverse process of atlas

Digastric, posterior belly

Accessory nerve

Internal jugular vein

Vagus nerve

Internal carotid artery

Sternocleidomastoid branch of occipital artery

Mastoid process

Articular disc of temporomandibular joint

Auriculotemporal nerve
Secretory branches

Styloid process
Styloglossus

External carotid artery

Stylohyoid

Glossopharyngeal nerve

Ascending pharyngeal artery

External carotid artery

Hypoglossal nerve

External carotid artery

Superior root of ansa cervicalis

Figure 7-73.   Dissection of the structures deep to the parotid bed. The facial nerve, the posterior belly of the digastric, and the nerve to it are retracted.

During its early development, the parotid gland lies between the two major branches of the facial nerve. As the gland enlarges, it envelops these branches and then superficial and deep parts of the gland fuse with each other between and around the branches of the facial nerve.

The *retromandibular vein* (Fig. 7-29), formed by the union of the temporal and maxillary veins, descends in the parotid gland, superficial to the external carotid artery but deep to the facial nerve (Fig. 7-73). Blood drains from the parotid gland into the retromandibular vein, which joins the posterior auricular vein to form the external jugular vein through its posterior branch and the facial vein through its anterior branch.

The *external carotid artery* enters the deep surface of the parotid gland (Figs. 7-28 and 7-71 to 7-73) where it lies deep to the facial nerve and the retromandibular vein. At the neck of the mandible, the external carotid artery divides into the superficial temporal and maxillary arteries.

Nerves Near the Parotid Gland (Figs. 7-22, 7-25, 7-27, and 7-71 to 7-73).

The Auriculotemporal Nerve (Fig. 7-71).   This is a branch of the mandibular division of the trigeminal nerve (CN V³). It is in the parotid space, but it passes superior to the superior part of the parotid gland. It communicates with the facial nerve, usually through two branches.

The Great Auricular Nerve (Figs. 7-22 and 7-72).   This is a superficial ascending branch of the cervical plexus. It passes external to the parotid gland where it divides into anterior and posterior branches, but it usually does not enter the parotid gland.

Vessels of the Parotid Gland (Figs. 7-22, 7-28, 7-29, 7-73, and 7-79).   This gland is supplied by the *external carotid artery* and its terminal branches (superficial temporal and maxillary arteries). They arise within the gland. The veins from the parotid gland drain into the retromandibular vein and then into the external jugular vein.

Innervation of the Parotid Gland (Figs. 7-22, 7-25, 7-27, and 7-74).   The parasympathetic component of the *glossopharyngeal nerve* (CN IX) supplies secretory fibers to the parotid gland through the *otic ganglion*. Stimulation of these fibers produces a thin, watery saliva. Sensory nerve fibers pass to the gland through the great auricular and auriculotemporal nerves. Sympathetic fibers also pass to the gland through a nerve plexus associated with the external carotid artery.

Lymphatic Drainage of the Parotid Gland (Figs. 7-22 and 7-32).   The parotid lymph nodes are located on or deep to the parotid fascia and within the gland. They receive lymph from the forehead, lateral parts of the eyelids, temporal region, lateral surface of the auricle, and the anterior wall of the external acoustic meatus. The lymph nodes within the gland also receive lymph from the middle ear. Lymph from the parotid nodes drains into the superficial and deep *cervical lymph nodes*.

The Temporal Region

This clinically important region contains the temporal and infratemporal fossae, which are located superior and inferior to the *zygomatic arch*, respectively (Fig. 7-76).

The Temporal Fossa

This oval fossa is bounded superiorly and posteriorly by the temporal lines and anteriorly by the frontal and zygomatic bones (Figs. 7-2 and 7-9). The *temporal fascia* stretches over the temporal fossa and the temporalis muscle (Figs. 7-23, 7-72, and 7-75). Inferiorly the temporalis fascia splits into two layers, superficial and deep. The superficial layer is attached to the superior margin of the zygomatic arch, and the deep layer passes medial to the arch to become continuous with the fascia deep to the masseter muscle.

The *floor of the temporal fossa* (Figs. 7-8, 7-9, and 7-76), which gives origin to the temporalis muscle, is formed by portions of four bones: parietal, frontal, greater wing of the sphenoid, and squamous part of the temporal bone. The area where these bones meet is called the *pterion* (p. 641). The temporal fossa contains the fan-shaped *temporalis muscle*, the "handle" of which passes deep to the zygomatic arch (Figs. 7-72 and 7-75). The temporal fossa is deepest where the temporalis muscle is thickest (*i.e.*, anteroinferiorly).

The Infratemporal Fossa

This is an irregularly-shaped space inferior and deep to the zygomatic arch and posterior to the maxilla (Figs. 7-71 and 7-

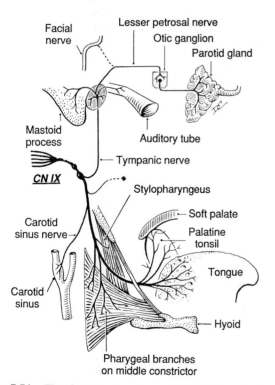

Figure 7-74.   The glossopharyngeal nerve (CN IX). Note that it sends secretomotor fibers to the parotid gland through its tympanic branch, the lesser petrosal nerve, the otic ganglion, and the auriculotemporal nerve.

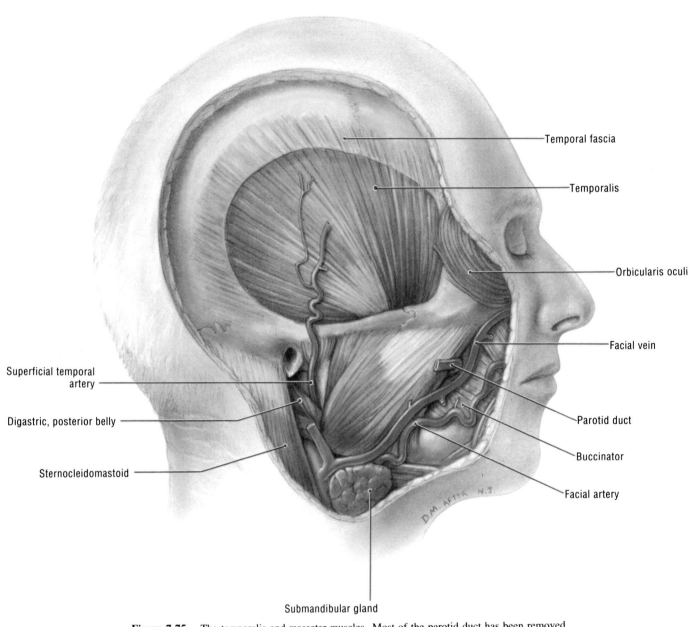

Temporal fascia

Temporalis

Orbicularis oculi

Facial vein

Superficial temporal
artery

Digastric, posterior belly

Sternocleidomastoid

Parotid duct

Buccinator

Facial artery

Submandibular gland

Figure 7-75.   The temporalis and masseter muscles. Most of the parotid duct has been removed (it is shown in Figs. 7-22A and 7-23).

76 to 7-78). It communicates with the temporal fossa through the interval between the zygomatic arch and the skull, which is traversed by the temporalis muscle and the deep temporal nerves and vessels.

Bones and Walls of the Infratemporal Fossa

The **lateral wall** of this fossa is formed by the *ramus of the mandible* (Fig. 7-9). The *lateral pterygoid plate* forms its **medial wall** (Fig. 7-76). Its **anterior wall** is formed by the *infratemporal surface of the maxilla*. This wall is limited superiorly by the *inferior orbital fissure* and medially by the *pterygomaxillary fissure*. The **posterior wall** of the infratemporal fossa is formed by the anterior surface of the *condylar process of the mandible* (p. 653) *and the styloid process* of the temporal bone (Figs. 7-15 and 7-76). The flat *roof of the infratemporal fossa* is formed mainly by the inferior surface of the greater wing of the sphenoid bone. It is separated from the temporal fossa by a ragged edge called the *infratemporal crest*. The *foramen ovale* and *foramen spinosum* open into the roof of the infratemporal fossa (Fig. 7-41). The inferior boundary of the infratemporal fossa is the point where the *medial pterygoid muscle* attaches to the medial aspect of the mandible near the angle (Fig. 7-77).

Contents of the Infratemporal Fossa

This fossa contains the inferior part of the temporalis muscle, the medial and lateral pterygoid muscles, the maxillary artery, the pterygoid venous plexus, the mandibular and chorda tympani nerves, the otic ganglion, and the inferior alveolar, lingual, and buccal nerves.

The Maxillary Artery (Figs. 7-28 and 7-77 to 7-79).   This vessel is the larger of the two terminal branches of the external carotid artery. It arises posterior to the neck of the condylar process of the mandible and passes anteriorly, deep to the neck, and *traverses the infratemporal fossa*. It passes superficial to the lateral pterygoid muscle and then disappears in the infratemporal fossa. The maxillary artery is divided into three parts by the lateral pterygoid muscle.

The branches of the first part of the maxillary artery are: (1) the deep *auricular artery* to the external acoustic meatus; (2) the *anterior tympanic artery* to the tympanic membrane (eardrum); (3) the *middle meningeal artery*; (4) the *accessory meningeal arteries* to the cranial cavity; and (5) the *inferior alveolar artery* to the mandible, gingivae (gums), and teeth. The largest of the meningeal arteries, the middle meningeal is the principal artery to the calvaria. It ascends between the two roots of the auriculotemporal nerve (Figs. 7-78 and 7-79) and enters the skull through the foramen spinosum to supply the dura and the interior of the calvaria (Figs. 7-42, 7-44A, and 7-76B).

The branches of the second part of the maxillary artery supply muscles through masseteric, deep temporal, pterygoid, and buccal branches (Fig. 7-79).

The branches of the third part of the maxillary artery, which arise just before and after it enters the pterygopalatine fossa (Fig. 7-79), are: (1) posterior superior alveolar, (2) middle superior alveolar, (3) infraorbital, (4) descending palatine, (5) artery of the pterygoid canal, (6) pharyngeal, and (7) sphenopalatine.

The Pterygoid Venous Plexus (Fig. 7-30).   This important venous plexus is located partly between the temporalis and lateral pterygoid muscles and partly between the two pterygoid muscles. It has connections with the facial vein through the cavernous sinus.

The Mandibular Nerve (Figs. 7-24, 7-25, 7-42, 7-78, and 7-80; Table 7-3).   All nerves in the infratemporal region except the chorda tympani, a branch of the facial nerve (CN VII), are derived from this inferior division of the trigeminal nerve. Descending through the foramen ovale into the infratemporal fossa from the medial part of the middle cranial fossa, the mandibular nerve (CN V³) divides into sensory and motor fibers. Branches of the mandibular nerve supply the four muscles of mastication (temporalis, masseter, and medial and lateral pterygoids), but not the buccinator, which is supplied by the facial nerve (CN VII). The branches of CN V³ are the auriculotemporal, inferior alveolar, lingual, and buccal nerves.

The Auriculotemporal Nerve (Figs. 7-22 to 7-25, 7-71, 7-73, and 7-77).   This nerve encircles the middle meningeal artery and breaks up into numerous branches, the largest of which passes posteriorly, medial to the neck of the mandible, and supplies sensory fibers to the auricle and temporal region. The auriculotemporal nerve also sends articular fibers to the temporomandibular joint and secretomotor fibers (parasympathetic) to the parotid gland.

The Inferior Alveolar Nerve (Figs. 7-21, 7-25, 7-77, 7-78, and 7-80).   This nerve enters the mandibular foramen, passes through the mandibular canal, and appears on the face as the mental nerve. While in the mandibular canal, it sends nerves to all teeth in the mandible on its side. One of its branches, the *mental nerve*, passes to the face through the mental foramen, which is generally in the region of the second premolar tooth. It supplies the skin and mucous membrane of the lower lip, the skin of the chin, and the vestibular gingiva of the mandibular incisors.

The Lingual Nerve (Figs. 7-25, 7-77, 7-78, 7-80, and 7-96).   This long nerve lies anterior to the inferior alveolar nerve. It is *sensory to the tongue* (L. *lingua*, tongue), the floor of the mouth, and the gingivae. It enters the mouth between the medial pterygoid muscle and the ramus of the mandible and passes anteriorly under cover of the oral mucosa, just inferior to the third molar tooth. *The chorda tympani branch of the facial nerve (CN VII) joins the lingual nerve in the infratemporal fossa* (Fig. 7-27), but the nerves maintain their separate identities. The chorda tympani is conducted by the lingual nerve to the tongue for its taste distribution.

The Buccal Nerve (Figs. 7-22, 7-25, 7-77, and 7-78).   This long nerve usually runs between the two heads of the lateral pterygoid muscle. It descends through the deep fibers of the temporalis muscle where its branches spread out over the lateral surface of the buccinator muscle. The buccal nerve supplies sensory fibers to the skin and mucous membrane of the cheek, as well as to the mandibular buccal gingiva in the molar region.

Mandibular Nerve Block.   As it emerges from the foramen ovale and enters the infratemporal fossa (Fig. 7-25), local anesthesia may be applied to the mandibular nerve. The injection needle is passed through the mandibular notch into

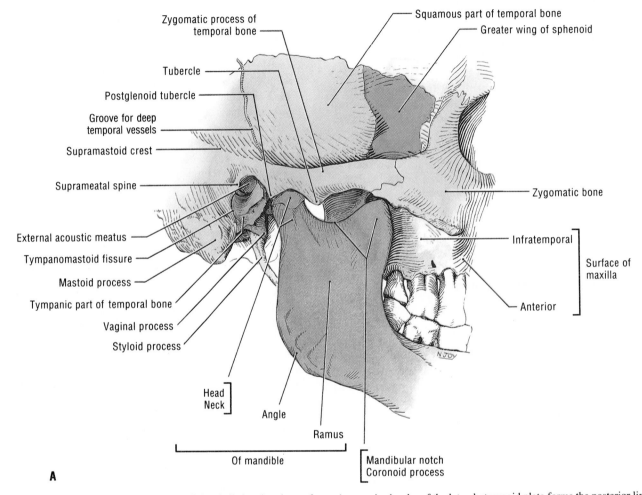

A

Figure 7-76. *A*, Part of the lateral side of the skull showing the roof and medial and lateral walls of the infratemporal fossa. Note that the posterior free border of the lateral pterygoid plate, when followed superiorly, leads to the foramen ovale in the roof of the fossa. Superiorly, the anterior border of the lateral pterygoid plate forms the posterior limit of the pterygomaxillary fissure, which is the entrance to the pterygopalatine fossa. *B*, The roof and medial and lateral walls of the infratemporal fossa.

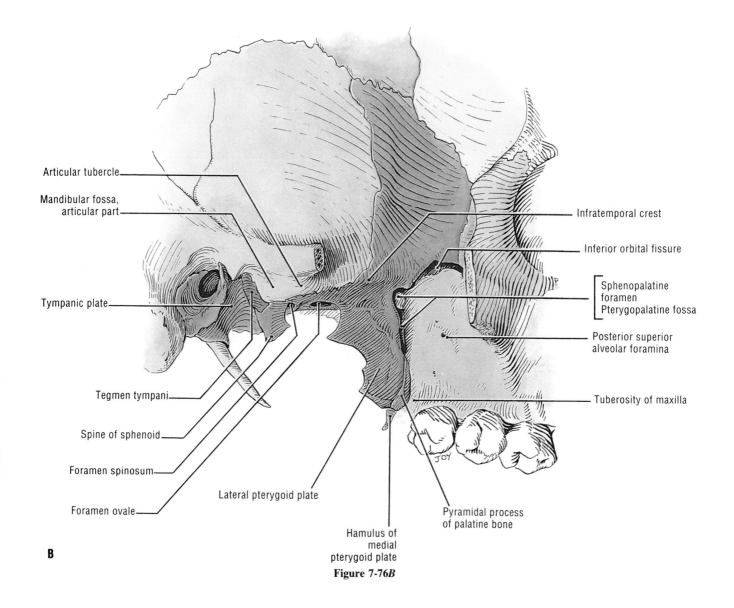

Articular tubercle

Mandibular fossa, articular part

Tympanic plate

Tegmen tympani

Spine of sphenoid

Foramen spinosum

Foramen ovale

Lateral pterygoid plate

Hamulus of medial pterygoid plate

Infratemporal crest

Inferior orbital fissure

Sphenopalatine foramen
Pterygopalatine fossa

Posterior superior alveolar foramina

Tuberosity of maxilla

Pyramidal process of palatine bone

B

Figure 7-76*B*

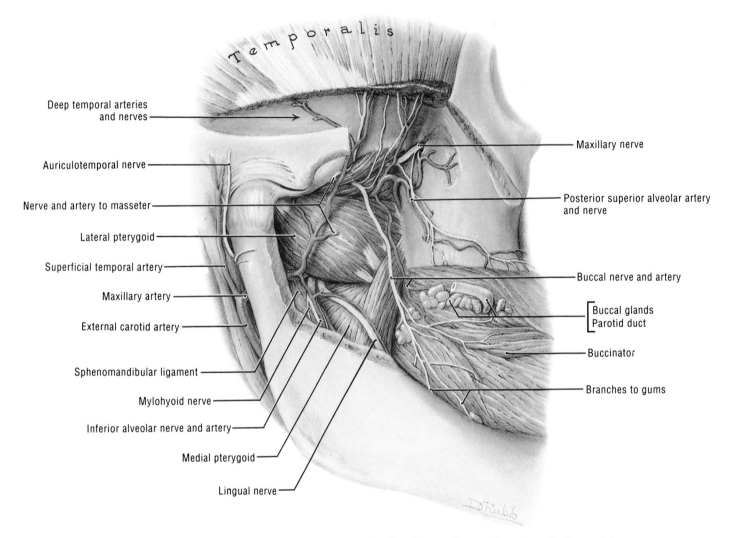

Deep temporal arteries and nerves

Auriculotemporal nerve

Nerve and artery to masseter

Lateral pterygoid

Superficial temporal artery

Maxillary artery

External carotid artery

Sphenomandibular ligament

Mylohyoid nerve

Inferior alveolar nerve and artery

Medial pterygoid

Lingual nerve

Temporalis

Maxillary nerve

Posterior superior alveolar artery and nerve

Buccal nerve and artery

Buccal glands
Parotid duct

Buccinator

Branches to gums

Figure 7-77. Superficial dissection of the infratemporal region. Observe the maxillary artery, the larger of the two end branches of the external carotid, running anteriorly, deep to the neck of the mandible.

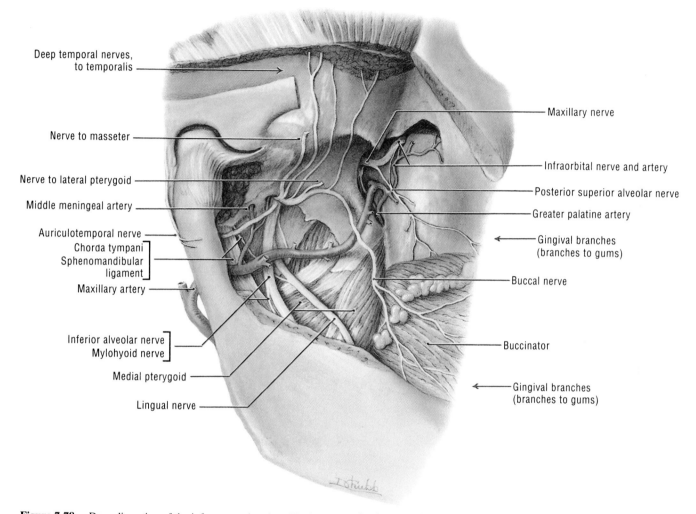

Deep temporal nerves, to temporalis

Nerve to masseter

Nerve to lateral pterygoid

Middle meningeal artery

Auriculotemporal nerve

Chorda tympani

Sphenomandibular ligament

Maxillary artery

Inferior alveolar nerve

Mylohyoid nerve

Medial pterygoid

Lingual nerve

Maxillary nerve

Infraorbital nerve and artery

Posterior superior alveolar nerve

Greater palatine artery

Gingival branches (branches to gums)

Buccal nerve

Buccinator

Gingival branches (branches to gums)

Figure 7-78. Deep dissection of the infratemporal region. The lateral pterygoid muscle and most branches of the maxillary artery have been removed. Observe the maxillary artery and auriculotemporal nerve passing between the sphenomandibular ligament and the neck of the mandible.

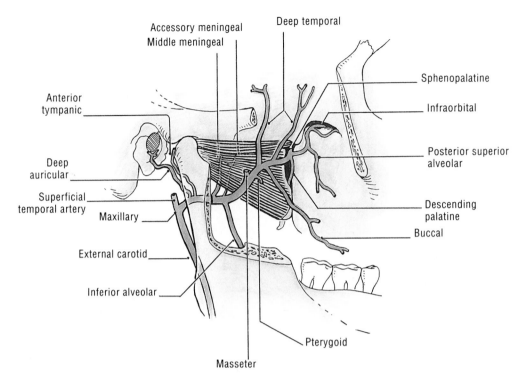

Figure 7-79. The maxillary artery, the larger of the two terminal branches of the external carotid. Observe that it arises at the neck of the mandible and is divided into three parts by the lateral pterygoid muscle.

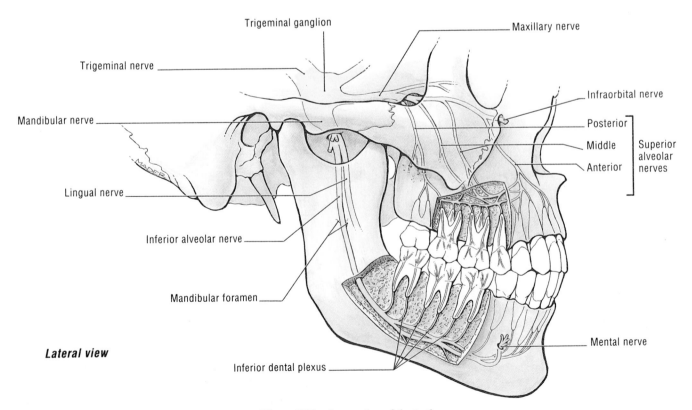

Figure 7-80. Innervation of the teeth.

Table 7-3.
The Branches of the Mandibular Nerve (CN V³)

Muscular Branches	Sensory Branches	Other Branches
Temporalis and masseter	Auriculotemporal	
Medial and lateral pterygoids	Inferior alveolar	
Tensor veli palatini Tensor tympani	Lingual	Articular
Mylohyoid and digastric (anterior belly)	Buccal	

the infratemporal fossa (extraoral approach), where the local anesthetic agent is injected. The following nerves are usually anesthetized: the auriculotemporal, inferior alveolar, lingual, and buccal (Table 7-3).

Inferior Alveolar Nerve Block. This procedure is performed regularly in dental offices. The inferior alveolar nerve is anesthetized by injecting the anesthetic fluid around the mouth of the mandibular canal (*i.e.*, the mandibular foramen). When an inferior alveolar nerve block is successful, all the mandibular teeth are anesthetized to the median plane. The skin and mucous membrane of the lower lip, the labial alveolar mucosa and gingivae, and the skin of the chin are also anesthetized because they are supplied by the mental branch of the inferior alveolar nerve (Figs. 7-25 and 7-80).

The Otic Ganglion (Figs. 7-25 and 7-74). This *parasympathetic ganglion* is located in the infratemporal fossa, just inferior to the foramen ovale, medial to the mandibular nerve, and posterior to the medial pterygoid muscle. Preganglionic parasympathetic fibers, derived mainly from the glossopharyngeal nerve, synapse in the otic ganglion. Postganglionic parasympathetic fibers, which are secretory to the parotid gland, pass from this ganglion to the auriculotemporal nerve.

The Temporomandibular Joint

This articulation is a *modified hinge type of synovial joint* (Figs. 7-81 and 7-82). The articular surfaces involved are: (1) the head or condyle of the mandible inferiorly and (2) the articular tubercle and the mandibular fossa of the squamous part of the temporal bone. An oval fibrocartilaginous **articular disc** divides the joint cavity into superior and inferior compartments. The disc is fused with the articular capsule surrounding the joint, and through this attachment, it is bound superiorly to the limits of the temporal articular surface and inferiorly to the neck of the mandible. The articular disc is more firmly attached to the mandible than it is to the temporal bone. Thus when the head of the mandible slides anteriorly on the articular tubercle as the mouth is opened (Fig. 7-82B), the articular disc slides anteriorly against the posterior surface of the articular tubercle.

The **articular capsule** of the temporomandibular joint is loose. The thin *fibrous capsule* is attached to the margins of the articular

area on the temporal bone and around the neck of the mandible. The **fibrous capsule** is thickened laterally to form the *lateral (temporomandibular) ligament* (Fig. 7-72). It reinforces the lateral part of the capsule. The base of this triangular ligament is attached to the zygomatic process of the temporal bone and the articular tubercle; its apex is fixed to the lateral side of the neck of the mandible. Two other ligaments connect the mandible to the cranium, but neither of them provides much strength to the joint (Figs. 7-77 and 7-78). The *stylomandibular ligament*, a thickened band of deep cervical fascia, runs from the styloid process of the temporal bone to the angle of the mandible and separates the parotid and submandibular salivary glands. The *sphenomandibular ligament*, a remnant of the first branchial or pharyngeal arch cartilage (Meckel's cartilage), is a long membranous band that lies medial to the joint. This ligament runs from the spine of the sphenoid bone to the lingula on the medial aspect of the mandible (Fig. 7-16). The *synovial membrane* lines the fibrous capsule and is reflected superiorly onto the neck of the mandible to the margin of the articular cartilage. In young people this membrane covers both sides of the articular disc (Fig. 7-81), but it gradually wears away and is absent in adults.

Movements of the Temporomandibular Joint (Fig. 7-82; Table 7-4). Two movements occur at this joint, an anterior gliding and a hingelike rotation. When the mandible is depressed during opening of the mouth, the head of the mandible and articular disc move anteriorly on the articular surface until the head lies inferior to the articular tubercle. As this anterior gliding occurs, the head of the mandible rotates on the inferior surface of the articular disc. This permits simple chewing or grinding movements over a small range. During this kind of chewing, both types of movement occur. The axes of these two movements are different. In simple opening of the mouth, the axis of the anterior gliding movement passes approximately through the mandibular foramen, whereas in hingelike rotation, the axis is through the head of the mandible. Because of these axes of movement, the vessels and nerves entering the mandible are not excessively stretched when the mouth is opened widely. In protrusion and retrusion of the mandible, the head and articular disc slide anteriorly and posteriorly on the articular surface of the temporal bone, with both sides moving together. The grinding movement occurs when protrusion and retrusion of the mandible alternate on the two sides.

Muscles Acting on the Temporomandibular Joint

The mandible may be depressed or elevated, protruded or retruded (Table 7-4); considerable rotation also occurs. These movements are controlled mainly by the muscles acting on the joint. The various movements result from the cooperative activity of several muscles bilaterally or unilaterally. Movements of the temporomandibular joint result chiefly from the action of the *muscles of mastication* (Fig. 7-83). The temporalis, masseter, and medial pterygoid muscles produce the biting movement (*i.e.*, elevate the mandible and close the mouth). The mandible is protruded by the lateral pterygoid muscles with help from the medial pterygoids and retruded largely by the posterior fibers of the temporalis muscle. Gravity is sufficient to depress the mandible, but if there is resistance (*e.g.*, when a person is wearing

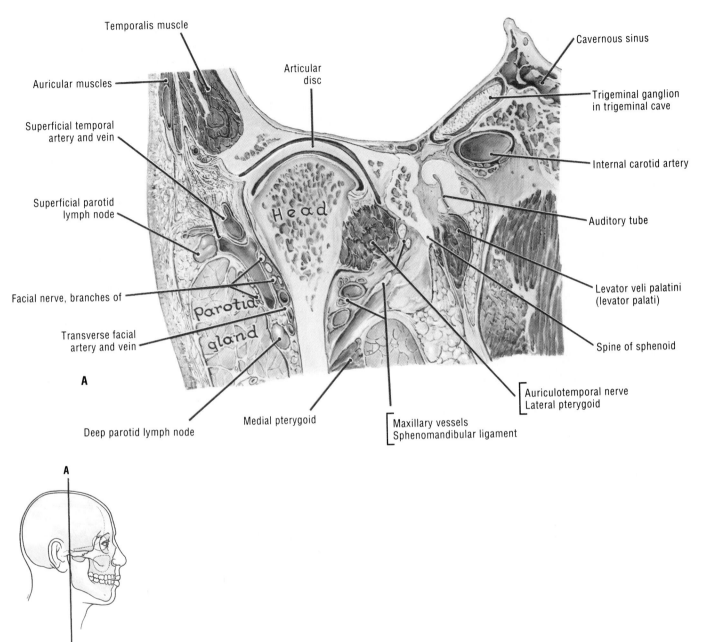

Temporalis muscle

Auricular muscles

Superficial temporal
artery and vein

Superficial parotid
lymph node

Facial nerve, branches of

Transverse facial
artery and vein

A

Articular
disc

Head

Parotid
gland

Cavernous sinus

Trigeminal ganglion
in trigeminal cave

Internal carotid artery

Auditory tube

Levator veli palatini
(levator palati)

Spine of sphenoid

Auriculotemporal nerve
Lateral pterygoid

Maxillary vessels
Sphenomandibular ligament

Deep parotid lymph node

Medial pterygoid

A

B

Figure 7-81. *A*, Coronal section of the temporomandibular joint. Observe the articular disc dividing the joint
cavity into superior and inferior compartments. *B*, Orientation drawing.

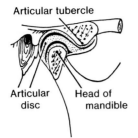

Articular tubercle

Articular disc   Head of mandible

A. Mouth closed

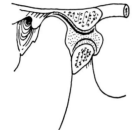

B. Mouth open

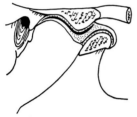

C. Dislocated joint

Figure 7-82.   The changing position of the head of the mandible and the temporal bone. *A*, Mouth closed. *B*, Mouth open. *C*, Dislocated temporomandibular joint.

The temporomandibular joint usually dislocates anteriorly (Fig. 7-82*C*). During normal opening of the mouth, the head of the mandible and the articular disc move anteriorly (Fig. 7-82*B*). During yawning or taking a large bite, contraction of the lateral pterygoid muscles may cause the heads of the mandible to dislocate, *i.e.*, pass anterior to the articular tubercle (Fig. 7-82*C*). In this position, the mandible remains wide open and the person is unable to close it. *Dislocation of the temporomandibular joint* can also occur during extraction of teeth if the mandible is depressed excessively. Most commonly, the joint is dislocated by a blow to the chin when the mouth is open (*e.g.*, when a person is laughing, gaping, or yawning). The displacement is usually bilateral and the mandible projects with the mouth fixed in an open position.

Fractures of the mandible (Fig. 7-19) may be accompanied by dislocation of the temporomandibular joint(s), a possibility that is not to be overlooked in treating these fractures. Because of the close relationship of the facial and auriculotemporal nerves to the temporomandibular joint (Figs. 7-71 and 7-81), care must be taken during operations on the joint to preserve the branches of the facial nerve overlying it and the articular branches of the auriculotemporal nerve that enter the posterior part of the joint. Injury to the articular branches of the auriculotemporal nerve supplying the temporomandibular joint, associated with traumatic dislocation and rupture of the articular capsule and the lateral ligament, leads to joint laxity and instability of the joint.

a chin strap), the muscles listed in Table 7-4 and the mylohyoid and anterior digastric muscles are activated.

The Temporalis Muscle (Figs. 7-72, 7-75, 7-77, 7-81, and 7-83*A*).   This extensive fan-shaped muscle covers the temporal region. It is a powerful masticatory (biting) muscle that can easily be seen and felt during closure of the mandible. Its attachments, nerve supply, and main actions are given in Table 7-5.

The Masseter Muscle (Figs. 7-22, 7-23, 7-71, 7-72, 7-75, and 7-83*B*).   This quadrangular muscle covers the lateral aspect of the ramus and the coronoid process of the mandible. The Greek word *maseter* means masticator or chewer. The attachments, nerve supply, and main actions of this muscle are given in Table 7-5.

The Lateral Pterygoid Muscle (Figs. 7-77, 7-79, 7-81, and 7-83*C*).   This short, thick muscle has two heads of origin. It is a conical muscle with its apex pointing posteriorly. Its attachments, nerve supply, and main actions are given in Table 7-5.

The Medial Pterygoid Muscle (Figs. 7-77, 7-78, 7-81, and 7-83*D*).   This thick, quadrilateral muscle also has two heads of origin that embrace the inferior head of the lateral pterygoid muscle. It is located deep to the ramus of the mandible. Its attachments, nerve supply, and main actions are given in Table 7-5.

The Oral Region

The oral region includes the mouth, teeth, gingivae (gums), tongue, palate, and the region of the palatine tonsil. The mouth or oral cavity is where the ingestion of food occurs and where it is prepared for digestion in the stomach and small intestine (see Fig. 2-26). Food is chewed by the teeth and *saliva* from the salivary glands is added to the food to facilitate the formation of a manageable *bolus* (L. a lump or choice morsel). Deglutition or swallowing is voluntarily initiated in the oral cavity (p. 836). This process pushes the bolus into the *pharynx*, the continuation of the digestive tract (p. 825).

The Oral Cavity

The oral cavity (mouth) consists of two parts: the *vestibule* and the *mouth proper*. The vestibule is the slitlike space between the lips and cheeks and the teeth and gingivae (Figs. 7-84, 7-85, and 7-99). It is the entrance to the digestive tract and is also used in breathing. The vestibule communicates with the exterior through the *orifice of the mouth*, the opening through which food and other substances pass into the oral cavity. The oral cavity is bounded externally by the cheeks and lips. The *roof of the oral cavity* is formed by the **palate** (Fig. 7-86). Posteriorly, *the oral cavity communicates with the oropharynx*, the oral part of the pharynx (p. 834), which is bounded by the soft palate su-

Table 7-4.
Movements of the Mandible and the Muscles Involved

Depression (Open Mouth)	Elevation (Close Mouth)	Protrusion (Protrude Chin)	Retrusion (Retrude Chin)	Side-to-Side Movements (Grinding and Chewing)
Lateral pterygoid	Temporalis	Masseter	Temporalis	Temporalis of same side
Suprahyoid	Masseter	Lateral pterygoid	Masseter	Pterygoids of opposite side
Infrahyoid	Medial pterygoid	Medial pyterygoid		Masseter

Table 7-5.
The Muscles Acting on the Temporomandibular Joint

Muscle	Origin	Insertion	Innervation	Main Actions
Temporalis (Fig. 7-83A)	Floor of temporal fossa and deep surface of temporal fascia	Tip and medial surface of coronoid process and anterior border of ramus of mandible	Deep temporal branches of mandibular n. (CN V^3)	Elevates mandible, closing jaws; its posterior fibers retrude mandible after protrusion
Masseter (Fig. 7-83B)	Inferior border and medial surface of zygomatic arch	Lateral surface of ramus of mandible and its coronoid process	Mandibular n. via masseteric nerve that enters its deep surface	Elevates and protrudes mandible, thus closing jaws; deep fibers retrude it
Lateral pterygoid (Fig. 7-83C)	*Superior head:* infratemporal surface and infratemporal crest of greater wing of sphenoid bone *Inferior head:* lateral surface of lateral pterygoid plate	Neck of mandible, articular disc, and capsule of temporomandibular joint	Mandibular nerve (CN V^3) via lateral pterygoid nerve from anterior trunk, which enters its deep surface	*Acting together*, they protrude mandible and depress chin *Acting alone* and alternately, they produce side-to-side movements of mandible
Medial pterygoid (Fig. 7-83D)	*Deep head:* medial surface of lateral pterygoid plate and pyramidal process of palatine bone *Superficial head:* tuberosity of maxilla	Medial surface of ramus of mandible, inferior to mandibular foramen	Mandibular nerve (CN V^3) via medial pterygoid nerve	Helps to elevate mandible, closing jaws *Acting together*, they help to protrude mandible *Acting alone*, it protrudes side of jaw *Acting alternately*, they produce a grinding motion

periorly, the epiglottis inferiorly, and the palatoglossal arches laterally (see Figs. 8-40 and 8-41).

The Lips

These mobile muscular folds surround the mouth, the entrance to the oral cavity. The lips (L. *labia*) are covered externally by skin and internally by mucous membrane (Figs. 7-21, 7-84, and 7-86). In between these layers are the muscles of the lips, especially the *orbicularis oris muscle*, and the superior and inferior labial arteries (Figs. 7-21 and 7-23). The *labial arteries*, branches of the facial arteries, anastomose with each other to form an arterial ring. The pulsations of these arteries can be palpated by grasping the lip lightly between the first two digits. *Labial salivary glands* are located around the oral vestibule, between the mucous membrane and the orbicularis oris muscle. These small glands resemble mucous salivary glands in structure; their ducts open into the vestibule.

The upper and lower lips are attached to the gingivae in the median plane by raised folds of mucous membrane, called the *labial frenula* (Fig. 7-84A). The junction of the upper lip and cheek is clearly demarcated by the *nasolabial sulcus*, running

laterally from the margin of the nose to the angle of the mouth. This sulcus is particularly obvious during smiling, especially in elderly people. The upper lip extends up to the nose medially and to the nasolabial sulcus laterally. The *mentolabial sulcus* indicates the junction of the lower lip and chin. The upper lip has a median, shallow, vertical groove called the *philtrum* (G. philtron, a love-charm). Each lip has the following four parts: a *cutaneous zone*, a *vermilion border*, a *transitional zone* (red area), and a *mucosal zone*. As the skin of the lip approaches the mouth, it changes color abruptly to red. This is the vermilion border. It is a transitional zone between the skin and the mucous membrane. The transitional zone appears red because of the presence of capillary loops close to the surface, which is composed of extremely thin and hairless skin. In dark skinned people the relatively higher concentrations of melanin deposits make the transitional zone brown rather than red.

The **sensory nerves** of the upper and lower lips are in the *infraorbital* and *mental nerves*, which are branches of the maxillary (CN V^2) and mandibular (CN V^3) nerves, respectively (Figs. 7-24, 7-25, and 7-80). **Lymph vessels** from both lips drain into the *submandibular lymph nodes* (Fig. 7-32). In addition, lymph from the central part of the lower lip drains into

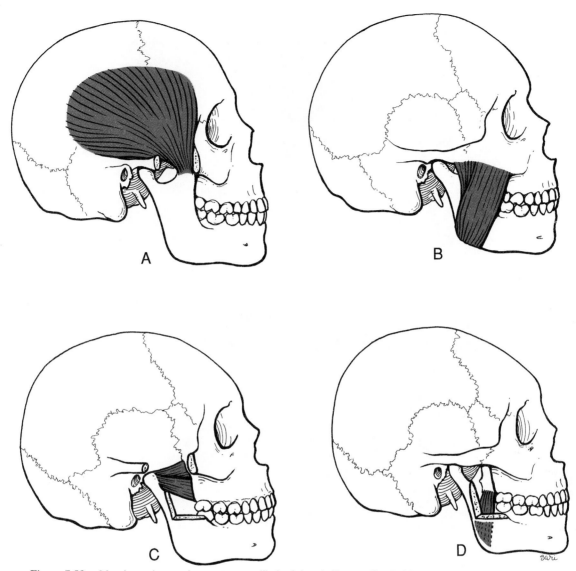

Figure 7-83. Muscles acting on the temporomandibular joint. *A*, Temporalis. *B*, Masseter. *C*, Lateral pterygoid. *D*, Medial pterygoid.

the *submental lymph nodes*. The submental nodes drain to either the submandibular lymph nodes or to the *jugulo-omohyoid node*. The submandibular lymph nodes drain to the deep cervical chain of lymph nodes.

> *Pustules of the upper lip* are potentially dangerous because the infection may extend intracranially through the superior labial vein, angular vein, and supraorbital vein (Fig. 7-23) and enter the *cavernous sinus* (Fig. 7-30 and p. 667).
>
> Persons with *facial palsy* (paralysis of the facial nerve) are unable to whistle because the air blows out through their paralyzed lips on one side. When persons with unilateral facial paralysis are asked to show their teeth, the nasolabial fold does not form on the injured side and the angle of the mouth does not rise. This results from paralysis of the facial muscles (p. 665), including the orbicularis oris, supplied by the facial nerve.
>
> **Cleft lip** (''Hare lip'') is a congenital malformation of the upper lip that occurs about once in 1000 births (Moore, 1988).

> The clefts vary from a small notch in the transitional zone and vermilion border to ones that extend through the lip into the nose. In severe cases the cleft extends deeper and is continuous with a cleft in the palate.
>
> **Carcinoma (Ca) of the lip** accounts for about 15% of all head and neck malignancies. It is more common in the lower lip and in males. A major etiologic (causative) factor of this type of Ca appears to be extensive exposure to intense sunlight.

The Cheeks

The lateral walls of the vestibule of the oral cavity, formed by the cheeks (L. *buccae*), have essentially the same structure as the lips with which they are continuous. In common usage, the cheeks include not only their movable parts, but also the prominences of the cheek over the zygomatic (cheek) bones and

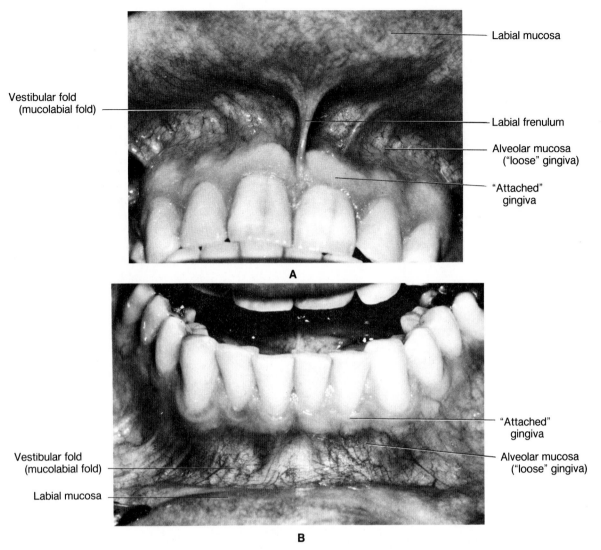

Figure 7-84. The vestibule of the mouth, showing the teeth and gingivae (gums) of the maxilla (*A*), and the mandible, (*B*). (Reprinted with permission from Liebgott B: *The Anatomical Basis of Dentistry*. Philadelphia, BC Decker, Inc., 1986.)

arches (Fig. 7-9). The principal muscular component of the cheeks is the *buccinator muscle* (Figs. 7-22*A*, 7-75, and 7-78). Superficial to the fascia covering this muscle is the *buccal fatpad* that gives the cheeks their rounded contour (Fig. 7-21), particularly in infants. This relatively large fat deposit is thought to reinforce the cheeks in infants and keep them from collapsing during sucking. The lips and cheeks function as a unit (*e.g.*, during sucking, blowing, eating, and kissing). They act as an *oral sphincter* in pushing food from the vestibule to the oral cavity proper. The tongue and buccinator muscle keep the food between the molar teeth during chewing. The buccinator compresses the cheek against the vestibular surfaces of the molar teeth and thereby pushes the food onto the occlusal surfaces of the teeth in opposition to the tongue that pushes the food in the opposite direction.

The buccal glands (Figs. 7-77 and 7-78) are small mucous glands that are situated between the mucous membrane and the buccinator muscle. There are groups of these glands around the terminal part of the *parotid duct*, which runs anteriorly from the

gland, external to the masseter muscle (Figs. 7-22*A* and 7-23). The parotid duct opens on a small papilla on the oral surface of the cheek, opposite the crown of the second maxillary molar tooth. *The sensory nerves of the cheeks* are branches of the maxillary and mandibular nerves (CN V^2 and CN V^3). They supply the skin of the cheek and the mucous membrane lining the cheek (Figs. 7-24, 7-25, and 7-80).

The Gingivae

The gingivae (gums) are composed of fibrous tissue that is covered with mucous membrane (Fig. 7-84). They are firmly attached to the margins of the alveolar processes (tooth sockets) of the jaws and to the necks of the teeth. The *alveolar mucosa* ("loose gingiva") is normally shiny red and nonkeratinizing. The *gingiva proper* ("attached gingiva") is normally pink, stippled, and keratinizing. It is firmly anchored to the underlying alveolar bone and to the necks of the teeth (Figs. 7-84 and 7-88). In addition to their nerve supply from the alveolar nerves supplying the teeth (Fig. 7-25), the gingivae receive nerve fibers from adjacent sensory nerves (buccal, infraorbital, greater palatine, and mental [Figs. 7-22, 7-27, 7-77, and 7-80]).

Improper oral hygiene results in food deposits in tooth and gingival crevices, which may cause inflammation of the gingivae called *gingivitis*. As a result, the gingivae swell and become red. If untreated, the disease spreads to other supporting structures, including the alveolar bone. This produces *periodontitis*, *i.e.*, inflammation and destruction of the alveolar bone and the *periodontal ligament* (p. 739). **Dentoalveolar abscesses** may drain to the oral cavity and lips.

The Teeth

This text does not contain the extensive description of the teeth that is essential for dental students. This information is traditionally taught in dental anatomy courses and excellent texts are available that deal with this area. This text gives a *general description of the teeth and their supporting structures*, as well as information that is essential for understanding their nerve supply and local anesthetic injection techniques.

Ten deciduous teeth (primary or "milk" teeth) begin to develop in the jaws before birth. The first tooth *usually* erupts at 6 to 8 months after birth and the last by 20 to 24 months of age (Table 7-6). As everyone knows, this eruption process is called "teething or cutting teeth." The deciduous teeth are usually shed between 6 and 12 years of age and are replaced by permanent teeth (Figs. 7-18 and 7-84). Eruption of the permanent teeth (normally 16 in each jaw) is usually complete by 18 years of age, except for the third molars ("wisdom teeth"). If they are malposed and/or impacted (Fig. 7-89), they may not erupt. Table 7-6 gives the average times for the eruption and shedding of deciduous teeth and the eruption of permanent teeth.

Parts and Types of Teeth (Figs. 7-84 and 7-87 to 7-89).   Each tooth consists of *three parts*, a crown, neck, and root. The **crown** is the part that projects from the gingiva and occludes (meets) with one or more teeth in the opposite jaw. It has a characteristic appearance that is distinctive for each type of tooth. The **neck** is the part of the tooth between the crown and the root. The **root** is fixed in the *alveolus* (tooth socket) by a fibrous *periodontal ligament*. The number of roots varies, *e.g.*, the incisors and canines have a single root each; the maxillary molars have three roots; the mandibular molars two. The 16 teeth

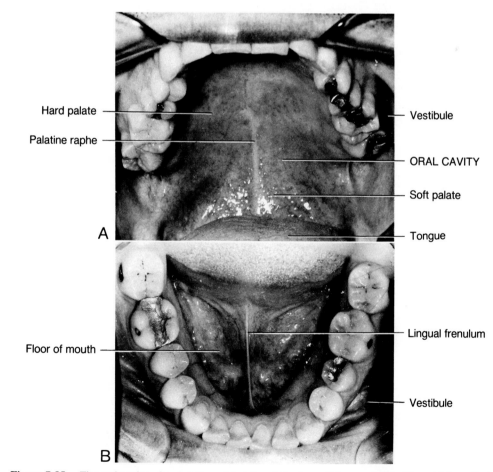

Figure 7-85.   The oral cavity, showing the palate (*A*) and the floor of the mouth (*B*). (From Liebgott B: *The Anatomical Basis of Dentistry*, Philadelphia, BC Decker, Inc., 1986.)

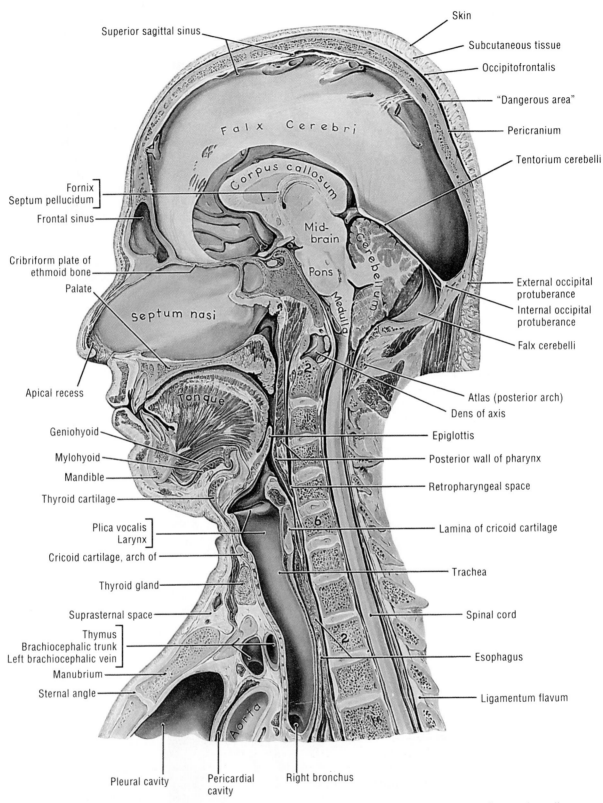

Superior sagittal sinus

Skin

Subcutaneous tissue

Occipitofrontalis

"Dangerous area"

Pericranium

Falx Cerebri

Tentorium cerebelli

Corpus callosum

Fornix
Septum pellucidum

Mid-brain

Cerebellum

Frontal sinus

Pons

Medulla

Cribriform plate of ethmoid bone

External occipital protuberance

Palate

Internal occipital protuberance

Septum nasi

Falx cerebelli

Apical recess

Tongue

Atlas (posterior arch)

Dens of axis

Geniohyoid

Epiglottis

Mylohyoid

Posterior wall of pharynx

Mandible

Retropharyngeal space

Thyroid cartilage

Plica vocalis
Larynx

Lamina of cricoid cartilage

Cricoid cartilage, arch of

Thyroid gland

Trachea

Suprasternal space

Spinal cord

Thymus
Brachiocephalic trunk
Left brachiocephalic vein

Esophagus

Manubrium

Sternal angle

Ligamentum flavum

Aorta

Pleural cavity

Pericardial cavity

Right bronchus

Figure 7-86. Median section of part of the head showing the nasal septum, palate, tongue, pharynx, jaws, lips, and other structures.

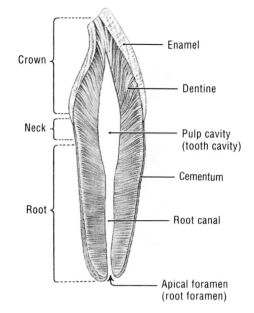

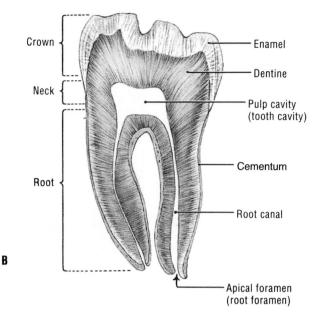

Figure 7-87.   Longitudinal sections of teeth. *A*, Incisor tooth. *B*, Molar tooth. The pulp, containing vessels and nerves, occupies the pulp cavity in vivo. The cementum is attached to the alveolar bone by periodontium to form a fibrous joint between the tooth and its alveolus (socket). The tooth is held firmly in its alveolus by the periodontal ligament.

in each adult jaw consist of four chisellike *incisors* ("cutters"); two tearing teeth called *canines* ("piercers"); four *premolars*; and six *molars* ("grinders").

Structure of a Tooth[20] (Fig. 7-87).   Most of the tooth is composed of *dentin* that is covered by *enamel* over the crown and *cementum* over the root. The *pulp cavity* contains connective tissue, blood vessels, and nerves. It is continuous with the periodontal tissue through the *root canal* and the *apical foramen* (root foramen). The root canals transmit the nerves and vessels to and from the pulp cavity.

Vessels of the Teeth (Figs. 7-31, 7-77, and 7-79).   The *maxillary artery* supplies both the maxillary (upper) and mandibular (lower) teeth. The **arteries** are derived from the posterior, middle, and anterior superior *alveolar branches of the maxillary artery* for the maxillary teeth and the inferior alveolar for the mandibular teeth. These arteries supply gingival, alveolar, and dental branches to the teeth. The dental branch enters the *root canal* through the apical foramen and forms a capillary plexus within the *pulp cavity*. **Veins** with the same names and distribution accompany the arteries and carry blood to the **pterygoid plexus** (Fig. 7-30). **Lymph vessels** from the teeth and gingivae pass mainly to the *submandibular lymph nodes* (Fig. 7-32). A vessel from the anterior part of the mandibular gingiva sometimes passes to a *submental lymph node*, and vessels from the posterior palatal gingiva pass to the *deep cervical lymph nodes*.

Nerves of the Teeth (Figs. 7-25, 7-77, 7-78, and 7-80).   The sensory supply of the maxillary teeth is from branches of the **maxillary nerve**, a division of the trigeminal nerve (*CN V²*). The *posterior superior alveolar nerve* supplies the molar teeth; the *middle superior alveolar nerve* supplies the premolars; and the *anterior superior alveolar nerve* supplies the canine and incisor teeth. The sensory supply of the mandibular teeth is from branches of the *inferior alveolar nerve*, a branch of the mandibular division of the **trigeminal nerve** (*CN V³*).

For purposes of description, incisor and canine teeth have lingual (L. *lingua*, tongue) and labial (L. *labium*, lip) surfaces and incisal (L. *incido*, to cut into) edges. The premolar and molar teeth have buccal (L. *bucca*, cheek), lingual, and occlusal (L. *occlusio*, to close up) surfaces. Teeth vary considerably in shape (Fig. 7-88). The incisors have chisellike edges (Fig. 7-89) and usually the maxillary teeth overlap the mandibular ones. The canines (L. *canis*, a dog) are conical, but they are poorly developed in man compared to carnivorous animals. Because premolars have two cusps (L. points) on the crown, they are often referred to as *bicuspids*, whereas maxillary molars have four cusps and mandibular molars five. The roots of the teeth also vary (Figs. 7-87 and 7-88). Each tooth has a single or multiple roots. The root of a tooth is separated from the cortical plate of its alveolus (socket) by a **periodontal ligament** (membrane). This membrane, composed of densely packed collagen fibers, attaches the tooth to its bony alveolus (socket).

Transverse ridges on the teeth indicate temporary arrests of tooth development. These localized disturbances of calcification can often be correlated with periods of illness, malnutrition, or trauma. *Discoloration of teeth* may follow the administration of **tetracycline antibiotics**. Tetracyclines become incorporated into the teeth and produce a brownish-yellow discoloration. As the enamel is not completely formed on the first and second molar teeth until 8 years of age, tetracycline therapy can affect the teeth of unborn infants and of children up to 8 years old.

[20]For more information on teeth, see Liebgott (1986) or Williams et al. (1989).

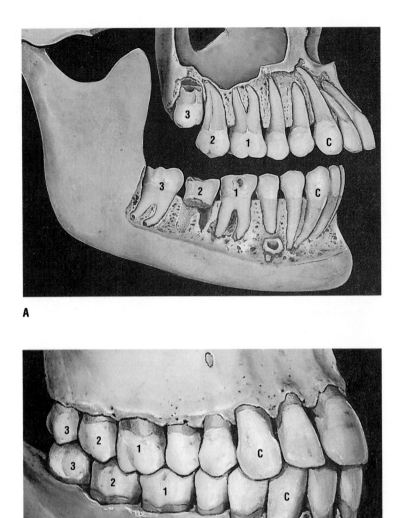

A

B

Figure 7-88. *A,* Permanent teeth on the right side of adult jaws. The alveolar bone has been ground away to expose the roots of the teeth. Note that the maxillary canine tooth (*C*) has the longest root and that the roots of the three maxillary molars are very close to the maxillary sinus. *B,* The teeth in occlusion.

Table 7-6.
The Usual Order and Time of Eruption of the Teeth and the Time of Shedding of the Deciduous Teeth[1]

	Deciduous Teeth				
	Medial Incisor	Lateral Incisor	Canine	First Molar	Second Molar
Eruption (months)[2]	6 to 8	8 to 10	16 to 20	12 to 16	20 to 24
Shedding (years)	6 to 7	7 to 8	10 to 12	9 to 11	10 to 12

	Permanent Teeth							
	Medial Incisor	Lateral Incisor	Canine	First Premolar	Second Premolar	First Molar	Second Molar	Third Molar
Eruption (years)	7 to 8	8 to 9	10 to 12	10 to 11	11 to 12	6 to 7	12	13 to 25

[1]Reproduced with permission from Moore, KL.: *The Developing Human: Clinically Oriented Embryology,* 4th ed. Philadelphia, WB Saunders, 1988.
[2]In some normal infants the first teeth (medial incisors) may not erupt until 12 to 13 months of age.

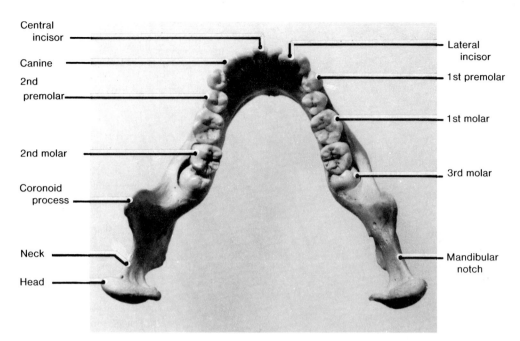

Figure 7-89. Superior view of the mandible showing the permanent teeth. Note that the third molar teeth are malposed (*i.e.*, pointing anteriorly). They probably did not erupt because they are also impacted.

Invasion of the pulp of the tooth by a deep carious lesion ("cavity") results in infection and irritation of the tissues in the pulp cavity. This condition causes an inflammatory process (*pulpitis*). Because the pulp cavity is a rigid space, the swollen pulpal tissues cause considerable pain (**toothache**). If untreated, the small vessels in the root canal may die owing to pressure and the infected material may pass through the apical foramen into the periodontal tissues. *Infection of the pulp* of a tooth may lead to infection of the periodontal ligament, destroying it and the compact layer of bone lining the alveolus. The **dentoalveolar abscess** that forms causes swelling of the adjacent soft tissues ("gum boil"). Pus from abscesses of the maxillary molar teeth may extend into the nasal cavity or the maxillary sinus (Fig. 7-88). The roots of the maxillary molar teeth are closely related to the floor of this sinus. As a consequence, *infection of the pulp cavity may also cause sinusitis*, or sinusitis may stimulate nerves entering the teeth and simulate a toothache. Pus from abscesses of the maxillary canine teeth often penetrates the facial region, just inferior to the medial canthus of the eye (Fig. 7-61). Such swelling may obstruct drainage from the angular vein (Fig. 7-29) and allow infected material to pass through the superior ophthalmic vein to the cavernous sinus (Fig. 7-30).

A tooth may lose its blood supply as a result of trauma. The blow to the tooth disrupts the blood vessels entering and leaving the apical foramen (Fig. 7-87). As people get older their teeth appear to get longer owing to the slow recession of their gingivae; hence, the expression "long in the tooth." Gingival recession exposes the sensitive cementum of the teeth (Fig. 7-87). This process occurs faster in persons who do not have the *tartar* (yellowish-brown deposit) removed from their teeth by a procedure called *scaling*. **Periodontitis** (inflammation of the periodontium) results in inflammation of the gingivae and may result in absorption of alveolar bone and *recession of the gingivae*. As a consequence of gingival recession, exposure of the periodontal ligament occurs allowing microorganisms to invade and destroy it.

The Palate

The palate forms the arched roof of the mouth and the floor of the nasal cavities (Fig. 7-86). Hence it separates the oral cavity from the nasal cavities and the nasal part of the pharynx or nasopharynx. The palate consists of two regions (Figs. 7-86, 7-90 to 7-93, and 7-99): the anterior two-thirds or *bony part*, called the **hard palate**, and the mobile posterior one-third or *fibromuscular part*, known as the **soft palate**. The palate is arched anteroposteriorly and transversely. The arch (vault) of the palate is more pronounced anteriorly in the region of the hard palate. The depth and breadth of the *palatine arch or vault* is subject to considerable variation.

The Hard Palate

The anterior bony part of the palate is formed by the palatine processes of the maxillae and the horizontal plates of the palatine bones (Figs. 7-6, 7-7, 7-85, and 7-86). Anteriorly and laterally, the hard palate is bounded by the alveolar processes and the gingivae. Posteriorly the hard palate is continuous with the soft palate. The *incisive foramen* is the mouth of the *incisive canal*. This foramen is located posterior to the maxillary central incisor teeth. This foramen is the common opening for the right and left incisive canals. The incisive canal and foramen transmit the nasopalatine nerve and the terminal branch of the sphenopalatine artery (Fig. 7-91).

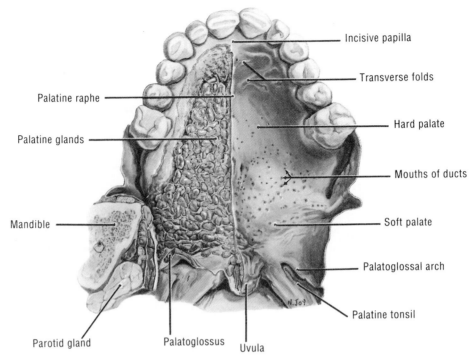

Incisive papilla

Transverse folds

Palatine raphe

Hard palate

Palatine glands

Mouths of ducts

Mandible

Soft palate

Palatoglossal arch

Palatine tonsil

Parotid gland   Palatoglossus   Uvula

Figure 7-90. The palate of an adult. Note that the second and third maxillary molars are missing and that the epithelium has been removed on the right side of the palate. Observe that the soft palate ends posteriorly in the uvula.

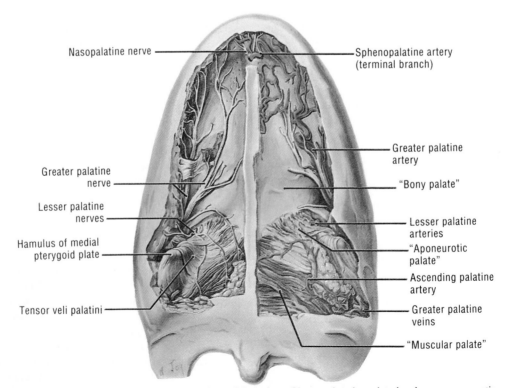

Nasopalatine nerve

Sphenopalatine artery (terminal branch)

Greater palatine nerve

Greater palatine artery

Lesser palatine nerves

"Bony palate"

Hamulus of medial pterygoid plate

Lesser palatine arteries

"Aponeurotic palate"

Ascending palatine artery

Tensor veli palatini

Greater palatine veins

"Muscular palate"

Figure 7-91. Dissection of the inferior surface of the palate. Observe that the palate has bony, aponeurotic, and muscular parts.

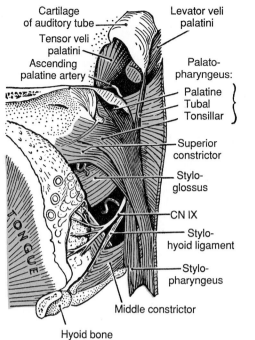

Cartilage
of auditory tube
Tensor veli
palatini
Ascending
palatine artery

Levator veli
palatini

Palato-
pharyngeus:

Palatine
Tubal
Tonsillar

Superior
constrictor

Stylo-
glossus

CN IX

Stylo-
hyoid ligament

Stylo-
pharyngeus

Middle constrictor

Hyoid bone

Figure 7-92. Dissection of the lateral wall of the pharynx showing the muscles of the palate and tongue.

In hard palates of young people, a suture line is visible between the premaxillary part of the maxilla and the palatine processes of the maxillae (Fig. 7-6A). This suture line represents the site of fusion of the median and lateral palatine processes during the 12th week of prenatal development (Moore, 1988). This suture line is absent in old people.

Medial to the third molar tooth, the *greater palatine foramen* (Fig. 7-6) pierces the lateral border of the bony palate. It is the inferior orifice of the *greater palatine canal*. The greater palatine vessels and nerve emerge from this foramen (Figs. 7-91 and 7-93) and run anteriorly in two grooves on the palate. The *lesser palatine foramina* transmit the lesser palatine nerves and vessels to the soft palate and adjacent structures. The hard palate is covered by mucous membrane that is intimately connected to the periosteum (Figs. 7-85, 7-90, and 7-99). Deep to the mucosa are mucus-secreting *palatine glands*. The orifices of the ducts of these glands give the mucous membrane of the palate an orange-peel appearance (Fig. 7-90). The mucous membrane of the anterior part of the hard palate has three or four transverse palatine folds (rugae) of mucous membrane. The *palatine raphe* runs posteriorly from the incisive papilla. It indicates the site of fusion of the palatine processes during the 12th week of development (Moore, 1988).

anesthetized by the same injection. The tissues anesthetized are: the palatal mucosa; the lingual gingivae of the six anterior maxillary teeth; the lingual plate of the alveolar bone; and the hard palate associated with the six maxillary teeth just mentioned. *Injections into the palate are painful* because the mucosa is tightly bound to the hard palate. Because of this, injections are given very slowly.

The Soft Palate

This posterior curtainlike part of the palate has no bony framework (Figs. 7-85, 7-86, 7-90 and 7-91); however, it contains a *membranous aponeurosis*. The soft palate is a movable, fibromuscular fold that is attached to the posterior edge of the hard palate. The soft palate or velum palatinum (L. *velum*, veil) extends posteroinferiorly to a curved free margin from which hangs a conical process, the *uvula* (L. *uva*, grape). At rest the soft palate hangs from the posterior aspect of the palate into the cavity of the pharynx. It separates the nasopharynx superiorly from the oropharynx inferiorly. During swallowing the soft palate moves posteriorly against the wall of the pharynx, thereby preventing regurgitation of food into the nasal cavities. Laterally the soft palate is continuous with the wall of the pharynx and is joined to the tongue and pharynx by the palatoglossal and palatopharyngeal arches, respectively (Fig. 7-90). The *palatine tonsil* is located in the triangular interval between the palatoglossal and palatopharyngeal arches (Figs. 7-90 and 7-95). Deep to the palatal mucosa are mucous glands (see Fig. 8-48). The soft palate is strengthened by the **palatine aponeurosis**, formed by the expanded tendon of the tensor veli palatini muscle (Figs. 7-91 and 7-92). This aponeurosis, attached to the posterior margin of the hard palate, is thick anteriorly and very thin posteriorly. The anterior part of the soft palate consists mainly of the aponeurosis, whereas its posterior part is muscular.

Muscles of the Soft Palate (Figs. 7-91 to 7-93 and 7-95). The five muscles of the fibromuscular soft palate arise from the base of the skull and descend to the palate. They produce various movements of this fibromuscular fold (Table 7-7). Its posterior part may be raised so that it becomes a horizontal continuation of the plane of the hard palate and is in contact with the posterior wall of the pharynx. The soft palate can also be drawn inferiorly so that its inferior surface is in contact with the posterior part of the superior surface of the tongue. During quiet breathing the soft palate is curved posteroinferiorly (Fig. 7-86; see also Fig. 8-50).

The Levator Veli Palatini (Levator Palati). The attachments, nerve supply, and main action of this cylindrical muscle are given in Table 7-7. It runs inferoanteriorly, spreading out in the soft palate, where it attaches to the superior surface of the palatine aponeurosis (Figs. 7-92 and 7-93). As its name indicates, this muscle *elevates the soft palate*, drawing it superiorly and posteriorly. It also opens the auditory tube to equalize air pressure in the middle ear and pharynx.

The Tensor Veli Palatini (Tensor Palati). The attachments, nerve supply, and actions of this thin, triangular muscle are given in Table 7-7. As it passes inferiorly, the tendon of this muscle hooks around the hamulus of the medial pterygoid plate (Fig. 7-76B) before inserting into the palatine aponeurosis (Figs. 7-91

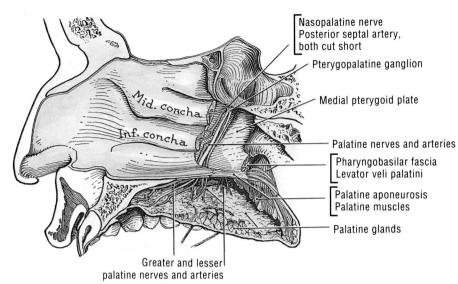

Nasopalatine nerve
Posterior septal artery,
both cut short

Pterygopalatine ganglion

Medial pterygoid plate

Palatine nerves and arteries

Pharyngobasilar fascia
Levator veli palatini

Palatine aponeurosis
Palatine muscles

Palatine glands

Mid. concha

Inf. concha

Greater and lesser
palatine nerves and arteries

Figure 7-93.   Dissection of the lateral wall of the nasal cavity and the palate. The mucous membrane, containing a layer of mucous glands, has been separated by blunt dissection. The posterior ends of the middle and inferior conchae are cut through; these and the mucoperiosteum are pulled off the side wall of the nose as far as the posterior border of the medial pterygoid plate. The perpendicular plate of the palatine bone was broken through to expose the palatine nerves and arteries.

Table 7-7.
The Muscles of the Soft Palate

Muscle	Superior Attachment	Inferior Attachment	Innervation	Main Actions
Levator veli palatini (Fig. 7-92)	Cartilage of auditory tube and petrous part of temporal bone	Palatine aponeurosis	Pharyngeal branch of vagus via pharyngeal plexus	Elevates soft palate during swallowing and yawning
Tensor veli palatini (Fig. 7-92)	Scaphoid fossa of medial pterygoid plate, spine of sphenoid bone, and cartilage of auditory tube		Medial pterygoid n. (a branch of the mandibular n.) via otic ganglion	Tenses soft palate and opens mouth of auditory tube during swallowing and yawning
Palatoglossus (Fig. 7-90)	Palatine aponeurosis	Side of tongue	Cranial part of CNXI through pharyngeal branch of vagus (CNX) via pharyngeal plexus	Elevates posterior part of tongue and draws soft palate onto tongue
Palatopharyngeus (Fig. 7-92)	Hard palate and palatine aponeurosis	Lateral wall of pharynx		Tenses soft palate and pulls walls of pharynx superiorly, anteriorly, and medially during swallowing
Musculus uvulae (Fig. 7-93)	Posterior nasal spine and palatine aponeurosis	Mucosa of uvula		Shortens uvula and pulls it superiorly

and 7-92). As its Latin name indicates, this muscle *tenses the soft palate* by using the hamulus as a pulley. It also pulls the membranous portion of the auditory tube open to equalize air pressure in the middle ear and pharynx. Acting together, the levator veli palatini muscles raise the soft palate as the tensor veli palatini muscles tense it. This forces the soft palate against the posterior wall of the pharynx, a movement that occurs during swallowing (p. 836).

The Palatoglossus Muscle (Fig. 7-90). The attachments, nerve supply, and actions of this slender slip of muscle are given in Table 7-7. It passes laterally and inferiorly in the palatoglossal arch and attaches to the side of the tongue. This muscle, covered

with mucous membrane, forms the *palatoglossal arch*. The palatoglossus elevates the posterior part of the tongue and draws the soft palate inferiorly on to the tongue. These movements close off the oral cavity from the oral part of the pharynx (*i.e.*, they close the oropharyngeal isthmus).

The Palatopharyngeus Muscle. The attachments, nerve supply, and actions of this thin, flat muscle are given in Table 7-7. This muscle, covered with mucous membrane, forms the *palatopharyngeal arch* (see Fig. 8-49). The palatopharyngeus muscle passes posteroinferiorly in this arch. It tenses the soft palate and pulls the walls of the pharynx superiorly, anteriorly and medially during swallowing.

The Musculus Uvulae (Fig. 7-93). The attachments, nerve supply, and actions of this delicate slip of muscle are given in Table 7-7. It passes posteriorly on each side of the median plane and inserts into the mucosa of the uvula. When the uvular muscle contracts, it shortens the uvula and pulls it superiorly. This movement assists in closing the nasopharynx during swallowing.

Nerves of the Palate

The sensory nerves of the palate, which are branches of the *pterygopalatine ganglion* (Figs. 7-25, 7-93, and 7-104), are the greater and lesser **palatine nerves**. They accompany the arteries through the greater and lesser palatine foramina, respectively. The *greater palatine nerve* supplies the gingivae, mucous membrane, and glands of the hard palate. The *lesser palatine nerve* supplies the soft palate. Another branch of the pterygopalatine ganglion, the *nasopalatine nerve*, emerges from the incisive foramen (Figs. 7-6, 7-25, 7-91, and 7-93) and supplies the mucous membrane of the anterior part of the hard palate.

Vessels of the Palate

The palate has a rich blood supply from branches of the **maxillary artery** (Figs. 7-28, 7-77, 7-79, and 7-91), chiefly the *greater palatine artery*, a branch of the descending palatine artery. This artery passes through the greater palatine foramen and runs anteriorly and medially. The *lesser palatine artery* enters through the lesser palatine foramen and anastomoses with a branch of the facial artery, the *ascending palatine artery* (Fig. 7-91).

The Tongue

The tongue (L. *lingua*; G. *glossa*) is a highly mobile muscular organ that can vary greatly in shape. It consists of three parts, a *root*, *body*, and *tip*. The root, its posterior part, is attached mainly to the floor of the mouth. At rest the tongue fills most of the mouth proper (Figs. 7-86, 7-94, and 7-99). The tongue is concerned with mastication, taste, deglutition (swallowing), articulation (speech), and oral cleansing, but its main functions are squeezing food into the pharynx when swallowing and forming words during speaking. The tongue is situated partly in the mouth and partly in the oropharynx. It is mainly composed of muscles and is covered by a mucous membrane on its dorsum, tip, and sides.

Gross Features of the Tongue (Fig. 7-95). The dorsum of the tongue is divided by a V-shaped *sulcus terminalis* into anterior oral (presulcal) and posterior pharyngeal (postsulcal) parts. The apex of the V is posterior and its two limbs diverge anteriorly. The *oral part* forms about two-thirds and the *pharyngeal part* or root about one-third of the dorsum of the tongue. At the apex of the terminal sulcus is a small median depression, the *foramen cecum* (Figs. 7-94 and 7-95); this pit is the remnant of the opening (point of origin) of the embryonic *thyroglossal duct* (Moore, 1988).

The oral part of the tongue is freely movable, but it is loosely attached to the floor of the mouth by the *lingual frenulum* (Fig. 7-85). On each side of the frenulum is a *deep lingual vein*, visible as a blue line. It begins at the tip of the tongue and runs posteriorly. The lingual veins lie close to the thin, transparent mucous membrane on the inferior or *sublingual surface* of the tongue. All the veins on one side of the tongue unite at the posterior border of the hyoglossus muscle to form the **lingual vein**, which joins either the facial vein or the internal jugular vein (Figs. 7-29 and 7-43). The inferior surface and sides of the tongue are covered with smooth, thin mucous membrane. On the dorsum of the oral part of the tongue is a *median groove* (Fig. 7-95). This groove, inconspicuous in some people, represents the site of fusion of the distal tongue buds during embryonic development (Moore, 1988). The dorsal lingual veins drain the dorsum and sides of the tongue and join the lingual veins, which accompany the lingual artery.

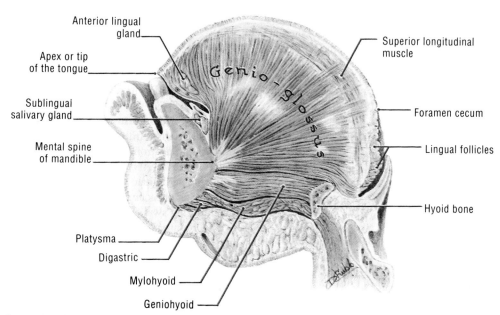

Figure 7-94. Median section of the tongue and the floor of the mouth. Observe that the tongue is composed mainly of muscle. The extrinsic muscles alter the position of the tongue and the intrinsic ones alter its shape.

to the pterygopalatine ganglion (Figs. 7-104 and 7-105). The artery breaks up into branches that accompany all the nerves in the fossa with the same names. The *posterior superior alveolar artery* arises from the maxillary artery as it enters the pterygopalatine fossa (Fig. 7-77). It descends on the infratemporal surface of the maxilla and supplies the molar and premolar teeth and the lining of the maxillary sinus (Figs. 7-88 and 7-103). The *infraorbital artery* (Figs. 7-78 and 7-105), often arising in conjunction with the posterior superior alveolar artery, enters the orbital cavity. It passes anteriorly in the infraorbital groove and canal (Fig. 7-105). While in the canal, it gives off the *anterior superior alveolar artery*. The infraorbital artery exits onto the face through the infraorbital foramen and sends branches to the lower eyelid, the lateral aspect of the external nose, and the upper lip.

The Nose

The nose is the superior part of the respiratory tract and contains the peripheral organ of smell (Fig. 7-106). It is divided into right and left nasal cavities by the *nasal septum* (Fig. 7-107). Each naris is divisible into an *olfactory area* and a *respiratory area*. The **functions of the nose** and nasal cavities are: (1) respiration, (2) olfaction, (3) filtration of dust, (4) humidification of inspired air, and (5) reception of secretions from the *paranasal sinuses* (Figs. 7-55 and 7-58) and the *nasolacrimal ducts* (Figs. 7-64 and 7-67).

The External Nose (Figs. 7-2, 7-6, 7-86, and 7-107).   Noses vary considerably in size and shape, mainly as a result of differences in the nasal cartilages and the depth of the glabella. The nose projects anteroinferiorly from the face to which its root is joined just inferior to the forehead. The dorsum of the nose extends from the root to its apex (tip). The inferior surface of the nose is pierced by two apertures, called the *anterior nares* (L. nostrils), which are separated from each other by the nasal septum (septum nasi). Each naris is bounded laterally by an *ala* (L. wing), *i.e.*, the side of the nose. The *posterior nares* apertures or *choanae* open into the nasopharynx (Fig. 7-6B).

Structure of the Nasal Septum (Figs. 7-2, 7-86, and 7-107). The partly bony and partly cartilaginous nasal septum divides the chamber of the nose into two narrow nasal cavities. The bony part of the septum is usually located in the median plane until 7 years of age; thereafter, it often deviates slightly to one or other side, more frequently to the right. The nasal septum has three main components: (1) the *perpendicular plate of the ethmoid* bone; (2) the *vomer*, and (3) the *septal cartilage*. The perpendicular plate, forming the superior part of the septum, is very thin and descends from the *cribriform plate* of the ethmoid bone. The vomer is a thin, flat bone that forms the posteroinferior part of the septum. It articulates with the sphenoid, maxilla and palatine bones.

The Skeleton of the Nose (Figs. 7-8, 7-9, and 7-107).   The immovable bridge of the nose, the *superior bony part of the nose*, consists of the nasal bones, the frontal processes of the maxillae, and the nasal part of the frontal bone. The movable *cartilaginous part of the nose* consists of five main cartilages and a few smaller ones. The nasal cartilages are composed of hyaline cartilage and are connected with one another and the nasal bones by the continuity of the perichondrium and periosteum, respectively. The U-shaped *alar nasal cartilages* are free and movable. They dilate or constrict the external nares when the muscles acting on the external nose contract (p. 660).

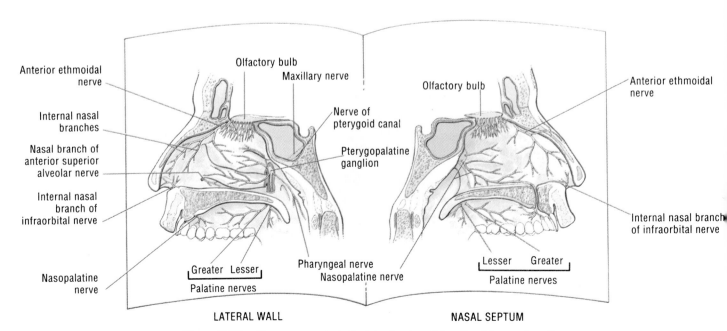

Figure 7-106.   The nerve supply to the lateral and medial walls of the nasal cavity.

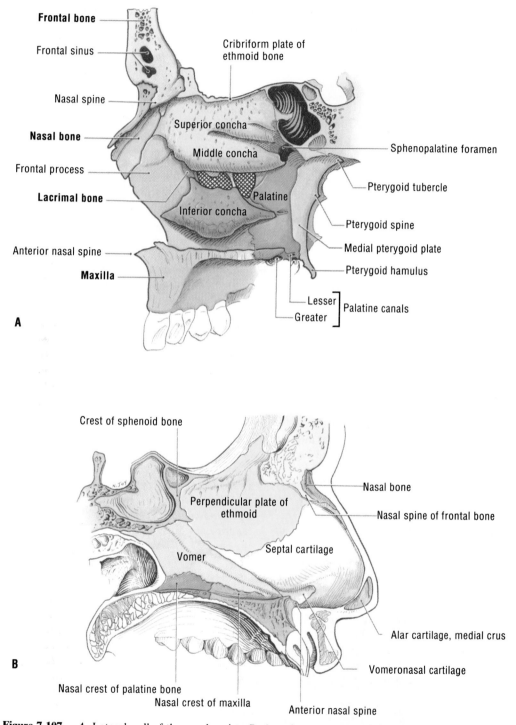

Figure 7-107. *A*, Lateral wall of the nasal cavity. *B*, Anterior part of the bisected skull showing the nasal septum. Note that the septum has a hard part (here partly bony and partly cartilaginous) and a soft or mobile part.

Fractures of the nose are quite common because its bony parts, the perpendicular plate of the ethmoid bone and the vomer, are thin bones. Usually the fractures are transverse (Fig. 7-20). If the injury results from a direct blow, the cribriform plate of the ethmoid bone may be fractured. The nasal septum may be displaced or deviate from the median plane as the result of a birth injury or a congenital malformation. More often, however, the deviation is caused by postnatal trauma (*e.g.*, during a fist fight). Sometimes the deviation is so severe that the nasal septum comes into contact with the lateral wall of the nasal cavity. As this obstructs breathing, surgical repair is usually necessary.

The Nasal Cavities

The nasal cavities are entered through the nostrils or *anterior nares*. They open into the nasopharynx through the *choanae* (Fig. 7-6*B*). Mucosa lines the entire nasal cavity, except the *vestibule* (Figs. 7-55 and 7-109). The entrance to the vestibule of the nose is lined with skin on both the nasal septum side and the lateral wall. The skin has hairs called *vibrissae*, a name derived from the French verb *vibro*, meaning to quiver.

The nasal mucosa (mucous membrane) is firmly bound to the periosteum and perichondrium of the supporting structures of the nose (Fig. 7-86). It is continuous with the lining of all the chambers with which the nasal cavities communicate: the nasopharynx posteriorly, the paranasal sinuses superiorly and laterally, and the lacrimal sac and conjunctiva of the orbit (Figs. 7-64 and 7-67). The nasal septum is covered with mucous membrane, except for the part in the vestibule. The inferior two-thirds of the nasal mucosa is called the *respiratory area* and the superior one-third is the *olfactory area* (Fig. 7-106). Air passing over the respiratory area is warmed and moistened before it passes through the rest of the upper respiratory tract to the lungs (see Fig. 7-14).

The Olfactory Area of the Nasal Mucosa (Figs. 7-106 and 7-108).   Yellowish in living persons, this area contains the peripheral organ of smell. Sniffing draws air to the olfactory area. The functional cells concerned with olfaction (L. *olfacere*, to smell) are called *olfactory receptor cells*. They are located in the mucosa of the olfactory area of the nose. The axons of the olfactory cells, constituting the *olfactory nerves* (CN I), are col-

lected into 18 to 20 bundles that pass through the foramina in the *cribriform plate of the ethmoid bone* in the anterior cranial fossa (Figs. 7-11, 7-12, 7-106, and 7-108). These nerves are unmyelinated, and after traversing foramina in the cribriform plate, they pierce the dura and arachnoid covering the brain. Surrounded by a thin tube of *pia*, the olfactory nerves cross the *subarachnoid space* containing CSF and enter the inferior aspects of the *olfactory bulbs* (Fig. 7-51 and 7-108). The olfactory nerves are described in more detail in Chapter 9.

The olfactory cell is a primitive type of sensory receptor. Its dendrite extends to the surface of the olfactory epithelium and ends as an exposed bulbous enlargement, which has cilia up to 100 μ long. Hence, an odoriferous substance has direct access to the neuron without the intervention of non-nervous tissue. It has been estimated that there are about 25 million olfactory cells in each half of the olfactory mucosa of young adults. The number of cells is reduced in old people.

Although nasal discharges are commonly associated with upper respiratory tract infections, a nasal discharge associated with a head injury may be **CSF**. This condition, known as *CSF rhinorrhea*, results from fracture of the cribriform plate, tearing of the meninges, and leakage of CSF (p. 701). This condition may be diagnosed by injecting a radioactive tracer into the CSF and later detecting it in pieces of cotton that have been inserted into the nostrils.

Anosmia (loss of the sense of smell) is not a serious handicap and occurs gradually with age. Olfactory cells degenerate at the rate of about 1% per year in most elderly people. Disorders of olfaction may result from conditions affecting: (1) the olfactory receptor cells in the nasal mucous membrane; (2) the secondary olfactory neurons in the olfactory bulb and tract; or (3) their intracranial connections (Fig. 7-108).

Nerves of the Respiratory Area of the Nasal Mucosa (Figs. 7-25, 7-27, and 7-106).   The inferior two-thirds of the nasal mucosa are supplied chiefly by the **trigeminal nerve** (*CN V*). The mucous membrane of the nasal septum is supplied chiefly by the *nasopalatine nerve*, a branch of the maxillary division of the trigeminal nerve (*CN V²*). Its anterior portion is supplied by the *anterior ethmoidal nerve* (a branch of the *nasociliary nerve* (Fig. 7-46), which is derived from the ophthalmic division of the trigeminal nerve (*CN V¹*). The lateral wall of the nasal cavity

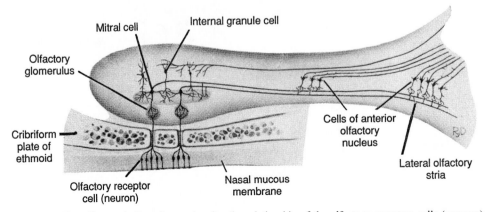

Figure 7-108.   The olfactory bulb and tract showing the relationship of the olfactory receptors cells (neurons) in the nasal mucosa with cells in the olfactory bulb of the brain (see also Figs. 7-51*A* and 7-106).

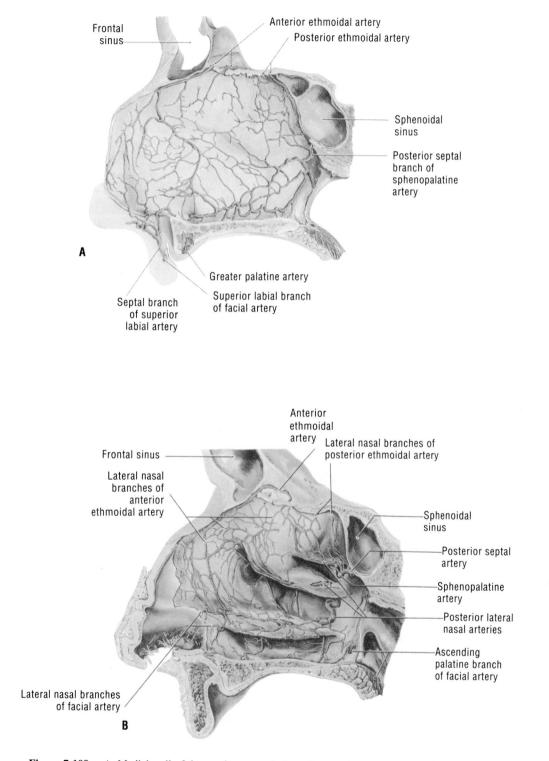

Figure 7-109. *A*, Medial wall of the nasal septum. *B*, Lateral wall of the nasal cavity. Observe that the sphenopalatine artery provides the main blood supply.

is supplied by nasal branches of the maxillary nerve (CN V²); the *greater palatine nerve*, and the *anterior ethmoidal nerve*.

Arteries of the Nasal Mucosa (Figs. 7-69A, 7-76B, 7-78, 7-79, 7-91, and 7-109). The rich blood supply of the mucosa of the nasal septum is derived mainly from the **maxillary artery**. The **sphenopalatine artery**, a branch of the maxillary artery, supplies most of the blood to the nasal mucosa. It enters by the *sphenopalatine foramen* and sends branches to posterior regions of the lateral wall and to the nasal septum. The *greater palatine artery*, another branch of the maxillary artery, passes through the incisive foramen to supply the nasal septum. The sphenopalatine and greater palatine arteries anastomose in the anteroinferior part of the nasal septum. The anterior and posterior **ethmoidal arteries**, branches of the *ophthalmic artery*, supply the anterosuperior part of the mucosa of the lateral wall of the nasal cavity and the nasal septum. Three *branches of the facial artery* (superior labial, ascending palatine, and lateral nasal) also supply anterior parts of the nasal mucosa.

Veins of the Nasal Mucosa (Figs. 7-30, 7-47B, and 7-69B). The veins of the nasal cavity form a rich venous network or plexus in the connective tissue of the nasal mucosa, especially over the inferior part of the septum. Some of the veins open into the sphenopalatine vein and drain to the *pterygoid plexus*. Others join the facial and infraorbital veins, and some empty into the ophthalmic veins and drain into the cavernous sinus.

The nasal mucosa becomes swollen and inflamed (rhinitis) during upper respiratory infections and with some allergies (e.g., hayfever). The Greek word rhis (rhin-) means nose; hence, rhinitis indicates inflammation of the nasal mucous membrane. Swelling of this membrane occurs readily because of its vascularity. **Infections of the nasal cavities** *may spread to: (1) the anterior cranial fossa through the cribriform plate of the ethmoid bone (Fig. 7-108); (2) the nasopharynx and retropharyngeal soft tissues (Fig. 7-86); (3) the middle ear through the auditory tube (Fig. 7-55); (4) the paranasal sinuses (Fig. 7-58); and (5) the lacrimal apparatus and conjunctiva (Figs. 7-64 and 7-67).*

Epistaxis (nasal hemorrhage; nosebleed) is relatively common owing to the richness of the blood supply to the nasal mucosa (Fig. 7-109). In most cases the cause is trauma and the bleeding is located in the anterior third of the nose. Mild epistaxis often results from nose picking, which tears the veins in the vestibule of the nose around the anterior nares. However, epistaxis is also associated with infections and hypertension. Spurting of blood from the nose results from the rupture of arteries, particularly at the site of anastomosis of the sphenopalatine and greater palatine arteries (Fig. 7-109). If nasal bleeding is so profuse that it cannot be stopped by usual treatments, the external carotid arteries are sometimes clamped in the neck. This is the source of the blood passing to the nose through the branches of the maxillary arteries (Fig. 7-79).

The Roof and Floor of the Nasal Cavity. The roof is curved and narrow, except at the posterior end. It is divided into three parts, frontonasal, ethmoidal, and sphenoidal, which are derived from the bones that form them. The floor is wider than the roof. It is formed by the palatine process of the maxilla and the horizontal plate of the palatine bone (Fig. 7-7).

The Walls of the Nasal Cavity (Figs. 7-55, 7-58, 7-106,

7-107, and 7-110 to 7-112). The medial wall is formed by the nasal septum; it is usually smooth. The lateral wall is uneven owing to the three longitudinal, scroll-shaped elevations, called **conchae** (L. shells) or turbinates (L. shaped like a top). These elevations are called superior, middle, and inferior nasal conchae, according to their position on the lateral wall of the nasal cavity. The superior and middle conchae are parts of the *ethmoid bone*, whereas the inferior concha is a separate bone. The inferior and middle conchae project medially and inferiorly, producing air passageways called the inferior and middle **meatus**[23] (L. passage). The short superior concha conceals the superior meatus. The space posterosuperior to the superior concha and into which the sphenoidal sinus opens is called the *sphenoethmoidal recess*.

The **superior meatus** (Fig. 7-55) is a narrow passageway between the superior and middle nasal conchae into which the posterior ethmoidal sinuses open by one or more orifices (Figs. 7-58 and 7-110). The **middle meatus** (Figs. 7-55 and 7-58) is longer and wider than the superior one. The anterosuperior part of this meatus leads into a funnel-shaped opening, called the *infundibulum* (Fig. 7-112), through which it communicates with the frontal sinus. The passage that leads inferiorly from the frontal sinus to the infundibulum of the middle meatus is called the *frontonasal duct*. There is one duct for each frontal sinus, and because there may be several frontal sinuses on each side, there may be several frontonasal ducts. The **inferior meatus** (Figs. 7-55 and 7-58) is a horizontal passage, inferolateral to the inferior nasal concha. The *nasolacrimal duct* opens into the anterior part of this meatus (Figs. 7-64 and 7-67). Usually the orifice of this duct is wide and circular (Fig. 7-111).

When the middle concha is raised or removed, a rounded elevation, called the *ethmoidal bulla* (L. bubble), is visible. This elevation is formed by the *middle ethmoidal cells* (Figs. 7-59 and 7-111) that constitute the *ethmoidal sinuses* (Fig. 7-58). The middle ethmoidal cells open on the surface of the ethmoidal bulla. Inferior to this bulla is a semicircular groove called the *hiatus semilunaris* (Fig. 7-111). The frontal sinus opens into this hiatus anterosuperiorly. Near the hiatus are the openings of the anterior ethmoidal sinuses (Fig. 7-110). The maxillary sinus opens into the middle meatus (Figs. 7-58 and 7-99).

The Paranasal Sinuses

These sinuses are *pneumatic or air-filled extensions of the respiratory part of the nasal cavity* into the following cranial bones: frontal, ethmoid, sphenoid, and maxilla (Figs. 7-13, 7-14, 7-55, 7-57B, 7-58, and 7-110). They are named according to the bones in which they are located as follows: frontal, ethmoidal, sphenoidal, and maxillary sinuses.

Relationship of the Paranasal Sinuses to the Orbit (Figs. 7-55, 7-57B, 7-58, 7-103, and 7-110). The frontal sinus is superior; the maxillary sinus inferior; the ethmoidal sinus medial; and the sphenoidal sinus posterior to the orbit. The sinuses are lined with mucous membrane that is continuous with that of the nasal cavities. However, the mucosa of the sinuses is thinner,

---

[23]The plural of meatus is the same as the singular, although it is not uncommon to see ''meatuses.''

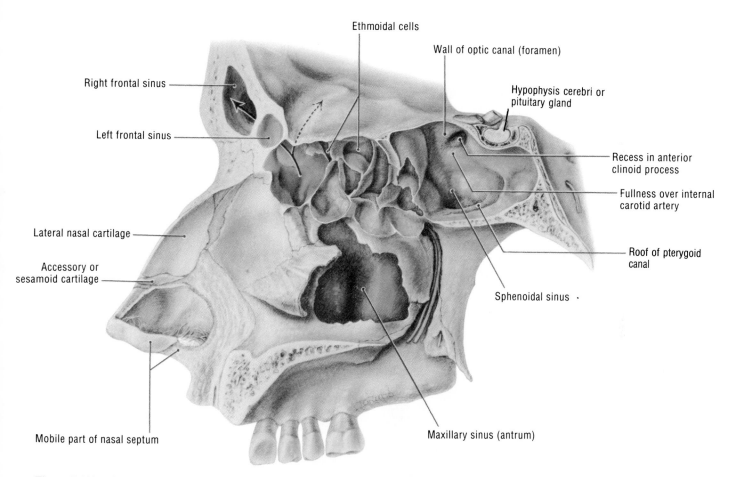

Ethmoidal cells

Wall of optic canal (foramen)

Right frontal sinus

Hypophysis cerebri or
pituitary gland

Left frontal sinus

Recess in anterior
clinoid process

Fullness over internal
carotid artery

Lateral nasal cartilage

Accessory or
sesamoid cartilage

Roof of pterygoid
canal

Sphenoidal sinus

Mobile part of nasal septum

Maxillary sinus (antrum)

Figure 7-110.   Sagittal section of the nasal cavity and palate. Bone has been removed to show the paranasal sinuses (opened). Collectively the ethmoidal cells are called the ethmoidal sinus. An anterior ethmoidal cell (*pink*) is invading the diploë of the frontal bone to become a frontal sinus. An offshoot (*broken arrow*) invades the orbital plate of the frontal bone. The sphenoidal sinus in this specimen is very extensive, extending (1) posteriorly, inferior to the hypophysis cerebri (pituitary gland), to the clivus (Fig. 7-40); (2) laterally, inferior to the optic nerve, into the anterior clinoid process; and (3) inferior to the pterygoid process but leaving the pterygoid canal and rising as a ridge on the floor of the sinus. The maxillary sinus is pyramidal in shape.

less vascular, and not so adherent to the bony walls as is the nasal mucosa. Mucus secreted by the glands in the mucosa of the sinuses passes into the nasal cavities through the openings or ostia (obscured by the conchae; p. 758) in their lateral walls (Fig. 7-111).

The paranasal sinuses develop as outgrowths from the nasal cavities mainly after birth (Moore, 1988). The original openings of the outgrowths persist as orifices in the nasal cavities. Hence, all the paranasal sinuses drain directly or indirectly into the nasal cavities. Secretions from the mucosa of the sinuses eventually drain through these openings into the nasal cavity (Figs. 7-110 and 7-111). The mucosal lining of the sinuses is also continuous with the nasal mucosa; this results from their origin as outgrowths from the nasal cavities.

The paranasal sinuses vary considerably in size and form in different people and in different races (*e.g.*, the frontal sinuses are generally small in Oriental people). Most of the sinuses are rudimentary or absent in newborn infants. No frontal or sphenoidal sinuses are present at birth, but there are usually a few ethmoidal cells and tiny maxillary sinuses. These sinuses enlarge during childhood. The frontal and sphenoidal sinuses develop during childhood and adolescence. Growth of the paranasal sinuses is important in altering the size and shape of the face during infancy and childhood, and in adding resonance to the voice during adolescence.

The Frontal Sinuses

These air chambers are located between the outer and inner tables of the frontal bone, posterior to the superciliary arches and the root of the nose (Figs. 7-1, 7-103, 7-107, and 7-109 to 7-112). The size of the superciliary arches varies in degree of development; however, the prominence of the superciliary arches

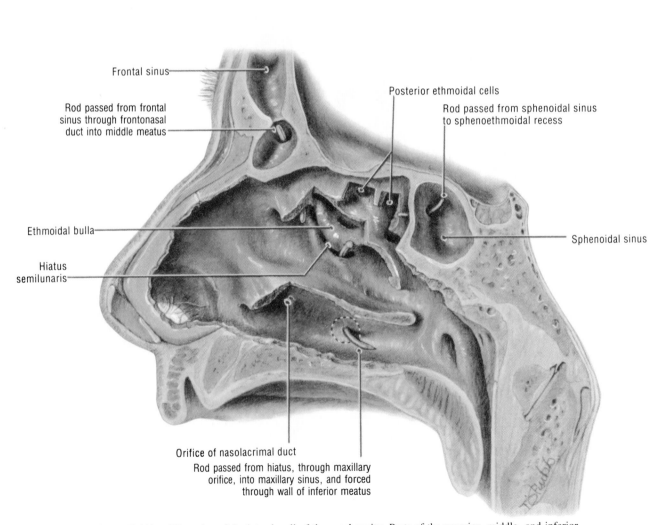

Frontal sinus

Rod passed from frontal sinus through frontonasal duct into middle meatus

Ethmoidal bulla

Hiatus semilunaris

Posterior ethmoidal cells

Rod passed from sphenoidal sinus to sphenoethmoidal recess

Sphenoidal sinus

Orifice of nasolacrimal duct
Rod passed from hiatus, through maxillary orifice, into maxillary sinus, and forced through wall of inferior meatus

Figure 7-111.   Dissection of the lateral wall of the nasal cavity. Parts of the superior, middle, and inferior conchae are cut away.

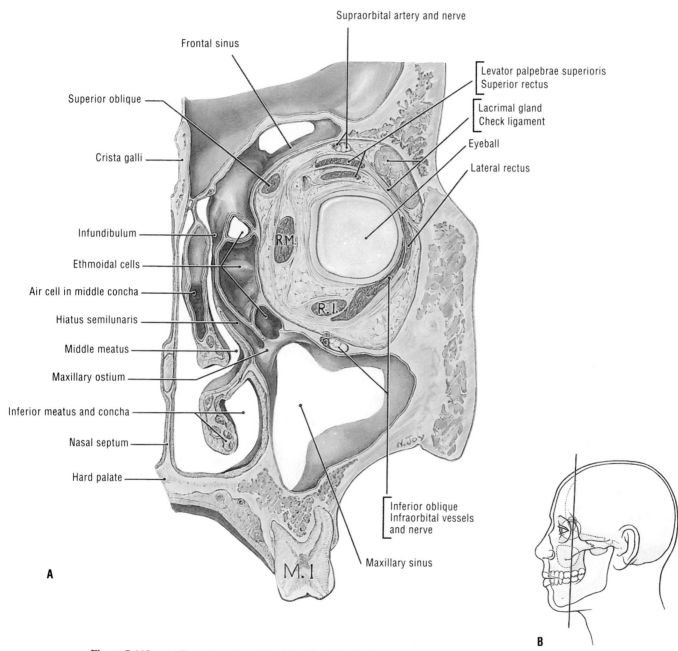

Figure 7-112. *A*, Coronal section of the left side of the head (posterior view), showing the orbital contents, nasal cavity, and paranasal sinuses. *B*, Orientation drawing.

is no indication of the size of the subjacent frontal sinuses. Usually the frontal sinuses are detectable in radiographs of children by 7 years of age.

The right and the left frontal sinuses are rarely of equal size in the same person and the septum between the right and left sinuses is rarely situated entirely in the median plane (Fig. 7-13). Often a frontal sinus has two parts: (1) a vertical part in the squamous part of the frontal bone and (2) a horizontal part in the orbital part of the frontal bone. One or both parts may be large or small. When the supraorbital part is large, its roof forms the floor of the anterior cranial fossa and its floor forms the roof of the orbit (Fig. 7-103). The frontal sinuses vary in size from about 5 mm (pea size) to large spaces extending laterally into the greater wings of the sphenoid bone.

The frontal sinuses may be multiple on each side and each of them may have a separate *frontonasal duct* (Fig. 7-111). Usually the frontal sinus drains through one frontonasal duct on each side. This duct drains inferiorly into the funnel-shaped *infundibulum* (Fig. 7-112), which eventually opens into the hiatus semilunaris of the middle meatus. The frontal sinuses are innervated by branches of the *supraorbital nerves* (Figs. 7-21, 7-24, and 7-42), which are branches of the ophthalmic division of the *trigeminal nerve (CN V¹)*.

The Ethmoidal Sinuses

These sinuses comprise several small cavities, called *ethmoidal cells*, within the ethmoidal labyrinth (G. *labyrinthos*, a maze) of the lateral mass of the ethmoid bone (Figs. 7-14, 7-58, and 7-110 to 7-112). The ethmoidal cells form the labyrinth of the ethmoid bone, which is located between the nasal cavity and the orbit. The number of cells varies from 3 to 18; the cells are larger when the number is small. Extremely thin septa of bone, covered with mucosa, form a variable number of interconnecting ethmoidal compartments or cells, which ultimately drain into the lateral wall of the nasal cavity (Figs. 7-58 and 7-112).

For purposes of description, the ethmoidal sinuses are divided into anterior, middle, and posterior groups. The *anterior ethmoidal cells* drain into the middle meatus, or indirectly into it through the infundibulum (Fig. 7-112). The *middle ethmoidal cells* open directly into the middle meatus (Figs. 7-55 and 7-111). There are several openings on the ethmoidal bulla and on the lateral surface of the middle meatus. The middle ethmoidal cells are sometimes called ''bullar cells'' because they form the ethmoidal bulla (L. bubble), a swelling on the superior border of the hiatus semilunaris (Fig. 7-111). The *posterior ethmoidal cells* open directly into the superior meatus (Fig. 7-111).

The ethmoidal sinuses are supplied by the anterior and posterior ethmoidal branches of the *nasociliary nerves* (Figs. 7-46, 7-59, 7-68, and 7-70), which are branches of the *ophthalmic nerves (CN V¹)*. The anterior and posterior ethmoidal branches of the *ophthalmic artery* supply blood to the ethmoidal sinuses (Figs. 7-69 and 7-109).

Usually the ethmoidal sinuses are not visible in plan radiographs before the age of 2 years, but they can be seen on

fine cut CT scans in newborns (Behrman, 1992). If nasal drainage is blocked, infections of the ethmoidal cells may break through the fragile medial wall of the orbit (Fig. 7-58). Severe infections in the orbit resulting from this source may cause blindness because some posterior ethmoidal cells lie close to the optic canal (Fig. 7-110). Spread of infection from these cells could affect the *dural nerve sheath* of the optic nerve and cause *optic neuritis*.

The Sphenoidal Sinuses

These pneumatic areas are in the body of the sphenoid bone. They occupy a variable amount of this bone and may extend into its wings. The sphenoidal sinuses are located posterior to the superior part of the nasal cavity (Figs. 7-14, 7-110, and 7-111). The two sinuses are separated by a bony septum that is usually not in the median plane. Owing to the presence of the sphenoidal sinuses, the body of the sphenoid bone is a fragile hollow structure. Only thin plates of bone separate the sinuses from several important structures: the *optic nerves* and optic chiasma, the *hypophysis cerebri* (pituitary gland), the *internal carotid arteries*, and the *cavernous and intercavernous sinuses* (Figs. 7-14, 7-30, 7-70, and 7-110). The posterior ethmoidal nerve (Fig. 7-68) and posterior ethmoidal artery (Fig. 7-109A) supply the sphenoidal sinuses.

Although it is sometimes stated that the sphenoidal sinuses are present at birth (although minute), this is not generally accepted because the sphenoidal sinuses are not observable in skull films of newborn infants. The current view is that the sphenoidal sinuses are derived from a posterior ethmoidal cell (Fig. 7-110) that begins to invade the sphenoid bone at about 2 years of age. In some people, several posterior ethmoidal cells invade the sphenoid bone, giving rise to multiple sphenoidal sinuses that open separately into the sphenoethmoidal recess (Figs. 7-55 and 7-111).

The Maxillary Sinuses

This is the largest pair of paranasal sinuses (Figs. 7-14, 7-58, 7-99, 7-103, 7-110, and 7-112). They are pyramidal-shaped cavities that may occupy the entire bodies of the maxillae. The apex of the maxillary sinus extends toward and often into the zygomatic bone (Figs. 7-13 and 7-110). The base of the maxillary sinus forms the inferior part of the lateral wall of the nasal cavity (Fig. 7-112). The roof of the maxillary sinus is formed by the floor of the orbit (Fig. 7-58), and its narrow floor is formed by the alveolar process of the maxilla (Figs. 7-1 and 7-13). The roots of the maxillary teeth, particularly the first two molars, often produce conical elevations in the floor of the maxillary sinus (Fig. 7-88).

The maxillary sinus drains into the middle meatus of the nasal cavity through the hiatus semilunaris (Figs. 7-58, 7-111, and 7-112) by an aperture in the superior part of its base. Because of the location of this opening, it is impossible for fluid in the sinus to drain when the head is erect until the sinus is nearly full. The *innervation of the maxillary sinus* is from the anterior, middle, and posterior superior *alveolar nerves*, branches of the maxillary nerve (Figs. 7-25C, 7-80, and 7-104). The *blood supply of the maxillary sinus* is from the superior alveolar branches of the third part of the *maxillary artery* (Fig. 7-79). Branches of the *greater palatine artery* contribute blood to the floor of the maxillary sinus (Figs. 7-91 and 7-109A).

The maxillary sinuses are very small at birth and grow slowly until puberty. They are not fully developed until all the permanent teeth have erupted (up to 25 years of age). The maxillary sinus is the one most commonly involved in infection, probably because its aperture is located superior to the floor of the sinus (Figs. 7-56 and 7-112), a poor location for its natural drainage. In addition, when the mucous membrane of the sinus is congested, the maxillary aperture may be obstructed. Gravity drainage from the maxillary sinus is best when one is lying on the side opposite the infected sinus. *The proximity of the maxillary molar teeth to the floor of the maxillary sinus poses potentially serious problems* (Figs. 7-88 and 7-112). During removal of a maxillary molar tooth, a fracture of one of the roots may occur. If proper retrieval methods are not used, the broken piece of the root may be driven superiorly into the maxillary sinus. As a result, a communication may be created between the oral cavity and the maxillary sinus. Infection can also spread to the maxillary sinus from an abscessed maxillary molar tooth.

As each paranasal sinus is continuous with the nasal cavity through an aperture that opens into a meatus of the nasal cavity (Figs. 7-55 and 7-112), infection may spread from the nasal cavities, producing inflammation and swelling of the mucosa of the sinuses (*sinusitis*) and local pain. Sometimes several sinuses are inflamed (*pansinusitis*), and the swelling of the mucosa may result in blockage of one or more openings of the sinuses into the nasal cavities. Because the superior alveolar nerves (Fig. 7-25), which are branches of the maxillary nerve (CN V^2), supply both the maxillary teeth and the mucous membrane of the maxillary sinus, inflammation of the mucosa of the sinus is frequently accompanied by the sensation of toothache, especially when bone is very thin in the inferior part of the wall of this sinus (Fig. 7-88).

Patients with fractures of the frontal, ethmoid, maxillary, or nasal bones (Fig. 7-20) are warned not to blow their noses because of the possibility of expelling air from their paranasal sinuses or nasal cavities into the subcutaneous tissues, cranium, or orbit. In addition, *CSF rhinorrhea* may occur (p. 701).

The Ear

The ear, containing the **vestibulocochlear organ**, consists of three main parts (Figs. 7-113 to 7-115): external, middle, and internal. It has two functions, *balance* and *hearing*. The external and middle parts are mainly concerned with the transference of sound to the internal ear, which contains the vestibulocochlear organ that is essential for equilibrium and hearing. The *tympanic membrane* (eardrum) separates the external ear from the middle ear. The *auditory tube* joins the middle ear or tympanic cavity to the nasopharynx (Figs. 7-55 and 7-113).

The External Ear

The external ear is composed of the oval *auricle* that collects sounds and the *external acoustic meatus*, (auditory canal) that conducts sounds to the tympanic membrane (Figs. 7-113 to 7-115), which separates it from the middle ear (p. 769).

The Auricle

The auricle (L. *auris*, ear) is the visible, shell-like part of the external ear (Figs. 7-114 and 7-115). It is connected to the skull by skin, the external acoustic meatus, ligaments, and muscles. The auricle (pinna) consists of a *single elastic cartilage* that is covered on both surfaces with thin, hairy skin. The external ear contains hairs, sweat glands, and sebaceous glands. The cartilage is irregularly ridged and hollowed, which gives the auricle its shell-like form. It also shapes the orifice of the external acoustic meatus. This cartilage is prolonged medially, where it is continuous with the cartilage of the external acoustic meatus. The auricle is connected with the fascia on the side of the skull by the *auricular muscles* (Fig. 7-22A). Many terms are used to describe the parts of the auricle, but only a few of them are commonly used. The auricle projects from the side of the head and acts as a trumpet for collecting sound waves and directing them into the relatively narrow external acoustic meatus. The shape of the auricle varies considerably in different people; this is of no clinical significance except when there are gross malformations, *e.g.*, as occurs in trisomies 13 and 18 (Moore, 1988).

The ear lobule (earlobe) consists of fibrous tissue, fat, and blood vessels that are covered with skin. The **arteries** are derived mainly from the posterior auricular artery and the superficial temporal artery (Figs. 7-22A, 7-23, and 7-71). The lobule contains no cartilage and is therefore easily pierced (*e.g.*, for earrings and for taking small blood samples). The skin of the auricle is supplied by the great auricular and auriculotemporal nerves (Fig. 7-24). The *great auricular nerve* supplies the superior surface and the lateral surface inferior to the external acoustic meatus with nerve fibers from C2. The *auriculotemporal nerve* supplies the skin of the auricle superior to the external acoustic meatus.

The **lymph vessels** of the lateral surface of the superior half of the auricle drain to the *superficial parotid lymph nodes*, which lie just anterior to the tragus (Figs. 7-32 and 7-115). Lymph from the cranial surface of the superior half of the auricle drains to the *retroauricular and deep cervical lymph nodes*. Lymph from the remainder of the auricle, including the lobule, drains into the *superficial cervical lymph nodes* (Fig. 7-32).

The External Acoustic Meatus

This auditory passage extends from the *concha* (L. shell) of the auricle to the tympanic membrane (L. *tympanum*, tambourine), a distance of about 2.5 cm in adults (Fig. 7-114). The lateral third of this *S-shaped canal* is cartilaginous, whereas its medial two-thirds is bony. In infants, the external acoustic meatus is almost entirely cartilaginous. The lateral third of the meatus is lined with the skin of the auricle and contains hair follicles, sebaceous glands, and *ceruminous glands*. The latter glands produce a waxy exudate called *cerumen* (L. *cera*, wax). The medial two-thirds of the meatus is lined with very thin skin that is

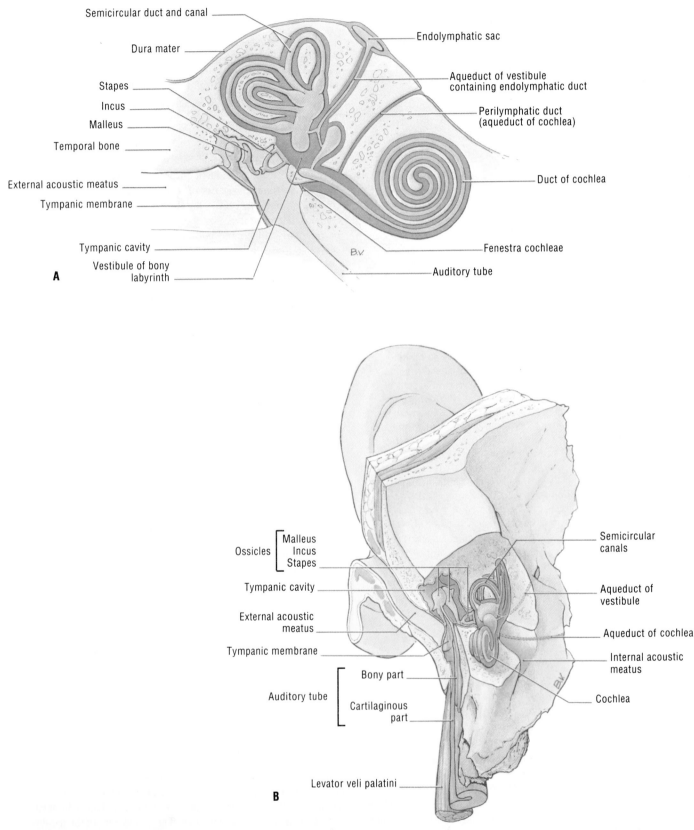

Semicircular duct and canal

Dura mater

Stapes

Incus

Malleus

Temporal bone

External acoustic meatus

Tympanic membrane

Tympanic cavity

Vestibule of bony labyrinth

Endolymphatic sac

Aqueduct of vestibule containing endolymphatic duct

Perilymphatic duct (aqueduct of cochlea)

Duct of cochlea

Fenestra cochleae

Auditory tube

A

Ossicles { Malleus Incus Stapes }

Tympanic cavity

External acoustic meatus

Tympanic membrane

Auditory tube { Bony part Cartilaginous part }

Levator veli palatini

Semicircular canals

Aqueduct of vestibule

Aqueduct of cochlea

Internal acoustic meatus

Cochlea

B

Figure 7-113. *A*, The three parts of the ear. The middle ear or tympanic cavity lies between the tympanic membrane and the internal ear. Three ossicles—malleus, incus, and stapes—stretch from the lateral wall to the medial wall of the tympanic cavity. Of these, the malleus is attached to the tympanic membrane; the stapes is attached by an anular ligament to an oval opening in the wall of the bony vestibule of the inner ear, called the fenestra vestibuli or oval window (Fig. 7-122); and the incus connects these two ossicles. *B*, Dissection showing the parts of the ear in situ.

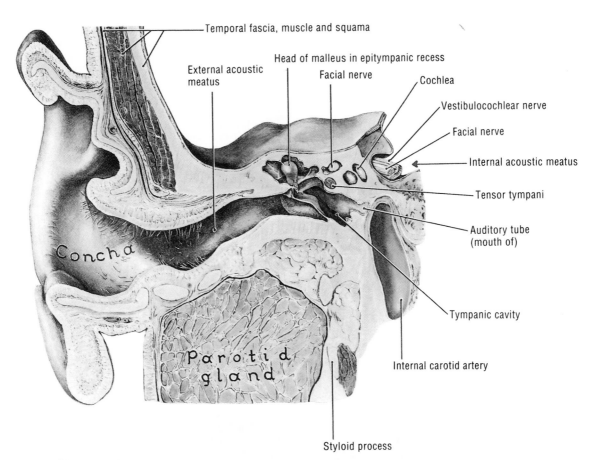

Temporal fascia, muscle and squama

External acoustic meatus

Head of malleus in epitympanic recess

Facial nerve

Cochlea

Vestibulocochlear nerve

Facial nerve

Internal acoustic meatus

Tensor tympani

Auditory tube (mouth of)

Tympanic cavity

Concha

Parotid gland

Internal carotid artery

Styloid process

Figure 7-114.   Anterior view of a coronal section of the right ear. Note that the bony part of the external acoustic meatus is narrower than the cartilaginous part.

continuous with the external layer of the tympanic membrane. The lateral end of the meatus, its widest part, is about the diameter of a pencil. The meatus becomes narrow at the medial end of its cartilaginous portion and in the bony part, about 4 mm from the tympanic membrane. The constricted bony part is called the **isthmus**. The inferior wall of the meatus is about 5 mm longer than its superior wall, owing to the obliquity of the tympanic membrane (Figs. 7-113 and 7-114).

The **arteries** supplying the external acoustic meatus are the posterior auricular branch of the *external carotid artery* (Fig. 7-22); the deep auricular branch of the *maxillary artery* (Fig. 7-79); and the auricular branches of the superficial temporal artery (Figs. 7-23, 7-28, and 7-71). The **veins** from the external acoustic meatus drain into the external jugular and maxillary veins (Fig. 7-29), and into the pterygoid plexus (Fig. 7-30). The *lymphatic drainage* of the external acoustic meatus is the same as described for the auricle (p. 763). **Innervation of the external acoustic meatus** is derived from three cranial nerves (Figs. 7-24, 7-25, and 7-27): (1) the auricular branch of the *auriculotemporal nerve*, derived from the mandibular nerve (*CN V³*); (2) the *facial nerve* (CN VII) by branches from the tympanic plexus (p. 771); and (3) the auricular branch of the *vagus nerve* (*CN X*).

The external acoustic meatus is directed somewhat anteriorly as well as medially; the tips of the *stethoscope* are

angulated to conform to this shape. The anatomy of the external acoustic meatus also has to be considered when using an *auriscope* to examine the meatus and tympanic membrane. The instrument can be inserted more readily when the auricle is pulled superiorly, posteriorly and a little laterally. *The meatus is relatively short in infants*, therefore extra care must be exercised so that the tympanic membrane will not be damaged. Foreign bodies inserted in the meatus by infants and children (*e.g.*, beans, candies) usually become lodged in the isthmus in the bony part, where its cartilaginous and bony parts meet (Fig. 7-114). Because the condylar process of the mandible is close to the external acoustic meatus (Figs. 7-15 and 7-73), a hard blow on the chin may drive the head of the mandible into the meatus and injure or fracture it.

The Tympanic Membrane

This thin, semitransparent, oval membrane is at the medial end of the external acoustic meatus. It forms a partition between the external and middle ears (Figs. 7-113, 7-114, and 7-116 to 7-118). It is a fibrous membrane that is covered with very thin skin externally and mucous membrane internally. In the adult it is oblique and slopes inferomedially. In living persons the tympanic membrane (eardrum) normally appears pearly gray and shiny. The tympanic membrane shows a concavity toward the meatus with a central depression, the *umbo* (Fig. 7-116*A*), which is formed by the end of the handle of the malleus, one of the three middle ear bones, called *auditory ossicles* (Fig. 7-119). From the umbo, a bright area referred to as the *cone of light*, radiates anteroinferiorly. The tympanic membrane moves in response to air vibrations that pass to it through the external acoustic meatus. The vibrations are transmitted from this membrane by the auditory ossicles through the middle ear to the internal ear (Figs. 7-113 and 7-117).

Blood Vessels of the Tympanic Membrane (Figs. 7-22, 7-28 to 7-30, and 7-79).   The *deep auricular artery*, a branch of the maxillary artery, supplies the external surface of the tympanic membrane. The internal surface is supplied by the stylomastoid branch of the *posterior auricular artery* and by the tympanic branch of the maxillary artery. The **veins** in the external surface of the tympanic membrane open into the external jugular vein. Those in the internal surface drain into the transverse sinus and the veins of the dura, and partly into the venous plexus of the auditory tube.

Nerve Supply of the Tympanic Membrane (Figs. 7-23 and 7-25).   The external surface of the tympanic membrane is supplied by the *auriculotemporal nerve*, a branch of the mandibular division of the trigeminal nerve (*CN V³*). Some innervation is supplied by a small auricular branch of the vagus nerve (CN X); this nerve may also contain some glossopharyngeal and facial nerve fibers (Fig. 7-27; see also Fig. 8-33).

Perforation ("rupture") of the tympanic membrane is one of several causes of middle ear deafness. *Perforation of the tympanic membrane* may result from foreign bodies, excessive pressure (*e.g.*, during scuba diving), or infection. Severe bleeding or escape of CSF through a ruptured tympanic membrane and the external acoustic meatus (*CSF otorrhea*) may occur following a severe blow on the head. Either of these conditions is indicative of a skull fracture and results from the close relation of the tympanic cavity, mastoid antrum, mastoid cells, and the bony external acoustic meatus to the meninges of the brain (Fig. 7-117). Sometimes a skull fracture passes through the bony part of the external acoustic meatus and causes bleeding or loss of CSF through the external acoustic meatus, even with an intact tympanic membrane. On rare occasions it is necessary to incise the tympanic membrane to allow pus to escape from the middle ear. Because the superior half of the tympanic membrane is much more vascular than the inferior half, incisions are made posteroinferiorly through the membrane (Fig. 7-116). This site also avoids the chorda tympani nerve and auditory ossicles (Fig. 7-118).

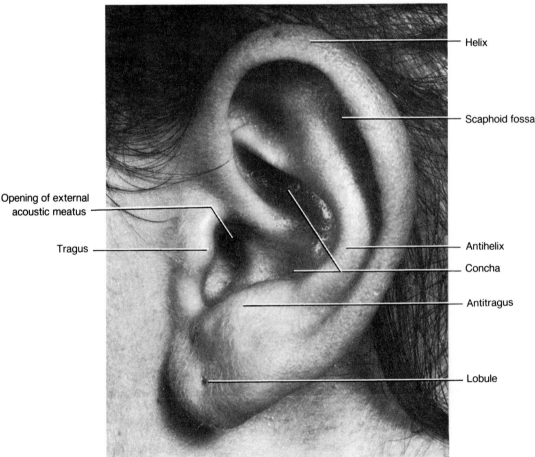

Figure 7-115.   The left auricle of a 12-year-old girl. The names given here are the ones commonly used in clinical descriptions.

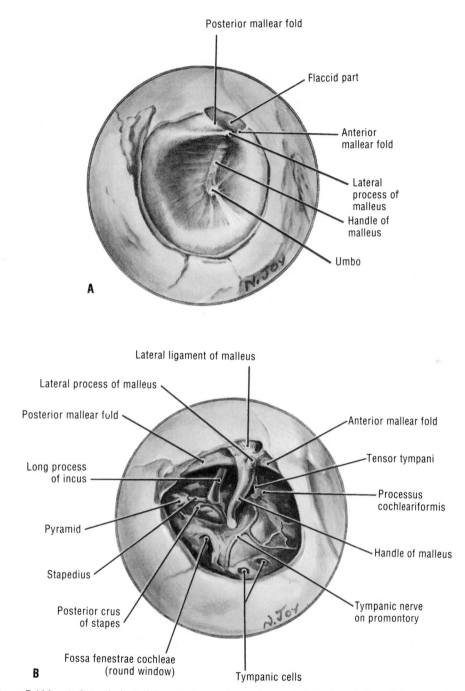

Figure 7-116. *A*, Lateral view of the right tympanic membrane. *B*, Inferolateral view of the tympanic cavity after removal of the tympanic membrane.

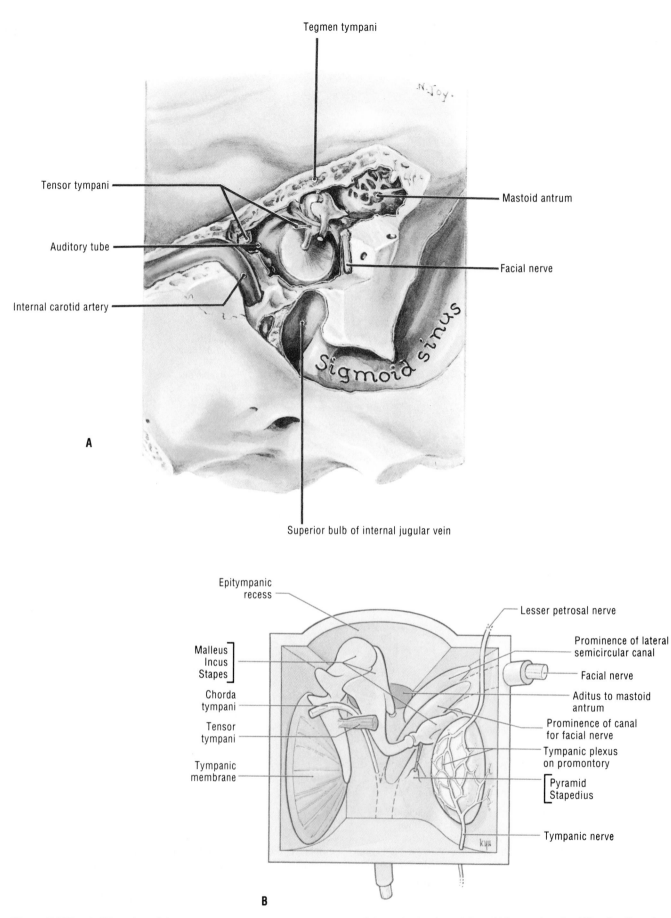

Tegmen tympani

Tensor tympani

Auditory tube

Internal carotid artery

Mastoid antrum

Facial nerve

Sigmoid sinus

A

Superior bulb of internal jugular vein

Epitympanic recess

Lesser petrosal nerve

Malleus
Incus
Stapes

Prominence of lateral semicircular canal

Facial nerve

Chorda tympani

Aditus to mastoid antrum

Tensor tympani

Prominence of canal for facial nerve

Tympanic membrane

Tympanic plexus on promontory

Pyramid
Stapedius

Tympanic nerve

B

Figure 7-117. *A,* Dissection of the middle ear showing the walls of the tympanic cavity. The mastoid antrum is a diverticulum of the tympanic cavity. It contains air and communicates with the mastoid cells. *B,* Schematic drawing of the middle ear that simplifies the dissection shown in *A* and emphasizes the relations of the tympanic cavity or middle ear. The anterior wall of the middle ear is removed.

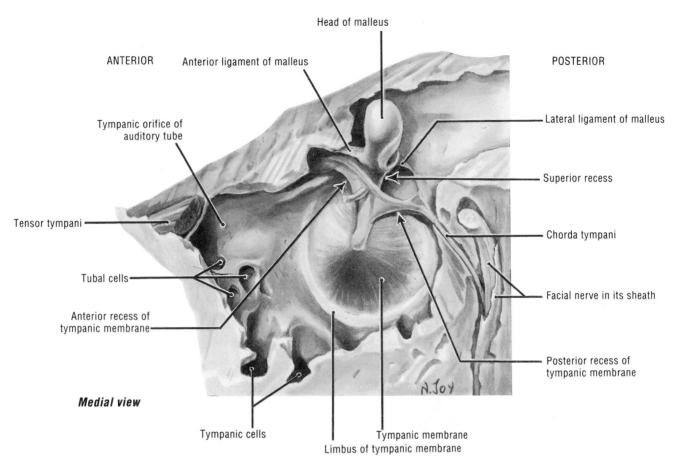

ANTERIOR

Head of malleus

Anterior ligament of malleus

POSTERIOR

Tympanic orifice of auditory tube

Lateral ligament of malleus

Superior recess

Tensor tympani

Chorda tympani

Tubal cells

Facial nerve in its sheath

Anterior recess of tympanic membrane

Posterior recess of tympanic membrane

Medial view

N. Joy

Tympanic cells

Tympanic membrane
Limbus of tympanic membrane

Figure 7-118. A dissection of the right middle ear to show the lateral wall of the tympanic cavity, which is formed almost entirely by the oval tympanic membrane. Note the head of the malleus in the epitympanic recess (Fig. 7-117*B*). Here its posterior surface articulates within the body of the incus.

The Middle Ear

This part of the ear is *a narrow cavity in the petrous part of the temporal bone*. It contains air, three auditory ossicles, a nerve, and two small muscles (Figs. 7-113, 7-116, and 7-117). The middle ear is separated from the external acoustic meatus by the tympanic membrane. It includes the *tympanic cavity* proper, the space directly internal to the membrane, and the *epitympanic recess*, the space superior to it. The middle ear is connected anteriorly with the nasopharynx by the auditory tube (Fig. 7-113). Posterosuperiorly the tympanic cavity connects with the mastoid cells through the *mastoid antrum* (Fig. 7-120). The tympanic cavity is lined with mucous membrane that is continuous with that lining the auditory tube, mastoid cells, and mastoid antrum.

Contents of the Tympanic Cavity or Middle Ear (Figs. 7-113, 7-114, and 7-116 to 7-118). This cavity contains the *auditory ossicles* (malleus, incus, and stapes); the stapedius and tensor tympani muscles; the *chorda tympani nerve* (a branch of the facial, CN VII); and the tympanic plexus of nerves.

Walls of the Tympanic Cavity or Middle Ear. This cavity is shaped like a narrow, six-sided box that has convex medial and lateral walls. It has the shape of a biconcave lens or red blood cell in transverse section. The general shape of the tym-

panic cavity is shown diagrammatically in Figure 7-117*B*. The *roof or tegmental wall* of the tympanic cavity or middle ear is formed by a thin plate of bone, called the **tegmen tympani** (L. *tegmen*, a roof). It separates the tympanic cavity from the dura on the floor of the middle cranial fossa. The tegmen tympani also covers the mastoid antrum (Fig. 7-117*A*). The *floor or jugular wall* of the tympanic cavity or middle ear is thicker than the roof, but it is formed by a layer of bone that may be thick or thin. It separates the tympanic cavity from the superior bulb of the internal jugular vein (Figs. 7-43 and 7-117). Because the internal jugular vein and internal carotid artery pass superiorly within the carotid sheath, they diverge at the floor of the tympanic cavity. The *tympanic nerve* (Fig. 7-74), a branch of the glossopharyngeal (CN IX), passes through an aperture in the floor of the tympanic cavity and its branches form the **tympanic plexus**.

The *lateral or membranous wall of the tympanic cavity* or middle ear is formed almost entirely by the tympanic membrane (Fig. 7-117). Superiorly it is formed by the lateral bony wall of the *epitympanic recess*. The handle of the malleus is incorporated in the tympanic membrane (Fig. 7-116), and its head extends into the epitympanic recess.

The medial or labyrinthine wall of the tympanic cavity or middle ear separates the tympanic cavity from the inner ear that consists of a membranous labyrinth (semicircular ducts and coch-

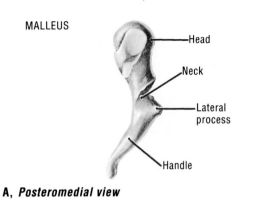

MALLEUS

Head

Neck

Lateral process

Handle

A, *Posteromedial view*

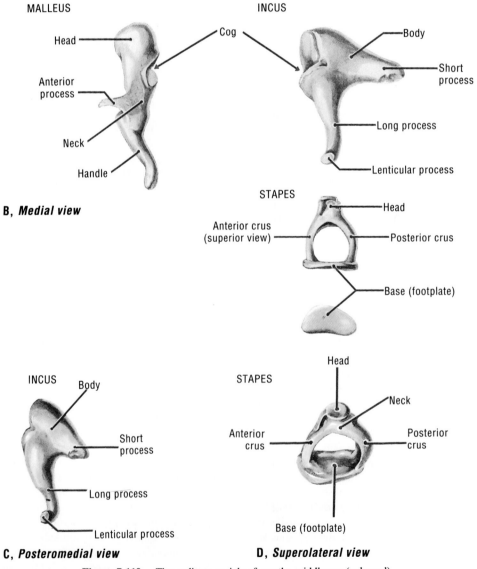

MALLEUS

Head

Anterior process

Neck

Handle

Cog

B, *Medial view*

INCUS

Body

Short process

Long process

Lenticular process

STAPES

Head

Anterior crus (superior view)

Posterior crus

Base (footplate)

INCUS   Body

Short process

Long process

Lenticular process

C, *Posteromedial view*

STAPES

Head

Neck

Anterior crus

Posterior crus

Base (footplate)

D, *Superolateral view*

Figure 7-119.   The auditory ossicles from the middle ear (enlarged).

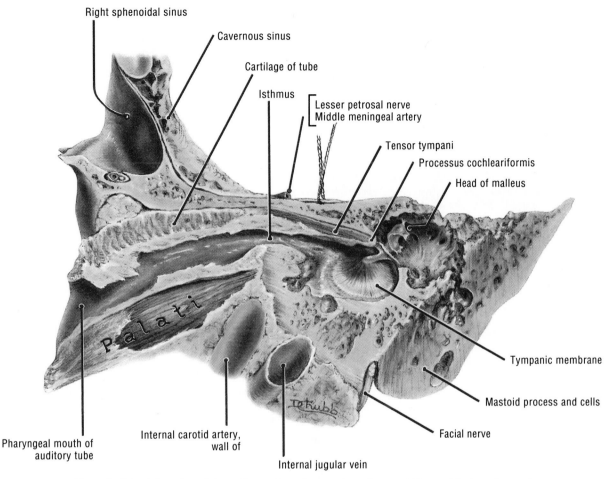

Right sphenoidal sinus

Cavernous sinus

Cartilage of tube

Isthmus

Lesser petrosal nerve
Middle meningeal artery

Tensor tympani

Processus cochleariformis

Head of malleus

Tympanic membrane

Mastoid process and cells

Facial nerve

Internal jugular vein

Internal carotid artery, wall of

Pharyngeal mouth of auditory tube

Figure 7-120.   Dissection of the auditory tube and tympanic cavity or middle ear. The lateral part of a longitudinally split specimen is shown.

lear duct) encased in a bony labyrinth (Figs. 7-113, 7-121, and 7-122). The medial wall of the tympanic cavity exhibits several important features. Centrally, opposite the tympanic membrane, there is a rounded promontory (L. eminence) formed by the large first turn of the cochlea. The **tympanic plexus of nerves**, lying on the promontory, is formed by fibers of the facial and glossopharyngeal nerves (CN VII and CN IX) that pass through an aperture in the floor of the tympanic cavity within the tympanic nerve (Figs. 7-74 and 7-117*B*). The medial wall of the tympanic cavity has two small apertures or windows (Fig. 7-122). The *fenestra vestibuli* is closed by the base of the stapes (Fig. 7-119), which is bound to its margins by an annular ligament. Through this window, vibrations of the stapes are transmitted to the perilymph within the bony labyrinth of the internal ear (Fig. 7-113). The *fenestra cochleae*, inferior to the fenestra vestibuli, is closed by a secondary tympanic membrane. This membrane allows the perilymph to move slightly in response to impulses from the base of the stapes. Thus both apertures are related to the tympanic cavity.

In *otosclerosis* (G. *ous* (ot), ear + *sklerosis*, hardening), there is a new formation of spongy bone around the stapes and fenestra vestibuli, resulting in progressively increasing

deafness. This bondy overgrowth may stop movement of the base of the stapes or the membrane of the fenestra cochleae. If either or both of these are immobilized, they must be freed to restore the hearing.

The *posterior or mastoid wall of the tympanic cavity* or middle ear has several openings in it (Fig. 7-117). In its superior part is the *aditus to the mastoid antrum*, which leads posteriorly from the epitympanic recess into the mastoid antrum and beyond into the mastoid cells (Fig. 7-120). Inferiorly is a pinpoint aperture on the apex of a tiny, hollow projection of bone, called the *pyramidal eminence* (pyramid). This eminence contains the stapedius muscle (Fig. 7-116*B*). Its aperture transmits the tendon of the stapedius, which enters the tympanic cavity and inserts into the stapes. Lateral to the pyramid, there is an aperture through which the chorda tympani nerve, a branch of the facial (CN VII), enters the tympanic cavity.

The *anterior wall or carotid wall of the tympanic cavity* or middle ear is narrow because the medial and lateral walls converge anteriorly. There are two openings in the anterior wall. The superior opening communicates with a canal occupied by the tensor tympani muscle (Fig. 7-120). Its tendon inserts into the handle of the malleus and keeps the tympanic membrane tense. Inferiorly the tympanic cavity communicates with the nasopharynx through the auditory tube. It runs inferomedially and anteriorly to open into the nasopharynx posterior to the inferior meatus of the nasal cavity (Fig. 7-55).

> Inflammatory conditions in the tympanic cavity or middle ear (*i.e.*, **otitis media**) may sometimes spread through the thin tegmen tympani (Fig. 7-117*A*). This causes inflammation of the meninges (meningitis) and brain (*cerebritis*). In infants and children the unossified *petrosquamous fissure* (Fig. 7-38) may allow direct spread of infection from the tympanic cavity to the meninges of the brain. In the adult, veins pass through the petrosquamous suture to the superior petrosal sinus (Figs. 7-47*B* and 7-49). Thus infection can spread through these veins to the dural venous sinuses (Fig. 7-30).
>
> *Earache is a common symptom that has multiple causes*, some in the ear and others at a distance. **Otitis externa** (inflammation of the external acoustic meatus) is one cause. Movement of the tragus results in increased pain because the cartilage in it is continuous with that in the external acoustic meatus (Figs. 7-114 and 7-115). Earache may be referred pain from distant lesions, *e.g.*, the mouth (dental abscess or cancer of the tongue) through the mandibular nerve (CN V³) or the pharynx and larynx through the vagus nerve (see Fig. 8-33). Some patients refer to *temporomandibular joint disease* as ear pain or an earache, because of the proximity of this articulation to the tragus of the ear (Figs. 7-81 and 7-82).

The Auditory Tube

The funnel-shaped auditory tube (pharyngotympanic tube) connects the nasopharynx to the tympanic cavity or middle ear (Figs. 7-113 and 7-120). Its wide end is toward the nasopharynx, where it opens posterior to the inferior meatus of the nasal cavity (Fig. 7-55). The auditory tube is 3.5 to 4 cm long; its posterior one-third is bony, and the other two-thirds cartilaginous. Its bony part lies in a groove on the inferior aspect of the base of the

skull, between the petrous part of the temporal bone and the greater wing of the sphenoid bone. The auditory tube is lined by mucous membrane that is continuous posteriorly with that of the tympanic cavity or middle ear, and anteriorly with that of the nasopharynx. A collection of lymphoid tissue, called the **tubal tonsil**, is located near the pharyngeal orifice of the auditory tube. The *function of the auditory tube* is to equalize pressure in the middle ear with the atmospheric pressure, thereby allowing free movement of the tympanic membrane. By allowing air to enter and leave the cavity, it balances the pressure on both sides of the membrane.

Vessels of the Auditory Tube (Figs. 7-28, 7-30, and 7-79).   The arteries of the auditory tube are derived from: (1) the *ascending pharyngeal artery*, a branch of the external carotid artery, and (2) the *middle meningeal artery* and the *artery of the pterygoid canal*, branches of the maxillary artery. The **veins** of the auditory tube drain into the *pterygoid plexus*.

Innervation of the Auditory Tube (Figs. 7-74, 7-104, and 7-117*B*).   The nerves supplying the auditory tube arise from the *tympanic plexus*, which is formed by fibers of the facial (CN VII) and glossopharyngeal (CN IX) nerves. The auditory tube also receives fibers from the *pterygopalatine ganglion*.

> The cartilaginous part of the auditory tube remains closed except during swallowing or yawning. The tube is opened by the simultaneous contraction of the tensor veli palatini and salpingopharyngeus muscles (Fig. 7-92), which are attached to opposite sides of the tube. When the cartilaginous part of the auditory tube opens (*i.e.*, during swallowing), it prevents excessive pressure in the tympanic cavity. We swallow during the descent of an aircraft to relieve the pressure in our middle ears.

> The auditory tube forms a route through which infections may pass from the nasal part of the pharynx or nasopharynx to the tympanic cavity. This tube is easily blocked by swelling

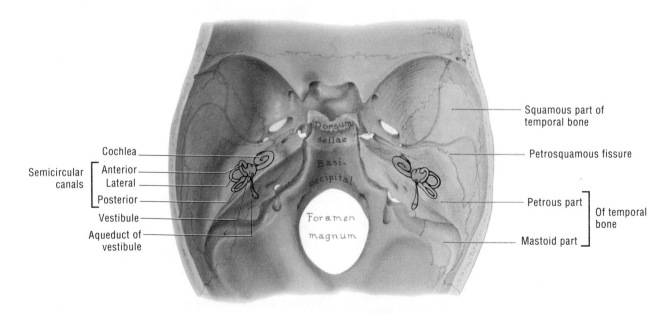

Figure 7-121.   Superior view of the interior of the base of the skull, showing the temporal bone and the location of the bony labyrinth.

of its mucous membrane, even by mild infections, because the walls of its cartilaginous part are normally in apposition. When the auditory tube is blocked, residual air in the tympanic cavity is usually absorbed into the mucosal blood vessels. This results in lowering of pressure in the tympanic cavity, retraction of the tympanic membrane, and interference with its free movement. As a result hearing is diminished.

The Auditory Ossicles

These little *ear bones* (malleus, incus, and stapes) form a chain across the tympanic cavity from the tympanic membrane to the fenestra vestibuli or *oval window* (Figs. 7-113, 7-114, 7-116, and 7-119). The malleus is attached to the tympanic membrane and the stapes occupies the fenestra vestibuli (Figs. 7-117 and 7-122). The incus is located between these two bones and articulates with them. The ossicles are covered with the mucous membrane that lines the tympanic cavity or middle ear.

The Malleus (Figs. 7-116, 7-117, 7-119, and 7-120). Its rounded superior part, the *head*, lies in the epitympanic recess. The *neck* lies against the flaccid part of the tympanic membrane. The handle of the malleus (L. hammer) is embedded in the tympanic membrane and moves with it. *The head of the malleus articulates with the incus*, and the tendon of the tensor tympani muscle inserts into its handle. The *chorda tympani nerve* crosses the medial surface of the neck of the malleus.

The Incus (Figs. 7-113, 7-117, and 7-119). Its large body lies in the epitympanic recess where it *articulates with the head of the malleus*. The long process of the incus (L. an anvil) articulates with the stapes, and its short process is connected by a ligament to the posterior wall of the tympanic cavity.

The Stapes (Figs. 7-113, 7-119, 7-121, and 7-122). The *base* (footplate) of the stapes (L. a stirrup), the smallest ossicle, fits into the fenestra vestibuli or oval window on the medial wall of the tympanic cavity. Its head, directed laterally, *articulates with the lenticular process of the incus*.

Functions of the Auditory Ossicles. The malleus functions as a lever with the longer of its two arms attached to the tympanic membrane. The base of the stapes is considerably smaller than the tympanic membrane. As a result, the vibratory force of the stapes is about 10 times that of the tympanic membrane. Thus the auditory ossicles increase the force but decrease the amplitude of the vibrations transmitted from the tympanic membrane.

Muscles Moving the Auditory Ossicles and Tympanic Membrane. Two muscles produce movements of the auditory ossicles and tympanic membrane: tensor tympani and stapedius.

The Tensor Tympani Muscle (Figs. 7-117 and 7-120). This muscle is only about 2 cm long.

Origin (Fig. 7-120). Superior surface of the cartilaginous part of the auditory tube, the greater wing of the sphenoid bone, and the petrous part of temporal bone.

Insertion (Figs. 7-120 and 7-121). Handle of the malleus. Its tendon turns around the processus cochleariformis before inserting into the handle of malleus.

Nerve Supply (Fig. 7-25). Mandibular nerve (CN V³) through fibers that pass through the otic ganglion.

Actions. The tensor tympani muscle pulls the handle of the malleus medially, tensing the tympanic membrane and reducing the amplitude of its oscillations. This tends to prevent damage to the internal ear when one is exposed to loud sounds.

The Stapedius Muscle (Fig. 7-116B). This tiny muscle is in the pyramidal eminence or pyramid.

Origin. Pyramidal eminence on the posterior wall of the tympanic cavity. Its tendon enters the tympanic cavity by traversing a pinpoint foramen in the apex of the pyramid.

Insertion. Neck of the stapes.

Nerve Supply (Fig. 7-27). Nerve to the stapedius muscle, which arises from the facial nerve (CN VII).

Actions. The stapedius muscle pulls the stapes posteriorly and tilts its base in the fenestra vestibuli or oval window, thereby tightening the anular ligament and reducing the oscillatory range. It also prevents excessive movement of the stapes.

The tympanic muscles have a protective action in that they dampen large vibrations of the tympanic membrane resulting from loud noises. Thus paralysis of the stapedius muscle (*e.g.*, resulting from a lesion of the facial nerve) is associated with excessive acuteness of hearing (a condition known as *hyperacusia*). This condition (G. *hyper*, over, above, + *akousis*, hearing) results from uninhibited movements of the stapes.

Mastoid Antrum and Mastoid Cells

This antrum is a spherical sinus, slightly smaller than the tympanic cavity or middle ear. It is located in the petromastoid part of the temporal bone (Figs. 7-49 and 7-121), posterior to the epitympanic recess (Fig. 7-117B). The mastoid antrum is connected to the tympanic cavity by the *aditus to the mastoid antrum* (aditus ad antrum), and is separated from the middle cranial fossa by a thin roof, the *tegmen tympani* (Fig. 7-117A). Its floor has a number of apertures through which the mastoid antrum communicates with the mastoid (air) cells (Figs. 7-49 and 7-120). Anteroinferiorly the mastoid antrum is related to the canal for the facial nerve. The lateral wall of the antrum is only 1 mm thick at birth, but it increases about 1 mm a year until it is about 15 mm thick. Unlike the paranasal sinuses, the mastoid antrum is well developed at birth and is almost adult sized.

No mastoid processes or mastoid cells are present at birth. As the *mastoid processes* develop, mastoid cells invade them from the mastoid antrum. By 2 years of age, these cells have bulged the temporal bones laterally and inferiorly, forming small mastoid processes, the internal structure of which resembles a honeycomb (Fig. 7-49).

Infections of the mastoid antrum and mastoid cells (*mastoiditis*) result from infection in the middle ear (*otitis media*). Infections may spread superiorly toward the middle cranial fossa through the petrosquamous fissure in young children (Fig. 7-38) or may cause *osteomyelitis* (bone infection) of the tegmen tympani (Fig. 7-117). Since the advent of antibiotics, *mastoiditis* as a complication of middle ear infection is uncommon. During operations for mastoiditis, surgeons have to be conscious of the course of the facial nerve (Figs. 7-117 and 7-118) so that it will not be injured. One access to the tympanic cavity is through the mastoid antrum. In a child, only a thin plate of bone needs to be removed from the lateral

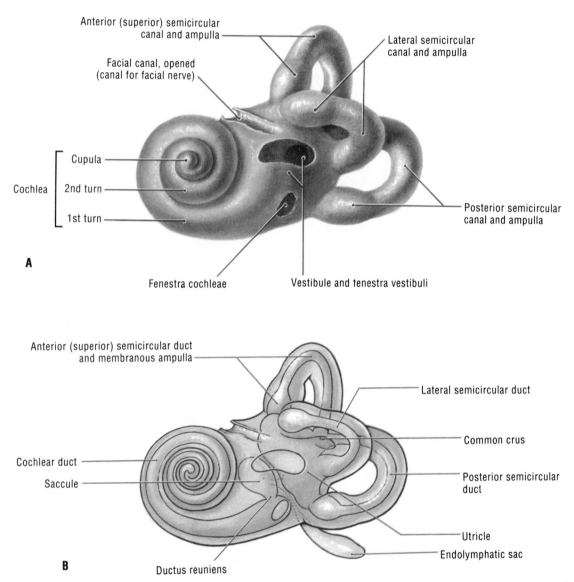

Anterior (superior) semicircular
canal and ampulla

Lateral semicircular
canal and ampulla

Facial canal, opened
(canal for facial nerve)

Cochlea

Cupula

2nd turn

1st turn

Posterior semicircular
canal and ampulla

A

Fenestra cochleae

Vestibule and tenestra vestibuli

Anterior (superior) semicircular duct
and membranous ampulla

Lateral semicircular duct

Common crus

Cochlear duct

Saccule

Posterior semicircular
duct

Utricle

Endolymphatic sac

B

Ductus reuniens

Figure 7-122. *A*, Lateral view of the left side of the bony labyrinth as it would appear if it was dissected from the petrous part of the temporal bone shown in Fig. 7-121. *B*, Similar view of the bony labyrinth with the membranous labyrinth superimposed on it. *C*, Section through the cochlea showing the structure of the cochlear duct and the spiral organ (of Corti).

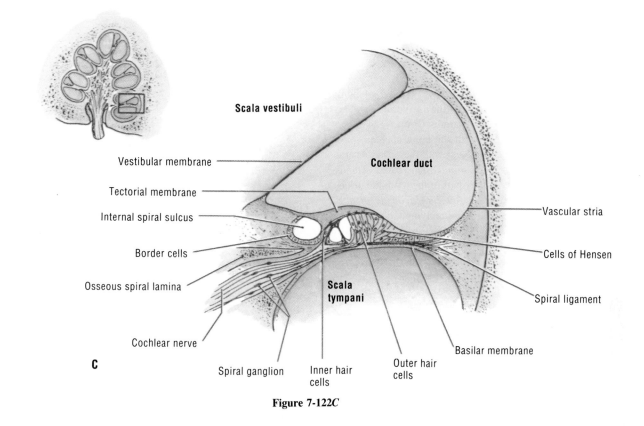

Scala vestibuli

Vestibular membrane

Cochlear duct

Tectorial membrane

Internal spiral sulcus

Vascular stria

Border cells

Cells of Hensen

Osseous spiral lamina

Spiral ligament

Scala tympani

Cochlear nerve

Spiral ganglion

Inner hair cells

Outer hair cells

Basilar membrane

C

Figure 7-122C

wall of the mastoid antrum in the suprameatal region to expose the tympanic cavity. However, in adults, bone must be penetrated for 15 mm or more to reach the mastoid antrum. At present, most *mastoidectomies* are endaural (*i.e.*, performed through the posterior wall of the external acoustic meatus).

The Internal Ear

This part of the ear contains the *vestibulocochlear organ*, which is concerned with the reception of sound and the maintenance of balance. It is buried in the petrous part of the temporal bone (Figs. 7-10, 7-113, and 7-121). The internal ear consists of the sacs and ducts of the membranous labyrinth. This membranous system contains **endolymph** and the end organs for hearing and balancing. The *membranous labyrinth*, surrounded by **perilymph**, is suspended within the *bony labyrinth* (Figs. 7-113 and 7-122*B*).

The Bony Labyrinth

The bony part of the internal ear is composed of three parts: cochlea, vestibule, and semicircular canals (Figs. 7-113 and 7-122). It occupies much of the lateral portion of the petrous part of the temporal bone (Figs. 7-10, 7-40, and 7-121).

The Cochlea (Figs. 7-113, 7-114, and 7-122). This shell-like part of the bony labyrinth contains the *cochlear duct*, the part of the internal ear that is concerned with hearing. The cochlea, so-named because of its shape (L. snail shell), makes two-and-one-half turns around a bony core, called the *modiolus* (L.

nave of a wheel), in which there are canals for blood vessels and nerves. It is the large basal turn of the cochlea that produces the *promontory* on the medial wall of the tympanic cavity. The bony cochlea measures about 5 mm from base to apex.

The axis of the modiolus is across the long axis of the petrous part of the temporal bone; thus the apex of the cochlea, called the *cupula*, points anterolaterally (Fig. 7-122*A*). A small shelf of bone, the *osseous spiral lamina*, protrudes from the modiolus similar to the thread on a screw (Fig. 1-122*C*). This starts at the vestibule and continues to the apex. The *basilar membrane* is attached to the osseous spiral lamina. The modiolus is pierced by several longitudinal channels that turn outward toward the spiral lamina and enter the spiral canal of the modiolus. This canal runs in the base of the spiral lamina and contains the sensory cochlear or *spiral ganglion* (Fig. 7-122*C*). Cells in this ganglion send their peripheral processes to the *spiral organ* (organ of Corti), which is concerned with auditory perception (hearing).

The *scala vestibuli* opens into the vestibule of the bony labyrinth (Fig. 7-113). Through its opening, perilymph can be exchanged freely between the vestibule and the scala vestibuli. The *scala tympani*, which also contains perilymph, is related to the tympanic cavity at the *fenestra cochleae* (Fig. 7-122*A*), which is closed by the secondary tympanic membrane. The scala vestibuli communicates with the scala tympani through a small aperture, the *helicotrema*, at the apex of the cochlea.

The **perilymph** in the scala vestibuli and scala tympani is *similar in composition to CSF*. This is understandable because there is a narrow connection, called the *perilymphatic duct*, between the perilymphatic spaces of the bony labyrinth and the subarachnoid space (Fig. 7-113A). This connection in the petrous part of the temporal bone runs from the scala vestibuli in the basal turn of the cochlea to an extension of the subarachnoid space around the glossopharyngeal, vagus, and accessory nerves (CN IX, X, and XI, respectively).

The Vestibule (Figs. 7-113 and 7-122). This oval bony chamber (about 5 mm in length), contains the *utricle* and *saccule*, which are parts of the balancing apparatus (Fig. 7-124). The vestibule is continuous anteriorly with the bony cochlea, posteriorly with the semicircular canals, and with the posterior cranial fossa by the *aqueduct of the vestibule*. This canal extends to the posterior surface of the petrous part of the temporal bone, where it opens posterolateral to the *internal acoustic meatus* (Figs. 7-40 and 7-114). It contains the endolymphatic duct and two small blood vessels (Figs. 7-113 and 7-124).

The Semicircular Canals (Figs. 7-113, 7-121, 7-122A and B, and 7-123). These bony canals communicate with the vestibule of the bony labyrinth. The semicircular canals (anterior, posterior, and lateral) lie posterosuperior to the vestibule into which they open and are set at right angles to each other. They occupy three planes in space. Each semicircular canal forms about two-thirds of a circle and is about 1.5 mm in diameter, except at one end where there is a swelling called the *ampulla*. The canals have only five openings into the vestibule because the anterior and posterior canals have one stem common to both.

The *anterior semicircular canal* (superior semicircular canal) lies at a right angle to the posterior surface of the petrous part of the temporal bone (Figs. 7-121, 7-122B, and 7-123). It is closely related to the floor of the middle cranial fossa (Fig. 7-12). It forms a transversely rounded elevation, called the *arcuate eminence* (Fig. 7-123), on the superior surface of the

petrous process, just medial to the tegmen tympani (Fig. 7-117A).

The *posterior semicircular canal* (Figs. 7-121 and 7-123) lies in the long axis of the petrous part of the temporal bone, immediately deep to the posterior or cerebellar surface, close to the *sigmoid sinus*.

The *lateral semicircular canal* (Fig. 7-121) is horizontal, and its arch is directed horizontally, posteriorly, and laterally. It lies deep to the medial wall of the aditus to the mastoid antrum (Fig. 7-117) and runs superior to the canal for the facial nerve.

The Membranous Labyrinth

This series of communicating membranous sacs and ducts are contained in the cavities of the bony labyrinth (Figs. 7-113, 7-122B and C, and 7-124). The membranous labyrinth generally follows the form of the bony labyrinth, but it is much smaller. It contains a watery fluid, called *endolymph*, which differs in composition from the perilymph around it in the bony labyrinth. *The membranous labyrinth consists of three main parts*: (1) two small communicating sacs, the *utricle* and *saccule* in the vestibule; (2) three *semicircular ducts* in the semicircular canals; and (3) the *cochlear duct* in the cochlea.

The parts of the membranous labyrinth are suspended in the vestibule of the bony labyrinth by trabeculae of connective tissue. A spiral thickening of the periosteal lining of the cochlear canal, known as the *spiral ligament*, secures the cochlear duct to the cochlear canal (Fig. 7-122C). Except at the *ampullae*, the semicircular ducts are much smaller than the semicircular canals. The various parts of the membranous labyrinth form a closed system

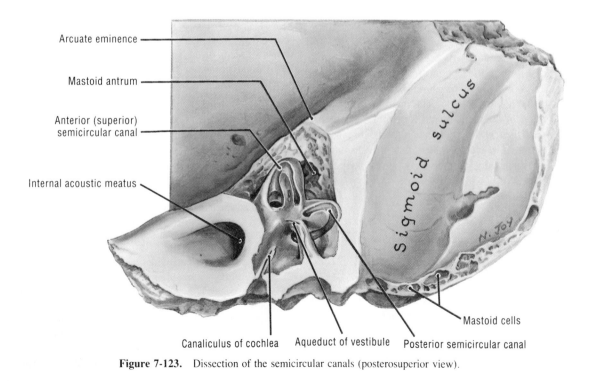

Arcuate eminence

Mastoid antrum

Anterior (superior) semicircular canal

Internal acoustic meatus

Sigmoid sulcus

Mastoid cells

Canaliculus of cochlea   Aqueduct of vestibule   Posterior semicircular canal

Figure 7-123. Dissection of the semicircular canals (posterosuperior view).

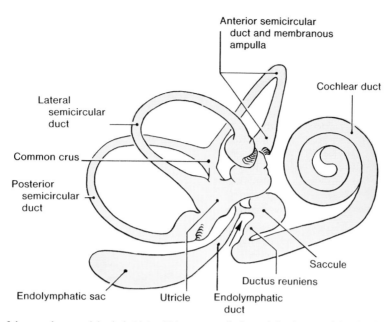

Figure 7-124. Lateral view of the membranous labyrinth (right side). In vivo, it is contained within the bony labyrinth (Fig. 7-122*B*). The membranous labyrinth is a closed system of ducts and chambers filled with endolymph and bathed by perilymph. Observe its three parts: the cochlear duct within the cochlea; the saccule and the utricle within the vestibule; and the three semicircular ducts within the three semicircular canals. Note that the utricle communicates with the saccule through the utriculosaccular duct (*arrow*). The lateral semicircular duct lies in the horizontal plane and is more horizontal than it appears in this drawing.

of sacs and ducts that communicate with one another (Figs. 7-113 and 7-124). The semicircular ducts open into the utricle through five openings, and the utricle communicates with the saccule through the *utriculosaccular duct* from which the *endolymphatic duct* arises. The saccule is continuous with the cochlear duct through a narrow communication known as the *ductus reuniens*.

The Utricle and Saccule (Figs. 7-113, 7-122, and 7-124). Each of these dilations has a specialized area of sensory epithelium called a **macula**. The *macula utriculi* is in the floor of the utricle, parallel with the base of the skull, whereas the *macula sacculi* is vertically placed on the medial wall of the saccule. The hair cells in the maculae are innervated by fibers of the vestibular division of the *vestibulocochlear nerve* (Fig. 7-125). The primary sensory neurons are in the *vestibular ganglion*, which is in the internal acoustic meatus. The *endolymphatic duct* emerges through the bone of the posterior cranial fossa and expands into a blind pouch, the *endolymphatic sac*. It is located under cover of the dura on the posterior surface of the petrous part of the temporal bone. The endolymphatic sac is a storage reservoir for excess endolymph formed by the blood capillaries within the membranous labyrinth.

> The **maculae** are primarily static organs for signaling the position of the head in space, but they also respond to quick tilting movements and to linear acceleration and deceleration. *Motion sickness* results mainly from prolonged, fluctuating stimulation of the maculae.

The Semicircular Ducts (Figs. 7-113, 7-122*B*, 7-124, and 7-125). Each duct has an ampulla or expansion at one end, containing a sensory area called a *crista ampullaris*. The cristae are sensors of movement, recording movements of the endolymph in the ampulla that result from rotation of the head in the plane of the duct. The hair cells of the cristae, like those of the maculae, are supplied by primary sensory neurons whose cell bodies are in the *vestibular ganglion* in the internal acoustic meatus (Fig. 7-123).

The Cochlear Duct (Figs. 7-113, 7-122*B* and *C*, and 7-124). This is a spiral, blind tube that is firmly fixed to the internal and external walls of the cochlear canal by the *spiral ligament*. It is triangular in transverse section and lies between the osseous spiral lamina and the external wall of the cochlear canal. The roof of the spiral cochlear duct is formed by the *vestibular membrane*, and its floor is formed by the *basilar membrane* and the external part of the osseous spiral lamina. The receptor of auditory stimuli is the **spiral organ** (organ of Corti), situated on the basilar membrane. It is overlaid by the gelatinous *tectorial membrane*. The spiral organ contains hair cells that respond to vibrations induced in the endolymph by sound waves.

Summary of the Pathway for the Conduction of Sound (Figs. 7-113, 7-117, and 7-124). Sound waves in the air are collected by the auricle and passed into the external acoustic meatus. They cause movements of the tympanic membrane and auditory ossicles. Stronger waves are transmitted to the perilymph at the fenestra vestibuli by the base of the stapes. Vibrations of the perilymph are transmitted to the basilar membrane, displacement of which (in response to acoustic stimuli), causes bending of the hair-like projections of the sensory hair cells of the **spiral organ**. These cells are in contact with the *tectorial membrane* (L. *tectum*, roof), which forms a roof over the hair cells. These cells are innervated by peripheral fibers or bipolar primary sensory neurons in the *spiral ganglion*, situated around the modiolus of the cochlea. The central processes of these nerve cells form the cochlear nerve, part of the vestibulocochlear nerve (Fig. 7-125).

subdural hemorrhage. These problems are discussed on p. 781.

Case 7-5

A 49-year-old woman developed a throbbing headache; it lasted for about 30 min. and then slowly faded away. Similar headaches occurred occasionally for the next week. One day as she was lifting a heavy chair, she experienced a sudden, severe headache that was accompanied by nausea, vomiting, and a general feeling of weakness. She decided to see her physician immediately. The physical examination detected *nuchal rigidity* and an elevation in blood pressure. Visualization of her optic fundus through an ophthalmoscope showed *subhyaloid hemorrhages* (bleeding between the retina and vitreous body). Her deep tendon reflexes were symmetrical and all modalities of sensation were normal. On the basis of these distinct signs and symptoms, the physician made a tentative diagnosis of a subarachnoid hemorrhage. *Arteriograms* showed a saccular aneurysm of the anterior communicating artery. A lumbar puncture was performed (see Fig. 4-55). Examination demonstrated bloody CSF. After centrifugation, the supernatant fluid was xanthochromatic (yellow-colored).

Diagnosis.   Subarachnoid hemorrhage owing to rupture of an aneurysm in the anterior communicating artery of the cerebral arterial circle.

> **Problems.**   Where would blood from the ruptured aneurysm most likely go? How do you explain anatomically the formation of subhyaloid hemorrhages? Why was the supernatant part of the CSF xanthochromatic? These problems are discussed on p. 781.

DISCUSSION OF PATIENTS' PROBLEMS

Case 7-1

Sudden facial paralysis may follow exposure to the cold; thus Bell's palsy was the most probable diagnosis in this case. The characteristic facial appearance results from a *lesion of the facial nerve* (CN VII). In the present patient, the motor supply to the muscles of the left face, forehead, and eyelids were most severely affected. Paralysis of the muscles of facial expression on the left side explains the expressionless look on that side of his face and his inability to whistle, puff his cheek, or close his left eye. When the facial nerve is paralyzed, the levator palpebrae superioris (acting unopposed) causes the eye to remain open, even during sleep. The drooling and difficulty in chewing result from paralysis of the orbicularis oris and buccinator muscles (Figs. 7-21 and 7-22*A*). Loss of taste sensation on the anterior two-thirds of the left side of his tongue is understandable anatomically because this region of the tongue receives taste fibers through the chorda tympani branch of CN VII (Fig. 7-96). This symptom also indicates that the nerve lesion is proximal to the origin of this nerve in the facial canal (Fig. 7-118).

Because of paralysis of the orbicularis oculi muscle, the lacrimal puncta are no longer in contact with the cornea. As a result, tears tend to flow over the left lower eyelid onto the cheek. In addition, the cornea may dry out during sleep (if an ointment is not used) because the eyelids on the affected side remain open. Drying of the cornea can also occur during the day owing to the inability to blink; this dryness could result in corneal ulceration. The site of the lesion was most likely in the facial canal in the petrous part of the temporal bone. The paralysis of the facial muscles is thought to be caused by inflammation of the facial nerve superior to the stylomastoid foramen (Fig. 7-7). The cause is generally thought to be a viral infection, which causes edema of the facial nerve and compression of its fibers in the facial canal or at the stylomastoid foramen. If the lesion is complete, all the facial muscles on that side are affected equally. Voluntary, emotional, and associated movements are all affected. In most cases the nerve fibers are not permanently damaged and nerve degeneration is incomplete. As a result, recovery is very slow but generally good. Some facial asymmetry may persist (*e.g.*, slight sagging of the left corner of the mouth).

Case 7-2

The area of skin and mucosa in which the stabbing pain was felt is supplied by the maxillary nerve (Figs. 7-24 and 7-25), the second division of the trigeminal nerve (CN V²). This wholly sensory nerve leaves the skull through the foramen rotundum (Figs. 7-41 and 7-42). At its termination as the infraorbital nerve, it gives rise to branches that supply the ala (side) of the nose, the lower eyelid, and the skin and mucous membrane of the cheek and upper lip. Branches of the maxillary nerve also innervate the teeth in the maxilla and the mucous membranes of the nasal cavities, palate, mouth, and tongue (Fig. 7-80).

The symptoms described by this patient are characteristic of *trigeminal neuralgia* (tic douloureux). It occurs most often in middle-aged and elderly persons. The pain may be so intense that the patient winces; hence, the common term *tic* (twitch). In some cases the pain may be so severe that mental changes occur; there may be depression and even suicide attempts. The maxillary nerve distribution, as in the present case, is most frequently involved, then the mandibular, and least frequently the ophthalmic. The paroxysms of sudden stabbing pain, as in the present case, are of sudden onset and are often set off by touching the face, brushing the teeth, drinking, or chewing. Often there is an especially sensitive "trigger zone," *e.g.*, the left upper lip in the present case. The complete cause of trigeminal neuralgia is unknown. Some persons believe the condition is caused by a pathological process affecting neurons in the trigeminal ganglion (Fig. 7-42), whereas others believe that neurons in the nucleus of the spinal tract may be involved (see Fig. 9-8). Medical or surgical treatment, or both are used to alleviate the pain. Only the anatomical aspects of these treatments are discussed here.

Attempts were made to block the nerve at the infraorbital foramen by using alcohol; this usually gives temporary relief of pain. The simplest surgical procedure is avulsion or cutting of the branches of the nerve at the infraorbital foramen (Fig. 7-21). Radiofrequency selective coagulation of the trigeminal ganglion by a needle electrode passing through the cheek and the foramen ovale is also used. To prevent regeneration of nerve fibers, the sensory root of the trigeminal nerve may be partially cut between the ganglion and the brain stem (*rhizotomy*). Although the axons may regenerate, they do not do so within the brain stem. Attempts are made to differentiate and cut only the sensory fibers to the division of the trigeminal nerve involved. The same result may be achieved by sectioning the spinal tract of CN V (*tractotomy*). After this operation the sensation of pain, temperature, and simple (light) touch are lost over the area of skin and mucous membrane supplied by the maxillary nerve (Fig. 7-24). This may be annoying to the patient who does not recognize the presence of food on the lip and cheek or feel it within the mouth on the side of the nerve section, but these disabilities are preferable to the excruciating pain.

Case 7-3

Carcinomas of the lip most commonly involve the lower lip. Overexposure to sunshine over many years, as occurs in outdoor workers such as farmers, is a common feature of the history in these cases. Chronic irritation from pipe smoking appears to be a factor also, which may be related to long-term contact with tobacco tar. The submental lymph nodes lie on the fascia covering the mylohyoid muscle between the anterior bellies of the right and left digastric muscles (Fig. 7-32). The central part of the lip, the floor of the mouth, and the tip of the tongue drain to these nodes, whereas lateral parts of the lip drain to the submandibular lymph nodes (Fig. 7-32). If cancer cells had spread further, metastases would have developed in the submandibular lymph nodes because efferents from the submental lymph nodes pass to them. In addition, lymph vessels from the submental lymph nodes pass directly to the jugulo-omohyoid node. As the submandibular nodes are situated beneath the deep cervical fascia in the submandibular triangle, the patient's chin may have to be lowered to slacken this fascia before these enlarged nodes can be palpated.

Because all parts of the head and neck drain into the deep cervical nodes, they might also be sites of metastases. As the jugulo-omohyoid node drains the submental and submandibular lymph nodes, it could be involved in the spread of tumor cells from a carcinoma of the lip. It is located where the omohyoid muscle crosses the internal jugular vein (Fig. 7-32).

Case 7-4

The temple is the area between the temporal line and the zygomatic arch (Fig. 7-8), where the skull is thin and is covered by the temporalis muscle and the temporal fascia. The blood vessels of the temple are very numerous. The *pterion* is a somewhat variable H-shaped area that lies deep to the temporalis muscle. Here, four bones approach each other or meet (frontal, parietal, temporal, and sphenoid). The pterion is an important bony landmark because it indicates the location of the frontal branch of the *middle meningeal artery*. The center of the pterion is 4 cm superior to the zygomatic arch and 3.5 cm posterior to the frontozygomatic suture. It lies in the anterior part of the temporal fossa. The thin squamous part of the temporal bone is grooved by the middle meningeal artery and its branches (Fig. 7-5). The temporal squama is easy to fracture, and the broken pieces may tear the artery and its branches as they pass superiorly on the external surface of the dura mater. This results in a slow accumulation of blood in the extradural space (Fig. 7-126), forming an *extradural hematoma* (epidural hematoma). The hematoma forms relatively slowly because the dura is firmly attached

to the bone by Sharpey fibers. These fibers resist stripping of the dura from the bone to a certain extent. A *subdural hematoma* is a localized mass of extravasated blood that is located on the surface of the brain, deep to the dura (Fig. 7-144).

The middle meningeal artery, a branch of the first part of the maxillary artery (Fig. 7-79), enters the skull through the foramen spinosum (Fig. 7-42). It divides within the first 4 or 5 cm of its intracranial course. The frontal branch passes superiorly from the pterion, more or less parallel to the coronal suture of the skull (Fig. 7-44A). The parietal branch passes posterosuperiorly, with its exact site depending on its point of origin. In the present case, the frontal branch of the middle meningeal artery was almost certainly torn. This artery is usually accompanied by a meningeal vein which may have also been torn. The lucid interval that followed the patient's recovery from the brief loss of consciousness (resulting from cerebral concussion) occurs because of the slow formation of the *extradural (epidural) hematoma*. In addition, this kind of a space-occupying intracranial lesion can be tolerated for a short time because some blood and CSF are squeezed out of the calvaria through the veins and subarachnoid space. However, as the cranium is nonexpansile, the intracranial pressure soon rises, producing drowsiness and then coma (G. *koma*, deep sleep).

The increased intracranial pressure forces the supratentorial part of the brain, usually the uncus, through the tentorial incisure (Fig. 7-45), squeezing the oculomotor nerve (CN III) between the brain and the sharp, free edge of the tentorium. Compression of CN III causes *third nerve palsy*, which results in a dilated, nonreacting pupil on the side of the lesion. An extradural hemorrhage in the characteristic position, illustrated by the present case, primarily causes compression of the temporal lobe underlying the pterion. Immediate surgical intervention is necessary to relieve the intracranial pressure so that further compression of the brain will not occur, which could cause death by interfering with the cardiac and respiratory centers in the medulla.

Case 7-5

Unruptured saccular aneurysms are usually asymptomatic. In the present case the initial headaches were probably caused by intermittent enlargement of the aneurysm or by slight bleeding from it into the subarachnoid space (the so-called warning leak). Her subsequent severe, almost unbearable headache was the result of gross bleeding from the aneurysm into the subarachnoid space (Fig. 7-44B). Blood in the CSF causes meningeal irritation, which produces a headache. As the anterior communicating artery is in the longitudinal fissure, rostral to the optic chiasma (Fig. 7-54B), blood escaping from the ruptured aneurysm would enter the chiasmatic cistern (Fig. 7-52) and other subarachnoid spaces around the brain and spinal cord. This explains why there was blood in the CSF obtained by lumbar puncture.

Some authorities recommend against a lumbar puncture in a case of subarachnoid hemorrhage that is so obvious as the present case because of the possibility of causing herniation of the brain. The lowering of CSF pressure in the spinal subarachnoid space by removing CSF might cause inferior movement of the brain resulting in herniation (*e.g.*, of the cerebellar tonsils).

Rupture of an aneurysm of the anterior communicating artery into the adjacent part of one frontal lobe may cause symptoms of a mass lesion in one hemisphere. In some cases, the *intracranial hematoma* may break into the ventricular system, causing an acute expansion of the ventricle, and may cause death. Blockage of subarachnoid spaces by large amounts of blood in the CSF could impair circulation of this fluid, resulting in a further increase in intracranial pressure. This could force the medial part of the temporal lobe (usually the uncus) through the tentorial notch and the cerebellar tonsils through the foramen magnum. *Herniation of the cerebellar tonsils* compresses the medulla containing the vital respiratory and cardiovascular centers and produces a life-threatening situation.

The subhyaloid hemorrhages observed during *funduscopy* resulted from the abrupt rise in intracranial pressure transmitted to the subarachnoid space around the optic nerve (Fig. 7-62). This compressed and obstructed the central retinal vein where it crosses this space. This results in increased pressure in the retinal capillaries and hemorrhages between the retina and vitreous body. After centrifugation of the CSF, the supernatant fluid was yellow because it contained serum bilirubin and products of hemolyzed red blood cells.

SUGGESTED READINGS

Barr ML, Kiernan JA: *The Human Nervous System: An Anatomical Viewpoint*, ed 5. Philadelphia, JB Lippincott, 1988.

Behrman RE: *Nelson's Textbook of Pediatrics*, ed 14. Philadelphia, WB Saunders, 1992.

Bertram EG, Moore KL: *An Atlas of the Human Brain and Spinal Cord*, Baltimore, Williams & Wilkins, 1982.

Carpenter MB: *Core Text of Neuroanatomy*, ed 3. Baltimore, Williams & Wilkins, 1985.

Haines DE: *Neuroanatomy. An Atlas of Structures, Sections, and Systems*, ed 2. Baltimore, Urban & Schwarzenberg, 1987.

Haines DE: On the Question of a Subdural Space, *Anat Rec*, 230:3-21, 1991.

Liebgott B: *The Anatomical Basis of Dentistry*, Philadelphia, BC Decker, 1986.

Moore KL: *The Developing Human: Clinically Oriented Embryology*, ed 4. Philadelphia, WB Saunders, 1988.

Mortimer CB, Kraft S: Ophthalmology. *In* Gross A, Gross P, Langer B (Eds): *Surgery: A Complete Guide for Patients and Their Friends*, Toronto, Harper & Collins, 1989.

Warwick R: *Nomina Anatomica*, ed 6. Edinburgh, Churchill Livingstone, 1989.

Williams PL, Warwick R, Dyson M, Bannister LH: *Gray's Anatomy*, ed 37. New York, Churchill Livingstone, 1989.

Many important structures are crowded together in the neck. It contains vessels, nerves, and other structures connecting the head, trunk, and upper limbs (*e.g.*, the carotid arteries, jugular veins, vagus nerves, lymphatics, trunks of the brachial plexus of nerves, vertebrae, muscles, esophagus, and trachea). The neck also contains important endocrine glands (*e.g.*, the thyroid and parathyroid glands). Several medical words are derived from the Latin word *collum*, meaning the neck. For example, **torticollis** (L. *tortus*, twisted + *collum*, neck) is the medical term for wryneck (stiffneck). *Cervix* is another Latin word for the neck; hence, the *cervical plexus* is a network of nerves in the neck (p. 792) and the *cervical triangles* are areas of the neck used for descriptive purposes. As lymphatic vessels in the head drain into *cervical lymph nodes*, their enlargement may indicate a tumor in the head, but the primary cancer could also be in the thorax or abdomen because the neck connects the trunk and head (*e.g.*, bronchogenic cancer, p. 77). There are several causes of *a pain in the neck* such as inflamed lymph nodes, muscle strain, and protrusion of a cervical intervertebral disc (p. 347).

Surface Anatomy of the Neck

The posterior aspect of the neck was described with the back (p. 327). Review this material, noting that the spinous process of the *axis* (C2 vertebra) is the first bony prominence that can be felt in the median plane inferior to the *external occipital protuberance* (Fig. 7-3). The spinous process of the **vertebra prominens** (C7) is easily palpable and is usually clearly visible when the neck is flexed (see Fig. 4-8). The *laryngeal prominence* ("Adam's apple")[1] is the important surface feature of the anterior part of the neck (Fig. 8-1). It is produced by the *thyroid cartilage*, the largest part of the laryngeal skeleton (Figs. 8-2 and 8-3). The superior part of the thyroid cartilage is most prominent and is more noticeable in men than in women and children, and in some men more than in others.

The sex difference in the angle formed by the laminae of the thyroid cartilage explains why the **laryngeal prominence** is more noticeable in men. In Figure 8-2C and D, note that the angle formed by the convergence of the laminae in the median plane is greater in women than in men. Also observe that each lamina has a greater anteroposterior breadth. Because of these anatomical differences, the superior border of the thyroid cartilage in most men projects anteriorly, producing a distinct laryngeal prominence. These sexual differences in the thyroid cartilage develop during *puberty* usually between 14 and 16 years of age.

The thyroid cartilage (Figs. 8-1 to 8-3 and 8-23) is located at the level of the fourth and fifth cervical vertebrae. It consists of two quadrilateral plates called *laminae*, which can easily be felt. Grasp the laminae between your first and second digits and move your thyroid cartilage from side to side. Also note that it rises when you swallow. The *vocal folds* (vocal cords) lie about level with the midpoint of the anterior border of the thyroid cartilage.

The hyoid bone[2] (Figs. 8-3, 8-14, and 8-17)) lies superior to the thyroid cartilage. This U-shaped bone is located at the level of the body of the third cervical vertebra. You can feel the body of the hyoid bone in the angle between the floor of the mouth and the anterior aspect of the neck. It is the first resistant structure that can be felt in the median plane inferior to the chin. It is easier to feel during swallowing. When the neck is relaxed, you can palpate the tips of the greater horns (L. *cornua*) of the hyoid bone, especially if you steady one side with your second digit (index finger).

The transverse processes of the atlas (C1 vertebra) can also be felt by deep palpation. Press superiorly with your second digit between the angle of your mandible and a point about 1 cm anteroinferior to the tip of your *mastoid process* (Fig. 8-3). As you do this, rotate your head slowly from side to side. During this movement, the skull and atlas (C1) rotate as a unit on the axis (C2).

The cricoid cartilage (Figs. 8-3, 8-36, and 8-60), part of the laryngeal skeleton, lies inferior to the thyroid cartilage. It can be felt inferior to the laryngeal prominence. Extend your neck and run your fingertip inferiorly from your chin over your hyoid bone and thyroid and cricoid cartilages. After you pass the cricoid, note that your fingertip sinks in because the arch of the cricoid cartilage projects anteriorly beyond the rings of the trachea. The cricoid cartilage lies at the level of C6 vertebra, where the pharynx joins the esophagus and the larynx and trachea join each other.

The tracheal rings (Figs. 8-3, 8-32, and 8-60) are palpable in the inferior part of the neck. These cartilaginous rings are usually not palpable just inferior to the cricoid cartilage because the isthmus of the thyroid gland lies anterior to them (Fig. 8-35). Grasp the trachea between your first and second digits, just superior to the jugular notch. Verify that it moves superiorly during swallowing.

The jugular notch (suprasternal notch) is a rounded depression in the manubrium of the sternum that is easily palpable between the medial (sternal) ends of the clavicles (see Fig. 1-6) and is clearly visible (Fig. 8-1). Put your second digit in your jugular notch and carefully press posteriorly until you feel your trachea.

---

[1]Supposedly the laryngeal prominence is where the forbidden apple stuck in Adam's throat. It is a fanciful way of accounting for the prominence of the thyroid cartilage in many men. The basis for this sex difference is in the anatomy of the thyroid cartilage of the larynx, which is illustrated in Figure 8-2.

[2]The term *hyoid* is derived from the Greek *hyoeides*, which means U-shaped like the Greek letter upsilon (v).

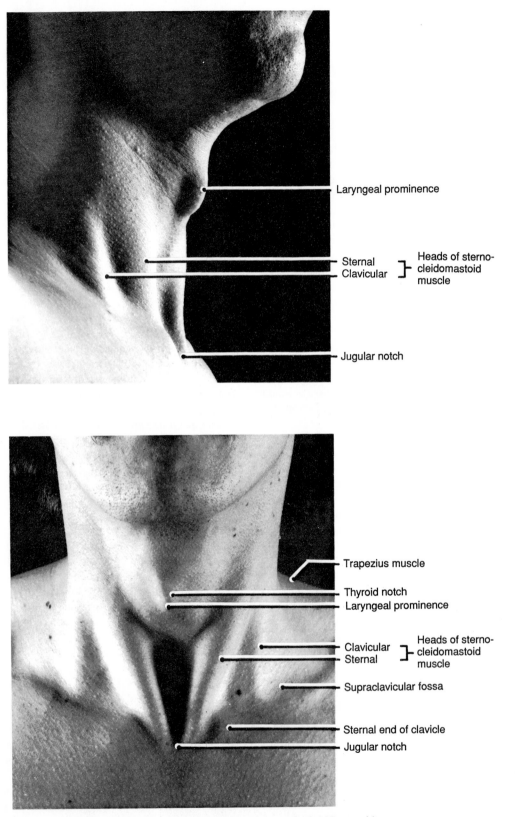

Figure 8-1.   Surface anatomy of the neck of a 28-year-old man.

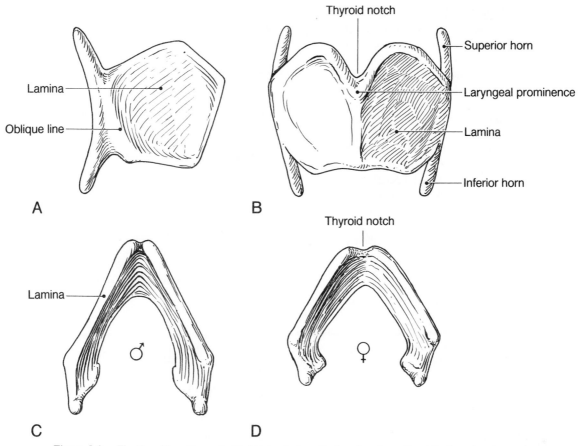

Figure 8-2.   The thyroid cartilage. *A*, Right lateral view. *B*, Anterior view. *C* and *D*, Superior views of thyroid cartilages from a man and a woman.

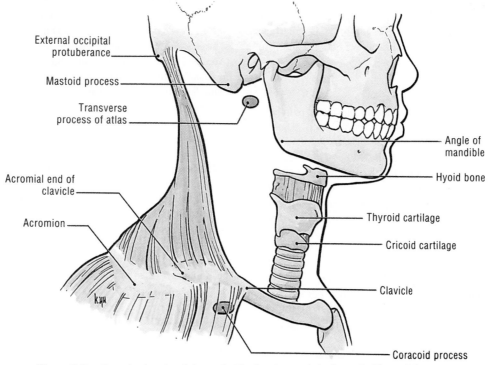

Figure 8-3.   Bony landmarks of the neck. The jugular notch is shown in Figure 8-1.

Turn your head to the right and move your digit to the left; it will encounter the tendon of the sternal head of the *sternocleidomastoid muscle* (Fig. 8-1). Let your digit pass over this tendon into a slight depression between the two heads of this muscle. Now run your digit over the clavicular head of this muscle and you will feel a triangular depression, called the *supraclavicular fossa*. This fossa is clinically important because it contains the *pressure point for the subclavian artery* (Fig. 8-12; see also Fig. 1-5 and p. 791). The medial ends of the clavicles are clearly visible at the root of the neck, particularly in thin persons (Fig. 8-1). Palpate one of your clavicles, verifying that it is subcutaneous throughout its length. The bony landmarks that form the *superior limit of the neck* are: the inferior margin of the *mandible* (lower jaw bone), the *mastoid process* of the temporal bone, and the *external occipital protuberance* (Fig. 8-3).

Put your hand on one of your shoulders and then shrug it as you feel the rounded edge of the large *trapezius muscle* (Figs. 8-1, 8-3, and 8-7). Verify by palpation that it extends from the external occipital protuberance to the bones of the *pectoral (shoulder) girdle*, which is composed of the clavicle and scapula (see Figs. 6-1 and 6-3). The trapezius, a muscle of the upper limb, extends over the posterior aspect of the neck and attaches the pectoral girdle to the skull and vertebral column. The anterior border of the trapezius marks the posterior limit of the side of the neck, and the anterior median line of the neck demarcates its anterior limit. The *sternocleidomastoid muscle* divides the lateral side of the neck into anterior and posterior triangles (Fig. 8-4) and forms an important landmark. When contracted, it forms a prominent ridge (Fig. 8-1). Although you can define the boundaries of these *cervical triangles* by palpation, they can be seen better during dissection after the skin and *platysma muscle* have been reflected (Figs. 8-6 to 8-8).

The *skin of the neck* is thin and pliable (Fig. 8-5). The subcutaneous connective tissue contains cutaneous nerves and superficial veins (Fig. 8-7). The flat platysma muscle, ascending on the face from the anterior aspect of the neck, is superficial to the main parts of these veins and nerves (Figs. 8-5 and 8-6).

Superficial Neck Muscles

There are three superficial muscles in the neck: platysma, sternocleidomastoid, and trapezius (Figs. 8-5 to 8-8). The last one is described in Chapter 6 (p. 530).

The Platysma Muscle (Figs. 8-5 and 8-6). This wide, thin subcutaneous sheet of striated muscle is located in the superficial fascia. It ascends to the face from the anterior part of the neck. The platysma (G. a flat plate) covers the superior part of the anterior cervical triangle and the anteroinferior part of the posterior cervical triangle. Its fibers blend superiorly with the facial muscles.

Inferior Attachment (Figs. 8-5, 8-6, and 8-8). Fascia and skin over the pectoralis major and deltoid muscles.

Superior Attachment (Fig. 8-6). Inferior border of the mandible and skin of the lower face.

Innervation (Fig. 8-7). Cervical branch of the facial nerve (CN VII).

Actions (Fig. 8-5). This thin muscle tenses the skin of the neck (*e.g.*, during shaving). It also draws the corners of the mouth inferiorly and assists in depressing the mandible. Acting superiorly, it produces skin ridges in the neck, releasing the pressure of the skin on the underlying veins. It thereby serves to ease, for example, a tight shirt collar. When the entire muscle contracts, it wrinkles the skin of the neck in an oblique direction and widens the mouth.

The platysma is one of several muscles of facial expression that we use to express sadness, horror, or fright. It also acts during violent deep inspiration (*e.g.*, after a long race). Although the platysma is thin and may be absent in some people, it tenses the skin of the neck in most people. Its fibers are usually apparent in thin and elderly people. This muscle is well developed in animals such as horses; they use it to shake flies off their necks.

When the platysma is paralyzed owing to injury of the cervical branch of the facial nerve, the skin tends to fall away from the neck in slack folds. Hence, *during surgical dissections of the neck*, care is taken to preserve the mandibular and cervical branches of the facial nerve (Figs. 7-22, 7-27, and 8-7). Injury to the mandibular branch of CN VII produces a noticeable facial deformity, whereas damage to the cervical branch produces unsightly postoperative defects (*e.g.*, the skin of the neck droops). Similarly, surgeons carefully suture the edges of the platysma after neck surgery, to prevent gaping of the skin incisions.

The Sternocleidomastoid Muscle (Figs. 8-1, 8-4, and 8-7). The sternocleidomastoid (sternomastoid),[3] a broad straplike muscle, is the *key landmark in the neck*. Running superolaterally from the sternum and clavicle to the lateral surface of the mastoid process, this muscle *divides the side of the neck into anterior and posterior triangles*, which are useful for descriptive purposes (Fig. 8-4). When it contracts, the sternocleidomastoid stands out as a well-defined prominence between the anterior and posterior triangles. The sternocleidomastoid is crossed by the platysma and external jugular vein. It covers the great vessels of the neck and the *cervical plexus* of nerves (Figs. 8-7, 8-12, and 8-16).

Inferior Attachment (Figs. 8-4 and 8-7). *Sternal head*: by a rounded tendon, which is attached to the anterior surface of the manubrium of the sternum, lateral to the jugular notch. *Clavicular head*: superior surface of the medial third of the clavicle.

Superior Attachment (Figs. 8-3, 8-4, and 8-7; see also Fig. 7-3). Lateral surface of the mastoid process of the temporal bone and the lateral half of the superior nuchal line of the occipital bone.

Innervation (Figs. 8-7 and 8-13). Spinal root of the accessory nerve (*CN XI*) and branches of the second and third cervical nerves (C2 and C3).

Actions. Acting alone, the sternocleidomastoid muscle *tilts the head to its own side* by drawing the mastoid process inferiorly (*i.e.*, it laterally flexes [bends] the neck and rotates it so the face is turned superiorly toward the opposite side). *Acting together*, these muscles flex the neck (*e.g.*, when raising the head from a pillow).

---

[3]The new name for this muscle is more informative because it indicates its complete inferior attachment to the sternum and the clavicle. The "cleido" part of its name is derived from the Greek work *kleis*, meaning clavicle.

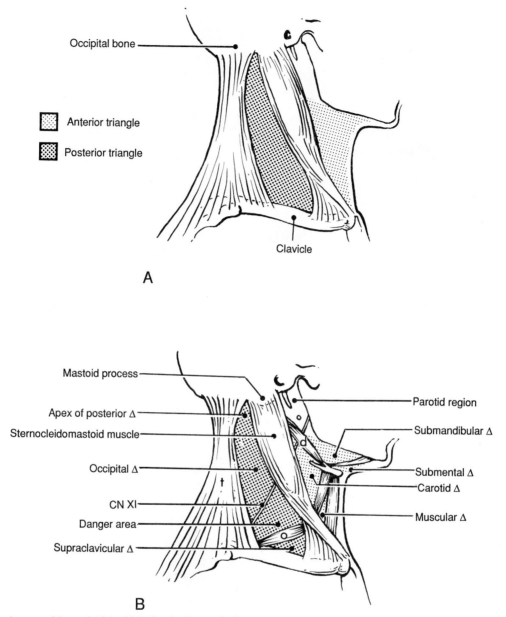

Figure 8-4.   Lateral aspect of the neck (right side), showing the cervical triangles that are used for descriptive purposes. *A,* The boundaries of the posterior triangle are: the clavicles, the posterior border of the sternocleidomastoid muscle, and the anterior border of the trapezius muscle. The boundaries of the anterior triangle are: the anterior border of the sternocleidomastoid muscle, the inferior margin of the mandible, and the anterior median line of the neck. *B,* Subdivisions of the cervical triangles. *O* indicates omohyoid; *D,* digastric; *T,* trapezius. Observe that the accessory nerve (CN XI) divides the posterior triangle into superior and inferior parts. Observe also that the inferior belly of the omohyoid muscle divides the posterior triangle into a large occipital triangle (superiorly) and a smaller supraclavicular triangle (inferiorly).

Occasionally the sternocleidomastoid muscle is injured at birth, resulting in a condition known as *congenital torticollis* or *wryneck* (Fig. 8-9). There is fixed rotation and tilting of the head owing to contracture of this muscle. Stiffness of the neck results from fibrosis and shortening of the muscle on one side. Because torticollis is a correctable condition, it is not usually seen in a more advanced form than that illustrated in Figure 8-9. Most cases of congenital torticollis result from tearing of fibers of the sternocleidomastoid when pulling the infant's head excessively during a difficult birth, particularly in a breech presentation.

Triangles of the Neck

Each side of the neck is divided into two triangles, anterior and posterior, by the sternocleidomastoid muscle (Fig. 8-4). These *cervical triangles,* which are useful for descriptive purposes, have a common boundary, the sternocleidomastoid.

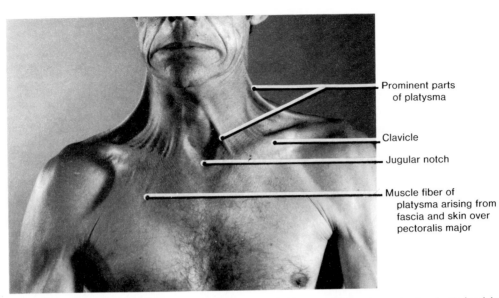

Figure 8-5. Surface anatomy of the platysma muscle. A 46-year-old man produced wrinkling of his skin in an oblique direction when he contracted his platysma. He also violently clenched his jaws, drawing down the corners of his mouth. Observe the fibers arising from the fascia and skin over the superior pectoral and anterior deltoid regions, particularly on the right side. Note that the sheet of muscle passes over the clavicle and the anterolateral aspects of the neck. Often, its fibers blend with the depressor anguli oris muscle (Fig. 8-6).

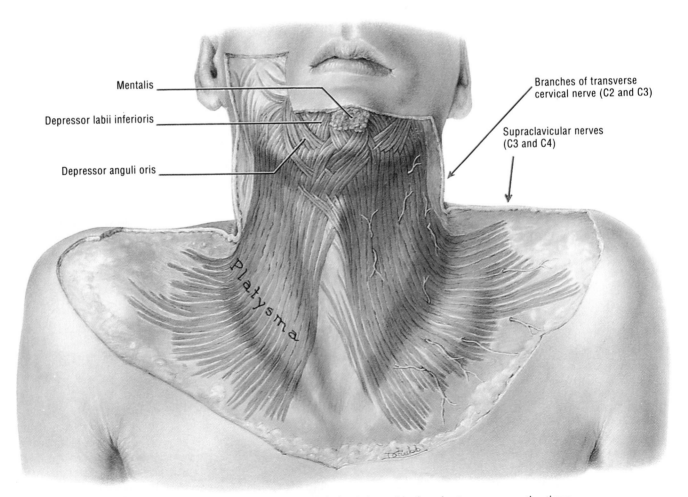

Figure 8-6. Dissection of the platysma muscle that is located in the subcutaneous connective tissue.

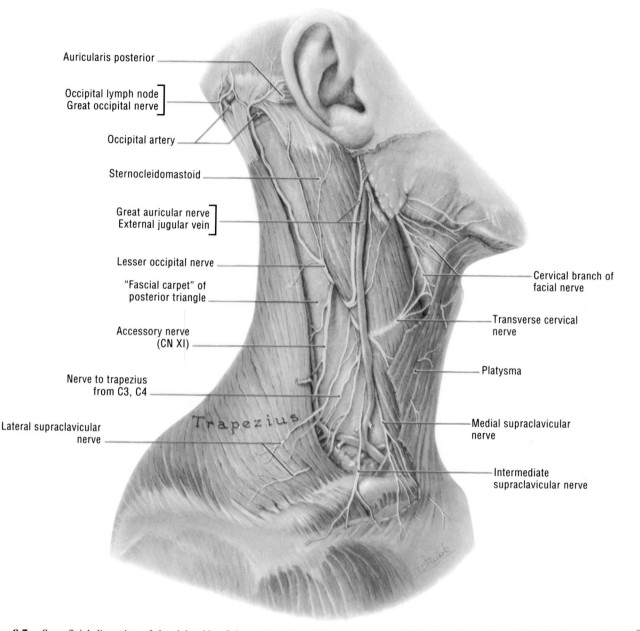

Auricularis posterior

Occipital lymph node
Great occipital nerve

Occipital artery

Sternocleidomastoid

Great auricular nerve
External jugular vein

Lesser occipital nerve

"Fascial carpet" of
posterior triangle

Accessory nerve
(CN XI)

Nerve to trapezius
from C3, C4

Lateral supraclavicular
nerve

Trapezius

Cervical branch of
facial nerve

Transverse cervical
nerve

Platysma

Medial supraclavicular
nerve

Intermediate
supraclavicular nerve

Figure 8-7. Superficial dissection of the right side of the neck, primarily to illustrate the posterior cervical triangle and its contents (Fig. 8-4). The superficial fascia and the superficial layer of the deep fascia have been removed. The deep layer of cervical fascia forms a "fascial carpet" for the floor of the posterior triangle. Note the accessory nerve, the only motor nerve superficial to this "carpet."

Posterior Triangle of the Neck

Boundaries of the Posterior Cervical Triangle (Figs. 8-4, 8-7, and 8-8). This triangle is bounded: *anteriorly* by the posterior border of the sternocleidomastoid; *posteriorly* by the anterior border of the trapezius muscle, and *inferiorly* by the middle third of the clavicle. The clavicle forms the *base* of the posterior triangle; its *apex* is formed where the borders of the sternocleidomastoid and trapezius muscles meet on the *superior nuchal line* of the occipital bone of the skull (see Fig. 7-3). The *occipital artery* passes through the apex of the posterior triangle before it ascends over the posterior aspect of the head.

Roof of the Posterior Cervical Triangle (Figs. 8-5 to 8-8). The posterior triangle is covered by deep fascia, which covers the space between the trapezius and sternocleidomastoid muscles. Superficial to the *deep fascial roof* are the superficial fascia, platysma, superficial veins, cutaneous nerves, and skin.

Floor of the Posterior Cervical Triangle (Figs. 8-7, 8-8, 8-10, 8-28, and 8-31). The *fascial and muscular floor* of this triangle is formed (superior to inferior) by the splenius capitis, levator scapulae, scalenus medius, and scalenus posterior muscles (Table 8-1). These muscles are covered by the prevertebral layer of *deep cervical fascia*. This "fascial carpet" is a lateral prolongation of the *prevertebral fascia*.

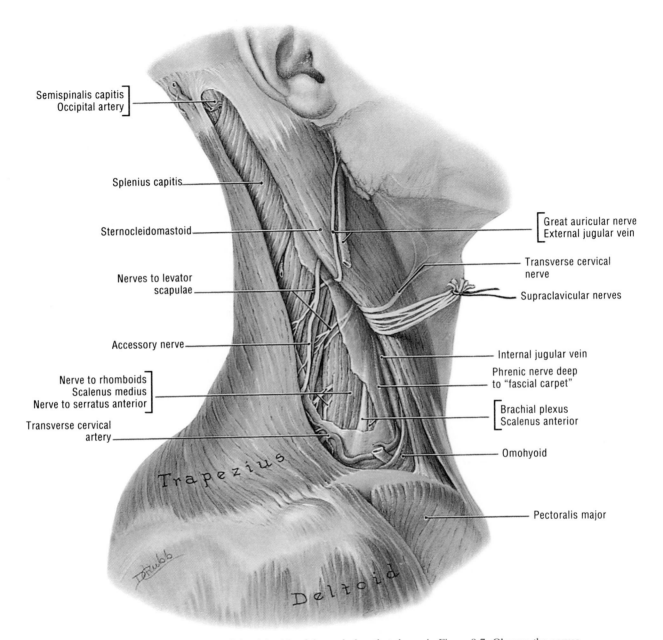

Semispinalis capitis
Occipital artery

Splenius capitis

Sternocleidomastoid

Nerves to levator
scapulae

Accessory nerve

Nerve to rhomboids
Scalenus medius
Nerve to serratus anterior

Transverse cervical
artery

Great auricular nerve
External jugular vein

Transverse cervical
nerve

Supraclavicular nerves

Internal jugular vein

Phrenic nerve deep
to "fascial carpet"

Brachial plexus
Scalenus anterior

Omohyoid

Pectoralis major

Trapezius

Deltoid

Figure 8-8.   Deeper dissection of the right side of the neck than that shown in Figure 8-7. Observe the nerves that are deep to the "fascial carpet" covering the muscular floor of the posterior triangle.

Contents of the Posterior Triangle

The posterior cervical triangle contains mostly vessels and nerves that connect the neck and upper limb (Fig. 8-12).

Veins in the Posterior Cervical Triangle (Figs. 8-7, 8-8, and 8-12).   The *external jugular vein* begins near the angle of the mandible, just inferior to the lobule of the auricle, by the union of the posterior division of the retromandibular vein with the posterior auricular vein. It crosses the sternocleidomastoid muscle in the superficial fascia and then pierces the deep fascial roof of the posterior cervical triangle at the posterior border of this muscle, about 5 cm superior to the clavicle. The external

jugular vein passes obliquely through the inferior part of the posterior triangle and usually ends by emptying into the subclavian vein about 2 cm superior to the clavicle. The external jugular vein drains most of the scalp and face on the same side.

The *subclavian vein* is the major venous channel draining the upper limb. It lies posterior to the clavicle and so is not really in the posterior triangle, but it may rise as far superiorly as the subclavian artery.

When venous pressure is within the normal range, the external jugular vein is either invisible or observable for only a short distance superior to the clavicle. However, when ven-

Not infrequently, the suprascapular artery arises from the third part of the subclavian artery, lateral to the scalenus anterior muscle, and passes anterior to the inferior and middle trunks of the *brachial plexus* of nerves and posterior to the superior trunk of this plexus (Fig. 8-12).

The *occipital artery* (Figs. 8-7, 8-8, and 8-24; see also Figs. 7-28 and 7-35), a branch of the external carotid artery, enters the apex of the posterior triangle before ascending over the posterior aspect of the head to supply the posterior half of the scalp.

Nerves in the Posterior Cervical Triangle (Figs. 8-4*B*, 8-7, 8-8, 8-11, and 8-12). The accessory nerve (CN XI) divides the posterior triangle into nearly equal superior and inferior parts. It runs between the sternocleidomastoid and trapezius muscles and supplies motor fibers to both of them. The **accessory nerve** enters the posterior triangle at or inferior to the junction of the superior and middle thirds of the posterior border of the sternocleidomastoid muscle. It passes posteroinferiorly through the triangle and then disappears deep to the anterior border of the trapezius muscle at the junction of its superior two-thirds with its inferior one-third (Fig. 8-5*B*). The superior part of the posterior cervical triangle contains only the *lesser occipital nerve*, which supplies the scalp (Fig. 8-7; see also Fig. 7-35), whereas the inferior part contains numerous important nerves, *e.g.*, the *ventral rami of the brachial plexus* (Fig. 8-12).

The accessory nerve (CN XI) is a motor nerve consisting of spinal and cranial roots (Fig. 8-13). The spinal root is composed of fibers that arise from cervical segments of the spinal cord (C1 to C5). They pass superiorly in the subarachnoid space to enter the posterior cranial fossa through the foramen magnum. Here they join the cranial root of the accessory nerve, the fibers of which originate in the medulla of the brain stem (see Fig. 7-51*A*). Both roots leave the skull through the jugular foramen (Fig. 8-27; see also Fig. 7-11). CN XI is described in more detail in Chapter 9 (p. 873).

The spinal root of the accessory nerve separates immediately from the cranial root and passes posteroinferiorly. It supplies the sternocleidomastoid muscle (Figs. 8-7 and 8-8) and then crosses the posterior cervical triangle, superficial to the deep fascia covering its floor, and supplies the trapezius muscle. It passes deep to this muscle about 5 cm superior to the clavicle.

Lesions of the accessory nerve are uncommon. CN XI may be damaged by: (1) traumatic injury; (2) tumors at the base of the skull; (3) fractures involving the jugular foramen; and (4) neck lacerations. Although contraction of one sternocleidomastoid muscle turns the head to one side, a unilateral lesion of CN XI usually does not produce an abnormality in the position of the head. However, weakness in turning the head to the opposite side against resistance can usually be detected in these patients. *Unilateral paralysis of the trapezius muscle* is evident by the patient's inability to elevate and retract the shoulder, and by difficulty in elevating the arm superior to the horizontal level. The normal ridge in the neck (nuchal ridge) formed by the trapezius is also depressed, and *drooping of the shoulder* is an obvious sign of injury to the spinal root of CN XI. During extensive *surgical dissections* in the posterior cervical triangle (*e.g.*, for removal of malignant lymph nodes), the accessory nerve is isolated to preserve it, if possible.

Figure 8-9. An 8-year-old boy with congenital torticollis (wryneck). Shortening of the right sternocleidomastoid muscle has tilted his head to the side of the contracture and rotated it so that his chin points to the left.

ous pressure is raised (*e.g.*, owing to *heart failure*), the external jugular vein becomes prominent throughout its course along the side of the neck (Fig. 8-7). Consequently, routine observation of this vessel during physical examinations may give diagnostic signs of heart failure, *obstruction of the superior vena cava* (*e.g.*, by tumor cells), enlarged supraclavicular lymph nodes, or *increased intrathoracic pressure*. Should an external jugular vein be lacerated where it pierces the roof of the posterior triangle, along the posterior border of the sternocleidomastoid (Fig. 8-7), air may be sucked into the vein during inspiration. This may occur because the vein does not retract (collapse) at this site owing to the attachment of its walls to the deep fascia (''fascial carpet'') of the posterior cervical triangle. A *venous air embolism* produced in this way fills the right heart with froth and practically stops blood flow through it. This results in *dyspnea* (G. bad breathing), cyanosis (G. blue color), and sometimes death.

Arteries in the Posterior Cervical Triangle (Figs. 8-8, 8-11, and 8-12). The *third part of the subclavian artery*, the large vessel supplying blood to the upper limb, begins about a fingerbreadth superior to the clavicle, opposite the lateral border of the scalenus anterior muscle. It is hidden in the inferomedial part of the posterior triangle, barely qualifying as one of its contents. The artery is in contact with the first rib posterior to the scalenus anterior muscle (Fig. 8-26). This *pressure point of the subclavian artery* can be compressed against this rib to control bleeding in the upper limb. The *transverse cervical artery* (Figs. 8-11 and 8-24) arises from the thyrocervical trunk, a branch of the subclavian artery. The transverse cervical artery runs superficially and laterally across the posterior triangle, 2 to 3 cm superior to the clavicle and deep to the omohyoid muscle to supply muscles in the scapular region (see Figs. 6-31 and 6-32). The *suprascapular artery* (Figs. 8-12 and 8-24), another branch of the thyrocervical trunk, passes inferolaterally across the inferior part of the posterior triangle, just superior to the clavicle. It then runs posterior to the clavicle to supply muscles around the scapula (see Figs. 6-31 and 6-32).

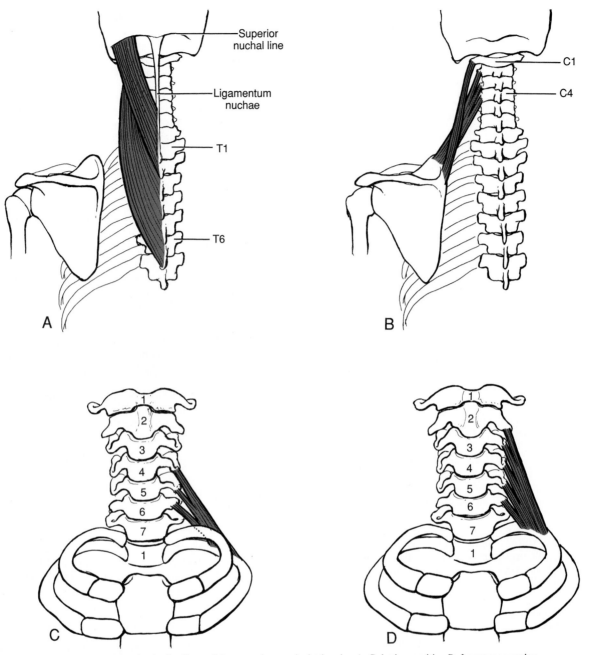

Figure 8-10.   Muscles in the floor of the posterior cervical triangle. *A*, Splenius capitis. *B*, Levator scapulae.
C, Scalenus posterior. *D*, Scalenus medius.

The Cervical Plexus of Nerves (Figs. 8-7, 8-8, and 8-26; see also Fig. 7-24*A* and *B*).   This network of nerves is formed by the communications between the ventral rami of the superior four cervical nerves. The plexus lies deep to the internal jugular vein and the sternocleidomastoid muscle. Cutaneous branches from the plexus emerge around the middle of the posterior border of the sternocleidomastoid to supply the skin of the neck and scalp, between the auricle and external occipital protuberance. The following nerves are derived from the ventral rami of C2 to C4 through the cervical plexus.

The *lesser occipital nerve* (C2, sometimes C3 also) ascends a short distance along the posterior border of the sternocleidomastoid muscle (Fig. 8-7) before dividing into several branches that supply the skin of the neck and scalp posterior to the auricle (ear) and the superior part of the auricle (see Fig. 7-24*A*).

The *great auricular nerve* (C2 and C3) curves over the posterior border of the sternocleidomastoid muscle and ascends vertically toward the parotid gland (Figs. 8-7 and 8-8; see also Fig. 7-22*A*). It supplies branches to the skin of the neck and then divides into anterior and posterior branches, which supply the

Table 8-1.
Muscles in the Floor of the Posterior Triangle of the Neck[1]

Muscle	Origin	Insertion	Innervation	Actions
Splenius capitis (Figs. 8-10*A* and 8-11)	Inferior half of ligamentum nuchae and spinous processes of superior six thoracic vertebrae	Lateral aspect of mastoid process and lateral third of superior nuchal line	Dorsal rami of middle cervical spinal nn.	Laterally flexes and rotates head and neck to same side; acting bilaterally, they extend head and neck
Levator scapulae (Figs. 6-40, 8-10*B*, and 8-11)	Posterior tubercles of transverse processes of C1 to C4 vertebrae	Superior part of medial border of scapula	Dorsal scapular n. (C5) and cervical spinal nn. (C3 and C4)	Elevates scapula and tilts its glenoid cavity inferiorly by rotating scapula
Scalenus posterior (Figs. 8-10*C* and 8-11)	Posterior tubercles of transverse processes of C4 to C6 vertebrae	External border of second rib	Ventral rami of cervical spinal nn. (C7 and C8)	Flexes neck laterally; elevates second rib during forced inspiration
Scalenus medius (Figs. 8-10*D* and 8-12)	Posterior tubercles of transverse processes of C2 and C7 vertebrae	Superior surface of first rib, posterior to groove for subclavian a.	Ventral rami of cervical spinal nn. (C3 to C8)	Flexes neck laterally; elevates first rib during forced inspiration

[1]Sometimes the *semispinalis capitis muscle* appears in the floor near the apex of the posterior cervical triangle (Fig. 8-8; see also Figs. 4-42 and 4-46). This muscle forms the largest mass in the posterior aspect of the neck. It extends the head and turns the face toward the opposite side. A portion of the inferior part of the *scalenus anterior muscle* oftens appears deeply in the inferomedial part of the posterior triangle (Fig. 8-11).

skin on the inferior part of the auricle on both surfaces, and an area extending from the mandible to the mastoid process.

The *transverse cervical nerve* (transverse nerve of the neck) from C2 and C3 curves around the posterior border of the sternocleidomastoid muscle near its middle, and then passes transversely across it. Its branches supply the skin over the anterior triangle of the neck (Figs. 8-4*A*, 8-7, and 8-8; see also Fig. 7-24*C*).

The *supraclavicular nerves* (C3 and C4) arise as a single trunk (Figs. 8-7 and 8-8), which divides into medial, intermediate, and lateral branches. They send small branches to the skin of the neck and then pierce the deep fascia just superior to the clavicle to supply the skin over the anterior aspect of the chest and shoulder. The medial and lateral supraclavicular nerves also supply the sternoclavicular and acromioclavicular joints, respectively (see Figs. 6-5, 6-21*A*, and 6-114).

The **phrenic nerve**, the sole motor nerve supply to the diaphragm (see Figs. 1-42 and 1-68), arises from the ventral primary rami of the C3 to C5. An important *muscular branch of the cervical plexus*, the phrenic nerve curves around the lateral border of the scalenus anterior muscle (Fig. 8-8). It then descends obliquely across its anterior surface, deep to the transverse cervical and suprascapular arteries. The phrenic nerve *enters the thorax* by crossing the origin of the internal thoracic artery (Fig. 8-24) between the subclavian artery and vein (Fig. 8-12). This nerve is also discussed on page 813.

The contribution to the phrenic nerve from the ventral ramus of C5 may be derived from an *accessory phrenic nerve*. It lies lateral to the main phrenic nerve (Figs. 8-11 and 8-12) and descends anterior and sometimes posterior to the subclavian vein. The accessory phrenic nerve, a part of the phrenic nerve, joins it in the root of the neck near the first rib, but it may not do so until it is in the thorax.

Close to their origin, the nerves of the cervical plexus receive *rami communicantes*, most of which descend from the large **superior cervical ganglion**, which is located in the superior part of the neck (Fig. 8-26). These rami may pass through the muscles and be difficult to identify. Injury to the sympathetic pathway on one side of the neck results in *Horner syndrome* (p. 815).

The Supraclavicular Part of the Brachial Plexus of Nerves (Figs. 8-8, 8-11, and 8-12).   The three trunks of the large and important **brachial plexus** are located in the posterior cervical triangle, superior to the clavicle. This *network of nerves*, derived from the ventral rami of C5 to C8 and T1, provides innervation for most of the upper limb (p. 512). The supraclavicular part of the plexus lies anterior to the scalenus medius muscle and the first digitation of the serratus anterior muscle. Its infraclavicular part is located in the axilla (see Fig. 6-16). Branches of the ventral primary rami of cervical nerves supply the rhomboid, serratus anterior, and nearby prevertebral muscles (Fig. 8-11; see also Fig. 6-40). Along the lateral border of the brachial plexus, the *suprascapular nerve* runs across the posterior triangle to supply the supraspinatus and infraspinatus muscles, which join the scapula to the humerus (see Fig. 6-44 and Table 6-6). The **nerve point of the neck**, a clinical designation, is located around the midpoint of the posterior border of the sternocleidomastoid muscle (Figs. 8-7 and 8-8). Several nerves lie superficially here, deep to the platysma, and are vulnerable during surgical dissections in the posterior cervical triangle.

Knife and other wounds in the neck may sever the relatively superficial nerves in the posterior triangle, resulting in *loss of cutaneous sensation in the neck* and the posterior part of the scalp (see Fig. 7-24*A* and *B*). For regional anesthesia before surgery, **nerve blocks**[4] are performed by injecting an anesthetic agent around the nerves of the cervical and brachial plexuses (Figs. 8-7, 8-11, and 8-12). The anesthetic agent *blocks nerve impulse conduction* similar to that which occurs when a dentist blocks (''freezes'') the nerves before performing dental work (see Fig. 7-80).

Cervical Plexus Block (Figs. 8-7, 8-8, 8-11, and 8-12).   The anesthetic agent is injected at several points along the posterior border of the sternocleidomastoid muscle. The main injection site is at the junction of its superior and middle thirds, *i.e.*, *around the nerve point of the neck*. Because the phrenic nerve, which supplies half the diaphragm, is usually

---

[4]Prevention of pain by temporarily arresting nerve impulses.

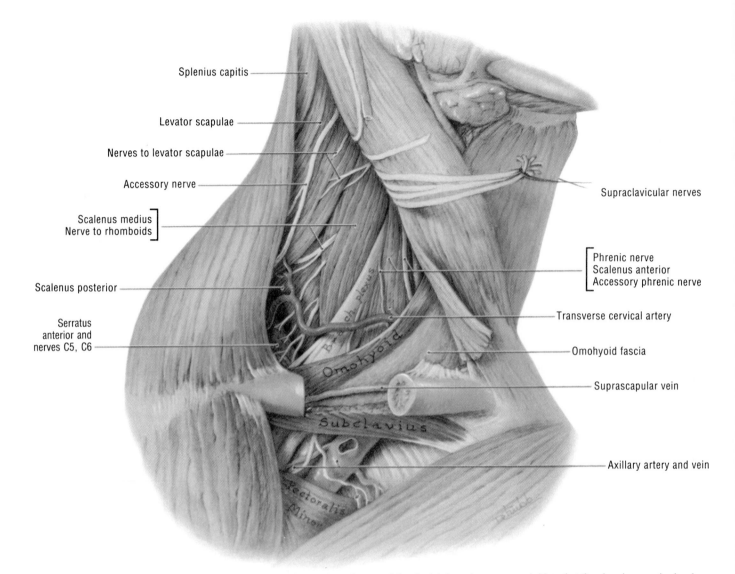

Splenius capitis

Levator scapulae

Nerves to levator scapulae

Accessory nerve

Scalenus medius
Nerve to rhomboids

Scalenus posterior

Serratus
anterior and
nerves C5, C6

Supraclavicular nerves

Phrenic nerve
Scalenus anterior
Accessory phrenic nerve

Transverse cervical artery

Omohyoid fascia

Suprascapular vein

Axillary artery and vein

Brach. plexus

Omohyoid

Subclavius

Pectoralis
Minor

Figure 8-11.   Dissection of the right side of the neck and superolateral part of the thorax, showing the muscles forming the floor of the posterior cervical triangle. The deep fascia (''carpet'') covering the floor and part of the clavicle have been removed. Note that the phrenic nerve is closely related to the anterior surface of the scalenus anterior muscle.

paralyzed by a cervical nerve block, this procedure is not performed on patients with pulmonary or cardiac disease.

Brachial Plexus Block (Figs. 8-8, 8-11, and 8-12).   For anesthesia of the upper limb, the anesthetic agent is injected around the supraclavicular part of the brachial plexus of nerves. The subclavian artery is located by palpation before making the injection to avoid entering it. The main site of injection is superior to the midpoint of the clavicle. The needle is directed inferomedially toward the first rib (see Fig. 1-5).

Muscles in the Posterior Cervical Triangle (Figs. 8-8, 8-10 to 8-12, and 8-16 to 8-18; Table 8-1).   As described (p. 789), four muscles form the floor of the posterior triangle. From superior to inferior they are: splenius capitis, levator scapulae, scalenus medius, and scalenus posterior.

The Splenius Capitis Muscle (Figs. 8-10*A* and 8-11; see also Fig. 4-41).   This is the larger superior part of the splenius mus-

cle. Its attachments, nerve supply, and main actions are given in Table 8-1. The splenius (G. *splenion*, a bandage) serves as a ''bandage'' that covers and holds the deeper muscles of the neck in place.

The Levator Scapulae (Figs. 8-10*B* and 8-11; see also Fig. 6-40).   This thick straplike muscle is composed of four loosely-bound slips. Its attachments, nerve supply, and main actions are given in Table 8-1. It is partially covered by the sternocleido-mastoid and trapezius muscles. As its name *levator* (L. lifter) indicates, this muscle elevates the scapula.

The Scalenus Medius Muscle (Figs. 8-10 to 8-12, and 8-34).   This is the longest and largest of the scalene muscles. Its attachments, nerve supply, and main actions are given in Table 8-1. The middle scalene muscle is *perforated by the dorsal scapular nerve* and the superior two roots of the long thoracic nerve. In Figure 8-11, note that the scalenus medius *lies posterior to the ventral rami (roots) of the brachial plexus* and the third part

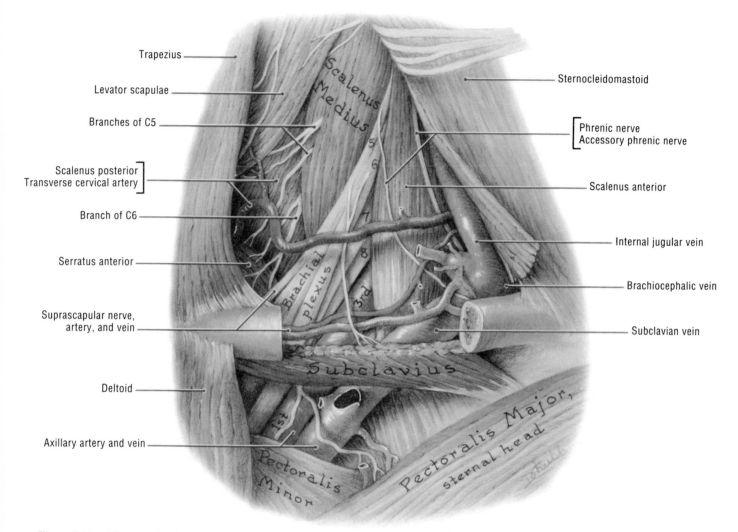

Trapezius

Levator scapulae

Branches of C5

Scalenus posterior
Transverse cervical artery

Branch of C6

Serratus anterior

Suprascapular nerve,
artery, and vein

Deltoid

Axillary artery and vein

Scalenus Medius

Brachial plexus

Subclavius

Pectoralis Minor

Sternocleidomastoid

Phrenic nerve
Accessory phrenic nerve

Scalenus anterior

Internal jugular vein

Brachiocephalic vein

Subclavian vein

Pectoralis Major, sternal head

Figure 8-12. Close up of a dissection of the right side of the neck and superolateral part of the thorax. Observe the brachial plexus of nerves passing to the upper limb and the parts of the subclavian vessels that are in the posterior cervical triangle. Compare with Figure 8-11, noting the the omohyoid muscle and its fascia have been removed to show the third part of the subclavian artery and the subclavian vein. Note that the internal jugular vein (usually the largest vein in the neck) is not in the posterior triangle, but it is very close to it.

of the subclavian artery. A detailed discussion of the brachial plexus appears in Chapter 6 (pp. 512–525).

The Scalenus Posterior Muscle (Figs. 8-11 and 8-12). This muscle is not completely separated from the scalenus medius muscle. Its attachments, nerve supply, and main actions are given in Table 8-1. This is the smallest and most deeply situated of the scalene muscles.

The slender inferior belly of the straplike **omohyoid muscle** (G. *omos*, shoulder), which is described with the infrahyoid muscles (Fig. 8-16; Table 8-3), also passes through the inferior part of the posterior triangle (Figs. 8-4*B* and 8-11). *The omohyoid muscle is an important landmark in the neck*; it can often be seen contracting when some people speak, especially those with thin necks.

Lymph Nodes in the Posterior Cervical Triangle (see Fig. 7-32). There is a superficial group of *cervical lymph nodes* along the external jugular vein and a deep group accompanying the internal jugular vein. These nodes receive afferent lymph

vessels from the parotid, occipital, and mastoid lymph nodes and from vessels in the muscles and skin. Efferent lymph vessels pass to the *supraclavicular lymph nodes* in the supraclavicular (subclavian) triangle of the neck (Fig. 8-4*B*).

Subdivisions of the Posterior Cervical Triangle

The inferior belly of the omohyoid muscle divides the posterior cervical triangle into a large superior occipital triangle (Fig. 8-4*B*) and a small inferior supraclavicular triangle.

The **occipital triangle** (Fig. 8-4*B*) was given its name because the occipital artery appears in its apex (Fig. 8-8). The most important nerve crossing the occipital triangle is the *accessory nerve (CN XI)*.

The **supraclavicular (subclavian) triangle** is the smaller subdivision of the posterior cervical triangle (Fig. 8-4*B*). Its location is indicated on the surface of the neck by the *supraclavicular fossa* (Fig. 8-1). The external jugular vein and suprascapular

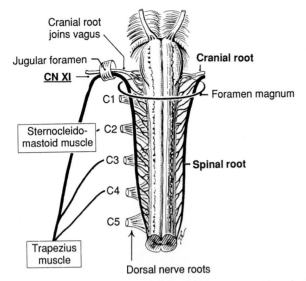

Cranial root
joins vagus

Jugular foramen

CN XI

Sternocleido-
mastoid muscle

Cranial root

C1 — Foramen magnum

C2

C3 — **Spinal root**

C4

C5

Trapezius
muscle

Dorsal nerve roots

Figure 8-13. The accessory nerve (CN XI). Observe that its spinal root is joined by fibers from the ventral ramus of C2 and supplies the sternocleidomastoid muscle. Also note that after it is joined by fibers from C3 and C4 it supplies the trapezius muscle (see also Fig. 8-27).

artery cross this triangle superficially (Figs. 8-7 and 8-26) and the subclavian artery lies deep in it (Fig. 8-12). These vessels are covered by the omohyoid fascia (Fig. 8-31). Because of the presence of the subclavian artery in this region, you will often hear the supraclavicular triangle called the *subclavian triangle*. The third part of the subclavian artery lies on the first rib and its pulsations can be felt on deep pressure.

Pressure in the supraclavicular triangle will occlude blood flow in the *subclavian artery*, where its third part passes over the superior surface of the first rib (Fig. 8-26). If there is a severe hemorrhage in the upper limb, pressure in this triangle will control the bleeding because the large subclavian artery supplies blood to all the arterial vessels in the upper limb (see Fig. 6-31). The *accessory nerve* may be used during surgery to divide the posterior cervical triangle into a carefree area superiorly and a danger area inferiorly (Fig. 8-4B). Care is essential during surgical dissections inferior to the accessory nerve because of the presence of many important vessels and nerves in this area (Figs. 8-7 and 8-12). Note that there are few nerves and vessels in the occipital triangle, which lies superior to this nerve (Figs. 8-4B, 8-7, and 8-12).

Anterior Triangle of the Neck

Boundaries of the Anterior Cervical Triangle (Fig. 8-4). The anterior triangle is bounded by the anterior median line of the neck, the inferior border of the mandible, and the anterior border of the sternocleidomastoid muscle. Its apex is at the jugular notch and its base is formed by the inferior border of the mandible and a line drawn from the angle of the mandible to the mastoid process. This triangular region is used for approaching many important structures in the neck (*e.g.*, larynx, trachea, and thyroid gland). Using the digastric and omohyoid muscles, it is common to divide the anterior triangle into smaller submandibular, submental, carotid, and muscular triangles for de-

scriptive purposes (Fig. 8-4B). These triangles are described subsequently (pp. 797–803).

Floor of the Anterior Cervical Triangle (Figs. 8-26, 8-28, 8-29, and 8-31). The floor of the anterior triangle of the neck is formed mainly by the pharynx, larynx, and thyroid gland. Deep to these structures is the prevertebral fascia covering the prevertebral muscles.

Contents of the Anterior Triangle

The anterior cervical triangle contains muscles, arteries, veins, nerves, lymph nodes, and viscera (*e.g.*, the thyroid and parathyroid glands).

The Hyoid Muscles (Figs. 8-14 to 8-17; Tables 8-2 and 8-3). The hyoid bone is held in place by several muscles that are attached to the mandible, skull, thyroid cartilage, manubrium of the sternum, and the medial end of the scapula. *The hyoid muscles are primarily concerned with steadying or moving the hyoid bone and larynx.* For descriptive purposes, they are divided into suprahyoid and infrahyoid muscles.

The suprahyoid muscles, as their group name indicates, are located superior to the hyoid bone and connect this bone to the skull. The group includes the mylohyoid, geniohyoid, stylohyoid and digastric muscles (Table 8-2).

The Mylohyoid Muscles (Figs. 8-14, 8-15, and 8-17). These thin, flat triangular muscles form a sling inferior to the tongue, which forms the *floor of the mouth*. The Greek word *myle* means "a mill" and denotes the role of these muscles in grinding food in the mouth. The attachments, nerve supply, and main actions of these muscles are given in Table 8-2.

The Geniohyoid Muscles (Figs. 8-14 and 8-15). These short narrow muscles contact each other in the median plane. They are located superior to the mylohyoid muscles, where they reinforce the floor of the mouth. The attachments, nerve supply, and main actions of these muscles are given in Table 8-2.

The Stylohyoid Muscles (Figs. 8-14 and 8-15). These muscles form a small slip on each side, which is nearly parallel to the posterior belly of the digastric muscle. The attachments, nerve supply, and main actions of these muscles are given in Table 8-2.

The Digastric Muscles (Figs. 7-71, 8-4, 8-15 to 8-17, and 8-22). Each of these straplike muscles has two bellies (G. *gaster*, belly) that descend toward the hyoid bone. They are joined by an *intermediate tendon* that is connected to the body and greater horn of the hyoid bone by a strong loop or sling of fibrous connective tissue. This fibrous pulley allows the tendon to slide anteriorly and posteriorly. Its attachments, nerve supply, and main actions are given in Table 8-2.

The difference in nerve supply of the two bellies of the digastric muscle results from the embryological origin of the anterior and posterior bellies from the first and second branchial or pharyngeal arches, respectively (Moore, 1988). CN V is the nerve to the first arch and CN VII supplies the second arch (Figs. 7-25 and 7-27).

The infrahyoid muscles, because of their ribbonlike appearance, are often called the *strap muscles* (Figs. 8-16, 8-17, and 8-23). As their name indicates, they are located inferior to the hyoid

Table 8-2.
The Suprahyoid Muscles[1]

Muscle	Superior Attachment	Inferior Attachment	Innervation	Main Actions
Mylohyoid (Figs. 8-14 and 8-15)	Mylohyoid line of mandible	Raphe and body of hyoid bone	Mylohyoid n., a branch of inferior alveolar n.	Elevates hyoid bone, floor of mouth, and tongue during swallowing and speaking
Geniohyoid (Figs. 8-14 and 8-15)	Inferior mental spine of mandible	Body of hyoid bone	C1 via the hypoglossal n. (CN XII)	Pulls hyoid bone anterosuperiorly, shortens floor of mouth, and widens pharynx
Stylohyoid (Fig. 8-15)	Styloid process of temporal bone	Body of hyoid bone	Cervical branch of facial n. (CN VII)	Elevates and retracts hyoid bone, thereby elongating floor of mouth
Digastric (Figs. 8-16 to 8-18)	*Anterior belly:* digastric fossa of mandible *Posterior belly:* mastoid notch of temporal bone	Intermediate tendon to body and greater horn of hyoid bone	*Anterior belly:* mylohyoid n., a branch of inferior alveolar n. *Posterior belly:* facial n. (CN VII)	Depresses mandible; raises hyoid bone and steadies it during swallowing and speaking

[1]These muscles connect the hyoid bone to the skull.

Table 8-3.
The Infrahyoid Muscles[1]

Muscle	Inferior Attachment	Superior Attachment	Innervation	Main Actions
Sternohyoid (Fig. 8-16)	Manubrium of sternum and medial end of clavicle	Body of hyoid bone	C1, C2, and C3 from ansa cervicalis	Depresses hyoid bone after it has been elevated during swallowing
Sternothyroid (Figs. 8-16 and 8-17)	Posterior surface of manubrium of sternum	Oblique line of thyroid cartilage	C2 and C3 by a branch of ansa cervicalis	Depresses hyoid bone and larynx
Thyrohyoid (Figs. 8-16 and 8-17)	Oblique line of thyroid cartilage	Inferior border of body and greater horn of hyoid bone	C1 via hypoglossal n. (CN XII)	Depresses hyoid bone and elevates larynx
Omohyoid (Figs. 8-5, 8-12, 8-16 to 8-18)	Superior border of scapula near suprascapular notch	Inferior border of hyoid bone	C1, C2, and C3 by a branch of ansa cervicalis	Depresses, retracts, and steadies hyoid bone

[1]These four straplike muscles anchor the hyoid bone (*i.e.,* they fix and steady it). They are concerned with the suprahyoid muscles (Table 8-2) in movements of the tongue, hyoid bone, and larynx in both swallowing and speaking.

bone. These muscles anchor the hyoid bone and depress the hyoid and larynx during swallowing and speaking (Table 8-3).

The Sternohyoid Muscle (Figs. 8-16, 8-17, and 8-23). This thin, narrow strap muscle is superficial, except inferiorly, where it is covered by the sternocleidomastoid muscle. Its attachments, nerve supply, and main actions are given in Table 8-3. It also helps to steady the hyoid during movements of the tongue, larynx, and pharynx.

The Sternothyroid Muscle (Figs. 8-17, 8-23, and 8-25). This thin muscle is located deep to the sternohyoid muscle, which it is shorter and wider than. Its attachments, nerve supply, and main actions are given in Table 8-3. It depresses the hyoid bone and larynx after it has been elevated by other muscles during swallowing and vocal movements. It also pulls the thyroid cartilage away from the hyoid bone, thereby opening the laryngeal orifice.

The Thyrohyoid Muscle (Fig. 8-17). This muscle appears as the superior continuation of the sternothyroid muscle. Its attachments, nerve supply, and main actions are given in Table 8-3. It is mainly responsible for closing the laryngeal orifice and preventing food from entering the larynx during swallowing.

The Omohyoid Muscle (Figs. 8-4, 8-11, 8-16 to 8-18, and 8-23). This muscle has two bellies that are united by an *intermediate tendon*, which is connected to the clavicle by a fascial sling. The prefix in this muscle's name is from the Greek word *omo*, meaning shoulder. *The omohyoid muscle is an important landmark in the neck* because it divides the anterior and posterior cervical triangles into smaller triangles (Fig. 8-4B). Its attachments, nerve supply, and main actions are given in Table 8-3. The omohyoid depresses, retracts, and steadies the hyoid during swallowing and speaking. In thin persons, the inferior belly of this muscle can often be seen contracting when they are speaking.

Subdivisions of the Anterior Triangle

In addition to the unpaired submental triangle (Fig. 8-4B), the anterior cervical triangle is divisible into three smaller triangles: submandibular, carotid, and muscular.

The Submandibular Triangle (Figs. 8-4B, 8-16, and 8-17). This glandular area on each side lies between the inferior border of the mandible and the anterior and posterior bellies of the digastric muscle. Because of this, some people refer to it as

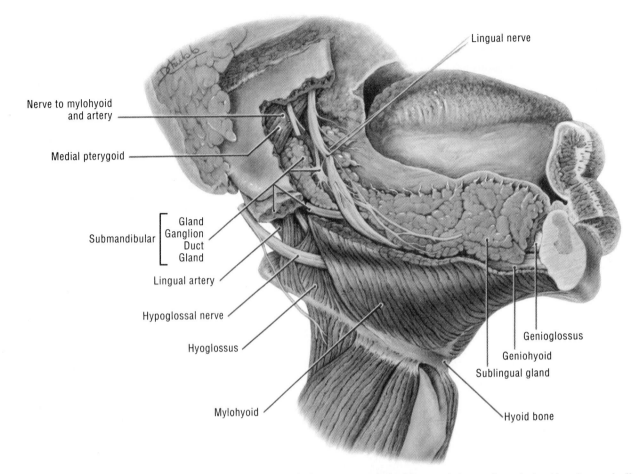

Lingual nerve

Nerve to mylohyoid and artery

Medial pterygoid

Submandibular {
Gland
Ganglion
Duct
Gland

Lingual artery

Hypoglossal nerve

Hyoglossus

Mylohyoid

Genioglossus

Geniohyoid

Sublingual gland

Hyoid bone

Figure 8-14.   Dissection of the suprahyoid region. The right half of the mandible and the superior part of the mylohyoid muscle have been removed. Observe that the cut surface of the mylohyoid muscle becomes progressively thinner as it is traced anteriorly. Also observe the lingual nerve (clamped) between the medial pterygoid muscle and the ramus of the mandible.

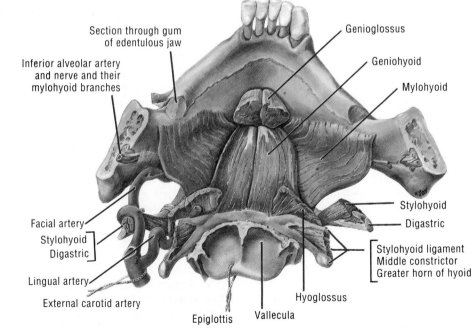

Section through gum of edentulous jaw

Inferior alveolar artery and nerve and their mylohyoid branches

Facial artery

Stylohyoid
Digastric

Lingual artery

External carotid artery

Epiglottis

Vallecula

Genioglossus

Geniohyoid

Mylohyoid

Stylohyoid

Digastric

Stylohyoid ligament
Middle constrictor
Greater horn of hyoid

Hyoglossus

Figure 8-15.   Dissection of the muscles of the floor of the mouth. Observe the geniohyoid muscles occupying a horizontal plane, with their apex at the mental spine of the mandible (Fig. 7-16), and their base at the body of the hyoid bone. The two mylohyoid muscles form the muscular floor of the oral cavity, which supports the tongue, sublingual glands, and other structures shown in Figure 8-14.

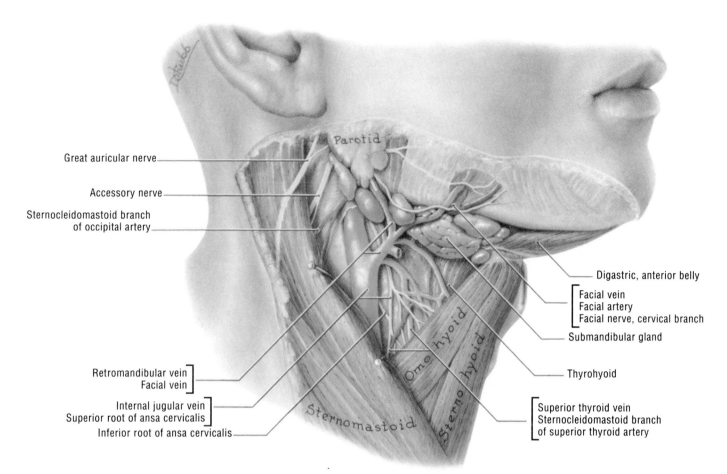

Great auricular nerve

Accessory nerve

Sternocleidomastoid branch
of occipital artery

Retromandibular vein
Facial vein

Internal jugular vein
Superior root of ansa cervicalis
Inferior root of ansa cervicalis

Digastric, anterior belly

Facial vein
Facial artery
Facial nerve, cervical branch
Submandibular gland

Thyrohyoid

Superior thyroid vein
Sternocleidomastoid branch
of superior thyroid artery

Figure 8-16.   Superficial dissection of the anterior triangle of the neck. Observe the submandibular gland, lymph nodes (*green*), and the cervical branch of the facial nerve in the subdivision of the anterior triangle called the submandibular triangle (Fig. 8-4*B*).

the "digastric triangle." The *floor of the submandibular triangle* is formed, from anterior to posterior, by the following muscles: mylohyoid, hyoglossus, and middle constrictor of the pharynx.

Contents of the Submandibular Triangle (Figs. 8-14, 8-16, and 8-17).   The *submandibular gland* nearly fills this triangle. It wraps itself around the free posterior border of the mylohyoid muscle, not unlike the capital letter U on its side. It sends a thin sheet of muscle superior to this gland. This deep process separates the superficial and deep parts of the submandibular gland. About half the size of the parotid, this salivary gland is usually palpable as a soft mass between the body of the mandible and the mylohyoid muscle. It is easily felt when the mylohyoid muscle is tensed by forcing the tip of the tongue against the maxillary incisor teeth. In Figure 8-16, note the position of the *submandibular lymph nodes*. They lie on the **submandibular gland** and along the inferior border of the mandible.

The *submandibular duct*, about 5 cm in length (Fig. 8-14), passes from the deep process of the gland, parallel to the tongue, to open by one to three orifices into the oral cavity. The submandibular duct opens on an elevation, the *sublingual papilla*, which is produced at the side of the lingual frenulum by the sublingual gland (see Figs. 7-85 and 7-102).

The **hypoglossal nerve (CN XII)**, which is motor to the intrinsic and extrinsic muscles of the tongue, passes into the

submandibular triangle, (Figs. 8-14 and 8-17). The *nerve to the mylohyoid muscle*, a branch of the inferior alveolar nerve (Figs. 7-25 and 8-17), and parts of the *facial artery and vein* also pass through the submandibular triangle (Fig. 8-16). The *submental artery* is a branch of the facial artery (Fig. 8-18).

The submandibular gland, along with the parotid gland, may become inflamed (*e.g.*, owing to mumps). The swelling formed by the enlarged gland can be seen easily and palpated. Because of the location of these glands (Fig. 8-16), it becomes painful to open the mouth and to eat.

The Carotid Triangle (Figs. 8-4 and 8-16 to 8-22).   This vascular area is bounded by the superior belly of the omohyoid, the posterior belly of the digastric, and the anterior border of the sternocleidomastoid muscle. *The carotid triangle is an important area because the common carotid artery and its branches ascend into it* (Fig. 8-17). Its pulse can be auscultated (L. to listen to) with a stethoscope, or it can be palpated by placing the digits in the triangle and compressing the artery lightly against the transverse processes of the cervical vertebrae. At the level of the superior border of the thyroid cartilage, *the common carotid artery divides within the carotid triangle* (Fig. 8-24) into the internal and external carotid arteries.

The Carotid Sinus (Figs. 8-19 and 8-24).   This is a slight

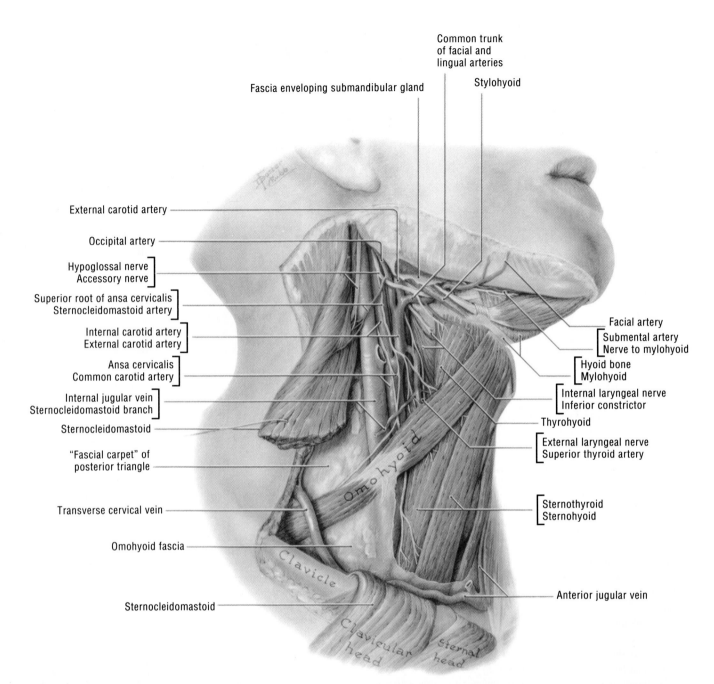

External carotid artery

Occipital artery

Hypoglossal nerve
Accessory nerve

Superior root of ansa cervicalis
Sternocleidomastoid artery

Internal carotid artery
External carotid artery

Ansa cervicalis
Common carotid artery

Internal jugular vein
Sternocleidomastoid branch

Sternocleidomastoid

"Fascial carpet" of
posterior triangle

Transverse cervical vein

Omohyoid fascia

Sternocleidomastoid

Fascia enveloping submandibular gland

Common trunk
of facial and
lingual arteries

Stylohyoid

Facial artery
Submental artery
Nerve to mylohyoid

Hyoid bone
Mylohyoid

Internal laryngeal nerve
Inferior constrictor

Thyrohyoid

External laryngeal nerve
Superior thyroid artery

Sternothyroid
Sternohyoid

Anterior jugular vein

Omohyoid

Clavicle

Clavicular head

Sternal head

Figure 8-17.   Deeper dissection of the right side of the neck than that shown in Figure 8-16. Observe that the facial and lingual arteries in this person arise by a common trunk that passes deep to the stylohyoid and digastric muscles to enter the submandibular triangle.

dilation of the proximal part of the internal carotid artery; it may involve the common carotid. A *blood pressure regulating area*, the carotid sinus is innervated principally by the glossopharyngeal nerve (CN IX) through a branch called the *carotid sinus nerve* (see Fig. 7-74). The sinus is also supplied by the vagus nerve and the sympathetic division of the autonomic nervous system (Figs. 8-22 and 8-28). The *carotid sinus reacts to changes in arterial blood pressure* and effects appropriate modifications reflexly.

> The carotid triangle provides an important surgical approach to the *carotid arterial system* (Fig. 8-17). It is also

important for approaches to: (1) the internal jugular vein; (2) the vagus (CN X) and hypoglossal (CN XII) nerves; and (3) the cervical sympathetic trunk (Figs. 8-21, 8-22, and 8-28).

The carotid sinus responds to an increase in arterial pressure by slowing the heart, owing to the parasympathetic outflow from the brain through the vagus nerve (Fig. 8-33). *Pressure on the carotid sinus may cause syncope* (fainting), and if the person happens to have a supersensitive carotid sinus, it may cause cessation of the heart beat (temporary or permanent). Although the common carotid artery can be occluded by compressing it against the carotid tubercle of the C6 vertebra (Fig. 8-29), *e.g.*, for control of hemorrhage in the neck, it is recommended that you not practice this on your

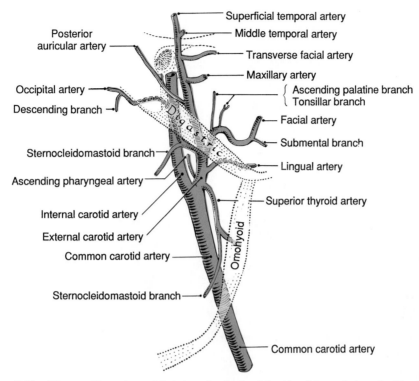

Posterior auricular artery

Occipital artery
Descending branch

Sternocleidomastoid branch

Ascending pharyngeal artery

Internal carotid artery

External carotid artery

Common carotid artery

Sternocleidomastoid branch

Superficial temporal artery
Middle temporal artery
Transverse facial artery
Maxillary artery
Ascending palatine branch
Tonsillar branch
Facial artery
Submental branch
Lingual artery
Superior thyroid artery

Common carotid artery

Digastric

Omohyoid

Figure 8-18.   The carotid arteries and their branches in the right side of the neck (see also Fig. 8-17).

colleagues because syncope may result (*i.e.*, the person may faint owing to deficiency of blood to the brain).

The carotid pulse (neck pulse) is easily felt in the side of the neck. It is routinely checked during cardiopulmonary resuscitation (**CPR**). The *common carotid artery* lies in a groove created by the trachea and strap muscles (Fig. 8-17; see also Fig. 7-28). The rescuer puts the tips of the digits gently on the victim's trachea, then slides them to the side nearest him/her, gently pressing posteriorly between the trachea and inferior larynx medially and the sternocleidomastoid muscle laterally. If there is a pulse, it can easily be felt. *Absence of a carotid pulse indicates cardiac arrest*, a life-threatening situation.

In elderly people with asymptomatic occlusion of one internal carotid artery owing to **atherosclerosis** (a common type of arteriosclerosis or hardening of the arteries), occlusion of the common carotid on the other side will deprive the brain of a major source of blood. As a result, unconsciousness may occur in 10 to 12 seconds. As soon as the pressure is released, the patient usually regains consciousness, but a permanent neurological deficit is likely to be produced if the artery is occluded for several minutes.

Carotid Endarterectomy (Figs. 8-17, 8-18, and 8-24; see also Fig. 7-28).   Atherosclerotic thickening of the intima of internal carotid arteries supplying the brain will cause partial or complete obstruction of blood flow. The resulting symptoms depend on the vessel obstructed, the degree of obstruction, and the amount of collateral blood flow from other vessels, *e.g.*, from those of the *cerebral arterial circle* (Fig. 7-54). The obstruction can be relieved partly or completely by opening the artery and stripping off the plaque with the adjacent intima. A common site for a carotid endarter-

ectomy is the internal carotid artery, just superior to its origin (Fig. 8-17). After the operation, drugs are used to inhibit clot formation at the operated area until the endothelium has regrown.

Two nerves in the carotid triangle are in danger of injury during a carotid endarterectomy: the vagus and recurrent laryngeal nerves (Figs. 8-17, 8-22, 8-25, and 8-63). Damage to these nerves may produce an alteration in the voice because they supply the larynx (p. 848). Temporary paralysis of these nerves may occur as the result of *postoperative edema*.

The Carotid Body (Figs. 8-19 and 8-20).   This small, reddish-brown, ovoid mass of tissue is located at the bifurcation of the common carotid artery, in close relation to the carotid sinus. *The carotid body is a chemoreceptor that responds to changes in the chemical composition of the blood*. It is supplied mainly by the carotid sinus nerve, a branch of CN IX (see Fig. 7-74), but it is also supplied by CN X and sympathetic fibers.

The carotid body responds to either increased carbon dioxide tension or decreased oxygen tension in the blood. A fall in the O_2 content or an increase in the CO_2 content of the blood circulating through the carotid body initiates reflexes through the glossopharyngeal and vagus nerves that stimulate respiration.

The Carotid Sheath (Figs. 8-21, 8-22, 8-32, and 8-34).   This tubular, thickly matted fascial condensation extends from the base of the skull to the root of the neck. It is formed by fascial extensions of the cervical fascia, which fuse with the prevertebral fascia. The inferior part of **the carotid sheath contains several clinically important structures**: (1) the *common carotid arteries*

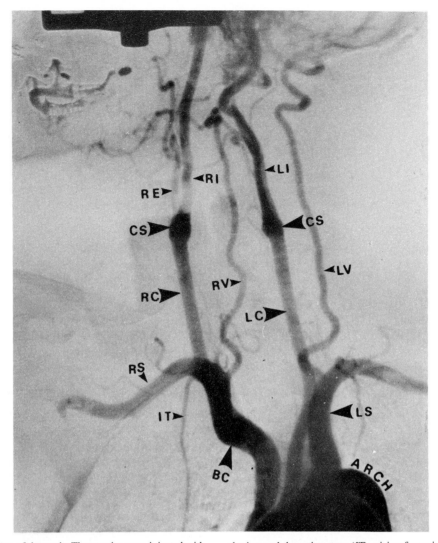

Figure 8-19. Arteriogram of the neck. The arteries were injected with a radiopaque material before the radiograph was taken. Note that the brachiocephalic artery (*BC*) divides into right subclavian (*RS*) and right common carotid (*RC*) arteries and that the common carotid artery divides into right external (*RE*) and right internal (*RI*) carotid arteries. Observe the internal thoracic artery (*IT*) arising from right the subclavian artery, as is the right vertebral artery (*RV*). On the left observe the common carotid (*LC*), internal carotid (*IC*), and the tortuous course of the left vertebral artery (*LV*). On both sides, observe the carotid sinus (*CS*).

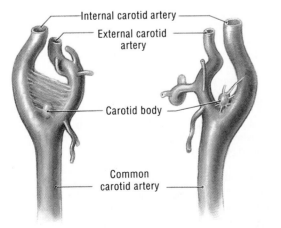

Figure 8-20. Posterior view showing two stages of a dissection of the carotid body. It responds to falling PO_2 or rising PCO_2 in the blood.

medially; (2) the *internal jugular vein* laterally; (3) the *vagus nerve* (CN X) posteriorly; and (4) the *ansa cervicalis* (Figs. 8-16 and 8-17). The superior root of the ansa cervicalis[5] descends between the common carotid artery and the internal jugular vein and is sometimes embedded in the carotid sheath. Many *deep cervical lymph nodes* lie along the carotid sheath and the internal jugular vein and between this vein and the common carotid artery. The cervical part of *the sympathetic trunk runs posterior to the carotid sheath* (Figs. 8-28 and 8-29).

The Muscular Triangle (Figs. 8-4*B*, 8-17, and 8-25).   This triangle is bounded by the superior belly of the omohyoid muscle (separating it from the carotid triangle), the anterior border of the sternocleidomastoid muscle, and the median plane of the neck. The muscular triangle *contains the infrahyoid muscles*

[5]Formerly called "descendens hypoglossi." The superior root of the ansa cervicalis contains C1 fibers that have "hitchhiked" along the hypoglossal nerve.

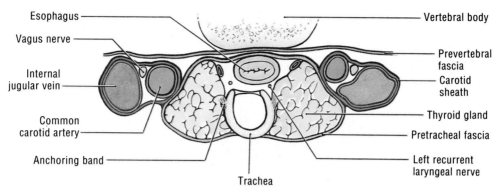

Figure 8-21. Transverse section of the neck showing, in particular, the relations of the thyroid gland and carotid sheath.

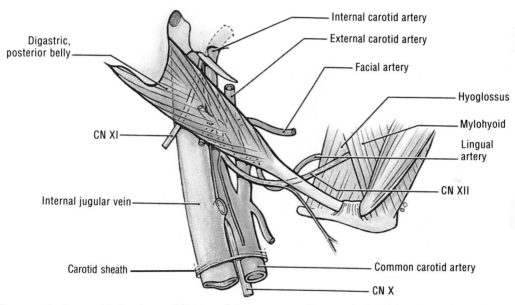

Figure 8-22. Structures in the carotid triangle, a subdivision of the anterior cervical triangle (see also Fig. 8-4*B*). Observe the carotid sheath and its contents (see also Fig. 8-21). Note the key position of the posterior belly of the digastric muscle, running from the mastoid process of the temporal bone to the hyoid bone. Note that the intermediate tendon of the digastric muscle is attached to the hyoid bone by a fascial sling. Observe that all the vessels and nerves cross deep to the posterior belly of the muscle (see also Figs. 8-17 and 8-18).

(Table 8-3) *and the neck viscera* (*e.g.*, the thyroid and parathyroid glands).

The Submental Triangle (Figs. 8-4*B*, 8-16, 8-17, 8-20, and 8-23). This unpaired *suprahyoid area* is bounded inferiorly by the body of the hyoid bone and laterally by the right and left anterior bellies of the digastric muscles. The *floor of the submental triangle* is formed by the two mylohyoid muscles, which meet in a median fibrous raphe. The *apex of the submental triangle* is at the symphysis menti (see Fig. 7-2), and its base is formed by the hyoid bone. The submental triangle contains the **submental lymph nodes**, which receive lymph from the tip of the tongue, the floor of the mouth, the mandibular incisor teeth and associated gingivae, the central part of the lower lip, and the skin of the chin. Lymph from the submental lymph nodes drains into the submandibular and deep cervical lymph nodes

(Figs. 8-16 and 8-31; see also Fig. 7-32). The submental triangle also contains small veins that unite to form the *anterior jugular vein* (Fig. 8-17; see also Figs. 6-33 and 7-29).

The muscular triangle (Fig. 8-4*B*) is used surgically for approaches to the thyroid and parathyroid glands and for exposure of the trachea, esophagus, and inferior levels of the carotid-jugular vascular system (Figs. 8-25, 8-32, and 8-36). Most cancers of the lip occur on the lower lip and tend to spread through the lymphatic system. Depending on the site of the lesion, metastases spread to the submental nodes from the central part of the lower lip and to the submandibular nodes from other parts of the lip. In advanced cancers of the central part of the lip, the submandibular and deep cervical lymph nodes would also be involved because they receive lymph from the submental nodes (see Fig. 7-32).

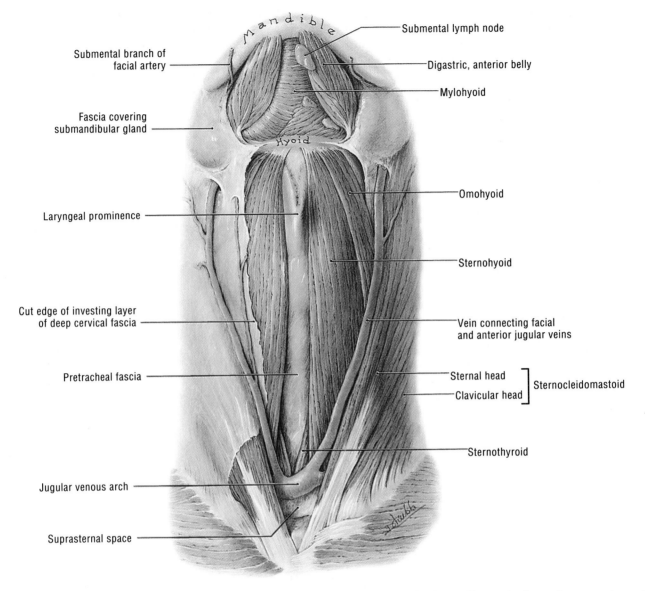

Submental branch of facial artery

Fascia covering submandibular gland

Laryngeal prominence

Cut edge of investing layer of deep cervical fascia

Pretracheal fascia

Jugular venous arch

Suprasternal space

Mandible

Hyoid

Submental lymph node

Digastric, anterior belly

Mylohyoid

Omohyoid

Sternohyoid

Vein connecting facial and anterior jugular veins

Sternal head ⎫
Clavicular head ⎬ Sternocleidomastoid

Sternothyroid

Figure 8-23.   Superficial dissection of the anterior aspect of the neck. Note the submental triangle, bounded inferiorly by the body of the hyoid bone and laterally by the right and left anterior bellies of the digastric muscles (see also Fig. 8-4*B*). Observe that it contains some submental lymph nodes and that the floor of the submental triangle is formed by the two mylohyoid muscles.

Most **arteries in the anterior cervical triangle** arise from the common carotid artery or one of the branches of the external carotid artery.

The Common Carotid Arteries (Figs. 8-17 to 8-22, 8-24, 8-25, 8-28, and 8-29).   The *right common carotid artery* begins at the bifurcation of the brachiocephalic trunk, posterior to the right sternoclavicular joint. The *left common carotid artery* arises from the arch of the aorta and ascends into the neck, posterior to the left sternoclavicular joint. Each common carotid artery *ascends within the carotid sheath* to the level of the superior border of the thyroid cartilage, where it terminates by dividing into the internal and external carotid arteries.

The Internal Carotid Artery (Figs. 8-17 to 8-22 and 8-24).   This is the direct continuation of the common carotid artery; it **has no branches in the neck**. As its name indicates,

it supplies structures within the skull. *The internal carotid arteries are two of the four major arteries that supply blood to the brain* (see Fig. 7-54). Each artery arises from the common carotid artery at the level of the superior border of the thyroid cartilage and passes superiorly, almost in a vertical plane, to enter the **carotid canal** in the petrous part of the temporal bone (see Figs. 7-6*B* and 7-42). A plexus of sympathetic fibers accompanies it. During its course through the neck, the internal carotid artery lies on the longus capitis muscle and the sympathetic trunk. The vagus nerve (CN X) lies posterolateral to it (Fig. 8-22). The internal carotid artery *enters the middle cranial fossa* beside the dorsum sellae of the sphenoid bone (see Fig. 7-42). Within the cranial cavity, the internal carotid artery and its branches supply the hypophysis cerebri (pituitary gland), the orbit, and most of the supratentorial part of the brain (see Fig. 7-54).

The External Carotid Artery (Figs. 7-28, 8-18 to 8-20, 8-22, and 8-24). This vessel begins at the bifurcation of the common carotid, at the level of the superior border of the thyroid cartilage. As its name indicates, it *supplies structures external to the skull*. The external carotid artery runs posterosuperiorly to the region between the neck of the mandible and the lobule of the auricle. It terminates by dividing into two branches, the maxillary and superficial temporal arteries. The stems of most of the *six branches of the external carotid artery* are in the carotid triangle (Fig. 8-24). The three important branches of this artery (printed in boldface) are described first (Table 8-4).

The superior thyroid artery (Figs. 8-18, 8-24, 8-25, and 8-32) is the most inferior of the three anterior branches of the external carotid. It arises close to the origin of the vessel, just inferior to the greater horn of the hyoid. The superior thyroid artery runs anteroinferiorly, deep to the infrahyoid muscles, to reach the superior pole of the thyroid gland. In addition to supplying the thyroid gland, it gives off muscular branches to the sternocleidomastoid and infrahyoid muscles (Table 8-2) and gives off the *superior laryngeal artery* (Fig. 8-25). This artery pierces the thyrohyoid membrane in company with the internal laryngeal nerve and supplies the larynx (Fig. 8-64).

The lingual artery (Figs. 8-22 and 8-24; see also Fig. 7-101) arises from the external carotid as it lies on the middle constrictor muscle of the pharynx.[6] It arches superoanteriorly,

---

[6]Sometimes the lingual artery arises with the facial artery by a common trunk, as shown in Fig. 8-17.

about 5 mm superior to the tip of the greater horn of the hyoid bone, and then *passes deep to the hypoglossal nerve* (CN XII), the stylohyoid muscle, and the posterior belly of the digastric muscle. It disappears deep to the hyoglossus muscle. At the anterior border of this muscle, it turns superiorly and ends by becoming the *deep lingual artery*.

The facial artery (Figs. 8-16 to 8-18, 8-22, and 8-24) arises from the external carotid, either in common with the lingual artery (Fig. 8-17) or immediately superior to it. In the neck the facial artery gives off its important *tonsillar branch* and branches to the palate and submandibular gland. The facial artery then passes superiorly under cover of the digastric and stylohyoid muscles and the angle of the mandible. It loops anteriorly and enters a deep groove in the **submandibular gland** (Fig. 8-16). *The facial artery hooks around the inferior border of the mandible and enters the face*; here, pulsations of the artery can easily be felt (see Fig. 7-23).

The *ascending pharyngeal artery* (Figs. 8-18 and 8-24) is the first or second branch of the external carotid. This small vessel ascends on the pharynx, deep to the internal carotid artery, and sends branches to the pharynx, prevertebral muscles, middle ear, and meninges.

The *occipital artery* (Figs. 8-8, 8-17, and 8-24; see also Figs. 7-28 and 7-35) arises from the posterior surface of the external carotid near the level of the facial artery. It passes posteriorly along the inferior border of the posterior belly of the digastric muscle and ends in the posterior part of the scalp. The occipital artery is often palpable as it crosses the superior nuchal line

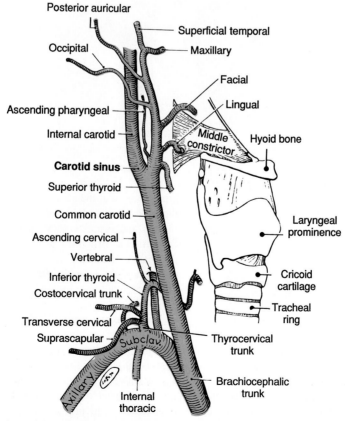

Figure 8-24. The subclavian and carotid arteries and their branches. Note that the terminal part of the right common carotid and the proximal part of the internal carotid artery are slightly dilated for about 1 cm to form the carotid sinus (see also Fig. 8-19).

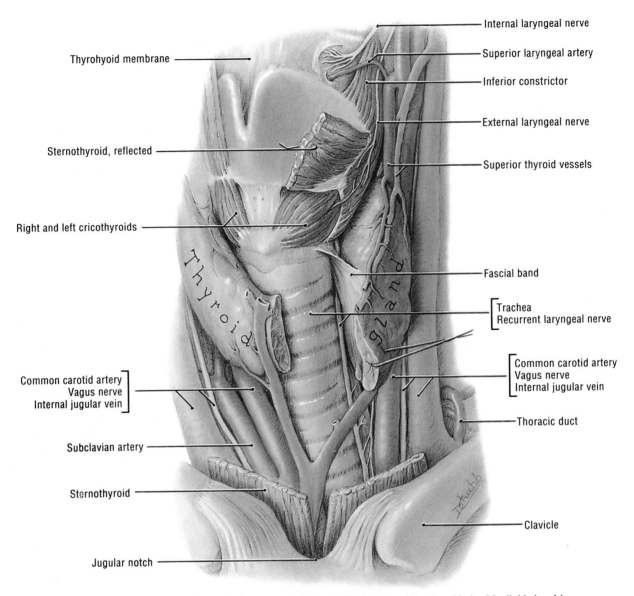

Internal laryngeal nerve

Superior laryngeal artery

Inferior constrictor

External laryngeal nerve

Superior thyroid vessels

Thyrohyoid membrane

Sternothyroid, reflected

Right and left cricothyroids

Fascial band

Trachea
Recurrent laryngeal nerve

Common carotid artery
Vagus nerve
Internal jugular vein

Common carotid artery
Vagus nerve
Internal jugular vein

Thoracic duct

Subclavian artery

Sternothyroid

Clavicle

Jugular notch

Figure 8-25. Dissection of the anterior aspect of the neck. The isthmus of the thyroid gland is divided and its left lobe is retracted.

Table 8-4.
Branches of the External Carotid Artery

Surface	Branches	Figure References
Anterior[1]	Superior thyroid a.	8-18 and 8-25
	Lingual a.	8-22 and 8-24
	Facial a.	8-17 and 8-18
Posterior	Occipital a.	8-17 and 8-24
	Posterior auricular a.	8-18 and 8-24
Medial	Ascending pharyngeal a.	8-18 and 8-24

[1]The three anterior branches of the external carotid artery are of major importance.

between the trapezius and sternocleidomastoid muscles. During this course, it passes superficial to the internal carotid artery and three cranial nerves (CN IX, CN X, and CN XI).

The *posterior auricular artery* (Figs. 8-18 and 8-24; see also Fig. 7-35) is a small posterior branch of the external carotid. It

arises from it at the superior border of the posterior belly of the digastric muscle. It ascends posterior to the external acoustic meatus and supplies adjacent muscles, the parotid gland, the facial nerve, structures in the temporal bone, the auricle, and the scalp.

Most of the **veins in the anterior cervical triangle** are tributaries of the large internal jugular vein.

The Internal Jugular Vein (Figs. 8-8, 8-12, 8-16, 8-17, 8-21, and 8-22; see also Fig. 7-43).   Usually the *largest vein in the neck*, the internal jugular drains blood from the brain and superficial parts of the face and neck. Its course corresponds to a line drawn from a point immediately inferior to the external acoustic meatus to the medial end of the clavicle. This large vein commences at the jugular foramen in the posterior cranial fossa, as the direct continuation of the sigmoid sinus (see Fig. 7-47A and C). From the dilation at its origin, called the *superior bulb* *of the internal jugular vein* (see Fig. 7-43), it runs inferiorly through the neck in the *carotid sheath*. It shares this sheath with the internal carotid artery (later with the common carotid) and the vagus nerve (CN X). The artery is medial, the vein lateral, and the nerve posterior, in the angle between these vessels (Fig. 8-22).

The internal jugular vein leaves the anterior triangle of the neck by passing deep to the sternocleidomastoid muscle (Fig. 8-16). Posterior to the sternal end of the clavicle, it unites with the subclavian to form the *brachiocephalic vein* (Fig. 8-12). Near its termination is the *inferior bulb of the internal jugular vein* (see Fig. 7-43), which contains a bicuspid valve similar to that in the subclavian vein. The internal jugular vein is usually larger on the right side than on the left side because of the greater volume of blood entering it from the superior sagittal sinus through the sigmoid sinus (see Figs. 7-30 and 7-47). *Deep cervical lymph nodes lie along the course of the internal jugular vein* mostly lateral and posterior. These nodes may be small and scattered; thus they are often difficult to identify in dissections.

Tributaries of the Internal Jugular Vein (Fig. 8-16; see also Figs. 7-30 and 7-43). This large vein is joined at its origin by the inferior petrosal sinus, the facial, lingual, pharyngeal, superior and middle thyroid veins, and often the occipital vein.

Pulsations of the internal jugular vein resulting from contractions of the right ventricle of the heart may be palpable and visible at the root of the neck (Fig. 8-31). Because there are no valves in the brachiocephalic vein or the superior vena cava, a wave of contraction passes up these vessels to the internal jugular vein. The *systolic venous pulse*, palpable in the internal jugular vein, is considerably increased in certain conditions (*e.g.*, disease of the mitral valve leading to increased pressure in the pulmonary circulation, the right side of the heart, and the great veins).

All lymphatic vessels from the head and neck drain into the deep cervical lymph nodes, many of which lie in the carotid sheath (Fig. 8-31; see also Fig. 7-32). They lie along the internal jugular vein, between it and the common carotid artery. The deep cervical lymph nodes are very important when there are metastases (*e.g.*, from a carcinoma in the mouth, larynx, or other structures in the head and neck). Removal of cancerous deep cervical lymph nodes is more difficult when they adhere closely to the internal jugular vein and the carotid sheath (Figs. 8-22 and 8-34).

Deep Structures in the Neck

Vertebrae, muscles, vessels and nerves are located in the deep aspects of the neck (Figs. 8-26, 8-29, 8-30 and 8-40).

Skeleton of the Neck (Figs. 8-26 and 8-40; see also Figs. 4-1, 4-16, and 4-37). The cervical vertebrae and joints of the neck are described in Chapter 4 with the section on the back (pp. 331 and 342, respectively). Revise your knowledge of these bones, particularly the atlas (C1) and axis (C2), and review the joints associated with them (see Figs. 4-30 and 4-34). The joints between the atlas and axis permit rotation, whereas those between the skull and atlas are structured to allow nodding movements of the head on the cervical region of the vertebral column (see Table 4-3 and p. 348).

Muscles of the Neck (Figs. 8-5 to 8-8, 8-10, and 8-12).

The superficial cervical muscles (platysma, trapezius, and sternocleidomastoid) have been described (p. 786), as were the hyoid muscles (p. 796) and the muscles in the floor of the posterior cervical triangle (p. 794).

The Prevertebral Muscles (Figs. 8-21, 8-26, 8-27, 8-29, 8-30, and 8-34). These deep anterior vertebral muscles are covered anteriorly by prevertebral fascia. They all *flex the neck and the head on the neck* and are supplied by ventral primary rami of the cervical nerves.

The longus colli muscle (longus cervicis), the longest and most medial of the prevertebral muscles, extends from the anterior tubercle of the atlas (C1 vertebra) to the body of T3 vertebra (Fig. 8-26). It is also attached to the bodies of the vertebrae between C1 and C3, and the transverse processes of C3 to C6 vertebrae.

The longus capitis muscle (Fig. 8-26), broad and thick superiorly, arises from the anterior tubercles of C3 to C6 cervical transverse processes and is attached superiorly to the base of the skull.

The rectus capitis anterior (Fig. 8-26), a short wide muscle, is attached inferiorly to the anterior surface of the lateral mass of the atlas (C1) and superiorly to the base of the skull, just anterior to the occipital condyle.

The rectus capitis lateralis (Figs. 8-26 and 8-27), a short flat muscle, is attached inferiorly to the transverse process of the atlas (C1) and superiorly to the jugular process of the occipital bone. In addition to flexing the head on the neck, this muscle and the rectus capitis anterior, help to stabilize the skull on the cervical region of the vertebral column.

The Root of the Neck

The *thoracocervical region* or root of the neck is the junctional area between the thorax and neck. It includes the *superior thoracic aperture* (see Figs. 1-1, 1-5, and p. 49), through which pass all structures going from the head to the thorax and vice versa.

Boundaries of the Root of the Neck (Figs. 8-25, 8-26, and 8-29). The thoracocervical region is bounded: *laterally* by the first pair of ribs and their costal cartilages; *anteriorly* by the manubrium of the sternum, and *posteriorly* by the body of the first thoracic vertebra (see Figs. 1-1 and 1-5).

Arteries in the Root of the Neck (Figs. 8-19; 8-25, 8-26, and 8-29; see also Fig. 1-70). The arteries in this junctional area *originate from the arch of the aorta*. They are the brachiocephalic trunk on the right side and the common carotid and subclavian arteries on the left side.

The Brachiocephalic Trunk (Figs. 8-24, 8-26, 8-28, and 8-36; see also Fig. 1-70). This trunk (artery) is the largest branch of the arch of the aorta. It is 4 to 5 cm in length and arises posterior to the center of the manubrium of the sternum. It passes superiorly and to the right, posterior to the right sternoclavicular joint, where it divides into the right common carotid and right subclavian arteries. The brachiocephalic trunk is covered anteriorly by the sternohyoid and sternothyroid muscles. At first, it lies on the trachea and then to its right side.

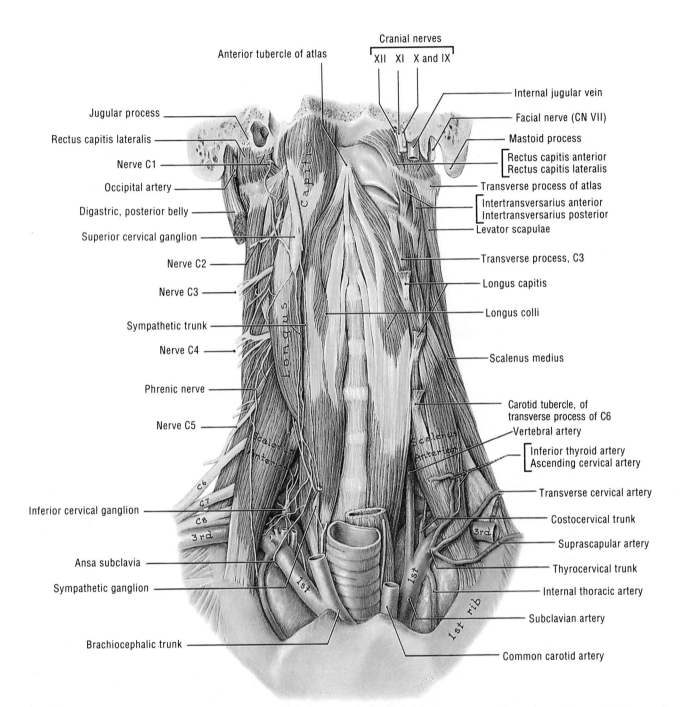

Figure 8-26. Dissection of the prevertebral region and root of the neck. The prevertebral fascia has been removed and the longus capitis muscle has been excised on the left side. Observe the cervical plexus of nerves arising from the ventral rami of C1 to C4 and the brachial plexus of nerves arising from the ventral rami of C5 to C8 and T1.

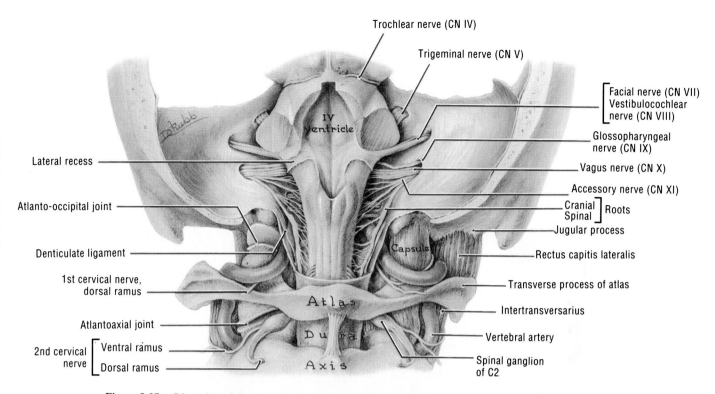

Trochlear nerve (CN IV)

Trigeminal nerve (CN V)

Facial nerve (CN VII)
Vestibulocochlear
nerve (CN VIII)

Glossopharyngeal
nerve (CN IX)

Vagus nerve (CN X)

Accessory nerve (CN XI)

Cranial ⎤
Spinal ⎦ Roots

Jugular process

Rectus capitis lateralis

Transverse process of atlas

Intertransversarius

Vertebral artery

Spinal ganglion
of C2

Lateral recess

Atlanto-occipital joint

Denticulate ligament

1st cervical nerve,
dorsal ramus

Atlantoaxial joint

2nd cervical
nerve ⎡ Ventral ramus
⎣ Dorsal ramus

IV
ventricle

Capsule

Atlas

Dura

Axis

Figure 8-27.   Dissection of the posterior aspect of the neck showing some of the cranial nerves and the second cervical nerve. Observe that this cervical nerve has a large spinal ganglion.

Usually the brachiocephalic trunk has no branches, other than its terminal ones (right subclavian and common carotid arteries), but sometimes a small vessel, called the *thyroid ima artery* (L. *ima*, lowest), arises from it and ascends anterior to the trachea to supply the isthmus of the thyroid gland. The clinical significance of this artery is discussed on page 820.

The Subclavian Arteries (Figs. 8-12, 8-25, 8-26, 8-28, and 8-29).   These are the *arteries of the upper limbs*, but they also supply branches to the neck and brain. The right subclavian artery, one of the terminal branches of the brachiocephalic trunk, arises posterior to the right sternoclavicular joint. *The left subclavian artery arises from the arch of the aorta* and enters the root of the neck by passing superiorly, posterior to the left sternoclavicular joint. Each subclavian artery arches superiorly, posteriorly, and laterally, grooving the pleura and lung, and then passes inferiorly, posterior to the midpoint of the clavicle. As these arteries rise 2 to 4 cm into the root of the neck, they are crossed anteriorly by the scalenus anterior muscles (Fig. 8-12). For purposes of description, *the scalenus anterior muscle divides the subclavian artery into three parts:* the first part is medial to the muscle, the second part posterior to it, and the third part lateral to it (Figs. 8-26 and 8-28).

Branches of the subclavian arteries are (Fig. 8-24): (1) the vertebral; (2) the internal thoracic; (3) the thyrocervical trunk; (4) the costocervical trunk; and (5) the dorsal scapular. *On the left*, all branches except the dorsal scapular arise from the first part of the subclavian artery. *On the right*, the costocervical trunk usually arises from its second part.

The vertebral artery (Figs. 8-26, 8-27, 8-29, and 8-30)

arises from the superior aspect of the first part of the subclavian and ascends through the transverse foramina of the cervical vertebrae, except for C7 vertebra (see Fig. 7-28). After winding around the lateral mass of the atlas, *the vertebral artery enters the skull through the foramen magnum.* At the inferior border of the pons, the vertebral arteries join to form a median vessel called the **basilar artery** (see Fig. 7-54), which is a very *important supplier of blood to the brain*.

The internal thoracic artery (Figs. 8-24 and 8-26; see also Figs. 1-19, 1-42, and 7-28) arises from the inferior aspect of the subclavian artery and passes inferomedially into the thorax (p. 59). It runs parallel to the sternum and gives off anterior intercostal branches to the first six intercostal spaces.

The thyrocervical trunk (Figs. 8-24, 8-26, and 8-29) arises from the first part of the subclavian artery, just medial to the scalenus anterior muscle. It gives rise to several branches, the largest and most important of which is the *inferior thyroid artery*, which passes to the inferior pole of the thyroid gland. Other branches of the thyrocervical trunk are: the *suprascapular artery* (except when it arises from the third part of the subclavian artery), supplying muscles around the scapula, and the *transverse cervical artery*, sending branches to the muscles in the posterior triangle of the neck (Figs. 8-11 and 8-12; see also Fig. 6-31).

The costocervical trunk (Figs. 8-24 and 8-26) arises from the posterior aspect of the subclavian artery and passes superoposteriorly over the cervical pleura. It divides into the superior intercostal and deep cervical arteries, which supply the first two intercostal spaces and muscles in the neck.

Veins in the Root of the Neck (Figs. 8-7, 8-17, and 8-23;

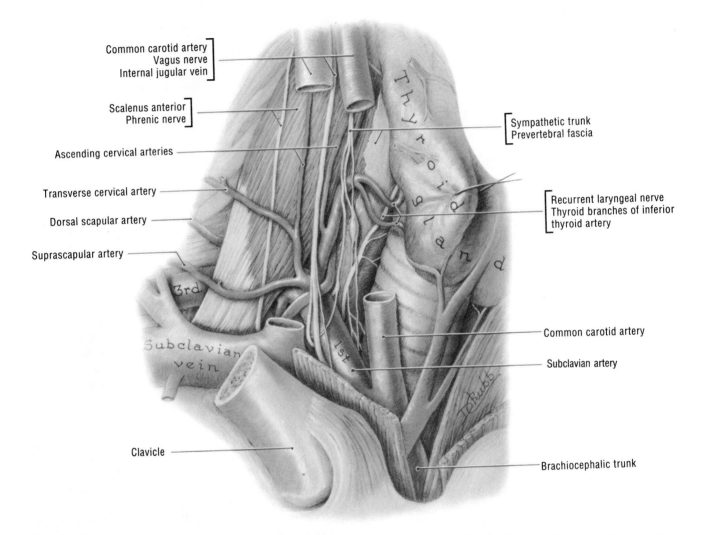

Common carotid artery
Vagus nerve
Internal jugular vein

Scalenus anterior
Phrenic nerve

Ascending cervical arteries

Transverse cervical artery

Dorsal scapular artery

Suprascapular artery

Clavicle

Thyroid gland

Sympathetic trunk
Prevertebral fascia

Recurrent laryngeal nerve
Thyroid branches of inferior
thyroid artery

Common carotid artery

Subclavian artery

Brachiocephalic trunk

3rd

Subclavian vein

1st

Figure 8-28.   Dissection of the right side of the root of the neck. The lateral part of the clavicle has been removed and sections have been taken from the common carotid artery and the internal jugular vein. The right lobe of the thyroid gland is retracted to expose the vagus nerve (CN X), crossing the first part of the subclavian artery and giving off the recurrent laryngeal nerve.

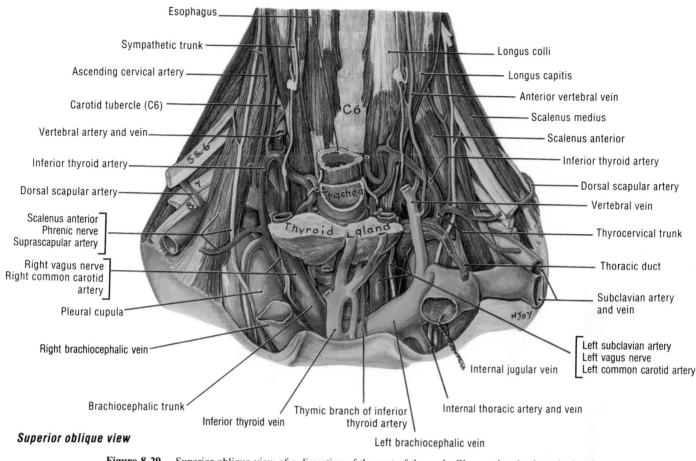

Superior oblique view

Figure 8-29. Superior oblique view of a dissection of the root of the neck. Observe that the thoracic duct is pulled inferiorly by the reflected internal jugular vein.

see also Fig. 7-29). The *external jugular vein*, receiving blood mostly from the scalp and face, is described on page 790.

The Anterior Jugular Vein (Figs. 8-17 and 8-23).   This is usually the smallest of the jugular veins. It arises near the hyoid bone from the confluence of the submandibular veins (see Fig. 7-29), which is sometimes called the *submental venous plexus*. It descends in the superficial fascia between the anterior median line and the anterior border of the sternocleidomastoid muscle. At the root of the neck, the anterior jugular vein turns laterally, posterior to the sternocleidomastoid muscle, and opens into the termination of the external jugular or directly into the subclavian vein. *There are usually two anterior jugular veins.* Just superior to the sternum, the right and left anterior jugular veins are united by a large transverse trunk, the *jugular venous arch*.

The anterior jugular veins have no valves. They may be replaced by a single trunk descending in the midline of the neck. These veins, like the external jugular veins, can often be made visible by ''blowing'' with the mouth closed.

The Subclavian Vein (Figs. 8-12, 8-28, 8-29, and 8-31).   This large vein is the *continuation of the axillary vein*. It begins at the lateral border of the first rib and ends at the medial border of the scalenus anterior muscle. Here it unites with the internal jugular vein, posterior to the medial end of the clavicle to form

the brachiocephalic vein. The subclavian vein, lying in the concavity of the subclavian artery superior to the clavicle, has a bicuspid valve near its termination. It usually has only one named tributary, the *external jugular vein*, which is often visible in the neck. The subclavian vein passes over the first rib parallel to the subclavian artery but is separated from it by the scalenus anterior muscle. It crosses the first rib anterior to the scalene tubercle (see Fig. 1-5). This tubercle separates the groove in the first rib for the subclavian vein from the groove for the subclavian artery.

The Internal Jugular Vein (Figs. 8-28, 8-29, 8-31, 8-32, and 8-34; see also Figs. 7-43 and 7-47A).   This vein ends posterior to the medial end of the clavicle by uniting with the subclavian to form the brachiocephalic vein. Throughout its course, the internal jugular vein is *enclosed within the carotid sheath* (Fig. 8-22; p. 801).

Nerves in the Root of the Neck (Figs. 8-25, 8-26, 8-28, 8-30, 8-32, and 8-34).

The Vagus Nerve (*CN X*) (Figs. 8-21, 8-22, 8-25, 8-28, 8-32 to 8-34, and 8-39; see also Figs. 7-40 and 7-49).   The vagus (L. wandering), so named because of its wide distribution, leaves the skull through the jugular foramen with the internal jugular vein and cranial nerves IX and XI. The vagus, the main parasympathetic nerve to the organs of the thorax and abdomen, passes inferiorly in the posterior part of the *carotid sheath* in the

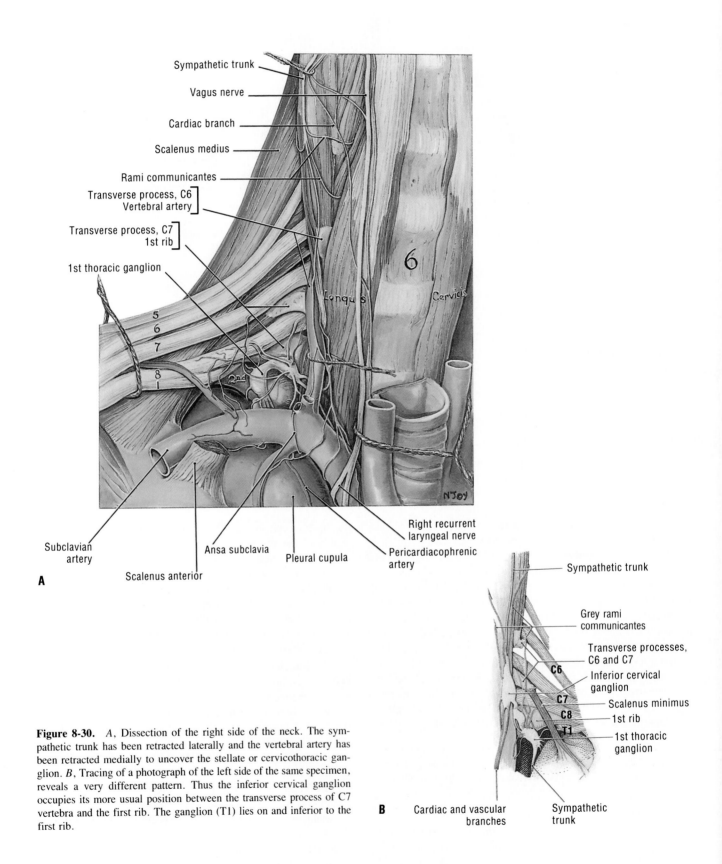

Figure 8-30. *A*, Dissection of the right side of the neck. The sympathetic trunk has been retracted laterally and the vertebral artery has been retracted medially to uncover the stellate or cervicothoracic ganglion. *B*, Tracing of a photograph of the left side of the same specimen, reveals a very different pattern. Thus the inferior cervical ganglion occupies its more usual position between the transverse process of C7 vertebra and the first rib. The ganglion (T1) lies on and inferior to the first rib.

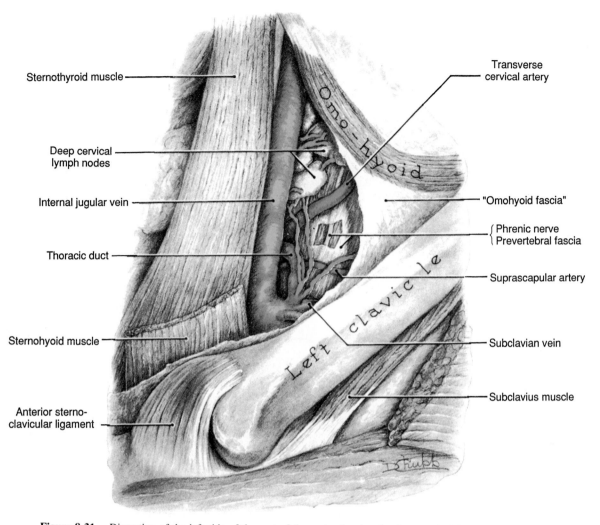

Figure 8-31. Dissection of the left side of the root of the neck, showing the deep cervical lymph nodes and the termination of the thoracic duct (see also Fig. 8-29).

angle between and posterior to the internal jugular vein and carotid artery. *On the right side*, CN X crosses the origin of the subclavian artery, posterior to the brachiocephalic vein and the sternoclavicular joint, to enter the thorax. The *recurrent laryngeal nerve*, a branch of CN X, loops around the subclavian artery on the right side and around the arch of the aorta on the left side. After looping, both recurrent laryngeal nerves pass superiorly to reach the posteromedial aspect of the inferior pole of the thyroid gland, where they ascend in the *tracheoesophageal groove* to supply all the intrinsic muscles of the larynx except the cricothyroid (Table 8-6). The *cardiac branches of CN X* (**cardiac nerves**) also originate in the neck and thorax (Fig. 8-33) and run along the arteries to the arch of the aorta and the cardiac plexuses of nerves (see Figs. 1-62, 1-68, and p. 106).

The Phrenic Nerve (Figs. 8-12, 8-26, 8-28, 8-29, 8-31, and 8-34; see also Fig. 1-42). This nerve, which is about 30 cm long, is the *sole motor nerve to the thoracic diaphragm* (p. 224). It arises chiefly from the fourth cervical nerve (with contributions from C3 and C5). The phrenic nerve is formed at the superior part of the lateral border of the scalenus anterior muscle, at the

level of the superior border of the thyroid cartilage and superolateral to the internal jugular vein. It descends obliquely with this vein across the scalenus anterior muscle, deep to prevertebral fascia and the transverse cervical and suprascapular arteries. *On the left*, the phrenic nerve crosses the first part of the subclavian artery, but *on the right*, it lies on the scalenus anterior muscle that covers the second part of this artery. The phrenic nerve crosses posterior to the subclavian vein on both sides and anterior to the internal thoracic artery to enter the thorax.

The unexpected innervation of the diaphragm by cervical nerve roots has an embryological basis and clinical significance (Moore, 1988). During the fifth week of development, the ventral rami from C3, C4, and C5 grow into the septum transversum, the primordium of the central tendon of the diaphragm when it is in the cervical region. As the developing diaphragm migrates caudally, it carries the phrenic nerves with it.

Injuries to the inferior cervical region of the spinal cord (*e.g.*, C7 segment) that are severe enough to cause paralysis

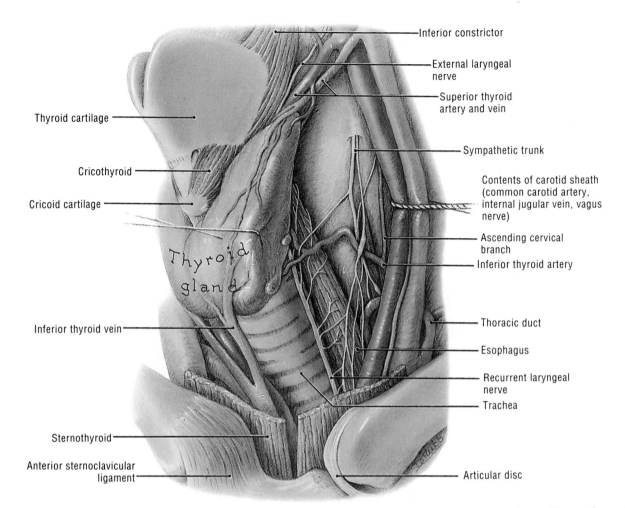

Thyroid cartilage

Cricothyroid

Cricoid cartilage

Thyroid gland

Inferior thyroid vein

Sternothyroid

Anterior sternoclavicular ligament

Inferior constrictor

External laryngeal nerve

Superior thyroid artery and vein

Sympathetic trunk

Contents of carotid sheath (common carotid artery, internal jugular vein, vagus nerve)

Ascending cervical branch

Inferior thyroid artery

Thoracic duct

Esophagus

Recurrent laryngeal nerve

Trachea

Articular disc

Figure 8-32.   Dissection of the left side of the root of the neck. Note the recurrent laryngeal nerve ascending beside the trachea, just anterior to the angle between the trachea and esophagus. Observe the contents of the carotid sheath (see also Figs. 8-21 and 8-22).

of the upper limbs have little effect on breathing because the phrenic nerves arise from more cranial segments (C3, **C4**, and C5). Breathing would not be normal however, because the intercostal muscles would be paralyzed.

Severance of a phrenic nerve in the root of the neck (or elsewhere) results in paralysis of the corresponding half of the diaphragm. To produce temporary therapeutic paralysis of one half the diaphragm (*e.g.*, to interrupt a severe case of *hiccoughs* [spasmodic sharp contractions of the diaphragm]), a **phrenic nerve block** is sometimes done. The anesthetic solution is injected around the phrenic nerve where it lies on the anterior surface of the middle third of the scalenus anterior muscle, about 3 cm superior to the clavicle (Fig. 8-31). To produce a longer period of paralysis of half of the diaphragm, (*e.g.*, for several months after the surgical repair of a diaphragmatic hernia), a *phrenic nerve crush* may be performed. In such cases, the phrenic nerve is crushed with a hemostat for up to 1 cm of its length. In other cases, a *phrenicotomy* is performed during which the phrenic nerve is sectioned.

An **accessory phrenic nerve** (Figs. 8-11 and 8-12) occurs in 20% to 30% of persons. It is frequently derived from C5 as a branch of the nerve to the subclavius muscle. It lies lateral to the main phrenic nerve and usually joins it in the root of the neck or in the superior part of the thorax. If an

accessory phrenic nerve is present, section or crushing of the phrenic nerve will not produce complete paralysis of the corresponding half of the diaphragm.

The Sympathetic Trunks (Figs. 8-26, 8-28, 8-30, and 8-32).   These longitudinal strands of autonomic nerve fibers and their associated *sympathetic ganglia* are located in the neck anterolateral to the vertebral column from the level of the first cervical vertebra. These trunks *receive no white rami communicantes in the neck*, but they contain **three cervical sympathetic ganglia** (superior, middle, and inferior). These ganglia receive their preganglionic fibers from the superior thoracic spinal nerves through white rami communicantes, the fibers of which leave the spinal cord in the ventral roots of thoracic spinal nerves (see Fig. 26, p. 30). From these sympathetic trunks, fibers pass to cervical structures as postganglionic fibers in cervical spinal nerves or leave as direct visceral branches (*e.g.*, to the thyroid gland). Branches to the head run with the arteries, especially the internal and external carotid arteries.

The Inferior Cervical Ganglion (Figs. 8-26, 8-30, and 8-34).   This collection of nerve cells lies at the level of the superior border of the neck of the first rib, where it is wrapped around the posterior aspect of the vertebral artery. It is usually fused with the first thoracic ganglion (and sometimes the second) to form a large ganglion, known as the **cervicothoracic (stellate) ganglion**. This ganglion lies anterior to the transverse process of the vertebra prominens (C7), just superior to the neck of the first rib on each side and posterior to the origin of the vertebral artery. Some postganglionic fibers from the ganglion pass into C7 and C8 nerves and to the heart; other fibers contribute to the vertebral plexus around the vertebral artery.

The Middle Cervical Ganglion (Fig. 8-28). This small collection of sympathetic nerve cells lies on the anterior aspect of the inferior thyroid artery, around the level of the cricoid cartilage and the transverse process of C6 vertebra, just anterior to the vertebral artery. Postganglionic branches pass from it to C5 and C6 nerves and to the heart and thyroid gland.

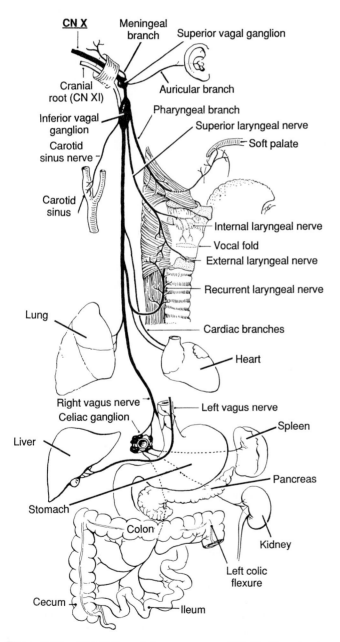

Figure 8-33. Distribution of the vagus nerve (CN X). Note that it receives fibers from the cranial root of the accessory nerve (CN XI).

The Superior Cervical Ganglion (Fig. 8-26). This large collection of nerve cells (2 to 3 cm long) is located at the level of the atlas (C1) and axis (C2). Because of its size, it forms a *good landmark for locating the sympathetic trunk in the neck*. Postganglionic branches from it pass along the internal carotid artery and enter the cranial cavity. It also sends branches to the external carotid artery and into the superior four cervical nerves. Other postganglionic fibers pass to the cardiac plexus (see Fig. 1-62).

Severance of a sympathetic trunk in the neck interrupts the sympathetic nerve supply to the head and neck on that side. Patients have a severe sympathetic disturbance known as the **Horner syndrome**, consisting of: (1) *pupillary constriction*, owing to paralysis of the dilator pupillae muscle in the iris; (2) *ptosis*, (lowering of the upper eyelid) owing to paralysis of the smooth muscle in the levator palpebrae superioris; (3) *sinking in of the eye*, possibly resulting from paralysis of the orbitalis muscle; and (4) *vasodilation and absence of sweating* on the face and neck, owing to lack of a sympathetic nerve supply to the blood vessels and sweat glands. *Hemisection of the spinal cord* in the cervical region also produces the Horner syndrome. In these cases, the disturbance results from interruption of descending autonomic fibers in the spinal cord.

Anesthetic fluid injected around the large cervicothoracic (stellate) ganglion blocks transmission of stimuli through the cervical and superior thoracic ganglia. A *cervicothoracic (stellate) ganglion block* may be performed to relieve vascular spasms involving the brain and upper limb.

Lymphatics in the Root of the Neck (Figs. 8-16, 8-29, 8-31, 8-32, and 8-34; see also Fig. 7-32). Several large lymph nodes are located in the *carotid sheath* along the blood vessels of the neck, particularly the internal jugular vein. Another group of nodes is found along the transverse cervical artery. All these are called *deep cervical lymph nodes* because they are deep to the deep cervical fascia. For descriptive purposes, the deep cervical lymph nodes are often divided into superior and inferior groups according to their relationship to the point of crossing of the omohyoid muscle over the internal jugular vein. These nodes receive lymph from the superficial cervical lymph nodes and from the entire head and neck. From the inferior end of the deep group of cervical lymph nodes, a *jugular lymph trunk* emerges and joins the venous system near the junction of the internal jugular and the subclavian veins. On the left side the jugular lymph trunk may empty into the thoracic duct (see Fig. 1-39).

The deep cervical lymph nodes, particularly those located along the transverse cervical artery (Fig. 8-31), may be involved in the spread of cancer from the thorax and abdomen. As their enlargement may give the first clue to cancer in these regions, they are often referred to as the *cervical sentinel lymph nodes*.

The Thoracic Duct (Figs. 8-25, 8-29, 8-31, and 8-32; see also Figs. 1-39 and 1-69). This large lymphatic channel, draining lymph into the venous system, passes superiorly from the thorax through the superior thoracic aperture at the left border of the esophagus (p. 115). It then arches laterally in the root of the neck, *posterior to the carotid sheath*, and anterior to the sympathetic trunk and the vertebral and subclavian arteries. The

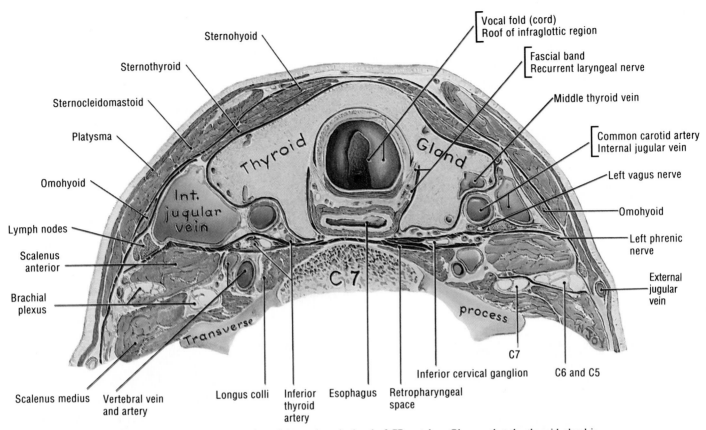

Figure 8-34.   Transverse section of the neck at the level of C7 vertebra. Observe that the thyroid gland is asymmetrically enlarged.

thoracic duct *enters the left brachiocephalic vein* at the junction of the subclavian and internal jugular veins. It drains lymph from the entire body, except the right side of the head and neck, the right upper limb, and the right side of the thorax (Fig. 24, p. 25). These areas drain through the *right lymphatic duct*, a vessel 1 to 2 cm long, which empties into the venous system at or near the junction of the right internal jugular and right subclavian veins.

> *Blockage of the thoracic duct* (*e.g.*, by tumor cells) usually produces no symptoms. The lymph apparently enters the venous system through other lymphatic channels. Malignant tumor cells pass (*e.g.*, from an abdominal cancer) through the thoracic duct into the root of the neck. Some tumor cells enter the venous system and others extend by retrograde permeation into the inferior deep cervical lymph nodes or *supraclavicular nodes*. The cancer cells proliferate here, forming metastases (new malignant tumors). The supraclavicular nodes, particularly on the left side, may be enlarged when there is a carcinoma of the bronchus (p. 77), stomach, or any other abdominal organ.

The Cervical Viscera

The cervical viscera are disposed in three layers (Figs. 8-21 and 8-34): (1) a *superficial endocrine layer* containing the thyroid, parathyroid, and thymus glands; (2) a *middle respiratory*

layer containing the larynx and trachea; and (3) a *deep alimentary layer* containing the pharynx and esophagus.

The Esophagus

This thick, distensible muscular tube extends from the pharynx to the stomach (about 25 cm). It begins in the median plane at the inferior border of the cricoid cartilage (Fig. 8-40) and ends at the cardiac orifice of the stomach (see Fig. 2-34). It lies between the trachea and the anterior longitudinal ligament on the surfaces of the vertebrae (Figs. 8-21, 8-29, and 8-34; see also Fig. 4-26). *On the right side*, the esophagus is in contact with the cervical pleura at the root of the neck (see Fig. 1-72), whereas *on the left side*, posterior to the subclavian artery, the thoracic duct lies between the pleura and the esophagus (see Fig. 1-73). The esophagus is also discussed in Chapter 2 (p. 160).

The Trachea

The trachea (windpipe) is a *fibrocartilaginous tube* that is supported by incomplete cartilaginous **tracheal rings**. These rings, which keep the trachea patent, are deficient posteriorly where the trachea is related to the esophagus (Figs. 8-21 and 8-34). Hence, the posterior wall of the trachea is flat. *The trachea extends from the larynx to the roots of the lungs*, a distance of about 12 cm (see Fig. 1-25). The isthmus of the thyroid gland usually lies over the second and third tracheal rings (Figs. 8-35

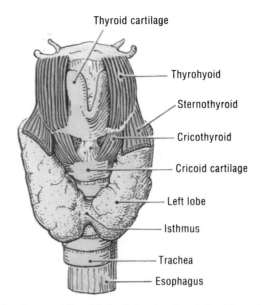

Thyroid cartilage

Thyrohyoid

Sternothyroid

Cricothyroid

Cricoid cartilage

Left lobe

Isthmus

Trachea

Esophagus

Figure 8-35. A normal thyroid gland showing its relationship to the trachea, esophagus, and cricoid cartilage. The sternothyroid muscles have been cut to expose the lobes of the gland. Note that the isthmus of the thyroid gland lies anterior to the second and third tracheal rings.

and 8-36). Inferior to the isthmus are the *jugular venous arch* (Fig. 8-23) and the inferior thyroid veins (Figs. 8-32 and 8-36). The *brachiocephalic trunk* is related to the right side of the trachea at the root of the neck (Figs. 8-26 and 8-36). Lateral to the trachea are the common carotid arteries and the lobes of the thyroid gland (Fig. 8- 21).

A surgical incision through the neck and the anterior wall of the trachea, called a **tracheotomy** (Fig. 8-37*A*), is often performed in *patients with laryngeal obstruction* to establish an adequate airway, either as an emergency lifesaving measure or as an elective procedure. Sometimes a round or square opening is made in the neck (Fig. 8-37*B*) rather than a slit, and the tracheal mucosa is brought into continuity with the skin. This operation is referred to as a **tracheostomy** (L. *ostium*, mouth). If the anticipated duration of endotracheal intubation is short, a tracheotomy is usually performed, but when supportive measures to maintain respiratory functions are expected to be longer than 72 hours (*e.g.*, in a comatose patient), a tracheostomy is usually performed.

Although emergency tracheotomies are occasionally performed without anesthesia and surgical instruments, *a tracheotomy is not a simple operation*. The common approach to the trachea through the skin, subcutaneous tissue, and deep cervical fascia is made by a transverse incision in the neck (Fig. 8-37*A*), usually midway between the laryngeal prominence and the jugular notch. Surgical opinions vary concerning the best site for making the incision, but it is commonly made through the second and third tracheal rings. The first ring is not cut because of the danger of narrowing of the trachea during healing. As the isthmus of the thyroid gland covers the second and third tracheal rings (Fig. 8-35), it is retracted inferiorly or superiorly or divided between clamps.

During a tracheotomy inferior to the thyroid gland, the following anatomical facts must be kept in mind to avoid possible damage to important structures: (1) the *inferior thy-*

roid veins form a plexus anterior to the trachea (Fig. 8-36); (2) a small *thyroid ima artery* is present in about 10% of people and ascends to the inferior border of the isthmus (p. 820); (3) the *left brachiocephalic vein* (Fig. 8-29), jugular venous arch (Fig. 8-23), and the pleurae (see Fig. 1-33), may be encountered, particularly in infants and children; (4) the *thymus gland* covers the inferior part of the trachea in infants and children; and (5) the *trachea is small, mobile, and soft in infants*, making it easy to cut through its posterior wall and damage the esophagus.

The Thyroid Gland

This important, highly vascular endocrine gland, brownish red and soft during life, *consists of right and left lobes*, which are united by a narrow **isthmus** that extends across the trachea, usually anterior to the second and third tracheal cartilages (Figs. 8-32 and 8-35). The largest of all the endocrine glands, it clasps the anterior and lateral surface of the pharynx, larynx, esophagus, and trachea like a shield (Figs. 8-32 and 8-34 to 8-36). The thyroid gland lies deep to the sternothyroid and sternohyoid muscles (Figs. 8-23 and 8-34). Each pear-shaped lobe (about 5 cm long) extends inferiorly on each side of the trachea, often to the level of the sixth tracheal cartilage posteriorly, and extends along the sides of the esophagus (Figs. 8-34 and 8-35).

The size of the thyroid gland varies greatly, but it usually weighs about 25 g. It is relatively larger and heavier in women, in whom it becomes slightly larger during menstruation and pregnancy. A conical *pyramidal lobe* is present in some people, which ascends from the isthmus of the thyroid gland, usually from the left side, toward the hyoid bone. This extra lobe may be attached to the hyoid bone by a fibrous band or a muscular slip, called the *levator of the thyroid gland* (levator glandulae thyroideae). The pyramidal lobe is derived from a persistent portion of the inferior end of the embryonic thyroglossal duct (Moore, 1988).

The thyroid gland is surrounded by a thin *fibrous capsule* of connective tissue. External to this, there is a false capsule formed by a sheath of *pretracheal fascia*, which is derived from the deep cervical fascia (Fig. 8-23). The gland is attached to the arch of the cricoid cartilage and to the oblique line of the thyroid cartilage. Because of these attachments, it moves up and down with the larynx during swallowing and oscillates during speaking.

Arterial Supply of the Thyroid Gland (Figs. 8-24, 8-25, 8-28, 8-32, 8-34, 8-36, 8-38, and 8-43). This highly vascular gland usually receives its blood from two rather large arteries: the superior and inferior thyroid arteries. These vessels lie between the capsule and pretracheal fascia (false capsule). All the thyroid arteries anastomose with each other on and in the substance of the thyroid gland, but there is little anastomosis across the median plane except for branches of the superior thyroid artery. The *superior thyroid artery*, the first branch of the external carotid artery, descends to the superior pole of the gland, pierces the pretracheal fascia, and then divides into two or three branches. The *inferior thyroid artery*, a branch of the thyrocervical trunk, runs superomedially *posterior to the carotid sheath* to reach the posterior aspect of the gland (Fig. 8-38). It divides into several branches, which pierce the pretracheal fascia to

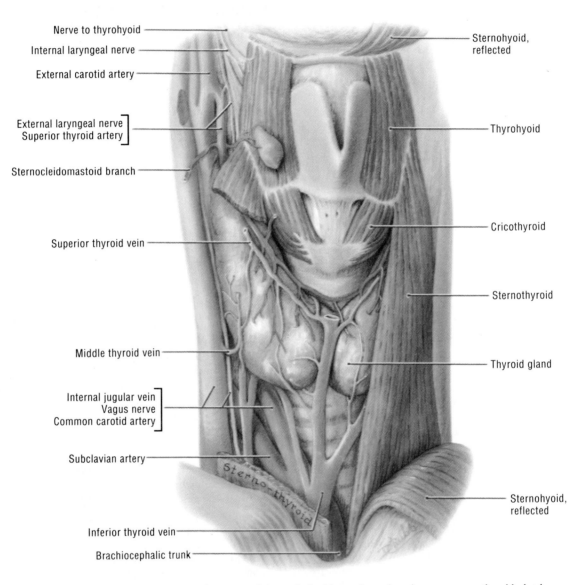
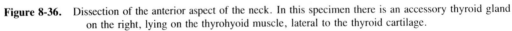

Nerve to thyrohyoid

Internal laryngeal nerve

External carotid artery

External laryngeal nerve
Superior thyroid artery

Sternocleidomastoid branch

Superior thyroid vein

Middle thyroid vein

Internal jugular vein
Vagus nerve
Common carotid artery

Subclavian artery

Inferior thyroid vein

Brachiocephalic trunk

Sternohyoid,
reflected

Thyrohyoid

Cricothyroid

Sternothyroid

Thyroid gland

Sternohyoid,
reflected

Sterno-thyroid

Figure 8-36.   Dissection of the anterior aspect of the neck. In this specimen there is an accessory thyroid gland
on the right, lying on the thyrohyoid muscle, lateral to the thyroid cartilage.

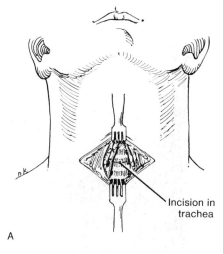

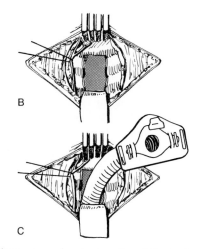

Incision in
trachea

A

B

C

Figure 8-37. *A*, Transverse incision in the neck and trachea (tracheotomy). *B*, The second and third trachea rings have been cut, creating a larger opening, and the tracheal flap has been sutured to the skin. This facilitates removal and reinsertion of the tracheotomy tube. *C*, A tracheotomy tube has been inserted into the trachea.

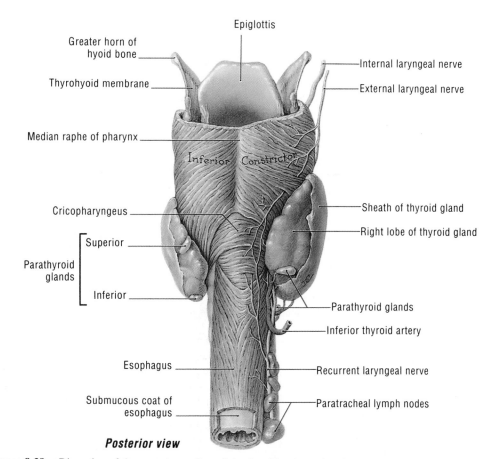

Epiglottis

Greater horn of
hyoid bone

Internal laryngeal nerve

Thyrohyoid membrane

External laryngeal nerve

Median raphe of pharynx

Inferior Constrictor

Cricopharyngeus

Sheath of thyroid gland

Right lobe of thyroid gland

Superior

Parathyroid
glands

Inferior

Parathyroid glands

Inferior thyroid artery

Esophagus

Recurrent laryngeal nerve

Submucous coat of
esophagus

Paratracheal lymph nodes

Posterior view

Figure 8-38. Dissection of the posterior surface of the thyroid and parathyroid glands. Both parathyroids on the right side are rather low, with the inferior gland being inferior to the thyroid gland.

supply the inferior pole of the thyroid gland. *The inferior thyroid artery has an intimate relationship with the recurrent laryngeal nerve* (Figs. 8-28 and 8-32).

In about 10% of people a third vessel, called the *thyroid ima artery*, supplies the thyroid gland. This small artery may arise from the arch of the aorta, the brachiocephalic trunk (most commonly), or the left common carotid artery. It ascends anterior to the trachea and supplies the isthmus of the thyroid gland.

During surgical operations on the thyroid gland, the inferior thyroid artery may have to be tied, but the recurrent laryngeal nerve, with which it is intimately related (Fig. 8-28), has to be preserved because it supplies all the intrinsic muscles of the larynx, except the cricothyroid (Table 8-6). The possible presence of a thyroid ima artery must be taken into account when incising the trachea inferior to the isthmus of the thyroid gland. As it runs anterior to the trachea, it is a potential source of serious bleeding. In addition to the named thyroid arteries, numerous small vessels pass to the thyroid gland from arteries supplying the pharynx and trachea. Hence, the thyroid gland still oozes blood during a *subtotal thyroidectomy*, even when the two main arteries supplying it are ligated.

Venous Drainage of the Thyroid Gland (Figs. 8-16, 8-25, 8-28, 8-29, 8-32, 8-34, and 8-36).   Usually three pairs of veins drain the venous plexus on the anterior surface of the gland. The *superior thyroid veins* drain its superior poles, and the *middle thyroid veins* drain its lateral parts. The superior and middle thyroid veins empty into the internal jugular veins. The *inferior thyroid veins* drain the inferior poles of the gland and empty into the brachiocephalic veins. Often they unite to form a single vein that opens into one of the brachiocephalic veins. Because they cover the anterior surface of the trachea, inferior to the isthmus of the thyroid gland, the inferior thyroid veins are potential sources of bleeding during a *tracheotomy* (Fig. 8-37).

Lymphatic Drainage of the Thyroid Gland (Figs. 8-29, 8-31, 8-34, and 8-38).   Lymph vessels run in the interlobular connective tissue, often around the arteries. They communicate with a capsular network of lymph vessels. From here, vessels pass to the *prelaryngeal lymph nodes* and to the *pretracheal* and *paratracheal lymph nodes*. Laterally, lymph vessels located along the superior thyroid veins pass to the inferior *deep cervical lymph nodes*. Some lymph vessels may drain into the brachiocephalic lymph nodes or empty directly into the thoracic duct.

Nerve Supply of the Thyroid Gland (Figs. 8-26, 8-28, 8-30, 8-32, 8-33, and 8-43).   The nerves are derived from the superior, middle, and inferior *cervical sympathetic ganglia*. They reach the thyroid gland through the cardiac and laryngeal branches of the vagus nerve, which run along the arteries supplying the gland. These postganglionic fibers are vasomotor and affect the gland indirectly through their action on the blood vessels.

Uncommonly, the thyroid gland fails to descend during development resulting in a *lingual thyroid gland* or a superior *cervical thyroid gland* in the region of the hyoid bone. Accessory thyroid gland tissue may also develop from remnants of the thyroglossal duct (Moore, 1988). Small detached masses of thyroid tissue may be present superior to the lobes or the isthmus as *accessory thyroid glands* (Fig. 8-36). Enlargement

of the thyroid gland, except for variable enlargement during menstruation and pregnancy, is called a **goiter**. The enlarged gland causes a swelling in the neck, which may exert pressure on the trachea or the recurrent laryngeal nerves (Fig. 8-34). There are various types of goiter, *e.g.*, *exophthalmic goiter* is a disorder caused by an excessive production of the thyroid hormone. One sign of this disease, bulging eyeballs, is called *exophthalmos*. It is sometimes necessary to remove the thyroid gland (**total thyroidectomy**), *e.g.*, during excision of a carcinoma of the gland. In the surgical treatment of *hyperthyroidism*, the posterior portion of each lobe of the enlarged thyroid gland is usually preserved (**subtotal thyroidectomy**).

During neck surgery, the risk of *injury to the recurrent laryngeal nerves* is ever present. Near the inferior pole of the thyroid gland, the right recurrent laryngeal nerve is intimately related to the inferior thyroid artery (Fig. 8-28). This nerve may cross anterior or posterior to the artery, or it may pass between its branches. Because of this close relationship, the inferior thyroid artery is ligated some distance lateral to the thyroid gland, where it is not so close to the nerve. Although the danger of injuring the left recurrent laryngeal nerve during surgery is not so great, it must be remembered that the artery and nerve are also closely associated near the inferior pole of the thyroid gland (Fig. 8-32). *Temporary aphonia* or disturbance of phonation (voice production) and *laryngeal spasm* may follow partial thyroidectomy. This is usually the result of bruising the recurrent laryngeal nerves during the operation or from the pressure of accumulated blood, serous exudate, or both after the operation.

Injury to the external laryngeal nerve, a branch of the superior laryngeal nerve (Fig. 8-33), is uncommon during thyroidectomy. However, as it is closely related to the superior thyroid artery during part of its course to the superior pole of the thyroid gland (Fig. 8-32), care should be taken not to damage it when the superior thyroid artery is ligated and sectioned. Injury to the external laryngeal nerve results in the voice becoming monotonous in character because the paralyzed cricothyroid muscle is unable to vary the length and tension of the vocal fold. To avoid injury to the external laryngeal nerve, the superior thyroid artery is ligated and sectioned near the superior pole of the gland, where it is not as closely related to the nerve as it is at its origin (Fig. 8-32). Because an enlarged thyroid gland may itself be the cause of impaired innervation of the larynx as the result of compression, it is good practice to examine the vocal folds before an operation. In this way, damage to the larynx or its nerves resulting from a surgical mishap may be distinguished from a preexisting injury resulting from compression.

The Parathyroid Glands

There are usually four parathyroid glands. Yellowish brown during life, these *small, ovoid endocrine glands* usually lie between the posterior border of the thyroid gland and its sheath (Figs. 8-38 and 8-39). They frequently lie just lateral to the anastomotic vessel joining the superior and inferior thyroid arteries. Usually there are two glands on each side of the thyroid gland, but the total number varies between two and six. The glands are named according to their positions as the *superior and inferior parathyroid glands*. The superior ones are more constant in position than are the inferior ones. They are usually

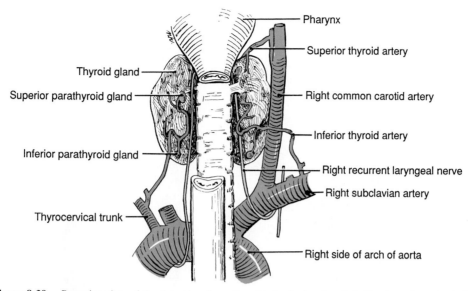

Figure 8-39. Posterior view of the thyroid and parathyroid glands showing their blood supply and relations.

located near the middle of the posterior surface of the lobes of the thyroid gland at the level of the inferior border of the cricoid cartilage. *The inferior parathyroid glands are variable in position.* They are usually located near the inferior surface of the thyroid gland, but they may lie some distance inferior to it (Fig. 8-38), even in the superior mediastinum (see Fig. 1-40). The best guide to their location is to follow their small arteries.

The variability in the number and position of the inferior parathyroid glands has an embryological basis (Moore, 1988). Their development is closely related to the formation of the thyroid and thymus glands. Usually the parathyroid glands are located outside the fibrous capsule of the thyroid gland, *i.e.*, between its capsule and sheath. However, in unusual cases, a parathyroid gland may be embedded in the thyroid gland.

Blood Supply of the Parathyroid Glands (Figs. 8-38 and 8-39). They are usually richly supplied by the *inferior thyroid arteries*, but they may be supplied from anastomoses between the superior and inferior thyroid arteries.

Venous Drainage of the Parathyroid Glands (Fig. 8-36). The veins from these glands drain into the plexus of veins on the anterior surface of the thyroid gland and trachea.

Lymphatic Drainage of the Parathyroid Glands (Figs. 8-31, 8-34, and 8-38). The lymph vessels drain with those from the thyroid and thymus glands into the *deep cervical lymph nodes* and *paratracheal lymph nodes*.

Nerve Supply of the Parathyroid Glands (Figs. 8-28 and 8-32). The nerves are derived from the *sympathetic trunks*, either directly from the superior or middle cervical sympathetic ganglia or from the plexus surrounding the superior and inferior thyroid arteries.

An awareness of the close relations between the parathyroid glands, the thyroid gland, and the recurrent laryngeal nerves is essential clinical knowledge. The possibility of injuring these nerves during *thyroidectomy* has been discussed

(p. 820). The parathyroid glands are also in danger of being damaged or removed during this operation, but they are usually safe during subtotal thyroidectomy because the posterior part of the thyroid gland is preserved. The variability in the position of the parathyroid glands, particularly the inferior ones, may create a problem during thyroid and parathyroid surgery. The superior parathyroids may be as far superior as the thyroid cartilage, and the inferior ones may be as far inferior as the superior mediastinum (see Fig. 1-40). Sometimes an inferior parathyroid gland is embedded within the inferior end of the thyroid gland. These possible aberrant sites are of concern when searching for an abnormal parathyroid gland (*e.g.*, when there is a *parathyroid adenoma* associated with hyperparathyroidism).

If the parathyroid glands atrophy or are inadvertently removed during surgery, the patient suffers from a severe *convulsive disorder* known as **tetany**. The generalized convulsive muscle spasms result from a *fall in blood calcium levels*. To safeguard these glands during thyroidectomy, the posterior portion of each lobe of the thyroid gland is usually not removed. In instances when it is necessary to do a total thyroidectomy (*e.g.*, owing to malignant disease), the parathyroid glands are isolated with their blood vessels intact before the thyroid gland is removed.

Fascial Planes of the Neck

Although the layers of cervical fascia have been mentioned previously, their surgical importance warrants that they be discussed in more detail. *The deep fascia of the neck is disposed in three layers*: investing, pretracheal, and prevertebral (Figs. 8-21, 8-23, 8-31, and 8-34).

The fascial planes of the neck are important because they form natural lines of cleavage through which the tissues may be separated and because they limit the spread of pus resulting from infections in the neck.

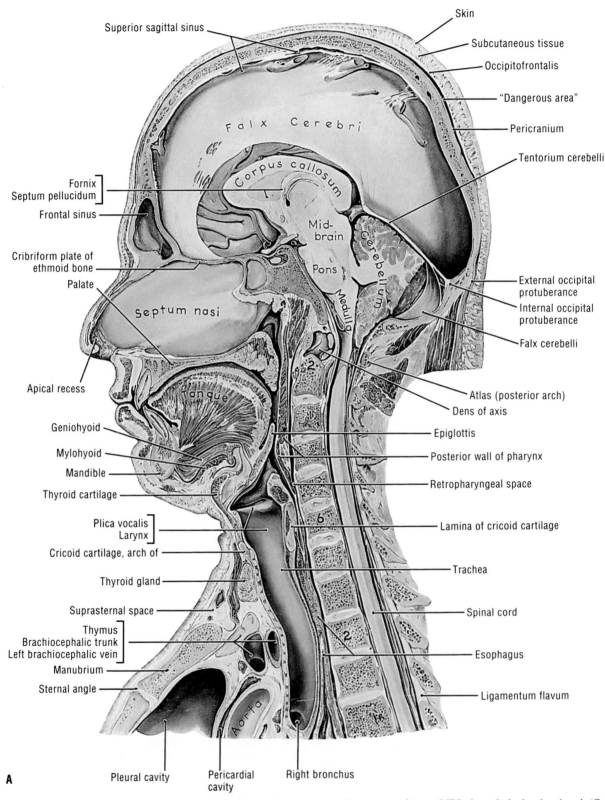

Superior sagittal sinus

Skin

Subcutaneous tissue

Occipitofrontalis

"Dangerous area"

Pericranium

Tentorium cerebelli

Falx Cerebri

Corpus callosum

Fornix
Septum pellucidum

Frontal sinus

Cribriform plate of
ethmoid bone

Palate

Mid-
brain

Pons

Cerebellum

Medulla

Septum nasi

External occipital
protuberance

Internal occipital
protuberance

Falx cerebelli

Apical recess

Tongue

Atlas (posterior arch)

Dens of axis

Geniohyoid

Mylohyoid

Mandible

Thyroid cartilage

Epiglottis

Posterior wall of pharynx

Retropharyngeal space

Plica vocalis
Larynx

Cricoid cartilage, arch of

Thyroid gland

Suprasternal space

Thymus
Brachiocephalic trunk
Left brachiocephalic vein

Manubrium

Sternal angle

Lamina of cricoid cartilage

Trachea

Spinal cord

Esophagus

Ligamentum flavum

Aorta

A

Pleural cavity

Pericardial
cavity

Right bronchus

Figure 8-40. *A,* Median section of the head and neck. Observe the pharynx extending from the base of the skull to the level of the body of C6 vertebra, where it is continuous with the esophagus. *B,* Median magnetic resonance image (MRI) through the head and neck (Courtesy of Dr. W. Kucharczyk, Clinical Director of Tri-Hospital Resonance Centre, Toronto, Ontario, Canada.)

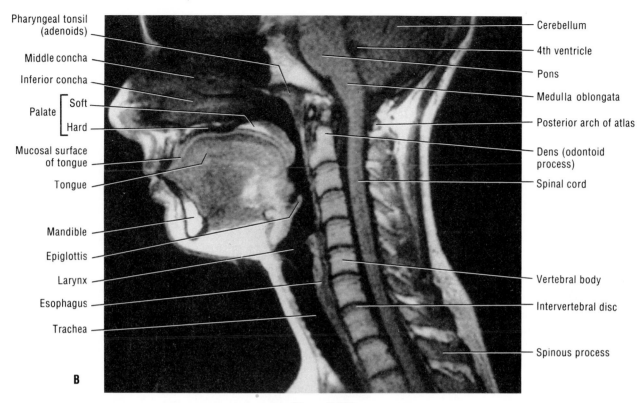

Pharyngeal tonsil (adenoids) —
Middle concha —
Inferior concha —
Palate { Soft —
Hard —
Mucosal surface of tongue —
Tongue —
Mandible —
Epiglottis —
Larynx —
Esophagus —
Trachea —

B

— Cerebellum
— 4th ventricle
— Pons
— Medulla oblongata
— Posterior arch of atlas
— Dens (odontoid process)
— Spinal cord
— Vertebral body
— Intervertebral disc
— Spinous process

Figure 8-40*B*

The Investing Fascia

This superficial or investing layer of deep cervical fascia encircles or *forms a collar around the neck*, surrounding the structures in it (Figs. 8-23, 8-31, and 8-34). It is located between the superficial fascia and the muscular layer. The investing fascia is attached superiorly to the *superior nuchal line* of the occipital bone (see Fig. 7-3), the mastoid process, the zygomatic arch, the inferior border of the mandible, the hyoid bone, and the spinous processes of the cervical vertebrae. Inferiorly, the investing fascia is attached to the manubrium of the sternum, the clavicle, and the acromion and spine of the scapula. Just superior to the sternum, *the investing fascia divides into two layers*, which are attached to the anterior and posterior surfaces of the manubrium of the sternum, respectively. The interval between these two layers of fascia is called the **suprasternal space** (Figs. 8-23 and 8-40). It encloses the sternal heads of the sternocleidomastoid muscles, the inferior ends of the anterior jugular veins, the jugular venous arch, and a few lymph nodes. The investing fascia also *forms the roof of the anterior and posterior triangles of the neck* (Figs. 8-4, 8-8, 8-16 and pp. 796 and 789).

The investing cervical fascia helps to prevent the extension of abscesses (collections of pus) toward the surface. Pus formed beneath it usually extends laterally in the neck, but if the abscess is in the anterior triangle, the pus may pass inferiorly and produce a swelling in the suprasternal space. It could also pass into the anterior mediastinum (see Fig. 1-40).

The Pretracheal Fascia

As its name indicates, this thin layer of deep cervical fascia is *anterior to the trachea* and limited to the anterior aspect of the neck (Fig. 8-23); however, it is more extensive than its name implies. It extends inferiorly from the thyroid cartilage and the arch of the cricoid cartilage into the thorax (Figs. 8-21, 8-23, and 8-34). T*he pretracheal fascia lies deep to the infrahyoid muscles* (Table 8-3). It splits to enclose the thyroid gland, trachea, and esophagus and blends laterally with the *carotid sheath* (Fig. 8-34). The main vessels of the thyroid gland are located deep to this layer. The pretracheal fascia descends into the thorax where *it blends with the fibrous pericardium* in the middle mediastinum (see Figs. 1-40 and 1-41).

Infections in the head or cervical region of the vertebral column can spread inferiorly, posterior to the esophagus, into the posterior mediastinum (see Figs. 1-40 and 1-74). They can also spread inferiorly, anterior to the trachea, and enter the anterior mediastinum. Air from a ruptured trachea, bronchus, or esophagus can pass superiorly in the neck. An unusual source of air in the face and neck may be caused by a dentist's drill passing through an alveolus (tooth socket). Similarly a small slit in the buccal mucosa, followed by hard blowing with the mouth closed, may result in subcutaneous *cervicofacial emphysema*. This unusual kind of swollen neck has been reported in glass blowers and in those who play wind instruments.

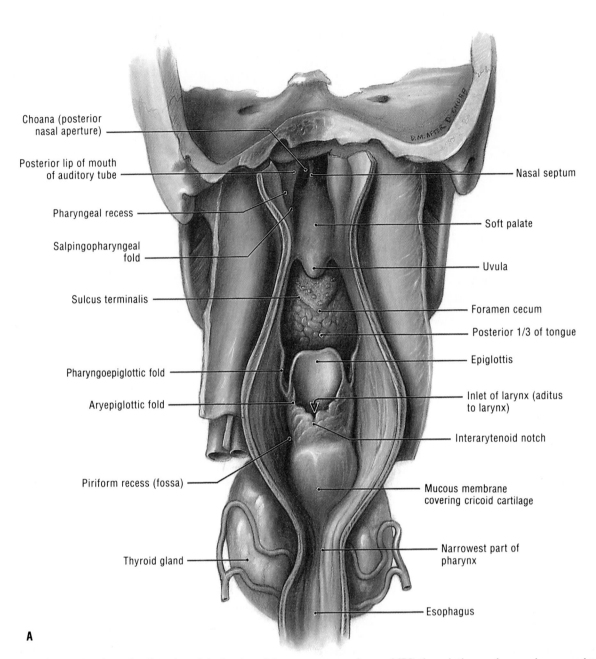

Choana (posterior nasal aperture)

Posterior lip of mouth of auditory tube

Pharyngeal recess

Salpingopharyngeal fold

Sulcus terminalis

Pharyngoepiglottic fold

Aryepiglottic fold

Piriform recess (fossa)

Thyroid gland

Nasal septum

Soft palate

Uvula

Foramen cecum

Posterior 1/3 of tongue

Epiglottis

Inlet of larynx (aditus to larynx)

Interarytenoid notch

Mucous membrane covering cricoid cartilage

Narrowest part of pharynx

Esophagus

A

Figure 8-41. *A*, Posterior view of a dissection of the interior of the pharynx. On each side of the inlet of the larynx and separated from it by the aryepiglottic fold, observe a piriform recess. *B*, Coronal magnetic resonance image (MRI) through the oropharynx, larynx, and trachea. (Courtesy of Dr. W. Kucharczyk, Clinical Director of Tri-Hospital Resonance Centre, Toronto, Ontario, Canada.)

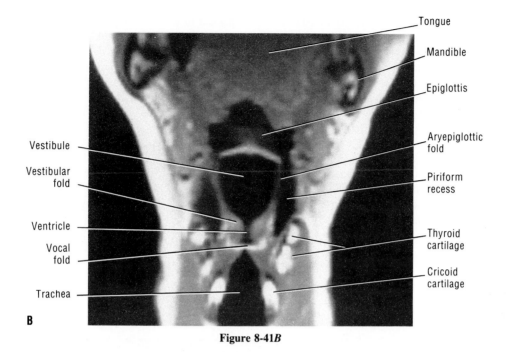

Tongue

Mandible

Epiglottis

Aryepiglottic fold

Piriform recess

Thyroid cartilage

Cricoid cartilage

Vestibule

Vestibular fold

Ventricle

Vocal fold

Trachea

B

Figure 8-41*B*

The Prevertebral Fascia

This layer of deep cervical fascia forms part of a tubular sheath (fascial sleeve) for the prevertebral muscles that surrounds the vertebral column.[7] It is also continuous with the deep fascia covering the muscular floor of the posterior triangle of the neck (Figs. 8-21, 8-31, and 8-34). The prevertebral fascia extends from the base of the skull to the third thoracic vertebra, where it fuses with *the anterior longitudinal ligament* in the posterior mediastinum (see Figs. 1-40 and 4-26). The prevertebral fascia extends inferiorly and laterally as the *axillary sheath* (see Fig. 6-34), which surrounds the axillary vessels and brachial plexus (Figs. 8-11 and 8-31).

Pus from an abscess located posterior to the prevertebral fascia is likely to extend into the lateral parts of the neck, deep to this fascia on the floor of the posterior triangle (Fig. 8-31). Here, it forms a swelling, posterior to the sternocleidomastoid muscle.

The Retropharyngeal Space (Figs. 8-34, 8-40, and 8-49). This *potential space* consists of loose connective tissue between the prevertebral fascia and the buccopharyngeal fascia. The *buccopharyngeal fascia* surrounds the pharynx superficially. This "space" permits movement of the pharynx, larynx, trachea, and esophagus during swallowing. It is closed superiorly by the base of the skull and on each side by the carotid sheath. It opens inferiorly into the superior mediastinum (Fig. 8-40; see also Fig. 1-40), which contains the thymus, great vessels of the heart, trachea, and esophagus (Fig. 8-40*A*; see also Fig. 1-65).

The retropharyngeal space is of considerable surgical interest because of the structures related to it. Pus located posterior to the prevertebral fascia may perforate this layer (Fig. 8-21) and enter the retropharyngeal space. This produces a bulge in the pharynx known as a *retropharyngeal abscess*, which causes difficulty in swallowing and speaking (*i.e.*, *dysphagia* and *dysarthria*). Infections in the retropharyngeal space may also extend inferiorly into the superior mediastinum (see Fig. 1-40).

The Pharynx

The pharynx is the continuation of the digestive system from the oral cavity. It is a funnel-shaped fibromuscular tube that is the common route for air and food. In its superior part, the pharynx receives the posterior openings of the nasal cavities, called *choanae* (Fig. 8-41*A*; see also Fig. 7-6*B*). The pharynx is located posterior to the nasal and oral cavities and the larynx (Figs. 8-40 and 8-41). It conducts food to the esophagus and air to the larynx and lungs. For convenience of description, the pharynx is divided into three parts: (1) the *nasopharynx*, posterior to the nose and superior to the soft palate; (2) the *oropharynx*, posterior to the mouth; and (3) the *laryngopharynx*, posterior to the larynx.

The pharynx, about 15 cm long, extends from the base of the skull to the inferior border of the cricoid cartilage anteriorly and to the inferior border of C6 vertebra posteriorly. It is widest (about 5 cm) opposite the hyoid bone and narrowest (about 1.5

---

[7]The adjective "prevertebral" is misleading because the prevertebral fascia surrounds the vertebral column and its associated musculature (Figs. 8-21 and 8-34).

cm) at its inferior end (Fig. 8-40), where it is continuous with the esophagus. The posterior wall of the pharynx lies against the prevertebral fascia, with the potential *retropharyngeal space* between them (Figs. 8-21, 8-40, and 8-49). In Figure 8-40 observe that the pathways for food and air cross each other in the pharynx. Hence, food sometimes enters the respiratory tract and causes choking, and air often enters the digestive tract and produces gas. Because this air is uncomfortable, it results in eructation (belching).

The pharyngeal wall is composed of five layers. From internal to external, they are: (1) a *mucous membrane* that lines the pharynx and is continuous with all chambers with which it communicates (Figs. 8-40 to 8-42); (2) a *submucosa*; (3) a fibrous layer forming the *pharyngobasilar fascia* (Fig. 8-43), which is

attached to the skull; (4) a *muscular layer* composed of inner longitudinal and outer circular parts; and (5) a loose connective tissue layer forming the *buccopharyngeal fascia* (Fig. 8-40). This fascia is continuous with the epimysium covering of the buccinator and pharyngeal muscles. The buccopharyngeal fascia permits movements of the pharynx. It contains the pharyngeal plexus of nerves and veins.

Muscles of the Pharynx

The pharyngeal musculature consists of three constrictor muscles (Figs. 8-43 and 8-44) and three muscles that descend from the styloid process, the cartilaginous part of the auditory tube, and the soft palate (Figs. 8-42 and 8-45; Table 8-5).

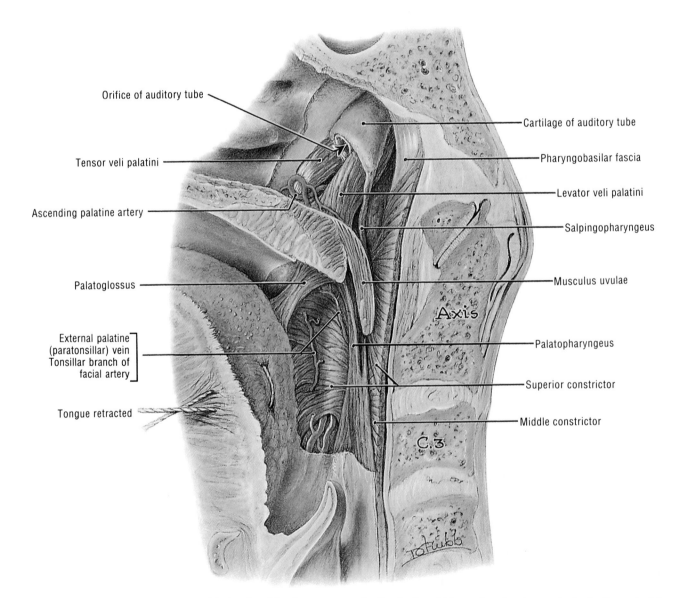

Figure 8-42. Lateral view of a dissection of the interior of the pharynx. The palatine and pharyngeal tonsils and the mucous membrane have been removed. Observe the remaining part of the submucous pharyngobasilar fascia, which attaches the pharynx to the basilar part of the occipital bone.

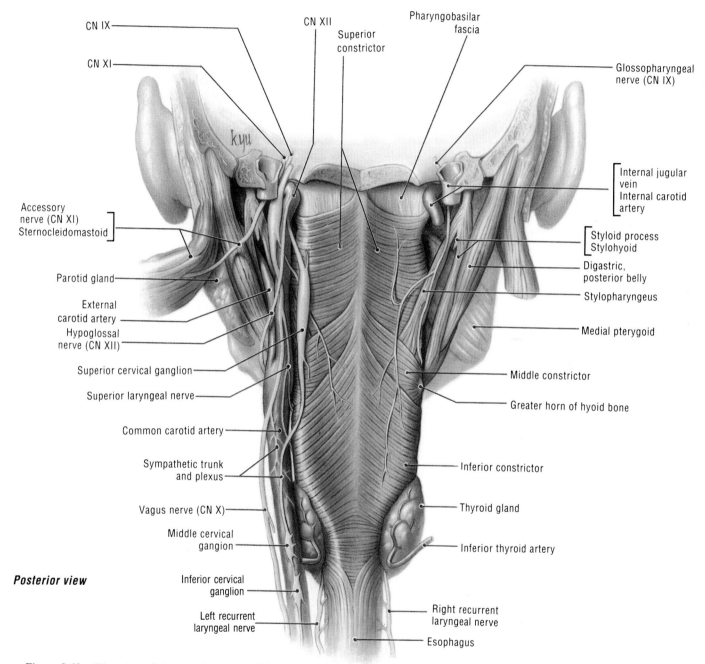

CN IX

CN XI

CN XII

Superior
constrictor

Pharyngobasilar
fascia

Glossopharyngeal
nerve (CN IX)

Internal jugular
vein
Internal carotid
artery

Accessory
nerve (CN XI)
Sternocleidomastoid

Styloid process
Stylohyoid

Digastric,
posterior belly

Parotid gland

Stylopharyngeus

External
carotid artery

Medial pterygoid

Hypoglossal
nerve (CN XII)

Superior cervical ganglion

Middle constrictor

Superior laryngeal nerve

Greater horn of hyoid bone

Common carotid artery

Sympathetic trunk
and plexus

Inferior constrictor

Vagus nerve (CN X)

Thyroid gland

Middle cervical
ganglion

Inferior thyroid artery

Posterior view

Inferior cervical
ganglion

Left recurrent
laryngeal nerve

Right recurrent
laryngeal nerve

Esophagus

Figure 8-43. Dissection of the posterior aspect of the pharynx and associated structures. Examine the three pharyngeal constrictor muscles.

Note that the inferior muscle overlaps the middle and the middle overlaps the superior.

External Muscles of the Pharynx

The external circular part of the muscular layer of the wall of the pharynx is formed by the paired superior, middle, and inferior **constrictor muscles**, which overlap one another (Figs. 8-43 to 8-45; Table 8-5). These muscles are arranged so that the superior one is innermost and the inferior one is outermost. As

their names indicate, they *constrict the pharynx during swallowing*. They contract involuntarily in a way that results in contraction taking place sequentially from the superior to the inferior end of the pharynx. This action propels the food into the esophagus. All three constrictors of the pharynx are supplied by the *pharyngeal plexus of nerves* (Fig. 8-45), which lies on the lateral wall of the pharynx, mainly on the middle constrictor muscle. This plexus is formed by pharyngeal branches of the glossopharyngeal (CN IX) and vagus (CN X) nerves (Fig. 8-33; see also Fig. 7-74).

The Superior Constrictor Muscle (Figs. 8-42 to 8-45; Table 8-5; see also Figs. 7-6 and 7-16). This broad muscle arises from the pterygoid hamulus, the pterygomandibular raphe, and the mylohyoid line of the mandible, posterior to the third molar tooth. A few fibers also arise from the side of the tongue. The *pterygomandibular raphe* is the fibrous line of junction between the buccinator and superior constrictor muscles. Fibers of the superior constrictor muscle curve posteriorly around the pharynx to insert into the *median raphe of the pharynx*. The most superior fibers reach the pharyngeal tubercle of the skull.

The Middle Constrictor Muscle (Figs. 8-42 to 8-45; Table 8-5). This muscle arises from the inferior end of the stylohyoid ligament and the greater and lesser horns of the hyoid bone. Its fibers curve posteriorly around the pharynx to insert into its median raphe.

The Inferior Constrictor Muscle (Figs. 8-42 to 8-45; Table 8-5). This muscle has a continuous origin from the oblique line of the thyroid cartilage, the fascia over the cricothyroid muscle, and the side of the cricoid cartilage. Its fibers pass superiorly,

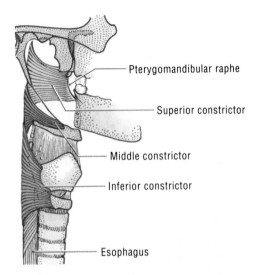

Pterygomandibular raphe

Superior constrictor

Middle constrictor

Inferior constrictor

Esophagus

Figure 8-44.   Right lateral view of the constrictor muscles of the pharynx, showing their attachments and borders. Note that each constrictor muscle is fan-shaped and that the narrow ends of the ''fans'' are fixed anteriorly.

Table 8-5.
The Muscles of the Pharynx

Muscle	Origin	Insertion	Innervation	Main Actions
Superior constrictor (Figs. 8-42 to 8-45)	Pterygoid hamulus, pterygomandibular raphe, posterior end of mylohyoid line of mandible, and side of tongue	Median raphe of pharynx and pharyngeal tubercle	Pharyngeal and superior laryngeal branches of vagus (CN X) through pharyngeal plexus	Constrict wall of pharynx during swallowing
Middle constrictor (Figs. 8-42 to 8-45)	Stylohyoid ligament and greater and lesser horns of hyoid bone	Median raphe of pharynx		
Inferior constrictor (Figs. 8-42 to 8-45)	Oblique line of thyroid cartilage and side of cricoid cartilage			
Palatopharyngeus (Fig. 8-42)	Hard palate and palatine aponeurosis	Posterior border of lamina of thyroid cartilage and side of pharynx and esophagus		
Salpingopharyngeus (Fig. 8-42)	Cartilaginous part of auditory tube	Blends with palatopharyngeus muscle		Elevate pharynx and larynx during swallowing and speaking[1]
Stylopharyngeus (Figs. 8-42 and 8-45)	Styloid process of temporal bone	Posterior and superior borders of thyroid cartilage with palatopharyngeus muscle	Glossopharyngeal n. (CN IX)	

[1]The salpingopharyngeus muscle also opens the auditory tube.

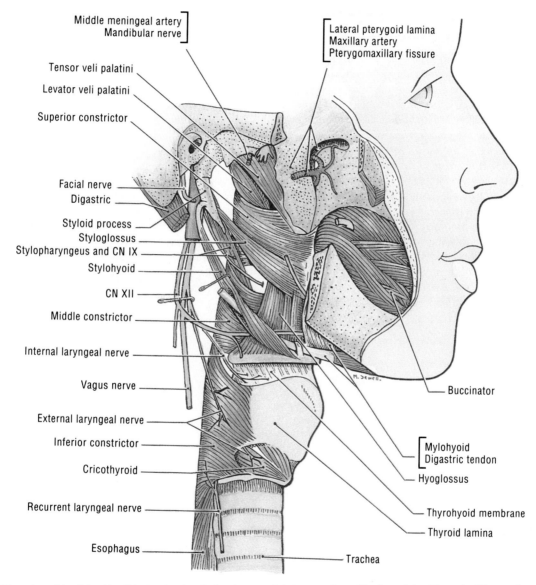

Middle meningeal artery
Mandibular nerve

Lateral pterygoid lamina
Maxillary artery
Pterygomaxillary fissure

Tensor veli palatini

Levator veli palatini

Superior constrictor

Facial nerve
Digastric

Styloid process
Styloglossus
Stylopharyngeus and CN IX
Stylohyoid

CN XII

Middle constrictor

Internal laryngeal nerve

Vagus nerve

External laryngeal nerve

Inferior constrictor

Cricothyroid

Recurrent laryngeal nerve

Esophagus

Buccinator

Mylohyoid
Digastric tendon

Hyoglossus

Thyrohyoid membrane

Thyroid lamina

Trachea

M. Jewell

Figure 8-45. Dissection of the right side of the head and neck showing the pharyngeal muscles and the buccinator muscle. Observe that the middle constrictor muscle is overlapped by the hyoglossus muscle, which is in turn overlapped by the mylohyoid muscle. Observe the gaps between the constrictor muscles through which vessels and nerves pass.

horizontally, and inferiorly, overlapping those of the middle constrictor, to insert into the median raphe of the pharynx. The fibers arising from the cricoid cartilage are believed to act as a sphincter, preventing air from entering the esophagus; they relax during swallowing.

The overlapping arrangement of the three constrictor muscles leaves the following four deficiencies or gaps in the pharyngeal musculature for structures to enter the pharynx (Figs. 8-44 and 8-45):

1. *Superior to the superior constrictor muscle*, the levator veli palatini muscle, the auditory tube, and the ascending palatine artery pass through the gap between the superior constrictor muscle and the skull (Figs. 8-44 and 8-45). Superior to the superior border of the superior constrictor, the pharyngobasilar fascia blends with the buccopharyngeal fascia to form, with the mucous membrane, the thin wall of the *pharyngeal recess* (Figs. 8-41 and 8-46).

2. *Between the superior and middle constrictor muscles* (Figs. 8-43 and 8-44) is the gateway to the mouth, through which pass the stylopharyngeus muscle, the glossopharyngeal nerve (CN IX), and the stylohyoid ligament.

3. *Between the middle and inferior constrictor muscles* (Figs. 8-43 to 8-45), the internal laryngeal nerve and the superior laryngeal artery and vein pass to the larynx.

4. *Inferior to the inferior constrictor muscle*, the recurrent laryngeal nerve and the inferior laryngeal artery pass superiorly into the larynx (Fig. 8-45).

Internal Muscles of the Pharynx

The internal, chiefly longitudinal muscular layer, consists of three muscles: *stylopharyngeus*, *palatopharyngeus*, and *salpingopharyngeus* (Figs. 8-42, 8-43, 8-45; Table 8-5). They all elevate the larynx and pharynx in swallowing and speaking.

The Stylopharyngeus Muscle (Figs. 8-43, 8-45, and 8-47). This long, thin, conical muscle descends inferiorly between the external and internal carotid arteries and enters the wall of the pharynx between the superior and middle constrictors muscles. Its attachments, nerve supply, and main action are given in Table 8-5. It elevates the pharynx and larynx and expands the sides of the pharynx, thereby aiding in pulling the pharyngeal wall over a bolus of food during swallowing.

The Palatopharyngeus Muscle (Figs. 8-42 and 8-47 to 8-50). This thin muscle and the overlying mucosa form the *palatopharyngeal arch*. Its attachments, nerve supply, and main action are given in Table 8-5. It elevates the pharynx and larynx, thereby shortening the pharynx in swallowing. It also constricts the palatopharyngeal arch.

The Salpingopharyngeus Muscle (Figs. 8-41, 8-42, 8-48, and 8-49). This slender muscle descends in the lateral wall of the pharynx where it is covered by the *salpingopharyngeal fold* of mucous membrane. Its attachments, nerve supply, and main action are given in Table 8-5. It elevates the pharynx and larynx and opens the pharyngeal orifice of the auditory tube during swallowing.

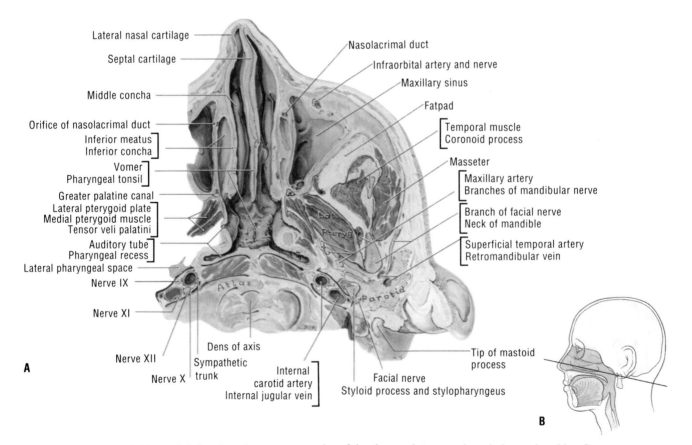

Figure 8-46. *A*, Inferior view of a transverse section of the pharynx that passes through the nasal cavities. *B*, Orientation drawing.

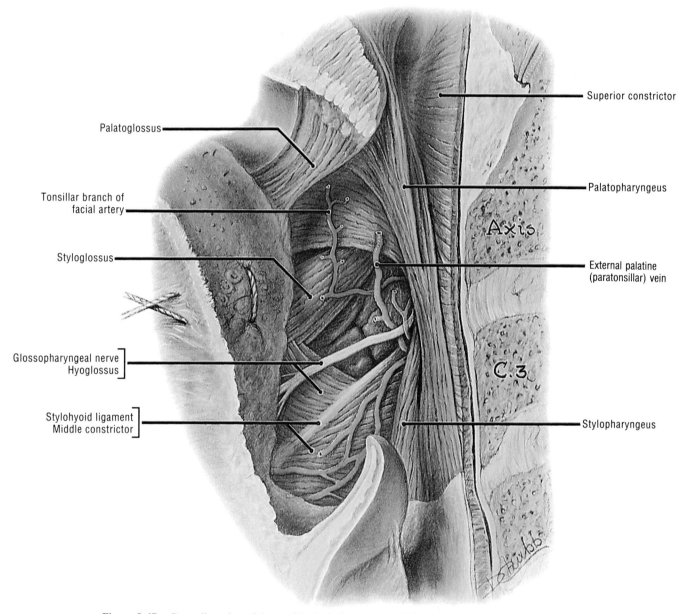

Palatoglossus

Tonsillar branch of
facial artery

Styloglossus

Glossopharyngeal nerve
Hyoglossus

Stylohyoid ligament
Middle constrictor

Superior constrictor

Palatopharyngeus

Axis

External palatine
(paratonsillar) vein

C.3

Stylopharyngeus

Figure 8-47.   Deep dissection of the tonsillar bed after removal of the palatine tonsil. The tongue is pulled
anteriorly and the inferior (lingual) attachment of the superior constrictor muscle is cut away.

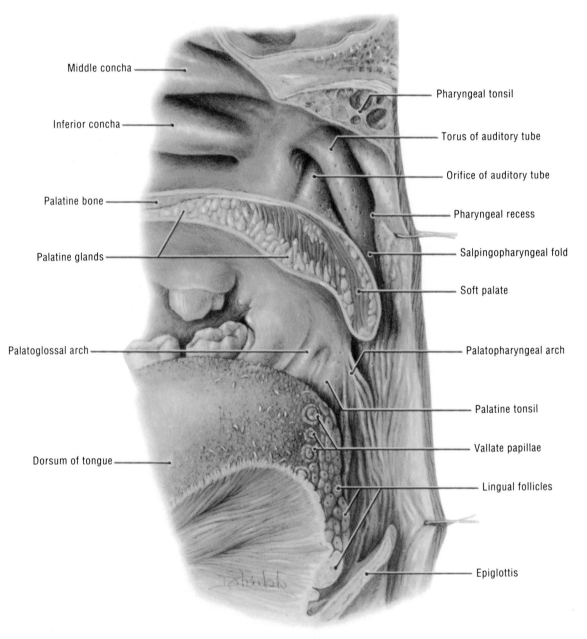

Figure 8-48. Lateral view of a dissection of the interior of the pharynx.

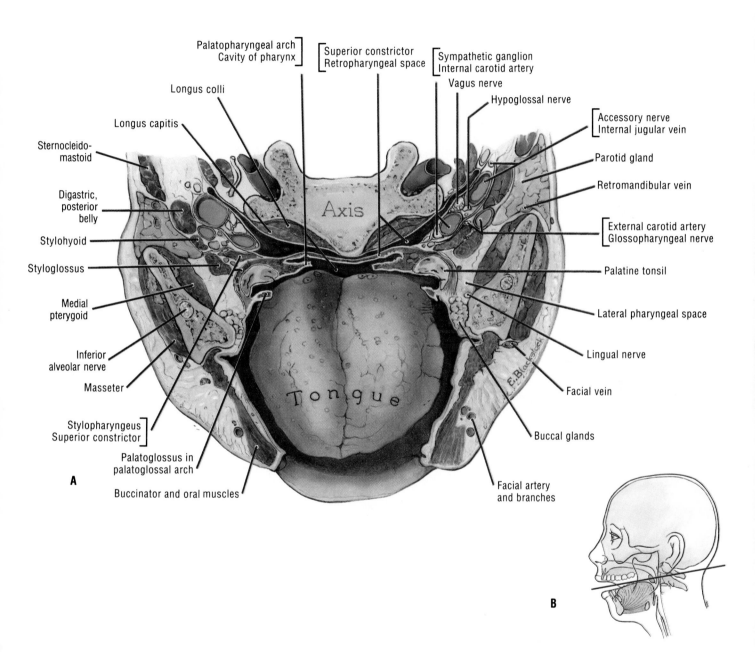

Figure 8-49. *A*, Transverse section of the head that passes through the mouth. Anterior to the ribbonlike palatoglossus muscle and its arch is the oral cavity; posterior to it is the pharynx. *B*, Orientation drawing.

Interior of the Pharynx

The pharynx communicates with three cavities: the nose, mouth, and larynx (Figs. 8-41, 8-42, 8-46, and 8-50).

The Nasopharynx

The nasal part of the pharynx has a *respiratory function*. It lies superior to the soft palate and is a posterior extension of the nasal cavities. The nose opens into the nasopharynx via two large posterior apertures called **choanae** or *internal nares* (Fig. 8-41; see also Fig. 7-6). They are separated by the bony nasal septum (see Fig. 7-107). The roof and posterior wall of the nasopharynx form a continuous surface that lies inferior to the body of the sphenoid bone and the basilar part of the occipital bone. In the mucous membrane of the roof and posterior wall of the nasopharynx is a collection of lymphoid tissue, known as the **pharyngeal tonsil** (Figs. 8-46 and 8-48).

The pharyngeal *orifice of the auditory tube* is on the lateral wall of the nasopharynx, 1 to 1.5 cm posterior to the inferior concha and level with the superior border of the palate (Figs. 8-42 and 8-48). The orifice, which is directed inferiorly, has a hoodlike *tubal elevation* over it called the *torus of the auditory tube* or the *torus tubarius* (L. *torus*, swelling). The torus is produced by the projecting base of the cartilaginous part of the **auditory tube**. Extending inferiorly from the torus is a vertical fold of mucous membrane, known as the *salpingopharyngeal fold* (Fig. 8-48). It covers the salpingopharyngeus muscle which opens the pharyngeal orifice of the auditory tube during swallowing (Fig. 8-42). The collection of lymphoid tissue in the submucosa of the pharynx, posterior to the orifice of the auditory tube, is known as the **tubal tonsil**. Posterior to the torus and the salpingopharyngeal fold, there is a slitlike lateral projection of the pharynx called the *pharyngeal recess* (Figs. 8-41, 8-46, 8-48, and 8-49). It extends laterally and posteriorly.

Hypertrophy or enlargement of the pharyngeal tonsils, commonly known as **adenoids**, can obstruct the passage of air from the nasal cavities through the choanae into the nasopharynx. This hypertrophy obstructs nasal respiration and makes mouth breathing necessary. In chronic cases the patient develops a characteristic facial expression called the ''adenoid facies.'' The open mouth and protruding tongue give the person a dull expression. Impairment of hearing results from nasal obstruction and blockage of the pharyngeal orifices of the auditory tubes. Infection from the enlarged pharyngeal tonsils may spread to the tubal tonsils (*tubal tonsillitis*), causing swelling and closure of the auditory tubes. Infection spreading from the nasopharynx to the middle ear or tympanic cavity (see Fig. 7-113) causes **otitis media** (middle ear infection), which may result in a temporary or permanent hearing loss.

The Oropharynx

The oral part of the pharynx has a *digestive function*. It is continuous with the oral cavity through the oropharyngeal isthmus. The oropharynx is bounded by the soft palate superiorly, the base of the tongue inferiorly, and the palatoglossal and palatopharyngeal arches laterally (Figs. 8-40, 8-41, 8-48, and 8-50). It extends from the soft palate to the superior border of the epiglottis.

The Palatine Tonsils (Figs. 8-47 to 8-51). Usually referred to as ''the tonsils,'' these collections of lymphoid tissue lie on each side of the oropharynx in the triangular interval between the palatine arches. They are called palatine tonsils because their superior one-third extends into the soft palate and because of their relation to the palatine arches. The palatine tonsils vary in size from person to person. In children the palatine tonsils tend to be large, whereas in older persons they are usually small and often inconspicuous. The visible part of the tonsil is

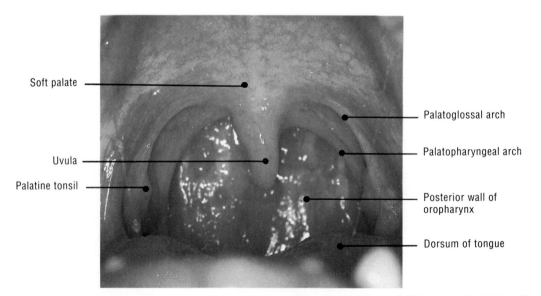

Soft palate

Uvula

Palatine tonsil

Palatoglossal arch

Palatopharyngeal arch

Posterior wall of oropharynx

Dorsum of tongue

Figure 8-50.   The oral cavity and palatine tonsils of a young adult man taken with the mouth wide open and the tongue protruding as far as possible. (From Liebgott, B: *The Anatomical Basis of Dentistry*. Philadelphia, BC Decker Inc., 1986.)

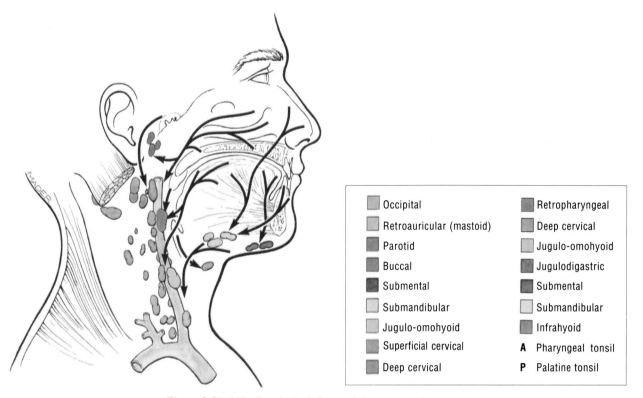

Figure 8-51. The lymphatic drainage of the tongue and tonsils.

no guide to its actual size because much of it may be hidden by the tongue and buried in the soft palate. Commonly the palatine tonsil is about 2 cm in its greatest dimension and usually does not fill the space between the palatine arches. Part of the remaining space, called the *intratonsillar cleft*, penetrates the tonsil and may pass into the soft palate. This cleft is the remains of the second pharyngeal pouch in the embryo (Moore, 1988). The exposed free surface of the palatine tonsil is characterized by the slitlike orifices of the mouths of the *tonsillar crypts*.

The **tonsillar bed**, in which the palatine tonsil lies, is between the palatoglossal and palatopharyngeal arches (Figs. 8-47 and 8-48). The thin, fibrous sheet covering the tonsillar bed is part of the pharyngobasilar fascia (Fig. 8-43). The tonsillar bed contains two muscles, the palatopharyngeus and the superior constrictor (Table 8-5), which are part of the muscular coat of the pharynx. The **tonsillar artery**, a branch of the facial artery, passes through the superior constrictor muscle and enters the inferior pole of the tonsil (Fig. 8-47). The tonsillar bed also receives small arterial twigs from the ascending palatine, lingual, descending palatine, and ascending pharyngeal arteries (Fig. 8-24). The **external palatine vein** (Fig. 8-47) is a large vessel that descends from the soft palate and passes close to the lateral surface of the tonsil before it enters the *pharyngeal plexus of veins*. One or more veins leave the inferior part of the deep aspect of the tonsil and open into this pharyngeal plexus and the facial vein (see Fig. 7-29).

The **nerves of the tonsil** are derived from the *tonsillar plexus of nerves* formed by branches of the glossopharyngeal and vagus nerves (Fig. 8-33; see also Fig. 7-74). Other branches are derived from the *pharyngeal plexus*, a network of fine nerve fibers in the fascia covering the middle constrictor muscle (Fig. 8-45).

The **lymph vessels from the tonsil** pass laterally and inferiorly to the lymph nodes near the angle of the mandible and to the jugulodigastric node (Figs. 8-16 and 8-51). Because of the frequent *enlargement of the jugulodigastric node* in tonsillitis, it is often referred to as the *tonsillar (lymph) node*. The palatine, lingual, and pharyngeal tonsils form an incomplete circular band of lymphoid tissue around the superior part of the pharynx that is called the **tonsillar ring** (Waldeyer's ring) of lymphoid tissue. The anteroinferior part of the ring is formed by the *lingual tonsil* (p. 746), a collection of lymphoid tissue in the posterior part of the tongue (Figs. 8-41 and 8-48). Lateral parts of the ring are formed by the palatine and tubal tonsils, and posterior and superior parts of the ring are formed by the pharyngeal tonsil (Figs. 8-46 and 8-49).

The tonsillar ring of lymphoid tissue does not form a strong defense system against the spread of infection from the oral and nasal cavities to the lower respiratory organs (bronchi and lungs). Infection of the tonsils (*tonsillitis*) is often associated with a sore throat and *pyrexia* (G. feverishness). Usually the jugulodigastric lymph node (*tonsillar lymph node*) in the deep cervical chain is enlarged and tender (Fig. 8-51).

Tonsillectomy is carried out by dissecting the tonsil from its bed, or by a *guillotine operation*. In each case the tonsil and the fascial sheet covering the tonsillar bed are removed. Considerable bleeding and other complications may follow removal of the tonsils. Owing to its abundant blood supply, bleeding may arise from the tonsillar artery or other arterial twigs (Fig. 8-47), but *bleeding commonly comes from the external palatine vein*. This vein, which descends from the soft palate, is immediately related to the lateral surface of the tonsil. The branches of the *glossopharyngeal nerve* to the tongue (see Fig. 7-74) accompany the tonsillar artery on the lateral wall of the pharynx (Fig. 8-47). Because this wall is thin, CN IX is vulnerable to injury. Also, edema around this nerve following tonsillectomy may result in a temporary loss of the sensation of taste in the posterior part of the tongue. *Careless removal of a tonsil could also injure the lingual nerve*. It does not supply the tonsil, but it passes lateral to the pharyngeal wall near the anterior part of the tonsil (Fig. 8-49). The *internal carotid artery* should be safe during tonsillectomy, but it can be injured if adjacent tissues are damaged in attempting to ensure that all tonsillar tissue is removed. The artery is especially vulnerable when it is tortuous and lies directly lateral to the tonsil (Fig. 8-43).

A *branchial fistula* (canal) opens internally into the intra-tonsillar cleft and externally on the side of the neck. This very uncommon, abnormal canal usually results from persistence of *remnants of the second pharyngeal pouch and second branchial groove* (Moore, 1988). The fistula ascends from its cervical opening through the subcutaneous tissue, platysma, and deep fascia of the neck to enter the carotid sheath (Figs. 8-21 and 8-34). It then passes between the internal and external carotid arteries on its way to its opening in the intratonsillar cleft.

The Laryngopharynx

The laryngeal part of the pharynx lies posterior to the larynx (Figs. 8-40 and 8-41). It extends from the superior border of the epiglottis to the inferior border of the cricoid cartilage, where it narrows to become continuous with the esophagus. Posteriorly the laryngopharynx is related to the bodies of C4 to C6 vertebrae. Its posterior and lateral walls are formed by the middle and inferior constrictor muscles, with the palatopharyngeus and stylopharyngeus internally (Figs. 8-42 and 8-43). The laryngopharynx communicates with the larynx through the aditus or *inlet of the larynx* (Fig. 8-41). The **piriform recess** is a small, pear-shaped depression of the laryngopharyngeal cavity on each side of the inlet of the larynx. This rather deep, mucosa-lined fossa is separated from the inlet by the *aryepiglottic fold*. Laterally the piriform recess is bounded by the medial surfaces of the thyroid cartilage and the *thyrohyoid membrane* (Fig. 8-53). The branches of the internal laryngeal and recurrent laryngeal nerves lie deep to the mucous membrane of the piriform recess (Fig. 8-52).

Innervation of the Pharynx

The motor and most of the sensory supply to the pharynx is derived from the *pharyngeal plexus of nerves* on the surface of the pharynx (Figs. 8-26, 8-38, 8-45, and 8-52; see also Fig. 7-74). This plexus is formed by pharyngeal branches of the vagus (CN X) and glossopharyngeal (CN IX) nerves and by sympathetic branches from the superior cervical ganglion. The *motor fibers in the pharyngeal plexus* are derived from the cranial root of CN XI, the *accessory nerve* (Fig. 8-13) and are carried by the vagus nerve to all muscles of the pharynx and soft palate (Fig. 8-33), except the stylopharyngeus (supplied by CN IX) and the tensor veli palatini (supplied by CN V3). The *sensory fibers in the pharyngeal plexus* are derived from the glossopharyngeal nerve (CN IX). They supply most of the mucosa of all three parts of the pharynx (see Figs. 7-74 and 9-10). The sensory nerve supply of the mucous membrane of the nasopharynx is mainly from the maxillary nerve (CN V²), a purely sensory nerve (see Fig. 7-25).

Foreign bodies (*e.g.*, chicken bones and "safety pins") entering the pharynx may become lodged in the piriform recess. If sharp, they may pierce the mucous membrane and injure the internal laryngeal nerve (Fig. 8-52). This may result in anesthesia of the laryngeal mucous membrane as far inferiorly as the vocal folds (Fig. 8-58). Similarly, the nerve may be injured if the instrument used to remove the foreign body accidentally pierces the mucous membrane.

Deglutition (Swallowing)

Although we swallow without thinking, deglutition is a complex process whereby food is transferred from the mouth through the pharynx and esophagus into the stomach. The term *bolus* (L. *bolos*, lump or choice morsel) is used to describe the mass of food or quantity of liquid that is swallowed at one time. Solid food is masticated (chewed) and mixed with saliva to form a soft bolus during chewing. Deglutition is described in three stages: in the (1) mouth, (2) pharynx, and (3) esophagus.

The first stage of swallowing is voluntary, during which the bolus is pushed from the mouth into the oropharynx, mainly by movements of the tongue. The tongue is raised and pressed against the hard palate by the intrinsic muscles of the tongue.

The second stage of swallowing is involuntary and is usually rapid. It involves contraction of the walls of the pharynx. Breathing and chewing stop, and successive contractions of the three constrictor muscles (Table 8-5) move the food through the oral and laryngeal parts of the pharynx. The bolus of food is prevented from entering the nasopharynx by elevation of the soft palate. The tensor veli palatini and levator veli palatini muscles tense and elevate the soft palate against the posterior wall of the pharynx (see Fig. 7-92; Table 7-7). These actions close the pharyngeal isthmus, thereby preventing food from entering the nasopharynx. Should a person happen to laugh during this stage, the muscles of the soft palate relax and may allow some food to enter the nasopharynx. In these cases the food, especially if it is liquid, is expelled through the nose. As the bolus of food passes through the oropharynx, the walls of the pharynx are raised. The contraction of the pharyngeal muscles elevate the pharynx and larynx (Table 8-5).

Watch someone swallow, particularly a thin man, and observe that the laryngeal prominence rises. The palatopharyngeus and stylopharyngeus muscles elevate the larynx and pharynx in swallowing. Palpate your hyoid bone with your thumb and second digit as you swallow and verify that it also rises. The hyoid bone is raised and fixed during swallowing by contraction of the geniohyoid, mylohyoid, digastric, and stylohyoid muscles (Table 8-2). Verify that elevation and anterior movement of the hyoid bone precedes elevation of the larynx.

During deglutition the *vestibule of the larynx* is closed

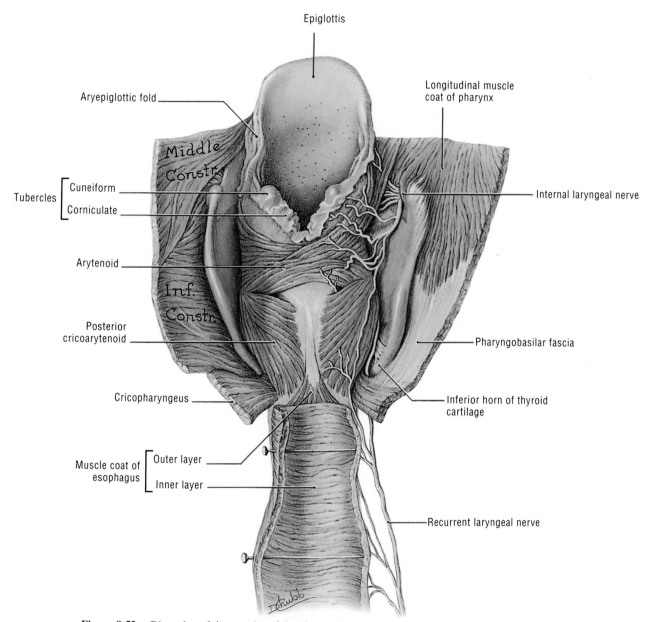

Epiglottis

Aryepiglottic fold

Longitudinal muscle
coat of pharynx

Middle
Constr.

Tubercles
Cuneiform
Corniculate

Internal laryngeal nerve

Arytenoid

Inf.
Constr.

Posterior
cricoarytenoid

Pharyngobasilar fascia

Cricopharyngeus

Inferior horn of thyroid
cartilage

Muscle coat of
esophagus
Outer layer
Inner layer

Recurrent laryngeal nerve

Figure 8-52.   Dissection of the muscles of the pharynx, larynx, and esophagus (posterior view). The mucous
membrane and the left palatopharyngeus muscle have been removed.

(Fig. 8-59*B*), the epiglottis is bent posteriorly over the inlet of the larynx, and the aryepiglottic folds are approximated (Figs. 8-40 and 8-41). These folds provide lateral food channels that guide the bolus of food from the sides of the epiglottis. The food now passes over the oral surface of the epiglottis and the closed inlet of the larynx. All these actions are designed to prevent food from entering the larynx.

The *third stage of swallowing* squeezes the bolus from the laryngopharynx into the esophagus. This is produced by the inferior constrictor muscle of the pharynx (Figs. 8-43 to 8-45).

Injury to the recurrent laryngeal nerves (*e.g.*, during a thyroidectomy) results in paralysis of the muscles in the aryepiglottic folds. As a result, the inlet of the larynx does not close completely during swallowing and food may enter the larynx. *Choking on food is a common cause of laryngeal obstruction*, particularly in persons who have consumed excessive amounts of alcohol, or who have *bulbar palsy* (degeneration of motor neurons in the brain stem nuclei of CN IX and CN X [p. 864] that supply the muscles of deglutition). Difficulty in swallowing is called **dysphagia**.

Young children swallow a variety of objects, most of which reach the stomach and pass through the gastrointestinal tract without difficulty. In some cases the foreign body stops at the inferior end of the laryngopharynx, its narrowest part (Figs. 8-40 and 8-41), or in the esophagus just inferior to the cricopharyngeus muscle, part of the inferior constrictor (Fig. 8-52). Foreign bodies in the pharynx are removed under direct vision through an *pharyngoscope* (an instrument for examining the interior of the pharynx). Radiographic examinations and magnetic resonance images (MRIs [*e.g.*, Fig. 8-41*B*]) will also reveal the presence of a foreign body in the pharynx (*e.g.*, a pin in the piriform recess).

The Larynx

The larynx (between the pharynx and trachea) communicates with the mouth and nose through the laryngeal and oral parts of the pharynx (Figs. 8-40 and 8-41). Although it is part of the air passages, the larynx normally acts as a valve for preventing swallowed food and foreign bodies from entering the lower respiratory passages. **The larynx is the phonating mechanism** (G. *phone*, voice), which is specifically *designed for voice production*. Phonation is defined as the utterance of sounds with the aid of the vocal folds (cords). Through movements of its cartilages, the larynx varies the opening between the vocal folds (Figs. 8-58 and 8-59), thereby varying the pitch of sounds produced by the passage of air through them. These sounds are translated into intelligible speech by articulatory and resonating structures (*e.g.*, the lips, tongue, and mouth).

The larynx is located in the anterior portion of the neck (Fig. 8-1). In adult males it is about 5 cm in length (Fig. 8-2) and is related posteriorly to the bodies of C3 to C6 vertebrae (Fig. 8-40). The larynx is shorter in women and children and is situated slightly more superiorly in the neck. This sex difference in the larynx normally develops at puberty in males, at which time all its cartilages enlarge.

The Skeleton of the Larynx

The laryngeal skeleton comprises nine cartilages that are joined by various ligaments and membranes (Figs. 8-53 to 8-56). Three of the cartilages are single (thyroid, cricoid, and epiglottis), and three are paired (arytenoid, corniculate, and cuneiform).

The Thyroid Cartilage (Figs. 8-1, 8-2, and 8-53 to 8-55). This is the largest of the laryngeal cartilages. It is composed of two quadrilateral *laminae* (L. thin plates). The inferior two-thirds of these laminae are fused anteriorly in the median plane to form a subcutaneous projection, called the *laryngeal prominence*.

The larynx is more prominent in postpubertal males than in females because the angle at which the thyroid laminae meet is smaller in males (Fig. 8-2*C*) and the anteroposterior diameter of the laminae is greater. These gender differences in the thyroid cartilage are usually evident by the 16th year.

Immediately superior to the laryngeal prominence, the two thyroid laminae diverge to form a V-shaped **thyroid notch** (Fig. 8-54). The posterior border of each lamina projects superiorly as the *superior horn* and inferiorly as the *inferior horn*. The superior border of the thyroid cartilage is attached to the **hyoid bone** by the *thyrohyoid membrane*. The lateral surface of each lamina is marked by an oblique line. It provides attachment for the inferior constrictor muscle of the pharynx and the sternothyroid and thyrohyoid muscles (Figs. 8-44 and 8-52). The inferior horns (L. *cornua*) of the thyroid cartilage articulate with the cricoid cartilage at special facets that allow the thyroid cartilage to tilt or glide anteriorly or posteriorly in a visorlike manner.

The Cricoid Cartilage (Figs. 8-53 to 8-56). The cricoid (G. ring) is *shaped like a signet ring* with its band facing anteriorly. The posterior (signet) part of the cricoid is called the **lamina** and the anterior (band) part is termed the **arch**. Although much smaller than the thyroid cartilage, the cricoid is thicker and stronger. Being the *most inferior of the cartilages of the larynx*, the cricoid forms the inferior parts of the anterior and lateral walls and most of the posterior wall of the larynx (Fig. 8-40*A*). The cricoid cartilage is attached to the inferior margin of the thyroid cartilage by the *cricothyroid ligaments* and to the first tracheal ring by the *cricotracheal ligament*.

The Arytenoid Cartilages (Figs. 8-41, 8-55, 8-56, 8-61, and 8-62). These paired cartilages, shaped like three-sided pyramids, articulate with the lateral parts of the superior border of the lamina of the cricoid cartilage. Each cartilage has an *apex* superiorly, a *vocal process* anteriorly, and a *muscular process* laterally. The apex is attached to the aryepiglottic fold, the vocal process to the vocal ligament, and the muscular process to the posterior and lateral cricoarytenoid muscles.

The Corniculate and Cuneiform Cartilages (Figs. 8-41, 8-55, and 8-56). These small cartilaginous nodules are in the posterior part of the aryepiglottic folds. These cartilages are attached to the apices of the arytenoid cartilages. The cuneiform (L. wedge-shaped) cartilages lie in the aryepiglottic folds and

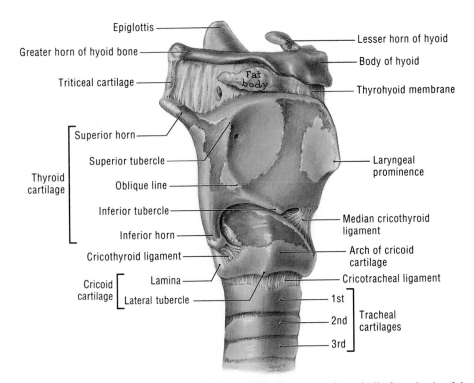

Figure 8-53. Lateral view of the skeleton of the larynx. The larynx extends vertically from the tip of the epiglottis to the inferior border of the cricoid cartilage. The hyoid bone is not part of the larynx.

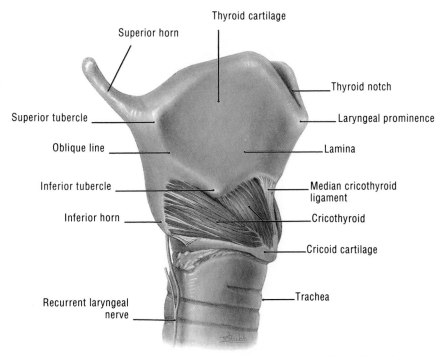

Figure 8-54. Lateral view of the thyroid cartilage, cricoid cartilage, cricothyroid muscle, and proximal part of the trachea.

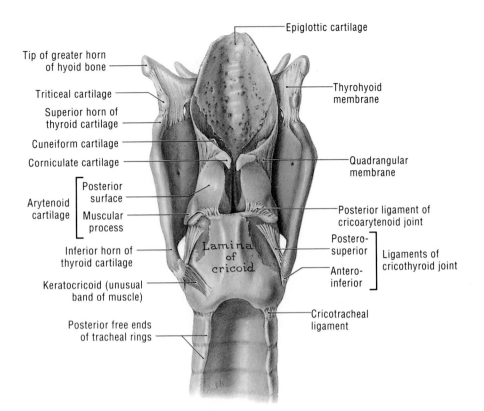

Epiglottic cartilage

Tip of greater horn
of hyoid bone

Triticeal cartilage

Superior horn of
thyroid cartilage

Cuneiform cartilage

Corniculate cartilage

Arytenoid
cartilage
  Posterior
  surface

  Muscular
  process

Inferior horn of
thyroid cartilage

Keratocricoid (unusual
band of muscle)

Posterior free ends
of tracheal rings

Thyrohyoid
membrane

Quadrangular
membrane

Posterior ligament of
cricoarytenoid joint

Postero-
superior

Antero-
inferior

Ligaments of
cricothyroid joint

Cricotracheal
ligament

Lamina
of
cricoid

Figure 8-55.   Posterior view of the skeleton of the larynx. Observe that the thyroid cartilage shields the smaller cartilages of the larynx. The hyoid bone, although not a part of the larynx, shields the superior part of the epiglottic cartilage.

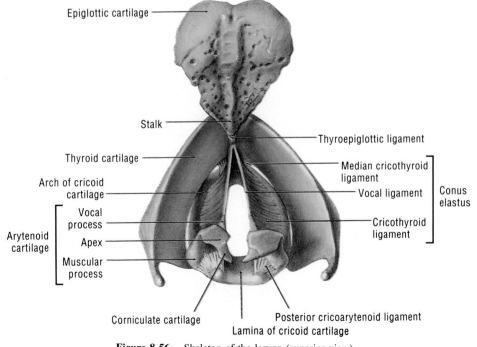

Epiglottic cartilage

Stalk

Thyroid cartilage

Arch of cricoid
cartilage

Arytenoid
cartilage
  Vocal
  process

  Apex

  Muscular
  process

Corniculate cartilage

Lamina of cricoid cartilage

Thyroepiglottic ligament

Median cricothyroid
ligament

Vocal ligament

Cricothyroid
ligament

Conus
elastus

Posterior cricoarytenoid ligament

Figure 8-56.   Skeleton of the larynx (superior view).

are approximated to the tubercle of the epiglottis when the inlet of the larynx is closed during swallowing.

The Epiglottic Cartilage (Figs. 8-40, 8-41, and 8-52 to 8-57). This thin cartilage, which is shaped like a leaf or bicycle saddle, gives flexibility to the epiglottis. Situated posterior to the root of the tongue and hyoid bone and anterior to the inlet of the larynx, the epiglottic cartilage forms the superior part of the anterior wall and the superior margin of the inlet of the larynx. Its broad superior end is free, and its tapered inferior end is attached to the *thyroepiglottic ligament*, located in the angle formed by the thyroid laminae. The anterior surface of the epiglottic cartilage is attached to the hyoid bone by the *hyoepiglottic ligament*. The inferior part of the posterior surface of the epiglottic cartilage that projects posteriorly is called the *epiglottic tubercle* (Fig. 8-59A).

> The thyroid, cricoid, and most parts of the arytenoid cartilages often calcify with age. *Fractures of the laryngeal skeleton* may result from blows received during boxing or karate or from compression by a shoulder strap during an automobile accident. Because of the frequency of this type of injury, most goalies in ice hockey have protective guards hanging from their masks that cover their larynges. **Laryngeal fractures** produce submucous hemorrhage and edema, respiratory obstruction, hoarseness, and sometimes an inability to speak, usually temporary.

Joints of the Larynx

Some of the laryngeal cartilages articulate freely, allowing them to move during voice production. There are two pairs of synovial joints in the larynx.

The Cricothyroid Joints (Figs. 8-53, 8-55, and 8-56). These articulations are located between the facets on the lateral surfaces of the cricoid cartilage and the inferior horns of the thyroid cartilage. Each joint has a fibrous capsule, which is lined by a synovial membrane. The main movements at these joints are rotation and gliding of the thyroid cartilage. These movements result in changes in the length of the vocal folds. They also slacken or tighten the vocal ligaments, which pass between the arytenoid cartilages and the thyroid cartilage.

The Cricoarytenoid Joints (Fig. 8-55). These articulations are located between the bases of the arytenoid cartilages and the superior sloping surfaces of the lamina of the cricoid cartilage. These joints permit the following movements of the arytenoid cartilages: (1) sliding toward or away from one another; (2) tilting anteriorly and posteriorly; and (3) rotary motion. These movements are important in approximating, tensing, and relaxing the vocal folds.

Laryngeal Ligaments and Membranes

The various laryngeal cartilages are united by several ligaments and membranes.

The Thyrohyoid Membrane (Figs. 8-53 and 8-55). This extrinsic ligament connects the thyroid cartilage and the hyoid bone, thereby suspending the larynx. It is separated from the posterior surface of the body of the hyoid by a bursa. Its thicker median part is called the *median thyrohyoid ligament*, and its thickened lateral parts are called the *lateral thyrohyoid ligaments*. The lateral ligaments connect the tips of the superior horns of the thyroid cartilage to the tips of the greater horns of the hyoid bone. They each contain a kernellike *triticeal cartilage* that helps to close the inlet of the larynx during swallowing.

The Cricothyroid and Cricotracheal Ligaments (Figs. 8-53 and 8-56). These ligaments connect the arch of the cricoid cartilage with the thyroid cartilage and the first tracheal ring, respectively. The fibrous *median cricothyroid ligament* produces a soft spot inferior to the thyroid cartilage. It is at this point that the airway is closest to the skin and is most accessible.

> If food or some object enters the larynx, the laryngeal muscles go into spasm; this tenses the vocal folds. As a result the *rima glottidis*[8] (Fig. 8-59) closes and no air can enter the trachea, bronchi, and lungs. Obviously the person is in danger of asphyxiation. If the foreign object cannot be dislodged, emergency therapy must be given to open the airway. The procedure used depends on the condition of the patient, the facilities available, and the experience of the person giving first aid. If available, a large bore needle is inserted through the median cricothyroid ligament to permit fast entry of air (Fig. 8-53). Later a **cricothyrotomy** may be performed, during which an incision is made through the skin and cricothyroid ligament for more adequate relief of the respiratory obstruction. This procedure may be followed by a **tracheotomy** and insertion of a metal tube into the trachea (Fig. 8-37).

---

[8]The space between the vocal folds (cords).

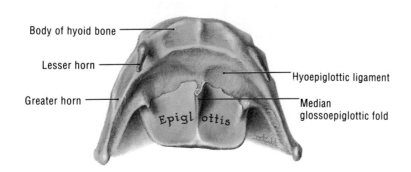

Figure 8-57. Superior view of the epiglottis and the hyoepiglottic ligament. Note that this ligament unites the epiglottic cartilage to the hyoid bone.

forced respiration, it is wide. The vocal folds are closely approximated during speaking so that the rima glottidis appears as a linear slit. Variation in the tension and length of the vocal folds, in the width of the rima glottidis, and in the intensity of the expiratory effort produces changes in the pitch of the voice. The lower range of pitch of the male voice results from the greater length of his vocal folds.

> The rima glottitis is the usual place where aspirated food or other material becomes lodged causing **laryngeal obstruction** (choking). Because the lungs still contain air, compression of the abdomen (*Heimlich maneuver*) will usually expel air from the lungs and dislodge the food or other material.

The Vestibular Folds (Figs. 8-58 to 8-60). These folds extend between the thyroid and arytenoid cartilages and *play little or no part in voice production*. They consist of two thick folds of mucous membrane enclosing the vestibular ligaments. Because they may be confused with the vocal folds (vocal cords), they used to be called the false vocal cords. The space between

the vestibular ligaments is called the *rima vestibuli*. The vestibular folds are part of the protective mechanism by which the larynx is closed during swallowing to prevent the entry of food and foreign particles into it.

Muscles of the Larynx

The muscles of the larynx are divided into extrinsic and intrinsic groups for descriptive purposes (Figs. 8-17, 8-36, 8-54, 8-61, and 8-62; Table 8-6). *The extrinsic muscles move the larynx as a whole.* The **infrahyoid muscles** (omohyoid, sternohyoid, and sternothyroid) are depressors of the hyoid bone and larynx (Table 8-3), whereas the **suprahyoid muscles** (stylohyoid, digastric, mylohyoid, and geniohyoid) and the stylopharyngeus are elevators of the hyoid bone and larynx (Tables 8-2 and 8-5). The thyrohyoid muscle draws the hyoid bone and thyroid cartilage together (Table 8-3), *i.e.*, it depresses the hyoid bone and elevates the thyroid cartilage.

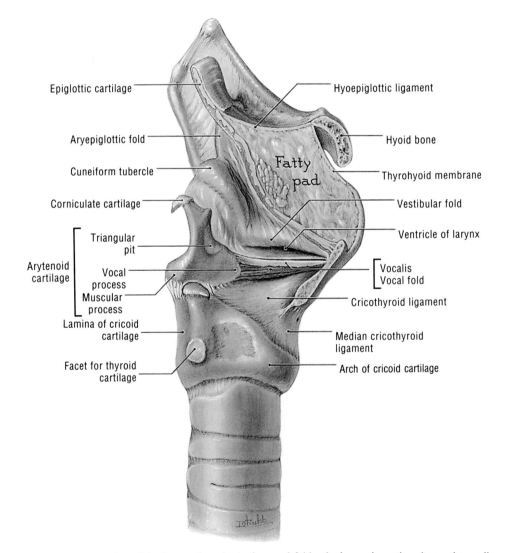

Figure 8-60. Lateral view of the larynx. Superior to the vocal folds, the larynx is sectioned near the median plane and the interior of its left side is seen. Inferior to this level, the right side of the larynx is dissected.

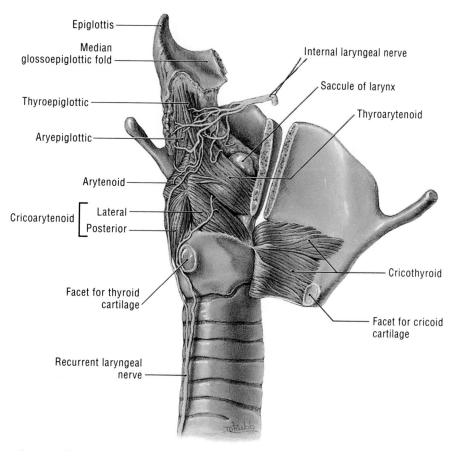

Epiglottis

Median
glossoepiglottic fold

Thyroepiglottic

Aryepiglottic

Arytenoid

Cricoarytenoid
Lateral
Posterior

Facet for thyroid
cartilage

Recurrent laryngeal
nerve

Internal laryngeal nerve

Saccule of larynx

Thyroarytenoid

Cricothyroid

Facet for cricoid
cartilage

Figure 8-61. Dissection of the muscles and nerves of the larynx. The thyroid cartilage is sawn through on the right of the median plane. The cricothyroid joint is laid open, and the right lamina of the thyroid car-tilage is turned anteriorly, stripping the cricothyroid muscles off the arch of the cricoid cartilage.

The intrinsic muscles are concerned with movements of the laryngeal parts, making alterations in the length and tension of the vocal folds and in the size and shape of the rima glottidis in voice production (Fig. 8-62). *All intrinsic muscles of the larynx are supplied by the recurrent laryngeal nerve*, a branch of CN X (Figs. 8-33, 8-61, and 8-63; Table 8-6), **except the crico-thyroid** muscle which is supplied by the *external laryngeal nerve*. The actions of the muscles of the larynx are easiest to understand if they are considered as functional groups.

Muscles of the Inlet of the Larynx (Figs. 8-41, 8-52, and 8-62). These muscles have a sphincteric action and *close the laryngeal inlet* as a protective mechanism during swallowing. Contraction of the transverse and oblique arytenoid muscles and the aryepiglottic muscles brings the aryepiglottic folds together and pulls the epiglottis toward the arytenoid cartilages. These movements help close the inlet.

The Transverse Arytenoid Muscle (Figs. 8-52 and 8-62*E*; Table 8-6). This is the only unpaired muscle of the larynx. It covers the arytenoid cartilages posteriorly and extends from the posterior aspect of one arytenoid cartilage to the same region of the opposite arytenoid.

The Oblique Arytenoid Muscle (Figs. 8-52, 8-61, and 8-62*F*; Table 8-6). This muscle, superficial to the transverse ar-ytenoid muscle, consists of two bundles that cross each other in

an X-like fashion. Some of the oblique fibers continue as the *aryepiglottic muscle*. The continuity of these muscles ensures that the arytenoid cartilages are brought together at the same time as the epiglottis is pulled inferiorly toward these cartilages. The inlet of the larynx is closed in two ways during swallowing, thereby preventing food from entering the larynx.

The Thyroepiglottic Muscle (Fig. 8-61; Table 8-6). This mus-cle arises from the internal surfaces of the laminae of the thy-roid cartilage and inserts on the lateral margin of the epiglottic cartilage. It constricts the inlet of the larynx.

Muscles of the Vocal Folds (Figs. 8-52, 8-56, 8-60, 8-61, and 8-62). These muscles open and *close the rima glottidis*, the space between the vocal folds.

Adductors of the Vocal Folds (Figs. 8-61 and 8-62*C* and *E*; Table 8-6). The lateral cricoarytenoid muscles arise from the lateral portions of the cricoid cartilage and insert into the mus-cular processes of the arytenoid cartilages. The **lateral cricoar-ytenoid muscles** pull the muscular processes anteriorly, rotating the arytenoids so that their vocal processes swing medially. These movements adduct the vocal folds and close the rima glottidis. This action is reinforced by the *transverse arytenoid muscle* that pulls the arytenoid cartilages together.

Abductors of the Vocal Folds (Figs. 8-52, 8-58, 8-61, and 8-62*B*; Table 8-6). The principal abductors of the vocal folds

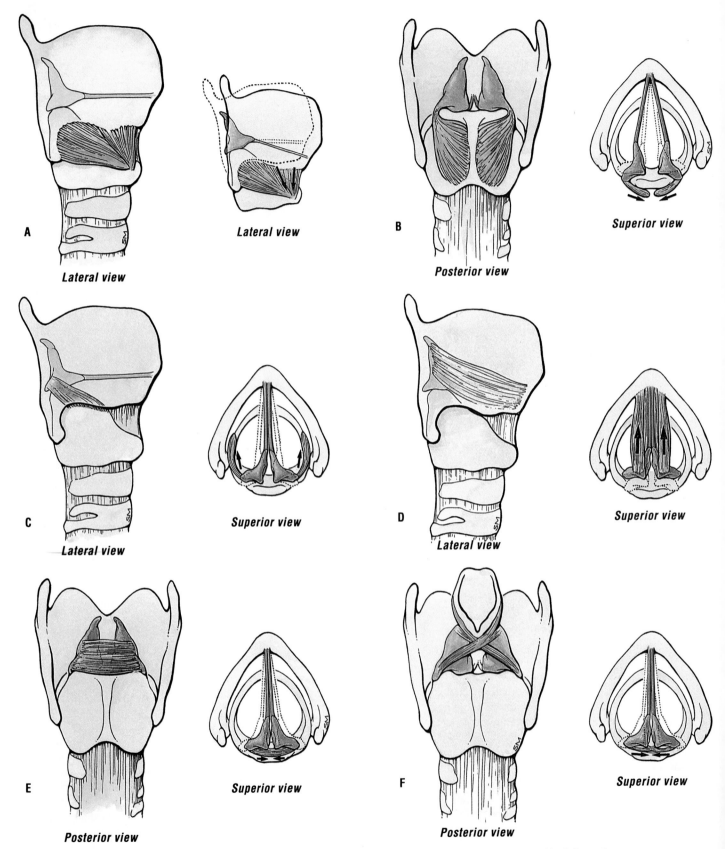

Figure 8-62. Actions of the laryngeal muscles. *A*, Cricothyroid. *B*, Posterior cricoarytenoid. *C*, Lateral cricoarytenoid. *D*, Thyroarytenoid. *E*, Transverse arytenoid. *F*, Oblique arytenoid.

are the **posterior cricoarytenoid muscles**. These muscles arise on each side from the posterior surface of the lamina of the cricoid cartilage and pass laterally and superiorly to insert into the muscular processes of the arytenoid cartilages. They rotate the arytenoid cartilages, thereby deviating them laterally and widening the rima glottidis.

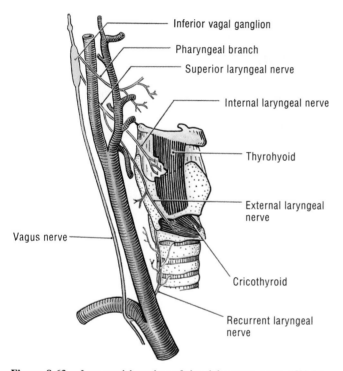

Figure 8-63. Laryngeal branches of the right vagus nerve (CN X). Note that the nerves of the larynx are the internal and external branches of the superior laryngeal nerve and the terminal branches of the recurrent laryngeal nerve.

Tensors of the Vocal Folds (Figs. 8-54, 8-61, and 8-62*A*; Table 8-6). The main tensors of these folds are the triangular **cricothyroid muscles,** which are located on the external surface of the larynx between the cricoid and thyroid cartilages. The muscle on each side arises from the anterolateral part of the cricoid cartilage and inserts into the inferior margin and anterior aspect of the inferior horn of the thyroid cartilage. These muscles tilt the thyroid cartilage anteriorly on the cricoid cartilage, increasing the distance between the thyroid and arytenoid cartilages. As a result, the vocal ligaments are elongated and tightened and the pitch of the voice is raised.

Relaxers of the Vocal Folds (Figs. 8-60, 8-61, and 8-62*D*; Table 8-6). The principal relaxers of these folds are the broad **thyroarytenoid muscles.** They arise from the posterior surface of the thyroid cartilage near the median plane and insert into the anterolateral surfaces of the arytenoid cartilages. One band of its inferior deeper fibers, called the **vocalis muscle,** arises from the vocal ligament and passes to the vocal process of the arytenoid cartilage. The thyroarytenoid muscles pull the arytenoid cartilages anteriorly, thereby slackening the vocal ligaments. *The vocalis muscles produce minute adjustments of the vocal ligaments* (*e.g.,* as occurs during whispering). They also relax parts of the vocal folds during phonation and singing.

Blood Supply of the Larynx

The superior and inferior **laryngeal arteries** supply the larynx. They are branches of the superior and inferior thyroid arteries, respectively (Figs. 8-24, 8-25, and 8-64). The *superior laryngeal artery* runs with the internal branch of the superior laryngeal nerve through the thyrohyoid membrane, and then branches to supply the internal surface of the larynx. The *inferior laryngeal artery* runs with the inferior laryngeal nerve and supplies the mucous membrane and muscles of the inferior aspect of the larynx.

Table 8-6.
The Muscles of the Larynx

Muscle	Origin	Insertion	Innervation	Main Action(s)
Cricothyroid (Figs. 8-54, 8-61, and 8-62*A*)	Anterolateral part of cricoid cartilage	Inferior margin and inferior horn of thyroid cartilage	External laryngeal n. (Fig. 8-63)	Stretches and tenses the vocal fold
Posterior cricoarytenoid (Figs. 8-52, 8-61, and 8-62*B*)	Posterior surface of laminae of cricoid cartilage	Muscular process of arytenoid cartilage		Abducts vocal fold
Lateral cricoarytenoid (Figs. 8-61 and 8-62*C*)	Arch of cricoid cartilage			Adducts vocal fold
Thyroarytenoid[1] (Figs. 8-61 and 8-62*D*)	Posterior surface of thyroid cartilage	Anterolateral surface of arytenoid cartilage	Recurrent laryngeal n. (Fig. 8-63)	Relaxes vocal fold
Transverse and oblique arytenoids (Figs. 8-52 and 8-62*E* and *F*)	One arytenoid cartilage	Opposite arytenoid cartilage		Close laryngeal aditus by approximating arytenoid cartilages
Vocalis[2] (Fig. 8-60)	Angle between laminae of thyroid cartilage	Vocal process of arytenoid cartilage		Alters vocal fold during phonation

[1]The superior fibers of the thyroarytenoid muscle pass into the aryepiglottic fold (Fig. 8-41), and some of them reach the epiglottic cartilage. These fibers constitute the *thyroepiglottic muscle,* which widens the inlet of the larynx.
[2]This slender muscular slip is derived from inferior deeper fibers of the thyroarytenoid muscle.

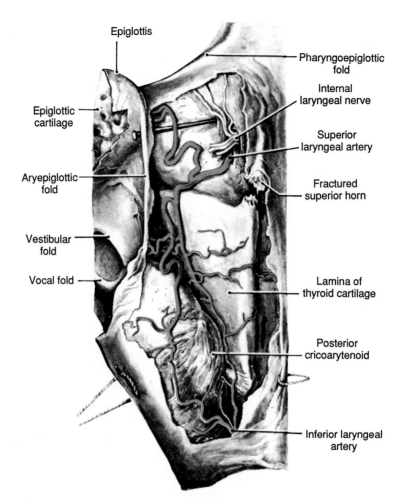

Epiglottis

Pharyngoepiglottic fold

Internal laryngeal nerve

Epiglottic cartilage

Superior laryngeal artery

Aryepiglottic fold

Fractured superior horn

Vestibular fold

Vocal fold

Lamina of thyroid cartilage

Posterior cricoarytenoid

Inferior laryngeal artery

Figure 8-64. Blood supply of the larynx. Observe the anastomoses between the superior and inferior laryngeal arteries. They are branches of the superior and inferior thyroid arteries, respectively (Fig. 8-32).

Innervation of the Larynx

The laryngeal nerves are derived from the **vagus (CN X)** through the internal and external branches of the *superior laryngeal nerve* and the *recurrent laryngeal nerve* (Figs. 8-52, 8-61, and 8-63). All the intrinsic muscles of the larynx, except the cricothyroid muscle, which is innervated by the *external laryngeal nerve* (Fig. 8-63; Table 8-6), are supplied by the recurrent laryngeal nerve with fibers from the accessory nerve (CN XI). The supraglottic portion of the *laryngeal mucosa* is supplied by the internal laryngeal nerve, a branch of the superior laryngeal nerve. The infraglottic portion of the laryngeal mucosa is supplied by the recurrent laryngeal nerve.

The Superior Laryngeal Nerve (Figs. 8-33 and 8-63). This branch of the vagus (CN X) arises from the middle of the inferior ganglion of the vagus at the superior end of the carotid triangle. It divides within the carotid sheath into two terminal branches, the internal laryngeal nerve (sensory and autonomic) and the external laryngeal nerve (motor).

The Internal Laryngeal Nerve (Figs. 8-52, 8-61, and 8-63). This is the larger of the two terminal branches of the superior laryngeal nerve. It pierces the thyrohyoid membrane with the superior laryngeal artery. It supplies sensory fibers to

the laryngeal mucous membrane superior to the vocal folds, including the superior surface of these folds.

The External Laryngeal Nerve (Fig. 8-63). This is the smaller of the two terminal branches of the superior laryngeal nerve. It descends posterior to the sternothyroid muscle in company with the superior thyroid artery. At first it lies on the inferior constrictor muscle of the pharynx and then pierces it to supply this muscle and the cricothyroid. The *cricothyroid muscle* is the only intrinsic laryngeal muscle that is not supplied by the recurrent laryngeal nerve (Table 8-6).

The Recurrent Laryngeal Nerve (Figs. 8-33, 8-38, 8-45, 8-52, 8-54, 8-61, and 8-63). This *clinically important nerve* ascends in the groove between the trachea and esophagus, where it is intimately related to the medial surface of the thyroid gland. It gives branches to the pharynx, esophagus, and trachea. The *recurrent laryngeal nerve supplies all intrinsic muscles of the larynx, with the exception of the cricothyroid muscle* that is supplied by the external laryngeal nerve. It also supplies sensory fibers to the mucous membrane of the larynx inferior to the vocal folds, including the inferior surface of these folds. The terminal part of the recurrent laryngeal is known as the *inferior laryngeal nerve*. It enters the larynx by passing deep to the inferior border of the inferior constrictor muscle of the pharynx and divides into anterior and posterior branches. These branches accompany the inferior laryngeal artery into the larynx (Fig. 8-64).

> The recurrent laryngeal nerve is vulnerable to injury during thyroidectomy (p. 820), carotid endarterectomy (p. 801), and other surgical operations in the anterior triangle of the neck. Bruising of the nerve causes *temporary aphonia*; crushing and sectioning of the nerve reduces the voice to a whisper.

Lymphatic Drainage of the Larynx

The lymph vessels *superior to the vocal folds* accompany the superior laryngeal artery through the thyrohyoid membrane and drain into the *superior deep cervical lymph nodes* (Figs. 8-31, 8-38, and 8-51). The lymph vessels *inferior to the vocal folds* drain into the inferior deep cervical lymph nodes (*supraclavicular nodes*) through the prelaryngeal, pretracheal, and paratracheal lymph nodes. The lymph vessels from the vocal folds do not communicate across the median plane, but vessels in the posterior wall of the larynx anastomose in the submucosa.

> The larynx may be examined visually by *indirect laryngoscopy* using a **laryngoscopic mirror**, or it can be viewed by *direct laryngoscopy* using a tubular, endoscopic instrument called a **laryngoscope**. The vestibular folds normally appear pink, whereas the vocal folds are usually pearly white in color. The size of the larynx varies somewhat from person to person and is not dependent on stature. This largely explains the difference in pitch of the voice in different persons. The larynx is larger in men than in women and the vocal ligaments are longer. The vocal folds in most men are also longer than in women; as a result the voice of most men is deeper than in most women.
>
> **Hoarseness** is the most common symptom of serious disorders of the larynx (*e.g.*, **cancer of the vocal folds**). In cases requiring *laryngectomy* (removal of the larynx), *esophageal speech* (regurgitation of ingested air) and other rehabilitative speech techniques can be learned. The mucosa of the larynx superior to the vocal folds is extremely sensitive, and contact with a foreign body immediately induces explosive coughing. This is a protective mechanism that is designed to keep foreign bodies out of the larynx.

Age Changes in the Larynx. The larynx grows steadily until about 3 years of age, after which little growth occurs until about 12 years of age. Before this, there are no major laryngeal sex differences. At puberty, particularly in males (13 to 16 years of age), the walls of the larynx become strengthened, the laryngeal cavity enlarges, the vocal folds lengthen and thicken, and the laryngeal prominence becomes conspicuous in most males (Fig. 8-1). The length of the vocal folds increases gradually in both sexes up to puberty. During puberty the increase in the length of the vocal folds is abrupt in males. The pitch of the voice lowers by an octave. The change in the length of the vocal folds is largely responsible for the voice changes that occur. The pitch of the voice of *eunuchs* (persons whose testes have been removed [*i.e., castrated males*] during childhood) or in whom testes have not developed (*agonadal males*) does not become lower unless male hormones are administered to induce lengthening of the vocal folds.

PATIENT ORIENTED PROBLEMS

Case 8-1

A 22-year-old woman consulted her physician about a swelling in the anterior aspect of her neck. Although painless, she was concerned because it seemed to be slowly getting larger. Physical examination revealed that the swelling was located just inferior to her hyoid bone and that it was cystic and freely movable. The physician grasped the swelling between his first and second digits and asked the patient to open her mouth and stick out her tongue. Feeling some movement of the mass, he asked the patient to stick her tongue out as far as possible and then retract it. The physician noted a definite superior tug on the mass as the patient's tongue protruded. The swelling also moved superiorly during swallowing. Fluid was aspirated from the swelling for laboratory investigation.

Diagnosis. Thyroglossal duct cyst (thyroglossal cyst).

> **Problems.** Explain the embryological basis of this cyst. Where are these cysts likely to be found? What is the anatomical basis for movement of the cyst superiorly when the patient protrudes her tongue and swallows? What would this condition be called if there had also been a midline cervical opening into the cyst? These problems are discussed on p. 851.

Case 8-2

A 27-year-old second-year medical student consulted her clinical instructor about a painless, plum-shaped swelling in the anterior triangle of her neck, inferior to the angle of her right mandible. As her mandibular third molar teeth (''wisdom teeth'') had not erupted, she thought the swelling might be caused by a dental abscess in the submandibular triangle. She also feared that the firm swelling might be caused by a tumor of the submandibular gland or of the jugulo-omohyoid lymph node. Radiographs of her mandible showed the crown of her left third mandibular molar tooth was in contact with the posterior surface of her second molar tooth. The instructor recommended that her impacted tooth be surgically removed after consulting her physician about the swelling in her submandibular region. He stated that her impacted molar tooth was not the cause of the swelling in the lateral aspect of her neck. On examination the physician found that the swelling was caused by a painless fluctuant cyst, located anterior to the superior one-third of her sternocleidomastoid muscle (Fig. 8-65).

Diagnosis. *Branchial cyst* (lateral cervical cyst). During excision of the cyst, it was discovered that a sinus tract passed superiorly from it.

> **Problems.** Explain the embryological basis of the branchial cleft cyst. Where did the sinus tract probably terminate? What nerve might be damaged during excision of this cyst? What signs would be present if this nerve were damaged? If the sinus tract had passed inferiorly, where would it probably open? These problems are discussed on p. 851.

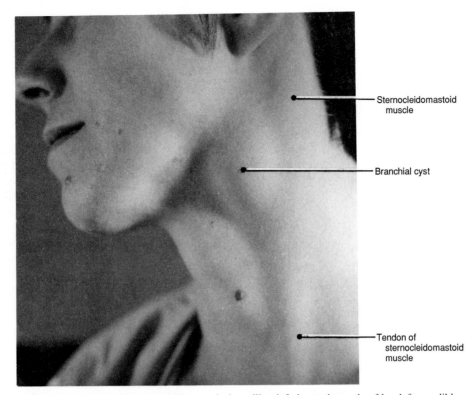

Sternocleidomastoid
muscle

Branchial cyst

Tendon of
sternocleidomastoid
muscle

Figure 8-65. A 27-year-old woman with a cervical swelling inferior to the angle of her left mandible and anterior to her sternocleidomastoid muscle. The swelling is produced by a branchial cyst.

Case 8-3

After completing your first anatomy examination, your father decided to celebrate and take you out for a steak dinner. Obviously he had had a few drinks before meeting you. You noted that his speech was slurred and that he was eating his steak very rapidly. Later you noticed your father's face change suddenly. He had a terrified look and then collapsed on the floor. At first you suspected that he had passed out from too much drinking, but as you examined him more closely you thought that perhaps he was having a stroke, a heart attack, or some other seizure. His pulse was strong and then his face began to turn blue (*cyanosis*). You then realized that your father was suffering from *asphyxia*. You opened his mouth widely and observed that a large piece of steak was caught in the posterior part of his throat. First you reached into his mouth with your second digit and tried to pull it out. On being unsuccessful, you rolled him into the prone position and, with your hands interlocked against his epigastrium, you gave him a forceful bearhug, exerting pressure on his abdomen inferior to his thorax. This increased his intraabdominal pressure and moved his diaphragm superiorly, forcing the air out of his lungs and expelling the piece of steak.

Problems. Where was the piece of steak most likely lodged? If the "Heimlich maneuver" had not been successful and a physician at another table had come to help you, what lifesaving measures do you think he might have taken? Discuss so-called "restaurant deaths." These problems are discussed on p. 851.

Case 8-4

A 30-year-old woman complained of a swelling in the anterior part of her neck, nervousness, and loss of weight. She told her physician that her family complains that she is irritable, excitable, and cries easily. During a physical examination a swelling was apparent on each side of her neck, inferior to the larynx. During palpation of the patient's neck from a posterior position, the physician felt an enlarged thyroid gland and noted that it moved up and down during swelling. The following signs were also detected: protrusion of the eyes, rapid pulse, tremor of the digits, moist palms, and loss of weight.

Diagnosis. Hyperthyroidism (*exophthalmic goiter*, Graves' disease). When the patient did not respond to medical treatment, a subtotal thyroidectomy was performed. After the operation the patient complained of hoarseness.

Problems. What is the anatomical basis for the swelling moving up and down during deglutition? Because the patient's thyroid gland was enlarged, what nerves might have been compressed or displaced? If a total thyroidectomy had been done, what other endocrine glands might inadvertently have been removed? What would result from this error? What was the probable cause of the patient's hoarseness? These problems are discussed on p. 851.

Case 8-5

A 10-year-old boy was admitted to hospital with a sore throat and earache. He had a high fever (temperature 40.5° C), a rapid pulse rate and rapid respirations. Examination of his throat revealed diffuse redness and swelling of the pharynx, especially of the palatine tonsils. His left tympanic membrane was bulging. The boy's medical history revealed that he had had chronic symptoms of inflammation of the nasal mucous membrane (*rhinitis*), including the pharyngeal tonsils (*tonsillitis*), resulting in persistent mouth breathing. On one occasion he had had a *peritonsillar abscess* (quinsy). Following antibiotic treatment, the boy's infection cleared up. In view of his history, it was decided to readmit him 3 or 4 months later for a **T&A** (tonsillectomy and adenoidectomy).

> **Problems.**   What is meant by the term tonsils? Explain the anatomical basis of the boy's earache. What lymph node in particular might be swollen and tender in this case? What is the probable source of hemorrhage during a tonsillectomy? Compression of what vessel would control severe arterial bleeding in the tonsillar bed? These problems are discussed on p. 852.

DISCUSSION OF PATIENTS' PROBLEMS

Case 8-1

A thyroglossal duct cyst develops from a remnant of the embryonic *thyroglossal duct*, which connects the thyroid gland with the base of the tongue in the embryo (Moore, 1988). Normally the thyroglossal duct atrophies and degenerates as the thyroid gland descends to its final site in the neck. Remnants of this duct may persist anywhere along the median plane of the neck between the foramen cecum of the tongue and the thyroid gland. These remnants may give rise to cysts in the tongue or neck, usually just inferior to the hyoid bone. Often the cyst is in intimate contact with the anterior part of this bone. It may be connected superiorly by a duct with the foramen cecum of the tongue, inferiorly with the pyramidal lobe or isthmus of the thyroid gland, or both. These connections explain why thyroglossal duct cysts move up and down during deglutition and when the tongue is protruded. Sometimes a thyroglossal duct cyst develops an opening onto the surface of the neck (*thyroglossal fistula*). This results from erosion of cervical tissues following infection and rupture of the cyst.

Case 8-2

All the conditions that came to the student's mind could have caused the swelling in the side of her neck. *Branchial cysts* may be derived from remnants of parts of the *cervical sinus*, the second branchial groove, or the second pharyngeal pouch. Although they may be associated with branchial sinuses, as in the present case, and may drain through them, these cysts often lie freely in the neck just inferior to the angle of the mandible. Branchial cysts may develop at any level in the neck, but they usually develop along the anterior border of the sternocleidomastoid muscle. The cyst usually extends deep to this muscle and involves other structures. In the present case the cyst was probably derived from a remnant of the embryonic **cervical sinus** (Moore, 1988). The sinus tract running superiorly from it was

probably derived from the second pharyngeal pouch. It probably passed between the internal and external carotid arteries, just superior to the hypoglossal nerve and the bifurcation of the common carotid artery. Very likely it terminated in or close to the *intratonsillar cleft*, the adult derivative of the cavity of the second pharyngeal pouch.

During the surgical excision of ascending sinuses associated with these cysts, the hypoglossal nerve may be bruised or injured, causing temporary or prolonged *unilateral lingual paralysis*. This would be indicated by hemiatrophy of the tongue and deviation of the tongue to the paralyzed side when it was protruded. This results from the unopposed action of the tongue muscles on the other side. If the sinus tract had passed inferiorly, it probably would have opened in the inferior third of the neck, along the anterior border of the sternocleidomastoid muscle. Branchial sinuses that open externally are derived from remnants of the second branchial groove (cleft).

Case 8-3

Probably the piece of steak was lodged in the *inlet of the larynx* (Fig. 8-41). Choking on food is a common cause of laryngeal obstruction, particularly in children, in persons who have consumed too much alcohol, and in persons with neurological impairment. Many "restaurant deaths," thought to be caused by heart attacks, have been shown to result from choking. Persons with dentures and/or who are drunk are less able to chew their food properly and to detect a bite that is too large.

The mucous membrane of the superior part of the larynx is very sensitive, and contact by a foreign body (*e.g.*, a piece of steak) causes immediate explosive coughing to expel it. However, if there is neurological impairment or the person is drunk, this response may be reduced or absent. Sometimes a foreign body enters the *piriform recess* (Fig. 8-41) or passes through the larynx and becomes lodged in the trachea or a main bronchus. Usually, as in the present case, the piece of steak is only partly in the larynx, but entry of air into the trachea and lungs is largely prevented. The patient would likely have died within minutes, almost certainly before there was time to get him to a hospital, if the piece of steak had not been dislodged using the **Heimlich maneuver,** enabling adequate respiration to be reestablished.

Had the emergency procedure not been successful, the physician would likely have first tried to get the piece of steak out of the patient's larynx with a long spoon or a slender fork. If these procedures had failed, he would likely have performed a lifesaving, emergency inferior *laryngotomy*. If he happened to have a large bore needle with him, he would have inserted it through the *median cricothyroid ligament* (Fig. 8-53). If not, he probably would have used a penknife or a steak knife to make an incision through the midline of the neck into the cricothyroid ligament (*cricothyrotomy*). Probably the physician would have inserted a large plastic straw or a tube of some sort (*e.g.*, the empty barrel of a ballpoint pen), to enable the patient to breathe while he was being taken to a hospital for removal of the piece of steak from his larynx and repair of the cervical wound.

Case 8-4

The tongue, hyoid, and larynx rise and fall during swallowing. Because the thyroid gland is attached to the larynx by pretracheal fascia, it also moves up and down during swallowing. Physiological enlargement of the thyroid gland is commonly seen at puberty and during pregnancy; otherwise, any enlargement of the thyroid is called a **goiter**. In the present case the patient's goiter resulted from *hyperthyroidism*. The association of hyperthyroidism with protrusion of the eyes (*exophthalmos*) was first described by an Irish physician, R.I. Graves. For many years his name has been associated with the disease.

The cause of exophthalmos is not precisely known; however, a considerable increase in the size of the orbital muscles is certainly a factor. In the surgical treatment of hyperthyroidism, part of each lobe of the thyroid is removed (**subtotal thyroidectomy**), thereby leaving less glandular tissue to secrete hormones. As the *parathyroid glands* typically lie on the posterior surface of this gland (Fig. 8-39), posterior parts of the lobes are left so that these glands will not be inadvertently removed. At least one of them is essential for secretion of parathyroid hormones that maintain the normal level of calcium in the blood and body fluids. If the parathyroid glands are removed, the patient soon develops a convulsive disorder known as **tetany**. The signs are nervousness, twitching, and spasms in the facial and limb muscles.

When the thyroid gland is being removed, there is danger that the important laryngeal nerves may be injured. Near the inferior pole of the thyroid gland, the *recurrent laryngeal nerves* are intimately related to the inferior thyroid arteries (Fig. 8-39). The nerves may cross anterior or posterior to this artery or between its branches before ascending in or near the groove between the trachea and esophagus. Because of the close relation between the recurrent laryngeal nerves and the inferior thyroid arteries, the risk of injuring them during surgery is ever present. These nerves supply all muscles of the larynx except the cricothyroids (Table 8-6). If one of the nerves is damaged or cut, there is likely to be a serious effect on speech (*e.g.*, hoarseness as in the present case) or a change in the quality of the voice (*e.g.*, a brassy sound). Some patients also have difficulty clearing their throats.

Temporary paralysis of the recurrent laryngeal nerves may also result from the effect of postoperative edema on them. It must be remembered also that a common cause of temporary hoarseness after surgery is trauma to the mucous membrane of the larynx by the endotracheal tube inserted as an airway by the anesthetist. If both nerves are severed, a very unusual occurrence, breathing will be severely impaired and speech will be difficult because the vocal folds remain partly abducted (the position of complete paralysis of the intrinsic muscles). Thus the rima glottidis is not fully open. If the nerves are compressed as a result of inflammation or the accumulation of fluid, the breathing and speech defects will normally disappear following healing and drainage of the operative site.

Case 8-5

Unless otherwise stated, reference to the tonsil usually refers to the *palatine tonsil*. The other tonsils are the lingual, pharyngeal, and tubal tonsils. Collectively, all the tonsils form the *tonsillar ring* around the isthmus leading from the oral cavity to the nasopharynx. Although it is often stated that the tonsillar ring acts as a barrier to infection, its function is not clearly understood. However, it is certain that this lymphatic tissue is important in the immune reaction to infection. The infection in the present case had spread along the auditory tube into the middle ear, producing **otitis media** and bulging of the tympanic membrane. This would be the chief cause of the boy's earache. The tonsils are supplied by twigs from the glossopharyngeal nerve (CN IX) and, because the tympanic branch of this nerve supplies the mucous membrane of the tympanic cavity, some of the pain related to the tonsillitis may have also been referred to the ear. When the opening of the auditory tube is closed, as it probably was in the present case, pressure changes in the middle ear can also cause earache.

The numerous lymphatic vessels of the tonsil penetrate the pharyngeal wall and terminate principally in the *jugulodigastric node* of the deep cervical chain of lymph nodes (Fig. 8-51). Because its enlargement is commonly associated with tonsillitis, it is often called the **tonsillar node**. *The external palatine vein is usually the chief source of hemorrhage following tonsillectomy.* This important and sometimes large vein descends from the soft palate and is immediately related to the lateral surface of the tonsil, before it pierces the superior constrictor muscle of the pharynx. In cases of severe and uncontrolled bleeding (*e.g.*, from the tonsillar branch of the facial artery), hemorrhage may be controlled by compressing or clamping the external carotid artery at its origin, because this vessel supplies blood to the tonsillar arteries.

SUGGESTED READINGS

Holinger LD: Tracheotomy. In Raffensperger JG (Ed): *Swenson's Pediatric Surgery*, ed 5. Norwalk, CT, Appleton & Lange, 1990.

Hew E: Anesthesia. *In* Gross A, Gross P, Langer B (Eds): *Surgery: A Complete Guide for Patients and Their Families*, Toronto, Harper & Collins, 1989.

Hung W: The growth and development of the thyroid. In Davis JA, Dobbing I (Eds): *Scientific Foundations of Paediatrics*, Philadelphia, WB Saunders, 1974.

Moore KL: *The Developing Human: Clinically Oriented Embryology*, ed 4. Philadelphia, WB Saunders, 1988.

Raffensperger JG: Congenital cysts and sinuses of the neck. In Raffensperger RG (Ed): *Swenson's Pediatric Surgery*, ed 5. Norwalk, CT, Appleton & Lange, 1990.

Solomon J, Rangecroft L: Thyroglossal duct lesions in children. *J Pediatr Surg* 19:555, 1984.

Tobias PV, Arnold M, Allan JC: *Man's Anatomy: A Study in Dissection*, ed. 4, vol. II, Johannesburg, Witwatersrand University Press, 1988.

The regional features of cranial nerves have been described in preceding chapters, especially those on the head, neck, thorax, and abdomen. The following descriptions bring into continuity the features already described. When reading this chapter, refer to the summaries of the cranial nerves given in Tables 9-1 and 9-2. *Cranial nerves are bundles of processes from nerve cells or neurons that either innervate muscles or glands or carry impulses from sensory areas.* The **twelve pairs of cranial nerves** are continuous with the brain and are numbered from anterior to posterior, according to their attachments to the brain as follows (Fig. 9-1): CN I is attached to the telencephalon or anterior part of the forebrain (mainly the cerebral hemispheres); CN II to the diencephalon or central core of the cerebrum (via the optic chiasma); CN III and IV to the midbrain; CN V to the pons; CN VI, VII, and VIII to the junction between the pons and medulla; CN IX, X, and XI (cranial root) and CN XII to the medulla. The spinal root of CN XI is attached to the superior part of the spinal cord.

The cranial nerves provide motor (or efferent) and sensory (or afferent) innervation for the head, neck, thorax and abdomen. They are called cranial nerves because they *emerge through foramina in the cranium* and are covered with tubular sheaths formed from the *cranial meninges* (see Figs. 7-11, 7-41, and 7-45; Table 7-1, p. 680). Some cranial nerves are purely afferent (sensory); others are entirely efferent (motor); and some are mixed (Table 9-1). Centrally the fibers of cranial nerves are connected to **nuclei** (groups of nerve cells or areas of gray matter) in which afferent (or sensory) fibers terminate and from which efferent (or motor) fibers originate. Except for the olfactory areas of CN I, the nuclei of cranial nerves are located in the brain stem and correspond to the functional columns of the spinal cord. Some belong to the *somatic motor (or efferent) column*; others to the *visceral motor (or efferent) column*; some to the *visceral sensory (or afferent column)*; and others to the *somatic sensory (or afferent) column*. All these columns, except the somatic motor column, are divided into special and general parts.

Cranial Nerve Nuclei. Most of these groups of neurons are *located in the brain stem* (Fig. 9-8). The nucleus of the spinal trigeminal tract extends into the spinal cord and the spinal nucleus of the accessory nerve (CN XI) is located in the superior part of the spinal cord. The general location of the nuclei is mentioned with the description of each cranial nerve. (For details about them, see Barr and Kiernan [1988] and/or Williams et al. [1989].) The cranial nerves that have **motor nuclei** and supply skeletal muscles send their fibers directly to the muscles concerned. Nuclei for cardiac muscle, for smooth muscle in the viscera, and for glands send preganglionic fibers to autonomic ganglia for peripheral relay to the structures concerned. The **sensory nuclei** of cranial nerves are groups of cell bodies of second sensory neurons. The cell bodies of the first sensory neurons are outside the CNS in the ganglia of the nerves (*e.g.*, in the trigeminal ganglion of CN V; Fig. 9-7B).

The Olfactory Nerve (CN I)

Type: Special Sensory (Fig. 9-2; Tables 9-1 and 9-2)

This is the **nerve of smell**. Its fibers are the unmyelinated axons of *olfactory neurosensory cells* that are located in the olfactory neuroepithelium covering the superior concha of the nasal cavity and the superior part of the nasal septum. The axons of these cells unite to form 18 to 20 small nerve bundles that are known collectively as the *olfactory nerve*. Each of these nerves is surrounded by the three layers of the cranial meninges (see Fig. 7-44). The nerve bundles that form the olfactory nerves pass through foramina in the *cribriform plate* of the ethmoid bone and enter the **olfactory bulbs** in the anterior cranial fossa (Fig. 9-1). Here they synapse with *mitral cells*. The central processes of these cells pass along the **olfactory tracts** to the olfactory areas of the brain in the region of the *anterior perforated substance* and uncus (Fig. 9-1).

Table 9-1.
Names and Types of Cranial Nerves[1]

Nerve and Its Number	Special Sense	Sensory (Afferent)	Motor (Efferent)	Para- sympathetic
Olfactory (CN I)	*			
Optic (CN II)	*			
Oculomotor (CN III)			*	*
Trochlear (CN IV)			*	
Trigeminal (CN V)		*	*	
Abducent (CN VI)			*	
Facial (CN VII)	*	*	*	*
Vestibulocochlear (CN VIII)	*			
Glossopharyngeal (CN IX)	*	*	*	*
Vagus (CN X)	*	*	*	*
Accessory (CN XI)			*	
Hypoglossal (CN XII)			*	

[1]Four sensory modalities may be carried by the cranial nerves, and three of them carry special sense only (CN I, II, and VIII) and have no motor component. Note also that four nerves (CN III, VII, IX, and X) carry parasympathetic fibers to smooth muscles and glands.

---

[1]Many students and colleagues have requested that a separate chapter giving a summary of the cranial nerves appear in this edition, similar to the one that was in the first edition. Obviously this has resulted in repetition of some facts and clinical comments, but these summaries serve to emphasize the main points of clinical interest in the cranial nerves. Persons who wish detailed information on the central connections of these nerves should consult one or more of the following references: Barr and Kiernan, 1988; Bertram and Moore, 1982; McMinn, 1990; Tobias et al., 1988; Williams et al., 1989; and Wilson-Pauwels et al., 1988.

Table 9-2.
Overview of the Cranial Nerves

	Efferent or Motor		Afferent or Sensory		
Nerve	Striated Muscles	Smooth and Cardiac Muscles and Glands	Skin	Mucous Membranes and Organs	Special Senses
CN I					Olfaction or sensation of smell
CN II					Vision or sight
CN III	Supplies all muscles of eyeball except superior oblique and lateral rectus	Muscles of lens and iris of eye		Proprioceptive fibers from eye muscles	
CN IV	Supplies superior oblique muscle of eyeball			Proprioceptive fibers from eye muscles	
CN V	Supplies muscles of mastication and tensors of tympanic membrane and palate	Carries parasympathetic ganglia for preganglionic nerve fibers of CNs III, VII, and IX	Face and anterior part of scalp	Teeth, mucous membrane of mouth, nose and eye	
CN VI	Supplies lateral rectus muscle of eyeball			Proprioceptive fibers from lateral rectus muscle	
CN VII	Supplies muscles of facial expression	Nervus intermedius; glands of mouth, nose and palate; lacrimal gland; submandibular and sublingual glands	External ear		Chorda tympani supplies taste sensation from anterior two-thirds of tongue
CN VIII					Hearing and equilibrium
CN IX	Supplies stylopharyngeus muscle	Parotid gland		Tympanic membrane, middle ear, pharynx, and tongue (posterior one-third)	Taste, posterior one-third of tongue
CN X	Supplies muscles of pharynx	Organs in neck, thorax, and abdomen	External accoustic meatus and tympanic membrane	Organs in neck, thorax, and abdomen	Taste, epiglottis
CN XI	Supplies muscles of soft palate, pharynx, larynx, and sternocleidomastoid and trapezius muscles				
CN XII	Supplies extrinsic and intrinsic muscles of tongue				

Elderly people usually have a reduced acuity of the sensation of smell that probably results from a progressive reduction in the number of olfactory neurosensory cells in the olfactory epithelium (Barr and Kiernan, 1988). To test the sense of smell, the patient is blindfolded and asked to identify common odors placed near the external nares (nostrils). Because loss of the sense of smell (**anosmia**) is usually unilateral, each naris has to be tested separately. *Chronic rhinitis*[2] is the most common cause of anosmia. In fact, most deficiencies of the sense of smell result from infections of the nasal mucosa rather than neurological disease. In severe head injuries the olfactory bulbs may be torn away from the olfactory nerves, or some olfactory nerve filaments may be torn as they pass through a fractured cribriform plate (Fig. 9-2C; Table 9-3). If all the nerve bundles on one side are torn, there will be a *complete loss of the sense of smell* on that side. Tumors and abscesses in the frontal lobes of the brain or tumors of the meninges (*meningiomas*) in the anterior cranial fossa, may also cause anosmia by compressing the olfactory nerve(s), tract(s), or bulb(s).

Tearing of the meninges around the olfactory nerves (*e.g.*, resulting from a *fracture of the cribriform plate*) results in **leakage of cerebrospinal fluid (CSF)** into the nose, a condition called *CSF rhinorrhea* (p. 701). This type of head injury also opens a route for an infection to spread from the nasal cavity to the meninges and brain. *Olfactory hallucinations*, which are usually unpleasant, arise from lesions in the *uncus* of the temporal lobe (Fig. 9-1).

The Optic Nerve (CN II)

Type: Special Sensory (Figs. 9-1 and 9-3; Tables 9-1 and 9-2)

This nerve is the **nerve of sight**. Its fibers arise from ganglion cells in the neuroepithelium of the *optic retina* and converge

---

[2]Inflammation of the nasal mucosa over a long period of time.

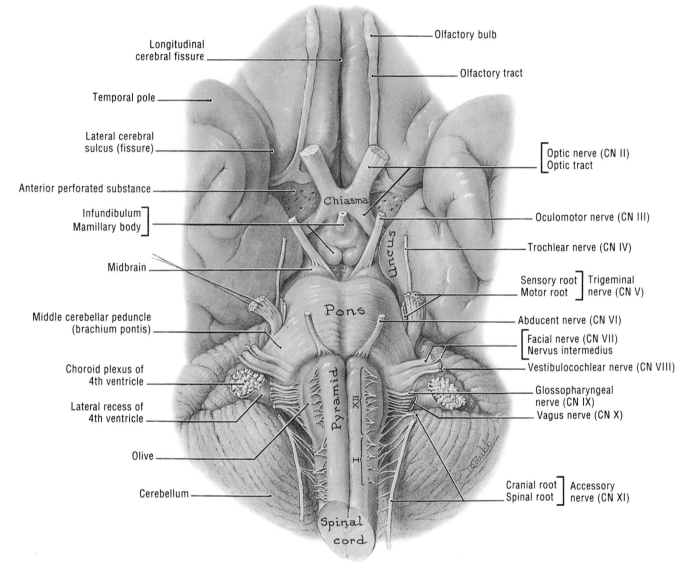

Longitudinal cerebral fissure

Temporal pole

Lateral cerebral sulcus (fissure)

Anterior perforated substance

Infundibulum
Mamillary body

Midbrain

Middle cerebellar peduncle (brachium pontis)

Choroid plexus of 4th ventricle

Lateral recess of 4th ventricle

Olive

Cerebellum

Olfactory bulb

Olfactory tract

Optic nerve (CN II)
Optic tract

Oculomotor nerve (CN III)

Trochlear nerve (CN IV)

Sensory root ⎤ Trigeminal
Motor root ⎦ nerve (CN V)

Abducent nerve (CN VI)

Facial nerve (CN VII)
Nervus intermedius

Vestibulocochlear nerve (CN VIII)

Glossopharyngeal nerve (CN IX)
Vagus nerve (CN X)

Cranial root ⎤ Accessory
Spinal root ⎦ nerve (CN XI)

Chiasma

Uncus

Pons

Pyramid

XII

II

Spinal cord

Figure 9-1.   Inferior surface of the brain showing its features and the attachment of the 12 cranial nerves.

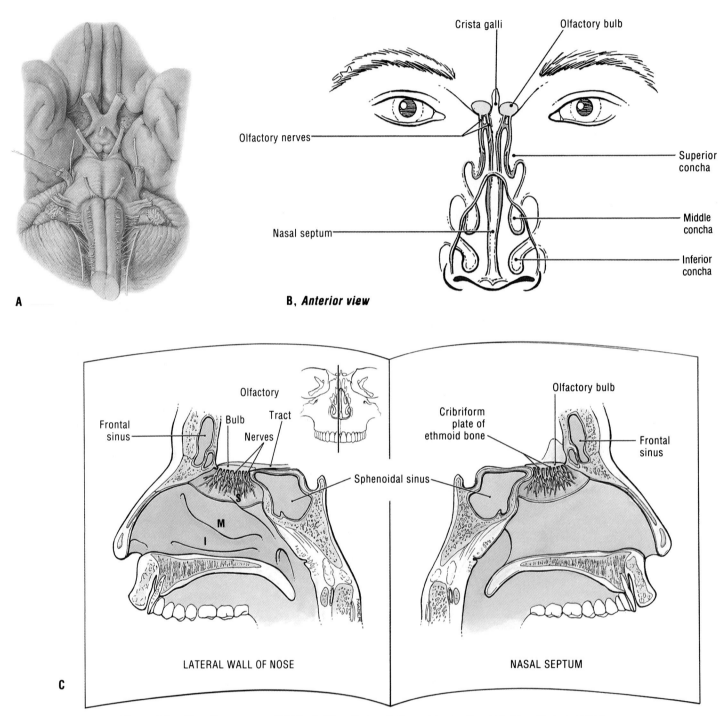

A

B, *Anterior view*

Crista galli

Olfactory bulb

Olfactory nerves

Superior concha

Nasal septum

Middle concha

Inferior concha

C

Olfactory

Frontal sinus

Bulb

Tract

Nerves

S

M

I

LATERAL WALL OF NOSE

Olfactory bulb

Cribriform plate of ethmoid bone

Frontal sinus

Sphenoidal sinus

NASAL SEPTUM

Figure 9-2.   The olfactory nerve (CN I). *A*, The olfactory bulbs and tracts. *B* and *C*, The olfactory system.

Table 9-3.
Summary of Cranial Nerve Lesions

Nerve	Type and/or Site of Lesion	Abnormal Findings
CN I	Fracture of cribiform plate	Anosmia (loss of smell); CSF rhinorrhea (p. 701)
CN II	Direct trauma to orbit or eyeball; fracture involving optic canal	Loss of pupillary constriction
	Pressure on optic pathway; laceration or intracerebral clot in the temporal, parietal, or occipital lobes of brain	Visual field defects
CN III	Pressure from herniating uncus (Fig. 9-1) on nerve; fracture involving cavernous sinus; aneurysms	Dilated pupil, ptosis, eye turns down and out; pupillary reflex on the side of the lesion will be lost
CN IV	Stretching of nerve during its course around brain stem; fracture of orbit	Inability to look down when the eye is adducted
CN V	Injury to terminal branches particularly CN V² in roof of maxillary sinus; pathological processes affecting trigeminal ganglion	Loss of pain and touch sensations; paraesthesia; masseter and temporalis muscles do not contract; deviation of mandible to side of lesion when mouth is opened
CN VI	Base of brain or fracture involving cavernous sinus or orbit	Eye fails to move laterally; diplopia on lateral gaze
CN VII	Laceration or contusion in parotid region	Paralysis of facial muscles; eye remains open; angle of mouth droops; forehead does not wrinkle
	Fracture of temporal bone	As above, plus associated involvement of cochlear nerve and chorda tympani; dry cornea and loss of taste on anterior two-thirds of tongue
	Intracranial hematoma ("stroke")	Forehead wrinkles because of bilateral innervation of the frontalis muscle; otherwise paralysis of contralateral facial muscles
CN VIII	Tumor of nerve (acoustic neuroma)	Progressive unilateral hearing loss; tinnitus (noises in ear)
CN IX	Brain stem lesion or deep laceration of neck	Loss of taste on posterior one-third of tongue; loss of sensation on affected side of soft palate
CN X	Brain stem lesion or deep laceration of neck	Sagging of soft palate; deviation of uvula to normal side; hoarseness owing to paralysis of vocal fold
CN XI	Laceration of neck	Paralysis of sternocleidomastoid and superior fibers of trapezius; drooping of shoulder
CN XII	Neck laceration; basal skull fractures	Protruded tongue deviates toward affected side; moderate dysarthria (disturbance of articulation)

toward the *optic disc* at the posterior pole of the eye (Fig. 9-3*C*; see also Fig. 7-62). These fibers unite to form the large *optic nerve* that passes posteromedially through the posterior half of the orbit (Fig. 9-3*B*; see also Fig. 7-57*A*). It passes through the *optic canal* into the middle cranial fossa to join its partner at the **optic chiasma**. The midregion of the optic chiasma is composed of crossed fibers from the medial or temporal half of each optic nerve (Fig. 9-3*C*).

Traditionally the optic nerve is described as forming at the posterior end of the eye and passing to the optic chiasma. However, it *begins in the retina* where its fibers form the internal layer of the retina. These fibers, which are axons of cells in the *ganglion cell layer*, converge on the **optic disc**, pierce the external layers of the retina, and join to form the large optic nerve (see Fig. 7-62).

Within the optic canal, the *ophthalmic artery* lies inferolateral to the optic nerve (see Fig. 7-70). From the optic chiasma the **optic tracts** continue dorsolaterally around the midbrain (Fig. 9-3*B*). The left optic tract carries nerve fibers from the temporal or lateral half of the left retina and the nasal or medial half of the right retina (Fig. 9-3*C*). The right optic tract contains fibers from the temporal half of the right retina and the nasal half of the left retina. Most fibers in the optic tracts terminate in the **lateral geniculate bodies** of the thalamus (Fig. 9-3*B* and *C*). From these nuclear complexes, axons form *geniculocalcarine tracts* (optic radiations) that are relayed to the **visual cortex**. These visual areas occupy the superior and inferior walls (lips) of the *calcarine sulci* on the medial surfaces of the occipital lobes of the brain (see Fig. 7-51*B*).

CN II is enclosed by three sheaths that are continuous with the three layers of the cranial meninges (dura, arachnoid, and pia). The thick, fibrous *dural sheath* of the optic nerve is continuous with the dura mater of the brain and blends with the sclera (see Figs. 7-59, 7-62, and 7-70). A **subarachnoid space containing CSF** is located between the intermediate and inner sheaths, which are derived from the arachnoid and pia, respectively (p. 681). The *central artery and vein* of the retina pass through these meningeal sheaths and are included in the distal part of the optic nerve.

The central artery and vein of the retina are enclosed in the optic nerve when the *optic fissure* closes during the sixth to eighth weeks of development (Moore, 1988). The retina and optic nerve develop from the *optic vesicle*, an outgrowth of the brain; hence, *CN II is actually a tract of the brain rather than a nerve.*

Because of the developmental and structural relations just described, infections of the meninges readily extend along the optic nerves. Because the optic vein is enclosed in the anterior part of the optic nerve, which is surrounded by CSF in the subarachnoid space, an increase in CSF pressure impedes the return of venous blood from the eye. As a result, *swelling of the optic disc* (papilla) occurs owing to edema (**papilledema**). This condition presents valuable clinical evidence of an increase in intracranial pressure.

In addition to the routine tests for visual acuity and color perception, the visual fields are tested for blindness. **Visual field defects** result from lesions that affect different parts of the visual pathway. The type of defect produced depends on

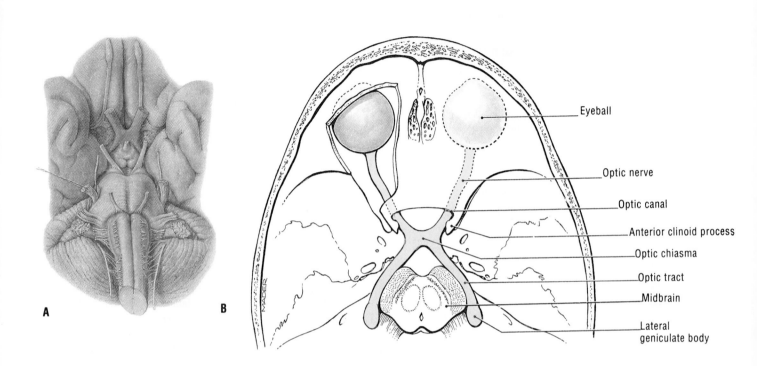

Eyeball

Optic nerve

Optic canal

Anterior clinoid process

Optic chiasma

Optic tract

Midbrain

Lateral geniculate body

A

B

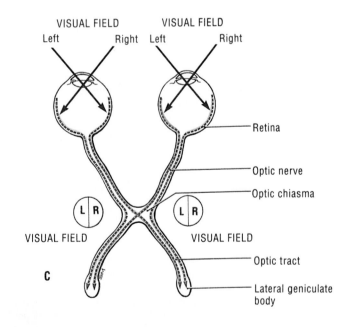

VISUAL FIELD

Left   Right

VISUAL FIELD

Left   Right

Retina

Optic nerve

Optic chiasma

L | R   L | R

VISUAL FIELD   VISUAL FIELD

Optic tract

C

Lateral geniculate body

Figure 9-3. The optic nerve (CN II). *A*, Its attachment to the diencephalon. *B*, The optic nerves pass from the eyes to the optic chiasma. From here, the optic tracts pass around the midbrain. *C*, Crossing of the optic nerve fibers in the optic chiasma is a requirement for binocular vision. Fibers from the nasal or medial half of the retina decussate in the chiasma and join uncrossed fibers from the temporal or lateral half of the retina to form the optic tract. The large arrows represent rays of light. A great majority of the nerve fibers in the optic tract end in the lateral geniculate body. Nerve fibers from this body pass to the visual cortex in the superior and inferior walls (lips) of the calcarine sulcus (p. 693).

where the pathway was interrupted (Fig. 9-4). Obviously, severe degenerative disease or trauma involving the optic nerve results in blindness in that eye. Clinically *the most common lesions are at the optic chiasma and in the optic radiations* (Fig. 9-3*C*; see also Fig. 7-53*B*). The most common lesion affecting the optic chiasma is a *pituitary tumor* that grows superiorly and compresses nerve fibers in the chiasma. For a detailed description of the various types of visual field defect, see Barr and Kiernan (1988).

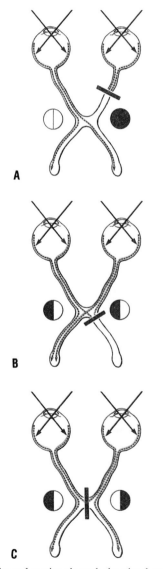

Figure 9-4. Horizontal section through the visual apparatus showing various visual defects caused by lesions affecting different parts of the optic pathway. *A*, Section of the right optic nerve results in blindness in the right eye. *B*, Section of the right optic tract eliminates vision from the left visual fields of both eyes. *C*, Section of the optic chiasma reduces peripheral vision.

The Oculomotor Nerve (CN III)

Type: Motor—To Striated and Smooth Muscle (Figs. 9-1 and 9-5; Tables 9-1 and 9-2)

As its name implies, this nerve **moves the eye**. It is somatic motor to striated ocular muscles and visceromotor to smooth muscles in the eye. CN III is attached to the base of the midbrain near the median plane (Fig. 9-1). *The oculomotor nerve supplies somatic efferent fibers to all the extraocular muscles of the eye except the superior oblique and lateral rectus* (Figs. 9-5 and 9-6). It also supplies the levator palpebrae superioris muscle and, through its *ciliary parasympathetic ganglion*, it supplies general visceral efferent fibers to smooth muscle in the sphincter pupillae and ciliary muscles (see Fig. 7-65).

When it emerges from the midbrain, CN III enters the subarachnoid space and passes anteriorly in the *interpeduncular cistern* (Fig. 9-1; see also Figs. 7-45, 7-46, and 7-52). It runs between the posterior cerebral and superior cerebellar arteries and lateral to the posterior communicating artery (see Fig. 7-54). It then pierces the arachnoid and internal layer of the dura and passes into the lateral wall of the **cavernous sinus** (see Fig. 7-48). Here, it lies superior to the trochlear nerve (CN IV) and lateral to the internal carotid artery. CN III enters the orbit through the *superior orbital fissure* within the *common tendinous ring* (Fig. 9-6; see also Fig. 7-66) and divides into superior and inferior branches between the two heads of the lateral rectus muscle. The *superior branch of CN III* supplies the superior rectus and levator palpebrae superioris muscles (Fig. 9-5; see also Figs. 7-60 and 7-65). The *inferior branch* supplies the medial and inferior rectus muscles and terminates in the inferior oblique muscle. It also supplies a motor parasympathetic root to the *ciliary ganglion* (Fig. 9-5). Postganglionic fibers pass in the *short ciliary nerves* to the sphincter pupillae and ciliaris muscles (see Fig. 7-68).

The Oculomotor Nerve Nuclei (Fig. 9-8).   There are two motor nuclei. The *somatic motor nucleus* is located in the periaqueductal gray matter of the midbrain, ventral to the cerebral aqueduct at the level of the superior colliculus (Fig. 9-8; see also Figs. 7-39 and 7-42). The *somatic visceral nucleus* (Edinger-Westphal or accessory oculomotor nucleus), lies dorsal to the rostral two-thirds of the somatic motor nucleus.

Section of the oculomotor nerve results in: (1) *ptosis* (drooping of the eyelid) owing to paralysis of the levator palpebrae superioris muscle; (2) *lateral strabismus* (squinting) owing to the unopposed action of the lateral rectus and superior oblique muscles (Fig. 9-5; see also Table 7-1); (3) *dilation of the pupil* owing to paralysis of the sphincter pupillae muscle; (4) *loss of accommodation* of the light reflex owing to paralysis of the sphincter pupillae and the ciliaris muscles; (5) *proptosis* (prominence of the eyeball) owing to relaxation of the ocular muscles; and (6) *diplopia* (double

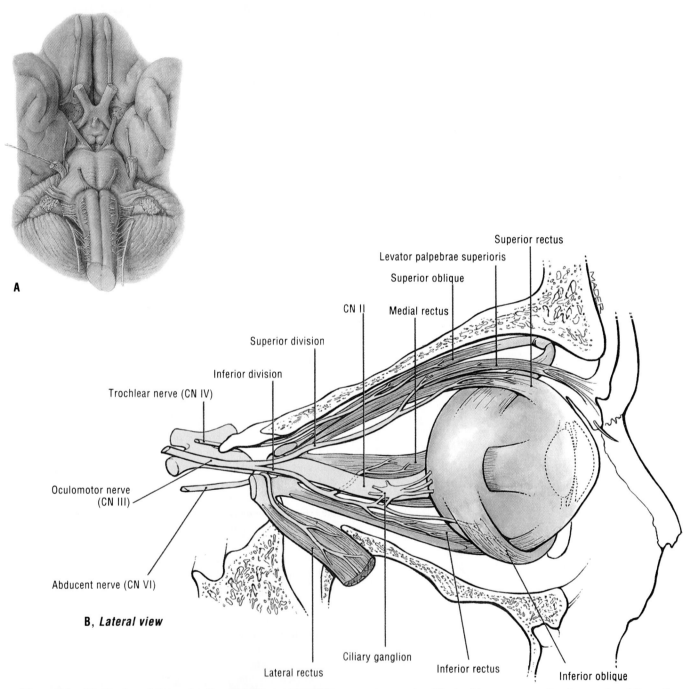

A

B, *Lateral view*

Figure 9-5. Distribution of the oculomotor (CN III), trochlear (CN IV), and abducent (CN VI) nerves. *A*, The attachment of these nerves to the midbrain and pons. *B*, Somatic motor and parasympathetic motor innervation of the orbit. CN III, CN IV, and CN VI supply the extra-ocular muscles. The trochlear nerve supplies the superior oblique; the abducent nerve supplies the lateral rectus; and the oculomotor nerve supplies the levator palpebrae superioris, superior rectus, medial rectus, inferior rectus, and inferior oblique muscles.

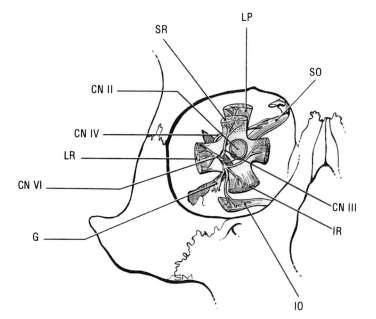

Figure 9-6. Muscles and nerves of the orbit. Movements of the eyes are produced by six extraocular muscles (four recti and two oblique). These muscles are supplied by three cranial nerves (CN III, CN IV, and CN VI).

vision). **Aneurysms** of the posterior cerebral or superior cerebellar arteries may exert pressure on the oculomotor nerve as it passes between these vessels (Table 9-3; see also Fig. 7-54). The effects produced depend on the extent of pressure exerted. Because the oculomotor nerve lies in the lateral wall of the cavernous sinus, it may be affected when injuries or infections of this sinus occur (see Fig. 7-48).

The Trochlear Nerve (CN IV)

Type: Motor (Figs. 9-1, 9-5, and 9-6; Tables 9-1 and 9-2)

This is the smallest cranial nerve. *It supplies only one eye muscle, the superior oblique.* The fibers of the right and left trochlear nerves decussate dorsal to the cerebral aqueduct (see Fig. 7-52) before emerging from the midbrain. *CN IV is the only cranial nerve to emerge dorsally from the brainstem,* and it has the longest intracranial course of all the cranial nerves (see Fig. 7-42). CN IV emerges at the medial border of the superior cerebellar peduncle and enters the lateral wall of the **cavernous sinus,** where it lies inferior to the oculomotor nerve and superior to the ophthalmic nerve (see Fig. 7-48). It leaves the cranium through the *superior orbital fissure,* superior to the common tendinous ring (Fig. 9-6; see also Figs. 7-65 and 7-66). CN IV runs medially, superior to the levator palpebrae superioris muscle, and supplies the superior oblique muscle from its superior aspect.

Trochlear Nerve Nucleus (Fig. 9-8). This motor nucleus for the superior oblique muscle of the eye is located in the

periaqueductal gray matter[3] immediately caudal to the oculomotor nucleus at the level of the inferior colliculus.

> Because of its long intracranial course and small size, the trochlear nerve may be torn when a *head injury* occurs. It is also vulnerable during surgical approaches to the midbrain (Table 9-3). Severance of the trochlear nerve paralyzes the superior oblique muscle, limits inferolateral ocular movements, and causes **diplopia** (double vision). The latter condition is most severe when the eye is directed inferomedially. Because the eye cannot look down as far as it should when turned in, persons with a trochlear nerve lesion often have difficulty walking downstairs. To counteract this, they hold their heads up and inclined to the other side.

The Trigeminal Nerve (CN V)

Type: Mixed—Motor[4] and General Sensory (Figs. 9-1 and 9-7; Tables 9-1 and 9-2)

This is the *motor nerve for the muscles of mastication* and several small muscles and the *principal general sensory nerve for the head.* CN V is a triple nerve (L. *trigeminus,* triplet) that **consists of three large sensory nerves** from the face (see Fig. 7-24): *ophthalmic* (CN V[1]), *maxillary* (CN V[2]), and *mandibular* (CN V[3]). These nerves form the large *sensory root of the tri-*

---

[3]The periaqueductal gray matter surrounds the cerebral aqueduct of the midbrain (see Fig. 7-39).
[4]The special visceral efferent or motor component of CN V is sometimes called the *brachial motor component* because it supplies muscles derived from the first branchial or pharyngeal arch in the embryo (Moore, 1988).

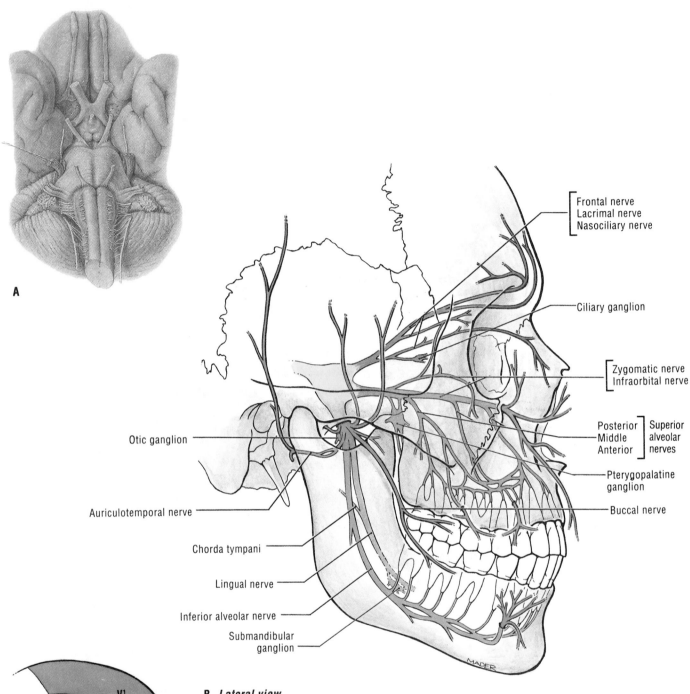

Frontal nerve
Lacrimal nerve
Nasociliary nerve

Ciliary ganglion

Zygomatic nerve
Infraorbital nerve

Posterior] Superior
Middle   ] alveolar
Anterior] nerves

Pterygopalatine ganglion

Otic ganglion

Buccal nerve

Auriculotemporal nerve

Chorda tympani

Lingual nerve

Inferior alveolar nerve

Submandibular ganglion

B, *Lateral view*

C, *Lateral view*

V¹
V²
V³

Figure 9-7. The trigeminal nerve (CN V). *A,* Inferior surface of the brain showing the attachment of the trigeminal nerves. *B* and *C,* Their divisions are: (1) the ophthalmic division (CN V¹), the nerve of the frontonasal prominence; (2) the maxillary division (CN V²), the nerve of the maxillary prominence; and (3) the mandibular division (CN V³), the nerve of the mandibular prominence. As shown in *C,* these prominences form most of the face and anterolateral parts of the scalp. All three divisions are sensory. Each division provides the sensory component to an autonomic ganglion as follows: CN V¹ to the ciliary ganglion; CN V² to the pterygopalatine ganglion; and CN V³ to the submandibular and otic ganglia. CN V³ is motor to four pairs of muscles: (1) the elevators of the mandible—temporalis and masseter; (2) medial and lateral pterygoids and (3) two tensors—tensor veli palatini and tensor tympani; (4) two muscles in the floor of the mouth—mylohyoid and anterior belly of digastric.

geminal nerve (Fig. 9-1); there is also a much smaller *motor root of the trigeminal nerve*. Together, these roots make CN V the largest cranial nerve. Fibers in the sensory root are mainly axons of neurons in the *trigeminal ganglion* (semilunar ganglion), which occupies a recess in the dura of the middle cranial fossa called the *trigeminal cave* (see Figs. 7-25 and 7-46). The peripheral processes of cells in this ganglion constitute the ophthalmic and maxillary nerves (CN V^1 and CN V^2) and the sensory component of the mandibular nerve (CN V^3). CN V is attached to the lateral part of the pons at the junction between the pons and the middle cerebellar peduncle (Fig. 9-1). Its two roots run anteriorly in the posterior cranial fossa, inferior to the tentorium cerebelli (see Figs. 7-37 and 7-45).

Trigeminal Nerve Nuclei (Fig. 9-8; see also Fig. 7-39). There are **four nuclei**: one motor and three sensory. The motor nucleus of CN V is located medial to the chief sensory nucleus in the dorsolateral area of the pontine tegmentum. More simply stated, it is in the superior part of the pons deep to the floor of the fourth ventricle. The central processes of the trigeminal ganglion cells enter the pons and terminate in the elongated *pontine trigeminal nucleus* and the *nucleus of the spinal tract*. The pontine trigeminal nucleus or **chief sensory nucleus** is in the dorsolateral area of the pontine tegmentum at the level of entry of the sensory fibers. The *mesencephalic nucleus* is located lateral to the cerebral aqueduct. The *spinal nucleus* is in the inferior part of the pons and throughout the medulla.

The Ophthalmic Nerve (CN V^1)

This is the smallest of the three divisions of the trigeminal nerve (Fig. 9-7). This *purely sensory nerve* supplies superficial and deep parts of the superior region of the face, including the eyeball, lacrimal gland, conjunctiva, nasal mucosa, and the skin of the scalp, forehead, upper eyelid, and nose (Fig. 9-7; see also Figs. 7-24 and 7-25). After it forms from dendritic processes of ganglion cells in the trigeminal ganglion, CN V^1 passes anteriorly in the lateral wall of the *cavernous sinus*, inferior to the trochlear nerve (see Figs. 7-46 and 7-48).

Branches of the Ophthalmic Nerve (Fig. 9-7; see also Figs. 7-24 and 7-46). CN V^1 *divides into three branches*: nasociliary, frontal, and lacrimal nerves. These branches enter the orbit through the superior orbital fissure (see Fig. 7-56).

The *frontal nerve* is the largest branch of the ophthalmic nerve (Fig. 9-7). It passes superior to the common tendinous ring and travels anteriorly, just inferior to the roof of the orbit. It proceeds between the levator palpebrae superioris and the periosteum and branches to form a small *supratrochlear nerve* and a larger *supraorbital nerve* (see Figs. 7-25 and 7-67). The supraorbital nerve provides sensory branches to the upper eyelid and the scalp, and the supratrochlear nerve supplies the upper eyelid and forehead.

The *nasociliary nerve* (Fig. 9-7; see also Figs. 7-46, 7-66, and 7-68), which is intermediate in size between the frontal and lacrimal nerves, enters the orbit through the *central tendinous ring*. After entering the orbit, the nasociliary nerve curves toward its medial wall and gives off several branches: (1) *long ciliary nerves* to the eyeball; (2) *ganglionic branches* to the ciliary ganglion; (3) posterior and anterior *ethmoidal nerves* to the eth-

moidal sinus and nasal cavity; and (4) the *infratrochlear nerve* to the medial aspect of the upper eyelid and the lacrimal sac. The nasociliary nerve crosses the optic nerve with the ophthalmic artery.

The *lacrimal nerve* (Fig. 9-7; see also Figs. 7-46 and 7-60), the smallest of the main branches of the ophthalmic nerve, enters the orbit through the *superior orbital fissure*. It runs along the superior border of the lateral rectus muscle with the lacrimal artery and supplies the lacrimal gland and the adjoining conjunctiva. It ends in the upper eyelid, where it joins filaments of the facial nerve.

The Maxillary Nerve (CN V^2)

This intermediate division of the trigeminal nerve is a *purely sensory nerve* (Fig. 9-7; see also Figs. 7-24, 7-25, 7-46, and 7-48). It arises from the trigeminal ganglion and runs anteriorly in the inferior part of the *cavernous sinus*, inferior to the ophthalmic nerve. CN V^2 leaves the middle cranial fossa through the *foramen rotundum* (see Figs. 7-41, 7-42, and 7-60). It then enters the *pterygopalatine fossa* (see Figs. 7-103 and 7-104) and bends laterally to leave this fossa through the *pterygomaxillary fissure*. It enters the *infratemporal fossa* (see Figs. 7-77 and 7-78) and then the orbit through the *inferior orbital fissure* (see Figs. 7-25 and 7-60). It continues as the *infraorbital nerve*, which exits through the infraorbital canal and foramen (Figs. 7-1, 7-21, and 7-24).

Branches of the Maxillary Nerve (Fig. 9-7; see also Figs. 7-25 and 7-104). *Meningeal branches* arise from CN V^2 while it is in the middle cranial fossa; they supply the dura. *Ganglionic branches* arise within the pterygopalatine fossa and enter the pterygopalatine ganglion as its sensory root. The *posterior superior alveolar nerves* descend on the infratemporal surface of the maxilla and enter the posterior superior alveolar foramina and supply the maxillary sinus and the roots of the maxillary molar teeth (Fig. 9-7). The *middle superior alveolar nerve* arises from the infraorbital part of the maxillary nerve to supply the mucosa of the maxillary sinus, the roots of the maxillary premolar teeth, and the mesiobuccal root of the first molar tooth (Fig. 9-7). The *anterior superior alveolar nerve* arises from the maxillary nerve just before the infraorbital nerve passes into the infraorbital foramen. It descends through its own canal in the anterior wall of the maxillary sinus and sends branches to this sinus, the nasal septum, and the roots of the maxillary central and lateral incisors and canine teeth (Fig. 9-7; see also Figs. 7-25 and 7-106). *Facial branches* arise from the infraorbital nerve, which supply the lower eyelid, nose, and upper lip (Fig. 9-7; see also Figs. 7-21 and 7-25).

The Mandibular Nerve (CN V^3)

This mixed nerve (sensory and motor) *contains all the motor fibers of the trigeminal nerve* (Fig. 9-7). After it arises from the trigeminal ganglion, CN V^3 descends to the *foramen ovale* and then passes through this opening in the middle cranial fossa with the motor root of the trigeminal nerve (see Figs. 7-41 and 7-42). Just outside this foramen, the motor and sensory roots of the mandibular nerve unite (Fig. 9-1). CN V^3 divides into anterior and posterior divisions, which supply the teeth and the gingivae

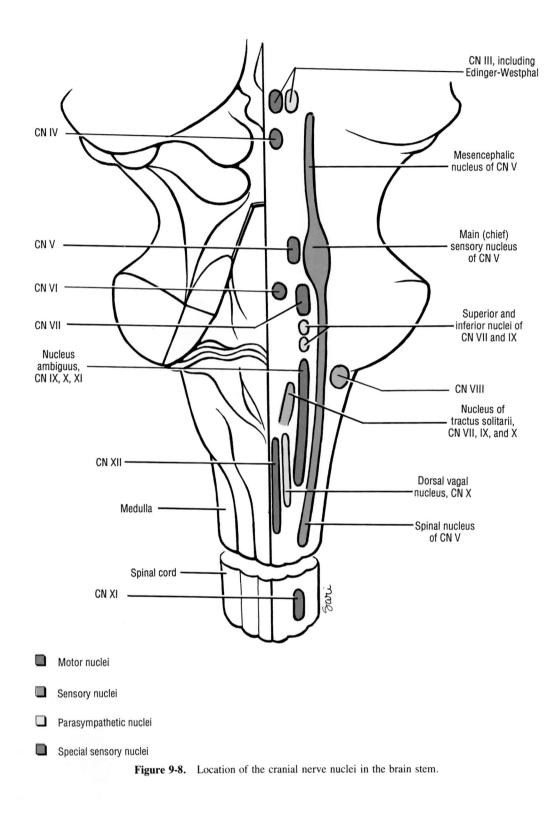

CN III, including
Edinger-Westphal

CN IV

Mesencephalic
nucleus of CN V

CN V

Main (chief)
sensory nucleus
of CN V

CN VI

CN VII

Superior and
inferior nuclei of
CN VII and IX

Nucleus
ambiguus,
CN IX, X, XI

CN VIII

Nucleus of
tractus solitarii,
CN VII, IX, and X

CN XII

Dorsal vagal
nucleus, CN X

Medulla

Spinal nucleus
of CN V

Spinal cord

CN XI

- ■ Motor nuclei
- ■ Sensory nuclei
- ☐ Parasympathetic nuclei
- ■ Special sensory nuclei

Figure 9-8.   Location of the cranial nerve nuclei in the brain stem.

of the mandible, the skin in the temporal region, part of the auricle, the lower lip, most of the lower part of the face, and the muscles of mastication (Fig. 9-7; see also Figs. 7-24 and 7-25). It also supplies the mucous membrane of part of the tongue and the floor of the oral cavity (see Fig. 7-96).

Branches of the Mandibular Nerve (Fig. 9-7; see also Figs. 7-24, 7-25, 7-71, 7-77, and 7-78). *The meningeal branch* (nervus spinosus) arises from the mandibular nerve just distal to the foramen ovale and then reenters the cranium through the foramen spinosum (hence, its alternate name) with the middle meningeal artery. This sensory branch helps to supply the dura covering that part of the brain that lies in the middle cranial fossa (see Fig. 7-42).

The nerve to the medial pterygoid arises from the stem of CN V^3 and supplies this muscle and sends fibers to the tensor tympani and tensor veli palatini muscles.

Motor branches from the anterior division of CN V^3 supply the temporalis, masseter, and lateral pterygoid muscles (see Fig. 7-77).

The buccal nerve (long buccal nerve) arises from the anterior division of CN V^3 (Fig. 9-7B) and pierces the tendon of the temporalis muscle (see Fig. 7-78). It supplies the cheek and mandibular buccal gingiva.

The auriculotemporal nerve, a sensory branch of the posterior division of CN V^3 (see Fig. 7-77), runs posteriorly as two roots that encircle the middle meningeal artery. It ascends onto the temple and the lateral part of the scalp (see Figs. 7-25 and 7-35) and sends sensory branches to the *temporomandibular joint*, auricle, external acoustic meatus, and skin of the temple and the lateral part of the scalp. Postsynaptic fibers of CN IX from the *otic ganglion* pass with this nerve to the parotid gland (see Fig. 7-25).

The lingual nerve (sensory) arises from the posterior division of CN V^3, deep to the lateral pterygoid muscle (Fig. 9-7B; see also Fig. 7-77). It enters the floor of the mouth and supplies sensory fibers to the submandibular gland (see Fig. 7-102). It ends by sending sensory fibers to the mucosa of the anterior two-thirds of the tongue (see Fig. 7-96), the floor of the mouth, and mandibular gingivae.

The inferior alveolar nerve (sensory) arises from the posterior division of CN V^3 (Fig. 9-7B; see also Fig. 7-25) and passes inferiorly, deep to the lateral pterygoid muscle (see Figs. 7-77 and 7-78). When it reaches the inferior border of this muscle, it turns laterally to enter the *mandibular foramen* and mandibular canal (see Fig. 7-80). When it leaves this canal, it sends sensory branches to the mandibular teeth and then emerges from the *mental foramen* as the *mental nerve* (Fig. 9-7B; see also Fig. 7-21). It supplies sensory branches to the skin of the chin, the mucous membrane and skin of the lower lip, and the labial mandibular gingivae.

Lesions of the motor part of the CN V are unusual. If present, contraction of the masseter muscle is affected. Lesions involving the entire sensory part of the trigeminal nerve cause anesthesia (lack of sensation) of the anterior half of the scalp, the face, cornea, conjunctiva, mucous membranes of the nose, mouth, and anterior part of the tongue (Fig. 9-7; Table 9-3). Lesions of the divisions of the trigeminal nerve

result in more limited anesthesia (Fig. 9-7C). Sensation is tested for light touch with a piece of cotton or with test tubes containing warm and cold fluid (see Fig. 7-26). Pain perception is tested with a pinprick.

Trigeminal neuralgia (tic douloureux) is the most common condition affecting the sensory part of CN V. It causes excruciating pain (p. 664). It is characterized by pain of sudden onset in the area of distribution of one of the trigeminal nerve divisions, usually the maxillary nerve. The ophthalmic division (CN V^1) is not commonly involved. The most common neuralgia associated with CN V^2 and CN V^2 divisions of the trigeminal nerve is the type associated with dental caries (cavities).

The Abducent Nerve (CN VI)

Type: Motor (Fig. 9-1; Tables 9-1 and 9-2)

As its name indicates, the abducent (L. abducens) nerve supplies the lateral rectus muscle that **abducts the eye**. CN VI supplies only this muscle. It leaves the ventral aspect of the brainstem between the caudal border of the pons and the superior end of the pyramid of the medulla. CN VI *ascends on the clivus* of the basilar part of the occipital bone (see Figs. 7-40 and 7-45) and then runs anterolaterally in the *pontine cistern* (see Fig. 7-52). It passes dorsal to the anterior inferior cerebellar artery and *enters the cavernous sinus* (see Fig. 7-54), where at first it lies lateral and then inferolateral to the *internal carotid artery* (see Fig. 7-48). It leaves the skull through the medial end of the *superior orbital fissure* and enters the orbit (Fig. 9-5; see also Figs. 7-46, 7-60, and 7-66). Here, it enters the medial side of the lateral rectus muscle and supplies it.

Nucleus of the Abducent Nerve (Fig. 9-8). This motor nucleus is located in the pons near the median plane. It lies deep to the facial colliculus in the floor of the fourth ventricle.[5]

Injury to the abducent nerve results in *medial strabismus* (convergent squint) and *diplopia* (double vision). There is an inability to direct the affected eye laterally. The long course of CN VI through the pontine cistern and its sharp bend over the petrous part of the temporal bone (see Fig. 7-42) make it vulnerable during neurosurgery and when increases in intracranial pressure push the brain stem caudally and stretch it (Table 9-3). CN VI may also be injured when fractures of the base of the skull occur.

---

[5]The facial colliculus is a slight swelling in the floor of the fourth ventricle. It is formed by fibers from the motor nucleus of the facial nerve looping around the abducent nucleus.

The Facial Nerve (CN VII)

Type: Mixed—Motor[6] and Sensory (Figs. 9-1 and 9-9; Tables 9-1 and 9-2)

This nerve is composed of two distinct roots: (1) the *facial nerve proper* (the motor root and larger trunk), and (2) the *nervus intermedius* (the sensory root[7]). The roots emerge from the brainstem at the caudal border of the pons, lateral to the superior end of the olive of the medulla (Fig. 9-1). Both roots enter the *internal acoustic meatus* (see Figs. 7-40 and 7-125), a passage that leads into the petrous part of the temporal bone and joins the *facial canal*. The facial nerve runs laterally in this canal and exits the skull through the *stylomastoid foramen* (see Figs. 7-27 and 7-117). CN VII enters the parotid gland and breaks up into its five terminal motor branches (Fig. 9-9B; see also Figs. 7-22 and 7-27).

The motor root of CN VII supplies the muscles of facial expression (p. 664), muscles of the scalp and auricle, the buccinator, platysma, stapedius, and stylohyoid muscles, and the posterior belly of the digastric muscle. *The sensory root of CN VII conveys (1) taste fibers from the chorda tympani nerve*, which come from the anterior two-thirds of the tongue (Fig. 9-8; see also Figs. 7-27 and 7-96); (2) taste fibers from the soft palate through the *palatine and greater petrosal nerves* (Fig. 9-9B; see also Figs. 7-27, 7-91, and 7-93); and (3) preganglionic parasympathetic (secretomotor) innervation to the submandibular, sublingual, and lacrimal glands and the glands of the nasal and palatine mucosae.

Branches of the Facial Nerve (Fig. 9-9B; see also Figs. 7-22 and 7-27). The *greater petrosal nerve*, a slender branch containing taste and parasympathetic fibers, starts from the *geniculate ganglion* and is joined by a branch of the tympanic plexus. It *arises in the facial canal* from the genu (bend) of the facial nerve and emerges from the anterior part of the petrous part of the temporal bone. It passes toward the foramen lacerum and enters the pterygoid canal. Here, it joins the deep petrosal nerve to form the *nerve of the pterygoid canal* that passes through this canal and enters the *pterygopalatine ganglion* (Fig. 9-9B; see also Figs. 7-93, 7-104, and 7-106). From here, taste fibers pass to the palate, and postsynaptic parasympathetic fibers go to the lacrimal gland and the mucosa of the palate, nasopharynx, and nasal cavity. The *nerve to the stapedius* muscle arises within the facial canal (Fig. 9-9B; see also Fig. 7-116B) and is motor to this muscle that moves the stapes (an auditory ossicle).

The **chorda tympani nerve** (parasympathetic and taste) arises from the facial nerve within the facial canal (Fig. 9-9B; see also Fig. 7-118). It passes through the middle ear and leaves the skull through the *petrotympanic fissure*. It joins the lingual nerve in the infratemporal region. Taste fibers are distributed to the anterior two-thirds of the tongue through the lingual nerve (see

Fig. 7-96). Parasympathetic fibers, after synapsing in the *submandibular ganglion*, supply the submandibular and sublingual salivary glands (see Figs. 7-27 and 8-14).

Suprahyoid motor branches arise from the facial nerve after it leaves the stylomastoid foramen. They supply the stylohyoid muscle and the posterior belly of the digastric muscle (see Figs. 8-15 and 8-16). *Facial branches* (motor) arise from CN VII within the parotid gland. There are five main groups: temporal, zygomatic, buccal, mandibular, and cervical nerves to the muscles of facial expression and the platysma (Fig. 9-8B; see also Figs. 7-22 and 7-27).

Nuclei of the Facial Nerve (Figs. 9-8 and 9-9; see also Fig. 7-27). Most fibers of the motor root arise from the *facial motor nucleus* that is located in the ventrolateral part of the tegmental area of the pons. The special sensory root (taste) arises from the rostral end of the *nucleus solitarius* in the medulla. The cell bodies of the primary sensory neurons are in the *geniculate ganglion* and end on the nucleus solitarius. Pain, touch, and thermal sensations from around the external ear end in the elongated *sensory nucleus of CN V*.

Lesions of the facial nerve cause *facial paralysis* that is usually unilateral (Table 9-3). There is also a *loss of taste sensation* in the anterior two-thirds of the tongue and *decreased salivation*. Paralysis of the lower face may result from occlusion of a blood vessel that supplies the *internal capsule* or the motor cortex (see Fig. 7-53B). When the paralysis also involves the muscles around the eye and in the forehead (Fig. 9-9B; see also Figs. 7-21 and 7-22), the lesion involves the cell bodies in the facial nucleus or their axons (Fig. 9-8).

In the most common condition, called **Bell's palsy** or paralysis, the facial nerve is affected as it passes through the facial canal (see Fig. 7-118). The paralysis results from infection and inflammation of the nerve and other tissues in this canal. This causes pressure on the nerve and paralysis of the facial muscles. The signs of Bell's palsy (p. 661) depend not only on the severity of the infection, but also on the site of the infection in the facial canal. All functions of the nerve are lost if the lesion is proximal to the geniculate ganglion (Fig. 9-9). In these cases there is no regeneration of nerve fibers. For details of facial nerve lesions, see Barr and Kiernan (1988). Patients with mild signs of Bell's palsy slowly recover completely because the nerve fibers are not usually so severely damaged to cause them to degenerate.

The Vestibulocochlear Nerve (CN VIII)

Type: Special Sensory (Figs. 9-1 and 9-10; Tables 9-1 and 9-2)

This nerve is composed of two parts: a *vestibular part*, called the **vestibular nerve**, is concerned with the **maintenance of equilibrium**; and a *cochlear part*, called the **cochlear nerve**, is concerned with **hearing**. Both parts emerge from the brainstem

---

[6]The special visceral efferent component is sometimes called the *brachial motor component* because it supplies muscles derived from the second branchial or pharyngeal arch in the embryo (Moore, 1988).

[7]The name "sensory root" of the facial nerve, often applied to the nervus intermedius, is not entirely accurate because the root also contains visceral motor fibers.

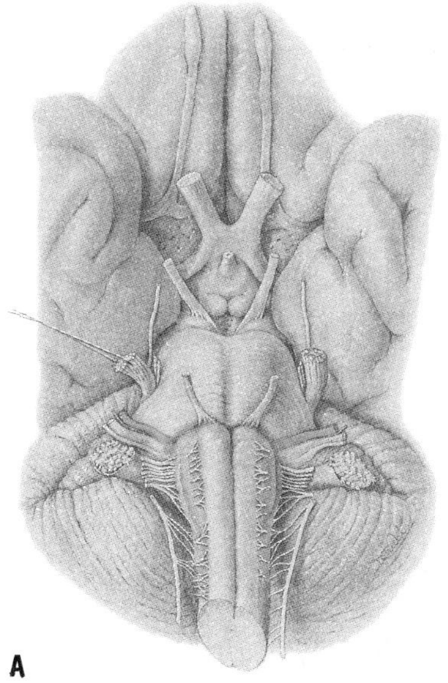

Figure 9-9. The facial nerve (CN VII). *A,* Its attachment to the junction between the pons and medulla. *B,* Its distribution in the head and neck. Observe that it is motor to the muscles of facial expression and the muscles around the eyes, mouth, and ears and of the scalp superiorly and the platysma inferiorly. It also supplies the stylohyoid and posterior belly of the digastric, as well as the stapedius. Also observe that it serves special sense. Taste fibers from the geniculate ganglion pass (1) from the palate through the pterygopalatine ganglion, the nerve of the pterygoid canal, and the greater petrosal nerve to the geniculate ganglion, and (2) from the anterior two-thirds of the tongue through the chorda tympani to the facial nerve and through it to the geniculate ganglion.

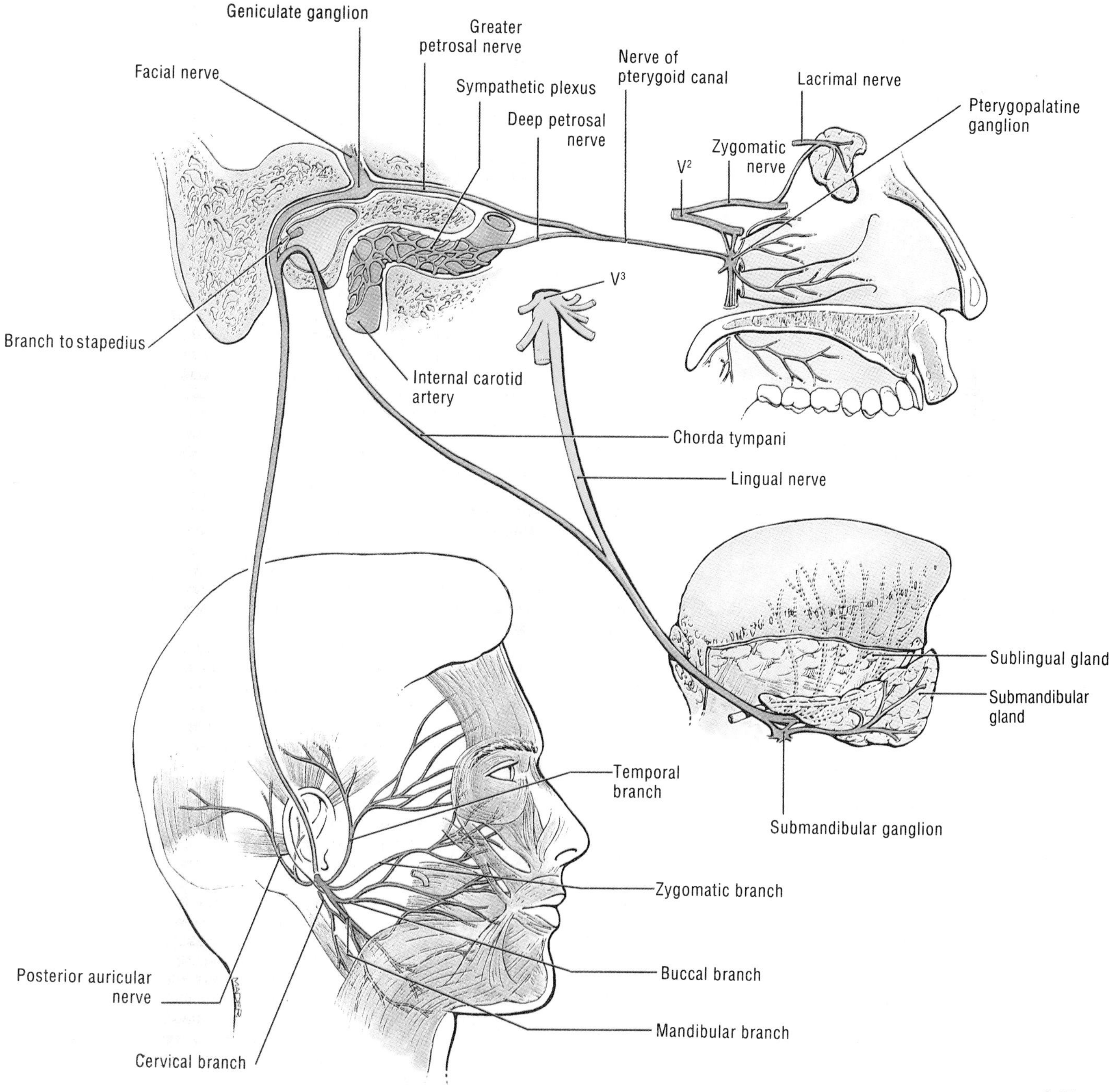

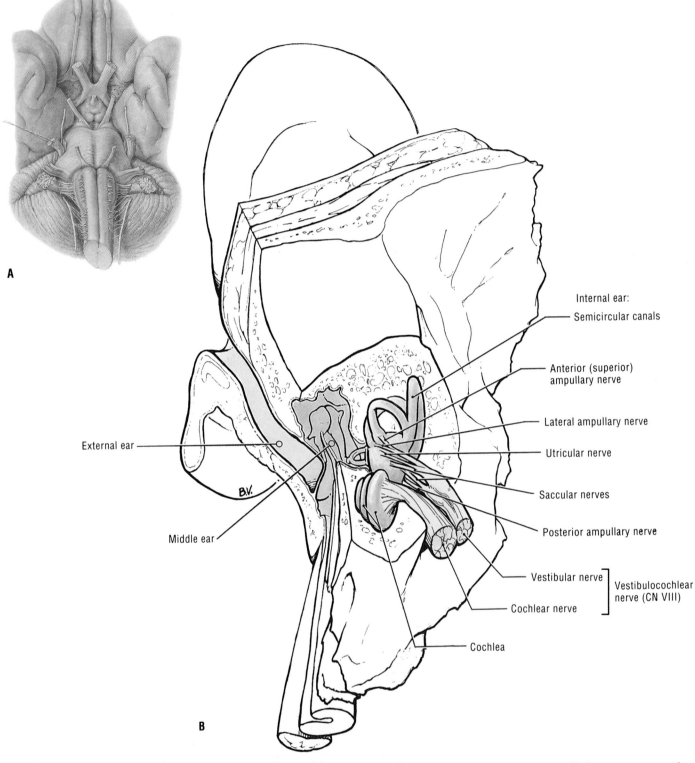

A

B

Internal ear:
Semicircular canals

Anterior (superior) ampullary nerve

Lateral ampullary nerve

Utricular nerve

Saccular nerves

Posterior ampullary nerve

Vestibular nerve ⎤
⎟ Vestibulocochlear
⎟ nerve (CN VIII)
Cochlear nerve ⎦

Cochlea

External ear

B.V.

Middle ear

Figure 9-10. The vestibulocochlear nerve (CN VIII). *A*, Its attachment to the junction of the pons and medulla. *B*, This nerve consists of two parts: (1) the cochlear nerve or nerve of hearing that conducts auditory information from the cochlea and (2) the vestibular nerve or nerve of equilibrium that conducts information about balance from the semicircular canals.

separately in the groove between the pons and medulla (see Fig. 7-125).

The Vestibular Nerve (Figs. 9-1 and 9-10*B*; see also Fig. 7-125). This nerve is the means whereby movements of the eyes and head are linked with the part of the membranous labyrinth that governs the appreciation of the body in space. It arises from cells in the *vestibular ganglion* that is located at the bottom of the *internal acoustic meatus*. The peripheral processes of the bipolar cells come from the utricle, saccule, and ampullae of the semicircular ducts (p. 776). The vestibular nerve passes through the internal acoustic meatus with the cochlear nerve, facial nerve, and labyrinthine artery.

The Vestibular Nuclei (Fig. 9-8). There are four vestibular nuclei located at the junction of the pons and medulla in the lateral part of the floor of the fourth ventricle. For their connections with the brain stem, cerebellum, and spinal cord, see Barr & Kiernan (1988).

The Cochlear Nerve (Fig. 9-10*B*; see also Figs. 7-122*C*, and 7-125). This nerve arises from cells in the *spiral ganglion* (cochlear ganglion) that is located in the *modiolus of the cochlea*. The peripheral processes of these bipolar cells travel to the spiral ganglion from hair cells in the *spiral organ* (of Corti). These processes reach this organ by passing through openings in the *osseous spiral lamina* that projects from the modiolus.

The Cochlear Nuclei (Fig. 9-8). There are two cochlear nuclei in the medulla. The dorsal and ventral nuclei are located superficially in the rostral end of the medulla, adjacent to the base of the inferior cerebellar peduncle. For a description of the pathway to the *auditory cortex* of the temporal lobe, see Barr and Kiernan (1988).

Lesions of the vestibulocochlear nerve may result in: (1) *tinnitus* (ringing or buzzing in the ears); (2) *impairment or loss of hearing*; and (3) loss of balance (*vertigo*). **Acoustic neuromas** on the extracerebral part of CN VIII are among the most common intracranial tumors (Table 9-3); other lesions are uncommon.

The Glossopharyngeal Nerve (CN IX)

Type: Mixed—Motor[8] and Sensory (Figs. 9-1 and 9-11; Tables 9-1 and 9-2)

This cranial nerve supplies one muscle, the *stylopharyngeus* (Fig. 9-11*B*; see also Fig. 7-74). It supplies the *parasympathetic secretomotor fibers to the parotid gland* (Fig. 9-11*C*), and it carries *sensory fibers from the pharynx, tonsil, and posterior part of the tongue* (Fig. 9-11*B*; see also Fig. 7-96). Its name is

derived from its supply to the tongue (G. *glossa*) and pharynx (G. throat). CN IX arises as three or four rootlets from the lateral aspect of the rostral part of the medulla (Fig. 9-1). These rootlets emerge from a groove between the olive and the inferior cerebellar peduncle and are closely related to the rootlets of the vagus nerve. The CN IX rootlets merge and leave the skull through the *jugular foramen* (see Fig. 7-40). The superior and inferior **glossopharyngeal ganglia** are located in the nerve as it traverses this foramen. The inferior ganglion is larger and more constant. The glossopharyngeal nerve descends and passes between the superior and middle pharyngeal constrictor muscles to enter the pharynx (Fig. 9-11*B*; see also Figs. 7-102 and 8-45). It passes through the *tonsillar bed* (p. 835) and ends by entering the posterior third of the tongue (Fig. 9-11*B*; see also Fig. 8-49). CN IX communicates with the *superior cervical sympathetic ganglion* and the vagus and facial nerves (see Figs. 7-74 and 8-26).

Branches of the Glossopharyngeal Nerve (Fig. 9-11*B* and *C*; see also Fig. 7-74). There are tympanic, carotid, pharyngeal, muscular, tonsillar, and lingual branches. The *tympanic nerve* (sensory and parasympathetic) arises from the inferior glossopharyngeal ganglion and reenters the skull through the *tympanic canal* in the temporal bone. It enters the middle ear where it divides into branches to form the *tympanic plexus* that supplies (1) a branch to the greater petrosal nerve (see Fig. 7-27); (2) branches to the mucosa of the tympanic cavity, auditory tube, and mastoid cells; and (3) a branch to the lesser petrosal nerve. The *lesser petrosal nerve* (Fig. 9-11*B* and *C*), a branch of the tympanic plexus, contains parasympathetic secretomotor fibers to the parotid gland. It leaves the middle ear through a small opening lateral to the hiatus for the greater petrosal nerve. It passes through the foramen ovale (see Fig. 7-41) and synapses in the *otic ganglion* (Fig. 9-11*C*). Postsynaptic fibers pass to the parotid gland.

The *pharyngeal branches* consist of three or four filaments that unite near the middle constrictor muscle with the pharyngeal branch of the vagus and laryngopharyngeal branches of the sympathetic trunk to form the *pharyngeal plexus* (Fig. 9-11*B*; see also Fig. 7-74). Through this plexus the glossopharyngeal nerve supplies sensory fibers to the mucosa of the pharynx. The *carotid branches* (sensory) pass to the carotid sinus and carotid body (Fig. 9-11*B*; see also Figs. 7-74, 8-20, and 8-24). The *nerve to the stylopharyngeus* is the only motor branch of the glossopharyngeal nerve (Fig. 9-11*B*; see also Figs. 7-74, 8-43, and 8-45). The *tonsillar branches* form a plexus around the tonsil by uniting with branches of the middle and posterior palatine nerves. Filaments from this nerve plexus supply the tonsillar and soft palate regions. The *lingual branches* (sensory) supply the mucosa of the posterior third of the tongue with general sensory and special sensory (taste) fibers (Fig. 9-11*B*; see also Fig. 7-96).

Glossopharyngeal Nuclei (Fig. 9-8). CN IX shares four nuclei in the medulla with CN X and CN XI (two motor and two sensory). The *nucleus ambiguus* is located deep in the superior part of the medulla. The *inferior salivary nucleus* is adjacent to the rostral part of the nucleus ambiguus in the inferior part of the pons. The *nucleus of the tractus solitarius* is lateral to the dorsal nucleus of the vagus in the superior part of the medulla. The elongated *sensory nucleus of the trigeminal* is located lateral to the nucleus ambiguus in the medulla.

---

[8]The special visceral efferent component of CN IX is sometimes called the *branchial motor component* because it supplies the stylopharyngeus muscle that is derived from the third branchial or pharyngeal arch of the embryo (Moore, 1988).

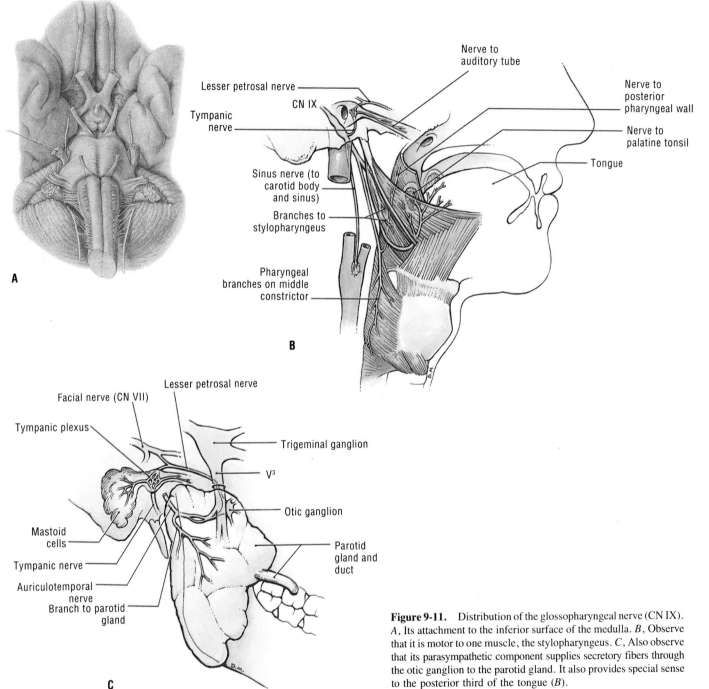

A

B

C

Lesser petrosal nerve

Tympanic nerve

CN IX

Sinus nerve (to carotid body and sinus)

Branches to stylopharyngeus

Pharyngeal branches on middle constrictor

Nerve to auditory tube

Nerve to posterior pharyngeal wall

Nerve to palatine tonsil

Tongue

Facial nerve (CN VII)

Tympanic plexus

Lesser petrosal nerve

Trigeminal ganglion

V³

Otic ganglion

Mastoid cells

Parotid gland and duct

Tympanic nerve

Auriculotemporal nerve

Branch to parotid gland

Figure 9-11. Distribution of the glossopharyngeal nerve (CN IX). *A*, Its attachment to the inferior surface of the medulla. *B*, Observe that it is motor to one muscle, the stylopharyngeus. *C*, Also observe that its parasympathetic component supplies secretory fibers through the otic ganglion to the parotid gland. It also provides special sense to the posterior third of the tongue (*B*).

Because CN IX, CN X, and CN XI share nuclei in the medulla, they have much in common functionally. These three cranial nerves innervate all the muscles of the larynx, pharynx, and soft palate. Consequently, an isolated lesion of CN IX is uncommon. Injury to the glossopharyngeal nerve alone could result in absence or diminution of taste from the posterior third of the tongue and *absence of the gag reflex* (Table 9-3). Normally this reflex is solicited by stroking the pharynx and may be induced unintentionally during dental treatments. **Glossopharyngeal neuralgia** results from a lesion to CN IX. Brief but severe attacks of pain usually begin in the throat and radiate along the side of the neck, near the anterior part of the auricle and the posterior aspect of the mandible. Often the pain is initiated by swallowing or protrusion of the tongue.

The Vagus Nerve (CN X)[9]

Type: Mixed—Motor[10] and Sensory (Figs. 9-1 and 9-12; see also Figs. 8-33 and 8-64; Tables 9-1 and 9-2)

This nerve has a more extensive course and distribution than any of the other cranial nerves. It traverses and supplies structures in the neck, thorax, and abdomen. *CN X arises as eight to ten rootlets that emerge from the medulla* in the groove between the olive and inferior cerebellar peduncle (Figs. 9-1 and 9-12A). These rootlets converge and pass through the *jugular foramen* (see Fig. 7-40). As the vagus nerve emerges, it accompanies the accessory nerve (CN XI). Just before it exits the cranium, CN X exhibits a swelling known as the rostral or *superior vagal ganglion.* It is joined to the cranial root of CN XI by one or two filaments (see Fig. 8-13). It is also joined to the inferior glossopharyngeal ganglion and to the sympathetic trunk by a filament from the *superior cervical ganglion* (see Fig. 8-26). Just after it leaves the jugular foramen, the vagus nerve exhibits another swelling called the caudal or *inferior vagal ganglion* (see Figs. 8-33 and 8-63). It is connected with the hypoglossal nerve (CN XII), superior to the superior cervical ganglion (see Fig. 8-26).

CN X descends through the neck within the *carotid sheath* (see Figs. 8-21 and 8-34) and enters the thorax. Here it joins the vagus nerve on the other side to form the **esophageal plexus** (Fig. 9-12B). This plexus follows the esophagus through the diaphragm into the abdomen. Here, the vagus nerves reform and are known as the anterior and posterior **vagal trunks** (see Fig. 2-42). The *anterior vagal trunk* supplies the anterior aspect of the stomach and gives branches to the lesser omentum that supply the liver, pyloric canal, the first two parts of the duodenum, and the head of the pancreas. The *posterior vagal trunk* supplies the

posterior aspect of the stomach and ends mainly in the **celiac ganglia** and adjacent plexuses (Fig. 9-12B; see also Figs. 2-82 and 8-33).

Branches of the Vagus Nerve (Fig. 9-12B; see also Fig. 8-64). These branches arise in the head, neck, thorax, and abdomen. The *meningeal branches* arise from the superior vagal ganglion and supply the dura in the posterior cranial fossa. *Auricular branches* (sensory) also arise from the superior vagal ganglion and pass to the ear, where they supply part of the auricle and the external acoustic meatus. The *pharyngeal branch* (motor) arises from the inferior vagal ganglion and passes to the pharynx (see Fig. 8-63). It supplies all the striated muscles of the pharynx and soft palate, except the stylopharyngeus (supplied by CN IX) and the tensor veli palatini (supplied by CN V[3]). The *carotid branches* (sensory) arise from the inferior vagal ganglion and pass to the walls of the *carotid sinus* (see Figs. 8-24 and 8-33). These fibers are auxiliary to those supplied by the glossopharyngeal nerve (see Fig. 7-74).

The *superior laryngeal nerve* (sensory and motor) arises from the inferior vagal ganglion and descends in the neck (Fig. 9-12B; see also Figs. 8-33 and 8-64). Here, it divides into two branches: (1) the *internal laryngeal nerve* that pierces the *thyrohyoid membrane* (see Fig. 8-45) and supplies the larynx superior to the vocal folds (cords) and (2) the *external laryngeal nerve* that is motor to the *cricothyroid muscle* of the larynx (see Figs. 8-54 and 8-64).

The *recurrent laryngeal nerve* (sensory and motor) turns superiorly around the subclavian artery on the right side (see Fig. 9-12B; see also Figs. 8-39 and 8-63) and around the arch of the aorta on the left side (Fig. 9-12B; see also Figs. 1-68 and 8-39). These nerves ascend in the neck and enter the larynx between the inferior constrictor muscle and the esophagus (see Figs. 8-38 and 8-52). The recurrent laryngeal nerve is *sensory to the larynx* inferior to the vocal folds, and *motor to the intrinsic muscles of the larynx* (see Fig. 8-61).

Cardiac branches (parasympathetic) leave the vagus nerve *in the neck* and pass to the **cardiac plexus of nerves** (Fig. 9-12B; see also Figs. 1-63 and 1-68). Here, the fibers synapse and postsynaptic fibers pass to the heart and act to slow it and to constrict the coronary arteries. *Pulmonary branches* (parasympathetic) are given off in the thorax that contribute to the anterior and posterior **pulmonary plexuses** (Fig. 9-12B; see also Fig. 1-72). The fibers synapse here, and postsynaptic fibers supply the smooth constrictor muscles in the bronchial tree. *Abdominal branches* arise from the anterior and posterior vagal trunks and synapse with ganglion cells in the walls of the viscera. They supply smooth muscle in the digestive tract as far as the left colic flexure.

Vagal Nuclei (Fig. 9-8). The location of the four nuclei (two motor and two sensory)—nucleus ambiguus, nucleus of the tractus solitarius, dorsal nucleus of the vagus, and the sensory nucleus of the trigeminal—is described with CN IX (p. 869). The vagus nerve shares these nuclei with CN IX and CN XI. For details about these nuclei, see Barr and Kiernan (1988).

Injury to the vagus nerve after it leaves the skull is uncommon, except during acts of warfare. Its functions are affected more often by damage (frequently of vascular origin)

---

[9]The Latin name *vagus*, meaning ''wandering,'' is an appropriate description of this nerve because it ''wanders'' from the medulla of the brain to the colon in the abdomen and pelvis.

[10]The special visceral motor component of CN X is sometimes called the *brachial motor component* because it supplies muscles derived from the fourth and sixth branchial arches in the embryo (Moore, 1988).

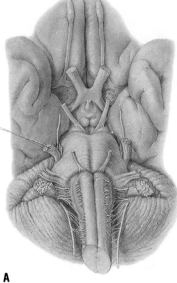

A

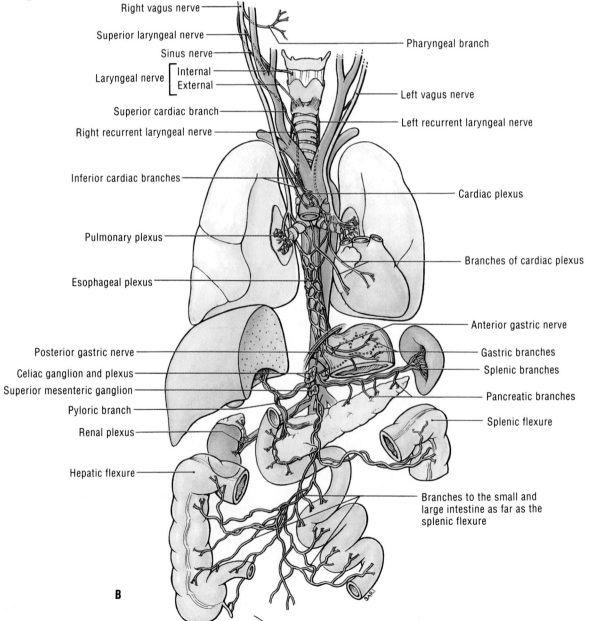

Right vagus nerve

Superior laryngeal nerve

Sinus nerve

Laryngeal nerve [Internal / External]

Superior cardiac branch

Right recurrent laryngeal nerve

Inferior cardiac branches

Pulmonary plexus

Esophageal plexus

Posterior gastric nerve

Celiac ganglion and plexus

Superior mesenteric ganglion

Pyloric branch

Renal plexus

Hepatic flexure

Pharyngeal branch

Left vagus nerve

Left recurrent laryngeal nerve

Cardiac plexus

Branches of cardiac plexus

Anterior gastric nerve

Gastric branches

Splenic branches

Pancreatic branches

Splenic flexure

Branches to the small and large intestine as far as the splenic flexure

B

to its nuclei in the medulla or by injury to it before it leaves the skull. The effects produced include: *palpitation* (forcible pulsation of the heart), *tachycardia* (rapid beating of the heart), vomiting, *slowing of respiration*, and a sensation of suffocation. A person with a vagal lesion inferior to the skull, although uncommon as stated, may have: (1) paralysis of the soft palate and larynx on the affected side, (2) hoarseness, and (3) anesthesia of the larynx (Table 9-3). These defects may result in the aspiration of foreign material. When one nerve is damaged, the vocal fold supplied by it is motionless. The person can speak but the voice is weak. The motor innervation of the soft palate can be tested by asking the patient to say "ah" and observing the movement of the palate. If paralyzed on one side, the unaffected side will rise further and pull the uvula toward the normal side. *Recurrent laryngeal palsies* often result from cancer (25%) and injury during surgery on the thyroid gland, the cervical part of the esophagus, the heart, and lungs (20%). Because of its longer course in the neck (p. 871), lesions of the left recurrent laryngeal nerve are much more common than those of the right (McMinn, 1990).

The Accessory Nerve (CN XI)

Type: Motor (Fig. 9-13; see also Figs. 7-51, 8-13, and 8-27; Tables 9-1 and 9-2)

This nerve has *cranial and spinal roots*. These roots are united for only a short distance. The **spinal root of CN XI** (spinal accessory nerve) arises from a column of motor neurons called the *spinal nucleus*, which is located in the lateral part of the gray matter of the cervical region of the spinal cord. The filaments arise from segments C6 to C1 and merge to form a trunk, which ascends through the *foramen magnum* (see Fig. 8-13). Here it joins briefly with the cranial root and exits then through the *jugular foramen* to enter the anterior and posterior triangles of the neck (see Figs. 8-11 and 8-16). The **cranial root of CN XI** arises from the *nucleus ambiguus* (Fig. 9-8). It really belongs to the vagus nerve and gives all its fibers to it to supply the skeletal muscles in the pharynx and palate. This root emerges from the lateral part of the medulla as four or five rootlets (Fig. 9-13A; see also Fig. 8-13). These rootlets merge and pass to the *jugular foramen*, where they unite with the spinal root of CN XI for a short distance. It is also connected with the superior vagal ganglion (see Fig. 8-13). After it leaves the jugular foramen, the cranial root separates from the spinal root and passes over the inferior vagal ganglion to which it is attached.

Accessory Nerve Nuclei (Fig. 9-8). There are two motor nuclei. The cranial root arises from neurons in the caudal part of the *nucleus ambiguus* in the medulla, and the spinal root arises from the *spinal nucleus*, a column of anterior horn cells in the superior five or six cervical segments of the spinal cord.[10]

---

[10]The motor neurons for the sternocleidomastoid and trapezius muscles differentiate near the cells that form the nucleus ambiguus. The neurons for these muscles migrate into the spinal cord (segments C1 to C6) and form the spinal nuceleus. This explains why part of a cranial nerve has its nucleus in the spinal cord (Fig. 9-8).

It would be simpler, although contrary to convention, to consider the cranial root of CN XI as part of the vagus nerve, leaving the spinal root as the definitive accessory nerve (Barr and Kiernan, 1988). Fibers of the cranial root are distributed mainly to the pharyngeal and recurrent laryngeal branches of the vagus nerve.

During dissections in the posterior triangle of the neck (p. 791) for the excision of a cancer, many lymph nodes and considerable connective tissue are removed. If the spinal root of the accessory nerve is damaged, paralysis of the superior part of the trapezius muscle occurs (Table 9-3). If the nerve is severed, the sternocleidomastoid muscle is also affected. This results in inferior and lateral rotation of the scapula and dropping of the shoulder. The patient also experiences a weakness during turning of the head to the opposite side. The functions of the accessory nerve may also be affected by fractures that involve the jugular foramen or lacerations in the neck or by gross enlargement of the cervical lymph nodes.

The Hypoglossal Nerve (CN XII)

Type: Motor (Figs. 9-1 and 9-14; Tables 9-1 and 9-2)

This is the *motor nerve to the tongue*. It supplies all the tongue muscles except the palatoglossus that is innervated by the vagus through the pharyngeal plexus (Fig. 9-14B; see also 7-100). A series of *10 to 15 rootlets emerge between the olive and pyramid* of the medulla (Figs. 9-1 and 9-13A). These rootlets unite to form one or two trunks that pass laterally, posterior to the vertebral artery. If there are two trunks, they unite after passing through the *hypoglossal canal* in the occipital bone (see Figs. 7-11 and 7-12A). Inferior to the skull, CN XII is medial to the CN IX, X, and XI nerves (Figs. 9-1). CN XII enters the anterior triangle of the neck (Fig. 9-14B; see also Fig. 8-17) and passes inferolaterally, close to the posterior surface of the *inferior vagal ganglion* (see Figs. 8-33 and 8-64). It comes to lie deep to the posterior belly of the digastric muscle, between the internal carotid artery and internal jugular vein. It then crosses lateral to the bifurcation of the common carotid artery. Thereafter it loops anteriorly, superior to the greater horn of the hyoid bone, and disappears between the mylohyoid and hyoglossus muscles (see Fig. 8-14). It enters the floor of the mouth and then divides to supply all the intrinsic muscles of the tongue (Fig. 9-13B; see also Fig. 7-102), as well as three of the four extrinsic muscles.

Branches of the Hypoglossal Nerve (Fig. 9-14B; see also Figs. 7-100, 7-102, and 8-17). There are meningeal, descending, thyrohyoid, and muscular branches. The branches of CN XII that occur before it reaches the tongue are all derived from C1 fibers (Fig. 9-14B). They join CN XII as it exits from the skull. *The fibers from the hypoglossal nerve have no supply outside the tongue.* However, its branches that are composed of C1 fibers are classified as parts of CN XII, although their origins are not from the hypoglossal nucleus. A very small *meningeal branch, derived from C1*, accompanies CN XII and is sensory. It leaves CN XII while it is in the hypoglossal canal. It returns

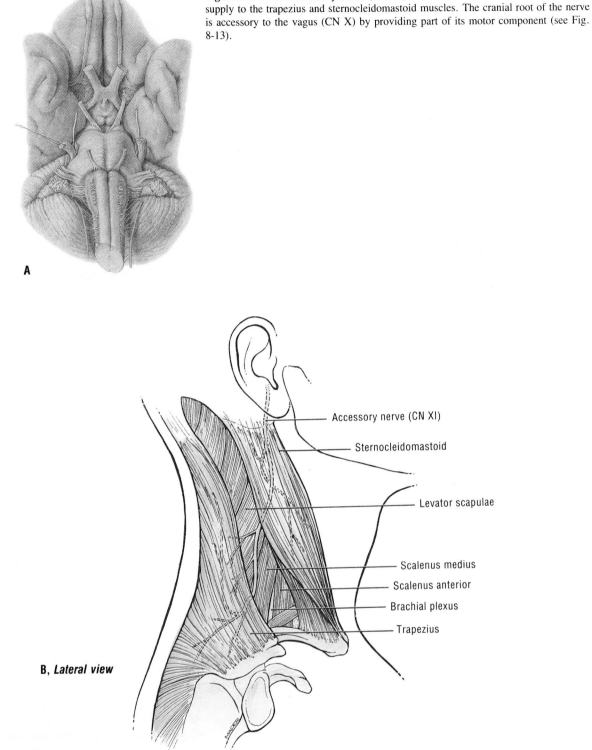

Figure 9-13. The accessory nerve (CN XI). *A*, Its attachment to the medulla. *B*, Its supply to the trapezius and sternocleidomastoid muscles. The cranial root of the nerve is accessory to the vagus (CN X) by providing part of its motor component (see Fig. 8-13).

A

Accessory nerve (CN XI)

Sternocleidomastoid

Levator scapulae

Scalenus medius

Scalenus anterior

Brachial plexus

Trapezius

B, *Lateral view*

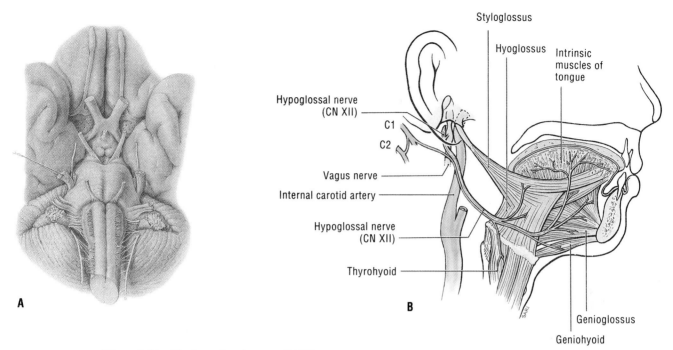

Figure 9-14.   The hypoglossal nerve (CN XII). *A*, Its attachment to the medulla. *B*, Its distribution to the muscles of the tongue.

to the cranium to supply the diploë of the occipital bone and a small ring of dura in the posterior cranial fossa beyond the foramen magnum that forms the walls of the occipital and inferior petrosal sinuses (see Fig. 7-30). The descending branch or *superior root of the ansa cervicalis* (formerly called the descendens hypoglossi) is also derived from C1 (see Fig. 8-16). It arises as the nerve crosses the internal carotid artery and runs inferiorly to join the *inferior root of the ansa cervicalis* (formerly called the descendens cervicalis), which is derived from C2 and C3. The two roots of the ansa cervicalis form the **ansa cervicalis**, a loop (L. *ansa*) in the cervical plexuses of nerves that supplies both bellies of the omohyoid muscle and the sternohyoid and sternothyroid muscles (see Fig. 7-100; Table 8-3). The *nerve to the thyrohyoid* (C1) crosses the greater horn of the hyoid bone to reach the muscle.

Muscular branches of CN XII supply the styloglossus, hyoglossus, genioglossus, and geniohyoid, as well as the intrinsic muscles of the tongue (see Tables 7-9 and 8-3).

Hypoglossal Nucleus (Fig. 9-8).   This nucleus is located between the dorsal nucleus of the vagus and the median plane in the floor of the fourth ventricle. It extends through most of the dorsal part of the medulla.

Complete section of the hypoglossal nerve causes *unilateral paralysis of the tongue* and, if of long standing, there is atrophy of the affected half. When the tongue is protruded, it deviates to the paralyzed side because of the unopposed action of the normal half. When it is retracted, the paralyzed side rises higher than the normal side. The larynx may also deviate toward the active side during swallowing owing to

paralysis of the infrahyoid muscles that depress the hyoid bone (see Table 8-2). If both nerves are sectioned the tongue is paralyzed. Taste and tactile sensations are unaffected, but speaking is slow and swallowing is difficult.

SUGGESTED READINGS

Barr ML, Kiernan JA: *The Human Nervous System: An Anatomical Viewpoint*, ed 5. Philadelphia, JB Lippincott, 1988.

Bertram EG, Moore KL: *An Atlas of the Human Brain and Spinal Cord*, Baltimore, Williams & Wilkins, 1982.

Brodal A: *Neurological Anatomy in Relation to Clinical Medicine*, ed. 3. New York, Oxford University Press, 1981.

Devinsky O, Feldman E: *Examination of the Cranial and Peripheral Nerves*, New York, Churchill Livingstone, 1988.

Fujimura I, De Souza RR, Ferraz De Carvalho CA, Rodriques, Jr AJ: A method for locating the marginal branch of the facial nerve in the neck. *Clin Anat* 3:143-147, 1990.

Haines, DE: *Neuroanatomy: An Atlas of Structures, Sections, and Systems*, ed. 2, Baltimore, Urban & Schwarzenberg, 1987.

McMinn RHM: *Last's Anatomy*, ed 8. New York, Churchill Livingstone, 1990.

Moore KL: *The Developing Human: Clinically Oriented Embryology*, ed. 4, Philadelphia, WB Saunders, 1988.

Tobias PV, Arnold M, Allan JC: *Man's Anatomy: A Study In Dissection*, ed. 4, vol. II. Johannesburg, Witwatersrand University Press, 1988.

Williams PL, Warwick R, Dyson M, Bannister LH: *Gray's Anatomy*, ed. 37, New York, Churchill Livingstone, 1989.

Wilson-Pauwels L, Akesson EJ, Stewart PA: *Cranial Nerves: Anatomy and Clinical Comments*, Toronto, BC Decker, 1988.

Page numbers in italics denote figures; those in boldface type
refer to major descriptions; those followed by "t" denote tables.